Nutrition

Encyclopedia of
ENVIRONMENTAL BIOLOGY

VOLUME I

A – E

Encyclopedia of
ENVIRONMENTAL
BIOLOGY

VOLUME I
A – E

Editor-in-Chief

William A. Nierenberg

Scripps Institution of Oceanography
University of California, San Diego

ACADEMIC PRESS

San Diego   New York   Boston   London   Sydney   Tokyo   Toronto

This book is printed on acid-free paper. ∞

Copyright © 1995 by ACADEMIC PRESS, INC.

All Rights Reserved.
No part of this publication may be reproduced or transmitted in any form or by any means, electronic or mechanical, including photocopy, recording, or any information storage and retrieval system, without permission in writing from the publisher.

Academic Press, Inc.
A Division of Harcourt Brace & Company
525 B Street, Suite 1900, San Diego, California 92101-4495

United Kingdom Edition published by
Academic Press Limited
24-28 Oval Road, London NW1 7DX

Library of Congress Cataloging-in-Publication Data

Encyclopedia of environmental biology / edited by William A.
   Nierenberg.
         p.   cm.
      Includes bibliographical references and index.
      ISBN 0-12-226730-3 (set). -- ISBN 0-12-226731-1 (v. 1). -- ISBN
   0-12-226732-X (v. 2) -- ISBN 0-12-226733-8 (v. 3)
      1. Ecology--Encyclopedias.   2. Environmental sciences-
   -Encyclopedias.   I. Nierenberg, William Aaron, date.
   QH540.4.552   1995
   574.5'03--dc20   94-24917
                                                              CIP

PRINTED IN THE UNITED STATES OF AMERICA
95   96   97   98   99   00   EB   9   8   7   6   5   4   3   2   1

Contents

C

D

E

Preface

My desire to compile an encyclopedia of environmental biology arose from my participation in the public arenas of acid rain and global warming debates. Governments continue to spend huge sums of money on research on both of these problems (the annual expenditure in the United States is 1.5 billion dollars per year), but the smallest fraction is spent in the area of greatest concern, the biological environment.

My first inkling of this state of affairs arose during my tenure as Assistant Secretary General of NATO for Scientific Affairs, when a colleague of mine, an experienced natural products chemist, became concerned at the decline in the support of taxonomy just when it was becoming a central focus in environmental issues. It can be surmised that the shifting emphasis to molecular biology that was taking place was at the root of the problem. The relative decline has continued unabated. Currently, there is a continuing neglect or abandonment of botanical museums and collections. However, by holding meetings on chemical taxonomy in the expectation of joining together two worlds that were growing apart, my colleague's efforts resulted in a modest budgeted effort in support of taxonomy.

Years later, when I served as Chairman of a White House committee on acid rain, the problem became much more focused. The purpose of the committee was to satisfy the new White House administration that the impending output of a major effort by a joint U.S./Canadian task force on acid rain represented the state of the art. We were able to assemble the best chemists, physicists, and environmental biologists, but we were unable to find freshwater ichthyologists familiar with the effects on fish of the acidification of their habitat. In fact, at the time, the reason for the negative effects on fish of a lowering of the pH of lakes and streams was not understood.

There also was the problem of forests, some of which seemed to show severe damage that many believed was caused by acid precipitation—the situation in Germany appeared to be most alarming. After ten years and a one-half billion dollar program in the United States, both the extent of the possible ecological damage and the underlying biological causes remain unclear. The same is true with respect to the German forests, which had been very much in the news. A joint U.S./German commission could find no clear relation between the observed damage and acid precipitation.

The severity of the problem of global warming that may be caused by the anthropogenic increase in the atmosphere of the so-called greenhouse gases, the most important of which is carbon dioxide, is widely debated. The effects on humans and the environment can be very complex, covering many aspects such as agriculture, sea level, and economics in the broad sense. However, the specifics of the possible biological effects on the environment are poorly understood, except in the agricultural area, for which one hundred years of excellent and sustained research have provided a good estimate of the effects of global change. The principal reason, here again, is the imbalance in research support between nonbiological programs such as high-powered computer modeling and satellite programs on the one hand and field biology and supporting bench biology on the other.

A third area of global concern is the stratospheric ozone problem. This serious problem remains controversial even though the nations of the world have agreed to substantially reduce chlorofluorocarbon emissions, the principal source of which is Freon, the air-conditioning fluid. Huge sums of money are spent monitoring research involving satellite missions and giant computer simulation programs. However, while the basic problem of the environmental and human physiological effects does get attention, the ratio of effort to expenditures is minimal. In particular, acceptable evidence that ultraviolet B is a major contributing cause of melanoma is still lacking.

This list of unbalanced support in environmental research where field biology receives inadequate attention and student support is minimal is very large and, in fact, is growing rather than diminishing. It is this complex of experiences that led to the concept of an encyclopedia dedicated to environmental biology. It is intended for use not only by the professional biologist, but also by the larger group of public officials—city managers and their staffs, city council members, county supervisors and staffs, and in fact, government at all levels. I hope that business executives and the staffs of environmental organizations will find these volumes useful as well in coping with the myriad local, regional, and global climate impacts that their activities touch upon.

This *Encyclopedia* could easily have been expanded if the time had permitted it. If this edition is as useful as anticipated, a second, much larger effort may be needed.

William A. Nierenberg

How to Use the Encyclopedia

The *Encyclopedia of Environmental Biology* is intended for use by both students and research professionals. Articles have been chosen to reflect important areas in the study of the environment, topics of public and research interest, and coverage of environmental issues vital to lawyers. Each article provides a comprehensive overview of the selected topic to satisfy readers from students to professionals.

The *Encyclopedia* is designed with the following features to allow maximum accessibility. Articles are arranged in alphabetical order by subject. A complete table of contents appears in the front matter of each volume. This list of titles represents topics carefully selected by our esteemed editorial board. Here, such general topics as "Acid Rain" and "Speciation" are listed.

A 10,000 entry Subject Index is located at the end of Volume 3. The index is the most direct way to access the *Encyclopedia* to find specific information. Although the article "Evolution and Extinction" is thorough and complete, additional text on evolution can be found in articles such as "Galápagos Islands" and "Bird Communities." Subjects in the index are listed alphabetically and indicate the

volume and page number where corresponding information can be found.

The Index of Related Titles appears in Volume 3 following the Subject Index. This index presents an alphabetical list of the articles as they appear in the *Encyclopedia*. Following each article title is a group of related articles that appear in the encyclopedia.

Articles contain an outline, a glossary, cross-references, and a bibliography. The outline allows a quick scan of the major areas discussed within the article. The glossary contains terms that may be unfamiliar to the reader, with each term defined in the *context of its use in that particular article*. Thus, a term may appear in the glossary for another article defined in a slightly different manner or with a subtle nuance specific to that article. For clarity, we have allowed these differences in definition to remain so that the terms are defined relative to the context of each article.

Each article contains cross-references to other *Encyclopedia* articles. Cross-references are found at the end of the paragraph containing the first mention of a subject covered more fully elsewhere in the *Encyclopedia*. By using the cross-references, the reader gains the opportunity to find additional information on a given topic.

The bibliography lists recent secondary sources to aid the reader in locating more detailed or technical information. Review articles and research articles that are considered of primary importance to the understanding of a given subject area are also listed. Bibliographies are not intended to provide a full reference listing of all material covered in the context of a given article, but are provided as guides to further reading.

Acid Rain

George Hidy

Electric Power Research Institute, California

I. Introduction
II. Air Chemistry
III. Effects on Human Systems
IV. Ecological Response
V. Summary

I. INTRODUCTION

Acidity in rainfall or precipitation has been a subject of considerable media attention since the 1970s. The reason for this scrutiny derives from concern about its pervasive geographical extent and its cumulative environmental effects on materials and cultural resources, on surface waters and aquatic ecosystems, and on terrestrial ecosystems. In this article, the phenomenon of acid rain is described in terms of its historical background, its origins in atmospheric chemical processes, and its biological implications.

A. Historical Background

The term *acid rain* first appeared in a remarkable 1872 book authored by the chemist Robert Angus Smith, "Air and Rain: Beginnings of Chemical Climatology." Smith's observations of atmospheric acidity in industrial Manchester, England, date back 20 years earlier. In the urban environment, he found that sulfuric acid was present in rainwater, supplementing a natural acidic carbonate chemistry. The results of Smith's work drifted into obscurity until the mid-1950s and later, when a modern awareness of acid rain emerged from an intersection of three seemingly unrelated scientific fields: limnology, agriculture, and atmospheric chemistry.

As early as the seventeenth century, Robert Hooke recognized atmospheric sources of nutrients for the growth of plants. In the meantime, the chemical composition of "bulk deposition" from the air (rain, snow, and dustfall) was characterized and documented in Europe, Scandinavia, and North America. By the twentieth century, atmospheric deposition of sulfur and nitrogen was accounted for as a part of the required fertilization for agricultural crops, as well as forest systems. Because of interest in nutrient deposition, large-scale deposition precipitation monitoring began by 1948 in Norway, Denmark, and Finland; by 1956, this monitoring was extended and coordinated as the European Air Chemistry Network by the Stockholm International Meteorological Institute.

The imaginative leadership of the Swedes Carl Gustaf Rossby and Erik Eriksson invigorated the embryonic science of atmospheric chemistry. Interpretation of the European monitoring data yielded insight about the origins of acidity in rainfall, particularly in relation to transport of air pollution over great distances exceeding 1000 km. Interest in the European monitoring data stimulated similar studies in the United States, initiated in the 1950s by Christian Junge, and later in the 1960s by the U.S. National Center for Atmospheric Research.

In the 1950s, a branch of limnological research began to follow up linkages between atmospheric chemistry, surface water chemistry, and aquatic biota stress. Early hypotheses about such linkages emerged from the work of the Minnesota ecologist Eville Gorham. In parallel, investigation of mysterious fish kills apparently associated with water acidification in Norway and Sweden led Scandinavian aquatic scientists to raise concerns about the origins of the acidity. Concern for the adverse environmental effects of acid rain was stimulated in the 1960s by the linkage between salmon and trout kills in Scandinavian lakes and streams and acid deposition. Sven Oden, a Swedish geologist, carried public expression for action to the United Nations in 1972, through the first Conference on the Environment. The reason for involving an international debate focused on the hypothesis that sulfur oxide pollution transport from Europe was acidifying Scandinavian waters.

By the mid-1970s, workers in eastern North America began to report observations of apparent acidification of remote lakes and parallel deterioration of fisheries. Lakes in southern Ontario and the Adirondack Mountains were of particular interest. A few years later, forest scientists in Europe and North America began to speculate about the role of acid deposition in forest dieback. As a result of international concerns, aggressive research programs to characterize the extent and severity of acid rain exposure were initiated in both Europe and North America. Mounting circumstantial evidence that pollution from industrialized areas could affect remote regions far distant from the pollutant sources resulted in pressure for international oversight of control strategies for sulfur and nitrogen oxide emissions in the 1980s.

In the 1980s, the United States initiated one of the largest publicly funded research programs ever attempted on a single issue—acid rain. After a 10-year Congressional authorization, the U.S. National Acid Precipitation Program (NAPAP) completed its work. This effort was closely paralleled by a Canadian research program. These two projects provided a large body of knowledge about acid deposition and its environmental effects in much of North America. NAPAP was reautho-

rized to investigate and report on the effectiveness of the acid rain reduction program legislated in 1990 for the United States.

B. Origins and Definition of Acidity

Acidity present in the atmosphere derives from the major acid-forming gases—carbon dioxide (CO_2), sulfur dioxide (SO_2), and nitrogen oxides (nitric oxide (NO) and nitrogen dioxide (NO_2). Small contributions derive from hydrogen chloride (HCl) and organic acids sometimes found in the air. In the atmosphere, chemical reactions take place either in the gas phase or in cloud water to oxidize SO_2 to sulfuric acid (H_2SO_4). Similarly, the nitrogen oxides are oxidized ultimately to form nitric acid (HNO_3). All the acid-forming gases in the air have both natural sources and anthropogenic sources. Most of the CO_2 is a product of the natural carbon cycle, involving, for example, plant respiration and forest fires. SO_2 can be produced as an intermediate oxidation product of biogenically emitted compounds such as hydrogen sulfide (H_2S) and dimethyl sulfide [$(CH_3)_2S$]. Nitrogen oxides are produced naturally from soil nitrification processes and respiration, as well as from lightning discharges. HCl derives naturally from volcanic emissions. Organic acids also may be traced back as potential oxidation products of hydrocarbon vapor emissions from vegetation, or the aerosolization of organic slicks on the oceans. Other acidic species in the atmosphere include acid-forming ammonium salts, for example, NH_4HSO_4, NH_4NO_3, and NH_4Cl.

Naturally occurring atmospheric acidity is substantially enhanced by air pollution, especially from SO_2, NO, NO_2 (NO_x), and HCl emissions, from the combustion of fossil fuels. There are also human-derived sources of CO_2, NH_3, and organic acids.

Acids from the atmosphere reach the earth's surface from precipitation (rain, snow, dew, drizzle, etc.) or by direct contact with the surface (dry processes). In the latter cases, atmospheric gases can be absorbed in soils, surface waters, and vegetation on contact. Nitric acid and hydrochloric acid, mainly in vapor form near or at the ground, may be absorbed directly on surfaces. Acids suspended as tiny particles (aerosols) in the condensed phase, for example, sulfuric acid, are deposited by fallout (sedimentation), impaction, or Brownian motion to underlying surfaces.

Acidity in the atmosphere depends on the mix of trace constituents suspended in the air that will form both anions and cations when dissolved in water. In addition to the principal anions involved, HCO_3^-, CO_3^{2-}, SO_4^{2-}, Cl^-, and NO_3^-, the cations derived from soil dust, K^+, Ca^{2+}, Mg^{2+}, and Na^+, are important adjuncts to H^+. Thus, the acid content of deposition cannot be determined from measurement of anionic species alone; it is the balance of cations and anions that determine acidity.

Acidity is normally measured in terms of the pH or negative logarithm (base 10) of the hydrogen ion concentration. An acid-neutral sample of water has a pH of 7. In the absence of alkaline anionic species in rainwater, equilibrium with gaseous CO_2 at atmospheric pressure and temperature would result in a pH of approximately 5.5. Because there are naturally occurring acid species in the air, the acidity of precipitation in a pristine unpolluted atmosphere is considered to range in pH from roughly 4.9 to 5.8 depending on the alkaline ion content. If the pH of precipitation lies well below 5.0, then the absorption of strong acids (mainly H_2SO_4 and HNO_3) derived from air pollution is usually a factor. Regions with low-pH rainfall are those that are of greatest concern for adverse environmental effects attributable to human activities.

Low pH in precipitation, or dry deposition of acid-forming species, does not necessarily create environmental problems. The underlying aquatic and terrestrial ecosystems need to be considered "susceptible" to acidification. Vulnerability of surface waters depends on watershed soil and rock conditions, as well as watershed vegetation interactions. If the alkalinity or acid neutralizing capacity (ANC) of the watershed surface water is low, then the capability to neutralize acid deposition is low. Examples of acidified water conditions are relatively limited and focus on regions of "thin" soils, silicaceous or granitic watersheds, and lakes or streams that reside in granite bowls, and where waters have an ANC near zero (or negative). Such areas are found, for example, in historically glaciated regions of Scandinavia and northeastern North

America. Similarly, terrestrial ecosystems in regions of thin-depth, low-alkalinity soils (podzols) are believed to be most susceptible to acid deposition effects. These have sometimes been identified with stands of conifers in parts of Europe and North America, such as fir, spruce, and pine.

Of course, acid deposition is often loosely linked with other environmental insults, including the effects of airborne sulfur and nitrogen oxides on humans, material corrosion, and visibility impairment. Through the linkages of air chemistry, nitrogen oxides are connected with environmental stress from elevated atmospheric ozone (O_3) levels near the ground.

II. AIR CHEMISTRY

Because high acidity in deposition has been attributed mainly to SO_4^{2-} and NO_3^- anions, the chemistry of acid rain has focused largely on their atmospheric behavior. For both species, the chemistry is quite complex and is blended with the broader aspects of meteorological processes that disperse pollution, as well as create cloud and precipitation interaction.

Although natural sources of sulfur and nitrogen compounds are important to global-scale air chemistry, these sources are overwhelmed by anthropogenic sources in major industrialized regions of the world, including parts of Europe, the Commonwealth of Independent States (former Soviet Union), North America, and Asia (particularly China, Japan, and Korea). The phenomenon of acid rain is largely confined to regions of 1000–2000 km surrounding and downwind of the industrialized regions. Acid rain is identified with these large regions and largely results from the accumulation of pollution in air that travels over sources that overwhelms the ability of the atmosphere to disperse and rid itself of the pollution. [See GLOBAL ANTHROPOGENIC INFLUENCES.]

The 1000-km scale of the phenomenon relates to the *residence time* of material in the air. This time scale is a measure of the effective lengths of time a chemical constituent will survive in the atmosphere under the influence of chemical transformations, dispersion, and dry and wet deposition loss at the ground. Residence times are largely determined by chemical reactivity and solubility in water. Some highly reactive chemical species are lost to transformation in seconds, whereas others remain in the air for years. The residence time for sulfur and nitrogen oxides generally ranges between 3 and 5 days; nitrogen oxides are more reactive and are estimated to be somewhat shorter. The "regional scale" of phenomena like acid rain is derived from these time durations, assuming air motion at about 10 km/hr.

The principal atmospheric reaction of SO_2 is its oxidation to sulfur trioxide (SO_3) followed by rapid reaction with water to form H_2SO_4. The oxidation of SO_2 occurs through homogeneous gas-phase reactions with the hydroxyl radical (OH), which is produced photochemically from NO_x reactions with organic species in the presence of sunlight. SO_2 also is oxidized in condensed water by reactions with dissolved hydrogen peroxide (H_2O_2), and to a lesser extent with O_3 or O_2. The last reactions are catalyzed by the presence of manganous ion and ferric ion. H_2O_2 comes from gas-phase photochemical reactions linked with OH formation. Sulfate in the atmosphere is in the condensed phase; reactions of H_2SO_4 with ammonia (NH_3) from biogenic sources form complex mixtures of H_2SO_4, ammonium sulfate, and ammonium bisulfate, with a variety of other material including soil dust, metal oxides and salts, and carbonaceous species.

Nitrogen oxide chemistry in the atmosphere is more complicated than that of the sulfur oxides. NO_x undergoes a series of photochemically stimulated gas-phase reactions with organic species (hydrocarbons and oxygenates) that produce a series of gaseous species, including NO_3, N_2O_5, N_2O_4, HONO, $HONO_2$, and organic nitrate (e.g., peroxacetyl nitrate of PAN). In urban areas like Los Angeles, this chemistry causes the well-known *photochemical smog*. Evidence suggests that nitric acid vapor is readily dissolved in condensed water or moist aerosol particles, rather than produced by aqueous reactions. Nitric acid also reacts with NH_3 to form ammonium nitrate. Some of the other nitrogen oxide species are quite reactive in air and

dissolve in water to form nitrate ion. [*See* Nitrogen Cycle Interactions with Global Change Processes.]

Acids can be collected in cloud or precipitation elements (on hydrometeors) by vapor absorption, chemical oxidation, and water vapor nucleation, or by aerodynamic processes involving collision with aerosol particles. If acid material is scavenged inside clouds, the resulting removal is sometimes called *rainout*. If scavenging occurs in hydrometeors falling below clouds, the process is called *washout*.

Dry deposition of gaseous species depends on their reactivity with the underlying surface, as well as fluid dynamic and molecular diffusion processes transferring the gas to the surface. Deposition of acid on particles depends on the stickiness or wetness of the surface and the particle, as well as the fluid dynamic process of particle capture. Wet deposition takes place from rain and snowfall, as well as collection of fog (cloud) particles, rime (ice particles), and dew formation on surfaces.

In most locations, except mountaintops or ridges, wet deposition is dominated by rain and snowfall. A typical range of strong acid deposition levels is shown for several sites in North America in Fig. 1. These locations were selected in the Integrated Forest Study (IFS) to represent a wide range of experience with different prevailing deposition regions. Aside from the processes of cloud and precipitation formation and diffusional processes near the surface, the range of distribution of acid deposition depends on the winds that transport suspended material and the dilution or dispersion properties of the turbulent atmosphere.

In general, concentrations of pollutants in the atmosphere tend to decrease exponentially with distance from a source to low levels within 100 km. However, if many sources are aligned roughly along the direction of prevailing winds, cumulative concentrations can remain elevated for substantial distances. The so-called long-range transport of air pollutants relates to this cumulative effect, as well as the atmosphere's ability at times to transport large volumes of air for long distances with relatively little mixing of (or dilution by) clean air. A tracing of polluted air by various direct and indirect means suggests that a "zone of influence"

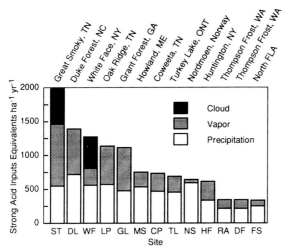

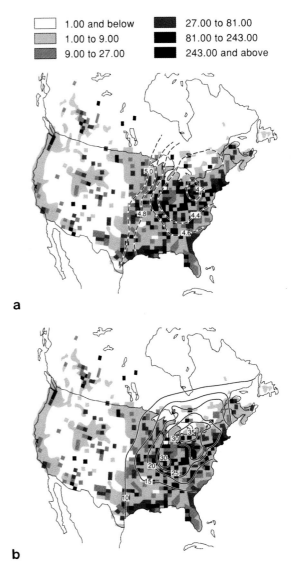

FIGURE 1 Atmosphere input fluxes of strong acidity to 13 IFS sites in equivalents per hectare per year (eq/ha/yr) based on multiyear annual averages at most sites and single-year fluxes at FS and MS. Vapor and aerosol fluxes are included in "Vapor"; "Cloud" was measured only at ST, WF, and LP. Sites are ordered according to total atmospheric deposition of strong acidity. [Reprinted from D. W. Johnson and S. E. Lindberg, eds. (1992). "Atmospheric Deposition and Forest Nutrient Cycling." Copyright 1992, Springer-Verlag, New York.]

FIGURE 2 Sulfur dioxide emissions in the United States and southern Canada in kg $SO^{2-}4$/ha/yr. (a) Superimposed on the emissions map is annual precipitation weighted pH and (b) sulfate deposition in kg/ha/yr. (Emissions from 1985 NAPAP inventory; pH and SO_4^{2-} deposition from NAPAP Interim Assessment, 1987.)

of large sulfur and nitrogen oxide sources can be inferred for hundreds of kilometers downstream. This long-range cumulative effect creates the potential for relatively high levels of acid deposition over the distances exceeding 1000 km, affecting pristine areas far from industrial-urbanized environments. Thus, the phenomenon of acid rain is generally not identified with localized air pollution problems in and around large sources or at urban areas but with larger scale regional effects.

Examples of widespread sulfate deposition observed in eastern North America are shown in the maps of Fig. 2. Superposition of SO_2 emission densities shows the coincidence between pH in precipitation water and regionally high emission densities. Similarly, there is a correspondence between high emission densities and SO_4^{2-} deposition. Emission–deposition patterns reflecting local effects and long-range transport of pollutants are also observed in Europe, Scandinavia, and Japan. Patterns are also similar for NO_x emissions and nitrate; NO_x emission distributions in eastern North America and elsewhere tend to coincide with SO_2 distributions. In Fig. 2, the international character of long-range pollutant transport is illustrated in the

distributions of $[H^+]$. Evidence has suggested that elevated acid deposition in southeastern Canada derives not only from local emissions, but also from transboundary transport northeastward from industrial areas of the United States. Similarly, Scandinavia is exposed to acid deposition from both local sources and long-distance sources in the United Kingdom and industrial Northern and Central Europe. Japan receives acid deposition exposure not only from its own industrial sources, but also from China and Korea westward across the Sea of Japan.

III. EFFECTS ON HUMAN SYSTEMS

Acid deposition is an important concern because of its potential to induce adverse environmental effects. Traditionally, effects to human systems are usually considered at highest priority because of the national and international laws and covenants protecting public health and welfare.

The direct effects of sulfur and nitrogen oxides on human health and welfare have been studied for many years. Knowledge of respiratory stress from high concentrations of SO_2, airborne particulate matter, and to a lesser degree NO_2 formed the basis of the U.S. National Ambient Air Quality Standards (NAAQS) in 1970. Welfare (economic or esthetic) effects associated with crop and forest damage and metal or stone corrosion and damage from high concentrations of SO_2 and NO_2, as well as visibility impairment from sulfate-containing haze, also were noted to support the NAAQS. These concerns resulted in similar standards being adopted in other countries, as well as international guidelines for pollution exposure. Since 1970, considerable effort has been made to reduce pollutant emissions to comply with NAAQS. The annual emissions of SO_2 have decreased in the United States by about 30% since 1970.

In the 1977 U.S. Clean Air Act Amendments, provision was made for the protection of remote, pristine areas from significant degradation in air quality and visibility from SO_2 pollution and airborne particles. This legislation has constrained industrial development in and around U.S. national parks, wilderness areas, and certain national forests.

The question of the effects of acidity on the human system has proven to be controversial. Studies of the effects of airborne sulfate in the 1970s suggested adverse respiratory response, but these were later disputed. More recently, the influence of strong acidity in aerosols has been considered. Experiments to date have not verified any respiratory effect of airborne acidity, except at concentrations far higher than found in the atmosphere. The occurrence of adverse health effects from either airborne SO_4^{2+} or acidity remains unresolved.

The adverse effects of high atmospheric concentrations of SO_2 and NO_2 on trees and agricultural crops have long been known. In fact, one of the early disputes in the 1930s on the effects of pollution exposure across national boundaries involved the fumigation of trees in the state of Washington from a nearby Canadian smelter operation. Plants sensitive to these gases show necrosis of leaves or needles at exposures in excess of $\sim$100 parts per million (ppm) SO_2 and NO_2 as well as sulfuric acid droplets for short periods. However, damage to agricultural crops has not been reported from acid or acid gases at the much lower concentrations normally found in rural, pristine areas. In most managed terrestrial ecosystems, the deposition of sulfur and nitrogen components acts as a fertilizer.

Possible indirect effects of acid deposition on drinking water supplies have been reported. These appear to derive from accelerated dissolution of heavy metals like lead from old pipes and cistern walls, but are not widespread as a hazard for most modern water supply systems. Of all the human system concerns, the impairment of visibility by various airborne sulfate appears to be the best documented. This esthetic effect is considered by some to merit marginal consideration. Nevertheless, the issue of visibility continues to be raised in conjunction with the environmental effects of acid deposition.

IV. ECOLOGICAL RESPONSE

Evidence of the ecological effects of acid deposition, though a strong motivating factor for research in the 1970s and 1980s, remains difficult to quantify conclusively, even for chemically vulnerable systems. Most of the early work concentrated on acidification of lakes and streams, resulting in possible decline or even loss of susceptible fish populations. However, to understand the behavior of aquatic systems, one has to know a great deal about acid deposition in terrestrial systems.

The processes by which acid is assimilated into natural ecosystems are illustrated in Fig. 3. When acid deposition contacts vegetation soil or bedrock, it results in chemical exchange processes and weathering as water runs off or percolates into the ground. Watershed runoff, as well as direct wetfall, maintains the lake water and the stream flow. Wetfall directly adds acidity to surface waters. However, runoff chemistry is affected by exchanges with vegetation, detritus, and soils. The surface waters continue to evolve chemically with integration with standing waters as well as bottom rock and sediments.

The chemistry of runoff is an integration of exchange processes in the tree canopy and the ground—a complex series of interactions that can be separated into the results of dry deposition and wet deposition. The interactions shown in Fig. 3 ultimately result in a runoff with a chemistry that is quite distinct from that of the precipitation water origins.

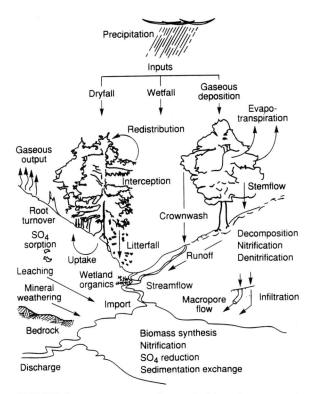

FIGURE 3 Important natural watershed/aquatic processes influencing the effects of acidic deposition on surface water acid–base chemistry. (From NAPAP SOS/T Report No. 10.)

The interactions begin at treetop level, where atmospheric gases and airborne particles, as well as hydrometeors, collect. As precipitation travels through the vegetation canopy to the ground, the acid solution reacts with the leaves and branches. Within the canopy, the throughfall may become more acidic from washoff of accumulated dry deposition or by leaching organic acid anions from the foliage, and from leaching the reaction products of dry-deposited SO_2. Throughfall, on the other hand, may become less acidic because of processes that may be thought of as exchanging H^+ for basic cations in the plant tissue. Of these processes, only the washoff of dry deposition represents a net change in the acidity of the soil-plant system. The other changes affect internal cycling only, although they may affect the vegetation by prematurely removing cations from active tissue, including a key nutrient element, magnesium.

On entering the soil, solutions undergo many reactions. When deposition dominated by sulfur enters a system previously unaffected by acid deposition, the SO_4^{2-} concentration of the soil solution begins to increase. Some soils have substantial capacity to adsorb SO_4^{2-} on the particle surfaces, and in such cases, this adsorption acts as a buffer, delaying the elevation of solution SO_4^{2-} concentrations. The SO_4^{2-} uptake and cycling capacities of the biotic component of the ecosystem also acts as buffers, slowing the increase in solution SO_4^{2-} concentration. Usually, the soil system will reach an equilibrium at which elevated SO_4^{2-} concentrations in solution become sufficiently high that the outgoing flux of SO_4^{2-} sulfur in the drainage water is approximately equal to the incoming sulfur in acid deposition. Because soils vary greatly in adsorption capacities, the time required before substantially elevated SO_4^{2-} concentrations are observed in the drainage waters may vary from a few weeks to many decades. [*See* SOIL ECOSYSTEMS.]

As SO_4^{2-} concentrations in soil solutions and water leaving the system increase, charge balance considerations dictate that these anions be accompanied by an equivalent amount of cations. In soils that are even moderately well supplied with bases (i.e., perhaps 15% or more of the negatively charged exchange sites are occupied by the cations

Ca^{2+}, Mg^{2+}, K^+, or Na^+), these base cations will comprise most of the increase in solution cation concentration. The remainder will be largely H^+ or aluminum species, particularly Al^{3+}, that comes into solution as a result of exchange reactions or the dissolution of soil minerals, although in some systems iron or manganese species may be significant. The increase in the rate of removal of basic cations will tend to acidify the system. Because the total supply of these cations in the soil is usually quite large relative to the annual input of H^+ in acid deposition, acidification of soils and waters by this cation export mechanism is likely to be a slow process involving decades or even centuries in deep soils or soils containing significant amounts of minerals that release basic cations upon weathering.

In low-base-status soils, a significant fraction of the increased cations in solution and in discharge waters may consist of H^+ and ionic aluminum species. This is generally undesirable because the Al^{3+} ion is toxic to many species of plant and animal life and because both aluminum and H^+ will reduce the alkalinity of the discharge water. In some cases, this reduced alkalinity and increased aluminum content may be sufficient to cause potentially substantial changes in the aquatic biota and loss of fisheries.

The changes in solution composition brought about by increased SO_4^{2-} concentrations resulting from acid deposition is mediated by a complex set of reactions with the soil system and is not simply a "wash through" of the acid impinging on the system. The effect is not only an increase in the cation concentration in solution but a change in proportions of the various cations as well through the soil cation-exchange capacity. Thus, upon reaching the ground, the acidic solution undergoes modification in contact with vegetation or detritus, and with soil particles and surface rock. Depending on the contact time of flow, the depth of penetration underground, and the nature of the soil–rock leachability of adsorption capacity, the flow will add acidity to streams and lakes. In other words, the surface waters are basically the repository for all the preceding interactions of precipitation with the terrestrial components.

A. Effects on Forests

Because there is substantial interaction between vegetation and acid deposition, it is reasonable to hypothesize that the forests or other terrestrial ecosystems potentially will be adversely affected by deposition. Concern regarding the potential impacts of acidic deposition on forests began to be raised about 15 years ago. Key factors motivating these concerns were the widespread reports of extensive forest damage in Europe and the frequent attribution of this damage to air pollution. In addition, as extensive research developed on the aquatic effects of acidic deposition, attention was drawn to the often-forested terrain surrounding lakes and ponds. With the significantly improved understanding of aquatic effects over the last decade, both public and scientific attention to forest effects has increased, which has led to more research on forest ecosystems and the possible impacts of air pollution in general and acidic deposition in particular.

Forests are highly complex ecosystems and trees have very long life cycles: Forests react to environmental changes and other stresses slowly, with subtle and often hard-to-identify changes. Assessing possible effects of acidic deposition has required the development of an improved knowledge of natural changes in forest systems, as well as knowledge of the impacts of a wide range of stress factors, including short-term weather cycles, long-term climate shifts, insects, pathogens, fire, wildlife, and human management practices, as well as air pollution.

A fundamental difference is that forest change is not the same as forest decline, and that not all forest decline is necessarily abnormal or unnatural. Maturation of trees and succession of different species are normal, natural processes that result in significant growth declines of particular species. Thus, identifying abnormal changes in forest systems, and associating such changes with particular causes, can be extremely difficult. [*See* FORESTS, COMPETITION AND SUCCESSION.]

The pathways through which deposition can act on plants are shown in Fig. 4. These begin either with foliar impact from uptake of gases or with

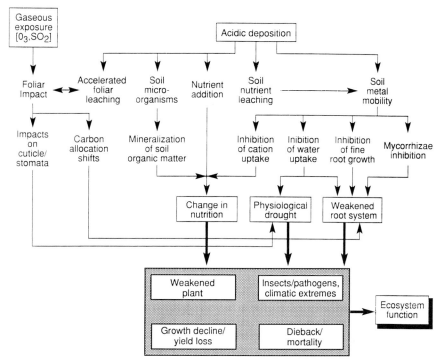

FIGURE 4 Principal pathways and mechanisms of plant growth response to atmospheric deposition and gaseous pollutants [From P. M. Irving (1991). "Acidic Deposition: State of Science and Technology." NAPAP, Washington, DC.]

leaching or ion displacement in the plant or the underlying soil. Potential impacts are observable on leaves and stomata, or in carbon allocation (allocation of carbohydrate to leaves, stems, and roots for growth); mineralization of soil organic matter is accelerated, accompanied by inhibition of nutrient and waste contact, as well as inhibition of root growth or mycorrhizal production. These in turn would produce changes in nutritional states, water status, and a weakening of the root system. This physiological impact hypothetically would result in weakened plants and increased susceptibility to pathogens and climate stress, potentially leading to decline and mortality.

During the 1970s and 1980s, forest declines have been observed in different locations in Europe, North America, and Japan. Some workers speculated that these declines were associated with air pollution and acid rain because they were found to coincide with the deteriorating forests. Perhaps most widely publicized was the forest deterioration

in Germany; this is where the first reports and warnings originated.

In Germany, an official forest damage survey has been conducted since 1982. The surveys in the early 1980s have included stands of Norway spruce (the most important timber tree in Germany), silver fir, and hardwoods such as beech. The overall extent of estimated damage involved up to 50% of the area surveyed. Within the damaged area, about 1% of the trees were seen to be dead or had greater than 60% foliage loss. The majority of affects trees experienced 11–25% needle or leaf loss. Because this loss involved older leaves and did not markedly affect growth, these changes are believed to be only marginally significant physiologically.

A significant feature of the apparent decline in Germany during the 1970s and 1980s is the relatively synchronous appearance of symptoms in different species and regions. This originally was a major reason for suspecting acidic deposition or other air pollutants as a major inciting factor. However, study of the decline in healthy appearance of the trees has indicated that no single factor can explain all the instances of the decline. The synchronizing factor may simply be weather, particularly drought or frost. A complicating factor, at least in the case of the Black Forest, may be related to the fact that both Norway spruce and beech are at the edge of the natural ranges of occurrences, and thus particularly susceptible to climate and other stress. Deforestation and land use changes throughout most of Europe and North America also affect watershed chemistry and forest health. Also significant are recent findings that even with the visible appearance of European forest decline, the *growth* of forests actually *increased* between 1970 and 1990. This was presumably associated with airborne nitrogen deposition.

Several hypotheses emerged from the German studies, depending on the specific location and species decline. Toxicity from aluminum mobilized from soils by acidic deposition is one; some workers believe heavy metal toxicity is involved. Stress on Norway spruce may involve ozone attack on the needle surface, followed by leaching of nutrients, such as Mg, by acid mist content. Magnesium deficiency may be from soil depletion as well. Needle reddening in other Norway spruce stands in southern Bavaria is a result of a fungal infection, which may have been exacerbated by weather extremes (frost shock), but this does not correlate with elevated levels of air pollution or acid deposition. A third type of disease of spruce appears to be restricted to the calcareous Alps of southern Bavaria. In this case, Mn^{2+} and K^+ deficiencies are involved, but acid deposition effects on soil nutrient levels appear unlikely.

The most widespread disease type in spruce is needle yellowing and senescence seen at higher elevations on more acidic soil substrates. This disease occurs in various regions, including the Bavarian Alps, the Black Forest, and the Harz and Hils mountains. Workers have determined that the symptoms reflect foliar nutrient deficiencies, mainly Mg^{2+}. The symptoms of the recent worsening decline appear consistent with Al^{3+} toxicity, Mg^{2+} deficiency, and ozone–acid mist foliar leaching of nutrients. The soil Mg^{2+} deficiency appears

to be the simplest explanation for the observations, but the others cannot be excluded. A variety of natural and anthropogenic factors have contributed to nutrient deficiencies, including the presence of poor parent rock and slow soil development, as well as inappropriate silviculture management practices.

Forest damage in other parts of Europe has been reported. Instances of severe decline in certain regions of Poland and (formerly) Czechoslovakia appear to be a result of exposure to very high SO_2 concentrations associated with nearby industrial emissions. A survey of forest damage in Great Britain has indicated a worsening forest condition, with damage levels similar to those found in Germany. In Scandinavia, recently initiated forest inventories record needle loss and reduced vitality in some locales. However, the geographic patterns of deterioration did not correlate with levels of acid deposition or other pollutants.

Damage and decline of cedars in Japan have been recognized since the 1960s. The known causes of cedar damage include winter drought, wind, salt spray, lightning strikes, insects, fungal disease, and rising groundwater. Circumstantial evidence from cedar conditions in low- and high-pollution areas also suggests that exposure to high ozone levels exacerbates tree dieback, but no concrete evidence of acid rain damage has been found.

The vast majority of forests in the United States and Canada do not suffer from widespread, visible health problems, nor do they exhibit any overall decline. Decline in specific species or areas that has been observed is typically a result of a combination of natural stresses and land use patterns. For example, historical forest use and changing land use patterns led to extensive cutting of eastern forests in the late 1800s and early 1900s. The normal processes of forest development and maturation on deforested lands could lead to some of the changes in forest condition that have been observed, independent of any additional stress factors.

Most forests in the eastern United States and Canada appear healthy and do not show decline symptoms that could be related to air pollution in general or acidic deposition in particular. Those cases in which acidic deposition has been hypothesized to play a significant role are discussed in the following. Forest decline in three species has been studied extensively in eastern North America; these are red spruce, sugar maple, and southern pine. Some cases of decline of certain forest species, particularly ponderosa pine, have been identified in the western United States. Research results strongly suggest that acidic deposition plays no role in western forest declines, although ozone may be a significant causal factor for deterioration in some cases.

Red spruce is showing various symptoms of decline over much of its range in the eastern United States. Although clear evidence indicates that red spruce declines have occurred before, dating back to the early 1800s, no systematic research records are available to allow historical comparisons. The extent of current red spruce decline has led to concerns that acidic deposition may be a major cause. Overall, a few thousand hectares of red spruce forest in the northeastern United States show damage from the multiple stresses that include acidic deposition. This area represents less than 0.1% of all eastern forest acreage. The most severely damaged stands are located at high elevations on thin soils, making them more vulnerable to stress of any kind. Taken as a whole, the body of research carried out to date demonstrates that no single factor accounts for the recent episodes of red spruce decline.

A wide range of possible explanations for observed red spruce decline have been suggested. Potential causal factors include weather cycles, disease, insects, natural stand development, and air pollution, including ozone and acidic deposition. Strong evidence suggests that weather patterns play a key role, with other stresses contributing to susceptibility to weather-related stress. Experimental evidence suggests that acidic deposition is not the primary cause of decline; rather, it contributes to processes that increase the susceptibility of trees to naturally occurring stresses such as climate and disease. (At high elevations, a combination of acid mist exposure and cold tolerance seems to be one important factor that adds to stress.) The relative importance of each stress factor may differ from site to site, making broad generalizations impossible.

Sugar maple decline has been observed in many stands in the northeastern United States and southeastern Canada. Decline has been reported periodically throughout this century and it is not clear that current rates of decline differ from historical averages. Research on sugar maple in Canada and the United States indicates that there is no single cause or even a predominant factor of decline. It is likely that different combinations of factors have been important in different locations. Natural stresses are certainly involved, but some effect from acidic deposition cannot be ruled out. For instance, acidic deposition could lead to reduced soil nutrient levels and consequently predispose trees to other stresses. However, little experimental evidence exists to suggest that acidic deposition plays a significant role in sugar maple decline. Rather, natural stresses, including drought and insects, are believed to be the predominant causes of decline. This conclusion applies to sugar maple stands in both the United States and Canada.

Widespread reductions in growth rates of natural stands of pine in the southeastern United States have been reported in recent years. Such reports have led to speculation that air pollution might be a cause of this decline, although research carried out to date has not demonstrated that air pollution is involved. Reported declines have been predominantly in natural stands; they have not been observed to a significant degree in intensively managed stands of similar species. The most likely causes of decline in growth rates in natural stands are a combination of natural stand development, historical land use practices, and other natural factors. In commercial stands, pines have been planted on lower-quality sites over the last 20 to 30 years. Thus, lower than average growth rates would be anticipated even in the absence of any other causal factors. Other evidence suggests that pests, drought, and competition from other species have played significant roles in the decline of southern pines in natural stands.

Controlled experiments have failed to identify any significant effects of acidic deposition on stands of southern pine. In contrast, controlled experiments have documented that ozone could affect the growth rates of some pine genotypes. Thus, if the observed decrease in growth rate in some natural stands of southern pine cannot be fully accounted for by natural and management factors, ozone, not acid rain, is the most likely contributor from air pollution.

B. Effects on Lakes and Streams

Perhaps the single most publicized environmental insult from acid rain is the acidification of surface waters. As illustrated in Fig. 3, the acid–base status and dynamics of lakes and streams, and their response to changes in composition of atmospheric deposition, are determined by several factors, including vegetation, soil properties, water movement, geologic characteristics, climate, in-lake processes, and deposition itself. The absolute and relative contribution of any single factor can vary substantially, both temporally and among lake watershed systems. For example, three neighboring lakes studied in the Adirondack Mountains showed markedly different response to acid deposition owing to variations in base status and watershed dynamics. [See LIMNOLOGY, INLAND AQUATIC ECO-SYSTEMS.]

Experience has indicated that the potential for acidification is most likely in areas where the watershed is poorly suited to neutralize acid deposition, and where the waters have low acid neutralizing capacity (ANC). Sensitive watershed characteristics are found in highly glaciated areas with thin soils and granitic (silicaceous) bedrock and uncompensated sediments very resistant to weathering. These include regions in the Laurentian Shield and alpine areas of North America, as well as much of Scandinavia. Also, one would expect acidification of runoff in areas of high rainfall, where soil cations have largely been leached away over centuries, or soils whose surface layers are naturally acidified from biological interaction with certain vegetation (e.g., conifers). ANC in clear natural waters is defined in terms of the aqueous bicarbonate and carbonate ion concentration:

$$[ANC] = [HCO_3^-] + 2[CO_3^{2-}] + [OH^-] + [H^+].$$

This equation is a way of stating the buffering ability of less dissolved carbonate in controlling the pH of natural waters. If [ANC] is very small (<50 μeq/liter) or zero, the natural waters are susceptible to acidification, with accompanying strong shifts in pH with small additions of acid. This aspect of change in acid status was of particular concern to limnologists, who feared that, with acid rain exposure, low-ANC lakes might acidify suddenly, without warning. Such extreme behavior has not been observed to date even in very high acid deposition areas.

The early estimates of the extent and severity of lake acidification in Scandinavia and North America relied heavily on limited historical observations based on chemical measurements of highly variable quality. The details of alleged widespread acidification of natural waters in Sweden and Norway, involving thousands of lakes, have not been verified because their survey and historical data have not been made available. Similar assertions of acidification of lakes in Ontario and Quebec, and in the northeastern United States, particularly the Adirondack Mountains, are clouded with ambiguity relating to inadequate historical records of water quality and incomplete accounting for naturally acidic waters,[1] as well as confounding causes of acidification related to land use and forest and management practices.

A thorough, one-time National Surface Water Survey (NSWS) of potential acid susceptibility was conducted during the course of NAPAP. This project focused mainly on low-ANC waters in the northeastern United States, the Appalachian Mountains region through the Smoky Mountains, and north and central Florida. Samples also were taken in the northern Great Lakes region, the alpine areas of the Sierra Nevada Range, the Rocky Mountains, and the Cascade Mountains.

Among lakes surveyed by the NSWS, 4% of some 28,000 lakes were acidic (ANC $\leq$ 0) and just over half had ANC $\leq$ 200 μeq/liter. Five percent had closed-system pH $\leq$ 5.5. Most acidic lakes

---

[1] Certain waters are naturally acidic from the presence of dissolved organic acids (brown waters) or negative alkalinity.

were found in the Northeast, Florida, and the Upper Midwest; < 1% of the lakes in the West, Minnesota, and the Southern Blue Ridge subregion were acidic. Total aqueous monomeric Al increased with decreasing pH; concentrations exceeded 50 μg/liter in 3.0% of the NSWS lakes.

Of the total stream length in the NSWS of over 20,000 km, 3% was acidic (ANC $\leq$ 0) and one-half had ANC $\leq$ 200 μeq/liter. Eight percent had closed-system pH $\leq$ 5.5. A greater percentage of streams was acidic at the upstream ends (6.1%) than at the downstream ends (2.3%). Most of the acidic stream length was in the mid-Appalachian and mid-Atlantic coastal plain regions. Inorganic monomeric Al concentrations increased with decreasing pH; 6–13% had $Al^{3+} > 50\mu$g/liter.

Atmospheric deposition is the dominant source of SO_4^{2-} in most NSWS surface waters. Among acidic NSWS lakes, 75% are inorganic dominated, with SO_4^{2-} as the dominant anion, 22% are organic dominated, and 3% are watershed-sulfur dominated. Among acidic, inorganic-dominated NSWS lakes and streams, acidic deposition is believed to be the main source of current acidity.

Among all acidic, inorganic-dominated NSWS lakes and streams, 95% of the lakes and 84% of the acidic upstream ends were found in six areas: southwest Adirondacks, New England, forested mid-Atlantic highlands, Atlantic coastal plain, northern Florida highlands, and low-silica lakes in the eastern Upper Midwest.

The NSWS did not fully survey acidic waters in the United States. Accordingly, the NSWS investigators believe that the survey tended to *underestimate* the low-pH surface waters in the acid-sensitive areas of the United States. Because the survey selectively focused on these areas, the NSWS findings tend to *overestimate* acid lakes and streams for the United States as a whole. Although surveys like the NSWS are invaluable in characterizing the chemistry of surface waters, in themselves they do not provide evidence of historical lake acid changes with the onset of elevated levels of acid deposition. Circumstantial evidence for lake acidification comes from correlations with high sulfate levels and high deposition rates of acid sulfate and nitrate. However, any history of change must be estimated from mathematical models of watershed chemistry.

Comparison of conditions in Scandinavia and North America suggest that low-ANC waters are particularly susceptible to acidification with acid sulfate deposition levels above about 15 kg/ha/yr.[2] For example, much of eastern North America falls within this category along the Appalachians and eastward from the Carolinas and Tennessee into southern Ontario and western Quebec. This threshold for potential damage from sulfate (acid) deposition has been used in Europe to set "critical loading" guidelines for sulfur oxide emissions management. To date, thresholds have not been adopted as a national standard or guideline in North America.

The best evidence for surface water acidification would come from historical records of lake water chemistry. Unfortunately, very few of these records are available in affected areas; those that are have substantial ambiguities relating to historical changes in acidity or alkalinity measurement methods.

Perhaps the least ambiguous direct historical evidence comes from analysis of lake bottom sediments. Tiny invertebrate species (diatoms, a type of plankton) are differentially sensitive to water pH, and their silaceous skeletons permit identification after death. Water pH can be inferred from the time variation in diatom population in undisturbed, layered sediments. Dating using sediment strata has been based on ^{210}Pb stratigraphics, supported by pollen and charcoal stratigraphics. Of twelve Adirondack lakes analyzed, four showed no decrease in pH in the last 150 years. The remaining eight lakes that indicated a decline in pH are currently less than 5.5. Of these, three had an inferred pre-1850 pH greater than 5.5 and five less than 5.5. Three of the lakes appear to have been acidified since 1850. These results suggest that lakes in the Adirondacks that are vulnerable to changes in pH at current acidic deposition levels have already

---

[2] Note that this "threshold" for lakes is not stated for acidity per se. This deposition level often is used more generally as a threshold for general ecological effects in acid-sensitive areas.

changed, because there is no indication of acidification after 1850 in any of the study lakes used with pH ≥ 4.8. Another study examined sediments in thirteen New England lakes. Six of these lakes with inferred pH less than 5.5 have been the same since the eighteeth century. Four of the remaining seven showed a small depression of 0.1–0.3 pH units, and the last three showed decreases of 0.5–0.7 pH units.

Paleolimnological studies of three low-pH lakes in Wisconsin show no recent acidification. Three acid lakes sampled in Norway show recent pH decreases and a fourth indicated no recent change. Three of four German lakes with present pH values less than 5 appear to have been acidified in the last 40 years, and the fourth seems to have been acid for a very long time. Sediment stratigraphy of six lakes in southwestern Scotland suggests highly variable levels and timing of acidification among lakes. The earliest date of one lake decline was 1840, whereas 1930 was the most recent date for most of the acidification in this group.

Thus, the results from paleolimnological studies indicate considerable intra- and interregional variability in pH decline, even in regions hypothesized to be sensitive to acid deposition. Even in high-deposition regions of the Adirondacks and near Sudbury, Ontario (Canada), some lakes have become less acidic since 1850. This is consistent with expectations based on current understanding of the biogeochemical, hydrologic, climatic, and deposition factors that influence surface water acid–base status.

I. Pulse Phenomenon

In acid-sensitive areas, lakes and streams commonly respond to the cumulative deposition of acidifying chemicals, including SO_4^{2-}, organic acids, and NO_3^-. In addition, a highly transient seasonal response has been observed. If acid accumulates in snowpack on a watershed, the spring meltwater can be acidic and cause a temporary pH depression in lakes and streams. This phenomenon has been observed in many alpine or equivalent climate conditions in Europe and North America. In the latter, nitrate appears to have a particularly important role. If the acid pulse creates a strong

pH depression (accompanied by decreased Ca^{2+} concentration and increased Al^{3+} concentration), fish kills have been observed, although these incidents appear to be quite rare.

C. Biotic Effects of Aquatic Acidification

Concerns for the adverse effects of anthropogenic acid deposition were stimulated in the early 1970s when Norwegian scientists attributed declines in salmon and trout populations in Norway to changes in surface water acidity. Reports claiming similar anecdotal effects in Sweden, Canada, and the United States followed, which led to scientific literature and popular press articles forecasting catastrophic consequences for aquatic ecosystems if national sulfur emissions were not reduced. These predictions have not materialized, but they did sensitize the public to the international and interregional consequences of unconstrained pollution growth.

Chemical changes that accompany lake and stream acidification can have deleterious effects on aquatic biota. This conclusion is supported by results of laboratory studies, field bioassays, and field observations. Organisms studied have ranged from bacteria and other decomposers through amphibia and other vertebrates that may be only partially dependent on water quality. However, acid lakes are definitely not barren of organisms; standing crop biomass of primary producer species and zooplankton may change little until lake pH approaches 4.

The primary impact of acidification appears to be on species composition and diversity. Unfortunately, few taxonomic groups have been studied sufficiently to predict species changes with the onset of acidification. Fish populations have been studied that are indigenous to acid-sensitive lakes. Fish are affected individually and at the population and community levels; some species are more sensitive than others. In lakes and streams experiencing acidification, fish appear to respond primarily to Al, pH, and Ca. Populations in low-Ca waters (an indicator of sensitivity to acid deposition) generally are less productive than those in hard waters. Ca has a protective effect for fish exposed to acidic conditions typified by low pH and high Al concentrations. Toxicities of fish to Al and pH depend on life stage and concentration in their environment. For example, the primary adverse effect on eggs seems to be related to pH, but toxicity to fry is primarily due to ionic Al.

The influence of acidification on the toxicity of metals other than Al is not well established. There is some evidence that mercury, for example is mobilized from sediments in acid waters. Adding solubilized mercury to natural waters can enhance its rate of accumulation in fish. Although accumulation at low levels does not seem to affect fish directly (i.e., toxicologically), it is believed to be a potential food contamination for humans who eat substantial quantities of freshwater fish. Much of the evidence suggests that recruitment failure from early life stage mortality is responsible for fish losses that have occurred with acidification. However, toxicity in lake waters depends on exposure as well as species sensitivity.

Fish populations that have been reduced in acidic lakes can recover and reproductive populations can be reestablished when they have been lost, if acid conditions are treated. Mitigation of acid waters is often accomplished through liming of the waters and the watershed. This treatment has been found to be quite effective in natural waters in both North America and Scandinavia.

Actual damage to fish populations from acidification is difficult to quantify. Estimates depend on judgments from evidence that (a) populations of species were once naturally reproducing in a water body, (b) acidity in the water body has increased, and (c) decline or absence of the population or community cannot be explained by other causal factors. Perhaps the best analysis of fish loss based on these three criteria was done in the 1980s for the Adirondack region. These studies suggested that 200–400 lakes may have lost fish populations. By area, this would amount to about 1.5–3.0% of the Adirondack lakes.

Estimates elsewhere in North America and Europe are much more subjective than the Adirondack work. Indirect evidence of the levels of acidic deposition and the acidity status of eastern Ontario waters suggest that some damage to fish population may have occurred. Loss or decline of Atlantic salmon populations in Nova Scotia is well documented through the early 1980s. However, their relation to acidification of lakes and streams in unresolved. Losses in fish populations in some regions of southern Norway and southern Sweden may have been substantially higher than in the Adirondacks. Because the three preceding criteria have not been reported, the level of damage or the degree of inflation of fish damage remains problematic in these countries.

Predictions for future losses of fish population have been attempted using mathematical models describing changes in water quality and fish population response. These models, of course, depend on knowledge of the complex biogeochemical interactions of acid deposition and watershed response and species-specific ecosystem response. For example, recent incorporation of organic acid behavior in mathematical models shows improvement in model performance. Because this knowledge is generally limited and incomplete, scientists' ability to estimate future ecological changes remains limited.

V. SUMMARY

Anthropogenically produced acid rain, or more precisely acid deposition, occurs from air pollution in the form of sulfur and nitrogen oxides. These species can be transformed into acids in the atmosphere or can form acids in direct contact with surfaces. Acid deposition occurs by dry processes of gas absorption at the ground or fallout of acidic aerosol particles or by wet processes involving precipitation from clouds, which have scavenged acids and other materials from the air.

The acid–base chemistry of dry and wet deposition is complicated and varies by region, depending on a balance between acid ions, mainly SO_4^{2-} and NO_3^-, and base cations including NH_4^+ and soil-derived material, Ca^{2+}, K^+, Mg^{2+}, and Na^+. Precipitation is naturally acidic with respect to pure water (pH $\approx$ 5.5) because of an aqueous equilibrium established with atmospheric CO_2, and becomes more acidic with incorporation of natural sulfur and nitrogen acid, as well as organic acids. Scavenging of strong acid from pollution in the air produces (anthropogenic) "acid rain" if pH levels reach approximately 4.8 or below.

Acid rain is observed in regions close to intensified industrial activity when pollution is prevalent. These regions can extend at greatly reduced concentrations up to 1000 km or more downwind of sources; they include eastern North America, Europe, and the Japan–China–Korea triangle.

Adverse environmental effects of acid deposition have been identified with human systems, as well as natural terrestrial and aquatic ecosystems. Acid deposition will tend to accelerate locally the corrosion of materials, and may locally influence agricultural crops if airborne concentrations are sufficiently high—well above current or anticipated levels. Humans are affected only indirectly by atmospheric acidity. Physiological effects of SO_2, NO_2, and aerosol particles on respiratory capacity at high concentrations in the air have been documented as part of developing U.S. and other National Ambient Air Quality Standards.

Acid deposition effects on natural ecosystems are commonly ascribed to damage to forests and to acidification of lakes and streams. These effects are expected in vulnerable regions where the deposition of acid is indexed to sulfate at an annual rate in excess of about 15 kg/ha/yr. Susceptibility is most likely in regions where the acid neutralizing capacity of soils and alkalinity of lakes is low. For susceptible lakes and natural waters, the ANC is ≤50 μeq/liter. Regions of this type are found in glaciated areas of North America, Scandinavia, and parts of Scotland where soils are thin and poorly weathered; the underlying bedrock is siliceous or granitic. A small fraction of lakes in these regions frequently have low alkalinity, and a subfraction appear to have been acidified in recent times.

Interactions of acid deposition with forest ecosystems are complex; toxicity appears to be related at least partly to Mg^{2+} nutrient leaching from leaves and soils and to mobilization of Al^{3+}. Most forests in North America are healthy, although long-term changes in soil chemistry may be of concern in a few areas.

Aquatic biota are affected by acidification of waters. Certain sensitive fish populations have been of principal concern. Effects on fish depend on species sensitivity and maturity; for eggs, direct exposure to acidity may be important. For fry and mature fish, aluminum toxicity and calcium imbalance are likely to be important. Acidification and consequent effects on biota can be mitigated effectively by liming of the affected waters.

Glossary

Acid emissions Air pollution emissions that are strong acids or produce acids from atmospheric reactions. The most common gases are sulfur dioxide (SO_2), nitrogen oxides (NO and NO_2), and hydrogen chloride (HCl).

Acid neutralizing capacity Capability of natural waters to neutralize acid deposition, usually measured in terms of alkalinity for a carbonate system or natural waters.

Acid pulse Pulse of acidic water released from melting snow and ice with spring runoff.

Acid threshold Level of deposition of acid above which sustained adverse effects on biota may occur; sometimes called *critical loading*.

Acidity Quality or state of being excessively or abnormally acid, as determined by aqueous hydrogen ion [H^+] or hydronium ion [H_3O^+] concentration. Hydrogen ion concentration is measured in terms of pH $= -\log_{10}$ [H^+].

Air quality models Mathematical constructs describing the transport, dilution, and chemical alteration of pollutant emissions as a function of spatial distribution and time.

Alkalinity In carbonate-bearing water, the equivalent sum of bases that are titratable with strong acid, such as H_2SO_4, HNO_3, and HCl.

Dry deposition Atmospheric gases or aerosol particles deposited on surfaces or the ground by absorption, impaction, or diffusion processes.

Hydrometeor Atmospheric precipitation element, including raindrops, hailstones, snowflakes, fog droplets, or graupel.

Long-range transport of air pollutants Movement of air pollution with persistent winds over distances in excess of 200 km from sources.

Major ions Principal anions in rainwater, SO_4^{2-}, NO_3^-, Cl^-, CO_3^{2-}, HCO_3^-, and cations, H^+, NH_4^+, Ca^{2+}, K^+, and Na^+.

Target loading Level of deposition of acidic or acid-producing species that is above a threshold for potential adverse effects on biota; proposed use for regulatory guidelines takes into account economic and political factors in relation to emission control strategy.

Throughfall Precipitation water reaching the ground after traveling through a vegetation canopy overhead.

Wet deposition Chemicals deposited on surfaces or the ground from rainfall, snowfall, or, sometimes, fog, dew, or rime.

Bibliography

Goldstein, R. A., *et al.* (1984). Integrated lake watershed acidification study (ILWAS): A mechanistic ecosystem analysis. *Philos. Trans. Roy. Soc.* (London) *Sci. B* **305,** 409–425.

Huckabee, J. W., *et al.* (1989). An assessment of the ecological effects of acid deposition. *Arch. Environ. Contam. Toxicol.* **18,** 3–27.

Irving, P. M., ed. (1991). "Acidic Deposition: State of Science and Technology." Washington, D.C.: National Acid Precipitation Assessment Program.

Johnson, D. W., and Lindberg, S. E., eds. (1992). "Atmospheric Deposition and Forest Nutrient Cycling." New York: Sringer-Verlag.

Seinfeld, J. (1986). "The Atmospheric Chemistry and Physics of Air Pollution." New York: Wiley (Interscience).

Agriculture and Grazing on Arid Lands

G. E. Wickens
Royal Botanic Gardens, United Kingdom

I. Introduction
II. Environmental Constraints
III. Management

I. INTRODUCTION

The objective of any agricultural or grazing system worldwide is for maximum sustainable productivity without environmental damage. The semiarid ecosystems present a particularly fragile environment where the main constraint to their utilization is water availability, with temperature a secondary constraint. Semiarid ecosystems are represented in both the subtropical and temperate continental biomes, generally flanking the major deserts of the world.

A. Historical Background

The rural population can be roughly classified as agricultural, in which the sole occupation is the cultivation of the land, or pastoral, in which tending livestock is the source of livelihood. The distinction is not always clear-cut, and there are various combinations of the two.

Depending on the stage of economic development, it is possible to recognize two broad groups of management systems: traditional and modern (Table I). The differences within and between the two systems are not, however, clear-cut; indeed, they are all subject to various combinations and permutations. The traditional system is largely dependent on land and labor, while the modern system has a large capital requirement and generally a lesser requirement of land and/or labor. The suggested sequence of evolution of management systems from hunter–gatherer to sedentary lifestyles must not be taken too literally; stages can be omitted or even reversed according to circumstances. Available water for the irrigation of minor crops, as either diet supplements or cash crops, can play an important role in the subsistence economy of traditional agriculture. [*See* AGROECOLOGY.]

B. Traditional Management Systems

The simplest and earliest form of utilizing semiarid lands is that of the hunter–gatherer, relying entirely on the wild flora and fauna for sustenance. The Australian aborigine typifies such a system, albeit with the use of fire as a management tool for creating fresh plant growth to entice game and thereby make hunting easier. [*See* TRADITIONAL CONSERVATION PRACTICES.]

1. Nomadism

The domestication of livestock and crops circa 7000 BC led to replacement of the hunter–gatherer

TABLE I
Agricultural and Grazing Management Systems[a]

Type	System	Management	Main production factors	Fertilizer/nutrient source
Agricultural Systems				
Traditional	Agropastoral	Transhumant/sedentary	Land/labor/water	Livestock/rotation
	Silvopastoral	Transhumant/sedentary	Land/labor/water	Livestock/tree crops
	Subsistence	Sedentary	Labor/land/water	Livestock/rotation
	Rain-fed cash cropping	—	—	Livestock/fertilizers/rotation
	Irrigated cash	Sedentary	Water/labor/land	Livestock/fertilizers/rotation
Modern	Agroforestry	Sedentary	Land/labor/capital/water	Green manure/fertilizer
	Plantation	Sedentary	Capital/water/labor/land	Fertilizer
	Cash cropping	Sedentary	Land/capital/labor/water	Fertilizer/rotation
Grazing Systems				
Traditional	Pastoral	Nomadic/semisedentary	Land	Range/browse
	Agropastoral	Transhumant/sedentary	Land/labor	Range/browse/crop by-products
	Agricultural	Sedentary	Labor/land	Crop by-products/forage
Modern	Ranching	Sedentary	Capital/land/labor	Range/browse/forage
	Game ranching	Sedentary	Land/capital/labor	Range/browse/forage
	Feedlot	Sedentary	Capital/labor/land	Feed/forage
	Dairy farming	Sedentary	Capital/labor/water/land	Feed/pasture/forage

[a] "Water" refers to requirements for irrigation [After Food and Agriculture Organization (1985) and Wilson (1991).]; the availability of potable water for humans and livestock is implied throughout.

by a nomadic or semisedentary management system. The nomadic herdsmen wandered freely over the range in search of good grazing and, where conditions permitted, sometimes stayed long enough to plant and harvest a crop. The horse-breeding Kirghiz of Siberia and the camel-owning Tuaregs of the Saharo–Sahel are examples of such a nomadic system. Although it is the only effective way to manage livestock in the arid and drier parts of the semiarid regions, the system rarely respects political boundaries, and consequently governments find it difficult to administer and approve.

2. Transhumance

The restriction in the freedom to wander, imposed by either natural or political restraints, gave way to an agropastoral system of transhumance in which part of the family unit, usually consisting of the old, infirm, and very young, remained behind in a permanent or semipermanent base camp to plant crops while the more active members of the family wandered along a vaguely predetermined circular route in search of grazing. The cattle-owning tribes of the Sahel are a typical example of such a semisedentary system of livestock manage-

ment. This is a very effective way of managing a semiarid range, provided that livestock numbers are kept within the carrying capacity of the range. Again, the system seldom meets the approval of governments and their requirements for taxation, administration, and education.

3. Silvopastoral Systems

A somewhat similar condition occurs when, instead of planting annual crops, reliance is placed on the seasonal produce from trees as a source of revenue for the purchase of food. Examples of such silvopastoral systems are the harvesting of cork from the cork oak forests of the Mediterranean region and the grazing of the ground cover by pigs, sheep, and goats; in the Sahel camels and cattle may browse the open *Acacia* savanna woodland while their nomadic or semisedentary herders supplement their income by harvesting gum arabic from *Acacia senegal*.

A more sophisticated development is one in which *A. senegal* is encouraged to dominate the bush fallow of a cropping system. This represents the transition to a more sedentary agropastoral system.

4. Sedentary Agriculture

A further stage in development is that of a purely sedentary agricultural system involving the establishment of permanent settlements for the entire family unit to cultivate the land and usually raise some livestock in the vicinity of the homestead. Any increase in livestock numbers beyond the carrying capacity of the local range and available feed and/or crop residues could be countered by either part of the family unit or a nominee adopting a transhumance role for the pastoral care of the livestock. Such agropastoral systems vary considerably in the degree to which either cropping or livestock becomes the major industry.

The indigenous Andean farmers of Chile, for example, cultivate the land in the immediate vicinity of their villages, growing under flood irrigation such subsistence crops as maize, potatoes, tomatoes, onions, and lima beans, plus lucerne for the livestock. The cultivated land is usually terraced, many of the terraces being of Incan construction, the water being brought from considerable distances along stone-lined channels. Large herds of llamas are kept for their wool, the wool being increasingly used to make articles for the tourist industry. From spring to autumn, various members of the village take turns at shepherding the llamas at grazing sites far from the village, enclosing them at night in small, sheltered, stone-walled paddocks. During the winter the llamas are brought down to stone-walled paddocks around the village, their diet being supplemented with lucerne hay and maize stover.

5. Traditional Irrigation Systems

Traditional irrigation systems can be relatively simple, such as the cropping of flood plains following seasonal inundation and simple water-spreading techniques, as practiced by the Papago Indians of Arizona, in which brushwood dams trap silt and retain and spread the occasional surface flow sufficiently long for the water to seep into the soil; tepary beans are then planted in the moistened soil. Growing rapidly and developing a very deep taproot, the crop is harvested 2 months later.

The Nabateans in the third century BC controlled the flow of water in the valleys of the Negev by means of a series of stone levees which slowed the flow of water sufficiently for the deposition of silt, thereby creating terraces where wheat could be sown after the floods. The Bedouin now cultivate these terraces and are able to obtain a yield of barley even in dry years.

More sophisticated irrigation of riverside lands involves the lifting of water from nearby lakes, rivers, or wells; the channeling of water from a more elevated source by means of canals or underground infiltration tunnels (*quanat* in Iran, *karez* in Pakistan, *foggara* in North Africa, *mina* or *laoumi* in Spain, or *karezes* in Mexico and Chile); or various combinations of these. Thus, in Egypt and northern Sudan in the irrigated lands bordering the Nile, the water may be lifted by means of water wheels, counterbalanced buckets (*shadufs*), or pumps and led by means of channels to the fields.

Attributed to the Persians circa 1000 BC, the *quanat* technique is widely used in Iran, Afghanistan, Pakistan, the Arabian Peninsula, Israel, Turkey, central Asia, Spain, Sicily, China (Sinkiang Province), and Central and South America. Often depending on community cooperation for their upkeep, the cost of their maintenance, or even construction of new *quanats,* this system is becoming increasingly more difficult and expensive. Many of these ancient irrigation systems are being abandoned in favor of tube wells.

Flood or furrow irrigation systems may be used, depending on the crop and the degree of sophistication of the farmer. The terrain is generally either a more or less level alluvial plain or terraced hill slopes and, depending on the seasonal availability of the water, the land may be cropped during the rainy or dry season, or both. For example, the mountain Fur of Jebel Marra, Sudan, grow irrigated wheat on terraced slopes during the dry season and cultivate rain-fed crops on the flatter lands during the summer. The water may be brought from considerable distances by means of stone-lined channels, with gullies crossed by hollowed-out palm trunks cut longitudinally. The lowland Fur, at the end of the rainy season, cultivate the lower terraces of the Wadi Aribo following the falling water levels, taking advantage of the moisture retained by the alluvial soils. Late in the dry season minor crops, such as tomatoes and peppers, may be grown in the dry river bed, irrigated from temporary wells there.

C. Modern Management Systems

The development of a sedentary agricultural or grazing system implies some form of long-term land tenure in order to safeguard the investment necessary for development. The traditional agricultural system of rotating cultivated land with a period under fallow until such time as the soil has recovered its fertility has been replaced by agroforestry, in which the fallow is replaced by the deliberate planting of fertility-restoring trees and shrubs. The planted species are often nitrogen fixing and preferably multipurpose species. Their planting ensures a more rapid recovery and longer maintenance of soil fertility than under traditional fallow, depending on natural regrowth and colonization.

1. Alley Cropping

A further development, especially in the more humid regions, is that of alley cropping, with the agroforestry species planted in rows (which also serve as windbreaks), leaving sufficient width for mechanized cultivation between the rows. Alternatively, the rows can consist of browse species with rough grazing or pasture between the rows.

2. Plantations

Plantations involve permanent cropping with perennial species. Such a system requires a considerable investment, since it could be 5 years or more before there is any return. Labor requirements for pruning, harvesting, packaging, and—for dates—pollinating could be high, but with seasonal fluctuations. Fruit crops such as guava, citrus, olives, almonds, and prickly pears, oil-seed crops such as jojoba, or even cut flowers can provide a reasonable cash return when grown in semiarid areas, although the use of supplementary irrigation, fertilizers, fungicides, and pesticides might be necessary to ensure high yields. Such soil and water conservation techniques as bunding, terracing, and contour planting may be required, although the system does

offer the least risk to erosion. However, care must be taken to ensure that stocking densities are not so high that transpiration losses exceed the water table recharge.

3. Cropping

A modern cropping system differs from traditional cropping in that the emphasis is placed on cash crops instead of subsistence crops. The crops may include, as in East Africa, species for the European flower market as well as cereals and pulses for export. The system usually involves a fairly high level of mechanization to reduce labor requirements, the growing of recognized cultivars, and the use of agrochemicals to ensure high yields. Appropriate soil and water conservation measures, including crop rotations, are necessary to ensure sustainable arable land use.

4. Irrigation

The two major requisites for irrigation are a first-class soil and an adequate supply of good-quality water of low salinity (i.e., an electrical conductivity of <7.8 dS/m), the quality of which is preferably higher than that considered suitable for human consumption. Adequate drainage is essential to prevent any build-up in soil salinity.

Increasing soil salinity is the major reason that irrigation schemes fail. Increasing salinity due to inadequate drainage is already a problem in the world's largest irrigation scheme in the Indus valley. The disposal of drainage waste can also create environmental problems, as witnessed by the build-up in salinity in lakes receiving drainage from the Nile delta and the serious increase in selenium toxicity of the Salton Sea, which receives drainage waters from the Colorado irrigation schemes.

5. Ranching

Ranching is the natural development from a semi-sedentary pastoral system to a sedentary system involving the tenure of extensive areas of land plus a heavy investment in boundary and internal paddock fences, watering points, shelter belts, etc. Ranching is widely practiced in the semiarid and savanna grassland regions of the world, especially the western and southwestern United States, Mexico, Argentina, southern Africa, and Australia. The system is widely used for the rearing of cattle for beef and sheep for their meat and/or wool; it can also be adapted for other forms of livestock, such as goats, ostrich, and camelids. The number of livestock kept is obviously limited by the carrying capacity of range, which may be as low as 8–12 ha per head of cattle. Provision must be made for winter or dry season fodder, such as hay and browse, plus access to valley bottom grazing and possibly irrigated pasture. [*See* RANGE ECOLOGY AND GRAZING.]

6. Game Ranching and Farming

A fairly recent development is the encouragement of wild animals, usually ungulates but also buffalo and vicuña, that graze either with or without cattle in a commercial ranching enterprise, generally in areas with more than 600 mm of annual rainfall. A further development is the domestication or semidomestication of the ungulates in fenced paddocks, with or without additional feeding. The eland has shown considerable promise for game farming, as it is amenable to handling, grows rapidly, produces high milk yields, and has a low water requirement, while the oryx gains weight on a diet that would not sustain cattle and requires only one-quarter of the amount of water. Game animals have a better physiological adaptation to high temperatures and limited water supply as well as achieving better weight gains and reproductivity than cattle, providing meat with a higher cutting-out percentage and a lower fat content. Although such animals are carriers of a number of diseases that are lethal to cattle, including trypanosomiasis, they are not susceptible. Game animals are therefore a source of meat in areas where the tsetse fly makes cattle ranching impractical.

7. Feedlots

First developed in the United States, the feedlot system of management is one in which a high density of livestock is accommodated in penned lots, with all their food brought to them. The system is dependent on a readily available source of feed and/or fodder. It is a suitable means of finishing livestock for slaughter that have been reared under more extensive management systems either locally or from afar. For example, in the United States beef cattle may be transported from the ranching areas to areas where cotton, groundnut, and maize by-products are readily available. In the desert regions of Arabia and North Africa, the system has been adapted to the feeding of dairy cattle.

8. Dairy Farming

The principal activity of a dairy farm is the production of milk from milch cows, sheep, goats, buffalo, or, in the drier regions, camels. Dairy farming requires either a nearby township requiring fresh milk and/or a milk factory for the manufacture of cheese, butter, etc. For fresh milk a high standard of stockmanship is necessary to ensure a constant supply of milk throughout the year, with any surplus being sent to the milk factory. If the milk is intended solely for manufacturing purposes, there is less emphasis on maintaining a constant supply throughout the year and the factory may even be closed when supplies of milk are low. Indeed, the lower monetary return to the farmer encourages seasonal production when forage is most readily available. For constant production throughout the year, some form of irrigated forage production, such as lucerne, grass pastures, or water meadows, will be necessary.

II. ENVIRONMENTAL CONSTRAINTS

There have been a number of attempts, by De Martonne, Thornthwaite, Miegs, UNESCO, and others, to define arid and semiarid climates by means of an aridity index involving calculations based on precipitation, temperature, and evapotranspiration. The high variability of rainfall distribution in arid regions, and to a lesser extent semiarid areas, unfortunately makes nonsense of mean annual rainfall statistics, so much so that in both regions the aridity index definition of a climatic zone would vary from year to year. Such systems also provide completely artificial delimitations, decided according to which side of a decimal point a particular index falls. Neither do such indices take into account the effect of the contribution by the soils in providing nutrients and available soil moisture to the plants in determining ecosystems. In practice, their usage is more for agroclimatic than ecological purposes. A more holistic approach involves the use of ecological climatic diagrams, by which it is possible to obtain a graphic representation of the climate as a whole, showing its seasonal course.

Mean temperatures and especially rainfall figures can be misleading when considering semiarid climates. Not only does the rainfall fluctuate from year to year, often quite dramatically in the more arid regions, there are also long-term cycles of relatively high and low rainfall. However, the annual pattern of temperature and rainfall should not be unduly affected. By the use of such climatic diagrams, nine major low-altitude climatic zones are recognized, within which the following three semiarid climatic diagram types, or zonobiomes, are recognized: a subtropical arid biome, a continental arid–temperature biome, and a Mediterranean biome with winter rain. Such divisions can have no distinct boundaries; they merge with the adjoining biomes. Thus, the more arid areas of the Mediterranean zonobiome, with winter rain and a long summer drought, are also considered, as well as the drier parts of the tropical summer rainfall and temperate zonobiomes. The relevant semiarid zonobiomes and their associated zonal soils and vegetation are briefly described below.

A. Subtropical Arid Zonobiome

In the subtropical arid zonobiome rain seldom falls. There is also very strong solar radiation by day, with considerable heat loss by irradiation at night, resulting in extreme daily fluctuations in temperature; infrequent night frosts may also occur. The effective annual rainfall is <200 mm and, under extreme desert conditions, below 50 mm. The zonal soils are characteristically sierozems or syrozems (raw desert soils); saline soils are also found. The landscape is characterized by rock, with subtropical desert vegetation. The most effective management systems are transhumant or semisedentary.

B. Continental Arid–Temperate Zonobiome

Low-rainfall continental areas with marked differences between summer and winter temperatures characterize the continental arid–temperate zone. The following degrees of aridity may be distinguished:

1. Semiarid steppe climate, with a dry summer period but only slight indications of drought, typically with frost-resistant steppe vegetation. This climate is generally adaptable to most forms of management systems.

2. Arid semidesert climate, characterized by a clearly marked drought period and a short wet season. This is similar to the subtropical desert zone with drought throughout the year, but distinguished by cold winters and by its frost-resistant desert vegetation. The zonal soils range from chernozems, castanozems, and burozems to sierozems. Transhumant and semisedentary are the appropriate systems in the lower rainfall regions, with sedentary systems in the higher rainfall regions.

C. Mediterranean with Winter Rain Zonobiome

Lying between latitudes 35° and 40° in the northern and southern hemispheres, this zonobiome is typified by winter rain and a long period of summer drought. The zonal soils are Mediterranean brown earths and fossil terra rossa. The vegetation consists typically of sclerophyllous woody plants that are sensitive to prolonged frost. The more arid parts of this zone merge with the subtropical deserts outlined above. This zonobiome is generally suitable for most forms of sedentary management systems.

The arid and semiarid mapping units using the above zonobiomes do not correspond entirely in outline or in climatic classification with those based on climatic indices. By their very nature climatic boundaries are never abrupt; likewise, vegetation boundaries are rarely abrupt, except where there are sharp changes in elevation or soil. Consequently, some differences in interpretation are to be expected; however, the more holistic approach using ecological climatic diagrams should provide more realistic mapping units on which the ecosystems can be based.

III. MANAGEMENT

Good management stategies are designed to ensure sustainable productivity within the restrictions imposed by good soil and water conservation practices and environmentally friendly crop and livestock husbandry.

A. Soil and Water Conservation

Soil conservation strategies are aimed at reducing soil losses due to wind and water erosion.

1. Wind Erosion

Bare soil without plant cover, depending on such factors as wind speed and duration and the size, roughness, and dryness of the surface particles, is liable to soil removal, which can soon strip the soil of its fertility and water-holding capacity. Measures required to reduce wind erosion include the reduction in wind velocity at the soil surface by means of a series of windbreaks, reducing the time during which bare soil is exposed by leaving crop residues on the soil surface, delaying cultivation, and ensuring a rough surface.

2. Water Erosion

Geological erosion by water is a normal process, while accelerated erosion reflects activities by humans and their livestock. Sheet erosion is caused by the accelerated movement of water across cultivated or overgrazed land. It results in the movement of topsoil from the hill slopes to the valley bottoms. The concentration of water along footpaths, cattle tracks, cultivation, ditches across the contour, etc., can result in gully erosion. The severity of water erosion is affected by such factors as the intensity and duration of the rainfall, slope, soil texture, and exposure of bare soil. Accelerated erosion of agricultural land can be controlled by reducing the water velocity downslope by avoiding

the cultivation of steep slopes, bunding, terracing, across-slope cultivation, strips of alternating cultivation and permanent vegetation along the contour and vegetation, mulching, check dams for existing gullies, etc. While the correct stocking rates to prevent overgrazing are essential, the alternation of watering points, feeding points, and gateways, preventing access to gullies and steep banks, etc., can also help to reduce erosion in grazing lands.

3. Water Conservation

Water conservation strategies are aimed at increasing the water available in the soil, reducing water losses through runoff and evaporation, and allowing the storage of water for future use. Strategies used to prevent water erosion by reducing surface velocity permit more time for water to infiltrate the soil. Mulching, including stone mulching, reduces losses from evaporation; plants, too, intercepting rainfall from leaf drip and stem flow, help to concentrate water in the vicinity of the roots while at the same time offering protection against evaporation.

4. Water Harvesting

This technique makes use of sheet flow by capturing water in a series of usually small chevron or arcuate catchments. By concentrating water within these catchments, the farmer is able to grow crops in an otherwise inadequate rainfall regimen. Water conservation measures are also required to provide potable and irrigation water, usually by means of dams and weirs across the river bed or by diverting water into reservoirs or tanks by means of channels or pumps. Dams and weirs can, in some instances, raise the local water table sufficiently to convert seasonally dry beds into permanent or semipermanent waterways.

B. Crop Management

The objective of good crop management is sustainable crop production without irretrievable loss of soil fertility and structure or introduction of toxic residues. In this context "sustainable crop production" includes both continuous cropping, which may often be possible on irrigated lands, as well as the more usual use of long and short fallow periods. Where good husbandry practices require the use of toxic chemicals for the control of crop weeds, insect pests, and diseases, these should consist of biodegradable chemicals that leave no lasting toxic residues.

I. Soil Fertility

Nutrients removed from the soil by harvested crops must be replaced so that soil fertility will not decline. This can be achieved by the use of artificial fertilizers, green cropping, or fallow. Increased yields may be obtained, provided that the use of fertilizers does not result in excessive foliar growth and associated delays in crop maturation and increased evapotranspiration. While nitrogenous fertilizers tend to increase foliar growth, phosphates may increase moisture utilization by hastening crop maturity.

The use of fertilizers is largely confined to modern farming systems or the more lucrative cash crops of the traditional farmers. Arid soils tend to be poor in organic matter, which is easily lost, destabilizes the soil, and increases erodibility. Thus, the continuous use of fertilizers without the intervention of a green crop or fallow period is not recommended. Unfortunately, the increasing population in many developing countries has increased the demand for land to such an extent that the period under cropping has been lengthened and the fallow period has decreased with disastrous results.

2. Cultivars

The matching of the crops grown to the available moisture is an obvious tenet of crop husbandry, for example, the growing of rapidly maturing varieties of dwarf maize instead of the more water-demanding traditional maize. However, when considering the inherent variability of arid and semiarid climates, for consistent food or income security the crop should be matched to the driest year, but where the producer can afford the risk, gambling on the wetter years could have greater long-term profitability. Some traditional farmers even cultivate "impure" crop varieties whose seed responds to different moisture levels, thereby ensuring a widely spread harvesting period and at least some harvestable crop irrespective of the rainfall.

3. Tillage

The short-term objectives for tillage are land preparation for sowing and planting, improvement of soil moisture, promotion of soil aeration, breakdown of organic matter, and improvement of root penetration and weed control, while the long-term benefits are prevention of erosion and maintenance of soil fertility. The minimum exposure of bare soil to the forces of erosion and the subsequent minimum of tillage operations consistent with good crop growth are essential features of arid land cropping. Although it is an inefficient method of saving moisture due to evaporation losses, alternating crop and fallow may be the only means of ensuring a crop every 2 years.

C. Range and Pasture Management

Good range and pasture management may be defined as maintaining a sustainable carrying capacity without deterioration of the pasture or range yet contributing to the nutritional requirements of the livestock. The nutritional requirements may require, for example, mineral and/or protein supplements due to seasonal or permanent deficiencies in the soil or forage.

I. Range and Pasture Evaluation

The value of a range or pasture can be evaluated from animal behavior trials and forage analysis, the three most important factors being availability, digestibility, and palatability of the forage. The availability varies with the plant and the type of animal. Thus, the camelids appear to be random grazers and browsers, eating even the most leathery or prickly vegetation, while sheep are selective grazers and normally do not browse. Goats graze and browse equally well, while cattle are primarily grazing animals, supplementing their diet by browsing. The cloven-hoofed animals are, to varying degrees, also tolerant of moist soils, whereas the camelids require dry conditions, so that the range of habitat also affects availability. These different grazing characteristics can sometimes be utilized in range improvement; a combination of cattle and goats, for example, might be used to control bush encroachment.

Seasonal availability of forbs and browse naturally determines what is available to be eaten and, given a choice of a range of forage, palatability will determine what is eaten first. Grasses and herbs are generally either not available or of low nutritional value during the dry season/winter months. Consequently, browse species or conserved fodder are essential food sources during such times of dearth. Browse may include evergreen species, such as many of the phyllodial *Acacia* species of Australia or species of *Atriplex,* deciduous and often leguminous trees and shrubs with tender shoots and edible fruits, or a few species, such as *Faidherbia albida,* whose life cycle is such that the nutritious pods and new leaves are produced at the start of the dry season, the tree remaining bare throughout the rainy season.

Other factors, such as the density and height of forage, determine how much can be gathered in one mouthful. This is an important consideration, since animal behavior studies in a temperate climate have demonstrated that cattle, for example, spend only a certain number of hours in feeding, irrespective of hunger. Thus, the traditional practice of penning livestock overnight could result in the animal's having insufficient time in which to feed unless supplied with a night feed. The digestibility of forage determines the available nutritional value to the beast. Tropical stock are, from necessity, more efficient digesters of fiber than temperate stock, while camelids are 25% more efficient in digesting coarse forage than other ruminants.

When forage is scarce either early in the season or as a result of drought, there is a danger of livestock eating toxic plants because of their availability or greedily eating plants that are not normally toxic if eaten in small quantities. Toxicity is a complex subject; in some cases animals can develop local immunity, as with *Morea* in South Africa, while introduced animals succumb. Toxicity may be species related; in the Andes llamas may safely graze *Stipa subaristata,* but the grass is lethal to donkeys and other livestock. While some plants may be eaten alone with impunity, they can produce a le-

thal cocktail when eaten with another species; for example, the pods of *Acacia georginae* are harmless but contain an enzyme that liberates hydrocyanic acid from a glucoside present in *Eremophila maculata.*

2. Range Improvement

Improvement in the vegetation is possible only if livestock numbers can be kept below the carrying capacity. This implies some form of control either by fenced paddocks or, in the case of free range and depending on ownership, some form of herding. The latter presents relatively few problems where the herder owns the land, but is extremely difficult where government land provides free grazing and personal gain is often put before national environmental needs. The latter is a problem in the Sahel, especially where desertification has destroyed many of the former traditional tribal grazing areas and their traditional tribesmen are now encroaching on neighboring tribal areas.

Ideally, the level of stocking requires careful seasonal regulation throughout the year to prevent overgrazing. With a paddock system some form of rotation is usually practiced to take into account any seasonal variation in availability and productivity. Any overgrazing will lead to an increase in unpalatable species at the expense of the more palatable; eventually, even these will decrease as the inevitable soil erosion increases. Overgrazing is also a problem with free range grazing, especially around watering points and night holding pens. Night holding pens on the open range can usually be moved elsewhere to remove the grazing pressure, but where these occur in villages a system of alternating access routes is desirable, along with an increase in the availability of supplementary feed to reduce overgrazing in the village vicinity.

The natural recovery of degraded range is extremely slow. More rapid improvement may be obtained by planting suitable browse trees and shrubs and grasses, especially in the winter rainfall areas with a Mediterranean-type climate, even in areas with as little as 200 mm rainfall. However, such plantings have proved to be slower in the summer rainfall areas. In the Sahel, for example, such plantings have proved to be unsuccessful in areas with <400 mm annual rainfall. The cost-effectiveness of improving free range is low, but the environmental benefits in controlling soil erosion, conserving water and wildlife habitats, etc., could justify the expense.

3. Fire

Fire is used to encourage grass growth at the start of the rainy season. Its regular use will, however, encourage fire-resistant species at the expense of more palatable species. In the higher-rainfall semi-arid regions fire can be an efficient means of controlling bush encroachment, inedible species, insect pests, and predators. [*See* FIRE ECOLOGY.]

4. Watering Points

The ability of the various breeds of livestock to survive without water affects the permissible distance between watering points that an animal may graze without loss of condition. The uncontrolled siting of watering points on open range can encourage overgrazing by increasing the numbers of livestock and attracting additional pastoralists. While access to wells and boreholes is easy to control, access to open water resources such as rivers and lakes is more difficult.

5. Livestock

The full potential of both primitive and improved breeds of livestock cannot be developed without adequate feeding. Provided that feeding and management are not limiting factors, the use of prime males is the usual method for upgrading livestock in order to take advantage of any range or pasture improvement.

Glossary

Agroforestry The deliberate planting of perennials on the same unit of land as agricultural crops and/or livestock, either in some form of spatial mixture or in sequence, thereby producing a significant interaction (positive and/or negative) between the woody and nonwoody components of the system, ecological and/or economical.

Browse The selective eating of the tender shoots of trees, shrubs, and perennial herbs and grasses by livestock and game. This contrasts with grazing, in which most of the aboveground parts of annual and perennial plants can be eaten.

Carrying capacity The number of livestock units the range or pasture is capable of maintaining without deterioration of the range or pasture.

Feed Animal food with a high food value relative to volume (also known as concentrates), low in fiber, some high in protein, others high in carbohydrates or fat.

Fodder Dried cured plant material such as hay, straw, and stover (also known as roughage).

Forage All browse and herbaceous food, including pasture, silage, and green feed, that is available to livestock or game animals.

Livestock unit The relationship of different classes of livestock (e.g., cattle, sheep, and goats) of varying ages, breeds (e.g., merino or karakul sheep), and levels of production to the food energy necessary for the maintenance and production of a standard animal, usually a bovine. Note: The requirements of a standard temperate bovine will differ from those of a bovine from the tropics, the rumen flora of the latter being more efficient in digesting fibrous forage.

Multipurpose species Plants that have more than one economic use, such as food, fuel, timber, medicine, nitrogen fixation, green manure, and browse.

Pasture Grazing lands whose composition consists of managed and sometimes selected indigenous or introduced herbs and grasses.

Range Natural grazing lands composed of the indigenous flora.

Stocking rate The number of animal units that can safely graze a degraded pasture or range yet permit improvement in the carrying capacity.

Bibliography

Arnon, I. (1991). "Agriculture in Dry Lands: Principles and Practice." Elsevier, Amsterdam.

Barrow, C. (1987). "Water Resources and Agricultural Development in the Tropics." Longman Scientific and Technical, Harlow, Essex, England.

Coupland, R. T. (Ed.) (1992). Natural grasslands. Introduction and Western Hemisphere. "Ecosystems of the World," Vol. 8A. Elsevier, Amsterdam.

Coupland, R. T. (Ed.) (1993). Natural grasslands. Eastern Hemisphere and résumé. "Ecosystems of the World," Vol. 8B. Elsevier, Amsterdam.

De Martonne, E. (1926). "Une nouvelle fonction climatologique: l'indice d'aridité." *La Meteorologie Nouvelle* **2,** 449–458.

Food and Agriculture Organization (FAO) (1985). "Tree Growing by Rural People," Forestry Paper 64. FAO, Rome.

Miegs, P. (1952). World distribution of arid and semi-arid homoclimates. *In* "Reviews of Research on Arid

Zone Hydrology," pp. 203–210. Arid Zone Research 1. UNESCO, Paris.

Singh, R. P., Parr, J. F., and Stewart, B. A. (1990). "Dryland Agriculture. Strategies for Sustainability." Springer-Verlag, New York.

Thornthwaite, C. W. (1948). An approach towards a rational classification of climate. *Geogr. Res.* **40,** 173–181.

United Nations Educational, Scientific, and Cultural Organization (1979). "Map of the World Distribution of Arid Regions." Explanatory note. MAB Technical Noters 7. UNESCO, Paris.

Walter, H. and Breckle, S.-W. (1985). Ecological Systems of the Geobiosphere. *In* "Ecological Principles in Global Perspective," Vol. 1. Springer-Verlag, Berlin.

Wickens, G. E. (1992). Arid and semiarid ecosystems. *In* "Encyclopedia of Earth System Science," Vol. 1, pp. 113–118. Academic Press, San Diego.

Wilson, R. T. (1991). "Small Ruminant Production and the Small Ruminant Genetic Resource in Tropical Africa," FAO Animal Production and Health Paper 88. FAO, Rome.

Agroecology

Miguel A. Altieri
University of California

I. INTRODUCTION

Agroecology is a scientific discipline that uses ecological theory to study, design, manage, and evaluate agricultural systems that are productive but also resource conserving. Agroecological research considers interactions of all important biophysical, technical, and socioeconomic components of farming systems and regards these systems as the fundamental units of study. Mineral cycles, energy transformations, biological processes, and socioeconomic relationships are analyzed as components of whole systems and are examined in an interdisciplinary fashion.

Agroecology is concerned with the maintenance of a productive agriculture that sustains yields and optimizes the use of local resources while minimizing the negative environmental and socioeconomic impacts of modern technologies. In industrial countries modern agriculture, with its yield-maximizing high-input technologies, generates environmental and health problems that often do not serve the needs of producers and consumers. Agricultural research has been designed to maximize the use of purchased inputs and technologies tailored for large farms, disproportionally favoring the agribusiness sector and large land owners. In developing countries, in addition to promoting environmental degradation, modern agricultural technologies have bypassed the circumstances and socioeconomic needs of large numbers of resource-poor farmers, and thus peasants in general have not benefited from the advance of the Green Revolution.

The contemporary challenges of agriculture have evolved from merely technical concerns to also include social, cultural, economic, and particularly environmental concerns. Agricultural production issues cannot be considered separately from environmental issues. In this light a new technological and development approach is needed to provide for the agricultural needs of present and future generations without depleting our natural resource base. The agroecological approach does just this, because it is more sensitive to the complexities of local agriculture and has broad performance criteria, which include properties of ecological sustainability, food security, economic viability, resource conservation, and social equity as well as adequate levels of production.

To put agroecological technologies into practice requires technological innovations, agriculture policy changes, and socioeconomic changes, but mostly a deeper understanding of the complex long-term interactions among resources, people, and their environment. To attain this understanding, agriculture must be conceived of as an ecological system as well as a human-dominated socioeconomic system. A new interdisciplinary framework to integrate the biophysical sciences, ecology, and other social sciences is indispensable. Agroecology provides a framework by applying ecological theory to the management of agroecosystems according to specific resource and socioeconomic realities, and by providing a methodology to make the required interdisciplinary connections.

II. WHAT IS AGROECOLOGY?

Initially, agroecology emerged as a loosely defined discipline that incorporated ideas about a more environmentally and socially sensitive approach to agriculture, one that focuses not only on production, but also on the ecological sustainability of the production system. This more normative use of the term "agroecology" implies a number of features about society and production and is aimed at understanding agricultural processes in the broadest manner.

At the operational/methodological level agroecology can be more narrowly defined as a scientific discipline that applies ecological principles to the study, design, and management of agricultural systems.

The agroecological approach regards a farming system as the fundamental unit of study, and within this system mineral cycles, energy transformations, biological processes, and socioeconomic relationships are investigated and analyzed as they relate to each other. Thus, agroecological research is concerned not just with maximizing production of a particular commodity, but rather with optimizing the agroecosystem as a whole. This approach shifts the emphasis in agricultural research and education away from disciplinary and commodity concerns toward complex introductions among and between people, crops, soil, and livestock, etc. At its most narrow level "agroecology" refers to the study of

interactions between people and food-producing resources within a farm, or to the investigation of purely ecological phenomena within the crop field, such as predator–prey relations, crop–weed competition, and nutrient cycling.

Whatever the level of analysis, the agroecological research and development approach differs from other conventional agricultural approaches in that it:

a. Is a holistic approach emphasizing system-level interactions, internalizing rather than externalizing environmental and social effects
b. Focuses on synergistic processes in the agroecosystem, resulting in complementarity and integration of biological/productive components
c. Focuses on the root causes of problems rather than on symptoms
d. Emphasizes the understanding of land conditions so that interventions are tailored to site-specific conditions
e. Builds on indigenous knowledge, as local farmers often possess a detailed understanding of local phenomena

III. THE INTERDISCIPLINARY NATURE OF AGROECOLOGY

Agroecology consists of an analytical framework integrating the numerous factors that affect agriculture, and it relies on insights from a number of biophysical, agronomic, and social science disciplines.

At the heart of agroecology is the idea that a crop field is an ecosystem in which ecological processes occur as found in other biological relationships; these processes include nutrient cycling, biological pest suppression, competition, commensalism, and succession. The purpose is to understand these processes in natural communities in order to build agroecosystems that mimic the structure and foundation of natural ecosystems. Natural systems serve as architectural and functional models for designing and structuring agroecosystems that meet environmental and social needs.

Implicit in agroecology is that when an understanding of the ecological processes is reached, a set of ecological principles can be derived so that the agroecosystems can be manipulated to stabilize production, with fewer negative environmental impacts, with fewer external inputs, and in a more sustainable manner. These principles can be summarized as follows:

a. Regeneration and conservation of local productive resources by
1. enhancement of soil fertility through organic matter and soil biology management
2. maximized soil health
3. soil conservation and erosion control
4. water conservation
5. enhancement of beneficial plant and animal diversity for optimal natural control, pollination, decomposition, etc.
b. Diversification of cropping systems in time and space through rotation, multiple cropping, and agroforestry
c. Recycling of biomass, organic matter, and nutrients through legume-based nitrogen fixation, green manuring, and use of farm residues
d. Optimal microclimate control through effective manipulation of solar radiation, moisture flow, and mechanical impact of wind, rain, and hail
e. Exploitation of plant–plant, animal–animal, and plant–animal interactions that result in the synergistic natural control of pests, recycling of nutrients and biomass, soil cover production, enhanced growth conditions for crops, etc.
f. Integration of plant and animal components for optimal biomass output and recycling

Agroecologists perceive people as part of the evolving local ecosystem and understand that each biological system has evolved to reflect the nature of the people and their social organization, knowledge, technologies, and values. In other words, social and biological systems have co-evolved and embody agricultural potential. This potential has been captured by traditional farmers through a process of trial, error, selection, and cultural learning. To grasp this complexity, agroecology has relied heavily on elements of several social science disciplines, such as sociology, anthropology, human ecology, and ethnobiology. Several agroecologists have explored indigenous production systems in developing countries and categories of knowledge about environmental conditions and agricultural practices, focusing on the native view of production systems. This emphasis on the human and social dimensions of production is an important basis for understanding the production logic or rationale of agricultural systems. Traditional agriculture is now increasingly studied by agroecologists to document indigenous systems and practices and to derive principles of sustainability for the design of yield-stable modern cropping systems. [See TRADITIONAL CONSERVATION PRACTICES.]

Agroecologists have also derived elements from economics, especially when evaluating the production efficiencies of alternative technologies when compared with conventional high-input technologies and when estimating the natural resource costs and benefits of alternative agriculture. A major concern is the need to generate and promote agroecological technologies that enhance economic viability but reduce ecological costs. The fact that most environmental costs (e.g., soil erosion and pesticide contamination of surface and ground water) of modern technological application are borne by society at large is not an acceptable proposition to agroecologists. [See SOIL ECOSYSTEMS.]

IV. AGROECOLOGY IN AGRICULTURAL RESEARCH AND EDUCATION

Research and education efforts focus mostly on individual components of a cropping system and rarely on the farming system as a whole. Agricultural systems are divided into various components (e.g., soil, water, crop species, and insect pests) and/or their functional components, such as soil fertility, ecophysiology, pest control, and crop management. Rarely are the linkages among the various components made obvious. Excessive attention is given to particular production constraints and to yield responses of a crop to a particular input. Understanding the agroecosystem as a whole and how it can benefit humans, other species, and the environment is rarely a goal. Agroecology provides a framework to better understand the complex interaction among various components, how they fit together in systems, and how this knowledge can be used for management in agriculture.

By viewing agroecosystems as coevolutionary products of human needs and biological systems, agroecology effectively links issues that lie at the interface of social, economic, and ecological sciences. At this interface we need to understand the short- and long-term impacts of human activities on agricultural and the natural environments. Although there may be many ways to view agroecosystems, the uniqueness of agroecological analysis is that it:

a. Utilizes a total farming systems approach rather than a discipline-specific analysis emphasizing one crop or one process
b. Incorporates anthropological and sociological methodologies in order to simultaneously analyze socioeconomic and biophysical production constraints
c. Uses the agroecosystem as the unit of study and concentrates on processes (e.g., nutrient cycling, water balance, energy flow, and population regulation) rather than on single-factor cause–effect relationships (e.g., the effect of fertilizer or an irrigation schedule on crop yield)

Agroecology brings together the concepts of ecologists, agronomists, and social scientists to look at agricultural systems in a more complete way. There is an urgent need to incorporate agroecological understanding and knowledge of resource use into the agricultural educational agenda and to prepare professionals to be able to seek strategies of sustainable management for agroecosystems. Professionals in the future will need skills to:

a. Introduce ecological rationality into agriculture in order to minimize the use of chemical inputs, implement watershed and soil conservation programs, and plan systems according to the land use capabilities of each region
b. Promote efficient use of water, forests, soil, genetic, and other nonrenewable resources and coordinate policies concerning pricing and taxation, natural resources accounting, land and resource distribution, and research and technical assistance to promote a more sustainable agriculture

V. AGROECOLOGY AND AGRICULTURAL DEVELOPMENT

In most developing countries it has become increasingly apparent that many conventional patterns of agricultural modernization have been unsustainable. Despite numerous internationally and state-sponsored rural development projects, poverty, food scarcity, malnutrition, health deterioration, and environmental degradation continue to be widespread problems. Such failures have stimulated a number of institutions, researchers, and development workers to promote agroecological technologies more sensitive to the complexities of local farming systems. Due to its novelty in viewing peasant agricultural development, agroecology has heavily influenced the agricultural research and extension work of many grass roots organizations. Several characteristics of the agroecological approach make it especially attractive and compatible:

a. Agroecology provides an agile framework for analyzing and understanding the diverse factors affecting small farms. It also provides methodologies that allow the development of technologies closely tailored to the needs and circumstances of specific farming communities
b. Agroecological techniques are socially activating, since they require a high level of farmer participation
c. Agroecological techniques are culturally compatible, since they do not question peasants' ratio-

nale, but actually build on traditional knowledge, combining it with elements of modern agricultural sciences

d. Agroecological techniques are environmentally sound, since they do attempt not to radically transform the peasant ecosystem, but rather to optimize the agroecosystem

e. Agroecological approaches enhance commercial viability, since they minimize (production) costs of predation by enhancing the efficiency of locally available resources

In practical terms the application of agroecology by grass roots organization in the Third World has translated into a variety of programs with important beneficial impacts:

- improvement of the production of basic foods, enhancing family nutritional intake and often increasing income
- conservation of native crop genetic resources and valuation of traditional crops such as *Amaranthus* and quinoa
- rescue, systemization, and application of peasants' knowledge and technologies
- promotion of efficient use of local resources (e.g., land, labor, and agricultural subproducts)
- increase of crop and animal diversity to enhance food security and minimize production risks
- improvement of the natural resource base through water and soil conservation, emphasizing erosion control, water harvesting, reforestation, etc.
- reduction of the use of external agrotechnical inputs to reduce pollution and economic dependency, but sustaining yields through the implementation of organic farming and other low-input technologies

VI. CONCLUSIONS

Agroecology emerges at a time when the world is searching for new and more sustainable methods of agricultural production and when technological issues require more environmentally sensitive man-

agement practices. Agroecology has been influenced by a broad range of concerns and bodies of thought, but this is the range of issues that impinges on agriculture. It is not surprising that agroecology finds congruencies with sociological, economic, agronomic, and development perspectives. A strong agroecological foundation can help prepare agricultural professionals to contribute to meeting long-term production, conservation, and socioeconomic goals that will ensure an acceptable quality of rural life.

Glossary

Commensalism A close association or union between two kinds of organisms, in which one is benefited by the relationship and the other is neither benefited nor harmed.

Ecological system A system made up of a community of plants, animals, and other organisms and their interrelated physical and chemical environment.

Ecophysiology Ecology of the functions and vital processes of living organisms.

Environmental degradation The deterioration of natural resources such as air, water, and soil.

Grass roots organizations Organizations that evolve from and/or serve the needs of indigenous peoples.

High-input technologies Technologies that require large amounts of products and resources that are not produced by the system but must be brought into it from outside sources.

Natural resource base Resources that are already present in an ecosystem, such as water, minerals, soils, and plants.

Nonrenewable resources Resources that, once used, cannot be regenerated.

Nutrient cycling The process by which nutrients are made available in an ecosystem.

Resource-poor farmers Farmers without access to the components necessary to generate a fair crop yield or income.

Socioeconomic Involving both social and economic factors.

Succession Replacement of one kind of ecological community by another.

Sustainable Capable of maintaining stable crop yields over an indefinite period of time despite external disturbances.

Bibliography

Altieri, M. A. (1987). "Agroecology, the Scientific Basis of Alternative Agriculture." Westview, Boulder, Colorado.

Altieri, M. A., and Hecht, S. B., eds. (1990). "Agroecology and Small Farm Development." CRC Press, Boca Raton, Florida.

Carroll, C. R., Vandermeer, J. H., and Rossett, P. (1990). "Agroecology," Biological Resource Management Series. McGraw-Hill, New York.

Conway, G. R., and Barbier, E. B. (1990). "After the Green Revolution: Sustainable Agriculture for Development." Earthscan, London.

Dover, M. J., and Talbot, L. M. (1987). "To Feed the Earth: Agroecology for Sustainable Development." World Resources Institute, Washington, D.C.

Gliessman, S. R., ed. (1990). "Agroecology: Researching the Ecological Basis for Sustainable Agriculture," Ecological Studies 78. Springer-Verlag, New York.

Pierce, J. T. (1990). "The Food Resource." Longman Scientific and Technical, New York.

Air Pollution and Forests

William H. Smith

Yale University

Forests cover one-third of the terrestrial surface of both the world and the United States. Humans value forest for the many products and services that forest ecosystems provide. Unfortunately, however, human activities can damage forest systems. One form of damage is associated with air pollutants produced by a variety of human activities. This article provides an introduction to the regional- and global-scale air pollutants that have the potential to adversely influence forests, with a focus on the developed countries of temperate latitudes.

I. FOREST VARIABILITY

An assessment of the complex interaction between regional-scale and global-scale air pollutants and forest systems can only be made in the context of full appreciation of forest variability. This variability encompasses differences in forest biology, forest management, and forest values.

The earth is covered by a mosaic of ecosystems. These ecosystems are connected and influence one another in a variety of ways. Temperate forest ecosystems occupy a position of prominence among all ecosystems. Temperate forests (1.8 billion ha) are second only to tropical forest ecosystems (2 billion ha) in size. The biomass of temperate

forest ecosystems (200–400 tons ha^{-1}) is second only to that of rain forest ecosystems (400–500 ha^{-1}). In terms of primary productivity (energy stored via photosynthesis), temperate forest ecosystems (5–20 tons ha^{-1} yr^{-1}) rank third behind tidal zone (20–40 tons ha^{-1} yr^{-1}) and rain forest (10–30 tons ha^{-1} yr^{-1}) ecosystems. Unfortunately, temperate forest ecosystems are also located in the zone of maximum air pollution because of their extensive distribution throughout the zone of primary urbanization and industrialization of the earth.

Within the temperate zone, forest systems have enormous variability. Forests may differ in soil type, climate, aspect, elevation, species composition, age, and health. Forests may be young, uneven aged, even aged, all aged, or overmature. Forests may be reproduced by seed, by sprouting, or by planting. Some forests have their structure completely shaped by natural forces, some may be influenced by human forces as well as natural forces, and other forests may be completely artificial in design and establishment.

Forest systems receive different amounts of human attention, which allows trees to be grouped along a continuum of human management efforts ranging from no management to intensive management. In part, the varied intensity of forest management reflects different values that people attach to forest systems and forest trees. These values (Table I) may be in the form of forest products, for example, wood (lumber and paper), wildlife, or water. In addition, and equally important, forests provide a variety of valued services to societies, including, among others: recreation; biological diversity; landscape diversity; several environmental amenity services, such as microclimate amelioration (shade for coolness, wind reduction for energy conservation), sound attenuation, visual attractiveness, and screening; water quality, flood and erosion management, soil and nutrient conservation; and persistent pollutant storage and detoxification. Some of these forest values are priced and some are unpriced. All values, however, whether in the form of forest products or forest services, whether quantified in economic terms or not, must be evaluated with reference to "adverse" air pollutant influence.

II. FUNDAMENTALS OF FOREST HEALTH

In addition to an appreciation of the elements of forest variability and of forest values, an under-

standing of fundamental forest health principles is essential for developing an accurate perspective on the influence of air pollutants on complex natural systems.

Factors capable of causing injury, disease, and mortality in forest systems are called stresses. Stress factors may be living or nonliving, natural or unnatural. Stresses recognized to have widespread and general importance in forest systems include climatic, pathologic, entomologic, anthropogenic, wildlife, fire, and stand dynamic elements (Table II). Trees are large and long-lived. Tree health integrates the influences of all stressors acting concurrently, sequentially, and interactively. Though there are numerous forest health problems, disease and mortality remain the exception rather than the rule. Temperate-zone forests combine the characteristics of relatively low biological diversity with relatively high stability. Tree mortality rates vary

TABLE I
Values Associated with Forest Systems

Products	Services
Wood	Existence value
Lumber	Recreation
Particleboard/plywood	Tourism
Paper	Habitat
Fuelwood	Biological diversity
Mulch	Genes
Wildlife (game)	Species
Water (quality)	Communities
Forage (wildlife, livestock)	Ecosystems
Other	Landscape diversity
Seeds	Amenity function
Edible nuts	Microclimate amelioration
Syrup (sugar maple)	Sound attenuation
Drugs (e.g., taxol)	Aromatic hydrocarbons
Pesticides (e.g., neem)	Visual attractiveness, screening
Christmas trees	Runoff/erosion management
Mistletoe	Soil/nutrient conservation
Mushrooms	Pollutant sequestration/ detoxification

TABLE II
Biotic and Abiotic Factors That Cause Disease and Injury of Forest Trees

	Living factors (biotic)	Nonliving factors (abiotic)
Cause disease[a]	Fungi	Air pollution
	Bacteria	Drought
	Viruses	Salt
	Mistletoes	Adverse soil chemistry or physical structure
	Nematodes	
	Mycoplasma	Nutrient deficiencies or excesses
	Insects	
	Mites	Competition
Cause injury[b]	Insects	Wind
	Mites	Fire
	Rodents	Temperature extremes
	Mammals	Moisture extremes
	Birds	Lightning
	Humans	Volcanic eruptions
		Landslides
		Avalanches
		Harvesting
		Pesticides
		Ice
		Snow
		Salt, other chemicals
		Radiation

[a] Disease = abnormal physiology from chronic interaction with stress.
[b] Injury = abnormal physiology from acute interaction with stress.

greatly with stage of forest development: they are high in seeding and early developmental stages, but medium to low in mature and old-age stages. Over a wide variety of forest histories, ages, species, and sites, in the northeastern United States the percentage of dead standing trees varies from 5 to 36% of total stem density.

Stress factor significance to forest systems exhibits great temporal and spatial variation. Numerous stresses have greatest importance at young tree ages, whereas others are more significant at old tree ages. Stresses vary with soil type, aspect, elevation, and microclimate. White-pine blister rust is an important disease in northern Wisconsin and northern New England, but is less significant in southern Wisconsin and southern New England because of climate. Management objectives further define the relative importance of specific stress factors. Unlike agricultural ecosystems, forest ecosystems do not have "absolute stresses" only "relative stresses." The significance of a particular stress is relative to the management goals and objectives that reflect the values imposed on the forest system or forest tree by the human manager. Complete or partial defoliation of hardwood trees caused by foliar pathogens or insects, in a single year, could cause significant economic loss in a forest being managed as a family campground. A single year of defoliation in this same forest, if it was being managed for wood production or as a wilderness area, would be largely without importance.

Forest ecosystems are dynamic and not static. Over time, forest systems are characterized by variability not constancy. Plant and animal diversity and forest productivity change with developmental stage of a forest. There is substantial evidence to support the conclusion that numerous living and nonliving forest stress factors play fundamental and natural roles in the regulation of species composition and productivity of forests. In natural forest ecosystems, disturbance or perturbation caused by environmental or biotic forces (living stress factors) may be necessary to maintain maximum diversity and productivity. The natural tendency in numerous temperate latitude forest ecosystems toward periodic perturbation at intervals of 50–200 years recycles the system and maintains a periodic wave

of peak biological diversity and a corresponding wave of peak production of living material. Insect outbreaks, fires, and wind-storms, even those events that cause massive destruction in the short run, may plan beneficial and essential roles in forest ecosystems in the long term. These roles are presumed to regulate tree species competition, species composition and succession, productivity, and nutrient cycling.

Ambiguous (uncertain cause) tree disease is the rule and unambiguous (certain cause) tree disease is the exception! Unambiguous stresses with catastrophic impact include volcanic eruption, such as Mount St. Helens in 1980, the 1938 hurricane in New England, and the 1988 wildland fires in Wyoming and California. Equally unambiguous and unique are wide-area defoliations by tussock moths, eastern and western spruce budworms, and gypsy moths, and wide-area mortality caused by chestnut blight disease, Dutch elm disease, fusiform rust, or white-pine blister rust. Though these stresses are dramatic, they are not the norm of forest health dynamics. More typically tree disease and death are the result of complex interactions of multiple stress factors and the relative importance of specific stresses is ambiguous. Air pollution impacts on forest systems generally fall into the ambiguous category! A special, and very important, type of "ambiguous" tree disease is called dieback/decline disease by forest health professionals. This disease occurs when a large proportion of a tree species population exhibits visible symptoms of stress or unusual and consistent growth decreases or death over an area of many square kilometers. [See FOREST PATHOLOGY.]

Forest dieback/decline diseases are irregular in distribution, discontinuous but recurrent in time, and the result of complex interactions of multiple stress factors. Decline stress is more characteristic of mature forests than immature forests. Generally declines result from the sequential influence of multiple stress factors. Stress factors are both nonliving and living in nature. Nonliving stress factors common to numerous declines include drought and low- and high-temperature stress. Living agents of particular importance include defoliating insects, borers and bark beetles, root-infecting fungi, and canker-inducing fungi. Typically declines are initiated by a nonliving stress, with mortality ultimately caused by a living stress agent. Quite commonly the stress responsible for decline initiation is the direct and indirect influence or change in some climatic parameter, for example, less than normal precipitation.

Dieback/decline stress is especially important currently in the United States in portions of the northern hardwood forest on sugar maple and on American beech. Very widespread decline and mortality of yellow birch occurred throughout New England, New York, and southeastern Canada from 1930 to 1960. An important dieback/decline of white ash was most severe in the Northeast during 1950–1960 and again may be increasing.

The most widespread and serious declines over the past several decades in the central hardwood forest have been associated with members of the red oak group (scarlet, pin, red, and black oak). The present distribution of oak decline is highly variable, but mortality can vary from 20 to 50% in severely impacted areas. Significant mortality has occurred in the George Washington and Jefferson National Forests in Virginia. There is evidence that similar declines occurred during earlier periods in this century in the southern Appalachians. Dieback/declines of lesser importance that have occurred in the central and southern hardwood forests involved dogwood, sweetgum, yellow-poplar, and magnolia.

Conifers also exhibit wide-area dieback/decline stress. Important North American examples of the recent past include Alaska cedar, western white pine, and white spruce in the West, and shortleaf pine, other southern pines, and high-elevation red spruce in the East. During the past decade, numerous forest species of Central Europe have also exhibited dieback/decline symptoms. In selected regions, symptomology has been especially dramatic on silver fir, Norway spruce, Scots pine, European beech, and various oak species. In Germany, the third national survey of forest health, conducted from July to September 1985, revealed that 55% of forest land was exhibiting foliar symptoms of stress. The inventory recorded that 33% of the symptomatic forests were classified in the lowest

category (1), 19% in category (2), and 3% in the heavily damaged category (3). What is the role of air quality in these declines?

Unfortunately our understanding of the interactive and sequential nature of stresses of significance in forest dieback/decline is very incomplete. A close examination of the literature related to North American and European dieback/decline phenomena of the past century reveals as much speculation as evidence relative to the cause and effect of decline phenomena. We have no formally accepted "rules of proof" for causality. The most conclusive documentations of cause associated with dieback/decline have been those concerned with specific stands or very restricted regions. The study of cancer in humans provides a useful analogy with the study of dieback/decline phenomena in forests. Cancer is recognized as a multistage process and numerous cancers involve multiple stress interactions. Confident assessment of the specific factors involved in certain cancers, and the relative importance of these factors, is not possible because of the current limited capabilities of the science of toxicology. Confident assessment of the specific factors involved in wide-area forest dieback/decline phenomena likewise exceeds the current limited capabilities of forest science.

III. REGIONAL-SCALE AIR POLLUTANTS AND FOREST HEALTH

Regional air pollutants of greatest documented or potential influence for forests include: oxidants, most importantly ozone; trace metals, most importantly heavy metals such as cadmium, copper, lead, manganese, chromium, mercury, nickel, vanadium, and zinc; and acidic deposition, most importantly the wet and dry deposition of sulfuric and nitric acids. Ozone and sulfuric and nitric acids are called secondary air pollutants because they are synthesized in the atmosphere rather than released directly into the atmosphere. The precursor chemicals, released directly into the atmosphere and causing secondary pollutant formation, include hydrocarbons and nitrogen oxides, in the case of ozone, and sulfur dioxide and nitrogen oxides, in the case of sulfuric and nitric acid. The combustion of fossil fuels by electric utilities, by industrial and manufacturing facilities, and in motor vehicles provides precursors to the atmosphere. Sulfur is oxidized from fossil fuels and the smelting of sulfur-rich ores. Heat of combustion causes nitrogen and oxygen to form nitrogen oxides. Incomplete combustion provides a rich assortment of hydrocarbons.

Forest damage from air pollutants is exposure related, and exposure–response thresholds for a specific pollutant are very different among the various tissues and organisms of a specific ecosystem. Ecosystem response is, therefore, a very complex process. In response to low exposure to air pollution, the vegetation and soils of an ecosystem function as a sink or receptor. When exposed to intermediate loads, individual plant species or individual members of a given species may be subtly and harmfully affected by nutrient stress, impaired metabolism, predisposition to insect or microbial stress, or direct induction of disease. Exposure to high deposition may induce acute disease or mortality of specific plants. At the ecosystem level, the impact of these various interactions is highly variable. In the first situation, the pollutant would be transferred from the atmosphere to the various elements of the biota and to the soil. With minimal physiological effect, the impact of this transfer on the ecosystem could be undetectable (innocuous effect) or stimulatory (fertilizing effect). If the effect of the pollutant exposure on some component of the biota is harmful, then a subtle adverse response may occur. The ecosystem impact in this case could include reduced productivity or biomass, alterations in species composition or community structure, or increased morbidity. Under conditions of high exposure, ecosystem impacts may include gross simplification, impaired energy flow and biogeochemical cycling, changes in hydrology and erosion, climate alteration, and major impacts on associated ecosystems.

A. Heavy Metals

Human activities introduce heavy metals to the atmosphere primarily in particulate form. Only cadmium and mercury are thought to enter the trophosphere in the vapor phase. During high-temperature combustion (electric power stations, metal smelters, steel mills, incinerators, and motor vehicles), metal elements and their oxides become volatilized. The elements of high volatility—for example, cadmium, chromium, nickel, lead, thallium, and zinc—show a pronounced concentrating effect as they condense on fine particle surfaces. Because of large surface-to-volume ratios, particles with diameters approximately 1 micron or less contain as much as 80% of their total elemental mass on the surface.

Preferential association of heavy metals with small particles is significant not only because these small particles may escape emission controls, but because these small particles have the longest atmospheric residence times and, therefore, can be carried long distances. Depending on climatic conditions and topography, fine particles may remain airborne for days or weeks and be transported 100–1000 km or more from their source. Ecosystems downwind of major power-generating, industrial, or urban complexes receive atmospheric deposition that has accumulated and integrated heavy metals from multiple sources. Heavy metal particles are deposited to forests by both wet and dry processes. Dry deposition (sedimentation and impaction between precipitation events) is presumed to be more effective for large particles and elements such as iron and manganese, whereas wet deposition may be more effective for fine particles and elements such as cadmium and lead.

Extensive evidence is available to support the suggestion that heavy metals deposited from the atmosphere to forest systems are accumulated in the upper soil horizons or forest floors. Because this accumulation is commonly in the soil horizons with maximum root activity and maximum activity of soil organisms, it is necessary to consider the potential for heavy metal toxicity to roots and soil organisms and the potential to interfere with nutrient cycling [See NUTRIENT CYCLING IN FORESTS.]

All the heavy metals, biologically essential or nonessential, can be toxic to forest trees at some threshold level of exposure. Only a few metals, however, have been documented to cause direct plant toxicity in actual field situations. Copper, nickel, and zinc toxicities have occurred frequently. Cadmium, cobalt, and lead toxicities have occurred less frequently and under more unusual conditions. Chromium, silver, and tin have not been demonstrated to be phytotoxic in field situations at high dose. Direct and acute heavy metal toxicities have been described for forest trees only in the immediate vicinity of point sources. Lower heavy metal exposures associated with regional-scale deposition are linked primarily with potential, long-term interactions with nutrient cycling processes. The high productivity of forest ecosystems is achieved and maintained through efficient nutrient cycling. For most forests, essential elements required to maintain productivity cannot be sustained by annual increments from precipitation and mineral substrates alone. Organic matter decomposition, release of nutrients from organic material, and efficient nutrient uptake by roots are also essential. Evidence indicates that heavy metals have the potential to interfere with nutrient cycling mechanisms in forest ecosystems. Some of the most important hypotheses are as follows:

1. The decomposition rate of forest floor organic matter by the soil biota may be retarded by heavy metal contamination of soil. Binding of heavy metal ions with colloidal organic matter may increase the resistance to decomposition and/or exert a toxic effect directly on an important decomposing microorganism or arthropod.

2. A large number of soil enzymes have been shown to participate in important extracellular soil processes, including free extracellular enzymes and enzymes bound to inert soil components, as well as active enzymes within dead cells and others associated with nonliving cell fragments. It is well established that heavy metals have potential for enzyme interference. Metal ions may inhibit enzyme reactions by complexing the substrate, by combining with the active group of the enzyme, or by reacting with the enzyme–substrate complex. Evidence has been presented that links at least 10 forest soil enzymes with heavy metal interference.

3. Evidence also indicates heavy metal reduction in soil processes. Numerous studies of large metal-emitting point sources have documented reduction in rates of soil respiration associated with cadmium, copper, mercury, nickel, lead, and zinc. The process of nitrification is important in forest soils and other soil types. Under experimental conditions, cadmium, chromium, iron, lead, mercury, nickel, silver, and tin have been demonstrated to be capable of reducing nitrification.

4. Symbiotic microorganisms have roles of great importance in nutrient relations in forest ecosystems. Forests frequently flourish in regions of low, marginal, or poor soil nutrient status. In addition to nutrient conservation and tight control over nutrient cycling, trees have evolved critically significant symbiotic relationships with soil fungi and bacteria that enhance nutrient supply and uptake. Numerous heavy metals, including cadmium, copper, nickel, and zinc, have been shown to adversely influence fungi, bacteria, and actinomycetes that serve as important forest tree symbionts.

Exposure of forest soil ecosystems to heavy metal stress, however, is a function of both amount (concentration in soil) and biological availability (i.e., largely solubility). Heavy metals not available for ready exchange from soil binding sites or in solution are not available for root or microbial uptake. More than 90% of certain heavy metals deposited from the atmosphere may be biologically unavailable. Heavy metals may be absorbed or chelated by organic matter (humic, fulvic acids), clays, and/or hydrous oxides of aluminum, iron, or manganese. Heavy metals also may be complexed with soluble low-molecular-weight compounds. Soluble cadmium, copper, and zinc may be chelated in excess of 99%. Heavy metals may also be precipitated in inorganic compounds of low solubility, such as oxides, phosphates, or sulfates. Adsorption, chelation, and precipitation are, however, strongly regulated by soil pH. As pH decreases and soils become more acid, heavy metals generally become more available for biological uptake. Natural forest soils generally become more acid as they mature. Acidification in excess of natural processes is possible, for example, via acid deposition from the atmosphere, especially in soils with a pH greater than 5. Under this circumstance, soil acidification associated with acid deposition may result in increased biological availability of heavy metals in the forest floor.

In addition, the soil immediately adjacent (within approximately 2 mm) to physiologically active roots—called the "rhizosphere"—may also transform heavy metals from "unavailable to available" forms. Typically, "rhizosphere" soil has lower pH, lower water potential, lower osmotic potential, lower redox potential, and higher bulk density than soil away from roots. The rhizosphere processes of pH regulation, protonation, solubilization, reduction, and complexation may transform unavailable metals to available metals. Rhizosphere processes that may be particularly important in lead uptake include pH, hydrogen ion availability, organic acid availability, and phosphate availability. Organic acids capable of complexing copper and zinc in the rhizosphere may be important in root uptake. Manganese reduction to the divalent form may allow greater uptake of this heavy metal. Complexing agents and reduced rhizosphere pH may also facilitate cadmium uptake. Again, any alteration of rhizosphere processes caused by chemical changes induced by other air pollutants, for example, acid deposition or oxidants, could alter availability and uptake of heavy metals.

B. Acid Deposition

The tropospheric oxidation of sulfur dioxide and nitrogen oxide emissions ultimately produces sulfuric and nitric acids that are deposited on the downwind landscape by both wet (precipitation events) and dry (between precipitation events) processes. Unless there is an unusual natural source of acidity, regional precipitation means of pH 5.0 or less generally result from large human emissions of sulfur and nitrogen oxides. The U.S. National Acid Precipitation Assessment Program has recently provided substantial detail on the patterns of acid deposition. The most complete deposition monitoring results are for wet deposition, with six major wet deposition networks operating in North America for many years. Information for dry deposition, unfortunately, is much less complete as only two networks, using different procedures, have been in operation for only a few years.

Several generalizations concerning the wet deposition of the strong mineral acids, sulfuric and nitric, can be made. The maximum deposition areas of sulfate (SO_4^{2-}) and nitrate (NO_3^-) in precipitation are located in the northeastern United States and southeastern Canada, and these locations have remained stable over the last 10 years. The mean annual pH of precipitation presently falling in the northeastern United States and in adjacent portions of Canada (also common in north and central Europe) is commonly in the range of 3–5.5. Individual storm events have been recorded with pH values between 2.0 and 3.0. The Mountain Cloud Chemistry Program (MCCP) of the National Acid Precipitation Assessment Program represents an effort to characterize the specific chemistry of mountain precipitation, which is largely in the form of cloud water (fog). The five MCCP monitoring sites have revealed that concentrations of H^+, SO_4^{2-}, NO_3^-, and NH_4^+ range from 5 to 20 times higher in high-elevation cloud water relative to nearby low-elevation precipitation and that cloud water concentrations of SO_4^{2-} and NO_3^- at high elevations in the eastern United States were greater than those in the western United States. This high exposure of eastern mountains to acid deposition has special implications for the forest ecosystems located at high elevations along the Appalachian Mountain chain. [*See* ACID RAIN.]

Hypotheses for adverse forest consequences associated with acid deposition have focused primarily on the greater than normal input of H^+, SO_4^{2-}, and NO_3^- ions delivered via wet or dry deposition processes. Some of the most important hypotheses that have been explored include the following:

- deposition impacts one or more processes of tree reproduction or seedling physiology;
- deposition impacts a critical tree metabolic process, for example, photosynthesis, respiration, water uptake, or evapotranspiration;
- deposition causes alteration of carbon allocation away from growth or storage processes to maintenance respiration/repair processes and/or from aboveground to belowground tissues or vice versa;

- deposition increases/decreases microbial symbiotic relationships (mycorrhizal fungi, nitrogen-fixing bacteria);
- deposition increases/decreases microbial pathogen activity;
- deposition increases/decreases insect activity;
- deposition increases/decreases nonliving (abiotic) stress influence (temperature, moisture, wind stresses);
- increased nitrogen/sulfur deposition alters nitrogen/sulfur cycle dynamics, including fertilization (forest growth is commonly restricted by nitrogen availability);
- increased hydrogen ion input to soil increases/alters the availability of heavy metal cations and results in nutrient uptake limitations, root toxicity, or soil organism toxicity;
- increased hydrogen ion input raises soil mineral weathering rates and alters cation (tree nutrient) availability;
- cations (tree nutrients) are leached from foliar tissues by throughfall (precipitation passing through the canopy) and by stemflow (precipitation running down tree stems);
- cations (tree nutrients) are leached from soil horizons of active root uptake to soil regions with reduced root uptake or devoid of roots; and
- deposition results in increased concentrations of biologically available soil aluminum, resulting in reduced nutrient cation uptake and/or fine root toxicity or soil organism toxicity.

Clearly all of these hypotheses cannot be reviewed in this article. Because there is some consensus that the latter two may be especially important to long-term forest health, they will be explored in more detail. Trees, like other higher plants, require adequate supplies of 16 elements to grow normally. Several of these elements, notably calcium (Ca^{2+}), magnesium (Mg^{2+}), and potassium (K^+), are taken up by tree roots as cations from the soil solution. If nutrient cations become depleted in soil solutions, trees may be subject to nutrient deficiencies and in turn predisposed to additional damage by other living or nonliving stress factors. Nutrient cation leaching in soils requires additional cations to dis-

place cations for movement. The deposition of strong mineral acids from the atmosphere to forest soils provides displacement cations in the form of hydrogen ions (H^+) and mobile anions in the form of sulfate (SO_4^{2-}) and nitrate (NO_3^-) ions. In most forest soils, rapid biological uptake may immobilize the nitrate anion as most forest systems are nitrogen limited. In a similar manner, sulfate anions may be immobilized in weathered soils by absorption to free iron and aluminum oxides. In other soils, however, especially those low in free iron or aluminum, or high in organic matter, which appears to block sulfate absorption sites, sulfate may readily combine with nutrient cations and leach nutrients beyond the rooting zone. Field evidence for acid deposition leaching has been obtained from diverse forest ecosystems subject to excessive atmospheric acid input and located in regions with soils vulnerable to anion migration.

Aluminum is the third most abundant in the crust of the earth, where it occurs primarily in aluminosilicate minerals, most commonly as feldspars in metamorphic and igneous rocks, and as clay minerals in weathered soils. Despite high concentrations of total aluminum in forest soils, aluminum interference with nutrient uptake or biological toxicity is generally presumed unimportant, as the element is present in tightly bound or insoluble form. Additions of strong mineral acids to forest soils, however, may mobilize aluminum by dissolution of aluminum-containing minerals or may remobilize aluminum previously precipitated within the soil during podzolization or held on soil exchange sites. If soil solutions become enriched in dissolved aluminum, principally as free ions (Al_3^+) or as monomeric hydroxides [$AlOH^{2+}$, $Al(OH)_2^+$, $Al(OH)_4^-$], the risk of aluminum interference with nutrient cation uptake and the risk of direct root and microbial toxicity are increased. Aluminum-induced calcium deficiency may be important in the health and productivity of red spruce in northeastern United States. Reduced calcium uptake will ultimately suppress tree growth and may ultimately predispose trees to insect and disease attack.

Numerous forest soils have been surveyed for inorganic aluminum concentrations. In general, results indicate the highest available aluminum concentrations in association with mineral (as opposed to organic) soil layers, where base saturation (percentage of soil exchange sites occupied with nutrient cations) is less than 10–15%, soil pH is less than 4.9, and soil solution sulfate is greater than 80 μmol liter^{-1}. The highest soil solution aluminum in North America has been documented at Big Moose in the Adirondack region of New York. The highest aluminum concentration detected in Europe is in the Sölling region (northeast of Bonn in the north-central portion) of Germany, which has four times the concentration recorded at Big Moose. At sufficient concentration, specific thresholds vary significantly by tree species, available aluminum is directly toxic to plants. Direct toxic effects, focused in root tissues, include reduced cell division associated with aluminum binding with DNA, reduced root growth caused by inhibition of cell elongation, and destruction of epidermal and cortical cells.

Nonetheless, the evidence generated by the National Acid Precipitation Assessment Program, as well as other integrated studies, has failed to provide conclusive evidence that acid deposition decreases forest growth.

C. Oxidants

In the presence of sunlight, nitrogen dioxide is dissociated in the troposphere (lower atmosphere) and forms equal numbers of nitric oxide molecules and oxygen atoms. The oxygen atoms rapidly combine with molecular oxygen to form ozone. This ozone then reacts with nitric oxide, on a one-to-one basis, to re-form nitrogen dioxide. The steady-state concentration of ozone that is produced by this cycle is small. When hydrocarbons, aldehydes, or other reactive atmospheric constituents are present, however, they can form peroxy radicals that oxidize the nitric oxide back to nitrogen dioxide or form other compounds. With reduced nitric oxide available to react with ozone, the latter may accumulate to relatively high concentrations. Numerous tropospheric oxidants, for example, peroxyacetyl nitrate, nitric acid, and hydrogen peroxide, are toxic to vegetation. Ozone, however, is judged to be the most important atmospheric oxidant owing to its

demonstrated importance for agricultural crops and forest tree health. The U.S. Environmental Protection Agency (EPA) has reported that more than 100 urban areas in the United States failed to meet the ambient air quality standard for ozone (120 ppb for daily maximum 1-hr average, not to be exceeded more than once per year) during the late 1980s.

Urban plumes are the major contributors to elevated ozone concentrations measured at nonurban locations, and the fact that ozone can survive at significant concentration levels for more than one day of transport means that large portions of agricultural and forested landscapes are exposed to elevated tropospheric ozone concentrations throughout the United States. During the last decade, rural ozone monitoring and data storage have been initiated by numerous organizations, including the EPA (National Dry Deposition Network, Storage and Retrieval of Aerometric Data), the Electric Power Research Institute (SURE and ERAQS programs), the Tennessee Valley Authority, the National Park Service, and others. This monitoring shows that elevated tropospheric ozone concentrations can occur over very wide areas of forest systems in the United States, with generally lower frequency in the Pacific Northwest, Upper Great Lakes, and northern New England and New York, and generally higher frequency in the Ohio River Valley and Piedmont/Mountain Ridge and Valley (portions of Virginia, North Carolina, South Carolina, Georgia, Alabama, and Tennessee) areas. Additional monitoring has revealed high ozone levels for southern New England, the mid-Atlantic states, and southern California.

Unlike heavy metal and acid deposition, whose adverse effects on plants are generally associated with chronic, extended-term effects of persistent chemical accumulation or gradual process modification, the interaction of ozone (a very reactive pollutant) with plants is judged to be very rapid. For this reason, there is substantial consensus that short-term, high-concentration (episodic peak) exposures are particularly important for acute effects. Long-term, low-concentration exposures are also significant for subtle adverse plant health effects (e.g., yield reduction). The specific vegetative stresses imposed by ozone are extremely varied and the key hypotheses for adverse tree health impacts are similar to those presented in the discussion on acid deposition. Ozone is capable of influencing forest tree reproductive processes, of altering (both simulating and restricting) insect and microbial pathogens of forest trees, and, at sufficient dose, of destroying forest tree foliar and fine root tissue. However, by far the most significant adverse impact of ozone on forest tree metabolism is the ability of this pollutant to interfere with carbon fixation and allocation.

Photosynthesis (carbon fixation) is the most fundamental metabolic process of trees and is the primary determinant of forest ecosystem growth and biomass accumulation. The rate of net photosynthesis of mature trees frequently is within range of 10–200 mg of carbon dioxide taken up per gram of dry weight per day. This photosynthetic rate is extremely variable and is influenced by genetic differences, season of the year, time of day, foliage position within the crown of the tree, age of foliage, solar radiation, climate, and soil factors. Unfortunately air quality, particularly ozone concentration, must also be added to this list. Evidence for ozone restriction of forest tree photosynthesis comes from extremely varied studies ranging from seedlings to large trees and from trees growing in both controlled and natural environments.

Some of the most interesting and powerful evidence for ozone-induced reduction in photosynthesis comes from studies employing "open-top chambers," which allow exposure of trees to known concentrations of ozone in field (natural) environments. The chambers are cylinders or rectangles constructed with lightweight metal frames covered with clear plastic film. Standard chambers are approximately 3 m in diameter and 2.4 m in height. They are equipped with blowers to introduce approximately 60 m^3 of air per minute through a manifold surrounding the lower 1.2 m of the chamber. The air entering "control" chambers passes through both particulate and charcoal filters. Nonfiltered chambers permit the use of ambient pollutants as a baseline for developing additional

exposures. Different exposures are obtained by either adding pollutants at various concentrations to the ambient pollution load or by partial filtration to reduce the ambient pollution load. In recent years, a large number of investigations have employed field chambers to study the influence of ozone on both hardwood (oak, maple, poplar) and softwood (eastern white pine, ponderosa pine, southeastern pines) forest tree species. In most cases, ozone-induced reductions in photosynthesis were related to declines in growth or yield. In many cases, these declines occurred in the absence of any visible symptoms of stress.

With regard to ozone and agricultural productivity, the U.S. National Crop Loss Assessment Program has indicated yield losses in numerous crops where the growing season 7-hr mean for ozone is 40 ppb (78 $\mu g\ m^{-3}$). Woody plant evidence available for selected species suggests that forest trees may also experience important growth reductions at the ambient exposures that adversely impact agricultural species. At low ozone concentration (approximately 50 ppb, 98 $\mu g\ m^{-3}$), evidence from agricultural crops, hardwood trees, and conifers suggests linear reductions in net photosynthesis and growth with respect to ozone uptake. When uptake is the same, agricultural crops are more sensitive than hardwoods, which in turn are more sensitive than conifers. These differences in sensitivity are presumed to be due to several factors, with some of the most important being that conifers generally have lower pollutant uptake than crop species and require longer exposure to ozone to have comparable dose, conifer foliage is less productive per unit time than crop foliage, and conifer needles have a higher capacity to resist stress (low nutrient supply, insect, or microbial stress) than crop foliage. Hardwoods are judged to be intermediate with regard to these characteristics. Ambient air over large portions of eastern North America has an average of 50–70 ppb (98–1372 $\mu g\ m^{-3}$) ozone on clear summer days (natural background is 20–30 ppb, or 39–59 $\mu g\ m^{-3}$), with frequent peak concentrations of 80–110 ppb (159–216 $\mu g\ m^{-3}$). Evidence suggests that following 1–2 weeks of typical growing season pollution, agricultural crops will exhibit significant declines in net photosynthesis and growth. Hardwood forest trees will begin to exhibit similar decreases following several additional weeks of elevated concentrations, that is, with an integrated exposure (concentration × time above background of approximately 10–20 ppm-hrs). The threshold for conifer impact may be several months of elevated ozone concentrations resulting in exposures approximating 25–100 ppm-hrs.

In addition to photosynthesis (fixation of carbon), the allocation and utilization of carbon within a plant is a critical regulator of plant health. Carbon is allocated within trees to a variety of processes requiring carbon resources, including growth (roots, leaves, wood-biomass accumulation), respiration (energy generation for growth and tissue repair), metabolism, reproduction (flowers, seeds), and defensive chemicals to reduce insect attack or microbial infection. In addition to the influence on photosynthesis, exposure to ozone may influence carbon allocation within plants. Although all evidence is not consistent, several studies have suggested reduced movement of carbon to roots following ozone exposure. In trees, between 15 and 50% of photosynthate is allocated to produce, maintain, and replace fine-root systems annually. A reduction in root growth could increase the risk of drought stress, as well as exacerbate nutrient deficiencies. In addition, a variety of studies have indicated an increased allocation to respiration following ozone exposure. More carbon to respiration means less available to be stored in the form of woody tissue and this would result in less growth. Reduced carbon allocation to defensive chemical synthesis could result in greater insect feeding or pathogen infection. In summary, ambient concentrations of ozone are probably sufficient to cause reductions in photosynthesis or adverse perturbations to carbon allocation that result in growth reductions of most vegetation in areas where mean daytime growing season concentrations average above 50 ppb. This includes vast areas of crop and forest land throughout the United States and other portions of the developed temperate zone. The classic studies conducted in the San Bernardino National Forest in southern California provide evidence for both tree mortality and growth reduction caused by ozone.

D. Which Regional-Scale Air Pollutant Is Most Important for Forest Health?

The answers to at least three additional questions would seem appropriate in order to address this main question. What is the strength of evidence linking adverse forest effects with pollutant exposures characteristic of regional-scale areas? How many forest hectares are subject to significant deposition of each pollutant? What is the probability of effective pollution management (or abatement) via emissions standard, ambient air quality standard, or other regulatory or process conversion strategies?

Recent large research efforts in the United States—notably the National Acid Precipitation Assessment Program; the Response of Plants to Interacting Stresses, Integrated Lake and Watershed Assessment Study, and Aluminum in the Biosphere programs of the Electric Power Research Institute; and the Air Quality/Forest Health Program of the National Council of the Paper Industry for Air and Stream Improvement—have added great amounts of data and evidence relevant to an expanded understanding of the interaction between forest systems and air pollution. If we inventory the evidence and weigh it in terms of quality and quantity, we see apparent support for ozone as the most significant regional-scale contaminant.

With regard to distribution of the three regional-scale pollutants, there is substantial evidence by which to rank them, in terms of amount of U.S. rural region subject to deposition, in the following order: ozone > heavy metals > acid deposition. Rural ozone monitoring, as previously indicated, exhibits elevated levels over wide areas of agricultural and forest regions. Heavy metals, owing to their association with extremely fine particles in the troposphere and extremely diverse and diffuse sources, are widely deposited over rural landscapes. The National Acid Deposition Assessment Program has clearly demonstrated that acid deposition is region-specific because of the concentrated distribution of primary sources (sulfur dioxide in particular) and the relative efficiency of specific deposition mechanisms (notably cloudwater deposition). As a result, the North American regions with greatest acid deposition risk include high-elevation sites in the eastern United States and in eastern Canada.

The management (reduction) of regional-scale pollutants varies greatly according to source strengths, source distributions, available technologies, mitigation costs, and regulatory effectiveness. Heavy metal emissions to the atmosphere have been greatly reduced by the requirement for electrostatic precipitators or baghouses on new industrial, utility, and incineration sources. The adoption of an ambient air quality standard for lead (1977) and the phase-down of the use of lead as a gasoline additive (1975–1990) have dramatically reduced the deposition of this toxin in rural landscapes. In 1973, Dr. Thomas Siccama (of Yale University) and I initiated a detailed study of the biogeochemistry of selected heavy metals at the Hubbard Brook Experimental Forest, in the White Mountain National Forest in central New Hampshire. The Hubbard Brook forest is approximately 120 km northwest of Boston and relatively distant from any major local sources of heavy metal emission. We currently have over 20 years of data on the bulk precipitation input (dry plus wet deposition, monthly for 12 months annually) and streamwater output of cadmium, copper, iron, lead, manganese, nickel, and zinc. Over the first 10 years of our study, lead exhibited the most impressive accumulation in our northern hardwood forest. Examination of annual lead input, however, reveals a dramatic decrease from project initiation to present. This decrease in lead input to the forest correlates extremely well with the decrease in national urban lead concentrations (ambient air) and with the decrease in lead consumed in gasoline over the same time period.

The revised Clean Air Act (1990) holds great promise to reduce the precursors of acid deposition over the next decade. The revised Act calls for a 10 million ton reduction in sulfur dioxide emissions and a more than 2 million ton reduction in nitrogen oxide emissions from 1980 levels by 2000. After 2000, sulfur dioxide emissions will be capped and no increase in total emission will be allowed. The

principal precursors for ozone synthesis, nitrogen oxides and hydrocarbons, are also specifically addressed in the revised Clean Air Act. The 1990 Act tightens emission standards for cars and trucks, requires reformulated gasoline in certain high impact areas, and requires use of unconventional fuels by fleet operations in designated areas. The ability of the revised Act to achieve compliance with the ambient standards for ozone in the near term, however, remains uncertain. As a result, ozone appears to be the regional-scale pollutant with the highest probability of remaining a significant forest health stressor.

IV. GLOBAL-SCALE AIR POLLUTANTS AND FOREST HEALTH

Increasing evidence emphasizes the ability of human beings to influence specific global-scale atmospheric processes. Among these, some of the most important include: trace gas loading of the troposphere with the radiatively active species carbon dioxide, methane, nitrous oxide, and halocarbons and the hypothesis for global warming; chlorine loading of the stratosphere via chlorofluorocarbons and the hypothesis for stratospheric ozone depletion and potential for increased UV-B radiation at the surface of the earth; and global circulation and ultimate deposition of trace pollutants, including chlorinated pesticides, polychlorinated biphenyls (PCBs), polynuclear aromatic hydrocarbons (PAHs), and heavy metals. The implications of these perturbations in atmospheric processes are potentially profound for the natural resource systems of the earth. This is particularly true for climatic alteration! [See GLOBAL ANTHROPOGENIC INFLUENCES.]

Climate is defined as the time-averaged value of meteorological quantities. Over time, climate, like a forest, is characterized by change not constancy. In geological terms, the climate of the earth is most typically characterized by extended *moderate periods* with equable weather the year-round, lack of ice caps, and generally warm seas. Humans evolved after the last *moderate period* and our development has been in a period of climatic revolution. This period of revolutionary climatic alteration has been characterized by a complex of *cycles within cycles.* Over the past 3000 years, for example, the general evidence suggests that the northeastern portion of North America has become cooler and more moist. Over the past several hundred years, however, and particularly during the first half of the present century, there is evidence for a moderation trend. There is general agreement that there has been a systematic fluctuation in recent global climate characterized by a net worldwide warming of approximately 0.5°C between the 1880s and the early 1940s. This warming trend in global temperature persisted to the early 1950s, was followed by temporary cooling through the 1960s, and resumed warming into the mid-1970s. Warming has generally continued over the past two decades.

The regulation of global climate is complex and incompletely appreciated. Numerous hypotheses have been proposed to explain the forces responsible for the variability of climate. The most plausible of these include: variations of the solar constant, changes in solar activity, passage of the solar system through an interstellar gas-dust cloud, variation in the velocity of the earth's rotation, gigantic surges of the Antarctic ice sheet, changes in the earth's orbital parameters, and alterations in the interactions between glaciers and oceans. Climate may respond rapidly and dramatically to small changes in these independent variables. Added to this complexity and uncertainty is the suggestion that the activities of human beings, particularly land use activities and atmospheric contamination, are currently influencing global and regional climates. Numerous trace gases of the atmosphere have strong infrared absorption bands. As a result, these gases can have a significant effect on the thermal structure of the atmosphere because they absorb within the 7- to 14-μm atmospheric window, which transmits most of the thermal radiation from the surface of the earth and troposphere to space. A primary result of more carbon dioxide, methane, nitrous oxide, ozone, water vapor, and halocarbons in the atmosphere will be warming. Though incoming solar radiation is not absorbed by carbon dioxide and these trace gases, portions of infrared radiation from the earth to space are. Over time, the earth could become warmer. So although the forces controlling global temperature are varied and complex, as suggested, the increase of 0.5°C since the mid-1800s is generally agreed to be at least partially caused by increased carbon dioxide. By 2000 it may increase an additional 0.5°C.

General Circulation Models, widely used to predict future climates, are numerical models of the earth–atmosphere system that solve the basic equations for atmospheric motion and provide boundary conditions of the earth and ocean. Using the results of General Circulation Models, the National Academy of Sciences and the Intergovernmental Panel on Climate Change have estimated a mean global average surface warming of 3 ± 1.5°C (5.5 ± 2.7°F) in the next century with a doubling of carbon dioxide concentration. Throughout most of the northern United States and Canada, the growing season could be increased by 20% or more. General Circulation Model predictions of the change in the hydrologic cycle with doubling of atmospheric carbon dioxide are less clear. Precipitation is generally estimated to increase, but warming will intensify evaporation. The movement of water into and out of the soil–vegetation system has a large influence on the hydrologic cycle and shows considerable regional variability. If carbon dioxide in the atmosphere doubles over current concentrations, it has been estimated that soil moisture will decrease (via evaporation/transpiration) over much of the Northern Hemisphere during the growing season of June through August.

The current state-of-science does NOT permit confident prediction of future climate. There is general consensus that continued loading of the atmosphere with trace gases will eventually result in warming, but there is NO agreement on the amount or timing of this warming. The uncertainty surrounding future changes in hydrologic cycles, especially cloud dynamics, precipitation patterns, and soil moisture, is particularly significant. We cannot, at present, provide probability estimates for future climates for specific forest regions. We recognize, on the other hand, that the condition of the long-term climate is the most important regulator of forest ecosystem distribution. Climatic variables, principally temperature and moisture, establish the range of biotic components of ecosystems. Climate, in interaction with regional geology, determines the physical and chemical character of the soil substrate. In the short term, weather is variable and highly interactive with other forces that regulate the structure and function of forest ecosystems. Forests are complex, long-lived systems that may be quickly responsive to weather stress, but slowly adaptive to climate change. Strategic planning for future forest management must include the possibility of future climate change even if the probability cannot be estimated! We hypothesize that the initial responses (*near term*) of forest systems to climate change will be mediated largely by changes in the *traditional* abiotic and biotic regulators (the stressors, see Table II) of forest health and development. Ultimate (*extended term*) forest responses to climate change will be importantly directed by evolutionary-scale forces associated with tree reproductive biology and interspecies competition.

V. THE FUTURE

The intense research efforts of the last 30 years concerning the interaction of air contaminants with forest systems have taught us a great deal. Assessment of our improved understanding allows us to focus our future research energy and resources on the most important questions and challenges:

- The vast size of our forest resource, great variability of forest systems, and huge differences in forest values to humans greatly restrict generalizations concerning air pollution damage to forests.
- We have learned that pollutant exposures in North America at the regional scale cause subtle, not acute, responses in some wildland ecosystems.
- Subtle perturbation in ecosystem-scale processes on the order of 10–15% caused by air pollution are extremely difficult to partition from changes in these processes caused by other natural or unnatural stressors.
- Short-term research programs on the order of 3 to 5 years cannot adequately document forest

change occurring on multiple-decade time scales.
- The various regional-scale air pollutants, and the subtle stresses they impose on forest ecosystems, are not independent of one another nor of other stresses and, in fact, are highly interactive.
- The values of forests to humans are numerous. Some values are related to forest products and some to forest services. Some values are priced and some are unpriced. All values are important and air pollution stress has different significance for different values.

To effectively address these challenges we must expand our efforts in two critically important areas, research and monitoring. With regard to research we must fully recognize that the applied sciences of forest health management are directly dependent on our basic understanding of tree and forest form and function. We desperately need greater fundamental understanding of tree biochemistry and physiology. We also need more appreciation of basic forest ecology and ecosystem dynamics. Research programs dedicated to specific stress factors must have both field and laboratory components. Studies of mechanisms of disease obtained in managed environments of the laboratory, growth chamber, and greenhouse must be combined with studies conducted in the field to obtain a comprehensive perspective. Efforts in controlled and natural environments must receive equivalent support and encouragement. Long-term, ecosystem-scale research must also be encouraged and supported. Research information must be synthesized and integrated, and models of risk processes and management options must be developed. Predictions of ecosystem toxicology and stress processes need to be provided to resource managers, regulators, and decision makers.

Finally, we must implement a comprehensive environmental monitoring program that will generate accurate information on the deposition of air pollutants in rural areas and that will allow assessment of change in wildland ecosystems. Air quality monitoring is critical for accurate determination of ecosystem exposure and for reliable indication of the efficacy of pollution reduction strategies. Monitoring ecosystem change is essential for improved understanding of the dynamic nature of complex systems and to allow correlation of change with environmental stressors. Documentation of change in the pattern of natural landscapes is critically important to differentiate natural stressors from human stressors. The objectives of the Environmental Monitoring and Assessment Program (EMAP) of the EPA and the National Biological Survey of the Department of the Interior are fully consistent with the latter need. The objectives of EMAP include: estimation of the current status, extent, changes, and trends in indicators of the condition of U.S. ecological resources; tracking indicators of pollutant exposure and habitat condition in an effort to reveal associations between human-induced stresses and ecological condition; and compilation of periodic reports on ecological status and trends.

Research indicates that regional-scale air pollution is one of the significant contemporary anthropogenic stresses imposed on some temperate forest ecosystems. Gradual and subtle change in forest metabolism and composition over wide areas of the temperate zone over extended time, rather than dramatic destruction in the immediate vicinity of point sources over a short period, must be recognized as the primary consequence of regional air pollutant stress in North America. Global-scale air pollution, including the "greenhouse" trace gases and chlorofluorocarbons, with their capability to alter fundamental radiation balances of the earth and potential to cause rapid climate change, has the potential to dramatically alter forest ecosystems in the next century. The integrity, productivity, and value of forest (and other wildland) systems are intimately linked to air quality. We must elevate consideration of forest resources in societal considerations of energy technologies. Management and regulation of air resources is shortsighted if it fails to recognize the critical linkage between the quality of human experience and the quality of our natural environment.

Glossary

Anions Negatively charged ions, usually nonmetals or acid radicals.

Cations Positively charged ions, usually hydrated hydrogen or metal ions.

Cloud water Mass of condensed water vapor particles or ice suspended above earth or deposited via impaction or surfaces at the earth's surface.

Dry deposition Deposit of gaseous or particulate pollutants, in between precipitation of events, to surfaces on the earth.

Ecosystem health Quantification of the intensity of ecosystem stressors (both living and nonliving) in the context of appropriate spatial, temporal, and human-value scales.

Forest stresses Living or nonliving factors (of human or nonhuman origin) that are capable of adversely impacting forest ecosystems.

Global-scale air pollutants Contaminants produced around the world or circulating around the world that are capable of adversely impacting large portions of the earth; examples include the greenhouse gases capable of influencing global climate, chlorofluorocarbons capable of influencing stratospheric ozone concentrations, and organochlorine chemicals capable of bioaccumulation and biomagnification in food webs.

Ozone Molecule consisting of three atoms of oxygen; in the stratosphere it occurs naturally and shields the earth from harmful ultraviolet radiation, and in the troposphere (lowest portion of the atmosphere) it is produced by the reaction of nitrogen oxides and hydrocarbons produced by human activity in the presence of sunlight and warm temperatures and is a principal component of "smog."

Regional-scale air pollutants Contaminants deposited hundreds to thousands of kilometers downwind from the source or point of precursor release; examples include heavy metals associated with small particles, acid deposition, and tropospheric ozone.

Wet deposition Deposit of gaseous or particulate pollutants, during precipitation events (rain, snow, sleet, cloud water), to surfaces on the earth.

Bibliography

Adriano, D. C., and Johnson, A. H., eds. (1989). "Acidic Precipitation. Vol. 2. Biological and Ecological Effects." New York: Springer-Verlag.

Barker, J. R., and Tingey, D. T., eds. (1992). "Air Pollution Effects on Biodiversity." New York: Van Nostrand Reinhold.

MacKenzie, J. J., and El-Ashry, M. T., eds. (1989). "Air Pollution's Toll on Forests and Crops." New Haven, Conn.: Yale University Press.

Smith, W. H. (1990). "Air Pollution and Forests," 2nd ed. New York: Springer-Verlag.

U.S. Congress, Office of Technology Assessment (1993). "Preparing for an Uncertain Climate," Vol. I, OTA-0-567; Vol. II, OTA-0-568. Washington, D.C.: U.S. Government Printing Office.

Antarctic Marine Food Webs

Coleen L. Moloney and Peter G. Ryan

University of Cape Town, South Africa

Despite the rigorous climate, the seas immediately around Antarctica are highly productive, supporting large populations of marine vertebrates such as sea birds, seals, and whales. Traditionally, it has been assumed that the large abundance of predators in this region is the result of short food chains in which large phytoplankton are eaten by Antarctic krill, which in turn serve as food for top predators. However, Antarctic marine food webs are much more complex than portrayed by this idealized short food chain. Much of the primary production is due to small, not large, cells, and a variety of microorganisms such as bacteria and protozoa play an important role in consuming energy and in recycling nutrients that are used for plant growth. There is great spatial and temporal variability in productivity. The greatest productivity occurs in the spring and summer in association with seasonal pack ice. However, other regions also contribute to overall productivity, and three distinct environments with unique biological–physical characteristics are recognized: the coastal zone, the seasonal ice zone, and the open ocean zone. Different species dominate the communities in these three environments, and the structures and functioning of the component food webs differ.

I. INTRODUCTION

Food webs represent the intricate feeding connections among the plants and animals that are the living components of ecosystems. We study food webs in order to understand how these components interact and the relative roles played by the different parts in distributing energy and material to living and nonliving forms. By improving our understanding of how food webs function, we are in a better position to assess how changes in one component may affect the whole community. This is particularly important in assessing the consequences of environmental variability and natural change, as well as estimating the impacts of past, present, and future human activities. [*See* PLANT–ANIMAL INTERACTIONS.]

Marine organisms of the Antarctic live in a variable environment, with extreme seasonal changes in day length and ice cover, although temperatures

We thank Michael Lucas of the University of Cape Town for helpful discussions. CLM acknowledges the support of the National Science Foundation, award No. OCE-9016721 of the US-GLOBEC program, while based in the Department of Wildlife and Fisheries Biology, University of California, Davis, CA 95616.

remain constantly low. The cold temperatures and large variations in incident radiation impose limits on primary production and growth rates. Relatively few marine species can cope with these harsh conditions, and many of the plants and animals have special adaptations. For example, Antarctic fish generally have few red blood cells, if any, and reduced hemoglobin concentrations, because their blood is so viscous in the cold temperatures in which they live. As a result, the fish tend to be sluggish and are unable to fill the niche of active pelagic swimmers. Most of the approximately 120 fish species are small, and are sedentary demersal dwellers. The most important pelagic species is the Antarctic silverfish *Pleuragramma antarcticum,* located in coastal regions. A striking feature of Antarctic marine food webs is the lack of significant numbers of epipelagic fish, especially when compared with pelagic food webs in other regions of the world.

The absence of pelagic schooling fish in Antarctic waters has allowed a euphausiid crustacean, Antarctic krill *Euphausia superba,* to exploit the epipelagic niche. Krill are relatively large (~5 cm in length) mobile animals, as are fish, but krill are well adapted to the extreme Antarctic conditions. In spring and summer they feed on large-celled phytoplankton, which grow rapidly in the well-illuminated and nutrient-rich waters at the edge of the pack ice. During winter krill are uniquely adapted to feed on algae found on the undersurface of the pack ice, which covers large parts of the ocean for more than half of the year. This dual mode of feeding (pelagic in summer and inverted benthic in winter) has allowed krill to become a very successful colonizer of this physically extreme environment. [*See* KRILL, ANTARCTIC.]

The low species richness of the Antarctic contrasts with the large abundance of a few dominant species. The massive biomass of Antarctic krill (on the order of 100–1000 million metric tons) in part supports the large numbers of whales, seals, and sea birds that are characteristic of the region, and in many cases unique to the Antarctic. Historically, Antarctic waters served as a summer feeding ground for large numbers of whales, before the populations of most were reduced drastically by

commercial exploitation. The Antarctic also is home to the most abundant seal species in the world, the crabeater seal *Lobodon carcinophagus,* which is an obligate krill feeder, and the region plays host as breeding area to many sea birds, most notable of which are the penguins. The emperor penguin *Aptenodytes forsteri* breeds under probably the most extreme environmental conditions of any bird. Adults attend their eggs and chicks throughout the Antarctic winter in temperatures as cold as −48°C. During this period of enforced fasting, the adults can lose up to 40% of their body mass. In addition to these large animals, there is also an active microbial food network in the Antarctic, which does not "switch off" during winter. Many small organisms live in the pack ice, a significant structuring force of Antarctic environments, which also provides shelter and resting areas for many seals and sea birds.

Food web representations of any ecosystem are simplifications and summaries of the components and the most important processes. The Antarctic marine food web is a continually changing mosaic, having seasonal pulses of biomass and production and communities that change in response to environmental and biological factors. In this article we first present details of parts of the system, then summarize these in a coherent picture of how the pieces fit together. Finally, some current environmental issues are explored as well as the ways in which they can impact Antarctic marine food webs.

II. FOOD WEB STRUCTURES

Food webs are composed of primary producers (mostly plants) and consumers (usually animals and bacteria). The primary producers are so named because they use the energy from sunlight to manufacture complex organic molecules from simple constituents in a process called photosynthesis. The resulting growth of the organisms provides food for the herbivore consumers, and these herbivores themselves may be eaten by other consumers. In marine pelagic food webs the primary producers are phytoplankton and cyanobacteria, small single-celled organisms occurring singly or in aggregates as chains and colonies. Some phytoplankton also may engulf other organisms, but consumers generally consist of microorganisms such as bacteria (bacterioplankton) and protozoa (protozooplankton), small invertebrates such as crustaceans, mollusks, and jellyfish (zooplankton), large strong-swimming organisms such as squid and fish (nekton), and air-breathing vertebrates such as sea birds and marine mammals.

The amount of primary production available to a food web ultimately determines the total productivity of the system, because all other living organisms derive their energy from the primary producers. The productivity of large organisms such as squid, fish, sea birds, and marine mammals is determined only in part by the total amount of primary production. Also important are the routes by which energy is transferred to these large consumers. It

is possible to trace the transfer of food or energy from the primary producers to "top" consumers, whether these top consumers are ciliates, squids, or whales. Such representations of step-by-step feeding transfers among individuals are known as "food chains." Whenever an organism is eaten by another, energy is transferred from prey to predator, but energy also is lost at each transfer, or step, in the food chain. Short food chains involve fewer transfers and thus are more efficient (i.e., waste less energy) than long food chains. It was believed that large phytoplankton (e.g., diatoms) at the base of short food chains, with Antarctic krill as the major link to vertebrate predators, were characteristic of the Antarctic pelagic ecosystem (Fig. 1). These short efficient food chains could account for the large populations of sea birds, seals, and whales which were observed by early visitors to the region.

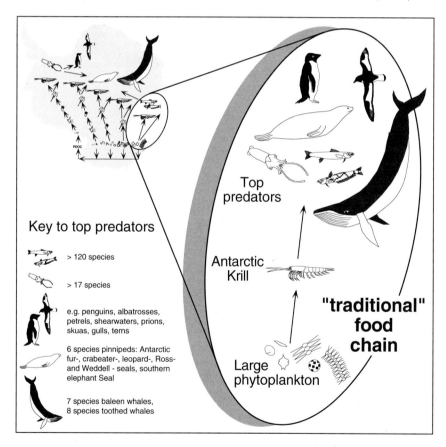

FIGURE I The "traditional" Antarctic marine food chain: large phytoplankton–Antarctic krill–top predators. In reality, this traditional food chain is only one of many food chains embedded within a more realistic Antarctic marine food web. The importance of different parts of the food web varies regionally and seasonally.

However, in reality no single food chain is representative of all feeding transfers in a community. Food webs are an intertwined mesh of all possible food chains.

The diatom–krill–marine vertebrate food chain (Fig. 1) is overly simplistic. Historical measurements of primary production and phytoplankton abundance were taken mainly in spring and summer and close to the Antarctic continent. It is now known that the results are not representative of the entire region or of the entire year. Large diatoms dominate the phytoplankton only in spring and summer and only in coastal regions, near the ice edge, and at oceanic frontal zones. The remainder of the region has relatively little primary production, carried out mainly by small phytoplankton cells. Krill are not able to feed efficiently on plant cells smaller than ~20 μm, and a large part of the primary production cannot be harvested directly by Antarctic krill. Instead, these small cells support a significant biomass of heterotrophic microorganisms throughout Antarctic seas, including free-

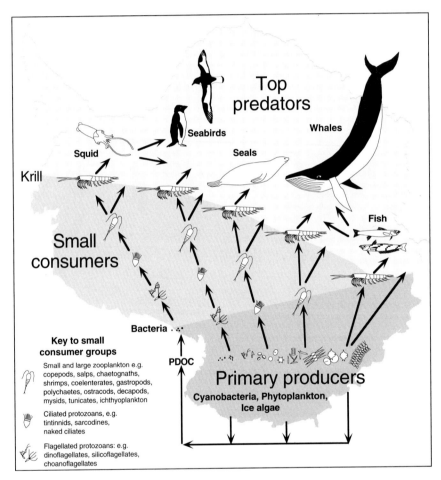

FIGURE 2 The Antarctic marine food web, showing all sizes of primary producers, small consumers, and top predators. Major groups are depicted by icons, explained in the keys of Figs. 1 and 2. All arrows represent transfers of carbon (or energy). Primary consumers lie at the interface between the primary producer and small consumer groups, and Antarctic krill are the major intermediary between small consumers and top predators. Phytoplankton secrete dissolved organic carbon (PDOC), which is used by bacteria. The weblike nature of interactions among small consumers has been "teased out" into a series of simplified food chains, each of which terminates in "top predators." In reality, these food chains may be truncated because of natural mortality within the small consumer groups. The detailed feeding interactions among the top predators are not shown.

living bacterioplankton and protozooplankton. These microorganisms are important consumers of primary production and also play a role in remineralizing nutrients for plant growth. The complex feeding interactions that take place among phytoplankton, bacterioplankton, and protozooplankton are often referred to as the "microbial loop."

The central role traditionally ascribed to Antarctic krill must be tempered by the fact that the species is not found throughout the region. Krill generally are absent from open-ocean regions and also do not occur in the productive shelf waters immediately fringing Antarctica; in these areas copepods, salps, and other euphausiid species dominate zooplankton assemblages. Current representations of Antarctic marine food webs have been made more realistic by including small phytoplankton and groups of microorganisms. This results in an increased num-

ber of possible food chains with marine vertebrates at the apex (Fig. 2). The relative importance of different food chains in different regions forms the focus of this article, reflecting the marked regional differences in ecosystem structure and functioning in the Antarctic.

III. REGIONAL DESCRIPTION

The Southern Ocean lies between the continental land mass of Antarctica and the Subtropical Convergence, but not all of the Southern Ocean falls within the Antarctic realm. The Antarctic marine province is restricted to the region south of the Polar Front (Fig. 3), covering an area of ~35 million km². The Subantarctic province extends north of the Polar Front to the Subtropical Convergence. The Polar Front or Antarctic Convergence is situ-

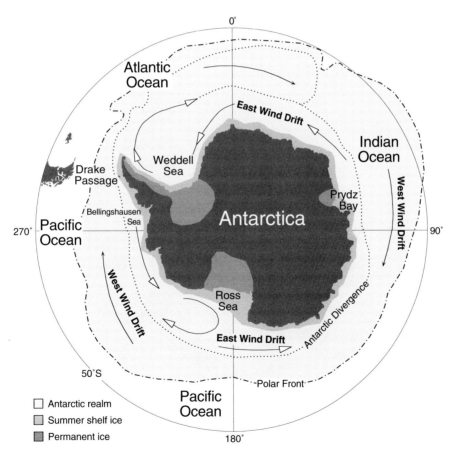

FIGURE 3 Antarctica and surrounding waters, showing the major oceanographic features and regions.

ated at approximately 50°S and bisects the major surface circulation in the region, the Antarctic Circumpolar Current. This current is a broad meandering eastward flow, often called the West Wind Drift after the west winds by which it is driven. South of the Antarctic Circumpolar Current is a region where water masses diverge and deep water is brought to the surface in a process termed "upwelling" (Fig. 4). This zone of upwelling is called the Antarctic Divergence. The surface circulation between the Antarctic Divergence and the coast of Antarctica is predominantly westerly and is known as the East Wind Drift.

Water that is brought to the surface at the Antarctic Divergence is rich in nutrients, giving rise to the characteristic chemistry of Antarctic surface waters, which have high concentrations of dissolved nutrients that are important for phytoplankton growth. The surface waters at the Antarctic Divergence have the highest concentrations of nitrates (25 μM), phosphates (2 μM), and silicates (60 μM) found anywhere in the world's oceans. However, despite the favorable nutrient status of these open ocean waters, the biomass and production of the phytoplankton are unusually low.

Superimposed on the large-scale circulation patterns described above are regional variations caused by coastal and bottom topography and the seasonal cycle of ice advance and retreat. Localized regions of high productivity occur in association with regional features. On the basis of physical, chemical, and biological factors, three subsystems have been identified: (1) the permanently open ocean zone, including frontal regions, (2) the seasonal ice zone, and (3) the coastal and continental shelf zone.

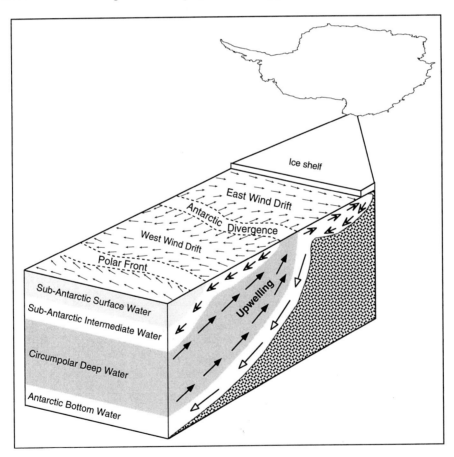

FIGURE 4 Schematic showing the main oceanographic features that influence the nutrient status of Antarctic surface waters. Nutrient-rich water is brought to the surface at the Antarctic Divergence.

IV. REGIONAL VARIATIONS IN ANTARCTIC MARINE FOOD WEBS

A. Food Web of the Permanently Ice-Free Zone

Much of the West Wind Drift is ice free year-round and is characterized by cold temperatures, abundant nutrients, strong wind stress, weak density stratification, deep vertical mixing, and little chlorophyll (an indicator of low primary production). This region includes the Indian, Atlantic, and Pacific Ocean sectors of the Southern Ocean, the Scotia Sea, and the Drake Passage (Fig. 3), covering an area of 14 million km². Upwelling at the Antarctic Divergence supplies large concentrations of nutrients to these surface waters. In low-latitude regions such high nutrient concentrations typically would be removed from the surface waters within weeks by phytoplankton blooms, but large phytoplankton blooms do not occur in the open ocean areas of Antarctica. In fact, chlorophyll concentrations in this part of Antarctica are as low (generally <1 mg of chlorophyll a·m⁻³) as those found in nutrient-poor tropical oceanic waters.

Given the abundance of nutrients, the most important factor limiting phytoplankton growth is probably light limitation through deep vertical mixing. The vertical structure of oceanic waters consists of a surface layer that is influenced by winds and mixed continually, so that organisms and particles suspended in the surface waters are transported between the top and bottom of the mixed layer (Fig. 5). Phytoplankton require sunlight to form energy-rich chemical compounds during photosynthesis. This energy is used by the cells during respiration. In order for net growth of an individual cell to occur, the energy fixed during photosynthesis should exceed the energy used during respiration. Respiration rates are relatively constant with depth, but photosynthesis decreases with depth as light levels decrease. At shallow depths photosynthesis exceeds respiration, but at greater depths respiration exceeds photosynthesis. The "compensation depth" (Fig. 5) is the depth at which the gains from photosynthesis are balanced by the losses due to respiration. The "critical

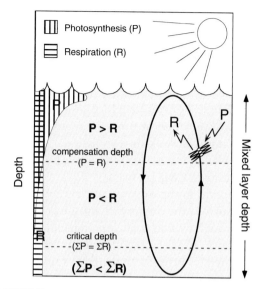

FIGURE 5 Hypothetical representation of the surface mixed layer of the open ocean zone, showing the relationship of photosynthesis and respiration with depth, and the compensation and critical depths. The phytoplankton follows the trajectory shown, moving between the surface and the bottom of the mixed layer.

depth" (Fig. 5) is the maximum mixing depth that permits net growth for a phytoplankton community. If phytoplankton are mixed below the critical depth during their vertical transport, they experience net energy losses. In much of the open ocean zone, the mixed layer is deep, ~50–200 m, and deep vertical mixing causes light to limit primary production. Most known areas of increased primary production in the Antarctic are associated with surface stratification of the water column, which results in a shallowing of the mixed layer.

In addition to light limitation, it has been suggested that micronutrients such as iron and other elements necessary for photosynthesis may limit phytoplankton growth. Iron is abundant in most coastal waters through river runoff, but Antarctica is a cold desert with virtually no surface runoff. Iron inputs through aeolian (windborne) contributions are necessarily low, given the mantle of ice that covers 98% of the continent. There have been serious, if controversial, proposals to increase primary production in the nutrient-rich waters of the Antarctic by supplying iron to the system.

The food web of the open ocean zone of the Antarctic is based on consistently low primary production, carried out mainly by small phytoplankton cells and cyanobacteria (Fig. 6). Because the dominant primary producers are small cells, they are eaten mainly by small consumers such as protozooplankton and small zooplankton, rather than by large zooplankton such as krill. Protozooplankton consist mainly of heterotrophic dinoflagellates and ciliates, and they may at times exert strong grazing control on phytoplankton populations. The dominant zooplankton are copepods and salps, but these do not occur in large biomasses. Antarctic krill are seldom found in the open ocean, except for a few regions such as the Scotia Sea and Drake Passage, areas through which krill are transported. As a consequence of the low zooplankton standing stocks, there is little food available for squid and vertebrate predators. Sea birds, seals, and whales do not feed extensively in the open ocean zone, except in regions of locally increased production, such as localized frontal zones near gyres and eddies and near islands.

B. Food Webs of the Seasonal Ice Zone

In winter the seas surrounding Antarctica are covered by ice. The ice consists of permanent ice and seasonal ice. Permanent ice is thick (>100 m) and derives from snow falling on the continent, so it consists of fresh water. The major regions of permanent ice occur in and around coastal seas and embayments, including the western parts of the Weddell and Ross Seas and the Prydz Bay region (Fig. 3). Seasonal ice is formed each year and comprises 90% of the ice cover. It is 1–2 m thick and derives from the freezing of seawater, so it is slightly saline [~5 parts per thousand (ppt)]. The major ice-forming season is the austral autumn

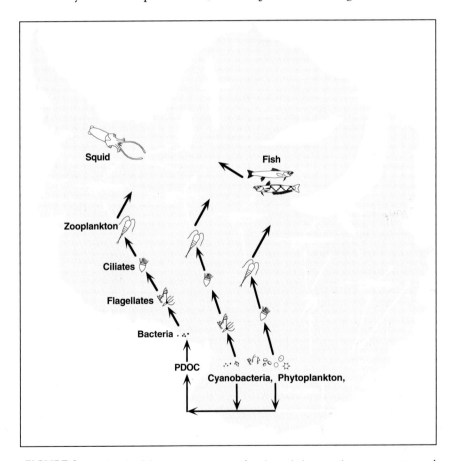

FIGURE 6 Food web of the open ocean zone, showing missing or minor components and linkages in white. Primary producers are small, and there is an active microbial component within the small consumers. Zooplankton are mostly small species and Antarctic krill are absent. Top predators occur in low numbers, and consist mainly of fish and squid, although some sea birds also are found.

(April), when there is rapid freezing and expansion of the ice cover northward. By late winter the pack ice can extend up to 1700 km offshore from Antarctica, covering ~20 million km². The region of the Antarctic which experiences the annual advance and retreat of the ice is known as the "seasonal ice zone" and covers ~16 million km². During summer the pack ice retreats toward the continent, and the ice-covered area shrinks to ~4 million km² (Fig. 3).

Within the seasonal ice zone three different environments are recognized. First is the ice itself, which has its own characteristic community and food web. Second is the water column underlying the ice. Third is the environment associated with the ice-edge zone, where the melting ice interacts with the open water. We explore seasonal changes in the food webs of each of these three environments.

I. Ice (Sympagic) Environment

The use of ice-breaking vessels in polar oceanographic research has greatly increased our understanding of the ice, or "sympagic," environment.

It is now known that considerable biological production occurs within the pack ice. Primary producers in the ice are known as ice algae, a diverse array of phytoplankton groups. During the freezing of seawater, ice crystals form in the water column and float up to the surface. Algae adhere to the crystals and become incorporated within the ice matrix as it freezes, or occur in brine pockets, channels, and pools in the ice floes. The brine pockets can vary in size from hundredths of millimeters to tens of centimeters, usually increasing in size during periods of ice melt. The salinity of the brine typically can reach 60–70 ppt, approximately twice that of normal seawater and greater than can be tolerated by most species. In addition to being tolerant to high salinities, ice algae such as ice diatoms and foraminifers also are well adapted to low light levels, because of light attenuation in the ice. The major factor controlling production of ice algae is probably light, and seasonal variations in primary production occur because of seasonal changes in incident radiation. Total primary production in the pack ice changes as the ice cover expands and contracts in area.

During "community founder" events when the ice freezes, predators frequently are excluded by chance or because they are too large for the small brine habitats. The ice algae are then protected from metazoan grazing, and this allows algal standing stocks to build up to very high levels. Concentrations exceeding 400 mg of chlorophyll a·m⁻³ have been measured in surface ice in late summer. This is two orders of magnitude greater than is commonly found in phytoplankton blooms in temperate regions, and three orders of magnitude greater than is found in ice-free waters in the Antarctic. When the pack ice melts, ice algae are released into the water column. It has been suggested that these cells may seed the phytoplankton blooms that develop in association with the ice edge in spring. However, experimental studies have demonstrated that many ice algae, upon release into the water, form dense glutinous aggregations which sink rapidly out of the euphotic zone. These aggregations may be caused in part by mass mortality when the algae encounter a sudden salinity decrease when they are released from the high salinity brine habitat into the seawater. Many ice algae have sticky cell coverings, which may facilitate their incorporation into the ice, as well as contribute to the formation of the aggregations. The characteristic species that dominate the ice algal communities are not the same as those that dominate in the ice-edge blooms, although some species are found in both habitats. There is probably some interchange between ice algae and the water column community, but the extent and importance of the relationship require further study.

In addition to ice algae, the sympagic community also consists of marine bacteria and protozoa (Fig. 7a). Biomasses and growth rates of sympagic bacteria can be 10 times greater than those of bacteria in the water column underlying the ice. Grazing by protozoan herbivores such as flagellates and ciliates may at times control the algal and bacterial ice communities, and protozoa in turn are eaten by larger zooplankton.

Possibly the most important aspect of the sympagic community is the role it plays in sustaining zooplankton, particularly Antarctic krill populations, during winter. Despite popular belief that the Antarctic pack ice is unfavorable for animals

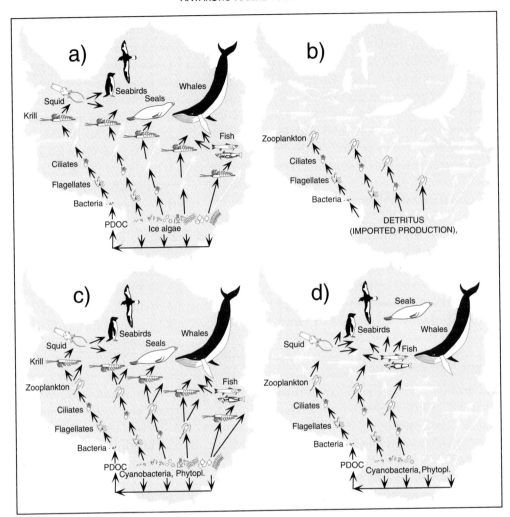

FIGURE 7 Food webs of the seasonal ice zone, showing missing or minor components and linkages in white. (a) Sympagic environment in winter. Primary producers are ice algae and there is an active microbial community. Antarctic krill use an inverted benthic mode of feeding, which supplies food for overwintering top predators. (b) Ice-covered water column. Secondary production is based on detrital imports rather than local primary production, and the community is sparse. (c) MIZ in spring and early summer. Primary producers are cyanobacteria and all sizes of phytoplankton. The zooplankton community is diverse, but dominated by Antarctic krill. Top predators are abundant. (d) MIZ in late summer and autumn. Primary producers are cyanobacteria and small phytoplankton, and small zooplankton dominate the community of consumers.

(krill were thought to survive the winter on stored lipid reserves), it is now known that extensive use is made of this seemingly harsh habitat throughout the year. Krill feed on the abundant algae that grow on the undersurface of the ice and in ridges and cracks, scraping off these cells with their filtering apparatus. They essentially switch from a pelagic to an inverted benthic feeding mode during winter, and in turn provide a food resource for vertebrate predators (Fig. 7a). Estimates of consumption of

krill by predators are variable, but on the order of 100 million metric tons per year. The most important krill predators are probably mesopelagic fish, not sea birds and marine mammals, as is commonly assumed.

The pack ice community of air-breathing vertebrate predators is composed of emperor and Adélie penguins; snow and Antarctic petrels; crabeater, Ross, Weddell, and leopard seals; and minke whales. These species are adapted to forage in association with pack ice, and they overwinter in the pack. Breeding occurs on the ice in winter (e.g., emperor penguins) or summer (e.g., Weddell seals) or in ice-free areas (e.g., other penguins and petrels and Antarctic fur seals). Other species of whales and sea birds migrate long distances to exploit the brief summer bonanza, but generally avoid dense pack ice (see below). In winter the diets of sea birds in the ice are more similar than at any other time of the year, probably because they are less opportunistic, because their preferred prey items are readily available near the undersurface of the ice. Krill are not the most important component of many marine vertebrates' diets during this time; mesopelagic fish and squid dominate in the diets of many sea birds. It is believed that the fish and squid move to the surface to feed on the krill associated with the ice, and thus become available to other predators. Any changes in the extent of pack ice, possibly caused by global warming, will affect the winter habitat of krill, and thus the sympagic community and food web.

2. Ice-Covered Water Column

The water column underlying the pack ice generally has very low primary and secondary production, much less than that occurring in the sympagic community and in the open water zone. This is due to light limitation caused by the overlying ice and is exacerbated by deep vertical mixing. Light levels under the ice may be reduced by 50–1000 times the intensity at the surface, and deep vertical mixing is caused by salinity changes. During ice formation salt is extruded from the freezing water, causing the salinity in the underlying water to increase. As a consequence of increased salinity, the water density increases and this dense surface water sinks, causing deep mixing in the water column and restricting primary production.

The relatively few animals in the water column under the pack ice feed mainly on detrital material and other animals, rather than plant life (Fig. 7b). An obvious exception is Antarctic krill feeding on ice algae, but krill are associated with the sympagic community (see above), not the water column. The ice-covered waters probably contribute little to overall productivity in the system, as long as

the ice remains in place overhead. In places the ice cover can be interrupted by leads and polynias. These openings are important for the air-breathing whales and seals that frequent the pack ice, and also provide penguins and seals with access to the water and resting areas on the ice. Despite increased incident light reaching the water column at these openings, chlorophyll concentrations in polynias are extremely low (<0.05 mg of chlorophyll $a \cdot m^{-3}$). Most feeding by the large predators is based on production from the sympagic community (including Antarctic krill), not from the water column.

3. Marginal Ice Zone

For most of the year the seasonal ice zone is a region of net ice melt. In winter local ice melt results from heating by warm (1–2°C) water from below, whereas in spring and summer the ice melt is caused mainly by increased solar/atmospheric heating from above. During the austral spring (October), the sea ice retreats southward, and as it retreats it forms an important biological zone known as the marginal ice zone (MIZ). The MIZ extends 50–250 km north of the ice edge, and is influenced by low-salinity meltwater from the retreating ice. The meltwater contains few dissolved salts and is less dense than the underlying seawater. As a consequence the well-mixed waters that are characteristic of the winter months are replaced by a stratified water column with a relatively shallow (30–80 m) low-salinity layer lying on top of a dense deep layer. Nutrient concentrations in the surface waters are high (as is typical for all Antarctic surface waters), average light levels become favorable for phytoplankton growth, and large phytoplankton blooms develop.

Phytoplankton blooms of the MIZ generally start in October–November but can occur as late as March. The blooms begin earliest in the northern regions, where the ice melt starts. December–January is the peak bloom season, and blooms have a duration of approximately 60 days. During the initial stages of the bloom in spring, large-celled centric diatoms and large long-chain-forming pennate diatoms dominate phytoplankton biomass (Fig. 7c), with relatively large chlorophyll concentrations of ~4 mg of chlorophyll a·m^{-3}. During summer the large-celled phytoplankton community is replaced by a small-celled one consisting of small pennate diatoms and flagellates (Fig. 7d). The food web changes from one based on "new" production (using upwelled nutrients) to one based on "regenerated" production (using nutrients recycled by consumers).

The microheterotrophic community of bacteria and protozoa is active throughout the year in the MIZ (Figs. 7c and d) and is responsible for much of the nutrient recycling. Bacterial biomass is typically 1–10 mg of carbon ·m^{-3} in late summer, ~10–20% of phytoplankton biomass. In spring bacterial production is ~15% of phytoplankton production, but in autumn it can exceed primary production in the water column. Protozooplankton reach their maximum biomasses in spring, but are also important in the MIZ in autumn, when they can exceed 20% of phytoplankton biomass. Protozooplankton are significant grazers of small phytoplankton and bacteria, as well as being important in recycling nutrients, excreting large concentrations of ammonia, which is used by plants for growth.

In spring and early summer the zooplankton biomass in the MIZ is ~150 mg·m^3 and is dominated by Antarctic krill (Fig. 7c). Krill are omnivorous and aggregate in regions of high summer productivity, to feed efficiently on the large diatoms and associated herbivores of the ice-edge blooms. Krill distribution is patchy, with major concentrations in regions such as Prydz Bay, the outer Weddell Sea, north of the Ross Sea, and the Bellingshausen Sea. During phytoplankton blooms in the MIZ sedimentation rates can be high, as diatoms and krill feces sink rapidly from the surface waters.

In late summer, when most phytoplankton in the MIZ are small cells, krill switch to feeding on ice algae and the zooplankton community is dominated by small zooplankton (Fig. 7d). Most of the more than 100 zooplankton species found in the MIZ are small-particle feeders, eating small phytoplankton and protozooplankton. These small zooplankton consist mainly of copepods (e.g., *Rhincalanus gigas, Calanoides acutus, Calanus propinquus,* and *Metridia gerlachei*) and the salp *Salpa*

thompsoni. Carnivorous zooplankton in the MIZ are mostly copepods and chaetognaths, and they mainly eat other copepods, not krill. In late autumn and early winter most zooplankton species descend to greater depths, where they overwinter as nonfeeding stages. Few fecal pellets are collected in sediment traps during winter, indicating that feeding is very much reduced.

Many sea birds and marine mammals are associated with the marginal ice zone. In terms of their feeding ecology, they can be divided into two main groups: those that feed within the pack ice (discussed above) and those that feed in open water, away from dense pack ice. However, both groups feed together at the ice edge. The open water community of sea birds includes a variety of penguins, albatrosses, petrels, shearwaters, prions, gulls, and terns, most of which breed farther north. Marine mammals associated with the MIZ include crabeater seals, many baleen whales, male sperm whales, and killer whales. In summer most sea birds and mammals prey on krill, fish, squid, and shrimp (Fig. 7c). In winter beaked whales, Antarctic fur seals, southern elephant seals, penguins, and some petrels occur at the ice edge, where their diets are composed mainly of fish and squid, although copepods also are eaten by right whales, prions, storm petrels, and diving petrels (Figs. 7a and d). The food items of these vertebrate predators are largely determined by what is available at different times.

The food web associated with the MIZ most closely represents the typical image most people have of Antarctic marine food webs. It is here that large phytoplankton blooms can be observed in the spring, and krill and krill predators are abundant.

C. Food Web of the Coastal and Continental Shelf Zone

The continental shelf of the Antarctic is very deep compared with continental shelves fringing other land masses. The coastal shelf regions are small in extent (~0.9 million km^2) and have their own characteristic biological communities. The coastal zone is separated from the seasonal ice zone by a thermal front over the shelf break. These shelf-break fronts are important feeding areas for top predators, such as minke whales and Adélie penguins. The major regions containing coastal zone environments are those found in the inner Ross and Weddell seas (Fig. 3). During winter these regions are ice covered. [*See* CONTINENTAL SHELF ECOSYSTEMS.]

Phytoplankton blooms develop in summer in coastal waters and are usually dominated by small pennate diatoms, such as *Nitzschia* spp. Antarctic krill are generally absent, and another euphausiid, *Euphausia crystallorophias,* is the major intermediary between phytoplankton and top predators, although other herbivores, such as the Antarctic silverfish, are also important (Fig. 8). Omnivores in the coastal zone include pagotheniid fishes and the squid *Psychroteuthis glacialis,* and important predators are nototheniid fishes; Adélie and emperor pen-

guins; Weddell, leopard, and crabeater seals; and minke whales.

Coastal regions have been the sites of superblooms, which are dense chlorophyll concentrations that develop in summer in a thin layer immediately under the ice, ~25–100 cm thick. These blooms are composed mainly of centric diatoms, which grow among the ice platelets in a stable protected environment. Superblooms have been noted in summer off the Antarctic Peninsula, in the eastern Weddell Sea, and in the Ross Sea. A single bloom can extend over areas up to 15,000–20,000 km^2, with typical chlorophyll concentrations of >300 mg of chlorophyll a·m^{-3}, an order of magnitude greater than is found in the most productive phytoplankton blooms in coastal regions elsewhere in the world. The seasonal pre-

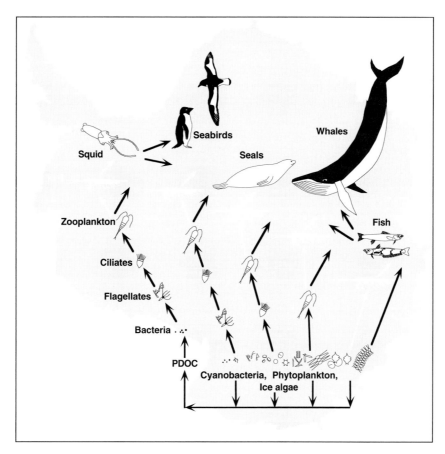

FIGURE 8 Food web of the coastal zone in summer, showing missing or minor components and linkages in white. Primary producers span a range of sizes and include species found in superblooms, which develop among ice platelets. Antarctic krill are absent. Other euphausiid species, copepods, and Antarctic silverfish dominate the consumer community. Vertebrate predators such as Adélie penguins and minke whales forage in the region.

dictability of superblooms has not yet been established. They eventually disappear because of a lack of nutrients; nitrate and phosphate concentrations are reduced to undetectable levels during the bloom, something that generally does not occur in phytoplankton blooms in the MIZ. However, heavy grazing may also contribute to bloom decay.

V. GENERALIZED ANTARCTIC MARINE FOOD WEB

Estimates of total primary production in Antarctic marine food webs are on the order of hundreds to thousands of millions of metric tons of carbon per year, <5% of the global total. Of this, ~25% occurs in the relatively small area of the MIZ in a burst of production in spring. Most of the remainder (~60%) is due to consistently low year-round primary production in the large open ocean areas of the West Wind Drift and the seasonal ice zone. Approximately 5% of total annual primary production occurs beneath the ice, ~8% is due to ice algae, and <1% is contributed by coastal phytoplankton blooms in spring and summer. In terms of production per unit area, the MIZ has the most productive Antarctic marine food web, and this region provides most of the energy for supporting top predators.

The food webs described above for the major regions of the Antarctic are not discrete entities. Figure 9 shows how the different systems grade into one another as the seasons change. The basic plant and animal groups occur in all systems, although actual species composition may differ. At different times and in different places different components and interactions dominate. A "background" food web occurs at all times in Antarctic waters. It consists of small-celled phytoplankton, bacteria, and microorganisms. Similar food web structures occur in practically all pelagic and freshwater ecosystems and are the dominant food webs in oligotrophic oceanic waters. At times these systems dominated by small organisms expand to include large phytoplankton and grazers, and at these times the systems are most productive.

Seasonal change is probably the most important component of the dynamics of the Antarctic ecosystem. In winter (Fig. 9a) pack ice forms a protected habitat for ice algae, with a constant, if low, light supply. Shade-adapted ice algae grow in this environment, providing food for protozooplankton and also for overwintering Antarctic krill. Resident sea birds and mammals feed on the mesopelagic fish and squid that move to the surface to exploit the krill. Most other zooplankton descend to great depths, where they overwinter as nonfeeding stages.

In spring and summer (Fig. 9b), increased stability of the water column due to ice melt allows physical conditions to occur in the MIZ which favor the growth of large-celled phytoplankton in addition to the small phytoplankton. The presence of large plant cells in the system provides food in sufficient quantities to allow large zooplankton (e.g., Antarctic krill) to grow. They in turn supply food for resident sea birds, seals, and whales, as well as for spring and summer visitors. The large standing stocks that develop in the MIZ represent the "traditional" Antarctic communities.

A similar type of food web to that of the MIZ is found in the coastal zone in spring and summer (Fig. 9b). Large phytoplankton blooms develop in response to stable conditions among ice platelets. However, in this system Antarctic krill are absent, and other zooplankton and fish are the main grazers.

In autumn (Fig. 9c) light levels begin to fade and the ice cover starts to expand. Ice algae are incorporated into the pack ice as it forms and sympagic communities develop, composed chiefly of microorganisms. In the water column small-celled phytoplankton at the ice edge support small zooplankton and protozooplankton, which form part of long food chains to the top predators. Nonresident sea birds and mammals return to more northerly regions, and Antarctic krill begin to feed in the under-ice habitat.

Over years large changes can occur in the physical environment, and these changes will affect the structure and functioning of the food webs depicted in Figs. 6–9. The numerous species in the ecosystem display their own specific peculiarities and adaptations; such details have been ignored in these representations, but may be important in affecting the manner in which the food webs respond to

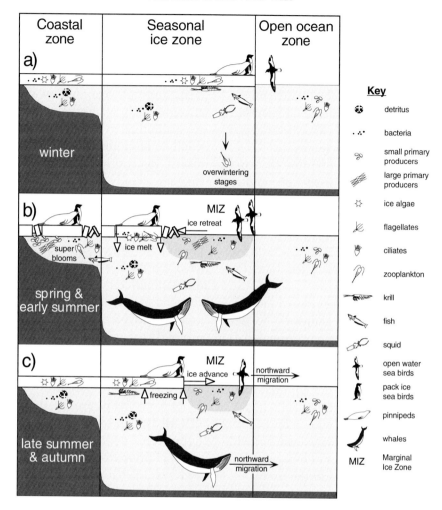

FIGURE 9 Schematic of generalized Antarctic marine food webs, showing the dominant food web components in different seasons for the three main environmental zones. Horizontal and vertical distances are not drawn to scale. Icons represent major feeding or taxonomic groups (see keys in Figs. 1 and 2), and are not all-inclusive.

change. For example, some species undergo extensive horizontal and vertical migrations, and different life stages of the same species can be found in different habitats, exploiting different food resources. The annual sequence depicting temporal and spatial changes (Fig. 9) is a simplification of many complex biological and physical interactions. It captures the essence of the variability in Antarctic marine food webs, but we still have much to learn.

VI. ENVIRONMENTAL ISSUES

A number of factors are currently causing environmental changes which impact Antarctic marine food webs. The "greenhouse effect" refers to the gradual warming of the earth as a result of increasing concentrations of "greenhouses gases" in the atmosphere during recent times. The major greenhouse gases are carbon dioxide (CO_2), nitrous oxide (N_2O), methane (CH_4), and chlorofluorocarbons (CFCs), and they act by trapping long-wavelength infrared reradiation from the earth in the atmosphere, reducing heat loss. CO_2 is the most important of the greenhouse gases, and the Antarctic Ocean is an important sink for CO_2 from the atmosphere, because the cold waters accelerate the process of CO_2 going into solution in seawater. Plants in the ocean take up the dissolved CO_2, which they use during photosynthesis. Grazing by animals and the resultant production of fecal pellets result in transfers of carbon to the sediments at the bottom of the ocean. It has been suggested that artificially increasing primary production in Antarctic waters (by supplying iron for photosynthesis) may help reduce concentrations of CO_2 in the atmosphere. While this suggestion has largely been discredited as a means for reducing global CO_2 concentrations, it does focus attention on the important role of Antarctic marine food webs in the biosphere. However, any large-scale artificial manipulations of food webs, if feasible, should be based on a sound understanding of the system and its dynamics, so that the consequences can be fully explored. [*See* GREENHOUSE GASES IN THE EARTH'S ATMOSPHERE.]

One of the possible consequences of global warming is a change in global climate. The ice fields of Antarctica play an important role in global heat balances. Conversely, any changes in global temperatures may affect the extent of Antarctic ice coverage, and consequently the communities associated with the pack ice. Antarctic krill populations are reduced during years with reduced ice cover, and sea bird populations have been shown to fluctuate in response to changes in breeding and feeding areas, caused by changes in the extent of the pack ice. [*See* GLOBAL ANTHROPOGENIC INFLUENCES.]

Another environmental problem facing Antarctic communities concerns the "ozone hole." Ozone (O_3) is a gas which occurs in the stratosphere. It plays a vital biological role in the atmosphere by absorbing harmful ultraviolet B (UV-B) radiation. A variety of factors, most of them induced by human activities, have contributed to deplete concentrations of ozone. The ozone hole over Antarctica allows increased UV-B radiation to reach the earth's surface and to penetrate below the surface of the ocean. UV-B rays may reach depths of 60–70 m, although they more commonly reach 10–20 m. [*See* ATMOSPHERIC OZONE AND THE BIOLOGICAL IMPACT OF SOLAR ULTRAVIOLET RADIATION.]

This increase in harmful UV-B rays may damage or kill plants and animals, and also has been shown to retard phytoplankton production by ~2%. This is not a very large amount, given that seasonal and interannual variability can cause differences of up to 25% in primary production. However, besides causing a decrease in overall productivity, UV-B radiation also may change the structure of the food webs, through changing the species and size composition of the plankton community. Species that are susceptible to UV-B radiation, such as prymnesiophyte phytoplankton *Phaeocystis* sp., may be replaced by more tolerant species. Indeed, such changes already may have occurred. The ozone hole has been recorded over Antarctica in spring since the mid-1970s, the period during which most measurements have been made on microbial communities in Antarctic waters. Microbial species have short generation times, and it is feasible that these rapid-response communities already have changed in response to UV-B radiation. The large grazers and predators would not be expected to respond as rapidly, although marine mammals and sea birds are exposed directly to the increased UV-B radiation because they spend a large part of their time out of the water. Body coverings of fur and feathers probably protect the animals' skins from UV-B damage, but there are concerns that their unprotected eyes and noses may be harmed.

In addition to these possible future changes, widespread changes have already occurred in Antarctic marine food webs as a result of human activities. Commercial sealing and whaling during the 19th and early 20th centuries caused major reductions in many whale and seal populations, some of which have still not recovered. With the decline in the stocks of marine mammals, it was suggested that more of their food, Antarctic krill, should be available to allow other populations to increase: the so-called "krill surplus" hypothesis. For example, increases in populations of chinstrap penguins and crabeater seals were attributed to the increased availability of food for these krill predators. Recently, however, this hypothesis has been challenged by detailed studies that show that krill is not the sole component of many predators' diets, and often is not the most important component. In addition, other environmental factors which impact breeding and feeding areas (e.g., changes in the extent of sea ice) have been shown to account for some of the population changes. Thus, it is still unclear exactly what impact human-induced population reductions have had or may have on other species in the Antarctic. In addition to past fisheries directed at whales and seals, in recent years there has been increased commercial fishing activity on krill, fish, and squid. Activities such as these can have far-reaching consequences for the Antarctic ecosystem and are cause for concern. We can attempt to address these issues only if we understand how Antarctic marine food webs function.

Glossary

Bacterioplankton Heterotrophic bacteria that are free-living in the plankton, as opposed to bacteria that are found attached to particles.

Benthic Living on the sea bottom.

Cyanobacteria Extremely small (generally <2 μm) photosynthetic prokaryotes, belonging to the order Chroococcales.

Demersal Living at the bottom of the sea.

Epipelagic Living in the surface layers of the water column, down to $\sim$200 m.

Euphotic zone The surface waters of the ocean which receive sunlight, and thus contain plants which can photosynthesize.

Food chain A linear flow of energy or material from primary producers through grazers and on to successive predators. A food chain is usually finite and culminates in the death of an organism, but it may be made infinite by including feedbacks.

Food web All possible feeding connections in a community. Few animals have diets restricted to one species or type of organism, so each species or group in a food web can receive food from many different sources, and can in turn be eaten by many other groups.

Marginal ice zone The region at the edge of the pack ice that is affected by meltwater.

Microbial loop The original definition referred to a food chain based on PDOC (see below), with energy or materials transferred to bacterioplankton and increasing sizes of protozooplankton, ultimately reaching small and large zooplankton and fish. The term is now widely used to refer loosely to the complex feeding and recycling processes that occur in the complete microbial food web.

New production Primary production in the oceans that is based on nitrogen in the form of nitrate (NO_3). Nitrate is usually supplied to surface waters from depth, and thus represents an external input to the euphotic zone. Over long periods and assuming equilibrium, the amounts of primary production based on nitrate should be balanced by sedimentation losses from the surface waters.

Pack ice The ice sheet covering large parts of the Antarctic marine ecosystem.

PDOC Photosynthetically produced dissolved organic carbon; small organic compounds which are secreted by healthy phytoplankton cells during normal growth and which can be taken up and used by bacterioplankton.

Pelagic Living within the water column.

Photosynthesis The process by which plants use the energy of sunlight to manufacture energy-rich organic compounds from carbon dioxide and water.

Polynia Expanse of open water in the midst of pack ice.

Protozooplankton Heterotrophic protozoa (single-celled organisms) that live in the plankton.

Regenerated production Primary production in the oceans that is based on reduced forms of nitrogen, such as urea or ammonia, excreted by heterotrophic organisms. Regenerated nitrogen is produced locally, and regenerated production represents recycling within the surface waters.

Sympagic Living in or associated with ice.

Bibliography

Hempel, G., ed. (1993). "Weddell Sea Ecology. Results of EPOS, European 'Polarstern' Study." Springer-Verlag, Berlin.

Kerry, K. R., and Hempel, G., eds. (1990). "Antarctic Ecosystems. Ecological Change and Conservation." Springer-Verlag, Berlin.

Legendre, L., Ackley, S. F., Dieckmann, G. S., Gulliksen, B., Horner, R., Hoshiai, T., Melnikov, I. A., Reeburgh, W. S., Spindler, M., and Sullivan, C. W. (1992). Ecology of sea ice biota. 2. Global significance. *Polar Biol.* **12**(3–4), 429–444.

Sahrhage, D., ed. (1988). "Antarctic Ocean and Resources Variability." Springer-Verlag, Berlin.

Siegfried, W. R., Condy, P. R., and Laws, R. M., eds. (1985). "Antarctic Nutrient Cycles and Food Webs." Springer-Verlag, Berlin.

Smith, W. O., Jr., ed. (1990). "Polar Oceanography." Academic Press, San Diego.

Aquatic Weeds

K. J. Murphy
University of Glasgow, Scotland

I. Natural Constraints to the Growth of Aquatic Plants
II. Aquatic Plants: Environmental Problem or
Environmental Asset?
III. Management of Aquatic Weeds
IV. Future Trends

Weeds can be defined as "plants which are not desired at their place of occurrence." In freshwater systems there are two main groups of plants: microalgae (simple plants too small individually to be seen by the unaided eye) and macrophytes (plants big enough to be visible to the naked eye). Although microalgae can cause serious problems (e.g., by forming dense, sometimes toxic, algal blooms in water supply reservoirs), the term *aquatic weeds* is reserved for macrophytes growing in sufficient abundance to cause nuisance in a body of fresh water.

Technically, any freshwater macrophyte species has the potential to cause an aquatic weed problem. In reality, only a few of the more than 400 known genera of freshwater macrophytes are a regular cause of nuisance severe enough to need active management. These few species, however, occupy large areas of freshwater systems worldwide. They may be conveniently grouped by growth form. *Free-floating* weeds have their photosynthetic tissue at the surface, and their roots dangling free in the water. *Macrophytic algae* are usually filamentous algae, which form tangled mats on or below the water surface. The remaining groups are all rooted in, or attached to, the substrate. *Emergent* weeds have their photosynthetic tissue above the water surface. *Floating-leaved rooted* weeds, like the free-floating species, have floating leaves on the surface, but are rooted in the sediment. *Submerged weeds* have all their photosynthetic tissue below the water surface.

I. NATURAL CONSTRAINTS TO THE GROWTH OF AQUATIC PLANTS

Freshwater ecosystems include natural habitats (e.g., lakes, rivers, streams, ponds, lagoons, temporary pools) and man-made or man-modified habitats (e.g., reservoirs, regulated rivers, canals, irrigation and drainage channel networks). All are potentially open to macrophyte colonization. The limiting factors to macrophyte success in a given freshwater system are the *stress* and *disturbance* which are imposed by *physical, chemical,* or *biological* constraints to plant survival, growth, and reproduction. Stress is defined in this context as any factor which limits photosynthetic carbon fixation. Disturbance is considered to be any factor which damages or destroys plant biomass.

Physical constraints to macrophyte growth include the stress caused by limited light availability and the disturbance associated with water and substrate movement. Like all plants, macrophytes can only survive if there is sufficient light to maintain a net photosynthetic carbon gain over and above the carbon expended by respiration. This is a particular problem for submerged macrophyte species. Light limitation excludes them from deep water, and if the water is turbid (often because of a high suspended silt load), the zone open to colonization may be restricted to only the shallowest parts of the water body. Fast-flowing waters, those with strong wave action, and systems where water and sediment are regularly disturbed by boat traffic are good examples of waters where the level of disturbance may be too great to permit successful exploitation by macrophytes.

An important *chemical* constraint is inorganic carbon supply. Although this is no problem for macrophytes with emergent or floating leaves, because they have direct access to gaseous CO_2 in the air, the availability of dissolved inorganic carbon (DIC) poses real problems for submerged macrophytes. Fresh water in equilibrium with air at 25°C contains about 10 μM of free CO_2 on average. This figure is highly variable because of the large resistance to diffusion of dissolved gases (about 10^4 times slower in water than in air)—giving rise to free CO_2 values which may range from zero to more than 350 μM. The upper-range value for CO_2 invasion rate from air into water is about 700 μM $CO_2/m^2/hr$. This would be typical of, for example, a soft water lake with low DIC. However, estimates of the DIC consumption rate of a dense submerged weedbed (e.g., *Hydrilla verticillata*) are as high as 10,000 μM $CO_2/m^2/hr$. Demand for DIC may therefore substantially exceed supply, and the water may become depleted of DIC, thereby potentially limiting carbon fixation, and hence growth, of submerged plants.

To overcome this stress problem, submerged macrophytes have evolved some interesting *physiological* and *morphological* adaptations. Some species can use bicarbonate ions (HCO_3^-) as their source of DIC. Others (so-called SAM plants) have modified photosynthetic pathways, involving production of C_4 acids, which help the plant to concentrate DIC internally for more efficient photosynthetic carbon

fixation. Morphological adaptations include the possession of thin or finely subdivided leaves in many submerged plants (to increase the surface area-to-volume ratio, and so ease the problems of slow diffusion). Some species of submerged plants also produce floating leaves (so-called heterophyllous species), which allows them access to atmospheric CO_2 and also increased light supply. Others forage for the CO_2 available in the sediment from the respiratory activity of microorganisms, by taking it up via their roots and piping it to the leaves through lacunal gas channels. [*See* PLANT ECOPHYSIOLOGY.]

The available supply of other ions, especially the nutrients N and P, within the aquatic system may also constrain the growth of aquatic plants. Whether water or sediment is the more important source of such nutrients, for rooted macrophytes, has been disputed in recent years. Many plants may be able to use both sources, but in general it now seems that the sediment is the more important source for most rooted aquatic plants.

Biological constraints to aquatic plant growth include competition from other plants, grazing (by invertebrates, fish, birds, and mammals), and disease. By definition, competitive effects are most important in crowded, productive habitats, where the plants experience low intensities of stress or disturbance from chemical or physical influences.

The role of herbivores in limiting the growth of freshwater macrophytes seems rather variable. There are few obligate grazers of aquatic plants in temperate fresh waters, one being the grass carp (*Ctenopharyngodon idella*), a native of rivers in northern China. There are more examples in warmer waters, including the manatee (*Trichecus manatus*) and fish species of the genus *Tilapia*. Many waterfowl, such as ducks, geese, and swans, consume macrophytes as an important component of their diet. Invertebrates which graze macrophytes include snails (e.g., *Marisa cornuaretis*) and some insects. One example of the latter is the mottled waterhyacinth weevil (*Neochetina eichhorniae*), which is an obligate feeder on water hyacinth (*Eichhornia crassipes*), one of the world's worst weeds. This insect is the basis of a successful biological control program used against this aquatic weed. In general, however, herbivory seems to play a less important role in regulating the abundance of freshwater macrophytes than is the case in, for example, terrestrial ecosystems such as grasslands.

A wide range of pathogenic organisms is known to infect aquatic macrophytes. For example, the fungal disease-causing organisms *Fusarium* and *Cercospora* cause rust, leaf spot, and leaf blight. Attempts to introduce disease as a means of controlling aquatic weed problems have met with only limited success. Most pathogens appear to be opportunistic in nature, being most effective when plants are already experiencing stress from other causes (e.g., a herbicide treatment).

II. AQUATIC PLANTS: ENVIRONMENTAL PROBLEM OR ENVIRONMENTAL ASSET?

A. Ecosystem Role of Freshwater Macrophytes

When present in small to moderate quantities, aquatic macrophytes play a positive and important role in ecosystem functioning. They provide habitat (e.g., nesting sites, substratum, and feeding sites) for a wide range of other organisms, from invertebrates to birds. Their importance as primary producers varies depending on the type of freshwater system in which they occur. They are least important in this context in deep, steep-sided, nutrient-poor lakes, such as Loch Ness, Scotland (Fig. 1a). They are very important as producers in shallow, nutrient-rich freshwater habitats such as Lake Sampson, Florida, U.S.A. (Fig. 1b). Moderate growths of aquatic macrophytes can help oxygenate the water, assisting the survival of fish and invertebrates. They are also important to processes such as nutrient cycling within freshwater ecosystems.

B. Problems Caused by Aquatic Weeds

In most cases, an aquatic weed problem is simply an excessively large growth of one or more aquatic macrophyte species within a given freshwater system which causes *blockage* or *amenity* problems within the water body or watercourse. The cause of such problems is usually directly attributable to human interference with the system (e.g., by nutrient enrichment) which relaxes one or more of the natural constraints to plant growth.

Aquatic weeds can seriously interfere with water movement and navigation, so the problems caused are particularly serious when they occur in navigable waterways and irrigation or drainage channel networks. In many cases the worst problems are caused by invasive nonnative macrophyte species. Examples include the tropical free-floating weeds *Salvinia molesta* and *Eichhornia crassipes*, which can be transported by water currents, or wind, to produce piles up to 5 m thick, extending over many hectares of lake or watercourse. The blockage problems produced by such enormous weed accumulations may cause flooding, break bridges, and obstruct or completely halt boat traffic movement.

Submerged weeds (e.g., *Hydrilla verticillata*) and filamentous algae (e.g., *Vaucheria dichotoma*) are another cause of problems in freshwater systems. In flowing waters, such as drainage channels, these plants increase the frictional resistance to water flow, which may increase the risk of flooding. Algae also increase maintenance costs in water-pumping and drip-irrigation systems. In navigable canal systems excessive submerged weed growth can obstruct boat movement. Although most submerged weed problems are found in lowland, often eutrophic waters, problems can also occur in upland, low-nutrient water bodies. For example, in recent years problem growths of the submerged acidophilous plant *Juncus bulbosus* have occurred in rivers regulated for hydroelectricity production in Norway.

In warmer parts of the world there may be risks to human or livestock health associated with aquatic weed growth. For example, certain disease-carrying mosquitoes make use of weed-infested waters for reproduction; the freshwater snails which act as a vector of the tropical disease schistosomiasis (bilharzia) are also encouraged by dense aquatic weed growth. [*See* PLANT–ANIMAL INTERACTIONS.]

FIGURE I (a) Loch Ness, Scotland: a nutrient-poor, deep lake providing a poor habitat for aquatic macrophyte growth. (b) Lake Sampson, Florida (U.S.A.): a moderately nutrient-rich, shallow lake providing a highly favorable habitat for aquatic macrophyte growth. Aquatic weed problems are much more likely to develop in the latter rather than the former type of water body.

III. MANAGEMENT OF AQUATIC WEEDS

A. Survival Strategies and Weed Control

Nearly all successful aquatic weeds have a *survival strategy* which combines high competitive ability with a good tolerance of disturbance (so-called CD strategists). Few occupy highly stressed habitats. In consequence, it is not surprising that aquatic weed control regimes which rely on disturbance of the plants (e.g., weed cutting) or competitive effects (e.g., introduction of competitive but non-nuisance plant species) tend to be less effective than those which impose severe stress on the target plant population (e.g., shade or herbicides).

The exception to this is weed control based on introduced grazers (e.g., grass carp), which sharply increase the disturbance experienced by the plant population. The reason why grazer-based aquatic weed control is effective may well be that aquatic macrophytes have a rather low evolutionary experience of combating grazing. The traits needed to stop an animal eating a plant, or to limit the damage caused by grazing (such as spines, protected meristems, hardened cuticle, or foul taste), are rarely found in aquatic plants.

B. Methods of Controlling Aquatic Weeds

Management measures for controlling aquatic weeds fall into four main categories: physical, environmental manipulation, chemical, and biological.

I. Physical Control

Physical control involves the use of manual or mechanical clearance measures. A simple, time-honored measure is scything, which is still widely used, for example, to control submerged *Ranunculus* growths in lowland chalk-streams in England which support high-value trout fisheries. In Argentine irrigation channels, "chaining" is a common technique. This method uses a heavy chain of sharp blades slung between two tractors, one on each bank. The chain is then dragged upstream, ripping out weed growth as it goes. Regrowth following such methods may be very fast: in Argentina chaining may be repeated up to seven times a year in main channels crucial for water supply to the irrigation network. Though simple, and of low capital cost, such weed control regimes are obviously expensive in fuel and labor costs. More advanced methods require purpose-built weed-clearing boats, weed buckets attached to hydraulic excavators, and other purpose-built machines (Fig. 2). Physical control remains the mainstay approach to aquatic weed control in most parts of the world. [*See* CONTROLLED ECOLOGIES.]

2. Environmental Manipulation

Environmental manipulation techniques alter the habitat in such a way as to discourage the growth of aquatic weeds. A good example is water-level manipulation, which is often used in reservoirs and irrigation channels to control submerged weed growth. Drawing down the water level exposes the plants to the air, killing them by desiccation.

3. Chemical Control

Chemical control involves the use of herbicides. Many different chemical compounds are used for aquatic weed control. Some countries do not permit the application of herbicides to water, but in many others herbicides are used against aquatic weeds. To minimize the potential risks to the environment and to human, crop, or livestock health, most countries impose strict controls on aquatic herbicide use. Some aquatic herbicides (e.g., diquat, acrolein) are fast-acting *contact* herbicides, which kill the weeds rapidly, then dissipate, break down chemically, or are otherwise deactivated within a short period of time. Others are slower-acting *residual* herbicides (e.g., terbutryn). These are designed to persist in the aquatic system in active form for longer periods, suppressing plant growth by interfering with one or more physiological processes needed for plant survival, for example, photosynthesis. Some of these herbicides are formulated in slow-release pellets (e.g., fluridone) which maintain the necessary phytotoxic effect over a long period of time.

FIGURE 2   Examples of machines developed for the mechanical control of aquatic weeds.

4. Biological Control

Insects, fish, and pathogenic fungi are the three groups of organisms that have had most success in the biological control of aquatic weeds. Some examples have already been mentioned. The principle of biological control is to increase the effect of an organism which acts as a natural constraint to the growth of the target weed population. Certain insect-based programs have been spectacularly successful against nonnative (i.e., introduced) aquatic weed problems. For instance, in the early 1980s, *Salvinia molesta* was a cause of serious weed problems on the Sepik River system in Papua New Guinea. In 1984 over 250 km^2 of this major transport route were infested by near-impenetrable floating mats of this weed. Following the introduction of a Brazilian weevil, *Cyrtobagous salviniae* (which is an obligate feeder on *Salvinia*), the infestation was reduced to less than 2 km^2 by the end of 1985. [*See* BIOLOGICAL CONTROL.]

Grass carp have been introduced in many parts of the world where conditions are suitable (they will not feed in cold waters, below about 10°C). Many programs have had good success, for example, in polder drainage systems in the Netherlands and in irrigation channels in Egypt. The unusual breeding requirements of the fish make natural spawning difficult outside their native river system (the Amur River in China). However, they are quite easy to breed artificially. This means that in many parts of the world there is little chance of the fish getting out of control when introduced to a new system—an important concern with most other biological control programs involving introduction of a nonnative control organism. In the United States, however, where there are river systems in which conditions resemble those of the grass carp's native habitat, fear of the carp escaping and breeding greatly limited the acceptability of the fish for aquatic weed control purposes. Here, the rapid expansion in use of grass carp in recent years has been due to the availability of sterile triploid grass carp artificially bred from normal parent fish.

Another fish species which is an effective destroyer of aquatic plants is the omnivorous common carp (*Cyprinus carpio*). This fish is a pest in some areas of the world, such as Australia, where it has destroyed native plant species and increased water turbidity by its feeding behavior: the fish root around in the sediments, uprooting plants and stirring up silt particles which increase water turbidity, thereby increasing shade stress on submerged plants. The same behavior makes the fish of interest as a weed control agent in weed-infested systems, and common carp are now being considered for this purpose in, for example, Argentine irrigation systems. If grass carp are comparable to sheep in their effects on plants, then by analogy common carp are the pigs of freshwater systems.

C. Environmental Impacts of Aquatic Weed Control

Much of the concern over possible undesirable environmental impacts of aquatic plant management programs has focused on the effects of herbicides, because these are likely to have effects on nontarget organisms as well as on the target weeds. In practice, however, the degree of impact of an aquatic weed control program on the aquatic ecosystem is closely related to the proportion of the plant community killed or removed, the speed with which this happens, and the length of time before regrowth occurs—irrespective of the type of control measure used.

For example, an inefficient physical clearance of an irrigation canal using a cutting method will rapidly remove some proportion of the weed biomass, but regrowth from the remaining plants is likely to be very rapid, thereby minimizing the effects of plant loss from the water body. Where this form of control is used, for example, in channels of the Río Negro irrigated area in southern Argentina, up to seven clearance operations per year may be needed to keep weed growth down to acceptable levels. In contrast, a single application of an efficient residual herbicide such as terbutryn in navigation canals and drainage channels in Britain is sufficient to control weed growth throughout the growing season.

Many of the environmental impacts associated with aquatic weed control are common to all control measures, including things like the loss of habitat for other organisms and reduction in primary production within the system. Herbicides are often considered to have a larger risk associated with their use than other measures, in part because of their inherent direct toxicity to nontarget organisms, though this is quite variable between different compounds and different groups of organisms. For example, acrolein is highly toxic to fish (with an LC_{50}—the concentration of herbicide that kills 50% of test organisms over 24 hr—as low as 0.08 mg/liter in bluegill: *Lepomis macrochirus*), whereas dalapon is very safe to fish (with an LC_{50} over 24 hr of 428 mg/liter for the same species).

The second important indirect effect of herbicide treatment, which is rarely encountered when other methods are used, is that the plants are killed and decay *in situ*. This can have deleterious effects on the oxygen regime of the water body, because a mass of rotting vegetation boosts the oxygen demand of decomposer organisms and can in the worst case completely deoxygenate the water for a long time, killing fish and other nontarget organisms by suffocating them. A summary of some of the interrelated impacts of a herbicide treatment on an aquatic ecosystem is shown in Fig. 3.

Although there is considerable concern over the side effects of aquatic herbicides, other weed control approaches may also damage the freshwater ecosystem. There is good evidence that weed-harvesting and weed-cutting programs may remove substantial quantities of fish and invertebrates from treated systems. Dredging also causes major destruction of bottom-dwelling plant and animal communities.

As a result of these potential effects, control programs for aquatic weeds have to be carefully managed. It is perfectly possible to use the available armory of aquatic weed control techniques to produce acceptable weed control without causing an unacceptable degree of environmental damage. However, poorly designed and implemented aquatic weed control programs can have serious consequences for the target freshwater ecosystem, including fish kills, water pollution problems, and potential hazards to human health. Aquatic weed control is a management tool which has to be used with respect in freshwater systems.

IV. FUTURE TRENDS

What can we predict for the future regarding aquatic weed problems? There are two important global trends which are particularly relevant. The first is the effects of increasing human population size and the increasing demands which this places on available freshwater supplies—for consumption, for crop irrigation, and for waste disposal. Eutrophication (nutrient enrichment) caused by nutrient runoff from croplands and sewage effluent disposal into freshwater systems is already decreasing the constraints on aquatic plant growth imposed by N and, especially, P limitation. This trend is very unlikely to be reversed in the foreseeable future and is likely to lead to increasing aquatic weed growth.

Allied to this, in its effects, is the increasing concentration of CO_2 in the atmosphere. If current hypotheses concerning the likely limitation of submerged weed growth by DIC limitation are correct, then any increase in atmospheric CO_2 is likely to have a potentially significant effect on CO_2 invasion rates into freshwater bodies. In this instance, the chemical constraint on submerged macrophyte production caused by carbon shortage for photosynthetic fixation may be relaxed. Worst-case scenarios which take into account both of these factors predict something not far short of an explosion of aquatic weed growth, mainly by submerged weeds, in fresh waters in many parts of the world. What is certain is that there is no evidence to suggest that the problems of aquatic weed growth are diminishing throughout the world, and plenty of evidence to suggest that they are increasing.

How then can we deal with these problems, preferably in an environmentally friendly way? At present there is a range of aquatic herbicides available. Because of the cost of safety-testing programs required for new herbicides designed for use in fresh waters, few or even no new compounds are expected to come onto the market for aquatic use in the foreseeable future. In addition, the wisdom of applying toxic chemicals to water is being questioned, so the use of herbicides is not increasing. In fact, in many parts of the world it is sharply decreasing or has been stopped altogether by legislation.

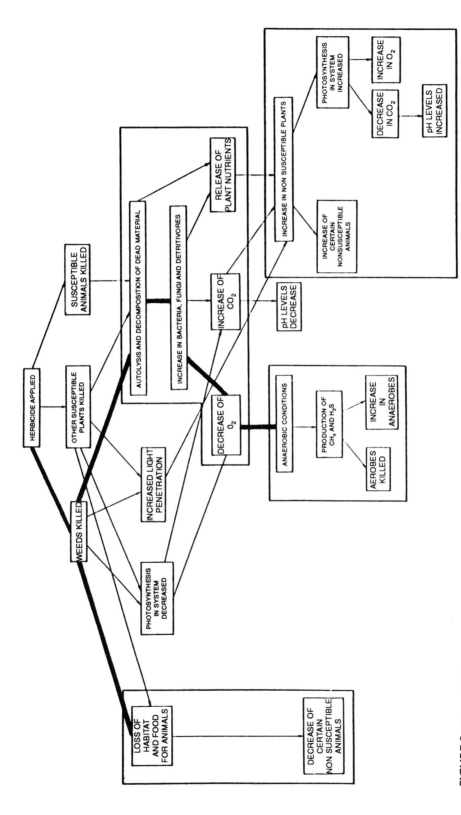

FIGURE 3 Potential effects of a herbicide on a freshwater ecosystem. Main effects (the importance of which is directly related to the biomass of weed plants present) are indicated by thick black lines.

Advances in physical control methods can be expected. There are already a number of "weed-eating" machines available, and research and development in this field is ongoing. The capital cost of such machines is quite high, and their efficiency is not usually as good as that of herbicides. In many areas, especially where labor costs are low, we may see old-fashioned manual clearance continue as the standard method of aquatic weed control, despite its inherent inefficiency (and dangers, especially in areas where waterborne diseases such as bilharzia pose a serious threat to workers in water).

Perhaps the most fruitful area of research to improve on current approaches to aquatic plant management lies in biological control. Many of the most intractable aquatic weed problems are caused by exotic weed species which have invaded or been introduced to waters away from their native range. This is exactly the sort of situation where classical biological control programs, which utilize control organisms brought in from the plant's native range, can be very effective. The spectacular success of the biological control of *Salvinia* demonstrates what can be achieved. Where the problems are not necessarily due to exotic weeds, then nonspecific biological control agents such as fish can be very successful, largely owing to the inherent susceptibility of aquatic weeds to grazing, and to the increased shading produced by the turbidity associated with the feeding activities of fish like the common carp.

No aquatic weed control measure is a panacea. All have their problems, environmental side effects, and limitations. Biological programs have the advantage of long-term sustainability and a general absence of major impact on nontarget components of the freshwater ecosystem. For these reasons, we might well expect to see an increasing dependence on such measures for the management of aquatic weeds.

Glossary

Aquatic herbicide Herbicide with an active ingredient molecule and carrier formulation suitable for use to control one or more species of aquatic weeds.

Aquatic weeds Freshwater plants, big enough to see with the naked eye, growing in sufficient abundance to cause nuisance to human use of a particular body of water.

Biological control agent Any organism which can be managed to increase the pressure on survival of an aquatic weed population, usually by introduction of an exotic species which is herbivorous on or pathogenic to the nuisance weed population or community.

Competition Interactive pressures on plant survival resulting from attempts by neighboring plants to capture the same unit of resources.

Disturbance External pressures on plant survival which affect vegetation by partially or totally destroying plant biomass.

Stress External pressures on plant survival which affect vegetation by reducing photosynthetic production.

Survival strategy Grouping of similar or analagous genetic characteristics which recurs widely among species or populations and causes them to exhibit similarities in ecology.

Bibliography

Charudattan, R. (1986). Integrated control of water hyacinth (*Eichhornia crassipes*) with a pathogen, insects and herbicides. *Weed Science* **34** (Suppl. 1), 26–30.

Cook, C. D. K. (1990). "Aquatic Plant Book." The Hague, The Netherlands: SPB Academic.

Fox, A. M., and Murphy, K. J. (1990). The efficacy and ecological impacts of herbicide and cutting regimes on the submerged plant communities of four British rivers. *Journal of Applied Ecology* **27**, 520–540.

Murphy, K. J. (1988). Aquatic weed problems and their management: A review. I. The worldwide scale of the aquatic weed problem. *Crop Protection* **7**, 232–248.

Pieterse, A. H., and Murphy, K. J., eds. (1990). "Aquatic Weeds: The Ecology and Management of Nuisance Aquatic Vegetation." Oxford, England: Oxford University Press.

Room, P. M., Harley, K. L. S., Forno, I. W., and Sands, D. P. A. (1981). Successful control of the floating weed *Salvinia*. *Nature* **294**, 78–80.

Varshney, C. K., and Rzóska, J., eds. (1976). "Aquatic weeds in south-east Asia." The Hague, The Netherlands: W. Junk.

Arboreta

Stephen A. Spongberg
Harvard University

An arboretum is a place where a collection of native and/or exotic woody or ligneous plants—trees, shrubs, and woody climbers—is obtained and cultivated for scientific, conservation, educational, and/or aesthetic purposes. By extension, an arboretum is an institution or organization involved in bringing together, establishing, utilizing, and maintaining such a collection. The term derives from the Latin *arboretum* (a place grown with trees) and *arbor* (tree).

An arboretum can be distinguished from a botanical garden as a botanical or horticultural institution that specializes in the development, cultivation, and utilization of a collection of ligneous plants to the exclusion of herbaceous plants. While a botanical garden may incorporate an arboretum (sometimes at a different site than the botanical garden itself) where collections of woody plants are concentrated, few arboreta in the strict sense actively seek accessions of, or devote space to, herbaceous plants other than for landscape effect. However, in practice no hard and fast distinction can be drawn between the two.

I. INTRODUCTION

Arboreta, as specialized gardens, share many fundamental features of operation and organization with their sister institutions, botanical gardens. Both are committed to devoting staff time and financial resources to bringing together collections of plants and cultivating and maintaining them for various combinations of scientific, educational, conservational, and amenity purposes. Land is one requirement which is also common to both, and land for arboreta is of primary concern inasmuch as the space requirements for the cultivation of comprehensive collections of mature examples of trees and shrubs far exceed those for herbaceous plants. Ideally, the acreage devoted to the purpose of an arboretum should have ecological as well as topographical diversity to ensure that a variety of habitats is available for plants with differing ecological requirements.

Another characteristic of arboreta and botanical gardens is the manner in which the collections are incorporated into the landscape. The most comcmon arrangement follows a systematic scheme, usually adhering to the outlines of a system of botanical classification whereby species of a given genus are grown in close proximity, and genera within a given botanical family are likewise grown in a common area. This systematic sequence is frequently extended to the family level as well, although ecological factors of the site may dictate departures from the adopted scheme. Intentional variations from a systematic arrangement of collections may be seen in many arboreta where collections are arranged on a geographic basis (e.g., all species of eastern Asiatic origin are grown together), a strictly ecological basis (e.g., alpine species are grouped together), or purely for landscape effect, in which species, regardless of their geographic origin, botanical relationships, or other considerations, are planted and grown in the landscape for their aesthetic appeal. Other planting schemes may be dictated by more utilitarian or horticultural concerns. Shrubs that adapt well to pruning for use as hedges may be grown in a demonstration area, woody climbers may be grown on a trellis or arbor without regard for generic or family groupings, smaller-growing trees adapted for use as street trees may be grown together in a demonstration plot, and dwarf or slow-growing conifers that adapt to a variety of landscape uses are commonly grown in a specialized display garden. Frequently, well-known, historic, or hardy horticultural varieties (cultivars) are also grown together for their educational value to home gardeners, landscape designers, and landscape architects. [*See* SYSTEMATICS.]

Another characteristic of both botanical gardens and arboreta relates to the fact that the individual accessions within the collection are labeled with their accession number and scientific and/or common names. Additionally, while it has not always been an established routine, today accurate and rigorous records of the source or origin and cultivation history are generally maintained—either by hand, in which case a card system is often utilized, or more frequently using a computerized data base—for each accession acquired and maintained in the collection. The record system, along with the labeling of plants in the collection, largely differentiates arboreta from park or other plantings of trees and shrubs that have been brought together for amenity purposes, and accession records provide the basis on which the collection, or any subset thereof, may be used for scientific purposes.

II. COLLECTION POLICIES

When received at an arboretum, plant materials destined to be accessioned and included in the permanent collections are normally scrutinized to ascertain that they are acceptable to the institution, following guidelines outlined in an officially accepted collection policy, which is often published and placed on public record. While historically many arboreta attempted to be comprehensive and all-inclusive by obtaining representatives of as many woody taxa as could withstand the prevailing climatic conditions of their sites, many institutions have found it necessary to limit the scope of their collections based on several criteria. Because acreage is limited, space may not be available to grow comprehensive collections of some or all large genera, and the institution may choose synoptic rather than comprehensive representation for these groups. Likewise, criteria may be established on the basis of provenance, whereby plants originating in commercial nurseries or from other cultivated sources are not admitted in preference to materials of known wild origin, an attitude that has become increasingly commonplace in recent decades as the scientific and conservation value of arboretum collections has become stressed. In a similar spirit some arboreta now restrict their accessions to botanical taxa with limited or no representation of modern rapidly eclipsed cultivars. Still other institutions limit their acquisitions to taxa of a given geographic region or, in a few instances, to species and cultivars of an individual plant family or genus. These latter examples represent policies that are usually dictated by a founding bequest and indicate the specialized interests of the benefactor.

While few arboreta can attempt to be comprehensive or all-inclusive by growing all those taxa of woody plants that could be grown, most institutions do specialize in one or a few groups of plants.

Consequently, most arboreta are well known for their collections in which they specialize, and they provide an important service to the botanical and horticultural communities by maintaining "archival" collections of botanical taxa as well as historic and modern cultivars and hybrids within these groups. Frequently these specialized collections also form the focal points of the arboretum and have wide public appeal during the season when they flower.

Some arboreta have direct ties with schools of forestry and serve as trial grounds for testing the general suitability of an assortment of species for the forestry industry, whereas other collections, especially in the humid tropics, trace their origins to the establishment of introduction stations for commodity crops. These latter institutions—the Jardim Botânico do Rio de Janeiro (Brazil), the Indian Botanic Garden (Calcutta, India), and the Botanical Gardens of Indonesia, Bogor (West Java, Indonesia) are primary examples—largely consist of woody plant collections. Arboreta linked with forestry schools frequently undertake provenance testing, in which comparisons of growth rates and productivity are made between accessions from a number of sources throughout the geographic range of the particular species. By these means superior selections are obtained for particular ecological and climatic conditions. Forest products are generally within the realm of arboretum collections, particularly in the tropics. The arboreta maintained by the Forest Research Institute (Kepong, Selangor, Malaysia) and the Botanical Garden at the Forest Research Institute and Colleges (Dehra Dun, India) are primary examples.

One of the immediate outcomes of the testing of new species in arboretum collections is data relating to plant hardiness and the adaptability of species to the climatic and edaphic conditions prevailing in the arboretum. As a consequence, arboreta can be viewed as long-term ongoing experiments focused on determining the tolerances and ecological amplitudes of woody plants; in addition to the other uses that the collection may facilitate, the data gathered over time help to answer the questions of what will grow where and under what climatic and cultural circumstances.

III. HISTORICAL BACKGROUND

The precursors of the modern-day arboretum are poorly documented, and none of the known examples fulfilled the goals and functions of arboreta as they exist today. The first examples did, however, consist of collections of trees and perhaps shrubs and woody climbers as well, and while their exact purpose is unknown, it is likely that these collections were established to satisfy the insatiable desire of early plantsmen to collect and grow unusual plants and to bring together and grow plants of medicinal and culinary importance. As long ago as 2800 BC, Chinese emperor Shen Ming gathered a garden of medicinal plants, and in Egypt Pharaoh Thotmes III established a systematic garden. Aristotle founded a botanical garden in Athens in 340 BC, and it was there that Theophrastus, the "father of botany," studied under Aristotle's tutelage.

In England the deer parks that belonged to the crown served as early examples of forest and game preserves, where both trees and wild game were protected from exploitation by the general populace. The trees, primarily oaks, were nonetheless available for the construction of ships for the royal navy, and as elsewhere in Europe, the naval stores drew heavily on forested regions for the timbers and other forest products necessary for the development of naval fleets. The demand for wood and wood products by society as a whole was enormous from the Renaissance on, and toward the end of the sixteenth century the forest resources of Europe had been largely depleted. In England John Evelyn's *Sylva, or a Discourse of Forest-Trees, and the Propagation of Timber* (1664) was written at the request of the Royal Society—of which Evelyn (1620–1706) was an early member—to address this calamity and to encourage the planting of trees on a large scale to avert a complete timber shortage. Evelyn's book went through numerous editions and had tremendous impact on raising the level of consciousness concerning forest resources and trees themselves as objects of scientific and aesthetic interest and as plants to be propagated and planted.

Earlier in France René du Bellay, Bishop of Man, had brought together a collection of trees at Touvoye during the sixteenth century. Included in this collection were some of the first western Asian trees to be cultivated in Europe, which had been introduced by physician and traveler Pierre Belon (1518?–1563). While they were never referred to as arboreta, the collections of European and North American trees grown on the estates of Veigny and du Monceau belonging to Henri Louis Duhamel du Monceau (1700–1782), head of the French department of naval affairs, serve as other early examples of arboretum collections. In these instances a definite purpose underlay the establishment of the collections: Duhamel had a direct interest in trees from a scientific standpoint, wanting firsthand information on their growth rates and the structural and physical properties of their timbers. Responsible for providing an ample supply of construction material for the French navy, Duhamel was keenly aware of the depletion of French forest resources and of the need to replenish those supplies through reforestation. Consequently, fast-growing species from foreign floras were sought for reforestation projects, and knowledge of the kinds of trees, the physical properties of their timbers, and their cultural requirements was investigated.

During the latter part of the seventeenth century and the early years of the eighteenth century the Reverend Henry Compton (1632–1713), Lord Bishop of London, gathered a diverse collection of trees, shrubs, and other plants, which he cultivated in the garden of Fulham Palace, his suburban London residence. Particularly interested in American species, Compton received seeds and plants of many New World species from travelers and settlers—many attached to the Anglican Church—in the American colonies. Notable among those who provided Compton with American plants was the Reverend John Banister (1650–1692), who provided valuable specimens and information concerning the natural history of the Virginia colony to naturalists in England. As a consequence of Banister's shipments of seeds, many eastern North American trees and shrubs were introduced into Europe by Compton, and Compton's horticultural activities served as another example for individuals interested in bringing together collections of woody plants.

The formal establishment of the Royal Botanic Gardens at Kew near London in 1759 was an event of great significance for botanical and horticultural enterprises, and the worldwide influence of the institution in the years that followed during the informal directorship of Sir Joseph Banks (1743–1820), and the official directorships of William Jackson Hooker (1785–1865), and Joseph Dalton Hooker (1817–1911) cannot be overestimated. All three directors encouraged plant introduction into Britain from the far reaches of the then-expanding British empire, and literally thousands of species became the objects of cultivation as the botanical exploration of the world's floras became the scientific focal point in the herbarium and library at Kew. As the world center for the study of botany, the Royal Botanic Gardens also served as the model for other institutions founded around the world during the eighteenth and nineteenth century.

In North America John Bartram (1699–1777) established a garden on the banks of the Schuylkill River in then suburban Philadelphia in 1728, and during the remainder of his life Bartram traveled extensively in eastern North America, collecting seeds and plants for cultivation in his own garden as well as for shipment to a consortium of plant enthusiasts in England. This undertaking marked the first botanical collection in North America, and Bartram's example was followed by his cousin, Humphry Marshall (1722–1801), who established a commercial botanical garden. Marshall also wrote *Arbustrum Americanum,* which was published in 1775 and stands as the first silva covering a portion of North America.

The short-lived Elgin Botanical Garden was established in New York City in 1801 by Dr. David Hosack (1769–1835), and the now-defunct Cambridge Botanical Garden adjacent to Harvard College in Cambridge, Massachusetts, was also founded in 1801 with support provided by the Massachusetts Society for Promoting Agriculture. While woody plants were included in the collections, these gardens properly fell into the category of botanical gardens, as did Shaw's Garden in St. Louis, Missouri, the predecessor of today's Missouri Botanical Garden, which had its beginnings in 1859.

The first arboretum to be established that incorporated the term "arboretum" in its name was the Derby Arboretum, an 11-acre tract that officially opened to the public on September 16, 1840. Planned as an arboretum from the outset, with individual specimens of a large number of tree and shrub species, each labeled and planted in a naturalistic manner on an artificially contoured landscape, the Derby Arboretum was the work of John Claudius Loudon (1783–1843) and constituted the first parkland in England specifically designed for public use. In shaping the landscape, bringing together the collections, and establishing the underlying philosophy for the Derby Arboretum, Loudon laid the foundations on which most subsequent arboreta would be based. In the *Catalogue of Trees and Shrubs* for the Arboretum, published in 1840, Loudon concluded

> . . . that the most suitable kind of public garden . . . was an arboretum, or collection of trees and shrubs, foreign and indigenous, which would endure the open air in the climate of Derby, with the names placed to each. Such a collection will have all the ordinary beauties of a pleasure-ground viewed as a whole; and yet, from no tree or shrub occurring twice in the whole collection, and from the name of every tree and shrub being placed against it, an inducement is held out for those who walk in the garden to take an interest in the name and history of each species, its uses in this country or in other countries, its appearance at different seasons of the year, and the various associations connected with it.

With the exceptions of the collections of trees brought together by John Evans and Minshall and Jacob Painter in the Philadelphia region earlier in the nineteenth century (the latter now the John J. Tyler Arboretum), the first North American collection to incorporate "arboretum" in its name was founded 32 years after the Derbyshire prototype had opened. This event occurred when the trustees of the will of James Arnold, a wealthy New Bedford, Massachusetts, merchant, transferred a portion of Arnold's residual estate to the President and Fellows of Harvard College. The income from this bequest was specified for the establishment, development, and maintenance of an arboretum to be known as the Arnold Arboretum and the appointment of a director who would serve as Arnold Professor of dendrology at Harvard College. Charles Sprague Sargent (1841–1927) was appointed the Arboretum's first director in 1872, and for 55 years Sargent was responsible for shaping and formulating the policies and programs of the Arnold Arboretum. The development, activities, and achievements of the Arnold Arboretum since its inception in 1872 have largely served as the standard against which similar institutions have subsequently been modeled and judged, both in North America and elsewhere. The successes of Sargent's administration were based in part on his ability to raise the necessary funds to implement his plans, and also due to a creative agreement that was forged between the City of Boston and Harvard in 1882, whereby the Arboretum acreage became part of the city park system but control of the collections and their development continued to reside with the Arboretum staff. Under this agreement the city would maintain the perimeter walls and gates as well as the roadway system and provide police surveillance. On its part the Arboretum, a department within Harvard University, agreed that the grounds would be open, free of charge, to the general public every day of the year from sunrise to sunset. Sargent was also fortunate to be able to enlist the help and assistance of Frederick Law Olmsted in planning the path and roadway system of the Arboretum as well as collaborating in allocating areas within the Arboretum for the various families of plants. As a consequence the Arnold Arboretum was incorporated into Boston's "Emerald Necklace," the city's network of parks and open lands that Olmsted was designing for the Boston Park Department.

Yet another aspect of Sargent's success as an administrator centered on his realization of the research potential of the institution he headed. The era's most distinguished dendrologist, Sargent authored the *Silva of North America* in 14 volumes, published between 1890 and 1902, and the *Manual of the Trees of North America* (1st Ed., 1905; 2nd Ed., 1922); both works remain as standard references today. Through the establishment of a comprehensive library devoted to the topics of botany, horticulture, and dendrology, an equally noted herbarium as a repository for specimens representing the ligneous plants of the world, and a series of scholarly and semipopular publications, Sargent established the Arnold Arboretum as a leading scientific institution. Moreover, the Arnold Arboretum's direct involvement with botanical and horticultural exploration of the floras of the world, with particular emphasis on those of eastern Asia, resulted in the introduction of many woody plants into cultivation and an expanded knowledge of the world's floras and the systematics of woody plants.

Today, over 120 years after its founding, the Arnold Arboretum, located on 265 acres (107 ha) of land in the Jamaica Plain section of Boston, maintains 11,277 accessions in its living collections, which represent 5270 taxa of woody plants. Additionally, its herbarium collection numbers in excess of 1.5 million specimens, and its library holdings are in excess of 100,000 volumes. The Arboretum also maintains an extensive photographic archive as well as general archival collections relating to its history and that of botany and horticulture in North America. Its programs in research, living collections maintenance and development, and public and education programs continue to be representative of those of a major university-affiliated arboretum, while its acreage is an integral part of the metropolitan Boston park system.

IV. NORTH AMERICAN ARBORETA

While the Arnold Arboretum serves as an example of a university-affiliated arboretum that is simultaneously part of a municipal park system, the majority of arboreta fall into three categories of ownership: private, public, or college or university affiliated. Today, many private arboreta operate under the terms of charitable trusts or constitute independent nonprofit corporations. In North America a sampling of notable private arboreta includes the Brooklyn Botanic Garden and Arboretum (Brooklyn, New York), the Bernheim Forest Arboretum (Clermont, Kentucky), the George Landis Arboretum (Esperance, New York), the Morton Arboretum (Lisle, Illinois), the Holden Arboretum (Mentor, Ohio), the Dawes Arboretum (Newark, Ohio), and the Walter Hunnewell Arboretum (Wellesley, Massachusetts). Examples of governmental or municipal arboreta include the Los Angeles State and County Arboretum (Arcadia, California), the Planting Fields Arboretum (Oyster Bay, New York), the Strybing Arboretum and Botanical Garden (San Francisco, California), and the United States National Arboretum (Washinton, D.C.). College- or university-affiliated arboreta include the Minnesota Landscape Arboretum (Chaska, Minnesota), the University of California Arboretum (Davis, California), the University of Guelph Arboretum (Guelph, Ontario, Canada), the Harold L. Lyon Arboretum of the University of Hawaii (Honolulu, Hawaii), Cornell Plantations (Ithaca, New York), the Morris Arboretum (Philadelphia, Pennsylvania), the University of Washington Arboretum (Seattle, Washington), and the Scott Arboretum (Swarthmore, Pennsylvania).

Outside of North America there are many notable arboreta, many combined with or incorporated into botanical gardens, and reference should be made to the *International Directory of Botanical Gardens V* for a more or less complete worldwide listing by country of arboreta and botanical gardens. Over 1400 gardens and arboreta are included.

V. PURPOSES AND FUNCTIONS OF MODERN-DAY ARBORETA

A. Research

As centers for the cultivation of comprehensive collections of woody plants, many arboreta include in their missions research into the botanical and horticultural aspects of woody plants. Studies in systematic botany are a natural extension based on the easy comparisons among species that are facilitated by living collections. These are frequently augmented by those made possible with herbarium specimens, and many arboreta maintain herbaria to document their living collections and to provide a means for plant identification as well as use in revisionary and monographic studies. Augmenting this and other aspects of the development and maintenance of their living collections, arboreta generally assemble reference and research library collections, which also facilitate research and education activities.

Of a more practical nature, arboreta are sometimes the sites of research laboratories for plant breeding, and many new selections of cultivars originate as a consequence of these activities. Cultivars are usually selected, named, and introduced into the horticultural marketplace for specific characteristics such as improved winter hardiness, pest and disease resistance, or superior horticultural attributes. Likewise, laboratories for plant pathology are sometimes established as part of an arboretum, as are laboratories for comparative plant morphology and anatomy, the latter disciplines requiring access to a ready source of fresh plant materials. Ongoing work in plant propagation in collections development also results in new propagation techniques and methods of producing large numbers of plants of taxa never before propagated or previously found difficult to propagate.

B. Conservation

While the numbers of individuals of a given taxon cultivated in arboreta may not suffice to ensure that the inherent genetic diversity of a rare or endangered species is preserved, long-term cultivation of these taxa in arboreta does serve as insurance that individuals of the taxon will be preserved in cultivation. Some woody plants (e.g., *Franklinia alatamaha* Marshall, or the Franklin tree) are extinct in their native habitats and are known only as cultivated specimens. Furthermore, research into cultural requirements and methods of propagation undertaken by arboreta may result in answers to some of the basic questions relating to why a particular plant ranks as rare and endangered in its native habitat. [*See* CONSERVATION PROGRAMS FOR ENDANGERED PLANT SPECIES.]

In North America the Center for Plant Conservation (CPC) based at the Missouri Botanical Garden in St. Louis has organized a program through which cooperating botanical gardens and arboreta are responsible for collecting, growing, and maintaining population samples of species native to the region surrounding the participating garden or arboretum. Through this cooperative network, and with financial support from the CPC, many of the rare and endangered species of North America are now represented in cultivation as a means of preventing their total loss should their native habitats be altered or destroyed. Other institutions have established seed banks in which seed samples of a wide variety of species are stored at cold temperatures (below $-17.8°C$ or $0°F$), thereby prolonging the dormancy of the seeds through desiccation for an indefinite period. [*See* SEED BANKS.]

With similar goals, the American Association of Botanical Gardens and Arboreta has established the North American Plant Collections Consortium (NAPCC). With six institutions currently participating, the NAPCC seeks to coordinate collection development between sister institutions to ensure that rare and endangered species are replicated in more than one collection and to prevent needless duplication of more commonplace taxa. Additionally, NAPCC seeks to increase overall genetic diversity in collections and to improve management practices and the documentation of significant collections.

On a worldwide basis, the Botanic Gardens Conservation International (BGCI, Descanso House, 199 Kew Road, Richmond, Surrey TW9 3BW, UK) fosters similar goals among arboreta and botanical gardens and promotes collaboration in basic research and both *ex* and *in situ* conservation of rare and endangered species. Established in 1987 as part of the World Conservation Union (IUCN), BGCI now has affiliations with more than 400 member institutions in 80 countries. In 1989, together with the World Wildlife Fund and IUCN, BGCI published *The Botanic Gardens Conservation Strategy* which outlines an overall program for the involvement of botanical gardens and arboreta in the conservation of natural diversity. The BGCI also publishes a series of journals, including *Botanic*

Gardens Conservation News, and in 1991 BGCI collaborated with the International Association of Botanical Gardens and the Worldwide Fund for Nature in the publication of the fifth edition of the *International Directory of Botanical Gardens.*

Conservation of many tree species in the tropics of both the Old and New Worlds is also achieved through a network of twelve research sites (eight more are pending) operated under the auspices of the Center for Tropical Science, Smithsonian Tropical Research Institute, with headquarters in Panama City, Panama. In these natural communities of tropical forests, the individual trees are named (sometimes provisionally), labeled, and monitored, and research is conducted on population structure, breeding systems, and other aspects of interest. Consequently, these protected sites fulfill the functions of arboreta.

C. Amenity

From the perspective of the general public, the amenity aspects of arboreta are perhaps paramount to all other functions. The open space made available and the beauty of the plantings are of inestimable value as community resources. It is likely that most visitors to arboreta classify these institutions as parks for passive recreation and fail to realize that additional purposes underlie the institution's activities. Adding to their amenity value, many arboreta represent period landscapes of historical significance in the evolution of landscape design, landscape architecture, and garden history. With the recent emphasis on landscape restoration for historic sites, these arboreta serve as key resources where historic landscapes may be viewed and studied.

D. Education

Many arboreta include on their staff individuals whose primary responsibility is the organization of courses, lectures, and other events of a horticultural or botanical nature that draw on the institution's living collections for the benefit and education of the public. These events may be free or available for a registration fee, and are intended for the interested lay public as well as gardeners at all levels and professional horticulturists, landscape designers, and arborists.

By virtue of the fact that the collections of arboreta are labeled, these institutions provide educational opportunities for the public who choose to inform themselves with the information made available. In addition, most arboreta provide maps and/or guide leaflets to the collections, many of which provide a listing of the sequence of flowering of the major collections throughout the seasons. Sales desks or gift shops also afford the opportunity for the public to purchase books on trees, gardening, and other aspects related to botany, horticulture, and arboriculture.

VI. ORGANIZATION OF ARBORETA

While smaller arboreta may be staffed by only a handful of people who spend the majority of their time caring for the plants in the arboretum itself, most larger institutions are divided into a number of departments that deal with different aspects of the day-to-day operations of the institution. A director normally heads the institution and is answerable to a board of trustees or other governing body. The director and his or her administrative staff are generally responsible for the financial aspects of the institution and its fiscal well-being. Overall policy decisions concerning the operation of the institution are also vested in the director and the administrative staff, which usually consists of the heads of the various departments.

The curatorial staff, generally consisting of a curator or keeper and individuals responsible for records, mapping, labeling, and the taxonomic and nomenclatural aspects of the collection, usually determines collections policy, that is, which plants (species, cultivars, etc.) will be admissible for the collection, and is also responsible for implementing this policy by a number of means, summarized below.

Acquisitions for an arboretum collection may be obtained either through purchase from commercial sources or, more often, through direct collection of propagating materials or the exchange of plant materials with sister institutions. Larger arboreta and some smaller institutions as well occasionally finance collecting trips to regions of floristic interest, usually to areas that are relatively less well known from a botanical standpoint and from which new taxa can be obtained for trial in cultivation. From a historical perspective many of the well-known trees and shrubs cultivated in the Temperate Zone of the Northern Hemisphere were obtained in eastern Asia during the last decades of the nineteenth century and the first decades of the twentieth century through expeditions sponsored by botanical gardens, arboreta, and commercial nurseries, notably, the Arnold Arboretum of Harvard University and the Veitch nursery firm of Chelsea, England. These expeditions usually involve the collection of propagating materials but also include the preparation of pressed and dried specimens, which will be deposited in an appropriate institutional herbarium at the conclusion of the expedition. These specimens vouch and allow for the identification of the living materials obtained; they also add documentary evidence of the flora of the region visited. Duplicate specimens may be shared with other institutions (often in exchange for identifications provided by specialists in the groups involved), and excess plant material raised from the seed collections is frequently shared with other arboreta as well.

Many arboreta and botanical gardens annually circulate an *Index Seminum* to sister institutions, and these lists of species for which seeds or other propagules can be provided on request form a primary means by which many arboreta augment their existing collections. Through such an exchange, which is worldwide in scope, an arboretum may receive these seed lists from sister institutions around the globe and request seeds of species that would otherwise be impossible for the staff of the arboretum to collect on their own. Seeds are sent gratis on a reciprocal basis, and data concerning the collecting locale (for wild-collected material), the collector's name, and the collection number usually accompany the shipment.

Many arboreta also prepare catalogs of their holdings, usually listing the accessions of each species and frequently indicating the provenance of each. Because of the risk of hybridization between species of a given genus when grown in close proximity in arboreta, these arboreta frequently offer to provide cuttings or scion materials from their collections to sister institutions. These catalogs form another basis of plant exchange between institutions.

Plant propagation at arboreta is usually centered in a well-equipped greenhouse and nursery complex where modern methods of propagation are used alongside time-tested methods, and new techniques are sometimes developed to overcome specific problems. Most new accessions are seed propagated, and frequently the requirements for successful germination must be determined through testing involving cold and/or warm stratification. Propagation of existing accessions is normally performed by rooting either soft- or hardwood cuttings, or the preferred method may involve grafting scion wood on an appropriate understock. These methods of asexual propagation ensure that an existing accession will be carried forward into a new generation without the risk of possible hybridization inherent in seed-propagated plants grown from seeds collected within the arboretum.

Grounds maintenance at most arboreta consists of several facets, which include both amenity aspects—such as lawn mowing, trash removal, and weeding—and horticultural curation, which includes the inspection of individual accessions. The latter process ensures that accessions within the collection receive corrective pruning as required, and also ensures that disease and insect infestations receive proper treatment, that fertilization is provided as required, and that overall cultural requirements for proper growth and development are satisfied. Plants that are declining are duly noted in this process and may be nominated for repropagation, if appropriate, in order to maintain the taxon represented by the plant in the collection. Deaccessioning of diseased or severely damaged plants or plants that do not meet collections criteria is also a routine aspect of grounds maintenance. The grounds staff is also usually responsible for the incorporation of new accessions into the collection and the aftercare required to ensure their success in the landscape. Periodic renovations of overgrown areas within the arboretum and the plant removals and new plantings that this work entails also fall within the purview of grounds maintenance.

Much of the work of grounds maintenance requires hand labor, although many aspects utilize a wide range of modern tools and machinery, and most arboreta attempt to implement ecologically sound methods of maintenance. This is particularly true in the areas of weed control and disease and pest management, in which integrated pest management stresses sound environmental standards with the use of chemical herbicides and pesticides. Other aspects of environmentally sound procedures include appropriate waste disposal and the recycling of materials through the production of compost and mulch for use within the collections.

The purpose of the plant records, mapping, and labeling department within arboreta is to maintain accurate and current information concerning the nature of the collection and its individual accessions. Generally, each plant in the arboretum carries a label with its accession number and name, which ties the plant to the more complete information concerning that accession in the system of plant records. In these records the source of the accession is maintained, as well as its provenance (if the accession was wild-collected), the means by which it was propagated, and whether or not it is a member of a lineage of accessions that has been repropagated from preexisting accessions. In addition, information concerning individual plants is frequently maintained in the record system. These data may relate to their growth rates, size, and habit; their flowering and fruiting dates; their horticultural attributes; and other information considered of interest or value (e.g., chromosome numbers, if determined).

While each accession carries its record label, these occasionally disappear (by accident or vandalism), and many arboreta maintain a system of maps showing the planting location of each accession, as well as features of topography and hardscape (buildings, roads, fences, gates, power and water lines, etc.). These maps serve as an important backup that allow each accessioned plant to be located and identified, regardless of whether the record label is present. Moreover, the system of maps provides an important tool whereby areas of the arboretum can be scrutinized and future developments can be planned and developed.

In addition to the record label, most plants in an arboretum collection carry a trunk or display label that provides the visiting public with the common and botanical names of the plant, its nativity, and frequently the botanical family to which it is assigned. Other information is sometimes included on the label which relates to an overall interpretive program designed to involve visitors to the arboretum in an educational manner. A series of map brochures that outline the road or pathway system and lead the visitor into the collections is sometimes made available.

During the past decade many arboreta and botanical gardens have utilized a computerized data base linked to a network of personal computers to maintain their plant records, and in some instances an interface between the records data base and a computer-assisted mapping program has also been implemented. These systems, of which BG-Base (BG-Base, Inc., The Holden Arboretum, 9500 Sperry Road, Mentor, Ohio 44060-8199, and/or Royal Botanic Garden, Inverleith Row, Edinburgh EH3 5LR Scotland, U.K.) is perhaps the most widely used, provide an easy means by which records can be maintained, the information can be manipulated in a variety of ways to facilitate collections curation, and the collection can be used for research and other specific needs. Through the link to a mapping program current, computer-generated maps can be produced in a short time, and the time-consuming process of hand drawing and lettering collections maps is alleviated. Furthermore, the possibilities of communication among institutions are greatly facilitated through electronic data base access.

VII. LINKAGE BETWEEN ARBORETA

Today's computer and telecommunications systems are facilitating direct "on-line" communication among arboreta. Additionally, the International Association of Botanical Gardens and

Arboreta was established at the eighth International Botanical Congress held in Paris in 1954, under the auspices of the International Union of Biological Sciences as a commission of the International Association of Botanical and Mycological Societies. The purpose of this organization has been to facilitate communication and collaboration among member institutions, to promote the study of taxonomy of plants to benefit the world community, to promote both *in* and *ex situ* conservation of plants, and to promote horticulture as an art and a science. These aims are sought through publications, meetings and symposia, and contact through regionally autonomous groups, which are represented at the IABG council.

All botanic gardens, arboreta, or other institutes and their staff are eligible for membership through various regional groups in Europe, Ibero-Macaronesia, Latin America, Australasia-Oceania, eastern Asia, and North America, where the American Association of Botanical Gardens and Arboreta provides corresponding service to North American institutions. The American Association of Botanical Gardens and Arboreta, with offices in Wayne, Pennsylvania, seeks to achieve similar goals through its sponsorship of an annual meeting as well as yearly regional meetings and a periodical publication, *The Public Garden*.

Given the great expense involved in many undertakings of collection development and maintenance, many arboreta have developed cooperative programs on a cost-share basis whereby the participating institutions work together to achieve common goals. Programs for plant exploration in several parts of the world have been conducted on this basis, and the North American China Plant Exploration Consortium, a group of North American arboreta, has plans for cooperative programs with colleagues in the People's Republic of China. Another example of cooperation among organizations to reduce duplication of effort and expense has been the establishment of "national collections" among a network of arboreta and gardens in Great Britain under the auspices of the National Council for the Conservation of Plants and Gardens. Through this scheme individual gardens and arboreta agree to develop and maintain comprehensive collections of taxa of a particular genus and to provide propagating materials, as required, to other cooperators. By this means conservation efforts are satisfied—particularly for old garden forms—needless duplication is avoided, and specialty collections are developed in one location, thereby enhancing the possibility for research.

The cooperation between sister arboreta frequently extends to the commercial nursery industry. The industry often relies on local arboreta for the identification of plants in commercial production and for the application of correct nomenclature. Arboreta also serve as the sources of new plants for introduction into commerce, whether they be newly introduced from another region of the world or new cultivars selected from within arboretum collections for particular horticultural attributes. Commercial nurseries on their part frequently support arboreta with financial contributions or in-kind donations of plants or helpful information.

VIII. THE FUTURE

While it is unlikely that many new arboreta will be founded in the future due to the high costs involved, established institutions will continue to serve the scientific community as a source of accurately named and correctly identified plant materials and to undertake research fundamental to our understanding of woody plant species. The role arboreta play in the protection and conservation of woody plants remains to be fully realized, as does their role in promoting an understanding by the general public of plants as key components in the world's economy and the global ecosystem. As cultural institutions arboreta will continue to fulfill their multifaceted roles to the many people who visit arboretum collections as plant professionals, urban or suburban gardeners, student naturalists, or city dwellers in search of a peaceful walk among the trees.

Glossary

Accession An individual plant, or a group of plants of the same taxon from a single gathering, that has been obtained for the collection and given an identifying number; an acquistion.

Arboriculture The science of tree planting, cultivation, and care.

Cultivar A cultivated variety, normally a plant selected and named for a particular desirable characteristic or attribute. Woody plant cultivars are propagated by asexual means, and the individuals composing the cultivar constitute a clone.

Horticulture The applied and theoretical science focusing on the cultivation of ornamental and comestible plants, both woody and herbaceous.

Ligneous Containing and producing lignin, the polymer that imparts the texture and qualities of wood to plant tissue. Woody or ligneous plants are differentiated from herbaceous plants that do not produce lignin.

Provenance The place of origin of an accession, normally data providing geographic and ecological information concerning the collecting site.

Silva A written account of the species of trees occurring naturally in a defined region.

Taxon A taxonomic group, or a number of groups taken together, of undesignated rank (e.g., family, genus, species, or subspecies). In the statement "Arboreta specialize in growing woody taxa," the term "taxa" can be taken to refer to any or all of the ranks listed.

Bibliography

Ashton, P. S. (1984). Botanical gardens and experimental grounds. *In* "Current Concepts in Plant Taxonomy" (V. H. Heywood and D. M. Moore, eds.), Systematics Association Special Volume 25, pp. 39–46. Academic Press, Orlando, Florida.

Heywood, C. A., V. H. Heywood, & P. W. Jackson. (1990). "International Directory of Botanical Gardens V." Koeltz Scientific Books, Koenigstein, Germany.

Jack, J. G. (1946). How to establish an arboretum or botanical garden. *Chron. Bot.* **10,** 405–418.

Raven, P. H. (1981). Research in botanical gardens. *Bot. Jahrb. Syst.* **102,** 53–72.

Stearn, W. T. (1971). Sources of information about botanic gardens and herbaria. *Biol. J. Linn. Soc.* **3,** 225–233.

Stearn, W. T. (1972). From medieval park to modern arboretum: The Arnold Arboretum and its historical background. *Arnoldia* **32,** 173–197.

Artificial Habitats for Fish

William Seaman, Jr.
University of Florida

Artificial habitats in the aquatic environment serve to enhance fishing, increase or manipulate assemblages of organisms, and mitigate ecological disruption. Artificial habitat technologies are used worldwide in freshwater and marine systems. These practices date from antiquity and use a variety of natural and man-made materials to create substrates that provide physical relief or otherwise modify habitat for aquatic life. In some cases large fisheries have been established. Only in recent years have advanced engineering and design principles been incorporated in this field, along with quantitative ecological and socioeconomic assessment of habitat structure and function. This subject area represents an emerging multidisciplinary branch of fishery and environmental science that is enabling practitioners to understand the basis for performance of artificial habitats, and thereby further describe fundamental ecological processes as well.

I. INTRODUCTION

Artificial habitats are deployed in aquatic environments by a variety of lay and technical interests for purposes that range from enhancing harvest of plants and animals to creating systems for ecological research. The earliest practices in this field were based on simple observations of the attraction of fishes to natural objects and have persisted for centuries. The historical goal of more successful subsistence fishing has been augmented in modern times by commercial, recreational, and environmental purposes. Artifical habitats are established worldwide in lakes, rivers, estuaries, and the ocean. Increasingly complex construction and deployment, especially at sea, now supplement traditional practices that rely on natural materials and simple procedures.

Ecologically, artificial aquatic habitats may influence either the behavior of organisms (particularly more mobile fishes) by directly attracting them to the vicinity of the structure or else some aspect of the life histroy of a species by enhancing an environmental variable that may be limiting. Persons who deploy these structures have used them to enhance fishing success by capturing fishes that aggregate at these sites, and also to provide the basis for growth of new biomass of plants, invertebrates, and fishes. This latter aspect exists when the artificial habitat truly mimics the ecological structure and function of natural habitats. Scientific inquiry has turned increasingly to examination of recruitment-limitation and habitat-limitation as factors influencing the abundance of plants and animals at artificial habitats. [*See* FISH ECOLOGY.]

Artificial habitats in the marine environment are built on a larger scale than in fresh water. Principal areas of the world where artificial habitats are utilized include coastal Southeast Asia and Japan, coastal Australia, some islands of the western tropical Pacific, the Carribbean and Mediterranean basins, and the United States (Table I). There is activity in at least 45 nations. Growing worldwide interest in this field in the last two decades has been accompanied by increased study, experimentation, and evaluation of performance of artificial habitats.

The use of artificial habitats in the management of natural resources has prompted some scientific and environmental controversy. Among fishery science practices, artificial habitats have been perhaps the most widely used by nontechnical audiences (e.g., sport fisheries) owing to the readily accessible and easily implemented technologies. Yet extensive use in some areas has raised concerns for overexploitation of fishery stocks. Particularly in the marine environment, deployment of artificial reefs as a means of solid waste disposal has caused debate. Also, the ecological mechanisms involved with some habitats are an issue.

II. CATEGORIES OF ARTIFICIAL HABITATS AND USES

For purposes of this discussion an artificial habitat is defined as a substrate, structure, or group of structures that are placed or built purposely in an aquatic

TABLE I
Principal Applications of Artificial Aquatic Habitat Technologies

Purpose of artificial habitat	Global areas of emphasis
Artisanal fishing	Africa, Caribbean basin, western Pacific
Commercial fishing	Japan, Taiwan, Philippines, western Pacific islands, Mediterranean Sea
Recreational fishing	Australia, United States
Environmental restoration and management	Mediterranean Sea, United States

ecosystem by humans. Although "artificial reefs" are most commonly associated with the term artificial habitat, this article takes a broader view by also addressing other types of structures and practices. Creation of entire new habitats or ecosystems, however, such as by digging and planting wetlands or excavating ponds, is beyond the scope of this article. Artificial habitats may of course be placed in a newly created system, such as in a reservoir, after its construction. Artificial habitats may be defined further according to their location, composition, or usage.

A. Artificial Habitat Location

In terms of placement, habitats are located in fresh and salt water and in flowing and standing waters, and may be suspended in the water column or placed on the bottom or floor of the aquatic system. Typically, artificial habitats are deployed individually to meet the purpose of a local interest group; coordination on a broader regional basis may be minimal, and national planning is not common. Figure 1 depicts representative structures throughout a hypothetical drainage system. In reality the number and diversity of artificial habitats in a given system may not be as great as in the illustration, but the purpose here is to portray the wide variety of situations in which placements are possible.

Placement of an artificial habitat may be restricted to a unique geographic site, as in the case of construction of a spawning channel in a degraded section of a particular stream or river as a means of increasing the survival of eggs. In Canada, for example, pink salmon survival in an "artificial" gravel bed greatly exceeded that from natural spawning areas. Commonly, though, there is some choice of sites for an artificial habitat, such as in creating a reef as a new recreational fishing site in an expansive ocean area of low physical relief so as to make it easily reached by anglers in small boats. In this case, variables such as distance from an access point, compass bearing from landmarks, or water depth would be considered in site selection.

B. Composition

Artificial habitats can be made of natural or manmade materials and deployed as individual items

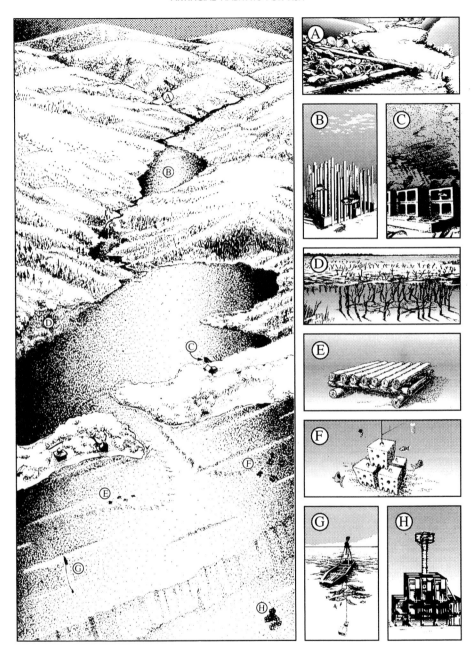

FIGURE 1 Artificial habitats are deployed in many different aquatic environments. Their scale matches the scale of the system in which they are placed, ranging from small structures of rocks and logs in freshwater streams to steel and concrete frameworks in the ocean comparable in size to multistory buildings. Representative habitats include (A) log deflector, (B) stake bed, (C) concrete block platform, (D) "acadja" or brush pile, (E) "casita" or log shelter, (F) concrete modules, (G) "payao" or bamboo raft, (H) steel reef.

or in an assembled structure. For example, logs can be assembled as a barrier and at an angle to the current in a stream to deflect water and increase its velocity, as is done commonly in North America

(Fig. 1A), and logs also can be assembled as a shelter for lobster such as in the Caribbean Sea (Fig. 1E). In the acadja or circular brush parks of African lagoons (Fig. 1D), individual tree branches are distributed, whereas bamboo is assembled as rafts in the Philippines (Fig. 1G). Rafts and other floating structures, plus those suspended in the water column, are commonly referred to as fish aggregating devices, or FADs.

Fabricated materials, either designed explicitly as habitat structure or acquired from surplus products, predominate in the ocean environment. Particularly concrete and steel are used, such as large concrete blocks in Italy (Fig. 1F) and many other areas, and steel frameworks in Japan (Fig. 1H). One popular natural item is rock, such as granite boulders. A common simple designed structure in freshwater lakes is the stake bed (Fig. 1B), and concrete building blocks may be assembled in structures in estuaries, for example, where they can be placed under docks (Fig. 1C) and the sea.

A greater diversity of structures and materials is used in marine environments in comparison to fresh waters. Freshwater artificial habitats tend to be on a smaller scale and emphasize the use of natural materials, such as wood, trees (Fig. 2D), gravel, stones, and rocks (Fig. 2F). Larger structures, for example, ships (Fig. 2I), and designed structures (e.g., Fig. 2K) appear at sea. A survey of worldwide marine practices determined that prefabricated concrete units were the most popular structure used in building artificial reefs. Both hollow and solid cubes (commonly 1–2 m) are deployed, typically in clusters of reef "modules," in areas such as Taiwan, Japan, Korea, Italy, and the United States. In contrast with prefabricated cubes, vehicle tires are the second most popular marine item and represent a surplus material adapted to artificial habitat construction on an opportunistic basis. Ordinarily tires are assembled in weighted modules intended for permanent deployment on the lake bottom or seafloor, although improper installation has caused environmental problems when tires break loose and wash onto coastal beaches or reefs.

Larger and denser surplus materials, including derelict vessels, obsolete petroleum platforms, and roadway or building construction rubble, are used worldwide in the coastal ocean. These items may be acquired at no cost to the reef-building interest, and they fulfill the most basic biological design

criteria of providing vertical relief, hard substrate, and shelter for organisms. Meanwhile, construction of reefs according to more extensive physical and ecological design guidelines is most common and diverse in Japan, where various fiberglass, steel, and concrete superstructures (e.g., Fig. 1H and 2K) are used.

C. Purposes for Artificial Habitats

From the earliest uses of artificial habitats to enhance subsistence harvest of fishes for food supply, a variety of other objectives developed since World War II and particularly in the last generation. Artificial habitats are used to achieve purposes related to artisanal fishing, commercial fishing, recreational fishing, other recreation (notably scuba diving), management of aquatic resource use patterns, restoration of environmental damage, and mitigation of environmental disruption. The earliest recorded description of artificial reefs appeared in the late eighteenth century in Japan.

I. Fishing Enhancement

The most intensively developed system of artificial habitats in the world is located in Japan's coastal waters, principally as a means of ensuring a reliable domestic seafood supply. It has been promoted as part of a systematic national fisheries development plan and has influenced about 10% of the sea bottom off Japan. As noted in Table I, other commercial fishing applications exist in Taiwan, Italy, the Philippines, and elsewhere. Artisianal fisheries have developed in less industrialized areas, including Mexico and other Caribbean nations, in Thailand, and in Africa. Recreational fishing applications are most common in coastal Australia and the fresh waters and coastal seas of the United States.

Artificial habitats to enhance artisanal (or subsistence) fishing are small and made of inexpensive and commonly available materials, such as brush, logs, bamboo, scrap tires, and concrete. In the Philippines, for example, 16,000 bamboo modules were deployed at depths of 15–25 m along one 40-km section of coastline by village fishermen who also planned and built them. Fish species captured included grunts, snappers, sea basses, and

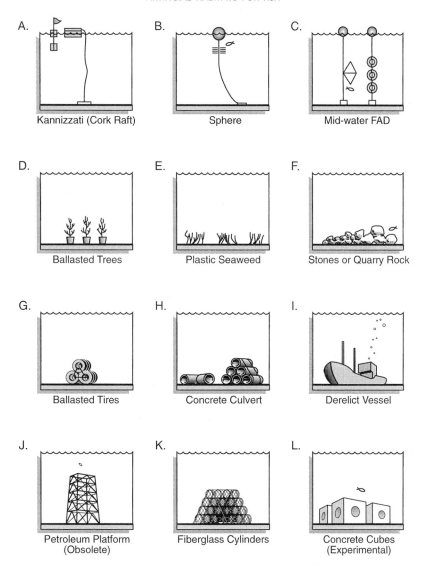

FIGURE 2 Common artificial habitat materials and designs.

others, with annual harvest equal to eight kg/m³. In Cuba and Mexico, shelters made of mangrove tree banches are used to attract spiny lobster. In one area, 120,000 habitat modules yielded 7000 metric tons of lobster. In the Gulf of Thailand, concrete cylinders were deployed over an area of 41 km² to establish a small-vessel, gill-net fishery for threadfin, which yielded over 5500 kg and an average of 8.3 kg/fishing trip during a season of at least six months.

Commercial fishing usually involves larger habitat structures and more extensive harvest operations such as larger vessels, inboard engines, and so on. Japan has spent well over U.S. $1 billion to build artificial reefs. As a result, a network of government investment and research interests, industrial reef-builders, and private fishing cooperatives has been established there to exploit a variety of seaweed, invertebrate (e.g., abalone, octopus), and finfish (e.g., sea bream, flounder) species. The Japanese program evolved over the last two centuries, from (1) local artisanal applications, to (2) enhancing fish catches near natural reefs after World War II, and most recently to (3) create new fishing grounds where none previously existed. In the Philippines, high yields of tuna (almost 280,000 metric tons in 1988) from purse seining around 3000 floating bamboo FADs were recorded. Concrete reefs placed in Italy enhanced the harvest of finfishes and mussels, so that profits were more than double those recorded for nonreef fishing.

Recreational fishing in Australia and especially the United States has benefitted from extensive use of artificial habitats. In the United States, for example, 32 state governmental freshwater fishery agencies had habitat installation programs as of the early 1980s, which created over 44,000 individual structures in nearly 1600 bodies of water. As of 1987, the 23 U.S. coastal states featured nearly 600 authorized ocean sites for artificial reefs. Marine reefs in the United States are built by both governmental and private interests and typically are created from surplus materials. Catch and effort data for sport fishing at ocean reefs are scant.

2. Environmental Management

The use of artificial habitats to modify the environment extends increasingly beyond the goal of enhancing fisheries and increasing harvest. Artificial habitats also are used in allocation, protection, and restoration of aquatic resources, and these actions may benefit fisheries indirectly. For example, placement of concrete modules along the seacoast of Italy as fishery habitat has the added result of establishing a physical barrier to prevent trawlers from entering a protected marine zone. Elsewhere on the northern Mediterranean coast (e.g., Monaco), concrete block structures are deployed specifically in coastal areas to establish reserves in which seagrass beds and juvenile fish habitat are protected. In Costa Rica, a reef of 5000 tires has been established, but not marked, to provide a habitat free of fishery exploitation. Placement of small concrete block units under docks (Fig. 1C) or in areas where seawalls and canals have destroyed natural habitat has been tested in the southern United States.

In recent years the use of artificial reefs as a way to mitigate damage of aquatic systems has been initiated. The most extensive experience has been gained on the temperate Pacific Coast of the United States, where new benthic habitat is created to replace other (natural) habitat that has been destroyed or polluted. Because a coastal electric power generating station killed organisms entrained in its cooling water system and also created turbidity that affected other organisms, government regulators decided that a local mitigation plan should require an artificial reef to restore the kelp forest system in southern California to account for the natural habitat that was lost. The degree to which artificial systems are functionally equivalent to natural systems, in terms of ecological structure and function, is not well documented.

3. Economic Development

Particularly in areas with clear coastal waters and an ambient environment that is attractive to tourism, coastal artificial reefs are deployed to provide new opportunities for recreation. In Miami, Florida, U.S.A., for example, reefs were created especially for scuba divers. Subsequently a small charter boat industry arose to cater to their needs. Tropical areas wishing to promote "eco-tourism" are considering artificial reefs to expand recreational fishing and diving as activities compatible with a philosophy of sustained use of natural resources.

III. SCIENTIFIC STUDY METHODS

Among the aquatic sciences, the study of artificial habitats has only recently become established as a branch of fishery science and the body of research knowledge still lags the application of various practices. Study of freshwater structures and systems began in the 1930s, while marine research accelerated in the 1970s. Three international conferences in the last decade have fostered an increased level of research and exchange of technical information, particularly for ocean artificial habitats.

Easier access to streams, lakes, and other inland waters enabled scientists to use traditional approaches to gathering data on so-called "habitat improvement" techniques. The emphasis on freshwater fish production dictated using assessment methods such as collection of fishes by seine, trap, electrofishing, and gill net, and environmental variables such as water clarity or oxygen content might also be measured. In shallower situations, direct observation is possible, or else some habitat structures or sampling plates on them might be retrieved from the body of water for inspection. Creel census of fishing success also is employed. Research in marine environments has taken a different approach.

The advent of scuba diving gear enabled scientists to make observations in some fresh waters and particularly in many marine systems. At sea, where artificial habitats may be at greater depths and access and surface working conditions may be physically more difficult than inland, scuba gear allows direct observation of both organisms and the habitat. Ecological study techniques developed for natural systems, such as coral reefs, have been readily applied to artificial habitats. Again, fishes have been emphasized; counting procedures according to various standard transects, areas, and times are used by divers. Where applicable, diving is the method of choice to study ocean fishes at artificial habitats, whereas traditional capture techniques (e.g., trawling) are used less commonly and advanced techniques such as remotely operated video systems are in the early stages of application.

Study of both physical processes at artificial habitats (e.g., upwelling caused by disruption of current) and primary biological production (i.e., photosynthesis) has lagged the study of fishes. However, as the interest of fishery science has evolved from a basic objective to describe the organisms present at an artificial habitat to a more holistic need to explain why a particular assemblage is present and how it interacts with the surrounding ecosystem, a broader multidisciplinary approach has developed. Nonbiological reasons for building an artificial habitat such as recreational diving dictate that socioeconomic assessments also be performed. A listing of the more common variables that would be addressed in an expansive multidisciplinary study of an artificial habitat appears in Table II. To date, however, published articles have focused on narrow aspects of the subject and a few variables, and generally do not attempt to describe the artificial habitat in the context of an ecosystem. Nonetheless, there is a growing interest among scientists, persons who use aquatic resources for other purposes, and those who finance construction to have reliable ecological, physical, and eco-

TABLE II
Variables Commonly Designated for Evaluation of Artificial Habitat

Factor	Characteristic measured
Abiotic conditions	Substrate, temperature, salinity, visibility, nutrients, sea state, current, tide, weather, lunar cycle, season, time, pollutants
Abundance of organisms	Number, biomass, diversity, colonization, succession, community similarity
Life history of species	Age, size, growth rate, condition, feeding, sex, reproduction, larval development, movement, behavior
Fishery and socioeconomics	Recruitment, catch, effort, catch per unit of effort, cost of project, expenditures of users, usage of project
Artificial habitat	Benchmark location, composition, location of material, area covered, volume, profile, structural complexity, site/vicinity map

nomic data for evaluating the performance of artificial habitats, so that a new body of knowledge is developing.

Whereas most of the research data base for artificial aquatic habitats is based on structures built for an applied purpose (and in marine systems often built by nonscientific personnel), in recent years experimental systems have been deployed. They are designed and built according to a study question and hypotheses, with sampling considerations dictated by statistical analysis requirements. Ultimately, a larger data base for both pre- and postdeployment conditions described by the variables in Table II will result.

IV. ECOLOGY OF ARTIFICIAL AQUATIC HABITATS

The distinction between artificial habitats and natural habitats blurs with time. As assemblages of aquatic organisms colonize and succeed on habitats created by humans, the physical appearance of the structure changes. Microbes, algae, periphyton, and invertebrates all may attach to the habitat surface, especially in the sea. The guiding biological principle for creation of artificial habitats is that biomass of one or more species will be increased owing to the influence of the habitat structure (Fig. 3).

Fishes commonly occupy an artificial habitat within hours or even minutes of its creation, whether it is a derelict vessel that is sunk as a "reef" or a carefully designed concrete, fiberglass, or steel module lowered from a barge. Clearly, the behavior of fishes to associate with introduced structure has made the deployment of surface, midwater and benthic devices popular as a means of enhancing fishing effectiveness, because the concentration of fishes at a structure makes capture easier.

A small body of literature has developed for the design, construction, and placement of artificial aquatic habitats. It is beyond the scope of this article to discuss the siting of habitats and their interaction with the physical environment, yet it is noteworthy that engineering practices range from elaborate in places such as the seacoast of Japan to less complex in many freshwater settings. The physical dynamics of artificial habitats are much better understood and documented than their biology. In this discussion, the biological influence of the habitat on the

adjacent ecosystem and as a substrate for organisms is emphasized. Just as a log deflector in a stream influences current flow, an artificial reef on the floor of a lake or the sea also has an effect by creating an upwelling that may carry nutrients up into the water column where phytoplankton can incorporate them and by providing a lee downcurrent where swimming organisms find shelter (Fig. 4).

The behavioral impact of artificial habitats is especially prominent for floating and midwater structures, for both freshwater and marine fishes that aggregate at them. These species—such as tunas and amberjacks in the ocean, and crappies in lakes—are considered as transients in terms of duration of dependence on the artificial habitat.

A variety of microbial, plant, and animal species also reside on artificial habitats, which provide a site for attachment, shelter, reproduction, and feeding both on-site and off-site (Fig. 4). The surface chemistry and texture of an artificial habitat influence the "fouling assemblage" that evolves. For example, certain plastics support algal and invertebrate assemblages similar to those on dead coral rock in terms of diversity and biomass. In Hawaii, concrete

FIGURE 3 Fishes are a highly visible component of the biological community that evolves on artificial aquatic habitats. Over time, plant and invertebrate species attach to the substrate of this marine reef. (Photograph courtesy of Frederic Vose, University of Florida Department of Fisheries and Aquatic Sciences.)

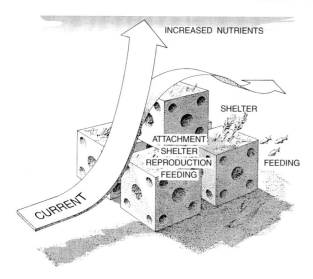

FIGURE 4 The physical impact of an artificial reef may divert prevailing current, create upwelling and lee areas, and provide surfaces and crevices for organisms.

supported assemblages that are most similar to those found on coral substrate. Larger attached organisms on marine reefs include plants and bivalve mollusks. In Japan, rocks in piles of 0.2–0.6 m are deployed to provide for settlement and protection of abalone larvae, and concrete blocks provide substrate for growth of kelp. Not all of these systems are for monoculture, and in some areas abalone and sea urchins that graze on the kelp are periodically transplanted to other areas for growth.

Fishes reproduce and feed at artificial habitats. Cold-water species such as lake trout spawn at large rocks placed at appropriate depth (e.g., 5 m), and smallmouth bass use gravel boxes for nesting in warm-water lakes. Feeding may occur directly at the habitat, as in grazing of algae and invertebrates by sheepshead on ocean reefs of concrete, or away from the habitat, as when species such as grunts nocturnally forage on the surrounding seafloor and then return to the shelter of the reef during daylight.

In spite of the common ecological principles that pertain to freshwater and marine systems, there are differences between artificial habitats placed in them. Unlike the marine environment, in fresh waters there are few reef-building organisms except for mussels. Hence freshwater artificial reefs are not physically enlarged over time by colonization of organisms, in contrast with the "fouling"

that takes place in the sea. Though a lake periphyton community can be diverse, the most complex community including algae, coelenterates, arthropods, echinoderms, and chordates is found on ocean reefs, where more organisms are adapted to utilizing crevices in the habitat structure.

V. APPLICATIONS AND IMPACTS OF ARTIFICIAL HABITAT TECHNOLOGIES

From a socioeconomic perspective, artificial habitats have contributed to increased or more efficient harvest of food and commercial products, satisfaction of recreational objectives, and the manipulation of the aquatic environment for purposes of conservation and management. Successful application of artificial habitat technology in certain areas have been emulated in other areas, so that at least 45 nations have artificial aquatic habitats in fresh and salt waters. Many of the experiences are recent, and relatively little analysis of impact or effectiveness has occurred.

The success of fishing at artificial habitats has raised concerns for the sustainability of the particular fishery stock that is exploited. With the introduction of bamboo rafts in the Philippines in 1976, catches of tuna rose from under 10,000 metric tons in the early 1970s, to 125,000 tons in 1976, to 279,000 tons in 1988, which contributed 20% of total fish production for the country that year. This made the Philippines the largest producer of tuna in Southeast Asia. However, concerns have been expressed that these FADs have altered the movement and feeding behavior of yellowfin and skipjack tuna, whereby juvenile (undersized) individuals are susceptible to heavy exploitation.

Three possible impacts of artificial habitats are depicted in Fig. 5, specifically for a fishery at a marine reef. These impacts may be among the most dramatic owing to the larger physical scale and impact of marine reefs, in contrast to the smaller and more geographically localized structures used in fresh waters. In scenario A in Fig. 5, the baseline or "prereef" condition of a fishery is depicted, whereas scenarios B, C, and D illustrate different

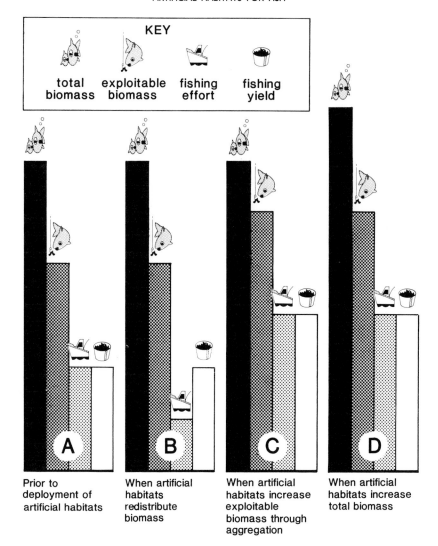

FIGURE 5   Possible influences of artificial habitat upon fishes subject to harvest. [Reprinted from W. Seaman, Jr. and L. M. Sprague, eds. (1991). "Artificial Habitats for Marine and Freshwater Fisheries." Copyright 1991, Academic Press, San Diego.]

situations after reef deployment. As a result of the artificial reef, total fish biomass, exploitable biomass, fishing effort, and yield might change. The intent of the reef, of course, is to have a positive impact, but it is possible that fish stocks might be damaged by intense, concentrated fishing (i.e., recruitment overfishing or growth overfishing). In scenario B, the artificial reef simply redistributes the exploitable biomass to make it easier to catch, so that less effort is needed to catch the same amount of fish. In scenario C, the exploitable biomass at the reef site is increased, and effort also goes up so that more fish are caught even though

total biomass in the system is not increased. Finally, in scenario D, the actual biomass of the overall fish stock is increased and thus all aspects of the fishery increase too.

Many artisanal, commercial, and sport fishing uses of artificial habitats lack reliable scientific data on levels of fishing effort and catch, and also the population dynamics of the target species. However, increasing concern for sustainability of harvest has prompted research into the aggregation and biomass production attributes of habitats. The most consistent documentation of increases of biomass at artificial habitats is for plants, including

Laminaria in Japan, sea urchins and abalone (Japan), mussels (Italy and the United States), and octopus (Japan). In the case of octopus, habitat could be limiting, and because one animal is typically associated with one den, an increase in the number of dens could contribute to increased population size. In brush parks of coastal Africa, the habitat initially serves as a refuge for fishes, but over time fishes reproduce there as well. Generally, an increase in the population of fishes would result from habitats that increase postlarval settlement and growth and survival of juveniles. Conversely, for recruitment-limited species the addition of habitat to the environment should not make a difference in abundance. The types of fishes that might be targeted for enhancement using artificial habitats are those that are demersal, territorial, and nonmigratory. By contrast, midwater and surface species, and those that are migratory and transient on reefs, would be less influenced by artificial habitat.

VI. CONCLUSION

Artificial habitats in the aquatic environment have direct and indirect effects on fishes, other taxa, and the physical dynamics of the system. The historical emphasis on fish production, though successful, has been tempered by concerns for long-term ecological impact and levels of sustained harvest and augmented by new objectives related to ecosystem restoration and conservation of biodiversity. A lack of data on the productivity and biological performance of artificial habitats in at least some settings may hamper the extensive application of these technologies, but research to evaluate artificial habitats has accelerated in recent years. Scientific study in this field is increasingly driven by management needs. Biological study is supplemented by some engineering and socio-economic research.

Increasing recognition of the types of fish species and other organisms that can be enhanced by artificial habitats is motivating the development of design criteria so that physical structure may be configured to meet the needs of individual species. Physical criteria for stability and longevity of structures are rea-sonably well known, especially for those built expressly as aquatic habitat. Biological criteria include environmental factors such as geographic location, surrounding substrate, isolation, depth, seasonality, temperature, and water conditions; physical variables of the habitat material include its design, composition, texture, shape, profile, size, void space, and dispersion/orientation.

As the objectives for which artificial habitats are created become more specific, it will be possible to establish a data base that will be useful to enhance their future performance. That is, objectives such as "to increase fishing" are increasingly rephrased to more quantified statements such as "increase catch of species A by X amount." As the fishery objectives of artificial habitats are better described, in turn there will be a more rational basis for newer applications in environmental management. From an art that has existed for centuries, a science that offers benefit to applied interests as well as basic environmental research is emerging.

Glossary

Artificial habitat Substrate, structure, or group of structures made of natural or man-made materials, placed in the aquatic environment to enhance harvest or abundance of organisms or in some other way manipulate assemblages of species or ecological processes.

Artificial reef Benthic artificial habitat in a lake, estuary, or the ocean, generally of relatively large size, which may include one or multiple pieces.

Fish aggregating device (FAD) Artificial habitat that is deployed in the water column or at the surface, which functions by influencing the behavior of fishes so they aggregate at it.

Fish attractor Nomenclature used for certain freshwater artificial habitats, particularly smaller benthic structures and any type of FAD.

Bibliography

Bulletin of Marine Science (1985). "Third International Artificial Reef Conference," Vol. 37, No. 1. Miami: Rosenstiel School of Marine and Atmospheric Science, University of Miami.

Bulletin of Marine Science (1989). "Fourth International Conference on Artificial Habitats for Fisheries," Vol. 44, No. 2. Miami: Rosenstiel School of Marine and Atmospheric Science, University of Miami.

Bulletin of Marine Science (1994). "Fifth International Conference on Aquatic Habitat Enhancement," Vol. 55, Nos. 2–3. Miami: Rosenstiel School of Marine and Atmospheric Science, University of Miami.

Everhart, W. H., and Youngs, W. D. (1981). "Principles of Fishery Science." Ithaca, N.Y.: Comstock Publishing Associates, Cornell University Press.

Mottet, M. G. (1981). "Enhancement of the Marine Environment for Fisheries and Aquaculture in Japan," Technical Report 69. Olympia: State of Washington, Department of Fisheries.

Seaman, W., Jr., and Sprague, L. M. eds. (1991). "Artificial Habitats for Marine and Freshwater Fisheries." San Diego: Academic Press.

Aspects of the Environmental Chemistry of the Elements

Nigel J. Bunce
University of Guelph, Canada

I. Introduction
II. Environmental Chemistry of Specific Substances

This account of environmental chemistry in the context of environmental biology must inevitably be very selective, as there is scarcely a point at which chemistry does not interact with biological systems. Comparison with the other planets of the solar system immediately reveals the enormous impact that life has wrought upon the chemistry of the Earth. The essential building blocks of life as we know it (the elements carbon, hydrogen, oxygen, nitrogen, phosphorus, and sulfur) are continuously cycled through the biosphere. Throughout geological time, living organisms have changed the face of the planet, depositing sedimentary rocks such as chalk and limestone, and through photosynthesis changing the atmosphere from a reducing medium to an oxygen-rich and hence oxidizing environment unlike that of any other planet. In recent recorded time, the activities of one particular species, *Homo sapiens,* have produced additional changes, such as the introduction of previously unknown substances into the environment and the large-scale alteration of the surface of the planet through agriculture, urbanization, mining, and other industrial operations.

Sunlight is an important energy source in the environment. Not only does incoming solar energy raise the temperature of the Earth's surface through physical absorption, but under appropriate conditions light energy can be harnessed to drive chemical reactions that would otherwise be energetically disfavored. Indeed, photosynthesis is an example of such a photochemical reaction; in the absence of light, the spontaneous reaction is respiration, the reverse of photosynthesis. Much of atmospheric chemistry is driven by solar energy.

I. INTRODUCTION

A. Compartments of the Environment

The environment may be divided into different "compartments" or "reservoirs": the solid land, or lithosphere; oceans, rivers, and lakes, or hydrosphere; and finally the atmosphere. Atmospheric chemistry may be modeled in the laboratory by chemical processes in the gas phase, while the chemistry of the hydrosphere is the chemistry of aqueous solutions. The chemistry of the lithosphere is the most complicated; it includes both reactions occurring in the solid state and reactions occurring on the surfaces of particles, at a gas–solid or a solid–liquid interface.

The physical and chemical processes that deposit a substance into a particular compartment are known as sources or source reactions. For example, the flow of rivers is a source of both water and dissolved salts into the oceans. Emissions from coal-fired electricity-generating plants are a source of acidic gases such as sulfur dioxide into the atmosphere. In the context of environmental pollution, distinction may be made between point sources, such as the emission of SO_2 from a power station, and nonpoint sources, which are by definition dispersed. For example, automobiles and trucks, residential woodstoves, and forest fires are nonpoint sources of nitrogen oxides into the atmosphere. [See RIVER ECOLOGY.]

Processes that remove a substance from a given reservoir are known as sinks. Deposition in the form of seashells (calcium carbonate) is an important sink for dissolved calcium in the oceans; oxidation to carbon dioxide and water is the chief sink for methane in the atmosphere.

The average abundance of the elements in the Earth's crust varies very widely (Table I), with oxygen and silicon considerably the most abundant by mass; among the metallic elements aluminum, iron, and calcium are the most common. Many elements of biological, toxicological, and economic importance are rare.

Minor elements include Cr, 100; Ni, 75; Zn, 70; Cu, 55; N, 20; Pb, 12; Br, 2.5; Sn, 2; I, 0.5; Cd, 0.2; Hg, 0.08; Ag, 0.07; and Au, 0.004 ppm. These data show that the familiarity of the elements has nothing to do with their relative abundance. Localization of elements in deposits of high concentration and their ease of isolation are more important than high average abundance. For example, biologically essential trace elements such as iodine must be carefully husbanded by organisms. Furthermore, the ease—or difficulty—of obtaining the various metals from their ores determined whether they came into human use early (gold, silver, lead) or late in history (aluminum, magnesium, titanium).

B. Residence Times

Residence times, τ, are defined as:

$$\tau = \frac{\text{amount of substance in the reservoir}}{\text{rate of inflow to, or outflow from, the reservoir}}$$

The numerator and denominator of this equation should have compatible units such as grams, grams per second; moles, moles per year; moles per liter; moles per liter per second, and so on. Residence times may cover the whole possible range, from very short to very long. For example, oxygen atoms, which are formed from nitrogen dioxide in the lower atmosphere by the action of sunlight, are extremely reactive and react with oxygen molecules to form ozone; the oxygen atoms have residence time in the atmosphere less than one second. At the other end of the scale, the residence time of dissolved ions such as Na^+ in the oceans is estimated to be as long as 70 million years, because their chief means of removal from the oceans is the formation of evaporites. [See ATMOSPHERIC OZONE AND THE BIOLOGICAL IMPACT OF SOLAR ULTRAVIOLET RADIATION.]

The residence time (also known as the lifetime) should not be used synonymously with the half-life. The half-life is by definition the time taken for the amount of substance entering the reservoir to reach half its maximum value, or the amount of substance leaving the reservoir to fall to half its initial value, in each case under the assumption that the reservoir is not being simultaneously depleted or replenished. In the language of chemical kinetics, the half-life ($t_{1/2}$) is defined as $\ln 2/(\Sigma k_1)$, where Σk_1 is the sum of all the first-order and pseudo-first-order rate constants for removal of the sub-

TABLE I

Occurrence of the Common Elements in the Earth's Crust

Metals		Nonmetals	
Element	c, ppm[a]	Element	c, ppm
Al	83,000	O	460,000
Fe	56,000	Si	280,000
Ca	42,000	H	1,400
Na	24,000	P	1,000
Mg	23,000	F	630
K	21,000	S	260
Ti	5,700	C	200
Mn	950	Cl	130

[a] Parts per million by mass.

stance from the reservoir. In kinetic terms, the residence time (lifetime) is defined as $1/(\Sigma k_1)$, a definition that is equivalent to that at the beginning of this section.

C. The Atmosphere

Unpolluted dry air at sea level (total pressure = 1 atm) has the following composition:

Major constituents (%)			
N_2	78.1	O_2	20.9
Ar	0.93	CO_2	0.035 ($\equiv$350 ppmv)

Minor constituents (parts per million by volume, ppmv)

Ne	18	He	5.2
CH_4	1.7	N_2O	0.3
H_2	0.5	CO	0.1

Even smaller amounts of NH_3, SO_2, Kr, Xe, O_3, and other gases are present. Water content is variable and ranges from close to zero to about 0.4%. Comparing the Earth's atmosphere with those of the other planets, the most notable feature is its high oxygen content. As a result, substances emitted into the atmosphere by both natural and anthropogenic processes tend to become oxidized; for example, methane and other hydrocarbons are oxidized to CO_2 and water.

The Earth's gravity causes most of the total mass of the atmosphere ($\sim 5 \times 10^{15}$ tonnes) to be concentrated close to the surface; 99% of the total mass lies below 30 km. The regions of the atmosphere of importance to environmental biology are the troposphere (lower atmosphere), extending from sea level to about 15 km altitude, and the stratosphere, extending from about 15 to 50 km. The issue of ozone depletion is a stratospheric problem, whereas acid rain, photochemical smog, and greenhouse warming are all tropospheric phenomena. These issues are discussed in Sections II, C, II, D, II, F, and II, B, respectively.

D. Gas Exchange, Henry's Law

The equilibrium solubility of any gas in any solvent is determined by the pressure of the gas in contact with the solvent. Quantitatively, the amount of gas that dissolves is proportional to the partial pressure of the gas. This statement is known as Henry's law, which can be written mathematically as

$$[\text{X, solvent}] = \text{constant} \times p(\text{X,g})$$

[X, solvent] represents the concentration of solute gas in the solvent and $p(\text{X,g})$ the partial pressure of the solute gas. The proportionality constant is an equilibrium constant, known as Henry's law constant, K_H. Common units for K_H are mol liter^{-1} atm^{-1} and mol m^{-3} atm^{-1}; alternatively, K_H may be defined with the equilibrium written in the opposite direction [X(aq) $\leftrightarrows$ X(g)], in which case K_H might have the units m^3 atm mol^{-1}. In the first case, a large value of K_H implies high water solubility; in the latter case a high value of K_H implies high volatility. Values at 298 K for K_H, defined as $K_H = [\text{X(aq)}]/p(\text{X,g})$, are given below for some common gases.

Gas	K_H (mol liter^{-1} atm^{-1})	Gas	K_H (mol liter^{-1} atm^{-1})
H_2	7.8×10^{-4}	CO	9×10^{-4}
N_2	6.5×10^{-4}	O_2	1.3×10^{-3}
CO_2	3.4×10^{-2}	O_3	1.3×10^{-2}
NO	1.9×10^{-3}	NO_2	6.4×10^{-3}
SO_2	1.2	NH_3	5.3

The solubilities of gases in water [for a given $p(\text{X})$] vary widely. Nonpolar substances such as CH_4, N_2, and O_2 have small Henry's law constants and hence have low solubilities in water, whereas very polar gases such as HNO_3 ($K_H \sim 10^5$ mol liter^{-1} atm^{-1}) and NH_3 are extremely soluble in water. All gases become less soluble in water as the temperature is raised.

The condition known by deep-sea divers as "the bends" is explicable in terms of Henry's law. The pressure under water increases by $\approx$1 atm for each 10 m depth. At 30 m, the diver's air supply must therefore be >4 atm (1 atm atmospheric pressure plus 1 atm per 10 m depth). If compressed air is breathed, then $p(\text{N}_2) = 0.8 \times p(\text{total}) = 3.2$ atm, instead of 0.8 atm at the surface. When the diver returns to the surface, where $p = 1$ atm, the excess nitrogen gas escapes from the blood, a condition that is both painful and dangerous. "The bends" are avoided by having divers breathe a mixture of helium and oxygen rather than air (because helium is less soluble in water than nitrogen), and by having the diver undergo gradual decompression after prolonged work at depth.

The differing solubilities of gases are of practical importance in connection with scavenging trace gases out of the troposphere by rainfall. "Wet deposition" is an important sink for gases such as HNO_3 and SO_2 that have large K_H. Nonpolar substances such as hydrocarbons and halogenated hydrocarbons such as chlorofluorocarbons have very low solubility in water and therefore may persist for long periods in the atmosphere, in the absence of other sinks such as chemical reaction.

A practical application of Henry's law is "air stripping" volatile contaminants from an aqueous industrial waste stream. Typically, the aqueous solution is sprayed as a fine mist down a tall cylindrical tower against a countercurrent of air. Equilibrium is reached, or approached, between the aqueous and gas phases, the efficiency of the process depending on Henry's law constant:

$$X(aq) \leftrightharpoons X(g)$$

Transfer to the gas phase is highly efficient for contaminants such as trichloroethylene, which have intrinsically low water solubility, but is even practical for water-soluble gases such as ammonia, provided that good contact is achieved between the gas and aqueous phases (a large tower and fine droplets). Transfer out of the aqueous phase is also promoted at higher temperatures; this is often practical at no cost if the waste process stream is already warm.

Air stripping is very cheap, but has the disadvantage that it transfers the unwanted solute from the aqueous phase into the atmosphere. Emerging technologies combine air stripping with treatment of the contaminated air stream, for example, by passing the warm air stream over an oxidation catalyst to convert contaminants into innocuous substances such as CO_2, H_2O, and N_2, or by "biofiltration," in which the exhaust gases are passed through a bed of biologically active material (often a slime of bacteria on an inert support) for final oxidation of the contaminant.

E. Acidity and Water Solubility

The water solubilities of many solutes, both organic and metallic, are pH dependent. In the case of organic substances, the pK_a of functional groups determines the charge status of the solute; when $pH < pK_a$ the solute will be in its acidic (protonated) form, whereas when $pH > pK_a$ the basic (unprotonated) form will predominate.

Functional group	Acidic form	Basic form	pK_a
Carboxylic acid	RCO_2H	RCO_2^-	3–5
Arylamine	$ArNH_3^+$	$ArNH_2$	3–5
Alkylamine	RNH_3^+	RNH_2	9–11
Phenol	$ArOH$	ArO^-	<10
Thiol	RSH	RS^-	≈10

Uncharged species are less water soluble, more lipophilic, and more easily able to penetrate cell membranes by permeation into the lipid bilayer. The uptake of xenobiotics from water, like the uptake of drugs in the body, therefore depends on the pH of the medium. Extraction of xenobiotics from environmental matrices such as soil and tissues requires prior adjustment of pH; for example, pentachlorophenol, $pK_a \approx 5$, cannot be extracted efficiently from soil into an organic solvent unless the pH of the soil slurry is <5.

Most metallic oxides, hydroxides, carbonates, phosphates, and sulfides are intrinsically very insoluble in water, and for this reason they occur as these salts in exploitable metal ores. The foregoing counterions are all examples of basic anions and have a tendency to associate with H^+ to form a (water-soluble) conjugate. Charge balance requires the associated metal ion to dissolve as well, as illustrated by the reactions

$$CaCO_3(s) + H^+(aq) \rightarrow Ca^{2+}(aq) + HCO_3^-(aq)$$
$$HgS(s) + H^+(aq) \rightarrow Hg^{2+}(aq) + HS^-(aq)$$
$$Fe(OH)_3(s) + 3H^+(aq) \rightarrow Fe^{3+}(aq) + 3H_2O(l)$$

The solubilities of these and similar salts therefore increase markedly with increasing acidity (lower pH). A typical environmental consequence is seen under conditions of acid mine drainage (see Section II, F), where biological oxidation of sulfide to sulfate (sulfuric acid) lowers the pH of mine tailings, thereby mobilizing the metal ions in the tailings.

A small number of metal oxides is amphoteric (able to dissolve in either acid or alkali). One example is aluminum oxide, whose solubility is at a minimum near pH 6.5 and rises in acidic solution [$Al(OH)_2^+$, $AlOH^{2+}$, Al^{3+}] and in alkaline solution [$Al(OH)_4^-$].

F. Lipophilicity and K_{ow}

Lipophilic substances are soluble both in nonpolar organic solvents and in the lipids of organisms. The "octanol–water partition coefficient" K_{ow} is a convenient quantitative measure of the lipophilicity of a solute. Octanol is chosen as a model organic solvent for typical lipids; a compound having a high value of K_{ow} (e.g., >1000) will tend to partition into lipids out of water and will thus be prone to bioconcentration (bioaccumulation), and hence to biomagnification through the food chain. Highly lipophilic substances such as DDT and the polychlorinated biphenyls have K_{ow} >10^6.

$$K_{ow} = \frac{\text{concentration of solute in octanol}}{\text{concentration of solute in water}}$$

Compounds with large K_{ow} can generally cross and accumulate in cell membranes easily. As a result, many toxic xenobiotics such as DDT, polychlorinated biphenyls (PCBs), dioxins, and Mirex are found at elevated levels in aquatic species taken from waters such as the Great Lakes and industrialized rivers such as the Hudson and Mississippi, the Danube, and the Rhine. The bioconcentration factor (BCF) is the ratio of the average concentration of the solute in the whole organism to the concentration of the solute in water:

$$BCF = \frac{\text{concentration of solute in organism}}{\text{concentration of solute in water}}$$

By taking octanol as a model solvent for fat, and assuming that the fatty tissues of an aquatic organism have reached equilibrium with the surrounding water, the BCF of a lipophilic solute can be related to K_{ow} according to the percentage by weight of fat in the organism:

$$BCF \approx K_{ow} \times \% \text{ by weight of fat}$$

K_{ow} can be used to predict the BCF where this has not been measured experimentally, often with the further assumption that the aquatic organism contains about 5% fat by weight (BCF $\approx 0.05 \times K_{ow}$).

Bioconcentration can cause the level of a potentially toxic substance to be much higher in the organism than in the water in which it lives. Frequently, however, equilibrium is not reached. For example, the weight-adjusted body burden of PCBs in lake trout captured from the Great Lakes increases with size and age; if equilibrium had been achieved, the experimental BCF would have been independent of these parameters.

The BCF can also be related to the rates (strictly, rate constants) of uptake and depuration (including metabolism and excretion) of a solute:

$$BCF = k(\text{uptake})/k(\text{depuration})$$

All substances prone to bioconcentration are taken up from the environment much faster than they are depurated. Indeed, the true criterion for bioconcentration is a faster rate constant for uptake than for depuration, rather than lipophilicity. This criterion explains the bioconcentrating properties of both lipophilic organic compounds and cumulative metallic toxicants.

G. Structure–Activity Relationships

A structure–activity or structure–reactivity relationship (SAR) is an empirical means of connecting the chemical reactivities or biological potencies of a series of chemically related compounds to aspects of chemical structure. As simple examples, a structure–activity relationship exists between water solubility and the chain length of aliphatic alcohols (the longer the carbon chain, the lower the water solubility) or between carcinogenicity and the structures of certain nitrogen-containing organic compounds [structures containing the N—N=O group (nitrosamines) tend to be carcinogenic]. Likewise, the correlation between BCF and K_{ow} of the previous section is an example of a structure–activity relationship.

The utility of SAR lies in predicting the properties of a new substance before measuring them experimentally. Ideally, a mathematical function is established between a structural parameter and the chemical or biological property to be estimated. For example, if the water solubilities of aliphatic alcohols of chain lengths C_{18}, C_{19}, and C_{21} had been measured in the laboratory, it would be feasible to estimate the water solubility of the C_{20} alcohol by interpolation of a graph of water solubility versus carbon number. Likewise, if the biological potencies of a series of polychlorinated biphenyl congeners had been related to their K_{ow} values, it would be reasonable to predict the biological potency of a different PCB congener from a measurement of its K_{ow}.

The SAR concept has been refined by the assignment of empirical numerical values to structural features of a molecule such as size, shape, electron distribution, aromaticity, polarity, hydrogen-bonding capability, hydrophilicity, and hydrophobicity. Multiple linear regression yields a quantitative structure–activity relationship (QSAR) of the form

$$\text{Property of interest} = a \times \text{length} + b \times \text{electron affinity} + c \times \text{hydrogen bonding ability} \dots$$

Examination of the magnitudes of the coefficients $a, b, c \dots$ and their standard deviations reveals which structural features are important in determining the intensity of the property of interest (large coefficient, important factor; small coefficient or wide confidence interval, relatively unimportant factor). To take a specific example, a QSAR between the structure of polychlorinated dibenzo-p-dioxins ("dioxins") and their ability to bind to a receptor protein known as the Ah receptor (see Section II, G) revealed that lipophilicity, size, and planarity were the key factors determining the strength of the receptor–ligand interaction. A further structure–activity relationship showed that

the most toxic members of the polychlorinated dibenzo-p-dioxin family were those with the strongest affinity for the Ah receptor, thus putting the toxicity of these important environmental toxicants on a molecular basis.

SAR and QSAR are of great value in predicting the properties of newly discovered or synthesized materials. For example, they guide synthetic chemists in the optimization of biologically active structures used as pharmaceuticals, pesticides and other agricultural chemicals, and environmental scientists in predicting the biological and toxicological properties of substances newly released or likely to be released into the environment. For example, the likely bioconcentration factor of a newly synthesized pesticide could be estimated following a simple measurement in the laboratory of its ability to partition between water and octanol.

H. Photochemistry

Photochemical reactions are those induced or accelerated by light. Sunlight is the radiation source of interest in the context of environmental biology. Sunlight-assisted reactions can only occur in locations where light can penetrate, namely, the atmosphere, the hydrosphere, and exposed surfaces such as leaves and soil. The two requirements for a photochemical reaction to occur are (i) that the radiation is absorbed by one of the reactants and (ii) that the photons are of sufficient energy to promote the reaction. The probability of light absorption can be determined from the absorption spectrum of the compound, whereas the relationship between wavelength and photon energy is given as

$$E(\text{photon}) = hc/\lambda = (2.0 \times 10^{-25})/\lambda \text{ J per photon, with } \lambda \text{ in meters}$$

More usefully:

$$E = (1.19 \times 10^5)/\lambda \text{ kJ mol}^{-1}, \text{ with } \lambda \text{ in nm}$$

Thus at wavelengths of 300, 400, and 700 nm, $E = 400$, 300, and 170 kJ mol^{-1} respectively. Energies >250 kJ mol^{-1} are sufficient to rupture many covalent bonds. In thermodynamic terms, light can be considered a source of internal energy, which can be used to drive an endothermic chemical reaction.

The solar spectrum at the Earth's surface ranges from a short-wavelength cutoff at ~300 nm through the near ultraviolet (UV-B, 300–320 nm; UV-A, 320–400 nm) and visible (400–700 nm) into the infrared. The upper atmosphere filters out shorter-wavelength radiation (UV-C, $\lambda < 300$ nm), so that it does not reach the Earth's surface. The filtering efficiency depends strongly on the depth of atmosphere through which solar radiation must travel to reach the Earth's surface. The short-wavelength cutoff moves deeper into the UV and the total UV intensity increases as the Sun rises higher in the sky (middle day at all latitudes, and in tropical versus temperate or polar latitudes) and at high elevation. Because stratospheric ozone is the chief UV absorber in the region 230 nm < λ < 320 nm, depletion of stratospheric ozone is anticipated to increase solar UV intensities.

Among biomolecules that can absorb solar radiation, DNA is of particular importance; its absorption maximum occurs near 280 nm, but the tail of its absorption band extends into the UV-B region. The absorption of radiation of these wavelengths is known to cause photochemical reactions in the nucleotides; such photochemical changes are presumed to be responsible for the elevated incidence of skin cancer which occurs upon excessive exposure to sunlight.

In atmospheric chemistry, relatively few substances of environmental concern absorb solar radiation directly. Instead, one of the commonest sinks for tropospheric trace constituents is attack by hydroxyl radicals (OH, not to be confused with hydroxide ion, OH^-). These reactions are indirectly photochemical, because hydroxyl radicals are formed in the troposphere by photolysis of ozone, which is a natural trace constituent of the troposphere (20–60 ppbv) and which occurs at higher concentration under conditions of urban pollution (see Section II, D). In the following equations, the asterisk (*) represents an excited state of the substance:

$$O_3 + h\nu \ (\lambda < 320 \text{ nm}) \rightarrow O_2^* + O^*$$
$$O^* + H_2O \rightarrow 2OH$$

The OH radical is very reactive (short residence time), and because it requires sunlight for its generation, its concentration rises and falls with the solar intensity, reaching essentially zero at night. The globally and time-weighted average concentration of OH is about 7×10^5 radicals per cm³ (~10^{-15} mol liter^{-1}), with a maximum concentration up to about 10-fold higher, depending on the place, season, and conditions. Despite its low concentration, the hydroxyl radical is so reactive that reactions with OH are a major sink for many atmospheric substances. Typical reactions of OH are abstraction of a hydrogen atom (often from a C-H bond) and addition to an unsaturated center such as a C = C bond, usually followed by interaction with molecular oxygen (see Section II, C). Chlorofluorocarbons are long-lived in the troposphere because they contain neither C-H bonds nor unsaturation and hence are completely unreactive with hydroxyl radicals.

The importance of photochemistry in the hydrosphere is limited by the ability of sunlight to penetrate. In many fresh and coastal waters, especially those that are turbid or colored, only solutes in the surface layer (a few cm to perhaps 1 m) are susceptible to photolysis, but in the open ocean sunlight penetrates several meters. As in the atmosphere, much of the photochemistry is indirect, rather than the result of direct absorption by the substrate. In this case, attack on the substrate frequently involves an excited state of oxygen (singlet O_2), which is formed by energy transfer from some other light-absorbing substance, a process called "photosensitized oxidation" (see Section II, C).

I. Environmental Modeling

Environmental models seek to predict the major sources and sinks of natural and anthropogenic substances in the environment. They vary greatly in their complexity and the extent to which they rely on first principles as opposed to the input of empirical data.

The goals of atmospheric models are frequently the prediction of the movements of air masses or of trace natural or anthropogenic constituents of air masses. Models may be two dimensional (lateral distribution) or three dimensional (including vertical distribution). The minimum parameters required are the source strengths to the atmosphere, both natural and anthropogenic, and sink strengths for wet and dry deposition. Modeling of wet and dry deposition rates also requires the Henry's law constant for partition of the solute between water and the gas phases, and the adsorption coefficients for distribution between the gas and adsorbed particle phases, all over a range of temperatures. These models may be further complicated if consideration of uptake at the Earth's surface or of chemical transformation in the atmosphere is required. Uptake at the surface requires knowledge of the fluxes for adsorption to vegetative surfaces and for adsorption at the ocean surface. In the case of chemical transformation, likely reaction partners are the hydroxyl radical, OH, the nitrate radical, NO_3, and ozone. Complete knowledge of all major source and sink chemical reactions is needed. The form of the rate law for each bimolecular reaction is given as:

$$Rate = k \cdot [substrate] \cdot [reaction\ partner]$$

Before modelling studies can be undertaken, the rate constants k for all relevant reactions must be measured in the laboratory over a range of temperatures, and the concentrations of all substrates and reaction partners must be obtained from field measurements. Because the rate expressions are all in the differential form, sophisticated methods of numerical integration are necessary to extract the desired information, namely, the temporal variation of the substrate.

Because reaction partners such as OH are products of solar photochemistry, an additional set of inputs to the models is the solar flux. Consider the information that is needed to calculate the rate of production of OH. As noted in Section I, H, OH is formed in a two-step reaction beginning with the photolysis of ozone (and for this discussion omitting completely the origin of the ozone). The absorption spectrum of ozone in the wavelength region 300–325 nm determines the efficiency of radiation absorption. This is wavelength dependent, as is the photochemical efficiency (quantum yield) of cleavage of O_3 to excited oxygen atoms. Tables of solar flux as a function of wavelength are available for different zenith angles ($Z = 90°$ − angle of solar elevation), and the zenith angle can be calculated from astronomical data for any point on the Earth's surface for any time and date. Combining this information with the concentration of ozone affords the rate of production of excited oxygen atoms, summed over all wavelengths at which ozone absorbs. The rate of formation of OH then follows from a knowledge of the intrinsic rate of deactivation of O★, which is temperature and pressure dependent, the concentration of water vapor, and the rate constant for reaction of O★ with H_2O.

Most present models for the movement of xenobiotics into the biosphere assume an equilibrium distribution of the xenobiotic between compartments such as biota, water, sediment, and air. In an aqueous ecosystem the following factors need to be recognized: BCF, adsorption coefficient (water/sediment), Henry's law constant (water/air), rates of chemical transformation such as oxidation and hydrolysis, rates of photochemical transformation (direct photolysis and singlet oxygen reactions), and rates of metabolism, along with source strengths and quantities of water, sediment, and biota. Structure–activity relationships are frequently useful for estimating missing experimental parameters where the corresponding information is available for structurally related compounds.

II. ENVIRONMENTAL CHEMISTRY OF SPECIFIC SUBSTANCES

A. Water

Two-thirds of the Earth's surface is covered with water. Most of this is ocean, as seen in the following figures:

Oceans	9.5×10^{19} mol (>99%)
Lakes and rivers	1.7×10^{15} mol
Atmosphere	7.2×10^{14} mol

There is also water deep underground, called "groundwater."

1. Water in the Atmosphere

Water is very unevenly distributed in the atmosphere, consistent with the variation in the weather with place and time. At any time about 7×10^{14} mol of $H_2O(g)$ are present, a tiny fraction of the 9.5×10^{19} mol present on the surface as $H_2O(l)$. Evaporation from the oceans (2.2×10^{16} mol yr^{-1}) and from lakes and rivers (3.5×10^{15} mol yr^{-1}) is balanced by precipitation over the land (5.5×10^{15} mol yr^{-1}) and the oceans (1.9×10^{16} yr^{-1}), leading to an average residence time of 3×10^{-2} yr (10 days) for water in the atmosphere.

Temperature strongly determines the capacity of the atmosphere to hold water vapor. The equilibrium, or saturated, vapor pressure of water increases from 0.8 at Torr $-20°C$ to 4.6 Torr at $0°C$ and 23.8 mm at $25°C$. Under most conditions, the atmosphere is subsaturated with respect to water vapor. The relative humidity is the prevailing $p(H_2O)$ as a percentage of the equilibrium value, and so a given value of the relative humidity does not represent a fixed water vapor content. Evaporation of water or of ice can occur whenever the relative humidity is $<100\%$.

2. Terrestrial Water

A vast array of environmentally important reactions takes place in aqueous solution—in raindrops, in rivers and lakes, and in the oceans. Almost all biochemical reactions occur in water, and water is the major constituent of all living matter (about 62% in the case of humans). The high melting point ($0°C$) and boiling point ($100°C$) of a small molecule such as water result from hydrogen bonding and are important for the existence of life, because water is a liquid at most places on Earth. Insulation by the troposphere warms the Earth's surface to an average $+15°C$, where water is a liquid, rather than the $-30°C$ that would be expected in the absence of an atmosphere, when all the water on Earth would be frozen and life as we know it would not be possible. The higher density of liquid water (1.0 g cm^{-3}) compared with the solid (0.9 g cm^{-3}) is also important for the existence of life, as otherwise ice would fall to the bottom of lakes in winter and would not melt completely in the summer. The turnover of lakes in fall and spring, which also has the effect of recycling nutrients, occurs because water has a maximum density at $4°C$.

The freezing point of water is lowered in the presence of dissolved solutes (the freezing point depression constant is $1.86°C$ kg mol^{-1}); for ionic solutes dissociation into ions must be considered to obtain the total concentration of solute species. $NaCl$ and $CaCl_2$ are used to melt ice and snow on roads and sidewalks by lowering the freezing point of the solution. Fish that live in polar waters produce natural antifreezes that are secreted into the bloodstream to prevent them from freezing. Many trees produce antifreezes to prevent their cells from freezing, and hence bursting, in winter. These trees may survive a hard winter, yet suffer severe frost damage during a late cold snap in the spring, when the antifreezes have been degraded, leaving the tree unprotected.

The amount of solid dissolved in natural waters varies widely. The values in Table II are typical of river and ocean water, although river water is quite variable in its mineral content. Dissolved solids in groundwater may exceed 1000 ppm (1 g per liter). These data indicate that the residence times of these ions in the oceans are long (10^3 years in the case of Ca, $>10^6$ years for Na and K). The oceans are the major reservoir of "soluble" ions such as Na^+ and Cl^-. Naturally occurring deposits of substances such as $NaCl$ (rock salt) are believed to have formed through the evaporation of ancient seas. Compared with other major ions, there is relatively more Ca^{2+} and HCO_3^- in river water because rivers dissolve

TABLE II

Typical Concentration of Ions in River Water and Seawater

Ion	c(river) (mmol liter^{-1})	c(ocean) (mmol liter^{-1})	Annual input to oceans (Tmol yr^{-1})
Cl^-	0.22	550	7.2
Na^+	0.27	460	9.0
Mg^{2+}	0.17	54	5.5
SO_4^{2-}	0.12	28	2.8
K^+	0.059	10	1.9
Ca^{2+}	0.038	10	12
HCO_3^-	0.095	2.3	32

ancient rocks containing $CaCO_3$, whereas the oceans precipitate $CaCO_3$ in the form of marine organisms' exoskeletons.

3. Dissolved Solids

Excessive quantities of dissolved solids cause problems when water is used for irrigation or in industry. Irrigation in a dry climate is inevitably accompanied by high rates of evaporation, leaving dissolved solids in the soil. Continued irrigation leads to a buildup of salts in the soil and may ultimately render the soil unfit for agriculture. Excessive salinity is a possible cause of the decline of the ancient civilizations in the Tigris and Euphrates valleys. Today the same problem is compounded by inorganic residues from fertilizers. In Pakistan, for example, loss of land to salinity threatens to wipe out the increases in agricultural production achieved through intensive farming. In North America, modern irrigation technology extends the working life of the land by "back-flushing" with a heavy application of water to wash out the accumulated salts with the runoff, which is then returned to the river from which the irrigation water was drawn. This leads to increased salinity of the river itself. For example, several tributaries of the Colorado River (U.S.A) have concentrations of dissolved solids in excess of 1000 ppm.

The rate of water use from underground aquifers may be orders of magnitude greater than the natural rate of recharge. By 1980, Texas had withdrawn over 20% of its share of the water in the Ogalalla Aquifer, which underlies parts of eight U.S. states. Because some parts of the aquifer contain high concentrations of dissolved salts, it is arguable that intensive agricultural production in the more arid parts of the U.S. Midwest may not be sustainable in the long term. Outside North America, the huge aquifer beneath the Sahara Desert is being consumed rapidly for agricultural irrigation in Libya. Because its rate of recharge is essentially zero, this amounts to mining a nonrenewable resource. Likewise, agricultural irrigation has so reduced the flow of rivers into inland seas such as the Dead Sea in Israel and the Aral Sea in Russia that their shores have each retreated many kilometers in the last few decades; these bodies of water may cease to exist in the foreseeable future.

In industrial and domestic use, the presence of "hardness cations" (Ca and Mg) causes precipitation of scum with soaps and deposits of scale in boilers, water heaters, and hot water pipes. Commonly, hard water is encountered in regions where the underlying rocks are gypsum ($CaSO_4$) or carbonates such as limestone ($CaCO_3$) and dolomite ($CaCO_3 \cdot MgCO_3$), which dissolve due to the presence of CO_2 in the water:

$$CaCO_3(s) + H_2O(l) + CO_2(aq)$$
$$\rightarrow Ca(HCO_3)_2(aq)$$

Scale is deposited when the water from limestone areas is heated, expelling CO_2 from the solution:

$$Ca(HCO_3)_2(aq) \rightarrow CaCO_3(s) + H_2O(l) + CO_2(g)$$

Water is softened in order to remove hardness cations, usually by ion exchange.

Naturally soft water contains only low concentrations of calcium and magnesium. Soft water is encountered commonly in regions such as New England, much of eastern Canada, Scotland, and Scandinavia, where the underlying rock is granite, which is very insoluble. From the environmental perspective, flora and fauna in soft-water areas are more likely to suffer from acidification (see Section II,F), because the water lacks bicarbonate ions to act as a natural buffer by reacting with added H^+.

4. Osmosis

If dilute and concentrated solutions are separated by a membrane that is permeable to solvent molecules but not to solute molecules, water will flow from the dilute solution into the concentrated one. The difference in pressure exerted on the semipermeable membrane by the two solutions is called the osmotic pressure. In the simplest case, where pure water is on one side of the membrane, the relationship is given by an equation whose form is reminiscent of the ideal gas equation:

$$\pi = c \cdot RT$$

When R expressed in liters atm^{-1} mol^{-1} K^{-1} and c is in mole liter^{-1}, π (the osmotic pressure) has the units of atm. Like the freezing point depression of water, the osmotic pressure depends only on the total number of moles of solute particles (molecules or ions) present in a fixed volume of solvent, and not on their identities.

Plant cells, unlike the cells of animals and bacteria, are constructed of cellulose and can withstand a difference in osmotic pressure. They preserve their rigidity by taking on water, maintaining a higher total concentration of solutes inside the cell than outside. In time of drought, the plant wilts because there is insufficient osmotic pressure to maintain rigidity. Animal and bacterial cells are lipid bilayers that have little structural strength and must be maintained in an isotonic solution (osmolar concentration ca. 0.3 mol liter^{-1}). In a hypotonic solution, entry of water causes the cell to swell or burst; conversely, in a hypertonic solution the cell becomes dehydrated.

5. Drinking Water

No other public health or medical innovation comes close in importance to access to a safe, clean supply of drinking water. Between 15 and 20 million babies are estimated to die every year worldwide as result of waterborne diarrheal diseases such as typhoid fever, amoeboid dysentery, and cholera. Surface water (drawn from a lake or river) almost always has a higher content of suspended materials and a higher microbial count than groundwater, and consequently requires more processing to make it safe to drink. Groundwater tends to be less contaminated than surface water both because organic matter in the water has had time to be decomposed by soil microorganisms and because the ground itself acts as a filter for pathogens. Many major waterways, such as the Great Lakes in North America and the Rhine and Danube rivers in Europe, are used for drinking water and for other purposes by a large number of communities. Communities lying downriver draw water that has been contaminated by sewage outfall and industrial use upstream.

Surface water is first subjected to primary settling to remove suspended particulate matter, followed by aeration to promote the oxidation of easily oxidizable substance that would otherwise consume the disinfectant to be added later. Colloidal mineral particles, bacteria, pollen, spores, and others are then removed using a filter aid such as filter alum $Al_2(SO_4)_3 \cdot 18H_2O$, which entrains these fine particles in a gelatinous precipitate of $Al(OH)_3(s)$. The clarified water is then disinfected using either chemical agents (chlorine, ozone, or chlorine dioxide) or by sterilization with UV-C radiation (254 nm) from mercury arc lamps. Chlorine is used most commonly, as it is cheap, effective, and retains a "chlorine residual" in the water, thus maintaining the water pathogen-free while it is in the distribution system. Chlorine is a more powerful disinfectant below pH 7.5 than above this value. Above pH 7.5, ClO^- is the principal disinfectant, whereas below pH 7.5 the principal disinfection agent is HOCl, which is more toxic to pathogens because it is uncharged and penetrates microbial cell membranes more easily than the charged ClO^-. Ozone is used only in large installations because of high capital costs, whereas chlorine dioxide is usually used on a small scale, and as an alternative to chlorine at times of pollution of the water supply. The advantage of ultraviolet disinfection is that it is quick acting and the water can be treated on a flow-through basis; in contrast, the chemical agents all require a large disinfection tank to be built so that the water will be in contact with the disinfectant for a residence time of 30–60 minutes.

6. Problem Substances in Drinking Water

Iron gives an unpleasant "metallic" taste to the water, but it is rarely present in amounts that might be toxic; it can be a problem with untreated well water, usually due to dissolution of the slightly soluble $FeCO_3$. Upon exposure to air, Fe(II) is oxidized to Fe(III), which precipitates as Fe_2O_3 (rust). In municipal water treatment, Fe(II) is oxidized during aeration and then precipitates.

Toxic metals such as lead, cadmium, and mercury may be present in water as a result of industrial activities, for example, groundwater contamination from municipal and chemical waste dumps, and in the vicinity of mines. Lead plumbing can be a source of lead in drinking water; with hard water a deposit of scale ($CaCO_3$) protects the surface of the lead piping from contact with water, but soft waters do not form scale and are often acidic, which promotes the dissolution of lead. Maximum allowable concentrations of lead in drinking water are typically 10 ppb.

High levels of nitrate ion (up to 100 ppm) are sometimes found in water on farms with shallow wells and in areas of intensive animal farming such as the Netherlands. However, even at the maximum recommended concentration of 50 ppm in drinking water, most people take in three or four times as much nitrate from food (mainly vegetables) as from drinking water. A specific toxic effect of nitrate ion is a condition of infants called methemoglobinemia, in which hemoglobin is converted to a form that cannot carry oxygen to the tissues. Severe cases of methemoglobinemia can result in mental retardation.

Organics in drinking water sources arise from agricultural runoff (e.g., pesticides), by the action of chlorine on natural organic compounds present in the water (trihalomethanes such as chloroform, $CHCl_3$), from industrial operations (e.g., phenols), and from leachage from waste dumps (e.g., trichloroethylene). Although there is evidence that chloroform may be carcinogenic at high doses in laboratory animals, the concentrations reached in drinking water (typically 10–20 ppb) are many orders of magnitude lower. Phenols react with chlorine during disinfection to produce powerfully odorous chlorinated phenols that can make the water undrinkable at only ppb concentrations. "Taste and odor" problems caused by phenols may be alleviated by switching temporarily from chlorine as disinfectant to chlorine dioxide.

B. Carbon and Hydrogen

Carbon and hydrogen transfer in and out of the biosphere together as a result of photosynthesis and respiration/decay. Water constitutes the largest reservoir of hydrogen, whereas carbon is found principally as the fossil fuels coal, petroleum, and natural gas, and as carbonate minerals, all of which are believed to have biogenic origin. [*See* GLOBAL CARBON CYCLE.]

I. Carbon Dioxide in the Atmosphere

The amount of carbon dioxide in the atmosphere is 1.4×10^{16} mol. In counterpoint to oxygen, a major source of atmospheric CO_2 is respiration, combustion, and decay, and an important sink is photosynthesis (about 1.5×10^{15} mol yr^{-1} each). Because CO_2 is somewhat soluble in water, exchange with the oceans must also be taken into account, some 7×10^{15} mol yr^{-1} being taken up and 6×10^{15} mol yr^{-1} being released by different regions of the oceans. The result is an atmospheric lifetime of about 2 yr, which makes the atmosphere moderately well mixed with respect to carbon dioxide.

Analytical data obtained over many years at the Mauna Loa Observatory in Hawaii, far from anthropogenic sources of CO_2, show an annual cycle in CO_2 concentration, with the peak about April and the trough around October (Fig. 1). The higher level of photosynthetic activity in the Northern Hemisphere in the period May to October causes CO_2 to be removed from the atmosphere a little faster than it is added, while sources dominate over sinks in winter. Superimposed on the annual cycle is a gradual increase in the partial pressure of CO_2, from about 315 ppmv in 1958 to over 350 ppmv by 1990, the cause of which is variously attributed to a combination of increased use of carbon-containing fossil fuels and decreased photosynthetic activity due to deforestation.

$p(CO_2)$ is expected to reach 600 ppmv during the latter part of the twenty-first century, and much debate surrounds whether the expected rise in $p(CO_2)$ will cause global warming. Carbon dioxide and other "radiatively active gases" absorb infrared radiation emitted from the Earth's surface, preventing the loss of this energy into space and thus warming the Earth's surface (cf. Section II,A.1). Only the minor atmospheric constituents are capable of absorbing infrared radiation, the requirement for which is that the initial and final vibrational states of the molecule must differ in dipole moment. This condition cannot be fulfilled for atoms such as argon, nor for homonuclear diatomic mole-

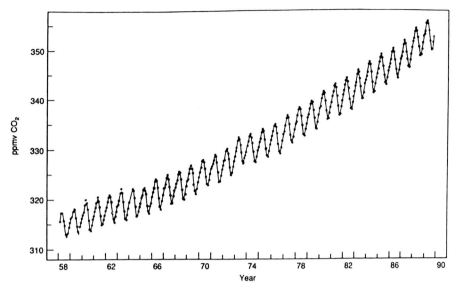

FIGURE I   Observations of p(CO$_2$) at the Mauna Loa Observation, Hawaii, for the period 1958–1990.

cules such as O$_2$ and N$_2$, which have zero dipole moment in all their vibrational states.

The conventional argument is that increased atmospheric concentrations of radiatively active gases (i.e., infrared absorbers) will trap infrared radiation more efficiently and cause a warmer troposphere. The major infrared absorbers, excluding water vapor, and their relative importance are CO$_2$, 50%; methane, 20%; chlorofluorocarbons, 20%; N$_2$O, 5%; and all others, 5%. A warmer troposphere would also increase the average p(H$_2$O) in the atmosphere through evaporation from the oceans because the equilibrium vapor pressure of water rises with temperature. The higher temperature would then further raise p(H$_2$O) and further increase the efficiency of infrared absorption.

The analysis of ice cores from Antarctica has shown a close correlation between the local temperature and p(CO$_2$) over the past 160,000 years, suggesting that the temperature of the trophosphere may be regulated by p(CO$_2$). The present and projected high CO$_2$ levels have led to predictions of a likely increase in average global temperature of 1–3°C over the next century. Because the difference in global temperatures between today and the last ice age is only some 4–5°C, temperature changes of this magnitude would be expected to have profound effects on the biosphere. These include

changes (both more and less) in average rainfall, thus affecting agricultural productivity (probably negatively in the North American grain belt; probably positively in sub-Saharan Africa and India), and changes in the species able to populate a given location, although the rate of change of temperature would influence whether species were able to modify their range quickly enough to escape extinction.

A minority of authors has argued that increased CO$_2$ levels in the atmosphere may be a consequence of atmospheric warming, rather than the cause, or that historical data correlating atmospheric CO$_2$ levels with temperature are insufficiently precise to determine which of the parameters p(CO$_2$) and temperature might be the cause and which the effect. These authors also note that the "fact" of global warming in the 1980s is biased by observations made at monitoring stations close to large Northern Hemisphere centers of population; other, remote sites showed evidence of cooling during the 1980s.

2. Methane and Other Hydrocarbons in the Atmosphere

The concentration of tropospheric methane is currently about 1.7 ppmv and the residence time about 10 years. Ice core data indicate that the historical level of CH$_4$ was stable at ≈0.7 ppmv until about two centuries ago, when the rate of increase (currently 1–2% annually) began accelerating. Although the chief sources of atmospheric methane all involve anaerobic decay, intensive agriculture is believed to be responsible for the increase, mainly from fermentation in the rumens of cattle and decay processes in rice paddies. A recent significant source, leakage from natural gas transmission, ought in principle to be controllable.

Sources of atmospheric methane (10^9 kg yr^{-1})	
Wetlands	150
Oceans, lakes, etc.	35
Cattle	120
Rice paddies	95
Others	150

Vast deposits of methane also exist in the Arctic tundra regions of Canada and Russia in the form of a frozen clathrate hydrate of approximate composition $CH_4 \cdot 6H_2O$. The methane is physically imprisoned in the framework of the ice structure and would be released if the temperature rose sufficiently to melt the permafrost.

The unpolluted atmosphere also contains low ppbv concentrations of many other low-molecular-weight hydrocarbons of biogenic origin. These include isoprene and monoterpenes such as the pinene isomers. Anthropogenic sources, principally in urban areas, include unburned hydrocarbons emitted from motor vehicles and lost from automobile gas tanks and service station spillage, and refinery emissions. Functionalized organic compounds include persistent substances such as chlorinated aliphatics and aromatics (see Section II,G).

The sink for CH_4 and other hydrocarbons in the troposphere is reaction with hydroxyl radicals (for formation, see Section I,H) by abstraction of a hydrogen atom, for example,

$$CH_4 + OH \rightarrow H_2O + CH_3$$

The CH_3 (or other alkyl radical) immediately reacts with atmospheric O_2 to form a peroxy radical:

$$CH_3 + O_2 \rightarrow CH_3OO$$

These two reactions have the effect of replacing a C-H bond by C-O; further reactions lead to CO_2 and H_2O. Nitrogen oxides are intimately involved in these reactions in the urban troposphere, and under these conditions ozone is a by-product of hydrocarbon oxidation (see Section II,D).

Space precludes giving detailed mechanisms for the oxidation of volatile organic compounds (VOCs), but structure–reactivity correlations exist for the initial attack of OH. Attack on saturated compounds such as alkanes, alcohols, and ethers involves abstraction from C-H bonds, with attack occurring in the order tertiary C-H > secondary C-H > primary C-H > methane. The long tropospheric residence time of methane compared with other aliphatic hydrocarbons is due to the slow attack of OH on the strong C-H bond of methane.

Attack on alkenes involves both abstraction from the reactive allylic positions and addition to the C = C bond. Both of these reactions occur with comparable efficiencies. The resulting free radicals then rapidly add oxygen, as above, to form peroxy free readicals:

$$C{=}C{-}C{-}H + OH \rightarrow C{=}C{-}C \cdot + H_2O$$
$$C{=}C{-}C \cdot + O_2 \rightarrow C{=}C{-}C{-}OO$$

or

$$C{=}C + OH \rightarrow \cdot C{-}C{-}OH$$
$$\cdot C{-}C{-}OH + O_2 \rightarrow OO{-}C{-}C{-}OH$$

The attack of OH on alkenes, and especially dienes, is very fast: close to the rate of encounter between the reaction partners. As will be shown in Section II,C,3, the atmospheric oxidation of organic substrates is accompanied by the formation of ozone, which causes a tropospheric pollution problem. In consequence, rapidly-oxidized organics such as alkenes and dienes have a high potential for producing ground-level ozone. This becomes environmentally significant when biogenic emissions such as isoprene and monoterpenes such as the pinenes are oxidized in a plume of urban air pollution.

3. Polycyclic Aromatic Hydrocarbons and Soot

Incomplete combustion of organic materials produces both polycyclic aromatic hydrocarbons (PAHs) and soot, which is an impure form of elemental carbon (graphite). Pure graphite has a layered structure consisting of infinite sheets of fused benzene rings, whereas soot also contains "fullerenes" (substances related to the C_{60} "soccer ball" molecule). Combustion of coal is especially prone to the formation of soot, as typified by industrial grime during Victorian times in Europe and northeastern North America, and today in industrializing nations such as India, China, and Eastern Europe. Smogs in these areas combine high concentrations of particles (mainly soot) and sulfur dioxide, which appear to act synergistically against the respiratory system, especially toward those already suffering from respiratory disease. During one particularly smoggy week in London, England in December 1952, the death rate was over 4000 more than normal, particularly among the elderly.

Polycyclic aromatic hydrocarbons are planar molecules containing fused benzene rings; they are formed whenever organic precursors are burned in a deficiency of oxygen. Sources include automotive exhausts, residential woodstoves, tobacco smoke, and emissions from the backyard barbecue. Environmental PAHs have attracted wide attention because some of them are carcinogenic; indeed dibenz[a,h]anthracene was the first pure chemical substance to be demonstrated to be carcinogenic. In the case of automotive sources, diesel engines are much dirtier overall than gasoline engines (3 g of soot per kg of diesel fuel vs. 0.1 g per kg of gasoline), but car exhaust tends to contain relatively more of the lower molecular weight PAH, which include some of the more carcinogenic members of the family.

In polluted urban air those PAHs with the smallest molar masses (naphthalene, phenanthrene, biphenyl, etc.) are found completely in the gas phase, while those with more than 5 fused rings are almost completely associated with solid particles. PAHs of intermediate size are found in both phases, the proportion depending upon the availability of solid surface area and upon the temperature. The gas

Anthracene
(not carcinogenic)

Dibenz[a,b]anthracene
(carcinogenic)

Pyrene
(not carcinogenic)

Benzo[a]pyrene
(carcinogenic)

phase PAHs undergo chemical transformation in the atmosphere, principally through the initial addition of a hydroxyl radical to one of the aromatic rings. Under strong sunlight this reaction is rapid; for example, the tropospheric lifetime of naphthalene under conditions of photochemical smog is only a few hours.

PAHs such as benzo[a]pyrene (B[a]P) are classical examples of substances which must be bioactivated to become carcinogenic. The mechanism of bioactivation of B[a]P involves the sequence of events shown on the next page, although it must be stressed that bioactivation at ring position 7 is only a minor metabolic pathway. Oxidation also takes place at other ring positions, following which conjugation to substances such as glutathione leads to successful detoxification and excretion.

The active carcinogen, benzo[a]pyrene-7,8-dihydrodiol-9,10-dihydroepoxide, forms covalent adducts with DNA, notably by nucleophilic attack of guanine-NH_2 residues at C_{10}, with concomitant opening of the epoxide ring. The resulting DNA adducts are hypothesized to lead to cancer through misreading of the genetic code during transcription.

4. Elemental Hydrogen in the Atmosphere

Hydrogen is a minor component of the atmosphere (0.5 ppmv), corresponding to a reservoir of 180,000

Benzo[a]pyrene

Benzo[a]pyrene
7,8-oxide

Benzo[a]pyrene
7,8-dihydrodiol
9,10-epoxide

Benzo[a]pyrene
7,8-dihydrodiol

tonnes. Sink strengths, principally uptake by soil, are estimated at about 90,000 tonnes per year, affording a lifetime of about 2 yr; therefore like CO_2, hydrogen is moderately well mixed in the atmosphere. Its concentration shows an annual cycle with peaks in April and troughs in October; unlike CO_2, the cycles between the Northern and Southern Hemisphere are in phase because of the much greater land area of the Northern Hemisphere.

In the upper atmosphere, where much of the hydrogen is present as atoms, up to one H atom in 10^6 has sufficient kinetic energy to have the chance of escaping the Earth's gravity (escape velocity 11.2 km s^{-1}). About 3×10^8 hydrogen atoms per second escape the Earth's gravitational pull for every square centimeter of the Earth's surface; conversely, low speed protons are captured from the space by the Earth's gravitational field.

5. Uptake of CO_2 by the Hydrosphere

The uptake and release of CO_2 into and from the hydrosphere is governed by a complex series of equilibria:

$$CO_2(g) \rightleftharpoons H_2CO_3(aq) \underset{}{\overset{-H^+}{\rightleftharpoons}} HCO_3^-(aq)$$
$$\underset{}{\overset{Ca^{2+}}{\updownarrow}} \quad {-H^+}$$
$$CaCO_3(s) \rightleftharpoons CO_3^{2-}(aq)$$

The Henry's law constant for the equilibration

between $CO_2(g)$ and $H_2CO_3(aq)$ is 3.4×10^{-2} mol liter^{-1} atm^{-1} at 298 K; combined with the acid dissociation constant of 4.2×10^{-7} mol liter^{-1} for H_2CO_3, the pH of pure water in equilibrium with tropospheric CO_2 is calculated to be about 5.6. Even completely "clean" rain is therefore slightly acidic. In more alkaline solutions (e.g., the oceans, pH 8.1) the foregoing equilibria lie further to the right, and the concentration of total dissolved carbonate ($[H_2CO_3] + [HCO_3^-] + [CO_3^{2-}]$) is higher. The rate of mixing of the oceans limits the rate at which increasing atmospheric loads of CO_2 can be absorbed by the oceans.

The oceans are close to equilibrium between $CaCO_3(s)$, $Ca^{2+}(aq)$, and $CO_3^{2-}(aq)$, as a result of which marine organisms are able to deposit exoskeletons made of $CaCO_3$. Deposition of these exoskeletons over geological time is the source of biogenic minerals such as chalk, limestone, and marble. In more recent times, the formation of caves and gorges in limestone areas is the result of gradual dissolution of the rocks by CO_2-laden river water. This reaction is responsible for the hardness and the high alkalinity of groundwater in limestone areas (see Sections II,A and II,F):

$$CaCO_3(s) + H_2CO_3(aq) \rightarrow Ca^{2+}(aq) + 2HCO_3^-(aq)$$

C. Oxygen

Major sources of oxygen in the lithosphere are water, oxo-anions such as silicates, carbonates, and phosphates. Atmospheric oxygen in the form of O_2 and oxygen dissolved in water are the chief forms of oxygen of interest in environmental biology.

I. Atmospheric Dioxygen

The total amount of O_2 in the contemporary atmosphere has been variously quoted as 3.8×10^{19} mol or 1.2×10^{18} kg. Oxygen enters the atmosphere through photosynthesis (5.0×10^{15} mol yr^{-1} or 4.0×10^{14} kg yr^{-1}), which is almost exactly balanced by respiration, decay, and the combustion of fossil fuels. These different sets of figures give the residence time of O_2 as 7600 yr and 3300 yr, respectively, both of which indicate that oxygen has a long residence time in the atmosphere and is well mixed, that is, its partial pressure does not vary with season or location, (unlike H_2O, CO_2, and H_2).

2. Stratospheric Ozone

Ozone (O_3) is produced photochemically by deep UV-C radiation in the upper atmosphere and is a minor (ppbv) component of the atmospheric at all altitudes:

$$O_2 + h\nu \ (\lambda < 240 \ nm) \rightarrow 2O$$
$$O + O_2 \rightarrow O_3$$

The concentration of ozone at any altitude in the stratosphere is determined by the source and sink strengths. Solar intensity regulates the stratospheric source strength. Of the sinks, mechanism (A) involves photochemical cleavage of ozone by ultraviolet radiation having $\lambda < 325$ nm:

(A1) $\quad O_3 + h\nu \ (\lambda < 320 \ nm) \rightarrow O_2 + O$
(A2) $\quad\quad\quad\quad O + O_3 \rightarrow 2O_2$

The absorption of radiation by ozone, and its concomitant destruction, is the process by which ozone shields the Earth's surface from high-energy radiation in the region 240–325 nm, which is capable of damaging DNA. It is believed that the emergence of life from the oceans of the early Earth became possible only following the development of an atmosphere sufficiently rich in oxygen to allow the formation of an ozone shield.

Mechanism (B) is a series of related sinks, all involving free radical chain reactions in which the production of a single free radical X can initiate many cycles of ozone destruction:

(B1) $\quad\quad\quad X + O_3 \rightarrow XO + O_2$
(B2) $\quad\quad\quad XO + O \rightarrow X + O_2$

The sum of reactions (B1) and (B2) is equivalent to reaction (A2). Because X is consumed in reaction (B1) and regenerated in (B2), it is a catalyst for the destruction of ozone.

Four such cycles are known, involving X as Cl, NO, H, and OH. All occur naturally, their relative importance varying with altitude. The cycle with X = Cl, and an analogous cycle having X = Br, are of environmental importance in the context of the release of chlorofluorocarbons (CFCs) and bromochlorofluorocarbons (halons) into the atmosphere. These substances are extremely resistant to chemical degradation in the troposphere, as they are unreactive with hydroxyl radicals. They gradually migrate to the stratosphere, where they undergo photolysis by deep ultraviolet radiation to give chlorine and bromine atoms, respectively, which initiate the chain reaction of (B1) and (B2), thus increasing the sink strength for the destruction of ozone, for example,

$$CF_2Cl_2 + h\nu \ (\lambda < 250nm) \rightarrow CF_2Cl + Cl$$

The resultant stratospheric ozone depletion lessens the effectiveness of ozone at screening the Earth's surface from UV-B and UV-C radiation. The control of chlorinated and brominated substances under the "Montreal Protocol" has the goal of limiting the emissions to the atmosphere of substances that contain chlorine or bromine and that are sufficiently unreactive in the troposphere to survive transport into the stratosphere. Because CFCs have atmospheric $t_{1/2}$ up to 100 years, their emissions are considered to be a global pollution problem; CFCs released anywhere on the planet become globally dispersed much faster than they are removed from the atmosphere.

3. Ground-Level Ozone

In remote areas, tropospheric $p(O_3)$ is usually 20–50 ppbv, supplied largely by incursion from the lower stratosphere. Whereas the toxicity of ozone is irrelevant at stratospheric altitudes, in the troposphere ozone is phytotoxic at concentrations <100 ppbv and toxic to humans at only slightly higher levels, causing respiratory distress. Its toxicity is linked to its high chemical reactivity compared with O_2 and reflects its endergonic nature, that is, $\Delta G^\circ_f = +163.2$ kJ mol^{-1} at 298 K.

Tropospheric ozone and the nitrogen oxides (collectively called NO_x) exist in a pseudoequilibrium, which is driven to the right by sunlight

($\lambda < 400$ nm) and which reverts to the left thermally:

$$NO_2 + O_2 \underset{\text{thermal}}{\overset{h\nu}{\rightleftharpoons}} NO + O_3$$

The phenomenon of photochemical smog may develop under conditions where urban air pollution by NO_x and oxidizable substrates such as CO and hydrocarbons accompanies bright sunlight and temperature $>18°C$. These conditions frequently occur in major cities such as Los Angeles, Mexico City, Sao Paulo, New Delhi, and so on. Photochemical smog is characterized by elevated levels of NO_x, haze, and the presence of an oxidizing atmosphere, containing ozone and organic nitrates such as peroxyacetyl nitrate (PAN, $CH_3CO \cdot OO \cdot NO_2$), in contrast with smogs due to coal, which contain soot and SO_2, and which are chemically reducing.

Two mechanisms account for the accumulation of ground-level ozone. High levels of NO_x drive the preceding pseudoequilibrium to the right. In addition, ozone is formed as a by-product of the oxidation of substrates such as CO and hydrocarbons.

The detailed mechanism for the oxidation of CO to CO_2 is

$$CO + OH \rightarrow CO_2 + H$$
$$H + O_2 \rightarrow HO_2$$
$$HO_2 + NO \rightarrow NO_2 + OH$$
$$NO_2 + O_2 + h\nu \rightarrow NO + O_3$$

The summary reaction for the multistep oxidation of CO is shown below; comparable reactions can be written for the oxidation of methane and more complex hydrocarbons:

$$CO + 2O_2 \rightarrow CO_2 + O_3$$

Photo-assisted oxidation of carbon monoxide is a net producer of tropospheric ozone when the ratio of concentrations [NO]/[NO_2] is constant, and sufficient NO_x and light are present to sustain the conversion of NO_2 to NO. However, in remote locations where [NO_x] is low, the oxidation of CO and other oxidizable constituents of the troposphere can be a net sink for ozone:

$$CO + OH \rightarrow CO_2 + H$$
$$H + O_2 \rightarrow HO_2$$
$$HO_2 + O_3 \rightarrow OH + 2O_2$$

The summary reaction is:

$$CO + O_3 \rightarrow CO_2 + O_2$$

The impact of photochemical oxidants on plants and human health has led to strenuous efforts to control their formation by the use of emission controls on automobiles, which are major sources of ozone precursors, that is, hydrocarbons and NO_x. However, recent evidence indicates that emission controls have had little success in improving air quality in the United States, because the production of ozone depends critically on the ratio [hydrocarbons]/[NO_x] as well as their absolute quantities. Underestimation of the emission inventories of hydrocarbons, and especially ignorance (until recently) of the presence of a significant concentration of hydrocarbons of biogenic origin, has encouraged a suboptimal strategy for the design of catalytic converters on automobiles.

4. Oxygen in Water

The atmosphere contains 0.21 atm O_2, and from the value of K_H at 25°C (1.3×10^{-3} mol liter^{-1} atm^{-1}), the equilibrium solubility of O_2 in water at this temperature is calculated to be 2.7×10^{-4} mol liter^{-1} (8.7 mg liter^{-1}, 8.7 ppm). Dissolved oxygen is essential for the survival of aquatic animals; for example, most fish require 5–6 ppm of dissolved oxygen. Fish kills occur if the oxygen supply is depleted, either due to thermal pollution, because the solubilities of all gases in water decrease with increasing temperature, or due to the presence of oxidizable substances in the water, because their oxidation consumes O_2. Excessive amounts of nutrients in water bodies promote the growth of algal blooms, which grow quickly but require oxygen when they die to oxidize their biomass to carbon dioxide and water. Untreated or partially treated sewage and factory effluents, especially from food processing (meat-packing plants, vegetable and fruit canneries) and animal feedlots (manure seepage), are common sources of oxidizable organic compounds that are readily utilized by microorganisms. Persistent water pollutants, such as polychlorinated biphenyls, do not pose a problem of oxygen depletion because they are so slow to undergo biological oxidation. Oxygen in water is replenished from the atmosphere; the concentration of O_2 (aq) falls if reoxygenation from the atmosphere cannot keep pace with the rate of oxygen depletion. Flowing water, especially waterfalls or rapids, promotes reoxygenation; stagnant, oxygen-depleted water tends to have a foul odor due to the presence of odorous nitrogen and sulfur compounds (NH_3, H_2S, and their organic derivatives) produced by anaerobic microorganisms.

Measures of the oxygen status of a water body include dissolved oxygen concentration (DO), total organic carbon (TOC), chemical oxygen demand (COD), and biochemical oxygen demand (BOD).

DO: usually measured using a portable oxygen sensor, which is a voltammetric device calibrated to read directly in ppm O_2.

TOC: an instrumental method in which all carbon-containing substances in the water are oxidized catalytically to CO_2, which is measured by gas chromatograph.

COD: a titration method in which the more easily oxidized organics are oxidized using $Na_2Cr_2O_7/H_2SO_4$ under stated conditions of time and temperature.

BOD: measured by incubating the water sample with aerobic microorganisms under stated conditions of time and temperature (usually 5 days, 25°C). The BOD is the difference in dissolved oxygen measured before and after the experiment.

High BOD is considered a source of water pollution because of its potential for oxygen depletion. The high BOD of sewage and industrial waste waters if reduced by treating them under conditions where microbial oxidation can take place rapidly, before release to the environment. Open lagoons or closed bioreactors may be used, with the water highly oxygenated to maximize microbial activity.

The microorganisms reduce BOD both by using organic compounds as an oxidizable energy source (producing CO_2 and water) and by incorporating them into microbial biomass. Incorporation of nutrients into microbial biomass is promoted by providing nutrients in the correct proportions: a ratio C : N : P of about 100 : 15 : 1. If the proportions of N or P in a given industrial waste stream are low, it may be necessary to supplement them by the use of fertilizers in the lagoon or bioreactor.

5. Singlet Oxygen

Singlet oxygen is an electronic excited state of molecular dioxygen, which has a triplet ground state (two unpaired electrons). Light-absorbing "sensitizers" transfer their photon energy to ground state O_2, thus producing singlet oxygen ($^1O_2$, all electrons paired), which is 90 kJ mol^{-1} more energetic than the ground state, and hence a powerful oxidant.

$$\text{Sens} + h\nu \rightarrow \text{Sens}^*$$
$$\text{Sens}^* + O_2 \rightarrow \text{Sens} + {}^1O_2$$

Whereas ground state oxygen, with its two unpaired electrons, tends to react as a free radical (e.g., its immediate coupling with other free radical species during atmospheric oxidations), singlet oxygen reacts chiefly as an electrophile, reacting preferentially at nucleophilic sites like the lone pair electrons of atoms such as sulfur and nitrogen. These reactions are called "photosensitized oxidations" because they require light, sensitizer, and O_2 in addition to the oxidizable substrate. Another mechanism of photosensitized oxidation, observed mostly for carbonyl compounds as photosensitizers, involves initial hydrogen abstraction from the substrate by the photosensitizer, followed by the addition of ground state O_2 to the resulting free radical. Besides oxidations under environmental conditions (provided that light can penetrate), photosensitized oxidations of biomolecules containing N and S, such as proteins, can cause the death of cells, a phenomenon recognized early in this century under the name photodynamic action for killing microorganisms with a combination of light, dye, and O_2. Photodynamic action has recently been exploited under the name photodynamic therapy for treatment of certain cancers. A light-absorbing dye such as hematoporphyrin is taken up by the tumor and the light is directed to the tumor site using optical fibers.

6. Ozone as Oxidant in Aqueous Solution

Ozone is intrinsically more soluble in water than O_2, having $K_H = 1.3 \times 10^{-2}$ mol liter^{-1} atm^{-1}. It is used as a disinfection agent for drinking water; its toxicity to microorganisms results from its oxidizing power. As an oxidizing agent, ozonation does not give traces of the halogenated by-products that are produced when chlorine is used. Ozone is produced by passing an electric discharge ($\sim$15,000 V) through dry air, and then dissolving the ozone formed in a large tank into which the raw drinking water flows. The capital cost of this equipment is higher than that needed for chlorination, and so ozonation is practiced principally by large regional water treatment facilities.

Because of its high oxidation potential, ozonation has been exploited for the treatment of aqueous streams of refractory industrial wastes. In this application, the ozonated waste is irradiated with deep ultraviolet (254 nm) light, promoting a highly unselective oxidation, probably involving the attack of aqueous-phase hydroxyl radicals on the substrate.

D. Nitrogen

The atmosphere contains some 3.9×10^{18} kg of elemental nitrogen, whose major natural sinks are biological nitrogen fixation (2×10^{11} kg yr^{-1}) and the production of NO in thunderstorms and through combustion; that is, any condition under which air is heated (7×10^{10} kg of N per year):

$$N_2(g) + O_2(g) \rightarrow 2NO(g)$$

In addition, about 5×10^{10} kg of nitrogen are fixed industrially each year by the Haber process, mostly for use as fertilizer:

$$N_2(g) + 3H_2(g) \rightarrow 2NH_3(g)$$

Because the rates of transfer of nitrogen in and out of the atmosphere are small in comparison with the size of the reservoir, the calculated residence time of N_2 is large ($\sim$10^7 years). [See NITROGEN CYCLE INTERACTIONS WITH GLOBAL CHANGE PROCESSES.]

Fixed atmospheric nitrogen is ultimately deposited as HNO_3 in rainwater. Microbial processes interconvert terrestrial and aquatic nitrogen (as NO_3^- and NH_4^+), which cycle through the biosphere as proteins and nucleic acids, and return to the atmosphere as N_2 and N_2O by the action of denitrifying bacteria. Increased fertilizer use increases the rate of biological denitrification, with up to half of all nitrogenous fertilizer being denitrified even before the crop takes it up. Denitrification is largely responsible for the gradual increase in $p(N_2O)$ from 0.29 to 0.31 ppmv since 1970. N_2O is rather unreactive in the troposphere, with residence time $\sim$20 yr; its chief sink is decomposition by UV-C radiation following migration to the stratosphere.

NO and NO_2 are naturally present in the troposphere at low ppbv levels. In urban atmospheres, combustion processes such as automobiles, electric power generation, and municipal incineration increase the local concentration of NO.

As noted in Section II,C, NO and NO_2 are interconverted in the atmosphere:

$$NO + O_3 \rightleftharpoons NO_2 + O_2$$

A complex series of reactions rationalizes the role of NO_x in the production of high ozone levels in polluted urban atmospheres. The oxidation of hydrocarbons and other oxidizable substrates is initiated by reaction with the hydroxyl radical, whose formation by photolysis of ozone was described in Section I,H. Peroxy radicals such as CH_3OO provide another route for the conversion of NO to NO_2 (cf. Section II,C):

$$CH_3OO + NO \rightarrow CH_3O + NO_2$$

The main sink for tropospheric NO_x is a rapid ($t_{1/2}$ typically several hours) reaction between NO_2 and OH to give gaseous nitric acid, which undergoes either wet deposition in rain or dew, or dry deposition adsorbed to particles:

$$NO_2(g) + OH(g) \rightarrow HNO_3(g)$$

High levels of NO_x in urban atmospheres depress the rate of oxidation of hydrocarbons by scavenging hydroxyl radicals. There is thus a complex interplay between NO_2 and OH, with NO_2 acting both as an ultimate precursor and as a sink.

Although NO_x is removed from the atmosphere within hours, ground-level ozone (see Section II,C,3), which is formed a by-product of the oxidation of hydrocarbons in the presence of NO_x, is much more persistent. Air pollution resulting from NO_x and hydrocarbon emissions in an urban area can therefore have a negative impact on agriculture well downwind of the source, and so photochemical oxidant formation is a regional, rather than a local, air pollution problem. The composition of the plume changes considerably as the plume migrates, with O_3 building up as NO_x abates.

E. Phosphorus

Although phosphorus is an essential element for life, it has limited bioavailability, because it occurs chiefly in highly insoluble forms such as fluorapatite, $Ca_5(PO_4)_3F$, and hydroxylapatite, $Ca_5(PO_4)_3OH$. A sedimentary form of hydroxylapatite, known as phosphorite, is the chief mineral mined, with the United States (mainly in Florida), Morocco, and Khazakhstan being major commercial producers. The solubility and hence the bioavailability of phosphate for use in fertilizers is increased by treating rock phosphate with concentrated sulfuric acid to give the more soluble calcium dihydrogen phosphates, for example, "superphosphate" $Ca(H_2PO_4)_2 \cdot H_2O$:

$$2Ca_5(PO_4)_3OH + 7H_2SO_4 + H_2O \rightarrow$$
$$3Ca(H_2PO_4)_2 \cdot H_2O + 7CaSO_4$$

"Triple superphosphate" is a higher-purity form of $Ca(H_2PO_4)_2 \cdot H_2O$ manufactured from fluorapatite.

In the presence of water, and hence upon application to soil, $Ca(H_2PO_4)_2 \cdot H_2O$ disproportionates:

$$Ca(H_2PO_4)_2 (+nH_2O) \rightarrow CaHPO_4 + H_3PO_4$$
$$(+n\text{-}1 \; H_2O)$$

Uptake into plants at typical soil pH occurs mostly as $H_2PO_4^-$, but much is lost through runoff before it can be assimilated.

Commercial exploitation of phosphorus has been of major importance in raising agricultural yields. Although phosphate minerals are unlikely to be exhausted in the near future, the natural biogeochemistry of phosphorus cycles this element between rocks, oceans, and biota on a very long time scale (millions of years); large-scale exploitation of phosphate deposits represents a considerable interference with natural processes and could significantly alter the balance between immobile phosphate in rocks and bioavailable phosphate in the oceans.

I. Eutrophication

Eutrophication is the accelerated aging of lakes through excessive inputs of nutrients into waterways. One manifestation of eutrophication is the growth of algal blooms, which deplete the water of oxygen when they decay (see Section II,C). [*See* EUTROPHICATION.]

The growth of algae is limited by the availability of nutrients in the water. The uptake of nutrient elements into biomass requires the nutrient elements carbon, nitrogen, and phosphorus in the approximate ratio C: N: P = 100: 15: 1. Carbon is never limiting in water because it can be resupplied from the atmosphere as CO_2, and blue-green algae can supplement dissolved nitrogen by fixing atmospheric N_2. Thus phosphorus is usually the limiting nutrient, even though it is needed in the smallest amount. Phosphates enter aquatic ecosystems through runoff of phosphate fertilizers from farms and through the use of phosphates such as sodium tripolyphosphate (STP, $Na_3H_2P_3O_{10}$), which are used as "builders" in detergent formulations. These serve the dual role of holding the pH of the wash above 7 and sequestering hardness ions such as Ca^{2+} (unlike the calcium salts of the monophosphate ion that are highly insoluble in water, complexes between calcium ion and both cyclic and acyclic polyphosphates are soluble in water). Polyphosphate anions are hydrolyzed to monophosphate during secondary sewage treatment, and a high proportion formerly passed out into streams and lakes, causing annual problems of algal blooms.

Eutrophication has declined sharply in North America during the past 20 years as a result of legislation to limit the phosphate content of detergents and through treatment of sewage to remove phosphate. In Canada, for example, the phosphate content of detergents was limited to 20% by weight in 1970 and 5% in 1973. As a result, the average phosphorus content of raw Ontario sewage dropped from 10 ppm in 1969 to 5 ppm in 1974 and has remained below 5 ppm since then. Today the chief source of phosphates in natural waters in North America is agriculture, in the form of runoff from fertilizer application to the land and runoff of manure from feedlots.

The improvement in water quality in the Great Lakes basin has been most evident in Lake Erie, which was generally regarded as "dead" in the 1970s, with stinking algal blooms being an annual occurrence and the commercial fishery no longer viable. With the reduction in phosphate loadings, Lake Erie is essentially free of algal blooms and once again supports commercial fishing. Lake Erie's health was restored relatively quickly because of the short residence time of the water in the lake (2.7 years), which allows pollutants to be flushed out quickly. Ecological damage to a larger body of water takes longer to be manifested, but also takes longer to reverse.

2. Phosphate Removal from Sewage

Conventional sewage handling involves primary treatment, a simple settling process to remove solids, followed by secondary treatment, which is mainly BOD reduction through microbial oxidation using either trickling filters or closed activated sludge reactors. Although secondary treatment incidentally removes some phosphate through incorporation into microbial biomass, much of the phosphate passes out in the effluent stream.

Tertiary treatment for phosphate involves chemical precipitation, using agents such as lime, filter alum, and ferric chloride. In treatment with lime $[Ca(OH)_2]$, the hydroxide ions bring the pH to about 9, while the elevation of the concentration of Ca^{2+} causes precipitation of calcium phosphate. Note that at pH 9, HPO_4^{2-} (not PO_4^{3-}) is the predominant form of phosphate ion:

$$5Ca^{2+}(aq) + 3HPO_4^{2-}(aq) + 4OH^-(aq) \rightarrow Ca_5(PO_4)_3OH(s) + 3H_2O(l)$$

Although hydroxylapatite $[Ca_5(PO_4)_3OH]$ is thermodynamically the most stable calcium phosphate, other, more soluble materials such as $Ca_3(PO_4)_2$ are the kinetic products of precipitation. Their formation must be avoided to optimize the removal of phosphate from the sewage stream. This is accomplished by "seeding" the solution with a small amount of finely powdered hydroxylapatite.

When filter alum, $Al_2(SO_4)_3$, is used as the precipitant, the product is $AlPO_4$. Precipitation is optimally carried out near pH 5, where $H_2PO_4^-$ is the major phosphate species. At higher pH, the added aluminum salts preferentially precipitate as $Al(OH)_3$ rather than as $AlPO_4$:

$$Al^{3+}(aq) + H_2PO_4^-(aq) + 2OH^-(aq) \rightarrow AlPO_4(s) + 2H_2O(l)$$

Correspondingly, $FeCl_3$ affords $FePO_4$, which is converted during sewage sludge digestion into $Fe_3(PO_4)_2$ under anaerobic conditions [reduction of Fe(III) to Fe(II)]:

$$Fe^{3+}(aq) + H_2PO_4^-(aq) + 2OH^-(aq) \rightarrow FePO_4(s) + 2H_2O(l)$$

Like $Al_2(SO_4)_3$, $FeCl_3$ is used at about pH 5 to minimize loss of the precipitant as $Fe(OH)_3$.

3. Organophosphorus Compounds

Many compounds having carbon-to-phosphorus bonds have pronounced physiological activity as anticholinesterases (ACE). ACEs inhibit the action of cholinesterase, thereby preventing the hydrolysis of acetylcholine, a neurotransmitter. The consequences of this inhibition may include paralysis and death. Many organophosphates are selectively more

$$(CH_3)_2N\overset{\overset{O}{\|}}{\underset{\underset{CN}{|}}{P}}OC_2H_5$$

Tabun

$$(CH_3)_3C\overset{\overset{CH_3}{|}}{\underset{}{CH}}O\overset{\overset{O}{\|}}{\underset{\underset{F}{|}}{P}}CH_3$$

Soman

$$C_2H_5O\overset{\overset{O}{\|}}{\underset{}{C}}\overset{}{\underset{\underset{\underset{O}{\|}}{C_2H_5O-C-CH_2}}{CH}}S\overset{\overset{S}{\|}}{\underset{\underset{OCH_3}{|}}{P}}OCH_3$$

Malathion

$$NO_2-\overset{R}{\underset{}{\bigcirc}}-O\overset{\overset{S}{\|}}{\underset{\underset{OCH_3}{|}}{P}}OCH_3$$

R = H; Parathion methyl
R = CH₃; Fenitrothion

toxic to insects than to mammals and are used as insecticides. Organophosphorus insecticides are preferred to the older organochlorines because of greater selectivity and reduced persistence (they are hydrolyzed relatively rapidly to nontoxic materials). Some organophosphates are too toxic toward mammals, including humans, to be considered for use as insecticides; some of this latter class have been investigated as "nerve gases" for military use. One mg of the military nerve gas Soman is a fatal human dose. No completely satisfactory antidote has been developed, and such antidotes that do exist must be administered within one minute of exposure.

The mechanism of action of ACE substances follows the mechanism of hydrolysis of acetylcholine. However, whereas the acetylcholine–esterase complex is rapidly recycled, the inhibitor occupies the enzyme site essentially irreversibly. Because cholinesterases are in limited supply and are resynthesized only slowly, inhibition of the cholinesterase may have fatal consequences for either insect or soldier.

F. Sulfur

Sulfur is found in the environment as the element, in subterranean deposits in Texas, and as metal sulfides, particularly of the transition and "heavy" metals (Ni, Cu, Zn, Hg, Pb, etc.), all of which are highly insoluble. These metals are recovered by mining the sulfides, and this leads to two specific environmental problems: acid rain, through air pollution by sulfur dioxide, and acid mine drainage, due to microbial oxidation of exposed sulfides in the presence of air. Pollution of the air by SO_2 is also associated with heavy industries that burn coal, which contains up to ~3% sulfur by mass. These include electricity generation and iron and steel production, as in the nineteenth-century Industrial Revolution in Europe and North America, and in developing countries today. In the absence of pollution control, the mass of SO_2 emitted is exactly twice the mass of sulfur in the coal. North American coals from southern West Virginia, Kentucky, and the Canadian and American Rockies contain <1% sulfur, whereas high-sulfur coals come from the American Midwest, northern Appalachia, and Nova Scotia.

Metal sulfide smelting involves roasting the ore in air to give the metal oxide, which is subsequently reduced to the element, usually with coke:

$$2MS \ (M = Ni, Zn, Pb) + 3O_2(g)$$
$$\rightarrow 2MO + 2SO_2(g)$$
$$MO + C(s) \rightarrow M + CO(g)$$

With Cu_2S and HgS, the metal is formed directly, for example,

$$Cu_2S(s) + O_2(g) \rightarrow 2Cu(s) + SO_2(g)$$

In the absence of pollution control, 1 mol of SO_2 is released for each mol of, for example, zinc produced. The situation is worse if the metal sulfide being mined is associated with other sulfides, for example, in Canada, where NiS is found along with a large excess of FeS_2. Even though some of the SO_2 is captured and converted into sulfuric acid at some of the largest smelters, thousands of tonnes of SO_2 are still released daily. The nickel smelter at Sudbury, Canada, remains the worlds's largest single point source emission of SO_2, despite pollution control measures that are discussed later in this section.

I. Acid Rain

Unpolluted rainwater has a pH close to 5.6, in consequence of the raindrops being in equilibrium with atmospheric CO_2 (See Section II,B). Acidic precipitation is generally defined as having pH lower than about 5.0; pH 4 to 4.5 is not uncommon, and isolated examples of rain and fog having pH lower than 2 have been recorded. Acid rain causes damage to crops, to forests, to environmentally sensitive lakes, and also to buildings and engineering structures made of stone (limestone) and metal (iron and steel). Acid rain is caused by atmospheric emissions of sulfur oxides and nitrogen oxides that are present in trace amounts (ppbv) in unpolluted air. Even in polluted air their concentrations rarely exceed 1 ppmv. Acid rain results when these gases are oxidized in the atmosphere and return to the ground dissolved in raindrops. [*See* ACID RAIN.]

SO_2 falls as H_2SO_3 and H_2SO_4:

$$SO_2 + H_2O \rightarrow H_2SO_3$$

$$SO_2 \xrightarrow{\text{oxidize}} SO_3 \rightarrow H_2SO_4$$

NO_x falls as HNO_3, as discussed in Section II,E.

In most areas, sulfur oxides are the major contributor to acid precipitation, an exception being the U.S. West Coast, where acid rain and photochemical smog are closely linked through the formation of HNO_3 in urban air (Section II,D). In a few regions of the world, notably parts of Alaska and New Zealand, highly acidic rain falls naturally as a result of the emission of HCl and SO_2 from volcanoes.

Even at concentrations <1 ppmv, NO_2 and/or SO_2 have greater impact on the pH of rain than the much larger ambient concentrations of CO_2: they are more soluble in water (larger K_H) and form stronger acids (HNO_3, H_2SO_3, and H_2SO_4). For example, 0.1 ppmv of $SO_2(g)$ will produce a pH <4.5 in equilibrium with rainwater when precipitation occurs as H_2SO_3, compared with the pH 5.6 produced by 350 ppmv of $CO_2(g)$. Deposition occurs either in the aqueous form (wet deposition) or in association with particulate matter (dry deposition), in which case much of the sulfur will deposit in the form of sulfite or sulfate ions rather than the free acids.

Oxidation of SO_2 to SO_3 occurs by several mechanisms. In dry air, homogeneous gas-phase oxidation is initiated by attack of the hydroxyl radical,

and hence only occurs during daylight ($t_{1/2}$ several days). Molecular oxygen converts the resulting HSO_3 radical to SO_3, which is then deposited as H_2SO_4:

$$SO_2 + OH \rightarrow HSO_3$$
$$HSO_3 + O_2 \rightarrow SO_3 + HO_2$$

Under moist conditions, a faster aqueous-phase oxidation occurs in clouds or raindrops. The oxidizing agent is hydrogen peroxide, which is formed by the disproportionation of HO_2 radicals. The oxidation of SO_2 may involve disproportionation of HO_2 in the gas phase followed by dissolution of the H_2O_2 in the water droplet ($K_H = 10^5$ mol liter^{-1} atm^{-1}); alternatively, the HO_2 radicals may enter the aqueous phase ($K_H = 2 \times 10^3$ mol liter^{-1} atm^{-1}) before disproportionation:

$$2HO_2(g) \rightarrow H_2O_2(g) + O_2(g)$$
$$SO_2(aq) + H_2O_2(aq) \rightarrow H_2SO_4(aq)$$

SO_2 is also oxidized heterogeneously on particles, catalyzed by salts of transition metals such as iron, manganese, and vanadium, which are present in the atmosphere as a result of burning coal and oil:

$$SO_2 + \tfrac{1}{2}O_2 \xrightarrow{\text{catalyst}} SO_3 \rightarrow H_2SO_4 \text{ (or } SO_4^{2-})$$

Because the half-lives of SO_2 and SO_3 in the atmosphere are only a few days, acid precipitation can be expected over whatever distance is traveled by the pollution plume during that time (several hundred kilometers). Like ground-level ozone, acidic precipitation is a regional pollution problem. Among the political problems it poses is transboundary pollution, where the citizens of one country or state are the unwilling recipients of pollution from a neighbor.

2. Biological Effects of Acidic Emissions

The effects of acidic precipitation on natural waters have been studied extensively. Data from the Great Lakes basin suggest annual wet deposition rates of sulfur and nitrogen of almost 400,000 and 200,000 tonnes, respectively. Acidification is mainly a problem in areas where the underlying rocks provide poor buffering capacity, that is, the water in contact with these rocks has low alkalinity. Rocks such as granite offer little protection by way of buffering, whereas carbonate rocks such as chalk and limestone are able to react with the acid and hence to neutralize it. Lakes and streams in limestone areas are therefore fairly insensitive to acidic precipitation:

$$CaCO_3(s) + 2H^+(aq) \rightarrow Ca^{2+}(aq) + H_2CO_3(aq)$$

Lakes in areas of carbonate rocks commonly have pH 7 to 7.5, and total alkalinity 10^{-3} mol H^+ liter^{-1} or higher. Their resistance to acidification is explained by higher total alkalinity together with replacement of alkalinity lost to H^+ by dissolution of the rock. In contrast, lakes in areas of granitic rocks, namely, northern and eastern Canada, the northeastern United States, and Scandinavia, are poorly buffered and have "normal" pH about 6.5 to 7 and total alkalinity in the range 10^{-4} mol H^+ liter^{-1}, which cannot be replenished by dissolution.

Acidified lakes do not support the variety of life that can be found in their nonacidified counterparts. Loss of game fish can be expected in lakes whose pH has already dropped below pH 5. By 1976, about half the lakes in the Adirondack Mountains of New York State had no fish in them, whereas 40 years earlier they almost all supported a population of sport fish. Correspondingly, 80% of Adirondack lakes were found to have lost alkalinity over the past 60 years, the median loss being 50 μmol H^+ liter^{-1}. Below pH 4, lakes become a suitable habitat for white moss, which forms a thick "felt mat" on the lake bottom, preventing the exchange of nutrients between the water and the bottom sediments, and also preventing the sediments from exerting any buffering action. The result is a lake whose waters are crystal clear but that support very few forms of aquatic life.

A special problem for aquatic species is that springtime spawning coincides with the worst "pulse" of acidity of the year, when the winter's accumulation of acid snow reaches waterways during the annual spring runoff. This decreases the rate of hatching and reduces the viability of the newly hatched fry.

Acidified lakes have been rejuvenated using powdered limestone sprayed from aircraft to neutralize the excess acid. The method is costly—about $20,000 to $50,000 to maintain a single 20-ha lake for 10 years—and relief is only temporary unless the source of acidity is controlled.

Acidic precipitation damages structural materials such as limestone by sulfation, which involves replacement of calcium carbonate by calcium sulfate, which is both more water-soluble than calcim carbonate and has less structural strength. This causes the outer layers of the stone to flake off.

$$CaCO_3(s) + 2H_2SO_4(aq) \rightarrow \\ CaSO_4(s) + H_2CO_3(aq)$$

Iron and steel also suffer faster corrosion in an acidic environment, and the protection of such structures with paint costs billions of dollars annually.

Plants are highly susceptible to the effects of acidic emissions, both the gaseous pollutants themselves and through lowered pH. Sulfur dioxide inhibits plant growth at concentrations less than 0.1 ppmv and causes observable injury between 0.1 and 1 ppmv after only a few hours of exposure. Combinations of gaseous pollutants such as SO_2/NO_2 and SO_2/O_3 act synergistically. Mist poses a special threat to forests in upland areas because acidic emissions at ground level make the base of the cloud the most acidic. Leaves show observable damage below pH 3.5. Excessive acidity in poorly buffered soils inhibits plant growth and seed germination.

Much controversy has been focused on whether acid rain is responsible for the reduced productivity of forests seen in recent years in Scandinavia, eastern North America, and Germany. The observed symptoms are consistent with long-term acidification, namely, chlorosis, with browning and dropping of the needles. Acid rain has been assumed to be the cause of these problems, but synergy with ozone or with volatile organic compounds of industrial origin is also possible.

Acidic gases are respiratory irritants, although atmospheric concentrations <1 ppmv of SO_2 are absorbed high in the respiratory tract and do not penetrate to the more sensitive alveoli. Respiratory irritation may occur at 1–2 ppmv, especially in elderly people and those already suffering from lung diseases such as emphysema. Respiratory distress is more pronounced if particulate matter is also present: this phenomenon (known as "London smog") was very prevalent in the United Kingdom in the days when coal was used for domestic heating (Section II,B,3). Penetration to the alveoli occurs when $p(SO_2)$ reaches $\sim$25 ppmv, which may be encountered in industries such as smelting, tanning, paper-making, and sulfuric acid manufacture. However, the irritant effects of SO_2 at these concentrations (wheezing, coughing, tearing) give a severe enough warning that actual injury is rare.

3. Abatement of Sulfur Oxide Emissions from Stationary Sources

Current trends may be illustrated by the large nickel smelter owned by Inco at Sudbury, Ontario, where three strategies are being followed to reduce emissions of SO_2. First, improved physical separation of the ore, using oil flotation, allows a higher proportion of unwanted sulfides such as FeS_2 to be rejected prior to roasting. Second, part of the SO_2 produced during roasting is converted to sulfuric acid before the gases escape to the atmosphere. Third, air in the roasting step is being replaced by pure oxygen; this affords a higher-strength SO_2 stream from which SO_2 can be converted to sulfuric acid more efficiently. Progressive improvements in the control of acidic emissions in the Sudbury area since the 1960s have allowed the return of plant life to an area which was once known for its barrenness, caused by acidic emissions from the nickel smelter.

The concentration of SO_2 in the exhaust gases from burning coal is too low for viable conversion to sulfuric acid. The options are to remove (much of) the sulfur prior to combustion by means of oil flotation, taking advantage of the higher density of sulfide-rich minerals compared with coal, or to scrub the exhaust gases with base (flue gas desulfurization) using an aqueous slurry of lime [$Ca(OH)_2$] or limestone ($CaCO_3$). Scrubbing is expensive; in addition, large volumes of an aqueous slurry of $CaSO_3/CaSO_4$ must be disposed of:

$$SO_2(g) + Ca(OH)_2(aq) \rightarrow CaSO_3(s) + H_2O(l)$$
$$SO_2(g) + CaCO_3(s) \rightarrow CaSO_3(s) + CO_2(g)$$

Fluidized bed combustion of coal admixed with limestone is an alternative to scrubbing, but it too leaves a mixture of $CaSO_3$ and $CaSO_4$ for disposal.

4. Acid Mine Drainage

Acid mine drainage involves the acidification of streams in the vicinity and downstream of a mine. Its manifestations are the mobilization of toxic metals and the deposition of iron, which forms an unsightly slimy orange precipitate of $Fe(OH)_3$ on the rocks of the streambed. Acid mine drainage is a biological phenomenon involving bacterial oxidation of waste sulfide-containing minerals at metal extraction sites and coal mines. These wastes ("tailings") are finely crushed minerals rejected from ore concentration and are discarded in tailings ponds, where *Thiobacillus* bacteria use the sulfide minerals to reduce molecular oxygen concurrent with oxidation of sulfide to sulfuric acid:

$$H_2S(aq) + 2O_2(g) \rightarrow SO_4^{2-}(aq) + 2H^+(aq)$$

or

$$S^{2-}(aq) + 2O_2(g) \rightarrow SO_4^{2-}(aq)$$

The large increase in the acidity of the solution can be conceptualized either as replacing the strongly basic sulfide ion by the feebly basic sulfate ion, or as replacing the weakly acidic H_2S by the strongly acidic H_2SO_4. Sulfide oxidation is a biological process below pH 4; in less acidic solutions a parallel chemical oxidation also occurs. Iron (II) in the form of FeS_2 is also oxidized, to $Fe^{3+}(aq)$:

$$4FeS_2(s) + 2H_2O + 15O_2 \rightarrow 4Fe^{3+}(aq) + 8SO_4^{2-}(aq) + 4H^+(aq)$$

Close to the mine pH values as low as 1–2 may be reached; under these conditions ferric oxide (in the tailings or in the abandoned mine itself) can dissolve:

$$Fe_2O_3(s) + 6H^+(aq) \rightarrow 2Fe^{3+}(aq) + 3H_2O(l)$$

Downstream from the mine site, dilution with uncontaminated water raises the pH, and the dissolved iron precipitates at pH > *ca* 3.5 as the insoluble $Fe(OH)_3$:

$$Fe^{3+}(aq) + 3H_2O(l) \rightarrow Fe(OH)_3(s)$$
$$[or \; ^1/_2Fe_2O_3 \cdot nH_2O] + 3H^+(aq)$$

Besides the precipitation of $Fe(OH)_3$, acid mine drainage also causes loss of acquatic life due to excessive acidity and mobilization of other (toxic) metals (See Section I,E), which may be present in solution at concentrations >100 ppm. Large abandoned mines may have thousands of tons of these metals in solution in tailings ponds covering thousands of hectares.

G. Chlorine

Chlorine occurs in nature chiefly as the chloride ion. Over geological time, soluble chloride salts have been leached from the lithosphere and concentrated as chlorides from the lithosphere into the oceans, which contain Cl^- at a concentration 0.55 mol liter^{-1}. Solid chloride deposits, mainly NaCl and KCl, are evaporites, formed by the evaporation of ancient seas. Most organochlorine compounds are of xenobiotic origin, although marine bacteria produce CH_3Cl, which escapes into the atmosphere, and some marine plants elaborate halogenated secondary metabolites. Organochlorines, especially aromatic organochlorines, are very resistant to both hydrolysis and oxidation and include some of the most persistent pollutants, such as polychlorinated biphenyls, "dioxins," chlorinated pesticides, and chlorofluorocarbons. Besides persistence, many organochlorines are hydrophobic, lipophilic, and hence prone to bioconcentration and biomagnification in the food chain.

I. Elemental Chlorine

Elemental chlorine is manufactured along with sodium hydroxide by the electrolysis of brine using the chlor-alkali process:

$$2NaCl + 2H_2O \rightarrow 2NaOH + Cl_2 + H_2$$

Elemental chlorine is used in the production of chlorinated solvents such as di-, tri-, and tetrachloromethane, chlorinated ethanes and chlorinated ethylenes, the former production of highly chlorinated pesticides such as DDT, Aldrin, and Mirex, former production of polychlorinated biphenyls, production of chlorinated phenols and hence chlorinated phenoxy herbicides, the production of chlorofluorocarbons, the bleaching of wood pulp for paper production, and the disinfection of drinking water (See Section II,A).

Bleaching pulp and paper is a declining use of chlorine. In the late 1980s concerns were expressed about small amounts of "dioxins" present in chlorine-bleached paper, and about the toxicity of chlorinated pulp mill effluents on the aquatic life of receiving waters. The chlorine content of pulp mill effluents is measured as adsorbable organic halogen (AOX), a parameter which is being used in several jurisdictions to regulate pulp mill effluents, with progressively stricter standards to be achieved by the period 2000–2005. Two measures, the installion of BOD reduction facilities and partial substitution of ClO_2 for Cl_2 as a bleaching agent, have both lowered AOX and reduced dioxin levels in the finished paper to below detection.

One measure of the toxicity of pulp mill effluents is its induction of mixed function oxidase enzymes in fish. Contrary to expectation, enzyme induction does not correlate with either total AOX or the lipophilic, solvent-extractable fraction of AOX, which had been assumed to be most toxic. Comparable effects are seen in fish downstream of both bleaching and nonbleaching mills. The substance responsible for enzyme induction therefore may not be associated with AOX, in which case AOX may be a poor parameter on which to regulate pulp mill effluents.

2. Chlorinated Aliphatic Solvents and Chlorofluorocarbons

Large volumes of chlorinated aliphatic solvents are used in applications such as metal degreasing and dry cleaning. Although solvent recycling is practiced, most of these solvents eventually reach the atmosphere. The same is true of chlorofluorocarbons such as CFC-11 ($CFCl_3$) and CFC-12 (CF_2Cl_2), blowing agents and refrigerants, respectively. CFCs and certain other chlorinated aliphatic solvents such as CH_3CCl_3 are of concern because of their potential for ozone depletion through the photochemical release of chlorine atoms in the stratosphere (See Section II,C).

The ozone-depleting potential (ODP) of a substance is governed by its chlorine content and its ability to reach the stratosphere. Compounds containing hydrogen are less likely to reach the stratosphere because they are oxidized by hydroxyl radicals in the troposphere (See Section II,B). CFC-11 and CFC-12 contain no hydrogen; they are virtually unreactive in the troposphere and thus pose the greatest threat to stratospheric ozone. Among chlorinated alkanes, 1,1,1-trichloroethane has ODP ~0.15 relative to CFC-11 (1.00), but most other high-volume chlorinated alkane solvents are more reactive and have negligible ODP. Suitable replacements for the "hard" CFC-11 and CFC-12 should have minimal chlorine content and sufficient hydrogen to make the substance reactive with tropospheric OH. Replacements for CFC-11 in foams include CFC-22 (CHF_2Cl), CFC-123 (CF_3CHCl_2), and CFC-141b (CH_3CFCl_2), all of which have ODP 0.02–0.05. CFC-134a (CF_3CH_2F), CFC-125 (CF_3CHF_2), and CFC-152 (CH_3CHF_2) are chlorine-free replacements for CFC-12 as refrigerants and hence have zero ODP.

Halons are brominated CFCs used principally in firefighting equipment for electrical installations: examples are H-1301 (CF_3Br) and H-1211 (CF_2ClBr). Their firefighting properties depend on the presence of weak C–Br bonds, which cleave at the high temperature of a fire and interrupt the chain reactions that support combustion. Like CFCs, halons are unreactive in the troposphere and have high ODP because they photolyze in the stratosphere to give bromine atoms that catalyze the destruction of ozone. CF_3I is a potential replacement for halons; it undergoes photochemical C–I bond cleavage in the troposphere and hence does not reach the stratosphere. Halons and the "hard" CFCs are scheduled for elimination under the terms of the "Montreal Protocol."

3. Chlorinated Pesticides

Many early insecticides were highly chlorinated organic compounds such as DDT, Aldrin, Mirex, and Toxaphene.

DDT (di-p-chlorophenyl-trichloroethane)

Although these substances were and still are inexpensive and more effective than previous pesticides, they suffer from low selectivity between target and nontarget organisms, persistence in the environment due to low chemical reactivity, and lipophilicity leading to bioconcentration and biomagnification. For example, a 1970s study on DDT in the Great Lakes showed concentrations of only 0.03 ppb in the water, but 300,000 ppb in the tissue of herring gulls at the top of the food chain.

Overexposure of mammals and humans to organochlorines causes liver lesions and long-lasting central nervous system symptoms such as aching joints, lethargy, and tremors, probably because the lipophilic organochlorines dissolve in the myelin sheaths surrounding the nerves. In predator birds, poor reproductive success in the 1960s was correlated with high body and egg burdens of organochlorines. Accordingly, the use of these insecticides was banned in North America and Europe around 1970, and since then levels of DDT, Aldrin, and others in fresh water and in wildlife have declined substantially. However, the efficacy and low cost of some organochlorines, for example of DDT against malaria mosquitos, has led to their continued use in some developing countries. Levels of DDT in fish taken from the North American Great Lakes reached a minimum in 1985 and have risen slightly since, presumably through atmospheric transport from other parts of the world. Modern pesticides such as organophosphates are more selective than organochlorines, nonbioaccumulative, and nonpersistent. However, their higher aqueous solubility makes them more mobile in the environment and potential water pollutants.

4. Halogenated Aromatic Pollutants

Halogenated aromatic pollutants include the polychlorinated biphenyls (PCBs), polychlorinated dibenzo-*p*-dioxins (PCDDs), and polychlorinated dibenzofurans (PCDFs). All these are complex families of pollutants; for example, the PCB family contains 209 "congeners" (that is, members differing in the number and/or arrangement of chlorine atoms on the biphenyl nucleus). Likewise, there are 75 PCDD congeners and 135 PCDF congeners. In most cases, all these families of halogenated aromatic pollutants occur in the environment as complex mixtures of many congeners, which makes their analysis extremely complex. All these families are highly lipophilic (with values of K_{ow} ranging to 10^6 and higher) and chemically unreactive, as a result of which they are persistent in the environment and prone to bioaccumulation in biota.

PCBs were manufactured between about 1930 and 1970 for use as dielectric fluids in electrical transformers, for which their electrical and thermal insulation and high chemical and thermal stability were desirable properties. In the late 1960s they were found to be ubiquitous in the environment, including locations as remote as the Arctic and Antarctic, where they migrate by atmospheric transport and deposition. Although electrical transformers are "closed" uses, PCBs have entered the environment due to careless disposal of discarded electrical equipment, and a trend in the latter years of their production to "open" uses such as plasticizers and deinking fluids. In most jurisdictions, PCBs may legally be drained from transformers and reused, even though no new PCBs may be manufactured.

PCDDs and PCDFs have no known uses, but are formed as trace contaminants in processes such as incineration and other combustion processes, as by-products in the manufacture of chlorinated phenols and their derivatives such as phenoxy herbicides, and as by-products of chlorine bleaching of wood pulp. In all of these reactions the actual yields of PCDDs and PCDFs are extremely low. The pattern of PCDD and PCDF congeners in air and sediment samples varies with the source. Incineration yields predominantly the less toxic hepta- and octachloro congeners, whereas the much more toxic 2,3,7,8-tetrachlorodibenzo-*p*-dioxin (TCDD) predominates in products derived from 2,4,5-trichlorophenol or the bleaching of paper. TCDD was a contaminant in the manufacture of the trichlorophenol derivative 2,4,5,-T (2,4,5-trichlorophenoxyacetic acid and its salts and esters), which was used as a brush killer and also as the chemical defoliant "Agent Orange" in the Vietnam War. Contamination by TCDD was responsible for withdrawing the registration of 2,4,5-T in North America. Incidents of pollution involving TCDD include the contamination of Times Beach, Missouri as a result of using "still bottoms" from the manufacture of 2,4,5-trichlorophenol for dust control on roadways, and the contamination of the town of Seveso, Italy following an explosion in 1976 of a 2,4,5-trichlorophenol manufacturing facility.

Structure–activity relationships show the most toxic PCDDs and PCDF congeners to be fully substituted with chlorine in the "lateral" (2,3,7, and 8) positions. Numerous toxic end points are seen in experimental animals treated with TCDD, including lethality at μg/kg doses via a wasting syndrome, carcinogenicity, teratogenicity, immunotoxicity, hepatic and dermal lesions, and reproductive complications, although not all symptoms are manifested in all species. In humans, the

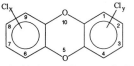

Polychlorinated dibenzo-*p*-dioxin

Polychlorinated biphenyl

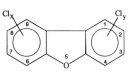

Polychlorinated dibenzofuran

2,3,7,8-Tetrachloro dibenzo-*p*-dioxin

perisistent skin lesion chloracne is well established; the question of whether TCDD is a human carcinogen upon long-term exposure remains controversial due to difficulties in eliminating confounding variables; small numbers of rare cancers have been seen recently in persons exposed at Seveso, Italy in 1976. At the time of writing, concern over the possible effects of TCDD as an environmental estrogen appears to be emerging as an issue as significant as the possible carcinogenicity of TCDD.

Because PCDDs and PCDFs frequently occur in the environment as complex mixtures rather than as single congeners, the toxic potential of a mixture is frequently estimated as TCDD equivalent concentration (TEQ) through the application of consensus "toxic equivalency factors (TEFs)," which attempt to assess the toxicity of a PCDD or PCDF congener relative to TCDD as a reference toxicant.

The toxicity of TCDD is initiated by its association with an intracellular protein known as the Ah receptor, which has a molar mass of about $300,000 \ g \ mol^{-1}$. It binds remarkably strongly to TCDD, following which it loses a subunit and is carried into the nucleus by a transporter protein. There it associates with DNA binding sites, triggering the production of a series of messenger RNA molecules and, ultimately, the toxic responses.

Human poisonings involving PCBs include the Yusho (Japan) and Yu-Ching (Taiwan) episodes in which cooking oil became inadvertently contaminated by PCBs. Symptoms of toxicity included severe chloracne and various liver lesions, as a result of which PCBs required the reputation of extreme toxicity. Later work showed that most of the toxicity of these PCB formulations was due to small amounts of PCDFs that had formed through long-term exposure of the PCBs to air. Among PCB congeners, those that contain chlorine substituents in the *ortho* positions are relatively nontoxic, whereas those chlorinated only *meta* and *para* have significant TCDD-like toxicity. The structure–activity relationship reveals that the more toxic PCBs can adopt a "coplanar" conformation whose shape resembles that of TCDD, whereas *o*-chlorinated PCBs are forced to adopt a nonplanar conformation on account of steric repulsion between *ortho* substituents.

PCBs are typically present at 1000 times the concentrations of PCDDs in the tissues of mammals and aquatic birds; although the "coplanar" PCBs comprise only a small proportion of the total, their TEFs are such that they may contribute a substantial fraction of the dioxinlike activity. The concentrations of organochlorines in human milk have declined over the period 1972–1985 in Sweden, but correlate linearly with the total fat content, reflecting their high lipophilicity. Organochlorine levels in the milk decline with the number of children nursed, that is, transport into the milk appears to deplete the reservoirs in the mother's body. The chief routes of exposure to adult humans are through meat and dairy products, total exposures being on the order of $1-4 \ pg \ kg^{-1} \ day^{-1}$ for dioxinlike compounds. Controversy exists concerning whether or not this exposure is above a "safe" level.

Fires involving PCBs either in use or in storage have been the cause of major environmental contamination, partly as a result of the formation of PCDFs through partial oxidation of PCBs. Intentional destruction of PCBs requires vigorous chemical conditions, such as combustion to CO_2, H_2O, and HCl in the presence of a large amount of fuel (PCBs do not burn by themselves); above 1000°C, destruction can be accomplished at "six nines" (99.9999%) efficiency, without significant PCDD/PCDF formation. In some jurisdictions, PCB destruction may be carried out within cement kilns, where the CaO formed during calcination acts as a trap for HCl. An alternative destruction technology exists for mineral oil-based transformer fluids that are contaminated with low levels of PCBs; these may be treated with reactive metals such as sodium or potassium, with the formation of the metal chloride and a polymer composed of the dechlorinated PCB. The recovered mineral oil can than be reused.

H. Metals

Metals of environmental interest includes elements that are macronutrients in the biosphere (Na, K, Ca) or micronutrients (Fe, Cu, Zn, etc.), and some that have no known biological funtion (e.g., Cd, As, Pb, Hg) and that are generally regarded as toxic. In the case of environmental pollution by metals, it is the element itself that is toxic, even though speciation may afford chemical forms of differing toxicity. For example, inorganic compounds of lead and mercury are less toxic than organomercurials and organoleads, but the opposite situation exists for arsenic, which enters the environment through burning coal and oil, in which it is a trace element, from mining operations, and from smelting, especially of copper. The order of toxicity is organoarsenics < arsenic (V), which predominates under aerobic conditions, < arsenic (III), which is formed by reduction of arsenic (V) in sediments. Assay methods for metals must therefore be able to differentiate the various chemical species that may be present rather than measuring the total concentration, for example, by atomic absorption spectroscopy. Additionally, there can be no "ultimate destruction" treatment for metals corresponding to incineration of chlorinated organics.

Many metallic elements are cumulative toxicants because they are excreted slowly. Limits on exposure through food and drinking water are set to ensure that continued daily exposure does not build up to a toxic dose. Where release of toxic metals into the environment is inevitable, disposal should involve chemical forms that are relatively immobile in order to minimize their dispersal and consequent contamination of groundwater.

I. Iron

In the oxygen-rich environment of planet Earth, almost all iron is present as insoluble Fe(III) oxides (Fe_2O_3 and Fe_3O_4, which are exploited commercially, and are biologically unavailable because Fe^{3+} is significantly soluble in water only below about pH 3.5). Iron also occurs as insoluble Fe(II) sulfides such as pyrite, FeS_2. Section II,F describes the role of iron in acid mine drainage.

Careful husbandry of iron stores is essential in the biosphere. Certain plants and microorganisms produce macrocyclic organic ligands called siderophores whose association constants for complexation with Fe^{3+} are so large that they can capture trace concentrations of Fe^{3+}(aq) from soil. Mammals have a complex storage and recycling system for the iron released from spent red blood cells; humans have a total body content of about 2 g of iron, but conserve this so strongly that only 1 mg is excreted daily, and excessive consumption of iron can be toxic.

2. Mercury

Mercury is a rare element, which occurs as HgS, cinnabar, or as the free metal. The metal is obtained by heating the sulfide in air (See Section II,F). Significant amounts of mercury reach the environment through burning coal and smelting metals such as lead, copper, and zinc, whose sulfide ores are often associated with mercury.

Mercury poses an environmental problem because its boiling point is relatively low (357°C) and its vapor pressure is significant even at room temperature, so that care must be taken to provide good ventilation in industries that use mercury [the expression "mad as a hatter" originated in the felt hat trade, in which significant worker exposure to $Hg(NO_3)_2$ occurred up to the 1940s]. Disposal of spent mercury batteries, which are used in watches and cameras, is a significant source of environmental mercury, especially in communities that incinerate garbage, because the mercury passes out with the flue gases.

A major complicating factor in the environmental chemistry of mercury is biological methylation, which converts inorganic mercury to lipophilic (and more toxic) forms that can cross the blood–brain barrier and that are responsible for neurological symptoms such as quarrelsome behavior, headache and depression, and muscle tremors. Studies with mice showed the unborn fetus to be particularly susceptible, with fetotoxicity and teratogenicity seen at <1 ppm of CH_3HgCl in the mothers' diet. Hg^{2+} is methylated to CH_3Hg^+ and $(CH_3)_2Hg$ in sediments by the attack of (microbial) methylcobalamin (vitamin B_{12}), which contains a nucleophilic CH_3-Co group that is transferred to electrophilic Hg^{2+}. Environmental CH_3Hg^+ occurs mostly as CH_3HgCl (CH_3HgSCH_3 is found in shellfish). CH_3Hg^+ derivatives predominate below pH 7; further methylation to $(CH_3)_2Hg$ occurs only at higher pH.

A significant past source of environmental contamination by mercury has been the chloralkali industry. When mercury cells were used, contamination of receiving waters by mercury was likely. Modern chloralkali plants use fluorocarbon membrane cells and hence are mercury-free. Data on losses of mercury from Swedish chloralkali plants using mercury cells (in 1970, 200 g ton^{-1} of chlorine; in 1975, 1–2 g; in 1980, 0.15 g) show the improvements due to prevention of physical losses into effluent waters. In some locations, past contamination was so great that pools of liquid mercury could be found in river sediments. Dredging would be prohibitively expensive and would disturb the river sediments, thereby resuspending the mercury in the biotic zone. Doing nothing allows the mercury to become ever more deeply buried in the sediment, in effect returning the mercury to the environment from which it was originally extracted.

The largest episode of mass poisoning due to mercury occurred in the 1950s in the Japanese fishing village of Minamata, where about 1300 people were afflicted with symptoms of mercury poisoning such as anorexia, irritability, and other psychiatric disorders, of whom about 200 died. The affected residents lived close to a chemical plant that used a mercury catalyst; they ate large amounts of local fish and shellfish, which were found to be contaminated with mercury. The mercury intoxication showed a dose–response behavior, with the most severe symptoms correlating with proximity to the chemical plant and with body burdens of mercury, which were monitored by analysis of the patients' hair for CH_3HgCl.

3. Environmental Regulation of Mercury

Clean air contains less than 10 ng m^{-3} of mercury, mostly from volcanic activity. Elemental mercury is mainly hazardous as the vapor; typical workplace air quality standards are 0.05 mg m^{-3} for inorganic mercury and 0.01 mg m^{-3} for the lipophilic and bioconcentratable organic forms of mercury. Standards for mercury in drinking water are typically 1.0 μg liter^{-1} (1 ppb). The Great Lakes joint water quality agreement between Canada and the United States sets a target of 0.2 μg of mercury per liter of unfiltered lake water. In Canada, fish should not be eaten if their mercury content exceeds 0.5 ppm.

4. Lead

Lead occurs in nature mostly as the sulfide, galena (PbS). Lead and zinc sulfides are commonly found and mined together. Lead has been used as a metal since the times of the Egyptians and the Babylonians. The Romans employed lead (plumbum) extensively for conveying water; there is speculation that a contributing factor to the decline of the Roman Empire was subclinical or sublethal lead poisoning among the ruling class. Through the Middle Ages and beyond, the malleability of lead and its resistance to corrosion encouraged its use as a roofing material for important public buildings such as the great cathedrals of Europe. Present-day production of lead is in the millions of tonnes annually; uses include the lead–acid storage battery and the addition of organolead compounds to gasoline as antiknock agents. Despite recycling, used car batteries are the major source of lead in municipal waste and cause local lead pollution when the old electrodes are redistilled (lead has a m.p. of 327°C and b.p. of 1740°C) to recover the metal. Smaller-scale uses of lead are in solder, leaded glass for ornamental purposes, and as a shielding material for radioactive sources. Compounds of lead are widely used as coatings: "red lead" (Pb_3O_4), used to undercoat steel; lead chromate ($PbCrO_4$), used in the paint on North American school buses; and "white lead" [$2PbCO_3 \cdot Pb(OH)_2$], which was formerly used as the pigment base for household paints (now replaced by zinc oxide and titanium dioxide). These varied uses explain why lead is widely dispersed in the environment.

Lead levels in humans are frequently 10% of toxic levels. Some authors have argued on the basis of analysis of ancient bones and ancient ice cores that relatively high levels of lead have always existed in the environment. Others claim that the apparent high lead content of ancient samples is the result of inadvertent contamination of the samples during their collection and analysis.

5. Toxicology of Lead

Lead causes neurological problems, including irritability, sleeplessness, and irrational behavior. The appetite is depressed to the extent that death can ensue due to starvation. Organolead compounds are more toxic than simple lead salts because they are lipophilic and can cross the blood–brain barrier. Both the fraction of lead absorbed from the intestinal tract and the proportion of organolead found to cross the blood–brain barrier increase in the order fetus > child > adult. Children exposed to lead, for example, by chewing objects covered with old paint, can suffer mental retardation, lower performance on I.Q. tests, and hyperactivity. The similar ionic radii of Pb^{2+} and Ca^{2+} suggest that lead can substitute chemically for calcium, and hence that lack of dietary calcium may predispose children to lead toxicity.

The half-life of lead in humans is estimated to be about 6 yr (whole body) and about 15–20 years (skeletal). Skeletal burdens of lead increase almost linearly with age, suggesting that the steady state with respect to uptake and excretion of lead is not normally reached. For patients who are clinically affected, chelation of Pb^{2+} with ethylenediaminetetraacetic acid (EDTA) is beneficial in reducing body burdens of lead.

6. Lead in Gasoline

Tetraethyllead (TEL) promotes smooth combustion of gasoline by decomposing evenly into ethyl free radicals, which act as free radical initiators for smooth fuel combustion:

$$Pb(C_2H_5)_4 \xrightarrow{\text{heat}} Pb(g) + 4\,C_2H_5$$

Lead emissions occur during both the manufacture of TEL and from cars using leaded gasoline. Lead atoms are formed when TEL decomposes in the car engine and are scavenged by the addition of halogenated ethanes to the gasoline, which convert the lead to inorganic lead halides. These pass out with the exhaust gases and hence prevent the condensation of lead in the engine.

Lead emissions from vehicles are a major source of environmental contamination by lead, and the nature of the source ensures wide despersion of the pollutant. Canadian reports indicate that 70% of lead emissions to the environment in the 1970s were due to leaded gasoline, this proportion falling as

leaded gasoline was gradually withdrawn. Reasons for the removal of lead from gasoline include both reduction of lead emissions and the incompatibility of TEL with the catalytic converters used to counter NO_x and hydrocarbon emissions (See Section II,D). Leaded gasoline is now unavailable in Canada and the United States, but is still used as an octane-enhancer in many parts of the world.

Glossary

Acid mine drainage Acidified water seeping from mineworkings or tailings that contain sulfides. Microbial oxidation in the presence of air converts these materials to sulfuric acid, thereby acidifying streams and increasing the concentrations of toxic metals in the water.

Acid rain Rainfall having acidity greater than (or pH less than) that of raindrops in equilibrium with atmospheric carbon dioxide (pH 5.6).

Bioconcentration Tendency of a substance to accumulate in biota. The bioconcentration factor is the ratio of the concentration of the substance in the biota to the concentration in the water, for aquatic species, or in the diet, for humans and other terrestrial species.

Eutrophication Accelerated aging of lakes caused by excessive quantities of nutrients, such as phosphates, increasing the rate of biomass production.

Octanol–water coefficient, K_{ow} Partition coefficient for the distribution of a solute between water and octanol. Octanol is used as a model solvent for biological lipid materials.

Residence time Time spent by a substance in a given compartment of the environment; defined in terms of the ratio of the amount of substance in the reservoir to the total rate of inflow to, or outflow from, the reservoir.

Structure–activity relationship Empirical correlation showing a trend between the molecular structures of a group of substances and their chemical, biological, or toxicological activities.

Xenobiotic Term to describe a substance not of natural origin; a synthetic or human-made substance.

Bibliography

Bunce, N. J. (1993). "Introduction to Environmental Chemisty." Winnipeg, Manitoba, Canada: Wuerz Publishing.

Bunce, N. J. (1994). "Environmental Chemistry," 2nd ed. Winnipeg, Manitoba, Canada: Wuerz Publishing.

Cunningham, W. P., and Saigo, B. W. (1990). "Environmental Science, A Global Concern" Dubuque, Iowa: W. C. Brown.

Finlayson-Pitts, B. J., and Pitts, Jr, J. N. (1986) "Atmospheric Chemistry. New York: Wiley-Interscience.

Francis, B. M. (1994). "Toxic Substances in the Environment." New York: Wiley–Interscience.

Manahan, S. E. (1991). "Environmental Chemistry," 5th ed. Chelsea, Mich.: Lewis Publishers.

U.S. National Research Council, Committee on Tropospheric Ozone Formation and Measurement (1991). "Rethinking the ozone problem in urban and regional air pollution," National Washington, DC: Academy Press.

Wayne, R. P. (1991). "Chemistry of Atmospheres," 2nd edition. Oxford, England: Oxford University Press.

Atmosphere–Terrestrial Ecosystem Modeling

Roger A. Pielke

Colorado State University

I. Atmospheric Modeling
II. Terrestrial–Ecosystem Modeling
III. Linkages between the Atmosphere and
Terrestrial Ecosystem
IV. Family of Modeling Solutions
V. Field Programs
VI. Conclusions

Terrestrial–ecosystem–atmospheric interactions refer to exchanges of heat, moisture, momentum, and trace gases and aerosols between vegetated land surfaces and the overlying air. These feedbacks represent a dynamic coupled system that would be expected to evolve differently as a result of the interaction between the two mediums. This article summarizes these interactions and discusses why they are important.

I. ATMOSPHERIC MODELING

Modeling of meteorological flows requires the use of conservation equations for fluid velocity, heat, mass of dry air, water substance in its three phases, and mass of other natural and anthropogenic atmospheric constituents. Spatial scales of simulation have ranged from high-resolution representations of the boundary layer where grid increments are on the order of tens of meters or less, to general circulation representations of the entire globe.

The characterization of biospheric processes in these models, however, has been limited to simple representations where most aspects of the soil and vegetation are prescribed. Stomatal conductance responds to atmospheric inputs of solar radiation, air temperature, air relative humidity, precipitation, and air carbon dioxide concentration, and to soil temperature and moisture. Up to the present, in meteorological models, these are the only meteorological variables to which the vegetation and the soil dynamically respond. A schematic of how the atmosphere is influenced by vegetation, as contrasted with bare soil, is shown in Fig. 1.

The names of these soil–vegetation parameterizations include BATS (Biosphere-Atmosphere-Transfer Scheme), SiB (Simple Biosphere Scheme), and LEAF (Land-Ecosystem-Atmosphere Feedback Scheme).

An illustration of the form in which these modeling components are used in atmospheric models is shown in Fig. 2. A recent version of SiB has the additional dynamic feedback between carbon as-

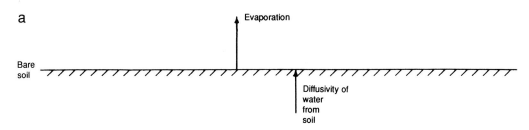

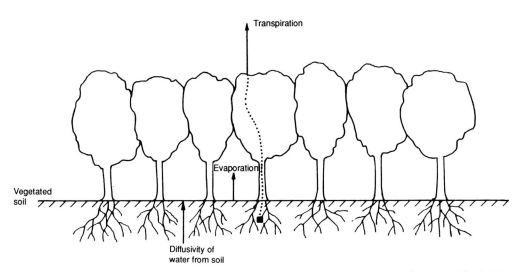

FIGURE 1   (a) Schematic illustration of the surface moisture budget over bare soil and vegetated land. The roughness of the surface (and for the vegetation its displacement height) will influence the magnitude of the moisture flux. Dew and frost formation and removal, and precipitation, will also influence the moisture budget. (b) Schematic illustration of the surface heat budget over bare soil and vegetated land. The roughness of the surface (and for the vegetation its displacement height) will influence the magnitude of the heat flux. Dew and frost formation and removal, and precipitation, will also influence the heat budget. [From R. A. Pielke and R. Avissar (1990). *Landscape Ecology* **4,** 133–155.]

similation in the vegetation and water vapor loss to the atmosphere. The concept is that plants have evolved such that they attempt to optimize maximum CO_2 intake with minimal H_2O loss. [*See* PLANT ECOPHYSIOLOGY.]

The utilization of these soil–vegetation schemes has been most extensive in general circulation models and in mesoscale models. In general, only one vegetation and soil characterization is used for each horizontal model grid interval. Such a representation, based on averaged conditions, is presumed to represent the net effect of the landscape within that grid cell. A newly considered approach is to use a *mosaic* formulation where subregions within a model grid are evaluated separately and the resultant grid-averaged heat, moisture, and momentum fluxes are obtained by a fractional weighting of the subregion fluxes.

It has been shown that heterogeneous landscape patterns, such as between irrigated and nonirrigated land, snow versus snow-free ground, etc., can produce atmospheric circulations that are as strong as sea breezes that develop as a result of the differential heating between land and lakes and oceans. The resultant fluxes of energy and trace gases into the atmosphere due to these mesoscale circulations have yet to be adequately considered in larger-scale atmospheric models.

II. TERRESTRIAL–ECOSYSTEM MODELING

The modeling of terrestrial ecosystems involves the short-term response of vegetation and soils

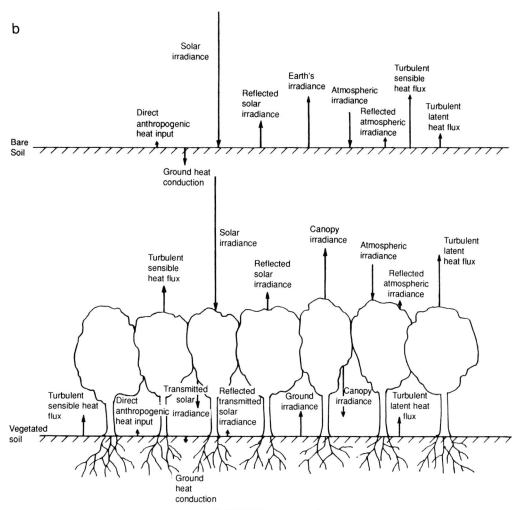

FIGURE 1   *Continued*

to atmospheric effects, and the longer-term evolution of species composition, biome dynamics, and nutrient cycling associated with landscape and soil structure changes. The assimilation of carbon resulting from the growth of vegetation, and its subsequent release during decay, has been a focus of these models. The modeling framework of these simulation tools has generally involved empirically based logic statements, which, though frequently based on fundamental biophysical concepts, are not expressed as differential equations. The spatial scales of these simulations have ranges from patch sizes to biome scales. These models, particularly when applied on the smaller spatial scales, include a stochastic component to represent unpredictable random inputs from the atmosphere

and interactions within the vegetation such as a falling tree, for example. [*See* ECOLOGICAL ENERGETICS OF ECOSYSTEMS.]

These models require atmospheric inputs such as temperature, relative humidity, net radiation, and precipitation to integrate their formulations forward in time. Output from nearby climatological stations have been used as the needed boundary conditions for these models when applied on the patch up to the regional scale. On the global scale, output from general circulation models has been used to estimate potential changes of biome type in response to hypothesized climate change scenarios using the concept of a Holdridge diagram. An example of a Holdridge diagram is shown in Fig. 3, where once annual characteristics

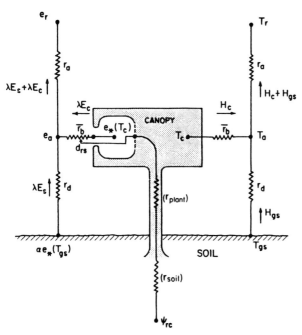

FIGURE 2 Framework of the Land Ecosystem Atmosphere Feedback (LEAF) model. The transfer pathways for latent and sensible heat flux are shown on the left- and right-hand sides of the diagram, respectively. The water vapor pressure is given by e and temperature is given by T. E refers to evaporation fluxes of water vapor, with H the heat flux. The quantities r and d refer to the resistance to water vapor flux and a parameter that is used to represent this flux, respectively. The resistance $\bar{r}_b$ is associated with flow between leaves and the air within the canopy, $\bar{r}_a$ is the resistance between the air within the canopy and the atmosphere above, and $\bar{r}_d$ is the resistance between the ground and the air within the canopy. The soil moisture availability is represented as a soil water potential, ψ. Subscripts r, a, c, g, and s refer to the air above the canopy, canopy air at its top, air within the canopy, the ground surface, and the soil surface. The saturation vapor pressure is expressed by e_* with α the fractional part of that water vapor level that is actually evaporated. The heat of vaporization is λ. Finally, in analogy to electrical diagrams, resistance is denoted by a zigzag line. [From T. J. Lee, R. A. Pielke, T. G. F. Kittel, and J. F. Weaver (1993). *In* "Environmental Modeling with GIS" (M. Goodchild, B. Parks, and L. T. Steyaert, eds.), pp. 108–122. Oxford University Press, Oxford, England.]

III. LINKAGES BETWEEN THE ATMOSPHERE AND TERRESTRIAL ECOSYSTEM

A. Short-Term Interactions

The vertical structure of the daytime atmospheric boundary layer is critically dependent on the partitioning of net radiation into sensible and latent turbulent heat flux and groundheat conduction. A deeper boundary layer, for example, results when more of this radiative energy is realized as sensible heat flux. When vegetation is present, the response of leaf conductance to atmospheric conditions represents a rapid feedback between the biosphere and the atmosphere. The passage of a cloud during daylight, for example, will significantly reduce net radiation with stoma apertures responding within minutes.

The drying of the near-surface soil and the depletion of deeper soil moisture as a result of transpiration represent another, somewhat slower feedback with the atmosphere that varies over days to a few weeks. When vegetation becomes water stressed, for instance, stoma will close to conserve the remaining water, so that a larger fraction of the net radiation is realized as sensible heat flux. Precipitation represents a short-term feedback that can quickly replenish the soil moisture, as well as provide shallow liquid water layers on the vegetation, which is referred to as *interception*.

B. Long-Term Interactions

Seasonal interactions include feedbacks between increases in biomass from spring into the summer that will modify the partitioning of latent and sensible heat fluxes. Nutrient limitations can also constrain biomass growth, particularly in moist environments. It has been shown that the increase of mean temperature in the spring in the eastern United States is interrupted as vegetation leafs out. Also, drought conditions over the eastern United States are apparently perpetuated by the reduced transpiration from the water-stressed vegetation.

Other important influences of the land surface on the atmosphere include the modification of the

of mean growing season temperature, mean annual precipitation, and mean annual evapotranspiration are known, one can estimate the biome of a region. More recent representations of equilibrium vegetation distributions are available, but cannot be presented graphically or succinctly as in this diagram.

An illustration of the framework of terrestrial-ecosystem models is shown in Fig. 4.

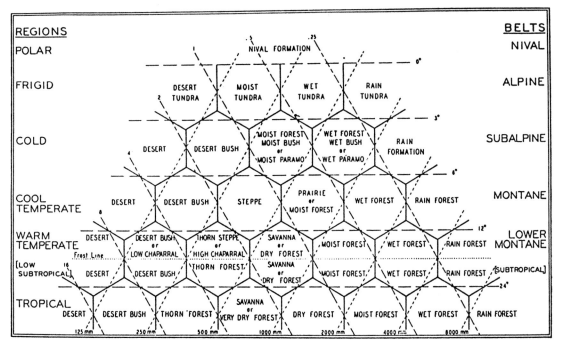

FIGURE 3 Chart that differentiates the vegetation of land areas of the world into 100 closely equivalent formations separated by mean growing season temperature (shown by the horizontal line), mean annual precipitation in mm (shown by the line that slopes up to the right), and the evaporation in terms of the number of times that the actual precipitation could be evaporated in one year (shown by the line that slopes up to the left). A positive value greater than one indicates that evapotranspiration exceeds precipitation. [From L. R. Holdridge (1947). *Science* **105**, 367–368.]

albedo of the surface. The albedo with respect to solar radiation has been shown to be a critical landscape–atmospheric interaction by which, for example, desertification of the Sahel region of Africa may have resulted from excessive grazing by domesticated goats of the darker vegetation such that a larger fraction of the solar radiation is reflected back into space. It has been suggested that the Rajastan Desert in India has resulted from the same mechanism. [*See* DESERTIFICATION, CAUSES AND PROCESSES.]

On time periods of years, species composition and soil characteristics, including nutrient turnover, can change in response to long-term atmospheric changes and through natural vegetation succession and human disturbances, as well as episodic events such as fire and severe windstorms. These terrestrial–ecosystem changes would be expected to then feed back to the atmospheric structure. Obvious prehistoric examples of these changes include the dynamic changes of the landscape as boreal and temperature biome regions

shifted poleward after the retreat of the maximum Pleistocene glaciation.

IV. FAMILY OF MODELING SOLUTIONS

The solution of nonlinear mathematical systems, such as represented by atmosphere–ecosystem models, can yield *equilibrium, periodic,* or *chaotic* solutions. In the past, ecologists have generally presumed that ecosystems would approach an equilibrium, which has been referred to as a *climax* state. Atmospheric scientists, in contrast, have recently recognized that the atmosphere is a chaotic system that is inherently unpredictable after a few weeks or so. On the longer time periods, from decades to hundreds of years to thousands of years, the available climate record based on such proxy data as pollen records, carbon radioisotope ratios, etc., also indicate a chaotic behavior. Such variability, however, is not yet represented by general circula-

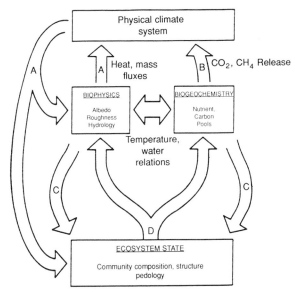

FIGURE 4 Important interactions between the vegetation and the atmosphere with respect to weather and climate. A, influence of changes in the physical weather and climate system on biophysical processes; these may feed back to the atmosphere through changes in energy, heat, water, and carbon dioxide exchange; B, change in nutrient cycling rates; release of carbon dioxide and methane from the soil carbon pool back to the atmosphere; C, changes in biogeochemical processes and water and nutrient availability influence community composition and structure; D, change in species composition results in changes in surface biophysical characteristics and biogeochemical process rates. (Adapted from Experiment Plan, Boreal Ecosystem-Atmosphere Study, July 1993, Version 1.0, pp. 1–3.)

tion modeling simulations of human-caused climate change scenarios (e.g., as shown by the monotonic response of global temperatures in climate models to increased greenhouse gases), perhaps as a result, at least in part of the neglect of atmosphere–terrestrial ecosystem interactions.

There are few existing coupled atmospheric–ecosystem models. The *Gaia hypothesis* had been proposed as a mechanism whereby biospheric–atmosphere interactions act to regulate climate within a narrow range of conditions. An idealized model referred to as *Daisyworld,* has been used to illustrate the Gaia concept. In that model, the preferential growth of white daisies when solar luminosity is elevated and black daisies when solar luminosity falls results in a stable equilibrium climate. However, if the interaction between the luminosity and the germination of daisies is sufficiently nonlinear, a chaotic response will occur with temperatures

varying unpredictably over a wide range (Fig. 5). Although this is a grossly simplified model, it does suggest that the nonlinear feedbacks between the atmosphere and terrestrial ecosystems do have the potential for a significant nonequilibrium response.

V. FIELD PROGRAMS

Recently, there have been several field programs that are designed in part to explore short-term (out to seasonal scale) atmosphere–terrestrial ecosystem interactions. These include the First ISLSCP Field Experiment in eastern Kansas (FIFE), the Hydrologic Atmospheric Pilot Experiment and Modélisation du Bilan Hydrique (HAPEX-MOBILHY) in southwest France, the HAPEX-Sahel in Niger in central Africa, the Anglo-Brazilian Amazonian Climate Observation Study (ABRACOS) in the Amazon region of Brazil, and the Boreal Ecosystem-Atmosphere Study (BOREAS) in Manitoba and Saskatchewan in Canada. A goal of these programs is to explore atmospheric–terres-

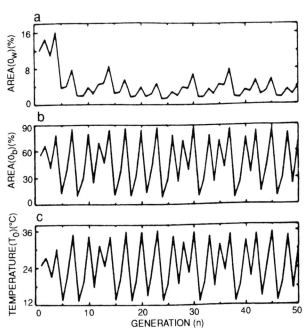

FIGURE 5 Chaotic behavior of the area of white daisies, black daisies, and temperature as a function of the generation of the daisies. [From X. Zeng, R. A. Pielke, and R. Eykholt (1990). *Tellus* **42B,** 309–318.]

trial ecological interactions for different major biome types.

VI. CONCLUSIONS

Although land covers only about 29% of the earth, and not all land is significantly covered by vegetation, the 5.6×10^{14} kg of carbon stored in terrestrial vegetation, with a net primary production of 6.0×10^{13} kg of carbon per year, result in a strong dynamic interface between the atmosphere and the land surface. The relative neglect of this interactive system by the scientific community is somewhat surprising because regional areas in the ocean (e.g., the El Niño) have been shown to influence global weather patterns. Thus we should expect that this topic will gain increasing attention, as its importance to local, regional, and global climate issues becomes known.

Glossary

Albedo Percentage reflection of radiation.
Biomass Mass of plant material.
Biome A major ecological community type.
Boundary layer Depth in the atmosphere to which direct exchanges of mass occur with the surface air.
Chaos Property of a system describing how, as a result of nonlinear interactions, the system becomes unpredictable after a finite period of time.
Net radiation Sum of the downward solar and atmospheric irradiance minus the reflected solar irradiance and the emitted land surface radiation.

Bibliography

André, J.-C., Bougeault, P., Mahfouf, J.-F., Mascart, P., Noilhan, J., and Pinty, J.-P. (1989). Impact of forests on mesoscale meterorology. *Philos. Trans. Roy. Soc. London Ser. B* **324**, 407–422.

Avissar, R., and Verstraete, M. M. (1990). The representation of continental surface processes in atmospheric models. *Rev. Geophys.* **28**, 35–52.

Bradley, R. S. (1985). "Quaternary Paleoclimatology: Methods of Paleoclimatic Reconstruction." Boston: Allen & Unwin.

Cotton, W. R., and Pielke, R. A. (1992). "Human Impacts on Weather and Climate," Geophysical Science Series Vol. 2. Fort Collins, Colo.: ASTeR Press.

Dickinson, R. E., Henderson-Sellers, A., Kennedy, P. J., and Wilson, M. F. (1986). "Biosphere–Atmosphere Transfer Scheme for the NCAR Community Climate Model," NCAR Technical Note, NCAR/TN–275 + STR. Boulder, Colo.: NCAR.

Henderson-Sellers, A. (1993). The project for intercomparison of land-surface parameterization schemes. *Bull. Amer. Meteor. Soc.* **74**, 1335–1350.

Lee, T. J., Pielke, R. A., Kittel, T. G. F., and Weaver, J. F. (1993). Atmospheric modeling and its spatial representation of land surface characteristics. *In* "Environmental Modeling with GIS" (M. Goodchild, B. Parks, and L. T. Steyaert, eds.), pp. 108–122. Oxford, England: Oxford University Press.

Lovelock, J. E. (1988). "The Ages of Gaia." New York: W. W. Norton.

Nielson, R. P. (1993). Vegetation redistribution: A possible biosphere source of CO_2 during climatic change. *Water, Air, and Soil Poll.* **70**, 659–673.

Pielke, R. A., Schimel, D. S., Lee, T. J., Kittel, T. G. F., and Zeng, X. (1993). Atmosphere–terrestrial ecosystem interactions: Implications for coupled modeling. *Ecological Modelling* **67**, 5–18.

Segal, M., and Arritt, R. W. (1992). Nonclassical mesoscale circulations caused by surface sensible heat flux gradients. *Bull. Amer. Meteor. Soc.* **73**, 1593–1604.

Sellers, P. J., Hall, F. G., Asrar, G., Strebel, D. E., and Murrphy, R. E. (1992). An overview of the First International Statellite Land Surface Climatology Project (ISLSCP) Field Experiment (FIFE). *J. Geophys. Res.* **97**, 18345–18371.

Atmospheric Ozone and the Biological Impact of Solar Ultraviolet Radiation

Dan Lubin and Osmund Holm-Hansen
University of California

The spectral distribution of solar electromagnetic radiation reaching the surface of the Earth is controlled by absorption and scattering of radiant energy by a variety of elements and compounds in the Earth's atmosphere. During the past two decades there has been a recurring concern about the integrity of the Earth's ozone layer, which is the primary regulator of solar ultraviolet-B (UV-B) radiation at the Earth's surface. Because UV-B radiation can be damaging to biological systems, there is widespread concern that ozone depletion might be accompanied by numerous adverse impacts to both ecology and human health. The concentration of ozone in the Earth's atmosphere shows considerable variability, on both a seasonal and latitudinal basis. Unusually low levels of ozone, however, have been recorded during the past three decades over Antarctica during October and November of each year, and there are indications that similar chemical reactions leading to a decrease in ozone levels may also occur in the northern hemisphere, but

with less severity. It is believed that the injection of anthropogenic compounds into the stratosphere, particularly chlorofluorocarbons (CFCs), is responsible for this seasonal loss of ozone. Although a decrease in stratospheric ozone abundance implies a theoretical increase in UV-B at the Earth's surface, there are several lower atmosphere phenomena that, if they result in an upward trend in tropospheric opacity, could offset a UV-B increase. The relationship between decreasing ozone abundance and increasing UV-B has yet to be established on a global basis by actual radiation measurement. An even more important (but also more complicated) subject is the role of solar ultraviolet radiation in biological processes. This must be considered at the cellular level, the physiological level, and in terms of the ecosystem as a whole. Once these roles are defined, they must be considered in conjunction with the global climatology of ultraviolet radiation to properly assess any impact of stratospheric ozone depletion.

I. ATMOSPHERIC OZONE

A. Ultraviolet Radiation from the Sun

To understand the existence of the ozone layer we must understand how solar energy interacts with the Earth's atmosphere. The rate of solar energy reaching the top of the atmosphere, adjusted to the mean Earth–Sun distance, is 1368 watts per square meter, and this is known as the *solar constant*. The annual variability in solar energy reaching the Earth's atmosphere is approximately 6.9%, due to the Earth's elliptical orbit about the Sun. There are also small variations in the solar constant associated with the 11-year sunspot cycle and the 27-day solar rotation period, and these affect primarily the shortest wavelengths of light emitted by the Sun. Solar energy propagates in the form of electromagnetic waves, and the vast majority of the Sun's energy has a wavelength less than 4000 nanometers (nm, 10^{-9} m). Approximately 40% of the Sun's energy lies within the wavelength range 400–700 nm, which is referred to as the *visible* part of the electromagnetic spectrum because it is detectable by the retina of the human eye. The wavelength range 700–4000 nm contains 51% of the Sun's energy and is referred to as the *near infrared*. Only about 9% of the Sun's energy output occurs at wavelengths shorter than 400 nm, and most of this is referred to as the *ultraviolet* part of the solar spectrum.

The ultraviolet spectrum is usually divided into several bands: the *UV-A* (315–400 nm), *UV-B* (280–315 nm, although some researchers prefer the definition 280–320 nm), *UV-C* (200–280 nm), the *far UV* (120–200 nm), and the *extreme UV* (10–120 nm). These divisions are not arbitrary; each band plays a relatively distinct role in both atmospheric and biological processes. For all practical purposes, the far and extreme UV are relevant only to the Earth's upper atmosphere. UV-C is the major driving force behind the creation of ozone in the stratosphere, but it does not reach the Earth's surface in any measurable amount. UV-B and UV-A are the only ultraviolet-wavelength regions relevant to biological processes at the Earth's surface.

B. The Earth's Ozone Layer

The Earth's lower atmosphere is divided into two distinct layers. The *troposphere* extends from the surface up to an average of 10 km (as high as 15 km in the tropics and only 7–8 km over Antarctica). Because of the exponential decrease in atmospheric density and pressure with height, about 80% of the atmosphere's mass lies in the troposphere. On a global basis, air within the troposphere is in nearly constant vertical motion (due ultimately to the lapse rates in air temperature with height), and hence the troposphere is where most observable weather occurs. The *stratosphere,* so called because of its stratified or vertically stable nature, extends from approximately 10 to 50 km. Air temperature in the stratosphere increases with height due to absorption of solar energy by ozone, and this explains the vertical stability. Above these two lower layers lie the *mesosphere* (50 to 80 km) and *thermosphere* (80 to 100 km), whose primary role in governing solar radiation is to absorb sunlight at the shortest wavelengths (generally 200 nm and shorter).

By mass, 75.5% of the Earth's atmosphere is nitrogen and 23.1% is oxygen. At atmospheric temperatures, both major gases generally exist as diatomic molecules (O_2 and N_2). The important exception is when an oxygen molecule is struck by a solar photon (fundamental quantum unit of light) having a wavelength shorter than 242 nm. The shorter the wavelength, the more energy an individual photon possesses: a photon having wavelength shorter than 242 nm imparts enough energy to the oxygen molecule to break it apart into two oxygen atoms. This process is known as *photodissociation:* its result is to make oxygen atoms available to recombine with oxygen molecules and form ozone (O_3). The term photodissociation applies not only to atmospheric oxygen but to the light-induced breakup of any molecule. Ozone is photodissociated by sunlight having wavelengths less than 1100 nm, although the vast majority of ozone photodissociation occurs at UV-B and longer UV-C wavelengths. When photodissociation occurs the photon is lost, as it has imparted its energy to the molecule. This process is therefore also known as *absorption,* whereby the solar energy has been absorbed by the molecule and transformed into the kinetic energy of the two constituent atoms. In an ideal "pure oxygen" atmosphere, an equilibrium ozone concentration is maintained by the suite of chemical reactions

$$O_2 + h\nu \rightarrow O + O \quad (\lambda < 242 \text{ nm})$$
$$O + O_2 + M \rightarrow O_3 + M$$
$$O_3 + h\nu \rightarrow O_2 + O \quad (\lambda < 1100 \text{ nm})$$
$$O + O_3 \rightarrow O_2 + O_2$$
$$O + O + M \rightarrow O_2 + M,$$

where λ is the wavelength and the symbol $h\nu$ denotes a photon (in physics, $h = 6.6262 \times 10^{-34}$ Joule seconds, Planck's constant; ν denotes the frequency of the light, $\nu = c/\lambda$, where $c = 2.998 \times 10^8$ meters per second, the speed of light). The symbol M refers to any third body, usually an N_2 molecule, which is necessary to conserve both energy and momentum in the particular reaction. [*See* ASPECTS OF THE ENVIRONMENTAL CHEMISTRY OF ELEMENTS.]

Ozone is thus both made and destroyed by solar energy. Its existence in a "layer" is due to the fundamental structure of the atmosphere. From the top of the atmosphere down, the number of oxygen molecules available for ozone production increases exponentially. However, with increasing atmospheric density the amount of available solar energy falls off as more and more photons are absorbed by oxygen atoms. This increase in available molecular oxygen with decreasing available solar energy results in a maximum ozone concentration between 20 and 30 km over most of the Earth. This applies to solar energy having wavelengths less than 242 nm. Solar energy at longer wavelengths destroys the ozone molecule, this photodissociation being most efficient at 250–260 nm. Between 260 and 320 nm, the *cross section* for ozone absorption (a measure of photodissociation efficiency) decreases by a factor of nearly 800, and the actual attenuation of solar energy varies roughly exponentially with the cross section. From the vantage point of the Earth's surface, this means that as wavelength decreases below 320 nm, the ultraviolet *irradiance* (flux of solar energy in watts per square meter) decreases rapidly by several orders of magnitude. At the Earth's surface, irradiance at wavelengths below 285 nm is basically immeasurable. Some sample measurements of ultraviolet irradiance are discussed in Section I,D.

C. Anthropogenic Destruction of Ozone

We have seen how an equilibrium exists in the stratosphere between production of ozone by solar energy having wavelengths shorter than 242 nm and photodissociation of ozone by solar energy having wavelengths between 242 and 1100 nm. The actual situation is more complicated than the "pure oxygen" atmosphere described earlier—in nature there are numerous sources and sinks for stratospheric ozone, including hydroxyl radicals and nitrogen oxides—but the fundamental phenomenon involves an equilibrium ozone amount maintained by the Sun's energy. During the mid-1970s, atmospheric scientists began to realize that some anthropogenic molecules, primarily CFCs, might be significant sinks for stratospheric ozone. Their significance arises from their breakup in the stratosphere (by photodissociation), which releases free chlorine atoms. The free chlorine can then destroy ozone by several possible reaction cycles, the most well known being

$$O_3 + Cl \rightarrow O_2 + ClO$$
$$O + ClO \rightarrow Cl + O_2.$$

The important thing to notice about this pair of reactions is that the chlorine atom has destroyed an ozone molecule, but emerges free again. Because it emerges unchanged, it is referred to as a *catalyst* for ozone destruction, and this pair of reactions is known as a *catalytic cycle*. In principle, the chlorine atom is free to continue destroying one ozone molecule after another, on into the millions, until it eventually settles out of the stratosphere. In practice, most free chlorine released from CFC photodissociation is immediately bound by various chemical processes into relatively inert *reservoir species,* such as HCl or $ClONO_2$ (but see Section I,F on the polar vortex and the significance for the

Antarctic ozone "hole"). At most latitudes, the chlorine that remains free is thought to be able to lower the equilibrium ozone abundance by an average of 2.7% per decade. [*See* GLOBAL ANTHROPOGENIC INFLUENCES.]

D. Measurement of Ozone

Atmospheric ozone plays a key role in defining the radiation environment of the biosphere, but its actual abundance is quite small. If the entire column of ozone over the United States, from the Earth's surface to the top of the atmosphere, were compressed to standard temperature and pressure (STP, 1013.5 millibars and 273.16 Kelvin), it would be approximately 0.35 cm thick. It is convenient to use this total column thickness as a measure of ozone abundance for two reasons: (1) the actual ozone density varies by an order of magnitude between the stratosphere and the troposphere, making a reference to ozone concentration somewhat ambiguous; and (2) the total column amount is the most relevant quantity in terms of UV-B radiation reaching the Earth's surface. The most common measure of total ozone abundance is the *Dobson unit* (named after the pioneering atmospheric physicist Gordon Dobson), which is the thickness of the ozone column (compressed to STP) in millicentimeters. Thus the typical ozone abundance over the United States is around 350 Dobson units. At STP, one Dobson unit is equal to 2.69×10^{20} molecules per square meter.

Atmospheric ozone is routinely monitored from the Earth's surface, *in situ,* and from space. From the Earth's surface, the total ozone abundance is most commonly measured using a *Dobson spectrophotometer.* This device works on the principle of *differential absorption* of radiation. The solar disk is viewed through a spectrograph, which is fixed to select discrete pairs of UV-B wavelengths, and the ratio of solar intensity between these two wavelengths is measured by the instrument. Because of the wide variability in the efficiency of ozone absorption throughout the UV-B (the solar intensity reaching the Earth's surface varies by more than four orders of magnitude between 280 and 320 nm), the ratio of solar intensity between any

two discrete wavelengths (separated by a few nanometers) is highly sensitive to the total atmospheric ozone amount. Given the solar elevation angle, the total ozone amount is readily calculated from the measured ratio. Currently there are 90 Dobson spectrophotometers in regular operation worldwide.

Several ozone-measuring spectrophotometers have been launched into orbit since the early 1970s. The most successful of these has been the Total Ozone Mapping Spectrometer (TOMS) aboard the NASA Nimbus 7 polar orbiting spacecraft. The Nimbus 7 TOMS operated reliably from October, 1978, until May, 1993, an extraordinarily long period of time for a satellite instrument. The Nimbus 7 instrument has subsequently been replaced by a new TOMS aboard the Russian Meteor-3 spacecraft, and there will be several generations of TOMS after Meteor-3. The TOMS operates by nearly the same principle as the ground-based Dobson spectrophotometer, but measures ultraviolet light backscattered to space by the Earth–atmosphere system. The purpose of TOMS is to provide once-a-day global maps of ozone abundance with a ground resolution of 1° in latitude by 1.25° in longitude. Other satellite spectrometers, such as the Solar Backscatter Ultraviolet (SBUV) instrument, have a coarser ground resolution but employ more sophisticated spectrographic techniques to measure part of the vertical distribution of stratospheric ozone.

Although ground-based Dobson spectrophotometers can make very precise measurements of the total ozone column, and orbiting sensors can map the global ozone distribution, a measurement of the entire vertical distribution of stratospheric ozone (from aircraft, balloon, or sounding rocket) is crucial for understanding the principles governing ozone abundance and possible depletion. *In situ* vertical profiling of ozone concentration is most commonly done using weather balloons (radiosondes) equipped with an *electrochemical concentration cell* (ECC). ECC-ozonesondes have made fundamental contributions to understanding ozone depletion. For example, vertical ozone profiles recorded over Antarctica have demonstrated that the springtime ozone depletion (the Antarctic ozone "hole," discussed later) often occurs at discrete altitude ranges throughout the stratosphere, whereas standard gas-phase models of anthropogenic ozone depletion predict that ozone depletion should occur starting at the top of the ozone layer. This evidence has supported *heterogeneous chemistry* as an explanation for the Antarctic ozone hole, as discussed in Section I,F.

E. Distribution and Trends in Stratospheric Ozone

The natural ozone column abundance varies by nearly a factor of two with season and latitude. One might expect that because ozone is produced mainly by sunlight, there should be greater ozone abundances where the solar irradiance is the highest (i.e., over low latitudes). The actual situation is nearly the opposite. The equatorial regions consistently have the smallest total ozone abundances, remaining nearly constant around 250 Dobson units throughout the year. Midlatitude ozone abundances (e.g., over the continental United States) can vary from 350 to 400 Dobson units during spring down to 300 Dobson units during autumn. The polar regions tend to have the largest total ozone abundances and the greatest seasonal variability, ranging from a nominal 400–450 Dobson units during spring down to 280–300 Dobson units during autumn.

This distribution in ozone is due primarily to air transport processes. The stratosphere is very stable against vertical motion, but there is enough mean horizontal air motion to affect the geographical distribution of ozone. Tropospheric weather does occasionally force air into the stratosphere, particularly during the intense convection that occurs in the tropics. This lower atmosphere air tends to fan out horizontally toward the poles, where it then descends and reenters the troposphere. The time scale for the poleward motion of such an air parcel is on the order of 6 months. Measurements of the vertical ozone profile tend to confirm these transport processes. In the lower part of the stratosphere, tropical ozone concentrations are the smallest whereas subpolar ozone concentrations are the largest, implying poleward ozone transport.

Motivated by theoretical predictions of ozone depletion, atmospheric scientists and statisticians have carefully analyzed the time history of global ozone reported by TOMS, using detailed statistical methods to account for the numerous natural cycles. Analysis of the TOMS data from 1978 through 1990 has suggested a globally averaged (for the region 65°N to 65°S) downward trend of $-2.7 \pm 1.4\%$ per decade. This trend shows considerable varibility with latitude and season. At Northern Hemisphere midlatitudes, the largest mean downward trends occur during winter and are on average -6 to -7% per decade. During summer (the period of maximum sunlight and UV exposure) the trends are smaller, between -2 and -3% per decade. At Southern Hemisphere midlatitudes, trends during winter are on average -3 to -6% per decade, whereas trends during summer are on average -2 to -5% per decade. At tropical latitudes the trends are zero or even slightly positive. Similar results have also been obtained from analysis of the worldwide Dobson spectrophotometer network.

F. Antarctic Ozone Hole

In 1985, the British Antarctic Survey reported a distinctly large trend of decreasing ozone abundance during the month of October, apparently beginning in the mid-1970s. This trend was evident at both Halley Bay station (76°S, 27°W) and the Argentine Islands (65°S, 64°W), although much more pronounced at Halley Bay. NASA subsequently confirmed this phenomenon in the TOMS data. ECC-ozonesonde data from Antarctica then revealed that this springtime ozone decrease typically begins just at the end of winter. Over much of Antarctica the ozone amount routinely falls below 200 Dobson units, and sometimes nearly below 100 Dobson units. Because these dramatic reductions go against the standard climatology for ozone, and because they were noticed to begin after CFCs were already in widespread use, the United States promptly deployed three National Ozone Expeditions (NOZE) to Antarctica during the middle to late 1980s. These expeditions resulted in a plausible explanation for this springtime ozone "hole."

The mechanism for springtime ozone depletion is set in motion during the austral winter with the presence of the polar vortex. Because of strong circumpolar air circulation, the stratosphere over Antarctica becomes isolated and cold, with temperatures dropping below 190K ($-83°C$) as the Antarctic stratosphere cools radiatively to space. In this isolated and cold environment, polar stratospheric clouds (PSCs) form, which consist mainly of nitric acid and ice particles. On the surfaces of PSCs, chemical reactions take place that break up the reservoir compounds in favor of photochemically active species (species that can be photodissociated), for example,

$$ClONO_2 + HCl \rightarrow Cl_2 + HNO_3$$
$$ClONO_2 + H_2O \rightarrow HOCl + HNO_3.$$

There is ample laboratory evidence that the reaction rates for these and similar processes are greatly enhanced in the presence of ice particle surfaces. Such reactions fall under the category of *heterogeneous chemistry,* or chemistry involving a mixture of phases of matter (gas, solid, liquid). When the Sun returns in the early Antarctic spring, photodissociation of Cl_2 and HOCl takes place, resulting in a large amount of free chlorine that then destroys ozone catalytically. The polar vortex begins to disappear during early to middle spring, and with it go the PSCs. During November and December, ozone-rich air from lower latitudes is transported poleward and the ozone hole disappears as normal ozone concentrations are reestablished.

The ozone hole mechanism is thought to involve a combination of anthropogenic chlorine and the unique seasonal meteorology over Antarctica. Its direct biological impact is on the Antarctic marine ecosystem; however, parcels of ozone-poor air from the springtime Antarctic are often observed to drift over the tip of South America and have been observed as far north as Australia. Occasionally the polar vortex itself extends over the tip of South America. Hence the ozone hole can occasionally perturb the ultraviolet radiation environment of inhabited areas.

There has also been much concern over a similar Arctic ozone hole, and precursors to the complete polar ozone depletion mechanism have been measured in the Arctic. However, the Arctic stratosphere generally does not get as cold as the Antarctic stratosphere, and Arctic PSCs are a much less frequent occurrence, due to wave motion induced in the stratosphere by northern hemisphere land surfaces. These complex atmospheric dynamics prevent the Arctic polar vortex from becoming as well-defined as the Antarctic polar vortex. It is thus not likely that a deep or seasonal ozone hole can or will be formed in the northern hemisphere in the same manner as over Antarctica. However, it should also be noted that heterogeneous ozone chemistry is not limited to PSCs. Aerosols propelled into the stratosphere by volcanic eruptions can play the same role. The record low ozone abundances above 45°N (up to 30% column reductions) during 1992–1993 are thought to be directly related to the 1991 eruption of Mt. Pinatubo in the Philippines.

G. Ultraviolet Radiation at the Earth's Surface

The spectral ultraviolet irradiance at the Earth's surface is governed by several physical factors. The most fundamental is simply the angle of incidence. The amount of solar energy striking the top of the atmosphere, per unit area and unit time, increases with the cosine of the *solar zenith angle* (the angle between the position of the sun and the local vertical direction). This factor alone is enough to make surface irradiance vary greatly with time and geographical location. The variation is even greater owing to the presence of the ozone layer. Not only does stratospheric ozone account for most UV-B attenuation, but as the solar zenith angle changes, the optical path length through the ozone layer changes. Attenuation of UV-B varies basically exponentially with optical path length through the ozone layer, so that the changing solar zenith angle brings about fluctuations in UV-B surface irradiance that are larger than one would expect from incidence angle alone. [*See* ATMOSPHERE–TERRESTRIAL ECOSYSTEM MODELING.]

In the troposphere, where the atmosphere becomes exponentially more dense, *scattering* (redirection) of solar photons by atmospheric molecules becomes an important regulator of both ultraviolet and visible radiation, as some radiation will be reflected back to space before reaching the surface. Scattering by atmospheric molecules, or *Rayleigh scattering,* has an efficiency that varies inversely with the fourth power of the wavelength. Rayleigh scattering is therefore noticeably more important in the ultraviolet than in the visible. If human vision was responsive to ultraviolet rather than visible light, the Sun would appear relatively dimmer and the entire sky would appear relatively brighter. At shorter wavelengths, Rayleigh scattering makes the radiation more *diffuse* or *isotropic* (arriving equally from all directions). Another factor affecting the amount of ultraviolet radiation received by a living organism is the reflectance of the Earth's surface, or surface albedo. The ultraviolet albedo of most earth and vegatation is less than 10%, although gypsum sand can have albedo as high as 30%. The albedo of water is around 5%, whereas that of pure snow can approach 100%. Surface albedo affects the ultraviolet radiation environment in two ways: (1) it helps establish how much of an organism's surface area is exposed to UVR (e.g., nearly double over snow) and (2) multiple reflection of photons between the surface and the Rayleigh scattering atmosphere or the bases of clouds can significantly increase the downwelling solar radiation at the Earth's surface (again, this is most important over a high-albedo surface such as snow).

Figure 1 shows three measurements of ultraviolet irradiance at the Earth's surface, as a function of wavelength, and these three irradiance spectra illustrate the contributions due to solar zenith angle and ozone abundance. The spectra denoted by the solid and dotted curves were measured at Ushuaia, in southern Argentina, under clear skies and the same solar zenith angle (44°). The spectrum of November 6, 1991, was recorded under a normal ozone column of 347 Dobson units. The spectrum of October 20, 1991, was recorded when the edge of the polar vortex had extended over Ushuaia, at which time the ozone column fell to 189 Dobson units. At wavelengths longer than 325 nm, the two irradi-

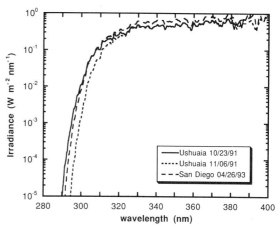

FIGURE I Three examples of spectral ultraviolet irradiance at the earth's surface, measured in populated areas. The solid curve depicts a measurement made at Ushuaia, in southern Argentina, when the edge of the Antarctic ozone "hole" was overhead (total column ozone equals 189 Dobson units). The dotted curve depicts a measurement also made at Ushuaia, but when total column ozone was normal at 347 Dobson units. The dashed curve depicts a measurement made at San Diego, California, when the total column ozone was normal at 325 Dobson units. The solar zenith angle for both Ushuaia measurements is 44°, whereas the solar zenith angle for the San Diego measurement is 19°. These measurements were made by spectroradiometers belonging to the U.S. National Science Foundation's Polar Network for Monitoring Ultraviolet Radiation, and the data have been provided courtesy of Biospherical Instruments, Inc.

ances are equal. As wavelength decreases into the UV-B, irradiance recorded on October 20 becomes dramatically larger than irradiance recorded on November 6. The difference is a factor of two at 307 nm, a factor of eight at 300 nm, and one order of magnitude at 298 nm. The spectrum denoted by the dashed curve was measured in San Diego, California, during the same season (midspring), but because of the lower latitude the solar zenith angle was smaller (19°). At UV-A wavelengths, irradiances are 1.3 times larger during the San Diego measurement than during the October 20 Ushuaia measurement. As wavelength decreases into the UV-B, this difference becomes smaller until the irradiances are equal at 306 nm. By 300 nm, the October 20 Ushuaia irradiance is 1.3 times larger than the San Diego irradiance, and this increases to a factor of three as wavelength decreases to 294 nm. Figure 1 indicates some complexity in the solar ultraviolet spectrum at the earth's surface, even under clear skies, but the most important fact is that UV–B radiation is attenuated by ozone whereas UV–A radiation is attenuated only by Rayleigh (and cloud) scattering.

Along with solar zenith angle and ozone abundance, a significant regulator of ultraviolet radiation (and in fact all solar radiation) is cloud cover. Clouds, because of the size and number density of their water droplets or ice crystals, are much more effective scatterers than the clear (Rayleigh scattering) atmosphere. Under an overcast sky the surface ultraviolet irradiance can be less than half that under clear skies, however, the actual extent to which clouds attenuate solar radiation depends on both the optical thickness of the clouds and the fraction of the sky covered by clouds. Because clouds have such widely varying morphology in nature, there are no simple numbers for cloud attenuation. One basic property of cloud scattering is that its efficiency is approximately the same for all wavelengths less than 1000 nm. In the near infrared, cloud droplets begin to absorb as well as scatter radiation, and the scattering efficiency varies more strongly with wavelength. In the ultraviolet, however, clouds may be thought of as a "neutral density filter" (to use a photographic analogy) for surface irradiance. Thus cloud cover will tend to reduce the surface irradiance at all ultraviolet wavelengths by nearly the same factor, but will have little effect on the ratio of one UV–B wavelength to another (or the ratio of total UV–B to UV–A and visible) as established by the stratospheric ozone abundance. Cloud cover is clearly a major factor, but as of this writing a global climatology of how clouds affect ultraviolet radiation has not yet been established.

Over urban or industrial areas, tropospheric pollutants can also reduce the amount of UV–B irradiance at the Earth's surface. The three most important absorbing gases in this case are SO_2, NO_2, and anthropogenic ozone (e.g., smog). Under the heaviest pollution episodes, UV–B irradiance can be reduced by up to 20%. One factor that can reinforce this absorption is the presence of clouds within the same atmospheric layer as these absorbing gases. The efficient cloud scattering greatly increases the effective optical path of sunlight through the absorbing gases.

II. BIOLOGICALLY ACTIVE ULTRAVIOLET RADIATION

A. Action Spectra and the Biological Dose

The response of most biological processes to ultraviolet radiation depends strongly on wavelength. This wavelength dependence must be measured in the laboratory or in nature before the impact of future changes in ultraviolet radiation can be assessed. This wavelength dependence is conventionally expressed in terms of an *action spectrum,* or relative effectiveness of ultraviolet light at any given wavelength in the biological process being studied. If such an action spectrum, $A(\lambda)$, is convolved with the instantaneous spectral irradiance, $F(\lambda,t)$, the result is a *biologically effective irradiance* or *dose rate,* $F_e(t)$, relevant to the particular life process being studied, that is,

$$F_e(t) = \int A(\lambda)F(\lambda,t)d\lambda \qquad \text{(weighted W/m}^2\text{)}.$$

Integration of the dose rate over the relevant period of time (day, year, duration of laboratory experiment) gives the *dose,* F_d:

$$F_d = \int F_e(t)dt = \int A(\lambda)F(\lambda,t)d\lambda dt$$

$$\text{(weighted J/m}^2\text{)}.$$

Two normalization conventions exist for action spectra: (1) normalization to unity at peak sensitivity and (2) that all action spectra should be normalized to unity at 300 nm. It is necessary to know the normalization convention so that doses and dose rates for various life processes can be readily compared.

Three examples of action spectra are shown in Fig. 2. One of the most often used is the action spectrum for damage to laboratory deoxyribose nucleic acid (DNA). When exposing solubilized DNA to UVR in laboratory experiments, the relative effectiveness of UVR in causing lesions to the DNA molecule increases by more than four orders of magnitude as wavelength decreases from 320 to 300 nm. Hence the dose and dose rate for DNA

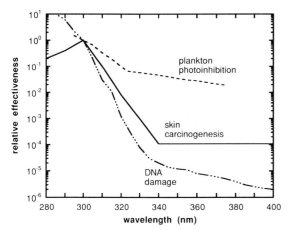

FIGURE 2 Three examples of biological action spectra. The solid curve depicts the MEE48 action spectrum for skin carcinogenesis in mammals, determined using experiments on laboratory mice. The dashed line depicts an action spectrum for the inhibition of photosynthesis in Antarctic phytoplankton, determined by experiments on organisms in the Southern Ocean. The dash-dotted line depicts the action spectrum for damage to unprotected DNA, determined in the laboratory without reference to any living organism.

damage are highly sensitive to atmospheric ozone abundance. In contrast, the action spectrum for inhibition of photosynthesis in Antarctic phytoplankton indicates that UV-A radiation plays an important role in this process, and the dose rate for photoinhibition is therefore far less dependent on atmospheric ozone abundance. Some researchers now caution against using the laboratory DNA action spectrum for assessing the biological impact of ozone depletion. The reason is that DNA inside living organisms is actually sheltered by the cell structure, which often includes protective UV-blocking pigments. Hence dose rates computed using the laboratory DNA action spectrum may in fact overstate the impact of ozone depletion. Most action spectra derived for a specific life process, such as the one shown in Fig. 2 for skin carcinogenesis in mice, have relatively less UV-B weighting than the laboratory DNA action spectrum.

It is generally possible to determine how a specific type of biological dose depends on the ozone column. This is expressed as a *radiation amplification factor*, RAF:

$$RAF = -(dF_d/F_d)/(dN_{O_3}/N_{O_3}),$$

where N_{O_3} is the total ozone abundance in Dobson units. For small ozone depletions (a few percentages), most biological dose rates increase approximately linearly with decreasing ozone, and one can speak of a "percent rule." The most well known is the so-called "two for one" rule, whereby each percent decrease in ozone abundance results in a 2% increase in DNA-damaging ultraviolet radiation (this rule was determined using the laboratory DNA action spectrum of Fig. 2 and so may be an overstatement). For larger ozone variations, such as the Antarctic ozone "hole," the RAF is not a simple percent but varies nonlinearly with total ozone abundance. Similarly, one can define a *biological amplification factor* (BAF) as the expected increase, in nature, of a given effect E in response to a change in the ultraviolet dose:

$$BAF = (dE/E)/(dF_d/F_d),$$

where, for example, E might be an incidence of skin cancer or a rate of photosynthesis. The total amplification factor, AF, for a given life process is the product of the RAF and the BAF.

B. Dependence on Latitude and Ozone Column

If the ozone abundance and distribution are known, along with other phenomena such as cloud optical thickness and surface albedo, then the spectral ultraviolet irradiance at any altitude can be calculated using the theoretical physics of *radiative transfer*. This branch of physics is fundamentally important to atmospheric science and climate study, and there are presently several advanced radiative transfer models available that can estimate UV-B fluxes. Although the basic physics of scattering and absorption is accurately represented in these models, the quality of model output depends on the validity of the parameters used as model input. If all other parameters are held constant, a theoretical radiative transfer model will always predict an increase in UV-B in response to a decrease in ozone. In the absence of cloud information, one can simply perform clear-sky radiative transfer calculations to get a first-order estimate of UV-B trends. Several such studies can be found in the recent literature, and these are useful to place ozone trends in perspective. Actual trends in surface UV-B, particularly in response to small ozone depletions, will be heavily influenced by the climatology of and any trends in cloud cover. This is still an unresolved issue from an observational standpoint.

One example of a theoretical radiative transfer calculation is shown in Fig. 3, which shows how the maximum dose rate for carcinogenically active radiation varies with northern hemisphere latitude, under clear-sky conditions. This dose rate refers to the action spectrum of Fig. 2, which is a recent development based on laboratory experiments with hairless mice and which is thought to give a fairly accurate representation of the potential for nonmelanoma skin cancer in humans. The calculations apply to local apparent noon on summer solstice and were performed using the actual climatology of stratospheric ozone as discussed in Section I,E, and for a hypothetical case where the ozone column is fixed at 325 Dobson units at all latitudes (the climatological value for 40°N, roughly the latitude of Philadelphia, Pennsylvania, or Boulder,

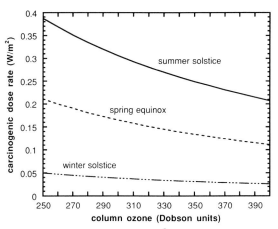

FIGURE 4 The dose rate of carcinogenically active ultraviolet radiation at 40°N at local noon under clear skies, for three different times of year.

Colorado). This illustrates how strongly a biologically effective dose rate depends on the natural climatology of ozone, particularly at lower latitudes.

Figure 4 shows how this same carcinogenic dose varies with total ozone abundance at 40°N at three different times during the year, according to a clear-sky radiative transfer calculation. Figures 3 and 4 allow us to put the ozone trend of −2.7% per decade in perspective (this is also approximately the trend for 40°N during summer). This trend implies a decrease in ozone abundance from 325 to around 316 Dobson units after one decade, which in turn implies (from Fig. 4) a dose rate increase from 0.275 to 0.285 W/m². The solid curve of Fig. 3 shows that this is the equivalent to moving 1.7° (roughly 100 nautical miles) closer to the equator. All other factors being equal, this downward trend in ozone is the equivalent of moving from Philadelphia to Washington, DC. Given the enormous geographical extent of biological systems thought to be at risk, whether they be the human population, staple crops, or marine phytoplankton, determining a realistic impact of these small ozone trends remains a formidable task.

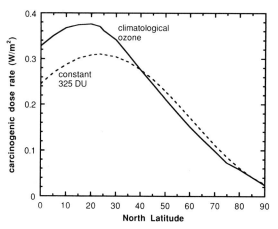

FIGURE 3 The dose rate of carcinogenically active ultraviolet radiation (the ultraviolet irradiance spectrum weighted by the MEE48 action spectrum shown in Fig. 2) as a function of Northern Hemisphere latitude at the summer solstice. The curves refer to local noon (minimum solar zenith angle) and clear skies and are theoretical dose rates from a radiative transfer model. The dashed curve refers to a hypothetical case where the total column ozone abundance is fixed at 325 Dobson units at all latitudes. The solid curve refers to the actual global ozone climatology with more ozone at high latitudes than at low latitudes.

C. Measuring and Monitoring Ultraviolet Radiation

UVR measuring devices fall into two general categories, *broadband radiometers* and *spectroradiometers*.

A broadband radiometer reports a single irradiance measurement that is the integral over most or all of the ultraviolet spectrum (watts per square meter). There are several ways to make this type of measurement, including filter and photodiode combinations, phosphorescent photodiodes, or chemical or biological actinometry. The most common broadband radiometer, the *Robertson–Berger meter* (RB meter), uses a phosphorescent cell and photodiode configuration to duplicate the action spectrum for sunburn in human skin. The RB meter therefore reports a specific biological dose rate. A spectroradiometer resolves ultraviolet irradiance into either discrete wavelengths or small-wavelength integrals (watts per square meter per nanometer). This can be done with either a grating monochromator or a prism, or by means of narrowband interference filters. Some of the more advanced spectroradiometers use a fixed grating and a charged-coupled device (CCD) array, so that they can record the entire irradiance spectrum instantaneously.

Scientifically, the spectroradiometer is a much more versatile device. Ultraviolet irradiance spectra (such as those shown in Fig. 1) can be interpreted using any action spectrum, and the wavelength resolution permits detailed investigation of atmospheric processes. The major drawbacks to the spectroradiometer are cost and complexity. For UV-B measurement, a relatively inexpensive single-grating monochromator is inadequate. Because UV-B irradiance at the ground varies by more than three orders of magnitude as a function of wavelength, a double monochromator must be used to reject stray light (light from longer wavelengths leaking through the first grating and corrupting the sensitive UV-B measurement). Also, the low energy levels one must measure when viewing one wavelength at a time usually necessitate thermoelectric cooling of the photodetector. Monochromators usually have delicate moving parts that must be protected from temperature variation and mechanical vibration to ensure wavelength accuracy. The use of a spectroradiometer in a long-term UV-B monitoring mode constitutes a challenging engineering task. The primary advantages of broadband radiometers are their low cost

and simplicity. Most biology or medical laboratories can afford a broadband radiometer as an ancillary measurement, and these devices can be left in unattended operation for long periods of time (although one should be careful not to take optical cleanliness, temperature stability, and radiometric calibration for granted). The main disadvantage of a broadband radiometer is the loss of information when making a single irradiance measurement integrated over all ultraviolet wavelengths.

At present there are few ultraviolet radiation monitoring programs having a data base longer than a decade. This situation of data scarcity is beginning to change as the United States, Canada, Australia, New Zealand, Great Britain, Norway, Sweden, Germany, Greece, and several other countries deploy robust networks including spectroradiometers. The best equipped ultraviolet monitoring program to date was established in 1988 by the U.S. National Science Foundation in Antarctica, in response to the ozone "hole." Presently this network consists of six automated spectroradiometers, three in Antarctica (at the South Pole, McMurdo Station, and Palmer Station), one at Ushuaia in southern Argentina, one in Barrow, Alaska, and one in San Diego, California.

There is also an RB meter network in North America that has been managed by the National Oceanic and Atmospheric Administration for the past few years. While the network was being managed by Temple University, the first analysis of a decade of this data actually suggested a downward trend in RB dose rate by an average of 7% per decade. There are some uncertainties with this analysis related to long-term stability and calibration, and also the location of many of the instruments near airports or urban areas. Some researchers are beginning to discount the usefulness of broadband measuring devices altogether, citing the uncertainties with the North American RB meter network's long-term record. Some of this criticism is unfair. Without proper attention to stability and calibration, a spectroradiometer will produce even worse data than a broadband instrument. Given current monitoring recommendations by the World Meteorological Organization, and technology and funding realities, a realistic ultraviolet monitoring strategy should probably involve a network of several broadband instruments supplemented by a few spectroradiometers at key sites.

In addition to the downward trend noticed in the North American RB meter data, several other long-term surface irradiance measurements can be summarized as follows (because of the scarcity of data, here "long-term" refers to any data set of a few years' duration). A spectroradiometer operated from 1975 to 1990 by the Smithsonian Institution recorded an increase in the monthly mean UV-B irradiance until 1986, with a subsequent decrease between 1987 and 1990. A recent report of the Toronto spectroradiometer data (1989–1993) showed basically steady levels of UV-B irradiance during the first four years, with 1993 showing noticeably higher irradiances in response to record low northern hemisphere ozone levels. The NSF ultraviolet monitoring network in Antarctica (1988–present) demonstrated a high degree of sensitivity in UV-B irradiance to the presence of the ozone "hole," with considerable interannual variability that can be directly correlated with the magnitude of the springtime ozone depletion. Scientists at the National Oceanic and Atmospheric Administration are presently attempting to document and catalog ultraviolet monitoring effects worldwide, and results of this effort should provide a useful global climatology.

D. Ozone and UVR Variation over Geologic Time

The question is often asked whether ozone "holes" (similar to the seasonal ozone depletions occurring over Antarctica at the present time) occurred before the use of commercial CFC products. To answer this question, one must look at the entire time span of life on Earth. It is believed that life arose more than 2.5 billion years ago when the Earth's atmosphere consisted primarily of reduced gases (e.g., ammonia, methane, nitrogen) with extremely low concentrations of oxygen and essentially no stratospheric ozone. Under these conditions the UVR at the Earth's surface would have been very high and might have extended in wavelength down to 200 nm. It is thought that life most likely arose in aquatic environments, where most of the UVR would have been absorbed by water molecules in the upper 50 m or so of the water column. Below that depth, organisms could grow without any significant amount of solar UVR. With the advent of photosynthetic organisms (which utilize the energy of sunlight to chemically reduce CO_2 and evolve molecular oxygen) about 500 million years ago, oxygen started to accumulate in the atmosphere to its present concentration of approximately 21%. With the increase of atmospheric oxygen, ozone would start to accumulate via reactions discussed previously. At some time in geologic history, the UVR incident upon the Earth's surface would be sufficiently reduced to permit colonization of terrestrial environments by plants and animals. The photobiological characteristics of present-day organisms should therefore be viewed in the evolutionary context, whereby organisms have evolved and adapted to changing conditions of incident sunlight, including both visible and UVR. There are no data to suggest the formation of seasonal ozone holes since the atmosphere became enriched with oxygen and ozone.

III. BIOLOGICAL EFFECTS OF ULTRAVIOLET RADIATION

For a photon of solar electromagnetic radiation to have any effect on an organism, the energy in the photon must first be absorbed by a molecule or substance (called a *chromophore*). The chromophore may be an important cell constituent that may be directly damaged (referred to as a *target*), or the absorbed energy may either damage target molecules via photodynamic reactions involving energetic intermediates or may be dissipated in nondamaging ways, such as heat loss or coupled chemical reactions (Fig. 5). The primary concern regarding UV-B radiation (particularly the shorter wavelengths from 290 to about 310 nm) is that DNA, which is the carrier of genetic information in all plants and animals, has significant absorption in this spectral region and thus may be a direct target, with the result that normal functioning of the DNA is impaired. [*See* BIOGEOCHEMISTRY.]

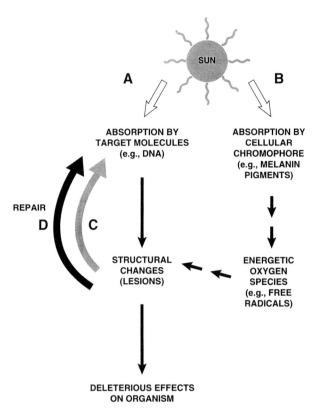

FIGURE 5 Alternate mechanisms for solar UVR to damage cellular material. Damaging radiation may be absorbed directly by the *target* molecule (pathway A) and result in molecular *lesions,* which are deleterious to the cell; the alternate pathway (B) involves absorption of the radiation by a chromophore that generates energetic oxygen species, which then react with the target molecule to produce similar lesions. Note that lesions in the target molecules can be repaired by either photochemical reactions (pathway C) or dark reactions (pathway D).

A. UVR, both Deleterious and Beneficial

It is often assumed that visible radiation is the "essential" component of sunlight, and that all UVR is damaging to biological systems. This misconception likely arose because most people are concerned primarily by events that can be "seen" by the human eye. In reality, however, the electromagnetic spectrum is a continuum, without any sharp demarcation in regard to beneficial or deleterious effects on organisms. Three examples that illustrate the beneficial aspects of UVR are: (1) vitamin D (deficiency of which results in rickets) can be synthesized in the skin tissue of humans via a photochemical reaction utilizing UV-B radiation; (2) all

green plants can utilize UVR (down to at least 350 nm) as an energy source to reduce CO_2 via photosynthesis (but with less efficiency than with visible radiation); and (3) lesions in DNA resulting from exposure to UV-B radiation can be repaired by photochemical reactions utilizing UV-A radiation. The deleterious effects of both UV-B and UV-A radiation are discussed in the following sections. It should be noted that visible radiation can also be very damaging to both plant and animal tissue if irradiances are sufficiently high.

B. Cellular Targets of UVR

The general photochemical effects induced by solar UVR are similar in all organisms, although various reactions are specific to various groups of organisms (e.g., damage to photosynthetic pigments in autotrophic plants and formation of carcinomas in skin tissue of animals). The end result of UVR-induced damage is fairly easy to document (e.g., death of an organism or a decrease in some specific metabolic rate), but the mechanism of cellular damage is usually difficult to determine. The major impacts of UVR on cellular processes that have been reported are considered in the following sections. Structural changes (called *lesions*) in DNA can be caused by UVR, leading to impairment in control of metabolic control, interference with mitotic processes, and possible death of the organism. Many types of lesions are possible, such as single- and double-strand breaks, DNA–protein cross-links, and dimer formation. Structural changes can be induced by UVR in messenger ribonucleic acid (RNA), which is involved with transcription and transmission of the coded genetic information contained in DNA. UVR can also induce structural changes and cross-linking of proteins or changes in binding of proteins to other cell constituents. This would include the network of microtubules and filaments that compose the "cytoskeleton" in cells and that are of much importance in growth and cell division processes. Such structural changes can thus interfere with normal cell functioning. Of particular importance here is the effect on membrane lipids and the functioning of membranes that control the rate at which inorganic or organic molecules are actively transferred between the cell and the environment or between organelles within the cell.

The electron transport system in cells can be impaired by UVR, with the result that respiration rates are decreased and metabolic control as regulated by cellular concentrations of adenosine mono-, di-, and triphosphate (AMP, ADP, and ATP, respectively) are seriously affected. Rates of nitrogen fixation by prokaryotic organisms can also be decreased by this type of impairment. As a result of excessive UVR exposure, photosynthetic rates may be reduced by a variety of mechanisms, such as damage to Photosystem II and limiting concentrations of CO_2. The molecular structure and function of essential photosynthetic pigments (e.g., chlorophylls and carotenoids) can be altered, resulting in loss of photosynthetic activity. High UVR irradiance can result in bleaching of pigments and loss of color. Impairment in motility of cells or cell components can result from UVR exposure, with subsequent loss of ability of the organism to respond to environmental stresses. In animal tissue, UVR can have serious effects on skin tissue components (e.g., tumor formation, destructive effects on connective tissue, diverse effects on immune-response cells) and on ocular tissue (e.g., corneal keratoses and cataracts on the lens). Finally, UVR can produce active oxygen species (e.g., singlet oxygen, hydroxyl free radical) that can result in oxidation of essential constituents throughout the cell.

C. Effects of UVR on Organisms

The bulk of data on UV effects on organisms has been obtained by laboratory experiments with lower plants or animals (e.g., unicellular algae, bacteria, mice) or with cultured animal tissues, usually using lamps to simulate possible changes in solar radiation. It is generally difficult and expensive to construct a lamp apparatus that closely approximates the full solar spectrum. Consequently, many results obtained under laboratory lamps pertain to UV doses higher than those found in nature, or to ratios of UV-B/UV-A/visible irradiance that are quite different from nature. Thus there is consider-

able uncertainty in extrapolating such data on biological impact of UVR to natural communities. There have been relatively few studies utilizing natural assemblages of organisms under natural environmental conditions.

I. Effects on Human Beings

The major impacts of UVR that have been described for humans involve those body parts that are most exposed to direct solar radiation. Thus there is less of a danger of enhanced UV-B impacts on humans than for other organisms, as the human population can potentially be educated to limit sun exposure.

a. Skin Tissue

The most important impacts on skin tissue are (1) the erythemal reaction (sunburn), which is caused mostly by UV-B but also in part by UV-A, and the tanning reaction (synthesis of melanin pigments), which is related primarily to UV-A radiation; and (2) damage to DNA in skin tissues, resulting in abnormal cell growth, which includes benign keratoses and skin cancers (both basal and squamous cell carcinomas), and possibly melanomas, which are more life-threatening than the carcinomas. Consequences of the sunburn and tanning reactions occur shortly after exposure to UVR; in contrast, development of skin carcinomas is related to accumulated UVR dose, so that the interval between exposure to UVR and initiation of tumor formation may be years to decades. It should be noted that the functional relationship between UVR exposure and melanoma formation is less obvious than that for carcinomas, leading some researchers to question the general belief that sun exposure is implicated in initiation of skin melanomas. Carcinomas generally develop at the site of UVR exposure, whereas melanomas often develop in areas of the body that have not received the greatest UVR exposure.

It should also be cautioned that epidemiological data showing the rate of increase of various kinds of skin cancers in different demographic regions cannot be attributed necessarily to effects of increasing UV-B resulting from lowered ozone concentrations. These data sets, which often extend from the 1960s or earlier to the present, must be considered in relation to improvements in diagnosis and recording of disease incidence and to a change in life-styles and increased leisure time, whereby many people in recent years have greatly increased their UVR dose absorbed by exposed skin. Also, it is known that epidemiologies of basal or squamous cell carcinomas can often be incomplete. Because these diseases often appear and persist only as minor skin irritations, they are often unreported or misdiagnosed by physicians. Epidemiologies of melanoma, a much more serious affliction, are usually more reliable.

UVR also has an effect on the aging and appearance of the skin. Sufficient UVR penetrates through the layer of squamous cells that both UV-A and UV-B are implicated in extensive histological and biochemical changes in both the germinative layer of the epidermis and many tissues of the dermis, including connective tissue, elastic fibers, blood and lymph vessels, glands, muscles, and nerves. The resulting damage is not reversible in laboratory tests and includes premature aging, sagging, and wrinkling of the skin.

b. Ocular Tissue

As the eye is the light-sensing organ, it cannot simply block the entry of solar UVR into the eye by synthesis of UV-absorbing compounds. Exposure of ocular tissues to UVR may result in seriously impaired vision by various types of damage. Short-wavelength UV-B radiation is known to be responsible for irritation of the cornea (keratitis) as well as for development of keratoses on the cornea. UVR is involved with formation of cataracts on the lens of the eye, but it is not known how much of this damage is due to UV-B as compared to UV-A radiation. Much of the UV-B radiation apparently is absorbed by the cornea, whereas most of the UV-A radiation penetrates to the lens, with the longer wavelengths also penetrating to the retina, where damage to visual acuity may occur. The mechanisms of damage to the structure and functioning of the lens by UVR are complex and may involve nucleic acids, histological changes, metabolic alterations, and opacities.

c. Systemic Effects

There is considerable evidence that UVR may have harmful effects on the body's immune system, which is a complex regulatory network that functions to protect the organism from harmful substances and also to help ward off infectious diseases. The site of UV-induced damage to the immune system involves a variety of cells located in the epidermis and dermis. The most damaging radiation is UV-B, which is largely absorbed by chromophores in the outer epidermis, but approximately 10% of UV-B and 30% of UV-A radiation can penetrate into the dermis where UVR can cause serious damage (ultrastructural changes, dysfunction, lysis) to blood cells circulating in the capillaries. The activity of the antigen-carrying Langerhans cells in the epidermis is impaired most by UV-B and the shorter UV-A wavelengths. Some data suggest that such immunosuppression reactions can be reversed by the longer UV-A wavelengths. The effects of UVR on the immune system do not cause a general systemic increase in immunosuppression activity, but rather are specific to certain responses. Adverse effects of UVR on skin tissue may be enhanced through photosensitized reactions involving both natural substrates (e.g., riboflavin, porphyrins) or exogenously added substances such as some antibiotics and diuretics. In addition to effects on skin tissue, some exogenously added therapeutic agents that are also photosensitizers may cause damage to vision when ocular tissue is exposed to UVR. Photosensitized reactions, however, can also be used effectively in chemotherapy to treat various disorders. Psoralen compounds, for instance, are used in conjunction with UV-A radiation to treat the skin disorder psoriasis. UV-A radiation is also useful in treating systemic lupus erythematosus.

Very little is known regarding the significance of UVR on orientation and migration of animals, or on general body physiology. Studies with rodents, however, suggest that UVR can affect both neuroendocrine function and circadian physiology.

It is not feasible to experimentally determine the action spectra for serious impacts of UVR on humans. Most action spectra data for carcinomas and melanomas, for instance, are obtained by experimentation with lower vertebrates (e.g., mice and fish), which introduces considerable uncertainty when trying to extrapolate the findings to humans. An indication of the lack of definitive knowledge regarding the action spectrum of UVR for tumor formation is a recent study with a mutant strain of fish showing that UV-A radiation is responsible for up to 95% of the inducible melanomas. The effect of UV-A radiation (which is not absorbed directly by DNA) on DNA function is apparently via photodynamic reactions involving melanin pigments as the initial chromophore.

2. Effects on Terrestrial Plants

Vascular plants utilize solar radiation not only to provide energy and reducing power to reduce CO_2 to carbohydrate and other organic molecules, but UVR, visible, and some near-infrared radiation is involved with many photochemical reactions regulating growth, cell division, flowering and fruiting response, and seed germination. Phototropic responses (movement of plant parts in response to light), which may be caused either by differential growth patterns controlled by photochemical interaction with plant hormones such as the auxins or by hydrostatic pressure changes in specialized tissues, might be affected by increased irradiance of UVR. Thus it is not surprising that changes in UVR exposure can alter general morphology and physiology, histology of tissues, and rates of photosynthesis, with concomitant impacts on productivity and yield of agricultural crops. Data indicate that rates of photosynthesis, which will affect overall growth and crop yield, can be decreased by UVR via effects of the enzyme involved with the initial carboxylation reaction (ribulose 1,5-diphosphate carboxylase), damage to the reaction center of Photosystem II, oxidation of chlorophyll and accessory photosynthetic pigments, and CO_2 limitation resulting from loss of integrity of the active uptake mechanism for CO_2, which is located on the plasmalemma.

Terrestrial plants that are exposed to bright sunshine may show adaptations that minimize the impact of UVR on plant tissue. The upper epidermis of leaves of angiosperms and particularly the needles of gymnosperms, for instance, absorbs much of the incident UV-B radiation, permitting mainly UV-A and visible radiation to penetrate to the inner portions where photosynthesis occurs. Many plants also show heliotropism movements, whereby they can orient the position of the leaf so that the leaf surface exposed to solar radiation can be varied to permit either minimal or maximal exposure to light.

The targets and mechanisms involved in UVR-induced damage in higher plants, as well as the adaptive mechanisms to counteract or minimize UVR, are basically similar to those mentioned earlier for animal tissue. It is important, however, to consider the impact of UVR on all stages of the plant's life cycle. Developmental processes that might show enhanced sensitivity to UV damage would include the periods of initiation of new leaves, flowering, growth of the pollen tube into the stigmatal tissue, seed germination, and subsequent growth of the young seedling. The adaptive capacity of plants to UVR stress is evidenced by the differential sensitivity of individuals of the same species when grown at high and low altitudes; the plants grown at mountain altitudes show more resistance to solar UVR than those plants grown at sea level. Some data suggest that this "acquired" resistance to UVR is transmitted to the seeds of the plants, although the mechanism for this is not known.

Most studies on the effects of UVR as described here, particularly effects of enhanced UV-B resulting from reduced ozone concentrations in the stratosphere, have utilized lamps to simulate solar radiation and have emphasized long-term effects. There is concern that enhanced UV-B radiation might significantly decrease agricultural productivity, with a resulting decrease in the world's food supply. However, considering the variability in plant sensitivity and adaptive responses to UVR, even within similar species and cultivars, it is not possible to make any quantitative prediction regarding the potential loss of yield from agricultural crops.

3. Effects on Aquatic Environments

The shorter UV-B wavelengths are attenuated rapidly in the aquatic environment by absorption and scattering by water molecules as well as by dissolved organic and particulate material. The longer UV-A wavelengths penetrate to much greater depths, but maximum transmission of solar radiation is in the visible region between 450 and 500 nm. Therefore the deeper an organism is in the water column, the less damage it will incur by UVR. Studies have shown that the threshold of UV-B radiation that results in detectable biological damage in phytoplankton corresponds to an irradiance equivalent to approximately 10% of incident UV-B radiation on a sunny day, which would correspond to a depth of approximately 10–15 m in clear waters. The biological impact of UV-A on phytoplankton can be detected down to 15–20 m; below that depth, it is unlikely that UVR will have any significant biological effect (Fig. 6). Organisms living close to the surface, or within the intertidal

zone, may thus be exposed to potentially damaging irradiances of both UV-B and UV-A. This would also apply to organisms living at or close to the surface of the bottom sediments. This would be of particular interest in regard to tropical coral reef environments, where there is a rich and varied community of plants and animals existing in relatively shallow waters.

Much concern has been expressed regarding the loss of primary production by phytoplankton in the oceans. The assimilation of CO_2 by phytoplankton, however, extends down to depths where the irradiance is about 0.5% of that incident upon the surface. As the threshold for photoinhibition of photosynthesis is at relatively shallow depths (Fig. 7), the impact of UVR on integrated primary production in the water column is moderate and

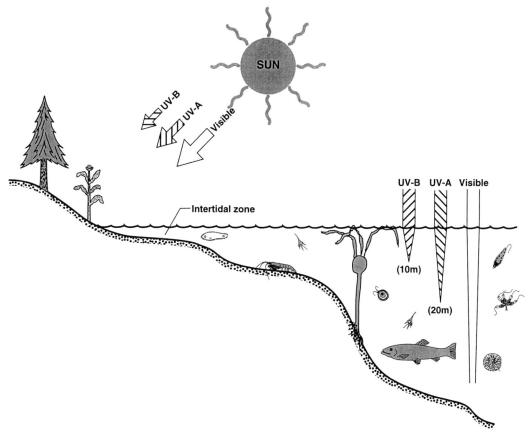

FIGURE 6   An illustration of the relative vulnerability of terrestrial plants and organisms to solar UVR in contrast to organisms living in marine or fresh waters. Terrestrial organisms are exposed to the full irradiance of sunshine, whereas organisms in water are partially or fully protected from UVR by attenuation of UVR by water molecules and suspended and dissolved materials. The arrows show the approximate depths to which damage by UV-B (about 10 m) and UV-A radiation (15–20 m) can be detected. Visible light penetrates to much greater depths.

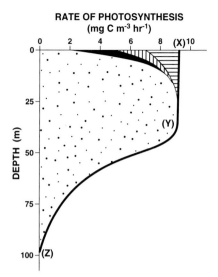

RATE OF PHOTOSYNTHESIS
(mg C m^{-3} hr^{-1})

FIGURE 7 Diagram to illustrate the inhibitory effects of UV-A and UV-B radiation on rates of primary production under normal ozone conditions, and the impact of enhanced UV-B radiation under an extreme "hole" in the Antarctic. These results were determined from photobiology experiments in the Southern Ocean in which either UV-B or UV-A radiation (or both) was screened from various phytoplankton samples by appropriate filters. The solid area depicts the loss in primary production due to enhanced UV-B under the ozone hole. The vertically lined area depicts the reduction in primary production resulting from normal levels of UV-B. The horizontally lined area depicts the reduction in primary production resulting from normal levels of UV-A (unaffected by atmospheric ozone). The dotted area represents the integrated primary production not affected by UVR. In the depth range from (X) to (Y), photosynthesis is described as being *light saturated,* meaning that there is more available sunlight than the phytoplankton need, and where excess irradiances of UV-A and UV-B can be photoinhibitory. In the depth range from (Y) to (Z), photosynthesis is described as being *light limited,* meaning that solar irradiance has been so attenuated by the water column that rates of photosynthesis decrease with depth.

(Fig. 7), the impact of UVR on integrated primary production in the water column is moderate and not calamitous. Over 50% of the loss due to UVR is generally caused by UV-A, which is not affected by stratospheric ozone depletion. Under an extreme ozone hole in the Antarctic (125 Dobson units over the Southern Ocean), the additional loss of integrated primary production due to enhanced UV-B radiation has been estimated to be in the range of 5–10% of primary production occurring during normal ozone conditions. Because the ozone hole is seasonal (maximum ozone depletion in October–November; during severe depletions the maximum UV-B irradiance may extend into

December) and the incident UV irradiance is usually decreased by cloud cover, the overall impact of stratospheric ozone depletion over the Southern Ocean (all waters south of the Polar Front) has been estimated to be less than 0.5% of the yearly production.

These impacts have regional variability associated with the sea ice. Antarctic sea ice exhibits a fivefold seasonal variability in area, and the springtime ozone depletion occurs when the sea ice is in retreat. The rate of sea ice retreat governs the rate at which phytoplankton are released into the open water column. Recent shipboard research in the vicinity of the Antarctic Peninsula has also shown that the melting of sea ice tends to stabilize the nearby upper water column. This promotes phytoplankton growth, but also limits the circulation of organisms to lower depths where they would be more sheltered from UVR. The greatest biological impact of the ozone hole would therefore be expected when the polar vortex places ozone-poor stratospheric air directly over the sea ice edge. Away from the sea ice edge, mixing by wind and currents increases the depth to which phytoplankton may be circulated, and there the ozone hole would be expected to have less impact on primary production. Near the sea ice edge and under the ozone hole, reduction in primary production has been estimated by some researchers to be as large as 6–12% relative to normal ozone conditions. It should be noted that the rate of sea ice retreat varies considerably from one year to the next, and the biological impact of the springtime ozone depletion is therefore linked to the dynamics of the Antarctic climate.

Another concern regarding the impact of UVR on the marine environment is that it may affect the nature and dynamics of the food chain (web) leading up to the species of fish or crustaceans harvested by humans for food. Data indicate that some phytoplankton groups are more sensitive to UVR than others, and also that there is differential sensitivity related to the size of the cells. Any change in the structure of the lower food web (e.g., the types of cells or their size) may have noticeable effects on the higher trophic levels as the intermediate trophic levels (zooplankton, etc.) show a preference for the type and size of food particle ingested. It is thus possible that changes in the structure of the lower food web in natural waters caused by UVR may change a food chain characterized by desirable fish species at the higher trophic levels to one dominated by jellyfish or salps, organisms that are not harvested by humans for food.

Rates of biological nitrogen fixation (the reduction of N_2 to ammonia by some bacteria and blue-green algae) have also been reported to be decreased by UVR. If the biological production of available nitrogen by nitrogen fixation were significantly impaired, this could have important consequences in the terrestrial environment and especially in rice paddies, where the growth of blue-green algae is of great importance in the maintenance of soil fertility.

Most other organisms living in the sea are heterotrophic (i.e., bacteria, zooplankton, and fish that utilize preformed organic compounds for food and are independent of solar radiation) and thus can live at depths where solar UVR would have little or no impact. Ambient levels of UVR are sufficient to increase mortality rates of zooplankton if they remain in the upper 5 m or so of the surface. Small heterotrophic protozoans (mostly ciliates and flagellates smaller than 100 microns in dimension) are very important in the "microbial loop" and generally are most numerous in the upper mixed layer of the water column. Damage to such organisms by UVR might affect the rate at which organic carbon is recycled between different groups of organisms and ultimately respired to CO_2. It should also be noted that the young stages of some fish larvae develop in the upper 10–15 m of the water column; studies have suggested that ambient levels of UVR may result in increased mortality rates, and hence there is some concern regarding the impact of ozone loss on important fisheries.

IV. MECHANISMS AND RESPONSES THAT MAY MINIMIZE INJURY BY UVR

All photosynthetic organisms require sunlight as a source of energy for reduction of CO_2 and hence cannot avoid concomitant exposure to UVR. Most organisms have adapted by various strategies and mechanisms to minimize the deleterious effects of UVR, but these adaptations do not seem to eliminate all cellular damage. It is generally believed therefore that most organisms exposed to sunshine are living under some degree of "UV stress," and that any additional UV-B resulting from ozone depletion will cause additional cellular damage. Common ways for organisms to minimize UV-induced damage include the following strategies.

Some organisms live in dim to dark conditions (e.g., organisms in the deep sea or shade plants growing under forest canopies), where they do not receive any significant doses of UVR. Other organisms exhibit diurnal movement into food-rich areas at night and return to darker areas during daylight (e.g., nocturnal animals and the diurnal migration of zooplankton and fish). Larger zooplankton (e.g., copepods and amphipods) and nektonic organisms (e.g., krill, fish) can move up and down in the water column and thus minimize their exposure to UVR. It is not known, however, if such organisms have sensory organs that can respond to UV wavelengths or if they respond only to visible light. If they cannot sense and respond to UVR, the organisms might be in danger from enhanced UV-B radiation resulting from decreased atmospheric ozone concentrations.

Some organisms can physically minimize their surface areas exposed to sunshine (e.g., small leaves on desert plants and leaves that orient vertically). Many organisms synthesize UV-absorbing compounds that dissipate UV energy without injury to the organism. Examples include melanin pigments in humans, mycosporinelike amino acids found in many marine plants and animals, and flavonoid compounds in terrestrial plants. Most higher organisms develop UV-absorbing layers to shield underlying sensitive tissues, for example, the epithelial layer in skin, the cuticle on the upper surface of plant leaves, and animal body covering such as hair, fur, or feathers. At the cellular level, some organisms exhibit movement of cell organelles to minimize the ratio of surface to volume. For example, when some plants such as large diatoms are transferred from low-light to high-light conditions, the numerous chloroplasts located around the periphery of the cell may coalesce and move to the polar region of the cell.

Enzymatic mechanisms can also repair cellular damage wrought by UVR. The best known of these are the many enzymatic reactions that repair damage to DNA resulting from exposure to UV-B radiation. These reactions, which are found in all plant and animal cells (except for a few mutant strains of bacteria produced in the laboratory), are either coupled to photochemical reactions (called *light* or *photorepair* mechanisms) or may take place in darkness following UV exposure (called *dark repair* mechanisms). Most of the photorepair reactions are activated by UV-A radiation or the shorter wavelengths in the visible portion of the spectrum. The importance of these repair mechanisms is shown by the extreme UV sensitivity of mutant strains that do not have any of these light or dark repair mechanisms. These mechanisms apparently are of primary importance for the survival of all organisms exposed to sunlight. Also found at the cellular level is radiation-induced synthesis of enzymes (e.g., photolyase) or antioxidants (e.g., vitamin E or glutathione) to protect against active oxygen species or free radicals.

V. PRESENT STATUS OF SCIENTIFIC UNDERSTANDING

A meaningful assessment of stratospheric ozone depletion involves three equally important issues: (1) determination of trends in stratospheric ozone and plausible explanations for these trends, (2) determination of whether trends in stratospheric ozone cause any changes in biologically active UVR that are significant in the context of the natural variability in surface UVR, and (3) understanding the role of UV-A and UV-B in biological processes, and how enhanced UV-B might affect living things, from cellular processes through physiology of the organism and ultimately the structure of the ecosystem. Despite the widespread concern about stratospheric ozone depletion, attention to each of these three issues has been historically uneven.

An enormous commitment of effort and resources has been made toward understanding the ozone layer. The global climatology of stratospheric ozone is now well known, and there is a good understanding of the photochemical and catalytic process that govern ozone production and loss in the stratosphere. Recent trends in ozone distribution have been identified on a global basis, and the unique mechanism of the Antarctic ozone "hole" has been explained in a surprisingly short period of time. Because of these research efforts, industrial countermeasures against ozone depletion are being implemented or seriously considered, such as the rapid phaseout of chlorofluorocarbons via the Montreal Protocol. The large expenditure of scientific resources on the study of stratospheric ozone is entirely justifiable. Even if the present anthropogenic ozone depletion issue were to prove for some reason unimportant, studies of stratospheric ozone constitute fundamental geophysical research and are important for a complete understanding of the climate and environment of our world.

However, given the concern about ozone depletion, the scientific community as a whole has been comparatively negligent in developing a proper understanding of the ultraviolet radiation environment at the earth's surface and in the biosphere. As of 1994 there exists no global baseline climatology of ultraviolet radiation established by robust measurements, and there exists no convincing evidence of increasing global trends in surface UV-B in response to trends in ozone. Much of the scientific community has simply taken for granted that any statistically significant downward trend in stratospheric ozone must result in a biologically significant upward trend in surface UV-B. After twenty years of concern, the most often quoted studies of surface ultraviolet radiation are still results from clear-sky radiative transfer models. These are useful theoretical exercises that inevitably show certain radiation amplification factors, but they usually do not account for any climatologies of (much less any estimate of trends in) cloud cover, aerosols, and tropospheric ozone and other anthropogenic gases that could easily offset an impact of the small global trends observed in stratospheric ozone. It was only with the discovery of the Antarctic ozone "hole" that government agencies made a serious commitment to deploy networks of spectroradiometers. Prior to 1988 the vast majority of surface UVR data came only from broadband instruments, although a handful of spectral instruments were in solitary use. This situation is changing, however. Resources are beginning to materialize for robust UVR-monitoring networks in many countries worldwide. Within the next ten years the scientific community should have access to robust and properly calibrated spectral and broadband UVR data from many locations.

There has been a respectable commitment on the part of the scientific community and various government agencies to study the biological effects of UVR, although our present understanding of UV photobiology is nowhere near as advanced as our understanding of the physics and chemistry of the ozone layer. Biology is by far the most difficult of the three major issues related to ozone depletion. Phenomena such as rates of skin cancer, crop yields, phytoplankton primary production, and the like are affected by a wide variety of environmental stresses in addition to UV-B, such as total sunlight, climate, temperature, water availability, water properties, nutrients, grazing, parasites, and (in humans) behavioral patterns and public awareness. A major difficulty involves generalizing relatively simple laboratory experiments to real (and complex) ecosystems.

As one example, we can consider impacts to marine phytoplankton, an area where considerable progress has been made. It is possible to place actual specimens at depth in the water column, screen out UV-B, UV-A, and other wavelength ranges of sunlight, and quantify the role of UVR in CO_2 fixation. It remains extremely difficult, however, to (1) simulate the roles of both vertical and horizontal mixing in the water column, and hence the rapid circulation of organisms throughout various depths and locations (thus the role of an important physical variable cannot be directly ascertained); and (2) monitor the vast tracts of ocean that contain phytoplankton blooms using surface research vessels (hence the generalization of specific findings to an entire marine ecosystem remains imprecise).

One can partially address these issues by (1) simultaneous and detailed hydrography measurements to supplement photobiology experiments placed at a wide variety of depths and (2) using oceanographic satellite images, such as those from the Coastal Zone Color Scanner (CZCS) or the forthcoming SeaWIFS instrument, to map the large-scale phytoplankton distribution. Even with such refinements, there is still the need for extrapolation.

With respect to skin cancer in humans, it is possible to perform useful studies with hairless mice or other animals in the laboratory to estimate action spectra and the relative importance of UV-B versus UV-A. It is also possible to study epidemiologies of skin cancer in terms of history and geographical distribution, and to try to draw conclusions from these records themselves. Establishing the link between laboratory action spectra and actual epidemiology remains to be accomplished. A far more difficult effect to quantify is suppression of the human immune system. It is possible to measure some reductions of immune system efficiency in laboratory animals, and to understand some of the basic physiology behind this immunosuppression, but to assign a role for enhanced UV-B in the actual epidemiologies of infectious diseases not directly caused by sun exposure is virtually impossible.

Empirical action spectra can be useful tools to relate macroscopic UVR effects to actual UVR irradiances, but by themselves they do not constitute a rigorous understanding of the role of UVR in nature. Much more work remains to be done on actual mechanisms of UV damage and repair. To genuinely understand a UVR-related effect on an organism, one must identify chromophores, targets, and repair mechanisms, and quantify the importance of the various candidates for each. The biochemistry behind these microscopic effects is still poorly understood.

Finally, the issue of stratospheric ozone depletion has the challenging aspect of being widely interdisciplinary. A proper assessment of ozone involves the interaction of atmospheric scientists, who have a background primarily in physical science, with photobiologists, ecologists, and medical doctors, who have backgrounds primarily in the life sciences. The scientific culture that expected most researchers to have a strong and current background in both the physical and life sciences came to an end sometime during the early nineteenth century, as human knowledge expanded rapidly and scientific specialization became necessary. Nevertheless, to fully understand the issue of ozone depletion, and to make defensible statements about what should be done, a scientist should be conversant in both atmospheric science and biology. The scientist or other interested individual must make the effort to supplement his or her own specific background. It is with this necessity in mind that we have prepared this article to include discussions of ozone, ultraviolet radiation, and biology under one title.

Glossary

Action spectrum Relative efficiency by which ultraviolet radiation impacts a particular biological process, as a function of electromagnetic wavelength. The action spectrum is generally determined by laboratory or field experiments on substances or organisms, and when convolved with the actual ultraviolet irradiance (multiplied one wavelength at a time and then summed over all wavelengths), it yields an ultraviolet *dose rate* relevant to the particular biological process.

Anthropogenic Generated or caused by human activity.

Chromophore Molecule or substance that is the initial absorber of the energy contained in electromagnetic radiation.

Dobson unit Convenient unit of measure for the total amount of ozone in a unit column of a planetary atmosphere, from the surface to the top of the atmosphere. The ozone column abundance in Dobson units is equal to 1000 times the hypothetical thickness (in centimeters) of all atmospheric ozone if it were compressed to standard temperature and pressure. One Dobson unit is equal to 2.69×10^{20} molecules per square meter.

Euphotic zone Zone from the surface of the water to the depths at which the rate of photosynthetic CO_2 fixation is equal to the rate of respiration. The lower depth of the euphotic zone is generally taken to be that depth where the irradiance is equal to 1% of the solar irradiance incident upon the surface of the water.

Irradiance Rate at which electromagnetic energy crosses a unit area in unit time (e.g., watts per square meter, or Joules per square meter per second); also known as *flux* (in physics) or *fluence* (in biology). One can refer to a *spectral irradiance* (one wavelength at a time, in units of watts per square meter per nanometer) or to an *integrated irradiance* (watts per square meter, referring to all electromagnetic energy in a specific wavelength range such as the UV-B).

Lesion Deleterious structural change in any of the biological macromolecules (such as DNA) resulting from absorption of ultraviolet radiation.

Primary production Rate at which autotrophic plants utilize the energy of solar radiation to fix CO_2 and reduce it to organic material. For terrestrial plants, the production is often expressed as "yield" of crop per acre, whereas in aquatic environments it is usually expressed as the amount of organic carbon produced per day (or per year) per unit volume (e.g., cubic meter) or per unit surface area (e.g., per square meter, integrated from the surface to the bottom of the euphotic zone).

Target Molecule or substance in biological cells that is the ultimate "acceptor" of radiant energy and that undergoes some chemical change. The result is that normal functioning of the cell or organism is generally impaired.

Bibliography

Herman, J. R., McPeters, R., Stolarski, R., Larko, D., and Hudson, R. (1991). Global average ozone change from November 1978 to May 1990. *J. Geophysical Res. (Atmospheres)* **96**, 17,297–17,305.

Niu, X., Frederick, J. E., Stein, M. L., and Tiao, G. C. (1992). Trends in column ozone based on TOMS data: Dependence on month, latitude, and longitude. *J. Geophysical Res. (Atmospheres)* **97**, 14,661–14,669.

Tevini, M., ed. (1993). "UV-B Radiation and Ozone Depletion." Ann Arbor, Mich.: Lewis Publishers.

Urbach, F., ed. (1992). "Biological Responses to UV-A Radiation." Overland Park, Kans.: Valdenmar Publishing.

Wayne, R. P. (1991). "Chemistry of Atmospheres." New York: Oxford University Press.

Young, A. R., Bjorn, L. O., Moan, J., and Nultsch, W., eds. (1993). "Environmental UV Photobiology." New York: Plenum Press.

Biodegradation of Pollutants

Durell C. Dobbins

BioTrol, Inc.

Pollutants are substances, especially wastes, that have deleterious effects on living organisms. Many pollutants, especially organic pollutants, may be modified by organisms in such a way that their negative effects are diminished. This biologically mediated modification of organic chemicals ("biodegradation") usually results in products that are more thermodynamically stable than their unmodified parent compounds. Microorganisms participate in biodegradation by producing enzymes—protein catalysts—that increase the rates of certain chemical reactions. By the action of these catalysts, even complex chemicals may be modified to simpler compounds that are used either as structural subunits for cell growth and maintenance or as fuel molecules for generation of energy.

Under most circumstances, biodegradation of a pollutant is considered complete when its carbon is oxidized to carbon dioxide with the production of water. Degradation is complete because, given the constraints of the environment, the carbon can-

not be further modified in such a way that microorganisms gain energy from the modification. Since the products are no longer organic molecules, the reaction is said to have resulted in the mineralization of the pollutant. Because these products are innocuous, mineralization reactions are ideal with respect to environmental and public health. Incomplete reactions, called biotransformations, do not necessarily result in detoxification of the parent molecule and sometimes even increase its toxicity.

I. INTRODUCTION TO ENZYME-CATALYZED REACTIONS

A basic principle of physical chemistry is that chemical decomposition occurs spontaneously over time because of the tendency of all organized systems toward randomness (entropy). For reactions (even spontaneous reactions) to proceed, the reactants must pass through an "activated" transition state, such that the rate of the reaction is dependent on the rate at which reactant molecules achieve the transition state. Catalysts decrease the activation energy required for a reaction to proceed, presumably by destabilizing chemical bonds through physical interactions with the reactants. In true catalysis, neither the catalyst itself nor the reaction equilibrium (relative concentrations of reactants and products) is affected; but the rate of the reaction is often dramatically affected.

Living organisms maintain a high level of order within their cellular composition. Because of entropy, construction and maintenance of that orderly state require that energy be obtained from their surroundings. Organisms are largely of two groups with respect to their source of energy. Autotrophs (or producers) synthesize organic molecules from atmospheric carbon dioxide using energy captured from sunlight. The primary organisms that participate in biodegradation of pollutants, however, are called heterotrophs (or consumers). These organisms obtain energy by catalyzing the decomposition of extraneous organic carbon. The rate at which energy is released from chemical decomposition in the absence of a catalyst is insufficient to support the relatively rapid consumption of energy required for cell growth and maintenance. Enzymes increase the rate at which chemical bond energy is released from reactant chemicals (substrates) within an environment where the released energy can be captured in a readily utilizable form. Released energy is harnessed in living systems by the conversion of adenosine diphosphate (ADP) to adenosine triphosphate (ATP), widely recognized among biologists as the "energy currency" of living cells. ATP gives up energy in its reconversion to ADP for driving biological reactions that require energy. [Both the phosphorylation (of ADP) and dephosphorylation (of ATP) reactions are themselves enzyme-mediated.]

When pollutants are released to the environment, their rates of spontaneous decomposition may be quite low, but in the presence of enzymes that are capable of catalyzing their destruction, their rate of decomposition may be essentially instantaneous relative to rates of other processes that affect their

fate. Biodegradation of complex chemicals takes place in several steps, involving combinations of enzyme-catalyzed and/or uncatalyzed chemical reactions. The reaction sequence, called a metabolic pathway, may be simple (having predictable, stoichiometric quantities of end products) or complex (having a variety of side reactions and/or parallel pathways, and sometimes resulting in the formation of organic by-products).

II. RELATIONSHIP BETWEEN ORDINARY CELL FUNCTION AND BIODEGRADATION OF POLLUTANTS

Catalysis of decomposition by living organisms is a feature of ordinary cell function. In fact, microorganisms are the earth's most vigorous consumers, accounting for the vast majority of organic chemical decomposition. Decomposition functions carried out by microorganisms include oxidation of both plant and animal products. However, the occurrence of microorganisms in polluted environments and pollutant degradation are considered extraordinary because they require organisms to survive in situations to which they had not adapted. Nevertheless, microorganisms in natural environments are capable of degrading most compounds using existing enzyme systems. [See NUTRIENT CYCLING IN FORESTS.]

The organisms involved in the degradation may or may not obtain energy or structural elements (especially carbon) from the pollutants as they are degraded. If no energy is obtained, then the microorganisms must continue their metabolism at the expense of another source of carbon and energy. When this happens, the carbon source that promotes metabolism is referred to as the primary carbon source and the pollutant is said to be co-oxidized or cometabolized.

The introduction of a pollutant to an environment can alter the relative abundance and activity of various naturally occurring strains of microorganisms. Microorganisms that are better prepared to exist in the presence of the pollutants (i.e., organisms having a selective advantage) thrive, whereas other (even dominant) organisms subside. Traits that confer a selective advantage in the presence of pollutants include (1) a tolerance for the particular pollutant, (2) a means of detoxifying the pollutant as it enters the microenvironment of the organism, and (3) the ability to derive benefit from the pollutant such as nutrients and energy.

Following exposure, some environments display an increase in the rate of decomposition of a pollutant over time. This results from one of a number of physiological events that are referred to broadly as microbial adaptations. Adaptations can result from several phenomena: first, microorganisms capable of a particular function can multiply, thereby increasing the overall metabolic capacity within that environment for carrying out the decomposition activity; second, genetic material may be exchanged among microorganisms within an environment, leading to an increase in the breadth of microorganisms that are physiologically equipped to conduct the activity; or finally, microorganisms may not express the critical genes until a suitable environment for their expression is present. The environmental factors that play a role in regulation of catabolism are many; specific factors that commonly intervene are discussed in later sections. They include concentrations of beneficial and inhibitory substances and interactions among microorganisms.

Pollutants may be classified as either naturally occurring compounds that enter extrinsic environments (e.g., open ocean spills of petroleum) or unconfined anthropogenic (i.e., "man-made") chemicals. Naturally occurring chemicals are often processed to alter the chemical composition and/or structure of the raw material, which may exacerbate the detrimental effects of the chemicals on living organisms. For example, distallation of petroleum can be used to concentrate the volatile aromatic compounds such as benzene, toluene, and simple alkylbenzenes. These compounds are readily biodegraded at low concentrations and in the presence of aliphatic petroleum constituents. However, when concentrated, they become highly toxic to the microorganisms that are otherwise capable of degrading them. Thus, for example, microorganisms proliferate at crude oil/water interfaces, but will not grow readily at benzene/water

interfaces. Anthropogenic chemicals are often referred to as xenobiotic because they are foreign to most organisms. [See GLOBAL ANTHROPOGENIC INFLUENCES.]

III. EFFECTS OF POLLUTANT STRUCTURE ON BIODEGRADATION

Most organic chemicals tested to date have proven to be biodegradable, whether synthetic or naturally occurring. However, several classes of compounds have proven resistant to biodegradation. This resistance is evidenced by their accumulation under environmental conditions. Examples of abundant, naturally occurring, organic compounds that persist under environmental conditions include humic substances, porphyrins, and terpenoid resins. There are also a number of anthropogenic chemicals that persist or accumulate under some environmental conditions. A famous example is DDT [(1,1'-(2,2,2-trichloroethylidene)) bis[4-chlorobenzene]], which is still present in agricultural areas where it was presumably last applied as a pesticide more than 30 years ago.

Accumulation of a chemical in the environment indicates that its decomposition rate is lower than its rate of introduction. Even for some readily biodegradable chemicals, pockets of accumulation may arise due to specific environmental conditions that do not favor decomposition. Thus, even though the bulk of crude oil is readily degradable in the presence of oxygen and other nutrients, it persists for long periods of time in its subterranean environment.

Petroleum is a particularly important pollutant because of its predominance as a source of raw materials and fuels. Although the constituents of petroleum that make up the bulk of its mass are relatively few in number, petroleum chemistry is extremely complex, and many of the less prevalent constituents degrade slowly under aerobic conditions. Common examples include high-molecular-weight polynuclear aromatic hydrocarbons (also called PAHs or PNAs). These compounds consist of multiple aromatic rings that are highly resistant to biodegradation compared to their low-molecular-weight counterparts (such as naphthalene and the substituted naphthalenes), which are readily biodegraded in aerobic environments.

A few anthropogenic compounds exist that are extremely resistant to biodegradation, including the highly chlorinated PCBs, polychlorinated dioxins and furans, and organochlorine pesticides. Among anthropogenic chemicals are compounds for which only a few, highly specialized degraders have been identified. An example is trichloroethylene and related, chlorinated aliphatic compounds, which are co-oxidized by a few known classes of organisms with enzymes that are of low substrate specificity and have other physiological functions. These enzymes include the methane monooxygenases of certain methane-oxidizing microorganisms, toluene oxygenases of some heterotrophs (of the genus *Pseudomonas*), and the ammonia monooxygenase of *Nitrosomonas europaea*.

Although the biochemistry of biodegradation is complex, several recurring themes have been identified that contribute to the resistance (or conversely, to the susceptibility) of chemicals to biodegradation. One such theme is polymerization (as in the case of plastics and cellulosic compounds). This may be a combination of the effects of molecular size on cellular uptake of the compounds and steric hindrance to the enzymatic action on them. A second recurring theme is a high degree of halogenation. This could be due to the high degree of energy investment required by an organism in breaking stable carbon–halogen bonds or to the paucity of structural analogues in nature. A third theme is that biodegradability has also been shown to be inversely related to compound hydrophobicity (expressed as the octanol/water partition coefficient).

The number, type, and location of substituents on aromatic rings affect their biodegradability. Typically, the presence of a hydroxyl or carboxyl substituent decreases the resistance of the compound to degradation, whereas other substituents (e.g., methyl, nitro, amino, sulfonyl, phosphonyl, and especially halogen substituents) tend to increase its resistance to enzymatic degradation. In general, the greater the number of obstructive substituents, the more resistant the compound is to degradation.

Ortho- and para-substituted compounds are generally degraded more rapidly than meta-substituted compounds. Aliphatic compounds are increasingly resistant to degradation as their molecular weight and degree of branching increase.

IV. COMMON METABOLIC MOTIFS

Because of the diversity of natural microbial communities, the multiplicity of enzymes expressed by a given microorganism, and the low substrate specificity of some enzymes, the metabolic diversity of natural microbial communities is enormous. However, there are only two main mechanisms for generating energy from heterotrophic decomposition—substrate-level phosphorylation and oxidative phosphorylation. There are also few identifiable substrates that contribute to substrate-level phosphorylation. Furthermore, relatively few compounds are recognized by microbial enzyme systems as precursors for cell structural materials.

Thus, at the end of the metabolic pathway where the substrates are complex, similarities in metabolic pathways are best described in general terms and are limited only by the types of enzymes available and their substrate specificities. Indeed, many parallel and analogous reactions may occur simultaneously that result in similar, albeit different, intermediates. Common reaction types at this end of the pathway include those reactions that result in the formation of smaller molecules that tend toward the gross oxidation level of cellular material. These include, most predominantly, oxidations and hydrolyses, many of which occur extracellularly. At this end of the reaction sequence, futile reactions might occur as products cannot be further metabolized by the organism that catalyzed the initial reaction. Indeed, other organisms may or may not continue the reaction sequence by further modifying the pollutant.

At the opposite end of the metabolic pathway, where the substrates are simpler and more sequestered by individual organisms, substrates are likely to be of a few recognizable forms. At this end of the pathway, analogous or even identical metabolic processes are shared by taxonomically diverse microbes. Some of these recognizable motifs in pollutant degradation include β-oxidation of aliphatic chain structures (similar to the oxidation mechanism for fatty acids) and *ortho-* and *meta*-fission pathways for aromatic rings. At this level, few reaction steps will be futile and a high proportion of the substrate atoms will participate in generation of energy or cell material.

V. RELATIONSHIP TO MICROBIAL TAXONOMY

A. Relationship to Metabolic Characteristics

Classification schemes for microorganisms are often based on metabolic characteristics. For this reason, the idea that taxonomically related groups of organisms are most often involved in metabolism of a specific group of pollutants should not be surprising. Nor should it be surprising that heterotrophic organisms displaying a high degree of metabolic diversity are involved in a variety of pollutant modifications. The latter description aptly fits organisms of the genus *Pseudomonas,* which are frequently involved in the degradation of pollutants. It also fits actinomycetes, which, in their usual habitat, degrade chemically complex soil organic matter.

Organisms possessing enzymes with a low degree of substrate specificity are also likely pollutant degraders. Examples from this group include the lignolytic fungi such as *Phanerochaete* spp. These organisms produce nonspecific ligninases that oxidize a wide variety of carbon substrates. Because the enzymes are nonspecific, the transformation products are varied.

Although the previously listed organisms are predominantly aerobic, this class of organisms is not unique in its pollutant-degrading abilities. Anaerobic microorganisms possess characteristics that are especially well adapted for elimination of certain types of pollutants. A classic problem in the biological treatment of high-carbon-concentration waste materials is the mass transfer of oxygen to aerobes for oxidation. As an alternative, anaerobic diges-

tion has become a popular treatment of such wastes prior to, or in lieu of, aerobic treatment. Like aerobes, anaerobes catalyze the oxidation of organic chemicals, but the terminal electron acceptor (oxidant) in the process is a compound other than molecular oxygen. Microorganisms are classified, based on the compounds used as terminal oxidants in respiration, as nitrate or sulfate reducers, as methanogens (carbon dioxide reducers) or fermenters (organisms that use oxidized organic molecules as terminal electron acceptors).

Anaerobic digestion is advantageous for high-carbon-concentration wastes for several reasons. First, the rate of oxygen transfer limits the rate of biodegradation of readily degradable organic material in high-concentration systems. Second, as a rule, anaerobic digestion processes produce relatively minor amounts of biomass relative to the amount of carbon digested. Instead, reduced carbon is converted to carbon dioxide, which in turn may be reduced to methane by methanogenic organisms. So, a substantial fraction of the carbon from high-concentration waste is converted anaerobically to gases that leave the system. Finally, methane that is generated during the digestion may be used as fuel, offsetting a fraction of the operating costs of the waste treatment process.

The environment of anaerobes is also conducive to reductive transformations. Most notably, reductive dehalogenation results in the displacement of halogen substituents on organohalogen compounds by hydrogen. This is an important reaction, because reductive dechlorination is one of the few known mechanisms for elimination of many highly chlorinated pollutants, such as PCBs and polychlorinated dioxins and furans, from the environment.

B. Role of Microbial Consortia

Biodegradation of complex organic chemicals often requires the metabolic capabilities of multiple organisms. Organisms that live in close association with one another that participate (either directly or indirectly) in a particular activity are often referred to as consortia. Microbial consortia have been implicated in the degradation of many common pollutants, including polynuclear aromatic hydrocarbons, pesticides, and chlorinated solvents.

VI. EFFECTS OF THE ENVIRONMENT

One of the fundamental characteristics of living organisms is their ability to respond to environmental stimuli. Environmental conditions play a pivotal role in the regulation of biological processes in general, and particularly of biodegradation processes. Many environmental factors have been identified that affect the biodegradation of pollutants, several of which are discussed next.

A. Pollutant Concentration

In general, biodegradation is affected by pollutant concentrations in the same way that all chemical processes are affected by their reactant concentrations. The more molecules of the reactant that are present in a given system, the greater is the number of molecules reaching its activated state at a given time, and thus the greater is the number of reactant molecules being converted to products at a given time. This simple scenario is somewhat complicated, however, by the catalytic aspects of biochemical reactions.

Although enzymes process reactants at extremely high rates, they are limited by the number of active sites available at a given time that accept reactants for processing. Therefore, at low substrate concentrations, the interactions of enzymes with substrate molecules are relatively infrequent, so the reaction rate increases approximately linearly with increasing substrate concentrations. At high substrate concentrations, however, the enzyme active sites become "saturated" with reactant molecules owing to the frequent interactions, and the rate becomes essentially constant, regardless of further increases in the substrate concentrations.

Figure 1 illustrates the Michaelis–Menten model (named for the biochemists that first reported it in 1913), which adequately describes most simple enzyme-catalyzed reactions. On an empirical level, this model has also been used successfully to compare rates of complex degradation reactions by

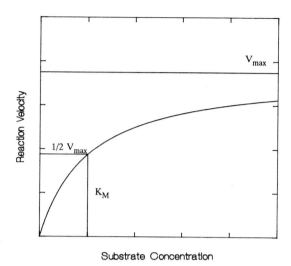

FIGURE 1 Michaelis–Menten curve showing the relationship between reaction velocity and substrate concentration. See text for details.

mixed microbial communities as well. The model is represented by a family of curves that are defined by the following equation in two parameters:

$$\nu = \frac{V_{max} \cdot S}{K_M + S,} \qquad (1)$$

where ν is the reaction velocity (mg · liter^{-1} · hr^{-1} reactant converted) and S is the substrate concentration (mg · liter^{-1}). The parameters are V_{max}, the maximum theoretical velocity of substrate degradation as the value of S becomes high, and K_M, (Michaelis–Menten half-saturation constant), the value of S that corresponds to one-half of V_{max}.

Other common kinetic models that have been used to describe pollutant degradation are the simple, first-order model

$$r = K_1 \cdot S, \qquad (2)$$

where r is the instantaneous rate of substrate degradation and K_1 is an experimentally derived first-order rate constant, and the second-order model

$$r = K_2 \cdot S \cdot X, \qquad (3)$$

wherein the rate of biodegradation is not only a function of the concentration of the contaminant, but also of the concentration of microbial biomass (X) involved in the conversion. K_2 is the experimentally derived, second-order rate constant. Although in concept, biodegradation would be expected to depend on the biomass concentration, few examples of biomass-dependent degradation of pollutants in natural environments have been demonstrated. This is probably because activities vary greatly among microorganisms within an environment.

A second mechanism by which the concentration of pollutants affects their degradation is toxicity. Though the biological effects of pollutants differ among various types of organisms (and indeed, quite significantly even among strains of bacteria), toxicity may either reduce the rate of biodegradation of pollutants or even prohibit their degradation.

B. Nutrient Concentrations

Sustained growth of microorganisms in the presence of a suitable source of carbon and energy also requires sources of other nutrients to balance carbon uptake. The relative weight proportions of vital nutrients as they occur in microbial biomass are carbon, 50%; oxygen, 20%; nitrogen, 10 to 15%; hydrogen, <10%; and phosphorus and sulfur, <5% combined. The balance of cell weight is contributed by a multiplicity of trace nutrients. To continually increase the concentration of biomass in an environment or convert a carbon-containing pollutant to microbial biomass, a continuous source of these nutrients is required. However, this rarely occurs. Most often, an established microbial community will be involved in metabolizing pollutants. Biomass creation is balanced to some degree by biomass recycling within the microbial community, and much of the organic carbon associated with the pollutant molecules is converted to carbon dioxide in energy-generating processes.

Alternatively, many biodegradation processes do not result in microbial assimilation of the carbon or in generation of energy. Instead, the microorganisms utilize another compound or set of compounds as their growth substrate while co-oxidizing the pollutant. In these cases, the amounts of other structural ingredients required by the microbial population will be related to the rate of biomass generation at the expense of the primary growth substrate and will be decoupled from the rate of pollutant utilization.

In either case, the weight of biomass generated relative to the weight of nutrient material incorporated is identified as the yield on that nutrient. These yields are often important operating parameters for engineered biological processes.

C. Temperature

Temperature affects biodegradation in at least two ways. First, most biological reaction rates (like other chemical reaction rates) are governed by physical principles that are described by the well-known Arrhenius relationship (named for Svandte Arrhenius, who described this relationship in 1887). Specifically, increasing the temperature of a reaction system results in an increase in the rate of reaction, because of the corresponding increase in the energies of the reactant molecules. Because biochemical reactions generally occur over a relatively narrow range of temperatures, temperature effects on reaction rates for pure enzyme systems are fairly predictable within the confines of the temperature stability of the enzymes involved. As a rule of thumb, rates of biochemical reactions increase approximately twofold per 10°C increase in temperature. For a reaction sequence involving several enzyme-catalyzed steps at a given temperature, the rate of the process is governed by the slowest (rate-limiting) step at that temperature.

The second effect of temperature on pollutant biodegradation results from physiological effects of temperature on microorganisms at the cellular level. Based on the observed temperature optima of cell replication, most microorganisms are empirically classified as psychrophiles (whose optimum temperature for growth occurs within a range from about 0° to 15°C), mesophiles (25° to 40°C) or thermophiles (40° to 70°C). Determinants of operable temperature ranges for biological activities include the concentrations of various intracellular temperature-regulating compounds, inclusion of variously saturated fatty acids to control fluidity of membranes, and the heat stabilities of cell components.

D. Other Factors

Similarly, microorganisms exhibit physiologically based activity optima for a variety of environmental factors that affect the rates of degradation of pollutants. These factors include concentrations of various ions (including pH), cofactors (e.g., riboflavin), toxic or inhibitory substances, and competing substrates, the redox potential of the environment, availability of suitable electron acceptors, and interactions among microorganisms. The complexity of the environment with respect to these factors and the interplay among them render the occurrence and rates of biodegradation unpredictable.

VII. APPLICATIONS OF BIOTECHNOLOGY TO ENVIRONMENTAL PROBLEM SOLVING

Because recent, international public interest has focused on the prevalence of manufacturing-derived pollutants in the environment, new interest has arisen in the use of engineering techniques to manage pollution problems. The following are types and descriptions of engineered processes for pollution abatement.

A. Biological Wastewater Treatment

The utility of microbiological techniques in the amelioration of pollution is illustrated by municipal wastewater treatment, which is based on the consumption by bacteria of human waste and carbonaceous substances in environmental runoff. Wastewater treatment involves a combination of engineered processes that typically include at least one biological process. Because the bulk of the carbon entering waste treatment systems is largely in the form of readily biodegradable compounds, the bacteria convert it to carbon dioxide and biomass. The motivators for municipal water treat-ment are primarily elimination of microbial pathogenesis and reduction of biochemical oxygen demand of water prior to discharge to receiving streams. [See WASTEWATER TREATMENT BIOLOGY.]

Typical treatment scenarios for water are divided into phases. Primary (or pre-) treatment refers to the removal of solids from suspension. Secondary treatment refers to digestion, usually by aerobic decomposition, of most of the remaining (substantially dissolved) organic material. Tertiary (or post-) treatment is occasionally used for a variety of problems on a plant-specific basis. Typical problems include overabundance of inorganic nutrients, toxicity in the effluent stream, or persistence of pathogens beyond the secondary treatment unit. Tertiary treatment may involve biological, chemical, or physical processes.

The following are typical uses of biological processes in treatment plant design. Solids from the primary treatment unit may be digested anaerobically, converting reduced organic carbon into (primarily) methane and carbon dioxide. The microorganisms responsible for the conversion include mixed populations of methanogens, hydrogen-producing anaerobes, and mixed acid fermenters. The reactions are characterized by close interdependence of the participating organisms. The overall carbon yields of the microorganisms are low, so the process results in a substantial decrease in the mass of the solids.

Secondary treatment may consist of a variety of aerobic or anaerobic processes. Anaerobic processes are substantially the same as the solids treatment processes just described. Aerobic processes are classified as either attached- or suspended-growth systems. Attached- growth systems include: submerged, fixed-film bioreactors, in which a reactor packing material (usually of plastic or ceramic construction) is used as a support for the growth of organisms; trickling filters, in which the solid support (often consisting of rocks or, increasingly, random or structured plastic support materials) is sprayed with the water to be treated; rotating biological contactors, in which the solid support is a disk suspended on a shaft that rotates, partially submerged, in the water undergoing treatment; and fluidized beds, in which solid support material such as sand or particulate organic carbon is fluidized by passage of water through the bed. Variations on these themes are common, each having its own design advantages and disadvantages.

Aerobic, suspended-growth systems, the most common form of secondary treatment, involve air diffusion into a continuously stirred tank reactor. The reactor often has a sequential, multichambered design to provide a plug-flow quality (i.e., reactions in separate chambers display adaptation to different influent concentrations of organic carbon, resulting in more complete carbon stabilization). The reactor is followed by a settling tank to remove suspended solids prior to discharge to receiving streams and a portion of the biomass is returned to the reactor to optimize carbon stabilization. A more passive approach to secondary treatment is an aerated lagoon. This is simply a surface impoundment for water, wherein the water is vigorously aerated and (perhaps) supplemented to promote biological activity and/or air-stripping of volatile compounds.

Typical biological processes used for tertiary treatment include, for example, nitrate removal systems wherein denitrifying (nitrate-reducing) bacteria convert dissolved nitrate into nitrogen gas, which exits the system by volatilization. Addition of disinfectants (such as hypochlorite or ozone) to destroy surviving pathogens may be considered a tertiary treatment.

B. Volume Reduction of Organic Material

A second type of pollutant minimization process involving microorganisms is the volume reduction of carbonaceous wastes. Such processes include anaerobic digestion (as described earlier) and solid waste composting. Solid waste composting is a process that involves the conversion of reduced carbonaceous materials to carbon dioxide and water under aerobic conditions. A portion of the metabolic heat that is generated in the process is retained by the compost, causing an elevation in its temperature. Microbial succession (i.e., a change in the microbial community favoring different groups of microbes in succession) can be observed during composting operations from predominantly mesophilic to thermophilic organisms. The latter may operate optimally at temperatures >60°C. In addition to volume reduction of the waste material, the high temperatures achieved during composting lead to the inactivation of viruses and destruction of other pathogens (microbial and invertebrate). Because of the increased temperatures involved in composting, conversions of potentially toxic chemicals may be more rapid than similar reactions at ambient temperatures.

C. Bioremediation

The need for remediation technologies has been created by decades of indiscriminate disposal of hazardous wastes primarily by commercial and military entities. Legislation passed largely since 1970 has brought polluting activities under stricter control and resulted in stiffer penalties for violations, and has thus served as the driving force for a growing remediation industry.

Bioremediation is environmental decontamination carried out by microorganisms, predominantly bacteria. Although the microorganisms involved possess the natural ability to conduct the process, the rate of the process itself is often limited by the unsuitability of the native environment for rapid decontamination. In many cases, the rate-limiting factors can be controlled to expedite biodegradation. The factors that most often limit the rates of biodegradation of readily biodegradable pollutants are concentrations of oxygen, nitrogen, and phosphorus. Although bacteria require other elements (such as sulfur, magnesium, potassium, and a variety of trace metals) for growth as well, these elements are generally present in sufficient quantities under ambient conditions, so their addition is rarely required to promote biodegradation. The ambient temperature of the contaminated environment also affects the rate of biodegradation. The cost of changing a contaminated environment in order to bring about an increase in the rate of pollutant biodegradation can be high, so a compromise is often accepted between the treatment cost and the time required to achieve clean-up goals.

The advantage of bioremediation over other forms (i.e., physical and chemical) of remediation is that microorganisms have the ability to degrade stable compounds without expensive inputs of energy or reactant chemicals, resulting in lower costs. These processes have been devised for treatment of a wide variety of contaminated materials, including soil, surface water, industrial process water, sludges, subsurface solids, and groundwater. Aside from uses of biological water treatment for elimination of industrial wastes, the first widely practiced biological remediation technique involved application of petroleum-laden sludge to agricultural soil. When amended with agricultural fertilizers, bacteria within the soil carry out the oxidation of petroleum hydrocarbons. This process is well established in the scientific literature and in industrial practice, particularly among petroleum refiners, and is known as land farming.

Several well-recognized technologies have been developed, and new technologies are being developed aggressively. The following are descriptions of existing technologies. They may be divided based on the materials treated (solids, water, or both) and the types of engineered systems used. A list of technologies with their associated advantages and disadvantages is provided in Tables I and II.

Remediation options for addressing soils and solids are divided into two categories and identified as either *in situ* (in place) or *ex situ* depending on whether excavation of the soil is required. Excavation is avoided, where possible, because it adds expense to the treatment, it is disruptive to site activities, and additional regulations often apply to excavated soil that do not apply to soil treated in place.

In situ treatment requires detailed analysis of the subsurface environment, including assessment of hydrogeological parameters (hydraulic conductivity, flow direction, stratigraphy, heterogeneity, etc.) prior to implementation of the treatment. This form of treatment is carried out by delivering required additives through an infiltration and hydraulic capture system designed to prohibit further

TABLE I

Developed Bioremediation Technologies for Water Treatment with Associated Advantages and Disadvantages

Technology	Advantages	Disadvantages
Oxidation pond	Inexpensive	Odor production Slow conversion of many toxic chemicals Long residence times for BOD reduction Poor containment Periodic sludge removal required
Aerated lagoon	Inexpensive Higher oxidation rates	Poor containment Periodic sludge removal required
Activated sludge	High oxidation rates Biomass recycle Short residence times Good containment	Toxicity causes biomass washout Sensitive to changes in flow and loading
Trickling filter, rotating contactor	Attached biomass withstands toxicity Good oxygenation at biomass surface Good containment	Anaerobic layers in biofilm create odors Mechanical parts subject to failure
Fixed film	Low maintenance Attached biomass withstands toxicity Withstands fluctuation in loading rate Good containment High biomass/water ratio High removal efficiencies	Incompletely mixed—mass-transfer resistance Longer residence times required Plugging if designed loading exceeded
Fluidized bed	Attached biomass withstands toxicity Withstands fluctuation in loading rate Good containment Reduced mass-transfer resistance	Solids circulation—wear and tear

TABLE II
Developed Bioremediation Technologies for Solid-Phase Treatment with
Associated Advantages and Disadvantages

Technology	Advantages	Disadvantages
In situ	No excavation Inexpensive Nondisruptive Nonseasonal	Not for heterogeneous sites Not for low-permeability sites Not for aquifers that are subject to plugging Little physical control Treatment times in years
Ex situ	Greater physical control	Disruptive to site activities Land ban restrictions for hazardous wastes
Land farming	Inexpensive	Little containment of pollutants Potential contamination of native soil One or more growing seasons required
Land treatment	Relatively inexpensive Runoff control Little cross-contamination potential	One or more growing seasons required Less effective for fine-grained soils
Heap treatment	High degree of physical control Runoff control Little cross-contamination potential Space-conservative Forced aeration	Not for low-permeability soils
Bioslurry	Applicable for fine-grained soil High degree of physical control Contained process	Relatively expensive Tankage required Long residence times Coarse-grained soils fail to suspend

spread of contaminants by movement in groundwater. The system consists of a series of injection wells or infiltration galleries and one or more extraction (recovery) wells. Water is extracted, amended, and reinfiltrated to the subsurface to promote bacterial action on the pollutants. Often, biological or other treatment of the water is required after extraction of the groundwater from the subsurface and prior to reinfiltration to avoid well-clogging by growth of bacteria.

For treatment of unsaturated zones, amended water is percolated through soil, bringing nutrients into contact with microorganisms at the site where biodegradation takes place. Alternatively, air can sometimes be used as the carrier for oxygen in a pair of related technologies called biologically enhanced soil vapor extraction, and bioventing. Oxygen can also be delivered into saturated subsurface materials by forcing it into injection wells: a process known as air sparging. This process can have limited influence on both the saturated and unsaturated subsurface environments.

There are several forms of treatment for excavated soils, including land farming, land treatment, heap treatment, and bioslurry treatment. Land treatment is identical to land farming, except that the treatment takes place within a treatment cell that is sealed with an impermeable liner to control runoff from precipitation, thus protecting underlying environments from the potential spread of contamination. Heap treatment involves similar treatment of soil under conditions of confined space. Because farm implements cannot be used for the incorporation of nutrients and aeration, treatment of a heap of soil requires some mode of transmission of the necessary elements, either by repeated addition and mixing or by infiltration through an irrigation/collection system.

Bioslurry treatment involves the suspension of soil in an aqueous bioreactor, which provides the highest level of engineering control over the microbiological environment. Mass-transfer limitations of oxygen and nutrient materials are maintained at a minimum because of high-intensity mixing within the reactor. Whereas other solid-phase systems are more applicable to coarse-textured soils because of their permeability to water and oxygen, bioslurry systems are more applicable to fine-textured materials (silts and clays) because they are more readily suspended than sands.

Composting systems have also been used in bioremediation, although the scope of their application has been limited. Composting requires the addition of a carbonaceous bulking agent that fuels its characteristic increases in temperature. This method is well suited to treatment of sludges (already high in organic carbon) and to scenarios where higher temperatures improve the process. Examples are pollutants that are mobilized, increasing their accessibility to microorganisms at higher temperatures, and pollutants with high activation energies that result in low rates of biodegradation at typical ambient temperatures. Although soils can be treated by composting, the soils themselves do not add substantial organic carbon to the treatment process. Less equipment- and labor-intensive processes are usually more cost-effective where feasible.

Water treatment systems were discussed in a previous section with particular regard to municipal waste treatment. These systems are also applicable for the treatment of industrial process water and for remediation of contaminated surface water and ground water. Though the design motifs are the same as discussed in previous sections, the goals of these types of water treatment are often different. Whereas the primary concerns related to the release of municipal waste to receiving streams are oxygen demand and pathogenicity, the primary concern associated with process streams and polluted waters is toxicity. So, while the goal of municipal water treatment is to stabilize reduced carbon by complete oxidation or incorporation into biomass, the goal of treatment for process/polluted water is elimination of toxic components, often without regard to the oxygen demand of the effluent water.

D. Role of Genetically Engineered Organisms

Genetic engineering describes the human-mediated transfer of genetic material between living organisms. The goal of genetic engineering is to transfer the metabolic capability of one organism into another organism. This is done for a variety of reasons. First, an organism having a valued metabolic trait may not be well suited to the environment where the expression of that trait would be valuable. Thus, if the trait is transferred to another organism that is well suited to that environment, the trait will be more effectively expressed there. Second, it is occasionally desirable to place a genetic trait under a specific form of metabolic control, or to remove existing metabolic controls. Third, it may be desirable to place more than one valued metabolic trait in a single organism. Fourth, specific genetic traits may be placed together such that one trait brings about a detectable signal that serves as an indicator that the other activity is taking place. These "reporter genes" are exemplified by the *lux* gene sequence, isolated from a fluorescent bacterium, which causes the bacterium to give off visible light when the gene sequence is active. This allows the level of activity to be estimated by the intensity of light given off by the culture.

Each of the aforementioned techniques has been used for solving pollution problems. However, to date these applications have been largely confined to the laboratory owing to regulatory constraints. These constraints have been issued based on concerns related to the release of engineered microorganisms to the environment. These techniques may eventually pose solutions to some of the toughest environmental pollution problems.

E. Bioaugmentation

An alternative to engineering organisms for a specific purpose is the much more commonly practiced selection and concentration of microorganisms with specific, desirable metabolic characteristics. Concentrated organisms can then be reintroduced into the environment for expression of their activities. The addition of microbial concentrates to improve a process is called bioaugmentation.

In its simplest form, bioaugmentation may be the simple addition of a shovel-full of anaerobic digester waste to a septic tank to "seed" the tank. In more complex forms it may involve rigorous procedures for isolation and enrichment of organisms that are specific degraders for a pollutant. These organisms may be added to a bioreactor to increase the rate of degradation. Because most organic chemicals are biodegraded by naturally occurring organisms, addition of concentrated organisms is usually superfluous, and increases in degradation activity due to added organisms are generally marginal and transitory. Increases are transitory because microorganisms cultured under laboratory conditions are often poorly suited to the environments to which they are released for expression of their activities. Instead, they are quickly out competed by natural organisms that are well adapted to that environment. So, unless a specialized organism can establish itself as a stable and active portion of the microbial community, continuous or intermittent addition of the organisms would be required to obtain long-term benefits.

Examples of bioaugmentation that have proven critical to bioremediation have been cited in rare cases where no natural microorganisms exist in a particular environment that carry out the specific activity of interest, and in cases where active bacteria have been essentially eliminated from the environment by toxicity. Where toxicity prevails, the source of the toxicity must be removed prior to the addition of bacteria or the added organisms will likely be rapidly destroyed as well. Under these circumstances, addition of specialized or pure cultures may be less effective for reestablishment of a native population than addition of a diverse culture, such as activated sludge or stabilized manure.

VIII. RELATIVE EFFECTS OF NATURAL AND INDUCED/ ENHANCED BIODEGRADATION ON ENVIRONMENTAL QUALITY

The earth is characterized by a carbon cycle of enormous proportions that involves fixation of atmospheric carbon dioxide into plant material, consumption of plant material by primary consumers, consumption of their biomass by secondary consumers, and so on. At each stage the biomass may die and enter the realm of the microbial decomposers—the ultimate consumers—that mineralize the carbon and release it, once again, as carbon dioxide. Although the amount of global carbon involved in organic pollution may be considered small, the negative effects of that small portion make it significant. Fortunately, the enzymes that participate in the global cycle are prepared to process most of these unusual compounds as well. The assimilative capacity of the earth for inputs of pollutants has long been held in awe, although the pollution resulting from some human practices has challenged the health and even the existence of some groups of living organisms. Thus, enlightened environmental stewardship is required to maintain the quality of environmental and public health. [See GLOBAL CARBON CYCLE.]

One need only examine the effects of wastewater treatment on environmental and public health to be convinced that engineering practices can have broad-scale health effects. However, the success of these efforts will depend on our ability to identify toxins and respond appropriately. Among polluted sites on record, microbiological techniques are currently being used to decontaminate only a small proportion of the total. However, there is no accounting for the large proportion of organic pollutants that have been biologically processed over the millennia without recognition of the benefits of the process. The advantages of bioremediation techniques will surely lead to their increased utilization, especially as regulatory agencies increase incentives for industry to destroy contaminants rather than transfer them to other environments.

Glossary

Bioremediation Biological decontamination of soil, water, or other environmental media.

Catabolism Portion of metabolism that relates to the breakdown of molecules and the release of energy from chemical bonds.

Catalyst Substance that increases the rate of a chemical reaction without itself being chemically modified, and without affecting the equilibrium of the reaction.

Cometabolism Metabolism of a substrate from which the participating organisms do not obtain either energy or structural elements.

Decomposition Breakdown of chemicals resulting in products that are nearer the thermodynamic equilibrium state of the elements.

Metabolic pathway Sequence of biochemical reactions that result in the formation of products from reactants.

Metabolism Total of all cell processes; or processes carried out on a particular substrate.

Mineralization Breakdown of organic chemicals to mineral (inorganic) end products (e.g., carbon dioxide water, chloride).

Terminal electron acceptor Final oxidant in a respiratory chain that ultimately accepts the electrons released during biochemical breakdown of reduced carbon compounds.

Transformation Biodegradation of organic chemicals that results in organic end products, typically the result of cometabolism.

Bibliography

Alexander, M. (1981). Biodegradation of chemicals of environmental concern. *Science* **211**(9), 132–138.

Atlas, R. M., and Bartha, R. (1981). "Microbial Ecology: Fundamentals and Applications." Reading, Mass.: Addison–Wesley.

Dagley, S. (1971). Catabolism of aromatic compounds by microorganisms. *Adv. Microb. Physiol.* **6,** 1–46.

Dobbins, D. C., Aelion, C. M., and Pfaender, F. K. (1992). Subsurface, terrestrial microbial ecology and biodegradation of organic chemicals: A review. *Crit. Rev. Environ. Contr.* **22**(1/2), 67–136.

Evans, W. C., and Fuchs, G. (1988). Anaerobic degradation of aromatic compounds. *Annu. Rev. Microbiol.* **42,** 289–317.

Gibson, D. T., ed. (1984). "Microbial Degradation of Organic Compounds." New York: Marcel Dekker.

Klug, M. J., and Reddy, C. A., eds. (1984). "Current Perspectives in Microbial Ecology." Washington, D.C.: American Society for Microbiology.

Reineke, W., and Knacknmuss, H.-J. (1988). Microbial degradation of haloaromatics. *Annu. Rev. Microbiol.* **42,** 263–287.

Tursman, J. F., and Cork, D. J. (1992). Subsurface contaminant bioremediation engineering. *Crit. Rev. Environ. Contr.* **22**(1/2), 1–26.

Biodiversity

G. T. Prance
Royal Botanic Gardens, United Kingdom

I. Introduction
II. Species Diversity
III. Genetic Diversity
IV. Habitat Diversity
V. Biodiversity under Threat
VI. The Tasks Ahead

Biodiversity is the total variability among organisms and the habitats in which they live; it includes diversity of species, genetic materials, and ecosystems.

Biodiversity is under unprecedented environmental threat. Species extinction is occuring at rates 10,000 times greater than natural rates, and yet fewer than 5% of the world's species have been described and named by systematists, so the unknown is being lost. Meanwhile, entire habitats are disappearing, preventing an understanding of their components' interactions, an understanding we need to maintain the quality of life on earth.

Those species which survive are threatened by diminished genetic variability. All components of biodiversity are menaced by the environmental destruction, which is marring the efficiency of the function of the ecosystem and reducing future options for human utilization of organisms.

I. INTRODUCTION

"Biodiversity" is an all-inclusive term to describe the total variation that occurs among living organisms of our planet, and it includes three main components: (1) the diversity of species that occurs in the world, from the familiar plants and animals to the less conspicuous fungi, bacteria, protozoans, and viruses; (2) the genetic variation that occurs within individual species that causes them to vary in their appearance (phenotype) or their ecological responses and allows them to react to the process of evolutionary selection; and (3) the diversity of habitats or ecological complexes in which species occur together, whether they be such well-known ones as rainforest, tundra, and coral reefs or the complex of bacteria that inhabit the human body or a gram of soil.

"Biodiversity" is a relatively recent term that has now become familiar because it has entered into the political arena, especially through its use in the Biodiversity Convention that was much debated and agreed on by the United Nations Conference on Environment and Development held in Rio de Janeiro in 1992. The word "biodiversity," coined by Walter G. Rosen in 1965 for the National Forum on Biodiversity held in Washington, D.C. in 1986. The proceedings of that forum were edited by sociobiologist Edward O. Wilson under the title *Biodiversity* which is a contraction of biological diversity.

It is a useful term because of the interrelatedness among species, genes, and habitats. To preserve life on Earth, it is important to consider the interactions among all three aspects, and so it is not surprising that the term "biodiversity" was developed at a time when the conservation of living organisms became a major concern of biologists. The interference with any part of biodiversity has an effect on the other components, therefore an adequate conservation policy takes into account all three aspects of biodiversity.

II. SPECIES DIVERSITY

Species are the fundamental units of biodiversity, because without them the other parts would not exist. Habitats are made up of groups of species, and they are lost or drastically changed when their component species become extinct. It is also the species that carry the genes that comprise genetic diversity, and so the conservation of species diversity is basic to the conservation of all biodiversity. [See SPECIES DIVERSITY.]

The number of species described to date is approximately 1.6 million (see Table I) but the num-

TABLE I

Numbers of Species of Major Groups of Organisms Estimated to Exist[a]

Organism	Species described	Estimated number of species
Viruses	5000	500,000
Bacteria	3058	400,000
Fungi	70,000	1 million
Protozoa	40,000	200,000
Algae	40,000	350,000
Lichens	13,500	17,000
Bryophytes	14,000	18,000
Pteridophytes	10,000	12,000
Gymnosperms	650	650
Seed plants	250,000	300,000
Insects	950,000	8–50 million
Arachnids	75,000	150,000
Mollusks	70,000	200,000
Nematodes	15,000	100,000
Birds	9881	9900
Other vertebrates	35,000	40,000
Total	1,601,039	

[a] Data were derived from Groombridge (1992).

ber of undescribed species is far greater. It is obviously hard to estimate how many unknown species remain to be described, and calculations vary from 5 to 100 million, mainly because of uncertainty about the numbers of insects, bacteria, fungi, and nematodes. About 3000 species of bacteria have been described, but a recent study in Norway indicated that there were between 4000 and 5000 species in a single gram of soil from a beech forest (Klug and Tiedje, 1994). The current estimate that there are 400,000 species of bacteria could be a considerable underestimate. Only 70,000 species of fungi have been described, but mycologist David L. Hawksworth (1991) estimated that there may be as many as 1.5 million species. For some groups of organisms, we can be confident with the statistics. For example, in 1990 there were 9881 known species of birds (Sibley and Monroe, 1990), and the estimate of 250,000 species of seed plants is reasonably accurate. We can use the statistics from these better-known groups to extrapolate conservation policies for many other groups.

Species do not occur evenly around the world, and therefore conservation planners need to know about the distribution and density of species in different habitats and different countries. McNeely et al. (1990) estimated that 70% of the world's total species diversity occurs in only 12 countries: Australia, Brazil, China, Colombia, Ecuador, India, Indonesia, Madagascar, Malaysia, Mexico, Peru, and Zaire, and these have been termed "megadiversity" countries. Table II shows the number of plant species in the 10 most species–diverse countries, seven of which are on McNeely's list, compared with the 10 countries with the highest number of restricted-range birds. These are obviously priority countries in terms of conservation.

Some habitats are much more species diverse than others, and rainforests are in first place for the quantity of species. The forest with the greatest known diversity of tree species is at Cuyabeno in Ecuador where Valencia et al. (1994) found 307 species of trees in a single hectare; in second place is Yanomono, near Iquitos in Peru, where botanist Alwyn H. Gentry (1988) found 300 species of trees and lianas with a diameter of 10 cm or more in a single hectare. Table III shows the tree species

TABLE II

Eleven of the Most Species-Diverse Countries in the World for Higher Plants and the 10 Countries with the Highest Numbers of Restricted-Range Bird Species[a]

Country	No. of plant species	Country	No. of restricted-range bird species
Brazil	55,000	Indonesia	410
Colombia	35,000	Peru	210
China	30,000	Brazil	200
Mexico	25,000	Colombia	190
South Africa	23,000	Papua New Guinea	170
Former Soviet Union	22,000	Ecuador	160
Indonesia	20,000	Venezuela	120
Venezuela	20,000	Philippines	110
United States	18,000	Mexico	102
Australia	15,000	Solomon Islands	98
India	15,000		

[a] Data were derived from Groombridge (1992) and Bibby *et al.* (1992).

diversity of some rainforest areas, which varies from 307 down to 81 species, depending on local conditions such as soil, topography, and rainfall. The more humid areas without a pronounced dry season tend to have greater species diversity than a seasonal forest. In addition to the wealth of tree species, rainforests contain a large number of herbs, shrubs, and epiphytic and parasitic species that usually more than double the total species count for any area. In the Río Palenque Reserve in western

TABLE III

Tree Species Diversity for Some Tropical Rainforests

Locality	No. of species per hectare	Minimum diameter (cm)
Cuyabeno, Ecuador	307	10
Yanomono, Peru	300	10
Mishana, Peru	295	10
Jatun Sacha, Ecuador	248	10
Añangu, Ecuador	244	10
Johore, Malaysia	227	10
Mulu, Sarawak	223	10
Cocha Cashu, Peru	189	10
Manaus, Brazil	179	10
Tambopata, Peru	168	10
Xingu River, Brazil	162, 133, 118	10
Breves, Brazil	157	10
Oveng, Gabon	131	10
Alto Ivon, Bolivia	81	10

Ecuador, Calaway Dodson and Gentry (1978) found 365 species of vascular plants in one-tenth of a hectare. The total number of species recorded for this tiny reserve, which covers 1.7 km² of rainforest, was 1033 species. In the rainforest of Costa Rica, Timothy Whitmore and colleagues (1985) found 233 species of plants in 100 m². An area half the size of a tennis court contained about one-sixth as many species as the total flora of the British Isles. Ten hectares of forest in Borneo studied by botanist Peter Ashton (1964) contained 700 species of trees, the same number as in the whole of North America.

For almost any type of organism studied, the rainforest areas usually have the greatest species diversity. For example, the Amazon basin has 2000 species of fish; second is the Zaire River, with 790. These two rivers account for just over 33% of the world's total of 8400 freshwater species. However, it is insects that contribute the greatest species diversity to the rainforest and many other ecosystems. Insects comprise 56.4% of all described species and are likely to eventually be a much higher proportion (~65%) as more are described. A canopy tree in the tropical rainforest hosts at least 10 times more insect species than a tree of the same size at the temperate latitudes. In recent years insect diversity in tropical forests has been studied by the technique of fogging, in which a warm smoke laden with an insecticide is released through the canopy of a tree. The insects drop into collecting traps placed under the tree (see, e.g., Adis *et al.*, 1984; Erwin, 1990; Stork, 1991). Nigel E. Stork of the Natural History Museum in London found that 90% of the insects sampled in this way were undescribed. It is this mass of insects in the rainforest canopy, as well as the unknown masses of species in the seabeds, that leads to the uncertainty about the total number of species in the world. It was extrapolations from various fogging experiments that led entomologist Terry L. Erwin (1982, 1983, 1988) to calculate the much-debated figure that there are 30 million species in the world. Erwin found a total of 1200 beetle species alone in 19 trees of the tropical linden tree *Luehea seemannii* in Panama and that stimulated him into a series of conclusions that led him to estimate a world species total of 30 million.

Most people know that rainforests are areas of high biodiversity, but few are aware that their complexity is rivaled by that of coral reefs, the richest of all marine environments. The largest group of coral reefs is the Great Barrier Reef of Australia, which contains over 300 species of coral and supports 1500 species of fish. This area, which covers only 0.1% of the ocean surface, is home to 8% of the world's total fish species. In addition, there are 4000 species of mollusks and 252 species of birds in the coral cays [International Union for Conservation of Nature and National Resources (IUCN)/United Nations Environmental Programme (UNEP), 1988]. Other marine habitats also foster biodiversity. The most recently discovered new phylum of animals, the Loricifera, was described only in 1983; since then other new species of this benthic group have been described.

Conservationists are especially interested in two aspects of species diversity. First, the areas of high species diversity, and second, areas of high endemism of species. Areas that fall into either of these categories are obviously of high priority for conservation. Endemic species are species of restricted range that occur over only a small area. Environmental analyst Norman Myers (1988, 1990) identified 18 areas of tropical rainforest and Mediterranean type of vegetation that are characterized by their high proportion of endemic species and that are also threatened by environmental destruction. These so-called "hot spots" are listed in Table IV and are a good example of the use of species diversity to make the case for conservation. The 18 hot spots contain approximately 49,955 endemic species of plants, or 20% of the world total in an area of 746,400 km² or only 0.5% of the total land area. It is obvious why these should be targeted for conservation, and their protection would certainly preserve many thousands of insect species.

The greatest problem for studies of species diversity is the fact that such a small percentage of the total has been described by systematists.

III. GENETIC DIVERSITY

Infraspecific genetic diversity is the variation that occurs within species. In natural populations of a species there are small differences among individuals and larger ones among populations. For exam-

TABLE IV
Numbers of Endemic Species Present in "Hot Spots"[a]

	No. of endemic species present		
	Higher plants	Reptiles	Amphibians
Cape region of South Africa	6000	43	23
Upland western Amazonia	5000	?	
Atlantic coastal Brazil	5000	92	168
Madagascar rainforest	4900	234	142
Philippines rainforest	3700	120	41
Northern Borneo	3500	69	47
Eastern Himalaya	3500	20	25
Southwestern Australia	2830	25	22
Western Ecuador	2500	?	?
Chocó, Colombia	2500	137	111
Malay Peninsula	2400	25	7
California floristic province	2140	15	16
Western Ghats of India	1600	91	84
Central Chile	1450	?	?
New Caledonia	1400	21	0
Eastern Arc Mountains, Tanzania	535	?	49
Southwestern Sri Lanka	500	?	—
Southwestern Ivory Coast	200	?	2
Total	49,955	892	737

[a] Data were derived from Myers (1988, 1990) and Groombridge (1992).

ple, humans are one species, *Homo sapiens,* but it is possible to recognize individuals because of genetic differences (except identical twins, who are also genetically identical) and to distinguish the more general features of certain populations, such as the different races. We, and all other organisms, are what we are because of the genes that we have inherited. The genes are discrete segments of DNA molecules made up of four different nucleotide bases that form the genetic code that determines the characteristics of an individual. Each gene is formed of up to several thousand nucleotide pairs. All of the biological variation in the world comes from the numerous possible permutations of this apparently simple genetic code. Variation is caused by mutations, which are changes in both the genes and their arrangement on the chromosomes. This is spread in a sexually outbreeding population by the process of recombination. Since an average organism contains about 10^4–10^5 genes, the possible combinations are infinite, and it is estimated that our planet contains some 10^9 different genes. Some genes that control basic processes, such as photosynthesis in plants, are extremely stable and show relatively little variation, while others are extremely variable, for example, the genes that control flower colors. [*See* MOLECULAR EVOLUTION.]

Variation also occurs in the amount of DNA per cell and in the number of chromosomes within an individual. Both of these factors also contribute to the genetic diversity of organisms. Only a small proportion of the DNA is coded for specific proteins, and it is still not known what the real purpose is of the surplus DNA. In all living organisms except the bacteria and blue-green algae (prokaryotes), most of the DNA is organized into chromosomes. Most organisms have two copies of each chromosome, one of which was received from each parent. The fusion of genetic material from two separate individuals of a species during reproduction means that new combinations are formed that generate diversity.

The greatest significance of genetic diversity is that it allows the process of evolution to take place. In natural populations selection occurs for the combination of characters that are best suited for the environment. Selection is acting on the mutations occurring by changes in the genes themselves, on the recombination of parental material within a new individual, and on rearrangement of genes on the chromosomes. If the expression of a new genetic combination is more favorable to an organism, then it is more likely to survive to reproduce and form part of the next generation than another individual that does not have this combination, if it confers, for example, resistance to a disease. Conversely, if a change is harmful or lethal, as most genetic changes are, then an organism is less likely to survive. Many of the genetic changes that occur in organisms are neutral in their effects and are likely to survive, neither influencing evolution nor reproduction, but adding to the genetic diversity within the species. Ultimately, all genetic diversity has been derived from the process of mutation. A species has an increased potential for evolution and consequently a greater potential for survival if it has more genetic diversity. Species with little genetic variation are less able to adapt to new conditions such as climate changes. Inbreeding with similar genetic material also soon produces deleterious effects. For instance, the entire world population of cheetahs was found to be almost genetically identical. It is therefore not surprising that in both wild and captive populations abnormal animals are frequent, sperm fertility is reduced, and they are easily susceptible to a number of diseases. At some stage in their history, the cheetah population must have been reduced to a small number of genetically similar individuals; the species has not yet recovered from this genetic bottleneck and could be much more prone to extinciton than other species of wild cat (see O'Brien, *et al.,* 1985). [*See* EVOLUTION AND EXTINCTION.]

An important part of genetic diversity, especially in plants, is the number of different breeding systems that exist. Plants may be dioecious, with male and female flowers on separate individuals; monoecious, with separate male and female flowers on the same individual; or hermaphroditic, with both sexes in the same flowers. In the latter case there is a great variety of self-incompatibility systems to prevent self-fertilization. Self-incompatibility is often accompanied by heterostyly, in which different flowers have either long styles or long stamens (distyly, as in the primrose, *Primula vulgaris*) or also an intermediate morph (tristyly, as in the water hyacinth, *Eichhornia crassipes*). The great variation which plants have evolved to ensure cross-pollination would indicate that the avoidance of selfing and its deleterious effects has selective advantage. However, there are also many plants that are predominantly selfing that have robust populations. It is probable that in these cases harmful alleles have been eliminated in the past. Barrett and Charlesworth (1991) compared a predominantly outcrossing population of *Eichhornia paniculata* with a predominantly inbreeding one and found no difference in their ability to survive.

Some plants and animals breed by asexual means, for example, by vegetative reproduction, such as the runners on strawberry plants, or by simple division in unicellular animals. Even in these cases small mutations do occur and the genetic diversity produced is often made use of by horticulturists who propagate many plants vegetatively by cuttings or by layering. Another means of reproduction is parthenogenesis, in which seeds are formed from embryos that are entirely derived from the mother plant in which the megaspore does not divide by meiosis. Many blackberries (*Rubus*) and hawthorn (*Crataegus*) have predominantly parthenogenic reproduction. The result is that many of the distinct morphological variations in these groups are clones, rather than true outbreeding species.

Many genes exist in several forms or alleles, each expressing a particular character in a different way, for example, hair or eye color in humans. Some alleles, usually the most favorable, are dominant over others, and an individual only shows the characteristic of a nondominant or recessive allele if it has inherited it from both parents. A good example of this is sickle-cell anemia. An individual is affected only if both parents carried the abnormal allele and contributed it to the offspring. The heterozygous condition of this trait, where an individual carries one normal and one abnormal allele, confers a slight advantage to the individual, since it increases resistance to malaria.

Humans have made much use of the genetic variation of plants and animals to change wild types of natural species into more useful ones by the gradual artificial selection of certain traits. For example, all of our domestic dogs (*Canis familiaris*) are derived from a single species, the wolf (*Canis lupus*). The Saint Bernard and the Chihuahua are very different in appearance and behavior, but breeders have been able gradually to select for the characteristics they wanted because of the genetic variation available in the original species. The vegetables cabbage, cauliflower, broccoli, kohlrabi, kale, and Brussels sprouts all come from the same original species. Breeders have produced six different vegetables by selecting for edible leaves in cabbage and kale, edible flowers in broccoli and cauliflower, edible buds in Brussels sprouts, and edible stems in kohlrabi, and again this is only possible because of the genetic variations within the original species, *Brassica oleracea*. As Mark Twain said, "Cauliflower is nothing but cabbage with a college education"!

In recent years the techniques of molecular biology have enabled us to "fingerprint" the genetic make-up of an individual or to study the genetic variation of a population. The study of allozyme variation or the differences among specific enzymes coded by different alleles was the first and simplest technique to be used. Currently, restriction fragment-length polymorphisms and DNA sequencing using the polymerase chain reaction are much more often used. For the conservation of species, these techniques have many applications. For breeding programs of rare species, it is vital to know the degree of genetic similarity of individuals so that unrelated individuals can be crossed to avoid the deleterious effects of close inbreeding, because many species of plants and animals will require careful genetic management if they are to survive. Inbreeding also tends to reduce genetic variation, and so it is to be avoided at all costs if the purpose of the work is to rescue an endangered species already on the brink of extinction. Fingerprinting can also be used to verify the origin of an individual. In plants this is already proving useful in the control of trade in economically important species. Genetic fingerprinting can sometimes be used to determine whether a retailer propagated the material in his nursery or whether it came from the wild.

Molecular studies are also enabling us to produce better evolutionary classifications of organisms. As an understanding of the relationships among species, genera, and families is developed, we can also make more rational decisions about which species are more important to conserve. Biologists are now examining cladograms to ensure that representatives of each major clade are conserved. Protecting biodiversity may require placing conservation priorities on different taxa, and, if a choice must be made, selection of taxic diversity from an evolutionary entity such as a genus will preserve a much greater genetic diversity than if only one part of it is conserved (Faith, 1994; Vane-Wright *et al.*, 1991; Williams *et al.*, 1991). [*See* BIODIVERSITY, VALUES AND USES.]

Genetic variation is the basis for the variety of life on earth. Without it evolution could not have taken place. Species with little genetic variation may be less competitive in the face of environmental changes. It is therefore not sufficient to conserve just a few individuals of any species alone. We must also examine it genetically and ensure that an adequate sample of its genetic variation is conserved. Ultimately, survival depends on the ability of a species to adapt to change, such as climatic change, and to have enough variation to overcome deleterious mutations. This means that it is necessary to maintain an adequate population size of a species with sufficient individuals to be genetically viable.

Within the genetic diversity of plants and animals, there is a great treasure of unexplored properties that could be of considerable use to humankind. A single gene can be worth large sums of money, yet we are allowing them to be destroyed. In 1962 botanists Donald Ugent and Hugh H. Iltis discovered a new weedy species of wild tomato (*Lycopersicum chmielewskii*). When this was crossed with the commercial tomato, the hybrid was found to contain a much higher sugar content (Rick, 1976). The commercialization of this new sugar-rich variety has benefited the tomato industry by $8 million per year (Iltis, 1988). This valuable gene could easily have been lost through habitat destruction in Peru and elimination of a weedlike plant of no obvious value. We are already losing much genetic diversity and potentially valuable genes for increasing disease resistance, improving food crops, or

broadening the range of cultivated species. This cannot be wise, because between 1976 and 1980 alone genetic material from wild relatives of crop species contributed $340 million per year to U.S. farmers in increased yield and disease resistance (Prescott-Allen and Prescott-Allen, 1988).

IV. HABITAT DIVERSITY

Biodiversity also includes the great variety of ecosystems which make up the habitats or communities in which the organisms live. Early work by conservationists concentrated on the preservation of species, and as a result many lists of threatened and endangered species of plants and animals were prepared. As the practical aspects of conservation of these species are better studied, gradually a greater emphasis has been placed on ecosystem conservation. It is obviously more cost effective to operate conservation at the coarse-filter or habitat level than at the fine-filter or single-species level. [*See* CONSERVATION AGREEMENTS, INTERNATIONAL.]

The most recent version of the World Conservation Strategy (IUCN, UNEP, WWF, 1991) states that "Conserving biological diversity equals conserving ecosystems." Species do not exist alone; in addition to being genetically diverse populations, they are members of a community and have a distinct niche and role within it. Each species is dependent on others within the community in some way; for example, many flowering plants depend on their animal pollinators for cross-pollination and the animals depend on the plants for food. A community is bound together in a loose web of mutualistic relationships on the one hand and the competition among species for survival on the other. We have already seen that genetically weakened species will have less ability to survive the competition within the community.

It is possible to examine DNA and define the genetic variability within a species and it is also possible to define a species by its molecular structure, outward morphology, and breeding capacity. It is much more difficult to define habitats, because it is hard to draw the line between them and they can be classified at many different levels. In addition, the ecosystem is also defined by various physical parameters, such as soil type and climate patterns. Rainforests occur in the warm tropical belt, where rainfall is not too seasonal; tundras occur in cold northern latitudes; and deserts occur where there is little or no rainfall. Different communities occur on acid or alkaline soils or on sandy and clay soils. At the macro level communities are most frequently defined by the type of vegetation and the component plant species that occur. However, much smaller units are also communities, such as the moths and algae that inhabit the fur of an Amazonian sloth or the colonies of bacteria that inhabit a human body.

A community depends on the ecological processes that occur within it, and these are performed by the species which it contains. Processes such as the recycling of nutrients, the maintenance of soil fertility and water quality, and even the regulation of local climate depend on the component species of a community. The removal of a few species might not affect these processes greatly, but the removal of certain keystone species on which the others depend will cause the breakdown or radical change of that community. There are communities and plant species in Africa whose survival is dependent on the disturbance caused by elephants. If the keystone species, the elephant, is removed, then many other species will gradually die out.

Since each community consists of a collection of species, it follows that the preservation of both species diversity and genetic diversity depends on the preservation of habitat diversity and the web of interactions that occur within the community, and so it is not surprising that ecosystem diversity is considered a component of the all-inclusive term "biodiversity." To formulate adequate conservation strategies, it is essential to understand the way in which species interact through pollination and dispersal relationships, predator–prey relationships, defense mechanisms, and many other connections.

V. BIODIVERSITY UNDER THREAT

The term "biodiversity" was coined to stress the link among species, their genetic diversity, and

their habitats when their component parts began to be seriously threatened. Although extinction is a natural process and the fossil record shows that there have been a number of major extinction spasms in the geological past, such as the meteor impact that eliminated the dinosaurs 65 million years ago, current rates are far in excess of natural ones. It has been established that extinction rates of birds and mammals have now reached 100–1000 times greater than under natural conditions before the advent of the human race, and total species extinction is about 10,000 times the natural rate. Based on worldwide deforestation rates, Wilson has estimated that 0.2–0.3% of rainforest insect species are now lost each year. We do not yet know how many of these species there are, but if they were 8 million, we could already be losing between 16,000 and 24,000 species a year. Species are being wiped out before they have been named or before we even know that they exist. This rate of extinction is far greater than even the five major phases of extinction in the geological past (Raup, 1988). A well-documented contemporary extinction spasm is the loss of a large number of endemic cichlid fish species in Lake Victoria following the introduction of the Nile perch.

The most recent report from the United Nations Food and Agriculture Organization (FAO) showed that current rate of tropical forest loss is 17 million hectares per year, compared with 11 million hectares at the beginning of the 1980s. In the case of rainforests, the most important centers of biodiversity in the world, it is apparent that their conservation is now a race against time and that urgent and rapid policy changes are needed. However, it is not just the rainforests that are in danger. Many other ecosystems that hold important biodiversity are also under threat; for example, one of the most diverse floras in the world, that of the Cape Province of South Africa, is being destroyed by the encroachment of agriculture. For example, 85% of the coastal renosterveld has been transformed by human actions (Moll and Bossi, 1984). Coral reefs are also under increasing threat through marine pollution and silt produced from deforestation of nearby areas. The destruction which is occurring today affects all aspects of biodiversity. Species are becoming extinct at an alarming rate, whole habitats are being destroyed, and the genetic variability of many surviving species is being seriously reduced.

VI. THE TASKS AHEAD

The approximately 1.6 million organisms that have been described to date represent only a small percentage of the total species diversity on Earth. Assuming that there are at least 30 million species, only 5.5% have so far been described, and if higher species estimates turn out to be true, it could be as little as 2%. Biodiversity cannot be properly conserved, used, or managed if it is still so poorly known. The first task is therefore to greatly increase the taxonomic effort to identify the species of organisms in the world. It is a tragedy that at a time when we are realizing the importance of biodiversity and when it is most under threat, the number of systematists is decreasing rather than increasing in many countries. If biodiversity is really important, we need to mobilize an adequate task force of systematists to cope with the problem.

It is likely that tropical rainforests of the world will be reduced to 10% of their original size during the next 30 years. As the world population increases, pressure increases on many different habitats of biodiversity, and it has been calculated that one-fifth or more of the species of all groups of organisms in the world are likely to disappear over the next 30 years. The second task is to make greater efforts to conserve biodiversity, placing the strongest emphasis on areas that are under the most threat such as the hot spots that contain large clusters of endangered species. A much greater unity of effort for conservation of biodiversity is needed if we are to slow the current rate of species and habitat loss.

The biodiversity of the world already supports human life in so many practical ways, such as for food, clothing, and shelter. It also provides the essential ecological services of maintaining our atmosphere, creating and maintaining soils, sustaining hydrological cycles, and controlling world climate patterns. However, we use only a small proportion of species in a direct way. Biodiversity is still a largely untapped source of new food crops, new medicines, new fibers, new sources of energy to substitute for petroleum, new flavors and scents, and countless other products of commercial value. At present only 20 species of plants supply 90% of our food and 50% is from the three crops wheat, maize, and rice (see Table V). About 7000 species

TABLE V
The 15 Most Important Food Plant Crop Species in Terms of Production and Area Cultivated[a]

Crop	Scientific name	Area (1000 ha)	Production (1000 metric tonnes)
Wheat	*Triticum* spp.	229,347	505,366
Maize	*Zea mays*	131,971	488,500
Rice	*Oryza sativa*	144,962	472,687
Potato	*Solanum tuberosum*	20,066	300,616
Barley	*Hordeum vulgare*	78,698	176,574
Cassava	*Manihot esculenta*	14,010	135,551
Sugar cane	*Saccharum officinarum*	23,676	121,524
Sweet potato	*Ipomoea batatas*	7880	110,651
Sorghum	*Sorghum* spp.	91,859	104,592
Soybean	*Glycine max*	52,683	100,809
Oats	*Avena sativa*	25,288	49,630
Rye	*Secale cereale*	16,738	32,288
Peanuts	*Arachis hypogaea*	18,728	20,708
Beans	*Phaseolus* spp.	25,665	14,909
Peas	*Pisum sativum*	8832	13,199

[a] Data were derived from the Food and Agriculture Organization.

of plants have been used as food in some way, but many more must be capable of use. One danger of using so few is that if a disease were to seriously attack one of the three major cereal crops, there would be great global famine. For a more certain future we must use a greater diversity of species for food and preserve as much genetic diversity of each crop as possible. The potato famine of the last century in Ireland occurred because of the use of a single variety of potato that was susceptible to the fungal disease *Phytophthora infestans*. In contrast, an Andean market in the region of origin of the potato will often contain 50 different varieties of several distinct species. The genetic diversity which these native peoples grew protects them from susceptibility to elimination of their entire crop. The Andean natives cultivate both species diversity and genetic diversity in different habitats at a wide range of altitudes. Their use of all aspects of biodiversity ensures the future of their crops. We are still not learning from this experience—of the 7098 varieties of apple used in the United States between 1804 and 1904, 6121, or 86.2%, are now extinct (Foyle and Winters, 1981).

The third task is to make a greater effort to release the potential of biodiversity and to use a much greater proportion of the species as well as a much broader genetic diversity in our crops. If world leaders were to have a greater idea of the potential uses of biodiversity, they would make more effort to ensure its preservation. To sustain a viable future, we must achieve a better balance between the maintenance of human welfare and the retention of biological diversity in the widest sense.

Glossary

Alleles Variant forms of the same gene that have different effects on the phenotype. A shortening of the term "allelomorph."

Biodiversity The total variability among organisms and the habitats in which they live, including three components: species diversity, genetic diversity, and ecosystem diversity.

Ecosystem An interdependent group of organisms that form a community with each other and with their abiotic environment.

Endemic A species that is restricted to a particular geographic region or habitat.

Gene The basic unit of heredity, comprised of a segment of DNA that codes for a particular function or several related functions within an organism.

Genetic diversity The genetic variation that occurs within an individual species that enables it to adapt to different environmental conditions and evolutionary processes and that causes variation within a species.

Heterostyly The polymorphism of certain flowers to ensure cross-pollination in which flowers have anthers and styles of different lengths, for example, the long-styled (pin-eyed) and short-styled (thrum-eyed) morphs of the common primrose (*Primula vulgaris*).

Hot spot An area of great species diversity and or endemism that is under particular threat of destruction.

Incompatibility A mechanism to prevent the self-fertilization in plants, whereby the pollen of an individual is only able to fertilize the ovary of a different plant.

Keystone species A species such as the elephant that affects the survival and abundance of other species of the habitat in which it lives. If it becomes extinct, the community structure will change and other species will also become extinct.

Recombination The mixing of genotypes that occurs as a result of sexual reproduction, causing the offspring to have a different combination of genes than either parent.

Bibliography

Adis, J. J., Lubin, Y. D., and Montgomery, G. G. (1984). Arthropods from the canopy of inundated and terra firme forests near Manaus, Brazil, with critical considerations of the Pyrethrum-fogging technique. *Stud. Neotrop. Fauna Environ.* **19,** 223–236.

Ashton, P. (1964). Ecological studies in the mixed dipterocarp forests of Brunei State. Oxford Forestry Memoirs 25.

Barrett, S. C. H., and Charlesworth, D. (1991). Effects of a change in the level of inbreeding on the genetic load. *Nature (London)* **352,** 522–524.

Bibby, C. J., Collar, N. J., Crosby, M. J., Heath, M. F., Imboeden, C., Johnson, T. H., Long, A. J., Stattersfield, A. J., and Thirgood, S. J. (1992). "Putting Biodiversity on the Map: Priority Areas for Global Conservation." International Council for Bird Preservation, Cambridge, England.

Dodson, C., and Gentry, A. H. (1978). Flora of the Río Palenque Science Center. *Selbyana* **4,** 1–628.

Erwin, T. L. (1982). Tropical forests: Their richness in Coleoptera and other arthropod species. *Coleopterists Bull.* **36,** 74–75.

Erwin, T. L. (1983). Tropical forest canopies: The last biotic frontier. *Bull. Entomol. Soc. Am.* **30,** 14–19.

Erwin, T. L. (1988). The tropical forest canopy: The heart of biotic diversity. *In* "Biodiversity" (E. O. Wilson and

F. M. Peter, eds.), pp. 123–129. National Academy Press, Washington, D.C.

Erwin, T. L. (1990). Canopy arthropod biodiversity: A chronology of sampling techniques and results. *Rev. Peruana Entomol.* **32,** 71–77.

Faith, D. P. (1994). Phylogenetic pattern and the quantification of organisational biodiversity. *Phil. Trans. Roy. Soc. London, Ser. B,* **345,** 45–58.

Foyle, M. W., and Winters, H. F. (1981). "North American and European fruit and tree nut germplasm resources inventory," Miscellaneous Publication No. 1406. U.S. Department of Agriculture, Beltsville, Maryland.

Gentry, A. H. (1988). Tree species richness of upper Amazonian forests. *Proc. Natl. Acad. Sci. U.S.A.* **85,** 156–159.

Groombridge, B., ed. (1992). "Global Biodiversity: Status of the Earth's Living Resources," report compiled by the World Conservation Monitoring Centre. Chapman & Hall, London.

Hawksworth, D. L. (1991). The fungal dimension of biodiversity: Magnitude, significance, and conservation. *Mycol. Res.* **95,** 441–456.

Iltis, H. H. (1988). Serendipity in the exploration of biodiversity: What good are weedy tomatoes? *In* "Biodiversity" (E. O. Wilson and F. M. Peters, eds.), pp. 98–105. National Academy Press, Washington, D.C.

Klug, M. J., and Tiedje, J. M. (1994). Response of microbial communities to changing environmental conditions chemical and physiological approaches. *In* "Trends in Microbial Ecology" (R. Guerrero and C. Pedros-Alio eds.), pp. 371–378. Spanish Society for Microbiology, Barcelona.

McNeely, J. A., Miller, K. R., Reid, W. V., Mittermeier, R. A., and Werner, T. B. (1990). "Conserving the World's Biological Diversity." International Union for Conservation of Nature and Natural Resources, Gland Switzerland. [Reprinted, in 1992.]

Moll, E. J., and Bossi, R. (1984). A current assessment of the extent of the natural vegetation of the fynbos biome. *S. Afr. J. Sci.* **80,** 355–358.

Myers, N. (1988). Threatened biotas: 'Hot spots' in tropical forests. *Environmentalist* **8,** 187–208.

Myers, N. (1990). The biodiversity challenge: Expanded hot-spot analysis. *Environmentalist* **10,** 243–256.

O'Brien, S. J., Roelke, M. E., Marker, L., Newman, A., Winkks, C. A., Meltzer, D., Colly, L., Evermann, J. F., Bush, M., and Wilot, D. E. (1985). Genetic basis for species vulnerability in the cheetah. *Science* **227,** 1428–1434.

Prescott-Allen, R., and Prescott-Allen, C. (1988). "The First Resource: Wild Species in the North American Economy." Yale University Press, New Haven, Connecticut.

Raup, D. M. (1988). Diversity crises in the geological past. *In* "Biodiversity" (E. O. Wilson and F. M. Peters, eds.), pp. 428–436. National Academy Press, Washington, D.C.

Rick, C. M. (1976). Genetic and biosystematic studies on two new sibling species of *Lycopersicon* from interandean Peru. *Theor. Appl. Genet.* **47,** 55–68.

IUCN, UNEP, WWF (1991). Caring for the earth: A strategy for sustainable living. The World Conservation Union, United Nations Environment Programme and World Wide Fund for Nature, Gland, Switzerland.

IUCN, UNEP (1988). Coral reefs of the world. Volume 3: Central and Western Pacific. UNEP Regional Seas Directories and Bibliographies. International Union for the Conservation of Nature and Natural Resources (IUCN), Gland, Switzerland and United Nations Environment Programme, Nairobi, Kenya.

Sibley, C. G., and Monroe, B. C. (1990). "Distribution and Taxonomy of Birds of the World." Yale University Press, New Haven, Connecticut.

Stork, N. E. (1991). The composition of the arthropod fauna of Bornean lowland rain forest trees. *J. Trop. Ecol.* **7,** 161–180.

Vane-Wright, R. I., Humphries, C. J., and Williams, P. H. (1991). What to protect–Systematics and the agony of the choice. *Biol. Conserv.* **55,** 77–90.

Whitmore, T. C., Peralta, R., and Brown, K. (1985). Total species count in a Costa Rican tropical rain forest. *J. Trop. Ecol.* **1,** 375–378.

Williams, P. H., Vane-Wright, R. I., and Humphries, C. J. (1991). Measuring biodiversity: Taxonomic relatedness for conservation priorities. *Aust. J. Syst. Bot.* **4,** 665–679.

Valencia, R., Balslev, H., Paz Y Miño, G. (1994). High tree alpha-diversity in Amazonian Ecuador. *Biodiv. and Conservation* **3,** 21–28.

Biodiversity, Economic Appraisal

John B. Loomis

Colorado State University

I. Introduction
II. Commodity Production Benefits of Biodiversity
III. Medicinal Value of Biodiversity
IV. Economic Value of Keeping Options Open
V. Economic Benefits from On-Site and Off-Site Uses
of Biodiversity
VI. Conclusion

The economic benefits of preserving biodiversity include improving the desirable characteristics of crops, making plants more pest resistant, providing medicines to treat and cure illnesses, supporting ecotourism, and providing enjoyment to individuals. The economic benefits of preserving the current biodiversity provide an option value to society to ensure that future generations will have the opportunity to use and enjoy biodiversity. Thus a reduction in biodiversity decreases the well-being of not only the current generation but of all generations to come. Economic theory requires that such irreversible decisions be made only when the benefits of development are quite large and that full information about the potential benefits foregone is available. However, many development projects that destroy biodiversity are uneconomic and must be subsidized by governments. Given the net loss of many development projects and the potential benefits to humans of biodiversity, the economic theory of option value suggests that there are benefits to maintaining our options and preserving biodiversity in all cases except where the benefits of development are dispropor-

tionately large relative to the known benefits of biodiversity.

I. INTRODUCTION

Biodiversity provides numerous services that directly or indirectly are of value to humans. The most obvious is the use of diverse species as inputs to produce the wide variety of products that people consume and enjoy. It is from the diversity of plants and animals that we are provided a splendid variety of foods and fabrics. For example, our lives are made both more delightful and healthful because of the availability of a wide variety of naturally available seafood. While biological diversity enriches the lives of people in the industrialized world, biodiversity provides the means of survival to people in developing countries. The natural diversity of plants and animals is directly used by individuals to feed, clothe, and house themselves. Preservation of biodiversity also supports agricultural productivity and ecotourism, as well as providing the basis for many importance medicines. [*See* BIODIVERSITY.]

II. COMMODITY PRODUCTION BENEFITS OF BIODIVERSITY

The natural genetic variability between different species and within a given species is the raw material for developing crops with higher yields or natural defense mechanisms against pests. The crop breeding industry, which depends on natural biodiversity, adds $1 billion annually to crop values. Specific examples of the successful use of wild genes to improve domestic crops are numerous. A gene found in a wild tomato allowed the development of new hybrid tomatoes with more soluble liquids, higher sugar content, and better flavor. The added value to tomato processors was worth $8 million annually. [*See* BIODIVERSITY, VALUES AND USES.]

Developing plants that have natural defense mechanisms against pests such as fungus and insects reduces the use of costly and environmentally damaging pesticides. The well-known discovery of a wild perennial, virus-resistant maize is likely to provide significant benefits to the world's third most important crop, one with a value of $50 billion. A gene found in wild tomatoes that can protect cultivated tomatoes from a disease called "bacterial speck" will reduce the use of pesticides against this disease.

Numerous plants also produce natural pesticides to make them less desirable to herbivorous animals. The calabra bean found in West Africa contributed to the development of methyl caramate insecticides. The roots of a South American forest vine used by Amazonian Indians to stun fish became a source of a widely used and biodegradable pesticide called rotenone. By extracting and analyzing these natural pesticides, humans can work with rather than against nature in control of pests.

Intact forest ecosystems provide a self-sustaining supply of many resources that local communities depend on. Many indigenous people have sustainably harvested fruits, nuts, and firewood from native forests for centuries. By harvesting just the natural surplus from the forest, the forest's productivity is maintained for future generations. This is analogous to living off the interest from a bank account, rather than consuming the principal of the account. Even when these native people gather commercial products for sale from the forest, it can be done on a basis that is both sustainable and profitable. Research has documented that sustainable harvests of fruit, latex, and timber can provide nearly three times the long-term revenue (referred to as net present value) as can turning the natural forest into a timber plantation or into livestock pastures.

Intact forest ecosystems also provide a valuable buffering service as watersheds. The forest cover intercepts the rain, thus reducing its erosive power. The root structure of the forest holds the topsoil in place. The forest floor better allows for deep percolation of rainfall and the gradual release of the water to rivers and streams. Without the forest cover, rainstorms often result in floods, landslides, and substantial siltation of downstream water supplies. Thus the services provided by the forest are critical to maintaining a steady and clean supply of water downstream. One of the original legislative mandates for creating the National Forests in the United States was to protect forest cover in watersheds. Dozens of communities in the United States protect the forests in their watersheds to protect water quality. Removal of forest cover by timber harvesting or slash-and-burn agriculture has contributed to the premature siltation of dozens of dams throughout the world, which has reduced their capacity to store water and generate hydroelectric power. In the Himalaya region, removal of forest cover is partly blamed for the increasingly large floods that sweep downstream areas. Flooding, reduced water storage capability, and the need for removing sediment from water supplies all impose large economic costs on society. Unfortunately, many times the costs are imposed on downstream landowners or countries. This illustrates one of the sources of the loss of biodiversity—externalities. Externalities arise when landownership is fragmented and the government does not enforce the rights of downstream residents. Thus upstream landowners often clear forests without regard to the downstream consequences. If governments held upstream landowners liable for the flood damages and higher water treatment costs caused by their timber harvesting or slash-and-burn agriculture, less biodiversity would be lost.

III. MEDICINAL VALUE OF BIODIVERSITY

One of the largest indirect benefits provided by biodiversity is the development of new medicines. More than half the medicines used by people in developing countries are derived from plants. Examination of these plant extracts has led to the development of numerous and widely used drugs. Over the last two decades, 25% of all prescriptions from community pharmacies contain active ingredients extracted from plants. This represents a value of $8 billion annually. More than 100 important drugs are derived from plants alone, with new discoveries made each year from a variety of life-forms from fungi to trees. A heat-resistant enzyme needed to greatly speed up the diagnosis of strep throat was found in a bacteria in hot springs in Yellowstone National Park. The more rapid diagnosis allows for quicker treatment and fewer lost work days. Many substances found in plants, such as the rosy periwinkle and the Pacific yew tree, have anticancer properties. Because typical estimates in developed countries of the value of saving a human life are in the $2–$10 million range, life-saving drugs can be quite valuable. New drugs typically have a net economic value of $100–150 million over and above their development cost. Ecologists estimate there may be as many as 200 new drugs that could be developed from plant species alone. Therefore the preservation of flora biodiversity just for their medicinal value may contribute between $20 billion and $30 billion dollars. [See PLANT SOURCES OF NATURAL DRUGS AND COMPOUNDS.]

IV. ECONOMIC VALUE OF KEEPING OPTIONS OPEN

The Pacific yew tree illustrates how quickly a species can go from being a nuisance to a life-saver.

Prior to the discovery that the anticancer compound taxol was present in Pacific yew trees, they were seen as a "trash" tree in Douglas fir forests of the Pacific Northwest. Foresters were glad to get rid of the trees and reforestation emphasized trees capable of producing lumber. This emphasizes an important source of the economic value of biodiversity: keeping your options open. Because current generations do not know what desirable properties are present in each plant and animal species, it is economically foolish to exterminate those species unless the costs of protecting them are astronomical. It is quite rare that preservation of a viable population of a species in its natural environment has astronomical costs. The protection program for the endangered whooping crane has cost about $7 million dollars over the last 20 years for water and habitat management, protecting 7 million acres of old-growth forests for the northern spotted owl, and incidentally several species of endangered salmon and the Pacific yew tree will cost about $500 million in lost value of timber harvesting over the next 30 to 50 years. Thus it is often an economically wise investment to preserve as many species as possible. Though it may be impractical to preserve every subspecies, those species with distinct genetic information provide an important option value for future agricultural and medicinal uses. The payment today to maintain species that we may not directly use until future decades, if at all, is similar to an insurance premium. Most people purchase life insurance, car insurance, and homeowners insurance. In a typical year, they rarely have to rely on these types of insurance. Yet most people voluntarily pay hundreds of dollars for this insurance "just in case" some unanticipated event arises and they need it.

So it is with maintaining biological diversity: maintaining the thousands of species of plants and animals represents "nature's insurance policy." If we maintain nature's insurance policy we can draw from it for plant breeding and medicines when the unexpected blight or disease arises. Much like regular insurance, this has an economic value that we referred to as option value. For society as a whole, the option value for unique and irreplaceable items such as plant species is likely to be much larger than that for replacing common items like cars, which can be rebuilt or manufactured. The one-of-a-kind, family heirloom is often insured for more than common household items. Distinct plant and animal species are collectively the larger society's family heirlooms and they warrent a similar degree of protection.

Maintaining biodiversity takes on added importance when we recognize the extinction of species is irreversible. In some cases, extinction of critical or keystone species will result in the collapse of entire ecosystems that depended on that species. We lose these species and ecosystems not only for our generation, but for all generations to come. Economically efficient decisions that involve irreversibility require that we are doubly sure the benefits of selecting the irreversible option outweigh the costs in all possible future states of the world. For once such an irreversible option is adopted, we have committed all future generations to the path we have chosen. Sound economic decision making often requires us to postpone making irreversible decisions if we lack knowledge about the potential benefits of preserving a particular species or area. The irreversible decision can be postponed while we invest in acquiring this information. Only then can we determine whether the benefits of development outweigh the costs. Postponing an irreversible decision usually involves only minor costs, for postponement normally allows us the same range of choices in the future as is available now. On the other hand, early adoption of the irreversible option, which is often development, never allows a different choice in the future. Thus preservation of biodiversity in the face of incomplete information about the utility and value of particular species or ecosystems provides the added benefit of keeping options open. This is a type of option value that can be quite important. [See KEYSTONE SPECIES.]

V. ECONOMIC BENEFITS FROM ON-SITE AND OFF-SITE USES OF BIODIVERSITY

The sights and sounds provided by the diversity in many natural environments are a major at-traction to people. Visitors to natural environments receive pleasure from viewing plants and animals different from those where they live. They may receive inspiration, as well as become more knowledgeable about life on planet Earth.

The desire to obtain these benefits motivates people to spend thousands of dollars to travel to natural environments from Alaska to the Amazon. Both developed countries such as the United States and undeveloped countries such as Costa Rica and Kenya have economically benefitted from the global inflow of visitors to see their natural wonders. Tourism can be a profitable "export" for many countries. Besides the actual dollars spent and resulting income and employment created, visitors frequently receive a benefit in excess of what they paid. For example, given the relatively low entrance fee to visit Grand Canyon National Park, many individuals are not forced to pay the maximum amount they would be willing to rather than forego a visit to the park. An individual might pay travel costs of $75 and an entrance fee of $10, but they might have been willing to pay $145. The difference between this maximum amount ($145) and the amount they actually have to pay ($85) is the personal profit or net benefit, in dollars ($60), of being able to visit Grand Canyon National Park. Thus there are two monetary measures of the on-site benefits from visitation at natural areas: (a) the direct benefit reflects the various forms of satisfaction the visitor receives—this is measured in the form of the *net* willingness to pay, an amount that they retain; and (b) the effect their expenditures have on local income and employment. This second effect is particularly important if employment involves the hiring of workers who were previously unemployed or underemployed in their current job.

On a broader scale, there are literally millions of people who, while never expecting to visit these natural areas, derive satisfaction and enjoyment from: (a) knowing they exist as self-sustaining and regulating ecosystems; (b) knowing that preservation today provides these ecosystems to future generations; and (c) learning about these ecosystems from magazines, books, and television programs, as well as from other individuals who have visited these areas. Although the first category is sometimes called existence value and the second is bequest value, collectively all three values are often referred to as passive-use values, off-site benefits, or preservation values. The common thread is that preservation of the biodiversity in these areas provides what are called public good values to the general public.

These public good values are quite tangible to both the individual and society as a whole. Conservation organizations such as the Nature Conservancy and the World Wildlife Fund receive millions of dollars each year from individuals to preserve threatened ecosystems all over the world. Nearly all the individuals contributing to preserve these areas never expect to visit them. Yet their sense of well-being and satisfaction with life is influenced by whether these areas and their diverse biota are preserved. Government agencies also tap into this tangible desire for preserving natural environments. In California, Colorado, and Missouri there have been successful referendum votes on bond issues or taxes to acquire and preserve wildlife habitat and open space. In 1987, California voters passed Proposition 70, which provided $700 million for the acquisition of land for parks, wildlife habitats, and open space.

It is often difficult to actually collect the monetary equivalent of the benefits that people derive from knowing biodiversity exists. For example, it would be quite difficult for the government of Brazil to collect the dollar benefits that United States citizens obtain from knowing that Brazilian rain forests exist. In some cases, an individual knows that if others contribute to organizations like the Nature Conservancy to protect rain forests, they too can enjoy knowing that these areas exist or watch television programs about these areas even if they do not pay. The large potential for what is called "free riding" behavior means that private protection of biodiversity is likely to fall far short of the amount that is socially optimal. Only if governments force all beneficiaries to pay can the optimal amount of biodiversity be protected worldwide. But this raises two issues. First, how do we measure the magnitude of this passive-use value of biodiversity? Second, how can governments collect

a portion of that value to transfer to the countries around the world that bear the cost of protecting biodiversity? Protection of biodiversity involves both direct costs of policing the areas against poachers of both plants and animals and compensation for foregoing the opportunity to extract timber or minerals from the area.

To estimate what the public's passive-use values are for biodiversity, governments often use interviews or surveys that simulate a market or referendum for protection of these areas. For example, in California the general public was asked whether they would vote in favor of costly programs to expand wetland acres to increase the numbers of resident and migratory waterbirds. The survey was done by mailing a survey booklet to a representative sample of California households and then conducting the interview over the telephone. The results showed that individuals would pay about $250 to increase the amount of wetlands by about 100,000 acres in California. Because every household in California could enjoy the diversity of plant and animal life supported by these wetlands, the total or aggregate value to the citizens of the state is $2.5 billion. This in turn yields an average value of $25,000 per wetland acre protected, a value often in excess of the land's development value. While these specific types of wetlands are not commonplace, neither are they unique. If one were examining the preservation of a unique habitat found nowhere else in the world, the appropriate benefits may be to the worldwide population. Thus even if one-quarter of the households in the world would pay $1 for preservation of a particular area of habitat that supports a unique mix of species found nowhere else in the world, this would amount to well over $1 billion dollars. Of course much more research is needed to understand the values that different cultures place on maintaining the biodiversity in their own countries and in other countries around the world.

One of the remaining challenges is the actualization of this willingness to pay into a transfer of economic resources to the country that is preserving the biodiversity. Various avenues are ecotourism arrangements and forgiving a portion of the country's outstanding foreign debt in exchange for their preservation of certain tracts of land (e.g., what are sometimes called debt for nature swaps). In attempting to slow the rise of carbon dioxide in the atmosphere, it may be cost-effective for developed countries to pay tropical countries not to engage in slash-and-burn agriculture. The reduction in carbon dioxide per dollar spent might be much greater from reducing slash-and-burn agriculture than by trying to raise automobile mileage standards.

VI. CONCLUSION

Collectively, humans derive many benefits from the preservation of biodiversity—it helps in improving our food supply and medicines and provides recreation opportunities. At present, the benefits of biodiversity are often limited by our lack of inventories of what is present in nature. As we collect this inventory data and improve our understanding of the wealth of possibilities, the current benefits of biodiversity will no doubt be multiplied severalfold. For this bounty of foods and medicines to be realized, we must maintain the biodiversity that exists. The promise of genetic engineering requires that we save the raw materials found in nature. At present we are literally discarding resources that may hold the key to producing foods without the need for chemical pesticides and for providing cures to many types of cancer. This is not only an inefficient use of natural resources today, but is a myopic selfishness that deprives all future generations of the opportunity to use and to view these species. Our destruction of biodiversity is irreversible. It cannot be corrected once the species is extinct. Economic efficiency requires that only investments that provide very large returns be undertaken when the consequences of those investments are irreversible. Unfortunately, many governments subsidize developments that are uneconomic and that destroy biodiversity. This is true in developed countries (e.g., U.S. Forest Service timber sales in Alaska, Colorado, and New Mexico) and in developing countries. In many, many instances, preserving biodiversity costs very little.

We must discard short-sighted policies and the transplantation of "inappropriate technology" from one culture to another. Future generations will be poorer in food supply, medicines, recreation opportunities, and spirit for our failure to protect the biodiversity we inherited from the previous generations.

Glossary

Bequest value Willingness to pay of the current generations to ensure that future generations will have a particular species or ecosystem available.
Economic value Dollar value of the satisfaction or enjoyment an individual receives that occurs regardless of whether an actual cash payment is made.
Existence value Willingness to pay to know that a species or ecosystem exists.
Irreversibility Change that once made cannot be undone.
Option Value Economic value of maintaining the opportunity to choose from a wider range of alternative courses of action, rather than foreclosing all but one choice in the future.

Bibliography

Krutilla, J., and Fisher, A. (1985). "The Economics of Natural Environments." Baltimore: Johns Hopkins University Press.

Loomis, J., Wegge, T., Hanemann, M., and Kanninen, B. (1989). "The Economic Value of Water to Wildlife and Fisheries in the San Joaquin Valley: Results of a Simulated Voter Referendum." Transactions of the 55th North American Wildlife and Natural Resources Conference. Washington, D.C.: Wildlife Management Institute.

Loomis, J. (1993). "Integrated Public Land Management." New York: Columbia University Press.

Lovejoy, T. (1994). People and biodiversity. *Nature Conservancy*, January/February 1994.

Panayotou, T. (1993). "Green Markets: The Economics of Sustainable Development." San Francisco: International Center for Economic Growth, ICS Press.

Pearce, D., Markandya, A., and Barbier, E. (1989). "Blueprint for Green Economy." London: Earthscan Publications Ltd.

Peters, C., Gentry, A., and Mendelsohn, R. (1989). Valuation of a tropical forest in Peruvian Amazonia. *Nature* **339**, 655–656.

Reid, W., and Miler, K. (1989). "Keeping Options Alive: The Scientific Basis for Conserving Biodiversity." Washington, D.C.: World Resources Institute.

Wilson, E. O. (1988). "Biodiversity." Washington, D.C.: National Academy Press.

Biodiversity, Processes of Loss

Richard B. Norgaard
University of California

Largely due to the types and intensities of human activity on the globe, biodiversity is being lost at rates unprecedented since the end of the Cretaceous era 65 million years ago. The data on rates of loss are far from precise. Current rates of extinction are thought to be 1000–10,000 times the natural rate. The estimates of the percentage of species subject to extinction if we do not change how we interact with the environment range from 20% to as high as 50% by the year 2050. The genetic diversity of surviving species is also decreasing, and diverse types of ecosystems are being simplified or radically transformed. These estimates have a large range because contemporary estimates of total species currently range between 5 and 30 million, because of the complexity of the relationships between the loss of one species and the subsequent likely loss of others, and because of the uncertainties with respect to the magnitude of the effects and interactions among the processes of loss. Regardless of the uncertainties, however, it is clear that stemming the processes of biodiversity loss presents a major challenge.

Biological diversity is far greater near the equator than near the poles. Thus, much of the concern with respect to the loss of biodiversity is being focused on the developing countries of the tropics.

While the consequences of human activity are greater in the developing countries, the processes of loss are similar across regions and have been going on far longer and more intensely in the developed countries. In addition, some of the processes, such as the emission of greenhouse gases, which are expected to induce climate change, occur locally but have global consequences. Therefore, the processes of loss in developed and developing countries are best addressed together.

I. BIOLOGICAL EXPLANATIONS AND SOLUTIONS

Biodiversity loss is occurring through a variety of direct processes which can be reduced by restricting particular types of human activities as well as activities in particular places. The major mechanism of loss is undoubtedly habitat transformation. The temporary conversion of moist tropical forests to agricultural uses and the destruction of coral reefs are thought to be especially critical because these habitats are among the most species rich and systemically vulnerable. Tropical forests are being deforested and converted to pasture for cattle production and agricultural settlement and are being logged for wood products very rapidly, while the new uses to which these lands are put are rarely sustainable. Coral reefs are being destroyed by dynamite in one-time fishing efforts and through pollution and siltation due to poor onshore land use practices. Many other types of ecosystems are being transformed into agricultural and urban uses, frequently driven by population growth. The relative importance of other direct mechanisms is less clear. Air, soil, and water pollution, including acid precipitation, resulting from modern agricultural and industrial processes as well as from the combustion of hydrocarbons for transportation, put direct pressures on species. These influences change habitats relatively slowly and in ways which are less dramatic to the human eye than, for example, tropical deforestation, but for the species sensitive to particular pollutants, the change is decisive. The introduction of exotic species constitutes another important process of biodiversity loss. Species are introduced deliberately to increase the productivity of agriculture and, without forethought, are transported from distant places, for example, in the ballast water of ships and through increased international travel and product trade generally. Introduced species frequently outcompete native species, driving them to extinction. Some species, usually of commercial importance, have been driven to extinction through excessive harvesting. Over the next century the potential for climate change to further reduce biodiversity is seen by many as an especially ominous threat because of its global nature. The greenhouse gases which are expected to induce future climate change are being emitted now and will be over the coming decades. Reducing this future loss entails reducing greenhouse gas emissions now and designing other conservation strategies with future climate change in mind.

Many peoples around the globe have long established sacred areas and royalty has held excess lands, both of which have had the side benefit of protecting biodiversity. The establishment of bioreserves deliberately for the purpose of protecting whole ecosystems and critical species, however, largely began in the early twentieth century in developed nations and largely during the last three decades in developing nations. To a significant ex-

tent, the philosophy of the conservation movement has been that if people are the problem, then isolating areas from human activity is the solution. In the industrialized countries of the north, many processes of loss have been controlled in bioreserves. Efforts to isolate biological systems from adverse human activity in developing countries has proved much more difficult because of the high costs of enforcing Western land use practices when they conflict with traditional beliefs and historic interactions with the land. In many cases those most directly affected have been indigenous peoples and others following traditional ways of life, for whom alternative livelihoods have been difficult to facilitate. [See ECOSYSTEM INTEGRITY, IMPACT OF TRADITIONAL PEOPLES.]

Beyond direct protection of biological resources from human influence, biologists have emphasized the need for better identifying existing genetic, species, and ecosystem diversity. They have argued that better understanding and the greater appreciation that understanding brings are critical to slowing and eventually halting the rate of loss. Appreciation of biodiversity for its aesthetic, scientific, material, and life-sustaining benefits is essential, but it is by no means sufficient. Halting the loss of biodiversity will depend on widespread understanding of the processes of loss, a political consensus to change those processes, and concerted action to implement the changes. International agreements have proved contentious and difficult to achieve precisely because people in developed and developing countries perceive the processes of loss differently, while many, North and South, also disagree on the consequences. A merging of understanding is under way through the joint efforts of nongovernmental organizations and international agencies. This cooperative effort among numerous scientific, development, and environmental organizations led to a global biodiversity accord in 1992. [See CONSERVATION AGREEMENTS, INTERNATIONAL.]

II. ECONOMIC EXPLANATIONS AND SOLUTIONS

The processes of biodiversity loss are frequently positively reinforcing (in a cybernetic sense). The degradation of any particular area increases economic pressure on other areas. The expected loss of woody species through climate change reduces the possibilities for carbon fixation and hence reduces the opportunities to ameliorate further climate change. To bring a system into equilibrium, negative feedbacks in the system are needed. Economists argue that prices can play an important negative feedback role. In market systems prices increase to reduce the quantity demanded when supplies are low and prices drop to increase the quantity demanded when supplies are high, keeping demand and supply in equilibrium. The problem, economists claim, is that most genetic traits, species, and ecosystems are being lost because they do not have prices on them acting as a negative feedback in the system to keep use in equilibrium with availability. When individuals of a species become fewer, there is no increase in price to decrease the quantity used. By putting economic values on species and by various other ways of including them in market signals, biodiversity loss would be reduced. Furthermore, the economic explanation and solution are systemic. Unlike bioreserves, which reduce human pressures on species within the protected area but typically increase them beyond it, including the value of biodiversity in the price system would affect decisions in every sector of the economy. [*See* BIODIVERSITY, ECONOMIC APPRAISAL.]

Biologists are also becoming increasingly interested in establishing economic values of species, arguing that if the economic value of species to society were understood, more species might be conserved. Clearly, if we knew the relative values of biological resources, we would be in a better position to manage them more effectively. Also, to the extent that these values could be included in the market system, markets themselves would assist in the conservation of biodiversity. So long as we do not assume that economic values can include all values, the situation could be improved through amending market signals.

A second economic explanation attributes biodiversity loss to interest rates. The rate of interest affects how, by economic reasoning, people *discount* the future. If the rate of interest is 10%, then $1.00 1 year from now is worth only $0.91, since one can put $0.91 in the bank today and, earning 10% interest, it will be worth $1.00 next year. The problem is that $1.00 one decade from now is only worth $0.34; two decades from now, a mere $0.11. Clearly, discounting at 10%, a species must have a very high value in the distant future to be worth saving today. With a lower rate of interest, it would be discounted much less and hence its future value would be worth relatively more today. Thus, lower discount rates appear to favor conservation. If a person can earn an 8% return per year by investing in industrial expansion through stock or bond markets, he or she has little incentive to invest in trees that only increase in value at 3% per year. By economic logic, biological resources that are not increasing in value as fast as the rate of interest should be exploited and the revenues should be put into industrial capital markets. Since even many economists find exploitation to extinction rather crass, there has been considerable interest as to whether the rate of interest observed in private capital markets might be significantly higher than the social rate of interest, the rate at which society as a whole would be willing to invest in the future. Some people have advocated using a 0% discount rate, that is, not discounting at all. While this would favor slow-growing trees, without other criteria being imposed on investments, it would also justify investing in many development projects, many of which would adversely affect biodiversity.

There are two problems with economic explanations and solutions as argued to date. The first is that economic values and other values, such as moral considerations of equity, are interactive. Moral values and economic values cannot simply be looked at separately or added together. Moral choices determine economic values. How economies operate and the prices they generate are dependent on how rights to resources are distributed initially. Imagine two countries with identical land resources, produced capital goods, population levels, and educational levels. In country A capital, land, and education are distributed relatively equally among the populace, whereas in country B they are distributed very unequally. Imagine that markets work perfectly in each country so that resources are efficiently allocated to produce the goods demanded in each country. But because of the differences in the distribution of the ownership of resources, levels of income vary more in country B, resulting in different goods being demanded. Land and labor, for example, might be allocated to the production of rice, chicken, and bicycles in country A, whereas in country B these resources are allocated to beans for those with few resources and to beef and cars for those with many. Both economies are efficient, but how resources are allocated to goods and services and their prices, including the rate of discount, depend on the initial distribution of resources among people. The problem, in short, is that economic values are not uniquely determined by markets.

Just as the prices of goods depend on the initial distribution of resources among people, the value of biodiversity depends on how rights to biological resources are assigned among peoples. Of particular importance are the rights of future generations. Economic valuation as advocated by economists implicitly assumes that the current assignments of rights within and across generations are morally acceptable and that only greater efficiency in the use of biologicals is necessary for the conservation of biological diversity. This assumption has no theoretical basis and merely follows from a tradition established while addressing questions with shorter-term consequences for which the rights of future generations were less at stake. When we decide to protect the rights of future generations which were not heretofore threatened by our activities, or decide to give them new rights, current economic behavior is constrained, more resources are transferred to future generations, and prices for goods and services, including environmental goods and services, change. Furthermore, when we choose to protect the future, the rate of interest goes down. How the economy efficiently operates reflects our moral decisions to protect our descendants. Currently, however, we are trying to adjust prices in order to justify protecting future peoples, falsely trying to derive a moral decision by efficiency reasoning.

The second problem with the dominant economic explanations and solutions is that they fail to address the underlying processes that have led to unsustainability. Economists are advocating the "fine-tuning" through negative feedbacks of a system which many sense went astray some time ago and is now far from where we would choose it to be if we could have foreseen the consequences of our decisions then. Economists seek negative feedbacks to achieve an equilibrium without questioning how the system went off course and how far we might be from where we would choose the new equilibrium to be. Unless we address the underlying processes which led the economy astray, no amount of fine tuning will reverse the trend of biodiversity loss.

III. UNDERLYING PROCESSES OF LOSS

Like the proximate mechanisms of loss, the underlying processes which drive the direct mechanisms also combine and interrelate, making a classification and separate descriptions of the processes inherently difficult. The underlying processes include those driving increasing materialism, those perpetuating economic inequity, those propelling population growth, those forcing increasing technological dependence on fossil fuels, and a variety of interrelated processes maintaining forms of social organization which prevent a corrective response to the underlying processes. It is important to keep in mind that these processes are interactive and mutually reinforcing. It is not possible to determine the relative importance of them separately because they cannot be separated.

A. Materialism

A sense of identity and self-worth for the vast majority of people throughout history was acquired through being a member of a community and living within its moral precepts. A small number of people in a few civilizations during the past three or four millennia acquired material wealth, but significant material possessions were not technically possible for most. Only with the rise of industrial technologies over the past two centuries has the idea occurred that material wealth could—and should—become available to all people. Materialism became a commonly held objective in both market and Marxist political rhetoric and in their respective economic theories. During the twentieth century, furthermore, in part because of modern urban–industrial living patterns, it became accepted that people not only required energy for transport to and from work but could expect to use energy for long-distance travel for the purposes of recreation. High levels of material and energy consumption drive environmental transformations and pollution, key proximate mechanisms of biodiversity loss, through the exploitation of mineral resources, the extraction and combustion of hydrocarbons for energy, the transformation of hydrocarbons into plastics, fertilizers, and chemicals, and the intensive use of agrochemicals. Travel, the heating and cooling of indoor habitats, and the production of material goods, almost all fueled by fossil hydrocarbons, contribute the bulk of the greenhouse gases forcing global warming and thereby a further potential loss of biodiversity. Efforts to increase the efficiency of material and energy use have been quite effective in industrialized nations since the energy crisis (1973–1974), but such efforts offer few long-term prospects, since they have been promoted for their ability to maintain lifestyle, thereby reinforcing materialism.

The idea of progress as it evolved since the renaissance was that science would provide technologies to control nature and thereby reduce the effort necessary to meet material needs. With abundance there would be no material conflicts and more time, thereby facilitating the conditions for artistic, community, moral, and spiritual development. In fact, material plenty bred materialist values and personal identity through material consumption and possession. Materialism bred greater materialism rather than the development of higher human faculties and identity through community. Maintaining identity through living within a community and its moral precepts is increasingly less important among either the well-to-do or those who aspire to greater material wealth. Critiques of individualism and materialism among social philosophers and searches for new bases of community and identity complement environmental critiques of modernity. Faith in progress is clearly declining at the end of the twentieth century. The former Soviet Union failed to deliver material well-being as promised and broke into numerous smaller nations divided along historic ethnic boundaries. Many poor nations are no longer working together to achieve a development which keeps eluding them and are suffering from ethnic strife. On the one hand, culture is replacing material consumption as a source of identity. On the other, social instability and ethnic war make environmental management and the conservation of biological diversity all but impossible. Nevertheless, many still hold to earlier Western beliefs that progress will solve all problems. The problems generated by the idea of progress and how we will learn to work together without it are not receiving adequate scholarly analysis and public discussion.

B. Inequity

The vast majority of the peoples on the globe still consume very little because they do not have sufficient long-term access to resources to meet their ongoing material needs. Furthermore, they are well aware that others consume far more than they do, know that their poverty is relative, and hence rightfully strive to improve their own relative condition. Striving to meet their material needs and aspirations without secure access to adequate resources, the poor have little choice but to use the few resources at their disposal in an unsustainable manner. The poor, excluded from the productivity of the fertile valleys or fossil hydrocarbon resources controlled by the rich, are forced to work land previously left idle because of its fragility and low agricultural productivity: the tropical forests, the steep hillsides, and arid regions. Tropical rainforests are difficult to convert to agricultural land partly because of their inherent species and ecosystemic diversity. Fragile lands, onto which the poor have been forced, are important reserves of diversity because they are the last to be directly affected by people.

The excessive material and energy consumption of the wealthy 20–30% of the total population is the other side of the equity coin. The middle classes and the rich in northern industrialized nations as well as the elite in middle-income nations and even in many poorer ones have global access to resources. Many of the environmental and resource impacts of their consumption decisions occur at a great distance, beyond their view, beyond their perceived responsibility, and beyond their effective control.

The relationships between unequal access to resources and the unsustainability of development generally, and the loss of biodiversity in particular, were major themes of the United Nations Conference on Environment and Development held in Rio de Janeiro in June 1992. Rich peoples and political leaders of northern industrialized countries generally have understandably had some difficulty participating in this discourse and even greater difficulty participating in the design of new global institutions to address the role of inequity in environmental degradation.

C. Population

Reverend-turned-economist Thomas R. Malthus argued early in the nineteenth century that the geometric potential of human population increase would periodically meet an arithmetic constraint in the rate of increase in food availability. When population exceeded food supply, Malthus asserted that population would decline through war, disease, and the ravaging of the land. Malthus' model is charmingly simple, and consequently demographic history generally does not support it. However, periodically in specific places Malthus' model is confirmed, and history may yet confirm it globally. Few question whether population must ultimately be stabilized in order to sustain human well-being at a reasonable level. The expansion of human populations into previously unpopulated or lightly populated regions, the intensity with which firewood is collected, and the push to increase food production through modern agrochemical monocultural techniques, so harmful to biodiversity, are driven over the long run by population increase. The continued rapid rates of population increase in the poorest nations threaten to keep them poor while diminishing the possibilities that all people in these nations will ever be able to consume at levels comparable to those in the rich nations using

current modern technologies without vastly accelerating the process of environmental degradation and biodiversity loss.

Without any people there would be no human-induced biodiversity loss. People working on reducing the rate of population growth, however, are not misanthropes. They are addressing the problem in the context of current population levels and are fully aware that population is only one of many interrelated variables. Unfortunately, the full complexities of the interaction of the causal factors are not consistently communicated, especially through the mass media. Due to incomplete communication and misrepresentation, people primarily concerned with the social processes maintaining inequities in the access to resources too simply label those primarily concerned with population as "neo-Malthusians." As a consequence, effective discourse and action on both population and inequity are being stymied by misunderstanding and divisive misrepresentation among those who are concerned about the problems. Such misunderstanding is one among many, suggesting that one underlying problem is that people "want" simple explanations and solutions and go into a collective paralysis when problems are complex; some favor one simplicity, others another. The multiplicative effects of population and materialism are especially important to consider together. People in countries with high birth rates and low consumption are now rightfully pointing out that a single person in a rich country consumes as much as 10 people in the poor countries. Also, activists in rich nations are now arguing that if the rich want to maintain their per-capita level of consumption, they should strive for negative population growth rates in order to improve the well-being and resource management options of the poor.

D. Technology

People have interacted with their environments over millennia in diverse ways, many of them sustainable over very long periods, many not. Some traditional agricultural technologies, at the intensities historically employed, probably increased biological diversity by creating new niches and opportunities for evolution. There is general evidence that traditional technologies, again at the levels employed, included biodiversity-conserving strategies as part of the process of farming. Technology today, however, is perceived as a leading culprit in the process of biodiversity loss. This is exactly contrary to the conventional progressive view that modern technologies would free us from the vagaries of nature. It is now becoming clear that modern agricultural technologies override and frequently destroy nature rather than free us from it. Modern agricultural technologies only override nature locally and temporarily at best. Pesticides kill pests, solving the immediate threat to crops. But the vacant niche left by the pest is soon filled by a second species of pest, the original pests evolve resistance, pesticides drift to interfere with the agricultural practices of other farmers, and pesticides and their by-products accumulate in soil and groundwater aquifers to plague production and human health for years to come. Each farmer strives to control nature but creates new problems beyond his or her farm and in subsequent seasons. With the adoption of new technologies creating new problems beyond in space and time, preharvest crop losses due to pests since World War II have remained around 35%, while pesticide use increased dramatically. Similarly, fertilizers can reduce the vagaries and limits of nitrogen-fixing bacteria and other soil microorganisms that assist in nutrient uptake. But in the process of overriding them, these natural properties of ecosystems are lost while the effects of nitrogen and phosphate pollution in groundwater aquifers and surface waters accumulate. While there is no doubt that modern genetic engineering has produced far more productive varieties of rice, wheat, corn, and other grains, these varieties depend on irrigation, fertilization, and pesticide technologies with substantial environmental costs beyond agriculture and over time. The accelerated evolution of plant diseases associated with the monocropping of a limited number of varieties makes further crop improvements in disease resistance necessary. Whether or not modern agriculture can be thought of as less dependent on nature, it is surely more dependent on continual advances in agricultural research and technology. [See AGROECOLOGY.]

The realization over the past few decades of the unexpected consequences of modern agriculture drives a new interest in how traditional agricultural technologies conserve biodiversity and the potentials for combining modern scientific understanding with traditional techniques. Clearly, people must still interact with the environment; the question is not whether, but how. While modern agriculture certainly causes environmental degradation, it is also clear that it feeds many who would otherwise have to be supported through the use of more land in agriculture, further threatening biodiversity. Traditional agricultural practices in areas of rapidly increasing population result in the expansion of farming into previously undisturbed regions. Furthermore, with the onset of development, the social relations and traditions which conserved resources break down, further hastening environmental degradation. Nevertheless, traditional agricultural technologies and social organization provide helpful insights into how we might live with nature more successfully. [See TRADITIONAL CONSERVATION PRACTICES.]

Beyond modern agriculture, fossil fuel-based technologies support industry, transportation, and thereby the concentration of people in urban areas. While concentrating people into cities reduces their direct impact on the land and hence on biodiversity, fossil fuel-based transport, residential heating and cooling, and industry produce the vast majority of carbon dioxide and other greenhouse gases driving potential climate change. Lesser, more immediate, threats from ongoing air, soil, and water pollution as well as from accidents such as oil spills are also inextricably linked to fossil fuel-based technologies. Improved technologies can increase energy and material efficiency, reducing energy and material flows and thereby the rate of environmental transformation. But technologies which merely reduce the consequences of existing approaches offer little hope in the longer run. The multiplicative effects of increasing population and materialism on resource use and environmental transformation will certainly outpace efficiency increases.

New technologies which work with natural processes rather than override them are sorely needed. During the past two centuries technologies have descended from physics and chemistry through engineering and agricultural science. Ecologists were never given the opportunity to systemically review such technologies, nor is it clear that our ecological understanding is sufficient to review them adequately now. A few agricultural technologies, such as the control of pests in agriculture through the use of other biologicals, have descended from ecological thinking. But research and technological development in biological control were nearly eliminated with the introduction of DDT to agriculture after World War II. Research and development of agricultural technologies requiring fewer energy and material inputs eventually received considerable support in industrial countries after the rise in energy prices during the 1970s and the farm financial crises in the United States during the early 1980s. However, support for agroecology, for technologies based on the management of complementarities among multiple species, including soil organisms, is still minimal. Learning how to use renewable energy sources will be long and difficult, since most of our knowledge has developed to capture the potential of fossil energy. Our universities and other research institutions are still structured around disciplinary rather than systemic thinking, and public understanding of the shortcomings of current technologies and possibilities for ecologically based technologies is generally weak. Scientists and technologists reproduce themselves and their institutions through direct control and education; hence, science and technology sometimes respond slowly to changes in the social awareness of environmental problems.

E. Social Organization

The foregoing comments on the organization of science and technology bring us to modern social organization more generally. Whether dominantly market or centrally directed, modern societies exhibit a historical process which is most simply and effectively understood as "distancing." Increasing specialization distances scientists from each other, impeding a collective scientific consciousness of how we are interacting with nature, let alone how we could. Increasing dependence on modern technologies shifts problems from the local and immediate to the distant and future. Increasing specialization in production processes distances capitalists

and workers from an overall consciousness of production technologies. Increasing urbanization distances people from the soil and water on which they depend. Increasing globalization distances people from the environmental impacts of their consumption. Distancing makes it more difficult for people to be environmentally aware and to design, agree to, implement, and enforce environmental management strategies. At the same time, global communication is now nearly instantaneous and is affordable for many. People individually are acquiring new perceptions of our global environmental predicament and are working through nongovernmental organizations, which are less prone than governmental agencies to the communication barriers of disciplinary and professional specialization. There is now discernible tension between bureaucratized but formal channels of power and less bureaucratized informal channels of global communication and agreement.

IV. CONCLUSIONS

Biological diversity is being lost through a variety of interrelated processes. Some of these processes are being halted relatively easily through changes in land use policies. The dynamiting of coral reefs and the subsidization of tropical forest conversion for cattle ranching were reduced during the 1980s through changes in national policies because they were obviously myopic and not in national or regional interests. Tropical rainforest habitats, however, continue to be transformed through other, interactive, processes. Pollution, especially in partially industrialized countries, continues to be a major threat to species. Also, climate change, driven largely by the production and consumption patterns of the richest nations, presents the greatest potential long-term threat to biodiversity. These proximate processes of loss are driven by underlying processes which reinforce materialism, global inequities, excessive procreation, fossil fuel-based technologies, and maladaptive social organization in both developed and developing societies. Stemming the loss of biodiversity ultimately will require a revisioning of the good life and of how to achieve it technologically, economically, and politically.

Glossary

Agroecology The study and use of agricultural practices that work with ecological processes, such as nitrogen fixation and nutrient uptake assisted by soil microbes and the control of pests through the introduction or management of predators and parasites.

Biodiversity The variation within species, the variety of species, and the combinations in which species interact, or genetic, species, and ecosystem diversity.

Bioreserves Areas in which human activity is restricted, except as a component of reserve management in order to protect specific species, groups of species, or ecosystem characteristics.

Exotic species Species which have evolved in one area and been introduced to and become established in another.

Fossil fuels Hydrocarbon energy sources, including coal, oil, and natural gas derived from fossil organic material.

Greenhouse gases Carbon dioxide, methane, and other gases in the upper atmosphere which reflect radiant heat from the earth back to the earth. An increase in these gases from human activity, largely the burning of fossil hydrocarbons, is thought to be increasing global temperatures.

Bibliography

McNeely, J. A. (1988). "Economics and Biological Diversity: Developing and Using Economic Incentives to Conserve Biological Resources." International Union for Conservation of Nature and Natural Resources, Gland, Switzerland.

Norgaard, R. B. (1992). "Sustainability and the Economics of Assuring Assets for Future Generations," Policy Research Working Paper 832. World Bank, Washington, D.C.

Norgaard, R. B. (1994). "Development Betrayed: The End of Progress and a Coevolutionary Revisioning of the Future." Routledge, London.

Orians, G. H., Brown, G. M., Kunin, W. E., and Swierzbinski, J. E., eds. (1990). "The Preservation and Valuation of Biological Diversity." University of Washington Press, Seattle.

SAAWOK Symposium (1992). Biodiversity: Challenge, science, opportunity. Proceedings of Science as a Way of Knowing (SAAWOK): Bodiversity Symposium. *Am. Zool.* (in press).

Wilson, E. O., ed. (1988). "Biodiversity." National Academy Press, Washington, D.C.

World Resources Institute, World Conservation Union, and United Nations Environment Programme in consultation with others (1992). "Global Biodiversity Strategy." World Resources Institute, Washington, D.C.

Biodiversity, Values and Uses

Margery L. Oldfield
The Seatuck Foundation

Biodiversity encompasses three major components correlated with different levels of biological organization: ecosystem diversity, or different types of natural ecosystems—commonly referred to as wildlands or wilderness areas; species diversity, or the variety of species within natural communities; and genetic diversity, or genetic variation found within species or distinct populations of living organisms. Biodiversity is highly valued at all three levels for providing products and commodities to meet basic human needs, amenities and services to promote human health and well-being, and aesthetic beauty and natural settings for moral contemplation and spiritually uplifting experiences. Thus, biodiversity provides the biotic raw materials—ecosystems, species, and genes—that underpin every major type of human endeavor at its most fundamental level.

I. INTRODUCTION

Without biodiversity, no human civilization could long endure, and modern industrialized society could not exist as we know it today. However, biodiversity is often undervalued and unappreciated in modern, technologically advanced nations and urban areas. In market economies, a biological resource, for example, wilderness used for recreation, pure air or water, or a disease resistance gene incorporated within a modern wheat variety, may have significant value to a society or economy. Yet, it may still not be considered economically important if there is no "spot market" or other mechanism within the marketplace to attach a monetary price to it. Biodiversity has been traditionally viewed as a source of "free goods"—goods, commodities, or services available in an unlimited supply that need not be allocated among users, or "public goods"—goods or services used collectively and consumed simultaneously by most people within a society. This means that even though biodiversity is not "valueless," it is often "priceless." [*See* BIODIVERSITY.]

This distinction between "value" and "price" is important. The definition of value employed here is an estimation of the general worth, usefulness, importance, *or* monetary price of a good, amenity, or service. In contrast, price specifically refers to the cost or amount of money (e.g., in dollar equivalents) set as a consideration for the sale of a good or service. Over the past decade, much progress has been made in the economic quantification of monetary values of biodiversity. However, it is important to acknowledge that intangible values of biodiversity that defy economic quantification may be just as significant for sustaining humanity as well as all life on earth. Thus, even though many examples of monetary values of biodiversity will be discussed here, the overall focus will be on intangible, nonmonetary as well as monetary and other tangible values. [*See* BIODIVERSITY, ECONOMIC APPRAISAL.]

II. INTRINSIC VALUE

Intrinsic value refers to the nonanthropocentric value of species, ecosystems, and genes, or value derived from the essential nature or constitution of biodiversity per se. The intrinsic value perspective acknowledges that other living entities have value independent of any human estimation of their worth. The most basic tenet of intrinsic value is a belief that *Homo sapiens* has no right to extinguish or jeopardize the existence of other life-forms. This is a very ancient and universal concept; it has been articulated by human cultures as diverse as the Sioux Indians of the North American Plains, the Kuna Indians of Panama, and the San of Botswana. In short, the intrinsic value perspective holds that all species have value in themselves, sometimes referred to as a "right to exist," and that they have a right to continued coexistence with us as fellow passengers on "Spaceship Earth."

In addition to these moral and ethical considerations, philosophical and evolutionary perspectives have been advanced in defense of the concept of intrinsic value. Philosophically, a species or an ecosystem represents an "end-in-itself;" therefore, it should be valued for itself rather than as a mere instrument for the fulfillment of human desires and needs, that is, as a means to meet human ends. Furthermore, our morality has been questioned on the grounds that the willful or inadvertent extinction of other species is tantamount to "playing God;" some consider it morally reprehensible for humans to destroy God's creations. The evolutionary perspective emphasizes the human arrogance inherent in viewing our own survival and continued existence as more important than that of other species. This view is particularly germane when one considers that *Homo sapiens* is one of the most

evolutionarily recent species on earth, and therefore nearly all wild species that exist today have evolved in the absence of human influence or interference.

Despite these philosophical, moral, and evolutionary arguments, the concept of intrinsic value has been surrounded by controversy, and its very existence has been questioned on philosophical grounds:

- Did wild species have any value before species capable of "conscious valuation" evolved on earth? Is *Homo sapiens* the only species capable of conscious valuation?
- Can wild species and entire ecosystems be "valued" by other species that are not "conscious valuers"? In other words, even though every species needs other species to sustain its existence, this may not validate the concept of intrinsic value in the absence of species capable of conscious valuation.
- Is the concept of intrinsic value a valid concept only if *all* "conscious valuers" *should* be capable of perceiving it (even if all do not)? If many or most "conscious valuers" cannot perceive the intrinsic value of other living things, does this invalidate the concept?

Although many conservationists consider intrinsic value as the "correct moral stance" toward nature, this concept has been criticized by others as a relatively ineffective rationale for promoting an understanding and appreciation of the necessity of biodiversity conservation, particularly within economic, political, and environmental policy arenas. It has been viewed as a highly impractical approach for promoting nature conservation in developing countries where many or most people do not have even the most basic necessities required to sustain life. In addition, many conservationists believe that a balanced, holistic approach to biodiversity values must include consideration of both intrinsic and extrinsic values.

III. EXTRINSIC VALUES

Extrinsic values are determined by human estimations of the worth or significance of biodiversity and, therefore, constitute anthropocentric uses that have an external origin in relation to biodiversity per se. Oftentimes referred to as "utilitarian" or "instrumental" values, extrinsic values constitute the diverse uses of genes, species, and ecosystems that serve the material and spiritual needs of human beings. In contrast, intrinsic value is an innate or inherent quality of biodiversity, even though appreciation or recognition of intrinsic value actually emanates from the moral, spiritual, or philosophical beliefs of human beings. Although various classification methods have been used to categorize extrinsic values, the method adopted herein is an adaptation of the "uses of wildlife" classification scheme (see Section IV). The major extrinsic value categories that will be discussed by use of pertinent examples are: (A) spiritual and religious, (B) psychological and therapeutic, (C) sociocultural and sociohistorical, (D) research and educational, (E) ecological and evolutionary, (F) recreational and tourism, and (G) monetary–economic or utilitarian.

Extrinsic values are sometimes classified into three broad categories that reflect the extent to which they are amenable to economic quantification: moral values, amenity values, and commodity values. "Moral values" are highly intangible, immaterial values; thus, they are very difficult to quantify (see Section IV). These include spiritual and religious values, and to some extent psychological values. In contrast, "commodity values" are much more amenable to quantification because they entail consumptive uses related to provision of commodities (goods) and services. By definition, these would include monetary–economic or utilitarian values of harvested resources, consumptive recreational values, and research, educational, or even spiritual and religious uses that require the death or sacrifice of animals or plants (see the following section). Finally, "amenity values," or nonconsumptive uses of species and ecosystems that meet nonmaterial human needs, possess both tangible and intangible components. Although these may be economically quantified to some extent, other components of value may be difficult to quantify. Amenity values include psychological, therapeutic, nonconsumptive recreational, educational,

ecological, and evolutionary values. In some cases, interrelationships exist among these broad value categories. As an example, the free ecological and evolutionary services of natural ecosystems oftentimes provide us with amenity values such as scenic beauty and field sites for education. However, ecological and evolutionary processes also generate biological resources that can be tapped to provide commodity values, for example, medicinally useful compounds and genes used for resistance to crop diseases and pests.

The distinction between "consumptive" and "nonconsumptive" uses of biodiversity in relation to amenity versus commodity values should be clarified. Traditionally, nonharvesting recreational activities, such as birdwatching and nature photography, have been considered as "nonconsumptive" uses. In contrast, recreational activities such as sport hunting, fishing, and trapping have been viewed as "consumptive" uses because they usually result in the death of animals. However, a significant proportion of hunters and fishermen do not kill the animals they capture, and there is mounting evidence that nonconsumptive users inadvertently kill plants and animals, for example, by overuse of hiking trails and disturbance of animals that forces them to abandon their eggs or young. During the past decade, much has been written about the "myth of the nonconsumptive user," as the nonconsumptive recreational literature on visitor impacts and human disturbance has steadily increased. Nevertheless, because many important studies on the values of biodiversity consider nonconsumptive and consumptive uses separately, the traditional distinctions between these two terms will be utilized in this article.

A. Spiritual and Religious Values

Human spiritual and religious uses of biodiversity are the most intangible of all extrinsic values of species and ecosystems. These values contribute to the spiritual growth and awareness of human beings, or are linked to the religious beliefs, doctrines, ceremonies, or rituals of various cultures. Examples related to the three major components of biodiversity include:

- *Ecosystem diversity*—Black Mesa mountain, Mount Everest, Mount Fuji, Mount Kenya, and other mountains that serve as the focal point for rituals or religious celebrations; some forests are also spiritually valuable to certain cultures, for example, the sacred forest groves in southwestern India, each of which is dedicated to a deity who protects all living creatures in the ancient natural sanctuary.
- *Species diversity*—Asian elephants, monkeys, birds, cobras, and trees are valued in the Buddhist, Hindu, and various tribal religions; many wild species have symbolic or religious significance, for example, *Ficus religiosa* is a sacred, protected tree in India.
- *Genetic diversity*—Traditional Papago corn varieties, the "festival rices" of Nepal, and the Mithan cattle breed in northeastern India and northern Myanmar are examples of the religious or ceremonial significance of genetic variation within species.

Aboriginal and tribal societies typically ascribe religious value to a variety of species. As an example, the Papago Indians of the Sonoran desert region of the U.S./Mexico borderlands utilize the following wild species for religious purposes: saguaro (*Carnegiea gigantea*), Goodding willow (*Salix gooddingii*), bulrush (*Scirpus olneya*), Golden Eagle (*Aguila chrysaetos*), Red-tailed Hawk (*Buteo jamaicensis*), turkey (*Meleagris gallopavo*), coyote (*Canis latrans*), ringtail (*Bassariscus astutus*), badger (*Taxidea taxus*), pronghorn antelope (*Antilocapra americana*), and the mule deer (*Odocoileus hemionus*), which is shown in Fig. 1. In general, Native Americans assigned special spiritual significance to impressive natural settings such as the Grand Canyon, as well as to special "totem" species—usually powerful predatory animals such as coyotes, bears, eagles, or hawks. [*See* SPECIES DIVERSITY.]

An excellent example of the religious value of an animal species is that of the endangered leatherback sea turtle (*Dermochelys coriacea*)—the largest living reptile on earth. The leatherback features prominently in the ceremonies and rituals of the Seri Indians, who refer to themselves as "The People" (*Cocaác*). The Seri are hunter-gatherers who

FIGURE I   A mule deer (*Odocoileus hemionus*) at Big Bend National Park in Texas; the Papago Indians ascribe religious value to this animal species. (Photo credit: M. L. Oldfield)

live along an extremely arid coast in Sonora, Mexico. In 1981, the capture of one of these gigantic sea turtles initiated a four-day ceremonial festival. A special shelter was constructed to shade the dead turtle, branches from the sacred elephant tree were scattered over the carcass, and a Seri woman painted powerful designs on the shell and flippers of the animal (Fig. 2).

Sometimes, a particular species may be simultaneously valued for spiritual or religious uses by different societies or cultures. In such cases, conflicts over the proper use and management of the species may result. As an example, the Port Orford cedar (*Chamaecyparis lawsoniance*) of northern California and southern Oregon is a sacred tree historically used in religious ceremonies of Native Americans of the Yurok, Karok, and Hupa tribes in northwestern California. Management of the dwindling stands of this cedar recently precipitated a major controversy, because it is also valued for use in Japanese temples as a substitute for the commercially extinct Japanese Hinoki cypress. In the 1980s, commercial U.S. logging operations were exporting Port Orford cedar lumber to Japan, where it commanded prices of up to $3000 per 1000

FIGURE 2   A Seri Indian woman painting a leatherback sea turtle (*Dermochelys coriacea*). In 1981, this turtle was the focus of a four-day religious ceremony in Sonora, Mexico. (Photo credit: Mary Beck Moser, "People of the Desert and Sea," University of Arizona Press, Tucson.)

board-feet. The controversy over domestic versus foreign uses of this species was fueled by criticism about the logging operations, which were implicated as a causal factor in the spread of a root-rot fungus that kills the cedars. American Indians attempted to reserve all the remaining cedars growing on tribal lands administered by the U.S. Forest Service for native ceremonial purposes. The logging industry and other citizens' groups reacted by seeking a management plan to restrict the industry's access to particular areas in order to reduce the spread of the deadly fungus.

In contrast to the relatively common religious uses of biodiversity by tribal and aboriginal peoples, spiritual and religious values are not frequently cited in relation to modern, technological societies. This may be due to a number of reasons. First, most people living in modern societies are clustered into urban/suburban environments and are generally disassociated from wild species and natural environments. Second, Judeo-Christian religions frown upon what they view as "pagan" worship of animal or plant species. Third, the U.S. Constitution mandates the separation of church and state. Because spirituality often connotes a relation-

ship to organized religion, the U.S. government, which funds much research on biodiversity values in the United States, may consider this area of research inappropriate. Finally, there is a general lack of understanding of the functional meaning and application of spirituality within a research context, in part because of a dearth of information on this topic.

Despite these obstacles, the spiritual value of wilderness in general, and of specific sites or species in particular, has been catalogued by studies of wilderness users in North America and is acknowledged by various religions. The Judeo-Christian view of the religious value of wildlife is evident in the parable of Noah's ark in the book of Genesis; this story describes how God's will for ensuring the continued existence of each species on earth was carried out by Noah in humanity's "first Endangered Species Project." Similarly, the religious value of "the wilderness" is evident from the scriptures that describe how Moses received the Ten Commandments from God in the wilderness, John the Baptist preached in the wilderness of Judea, and the Spirit of God led Christ into the wilderness to be tempted by the devil and obtain spiritual inspiration. Wilderness—in the form of natural ecosystems—has held special symbolic significance in relation to Christian spiritual experiences. Christian theologians have always viewed the earth as God's creation and many argue that an understanding of the workings of natural ecosystems can aid in our understanding of the nature of God. Other religions also refer to the spiritual value of natural areas or "sacred places." For example, Buddha retreated to the forest for six years to find enlightenment. In many modern religions, there is a religious interpretation of the earth's "least modified environments" as evidence of God's power and glory.

The spiritual significance of "the wilderness experience" underlies much of the wilderness preservation philosophy of modern societies. The language of the Wilderness Act of 1964 implies spiritual values of wilderness, and the U.S. Code of Federal Regulations states that the wilderness resources of our National Forests shall be managed to promote and perpetuate the values of inspiration and solitude. Preservation rationales offered by Americans emphasize spiritual and mystical dimensions as well as freedom, solitude, and nature appreciation. Notions of spirituality appear to emanate from feelings of grandeur and the overpowering elements of wilderness areas—forces and elements larger than the human mind can easily comprehend (Fig. 3). Interestingly, the construction of many cathedrals and other formal places of worship appears to mimic wilderness in the sense of vast open spaces that reach toward the sky.

Wilderness experiences have been frequently equated with spiritual endeavors and experiences of extreme states of consciousness. For example, students at the University of Colorado who were polled about such experiences cited an opportunity for "spiritually uplifting experiences" as a major perceived benefit of wilderness areas. Many renowned conservationists and wilderness writers, such as John Muir, Henry David Thoreau, and Aldo Leopold, used language that suggested wilderness experiences close to the notions of "peak" or "transcendent" experiences that may catalyze higher states of consciousness. John Muir wrote that "wild nature" should be considered as the "conductor of divinity"; he believed that wilderness provides humanity with spiritual and psychic

energy. Joseph Wood Krutch thought of wilderness as the "permanent home" of the human spirit.

Another common theme in the American wilderness experience is that of achieving a greater appreciation of "the relatedness of all living things" as well as an enhanced understanding of "the unity of all life." That people in modern societies aspire to such spiritual awareness is refreshingly similar to the beliefs espoused by many tribal societies that have depended on natural environments for their survival. For example, Black Elk, an Oglala Sioux Indian of the North American Great Plains, believed that "all life is holy" and all species "are the children of one mother and their father is one spirit." Similarly, the Kuna Indians of Panama believe that "the land is our mother" and "all living things that live on her are brothers." In sum, the wilderness is one of the few places where people living in technologically advanced societies can retreat to discover their own "spiritual riches," to find spiritual freedom and solitude, and to contemplate "what we are" rather than "what we have."

B. Psychological and Therapeutic Values

Psychological and therapeutic values of biodiversity relate to the perceived psychological and intellectual benefits derived from the direct use of wilderness or other natural areas, or from observing or interacting with wild and domesticated species. Examples related to the three major components of biodiversity include:

- *Ecosystem diversity*—Feelings of mental health and well-being, the growth and development of self-esteem, and other psychological and therapeutic values can be derived from the use of wildlands and natural areas.
- *Species diversity*—Some people are inspired by watching a soaring eagle or hawk or by observing powerful predatory animals such as bears or lions; psychological benefits derived from observing wild species include intellectual achievement, a sense of well-being, aesthetic appreciation, and enjoyment of the "wonders of nature."

FIGURE 3A The majestic Grand Canyon National Park. (Photo credit: L. R. Lawlor)

FIGURE 3B The unusual hoodoo formations of Bryce Canyon National Park provide an awe-inspiring vista. (Photo credit: L. R. Lawlor)

• *Genetic diversity*—The psychological and therapeutic benefits of owning and caring for horticultural varieties of plants, garden crops, or domesticated breeds of pets have been demonstrated, for example, for emotionally disturbed children, the elderly, and people who live alone.

The psychological benefits of uses of wilderness and natural areas have been assessed by a variety of methods. Two similar approaches using two different types of leisure benefit inventories—PAL (Paragraphs About Leisure) and REP (Recreation Experience Preference)—have indicated the relative importance of more than 40 psychological "benefit themes" as they were subjectively evaluated by wilderness users. The PAL scores of recreationists at Huron River parks in Michigan were analyzed for eight psychological benefit factors. "Compensation" (satisfaction of the need for unusual or new experiences, i.e., needs not satisfied by daily routines) was the only factor that consistently received high scores (PAL = 63–65) in association with the following wilderness-related activities: camping, canoeing, hiking, and lake fishing. Above average scores (PAL = 56–60) were observed for camping and hiking in relation to the factors of "self-expression" (satisfaction of the need to initiate novel activities or express oneself through creative use of talents, or to enjoy power or recognition for these activities) and "companionship" (satisfaction of the need for playful, supportive relationships with others that serve to enhance one's self-esteem and acceptance).

The REP scores of almost 3000 backpacking hikers and campers who utilized eight designated wilderness areas and four undesignated wilderness sites indicated that satisfaction with their recreation experience was:

• Strongly associated (100%) with a preference to "reduce tensions";

- Strongly associated (92%) with a preference to "enjoy nature";
- Strongly (75%) or moderately (25%) associated with a preference to "escape noise" and crowds;
- Strongly (75%) or moderately (25%) associated with a preference for "outdoor learning" experiences;
- Moderately associated with a preference for experiencing "independence" (100%), "sharing similar values" with others (92%), "introspection" and spiritual experiences (83%), and personal achievement/stimulation (58%).

In this context, backcountry wilderness users seemed to more often appreciate the psychological benefits of enjoying nature, reducing tensions, escaping crowds, and having opportunities for outdoor learning experiences than psychological benefits associated with, for example, spirituality, personal achievement, or interactions with others. Moreover, all of these psychological benefits were highly ranked in comparison with other benefits judged to be of moderate to low importance—physical rest, teaching or leading others, and risk-taking.

Psychotherapeutic camping programs located in either backcountry wilderness or mountainous, forested areas within parks, national forests, and forest reserves have also been cited for their psychological and psychiatric benefits for treating emotionally disturbed adolescents and the mentally ill. A variety of wilderness challenge and adventure programs, such as Outward Bound, and wilderness and outdoor camping therapy programs have been acknowledged for:

- Assisting adolescents in processing their feelings and communicating their feelings to others;
- Reducing pathological symptoms and emotional problems;
- Enhancing the self-esteem and self-confidence of emotionally disturbed, abused, neglected, as well as behaviorally "normal" adolescents;

- Improving students' attitudes and behaviors about school;
- Reducing rates of recidivism of juvenile delinquents;
- Increasing the quantity and quality of social interactions;
- Improving patient–staff relationships;
- Increasing patient "fun" and enthusiasm.

Although future research will be necessary to clarify the role and nature of psychological and psychotherapeutic benefits of biodiversity, preliminary studies indicate that these intangible benefits may contribute substantially to human mental health and well-being.

C. Sociocultural and Sociohistorical Values

Sociocultural and sociohistorical uses are relatively intangible values of biodiversity because they are linked to the cultural or historical significance of species or natural areas. These include national or state symbols and other symbols of civic pride, "totem" or symbolic animals of tribal cultures, and biodiversity that promotes social cohesion by serving as a focal point of social activities. Examples related to the three major components of biodiversity include:

- *Ecosystem diversity*—Many national parks, monuments, and historic sites are culturally or historically significant to Americans because they demonstrate close ties between U.S. history and the American landscape, for example, the Great Smoky Mountains National Park, Mesa Verde National Park, Voyageurs National Park, Kobuk Valley National Park, and Canyon de Chelly National Monument (Fig. 4).
- *Species diversity*—Wild species are important as national, state, or cultural symbols, for example, the endangered Bald Eagle (*Haliaeetus leucocephalus*), chosen by the Continental Congress in 1782 as a symbol of our American ideals of freedom, the Texas bluebonnet (*Lupinus texensis* and the rare *L. subcarnosus*),

FIGURE 4 Canyon de Chelly National Monument is important for its archaeological sites as well as its scenic beauty. (Photo credit: M. L. Oldfield)

the official state flower of Texas, and the Eastern Bluebird (*Sialia sialis*), the official state bird of New York.

- *Genetic diversity*—Strong cultural preferences still persist for native, traditional crop varieties, for example, potatoes in the Peruvian Andes and corn in the U.S. Southwest and Mexico. Many rare or unusual livestock breeds also possess sociohistorical or sociocultural value: the Texas Longhorn of the Southwest, the endangered Kuri cattle of Lake Chad, and the native, nonhumped cattle breeds of Africa such as the endangered trypano-tolerant N'Dama and West African Shorthorn.

Sociocultural values arise when wild species serve as a focal point in community activities or promote social cohesion and unity. An excellent example is the sociocultural role of the bowhead whale (*Balaena mysticetus*) in several Inupiat and Yupik Eskimo villages in Alaska. Bowhead whal-

ing is the most significant activity in the subsistence cycle of these cultures. The bowhead dominates in the legends and festivals of these people, because they literally owe their way of life and very existence to this endangered species. Whalers are the most respected members because they provide the entire community with most of its annual protein needs. They help to maintain social stability within the villages, which are actually loose groups of distantly spaced homes. The entire community divides and shares the meat and skin of the first bowhead whale caught each year—a custom that strengthens communal and kinship bonds. Moreover, without the meat and oil from a single whale, the entire community may risk starvation or spend the year living on the brink of survival, relying instead on daily hunts to obtain meat from much smaller animals. After the initial ceremony, the remainder of the meat and other parts of the animal are shared at subsequent feasts and ceremonies—most notably, the Captain's Feast, during which the section from the belly to the tail is divided, and the Whale's Tail Feast. Bowhead whale meat is also provided for Thanksgiving and Christmas feasts. Thus, the bowhead plays a major sociocultural role in these Inupiat and Yupik villages: it provides community identity and ensures that each village has continuity with its past. Other whale species have sociocultural significance as well. The endangered humpback whale (*Megaptera novaeangliae*) has long been socioculturally significant to many aboriginal whaling cultures, such as the Tonga Islanders. Two other endangered species, the black right whale (*Eubalaena glacialis*) and sperm whale (*Physeter catodon*), were once the mainstay of the Yankee whaling industry. These great whales still play important roles in the sociocultural history of Long Island and other coastal communities in the U.S. Northeast.

Sociohistorical values of biodiversity involve contributions to a culture's history, folk traditions, or legends. An interesting example of the sociohistorical value of wild species is a folk legend of the Cocopa Indians that involves the horned lizard (*Phrynosoma platyrhinos*) (Fig. 5) and an endangered, semidomesticated saltgrass species (*Distichlis palmeri*), also known as Wildwheat to scien-

FIGURE 5 The horned lizard (*Phrynosoma platyrhinos*) plays an important sociohistorical role in the legends of the Cocopa Indians. (Photo credit: W. C. Sherbrooke)

tists. The Cocopa tell of a time when their ancestors lived in the Cocopa Mountains and the foothills were surrounded by a great freshwater lake that once covered the Mexicali and Imperial valleys. After the climate warmed, the waters of the surrounding lake receded and left a dry lower delta cut by rivers, including the Hardy River where the Cocopa now live. The legend tells the story of how the Cocopa discovered the saltgrass (*nypá*), a salt-tolerant food grain plant, and determined that it was time to move from the mountains to the river delta. At the time that the vast lake was receding, a mountain person named Kinakul sent the horned lizard, Hiesh, down to the delta to see if the land was dry. Hiesh traveled as far as the Laguna Salada, where he found some ripe *nypá* spikelets, which he gathered and stuck into the top of his head. When he returned to the mountains, Hiesh's new spikelets were a sign to the Cocopa that the waters had sufficiently receded, and that there was a suitable food plant on the lower delta for the people to eat. This is also their explanation of the origin of the spikes on the head of this ant-eating reptile.

D. Research and Educational Values

Research and educational values refer to uses of biodiversity for advancing scientific knowledge and research, or for promoting education. Examples related to the three major components of biodiversity include:

- *Ecosystem diversity*—Natural research areas, such as Olympic National Park and the Great Smoky Mountains National Park, have been used for baseline monitoring in long-term studies of, for example, the effects of pollutants on natural ecosystems and the role of forests and watersheds in nutrient cycles. Natural ecosystems also serve as "living classrooms" for teaching students about paleontology, archeology, geology, botany, zoology, microbiology, population biology and ecology, ethology, population genetics, entomology, wildlife management, and the recreation and leisure sciences.
- *Species diversity*—Wild species contribute to the education and training of anthropology, biology, geology, and medical students; they also serve important roles as "research models" that contribute to our understanding of human pathology, neurophysiology, developmental biology, structural chemistry, genetics, ecology and evolutionary biology, and climatology.
- *Genetic diversity*—Genetic variation within particular species has been invaluable in advancing research and education in genetics (e.g., fruit fly, *Drosophila melanogaster,* and corn, *Zea mays*), virology (tobacco, *Nicotiana* spp.), and other areas of scientific research. The Yucatan pig, a small, hairless breed of swine imported from Yucatan in 1960, has been used to train surgeons and other medical personnel in surgical procedures.

The research and educational role of intact natural ecosystems is extremely important for students of evolutionary biology, ecology, population genetics, paleobiology, and many other scientific disciplines. However, the research and educational significance of particular species or genetic strains or breeds of plants and animals has also been well documented. Evolutionarily ancient species considered to be "living fossils" provide interesting examples. These species possess many primitive traits or characteristics, and therefore they often play pivotal roles in determining the evolutionary histories (i.e., phylogenies) of living and extinct organisms. Species known as "living fossils" are oftentimes classified as a "sister taxon" or "outgroup"; an outgroup is a distantly related taxon that possesses both primitive (ancestral) and some evolutionarily "derived" characteristics that link it to the "ingroup" taxon that is considered more evolutionarily advanced. Table I lists some important examples of living species considered to be "outgroups" in vertebrate systematics (the taxonomic classification and study of the anatomical, morphological, or other traits of living organisms). Many of the "living fossils" that are classified as outgroups are endangered or rare species, for example, the coelacanth, tuatara, saltwater crocodile, and duck-billed platypus.

Perhaps the most important research value of wild animal species has been as desirable, unique, or even indispensable "animal models." Research on nonhuman primates has played an essential role in our ability to understand, prevent, and treat many human diseases, including poliomyelitis, hepatitis, measles, yellow fever, rubella, Parkinson's disease, AIDS, and virally induced human cancers. However, biomedical research uses of animals, especially mammals and other vertebrates, have been frequently opposed by animal rights activists. At the heart of the animal rights controversy are species used as domestic pets (e.g., cats and dogs) and the nonhuman primates. People often object to research uses of pet animals because they are so familiar to us and are the focus of so much human affection. Objections to research uses of nonhuman primates may also emanate from their observed similarity to humans. In fact, it is their very genetic similarity to humans that has made nonhuman primates the most important and perhaps most indispensable taxonomic group of wild species currently used for studying ethology, human evolutionary history, and human pathology. For example, the DNA (gene) sequences of the chimpanzee (*Pan troglodytes*) are 98% identical to those found in human DNA. Therefore, it is not surprising that chimpanzees exhibit such marked biological and behavioral similarities to humans—similarities that make them invaluable as research subjects for biomedical and psychobiological research.

The chimpanzee has recently served as the principal animal model for studying infectious diseases such as hepatitis and AIDS and for developing and testing antiviral compounds and vaccines to treat

TABLE I

Some Taxonomically Important Vertebrate Outgroups with Living Representatives[a]

Outgroup taxon	Living species	Ancestral traits	Ingroup taxon	Derived traits
Class Agnatha (jawless fishes)	Atlantic hagfish (*Myxine glutinosa*)	Gill bars, cranium, and brain	Vertebrata (vertebrates)	Presence of vertebrae, 2–3 semicircular canals
	Sea lamprey (*Petromyzon marinus*)	Absence of jaws, presence of vertebrae	Gnathostoma (jawed vertebrates)	Presence of jaws and teeth, paired fins
Subclass Sarcopterygii (lobe-finned fishes)	Coelacanth (*Latimeria chalumnae*) South American lungfish (*Lepidosiren paradoxa*)	Presence of gills and lungs, fleshy fins with supporting (unique) skeletal structure	Tetrapoda (tetrapods)	Paired pectoral and pelvic limbs with one proximal and two distal bones
Order Sphenodontia (sphenodontid reptiles)	Tuatara (*Sphenodon punctatus*)	Diapsid skull, oviparity (amniotic egg), determinant growth	Squamata (squamate reptiles)	Fused snout bones, unique palate characters
Order Crocodilia (crocodiles and alligators)	Saltwater crocodile (*Crocodylus porosus*)	Diapsid skull, fenestra in front of eye orbit, advanced parental care	Aves (birds)	Wings and feathers, loss of teeth, endothermy
Order Monotremata (monotremes)	Duck-billed platypus (*Ornithorhynchus anatinus*) Australian spiny echidna (*Tachyglossus aculeatus*)	Therapsid skull, hair, and mammary glands, oviparity (lays eggs), reptilian skeletal elements	Metatheria and Eutheria (marsupial and placental mammals)	Marsupium or placenta, functional teats, viviparity

[a] Source: F. H. Pough, J. B. Heiser, and W. N. McFarland (1989). "Vertebrate Life." Macmillan, New York.

these diseases. Before the AIDS epidemic, hepatitis B vaccine was developed in the United States by using the chimpanzee as the principal animal research model. This is the first human vaccine that prevents both a chronic infectious illness and a type of fatal cancer induced by hepatitis B virus–hepatocellular carcinoma. In 1988, wild African chimpanzee populations were added to the U.S. endangered species list because of the combined effects of habitat loss, the live trade for pets, research subjects, and zoological specimens, and their use as a source of food. A particularly destructive aspect of the live trade was the common practice of capturing juvenile chimpanzees by shooting their mothers and other protective adults. Today, biomedical research institutions in the United States use captive-bred chimpanzees (Fig. 6).

Other species of nonhuman primates have played an essential role in the study or treatment of various human diseases. Examples include the squirrel monkey (*Saimiri sciureus*) for the study of cardiovascular diseases, the woolly monkey (*Lagothrix lagothricha*) for leukemia studies, leaf monkeys (*Presbytis* spp.) for studies of bubonic plague, the owl moneky (*Aotus lemurinus lemurinus = trivirgatus*) for treating drug-resistant malaria strains, the rhesus macaque (*Macaca mulatta*) for studying the fatal dengue viruses that infect Southeast Asian children, the Celebes macaque or black ape (*Macaca nigra*) for diabetes research, the African green (vervet) monkey (*Cercopithecus aethiops*) for hypertension and toxicology studies and production of virus-free polio vaccines, and marmosets (*Callithrix* spp.) and tamarins (*Saguinus* spp.) for cancer and hepatitis research. In 1979, the discovery of leprosy in a naturally infected sooty mangabey monkey (*Cercocebus atys*) was a valuable addition to the only other animal research model—the nine-banded armadillo (*Dasypus novemcinctus*). Prior to that time, researchers had been unable to locate a suitable animal model for study of leprosy in a species more closely related to humans. Use of the sooty mangabey, rhesus macaque, and African green monkey may now enable researchers to better understand the transmission of leprosy and to develop a protective vaccine as well as new therapeutic regimens for treating dapsone-resistant leprosy strains. More recently, owl monkeys and captive-bred specimens of the endangered cotton-top tamarin (*Saguinus oedipus*) have been used to experimentally demonstrate that Epstein–Barr virus induces the cancer known as Burkitt's lymphoma.

FIGURE 6 A female chimpanzee (*Pan troglodytes*) and her offspring look out over the corral walls at the University of Texas M. D. Anderson Cancer Center chimpanzee breeding facility in Bastrop, Texas. This facility is one of five dedicated breeding colonies funded by the National Institutes of Health as part of a national plan to a manage self-sustaining, captive chimpanzee population. (Photo credit: Department of Veterinary Sciences, U.T. M. D. Anderson Cancer Center)

E. Ecological and Evolutionary Values

Ecological and evolutionary values of genetic diversity, species, and natural ecosystems are related to the free services of nature that support human civilizations and sustain the biosphere. Examples related to the three major components of biodiversity include:

- *Ecosystem diversity*—Natural processes within ecosystems provide the following benefits: fixation of solar energy and organic carbon by primary producers; nutrient cycling; soil production; absorption and breakdown of toxic substances, pollutants, and organic wastes; regulation of climate and temperature;

maintenance of the atmosphere's carbon dioxide–oxygen balance; production and accumulation of living biomass; and provision of *in situ* (on-site) reservoirs of genetic diversity that are used to establish *ex situ* (off-site) genetic resource conservation facilities.

- *Species diversity*—Wild plants, microbes, and animals play important roles in nutrient cycles, forest regeneration, and biomass production in natural ecosystems; others, such as wild pollinators and biological control organisms, provide useful ecological services to support agriculture.
- *Genetic diversity*—Genetic variation provides the necessary "raw materials" for evolutionary processes and for the continuing adaptation of species or populations to changing environmental conditions.

Ecological interactions among organisms and between living organisms and their abiotic environment are important factors that influence evolutionary processes in natural ecosystems. For example, predators exert natural selection pressures that cause evolutionary changes in their prey populations, and disease pathogens select for disease-resistant (and select against disease-susceptible) individuals within their host populations. Predator–prey, parasite–host, and mutualistic associations, such as those that exist between plants and their pollinators or seed-dispersers, provide examples of coevolution or the joint evolution of two or more species that share close ecological associations over long periods of time.

In some cases, commodity values accrue from the "free" ecological and evolutionary services of nature. Examples include the monetary contributions of wild pollinators and biological control organisms to modern agriculture. For instance, natural areas contain essential habitat and resources, such as wildflowers, that sustain wild pollinators and many parasitic and predatory insects that aid the agricultural sector of the U.S. economy. In California, grape growers have saved upward of $40–$60 per acre in reduced pesticide applications because the parasitic wasps that attack grape leaf-

hoppers use wild berry (*Rubus* sp.) brambles to locate alternative hosts during the off-season. Natural ecosystems also provide cover, food, and breeding sites for the many species of wild pollinators that contribute to the production of more than $1 billion worth of crops either imported by or grown in the United States.

Evolutionary processes occurring within natural ecosystems are also primarily responsible for much of the useful genetic variation valued by humanity. As an example, useful genes from wild crop relatives confer resistance to disease pathogens and insect pests and tolerance to salt, drought, flooding, or other harsh environmental conditions. These genetic traits evolved as a consequence of selection pressures within natural ecosystems over very long periods of time. Long-term ecological interactions between predators and their prey and parasites or pathogens and their hosts have also resulted in the evolution of economically valuable medicinal and industrial compounds (see Section III,G). Evolutionary ecologists have documented the important "defensive" role that many toxic, pharmacologically active or industrially useful compounds play in protecting wild species from their predators, parasites, and disease pathogens. [*See* PLANT SOURCES OF NATURAL DRUGS AND COMPOUNDS.]

Cardiac glycosides—a group of pharmacologically active compounds—provide an excellent example of toxic, defensive compounds. Both digitalin and digitoxin from the purple foxglove (*Digitalis purpurea*) have been used to treat hypertension, congestive heart failure, and other heart afflictions. When monarch butterfly (*Danaus plexippus*) larvae are reared on milkweeds that contain cardiac glycosides, they sequester these toxic compounds in their wings and other body tissues, and thereby gain protection against insect-eating birds. Another example is that of plant-derived pharmaceutical and industrial chemical compounds that also play a defensive role in protecting plant species. The principal chemical feeding deterrents avoided by snowshoe hares, moose, white-tailed deer, and other browsing mammals include common forestry products such as tannins, turpentine, and resins as well as medicinal compounds such as alkaloids and antibiotics. The presence of camphor, a medicinal compound that has antimicrobial properties, deters feeding by snowshoe hares. Because browsing mammals depend on symbiotic microorganisms in their digestive system to process and digest their food, it is not surprising that they avoid consuming plants that contain antibiotics or other antimicrobial compounds.

In addition to commodity values, ecological and evolutionary processes can also be translated into amenity values, for example, the research and educational value of "living fossils" and other evolutionarily ancient species. It has been argued effectively that evolutionary relicts and other taxonomically unique species should be valued more highly, and therefore afforded greater protection by conservationists, than taxonomic groups that contain many species. Thus, a unique "outgroup" species should be considered more valuable than any of the "ingroup" species to which it is distantly related (see Table I). Using this "calculus of biodiversity" argument, it would be more important to save, for example, New Zealand's endangered tuatara (*Sphenodon punctatus*) than the many endangered, but nonnative bird species that prey on the eggs and young of this taxonomically unique reptile.

The values associated with "keystone" predators and other "ecologically unique" species are comparatively difficult to conceptualize, yet they deserve special consideration. An "ecologically unique" species or geographically distinct population is one whose removal would cause extinctions or major population shifts among other species within a natural community. Obviously, if the removal of a single "key" species produces a cascade of extinctions, such a species would have immense ecological and evolutionary value. Examples of ecologically unique species include: (1) primary producers, such as photosynthesizing plants in terrestrial and most aquatic ecosystems; (2) "keystone" predators that foster coexistence among competing species; and (3) coevolved mutualists, or two or more species that depend on each other to survive and/or reproduce. The latter group includes species involved in plant–pollinator, plant–seed disperser, plant–fungus (lichen), and plant–microorganism mutualisms. There are many examples of keystone predators, such as the starfish *Pisaster,* whose presence in nearshore marine communities enhances overall species diversity. Bacteria that serve as primary synthesizers of organic molecules are keystone producers in deep-sea hydrothermal vents, and approximately ten "keystone" plant producer species support about 85% of the biomass of mammals in a Peruvian rain forest during the dry season. [*See* KEYSTONE SPECIES.]

F. Recreational and Tourism Values

Nature-based recreation and tourism values of biodiversity relate to amenity resources that provide opportunities for people to relax and enjoy the aesthetic beauty of nature, the challenges of wilderness environments, and other recreational or leisure pursuits. These amenity resources include natural settings such as parks, reserves, wilderness and scenic areas, forests, streams, and lakes as well as wildlife, fish, and plant species. Examples related to the three major components of biodiversity include:

- *Ecosystem diversity*—Each year, more than 700–800 million people visit our national and state parks to enjoy the varied landscapes and forests protected at these sites; tourism at parks and reserves has become increasingly important to the developing economies of countries such as Costa Rica, Venezuela, Tanzania, and Kenya.
- *Species diversity*—In 1991, 76 million Americans observed, fed, and/or photographed hundreds of species of birds, mammals, reptiles, amphibians, butterflies, and other insects, and approximately 40 million recreational fishermen and hunters pursued more than 80 species of fishes and 90 species of wildlife in the United States.
- *Genetic diversity*—Recreational benefits are enjoyed by millions of members of horticultural associations and home gardeners who cultivate unusual varieties of plants and by millions of recreational fishermen who take advantage of fish enhancement and stocking programs; each year, approximately 100,000 tourists visit the Rare Breeds Survival Trust in

the United Kingdom and about 10 million tourists visit "living historical farms" in North America.

Recreation and tourism activities that typically occur within wilderness areas include: hiking, camping, whitewater rafting, canoeing, rock climbing, birdwatching, sightseeing, nature photography, hunting, and fishing. A 1987 study assessed the net economic value (consumers' surplus) of outdoor recreation activities in parks, forests, and natural areas in the United States. Table II shows the aggregate consumer demand for the 12 outdoor recreation activities most closely associated with wild species and wilderness areas. These activities totaled about $47 billion in U.S. consumers' surplus (for 1.23 billion trips or 3.3 billion recreation days) out of a total of $122 billion net economic value for all 37 recreation activities that took place at parks, forests, and other outdoor recreation sites.

In the United States in 1991, more than 35 million recreational fisherman expended nearly $24 billion—and 14 million recreational hunters spent $12 billion—on equipment, trip-related expenses, licenses and permits, and other expenditures. From 1975 to 1991, the principal animal species captured by U.S. recreational fisherman and hunters included:

- *Freshwater fishes*: Northern pike (*Esox lucius*), muskie (*E. masquinongy*), walleye (*Stizostedion vitreum*), white bass (*Morone chrysops*), white perch (*M. americana*), yellow perch (*Perca flavescens*), largemouth bass (*Micropterus salmoides*), cutthroat trout (*Salmo clarki*), Atlantic salmon (*S. salar*), brook trout (*Salvelinus fontinalis*), lake trout (*S. namaycush*), sockeye salmon (*Oncorhynchus nerka*), pink salmon (*O. gorbuscha*), steelhead trout (*O. mykiss*), and catfish (*Ictalurus* spp.).
- *Saltwater fishes*: Atlantic bonito (*Sarda sarda*), Atlantic mackerel (*Scomber scombrus*), red snapper (*Lutjanus campechanus*), bluefish (*Pomatomus saltatrix*), spot (*Leiostomus xanthurus*), Atlantic croaker (*Micropogon undulatus*), red drum (*Sciaenops ocellata*), white perch (*Morone americana*), summer flounder (*Paralichthys dentatus*), Gulf flounder (*P. albigutta*), Atlantic cod (*Gadus morhua*), and pollock (*Pollachius virens*).
- *Birds*: Bobwhite Quail (*Colinus virginianus*), Mourning Dove (*Zenaida macroura*), Turkey

TABLE II

Net Economic Value of Recreational Activities Related to Wild Species and Wilderness Areas in the United States in 1987[a]

Recreational activity or category	Total U.S. consumers' surplus (billion $)	Total U.S. trips (× 1,000,000)	Total recreation days (× 1,000,000)
Fishing	12.81	423	750
Game hunting	2.53	114	211
Collecting berries	0.78	19	32
Cutting firewood	0.95	30	64
Wildlife observation	3.21	70	249
Nature study	2.45	71	250
Sightseeing	16.00	292	1,124
Primitive camping	1.30	38	157
Backpacking	2.09	26	81
Day hiking	2.38	91	193
Cross-country skiing	0.27	10	28
Photography	1.90	42	180

[a] Source: J. C. Bergstrom and H. K. Cordell (1991). *J. Leisure Res.* **23**(1), 67–86. All figures are rounded to nearest whole number.

(*Meleagris gallopavo*), Ruffed Grouse (*Bonasa umbellus*), Sharp-tailed Grouse (*Pediocetes phasianellus*), Prairie Chicken (*Tympanuchus cupido*), American Woodcock (*Scolopax minor*), Canada Goose (*Branta canadensis*), White-fronted Goose (*Anser albifrons*), Snow Goose (*A. caerulescens*), Greater Scaup (*Aythya marila*), Canvasback (*A. valisineria*), Wood Duck (*Aix sponsa*), Mallard (*Anas platyrhynchos*), Green-winged Teal (*A. crecca*), Pintail (*A. acuta*), American Wigeon (*Mareca = Anas americana*), and Black Duck (*A. rubripes*) (Fig. 7).

- *Mammals*: white-tailed deer (*Odocoileus virginianus*), mule deer (*O. hemionus*) (see Fig. 1), elk or wapiti (*Cervus canadensis*), moose (*Alces alces*), pronghorn (*Antilocapra americana*), mountain goat (*Oreamnos americanus*), collared peccary (*Tayassu tajacu*), snowshoe hare (*Lepus americanus*), Eastern gray squirrel (*Sciurus carolinensis*), woodchuck (*Marmota monax*), and American black bear (*Ursus americanus*).

Numerous economic studies have attempted to quantify the monetary value of recreational hunting and fishing. For example, the value of elk sport hunting in the Gallatin National Forest in Montana was estimated at about $12 million over a 50-year planning period. And a 1989 study of the economic benefits of deer in California determined their value for recreational hunting at approximately $229 million as opposed to $50 million per year for nonconsumptive recreational uses.

Even though more Americans engage in nonconsumptive recreational uses of wildlife, they spend less per capita than anglers, hunters, and other consumptive users. In the United States in 1991, 76 million nonconsumptive wildlife users spent $18 billion on equipment, trip-related expenses, society membership dues, and other expenses. The animal species most commonly observed and photographed by nonconsumptive users include: shorebirds, raptors (birds of prey), songbirds, upland gamebirds, waterfowl, squirrels, chipmunks, deer, elk, rabbits, hares, coyotes, wolves, bears, raccoons, marine mammals, trout, salmon, butterflies, beetles, and spiders.

Whale watching (Fig. 8) is a rapidly growing nonconsumptive recreational industry focused largely on endangered species of baleen whales, such as the humpback whale (*Megaptera novaeangliae*), black right whale (*Eubalaena glacialis*), fin whale (*Balaenoptera physalus*), blue whale (*Balaenoptera musculus*), and California gray whale (*Eschrichtius robustus*). Early estimates (1981) of total income generated by the U.S. whale watching industry totaled roughly $5–$6 million per year. However, this industry now supports more than 250 individual enterprises and is worth more than

FIGURE 7 Although the Black Duck (*Anas rubripes*) is still prized by recreational hunters in the Northeast, this species experienced significant population declines from the 1950s until the early 1980s. (Photo credit: Glen Smart, U.S. Fish & Wildlife Service)

FIGURE 8 Each year, thousands of whale-watchers observe humpback whales (*Megaptera novaeangliae*) and other cetaceans in the North Atlantic Ocean off Montauk Point, Long Island, New York. (Photo credit: Okeanos Ocean Research Foundation)

$100 million annually. A new study by the National Marine Fisheries Service has estimated the total income or revenues generated from whale watching activities in various regions:

- *Northeast*: In 1993, the Director of the Massachusetts Office of Tourism estimated total revenues in excess of $100 million per year; the New England Whale Watching Association recently estimated that the industry generates direct income of $22.5 million and indirect income (hotels, transportation, etc.) of $60 million annually. Whale watching companies in New Hampshire and Maine currently report annual income generated from ticket sales, food, and souvenirs at about $3 million.
- *Southeast*: Ticket sales generate annual gross revenues of about $3.2 million.
- *Southwest*: During the 1983-1984 season, whale watching trips in California generated gross income of $2.6 million; the number of people participating in this form of recreation more than doubled (from 115,000 to 255,000) in a single decade. In Hawaii, day-trips average $55 (half-day) and $90 (full-day), and cruises range from $895 to $2950 (excluding airfare).

Birdwatching or "birding" is probably the most popular of all nonconsumptive recreational uses of wildlife. In 1991, 63 million Americans fed or observed wild birds at or near their homes, whereas only 37 million fed or observed deer or other mammals. "Birders" spent an estimated $2 billion for birdseed and $468 million for bird feeders, bird baths, bird houses, and nest boxes. In the same year, about 24.5 million Americans observed, photographed, or fed birds of prey, shorebirds, waterfowl, and other birds in their home state or other states. "Avitourism," or travel specifically undertaken to engage in activities related exclusively to wild birds, has increased in popularity over the past few decades. In recent years, avitourists have spent up to an average of $2000 annually on birdwatching trips. In 1984, the annual value of the endangered Whooping Crane (*Grus americana*) (Fig. 9) to visitors at the Aransas National Wildlife Refuge was

FIGURE 9 Whooping cranes (*Grus americana*) take flight from the Aransas National Wildlife Refuge in Texas; thousands of avitourists visit Texas each year to observe this endangered bird species. (Photo credit: Steve Hillebrand, U.S. Fish & Wildlife Service)

estimated at approximately $100,000 (or $1.40 per visitor). More recently, avitourism at High Island, Texas, has generated total revenues in excess of $2.5 million per year for the regional economy (for hotels, restaurants, airlines, auto rental agencies, etc.).

G. Monetary-Economic or Utilitarian Values

Monetary-economic or utilitarian values are related to biological resources and the tangible commodities, goods, and services obtained from the use of natural ecosystems, wild species, and genetic resources. Examples related to the three major components of biodiversity include:

- *Ecosystem diversity*—Terrestrial ecosystems provide habitats for wild species that are sources of novel foods, medicines, and industrial products or that serve as pollinators or biological control organisms for agriculture. Natural forests supply us with timber, plywood, paper, and other forest products; in 1990, U.S. exports of forest products derived from wild trees were valued at $239 billion. Natural rangelands and pasturelands also support wild grazing animals and 3 billion head of domesticated livestock; these animals yield meat, milk, and other food products. Marine and other aquatic ecosystems provide novel sources of medicines and support commercial fishing operations. Between 1976 and 1980, U.S. fisheries imports and production averaged $4 billion annually, and global fisheries productivity totaled almost $12 billion per year.

- *Species diversity*—In addition to the contributions of wild fish, shellfish, and timber-producing species, the annual value of other commodities derived from wild species between 1976 and 1980 included: U.S. imports and production of wild-derived industrial and food products valued at an average of $229 million, North American fur and skin exports averaging about $122 million in the United States and $54 million in Canada, and live specimens of wild plants and animals imported by the United States valued at more than $90 million.

- *Genetic diversity*—Between 1976 and 1980, the contribution of wild-derived genes to crop productivity in North America was estimated at more than $340 million annually. Useful genes worth hundreds of millions of dollars are also obtained from "landrace" crop cultivars, or ancient cultivated varieties of crop plants maintained by traditional agriculturalists.

Ecological and evolutionary amenity values provide valuable genetic resources that support our modern agricultural production systems (see Section II,E). A specific example is that of the genetic improvement of modern tomato (*Lycopersicon esculentum*) cultivars. Since the 1980s, the U.S. tomato crop has averaged more than $800 million annually. Each year, growers save millions of dollars owing to increased yields and enhanced quality factors obtained from wild-derived genes. The first disease-resistant tomato cultivars, released in the 1940s, carried the dominant *I* gene for wilt resistance. The *I* gene was obtained from a wild progenitor, or one of the wild ancestors of the tomato—the wild currant tomato, *L. pimpinellifolium*. Since then, other wild tomatoes have provided useful genes for genetic improvement of our modern varieties; for example, *L. peruvianum* has provided other disease-resistance genes, and *L. hirsutum* has been used as a source of genes for resis-

tance to nine tomato insect pests. An endemic tomato species of the Galapagos Islands, *L. cheesmanii,* has been investigated as a source of salt-tolerant genes to facilitate the adaptation of modern cultivars to arid, alkaline soils. In addition, many "landrace" tomato (*L. esculentum*) varieties have been tapped as sources of useful genes for increased vitamin C content and for resistance to leaf mold, early blight, late blight, and gray leafspot. Between 1976 and 1980, the productivity or yields of other crops were also enhanced by the use of wild-derived crop genetic resources. The average annual monetary value of these genetic resources included those used to improve bread wheat ($35 million), sugarcane ($34 million), sugar beet ($3 million), sunflower ($88 million), and cacao ($49 million).

Future sources of wild-derived genes for use in modern agriculture are primarily maintained in genetic reservoirs within natural ecosystems, whereas future sources of genes from "landrace" crop varieties are being maintained on-site in traditional agricultural systems. During the past half century, however, extinction of wild crop relatives and landrace crop varieties has accelerated dramatically. Today, many wild crop relatives are endangered in the United States; these include Walker's manihot (*Manihot walkerae*), Texas wild-rice (*Zizania texana*), steamboat buckwheat (*Eriogonum ovalifolium* var. *williamsiae*), clay-loving buckwheat (*E. pelinophilum*), Schweinitz's sunflower (*Helianthus schweinitzii*), Okeechobee gourd (*Cucurbita okeechobeensis* subsp.), and the scrub plum (*Prunus geniculata*). The situation for endangered, native landrace crops is even more serious, because the maintenance of crop varieties is often dependent on the survival of disappearing human cultures. In the southwestern United States and Mexico, a number of native crop varieties are rare or endangered, including the Pima mottled lima bean (*Phaseolus lunatus*), Yaqui red tepary bean (*P acutifolius*), Warihio Indian panic grass (*Panicum sonorum*), Warihio green amaranth (*Amaranthus hypochondriacus*), Hopi red dye amaranth (*A. cruentus*), and both the Hopi and the Pima striped cushaw squashes (*Cucurbita argyrosperma*).

In the United States, the U.S. government and nongovernmental organizations such as Native Seeds/SEARCH are focusing greater attention on both on-site and off-site conservation of crop culti-vars and wild relatives of native U.S. crop species. An example is the wild cranberry (*Vaccinium macrocarpon* and *V. oxycoccus*). The Agricultural Research Service and U.S. Forest Service of the Department of Agriculture are cooperating to establish a series of on-site "Genetic Resource Management Areas" in the mid-Atlantic region. In addition, stolons or "runners" have recently been collected from wild cranberry populations in Massachusetts, New York, and other locations in the East. These wild population samples are being propagated for off-site conservation at the Blueberry and Cranberry Research Center of Rutgers University in Chatsworth, New Jersey (Fig. 10); it is hoped that these cranberry "breeding bogs" in New Jersey will yield progeny of modern varieties that contain wild-derived genes for resistance to cranberry diseases and insect pests.

The pharmaceutical industry has also benefited immensely by tapping the pharmaceutical wealth of natural ecosystems. In 1985, $8 billion worth of drugs containing naturally derived ingredients were dispensed from community pharmacies. The pharmaceutical industry capitalizes on the free ecological and evolutionary services of nature, because phar-macologically active compounds evolved in natural ecosystems, where they play an important role in protecting species from predators, parasites, or disease pathogens (see Section III,E). The defensive or protective role of plant alkaloids—one of the most important and economically valuable types of medicinal compounds—have been demonstrated against herbivorous (plant-feeding) species of insects and mammals, including *Colobus* monkeys, mountain gorillas, sheep, and other livestock.

Two of the most famous plant-derived alkaloid drugs—the vinca alkaloids, vincristine (Oncovin) and vinblastine (Velban)—were obtained from the Madagascar periwinkle (*Catharanthus roseus*). These drugs, often referred to as the first modern anticancer drugs, have proved effective against a variety of cancers, including acute childhood leukemia, Hodgkin's disease, and malignant melanoma. In 1985, Eli Lilly sold approximately $100 million worth of Oncovin and Velban worldwide, at a profit of about $88 million. Similarly, the anticancer drugs etoposide and teniposide have been produced from podophyllotoxins obtained from the Asian (Himalayan) mayapple (*Podophyllum emodi* = *hexandrum*) since 1984. Bristol-Myers/Squibb sold $100 million worth of etoposide (Vespeside) to

FIGURE 10 A cranberry breeding bog at Rutgers University's Blueberry and Cranberry Research Center in Chatsworth, New Jersey. (Photo credit: N. Vorsa)

FIGURE 11   Horseshoe crabs (*Limulus polyphemus*) mating on a Massachusetts beach; wild horseshoe crabs are used to produce an assay that can detect the presence of endotoxic bacteria in intravenous drug products and vaccines. (Photo credit: Associates of Cape Cod, Inc.)

treat testicular cancer in 1990, and approximately $275 million in 1992, even though the Asian mayapple was listed on Appendix II of the Convention on International Trade in Endangered Species (CITES) in 1990.

Finally, the medical/biomedical supply industry also benefits from uses of wild biota. The horseshoe crab (*Limulus polyphemus*) of Atlantic seashores (Fig. 11) currently supports a diagnostic reagent products industry now worth about $20 million annually ($10 million in 1988). The horseshoe crab is not a "true crab," but instead is a distant relative of spiders. Once widely persecuted as shellfish predators, horseshoe crabs were later valued by scientists as "living fossils." Today, they are routinely harvested from the Atlantic Ocean and bled for their blood cells or "amebocytes," which are used to produce the *Limulus* amebocyte lysate (LAL) assay (Fig. 12). Later, they are returned to the ocean near the site where they were collected. The LAL assay is used primarily to detect the presence of endotoxins produced by gram-negative bacteria, such as those that cause spinal meningitis and other life-threatening infections. Intravenous drugs or drug products that are contaminated with endotoxins can cause shock, high fevers, and/or death in humans, their pets, or other animals. The LAL assay has revolutionized the drug and biomedical products safety testing industry because it reacts much more strongly and accurately to the presence of endotoxins than the rabbit pyrogen test of the 1970s and 1980s. Moreover, the LAL assay is 10–15 times more cost-effective than the rabbit assay, and it is an *in vitro* (test tube) rather than an *in vivo* (live animal) test.

FIGURE 12   A technician draws blood from horseshoe crabs during the Limulus Amebocyte Lysate (LAL) Assay production process. (Photo credit: Associates of Cape Cod, Inc.)

IV. ALTERNATIVE METHODS OF VALUATION

Many different systems of classification have been used to categorize values and uses of biodiversity. Harold Steinhoff and his colleagues identified four major methods used to classify wildlife values: (1) major uses; (2) the attitudes or motives of value holders; (3) commodity types purchased; and (4) exercised and option values, that is, economic methods of valuation. The classification system adopted by various scholars often reflects their academic orientation, goals, or attitudes. The intrinsic versus extrinsic valuation methodology adopted in this article is a variation on the theme of "major uses" of wildlife and wilderness, with the exception that intrinsic value is not "use-oriented." This classification methodology, which is most commonly used by noneconomists, dates back to R. T. King's work in 1947. Because this approach focuses on values that can be attached to biodiversity per se, it has been frequently employed by biologists, wildlife managers, and outdoor recreation analysts since the 1960s. Even though the concept of intrinsic value was acknowledged prior to that time, it was first articulated in the form of existence and bequest values by the economist John Krutilla in 1967.

The other three classification systems deserve further attention. The second method classifies values in relation to the attitudes and motivations of people. This approach focuses primarily on the perceptions and attitudes of "value holders" (people) per se, rather than on specific uses or values that emanate from biodiversity, and therefore, it is difficult to compare value categories developed by use of this method with those utilized here. The third classification system is narrowly focused on commodity values in relation to consumptive recreational activities (see Section III,F), for example, hunting and fishing. The final method, that of exercised and option values, is the methodology most commonly used by economists. This approach attempts to quantify or attach monetary-economic values to both intrinsic and extrinsic values of biodiversity; it is exemplified by many of the contributions found in "The Economic Value of Wilderness." [See BIODIVERSITY, ECONOMIC APPRAISAL.]

Considerable information and data obtained from the economic valuation approach have been included in the preceding sections because some value categories utilized by economists are comparable to those utilized by proponents of the major uses or intrinsic versus extrinsic values approach. Direct economic benefits derived from "exercised values" are basically synonymous with "commodity values" related to consumptive and nonconsumptive uses of biodiversity, for example, monetary-economic and recreational/tourism values. In contrast, indirect economic benefits known as existence values and bequest values accrue from "nonuse" activities and incorporate the concept of intrinsic value. Pure existence values arise when nonusers place a value on the knowledge that a wild species or natural ecosystem still exists, or when they believe that it has a "right to exist." Bequest values arise when nonusers wish to leave wild species or natural ecosystems as a "bequest" or endowment for the benefit of future generations. Finally, option values provide a means of calculating the potential value of biological resources, because they can be viewed as what someone would be "willing-to-pay" (WTP) to keep their future use and existence options open.

Indirect economic benefits are usually quantified by the use of "contingent valuation method" (CVM) surveys. Typically, option demands of consumers are estimated by "willingness-to-pay" (WTP) surveys or, alternatively, by "willingness-to-be-paid" (WTBP) surveys. WTP surveys attempt to determine what someone would be willing to pay to save an endangered species or preserve a tract of a tropical rain forest, for example. In contrast, WTBP surveys might attempt to determine how much someone would be willing to accept in payment to allow the endangered species to go extinct, or the rain forest to be destroyed.

Very few economic studies have attempted to quantify intrinsic and highly intangible values along with direct uses of biodiversity. As an example, the total value of the Whooping Crane (Fig. 9) to U.S. consumers was recently estimated at approximately $5 billion annually. Although "total valuation studies" such as these include consideration of some aspects of intrinsic and extrinsic values, they have thus far failed to adequately incorporate many value categories. Some notable exceptions include research (scientific) and educational values, ecological and evolutionary values, and spiritual and religious values. Moreover, the survey techniques utilized by economists in CVM, WTP, and WTBP studies have been frequently criticized for either underestimating or overestimating values of biodiversity. These criticisms have ranged from discussions of the methodological problems in surveys, for example, "nonresponse biases," to general skepticism about the ability of people to make rational choices that involve moral considerations and ethical commitments about biodiversity or other public goods. In 1985, Holmes Rolston provided a useful synthesis of intrinsic and extrinsic or instrumental values, and a critique of economic methods of valuing wildlands in relation to this approach.

V. CONCLUSIONS

In all of its various forms and at all levels of biological organization, biodiversity is valuable to humanity: it supports our modern societies, provides the biotic raw materials necessary to sustain economic productivity, contributes substantially to human health and well-being, and satisfies both essential and nonessential human needs. In the United States, the monetary-economic contributions of biodiversity total hundreds of billions of dollars annually. However, the indirect benefits and intangible contributions of biodiversity to our daily lives are perhaps even more important, even though they cannot be easily transformed into dollar equivalents.

It is important to consider the full range of biodiversity values—spiritual and religious, psychological and therapeutic, sociocultural and sociohistorical, research and educational, ecological and evolutionary, recreational and tourism, and monetary-economic—each time we make a decision that will adversely affect the survival of an endangered species or the earth's remaining wilderness areas. Whenever possible, we should err on the side of conservation, because our future—and the future world we will leave for our children and grandchildren—literally depends on the biodiversity conservation decisions of the present generation.

Glossary

Amenity values Nonconsumptive uses of genetic materials, species, or ecosystems to meet primarily nonmaterial human needs, for example, nonconsumptive recreational and sociohistorical values of biodiversity.

Biodiversity Diversity or variety of ecosystems, species, and genetic materials (genetic information within species or populations of living organisms).

Biological resource Living, renewable natural resources, including microorganisms, animals, plants, fungi, and natural communities or assemblages of these living entities found within natural or human-modified ecosystems. Unlike nonliving, nonrenewable resources, biological resources are potentially inexhaustible as a consequence of biological reproduction if the natural processes and habitats necessary for their replenishment are not destroyed; however, they can become exhaustible resources if driven to extinction.

Biomass Total weight of living material, usually expressed in terms of dry weight of an organism, population, or community per a designated unit area.

Commodity values Consumptive uses of biodiversity related to the provision of commodities or goods and services to meet primarily material human needs, for example, monetary-economic and consumptive recreational uses of biodiversity.

Extrinsic value Anthropocentric values or uses of biodiversity that have an external origin in relation to biodiversity; oftentimes referred to as "utilitarian" or "instrumental" values, extrinsic values are determined by human estimations of worth or significance.

Intrinsic value Nonanthropocentric value, that is, value independent of human estimations of worth, or value(s) derived from the essential nature or constitution of biodiversity per se.

Moral values Highly intangible values that meet immaterial or nonmaterial human needs, for example, religious and spiritual uses of biodiversity.

Value Any estimation of general worth, usefulness, importance, or monetary price of a good or commodity, amenity, or service.

Bibliography

Alvarez de Williams, A. (1987). Environment and edible flora of the Cocopa. *Environ. Southwest* **519**, 22–27.

Bergstrom, J. C., and Cordell, H. K. (1991). An analysis of the demand for and value of outdoor recreation in the United States. *J. Leisure Res.* **23**(1), 67–86.

Bryant, J. P., Provenza, F. D., Pastor, J., Reichardt, P. B., Clausen, T. P., and du Toit, J. T. (1991). Interactions between woody plants and browsing mammals mediated by secondary metabolites. *Annu. Rev. Ecol. Systematics* **22,** 431–446.

Decker, D. J., and Goff, G. R., eds. (1987). "Valuing Wildlife: Economic and Social Perspectives." Boulder, Colo.: Westview Press.

Driver, B. L., Brown, P. J., and Peterson, G. L., eds. (1991) "Benefits of Leisure." State College, Penn.: Venture Publishing.

Duke, J. A. (1992). "Handbook of Biologically Active Compounds and Their Activities." Boca Raton, Fla.: CRC Press.

Eubanks, T., Kerlinger, P., and Payne, R. H. (1993). High Island, Texas: Case study in avitourism. *Birding* **25**(6), 415–420.

Felger, R. B., and Moser M. B. (1991). "People of the Desert and Sea: Ethnobotany of the Seri Indians." Tucson: University of Arizona Press.

Klorer, P. G. (1992). Leaping beyond traditional boundaries: Art therapy and a Wilderness Stress Challenge Program for adolescents. *The Arts in Psychotherapy* **19,** 285–287.

Krutilla, J. V. (1967). Conservation reconsidered. *Am. Econ. Rev.* **58,** 777–786.

May, R. M. (1990). Taxonomy as destiny. *Nature* **374,** 129–130.

Nabhan, G. P., Rea, A. M., Reichhardt, K. L., Mellink, E., and Hutchinson, C. F. (1982). Papago influences on habitat and biotic diversity: Quitovac oasis ethnoecology. *J. Ethnobiol.* **2**(2), 124–143.

Norton, B. G., ed. (1986) "The Preservation of Species: The Value of Biological Diversity." Princeton, N.J.: Princeton University Press.

Novitsky, T. J. (1991). Discovery to commercialization: The blood of the horseshoe crab. *Oceanus* **27**(1), 13–18.

Oldfield, M. L. (1989). "The Value of Conserving Genetic Resources." Sunderland, Mass.: Sinauer Associates.

Oldfield, M. L., and Alcorn, J. B., eds. (1991). "Biodiversity: Culture, Conservation, and Ecodevelopment." Boulder Colo.: Westview Press.

Orians, G. H., Brown, Jr., G. M., Kunin, W. E., and Swierzbinski, J. E., eds. (1990). "The Preservation and Valuation of Biological Resources." Seattle: University of Washington Press.

Payne, C., Bowker, J. M., and Reed, P. C., eds. (1992). "The Economic Value of Wilderness." Asheville, N.C.:

U.S.D.A. Forest Service, S.E. Forest Experiment Station.

Pough, F. H., Heiser, J. B., and McFarland, W. N. (1989). "Vertebrate Life." New York: Macmillan.

Prescott-Allen, C., and Prescott-Allen, R. (1986). "The First Resource: Wild Species in the North American Economy." New Haven, Conn.: Yale University Press.

Rolston, III, H. (1985). Valuing wildlands. *Environ. Ethics* **7,** 23–48.

Rolston, III, H. (1994). God and endangered species. "Ethics, Religion, and Biodiversity," (L. S. Hamilton, ed.) pp. 40–64. Cambridge, U.K.: White Horse Press.

Rushing, K. K. (1978). The bowhead or the Eskimo: Must we choose? *National Parks and Conservation Magazine* **52**(9), 10–14.

Steinhoff, H. W., Walsh, R. G., Peterle, T. J., and Petulla, J. M. (1987). Evolution of the valuation of wildlife, "Valuing Wildlife: Economic and Social Perspectives," (D. J. Decker and G. R. Goff, eds.) pp. 34–48. Boulder, Colo.: Westview Press.

Stevens, T. H., Echeverria, J., Glass, R. J., Hager, T., and More, T. A. (1991). Measuring the existence value of wildlife: What do CVM estimates really show? *Land Economics* **67**(4), 390–400.

U.S. Congress, Office of Technology Assessment (1987). "Technologies to Maintain Biological Diversity." Washington, D.C.: U.S. Government Printing Office.

U.S. Department of Agriculture, Agricultural Research Service (1992). Proposal for a comprehensive management plan to conserve wild cranberry (*Vaccinium macrocarpon* Ait. and *V. oxycoccus* L.) genetic resources in the mid-Atlantic states. USDA–ARS Plant Exploration Office (BARC-West, Bldg. 003), Beltsville, Md.

U.S. Department of Commerce, National Marine Fisheries Service (in press). An assessment of whale watching in the United States. *Reports of the International Whaling Commission* **44.**

U.S. Department of the Interior, Fish and Wildlife Service and U.S. Department of Commerce, Bureau of the Census (1993). "1991 National Survey of Fishing, Hunting, and Wildlife-Associated Recreation." Washington, D.C.: U.S. Government Printing Office.

Western D., and Pearl, M. C. (1989). "Conservation for the Twenty-first Century." New York. Oxford University Press.

Wilson, E. O., and Peter, F. M., eds. (1989). "Biodiversity." Washington, D. C.: National Academy Press.

Witt, S. C. (1985). "Briefbook: Biotechnology and Genetic Diversity." San Francisco: California Agricultural Lands Project.

Biogeochemical Cycles

M. Meili

University of Uppsala, Sweden

I. Introduction
II. Basic Aspects of Biogeochemical Cycles
III. Elements of Biogeochemical Significance
IV. Selected Biogeochemical Cycles
V. General Biogeochemical Concepts, Dimensions, and Models
VI. Potential and Limitations of Compiling Biogeochemical Cycles

The concept of biogeochemical cycling is used to describe the distribution and transport of materials and the interaction of biological, chemical, physical, and anthropogenic processes controlling their turnover and transformation in the aquatic, terrestrial, and atmospheric environment. Usually, the focus is on biologically important elements or compounds such as carbon, nitrogen, phosphorus, or toxins, often on a global scale and on an annual basis. Cycles are often considered as closed with regard to mass conservation, but open with regard to energy. They are used to assess turnover dynamics by comparing magnitudes of pools and fluxes in different ecosystem compartments. Of particular interest are spatial and temporal scales of transformations and phase transitions.

I. INTRODUCTION

Biogeochemical cycles are compiled to provide a comprehensive representation of current knowledge on the distribution of substances between different reservoirs and on the movement of material from one region to another. Such a holistic perspective is useful for viewing proportions of reservoir sizes, flow rates, and processes that control dynamic equilibria or temporal changes in life conditions. This may improve our understanding of the natural environment, stimulate interdisciplinary thinking, and assist in environmental management.

Differing from geochemistry, biogeochemistry emphasizes the interactions of biological entities with their environment, either as agents or as products. Organisms are adapted to more or less narrow windows of biogeochemical conditions. Man-made changes in flux patterns change these niches and can lead to the extinction of species or whole habitats. Human intervention with natural biogeochemical cycles occurs by exploitation of resources (removal of materials) or by pollution (addition of materials). Both imply a redistribution of material through a redirection of fluxes. Small relative changes in some material fluxes may have dramatic effects on the natural environment if enhanced by cascading processes. An example is the potential damage of a minor increase in atmospheric carbon dioxide to life forms and habitats as a consequence of global warming with all its effects mediated by hydrological and biochemical processes. Further examples of global scale environmental issues are the terrestrial and aquatic production of food and its dependence on the climate, the availability of nutrients, and the presence of toxic agents; the release of sulfuric acid and its effects on atmospheric, terrestrial, and aquatic systems; the release of trace gases on the stratospheric ozone and the radiation climate; and the dispersal of synthetic chemicals such as pesticides into the biosphere. The understanding of natural biogeochemical cycles may help minimize the human impact on natural cycles. [*See* BIOGEOCHEMISTRY.]

II. BASIC ASPECTS OF BIOGEOCHEMICAL CYCLES

Basic aspects of biogeochemical cycles are the distribution of materials (location and size of reservoirs), their transport (pathways and rates of flow), and their turnover, given either as turnover rates, (ratio of flux/pool) or as residence times (ratio of pool/flux). The starting point is a qualitative model including the transformation between material species and the reactions involved. The ultimate goal is a quantitative model including the magnitude of reservoirs, fluxes, residence times, and concentrations of material. From such models, system kinetics can be derived and predictions made about the response of a system to perturbation or remediation.

Biogeochemical cycles are often conceptualized as box models and are conveniently visualized by boxes and arrows. Depending on the perspective and purpose, diagrams of biogeochemical cycles can be focused on the spheres of the Earth (geocentric view), on compounds (chemocentric view), or on organisms (biocentric view). Environmental issues are often viewed from the role and exposure of man (anthropocentric view) (Fig. 1).

III. ELEMENTS OF BIOGEOCHEMICAL SIGNIFICANCE

Among approximately 80 elements found in soils, only about one-third are essential components of

plants and animals. Most elements of biogeochemical importance have low atomic numbers. Some of the major elements are the key constituents of organic matter (C, H, O, N, P, S), while others serve as ionic matrices or contribute to supporting structures (Ca, Mg, Si, K, Na, Cl, F). Nitrogen and phosphorus are usually the limiting nutrients for the production of biogenic matter. Essential trace metals, often occur in coenzymes (Fe, Mn, Co, Cu, Zn, Se, Mo). Some important elements are not used by organisms: one that is very abundant (Al) and others that are highly toxic even at low concentrations (Hg > Cd > Pb). The elemental composition of biomass, seawater, and the earth's crust is compiled in Table I. [*See* ASPECTS OF THE ENVIRONMENTAL CHEMISTRY OF ELEMENTS.]

Carbon, sulfur, and mercury are elements that have experienced significant perturbations of their cycles over a recent period of about 10 human generations. These cycles are of particular interest also because they cover large spatial scales and include an interaction of all major spheres (atmosphere, hydrosphere, sediments, biosphere, pedosphere, lithosphere) and phases (gas, liquid, solid) over a wide range of time scales. Furthermore, these cycles are tightly coupled to each other and may serve as examples to demonstrate the complexity of biogeochemical cycling in general and of the human impact of natural cycles causing important environmental issues.

IV. SELECTED BIOGEOCHEMICAL CYCLES

A basis for the understanding of most biogeochemical cycles is the water cycle. Among the elemental cycles, the most important in biogeochemistry are those of the biological macronutrients: carbon, nitrogen, phosphorus, and sulfur. Pollutants are represented here by mercury, the most toxic heavy metal. Coupling of cycles must be considered for understanding a single cycle. A good example is the mercury cycle, which is tightly coupled to the cycling of organic matter and nutrients.

Physical, chemical, and biological transformations are elucidated, and spatial as well as temporal

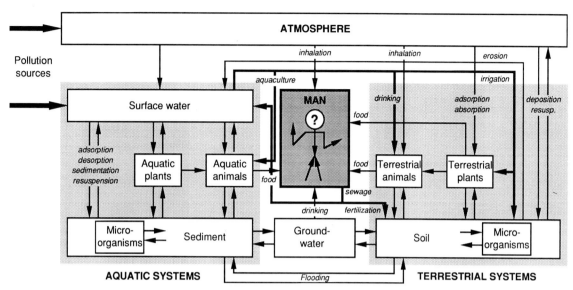

FIGURE I Anthropocentric view of the biogeochemical cycle of toxic trace pollutants such as heavy metals and persistent organic compounds. New fluxes caused by human activities are denoted by bold arrows.

scales are compared. Although environmental pools and fluxes often are quite uncertain for most elements and compounds, many diagrams display apparently accurate magnitudes. Such numbers often originate from simple assumptions or from balance calculations, but despite their uncertainty they provide a "best estimate" of internal relationships.

A. The Hydrological Cycle

Water is most important as a carrier and chemical medium for other substances. The relevant properties of water include its large dipole moment and ability to form hydrogen bonds. These determine its capacity to dissolve and stabilize charged species (ions) and account for the large number of reactions involving a proton transfer between compounds, such as acid–base reactions or the hydrolysis of biotic macromolecules.

The water cycle is of greatest significance, as it interacts with most other biogeochemical cycles. Water is very important as a medium for chemical reactions, in which the water itself may not participate at all. The evaporation of water leads to a concentration of dissolved compounds and eventually in a precipitation of dissolved solids, whereas rainfall leads to a dissolution and a dilution of materials. Both processes lead in the long run to the

formation and weathering of rocks. Water is also required to ensure any biotic growth and reproduction, by serving as a solvent, as a chemical reactant, and as a physical agent.

The hydrological cycle describes the distribution of saline and fresh water among major reservoirs and the rate at which water is moved from one region to another (Fig. 2). By far the largest pool of water is contained in the oceans, which cover 71% of the Earth's surface to an average depth of almost 3800 m. Inland waters cover less than 2% of the land surface. Fresh water constitutes only a few percent of the global water. A large freshwater resource is stored in the polar ice caps, of which Antarctica holds about 75%. A similar quantity of fresh water is found in groundwaters. Surface waters only constitute a minor fraction, of which about 20% is stored in Lake Baikal alone, with a similar amount in the Laurentian Great Lakes.

The fluxes are driven by solar radiation pumping water from land and oceans into the atmosphere. Evaporated water is distributed and redeposited as precipitation on land and oceans. Land surfaces receive on average about one-third less water per unit area than the oceans. From land, water is returned to the oceans as runoff or to the atmosphere as evapotranspiration. River runoff is

TABLE I

Selected Elements of Biogeochemical Interest and Their Abundance in Different Environmental Matrices[a]

Element			All biomass			The Earth's crust		Seawater	
Symbol	Atomic number	Atomic mass	ppm mass	ppm atoms	Essential to biota	ppm mass	ppm atoms	ppm mass	ppm atoms
H	1	1.0	65,900	496,800	·	1,400	28,800	111,000	110,000
B	5	10.8			·			4.5	0.410
C	6	12.0	393,460	248,900	·	200	350	27.8	2.387
N	7	14.0	5,020	2,720	·			0.4	0.030
O	8	16.0	524,290	249,000	·	466,000	604,000	883,000	55,200
F	9	19.0			·	625	680	1.3	0.068
Na	11	23.0	190	63	·	28,300	25,500	10,770	468
Mg	12	24.3	980	307	·	20,900	17,800	1,290	53.2
Al	13	27.0	560	157		81,300	62,500		
Si	14	28.1	1210	327	·	277,200	205,000	2.9	0.103
P	15	31.0	520	128	·	1,050	700	0.07	0.002
S	16	32.1	710	169	·	260	170	904	28.2
Cl	17	35.5	500	106	·			19,353	546
K	19	39.1	2290	444	·	25,900	13,700	399	10.2
Ca	20	40.1	3780	717	·	36,300	18,800	412	10.2
Ti	22	47.9				4,400	1,900		
Mn	25	54.9	210	29	·	950	360		
Fe	26	55.9	390	53	·	50,000	18,600	0.03	
Co	27	58.9			·				
Cu	29	63.6			·				
Zn	30	65.4			·				
Se	34	79.0			·				
Br	35	79.9						67	0.84
Sr	38	87.6				375	89	8	0.091
Mo	42	95.9			·				
Ba	56	137.3				425	64		

[a] ppm = parts per million; only the dominant elements are listed; seawater values do not include gases.

about one-third of the precipitation on land. Per unit area both are greatest in South America (0.6 and 1.65 m year^{-1}, respectively), nearly twice that of other continents. The relative contribution of runoff and precipitation to the oceans differs widely between regions. Runoff contributes annually about 6 to 7 cm of water to the Indian and Pacific oceans and about 20 to 25 cm to the Altantic and Arctic oceans. The contribution from precipitation is of similar magnitude in the Arctic oceans, but is 4-fold in the Atlantic and 14- to 20-fold in the Indian and Pacific oceans. Evaporation is about one-tenth in the Arctic compared to the other oceans.

The water renewal time is about 9 days in the atmosphere and about 2 to 3 weeks in large rivers. The other major reservoirs are on average much more stable, with renewal times of years (surface fresh water) to millennia. The renewal time of the oceans is about 3500 or 37,000 years, depending on the consideration of gross or net evaporation. The shorter time is of interest for the cycling of the water itself, the longer is more relevant for the fate of dissolved substances.

The man-made fraction of continental evaporation due to environmental modifications such as irrigation and deforestation is about 3%, but is projected to increase to 10% or even 50% in the near future.

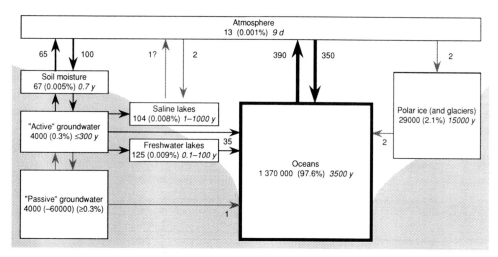

FIGURE 2 The global hydrological cycle. Principal resources and approximate annual fluxes of fresh and saline water in units of 10^3 km$^3 \approx 10^{15}$ kg H$_2$O. Pool sizes are also given as a percentage of the total. Italic numbers show approximate water residence times in years or days.

B. Carbon: The Key Element of Life on Earth

Carbon atoms constitute the backbone of an enormous variety of organic molecules because of the unique ability to form long covalent chains and rings. Moreover, large deposits of carbonates interact with water where they exert a major control of acid-buffering capacity and salinity. Relevant time scales vary from seconds (gas exchange, biochemical transformations) to millions of years (from formation to weathering of carbonate rocks). Accordingly, the biogeochemical carbon cycle is very complex and includes most physical, chemical, and biological dimensions. At the same time, it is probably the most studied and the best understood elemental cycle (Fig. 3). [*See* GLOBAL CARBON CYCLE.]

There are eight isotopes of carbon (molar weights of 10 to 16), two of which are stable (^{12}C and ^{13}C) with abundances of 98.89 and 1.11%, respectively. Natural differences in isotopic composition are used to trace carbon sources and to estimate exchange rates between reservoirs. Among the radioactive isotopes, which are continuously created in the atmosphere by cosmic radiation, the best-known is ^{14}C (often referred to as radiocarbon, half-live 5726 years) because of its widespread use in dating organic matter and carbonates.

Insignificant amounts of elemental carbon occur naturally in the three forms of amorphous carbon, graphite, and diamond. Most carbon is oxidized and occurs as carbonates, either in solid rocks or dissolved in the oceans. Carbonate rocks such as limestone and dolomite have formed as marine deposits, either from the shells of organisms or by the chemical precipitation of carbonate ions with calcium and magnesium into minerals such as calcite, aragonite, or dolomite. Dissolved carbonates equilibrate with gaseous carbon dioxide (CO_2), which is the dominant form of carbon in the atmosphere. Organic matter from biotic production (mainly photosynthesis) forms the large and complex pool of reduced carbon (simplified as CH_2O representing the chemical composition of primary carbohydrates). The most reduced form of carbon is methane (CH_4), a decay product of organic matter.

The cycling of gaseous carbon is closely linked to biotic activity. Methane is produced by anaerobic bacteria that derive their energy from the oxidation of simple organic molecules such as methanol and acetate or from molecular hydrogen. Important sites of methane production are rice paddies, lacustrine sediments, wetlands, and the rumen of cattle and hindguts of termites. A variety of autotrophic organisms fix large amounts of carbon dioxide or bicarbonate into organic molecules, either in photosynthesis or in chemosynthesis. Carbon dioxide

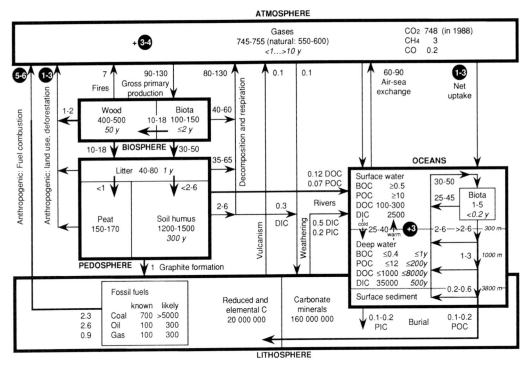

FIGURE 3 The global carbon cycle. Approximate pool sizes and annual fluxes in units of Gton = Pg = 10^{12} kg C. Bold arrows denote organic carbon fluxes. Annual net changes or fluxes caused by human activities are highlighted with black circles. Italic numbers show approximate carbon turnover times or ocean depths. DIC, dissolved inorganic carbon; BOC, POC, DOC, biotic, particulate, and dissolved organic carbon.

is released in the aerobic and anaerobic respiration of living organisms and in the decomposition of dead organisms by fungi and bacteria. The gross fluxes between the terrestrial system and the atmosphere are similar to those of the oceans, but the considerably smaller terrestrial pool size leads to a much faster turnover than in the ocean.

Organic acids and inorganic carbon dioxide, which is 10–100 times more abundant in soils than in the atmosphere, both contribute significantly to the weathering of rocks and minerals and control in this way the biogeochemical cycling of many other elements. Organic matter in soils is made up of plant and animal remains at various stages of decomposition, microbial cells, and substances produced during the process of decomposition. The decomposition of organic matter is selective and often incomplete, especially in cold, wet, and acidic soils. As a result, resistent organic compounds tend to accumulate in soils, often as colloidal aggregates. Because of this fractionation, observation of a soil in a steady state shows a

biogeochemical transformation on the way from living biota into detrital or humic matter. Humic substances are chemically complex acidic dark-colored molecules that range in molecular weight from a few hundred to a few hundred thousand in humic aggregates and are classified according to their solubility in acid and base: fulvic acid is soluble in both, humic acid is insoluble in acids, and humin is insoluble in both. Fulvic acids contain less H, N, and S, but more O, more carboxyl groups, fewer phenolic hydroxyl groups, and have a higher acidity. Humic substances form complexes with most metals and play an important role in the mobilization and transport of micronutrients and toxins from land to water, either by enhancing mineral solubilization and acting as a dissolved carrier or through inactivation and immobilization of the ligands in colloids. [See SOIL ECOSYSTEMS.]

Burning of fossil fuels is causing a steady increase in atmospheric CO_2 (about 0.5% per year) and CH_4 (>65% since the preindustrial period). These increases are expected to contribute significantly to a global warming by absorbing infrared radiation in the atmosphere and changing the global heat balance ("greenhouse effect"). [*See* GREENHOUSE GASES IN THE EARTH'S ATMOSPHERE.]

C. Sulfur: An All-Round Element

The sulfur cycle is one of the most heavily perturbed among the major elemental cycles (C, N, O, P, S). Human activities, mainly coal burning, have approximately doubled the emissions to the atmosphere. Sulfur occurs naturally in several oxidation states. In its reduced oxidation state, sulfur is a key constituent of life, providing structural integrity to proteins. As sulfate, in its fully oxidized state, it is the second most abundant anion in fresh water (after bicarbonate) and in seawater (after chloride), and it is the major cause of acidity in both natural and polluted rainwater ("acid rain"), thereby influencing the weathering of rocks. Sulfate in the atmosphere influences the hydrological cycle by constituting the dominant component of cloud condensation nuclei even in unpolluted areas.

There are eleven known isotopes of sulfur of which four are stable. The second most abundant isotope (^{34}S, 4.22%), along with the fractionation known to occur in many biogeochemical processes, has been successfully used to trace sources and sinks of sulfur.

In natural waters, sources of sulfur compounds are rocks (weathering and mining), soils (decomposed biota and fertilizers), and, at present, dominating, atmospheric transport as precipitation and dry deposition (including windblown sea salts as well as soot, gases, and sulfuric acid from fuel combustion). Oxic waters contain mainly sulfate, whereas sulfide tends to accumulate in anoxic zones of intensive decomposition where the redox potential is reduced below about 100 mV, such as wetlands, deep waters, and sediments. Mineral precipitation occurs as gypsum ($CaSO_4$) or as metal sulfides such as pyrite (FeS).

In organisms, the amount of sulfur varies from 0.02 to 5% in some sulfur-oxidizing bacteria, but is typically about 0.25% by dry weight, similar to phosphorus. Sulfur is nearly always present in

quantities adequate to meet the requirements for protein synthesis, which is usually limited by the availability of nitrogen. While the fraction used by organisms does not significantly influence the sulfur cycle, they create conditions that directly or indirectly influence the cycle. Numerous pathways of biotic transformations among sulfur species of different oxidation state have been identified (Fig. 4). Hydrogen sulfide (H_2S) is produced by microorganisms, either in the decomposition of organic material (protein) by heterotrophic bacteria or by sulfate-reducing heterotrophic and chemosynthetic anaerobic bacteria using sulfate and other oxidized species (instead of oxygen) as an electron acceptor in the oxidative metabolism. Sulfide is oxidized to sulfate directly or via other species, either by chemosynthetic aerobic bacteria gaining energy from this process or by photosynthetic anaerobic bacteria using reduced sulfide (instead of water) as an electron donor in the photosynthetic reduction of CO_2. Oxidation also occurs chemically without bacteria involved. Although the requirements of photosynthetic sulfur bacteria are rather specific and their distribution is restricted to zones with both light and steep redox gradients, they can significantly contribute to the annual bioproduction in lakes and estuaries.

The dominant source of gaseous sulfur emitted from the open oceans is dimethyl sulfide (DMS). Reduced sulfur, as hydrogen sulfide, is added in

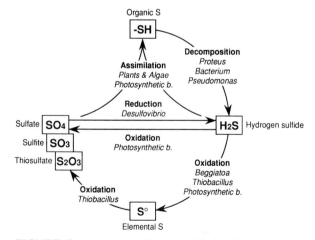

FIGURE 4 A complex microcycle: Biochemical transformations among various sulfur species by selected important organisms, mainly bacteria.

large quantities to the atmosphere from volcanic gases and from biogenic and industrial sources. H_2S undergoes several oxidative reactions to sulfur dioxide (SO_2) and sulfur trioxide (SO_3). These are rapidly converted into sulfuric acid (H_2SO_4) when dissolved in atmospheric water. As a result, the traveling distance and the residence time of sulfur gases are rather short (one or few days). The residence time is even shorter for windblown sea salts, of which only 10% are deposited on land (Fig. 5).

Sulfur dioxide constitutes about 95% of the sulfur compounds emitted by the combustion of fossil fuels. Over 90% of the man-made emissions to the atmosphere are produced in the northern hemisphere, with 80% of these emissions being deposited within the source continent. Deposition maps suggest that the spatial scale is about 1000 km. The flux of sulfate in rainwater over polluted industrial regions is at least 1 g S m^{-2}. This is about 10 times larger than the remote marine flux, which is higher than the natural continental flux, and illustrates the massive man-made perturbation of the sulfur cycle.

Increased emissions of sulfur have caused a considerable acidification of precipitation as well as soils and surface waters, especially in large acid-sensitive areas as are common in Scandinavia and Canada. This has perturbed other important cycles (Al, heavy metals, nutrients) and caused severe ecological damage in forest and freshwater ecosystems. On a global scale, sulfur emissions may influence the climate by an increased production of aerosols acting as cloud condensation nuclei. [*See* ACID RAIN.]

D. Mercury: Coupled Cycles of Pollutants and Nutrients

Mercury is not only one of the most toxic metals but it is also one of the most intriguing metals. It is not known to be essential for any metabolic process, yet it is readily accumulated by most organisms. Mercury occurs naturally in a variety of inorganic and organic compounds, not only in solid or dissolved states, but also in true liquid and gas phases. The transition of mercury between these compounds and phases is controlled by a multitude of environmental processes, including photochemical reactions, chemical oxidation and reduction,

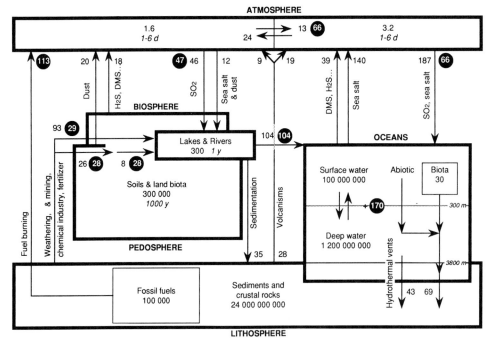

FIGURE 5 The global sulfur cycle. Approximate pool sizes and annual fluxes in units of Mton = Tg = 10^9 kg S. Annual net changes and additional fluxes caused by human activities are highlighted with black circles. Italic numbers show approximate sulfur turnover times.

microbial transformations, and physiological fractionation. Mercury differs from other metals by its "organic" character: it occurs naturally in organometallic compounds and has a very high affinity for most types of organic matter, especially the proteins of organisms. Furthermore, the natural cycling of mercury has been significantly disrupted and accelerated by anthropogenic activities. Evidently, the biogeochemical behavior of mercury is complex, and its understanding requires a holistic study of industrial, atmospherical, geological, hydrological, chemical, microbial, physiological, and ecological processes.

The toxic effects of mercury vapor to man have been known for centuries from the symptoms observed in mine workers ("mad hatters"). In the middle ages, mercury was used to treat syphilis. The first well-documented case of mercury poisoning through food is the disaster of Minamata (Japan) in the 1950s when dozens of people died from contaminated fish. Far from industrial pollution, mental deficiencies have been observed in children after prenatal exposure to methylmercury as a result of regular consumption of certain marine fish by their mothers. In thousands of remote forest lakes of the boreal zone, the maximum daily intake of methylmercury according to WHO guidelines (<30 μg) is contained in a single mouthful of flesh from the largest fish. This widespread environmental problem with direct implications for human health and the puzzling discrepancy of dose and response (high levels at low input) were discovered in the early 1960s.

The chemistry of mercury is very complex. Mercury occurs naturally in different oxidation states [Hg(0), Hg(I), Hg(II)] and in both inorganic and organic compounds. It can occur in the gas phase (elemental Hg, dimethylmercury), as a liquid (elemental Hg), in the solid state and in solution in a variety of forms. In the atmosphere, elemental Hg(0) is the primary form. In bedrock, mineralized soils, and anaerobic sediments, mercury occurs as cinnabar (HgS). In natural waters, Hg(II) compounds and complexes prevail (mainly with hydroxide, chloride, or humic matter), whereas the dominant form in animals is usually methylmercury (CH_3–Hg^+). Inorganic substances such as chloride, iron hydroxides, and sulfide affect the aqueous mercury cycle either by removing aqueous mercury in precipitates or by forming stable soluble complexes. In contrast to most other metals, mercury forms stable organometallic compounds under natural conditions. These compounds have many chemical and physiological characteristics that are typical of purely organic substances such as persistent organic chemicals (DDT, PCB, dioxin). The most abundant is methylmercury, which is formed by microorganisms. Microorganisms are also involved in the reduction and volatilization of oxidized mercury forms.

Mercury differs from most other metals not only by the complexity of its chemistry, but also by its high affinity for organic matter. As this is mainly due to the strong binding to reduced sulfur species, the biogeochemical cycling of mercury is closely linked to the sulfur cycle. Atmospheric deposition is largely proportional to the deposition of sulfuric acid. Quantitative relationships with organic matter are also found in lake and marine sediments, lake water, stream water, groundwater, and soils. In freshwater systems, most mercury is associated with the organic matrix formed by living organisms, detrital particles, and dissolved humic substances. Among the more abundant metals in these systems, iron, lead, copper, and aluminium show a similar behavior, but mercury has the strongest binding to humic substances. High mercury concentrations are also found as sulfide in dry fossil fuels originating from the burial of incompletely degraded organic matter.

Unlike most other metals, mercury has a cycle with a dominating atmospheric component, caused by the high vapor pressure of the elemental species. Mercury is naturally volatilized in significant quantities from land and oceans and is released by volcanic activity, chemical and physical processes in the Earth's crust, and by photoreduction and microbial activity in the biosphere. On a global scale, roughly one-third of the mercury input to the atmosphere originates from natural sources (Fig. 6). Atmospheric Hg(O) travels over large distances. It is slowly transformed by complex photooxidation processes into soluble species that enter the biosphere with wet precipitation. In addition, signifi-

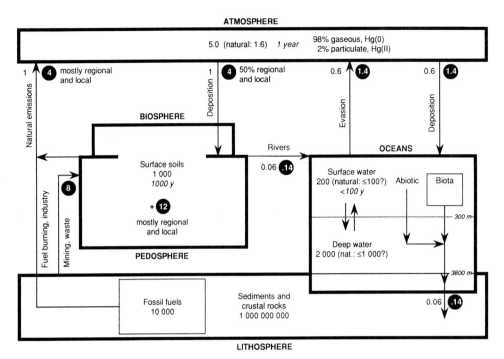

FIGURE 6 The global mercury cycle. Approximate pool sizes and annual fluxes in units of kton = Gg = 10^9 g Hg. Annual net changes and additional fluxes caused by human activities are highlighted with black circles. Italic numbers show approximate mercury turnover times.

cant dry deposition of aerosol particles and gaseous mercury occurs on vegetated surfaces.

The atmospheric pool has globally and regionally increased 2- to over 10-fold during this century due to the anthropogenic contribution of mercury in certain regions of the world. Main sources are coal combustion, mining and smelting of sulfidic minerals, metallurgic industry, waste incineration, and chlor-alkali production. The dramatically increasing pollution of the atmospheric, terrestrial, and aquatic environment during the past decades has reached such dimensions that in recent years, the contamination of wide areas far from any anthropogenic sources of pollutants has become a major concern. One of these areas is the boreal forest zone, where not only the acidification of surface waters has entered public awareness by threatening the fish stock in thousands of lakes, but also a widespread contamination of lacustrine fish with mercury even in remote areas.

Because of its particular chemical character, mercury is naturally ubiquitous in nature. As a result of the wide distribution, virtually all organisms contain measurable quantities in their tissues. Mer-

cury is readily accumulated by biota, and the highest concentrations in aquatic systems are usually found in fish. Natural concentrations in marine and freshwater fish are highly variable (<0.01 to >5 mg kg^{-1} wet weight) and depend on water quality, fish species, size, sex, and behavior. Average concentrations in natural waters are about 1 to 5 ng liter^{-1}, and the proporation of methylmercury (MeHg) is usually < 10%. In contrast, most mercury in animals is MeHg because of a selective bioaccumulation, and the concentration as well as the proportion of MeHg increases with the trophic level in food webs. Consequently, MeHg is the compound of interest from a toxicological point of view. Both ionic inorganic Hg^{2+} and MeHg form stable complexes with anionic sulfur ligands. Accordingly, both have a high affinity for biogenic matter and bind strongly to the sulfhydryl (thiol) groups in proteins and other tissue constituents of living cells. Compared to inorganic mercury, the MeHg uptake is about 10 times more efficient, both from food and from water. This is because of the more lipophilic character of MeHg favoring passage across cell membranes and resulting in a rather even distribution of mercury within animal bodies, which is in contrast to other heavy metals that have a higher affinity for proteins but are lipophobic and accumulate in visceral organs. Consequently, the elimination of MeHg from animals is very slow, partly due to the strong association with biotic tissues and partly as a result of the large pool of tissues such as muscle and brain that need to be depurated. In small lakes, the fish community contains about half or more of the methylmercury pool in the whole water column.

The strong interaction between mercury and biogenic matter largely determines the fluxes of mercury. The load of mercury on a lake and the concentration in the water at a given atmospheric deposition are highly dependent on the inflow of humic carrier substances from the watershed and on the level of soil contamination. In boreal areas, where surface waters have a high content of humic substances, mercury concentrations are generally high, both in water and in biota. In brownwater lakes, input of mercury is dominated by the flux from the surrounding soils whereas in clearwater lakes, concentrations are lower and are controlled by the rate of direct deposition on the lake surface and by the sedimentation rate of mercury. Cold climate may enhance the susceptibility of soil and lake ecosystems to mercury contamination in boreal regions, both by depressing productivity of terrestrial and aquatic biota and by influencing mercury fluxes among atmosphere, pedosphere, and hydrosphere. Large problems with elevated mercury concentrations in fish have arisen in remote areas after the creation of hydroelectric reservoirs, where mercury is released from flooded soils and enters the aquatic food web after transformation into methylmercury.

Organic interactions also control the ecotoxicology of mercury, including its bioavailability, its bioaccumulation by microbes and invertebrates, and its biomagnification in food chains. Bioavailability of mercury in lake waters can be described as a function of the concentration and the origin of organic matter. The accumulation in biota is pronounced in nutrient-poor ecosystems where the available mercury is distributed within a low total biomass. While primary bioproduction results in a

biodilution of mercury, humic matter with a low nutritive value results in a high mercury exposure of detritus food chains. Accordingly, detritivorous animals can contain up to 100 times more mercury than predators within the same habitat. The same applies across sysstems, as animals have high mercury levels in humic lakes but low levels in nutrient-rich lakes. Bioaccumulation patterns in animals suggest a dynamic equilibrium between animal and food mercury concentrations, probably controlled by food conversion efficiency. Accordingly, mercury concentrations usually increase with animal age or size, but can also decrease. Considerable seasonal variations in fish occur that can be attributed to changes in physiology rather than changes in the environment. Biomagnification of methylmercury in food chains is characterized by an enrichment between predator and prey, irrespective of ecosystem type, trophic level, animal size, or age. Consequently, mercury concentrations in animals are basically determined by their trophic level and by the bioavailability of methylmercury at the base of the food web.

Obviously, the mercury concentration in many ecosystem compartments is to a large extent controlled by biotic processes. Accordingly, mercury cycling is tightly coupled to the cycling of elements characteristic of biogenic matter (C, N, P, S) all the way from atmospheric emission to predatory animals. Fractionation (enrichment or depletion) between these elements and mercury occurs when mercury is transferred into another matrix or when organic matter is produced, decomposed, or reconstructed (dilution or accumulation).

V. GENERAL BIOGEOCHEMICAL CONCEPTS, DIMENSIONS, AND MODELS

The value of biogeochemical models in answering questions depends very much on appropriate and consistent definitions. The structure of the system and the transfer processes need to be considered, including the compartment interfaces, the type of reactions, transformations and phase transitions, the direction of dispersal and reactions, and all associated rate constants or variables (Table II). Ecosystems may theoretically be described by defining the biological, chemical, and physical states of all its components. By including the position and energies of these parts, densities or concentrations could be expressed as a function of time. For practical reasons, this approach needs to be simplified by breaking down large systems into manageable entities. This is achieved by defining a limited number of compartments with a rather homogeneous content differing from the surroundings. Definition of such aggregates may be biological (species, trophic levels, communities), chemical (isotopes, elements, classes of compounds), or physical (state of aggregation: gas/liquid/solid, temperature). Aggregate variables such as mass, density, concentration, and temperature are used because they simplify the description of a system and have applications in predicting its behavior through the conservation of mass and energy. Entropy and free energy may be useful variables for other purposes, as well as trophic levels within food webs (herbivores, carnivores, detritivores) or horizons within soils. [See ECOLOGY, AGGREGATE VARIABLES.]

A prerequisite of biogeochemical models is the conservation of mass. At steady state, fluxes to and from a compartment must balance. Imbalance indicates either the omission of one or several important fluxes (sources or sinks) or a net change (loss or accumulation) of material over time. Either of these alternatives can provide important knowledge, or at least an impetus to collect the missing information.

Traditionally, concentrations of contaminants are related to volumetric or weight units. In natural waters, this is very convenient for flux calculations, but is not always satisfactory when studying biogeochemical mechanisms. The environmental fate and effect of pollutants may not be determined by their concentrations but instead by the abundance, characteristics, and source of their matrix or of functionally related compounds that are cycling in the system simultaneously. For instance, most micropollutants have a high affinity for inorganic or organic particles, and only a minor proportion is truly dissolved in natural waters. In biota, many contaminants accumulate selectively in certain body tissues or interact with specific compounds. Many disadvantages of conventional concentration units can be eliminated by using mass or element (isotope) ratios instead of concentrations related to volume or weight.

The choice of spatial and temporal scales depends on the system dynamics and on the observer's perspective and interest. Spatial scales of biogeochemical cycling range from atomic, cellular, and individual to local, regional, global, and cosmological. The term "cycle" is not only used for closed, steady-state systems, but also for open or partial systems of interest where substances are processed, such as a lake or a forest. This approach requires a careful definition of system boundaries and includes unidirectional fluxes in addition to cycles.

Temporal scales of reactions within a system usually range over several orders of magnitudes, from very rapid processes (e.g., bio- and geochemical

TABLE II

Aspects and Components of Biogeochemical Cycles and Models

Aspects	Static	Dynamic	Determination
Entities:	Pools or reservoirs	Fluxes and transformations	Observation
What?	Element, compounds...	Transf.: Reagents, products	Definition
Where?	Compartments, spheres	Fluxes: Interfaces, directions	Definition
How?	Status	Transformations	Observations
Physical	Phases	Transitions	
Chemical	Forms, species	Reactions	
Biological	Organisms	Transformations	
How much?	Burdens	Rates of flow and transformation	Measurement
How fast?		Turnover, residence, age	Calculation

reactions or microcycles) to very slow processes (e.g., sediment diagenesis or geophysical macrocycles). Most models cover only few orders of magnitude. Very slow processes are treated as negligible or constant, for example, the movement of the earth's crust needed for the regeneration of marine sedimentary material. Correspondingly, very fast processes may be averaged over time and treated as quasi-stationary, for example, chemical equilibria. Temporal scales are also affected by the spatial scaling: turnover times of environmental compartments tend to increase with their size, since compartments are aggregated entities that are assumed to be well-mixed. The selection of time scales considered in models depends very much on the questions being asked. Depending on this selection, temporal cycles may need to be considered: diurnal cycles such as the light-dependent alternation between net production and respiration, or annual cycles including temperature effects on hydrological and biological processes.

VI. POTENTIAL AND LIMITATIONS OF COMPILING BIOGEOCHEMICAL CYCLES

Biogeochemical diagrams and models provide a useful compilation and synthesis of extensive quantitative knowledge about the natural environment. However, the approach also has drawbacks and limitations:

- Spatial and temporal resolution is limited because model structures necessarily are simplified (both between and within compartments) and because models usually include processes that work on similar time scales (very rapid or very slow reactions are rarely considered).

- Definition of a limited number of compartments necessarily leads to inhomogeneities within compartments, even when their components appear to be well mixed. This may significantly affect quantitative interpretations, which is illustrated with an example of potentially misleading

turnover times: In soils, which contain a small amount of rapidly decomposing litter (say 9%, turnover time 10 years) and large quantities of recalcitrant humic substances (91%, 1000 years), the average turnover time of organic matter is about 100 years, a number that fits neither fraction and provides little information about actual processes or about the response to events such as deforestation. Similar considerations apply to the dissolved organic matter in the surface water of lakes and oceans.

- Current relationships cannot be extrapolated forward or backward over time scales without caution. This is partly because many models only include processes working or responding during the period of observation, often a few decades of modern science. Long-term changes in system behavior, such as a slow saturation of compartments, may be difficult to predict. For instance, man-made changes of environmental fluxes are often compared to the preindustrial situation, although a consideration of geological time scales may change interpretations. Over centuries, the CO_2 concentrations in the atmosphere has been influenced by changes in the exchange rate between the atmosphere and the biosphere, but ultimately it is determined by geological processes: the release of carbon from the lithosphere and the sedimentation of carbon in the deep oceans, which are balanced by the movement of the earth's crust.

- Natural variability is usually neglected: cycles are usually based on averages and rarely account for variability in time (diurnal or seasonal cycles) or space (fluxes and processes may differ regionally), which may significantly influence pathways as well as magnitudes of material transport.

- The necessary setting of priorities potentially results in a certain degree of subjective bias with respect to the selection of compartments and regarding spatial and temporal resolution and detail. Depending on the system dynamics, pool sizes, and the time scale in focus, reservoirs are regarded as negligible entities, as dynamic entities, or as "passive" sources and sinks. For instance, marine scientists may distinguish a large number of different marine compartments and processes, whereas others may treat the oceans as one passive sink compartment.

- The predominantly materialistic view of biogeochemical cycles rarely considers important biological aspects such as biodiversity, habitat values, behavior, survival, or genetic evolution. Similarly, the focus is on phenomena rather than mechanisms.

Despite these considerations, the reconstruction and compilation of biogeochemical cycles are invaluable for many reasons and can serve scientists and laypersons at several levels:

- In applications (management, risk assessment): for dealing with environmental issues important to society, since problems are more likely (although not guaranteed) to be avoided or solved by understanding, by knowing the relative and absolute magnitude of dominant pools and fluxes, and by ranking processes by their importance.
- In communications (teaching, public information): for visualizing complicated and large-scale interactions in the global or local environment by means of instructive diagrams.
- In investigations (research and development): for identifying important gaps in our knowledge.

From an appropriate diagram, both structure and dynamics of a system emerge simultaneously. Pathways and rates of cycling can be compared and ranked by importance. Missing information such as the magnitude of diffuse fluxes can be obtained from mass balance calculations if all but one of the entities around a compartment are known. Evidently, estimates from such comparisons cannot be more accurate than the known entities. These may continue to grow in number along with the development of science.

Glossary

Biotic transformation Conversion of compounds by organisms, for example of inorganic to organic molecules or vice versa; can be associated with a phase transition.

Flux Rate of flow of compounds between compartments, useful for calculating turnover (residence) times of materials in the environment.

Phase transition Physical transition of compounds between gas, liquid (dissolved), and solid state, usually mediated by chemical processes or biotic transformations.

Bibliography

Butcher, S. S., Charlson, R. J., Orians, G. H., and Wolfe, G. V. (1992). "Global Biogeochemical Cycles." International Geophysics Series, Vol. 50. San Diego: Academic Press.

Fleischer, S., and Kessler, E., guest eds. (1993). Surface water acidification; effects and remedial measures. *Ambio* **22**, 257–337.

Heimann, M., ed. (1993). "The Global Carbon Cycle." NATO ASI Series I: Global Environmental Change, Vol. 15. Hamburg: Springer-Verlag.

Meili, M. (1991). The coupling of mercury and organic matter in the biogeochemical cycle—towards a mechanistic model for the boreal forest zone. *Wat. Air Soil Pollut.* **56**, 333–347.

Rundel, P. W., Ehleringer, J. R., and Nagy, K. A., eds. (1989). "Stable isotopes in ecological research." Ecological Studies, Vol. 68. New York: Springer-Verlag.

Williamson, S., guest ed. (1994). Integrating Earth System Science. *Ambio* **23**, 4–103.

Biogeochemistry

Fred T. Mackenzie

University of Hawaii

Biogeochemistry is the discipline that links various aspects of biology, geology, and chemistry to investigate the surface environment of Earth. This environment, the ecosphere (Fig. 1), encompasses the biosphere and parts of the other large subdivisions (reservoirs) of Earth's surface of atmosphere, hydrosphere, and shallow crust (soils, sediments, and crustal rocks). Processes controlling the concentration, distribution, and cycling of elements in and above Earth's crust are a focus of the field of study of biogeochemistry. In particular, studies of the biogeochemical processes of plant photosynthesis and respiration, bacterial decomposition, nitrogen fixation, nitrification and denitrification, methanogenesis and methanotrophy, fermentation, and sulfur oxidation and reduction are a critical component of biogeochemistry. These processes involve several biochemical pathways, including (1) autotrophy in which inorganic carbon in the environment is converted into organic materials, (2) heterotrophy in which organic substrates are used to make organic matter, and (3) mixotrophy in which both organic and inorganic substrates are used by an organism to manufacture organic tissue. The term biogeochemistry was first coined by V. I. Vernadsky in 1926 as a subdiscipline of geochemistry, a term first proposed by C. F. Schonbein in 1838.

The influence of life, including humans, on Earth is so pervasive that it is difficult to isolate a naturally occurring chemical reaction in the surface environment of the planet that is not affected by the biota. For more than 3.5 billion years, the biota has played a major role in the cycling of elements about Earth's surface. From the perspective of some scientists, the influence of the biota in the evolution of Earth and its present and future state has become the major factor in controlling the composition of the oceans and atmosphere, the rate of weathering of continental crust, and the deposition of marine sediments. The concept of Gaia, in which the surface environment of Earth is regulated by and for the biota, is a direct outgrowth of the strong control exerted by life on the Earth's environment.

Human-induced global environmental change is a consequence of direct and indirect rapid modification of Earth's surface environment by human activities. The distribution and rates of growth of the human population and the demand for economic growth, with concomitant utilization of resources, are forces that are acting as agents of global environmental change. Human activities resulting from these forcings include fossil fuel and biomass combustion, land-use change, agricultural practices, and halocarbon and other synthetic chemical production and release. All of these activities lead to changes in the composition of terrestrial, aquatic, and atmospheric systems and changes in the cycling of elements within the ecosphere. Therefore, human activities and their rates have

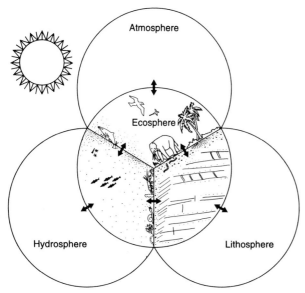

FIGURE I The ecosphere, our life-support system, showing its relationship to the other spheres of the Earth.

become a biogeochemical force of major and global importance and as such constitute a part of the study of biogeochemistry.

I. DEVELOPMENT OF THE FIELD OF BIOGEOCHEMISTRY

The development of the field of biogeochemistry involves a series of conceptual advances covering a 300-year span of time. The advances of greatest significance in chronological order include:

1. the nature of the hydrologic cycle;
2. the role of carbonic acid in rock and soil weathering;
3. the interactive roles of plants and animals in gaseous metabolism at Earth's surface driven by solar radiation;
4. the unification of theories of plant nutrition;
5. the wholly inorganic nutrition of green plants;
6. the role of microbes in organic decomposition;
7. the global significance of the biogeochemical cycles of carbon and nitrogen powered by solar radiation;

8. the recognition that the primitive atmosphere of Earth was anaerobic and that the gradual evolution of plants and storage of their decayed residues as kerogen in sediments led to an oxygenated atmosphere;
9. the major role of microbial decomposers in biogeochemical cycles;
10. the recognition of the biosphere as a major reservoir in Earth's system, equivalent in importance to the lithosphere, atmosphere, and hydrosphere;
11. the involvement of specific microorganisms in the cycles of elements like nitrogen, sulfur, and iron;
12. the fitness of the environment for living organisms;
13. the possibility of the origin of life by prebiotic synthesis of organic molecules in a reducing environment and their consumption by heterotrophic organisms;
14. large-scale regulation of biogeochemical cycles by the biota;
15. the view that humans are a major agent of change in the environment; and
16. the quantification of the view of the Earth as a biogeochemical system with interactive reservoirs of atmosphere, ocean, crust, and biota—the ecosphere. This system has various names, including Earth surface system and exogenic system. If it is assumed that the biota actually regulate the composition of the atmosphere and ocean, the term Gaia has been applied to this interactive system.

II. THE GLOBAL BIOGEOCHEMICAL SYSTEM

A. Parts of the Ecosphere

The ecosphere is the planetary life-support system. It is the thin film around the planet where the biota interact with the atmosphere, hydrosphere, and lithosphere in a complex system involving biogeo-chemical processes and cycles. [*See* BIOGEOCHEMI-CAL CYCLES.]

The major source of energy for driving processes in the ecosphere comes from the Sun. The surface temperature of the Sun is about 5480°C, and the light energy or radiation from this stellar body easily reaches the atmosphere. Some of the incoming solar radiation is reflected back to space in the outer reaches of the planetary atmosphere. Most of the energy reaching the outer boundary of our atmosphere is visible, shortwave, and ultraviolet radiation. The latter is largely absorbed by ozone. Of the incoming solar radiation, approximately 30% is reflected back to space by clouds and Earth's surface, 25% is absorbed by the atmosphere and reradiated back to space, and 45% is absorbed by the surface of land and water. The energy reaching the surface is used to evaporate water, photosynthesize organic matter, drive winds, currents, and waves, generate rising thermals, and is absorbed by greenhouse gases before being radiated back toward space. Only a small amount of solar energy, 0.08%, is involved in the critical process of photosynthesis that is the basis of life on Earth. [*See* ATMOSPHERIC OZONE AND THE BIOLOGICAL IMPACT OF ULTRAVIOLET RADIATION.]

I. The Atmosphere

The atmosphere is the gaseous envelope surrounding the planet and extending 500 kilometers above the Earth's surface. It is a dynamic system but one that is well balanced. It is the heat engine, along with the oceans, that distributes heat throughout the globe and drives the climate system of the planet. The atmosphere is a mixture of gases constituting air and is composed of 78.1% diatomic nitrogen, 20.9% diatomic oxygen, 0.036% carbon dioxide, and a number of trace gases (Table I). The water vapor content of the atmosphere varies from several percent in hot, humid environments to tens of parts per million in cold, dry conditions. The trace gases, although low in concentration, are of importance in many phenomena affecting the ecosphere. For example, the gases of carbon dioxide, methane, nitrous oxide, and tropospheric ozone contribute to the natural greenhouse effect that provides an equable climate for the biota. Stratospheric ozone shields the planetary surface and life from harmful ultraviolet solar radiation. Some sulfur

TABLE I

Gaseous Composition of Dry Air[a]

Constituent	Chemical symbol	Mole percent
Nitrogen	N_2	78.084
Oxygen	O_2	20.947
Argon	Ar	0.934
Carbon dioxide	CO_2	0.0360
Neon	Ne	0.001818
Helium	He	0.000524
Methane	CH_4	0.00017
Krypton	Kr	0.000114
Hydrogen	H_2	0.000053
Nitrous oxide	N_2O	0.000031
Xenon	Xe	0.0000087
Ozone[b]	O_3	trace to 0.0008
Carbon monoxide	CO	trace to 0.000025
Sulfur dioxide	SO_2	trace to 0.00001
Nitrogen dioxide	NO_2	trace to 0.000002
Ammonia	NH_3	trace to 0.0000003

[a] After Warneck (1988); Anderson (1989); and Wayne (1991).

[b] Low concentrations in troposphere; ozone maximum in the 30- to 40-km regime of the equatorial region.

gases emitted to the atmosphere from Earth's surface by biological processes tend to cool the planet, helping to regulate the Earth's surface temperature. [*See* GREENHOUSE GASES IN THE EARTH'S ATMOSPHERE.]

2. The Hydrosphere

The hydrosphere is the part of Earth that contains water in liquid, vapor, and frozen form. It includes water in the oceans and seas, glacial ice, groundwater, lakes, soils, atmosphere, and rivers (Table II).

The planet has had a hydrosphere for over four billion years. Throughout geologic time, the size and shape of the ocean basins and ocean circulation patterns have changed because of seafloor spreading and continental drift, which rearrange the configuration of land and ocean. There have been continuous changes in the composition of the atmosphere, including its water vapor and carbon dioxide content, and there have been changes in the type and distribution of clouds and land vegetation. All of these changes affect the temperature and climate of Earth and indirectly affect the volume of the cryosphere (ice). As a result, sea level has risen and fallen during Earth's history because of changes in ice volume, and for other reasons.

TABLE II

Distribution of Water in the Ecosphere[a]

Reservoir	Volume (10^6 km^3)	Percentage of total
Ocean	1370	97.25
Cryosphere (ice caps and glaciers)	29	2.05
Groundwater	9.5	0.68
Lakes	0.125	0.01
Soils	0.065	0.005
Atmosphere	0.013	0.001
Rivers	0.0017	0.0001
Biosphere	0.0006	0.00004
Total	1408.7	100

[a] After E. K. Berner and R. A. Berner (1987). "The Global Water Cycle: Geochemistry and Environment." Prentice–Hall, Englewood Cliffs, N.J.

Water circulating through the ecosphere is part of a continuous hydrologic cycle that makes life on Earth possible. It is a dynamic system with water stored in many places at any one time. The water cycle involves the transfer of water in various forms of liquid, vapor, and solid (ice and snow) through the land, air, and water environment. Matter and energy are involved in the transfer. Heat from the Sun warms the ocean and land surfaces and causes water to evaporate. The water vapor enters the atmosphere and circulates with the air. Warm air rises in the atmosphere and cooler air descends. The water vapor rises with the warm air. The farther from the warm planetary surface the air travels, the cooler it becomes. Cooling causes water vapor to condense on small particles (cloud condensation nuclei) in the atmosphere and to precipitate as rain, snow, or ice and fall back to Earth's surface. When the precipitation reaches the land surface, it is evaporated directly back into the atmosphere, runs off or is absorbed into the ground, or is frozen in snow or ice. Also, and perhaps most importantly for the ecosphere, plants require water and absorb it, retaining some of the water in their tissue. The rest is returned back to the atmosphere through transpiration. The net amount of water evaporated annually from the sea surface is equal to the yearly flow of rivers and groundwater to the ocean.

3. The Lithosphere

The lithosphere is the outer region of the solid Earth extending to a depth of about 100 kilometers. The lithosphere consists of naturally occurring inorganic compounds, called minerals, found in a variety of rock types and organic matter found in discrete coal, oil, and gas deposits or dispersed throughout sedimentary rocks as kerogen. The lithosphere and its outer skin, the oceanic and continental crust, interact strongly with the interior of the planet. The outer skin comes in contact with the atmosphere, hydrosphere, and biosphere and is in a state of continuous dynamic change.

After Earth formed, the outer crust was subjected to many forces that altered the exterior of the planet. Internal heat creates molten material (magma) within the inner Earth. The hot magma is transported toward the surface of the planet, where it is intruded into the crust or flows out as lava on its surface. Over time the surface rocks are subjected to weathering and erosion and give rise to soil and detritus for rivers and the ocean. Some of the rocks and their alteration products eventually recycle back into the interior of the planet. These and other forces are all part of the rock cycle and continually reshape Earth's surface, providing the environment for life to evolve.

Soils are an important biogeochemical system of the crust and are an integral component of the ecosphere. Soils are made up of matter existing in solid (organic and inorganic), liquid (water), and gaseous states. Gases, derived from the chemical reactions and biological activity in soils, accumulate in air spaces within the soil. Much of the carbon dioxide in soils is produced by the biological decay of organic matter in soils. Other gases, such as nitrogen and oxygen, come directly from the atmosphere. It may take hundreds to thousands of years for topsoils to form that are capable of supporting substantial plant life. Depending on the climate and type of bedrock, 200 to 1500 years are required to form just 2.5 centimeters of topsoil from bedrock. [See SOIL ECOSYSTEMS.]

Although a number of factors affect soil type, Russian scientists before the First World War demonstrated that given sufficient time for soil development, the principal factor determining soil type is the climate of a region. There is a strong correlation between climate, soil type, and vegetation. For example, in areas of heavy tropical rainfall, like the Amazon and Congo river basins, the soils are well leached, lack nutrients, and are deep red in color because of the oxidation of iron-bearing soil minerals to iron oxides. In such systems, the tropical forest trees obtain much of their nutrients from decaying vegetation on top of the soil and from the dead understory below the forest trees. The vegetation is characterized by diverse communities of tall trees and epiphytes. The land is generally very infertile for any vegetation except the tropical forest itself. In contrast, tundra soils of high latitudes are frozen except for a thin upper horizon that thaws in the summer. Because of the frozen subsoil, leaching of materials is restricted to the thin upper layer of the topsoil, and the subsoil has the same dirty gray color as its substratum. Grasses, mosses, lichens, and rushes characterize the tundra. [See SOIL MANAGEMENT IN THE TROPICS.]

4. The Biota

The biota, life itself, is the major component that defines the ecosphere and is the essence of the discipline of biogeochemistry. Mars and Venus do not have an ecosphere because there is no life on these planets. Only the Earth supports an ecosphere. In the narrow band of the ecosphere, the biota sequesters carbon and nutrients and, along with the atmosphere, hydrosphere, and lithosphere, helps to maintain the balance of carbon and other biologically reactive elements on the planet. Because of the importance of the biota in biogeochemistry, it is necessary to consider organisms and their relationship to each other and to their physical environment in some detail.

Where an organism lives is called its habitat, and what it does or how it interacts with its habitat is called its niche. There are specific niches and habitats for all organisms. Organisms of the same species living in a specific area are populations of that particular species. Populations never live in isolation but always interact with other populations.

A group of plant and animal populations living together in the same region is a community. Specific types of plants and animals live together within a particular community. For example, deer and mountain lions coexist. In contrast, cacti and reindeer never share the same community.

Communities of organisms that interact with one another, as well as with their physical and chemical environment, in such a way as to sustain a system are collectively called an ecosystem. These dynamically balanced systems are found from the tallest peaks to the floors of the deepest oceans. Ecosystems come in many sizes, exhibit variations in life-forms, and have distinct chemical and physical properties, such as those found in desert, forest, grassland, tide pool, stream, or pond. There are a multitude of dynamically balanced ecosystems on Earth. As one progresses toward the equator, the complexity of ecosystems increases from a few species and communities in the polar regions to multiple, diverse communities at the equator. The many ecosystems coupled together sustain the larger, complex, and intricately interlinked global ecosystem—the ecosphere.

a. Components of Ecosystems

Every ecosystem, in order to function properly, must have an energy and nutrient source and maintain a relationship between its biotic and abiotic components that allows the processes of energy flow and nutrient cycling to go on efficiently. These dynamically balanced components sustain the ecosystem.

i. Energy Flow Energy is defined as the ability or capacity to do work. There are many forms of energy, such as kinetic (energy of motion) and potential (stored energy). The movement of energy is described by basic scientific observations called the Laws of Thermodynamics. The first law states that energy can neither be created nor destroyed, but is only transferred from one form to another. The second law states that when energy changes from one form to another, part of it is always converted into unusable waste heat. As a result, a constant supply of energy is needed to sustain a system. The waste heat lost in each transformation must be replaced.

Energy flows through an ecosystem in a series of transformations. The light energy of the Sun enables plants to make organic tissue from carbon dioxide, water, and inorganic nutrients through the process of photosynthesis. The light energy is changed to chemical energy in plant cells and used for growth or stored in the plant tissue. When the plant dies and decomposes, or is eaten by a consumer, energy stored in the plant is transferred.

The source of energy for animals is the plant or other animals. Animals require energy to convert nutrients from their food into body tissue, because they cannot make body tissue directly from the energy of the Sun. When plants are eaten, a small portion of the energy stored in plants is transferred to animals for growth, maintenance, and performance of activities. When animals are food for other animals, another transfer of energy occurs. With each transfer, energy is lost to waste heat and ultimately radiated back into space as infrared radiation.

When animals use energy stored in their bodies, inorganic compounds are released through body excretions and eventually through the death of the animal. These inorganic compounds are one source of nutrients that are used by plants.

ii. Nutrient Cycling
Elements and compounds are cycled through the organic and inorganic systems. Their cycling is essential for life to exist. The complexities of the biogeochemical cycles and their involvement in the sustainability of life are topics of great scientific interest and practicality. Knowledge of these topics is necessary to evaluate the effects of global change on the ecosphere.

Of the over 100 elements, 17 are essential for the growth of most plants (Table III)—life could probably not exist without each of these elements. Six of the 17 elements, nitrogen (N), carbon (C), hydrogen (H), oxygen (O), phosphorus (P), and sulfur (S), compose approximately 95% of all matter in plants, animals, and microorganisms. In addition, the essential elements must also be present in a specific compound that the organism can use. These compounds are called nutrients. For exam-

ple, the carbon (C) in carbon dioxide (CO_2) is the only major form of carbon that plants can use. In general, they cannot use carbon directly from other compounds. Water (H_2O) provides the only form of hydrogen (H) that a plant can use. The deficiency or lack of any one of these 17 nutrients can be a limiting factor for organic growth in an ecosystem. The one nutrient that is not provided or that is deficient—the limiting nutrient—may limit the development of an organism in some way. [See Aspects of the Environmental Chemistry of Elements.]

iii. Relationship between Abiotic and Biotic Components
An ecosystem has abiotic and biotic components that interact with each other (Fig. 2). The abiotic components of an ecosystem are all the chemical and physical parts of the environment, such as nutrients and their availability, soils, temperature, water, and sunshine. Every organism must adapt to the abiotic factors of an area or it cannot live in that area. The loss or change in a single abiotic factor may lead to collapse of the system. This factor may be a limiting factor for an ecosystem. For example, the availability of water determines the types and extent of plant life in a desert, while the supply of nitrogen or phosphorus as nutrient compounds determines the extent of plant growth in the oceans and in many terrestrial environments. Disruptions in the supply of these life-essential materials can lead to ecosystem degradation and ultimately to collapse.

The members of the biota may be categorized by their function in an individual ecosystem and include producers, primary and secondary consumers, and decomposers. These organisms form part of a cycle—a food chain. A food chain results in the transfer of energy and nutrients from one organism to another within a particular ecosystem. A more complex intertwining of individual food chains in an ecosystem is a food web. The feeding level occupied by an organism in a food chain is its trophic level—who eats whom. The first trophic level contains the producers—the plants. Primary consumers, the plant eaters (herbivores), occupy the second trophic level, and the secondary consumers, the meat eaters (carnivores), occupy higher

TABLE III
The Essential Elements Necessary for Plant Growth[a]

Percentage of plant matter	Essential element	Available from nutrients	Present in	Role in plants
95%	Carbon, C	Carbon dioxide, CO_2	Air or dissolved in water	Essential in the structure of all organic molecules
	Hydrogen, H	Water, H_2O	Water	Essential in the structure of all organic molecules
	Oxygen, O	Carbon dioxide, CO_2	Air	Essential in the structure of most organic molecules
	Nitrogen, N	Nitrate, NO_3^- Ammonium, NH_4^+ Nitrogen gas, N_2, but only by nitrogen fixation	Some soil minerals or dissolved in water / Air	Essential in the structure of all proteins, nucleic acids, and some other organic molecules
5%	Sulfur, S	Sulfate, SO_4^{2-}		Essential in the structure of most proteins
	Phosphorus, P	Phosphate, PO_4^{3-}		Essential in the structure of nucleic acids and in energy transfers
	Potassium, K	Potassium ions, K^+		Essential in maintaining water balance; necessary in certain enzymes
	Calcium, Ca	Calcium ions, Ca^{2+}		Essential in maintenance of membrane function; essential in cell walls of most plants
	Magnesium, Mg	Magnesium ions, Mg^{2+}	Some soil minerals or dissolved in water	Essential in chlorophyll molecules
	Iron, Fe	Iron ions, Fe^{2+}, Fe^{3+}		Essential in photosynthesis and in energy-releasing reactions
	Manganese, Mn	Manganese ions, Mn^{2+}		All essential for the function of certain enzymes
	Boron, B	Boron ions, B^{3+}		
	Zinc, Zn	Zinc ions, Zn^{2+}	(Mn, B, Zn, Cu, Mo, Co)	
	Copper, Cu	Copper ions, Cu^{2+}		
	Molybdenum, Mo	Molybdenum, Mo^{2+}		
	Cobalt, Co	Cobalt ions, Co^{2+}		
	Chlorine, Cl	Chlorine ions, Cl^-		Essential in photosynthesis

[a] After Nebel (1981).

trophic levels. Humans occupy the higher trophic levels, existing both as carnivores and vegetarians.

First, there are the primary producers (autotrophs, e.g., green plants). The autotrophs evolved before animals because these organisms can get nourishment from inorganic compounds to form their tissues. Animals generally cannot extract sufficient nutrients directly from soil and water in order to survive. Land and aquatic plants have this ability in the process of photosynthesis. Land plants are able to make their own food (carbohydrates–organic carbon compounds) with water (H_2O), nutrients [nitrogen (N), phosphorus (P), sulfur (S), and trace quantities of other essential elements] removed from the soil by the roots of plants, carbon dioxide (CO_2) from the air, and light energy from the Sun that is trapped by chlorophyll. In the process, plants discharge oxygen (O as O_2), heat energy, and water vapor into the air while storing carbon in their tissue as glucose and other carbohydrates. In the aquatic realm, the photosynthetic process is similar, except in this case the plants, the microscopic phytoplankton of the ocean, take up carbon and nutrients directly from dissolved forms of these substances in their water environment. The organic compounds produced by photosynthesis,

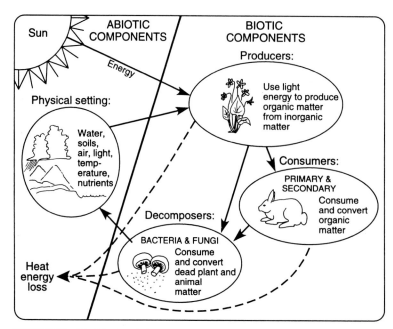

FIGURE 2   The biotic and abiotic components constituting an ecosystem structure and their interactions. The relationship between producers, consumers, and decomposers is critical to the functioning of the system. Light energy is degraded as it flows through the system.

and the energy stored in them, are the basis for the food chains of the land and oceans. This process is the same that initially provided Earth with its oxygenated environment.

Respiration is the reverse of the process of photosynthesis. It is the burning, through the use of oxygen, of the stored carbohydrates in plants and animals. Respiration produces energy for organisms to move, grow, and maintain existing body structures. Respiration has as its by-products CO_2 and H_2O. In plants, these by-products are released into the atmosphere through the stomata of leaves. Transpiration is the process whereby water is released and evaporated. The processes of photosynthesis and respiration are limited mainly to plants, algae, and cyanobacteria. They are the organisms that form the foundation for the entire living world.

The consumers (heterotrophs) are organisms living on land or in water that feed directly or indirectly on the producers. They cannot make organic compounds within their bodies from inorganic materials. They must feed on plants or other animals, who may have initially fed on plants, to get their nutrients and energy. Primary consumers are or-

ganisms that feed directly on plants and include deer, cows, or horses on land and crustaceans in the sea. Secondary consumers feed on the organisms that feed on the primary consumers. Humans, sharks, and cougars are perfect examples of secondary consumers. The secondary consumers keep a system in balance by maintaining the primary consumers at a population level such that the system has sufficient space, food, and other materials to sustain it.

Decomposers (heterotrophs) are organisms that are essential in the process of decay of dead organisms. The decomposers (fungi and bacteria) derive energy and nutrients from dead organisms and change organic compounds back into inorganic compounds. This decomposition completes the cycle of organic matter, because inorganic compounds are now available for plants to use again. The decomposers seem like a gross lot, but without them organic materials would rot exceedingly slowly (in outer space, where bacteria do not live, organic materials may persist and not decay), nutrients would be consumed and not replenished, and life would cease.

b. Ecological Pyramids

The type, variety, and number of the members of the biota must be in balance to sustain an ecosystem. There can only be a certain population of each type of organism, fulfilling the roles of producer, consumer, or decomposer in an environment.

Biomass is the total combined dry weight of any specific group of organisms. The biomass of the first trophic level is the total dry weight of all producers in an area. The biomass of the second trophic level is the total dry weight of all herbivores, and the third trophic level is the total dry weight of secondary consumers, the carnivores, and so on. A phenomenon common to many ecosystems is the decrease of biomass with an increase in the trophic level.

Energy enters an ecosystem as light from the Sun. Only about 0.08% of this energy is used by plants. However, all the energy absorbed by plants is not available to each of the following trophic levels, because most plant matter produced is not consumed by second trophic level feeders, the herbivores, but is consumed and digested by decomposers and detritus feeders feeding at the second trophic level. Furthermore, each time energy from food is transferred from one organism to another, some energy is lost through bodily excretions and through waste body heat as work is performed.

An ecological pyramid represents the decreasing amount of energy available for use by each subsequent trophic level. Only about 10 to 20% of biomass is actually transferred upward at each level, resulting in an approximately 90% decrease in potential energy available for each succeeding trophic level. Thus, the organisms at the bottom of the food chain, the plants, are more abundant. There are more plants than herbivores and more herbivores than carnivores.

An ecosystem remains in balance when the type and number of members of the biota are in certain proportions. The balance of nature is this stability of plants and animals in relationship to their environments, and humankind is capable of interfering with this natural balance. An example of how the balance of an ecosystem can go awry and collapse is that of plant–deer–cougar relationships in the southwestern part of the United States. In the early 1900s, Arizona placed a bounty on large predators (wolves, cougars, and coyotes) in an effort to protect and increase the deer population. This action led to massive extermination of these secondary consumers. The deer population (primary consumers) swelled without any natural force to cull the herd. As the bulging deer population searched for food, vegetation (producers) in the area was stripped. Eventually, many deer starved to death after their food source was depleted. The deer population ultimately decreased in number to less than that prior to the extermination of the predators. However, since then, predators have been reintroduced into their natural habitat and they, the deer, and the vegetation are slowly making a comeback. [See PLANT–ANIMAL INTERACTIONS.]

B. Biogeochemical Processes

There are innumerable biological, geological, and chemical processes operating within the ecosphere. To list them all here would be difficult. A few will be discussed to give the reader an idea of the variety and complexity of biogeochemical processes. Table IV is a summary of processes involved with prokaryotic organisms, that is, organisms like bacteria and cyanobacteria lacking a membrane-bound nucleus.

We begin with perhaps the most important biogeochemical process of all, that of photosynthesis. Life began in the absence of oxygen, and it is likely that oxygen was a very powerful poison for the early anaerobic organisms. These organisms must have initially fed off the organic material synthesized by nonbiological processes early in Earth's history. The process of synthesizing complex organic molecules from simple inorganic ones was a critical step in evolution, enabling the biomass to grow to the level of today.

In the overall photosynthetic process, carbohydrates are synthesized from atmospheric carbon dioxide and a hydrogen donor (H_2A). The reaction is

$$nCO_2 + 2nH_2A = (CH_2O)_n + nH_2O + 2nA \tag{1}$$

TABLE IV

Biogeochemical Reactions Involving Prokaryotes[a]

Element	Process	Examples of organisms and summary of partial reactions
Carbon	CO_2 fixation	$CO_2 + H_2 \rightarrow (CH_2O)_n + A_2$ (A = O, S)
		Photoautotrophs: cyanobacteria, purple and green sulfur bacteria
		Chemoautotrophs: sulfur- and iron-oxidizing bacteria
	Methanogenesis	$COO^- + H_2 \rightarrow CH_4$
		Methanogenic bacteria
	Methanotrophy	$CH_4 + O_2 \rightarrow CO_2$
		Methanotrophic bacteria
	Fermentation	$(CH_2O)_n \rightarrow CO_2$
		Anaerobic heterotrophic bacteria
	Respiration	$(CH_2O)_n + O_2 \rightarrow CO_2$
		Aerobic heterotrophic bacteria
Sulfur	Sulfur reduction	$SO_4 + H_2 \rightarrow H_2S$
		Sulfur-reducing bacteria
	Sulfur oxidation	$H_2S \rightarrow S°$
		Purple and green sulfur phototrophs
		$S° + O_2 \rightarrow SO_4$
		Sulfur-oxidizing bacteria
Nitrogen	N_2 fixation	$N_2 + H_2 \rightarrow NH_4$
		Phototrophic bacteria, nitrogen-fixing heterotrophic bacteria
	Nitrification	$NH_4 + O_2 \rightarrow NO_2, NO_3$
		Nitrifying bacteria
	Denitrification	$NO_2, NO_3 \rightarrow N_2O, N_2$
		Denitrifying heterotrophic bacteria

[a] After Stolz *et al.* (1989).

where H_2A is H_2O in the case of the cyanobacteria and all higher plants, H_2S in the case of photosynthetic sulfur bacteria, and organic compounds in the case of nonsulfur purple bacteria.

Atmospheric carbon dioxide is fixed in eukaryotic organisms of the kingdom Plantae through a process of oxygen-producing photoautotrophy. As the term photoautotrophy implies, the source of the carbon is inorganic carbon and the reaction occurs in the presence of light. Nutrients of nitrogen, phosphorus, and bioessential trace elements are required for the production of plant matter. The net photosynthetic reactions that produce organic matter on land and in the marine environment differ because the stoichiometric proportions of carbon, nitrogen, phosphorus, and sulfur in organic matter from land vegetation and marine plankton are different. The reactions representing organic matter production in the two environments are

Marine plankton

$$106CO_2 + 16HNO_3 + 2H_2SO_4 + H_3PO_4 + 120H_2O = C_{106}H_{263}O_{110}N_{16}S_2P + 141O_2 \qquad (2)$$

Terrestrial vegetation

$$882CO_2 + 9HNO_3 + H_2SO_4 + H_3PO_4 + 890H_2O = C_{882}H_{1794}O_{886}N_9SP + 901.5O_2 \qquad (3)$$

It can be seen that biological productivity represents (among other things) a coupling mechanism linking the biogeochemical processes and cycles of the individual organic elements of carbon, nitrogen, phosphorus, and sulfur. These elements plus those of hydrogen and oxygen are the major elements constituting organic matter. These elements plus a dozen or so minor elements are necessary to the maintenance of organic structures and physiological functions of living organisms.

The reverse of reactions (2) and (3) is the process of respiration/decay. The balance of the life cycle is maintained by the complementary processes of photosynthesis in plants and respiration in animals. Biochemically, respiration breaks down complex organic molecules synthesized during the process of photosynthesis and energy is released, about 686 kcal mole^{-1} of glucose, in the generalized reaction

$$C_6H_{12}O_6 + 6O_2 = 6CO_2 + 6H_2O + energy \quad (4)$$

The respiratory oxidation of foods by animals provides energy for an animal to use in a variety of ways, including maintenance of body temperature, muscular movement, and synthesis of complex organic compounds. Oxygen is the preferred oxidant; however, in anoxic waters, sediments, and soils, other oxidants are used in the respiration/decay of organic materials. These include nitrate, iron and manganese oxides, and sulfate. Representative reactions for the oxidation of organic matter of the composition of average marine phytoplankton are

$$(CH_2O)_{106}(NH_3)_{16}H_2PO_4 + 138O_2 = 106CO_2$$
$$+ 16HNO_3 + H_3PO_4 + 122H_2O \quad (5)$$

$$(CH_2O)_{106}(NH_3)_{16}H_2PO_4 + 84.8HNO_3$$
$$= 106CO_2 + 42.4N_2 + 16NH_3 + H_3PO_4$$
$$+ 148.8H_2O \quad (6)$$

$$(CH_2O)_{106}(NH_3)_{16}H_3PO_4 + 53SO_4^{2-} = 106CO_2$$
$$+ 16NH_3 + H_3PO_4 + 53S^{2-} + 106H_2O \quad (7)$$

In these reactions, carbon dioxide, nitrogen- and phosphorus-bearing nutrients, and bioessential trace elements are returned to the environment to be used again in bioproductivity. Reaction (5) represents the processes of oxic respiration and decay of dead organic matter in an oxygenated environment. Reaction (6) is the process of denitrification, returning diatomic nitrogen originally taken from the atmosphere in nitrogen fixation reactions back to this reservoir. In reaction (7), the reduced sulfide produced is often precipitated as iron sulfide miner-

als in marine sediments. When this reaction takes place in the pore waters of sediments during the early stages of the process, carbonate ($CaCO_3$) sediments may be dissolved by the acid produced by this anaerobic process. However, in the latter stages of the reaction, the carbonate alkalinity (HCO_3^- + $2CO_3^{2-}$) of the pore waters builds up, and carbonate minerals may precipitate in the sediments.

Another set of biogeochemical processes of major importance is that of weathering, which prepares rock for erosion. The products of weathering are dissolved chemical species and solids derived from alteration of primary rock minerals. The solid alteration products are dominantly clay minerals, like kaolinite [$Al_2Si_2O_5(OH)_4$] and montmorillonite [e.g., $Ca_{0.165}Al_{2.33}Si_{3.67}O_{10}(OH)_2$]. The clay minerals and particles of original rock minerals form the suspended and bed load of streams. The dissolved products of weathering, predominantly calcium as Ca^{2+}, carbon as HCO_3^-, and silicon as H_4SiO_4, constitute the dissolved load of rivers. Ultimately, the dissolved and solid products of weathering reach the ocean via transport by rivers, as particulates in atmospheric dust fallout, in groundwater flows, and as solids transported by glaciers. Of the approximately 20 billion tons of solids and dissolved materials reaching the ocean annually from the land, more than 80% is delivered by rivers. The rivers are the main purveyors of materials to the ocean, although hydrothermal reactions at midocean ridges are significant sources of dissolved calcium, silica, and iron for the oceans.

An example of a chemical weathering reaction is that of the weathering of albite ($NaAlSi_3O_8$) to kaolinite:

$$2NaAlSi_3O_8 + 2CO_2 + 11H_2O =$$
$$Al_2Si_2O_5(OH)_4 + 2Na^+$$
$$+ 2HCO_3^- + 4H_4SiO_4 \quad (8)$$

The weathering takes place principally in the presence of carbon dioxide-charged soil and groundwaters. The ultimate source of the CO_2 is the atmosphere. However, much of the CO_2 does not come directly from the atmosphere but is produced in soils by the respiration of plants and the oxidative decay of dead plants and animals. Rainwater in equilibrium with the partial pressure of CO_2 in the atmosphere of $10^{-3.5}$ atm. contains about $10^{-5.67}$ moles of hydrogen ion per liter of water (a pH of 5.67). In soils the PCO_2 may be two or more orders of magnitude greater than that of the atmosphere because of respiration/decay processes, giving rise to acidic soil solutions with a pH of 4 or lower. It is this aggressive soil solution that is responsible for the weathering of primary rock minerals like albite. In regions of important inputs of sulfur and nitrogen oxide gases to the atmosphere from the combustion of the fossil fuels of coal, oil, and gas and other anthropogenic activities, the pH of rainwater and consequently soil water may be lower than natural values. This is the result of the oxidation and hydrolysis of the sulfur and nitrogen oxide gases in the atmosphere and their rainout as sulfuric and nitric acids, respectively.

When the products of weathering reach the ocean, the solids settle out because of gravity and are deposited on the seafloor as gravel, sand, silt, and mud. The dissolved constituents are removed from the ocean on various time scales that depend on the reactivity of the constituent. Many of the removal processes involve marine organisms. In the present oceans, dissolved calcium and bicarbonate are precipitated as carbonate minerals [(Ca, Mg)CO_3] in the skeletons of planktonic foraminifera, pteropods, and Coccolithophoridae and benthic corals, echinoids, mollusks, and coralline algae. Of the total production of skeletal carbonate in the oceans equivalent to about one billion tons of carbon per year, 80% of it is recycled in the ocean. This efficient recycling is due to the fact that although the surface ocean is oversaturated with respect to most skeletal carbonate minerals, the deep sea at varying depths is undersaturated. Much of the sinking skeletal debris dissolves in transit to the seafloor or on settling on the sea bottom. Only about 20% of the production accumulates in shallow-water and deep-sea sediments, an amount of carbon approximately equal to one-half of that brought to the oceans annually by rivers. The remaining carbon is released to the ocean/atmosphere system on precipitation of skeletal carbonate minerals according to

$$Ca^{2+} + 2HCO_3^- = CaCO_3 + CO_2 + H_2O \quad (9)$$

In actual fact, the stoichiometry of this reaction is not quite one carbon precipitated as $CaCO_3$ and one released as CO_2. In a solution of seawater pH and composition, the ratio of carbon in $CaCO_3$ to carbon in CO_2 is about 1.5:1.

Dissolved silica is also removed from the oceans in the skeletons of marine organisms. Planktonic diatoms, radiolarians, and dinoflagellates and benthic sponges utilize dissolved silica to form their tests of opaline silica (opal-A). After death of the organisms, most of the opal-A dissolves, because the oceans throughout their extent are undersaturated with respect to this phase. Only about 40% of the total annual production of skeletal silica settles out of the euphotic zone of the ocean, and of this flux, only 5% accumulates in marine sediments. The remainder recycles in the ocean. The accumulation is about equivalent to the annual input of dissolved silica to the oceans by rivers and hydrothermal solutions at midocean ridges.

In contrast to carbon and silica, which are involved primarily in biological removal processes, river-borne dissolved sodium and magnesium are removed to a significant extent by inorganic reactions. Both of these elements enter into hydrothermal reactions involving circulating seawater and basalt at midocean ridges and in the process are removed from the seawater. Sodium is also removed from the ocean by the precipitation of halite (NaCl) from seawater. This process is very important as a removal mechanism for sodium *and* chlorine but occurs only when the right set of climatic and tectonic conditions are met, and seawater in relatively isolated arms of the sea can be sufficiently evaporated to reach halite saturation. Thus, because such environments are scarce today, it is likely that sodium and chlorine are accumulating in seawater. Some magnesium is also removed from seawater by diagenetic reactions in the pore waters of sediments. The relative importance of diagenetic and hydrothermal reactions for the removal of magnesium from seawater is a topic of current scientific research and debate.

It can be concluded from the foregoing discussion of some of the processes involved with material flows in the ecosphere that the circulation is complex. The flow of materials involves a myriad of chemical, biological, and geological processes. The system is truly biogeochemical in nature. On all time and space scales, if the composition of the ecosphere is regulated, the regulation is controlled by a complex of interwoven inorganic and organic processes. Maintenance of an equable environment, including that of climate, for life on Earth is a product of this interacting and interwoven complex of biogeochemical processes and cycles, which are discussed in the next section.

III. BIOGEOCHEMICAL CYCLES

The ecosphere is a highly interactive system with matter and energy flowing between and within individual ecosystems in interconnected biogeochemical cycles. For various reasons, including those involving nutrient cycling and limitations and climate control, it is essential that the integrity of all biogeochemical cycles on Earth is preserved. Most elements and compounds on Earth are involved in these cycles. For example, Fig. 3 shows the many gaseous compounds that are produced biologically in terrestrial ecosystems and exchange with the atmosphere. These compounds enter into

biogeochemical cycles. In their transport through the atmosphere, they may react to form other compounds before returning to Earth's surface.

Biogeochemical cycles of elements and compounds are generally portrayed in the form of box models. In biogeochemical cycling models, the major compartments or reservoirs (boxes) that contain the substance of interest are defined, and the mass of the substance in each box is determined. Examples of reservoirs on a global scale are the atmosphere, ocean, and biosphere. After the reservoirs in the cycle are defined, the processes or mechanisms that lead to transport of the substance between reservoirs, that is, transport paths, are determined. An example is the process of production of organic matter by terrestrial plants that results in removal of carbon dioxide from the atmosphere and uptake in land vegetation, a reservoir in the biogeochemical cycle of carbon. Finally, the rates (fluxes) at which the substance of interest moves between boxes must be known to develop a quantitative model of a biogeochemical cycle. An example is the rate at which carbon as carbon dioxide is removed from the atmosphere and incorporated in land vegetation. This flux is 63 billion tons of carbon per year.

Individually, or in a complex web of interaction, biogeochemical cycles serve many purposes. For example, processes occurring in the nitrogen and phosphorus cycles shuffle these elements between organic and inorganic materials and maintain production of organic materials in aquatic and terrestrial ecosystems. The carbon and sulfur cycles are an important part of Earth's regulatory system for climate; carbon dioxide is a greenhouse gas and sulfur, as sulfate (SO_4) aerosol, in the troposphere and stratosphere may help to cool the planet. To provide some feeling for how cycles operate, a few biogeochemical cycles, or parts thereof, are discussed here. Emphasis is on the biologically important element carbon and the elements nitrogen, phosphorus, oxygen, and sulfur that are tied intimately to it.

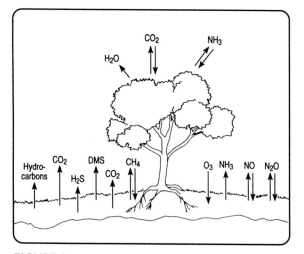

FIGURE 3 Exchange of gases between the terrestrial environment and the atmosphere. These gases in their transit through the atmosphere may react to form other compounds before returning to the Earth's surface.

A. Carbon

Carbon comprises approximately 50% of all living tissues and in the form of carbon dioxide is necessary for plants to grow. Another function that carbon dioxide serves is helping to sustain an equable climate on Earth. The concentration of carbon dioxide in the atmosphere has varied during the geologic past, but has remained within limits that permit life to exist on Earth. Carbon dioxide is cycled throughout the spheres of Earth on different time scales. For convenience, we can refer to these scales as short, medium, and long term (Fig. 4). Figure 5 is a box model of the modern biogeochemical cycle of carbon. A discussion of the behavior of the carbon cycle over a range of time scales follows. [See GLOBAL CARBON CYCLE.]

I. Short-Term Cycling: Photosynthesis and Respiration/Decay

Photosynthesis is a process that is part of the short-term carbon cycle (Figs. 4 and 5). We can look at the short-term cycling of carbon as carbon dioxide by beginning with the producers of organic carbon, the plants. Carbon in the form of atmospheric carbon dioxide is removed from the air by plants. This

removal occurs on land, for example, in forests, grasslands, and aquatic systems, and in the surface waters of the oceans. The primary producers, the photosynthetic phytoplankton in the oceans and plants on the terrestrial surface, transform inorganic carbon as carbon dioxide into organic carbon within their tissue. A simplified chemical reaction representing the photosynthetic reaction is

$$6CO_2 + 6H_2O = C_6H_{12}O_6 + 6O_2$$
$$\text{(carbon dioxide)} + \text{(water)} =$$
$$\text{(organic matter)} + \text{(oxygen)} \qquad (10)$$

Energy and nutrient substances, like phosphate (PO_4^{3-}) and nitrate (NO_3^-), are necessary for this reaction to occur. In the photosynthetic process, some of the energy from sunlight is used in the growth of plants, and some energy remains stored in the tissue of plants as carbohydrates.

Plants remove about 100 billion tons of carbon as carbon dioxide from the atmosphere each year, which is about 14% of the world's total atmo-

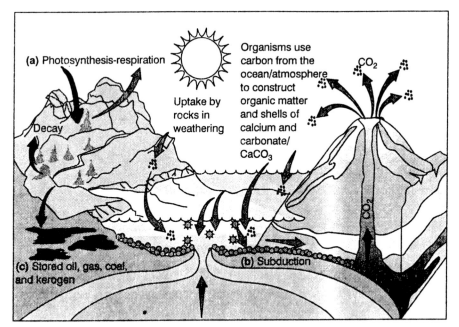

FIGURE 4 Diagram of the biogeochemical cycle of carbon prior to human interference showing the short-term photosynthesis–respiration part of the cycle (a); the long-term cycle involving accumulation of organic carbon and $CaCO_3$ in marine sediments, their subduction, alteration, and return of carbon dioxide via volcanism (b); and the medium-term situation involving storage of carbon in organic materials in sedimentary rocks (c). The diagram does not show the important anthropogenic perturbations of the carbon cycle, those of fossil fuel burning, cement manufacturing, and deforestation.

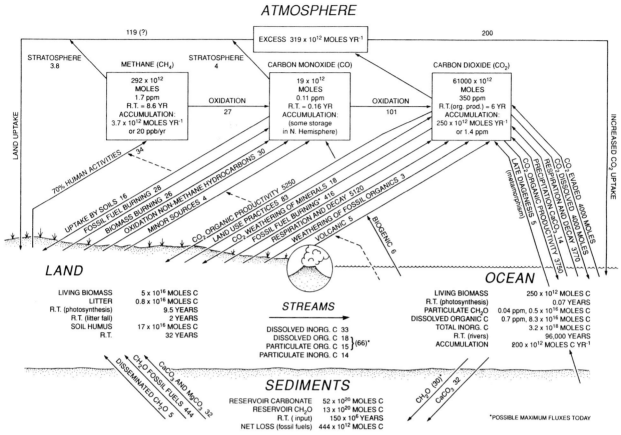

FIGURE 5 Global biogeochemical cycle of carbon. Fluxes in units of 10^{12} moles of carbon (C) per year. One mole of carbon = 12 grams.

spheric carbon. Much of the carbon dioxide taken from the atmosphere during photosynthesis in the terrestrial and oceanic realms is returned to the atmosphere during respiration and decay of plants. The reaction is the reverse of that of photosynthesis; oxygen is consumed, and energy, carbon dioxide, and nutrients are released. The yearly removal rate of atmospheric carbon dioxide in photosynthesis is slightly larger on land than in the ocean.

After photosynthesis, carbon may be next transferred to a consumer organism when plants are eaten for food. The carbon stored in the tissue of the plant enters an animal's body and is used as energy or stored for growth. Land animals, such as cows and deer, are primary consumers of this stored energy. Aquatic plants are eaten by zooplankton and larger consumer animals. Whales, for example, feed directly on the animal plankton krill.

When an animal breathes, some of the carbon taken up is released from the body as carbon dioxide gas in the process of energy expenditure (respiration). Aside from carbon stored in aboveground living and dead vegetation, carbon is also present in the root systems of terrestrial plants. When plants die, some of this carbon may be released as carbon dioxide or methane gas to the soil atmosphere, or may accumulate in the soil as dead organic material. This dead organic matter may be ingested by consumers, such as insects and worms living in the soil.

Some of the organic carbon generated in land environments is weathered and eroded, and the organic debris is transported by streams to the ocean. In the ocean, some of this debris, along with the organic detritus of marine plants, is sedimented on the ocean floor and accumulates in the sediments. However, some of the terrestrially derived organic matter is respired in the ocean to carbon dioxide; this carbon dioxide leaves the ocean and is transported over the continents, where it is utilized again in the production of land plants.

2. Long-Term Cycling: The CaCO₃–SiO₂ Connection

Long-term cycling of carbon dioxide (Figs. 4 and 5) involves a series of processes dating back to the beginning of plate tectonics and includes not only the land and ocean reservoirs but also that of limestone rocks. Limestone rocks are mainly composed of calcium carbonate ($CaCO_3$) and are the fossilized skeletal remains of marine organisms or, less commonly, chemical precipitates of calcium carbonate. Limestones are great storage containers for carbon; most of the carbon near Earth's surface is found in these rocks or in fossil organic matter in sedimentary rocks. Weathering and erosion of the surface results in the leaching of dissolved calcium, carbon, and silica (SiO_2) from limestones and rocks containing calcium silicate ($CaSiO_3$). The chemical weathering reactions are

$$CaCO_3 + CO_2 + H_2O = \\ Ca^{2+} + 2HCO_3^- \qquad (11)$$
$$\text{and}$$
$$CaSiO_3 + 2CO_2 + H_2O = Ca^{2+} + \\ 2HCO_3^- + SiO_2 \qquad (12)$$

The dissolved substances produced by these weathering reactions are transported to the ocean by rivers. In the ocean, the dissolved compounds are used to form the inorganic skeletons of benthic organisms and plankton, which are composed of calcium carbonate and silica. During formation of the calcium carbonate skeletons, the carbon dioxide derived from the weathering of limestone is returned to the atmosphere in a chemical reaction that is the opposite of that for the weathering of limestone.

When marine animals and plants die, their remains settle toward the seafloor, taking the carbon stored in their bodies with them. En route, their organic matter is decomposed by bacterial decomposers, just as on land, and some shells may dissolve, thus turning animal and plant organic and skeletal matter back into dissolved carbon dioxide, nutrients, calcium, and silica in the ocean. This carbon dioxide is stored in the deeper waters of the oceans for hundreds to a thousand or so years before being returned to the atmosphere by the upwelling of deep ocean waters. [*See* OCEAN ECOLOGY.]

Some of the animal and plant plankton sink to the bottom, where the carbon in the organic matter and shells escapes degradation and becomes part of the sediment. As the seafloor spreads through plate tectonics, the sediments containing the remains of marine plants and animals are carried along to subduction zones, where they are transported down into the mantle. At the severe pressures and high temperatures in the subduction zones, organic matter is decomposed and calcium carbonate reacts with the silica found in the subducted rocks and forms rocks containing calcium silicate. This metamorphic reaction (the "Urey reaction") is the reverse of that of weathering of calcium silicate. During this process of metamorphism, carbon dioxide is released and makes its way into the atmosphere in volcanic eruptions and via hot spring discharges. Once in the atmosphere, carbon dioxide can then combine with rainwater. The rainwater falls on the land surface and seeps down into the soils, where it picks up more carbon dioxide from decaying vegetation in the soils. This water, enriched in carbon dioxide, is very aggressive. It weathers and dissolves the compounds of calcium and silica found in rocks of the continents, and the cycle begins again. This series of processes has been active for at least 600 million years, since the advent of the first organisms that made shells, and was significant even earlier in Earth's history when calcium carbonate was deposited in the ocean by inorganic processes.

One outcome of changes in the rates of processes in the long-term biogeochemical cycling of carbon is that atmospheric carbon dioxide has varied in a cyclic fashion during the last 600 million years of Earth's history. Robert A. Berner of Yale University and colleagues have developed models of the carbon cycle to calculate these variations. Figure 6 shows one such calculation.

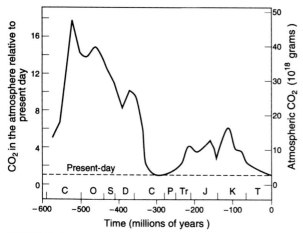

FIGURE 6 Model calculation of atmospheric carbon dioxide during Phanerozoic time.

The high atmospheric carbon dioxide levels of the Cretaceous and middle Paleozoic are a result mainly of intense plate tectonic activity, a time of increased metamorphism of limestone and release of carbon dioxide to the atmosphere from volcanoes. These high carbon dioxide periods are referred to as Hot Houses ("greenhouses"). The lower atmospheric carbon dioxide levels of the early Phanerozoic, the Carboniferous through Triassic, and much of the Cenozoic are an outcome of less intense plate tectonic activity and enhanced removal of carbon dioxide from the atmosphere by weathering. These three intervals in time are called Ice Houses. Other factors, like the evolution of plants and their effect on weathering, also play a role in regulating atmospheric carbon dioxide levels over the long term.

An important point is that atmospheric carbon dioxide has varied by a factor of about 10 during the last 600 million years. This variation certainly has had climatic implications, because carbon dioxide is an important greenhouse gas. In fact, the planet for much of its Phanerozoic history had a different atmospheric composition and a more equable climate from that of today.

3. Medium-Term Cycling: The Organic Matter–Oxygen Connection

Medium-term cycling of carbon dioxide (Figs. 4 and 5) involves organic matter in sediments, coal, oil, and gas and atmospheric oxygen. It commences

as with short-term cycling, with the removal of carbon dioxide from the atmosphere by its incorporation into plants and the accumulation of the dead plant and animal carbon in sedimentary organic matter. Sedimentary organic matter is the dead and fossilized remains of plants and animals. When dispersed throughout a sedimentary rock, it is called kerogen. Shales are very fine-grained sedimentary rocks that are often rich in kerogen. Coal, oil, and gas deposits are also the remains of the soft tissues of plants and animals, but they represent large and segregated accumulations of these materials in an altered condition.

Coal is derived mainly from terrestrial plant material that is often deposited in swampy environments. The plant material is altered during burial of the swamp sediments, and if buried deep enough it may be substantially changed owing to the increased temperatures and pressures at depth. Different types of coal are formed because of varying conditions of temperature and pressure. Anthracite is a hard, dense coal formed by alteration of plant material at a relatively high temperature and pressure, whereas bituminous coal is formed under less intense conditions. Peat used as a fuel for fires is little altered plant material, containing high amounts of carbon, that has not been buried deeply. Peat is an important component of tundra areas in the Northern Hemisphere.

Oil and gas represent highly altered organic matter, principally microscopic plant matter, the altered remains of marine phytoplankton that have been sedimented on the seafloor. During burial, these organic materials are broken down at elevated temperature and pressure to form oil and gas. The oil and gas may travel many miles in the subsurface before coming to rest in large accumulations in the voids of rocks. Often oil and gas are formed in shales, but leave these rocks and migrate to more coarse-grained rocks like sandstones and limestones. It is in these latter rocks that the great Cretaceous and Cenozoic oil and gas reservoirs of the world are found, like those of the Persian Gulf.

These deposits of coal, oil, and gas represent organic carbon that has escaped respiration and decay and thus carbon dioxide that has been removed from the atmosphere. The same is true of the kerogen dispersed as fine-grained materials in the sedimentary rocks. Because of their burial, the oxygen normally used to decay these materials remains in the atmosphere. The carbon in the coal, oil, and gas deposits and the kerogen are recycled back into the atmosphere to return carbon dioxide to that reservoir. At the same time, the oxygen that accumulated in the atmosphere is removed. All of this is accomplished when these fossil fuel deposits and kerogen are uplifted by plate tectonic forces after millions of years of burial and are exposed to the atmosphere. When this occurs, the oxygen that previously accumulated in the atmosphere reacts with the coal, oil, gas, and kerogen. The reaction involves the oxidative decay of these organic materials, which results in the removal of oxygen from the atmosphere and the return of carbon dioxide to the atmosphere. The ongoing dynamic cycle is complete.

Fossil fuel is a nonrenewable energy source, because coal, oil, and gas deposits take millions of years and specific environmental conditions to form. The mining of these deposits brings these materials back to the surface much more rapidly than would natural processes. The stored energy from the long dead organisms is released in the form of heat when the coal, oil, and gas are burned. This fossil fuel energy keeps us warm, powers our cars, and moves the machinery of industry. It is also a main cause of environmental pollution, because a by-product of fossil fuel burning is the release of gases and particulate materials into the environment. Climatic change is an important potential global environmental effect of the release of carbon and other gases to the atmosphere by combustion activities.

4. Summary

Carbon is found in all four spheres or reservoirs. It is essential to every life-form and occurs in all organic matter in the ecosphere. It is found as the gases carbon dioxide, methane, and several other compounds in the atmosphere, and it occurs as carbon dioxide dissolved in lakes, rivers, and oceans in the hydrosphere. In the crustal part of the lithosphere, it is found as calcium carbonate originally deposited on the seafloor, as kerogen

dispersed in rocks, and as deposits of coal, oil, and gas. It is because carbon is stored in the large sedimentary reservoirs of limestone and fossil organic carbon, and not in the atmosphere, that life on Earth is possible. If all this carbon were stored in the atmosphere, the result would be a heating of the atmosphere because of the increased absorption of the Sun's energy; the greenhouse effect would be very strong, and Earth would probably have a temperature like that of Venus!

B. Oxygen

The oxygen biogeochemical cycle is diagrammed in Fig. 7. Oxygen comprises 20.9% of the gases of the atmosphere, and its cycling is strongly coupled to that of carbon. Oxygen is produced by plants during photosynthesis, when carbon dioxide is consumed, and it is removed by respiration and decay, when carbon dioxide is produced. This is a short-term, nearly balanced cycle on land, because the amount of oxygen produced yearly by land plants is about equivalent to the amount used in the process of respiration and decay in the terrestrial

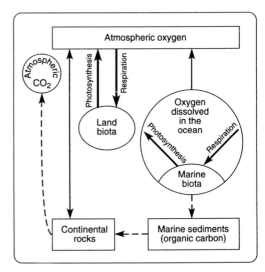

FIGURE 7 The biogeochemical cycle of oxygen. The cycle is strongly coupled to that of carbon. The arrows represent the flows, or fluxes, of oxygen from one box (reservoir) to another. The heavier the arrow, the larger the flux. The dashed line represents the flow of carbon in sedimentation on the ocean floor, burial in sediments, and uplift by plate tectonic processes. When uplifted, carbon is oxidized by atmospheric oxygen, and carbon dioxide is released back to the atmosphere.

realm. It takes about a decade for oxygen and carbon dioxide to recycle through living plants. However, there is a little leakage of nonrespired organic matter from the land to the ocean. This organic matter, the leaves and trunks of trees and smaller-sized organic debris, is carried to the oceans by rivers, where some of it is deposited in the sediments accumulating in the ocean.

In the oceans, the production of oxygen by phytoplankton slightly exceeds consumption of this gas in respiration and decay. As a result, oxygen is released to the atmosphere. The organic carbon not decayed by this oxygen, as well as some of that brought to the oceans by rivers, is deposited on the seafloor and accumulates in the ocean sediments. If this accumulation process continued too long, all the carbon dioxide in the atmosphere would disappear in less than 10,000 years, and the oxygen content of the atmosphere would double in less than several million years. Both of these changes are not likely to have occurred, at least on these time scales, because they would imply a cessation of life on Earth. Loss of atmospheric carbon dioxide would inhibit the photosynthetic process, and a large increase in oxygen would lead to the massive burning of plant life in forest and grassland fires. Fortunately, as mentioned earlier, oxygen is removed by weathering of fossil organic carbon and other materials found in rocks exposed on land, and during this process, carbon dioxide is returned to the atmosphere. The overproduction of oxygen in the oceans and uptake by rocks on land balance the cycle.

One outcome of changes in the rates of biogeochemical processes involved in the medium-term cycling of carbon is that enhanced burial of organic carbon in sediments implies the possibility of accumulation of oxygen in the atmosphere. Times of high organic carbon burial in the past should have given rise to high atmospheric oxygen levels. This seems to be the case. Figure 8 is a model calculation of atmospheric oxygen concentration variations during the Phanerozoic. Just before the Carboniferous, vascular plants evolved and spread over the continents. Their organic remains were a new source of organic matter resistant to degradation by atmospheric oxygen. During the Carboniferous

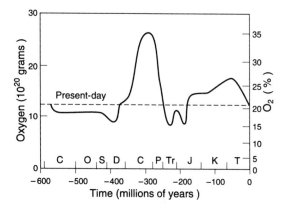

FIGURE 8 Model calculation of atmospheric oxygen during Phanerozoic time.

and Permian, large quantities of vascular plant organic matter were buried in the vast coastal lowlands and swamps of the time. This material became the coal deposits mined from rocks of Carboniferous and Permian age today. This large accumulation of organic matter gave rise to the high atmospheric oxygen levels of the late Paleozoic. Coal deposits are also important in Cretaceous and early Cenozoic rocks, another time of high atmospheric oxygen concentrations.

C. Nitrogen

Nitrogen is important to life, because it forms part of the molecules that make up living things, such as amino acids, the building blocks of proteins, and DNA. The nitrogen in proteins acts to bond together various amino acids to form the protein structure. The atmospheric abundance of nitrogen is very large compared to that found in the oceans or in rocks. Approximately 78% of the atmosphere is composed of diatomic nitrogen gas (N_2). Plants and animals cannot use nitrogen directly from the air; it must first be changed into the compounds ammonia (NH_3) or nitrate (NO_3^-) by certain bacteria that live in soils or water before it can be used by plants and incorporated into their tissue. Nitrogen fixation is the process of converting atmospheric nitrogen to usable nitrogen. One reaction representing the nitrogen-fixation process in which the bacterium *Azobacter* plays a role is

$$2N_2 + 6H_2O = 4NH_3 + 3O_2 \qquad (13)$$

The ammonia (NH_3) reacts with water to make ammonium ion (NH_4^+), which is then incorporated into plant tissue. Ammonium may also take part in another bacterially mediated process, that of nitrification. In nitrification, ammonium is converted to nitrate (NO_2^-) and then to nitrate (NO_3^-) by bacteria. The nitrate can then be used in plant production.

The roots of land plants take nitrogen from the soil, and the nitrogen is used as a nutrient element for growth. In the ocean, phytoplankton obtain their nitrogen for growth from the water and have complicated methods to use the dissolved nitrogen. Animals receive the nitrogen they need when plants are eaten. When plants or animals die, bacteria and fungi decompose the remains. In the process of denitrification, microbes use nitrogen as a food source and produce nitrogen gas (N_2) and nitrous oxide (N_2O) as by-products. An example of a denitrification process is that of the production of diatomic nitrogen gas by the bacterium *Pseudomonas*

$$4NO_3^- + 2H_2O = 2N_2 + 5O_2 + 4OH^- \quad (14)$$

Decomposition releases nitrogen back into soils or waters, which, if converted to ammonia by nitrogen-fixing bacteria, can be a fertilizer for plant growth. Some of the nitrogen is transferred back to the air as nitrogen and nitrous oxide gases, and the cycle is complete.

Much of the nitrogen used in productivity in both terrestrial and aquatic systems is recycled. Little new nitrogen is added via streams and the atmosphere. Of the total nitrogen utilized annually in terrestrial production, only about 25% of it is derived directly from the atmosphere in nitrogen fixation; the rest is recycled. In the ocean, recycling is even more efficient, with only about 1% of the nitrogen consumed yearly coming from river runoff or the atmosphere. The rest enters the productive shallow waters of the ocean through the upwelling of deep nitrogen-enriched water. These deep waters have been enriched in nitrogen by the sinking of dead organic matter out of the euphotic zone of the ocean and its decomposition in deep waters.

The modern global biogeochemical cycle of nitrogen is very complex, involving more than seven major gaseous and aerosol species of nitrogen in the atmosphere, three major species in aquatic systems, and nitrogen in organic matter and absorbed to minerals in rocks (Fig. 9). Bacterial processes in soils and aquatic systems play an important role in the transfer of nitrogen about Earth's surface.

The natural biogeochemical cycle of nitrogen is being heavily impacted by anthropogenic activities (Fig. 9). Nitrogen is stored in Earth in the form of dispersed organic matter, coal, oil, and gas. When the fossil fuels are burned, nitrogen gases may be released to the environment. In the exhaust streams of cars, nitrogen oxide gases are generated from atmospheric nitrogen and constitute an important component of urban smog and contribute to acid rain. Agricultural crops use nitrogen as a nutrient in growth. When these crops are harvested and taken to stores for our consumption or utilized to feed domestic animals, the nitrogen is removed with the crop. The nitrogen taken up during crop growth is not directly returned to the soil during decomposition of the crop plants. This nitrogen is not available for future plant growth. The missing nitrogen is replaced by artificial fertilizers of urea and ammonium nitrate made in an industrial process, or in some cases by sludge from waste treatment plants or by animal and human manure, and spread on the land surface. Frequently, excessive nitrogen is used on the fields, and this excess may get washed into streams and rivers causing pollution problems. Also, the decomposition of nitrogen fertilizers applied to the land surface produces nitrous oxide, a greenhouse gas. Production of nitrous oxide also occurs during the decomposition of sewage released into the environment and in areas of eutrophication.

It is possible that the additions of nitrogen to the landscape from fossil fuel combustion and agriculture are stimulating plant production in terrestrial and aquatic systems and leading to accumulation of biomass. This accumulated biomass may represent a significant fraction of the CO_2 that humans are adding to the atmosphere from land use activi-

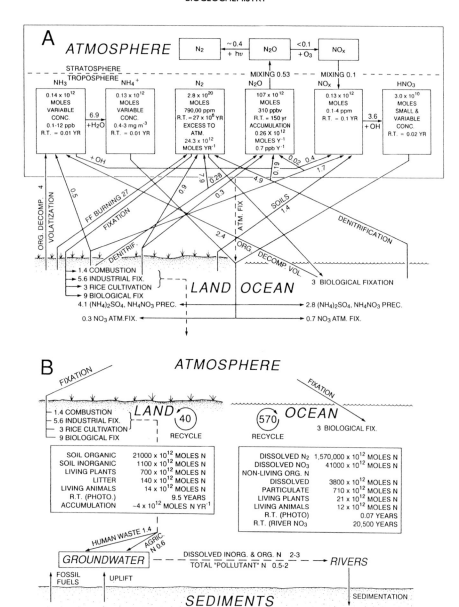

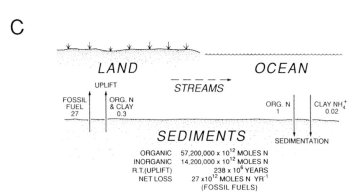

FIGURE 9 Global biogeochemical cycle of nitrogen. Fluxes in units of 10^{12} moles of nitrogen (N) per year. One mole of nitrogen = 14 grams. (A) Earth surface–atmosphere exchange; (B) land–ocean reservoirs; (C) sediment–land–ocean exchange.

ties (e.g., deforestation) and the burning of coal, oil, and gas. It may emerge, ironically, that humankind is inadvertently engaging in activities that help to balance out our emissions of the greenhouse gas CO_2 to the atmosphere. [*See* NITROGEN CYCLE INTERACTIONS WITH GLOBAL CHANGE PROCESSES.]

D. Phosphorus

Phosphorus is important to life because it is an integral part of the RNA and DNA of living organisms. It is found in both the organic tissue and skeletons of organisms as phosphate (PO_4). Unlike carbon and nitrogen, phosphorus is not present in the atmosphere as a gas and is not transferred about Earth's surface in gaseous form. The nutrient is leached by rain from rocks containing phosphate, absorbed by plants from soil waters, and then passed on to animals when they feed. It may be recycled from excretions of animals and from dead plants and animals through organic decay processes and returned to the soil. Phosphate is also transported to the oceans by rivers, where it is used by benthic organisms and marine plankton. Some phosphate settles to the ocean bottom in dead organic matter and inorganic particles and becomes part of the sediments. These sediments are eventually uplifted and phosphorus is returned to the land surface to begin its trip again. Figure 10 shows the modern global biogeochemical cycle of phosphorus involving the land, ocean, and sediments.

Similar to nitrogen, most of the phosphorus consumed in organic production in terrestrial and aquatic ecosystems is recycled. Phosphorus is considered to be the limiting nutrient in many continental ecosystems because it must be ultimately derived from the slow process of weathering of surface rocks. There is no large atmospheric reservoir of phosphorus as there is for nitrogen. Whether phosphorus is the limiting nutrient in the ocean is controversial, as nitrogen, iron, and other elements have also been proposed as limiting. For the global ocean over a long time, it is likely that phosphorus is the limiting nutrient, if for no other reason than that nitrogen can be obtained from the atmosphere by nitrogen fixation.

Phosphorus as inorganic phosphate compounds is an important element used in fertilizing agricultural crops. When the crops are harvested, the nutrient must be returned to the soils for the growth of future crops. Phosphorus is mined from the ground where it occurs in commercial deposits of calcium phosphate rock and is added to cropland as fertilizer. Generally an excess of phosphorus is added to agricultural land to stimulate production of crops. Like fertilizers of nitrogen, it eventually may be washed into aquatic systems. Because phosphorus can be a limiting nutrient in rivers, lakes, and coastal marine environments, phosphorus addition to these systems can act as a potent fertilizer. Rapid growth of plants in both freshwater and marine environments may occur and lead to problems of eutrophication.

E. Sulfur

The global biogeochemical cycle of sulfur involves three major atmospheric components, those of reduced gaseous forms of sulfur, like dimethylsulfide (DMS) and carbonyl sulfide (OCS), sulfur dioxide (SO_2) gas, and sulfate (SO_4) aerosol. In aquatic systems, dissolved sulfate (SO_4^{2-}) is the major chemical compound found, and in sediments, sulfur occurs in the minerals pyrite (FeS_2) and gypsum ($CaSO_4 \cdot 2H_2O$, or its dehydrated form, anhydrite, $CaSO_4$) and in organic matter.

The modern global biogeochemical cycle of sulfur is shown in Fig. 11. Dimethylsulfide is produced by algae in the surface waters of the ocean. The DMS concentration in the ocean is very low, but the dissolved gas is found nearly everywhere near the sea surface where it may escape into the atmosphere. It is a trace gas comprising much less than 1% of the gases in the atmosphere. Once in the atmosphere, it takes only a few days before DMS is oxidized to sulfate and, along with other chemical species in the atmosphere, condenses into small aerosol particles. These atmospheric sulfate aerosols act as nuclei on which rain droplets form, thereby facilitating formation of clouds and rain. The aerosol particles formed in the troposphere produced by oxidation of dimethylsulfide by themselves and acting as seeds for cloud formation affect

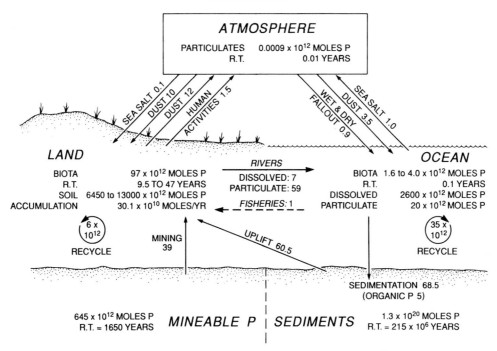

FIGURE 10 Global biogeochemical cycle of phosphorus. Fluxes in units of 10^{10} moles of phosphorus (P) per year. One mole of phosphorus = 31 grams.

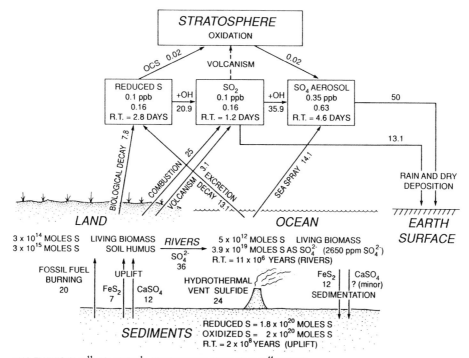

FIGURE 11 Global biogeochemical cycle of sulfur. Fluxes in units of 10^{11} moles of sulfur (S) per year. One mole of sulfur = 32 grams.

the radiant energy balance of the oceanic atmosphere. Cloud formation over the oceans may lead to reflection of incoming solar radiation and have a cooling effect on the troposphere and surface of Earth. Furthermore, these aerosol particles interact with sunlight and water and produce some acidity in rain.

As seen in Fig. 11, another reduced sulfur compound, carbonyl sulfide, is produced in the ocean from dissolved organic sulfur compounds and escapes the sea surface to the atmosphere. This gas is chemically inert in the troposphere, but in the stratosphere it too is oxidized to sulfate. This process has led to a blanket of sulfate aerosols, the sulfate veil or Junge layer, in the stratosphere surrounding Earth. This veil reflects incoming solar radiation and helps to cool the planet, thus maintaining an equable climate.

Two more features of the biogeochemical cycle of sulfur are important to climatic change. Volcanic explosions, like that of Mount Pinatubo in the Philippines in 1991, can lead to the injection of sulfate aerosol into the stratosphere, resulting in a short-term cooling of Earth's climate. Also, sulfur dioxide gas released from fossil fuel burning can form atmospheric sulfate aerosol, particularly in the Northern Hemisphere where most heavily industrialized nations are located, and potentially act as a cooling agent for the planet. Thus, aside from the role it plays in the acidity of rain, sulfur is an important player in regulating climate.

Eventually sulfur that is released into the atmosphere from the surface of Earth is returned to the ocean and land surface in precipitation. This flux completes the cycle.

The global sulfur cycle has been dramatically perturbed by the fossil fuel burning activities of human society. Prior to human intervention in the global sulfur cycle, the net transport of sulfur in the atmosphere was from ocean to land. Because of the combustion of fossil fuels and smelting of sulfide ores, the global situation has been reversed, and net transport is now from land to the ocean through the atmosphere. The flux of sulfur dioxide to the atmosphere from burning of fossil fuels in certain regions of the world greatly exceeds natural fluxes of the gas. On a global scale, the principal flux involved in the exchange of sulfur between Earth's surface and the atmosphere is that of fossil fuel burning (Fig. 11). One environmental problem related to this large flux is that of acid deposition.

F. Concluding Remarks

Chemical, biological, and biogeochemical cycles are found in all "spheres" of the planet Earth. These cycles interact with one another and are essential for life on the planet. The activities of people have significantly impacted these cycles and the Earth surface system. These human-induced changes in biogeochemical cycles can be the root causes of global environmental change on the time scale of human societies. The extent to which these cycles can accommodate change brought about by fluxes from human activities and their influence on and response to global environmental change are subjects of considerable scientific interest and practical importance today. The following discussion briefly explores in greater detail these human impacts on biogeochemical cycles and global environmental change induced by human activities. The material has become an important part of the discipline of biogeochemistry.

IV. HUMAN FORCING AS A FACTOR IN BIOGEOCHEMISTRY

Human forcing, particularly in the twentieth century, has become an important factor in biogeochemical studies. There is no doubt that the global environment of Earth has changed in the past. Change is more the nature of the planet than constancy. During the geologic past, continents have broken apart to drift across climatic zones, and they have rejoined to form great mountain ranges. Ocean basins have been created and then have been destroyed by subduction. The planet has been warmer and colder than present. Continental ice sheets have advanced and retreated. Atmospheric composition has changed over time, as have atmospheric and oceanic circulation patterns. The Sun's energy output has varied and thus the amount of radiant energy reaching Earth's surface. River water discharge and dissolved and suspended loads have not been constant over geologic time. Sea level has risen and fallen. Ecosystems and species have come and gone and migrated with climatic change. All aspects of the surface environment of Earth and the type and intensity of biogeochemical processes have changed over geologic time. [*See* GLOBAL ANTHROPOGENIC INFLUENCES.]

The physical and chemical changes in the environment have led to and constrained biological evolution. In turn, the evolution of species and ecosystems has interacted with the physicochemical system of the planet to produce the surface environment of the Earth of the past and as we know it today. The biota and environment are a product of coevolution, although the properties of the environment also exhibit cyclical characteristics. Gaia is a term applied to this interactive set of organic and inorganic processes that regulate the natural environment of Earth, and Earth system science or geophysiology is the study of this interactive system.

Table V lists changes in the conditions of Earth's surface environment owing to natural causes during late Pleistocene glacial–interglacial stages. It documents the geologic state of change in some environmental conditions during human evolution. Recent information on rates of natural change shows that some dramatic global change can be rapid, on the time scale of several decades to centuries. Thus it is very unlikely that the surface environment of the planet will not change in the future. The natural forcings of environmental change, such as plate tectonic motions and intensity, bolide impacts, and variations in the orbital parameters of the planet, can cause changes in the type and intensity of biogeochemical processes. In turn, the processes can act as positive or negative feedbacks on the forcing and enhance or ameliorate, respectively, change brought about by changes in forcing.

Table V also gives historical changes over the past century in some surface environmental parameters of the planet to compare with glacial–interglacial change in these parameters. It can be seen that these many parameters have changed more rapidly over the last several centuries than in more recent geologic time. The more rapid change is to a sig-

nificant degree a result of human forcings on biogeochemical processes in the ecosphere. Aside from historical data, the table also shows qualitatively the direction of change of environmental parameters in the late twentieth century and projections for the early part of the twenty-first century. For example, fossil fuel combustion and land-use activities are adding to the atmosphere on the order of 7 billion tons of carbon per year, resulting in a 0.4% per year increase in atmospheric CO_2 concentration in the early 1990s. In the early twenty-first century, because of these anthropogenic emissions, atmospheric CO_2 concentration will continue to increase and the probability of global planetary warming will be increased. Likewise, the global sulfur emissions to the atmosphere of nearly 90 million tons per year are leading to acidification of rain. Despite controls on these emissions by many developed countries of the world, the continuous strong global dependence on fossil fuels as an energy source on into the twenty-first century implies an increase in acidification of precipitation. Furthermore, because of point and nonpoint sources of pollutants, the coastal zones of the world are likely to receive increased burdens of organic matter, nutrients, and chemical pollutants resulting in continued stresses on these ecosystems.

Although the interactions between human activities and the total Earth surface system and its physical environment that harbors life are not well understood, these interactions are substantial, may be cumulative, and in many cases are accelerating. For all the environmental conditions of Table V, there is still much to be learned about their future course and the effects they have on the environment and biogeochemical processes worldwide.

Human-induced global environmental change is a consequence of direct and indirect rapid modification of the environment by human activities of urbanization, cultivation, and industrialization. The distribution and rates of growth of the human population and the demand for economic growth with concomitant utilization of resources are forces that are acting as agents of global environmental change. The growth of the human population, for some scientists and policy makers, is the most im-

TABLE V
Historical and Present Global Environmental Conditions of the Earth's Surface Environment and Recent and Future Changes in These Conditions. (y) Indicates a Quantity Not Accurately Known at Present[a]

Component	Conditions			Change in rates	
	Glacial	Interglacial	Present	Last 100 years (average)[b]	Future[b]
CO_2 concentration	180 ppmv	280 ppmv	356 ppm	↑ 0.4% y^1	↑
CH_4 concentration	0.35 ppmv	0.7 ppmv	1.7 ppmv	↑ 0.6% y^{-1}	↑
N_2O concentration	185 ppbv	275 ppbv	310 ppbv	↑ 0.02% y^{-1}	↑
SO_2 emissions	—	—	90×10^9 kg y^{-1}	↑ 1% y^{-1}	↑
DMS emissions	$60–400 \times 10^9$ kg y^{-1}	40×10^9 kg y^{-1}	40×10^9 kg y^{-1}	↑ ?	↑ ?
Temperature	284 K	288 K	288 K	↑	↑
Mineral aerosol flux	$5–10 \times$ (y)	(y)	2×10^{12} kg y^{-1}	↑	↑
Land runoff (H_2O) flux	2×10^{16} kg y^{-1}	3.7×10^{16} kg y^{-1}	3.7×10^{16} kg y^{-1}	↓ 0.1% y^{-1}	↑ [c]
Particulate erosion products in runoff flux	$\sim 1 \times 10^{13}$ kg y^{-1}	$\sim 7 \times 10^{12}$ kg y^{-1}	1.5×10^{13} kg y^{-1}	↑	↑ [c]
Dissolved salts in runoff flux	—	—	4×10^{12} kg y^{-1}	↑	↑
N_{total} riverine flux	—	1.4×10^{10} kg N y^{-1}	3.5×10^{10} kg N y^{-1}	↑	↑
P_{total} riverine flux	—	1.4×10^9 kg P y^{-1}	3×10^9 kg P y^{-1}	↑	↑
$C_{organic}$ riverine flux	—	4×10^{11} kg C y^{-1}	8×10^{11} kg C y^{-1}	↑	↑
Total marine net primary production	>(y)	(y)	3.8×10^{13} kg y^{-1}	↑	↑

[a] Modified from Mackenzie *et al.* (1991).
[b] ↑ -increase; ↓ -decrease.
[c] Damming will decrease runoff and particulate erosion flux; temperature increase will increase runoff.

portant factor involved in human-induced environmental change.

Human activities stemming from population growth and the demand for economic growth include fossil fuel and biomass combustion, land-use change, agricultural practices, and halocarbon and other synthetic chemical production and release. All of these activities result in emissions of chemicals to soil and aquatic systems and to the atmosphere. Furthermore, fossil fuel burning depletes a nonrenewable resource. Biomass burning, as presently done in most countries, depletes a potential renewable resource. Land-use changes lead to the conversion of forests to cropland and rangeland, wetlands to agricultural and urban uses, and the loss of habitat and biological diversity. Agricultural activities lead to the release of synthetic pesticides and nitrogenous and phosphorus-bearing fertilizers to the environment. Synthetic chemical production and use in industrial practices result in the venting of these chemicals or their by-products to aquatic systems and to the atmosphere. The production and utilization of chlorofluorocarbons and other halocarbons are promoting growth of the atmospheric concentration of greenhouse gases and stratospheric ozone-depleting chemicals. All of these anthropogenic activities are modifying the biogeochemical processes and cycles that were part of the ecosphere prior to human intervention in this system. Thus human forcing has become an important part of the discipline of biogeochemistry. A summary of global environmental change problems owing to human activities is given in Table VI.

TABLE VI

Some Problems of Global Environmental Change Owing to
Human Activities[a]

- Climatic changes from anthropogenic inputs to the atmosphere of CO_2 and other greenhouse gases, and SO_2 and its fate
- Disruptions in biogeochemical cycles of C, N, P, S, trace metals, and other elements
- SO_2 and NO_x emissions and acid deposition
- NO_x and VOC emissions and development of photochemical smog and tropospheric ozone
- Emissions of halocarbons and alterations in the stratospheric ozone layer and associated effects on ultraviolet radiation
- Increasing rates of tropical deforestation and other large-scale destruction of habitat, with potential effects on climate
- Disappearance of biotic diversity through explosive rates of species extinctions
- The global consequences of distribution and application of chemicals potentially harmful to the biota, e.g., pesticides
- Cultural eutrophication from agricultural runoff and municipal and industrial sewage disposal
- Exploitation of natural resources and consequent waste disposal and chemical pollution problems
- Water quality and usage
- Waste disposal: municipal, toxic chemical, and radioactive

[a] Population growth at 1.6–2.0% per year in the last 40 years is a factor common to all these problems.

Glossary

Autotrophy Biogeochemical pathway by which organisms convert inorganic into organic materials.

Biogeochemical cycle Representation of biological, geological, and chemical processes that involve the movement of an element or compound about the surface of the Earth.

Biogeochemistry Discipline that links various aspects of biology, geology, and chemistry to investigate the surface environment of Earth.

Biosphere The living and dead organic components of Earth.

Ecosphere System that includes the biosphere and its interactions with the physical systems of the hydrosphere, atmosphere, and lithosphere.

Ecosystem Complex of a community or group of communities and the environment that functions as an ecological unit in nature.

Eukaryote Organism having as its fundamental structural unit a cell that contains specialized organelles in the cytoplasm, a membrane-bound nucleus, and a system of cell division by mitosis or meiosis.

Exogenic system Outer sphere of Earth that includes the atmosphere, hydrosphere, biosphere, and shallow lithosphere.

Heterotrophy Biogeochemical pathway in which organic substrates are used by organisms to make organic matter.

Photosynthesis Process of synthesis of complex organic materials (e.g., carbohydrates) from carbon dioxide, water, and nutrients, using sunlight as a source of energy and with the aid of chlorophyll and associated pigments.

Prokaryote Any cellular organism that has no membrane about its nucleus and no organelles in the cytoplasm except ribosomes. Its genetic material is in the form of single, continuous strands forming coils or loops, characteristic of all organisms in the kingdom Monera, such as bacteria and cyanobacteria.

Bibliography

Anderson, D. L. (1989). "Theory of the Earth." Blackwell Scientific Publications, Boston, MA.

Berner, E. K., and Berner, R. A. (1987). "The Global Water Cycle: Geochemistry and Environment." Englewood Cliffs, N.J.: Prentice–Hall.

Bolin, B., and Cook, R. B., eds. (1983). "The Major Biogeochemical Cycles and Their Interactions." New York: John Wiley & Sons.

Brimblecombe, P., and Lein, A. Y., eds. (1989). "Evolution of the Global Biogeochemical Sulfur Cycle. SCOPE 39." New York: John Wiley & Sons.

Buat-Menard, P., ed. (1986). "The Role of Air–Sea Exchange in Geochemical Cycles." Dordrecht: D. Reidel.

Butcher, S. S., Charlson, R. J., Orians, G. H., and Wolfe, G. V., eds. (1992). "Global Biogeochemical Cycles." San Diego, Calif.: Academic Press.

Degens, E. T., Kempe, S., and Richey, J. E., eds. (1991). "Biogeochemistry of Major World Rivers." New York: John Wiley & Sons.

Gorham, E. (1991). Biogeochemistry: Its origins and development. *Biogeochem.* **13**, 199–239.

Gorham, E. (1992). Atmospheric deposition to lakes and its ecological effects: A retrospective and prospective view of research. *Japan J. Limnol.* **53**(3), 231–248.

Graedel, T. E., and Crutzen, P. J. (1993). "Atmospheric Change: An Earth System Perspective." New York: Freeman.

Holland, H. D. (1984). "The Chemical Evolution of the Atmosphere and Oceans." Princeton, N.J.: Princeton University Press.

Mackenzie, F. T., et al. (1991). What is the importance of ocean margin processes in global change? *In* "Ocean Margin Processes in Global Change," (R. F. C. Mantoura, J.-M. Martin, and R. Wollast, eds.), pp. 433–454. John Wiley & Sons, NY.

Mackenzie, F. T., and Mackenzie, J. A. (1995). "Our Changing Planet." New York: Macmillan.

Mantoura, R. F. C., Martin, J.-M., and Wollast, R., eds. (1991). "Ocean Margin Processes in Global Change." New York: Wiley–Interscience.

Nebel, B. (1981). "Environmental Science." Prentice-Hall, Englewood Cliffs, NJ.

Oremland, R. S., ed. (1993). "Biogeochemistry of Global Change." New York: Chapman & Hall.

Rambler, M. K., Margulis, L., and Fester, R., eds. (1989). "Global Ecology: Towards a Science of the Biosphere." San Diego, Calif.: Academic Press.

Schlesinger, W. H. (1991). "Biogeochemistry: An Analysis of Global Change." San Diego, Calif.: Academic Press.

Stevenson, F. J. (1986). "Cycles of Soil." New York: John Wiley & Sons.

Stolz, J. F., Botkin, D. B., and Dastoor, M. N. (1989). The integral biosphere. *In* "Global Ecology," (M. B. Rambler, L. Margulis, and R. Fester, eds.), pp. 31–50. Academic Press, San Diego, CA.

Warneck, P. (1988). "Chemistry of the Natural Atmosphere." Academic Press, Inc., San Diego, CA.

Wayne, R. P. (1991). "Chemistry of Atmospheres." Oxford University Press, New York, NY.

Wollast, R., Mackenzie, F. T., and Chou, L., eds. (1993). "Interactions of C, N, P and S Biogeochemical Cycles and Global Change." Berlin: Springer-Verlag.

Woodwell, G. M., ed. (1990). "The Earth in Transition: Patterns and Processes of Biotic Impoverishment." New York: Cambridge University Press.

Biological Communities of Tropical Oceans

Charles R. C. Sheppard

University of Warwick, United Kingdom

Important tropical marine communities include the three classical ones of coral reefs, mangroves, and seagrasses, but also the less spectacular and sometimes inconspicuous mud flats and algal beds. These names are all generic labels, however, and many variants of each type occur in different parts of the world. All of them are highly productive. Equally important is the pelagic zone which links all of the coastal ecosystems and which, because of its size, also has a high total productivity and which is especially important in regions where upwelling occurs. None of these ecosystems occur or function independently from the others. Biomass, nutrient, and species exchanges are extensive among many of them, and strong physical linkages mean that many are dependent for their existence on some of the others. Globally, many of the tropical ecosystems have been subjected to damage either directly, through overextraction of their products, or indirectly and sometimes inadvertently. Often, this is due to direct impacts to the ecosystem itself, but sometimes it is due to disruption of the more poorly understood linkages among them.

I. INTRODUCTION

The warm sea temperatures required by tropical marine communities have an irregular global pattern. The common reference to waters between the tropics of Capricorn and Cancer is an oversimplification which conceals many of the most interesting features of the region. Warm currents flowing poleward out of the region carry with them the conditions, spores, and larvae which allow tropical marine communities to flourish well beyond these geographical boundaries, and conversely, patches of cooler water within the tropics also permit communities of temperate character to flourish nearer the equator than do many coral reefs and mangroves. Visually spectacular examples of this are found in areas with upwelling, such as the Arabian Sea and the eastern Pacific Ocean, or where cooler surface currents flow toward the equator. It is possile in some of these areas, such as Oman and western Australia, to observe simultaneously both typical coral communities and communities of kelp or other temperate macroalgae. The upwelling condition that cause these cooler water patches may also ring substantial nutrient enrichment, and this in turn generates rich fishing grounds. [See OCEAN ECOLOGY.]

Another characteristic of warm seas is that their warm water rarely extends more than 50 or 100 m deep. Even where surface waters are over 25°C, marked thermoclines generally exist near the lower end of the photic zone. There are cases, such as the Red Sea, in which warm water does extend much deeper, but these are exceptions. Most commonly, therefore, tropical conditions extend only in a relatively thin film over the surface of the low-latitude oceans.

Important coastal tropical marine communities occur where this warm film laps over continental shelves and onto shorelines. Communities which have historically been of greatest interest are coral reefs, mangroves, and seagrasses. Equally important in environmental terms, and in terms of productivity and nitrogen fixation, are the lesser known and less spectacular communities of mud flats and blue–green algal flats, the latter being complex assemblages of blue–green algae (Cyanophyta), diatoms, and bacteria. Offshore, the thin layer of warm water that supports the tropical pelagic community is the largest in terms of area.

Traditionally, each community has been compartmentalized into separate self-contained subjects of study. This reflects the preferred interests of researchers more than biological reality, and in the broadest sense the tropical marine community does not function in this compartmentalized way at all. First, many of the classical communities, such as "coral reefs" or "mangroves," are only generic terms, whose components differ greatly, depending on geographical region or environmental condition. Second, most shorelines are mixed. For example, a coast may support mangroves, beside which are vast blue–green algal flats, seaward of which are seagrass beds, all of which can exist only because wave energy is reduced due to a coral reef further seaward. In such a situation the energy flows, nutrient fluxes, and physical interactions between and within communities operate over considerable distances.

This article describes biological communities of tropical oceans in terms of distributions, interactions, and comparative productivity rather than providing detailed descriptions of each. Also, the substantial human influence currently affecting tropical marine communities is addressed, and this especially requires that all tropical marine communities are considered together as components of an integrated whole.

II. THE MAIN COMMUNITIES

A. Coral Communities and Reefs

There are two main coral provinces: the Atlantic and the Indo-Pacific, which have no coral species (or, arguably, one) in common. In the Atlantic the Caribbean is the focus of diversity, with about 75 coral species. South of a sediment and freshwater barrier at the mouth of the Amazon River is an assemblage in Brazil which has several endemic species, and there is another smaller assemblage with poor diversity scattered across Atlantic islands to the central African coast. [See CORAL REEF ECOSYSTEMS.]

In the Indo-Pacific the highest coral diversity occurs in the Indonesian–north Australian region, where there are about 500 species. Further east, diversity declines remarkably smoothly, falling off both eastward and with increasing latitude. Westward across the Indian Ocean, the pattern is much more complex. There is a slight westerly decline, but the appearance of many endemics complicates this so that there are no regular "contour lines" of decreasing diversity. There is a gross reduction with latitude, but here too the pattern is not simple, as the highest-diversity areas are western Australia and the Red Sea. At the poorest extremes of this range, coral diversity still generally exceeds 50 species.

The broad global controls of coral communities are well known, in that they occur in well-lit waters where annual temperatures exceed ~20°C. However, this broad outline, which has been held from Darwin to the 1970s, is too simplistic. Reef corals do generally flourish in waters over 20°C, but may be found between 13°C and 38°C in some areas, provided that (1) there is no significant freshwater input, (2) the water is well illuminated, (3) sedimentation is not too severe, and (4) there is no substantial nutrient enrichment. In contrast to most other marine ecosystems, exposure is not limiting to coral communities generally, although many species may be excluded in exposures at either extreme; different coral assemblages are found from embayments with very restricted water movement to heavily exposed windward sides of oceanic atolls. Reef-building corals, and the assemblages of reef fishes and invertebrates which accompany them, may also occur among macroalgae commonly associated with temperate waters.

The term "hermatypic," meaning reef building, is still in common use to denote the species of corals which contain symbiotic algae called zooxanthellae, which grow rapidly and which form reef (the opposite is ahermatypic). Under condition in which sedimentation, temperature, or nutrient enrichment has a constraining influence, reefs may not develop, even where hermatypic corals grow abundantly. In these areas coral growth becomes uncoupled from reef growth. The nature of the coupling itself is largely unknown, but may involve microbiological and chemical processes causing recementation of the limestone sand. Generally, today the terms "zooxanthellate" and "azooxanthellate" have replaced the older terms of "hermatypic" and "ahermatypic," and this better reflects the many conditions under which corals thrive even where active reef accretion is not occurring. Visually, it may be difficult to distinguish between true reefs and identical assemblages of corals and accompanying species on nonaccreting substrates.

The larger faster-growing zooxanthellate corals dominate hard substrate in warm waters, but their symbiotic association with zooxanthellae limits their distribution to shallow well-lit areas. Azooxanthellate forms have no such limitation, and are found in darker recesses of reefs, both to abyssal depths and to the poles. The symbiosis of corals and zooxanthellae forms the key to the success of this group and of the reef ecosystem supported by it, due to its efficient energy and nutrient-recycling mechanisms. The gross primary productivity of coral communities is very high (see Section II,B).

Reef-associated fish, whose abundance and colors are arguably two of the most dramatic and characteristic aspects of coral reefs, have diversity and abundance patterns broadly similar to those of the corals. Total diversity is considerably greater, however, and commonly there are about four to five times more fish species than corals in any one location. Similar patterns are found with most invertebrate groups, although with some groups, such as mollusks, the total numbers of species are much greater still. A major difference between the Atlantic and Indo-Pacific shallow hard-substrate assemblages is the absence of alcyonarian "soft corals" from the Atlantic. Instead, Gorgonian soft corals replace them, and although the latter do occur on Indo-Pacific reefs, especially deeper, they are not as abundant.

Most true reefs have a characteristic profile of a reef flat and a reef slope (Fig. 1). The former results from previous upward growth to the low water level, followed by continuing outward growth seaward, and sometimes there are complex energy-dissipating structures such as ridges, spurs, and grooves at the seaward edge. The reef slope then dips downward, sometimes with steps or terraces which may mark earlier phases of low sea level. Fish and invertebrate diversity on the reef flat is generally relatively low, since it is exposed to intense insolation, including infrared and ultraviolet. This is the least diverse part of coral reefs and lies at an extreme end of the total range of environmental conditions experienced by them. Maximum diversity is found on the shallow reef slope, commonly just below the wave base, and both diversity and productivity then decline with increasing depth to 40–60 m, depending on overall water clarity. At higher taxonomic levels (i.e., at the level of orders and phyla) coral reefs as a whole are the most diverse ecosystem on earth, although at lower taxonomic levels the presence of insects allows the tropical forest to occupy this "exalted" position.

B. Mangroves and Estuarine Communities

Mangroves from important ecological transition zones between the terrestrial and marine provinces.

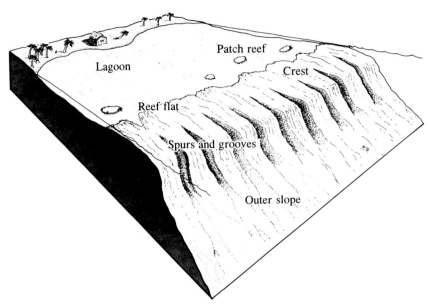

FIGURE 1 Structure of a typical coral reef showing the characteristic zones. The most vibrant coral growth and the most diverse part of the reef is seaward in the outer slope zone. Spurs and grooves develop in areas of high wave energy and help dissipate the power of waves, creating a sheltered lagoon behind the reef crest. Sediments build up in the lagoon and this area is often colonized by seagrasses.

Here, too, there are two distinct biogeographical areas. Over 40 species occur in the Indo-Pacific, where, in common with most groups, the highest diversity is in east Asia, reducing markedly to east and west. About 10 species occur in the tropical Atlantic and central America. Mangroves are restricted to warm waters, and there seem to be few locations which are too hot for them to thrive. However, they are clearly limited by low winter temperatures. Species have different salinity tolerances, and the most tolerant is *Avicennia marina,* which exists in waters with salinities up to 42 parts per thousand (ppt).

Mangrove trees are the visually dominant elements of diverse and highly productive communities; they are halophytic and flourish especially well in estuaries where salinity is lower than that of the open ocean, while in embayments with high evaporation and elevated salinity they are generally absent. Although they are predominantly intertidal, the width of their communities (the mangal) may nevertheless extend up to 20 km, such as in Malaysia or Yucatan, and even to 80 km in Bangladesh. Landward of the tidal influence they may grade into almost any terres-

trial habitat, from rain forest to desert, while seaward, mangrove communities generally have abrupt boundaries, commonly giving way to seagrasses or coral reefs.

The main influence on the associated biota, however, is marine, and although the mangal is generally regarded as a distinctive ecosystem, its associated flora and fauna show strong affinities with adjacent intertidal habitats that lack mangrove trees. Typically, this is a soft-substrate biota, but with attached biota also finding a foothold on stems and roots. It is mainly a question of physical attachment, shade, and shelter which separates a mangrove-associated biota from that of open estuarine mud flats. The mud they inhabit is almost always fine, organically very rich, and sustained and built up partly by production from the trees themselves. Mangroves require sheltered water, and the tangles of trunks and various kinds of aerial roots maintain low water energy in self-enhancing ways. This often leads to a seaward progradation of the mangrove stand. In this sense mangroves make efficient pioneering species, as they accumulate detritus, stabilize sediment, and, in many areas, convert sea to land. The zonation patterns of mangroves and the timing of the progradation sequences have been well documented.

A characteristic feature of mangroves is the canals and waterways through which river water descends or tidal sea water influxes. Aeration of this water is generally good. In the muds, however, anaerobic conditions commonly prevail. This condition is overcome by the mangrove trees by respiratory mechanisms such as pneumatophores in the roots. Associated biota may be placed into different groupings for convenience: (1) the canopy biota, including orchids, birds, and typically terrestrial insects which have little connection with the marine conditions below; (2) the intertidal mud flora and fauna, including species transitional between the terrestrial and marine such as mudskippers (small fish) and many crabs; (3) freshwater communities, usually limited to rain traps in the branches of the trees; (4) permanent pools of seawater, including the waterways and creeks; and (5) biota attached to trees and roots which is mainly filter feeding and dominated by bivalves. Of these, the intertidal biota may provide the most dramatic secondary productivity, since densities of gastropods, nematodes, and polychaetes commonly exceed 5000 m^{-2}. These are supported partly by the mangrove detritus itself, but also by algae and cyanophytes (blue–green algae) which grow on the mud. This area also supports large numbers of birds.

C. Seagrass and Other Soft-Substrate Communities

1. Seagrasses

Seagrasses are monocotyledons related more to pond weeds than to true grasses (they colonized the sea from freshwater), and are rooted in sands or muds by extensive underground roots and rhizomes. Worldwide there are about 50 species, and unlike reef and mangrove communities, some seagrasses occur in high latitudes. In the tropics stands are commonly monospecific, although it is common to observe three or four species mixed in

patchy mosaics. Dense stands generally are found in shallow sheltered water, but they have been observed down to 90 m. Seagrass stands are fairly dynamic, their position being determined by exposure, substrate grain size, and sediment movement, accumulation, and loss.

Important ecological attributes include (1) providing nursery and refuge areas for resident and migratory fauna, including fish and crustaceans; (2) stabilization of otherwise mobile sediments and protection against erosion; (3) accumulation of sediments; and (4) accumulation of organic matter and provision of an important amount of detritus.

Seagrass blades are generally coated by biota, with some beds providing up to 12 m^2 of attachment substratum for 1 m^2 of seagrass bed. Attached primary producers are dominated by nitrogen-fixing Cyanophyta and diatoms. Nutrient transfer between these and the host plants suggests that this may be a symbiotic interaction rather than a merely mechanical attachment, which helps explain the high productivity from these small plants. Many colonial animals, such as ascidians, hydroids, and bryozoans, attach to the blades, and with the substantial number of grazing and browsing species attracted to seagrass beds to feed on them, seagrasses are an important marine habitat.

2. Mud and Blue–Green Algal Flats

In the tropics mud flats may remain bare of larger plants to a degree rarely seen in higher latitudes. The total area covered by mud flats has not been estimated, but commonly matches or exceeds that covered by mangroves. These apparently bare flats are common in the high intertidal regions, but may also extend through to the sublittoral areas. High salinity, resulting from high evaporative rates, and high temperatures are the principal factors associated with these flats. In these areas surface water temperatures of well over 40°C and salinities exceeding 80 ppt often arise in tidal water, while in ponded areas with low water exchange, salinities commonly exceed twice this value.

Mud flats inundated with water of salinity up to 70 ppt support abundant invertebrates, often with spectacular bird life. This high productivity is based on a microflora, prominent among which are diatoms, filamentous algae, and cyanophytes (blue–green algae). As conditions become more extreme, diversity falls but abundance and productivity may not; a large bay in Arabia, for example, with salinities of 335 ppt each summer, supports the dinoflagellate *Dunaliella* in densities of several million per liter, creating "red tides." Invertebrates surviving throughout the year in abundance include copepods, flatworms, and nematodes, with the latter even found among the salt crystals which persist on the floor of the bay.

In intertidal regions of very high salinity, sabkha develops. This term comes from the Arabian region where this community occupies tens of thousands of square kilometers, but it is applicable to any region where desiccation, salinity, or heat is intense. It refers to low-lying, intertidal, or only seasonally inundated areas. Even when not immersed, moisture usually remains on the surface of such areas, either from capillary action or from hygroscopic surface salts which can attract water from relative air humidities of only 10%. This moisture is sufficient to support a specialized and very productive "algal mat." This is a complex assemblage of cyanophytes, nonphotosynthetic bacteria, and diatoms, several of whose components fix nitrogen. Where salinities are 50–70 ppt, diatoms dominate the mat, with abundant filamentous green algae. At greater salinity cyanophytes increase to provide over 80% of the biomass. The enormous areas covered by this community indicate that it is important, although it is rather under-studied. [*See* INTERTIDAL ECOLOGY.]

D. Macroalgae Beds

Beds of macroalgae (large seaweeds) on hard substrate are unusual in tropical waters, and suggest the presence of environmentally unusual conditions. (Heavily grazed but very productive filamentous and small leafy algae are also a ubiquitous part of coral reefs. These are not included in this category but are an important part of the coral reef community.) Areas of upwelling, cool water incursions, or nutrient enrichment contain macroalgae such as the kelp *Ecklonia* (Arabian Sea) and the *Sargassum* group (many locations). The latter forms dense

thickets on shallow substrates and dominates many reef crests of the Red Sea and the Indian Ocean. Under many of these conditions, corals appear to be out-competed by algae, perhaps due to nutrient enrichment and lowered temperatures. In other locations there is apparently no nutrient enrichment, and the algal domination appears to be a consequence of lower grazing pressure.

The green alga *Halimeda* dominates many sandy areas, especially in the Caribbean, and has been considered important also between reefs of the Great Barrier Reef and in lagoons of many oceanic atolls. This calcareous algae may develop into mounds up to 20 m high and 1 km broad off Australia, while in the Caribbean mounds over 140 m tall have been reported. These generally occupy low-energy environments, often in relatively deep water of over 100 m. The fast growth rate of the calcareous *Halimeda* is the reason for the development of these "soft" reefs, which harbor dense populations of polychaetes, microcrustacea, and micromollusks.

E. The Near-Shore Pelagic Zone

The tropical pelagic zone is clearly the most extensive "community," even though it is confined to a relatively thin surface layer over most of the tropical ocean. Upwelling zones, shallow seas, and areas of continental shelf whose substrate lies within the warm surface layer are generally the most important in terms of primary productivity and fisheries. [*See* MARINE MICROBIAL ECOLOGY.]

The least-studied pelagic component are the bacteria. Their contribution to pelagic biomass and productivity remains unknown, although their importance in terms of detrital breakdown and freeing nutrients and minerals from dead phyto- and zooplankton is considerable. They accumulate along the thermocline in stratified water, which corresponds to the location of dead plankton, whose fall through the water is generally arrested at this layer. Saprophytic bacteria have been recorded at densities of up to 470 ml^{-1} in such locations, where they have a high turnover. These numbers are small, however, when compared to the values of up to 200,000 organisms per gram of sand, and it is likely that densities of "pelagic" bacteria from resuspended mobile substrates at the bottom of shallow water are very much higher than under true pelagic conditions.

Phytoplankton distribution in tropical surface waters is not homogenous; eight planktonic biogeographical regions have been described for the tropical part of the Indian Ocean alone, for example, with regions being determined by relative dominance of diatoms, dinoflagellates, coccolithophores, and pelagic cyanophytes. Regions are demarcated to a large degree by proximity to upwelling, or by the degree of nutrient starvation over deep oceanic areas rather than by geographical location per se. The location and nature of the different deep benthic oozes formed from dead plankton, which are well mapped globally, reflect this. Phytoplankton diversity shows geographical patterns similar to those of benthic species; known dinoflagellate species in the Indian Ocean, for example, currently number 452, and this diversity shows a marked fall away from the Far East region, to 130 in the Arabian Sea. A similar fall occurs eastward of the southeast Asian region across the Pacific.

Equatorial regions commonly have fewer than 400 phytoplankton cells per cubic meter, while upwelling areas have offshore pelagic values of 1000 to >6000 cells per cubic meter over large areas, and densities can be more than 1000 times greater still in upwelling cores and on shallow continental shelves. These numbers, of course, are still considerably lower than densities found in blooms of the dinoflagellate *Oscillatoria,* when 10^9 cells per cubic meter may occur in surface waters. These cells also fix nitrogen, and so they greatly increase the local primary productivity in nutrient-poor waters. Outside such red tides, some estimates in areas of upwelling indicate that 10–20% of the total suspended carbon is in the form of living plankton, and that the remainder therefore is dead cells undergoing solution and remineralization by, for example, bacteria.

The gross pattern of zooplankton numbers generally follows that of phytoplankton, but with a time lag of a few weeks, since numbers of many zooplankton respond to the abundance of phytoplankton they feed on. Numbers are variable, and

in shallow water are complicated by several spawning patterns. In winter larvae may form <5% of the zooplankton, but this may increase to >90% each spring. "Demersal" plankton also complicate the picture. This is a complex group which is abundant over reefs, sand, and seagrass beds, which rises from refuges in these benthic substrates each dusk, in densities of up to 1.5 million per cubic meter, providing greatly enhanced pelagic food supply in such areas.

III. PRODUCTIVITY OF THE MAIN COMMUNITIES

In most tropical marine ecosystems gross production may be high, but net production (gross production, P, minus the species' or system's respiration, R) may be very low because the P/R ratio is very close to 1. Several studies have examined gross production or net production, and a summary of several different sets of measurements is given in Table I. Average or common values for the different ecosystems (and for different parts of some of them) are shown in the first column, while the second column gives values which are greater or smaller than most, but which nevertheless have been obtained from realistic field measurements. The P/R ratio gives an indication of the surplus production available, possibly for export to areas outside the ecosystem. When this ratio is >1, there is surplus of production over the respiration and that required for "internal consumption." [*See* MARINE PRODUCTIVITY.]

It is seen that reefs, several kinds of algae, mangroves, and seagrasses have high rates of production compared with the pelagic and bacterial systems, and that values over 5 g of C m^{-2} d^{-1} are common. For algae, mangroves, and seagrass there is clearly surplus for export. Seventy-five percent of organic carbon fixed in mangrove stands is exported into the surrounding waters, and in the case of seagrasses, some estimates suggest that 50% of the production is exported in the form of dissolved organic carbon, which is rapidly assimilated. The remainder enters the detrital food chains. Detritus is an often overlooked component, but lies at the

TABLE I

Values for Productivity of Tropical Marine Ecosystems[a]

Ecosystem	Average values (g of carbon/m²/day)	Extreme records (g of carbon/m²/day)	P/R ratio
Reefs			
Whole reef systems	3.2–4.0	2.3–6.0	Usually near 1
Outer slopes	2.0–7.1		0.7–1.1
"High-activity areas"	9.0–14.0	8.0–23.0	0.6–1.7
Reef flats	6.0–7.0	4.0–19	0.7–2.5
Shallow lagoons	3.0–5.0	2.9–12.9	0.7–1.2
Algae			
Algal turfs	2.0–4.0	1.0–14.7	0.5–13.7
Algal pavement	2.0–7.0		1.0–4.0
Coralline-encrusting algae	0.8–1.0	7.0	1.3–1.4
Sargassum, Sargassopsis	60.0 (dry organic weight)		
Tropical kelp (Oman)	4.0 (dry organic weight)		
Halimeda	0.2–6.0		2.3 (for value = 6)
Mangroves			
Global averages	10	7–14	1.25
"Dwarf," stressed	<6		
Sandy areas near reefs	0.9–1.5	0.6–2.7	0.6–1.1
Seagrasses			
Global averages	6.0–15.0		1.2
Arabian region	3.0		1.2–2.0
Bacterial			
On coral rubble	0.01–0.1		
On sediments	1.2		
Pelagic	0.007–0.11		
Phytoplankton			
Open oceanic	0.04–0.06	0.003–0.4	0.3–1.4
In upwelling areas	0.8–3.0		
Near shore or reefs	0.2–0.9	0.1–1.0	

[a] Average values are given, with extreme ranges recorded. P/R ratios are given, showing an excess of production over respiration.

basis of a substantial part of the total food web of tropical benthic as well as pelagic systems.

In the case of the coral reefs, the P/R ratio is commonly close to 1, so that much less is exported. It is now well known that coral reefs have an efficient recycling mechanism which enables them to thrive in waters low in organic matter and nutrients. Gross pelagic productivity of tropical oceanic water is commonly 0.01–0.03 g of C m^{-2} d^{-1} or less. In upwelling areas this rises dramatically, and may reach that of some benthic communities. In many cases production has a strongly seasonal nature, depending on monsoon periods. Rich upwelling areas, for example, may have a productivity only a little greater than normal oceanic values for parts of each year, but which rises 10- or 100-fold, to over 1–3 g of C m^{-2} d^{-1} when upwelling occurs.

The range of productivity seen in tropical marine ecosystems thus varies considerably, but commonly is substantial. The fact that gross pelagic productivity per unit area is substantially lower than benthic productivity, perhaps by a factor of 10 or 100, has led to many discussions about the relative importance of benthic compared to pelagic productivity, usually in connection with expressions pointing out the importance of one of the benthic ecosystems. But this debate is of little value because it generally ignores the fact that the total area of pelagic productivity greatly exceeds that of benthic production by a factor very much greater than 10 or even 100. Of course, the two main groups of ecosystem are complementary. Evidence is still in short supply, but it is increasingly understood that pelagic and benthic systems interact considerably in terms of energy and nutrient exchange. What is clear is that in the pelagic situation, productivity is greatly submaximal and is increased substantially given upwelling conditions, river runoff, or sewage inputs. This contrasts with the typical tropical benthic systems, in which these additional inputs may lead to markedly deleterious consequences, such as increasing algal domination on coral reefs.

IV. INTERACTIONS BETWEEN COMMUNITIES

A. Physical Interactions

Very few stretches of tropical coast support only a single ecosystem. Mangroves and seagrasses commonly co-occur in mosaics, and seagrasses commonly extend between coral reefs. Mud flats always occur beside mangroves, and even coral reefs, which are sometimes regarded as the opposite of mangroves and mud flats in their requirements, may occur in close proximity to them. Fringing reefs commonly provide sheltered back-filled lagoons which form ideal mangrove habitats. In many locations the physical interactions among these components are essential to the existence of each, so that they should be regarded as interdependent (Fig. 2).

Most important is the modification of wave energy. Coral reefs thrive under conditions of high wave energy, growing spur, groove, and algal ridge structures in response to the energy, which

has the result of dissipating it. These structures provide effective physical barriers which give rise to calm back-reef conditions, under which sediments can accumulate and then be colonized by seagrasses and mangroves (Fig. 2). Both groups of plants may also develop on this mud where this accumulates on shallow reef flats, although when mangroves grow under such conditions (commonly called "hard-substrate mangroves"), they rarely become tall or well developed.

Once colonization of sediments by seagrasses or mangroves has begun, sediment stabilization and further colonization become self-enhancing. Seagrasses promote sediment trapping and decrease sediment resuspension. Networks of roots and rhizomes prevent sediment mobilization and longshore drift, providing further improved conditions for colonization. Over a period there is a build-up to a point at which the surface becomes increasingly shallow and then exposed to air. A few species of seagrasses may survive short intertidal exposure, but generally beds will not survive above the low water level. In the case of mangroves, the end result of this is dry land, although new colonization at the seaward edge will probably also result in progradation of the intertidal mud seaward. Coral reefs also directly provide substantial amounts of carbonate sediment for these soft-substrate communities.

Conversely, both mangroves and seagrass beds also provide protection for reefs from terrestrial influences. Freshwater influxes and their accompanying high loads of nutrients and sediments are

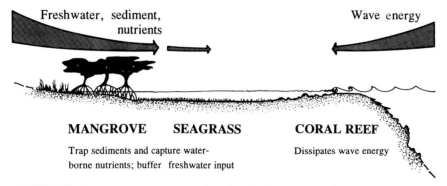

FIGURE 2 Mangroves, seagrasses, and coral reefs often grow in close proximity. By protecting the shoreline from wave action, reefs can allow mangrove stands to develop and seagrasses to flourish in shallow lagoons. Mangroves in turn protect reefs from sediment and freshwater runoff from the land.

tolerated well by mangroves, but not by reefs. The mangroves here act as physical sediment traps and as freshwater buffers, reducing the effects of these terrestrial influences on the reefs (Fig. 2). This role is becoming increasingly important as terrestrial degradation results in more severe flash flooding, carrying ever larger discharges of soil, nutrients, and even pesticides into the sea.

B. Biomass and Nutrient Interactions

A substantial proportion of reef species are found only on reefs, although some move across reef boundaries. In contrast, only about 10% of species found in seagrass beds are confined to that habitat, and probably a smaller proportion still are confined to mangroves; in the case of the last two habitats, migration to sandy or muddy habitats is common, and these communities are very similar to sublittoral and intertidal soft-substrate habitats not colonized by these larger plants. The differences relate rather to the increased abundance of some of the species when the plants exist, and it is thought that the increased physical or topographical complexity the plants provide is partly responsible for this congregating effect.

For some noteworthy herbivores, dependence on seagrasses is great, such as when they form the bulk of their diet (conspicuous examples are green turtles and the dugong). Seagrass beds also appear to be important to certain phases of the life cycles of some species (e.g., pearl oysters and shrimp) which may migrate annually or at one point in the life cycle, or are important to species which migrate diurnally (e.g., snappers) between seagrass beds and coral reefs (Fig. 3). In some of these cases, substantial nutrient transfer has also been claimed to occur from the feeding–defecation cycle, and notable examples are the diurnal movement of snappers between seagrass meadows and reefs (Fig. 3) and even of sea birds which feed in the pelagic zone and roost in mangroves.

In seagrass beds average leaf turnover, or renewal time, is commonly every 35 days, thus continually providing new detritus and a continuous supply of new substrate for sessile biota to colonize. Some of these attached biota are calcareous, and these

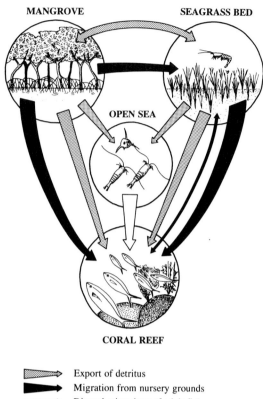

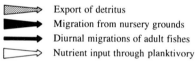

	Export of detritus
	Migration from nursery grounds
	Diurnal migrations of adult fishes
	Nutrient input through planktivory

FIGURE 3 Some of the main pathways of biomass and nutrient transfer among mangroves, seagrass beds, coral reefs, and open sea. Arrows show the direction of transfer, but their sizes do not indicate the magnitude of flows. Movement of detritus (e.g., leaf litter) forms one of the most important routes of biomass transfer. However, migrations of fish and shrimp from nursery grounds to adult habitats and daily feeding migrations of fishes among habitats are also important. The "wall of mouths" of plankton-feeding fishes on coral reefs has been found to effectively remove the plankton from seawater passing over reefs, providing important inputs of trace nutrients such as phosphorus.

epiphytes add significantly to the substrate itself. Detritus from seagrasses is >2% useful organic material, which significantly increases the food supply in localized areas such as the high intertidal region and the strand line and may also be important as far afield as ocean trenches.

Mangroves export larger quantities of dissolved and particulate organic matter. Breakdown of leaf material from the common Indo-Pacific mangrove species *Avicennia marina* occurs at ~80–100% per 100 days, so that export takes place well beyond the boundaries of the mangrove stand. This detritus has been detected across continental shelves and into deeper water, well beyond the usually accepted limits of influence of these coastal ecosystems (Fig. 3). In general terms organic carbon and phosphorus compounds are the critical components in this respect; nitrogen is fixed by components within many of the benthic systems, so that in reefs, seagrass beds, and a wide range of rubble and soft substrates, nitrogen is unlikely to be limiting.

Species such as shrimp and several commercially important fish are known to use mangroves as physical refuges at breeding and spawning times. For this reason one or more of these habitats may be critical to a species, even though it does not reside there permanently. This is important in considering effects of the disruption of these ecosystems by humans.

V. HUMAN INTERVENTIONS AND INTERACTIONS

The ecosystems described above all commonly fall into the category of "critical habitats." This term is used to mean either that they are vulnerable to disturbance to a degree which impairs their full ecological function or that they are easily impacted. It may also mean that they are in limited supply, and the remaining mangroves of a certain coastline, for example, may become increasingly important as more of their parts are removed. Today, the global extent of tropical marine community modification is becoming increasingly extensive and widespread. [*See* MARINE BIOLOGY, HUMAN IMPACTS.]

Many "traditional" uses of tropical marine communities have continued for hundreds of years in ways which, although not necessarily intrinsically sustainable, have managed to remain sustainable in practice because the numbers of people extracting goods from the ecosystems have remained low. Misconceptions persist concerning the sustainability of traditional uses. Many of the traditional practices themselves may not be sustainable beyond a very limited intensity of use, and it is the increased pressure from rapidly growing coastal populations which provides the main causes of degradation seen

increasingly over the last decade or two. The main extractive use from all of them has been seafood, although building materials from mangroves have also been important. However, in addition to these, many lists have also been made of numerous, sometimes more minor, uses of these habitats, often published in support of attempts to demonstrate the importance of protecting them and leaving them intact.

Some of the main impacts which have had profound importance are given here as examples. Mangrove clearance for aquaculture is especially damaging. Here, mangroves are cleared for pond construction, but such ponds may have limited life due to soil acidification problems, in which case they become unusable after a very few years. The scale of mangrove clearance is high. Globally, 56% of mangroves present in preindustrial times have been cleared for all purposes, although in many individual countries the degree of clearance is 60–85%. Added to this, where mangrove clearance has been as extensive as in areas such as central and south America, natural production of, for example, shrimp larvae, which are required for initial pond stocking, has also disappeared, since shrimp required the mangroves for breeding. Societal displacement is also severe with this activity, since local fishing communities find themselves unable to find food, and the (sometimes short-lived) profits from the shrimp aquaculture seldom benefit the local communities.

Mangrove stands, near-shore seagrass beds, and coral reefs have also suffered from coastal developments which require that landfill be placed on the habitat for building. Sometimes this is extensive, such as in the Saudi Arabian Gulf coast, where infill now comprises 40% of the coastline. The infill in this and many other cases comes from dredge material taken from shallow but nearby offshore regions, thus increasing the area which is destroyed. Landfill and most shoreline construction results in greatly increased sedimentation, to which reefs are particularly vulnerable. Expanding human settlement also results in increased discharges of sewage, to which reefs are very vulnerable, as well as increases in many industrial pollutants and pesticides. In this respect it is not only coastal developments which are important; increasingly, land use practices well inland have severe impacts on the marine communities. Deforestation, and the inappropriate agriculture which usually accompanies it, cause greatly increased discharges of freshwater from flash flooding, which carries to the sea massive loads of sediment, fertilizer, and pesticides. In many cases this has swamped the sediment-trapping ability of mangroves (where they have not been destroyed) and has resulted in widespread obliteration of coral reefs as well. Areas of high productivity have thus been converted into areas of muds, which are generally of very low productivity, since they remain very mobile and are subject to further blanketing.

Global climate change may also have impacts on these coastal marine ecosystems, although this is a poorly developed area of research at present. The phenomenon of coral bleaching, which is now increasing worldwide, is not well understood, but has been linked with temperature changes. In this and other cases it is probably not the very small average global temperature change which is important, but rather the increase in extreme values which underlies it. Further, the predicted increase in storm intensity and frequency is likely to have marked effects on the stability of some of the more exposed coastal mangrove and seagrass stands. Some research has been done on the effects of sea level rise on intertidal habitats such as mangroves and on the ability of coral reef growth to keep up with the rise, although very few clear conclusions have emerged to date.

In most areas of study, however, there is a sound and increasing understanding of the value provided to humans by these tropical marine ecosystems, but this understanding is not being conveyed adequately by scientists to the decision makers. Problems also arise not because of a lack of understanding, but because there are at present no adequate mechanisms for costing damage which may be done to an ecosystem or of appropriating its cost to any individual, operation, or organization in particular. The tragedy of the tropical marine eco-

systems is that the "tragedy of the commons" is currently being reenacted, because, both regionally and within countries, there is generally a very poor sense of guardianship of their present and future worth. In many countries there is no capacity, for example, for costing a lost fishery when a new golf course is built on the only, and critical, mangrove in the region, or of costing the required importation of fish to replace those lost because the reef nurseries have been dynamited. [See TRAGEDY OF THE COMMONS.]

Against this, however, in more and more areas, measures are being implemented to safeguard each ecosystem, in marine parks and reserves, for example. However, there are numerous case histories which have demonstrated that damage to one of the ecosystems will cause unintended damage to another, perhaps because of resulting sedimentary changes, or perhaps because of elimination of a habitat which is required for part of a life cycle. Many failures of management have occurred because these interactions have not been adequately considered. Increasingly, coastal zone management is recognizing the need to address the interactions among the tropical marine ecosystems as much as the ecosystems themselves, but much more effort is needed in conveying these principles to the authorities which have controlling influences on habitat use.

Glossary

Biogeography The global patterns of species distributions, or of ecosystem distributions, and the study of reasons why such patterns occur.

Cyanophyta A group of prokaryotic organisms which fix nitrogen; commonly called blue–green algae.

Endemic Occurring only in a particular circumscribed area or region.

Halophytic Tolerant of saline conditions.

Hermatypic Reef forming; refers to corals which contain zooxanthellae and which are major contributors to reef growth.

Productivity A measure of the amount of organic or living matter produced per unit time and per unit area, often measured in terms of carbon (sometimes chlorophyll or other substance). Gross productivity is the total amount made, while net production reflects the former minus the amount consumed *in situ*.

Zooxanthellae Single-celled algae that live in symbiotic association with several coelenterate groups, including corals, alcyonarian soft corals, and gorgonian soft corals.

Bibliography

International Union for Conservation of Nature and Natural Resources (1991). "Oceans, a World Conservation Atlas." Beazley, London.

Lewis, J. B. (1982). Coral reef ecosystems. *In* "Analysis of Marine Ecosystems" (A. R. Longhurst, ed.), pp. 127–158. Academic Press, New York.

Longhurst, A. R., and Pauly, D. (1987). "Ecology of Tropical Oceans." Academic Press, Orlando, Florida.

Sheppard, C. R. C., Price, A. R. G., and Roberts, C. M. (1992). "Marine Ecology of the Arabian Region: Patterns and Processes in Extreme Tropical Environments." Academic Press, London.

Biological Control

Alan A. Berryman

Washington State University

Biological control, in its broadest sense, is the control or regulation of a biotic variable at, near, or around a fixed point or equilibrium by a negative feedback mechanism. Examples can be found in the regulation of body temperature, the control of pest populations, and the natural balance of ecological communities. However, biological control is commonly used in a narrow sense to describe the regulation of pest populations at innocuous densities by their natural enemies. This article, with its focus on environmental biology, will mainly be concerned with the control of populations and communities of living organisms and their effluents by natural and engineered feedback mechanisms.

I. INTRODUCTION

The idea of a "balance of nature" or "homeostasis" has been a central theme in ecology for more than a century. Yet these notions are often clouded in mysticism, confusion, and controversy (see the controversy over population regulation discussed by Krebs in this volume). Ecologists often have no clear picture of what they mean by population control and ecological stability. Engineers, on the other hand, seem to have a very clear and formal view of the same concepts applied to automobiles and spaceships. We can argue, of course, that natural ecosystems are completely different from machines and, therefore, that the engineer's viewpoint is irrelevant. It may be more productive, however, to search for similarities between these disciplines, for in so doing we may learn how to better manage our endangered planet. [*See* POPULATION REGULATION.]

The most obvious thing that ecosystems and machines have in common is that they both change with time. In other words, they are *dynamic systems*. In addition, they both utilize and dissipate energy. Because of these similarities, changes in ecosystems are governed by the same basic rules as machines—the fundamental laws of motion of dynamic systems.

II. DYNAMIC SYSTEMS CONTROL

Ecologists who brave the literature on engineering control theory, cybernetics, and dynamic systems theory can become intimidated or confused by the complicated jargon and mathematics. Yet the basic rules of dynamics are quite simple and intuitive. Consider a *variable* of interest, say the speed of an automobile, the temperature of your body, or the number of individuals in a population of organisms. The value of such a variable, measured at a particular point in time, can be used to identify the condition, or state, of the system at that point of time. The variable is then called a *state variable*. [*See* ECOLOGY, AGGREGATE VARIABLES.]

Changes in the value of a state variable (*changes of state*) can be brought about by either exogenous or endogenous forces. *Exogenous forces* are those that affect a change in state but are themselves unaffected by current or past values of the state variable. Exogenous forces can act either as *forcing functions*, in which case the force is applied in a predictable manner (e.g., as a continuous or discontinuous function of time), or as *random functions*, in which case the force is applied in an unpredictable manner (e.g., as a random function of time). Weather variation from the climatic norm is a good example of a random function, whereas gradual climatic warming would act as a forcing function. Exogenous forcing functions cause state variables to move in particular directions and random functions cause them to fluctuate around their average values.

Endogenous forces, on the other hand, not only affect changes in state but are also affected by current or past values of the state variable. In other words, endogenous forces result from the action of *feedback loops.* For example, suppose that warm spring weather (an exogenous force) causes a population of pests to produce more offspring than normal in a particular year. This large pest population might then provide more food for predators, resulting in increased predator reproduction. More predators would then eat more pests, and this could cause the pest population to move back toward its original density. A chain of events such as this is called a *negative feedback* loop because the original change in state (a population increase) is opposed by the action of the endogenous force (predation). Note that predation is an endogenous force because the magnitude of its effect (the number of pests killed) depends on a previous value of the state variable (pest population density). (This is also known as *density dependence* and is discussed in more detail in Krebs' article in this volume.) Because negative feedbacks cause state variables to return toward their original values, they act as stabilizing forces in dynamic systems. Over time, therefore, negative feedback forces usually lead to *equilibrium* or balance in mechanical and ecological systems. In other words, they *regulate* or *control* the state variable. [*See* EQUILIBRIUM AND NONEQUILIBRIUM CONCEPTS IN ECOLOGICAL MODELS.]

Endogenous dynamics can also result from *positive feedback.* In contrast to stabilizing negative feedback, positive feedback induces instability in dynamic systems because they accentuate or amplify changes in state. For example, an increase in a pest population causes even greater changes in the future because large populations can produce more offspring. Hence, state variables dominated by positive feedback tend to move away from their previous values. For this reason, positive feedback is the force behind inflation spirals, arms races, organic evolution, and the extinction of species. As we will see later, positive feedback can also create breakpoints or thresholds in dynamic systems.

Engineers also recognize another kind of endogenous force called *feedforward.* Here, predicted values of a state variable are fed forward to affect future changes in state. For example, a manager might use the predicted density of pests to plan a spray project. In this case the magnitude of a negative feedback force (a pesticide spray) is determined by a predicted value of the state variable rather than its current or past values. Because feedforward involves the anticipation of future states, it requires a mathematical or mental model of the dynamic system. Thus, modern pest managers often employ population dynamics models, which include exogenous factors like weather and endogenous factors like parasites and predators, to predict pest densities and to make control decisions.

Engineers use feedback mechanisms such as thermostats, autopilots, governors, and amplifiers to control the dynamic behavior of machines. In particular, negative feedback mechanisms are employed to stabilize and control the trajectories of planes and rockets, and the temperatures of buildings. However, although negative feedback is necessary for stability in complex systems, it is not sufficient to ensure stability. To guarantee stability, the negative feedback force must act rapidly and gently (Fig. 1). If they do not, state variables may oscillate around their equilibrium points. For example, negative feedback loops may involve the sequential reactions of several components, in which case the feedback may take some time to return to its origin. Under these conditions, the state variable can overshoot and undershoot its equilibrium, giving rise to *periodic* or *cyclical* dynamics (Fig. 1, middle). Periodicity is amplified by strong feedback forces and, if the negative feedback force (called the *loop gain*) is strong enough, cyclical dynamics can change into aperiodic *chaos* (Fig. 1, bottom). Thus, negative feedbacks become increasingly unstable, creating periodic or aperiodic oscillations in the state variable, as longer *time delays* are introduced into the feedback loop and as the loop gain becomes stonger. Feedforward can be used to reduce time delays in negative feedback loops, and thereby enhance their stability.

III. NATURAL CONTROL OF ECOSYSTEMS

Although machines and natural ecosystems are similar in some respects, they also have their differences. Engineeers build negative feedback mechanisms to control dynamic systems at or near *desirable* equilibrium points or steady states, for example, to keep an office at a comfortable temperature or the guns of a battleship or target. In such cases the equilibrium point is set by humans for human purposes (this is why the equilibrium is often called the *set point*). Similar purposes and well-defined set points can be imagined in the control of body temperature, blood pressure, respiration, and other organismic functions. In ecological systems, however, it is difficult to see set points or purpose. Here, negative feedbacks tend to arise spontaneously, through chance interactions between organisms, and the equilibria they create have no obvious predetermined purpose—they merely emerge (or fail to emerge) as a result of the structure of the ecosystem. In other words, the equilibria are *emergent* rather than designed properties of the system. Engineers define these two kinds of negative feedbacks as *active* when the state vari-

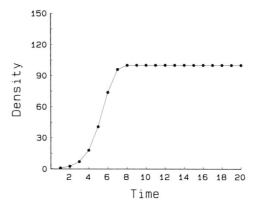

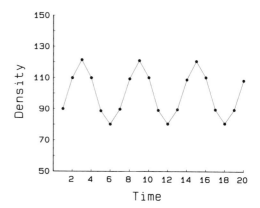

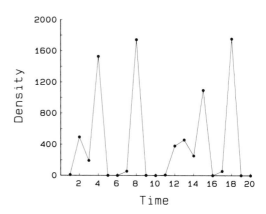

FIGURE 1 Dynamics of a single-species population model, the discrete "logistic" or "Ricker" equation $\frac{\Delta N_t}{\Delta t} = N_t e^{r - sN_t - d}$, where $\Delta N_t / \Delta_t$ is the change in population density with change in time (the change of state), N_t is the density of the population at time t (the state variable), r is the maximum per capita rate of increase of the population, s is the strength of the negative feedback force, and d is the delay in the feedback loop. (Top) Fixed-point attractor when $r = 1$, $s = 0.01$, and $d = 0$. (Middle) Periodic attractor when $r = 1$, $s = 0.01$, and $d = 1$. (Bottom) Chaotic attractor when $r = 4$, $s = 0.01$, and $d = 0$.

able is controlled at or near a preset equilibrium and as *passive* when the equilibrium emerges from the structure of the system.

Ecological systems can be influenced by a large number of potential feedback loops. For example, if we are concerned with changes in the number of individuals in a population of organisms (the state variable could be population density), then we might expect the following feedback loops to emerge as a result of *intra*- and *interspecific* interactions between individual organisms:

Biological Interaction	Feedback
Mating frequency	+
Competition *within* the species (food, space)	−
Cooperation *within* the species (hunting, defense)	+
Competition *between* species (food, space)	+
Cooperation *between* species (symbiosis)	+
Cannibalism	−
Predation on or by other species	−
Parasitism on or by other species	−
Pathogen epizootics	−

Theoretically, a population of organisms can be affected by any or all of the negative feedback loops specified here (this is another difference from machines, where a state variable is rarely affected by more than one feedback mechanism). Of all these potential loops, however, only one will normally regulate a particular state variable at any given time and place. To illustrate this, consider a pest population that has been subjected to a pesticide spray. Following this catastrophe, the dynamics of the population will be dominated by positive feedback (mating frequency) and the populations will grown back toward its previous density. As it grows, however, the enlarging pest population may come to the attention of generalist predators, which may then concentrate their feeding on this rising food supply. Should the force of predation be sufficient to overcome that of pest reproduction, the population could be regulated by these generalist predators at a fairly sparse density. On the other hand,

if the effect of predation is too weak to overcome the growth potential of the population, another force like starvation or disease will eventually control growth at some higher density. Populations of organisms, therefore, are often subjected to a *hierarchy* of feedback loops that can regulate density at or around several different equilibrium points. For example, the following dominance hierarchy has been observed in the feedbacks acting on insect populations:

Density	Dominant feedback	Time lag
Extreme	(−) Competition for food (starvation)	Short
High	(−) Pathogen epizootics	Medium
Moderate	(−) Numerical responses of specific parasitoids	Long
Low	(+) Cooperative attack or defense	None
Sparse	(−) Switching/aggregation of general predators	Short
Very sparse	(+) Mating frequency	None

With this dominance hierarchy, it is possible to have populations in which negative feedbacks dominate at both high and low densities but where unstable positive feedbacks dominate in between. In such cases, the positive feedback can create a breakpoint or threshold separating two potentially stable equilibria, a situation known as *metastability* (Fig. 2). Metastable systems can display highly unpredictable patterns of behavior as random forces shift their state variables from one *basin of attraction* to another. Because unpredictable dynamics are rarely desired in machines, this is another difference in ecological systems.

Finally, because ecological feedback loops may involve several reproducing species, time delays often occur in their negative feedback loops, and this can cause periodic or aperiodic oscillations (Fig. 1). For instance, an increase in a pest population could result in a larger population of specific parasitoids (through their reproductive or numerical responses) and this could feed back to suppress

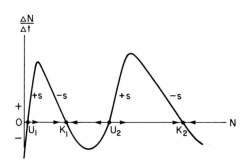

FIGURE 2 A complex "change of state" function $\frac{\Delta N_t}{\Delta t} = f(N_{t-d})$ for a population system in which feedback switches from one dominant mechanism to another at different population densities: $+s$ indicates the dominance of positive feedback (say mating frequency when density is low and cooperative defense when density is intermediate), $-s$ indicates negative feedback dominance (say switching/aggregation of generalist predators at medium density and competition for food at high density), U's are unstable equilibria, and K's are potentially stable equilibria. The basin to attraction to 0 (the extinction basin) is $N \in 0 \rightarrow U_1$, the low-density basin of attraction is $N \in U_1 \rightarrow U_2$, and the high-density basin is $N \in U_2 \rightarrow \infty$. Directions of changes in the state variable N are shown by arrows on the abscissa.

pest population growth. Because it takes time for food to be changed into offspring, and for new parasitoids to attack their prey, destabilizing time lags can be introduced into the negative feedback between parasitoid and prey populations. This is why cycles are frequently observed in such situations (Fig. 3).

Time delays are created in predator–prey interactions because each species needs time to react (through reproduction) to the abundance of the other. (Note that reactions not involving reproduction, such as predators switching to more abundant prey species, do not create long time lags.) Of course, feedback loops that involve many reproducing species can give rise to longer delays and even more complex dynamics. It is interesting, however, that lags greater than three years or three generations are rarely detected in natural populations, suggesting that long feedback loops do not normally dominate ecological systems. One reason may be that the strength of negative feedback loops, and hence their dominance, depends on the product of all the interactions in the loop. One weak interaction will weaken the entire loop and, because long feedback loops are likely to contain at least one weak link, they are less likely to dominate in natural systems.

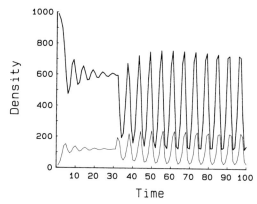

FIGURE 3 Dynamics exhibited by a predator–prey interaction simulated with the discrete logistic trophic chain equation $\frac{\Delta N_i}{\Delta t} = N_i e^{r_i - s_i \frac{N_i}{N_{i-1}} - p_i \frac{N_{i+1}}{N_i}}$, where N_i is the biomass density of the ith species in a trophic chain, s_i is the negative feedback effect of competing for food (in the $i-1$ trophic level), and p_i is the effect of predators (in the $i+1$ trophic level). The time subscript (t) is omitted for clarity. The simulation is performed on a two-species food chain: N_1 = prey (thick line), $r_1 = 2$, $N_{i-1}/s_1 = K_1 = 1000$ (assumes $N_{i-1} = constant$), $p_1 = 4$; N_2 = predator (thin line), $r_2 = 1$, $s_2 = 5$, $p_2 = 0$. Dynamics are simulated for 28 time steps and then the per capita rate of increase of the predator is increased to $r_2 = 1.5$.

IV. ECOLOGICAL ENGINEERING

Knowledge of the basic laws of dynamics can help us understand the causes of change in ecological systems and, perhaps, to control their state variables at desirable levels (desirable from the standpoint of human beings and their environments, or from the more general standpoint of the global ecosystem). Activities such as this fall under the heading of *ecological engineering*. There are three general ways to control the dynamics of ecological systems:

1. *Construct new feedback loops*. This can be done by introducing new organisms into an environment, or by creating feedbacks that involve human actions. For example, pests can sometimes be controlled at low densities by exposing them to natural enemies introduced from other lands. We could also use genetic engineering techniques to construct new organsims that feed on pests or pollutants, like bacteria that feed on oil spills. Here new negative feedbacks are created in an attempt to regulate pests or pollutants at desirable steady states (low levels).

The ecological engineer should realize, however, that new feedbacks can also destabilize the dynamics of previously stable systems. In Fig. 3, for example, I simulated a pest–predator system that originally has a stable equilibrium point, but then I increased the reproductive potential of the predator, say through genetic engineering. Although this "genetically engineered" predator was able to create a lower pest equilibrium, it also destabilized the equilibrium point, so that the pest now exhibits periodic outbreaks. Ecological engineers must be concerned not only with the equilibrium points of the system, but also with their stability properties.

Feedback loops can also be built by directing human action in response to changing levels of a state variable. For instance, if farmers respond to increasing pest damage by spraying a pesticide whenever the damage exceeds a predefined economic damage level, then a negative feedback control loop is created. The farmer might also use information about natural enemy abundance before deciding to spray. In this case feedforward is being used to predict whether natural enemies are likely to prevent pest damage in the future. Operations that integrate economic damage levels, natural enemies, and predictions of future damage into pest control decisions are usually known as *integrated pest management*.

2. *Alter feedback dominance.* The notion of feedback dominance can be employed to alter the equilibrium levels of state variables. For example, a pest equilibrium could be changed from high to low density by helping the negative feedback from generalist predators dominate over that from competition for food. This could be accomplished, for example, by providing nesting places, alternative food, and/or cover for the predators. In metastable systems, feedback dominance can be altered by changing the level of the state variable. For example, suppose that the function in Fig. 2 represents a pest population, then the pest could be moved from the outbreak basin of attraction (in the vicinity of K_2) into the low-density basin of attraction (in the vicinity of K_1) by reducing pest density below the outbreak threshold U_2. Conversely, if the species was beneficial, it could be moved into the higher basin of attraction by stocking to a density above U_2.

Of course, the feedback control can also be changed by weakening or removing dominant feedbacks. For example, we could weaken the feedback between a pest and its food plant by breeding resistant strains of the plant. If one produces a completely resistant strain, then the feedback between plant and pest is broken, and the pest population will not be able to exist on that strain of plant. However, even if resistance is incomplete, weakening the plant–pest interaction may enable another feedback loop, like predation, to become dominant.

3. *Alter the time delay.* Whenever we change the dominance of feedback loops by manipulating the strengths of interactions between organisms or their reproductive rates, we are faced with the danger of oscillatory (cyclical) instability if time delays are present in the feedback response (Fig. 3). Delays are particularly evident in loops containing two or more reactive elements, like reproducing species of prey and predator. Instabilities resulting from such delays can be minimized by manipulating the densities of predator or prey populations as illustrated in Fig. 4.

There are, of course, innumerable ways in which new feedbacks can be built into ecosystems, or by which feedback dominance or delays can be altered. The only limitation is human imagination. What is essential is that the human clearly understand the rules of the game—the laws of motion of dynamic systems. In other words, we should know how positive and negative feedback loops affect the dynamic behavior of complex systems, how time lags and feeback dominance operate to change equilibrium points and their stability, and how exogenous and endogenous forces can be manipulated to change the level and stability of emergent equilibria. With these rules in mind, it may be possible for ecologists to engineer stable, sustainable ecosystems that can support humans and the creatures on which they depend indefinitely. Without an intimate understanding of these rules, humans will make decisions that disrupt feedback loops, or their stability, resulting in unpredictable and catastrophic dynamic events.

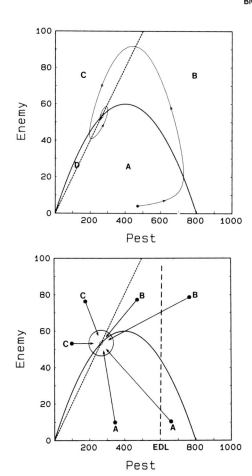

FIGURE 4 Zero growth isoclines for a pest–enemy interaction obtained by setting the left-hand sides of the logistic trophic chain equation (see Fig. 3) to zero and solving for N_1^* and N_2^*, and N_i^* is the equilibrium density of species i. (Top) The pest–enemy phase space is divided into four regions; A = region where both populations grow, B = pests decrease and enemies increase, C = both populations decline, and D = pests increase and enemies decline. The thin line shows a typical trajectory to the community equilibrium and illustrates how the orbit narrows as it nears the equilibrium point. (Bottom) Maintaining pest–enemy numbers within the target area minimizes the likelihood of the pest population exceeding the economic damage level (EDL) because wide orbits are avoided. This can be done by: in region A, spray with a selective pesticide that kills only the pest and/or supplement the enemy population; in B, spray with a general pesticide that kills both species; and in C, spray with a specific pesticide that kills enemies and/or supplement pests. Note that pesticides should be employed to effect optimal kill, *not* maximum kill, because overkill can result in a wide outbreak orbit—what has been called "the boomerang effect" or "pest resurgence."

Glossary

Active control Maintenance of a state variable at or near a preset equilibrium point or periodic orbit by means of a negative feedback mechanism.

Basin of attraction Definable region that attracts dynamic trajectories toward an equilibrium point or a periodic or chaotic orbit.

Chaos Trajectory that has a definable, bounded orbit but does not have a predictable periodic oscillation.

Community Assemblage of organisms living together in a given area.

Control Maintenance of a state variable at or near an equilibrium point or a periodic orbit by means of a negative feedback mechanism (see also active control and passive control).

Dynamic system System whose state varibles change with time.

Equilibrium Condition in which a state variable remains constant due to balance between opposing forces.

Endogenous From or of the inside.

Exogenous From or of the outside.

Feedback Stimulus, message, or effect that returns, after a period of time, to its beginning or source.

Feedback dominance One of many potential feedback loops that actually controls a particular state variable at a particular time and place.

Feedback hierarchy Series of feedback loops that dominate at different values of the state variable.

Feedforward Stimulus, message, or effect that is fed forward to a future time.

Forcing function Force that acts on state variable(s) in a predictable time-dependent manner.

Homeostasis State of balance or equilibrium in a system.

Loop gain Value by which a stimulus, message, or effect is multiplied by traveling around a feedback loop.

Metastability Bounded region of stability, such that a state variable returns only toward equilibrium for a limited set of inital conditions, and therefore has a restricted basin of attraction.

Natural enemy Organism that feeds upon or somehow harms another organism.

Negative feedback Loop gain is negative so that it acts in opposition to the original stimulus, message, or effect in a feedback loop.

Passive control Maintenance of a state variable at or near an arbitrary equilibrium point or periodic/chaotic orbit by means of a negative feedback mechanism.

Periodic Event or phenomenon that repeats itself at regular (or fixed) intervals of time.

Positive feedback Loop gain is positive and acts to accentuate or reinforce an original stimulus, message, or effect in a feedback loop.

Random force Force that acts on a state variable in an unpredictable way.

Regulation See control.

Stability Tendency for a state variable to return to its previous state following a disturbance.

State variable Observation or measurement that defines the state or condition of a system, e.g., the density of a population, your blood pressure, or an air pollution index.

System Structure made up of many interacting elements or components.

Bibliography

Berryman, A. A. (1981). "Population Systems: A General Introduction." New York: Plenum.

Berryman, A. A. (1989). The conceptual foundations of ecological dynamics. *Bull. Ecol. Soc. Amer.* **70,** 234–240.

Berryman, A. A. (1993). Food web connectance and feedback dominance, or does everything really depend on everything else? *Oikos* **68,** 183–185.

Berryman, A. A., Valenti, M. A., Harris, M. J., and Fulton, D. C. (1992). Ecological engineering: An idea whose time has come? *Trends Ecol. Evol.* **7,** 268–270.

Berryman A. A., Stenseth, N. C., and Isaev, A. S. (1987). Natural regulation of herbivorous forest insect populations. *Oecologia* **71,** 174–184.

Logan, J. A., and Hain, F. P. (1991). "Chaos and Insect Ecology." Blacksburg: Virginia Agricultural Experiment Station.

Milsum, J. H. (1986). "Biological Control Systems Analysis." New York: McGraw–Hill.

Milsum, J. H. (1968). "Positive Feedback: A General Systems Approach to Positive/Negative Feedback and Mutual Causality." Oxford, England: Pergamon.

Mitsch, W. J., and Jorgensen, S. E. (1989). "Ecological Engineering: An Introduction to Ecotechnology." New York: John Wiley & Sons.

Puccia, C. J., and Levins, R. (1985). "Qualitative Modeling of Complex Systems: An Introduction to Loop Analysis and Time Averaging." Cambridge, Mass: Harvard University Press.

Biological Engineering for Sustainable Biomass Production

Arthur T. Johnson and Patrick Kangas
University of Maryland

Biological engineering is that discipline of engineering using or applied to biological systems. An understanding of biology can help engineers address environmental problems, including sustainable production of biomass. To be truly sustainable, utilization of biomass requires a knowledge of ecological interactions in the biomass stages of production, harvesting, and processing.

I. DEFINITIONS

A. Biological Engineering

Engineers are those professionals who apply scientific principles for practical purposes such as the design, construction, and operation of efficient and economical structures, equipment, and systems. Engineers use their imaginations to create things that did not exist before, they calculate and compute to assure the workability and safety of their designs, and they oversee the fabrication or manufacture of the products used by the public. Once products are manufactured, they test, evaluate, and redesign to assure that the product incorporates the latest technology, is safe to use, and that it can endure many cycles of use.

Biological engineers specialize in products made from, used with, or applied to biological organisms. They study biochemistry, microbiology, genetic manipulations, physiology, ecology, biotechnology, bioinstrumentation, medicine, biorheology, and agriculture. They are as familiar with this material as they are with the engineering sciences, and this symbiosis between biology and engineering gives them unique capabilities in the modern world.

The "biological revolution" is the term used to describe the explosion of useful new knowledge about genetic codes, new techniques to maintain life and health, and new ways of looking at the environment on a global scale. A new technology is developing that will result in new products and new solutions for problems involving biological organisms. Biological engineers will apply this technology in the ways described here.

B. Biomass Production Systems

The production of biomass is a very important application for the attention of biological engineers. Biomass is the material that exists as part of living organisms or that once existed as part of living organisms. Biomass is primarily produced with energy from the sun by photosynthesis or with thermal energy from the earth's core by chemosynthesis. It consists primarily of carbohydrates (including carbon, hydrogen, and oxygen), in contrasts to the hydrocarbons (carbon and hydrogen) produced by physical processes and that are stored in the earth.

Biomass may be used for many life-sustaining or industrial uses. Chief among these are its uses as food for humans, as feed for animals, and as fertilizer for plants. Biomass acts as a source of energy and nutrients for its consumers.

It may also be used for fiber, as in textiles such as cotton, wool, silk, and burlap. Building materials often contain biomass in such forms as lumbar, thatch, animal skins, and various cellulosic insulation materials. Biomass can also be used as a chemical feedstock. Examples of this are in the production of alcohol from corn, in the extraction of rare biochemicals from genetically altered microorganisms, or in the production of pharmaceutical compounds for human, animal, or plant use.

Finally, biomass may be used as lubricants or directly as fuel. Sources of biomass for energy production are industrial wastes, agricultural wastes and residues, forest residues, specific energy crops on land, energy crops from the ocean, and urban wastes. Because biomass consists of carbohydrates rather than hydrocarbons, less carbon dioxide is produced and more water vapor is produced from burning biomass (per unit weight) as compared to hydrocarbon energy sources. To illustrate this, petroleum (a hydrocarbon) contains 85–87% carbon, whereas wood contains 50–60% carbon (based on dry weight). For the same amount of energy produced, the petroleum will produce about 40% more carbon dioxide than will the wood.

There are about 2×10^{14} kg of dry matter (biomass without the water) presently produced per year by photosynthesis. Of this total, 35% is produced in the oceans, 22% in tropical rain forests, 20% in other forests, 10% in grasslands, 6% in cultivated land, 3% in fresh water, and 4% in other locations.

Biomass has a number of characteristics that influence how biological engineers work with it.

First, biomass exhibits extreme variation in its characteristics, much of which is related to its source. Second, biomass tends to be distributed over wide areas and must be collected to form concentrations high enough to be of practical use. Third, much of the weight of biomass is due to its internal water component, which lowers its value for food, fiber, or fuel when considered on a weight or volume basis. Fourth, it is in a form that tends to decompose readily. Energy must often be expended to preserve biomass by drying, freezing, sterilizing, or otherwise processing so that it does not spoil. Thus, the processing of biomass, especially biomass used for food, often consumes up to twice as much energy as the production of the biomass. Fifth, using biomass for fuel often produces less damaging gaseous emissions than does burning of fossil fuels. Less carbon dioxide is produced for the same energy output for the two types of fuels (biomass and fossil fuels), less sulfur emissions come from biomass (burning wood produces only 7% of the sulfur emissions as low-sulfur coal), and lower amounts of nitrogen compounds are emitted from burning biomass compared to burning coal because the pyrolytic temperature is higher for coal. On the other hand, burning biomass produces more carbon monoxide (about twice that of coal) and particulates (about one and one-half times that of coal). Also, larger volumes of biomass need to be handled to provide the same heat.

C. Sustainability

Sustainable biomass production is the term used to describe growth of biomass with a very limited energy or nutrient supply. Some people argue that the energy or nutrient input to the system must be small enough so that biomass production can be maintained indefinitely without environmentally damaging importation of nutrients or addition of energy. Such a view must be considered the ideal in most cases, as removal of biomass from any particular location also removes nutrients and additional energy is usually required to concentrate and process the biomass for useful purposes. A major question is whether sustainability can ever be achieved if harvesting is to be done. Except in special cases where periodic enrichment occurs naturally, as in riverine floodplains, sustainability may only be a goal to work toward.

Biomass production based on photosynthesis is an environmentally benign means to transfer energy from a limitless source (the sun) to a useful material form (plant matter or animal tissue). The conversion of sunlight energy into carbohydrates and hydrocarbons is only 1–2% efficient. Solar cells by contrast, are 10–25% efficient and wind conversion is 10–22% efficient. Sunlight energy that is not used often ends up as heat.

These efficiencies are not directly comparable because of the nature of the energy inputs. For instance, if the energy required to fabricate the solar cell is included in the calculation, its energy conversion efficiency would be lower than 10–25%. One of the roles of the biological engineer is to conduct an energy analysis or audit, which relates total inputs to total outputs, resulting in an assessment of the net energy yield of a process or system.

All living tissue requires additional elemental nutrients to thrive. Among these are nitrogen (or protein in animal tissue), phosphorus, potassium, calcium, sulfur, iron, and many others. These must be supplied either from concentrated sources such as fertilizer or from diffuse sources such as soil or ocean water. Concentrated sources usually carry some environmental burden, such as energy of production, pollution potential, or nutrient mobility. Diffuse sources may not be able to support sufficient biomass production in any given area. [*See* ASPECTS OF THE ENVIRONMENTAL CHEMISTRY OF ELEMENTS.]

Thus, sustainable biomass production is a goal to produce biomass without untoward environmental effects from either the energy or nutrient requirements of the biomass.

II. GENERAL MODEL OF BIOMASS PRODUCTION

As we have seen, the usefulness of biomass for any of the purposes previously named requires the stages of harvest, concentration, and processing in addition to production methods. A general model of biomass production and processing appears in Fig. 1. Consideration of such a model should clarify the role of the biological engineer in the production and processing of biomass and show where strides can be taken toward the goal of sustainability.

Biomass production requires several general inputs. In Fig. 1 these are labeled as the seed source, carbon source, energy, nutrients, oxidative medium, biological interactions, cultivation, and moisture. Each of these may take many forms, and the role of the biological engineer is to supply one or more of these forms to meet the objectives of biomass production.

Take the case of seed supply, for instance. Seed may be naturally occurring in the environment from some indigenous reproductive population of the desired species. Such a situation is often too haphazard and inefficient for modern biomass production. The first attempts to improve natural seeding consisted of gathering seeds from specially grown crops of the desired biomass species. It was soon found, however, that these seeds had to be cleaned, preserved, carefully stored, and just as carefully planted to produce biomass in the quantities required. Biological engineers had to produce machines and processes to perform each of these functions. Plant seeds, of course, required different techniques from animal germ cells. [*See* SEED BANKS.]

With time, certain individuals of the biomass species were found to grow either faster or more efficiently or with less care, or to produce products with features more desirable than those from other individuals. Ways were sought to reproduce these exact individuals by asexual reproduction. Grafting and budding as asexual reproductive techniques have long been known, but newer technologies are developing in tissue culture and micropropagation. Using these technologies, somatic plant and animal cells can be used as sources for new clones of the original individual. In an exciting development, coniferous plants are being reproduced by somatic embryogenesis. These somatic embryos are then coated with gelled nutrient solutions and rigid seed coats to form artificial seeds that can be handled about the same as natural seeds. Biological engineers worked with plant scientists to develop the processes to produce these artificial seeds and thus contributed to more efficient biomass production.

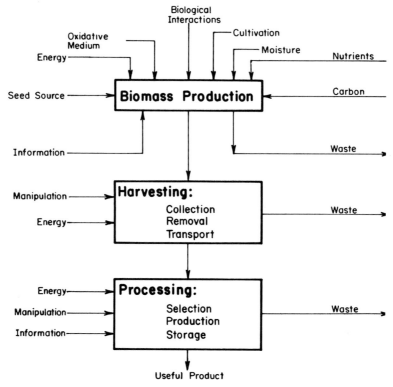

FIGURE I   General schematic for biomass production and processing.

Carbon is required by all living things. Autotrophic organisms (some bacteria and plants) can use inorganic sources of carbon such as atmospheric carbon dioxide. Heterotrophic organisms (some bacteria and plants, and animals) must have organic carbon sources such as carbohydrates. The commercial objective of biomass crops for energy is to use solar energy to capture atmospheric carbon in a form that can be used to supply energy at some point and again release the carbon to the atmosphere. The result is one form of the carbon cycle that follows carbon through its various forms. [*See* GLOBAL CARBON CYCLE.]

Carbon (and other nutrients) is often supplied from a "green manure" process in which other biomass is grown to improve the organic matter content of soils. Crops like sorghum are grown until eight or more feet tall, then chopped and plowed under the soil surface. All important biomass species require an oxidative medium to supply energy on demand to carry on natural processes. Usually, this medium is atmospheric or aquatic-dissolved oxygen. Sometimes other elements, such as sulfur in the ocean depths, supply energy needs.

Biological interactions include disease, parasitism, symbiosis, predation, biochemical interactions, and, for higher-level animals, social interactions. Control of these factors is extremely important for efficient biomass production. High-technology solutions to some of these problems include inoculation, vaccination, pheromone disruption, genetic manipulation, and drug therapy.

Whether the biomass is plant or animal matter, some type of cultivation is required. Cultivation is the process that seeks to maximize biomass production and minimize competition, and hence it encompasses biological and physical interactions. The cultivation input overlaps other inputs to biomass production, but is meant to convey specific knowledge about the species to be produced. Moisture is also required and, for most practical biomass production, the moisture must be abundant. Moisture supplies the medium for addition of nutrients and movement of metabolites; it also supplies hydrogen for carbohydrates.

An understanding of biological engineering principles is important in all of these crops and economic uses of biomass. In many cases the objective is conversion of solar energy and soil nutrients to plant tissue with minimum fossil fuel consumption. The biologically trained engineer understands the plant growth limiting factors whether they be plant physiology, genetics, moisture, temperature, wind, light interception, or soil matrix. The engineer must also understand the rheology and other biomaterials properties as well as energy requirements during harvesting and in-field processing.

An understanding of the various needs of biomass production, harvesting, and processing is important in the quest to achieve sustainability. Sustainability requires minimizing inputs by recycling waste as much as possible and adding as little external energy (except solar) as possible. The linking of the three steps of production, harvesting, and processing is shown in Fig. 1. Sustainability requires that wastes that could be environmentally damaging are reused as much as possible in a way that sustains the overall process.

The biological engineer must understand the entire cycle and the inputs, outputs, operations, and processes as well as the biological principles upon which the cycle is based. The biological engineer should realize that true sustainability can only be achieved as long as the energy equivalent of the dry matter removed as useful product or wastes does not exceed the solar energy captured during production. True sustainability requires complete recycling of nutrients and moisture and minimizing the effects of disease. The energy required for cultivation, manipulation, transportation, freezing, sterilization, packing, or other processing reduces the amount of useful product that can be obtained on a sustained basis. This is the challenge for the biological engineer: produce the maximum possible biomass-based product with the minimum use of nonrenewable resources.

III. APPLICATIONS AND CASE STUDIES

Three case studies are considered to illustrate some of the concepts described here. The case studies represent different types of biomass production with different challenges for engineering. They range from an open system with minimal management to a closed, highly managed system and were chosen to cover the spectrum of biomass production systems.

A. Tropical Tree Plantations

The plantation form of agriculture in the Western world developed and expanded during the period when European countries were exploring the globe and establishing colonies. Plantations can be extremely productive and they often became the dominant source of income for colonies as they exported valuable production products back to the mother countries. This was especially true in tropical regions that provided specialized products, often luxury items such as sugar or coffee that were not otherwise available. Since early colonial times a great deal of knowledge has accumulated on plantations and production can approach sustainable levels with proper management. Discussion is limited here to tropical tree plantations but other examples include sisal (harvested for hemp), pineapples, bananas, and tobacco.

Tree plantations in general are managed stands in which artificially regenerated species comprise a significant percentage (25% or more) of the basal area or volume of stands (Fig. 2). Although they can be highly ordered monocultures, they also can contain a diversity of species with an understory of shrubs and saplings. The tropics have some advantages for plantations in terms of long growing season, high temperature, and, at least in some areas, high moisture. They also have significant problems with pests, including disease organisms and insect herbivores. Trees within plantations are harvested for a variety of products such as rubber, fruits, nuts, lumber, pulpwood, and fuelwood. However, principles of plantation operation are often similar for different species. Emphasis focuses more on biological interactions than on traditional engineering, and human inputs can be minimal. In many cases, the main energy and water sources come from natural inputs and human management is limited to species selection and planting, pest

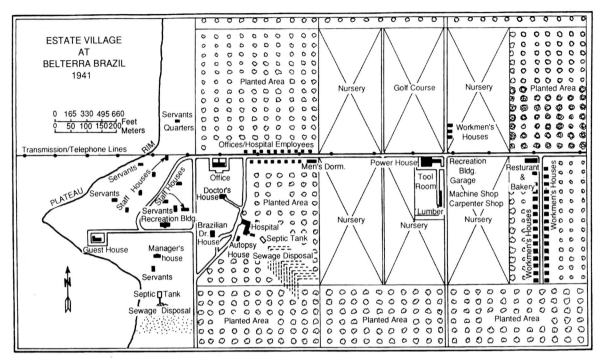

FIGURE 2 A typical tropical plantation design, showing the complete support services that were included. The rubber plantation at Belterra Village in Brazil included all required urban functions: hospital sewage disposal facilities, dwellings, offices, radio station, schools, recreational facilities, stores, and fire protection. [Redrawn from J. A. Russell (1942). *Economic Geography* **18**, 125–145.]

management, harvesting, and processing. Trees in this kind of plantation are analogous to trees in a natural forest that rely primarily on ambient energy inputs for generating production. Engineering is subtle here but can be important in terms of plant spacing, rotation cycles, economic water use, disease control, and harvest technologies. Management is probably more important than engineering in tree plantations but distinctions between these human activities are sometimes difficult to make.

Tree plantations in the tropics presently comprise about 1% of the total tropical forest area but they provide about 12% of the wood production. Thus, they are productive and have the potential for expansion (Table I). If this were to occur, it could release pressure for harvest from natural forests and provide opportunities for conservation and preservation.

However, plantations are not simple systems and there are many examples of failure. Two such examples come from Amazonia. In 1927 Henry Ford set up a large plantation (called Fordlandia) on the

Tapajós River to produce rubber for his automobile factories in the United States. The operation was beset by many problems, including labor relations and severe disease impacts, and never produced significant amounts of rubber. In 1945 the plantation was sold to the Brazilian government at great loss of initial investment. Another example is the large pulpwood project started by Daniel Ludwig in 1967 on the Jari River. Here the plan was to harvest fast-growing species, such as *Gmelina arborea,* for pulpwood and other species for fuelwood to operate a large floating paper mill. The original intention was for sustainable operation of the whole system but this does not seem to have been achieved. The plantation was sold to a consortium of Brazilian companies in 1982 and has become more of a settlement site than a productive pulping operation. Failures such as these can be lessons that lead to improvements, and as such they represent the basis for trial and error learning, which historically has been the precursor for engineering developments. Of course, it must be noted that there

TABLE I

Annual Productivity and Fuel Production Potential of Selected Biomass Crops[a]

Crop	Annual productivity (kg/ha)	Fuel production potential (kwh/ha)
Wood crops		
Loblolly pine	22,000	120,000
Aspen	9,000	43,000
Poplar	45,000	23,000
Eucalyptus	58,000	300,000
Agricultural crops		
Corn	36–58,000	200,000[b]
Wheat	40–67,000	220,000[b]
Potatoes	50,000	190,000
Rye grass	52,000	200,000
Sugarcane	143,000	560,000
Sugar beets	69–94,000	330,000

[a] Adapted from D. A. Tillman (1978). "Wood as an Energy Source." Academic Press, New York.

[b] This figure includes total biomass production rather than merchantable wood production. It includes branches, foliage, and other elements of the complete plant.

are successful tropical tree plantation systems, such as with species of *Pinus* and *Eucalyptus*, and the future is positive for their development. Using newer techniques from biotechnology, biological engineers can increase the productivity of tropical plantations and enable them to survive the economic challenges of an increasingly competitive marketplace.

B. Ethanol Production

Production of ethanol from crop plants as a fuel for automobiles developed in the mid-1970s as a consequence of the first global energy crisis. Ethanol was one of many alternative energy sources that was investigated to be a substitute, or at least a supplement, for petroleum. This represents an intermediate scale of biological engineering because aspects of conventional crop production and the technological process of conversion of biomass into ethanol have involved much more engineering input compared to the tree plantations.

The system of ethanol production includes crop production, harvest and pressing, and fermentation and distillation of juices to form ethanol. There are opportunities for sustainable use of by-products as

a source of heat for distillation with bagasse and as a fertilizer with liquid wastes. Quality control, conversion efficiency in distilleries, and by-product utilization are important considerations that require engineering input along with agronomic efficiency of the crop.

Overall energy analyses of pilot projects and commercial facilities have shown that there is little net energy yield, and sometimes a new energy loss, from ethanol production. However, there is continuing interest in the potential of this energy source. Strong programs exist in Iowa in the United States using corn as input and in Brazil using sugarcane. In both cases the production is subsidized by the government, which allows the operations to continue. If petroleum prices rise, ethanol may become more viable. An important advantage of this energy source is that it provides a country with an internal source of fuel and buffers the country from the world fuel markets. Also, new enzyme and recovery technologies are being developed that may improve efficiency and bring on large-scale commercial production. Biological engineers contribute in many ways to biomass growth, harvesting technologies, and production of ethanol from biomass and their creative uses of by-products may help lead to more sustainable production.

C. Biosphere 2

The most highly engineered example of sustainable biomass production is the Biosphere 2 project north of Tucson, Arizona, which consists of a large, closed mesocosm (1.3 ha) designed to test life support technologies for space travel. The project was started in 1984 through a private company called Space Biosphere Ventures. After experimentation with a smaller test module, the large mesocosm was constructed and on September 26, 1991, eight people were sealed inside for two years. A second mission with a crew of six was run from March 6 to September 17, 1994. The system was initially closed to gas and material cycles with the only energy input from outside in the forms of sunlight and electricity, but additional oxygen was injected since the first year of operation and some food was imported.

The facility is called Biosphere 2 because it simulates the Earth's biosphere in striving to be a closed system with human occupants. The facility contains a human living area, an agricultural biome, and five natural biomes (tropical rain forest, savannah, marsh, desert, and ocean) (Fig. 3). The structure consists of a space frame supporting special laminated glass and mounted on a belowground stainless-steel liner. Two "lungs," or large empty chambers, are connected to the facility to accommodate contraction and expansion of the atmosphere with daily heating and cooling. The facility is similar in appearance to a conventional greenhouse but is much larger and is sealed to gas exchange. It was designed to recycle air and water inside the structure.

Except for input of electricity and outside maintenance, Biosphere 2 is designed to be a sustainable system in that material cycles maintain the atmosphere and the agricultural biome provides food for the humans occupying the system. A true type of sustainability can be achieved if income gener-

ated by the project through tourism and technology development and sales can balance the costs of electricity and maintenance. However, especially at this early stage of the project (it was designed to last 100 years), the most important feature of the experiment is that it can maintain a fairly large number of humans in a nearly closed system for several years. The two major achievements are sustainable agricultural production to support the people and the maintenance of a regenerative atmosphere.

The agricultural biome consists of polycultural plots of many plant species and several species of fish and domestic animals that are also included in the system. The atmosphere has been maintained in large part by the inclusion of the large natural biomes. These systems contain a variety of microbial species that actively regulate the cycles of gases. Thus, the philosophy has been to include a high diversity of species in the form of ecosystems to produce sustainability. This is a major accomplishment that has required much biological engineering. Creative design of Biosphere 2 used

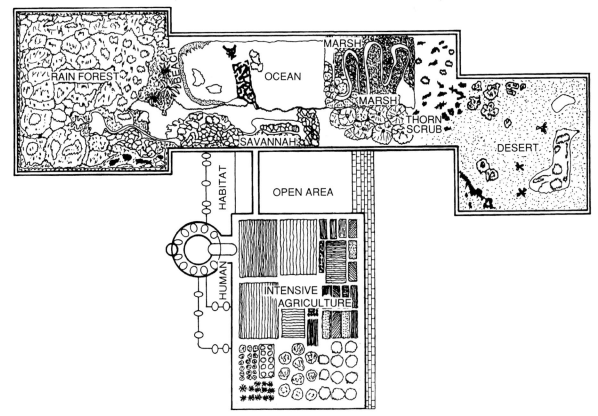

FIGURE 3   Layout of Biosphere 2, a nearly closed sustainable experiment in human habitation.

knowledge of gas exchanges, nutrient requirements, bioenergetics, and compatibilities of many living species. In fact, because of the use of ecosystems in the design, Biosphere 2 is actually an example of a subset of biological engineering called ecological engineering. This is a new discipline with a focus on utilizing multispecies systems to solve engineering problems.

Glossary

Biological engineering Engineering applied to or used with biological systems.
Biomass Organic matter created by living organisms.
Biome Ecological unit of the landscape that includes a major vegetation type.
Energy analysis Energy input–output balance calculation.
Mesocosm Medium-scale, constructed ecological system.
Sustainability Perpetual production with limited energy and nutrient inputs.
Waste Term here representing those products not of primary interest, such as by-products, coproducts, losses, and other wastes.

Bibliography

Allen, J. (1991). "Biosphere 2, The Human Experiment." New York: Penguin Books.

Hohmann, N., and Rendleman, C. M. (1993). "Emerging Technologies in Ethanol Production," Agriculture Information Bulletin No. 663. Washington, D.C.: USDA Economic Research Service.

Johnson, A. T., and Rehkugler, G. E. (1990). Career opportunities in biological engineering. *Careers and the Engineer* Spring, 30–34.

Kitani, O., and Hall, C. W. (1989). "Biomass Handbook." New York: Gordon & Breach Science Publishers.

Parnell, C. B., Jr. (1985). Systems engineering of biomass-fueled energy alternatives. *In* "Biomass Energy, A Monograph" (E. A. Hiler and B. A. Stout, eds.), pp. 213–248. College Station, Texas: Texas A&M University Press.

Russel, J. A. (1942). Fordlandia and Belterra, rubber plantations on the Tapajós River, Brazil. *Economic Geography* **18**, 125–145.

Shen, S. (1988). Biological engineering for sustainable biomass production. *In* "Biodiversity" (E. O. Wilson and F. M. Peter, eds.), pp. 377–389. Washington, D.C.: National Academy Press.

Takakura, T. (1989). "Climate Under Cover: Digital Dynamic Simulation in Biological and Agricultural Sciences." Tokyo: Laboratory of Environmental Engineering, University of Tokyo.

Tillman, D. A. (1978). "Wood as an Energy Resource." New York: Academic Press.

Vitousek, P. M., Ehrlich, P. R., Ehrlich, A. H., and Matson, P. M. (1986). Human appropriation of the products of photosynthesis. *BioScience* **36**, 368–373.

countries, it is still a common occurrence in agricultural areas in industrialized countries as well. It is most often used in conjunction with grain crops such as wheat, rice, and corn and in sugar cane cultivation. Burning the stubble after harvest reduces the amount of preparation time required for the next planting. It is also used to eliminate pests and is thought to reduce the carryover of disease.

The vast majority of biomass burning occurs in the savanna regions of Africa and South America. Because most of the savanna regions in the world are used for grazing, the primary incentive for burning in the savanna is to remove the dry grass and shrubs and encourage the growth of new grass that is more palatable to livestock. In addition, fire is used to eliminate shrubs, trees, pests, and for ash fertilization.

C. Reasons for Concern

Although biomass burning applications are not unique to the present, there is a great deal of concern surrounding the amount of fire activity especially in the tropics and subtropics and the possible effects this may have on our global environment. Biomass burning is a distinct biogeochemical process with implications in many different arenas within the biosphere, atmosphere, and geosphere. Very little is known concerning the effects of biomass burning within each of these systems. Even less is known about the interactions between the systems. Some of the more immediate areas of concern, primarily regarding biomass burning associated with deforestation and grassland management, include effects on species diversity; the gene pool; erosion of soil; increased water runoff, changes in the surface albedo; decreases in evapotranspiration; higher concentrations of CO_2, other trace gases, and aerosols; effects on atmospheric chemistry and the radiation budget; and possible implications for both regional and global climate change.

II. GLOBAL BIOMASS BURNING

A. Geographic Distributions and Seasonal Variations

Current estimates suggest that biomass burning consumes from 6 to 9 Pg (1Pg $= 10^{15}$g) of dry material each year. The overwhelming majority of dry biomass burned each year is concentrated within the developing countries in the equatorial and subtropical regions, with maximum burning occurring during the dry season. Current estimates suggest that over 85% of all biomass burning emissions are produced in the tropics. Nearly one-third of the global biomass burning emissions are believed to be concentrated in Africa in the region lying between 15°N and 25°S. Much of Africa is covered with savanna-type vegetation, consisting of grassland with varying amounts of trees and shrubs. Although very little is known about the actual amount of burning in Africa, estimates suggest that nearly 90% of the biomass burning emissions in Africa are related to burning of the savanna. The remaining fires are primarily linked to deforestation activities, burning of wood for fuel, and disposal of agricultural waste. Savanna fires follow a seasonal cycle, with the maximum occurring in the middle of the dry season, and are thought to burn on a 1- to 3-year cycle. The dry season can last from 8 to 9 months in the semiarid regions and from 4 to 5 months in the more humid regions near the perimeter of tropical forests. In the northern hemisphere, the dry season lasts from approximately December to March, with maximum burning occurring during January and February. In the Southern Hemisphere, the dry season extends from June to November with the peak burning period in September and October.

The majority of fires associated with deforestation activities in Africa are located in the tropical forests of the Ivory Coast, Cameroon, Nigeria, Zaire, and Madagascar. These biomass burning activities account for only a fraction of the total biomass burned each year in Africa, but they represent a serious threat to the future of the closed tropical forests in some of these countries. Some of the most aggressive deforestation in Africa is taking place in Nigeria and the Ivory Coast, where each year an estimated 10% of the remaining forest is cleared for agricultural applications. In the Ivory Coast approximately 70% of the forest present at the turn of the century has been removed.

Biomass burning from deforestation activities is most often associated with clear-cutting in the Amazon, which represents approximately 30% of all tropical forests. The United Nations Food and Agricultural Organization has suggested that during the 1980s roughly 154,000 km^2 of tropical forest were lost each year; nearly 40% was attributed to deforestation activities in South America, primarily in the Brazilian Amazon Basin. More conservative estimates based on satellite analyses suggest that the annual rate of deforestation in the Brazilian Amazon Basin lies between 15,000 and 30,000 km^2/ year. In Brazil, the dry season lasts from approximately June to October with peak burning during the months of July, August, and September.

The majority of biomass burning in South America occurs in the cerrado regions south of the Amazon. The cerrado is composed primarily of grass and scrub land. Preliminary estimates suggest that each year during the dry season approximately 400,000 km^2 or 20% of the cerrado is burned to rid the land of weeds, pests, disease, and dry vegetation. Burning in the cerrado often begins in May, steadily increasing in June and July, and reaches a maximum in August. The burning begins to taper off during the latter part of September.

Although biomass burning in Africa and South America dominates global fire activity, biomass burning in other locations should not be ignored. For example, clear-cutting in South America may represent the largest annual deforestation effort in the world, but other countries in Southeast Asia and Central America are clear-cutting a greater proportion of their forests each year, threatening the very existence of unique regional tropical rain forests. Although their contribution to global biomass burning emissions may not be substantial, these biomes contain some of the most diverse environments on earth. The loss of biodiversity and exotic species may prove to be of equal or greater significance to humankind. [See BIODIVERSITY, VALUES AND USES.]

The fires common to the eucalyptic forests and woodlands of Australia provide yet another example of fire activity. They represent a unique application of fire used to reduce accumulated forest fuels through a system of prescribed burns where the fires burn under controlled weather conditions. If a wildfire is subsequently ignited in these areas, the vegetation will burn with a much lower fire intensity, reducing the devastating effects of high intensity wildfires. Globally, emissions associated with fires in the temperate and boreal forests of the northern hemisphere represent only a very small fraction of the total biomass burned each year, yet they include some of the largest fires ever recorded. Although a system of prescribed burning has reduced the severity of wildfires in North America, they remain a continual threat to areas where wildlands and urban sprawl intermix, especially in California, as evidenced in the devastating losses in the Oakland fires in the fall of 1991 and the fires in southern California that were fanned by the Santa Ana winds in the fall of 1993.

B. Burning in Three Distinct Ecosystems

The following sections take a closer look at biomass burning activities in three distinct ecosystems where fire has played an important role in the past and that represent active areas of burning today, including the boreal forests of the Northern Hemisphere, the tropical forests in the Amazon Basin of South America, and savannas in South America and Africa.

I. Boreal Forests of the Northern Hemisphere

Although the tropics represent the current hot spot for biomass burning, fires are quite common in the boreal forests of the Northern Hemisphere. The largest areas of boreal forests are found in two major belts extending across North America and Eurasia between 45°N and 70°N, primarily in Alaska, Canada, Norway, Sweden, Finland, and the former Soviet Union. The majority of boreal forest fires occur in Canada and the northern portion of Russia and in Siberia. Over the past 50 years, Alaska has experienced a steady increase in the number of fires corresponding to an increase in human access to the forests, but there has been a decrease in the area burned, primarily attributed to quick response and fire management and control practices. The story is somewhat different for Canada, which has experienced both a steady increase in the incidence of fires primarily in response to human activity and also a significant increase in the area burned, especially during the 1980s. This increase was associated with an extreme interannual variability in burned area and may represent a temporary trend associated with unusually severe fire weather conditions throughout central and western Canada. Statistics concerning fire activity for Scandinavia are limited but suggest that there has been an overall decrease in the area burned throughout the region, with the majority of the fires attributed to anthropogenic sources. National fire statistics are virtually nonexistent for the former Soviet Union, but satellite analyses and other estimates suggest that forest fires in northern Russia and Siberia account for the largest areas of boreal forest burned each year.

Boreal forests are composed primarily of the coniferous trees pine, spruce, and fir and deciduous hardwoods such as birch, alder, willow, and poplar. The boreal forests provide an excellent example of a fire-dependent ecosystem where wildfires have played a major role in their development and very existence throughout history. The vegetation species of the boreal forests are uniquely suited to survive during years of severe wildfire activity usually associated with very dry conditions.

The annual combined contribution of biomass burned in both temperate and boreal forests represents less than 4% of the total biomass burned globally each year, but the area consumed by a single wildfire episode in the boreal forests can be equivalent to the work of thousands of colonists in the tropical forests. In May of 1987, nearly 100,000 km² of boreal forests were destroyed by fire in the People's Republic of China and Siberia over a three-week period. The fires were attributed to both lightning strikes and anthropogenic sources and occurred in the midst of a prolonged dry period.

Most fires in the boreal forest are the result of anthropogenic activities, although the largest fires are often associated with lightning strikes in remote wilderness areas not easily accessible to humans. In Canada, lightning accounts for roughly 35% of all boreal forest fires, but they represent approximately 85% of the total area burned. Although

the incidence of fires has increased with increased human access to the forest, most anthropogenic fires are located near fire control technology and can usually be contained before the fire spreads out of control. Lightning-induced fires, on the other hand, are often located in remote areas. The fires are not detected as quickly and cannot be easily reached with conventional fire suppression equipment and can easily spread over large areas. Over 95% of the area burned is associated with 2 to 3% of the fires. In recent years, prescribed fires have become more frequent as part of fire management policies in North America aimed at reducing the possibility of wildfires, especially in regions located near valuable resources or population centers.

2. Tropical Forests in the Amazon Basin of South America

In recent years the world has come to appreciate the importance of the continental tropics, yet it remains one of the least understood areas on earth. The tropical forest biome is believed to contain between 50 and 90% of all plant and animal species, over several million species, of which only one-half million have been identified. Tropical forests comprise the most productive and variable ecosystems. In addition, the tropics receive the maximum amount of visible and ultraviolet radiation, thus acting as a primary driving force for atmospheric circulation and conceivably playing a major role in atmospheric photochemistry. For many years the tropics were relatively undisturbed. This is no longer the case. Most of the tropical forests are located in developing countries where economic pressures and increased population demands are providing the impetus for expansion into the forest. Each year tropical forests are being converted to agriculture, rangeland, and abandoned degraded secondary forest. Much of this deforestation is taking place in the Amazon Basin of South America.

In the Amazon Basin, the forest is slashed and burned to use the cleared land for farming, grazing, and mining. To a lesser degree the forest is also cut for timber in response to a growing demand in the developed countries for beautiful and exotic hardwoods. During the past decade, one of the most intensive clearing sites was located in the province of Rondonia, Brazil, along the newly paved BR-364 highway. As part of a government-sponsored colonization project, immigrants to Rondonia were sold 100-hectare plots of forest to be cleared and used for agriculture and pasture. The photograph in Fig. 1 depicts an example of the slash-and-burn technique used throughout the Amazon Basin. As soon as the rainy season has ended, the vegetation is cut and left to dry. Sometimes bulldozers are used to collect the material into a pile. Once the material has had a sufficient time to dry, it is set on fire. Usually less than 30% of the material is burned in the first burn. Through successive burns and the process of decay, the land is cleared for agriculture. Most soils in the Amazon are not suitable for traditional farming, lacking necessary nutrients and elements. In the humid tropics the decaying organic matter associated with the

tropical rain forest is often the only source of soil nutrients because of advanced laterization. On average it is economically feasible to produce crops on a given parcel for only a few years owing to the high cost of fertilizers, disease, and loss in soil fertility. As the nutrients in the soil are consumed, the farmer clears more land and abandons the land cleared earlier. In this type of shifting agriculture, the abandoned land may be able to recover into a modified or secondary form of tropical forest vegetation, especially in areas along the perimeter of virgin undisturbed forest. In other instances, through the use of fire, the tropical forest vegetation is permanently eliminated. Land that is no longer able to be used for cultivating crops is converted to pasture and used for grazing cattle. Because of erosion, poor soil, repeated burns, and inappropriate management practices, eventually much of this land becomes incapable of supporting

FIGURE 1 A typical biomass burning scene in a tropical forest associated with the slash-and-burn techniques used throughout the Amazon Basin. This small fire was photographed in the Brazilian state of Rondonia in 1986. [From J. P. Malingreau and C. J. Tucker (1988). *AMBIO* **17,** 49–55.]

any type of agricultural application. [*See* FOREST CLEAR-CUTTING, SOIL RESPONSE.]

3. Savannas of Africa and South America

Biomass burning associated with the management of savanna ecosystems is found throughout the earth, although most of the activity is centered in Africa, South America, tropical Asia, and along the northern coast of Australia. Savanna-type vegetation can most easily be described as grasslands with varying concentrations of shrubs and trees. Approximately 40% of all savannas can be classified as arid savanna with an annual rainfall of less than 700 mm. The arid savanna consists mainly of shrub-type vegetation. Because of a lack of adequate precipitation, the grass layer is not very well developed, resulting in minimal fuel load during the dry season and low fire frequencies. Humid savannas, which comprise approximately 60% of all savanna-type vegetation, contain a considerable concentration of woody species. The annual rainfall in humid savannas is in excess of 700 mm, which allows the grass layer to flourish. Savannas can be characterized as a fire-dependent ecosystem that relies on recurrent fire activity to maintain the balance between grass, shrubs, and trees. Without fire the vegetation would soon be dominated by shrubs and brush. Fire generally consumes the grass, shrubs, and brush, but unless they are cut down first, the mature trees remain intact and have become resistant to repeated burns.

Some savanna fires are lightning initiated, but most are associated with human activity. Although fire has always been an integral part of the savanna landscape, and is subject to considerable interannual variability, there are indications that fire may be on the increase in response to population pressures, especially in the developing countries of Africa and South America. The largest concentration of savannas is found in Africa extending across the continent in two separate belts north and south of the tropical forests located along the equator. Figure 2 is a photo of biomass burning taken in Zambia, Africa in August, 1992. It represents a typical biomass burning scenario in the African savannas. Although very little is known about the actual ex-

tent of burning in the tropics, recent estimates suggest that on average 30–50% of the African humid savannas are burned each year. Other studies indicate that on average the interval between burns is on the order of 1 to 3 years. Based on the current knowledge of global burning practices, African savanna fires may be responsible for up to a third of all global biomass burning emissions. In South America the largest expanse of savanna (cerrado) is located south of the Amazon and east of the Andes in southern Brazil, Bolivia, Paraguay, and Argentina. Each year during the dry season the region is covered with smoke associated with biomass burning activities in the cerrado and from tropical deforestation efforts in the Amazon. Previously the cerrado was subject to fire on a 3- to 4- year cycle. In the past several decades, however, the interval between burns has steadily decreased.

The primary agricultural application for the savannas in Africa and South America is cattle grazing. As the dry season progresses, the grass begins to wither and becomes unpalatable to livestock, resulting in weight loss and decrease in milk production. To stimulate the growth of new grass, the savanna is burned. Within a matter of days to weeks the scorched landscape shows signs of a fresh layer of green grass. In some regions the savannas undergo repeated burns during a single season to keep up with the demand for fresh grass. This represents the largest component of biomass burning activity in the savannas. In recent years there has also been an increase in the conversion of savanna into farmland, primarily in response to the rapidly increasing population in developing countries and the possibility of selling surplus goods at home and abroad. By the latter part of the last decade the cerrado regions of southern Brazil were producing nearly one-third of the country's grain crops (e.g., wheat, rice, corns, beans, and soybeans). Biomass burning plays a significant role in cultivation practices as part of the initial clearing campaign and subsequently it is used to rid the land of weeds, pests, and disease. Following harvest, fire is used to eliminate the remaining debris in preparation for the next crop. Accidental fires are also a common occurrence.

FIGURE 2 This photo was taken in Zambia, Africa in August, 1992. It represents a typical biomass burning scenario in the African savannas. [Photo by Chris Justice UMD/NASA/GSFC.]

III. ENVIRONMENTAL IMPACTS OF BIOMASS BURNING

Although biomass burning is virtually a global occurrence, the overwhelming majority occurs in the tropics. The following sections discuss some of the principal concerns involving burning practices in the tropical forests and savanna regions and possible implications for both regional and global climate change.

A. Ecological Implications

One of the most serious implications of deforestation and associated biomass burning in the tropical forests is the impact on species diversity. Plant and animal life flourish in the constant warmth and humid conditions of the tropics. The tropical forests cover only 7% of the terrestrial surface of the earth, yet they contain over half and possibly as many as 90% of all plant and animal species. Only

a small fraction of these have been identified, and some have proven to be extremely beneficial to industry and have contributed to medical breakthroughs. Deforestation activities impact biological diversity by destroying habitats directly through clear-cutting, but also by isolating portions of a habitat and by increasing edge effects in the boundary regions between the primary forest and deforested regions. In terms of biological diversity, the area adversely affected by deforestation activities may be over twice as large as the actual deforested area. Maintaining biodiversity also plays a large role in determining whether an abandoned plot will be able to recover. Once a plot has been deforested, cultivated, and abandoned, the fate of that land is largely dependent on the availability of suitable species to colonize the area. Through successional development it may be possible for the land to return to a secondary type of tropical forest. Unfortunately, when the land is subjected to repeated burns, tropical forest species that are less resistant to fires are often eliminated and replaced by those that have developed adaptations to fire.

Of course there are many other ecological implications of biomass burning associated with deforestation and the successional process. The majority of the soil in the tropical forests is poor in nutrients. The vegetation in primary forests compensates for this through a shallow root system that obtains nutrients from decaying organic matter. Initially when the land is cleared and burned, there is a larger increase in key plant nutrients from the ash and decomposing plant material as the nutrients that were stored in the standing vegetation become available and can temporarily accelerate the nutrient cycle. Unfortunately this process is short-lived. In the absence of protective tree cover and a well-developed understory, the top 10–20 cm of organic matter is vulnerable to accelerated erosion and loss of soil nutrients as a result of weathering. As the nutrients are lost without a replenishment source, the land quickly loses its ability to sustain agriculture after only 2 to 3 years of cultivation. Furthermore, in some cases, as a result of the erosion process, large quantities of silt are deposited in the river channels and reservoirs affecting water quality and endangering aquatic life.

Although fire is a vital component of the savanna ecosystem, the increased frequency of fire activity is a cause for concern. In the short term, burning the dry vegetation may accelerate the mineral nutrient cycle by freeing up the nutrients tied up in dry vegetation, but repeated burns can lead to increased erosion, leaching of nutrients to groundwater, and eventually result in nutrient-deficient soils. By experience, some of the farmers of the cerrado region in Brazil have learned that if the grassland is burned too often, the pasture weakens and the productivity is greatly reduced.

B. Hydrologic Cycle and Climatological Implications

The composition of the earth's surface plays a major role in both the regional and global heat and water budgets and depends on factors such as the surface albedo, soil moisture, and surface roughness. The regional water budget in tropical forests is a distinct example of the close relationship between vegetation and the hydrologic cycle. Although numerous modeling studies have attempted to address the implications of deforestation on the local and regional hydrologic cycle, complicated interactions involving the geosphere, atmosphere, and biosphere have left the scientific community with only a marginal understanding of the impact of biomass burning associated with deforestation activities. The myriad of often contradictory results reveal the complexity of the interactions on a regional and global basis.

Some possible hydrological and climatological effects include changes in surface albedo, increase in diurnal temperature ranges, increased surface temperatures, decreased soil moisture, extended dry seasons, increased surface runoff during rainy periods, reduced evapotranspiration, and associated reductions in the sensible heat flux and latent heat release to the atmosphere. Approximately half of the precipitation in the tropical forests of South America is the result of moisture that is directly recycled through evapotranspiration associated with the dense array of trees and highly developed understory. When the land is deforested and replaced with savanna-type vegetation, which has a lower rate of evapotranspiration, there is a net loss in the amount of water that is directly recycled to the atmosphere. A reduction in the amount of water that is recycled through the atmosphere also reduces the latent heat available to the atmosphere, which ultimately affects one of the major global atmospheric heat sources. The majority of incoming solar radiation in the tropics is used in the evaporation process. Upon condensation, the stored latent heat becomes available as sensible heat and is transported to the temperate and polar regions by way of various atmospheric circulation cells. The consequences of reducing the amount of available latent heat and sensible heat in the tropics in terms of global climate change are not very well understood, but some preliminary results indicate that the reduction in the poleward transport of heat and moisture could lead to cooling in the extratropics, displaced storm tracks and a reduction in the length of the growing season in the midlatitudes.

C. Atmospheric Emissions

Studies have shown that intensive biomass burning associated with naturally occurring forest fires, deforestation practices, and savanna management is a major source of trace gases such as NO, CO_2, CO, O_3, NO_x, N_2O, NH_3, SO_x, CH_4, and other nonmethane hydrocarbons, as well as an abundant source of aerosols. Current estimates of trace gas emissions associated with global biomass burning are based on techniques that relate ratios of trace gas concentrations and particulate matter to the concentration of CO_2 produced by biomass burning as a function of ecosystem and fire conditions. Preliminary global estimates indicate that annual biomass burning may be associated with 38% of the ozone in the troposphere; 32% of global carbon monoxide; more than 20% of the world's hydrogen, nonmethane hydrocarbons, methyl chloride, and oxides of nitrogen; and approximately 39% of the particulate organic carbon. Although these estimates include a wide range of uncertainty, it is becoming evident that these emissions may be as important to global air chemistry as industrial activities in the developed world. The tropics receive the maximum amount of visible and ultraviolet radiation, thus acting as a primary driving force for atmospheric circulation. Because most biomass burning in the tropics is limited to a burning season, the temporal and spatial concentrations of emissions are expected to have a noticeable impact on atmospheric chemistry and climate in the tropics and globally as a result of increased trace gas emissions and the direct and indirect radiative effects of smoke aerosols.

In the past, much attention has centered around increased CO_2 and trace gases associated with biomass burning and the possible influence on net greenhouse warming. The exact contribution of biomass burning to increased global CO_2 concentrations is unknown but may lie somewhere between 25 and 40% and must be considered in evaluations of sources and sinks within the global carbon cycle. In the savanna regions, the CO_2 released when the biomass is burned is promptly reincorporated in regrowth through photosynthe-

sis, often in the same growing season. The CO_2 released in the process of clearing the tropical forests through burning and also in the subsequent decay of the remaining biomass, above and below ground, is much greater than the amount that is reincorporated in the vegetation that replaces it, resulting in a net flux to the atmosphere. Other pollutants associated with biomass burning such as ozone, methane, and nitrous oxide also contribute to the greenhouse effect. Unlike CO_2, there is no biological sink for methane once it is injected into the atmosphere. Although the amount of methane released is small in comparison to CO_2, methane is a much more powerful greenhouse gas.

Recently, analyses have indicated that the direct and indirect radiative effects of aerosols from biomass burning may be equally significant and may be a major factor in climate change. The two primary radiatively active components of aerosol emissions from biomass burning are particulate organic and graphitic carbon. Particulate organic carbon is the largest constituent and generally serves to scatter incoming solar radiation, whereas graphitic carbon particles can increase the solar radiation absorption in the atmosphere and clouds. The primary direct effect of particulate organic carbon aerosols is that they reflect incoming solar radiation back to space and act to cool the climate. The indirect effects of particulate organic carbon are associated with the ability of these aerosols to act as cloud condensation nuclei (CCN). Because clouds are one of the most important controls on the heat balance of the earth, significant changes in cloud type and cover could impact regional and global climates. Increased CCN loading generally results in increased droplet formation of a smaller size given a constant amount of available water. Clouds that are made of smaller droplets are usually assumed to be whiter, reflect more solar radiation, and may be less likely to produce rain. In the case of clouds that are optically thick and very bright to begin with, the effect of increased cloud drops is dominated by increased absorption and results in a darkening of the clouds.

IV. BIOMASS BURNING MONITORING EFFORTS

A. Remote Sensing Applications

Before we can assess the impact of global biomass burning we must be able to monitor the extent of burning. In the past decade many techniques have been developed to monitor biomass burning activities associated with deforestation and savanna fires. The vast majority of these methodologies involve the use of satellite remote sensing technology, which currently offers the most efficient and economical means to monitor biomass burning on both a global and regional scale. Table I provides an abbreviated list of current satellite sensors that have been utilized to monitor fires.

Fires exhibit a variety of attributes that can be monitored from space. Postfire char and scar can serve as an indication of biomass burning. Char refers to the blackened earth surface that is evident following a burn, whereas scar more specifically refers to the changes in vegetation that occur as a result of biomass burning. High-resolution Landsat and SPOT multispectral data have been used to locate fires by identifying char, which is black across the electromagnetic spectrum and can be distinguished from both green and senescent vegetation. Immediately following a burn, the ground is covered by a layer of black carbon soot, which typically has a lower albedo than surrounding unburned vegetation and is accompanied by a sharp decrease in the near-infrared (IR) wavelength (0.7–1.1 μm). A combination of the visible (0.58–0.68 μm) and near-IR data available from the NOAA AVHRR platforms have been used to derive vegetation indices to investigate biomass burning and its impact on the surrounding vegetation. The most widely used index is the normalized difference vegetation index (NDVI), which represents a measure of green leaf density. It is a contrast ratio between the visible (VIS) and near-IR (NIR) channels of the AVHRR instrument and is expressed as

$$NDVI = \frac{NIR - VIS}{NIR + VIS}.$$

The AVHRR visible channel corresponds well with the strong chlorophyll absorption band, whereas the near-IR channel corresponds to the highly reflective region in the near-IR associated with healthy green vegetation. Studies have shown that the NDVI results in high values for vegetated areas and low values for bare ground, clouds, water, and recently burned areas.

A decrease in the surface albedo also results in increased absorption of solar energy at the surface. If the burned areas are large enough, there will be a noticeable increase in the surface brightness temperature as observed in the shortwave IR window (4 μm), which is sensitive to both emitted energy from the earth's surface and reflected solar radiation. Figure 3 represents NOAA AVHRR channel 3 shortwave IR window imagery over the state of Rondonia, Brazil, for July 1982 and August 1985. The lighter areas within the light gray outline are deforested areas that appear warmer than surrounding undisturbed primary forests. The linear features are associated with forest-clearing roads that emanate from the paved highway BR-364. A comparison of the two years shows a dramatic increase in deforestation activities along BR-364 from 1982 to 1985. The lighter areas outside of the gray outline southwest of Rondonia represent naturally occurring savannas in the country of Bolivia.

One of the limitations of using vegetation changes to identify biomass burning associated with deforestation is in the difficulty of distinguishing between primary and secondary growth. Furthermore, in the cerrado regions, regrowth occurs within days of the burning episode. An alternative is to monitor the actual fire events. The shortwave (4 μm) and longwave (11 μm) infrared window channels available from the GOES VAS and NOAA AVHRR satellite sensors have been used in the detection of active fires. Although both spectral regions can be used to sense the earth's surface, the shortwave IR region is especially useful in detecting subpixel fire activity. As the surface temperature increases, the peak of the Planck function shifts toward shorter wavelengths, so the radiance increases more rapidly at 4 μm than 11 μm. Fires can be located by identifying anomalously high temperature sources in the shortwave IR channel.

TABLE I

Examples of Satellites Used for Biomass Burning Investigations

Satellite sensor	Biomass burning monitoring capabilities	Spatial resolution	Temporal resolution
Landsat			
Multispectral Scanner (MSS)	Visible (multiband), near-IR (smoke, char, scar)	80 m	16 days
Thematic Mapper (TM)	Visible (multiband), near-IR mid-IR, thermal IR (smoke, char, scar, fires)	30 m 120 m (thermal)	16 days
Systeme Probatoire pour l'Observation de la Terre (SPOT)	Visible (multiband), near-IR (smoke, char, scar)	10–20 m	variable (days–weeks)
NOAA Polar Orbiting Satellites Advanced Very High Resolution Radiometer (AVHRR)	Visible, near-IR, shortwave IR window, longwave IR window (smoke, char, scar, fires)	1 km LAC[a] 4 km GAC[b]	1 daytime 1 night-time
Geostationary Operational Environmental Satellite (GOES)			
GOES-7 Visible Infrared Spin Scan Radiometer (VISSR) and Atmospheric Sounder (VAS)	Visible, shortwave IR window, longwave IR window (smoke, scar, fires)	1 km visible 7 km or 14 km IR	half-hourly
GOES-8 Imager		1 km visible 4 km IR	half-hourly
European Space Agency (ESA) Remote Sensing Satellite (ERS-1)			
Synthetic Aperture Radar (SAR)	C-band microwave (char, scar)	30 m	3 days
Along-Track Scanning Radiometer (ATSR)	Shortwave IR window, longwave IR window (char, scar, fires)	1 km	3 days
Defense Meterological Satellite Program (DMSP)	Night-time low-light visible (active fires)	2.7 km global 0.6 km limited	1 daytime 1 night-time

[a] LAC-Local Area Coverage.
[b] GAC-Global Area Coverage.

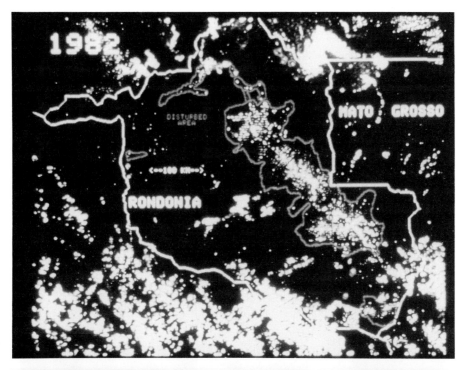

FIGURE 3 NOAA Advanced Very High Resolution Radiometer (AVHRR) shortwave IR window (4 μm) imagery for the state of Rondonia, Brazil, in July 1982 (top) and August 1985 (bottom). The lighter areas within the light gray outline are deforested regions that appear warmer than the surrounding undisturbed forests. A comparison of the two years shows a dramatic increase in the deforestation activities along BR-364 from 1982 to 1985. The lighter areas outside of the outline southwest of Rondonia represent naturally occurring savannas in the country of Bolivia. The image covers an area of approximately 700 km × 600 km. [From J. P. Malingreau and C. J. Tucker (1988). *AMBIO* **17,** 49–55.]

Figure 4 is a composite image indicating the locations of hot spots primarily associated with savanna fires in southern hemisphere Africa as detected in NOAA-11 AVHRR imagery from May–October, 1989. Various algorithms have been developed that use a combination of the shortwave and longwave IR data to create thresholding schemes that eliminate other hot but nonfire pixels and attempt to determine the subpixel area on fire. The light emitted from fires can also serve as a signal to locate biomass burning activities as they occur. The low-light visible sensor available on the DMSP satellite is sensitive to terrestrial light sources at night and has been used to monitor global fire activity. By identifying heat or light sources, remote sensing technology provides a means of locating fires and thus estimating the relative amounts of burning, as well as infering the distribution patterns and trends on various time scales.

Often the fire event is too small to be distinguished by satellite sensors and can be identified only by the associated smoke plume. During daylight hours, smoke plumes extending from the fire pixels can be detected in the visible portion of the electromagnetic spectrum. Figure 5 represents a GOES visible image of South America and the Atlantic Ocean at the peak of the burning season on the morning of 29 August 1988. Over a period

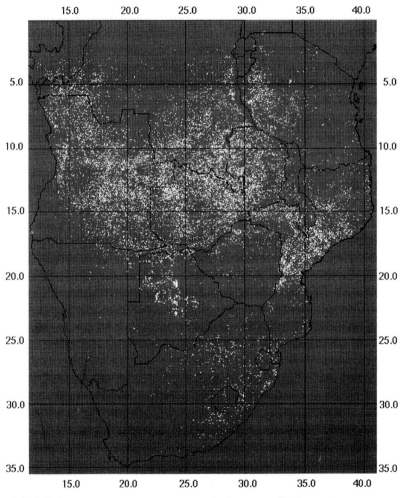

FIGURE 4 This composite image displays the locations of high temperature sources (active fires) primarily associated with savanna fires in southern hemisphere Africa as detected in NOAA-11 AVHRR imagery from May through October, 1989. [Produced by NASA/GSFC/GIMMS.]

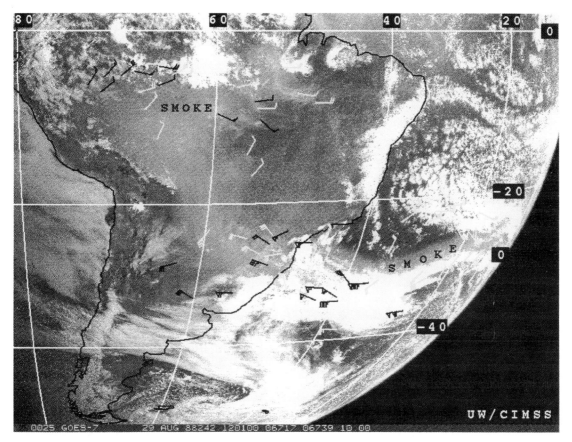

FIGURE 5 GOES visible image of South America and the Atlantic Ocean on the morning of 29 August 1988. The smoke pall covers the area east of the Andes Mountain from the equator to 30°S and extends out over the Atlantic Ocean at its southern boundary to the edge of the image. The prevailing circulation is inferred from looping satellite images and tracking cloud or smoke tracers. The wind barbs display counterclockwise flow east of the Andes channeling the smoke over the Atlantic Ocean. The darker barbs represent upper level winds and the light barbs depict low level flow. The smoke plume is clearly evident over the ocean as it contrasts with adjacent clouds. [From E. M. Prins and W. P. Menzel (1992). *Int. J. Remote Sensing* **13(15),** 2783–2799.]

of five days, the region from 50°W to the Andes Mountains was covered with smoke and haze. The smoke pall depicted in Fig. 5 is one of the largest ever documented in South America. It covers the area east of the Andes Mountains from the equator to 30°S and out over the Atlantic Ocean at its southern boundary to the edge of the image. The smoke represents a combination of emissions from burning in the selva (tropical forest) and cerrado (grassland) regions as well as natural forest fires in the foothills of the Andes Mountains west of the cerrado. The temporal resolution of the GOES data makes it possible to determine the prevailing circulation and transport of aerosols in South America by examining a series of half-hourly visible and infrared images and tracking the motion of smoke, haze, and adjacent clouds. The wind barbs depict an anticyclonic flow throughout the Amazon Basin. The easterly winds in the northern portion of the Amazon Basin (5°S) transport the aerosols westward. The Andes Mountains act to deflect the material southward, where westerlies transport the material over the Atlantic Ocean.

In addition to monitoring plume activity, satellite remote sensing has been used to provide information on the effect of biomass burning on aerosol budgets and global trace gases. Although satellite imagery cannot provide detailed measurements of atmospheric aerosol constituents at this time, they can provide information concerning aerosol loading. Visible and near-IR data available from the NOAA AVHRR satellites have been used to determine aerosol optical thickness, particle size, and single-scattering albedo, which can be used to investigate the radiative impact of biomass burning activities. By utilizing known relationships between trace gases and aerosol loading, investigators have used satellite data to infer trace gas emissions as well.

B. Current Limitations and Future Possibilities

Clearly, remote sensing technology allows us to monitor biomass burning in ways never before possible on a global scale, but there are many limitations. Remote sensing investigations of biomass burning are constrained by the trade-off between spatial and temporal resolution and the inability to monitor high-temperature sources. No instrument currently available to the general remote sensing community is capable of providing high-resolution data on an hourly basis. High-resolution data available from the Landsat and SPOT series satellites can only be obtained for a specific location on a biweekly schedule at considerable cost. Furthermore, there is no consistent archive for the tropical forests. Most biomass burning occurs in the tropics, where clouds often interfere with effective monitoring when using these high spatial but low temporal resolution sensors. The GOES, on the other hand, can provide visible and IR full-disk coverage of the earth every half-hour but at much lower resolutions (see Table I), which limits its use for detecting small localized fire activity.

To date, remote sensing studies of biomass burning have relied on data from sensors that were not designed to monitor fires. Active fire detection techniques have utilized infrared data available from meteorological satellites. The sensors on these platforms were explicitly designed to monitor temperatures ranging from very cold cloud tops to typical land surface temperatures, not the very hot temperatures associated with biomass burning. Designed for meteorological application, the shortwave IR window channel on the AVHRR instrument often saturates when it detects fire activity.

Future satellites offer promise for improved biomass burning monitoring capabilities. The GOES NEXT series of satellites, beginning with the launch of GOES-8 in 1994, will continue to provide diurnal visible and infrared data but with greater radiometric sensitivity, higher spatial resolution (4 km at nadir), and improved signal-to-noise ratios. The METEOSAT Second Generation satellite to be launched in 2000 will offer a major improvement in the ability to monitor diurnal burning activities in Europe and Africa, with a suite of channels similar to those currently available on the AVHRR instrument at a resolution of 3 km at the subsatellite point. The Moderate-resolution Imaging Spectroradiometer (MODIS) scheduled for launch in 1998 will provide global coverage every one to two days in 36 spectral channels at spatial resolutions ranging from 250 to 1000 m. It will include two high-gain channels at 4 and 11 μm with saturation levels at 500 and 600K, respectively, which will enable calculations of subpixel fire size and temperature. Visible and near-IR channels will be available at 250-m resolution to provide the capability for monitoring burn scars. The Canadian Radarsat, launched in 1994, offers a unique opportunity to investigate changes in vegetation associated with biomass burning using the C-band (5.6 cm), which can penetrate clouds and provide global coverage at a resolution of 10 to 100 m. Each of these platforms represents improved capabilites and opportunities for monitoring biomass burning in the future.

V. CONCLUSIONS

As we attempt to understand the extent and implications of biomass burning on our global environment, we must bear in mind that biomass burning applications are not a new phenomenon and remain in a constant state of flux. In the developed countries of the Northern Hemisphere, fire management practices over the past several decades, as well as an increase in the use of fossil fuels instead of wood, are contributing to a reduction or leveling off in biomass burning activities but a steady increase in trace gas emissions from fossil fuel consumption. For the majority of the tropics and subtropics, biomass burning associated with deforestation and grassland management seems to have increased over the past 20 years. With a gradual shift from subsistence to market economy in the developing countries, we can expect to see continued increases in biomass burning associated with agricultural practices in the tropical forests and savannas. Although fire has always played a major role in the tropical savannas, there seems to be a significant increase in the fire return interval associated with agricultural applications. Biomass burning associated with the conversion of tropical forests to agriculture constitutes a major disturbance in an ecosystem that is not adapted to fire activity. Biomass burning involves complex biogeochemical processes. The possible impacts of these processes on the global environment and on regional and global climate change have prompted much discussion and numerous investigations, but many questions remain unanswered.

Glossary

Albedo Fraction of the total incident solar radiation that is reflected back to space without absorption.

Biogeochemical processes Those interactions and relationships that involve the circulation of chemical components through the biosphere from or to the lithosphere, atmosphere, and hydrosphere.

Carbon cycle Green plants incorporate CO_2 during photosynthesis and release it in association with respiration, decay, and combustion. The oceans play a large role in the carbon cycle, serving as a regulator for the amount of CO_2 in the atmosphere by absorbing and releasing CO_2.

Greenhouse warming The greenhouse effect suggests a warming of the earth in response to increased CO_2 and other greenhouse gases in the atmosphere that allow solar radiation to freely enter the atmosphere but that absorb and reemit terrestrial radiation back to the earth's surface contributing to increased surface temperatures.

Hydrologic cycle Complex cycling of water in the environment from rainfall to runoff to evaporation and back again.

Infrared window Those wavelengths in the infrared portion of the electromagnetic spectrum in which the atmosphere is particularly transmissive of electromagnetic energy. Infrared windows exist from 3 to 5 μm and 8 to 14 μm.

Remote sensing The process of gathering and recording information about a phenomenon, area, or object using a device that is not in physical contact with the object, area, or phenomenon being studied.

Bibliography

Aldhous, P. (1993). Tropical deforestation: Not just a problem in Amazonia. *Science* **259,** 1390.

Crutzen, P. J., and Andreae, M. O. (1990). Biomass burning in the tropics: Impact on atmospheric chemistry and biogeochemical cycles. *Science* **250,** 1669–1678.

Crutzen, P. J., and Goldammer, J. G., ed. (1993). "Fire in the Environment: The Ecological, Atmospheric, and Climatic Importance of Vegetation Fires." New York: John Wiley & Sons.

Goldammer, J. G., ed. (1990). "Fire in the Tropical Biota." New York: Springer-Verlag.

Justice, C., and Dowty, P., ed. (1994). "IGBP-DIS Satellite Fire Detection Algorithm Workshop Technical Report." IGBP-DIS Working Paper #9. Paris: IGBP-DIS.

Levine, J. S., ed. (1991). "Global Biomass Burning: Atmospheric, Climatic, and Biospheric Implications." Cambridge, Mass.: MIT Press.

Malingreau, J. P., and Tucker, C. J. (1988). Large-scale deforestation in the southeastern Amazon Basin of Brazil. *AMBIO* **17,** 49–55.

Prins, E. M., and Menzel, W. P. (1992). Geostationary satellite detection of biomass burning in South America. *Int. J. Remote Sensing* **13(15),** 2783–2799.

Prins, E. M., and Menzel, W. P. (1994). Trends in South American biomass burning detected with the GOES visible infrared spin scan radiometer atmospheric sounder from 1983 to 1991. *J. Geophys. Res.* **99(D8),** 16,719–16,735.

Skole, D., and Tucker, C. J. (1993). Tropical deforestation and habitat fragmentation in the Amazon: Satellite data from 1978 to 1988. *Science* **260,** 1905–1910.

Biosphere Reserves

M. I. Dyer

University of Georgia

I. Introduction
II. Background and History
III. Purpose and Design
IV. Related Programs
V. Worldwide Network
VI. Needs for the Future

Biosphere Reserves serve ecologists and environmental scientists with a global network of "outdoor laboratories" in which monitoring and basic research provide long-term background information about ecosystem productivity and sustainable development. Two major efforts currently exist that point the way for future research and development: the Biosphere Reserve project of MAB [Man and the Biosphere Programme within UNESCO (United Nations Educational, Scientific and Cultural Organization)] and the Long-Term Ecological Research (LTER) program of the U.S. National Science Foundation.

I. INTRODUCTION

As human populations continue to expand both their numbers and their impacts across the earth, politicians, scientists, and resource managers increasingly require more stringent and accurate land-use plans. This continued growth places extreme demands on our natural resources, resulting in crowding, changes in land fertility, new and additional needs for foods and fiber, and alterations in the quality of life, all of which affect various parts of the Earth differently. On a global scale, needs in arid cold regions, for example, obviously do not match those in warm tropical regions. Yet, essentially all societies strive to utilize what they have available to them in an optimal, sustainable manner. The question then arises—"what constitutes optimality, and how will we know when resource use becomes sustainable?" To address these questions on local to global scales, ecologists, land-use planners, and politicians have developed a plan of work that involves a network of sites, or reserves, in a large variety of ecosystems or landscapes where short-term to long-term experimental studies, and monitoring of controls over ecosystem productivity, can help provide definitive answers. These reserves, referred to in general as Biosphere Reserves, provide ecologists and environmental resource managers with the equivalent of the "laboratories" of the cell biologist or the "observatories" of the astrophysicist. As such, they have become one of the more important developments in environmental sciences of the past two decades and will continue to grow in their potential contribution to developing responsible management strategies. [See NATURE PRESERVES.]

II. BACKGROUND AND HISTORY

Even though historically royalty and many wealthy individuals (and in some instances, native societies) set aside reserves that last to this day in Europe and North America, and a few areas in Asia, the establishment of Yellowstone National Park in the United States in 1872 stands out as one of the earliest contributions to preserving landscape dynamics and quality. Since then, many nations have developed their own reserve systems, mostly to preserve places with unique scenic beauty for tourism. In the late 1950s and early 1960s, after the successful International Geophysical Year in 1957 designed to examine earth processes, world leaders in ecology and environmental sciences met in London and other world capitals to plan a longer-term project designated the International Biological Programme (IBP). IBP, whose duration spanned from 1965 to 1974, became the first worldwide effort to describe controls over plant and animal populations, communities, and ecosystem patterns and processes. IBP focused on a large number of topics, including events controlling pattern and process in biogeographic areas representing specific forests, grasslands, tropics, and the Arctic. [*See* CONSERVATION AGREEMENTS, INTERNATIONAL.]

As IBP came to a close, Man and the Biosphere Programme (MAB) within the Science Sector of UNESCO emerged with its central focus of cataloging and helping to mediate global environmental problems. One of the main efforts within MAB became known as the Biosphere Reserve project (Project 8), an effort that persists to the present, even though it, like many MAB projects, has undergone radical change with time.

Early in the project development UNESCO attempted to develop a strong centralized science program, but that changed in 1976 when the focus turned to developing a large network of sites among the member nations. The program planners used Udvardy's classification of world biogeographical provinces as a basis for making decisions about the distribution of Biosphere Reserves. In 1994 the total number of Biosphere Reserves rose to 324 in 82 countries (Fig. 1), with the largest number (ca. 20%) defined in the United States and the former Soviet Union. With the exception of a few countries, to this point both the MAB secretariat and the nations contributing to the Biosphere

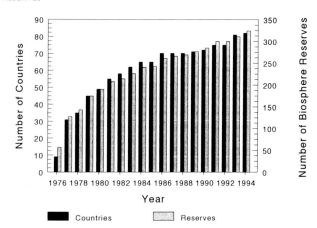

FIGURE I Histogram showing cumulative growth of member countries in the Biosphere Reserve concept and allocation of reserves by the MAB/UNESCO program since inception of the project in 1976. As of 1994 (which includes the most up-to-date data available from U.S. MAB offices), UNESCO has defined 324 Biosphere Reserves located in 82 countries (taking into account the changes in Germany and the former Soviet Union).

Reserve network have concentrated on the identification and basic characterization of their sites, with less emphasis on a research and monitoring design. This promises to change in the future as funds become more available to carry out the basic research and development plan.

The second major environmentally oriented network now operational also arose from the IBP experience. In the United States, ecosystem research funded by the National Science Foundation (NSF), the primary support for IBP during its tenure, took on an additional task in the late 1970s when it developed a new program called the Long-Term Ecological Research program. The focus of the LTER program centered on examining ecological processes and problems with temporal and spatial scales that could not be addressed by the more traditional research projects with a single focus and inherently shorter time frames. In 1979 and 1980, NSF started up five peer-reviewed projects, which have now grown to 18 sites that range from Arctic tundra and boreal forests; to temperate forests, grasslands, agriculture, lakes, streams, rivers, and coasts; to tropical forests; and to Antarctic marine systems and dry valleys. Although each project has a stand-alone design, efforts have emerged to develop a network of synthesis activities. Several governmental agencies own or operate the sites for their own specific purposes; however, through a peer-review process, NSF allocates annual research and operating funds for periods of up to 5 years, at which time each study undergoes an intensive review process in preparation for the next 5-year block of work.

A partial interlinking design exists between Biosphere Reserves and the LTER sites. Nine of the LTER research sites also have the designation of UNESCO Biosphere Reserves. Recently, with establishment of global change studies, and a major interest in ecosystem redevelopment of severely damaged environments, thought has turned to developing a more intensive research network patterned on both the Biosphere Reserve and LTER efforts that would serve an even wider research and management audience in ecological and environmental sciences.

III. PURPOSE AND DESIGN

A. UNESCO Biosphere Reserves

The main goals of Biosphere Reserves have evolved slowly since the inception of the idea, but currently involve three separate, but interacting objectives:

1. to provide stable areas set aside to conserve important ecosystems or those with unique genetic materials;
2. to provide for long-term local research and monitoring projects that address specific temporal processes, and from this local experience to develop an international network that covers spatial processes or problems; and
3. to provide areas where resource planners and managers can use the collective local and network data or experiences for education and sustainable environmental development strategies and tactics that serve the local area or region.

In many instances the current Biosphere Reserves have served to provide one or more of the main objectives; however, few countries have put such an ambitious plan into action, and almost nothing but the naming of the Biosphere Reserves exists

on a global basis at this time, despite the promotion of these goals for more than two decades.

The Biosphere Reserve, though simple in its main context, maintains a complex design. The overall design may work well for local matters, or for some relatively simple problems, such as establishing coastal zone monitoring, and it may even prove tractable for regional problems; however, thus far, the effort to carry out the design in its entirety into a global context has proved unworkable. Basically, as UNESCO has defined it, a Biosphere Reserve consists of several basic elements: (1) a permanently and strictly protected core area allowing only nondestructive monitoring projects; (2) a so-called buffer zone, usually in close proximity to the core area, where scientists and managers can conduct basic research and monitoring projects in addition to carrying out experiments or pilot programs on environmentally related development projects; and (3) a transition area beyond the buffer area in which a society conducts its normal day-to-day business, and where landscape planners may apply the information derived from the core and buffer zones (Fig. 2). In an even grander scheme, leaders in the Biosphere Reserve project envisaged using a large number of reserves across a complex landscape as a network to develop optimal or sustainable usage of environmental resources within a region or a nation. As one might expect, especially in highly variable biogeographic regions, the expense and logistical organization necessary to undertake such a project has effectively precluded this development so far, but it certainly holds potential for the future.

B. NSF LTER Network

As the NSF program dedicated to funding long-term process-level research at areas of excellence has grown in size from the original 5 LTER sites funded in 1980 to the 18 that now exist, attention has shifted toward attempts to build a network concept. Again, this network design vaguely follows some of that initiated by IBP over two decades ago. The original five "core areas" or principal research topics identified to guide the program include:

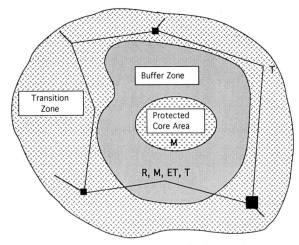

Core area representative of the region with only monitoring activities

Comparable buffer zone with research, monitoring, education and training, and in some areas, tourism

Transition area where local or regional society conducts its normal business

■ Human settlements

—— Roadways used throughout area

M Monitoring programs to record climate and biotic changes

R Research activities, including stress manipulations to examine cause and effect relationships in local environment

ET Education and training for local populace as well as those invited from other regions

T Tourism

FIGURE 2 Idealized concept of Biosphere Reserve and its functions. A hierarchical scheme contains a central core to provide background monitoring; a buffer zone with monitoring and experimental procedures for research, education, and training; and a transition zone that consists of normal landscapes and human activities indigenous to that region. The so-called transition zones occupy the terrain between any given number of Biosphere Reserves placed in a network design. [Redrawn from M. Batisse (1990). *Environ. Conservation* **17**(20), 111–116.]

1. patterns and control of primary production;
2. spatial and temporal distribution of populations selected to represent trophic structure;
3. pattern and control of organic matter accumulation in surface layers and sediments;
4. patterns of inorganic inputs and movements of nutrients through soils, ground-water, and surface waters; and,
5. patterns and frequency of site disturbance.

The spatial–temporal scales used as constraints for this array of topics range from daily events to 10 millennia for temporal events, and for space-related events from 1 m² plots for experimental

work to continent-size areas in syntheses involving topics such as global change (Fig. 3). For intersite comparisons, important elements from each site include inter- and intraannual weather patterns, the dominant plant and animal species—essentially the biotic community, organic matter and nutrient pathways and transfer rates—and the construction of mathematical or computer simulation models to link site-specific dynamics and to build intersite comparisons.

IV. RELATED PROGRAMS

Most countries have programs that address environmental problems, and in some instances these countries have developed networks of sites to attack specific problems. For decades in the United States the U.S. Department of Agriculture has developed agricultural practices through the use of field stations scattered across the country. Much of this work has developed in collaboration with state universities throughout the nation. Other USDA programs include watershed management and forest practices research administered by the U.S. Forest Service.

The U.S. Department of Interior, besides having perhaps the world's largest single network of reserves in the National Park system (largely to preserve scenic or historic features), also has a large network of wildlife refuges administered by the U.S. Fish and Wildlife Service and a variety of other agency programs that require field research and development at field sites. The U.S. Department of Energy has a small program called National Environmental Research Parks that focuses on environmental problems at National Laboratory facilities. Other federal and state agencies and private foundations also maintain a large number of similar reserves.

Countries elsewhere have similar national programs, for example, Canada, China, and the United Kingdom. Thus, around the world we can see that a great deal of work has focused on the idea of developing reserves for specific purposes. However, integration of their data resources has not developed to the same degree, and this awaits

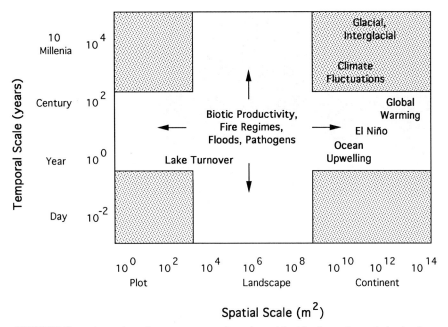

FIGURE 3 Relationships between temporal- and spatial-scale dynamics and the level at which representative research questions operate [J. F. Franklin, C. S. Bledsoe, and J. T. Callahan (1990). *Bioscience* **40**(7), 509–523. Copyright 1990 American Institute of Biological Sciences.]

new and critical future national and international management agreements.

V. WORLDWIDE NETWORK

A functional, well-designed worldwide network would contribute a great deal to our knowledge of environmental sciences. As the reader can easily imagine, the logistics would require a strong, smoothly operating, centralized office capable of interacting with both the scientific community and world leaders in resource management. Unfortunately, at the current time the international natural resource community lacks agreement on this effort. How should the world community organize such a critical operation? Positions expressed during two meetings structured by SCOPE [Scientific Committee On Problems in the Environment, a part of ICSU (International Council of Scientific Unions)] suggested that for many programs the design perhaps should have a "top-down" structure, one that could organize both long-term and short-term studies across a variety of defined problems. However, more often than not, the basic science needed to answer the questions comes about with a reductionist design, and integration of these data into the larger picture(s) can only proceed from a "bottom-up" approach. Coupled with the need for peer review that keeps basic science focused, a management approach that controls funding of science in the United States and other Western countries, this dichotomy presents a countervailing organizational problem for which no solution presently exists.

The developing International Geosphere-Biosphere Programme (IGBP) has planned for a large, integrated network of sites to complement its core research projects. Originally considered as a hierarchical or tiered design of Geosphere-Biosphere Observations distributed widely, the IGBP plans now call for a restricted number of Regional Research Centres (RRCs) placed strategically in developing countries of Asia, Africa, and South America. IGBP expects to draw from a nucleus of ongoing projects, such as the UNESCO Biosphere Reserve network, for at least part of its information base, but it expects that RRCs will bring together the main focus to support the IGBP core projects. In main, IGBP expects that the RRCs will:

"1. Promote cooperation among scientists of the region for. . .:
   i. facilitating coordination of Core Project research; and
   ii. defining regional research priorities and identifying regional questions that have global significance.
2. Provide facilities for data management of regional and global data sets.
3. Develop synthesis and modelling activities.
4. Distribute research results to scientists within the region and establish mechanisms for sharing the results with scientists from other regions.
5. Establish training and exchange programmes, especially in the use of new technologies and in the area of data analysis synthesis and modelling" [IGBP (1990). "The International Geosphere-Biosphere Programme." ICSU, Paris].

This plan, neither funded nor yet in place anywhere, will provide the basis for much of the analytical work that the global change program will need in the twenty-first century.

VI. NEEDS FOR THE FUTURE

No one really doubts that out global society needs to establish a stable, sustainable environment in which to live and maintain itself. To achieve such a goal, scientists and managers must know how to design reasonable objectives and to structure syntheses that lay out strategies and tactics for managing our environment. As the twentieth century ends, technology has given us many of the tools required to do the work, both in the field and in the office and laboratory, with new, easy-to-use personal computer and software systems that have great power. However, having the tools with which to do the work does not provide us with the entire picture.

To address such large-scale problems, such as developing and integrating environmental information, scientists and managers must work together to construct a mandate and a feasible plan.

As part of this overall plan, scientists need to have basic knowledge of their local regional resources, and managers must learn how to husband them. For this enormous undertaking, the world environmental society has turned to designing a reserve network, and by designating key reserves as components of that network we can lay the basic foundation and specify the operational steps for this overall environmental plan. Two excellent examples lie before us: the work that has established the UNESCO Biosphere Reserves and the scientific thought and work that has brought about the NSF LTER network can serve as a unified significant global effort. Several national and international governmental and nongovernmental organizations continue to work toward this goal.

Glossary

Biosphere reserve A reserve in the worldwide UNESCO/MAB design set aside to preserve ecosystem or genetic diversity, to provide for research, monitoring, and the definition of sustainable environmental development for a region, and to promote education and training.

Ecosystem Basic ecological unit, usually containing plant, animal, and microbial populations, that, in conjunction with local climate, regulates itself over long periods of time and maintains a unique identity.

ICSU International Council of Scientific Unions, an international parent body for scientific societies, with headquarters located in Paris.

IGBP International Geosphere-Biosphere Programme, a program set up by ICSU in Berne, Switzerland, in 1986 to oversee the international research work on global change. Secretariat offices are located in the Royal Swedish Academy of Sciences, Stockholm.

Long-term ecological research (LTER) U.S. National Science Foundation program established in 1979–1980 to investigate site-specific long-term ecological processes.

MAB Man and the Biosphere Programme located in the Science Sector of UNESCO to deal with worldwide problems involving humans and the environment.

Reductionist science Approach to a research question that often involves testing of basic ideas by developing experiments that consist of smaller components of the problem to address a specific hypothesis.

RRCs Regional Research Centres, regional centers designated by IGBP to examine global change phenomena.

SCOPE Scientific Committee on Problems in the Environment, an ICSU program to provide for intensive review and study of specific ecological problems around the world, with the Secretariat located in Paris.

Sustainable development Concept made popular by the Brundtland Commission Report and developed to address how economic development should proceed in terms of environmental productivity and quality. The main thought centers around doing nothing that will damage the environment, nor require a subsidy from outside the developed area to maintain its structure and utility over the long run.

"Top down" versus "bottom up" Concepts applied to how basic research and monitoring data apply to a scientific problem. A top-down scheme applies reductionist thinking in which the scientist or manager controls each step by disaggregating a problem; a bottom–up scheme takes highly detailed information and aggregates it into a whole to solve a problem. In actual practice, particularly in ecological systems, both approaches complement one another.

UNESCO United Nations Educational, Scientific and Cultural Organization, located in Paris.

Bibliography

Batisse, M. (1990). Development and implementation of the Biosphere Reserve concept and its applicability to coastal regions. *Environ. Conservation* **17**(2), 111–116.

Dyer, M. I., and Holland, M. M. (1991). The Biosphere-Reserve concept: Needs for a network design. *BioScience* **41**(5), 319–325.

Dyer, M. I., di Castri, F., and Hansen, A. J., eds. (1988). "Geosphere-Biosphere Observatories: Their Definition and Design for Studying Global Change," Biology International Special Issue 16. Paris: International Union of Biological Scientists.

Franklin, J. F., Bledsoe, C. S., and Callahan, J. T. (1990). Contributions of the long-term ecological research program. *BioScience* **40**(7), 509–523.

IGBP. (1990). "The International Geosphere-Biosphere Programme: A Study of Global Change. The Initial Core Projects," Report No. 12. Paris: ICSU.

LTER (1991). "Long-Term Ecological Research in the United States. A Network of Research Sites, 1991," LTER Publication No. 11, 6th ed. revised. Seattle: Long-Term Ecological Research Network Office.

MAB Information System (1986). "Biosphere Reserves. Compilation 4, October 1986." Prepared for UNESCO by the IUCN Conservation Monitoring Centre. Paris: UNESCO.

Risser, P. G., ed. (1991). "Long-Term Ecological Research: An International Perspective," Scope 47. New York: John Wiley & Sons.

Bird Communities

Raimo Virkkala

National Board of Waters and the Environment, Finland

Bird communities can be defined as the co-occurrence of individuals of several species in time and space. Bird communities have often been divided into different guilds consisting of species with similar ecology according to, for example, foraging patterns. During the 1960s and 1970s interspecific competition within guilds was emphasized in regulating communities, which were regarded as being at equilibrium. At present communities are considered more open systems than was previously believed, and spatial and temporal variations of species' populations are considerable. Migratory patterns of birds make bird communities highy different from communities in other animal or plant groups. Different bird species in a community operate at different spatial and temporal scales. Therefore, bird communities should be considered in a landscape context, in a mosaic of habitat patches. Due to the migratory habits of most birds species, preserving bird communities is clearly a global issue.

I. INTRODUCTION

Bird communities consist of individuals of several species varying in numbers in time and space. The concept of bird community refers to a local-scale phenomenon. However, communities are affected both by higher and lower scale patterns. The regional species pool determines the conditions for the composition of communities. Habitat selection patterns of individuals, in turn, are important at the lower scale defining the structure of bird communities. Physical environment and vegetation structure and composition affect the occurrence of individuals.

Communities are generally divided into different guilds. A guild consists of species of similar ecology according to, for example, foraging patterns or situation of a nest site. We can speak about a foliage-gleaning or a nectarivore guild. Species within guilds utilize a resource in a similar way, and interspecific competition occurs between species within a guild.

Bird communities can be considered in a hierarchical context: Communities are affected by processes operating at different spatial and temporal scales. Earlier bird communities were considered to be at equilibrium or the community variation closely tracked the variation of resources. Communities were considered to consist of species' populations restricted largely by interspecific competition. Interspecific competition was emphasized, as the structure and composition of bird communities were considered to be largely constrained by the availability of resources. Bird communities were also assumed to be rather closed systems and consequently the structure of communities could be defined.

During the past 15 years this approach has been criticized, as both temporal and spatial variations of bird communities are considerable and dynamics based on close tracking of resources are, in many cases, unprobable. In addition to interspecific competition, there are several other factors affecting the structure of communities (Fig. 1). These factors should also be taken into account in analyzing the observed patterns in bird communities.

Space boundaries of a bird community are not strict, for example, compared with a fish community of a lake. Most of the bird species in temporal and boreal areas are migratory, so the nesting area is recolonized every year. Even in the tropics there is small-scale migration among habitats as a consequence of, for example, drought periods. Migratory patterns of birds make the communities more open systems than in many other animal groups.

Time lags in the response of individuals make the close tracking of resources highly difficult. Birds usually reproduce only once or twice a year, and resources may vary in a shorter time scale. The effect of different factors for the structure of bird communities (see Fig. 1) is highly variable, for

example, if we consider the consequences of predation for the preyed individual's fitness (= 0) and the negative effect of interspecific competition on species (fitness >0).

The structure of bird communities is thus far more complex than was previously believed. Spatial and temporal scales in defining and studying communities are highly important: How close does the study scale match the scale at which different species operate? As bird ecologist John A. Wiens has stated, bird communities have been studied from the viewpoint of the human observer largely thinking that birds view habitats as we do. Different study scales should be taken into account in order to have a more objective view of bird communities.

In this article I concentrate on the following aspects of bird communities: the distribution of species in a bird community, spatial and temporal dynamics, the hierarchical structure of bird communities, and the conservation of bird communities.

Individuals of the same species breeding nearby form a colony, which is a highly different concept than a community. However, sea bird colonies are often mixed, so that several colonies of different species occur in the same cliff. I concentrate here on discussing bird communities composed mainly of noncolonial species, as a community consisting of colonial species is affected by slightly different processes, as, for example, social organization is important for these species.

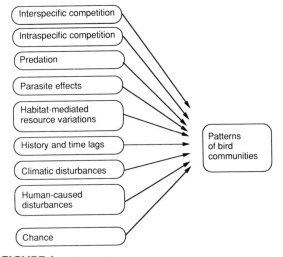

FIGURE I   Selected factors affecting observed patterns in bird communities.

II. DISTRIBUTION OF SPECIES IN A BIRD COMMUNITY

A. Habitat Selection Patterns of Birds

Birds do not distribute themselves randomly in a landscape. Thus, they clearly select their home ranges and nest sites. The habitat selection of birds is a process in which several factors are important. They can roughly be divided into two interacting categories: population biological and landscape factors. By "population biological factors" I mean such issues as population density, interspecific interactions, and the site tenacity of individuals. Landscape factors refer to habitat structure and the spatial scale of habitat selection.

Habitat selection can be considered a hierarchical process having a ranking order. At first species select the geographical range and breeding region. Second, individuals select their home ranges in a landscape. Third-order selection is connected with the use of various habitat components (e.g., habitat patches) within the home range. Microhabitat selection at the fourth level is correlated with the use of food items at feeding sites. Thus, a species might be absent from an area although there are suitable habitat patches (components) for the species, but the home range requirements for the species are not fulfilled, that is, the higher order of selection.

B. Habitat Structure and Composition

Habitat structure is often considered a key factor affecting bird species' habitat selection, as, for example, bird communities in clear-cut areas differ almost totally from those in mature or old-growth forests. Habitat structure includes tree species' density, diversity, and age distribution in a forest. The structure of habitat has been analyzed by measuring several habitat variables, such as trunk numbers of different tree species, and using multivariate analyses to correlate the habitat variables and the species' occurrence. It has been observed that in many cases individuals respond to structural features of habitats mainly independently of other species, because habitat features are correlated with species-specific nesting and feeding sites. It may well be that birds select habitats by using structural habitat cues to estimate the prey abundance. This structural cue hypothesis implies that habitat structure reflects habitat quality.

In addition to vegetation physiognomy, the floristic composition of vegetation is also important in the habitat selection of birds, particularly in open areas, such as agricultural areas and grasslands.

However, it is often difficult to connect the structure and composition of a habitat with the distribution of birds due to the apparent hierarchy of habitat selection (see above). In studying bird–habitat rela-

tionships there should be information on several spatial scales.

C. Population Density

Population density affects species distribution within space. Fretwell and Lucas introduced the concepts of ideal free and ideal despotic distribution of individuals, which reflect the importance of population density in the habitat selection of birds. In the model of ideal free distribution, habitat types have different quality values for individuals. The best and optimal habitat is inhabited first and when this habitat is fulfilled with individuals, the remaining individuals manage better in the less optimal habitat. Again, when this suboptimal habitat is saturated by birds or a certain threshold of habitat quality is reached, it is advantageous for the remaining individuals to settle in the more marginal habitat. The quality of a habitat type declines as the density of individuals increases. Noncolonial birds usually have territories which they defend, and therefore this model might be relevant for birds. The model predicts that the density is highest in the best habitat, second highest in the second best habitat, and so on. [See TERRITORIALITY.]

In the ideal despotic model the most dominant individuals occupy the best habitat and the subdominants occur in the worse habitats. In this model the density of dominant individuals in the best habitat can be even lower than that of subdominants in the lower-quality habitats.

These models of Fretwell and Lucas emphasize the role of density-dependent factors in the habitat selection of birds. There are, however, restrictions to these models, as density cannot necessarily be used as an indicator of habitat quality. In addition to density, habitat quality should be measured by offspring production and the survival of individuals based on fitness parameters. In some studies there has been a positive correlation between the fitness parameters and density; in others there has not. In spite of the limitations, the habitat distribution models of Fretwell and Lucas give a general idea of the effects of population density and intraspecific competition on the distribution of individuals.

D. Interspecific Competition

Species interactions might limit the occurrence of individuals in a given area, such as interspecific competition or predator–prey relationships. Interspecific competition can be defined as the negative effect of one species on the population size of another. Interspecific competition has generally been divided into two categories: exploitation and interference competition. Exploitation competition means that individuals have free access to the resources, but the use of resources by individuals of one species decreases their availability to individuals of another species. In interference competition the access of individuals is prevented directly by the aggressive actions of other species' individuals.

Interspecific competition was emphasized earlier as the main process regulating bird communities. Competition occurs when individuals share a similar resource that is within specific guilds. Resource limitation is a prerequisite of interspecific competition. Guilds and interspecific competition have been studied in connection with the niche concept. A niche can be regarded as including the specific features of the environment in which the species lives. Niche differences and overlaps among species have been interpreted as evidence of interspecific competition. However, reasons for changes in species' niches are difficult to estimate, and they cannot be regarded solely as a consequence of interspecific competition. For example, interspecific competition predicts niche contraction of a species, whereas intraspecific competition, such as in the models of Fretwell and Lucas, predicts niche enlargening. So the effects of inter- and intraspecific competition are contradictory.

Interspecific competition is probably important in closed systems such as isolated oceanic islands and bird groups strictly restricted by resources, such as species of resident titmice *Parus* spp. in boreal and temperate areas and nectar-feeding species. Resident titmice foraging in trees in boreal forests can utilize a considerable proportion of the food resources (50% or more of spiders and insects), so for these species resources can evidently be considered a limiting factor. Also, species of the nectar-feeding guild, such as hummingbirds in the New World or sunbirds in Africa, are clearly limited by their food resources: nectar in flowers.

However, most species in bird communities are not strictly restricted by resources. In addition, bird communities are not closed but open systems consisting largely of migratory species. Therefore, interspecific competition cannot be regarded as the main process affecting the habitat selection of birds and community structure, as several other factors are also important and should be taken into account in defining the structure of bird communities (see Fig. 1).

The evidence for species' total competitive exclusion from an area by another species is weak. However, the interference competition for limited resources can be important. For example, natural holes are a limiting factor for hole-nesting species, particularly in managed forests in which dead and dying trees, important as a hole supply, are removed. Competition for nest holes is often recognized in studies of hole-nesting birds.

E. Predation and Parasitism

The predator–prey relationship is another important form of species interactions. For a small passerine it is highly relevant whether its territory is situated in or outside a sparrow hawk's predation area. The effects of predation on the distribution of birds, however, are not well documented. Predation rates are high near the nests around which the activities of birds are centered. Therefore, an important aspect of predation is nest losses. In experimental studies nest losses have been documented to be extremely high near forest edges. Also, in the tropics predation rates causing nesting failures are apparently very high. Even a low predation rate on the nests might be important due to the devastating effects of predation on the offspring production of individuals. Thus, predation has considerable and direct influence on the fitness (i.e., offspring production and survival) of preyed species.

The other factor of predation is that avian predators highly fluctuate in numbers according to their food resources. This is confirmed in studies dealing with vole-eating birds of prey, as both the numbers of avian predators and their food (voles) have been quantified.

Parasites form a possibly much more important factor than we presently know about in affecting bird species distribution in communities. Two groups of parasitism can be separated: (1) internal or external parasites infecting all vertebrates and (2) brood parasites especially typical for birds. The former parasitism is important, for example, in nest site selection, as birds must often use nests from the previous year (e.g., birds of prey, sea birds, and hole-nesters) which might contain a heavy parasite load. The effect of internal and external parasites in the habitat selection of birds is poorly known. However, the occurrence of internal parasites has been observed to affect the mate selection of birds considerably, and thus their influence on the distribution of birds is probably much more important than has been thought so far.

Brood parasites such as cuckoos and cowbirds have a clear effect on their host species: Brood parasites remove host eggs and deposit their own eggs in host nests. In tropical areas in certain species groups (e.g., passerines), even 50% of all species may be subject to brood parasitism. Human-caused habitat fragmentation increases the numbers of brood parasites, as they commonly occur near edges of forest and open areas. Thus, landscape and biotic factors interact and are essential in the distribution of species being parasitized.

F. Time Constraints

Time lag and history affect bird species' present distribution and habitat selection. A bird species might be present in an atypical habitat for the species, in a habitat that has been totally altered as a consequence of, for example, forest cutting or herbicide poisoning. This might be due to the fact that adult birds are usually site tenacious: They breed in the same area in consecutive years. Although the habitat has been totally altered, the same individuals might still be present as before the alteration as a consequence of the site fidelity of birds.

On a longer time scale the distributional history of a species affects its present occurrence in bird communities. For example, a species might occur in a given area although its preferred habitat type is scarce. This habitat type, however, could have been common in earlier times but decreased considerably as a consequence of human activities. At present, the habitat type might be more common in some other areas nearby, but the species preferring this habitat type might not be there due to poor dispersal ability. Dispersal and recruitment abilities strongly affect species' present distribution in bird communities.

G. Chance

Chance is an important factor in nature and also in defining the composition of bird communities. Species colonization patterns of oceanic islands are affected by chance, but of course the dispersal ability of a species also has a great influence on this colonization. In the process of habitat selection of individuals, there is always the stochastic component, for example, in the yearly redistribution of migratory birds. Although environmental variation and population density of a species remain at the same level from a given year to the next, there might still be observable changes in the distribution of birds between the two years due to chance only.

Bird communities are not so strictly ordered as we think and they are largely affected by chance. Small populations are particularly highly susceptible to both environmental and demographic stochasticities. A small population may easily become extinct as a consequence of environmental stochasticity, such as an extremely cold winter or a storm. In a small population demographic stochasticity easily causes a skewed age distribution or an unequal sex ratio, for which reasons the population ultimately disappears. In Sweden, the middle spotted woodpecker (*Dendrocopos medius*) remained for decades as a small isolated population of 15–20 pairs, but became extinct in 1982 after exceptionally cold winters (environmental stochasticity) and low reproductive success. The consequences of environmental and demographic stochasticities should be taken into account in preserving small populations [*See* ENVIRONMENTAL STRESS AND EVOLUTION.]

III. COMMUNITY PARAMETERS

Species' numbers, total density, and biomass are often considered the main parameters by which bird communities are measured. However different community parameters measure different aspects of community composition. High total density does not necessarily mean high biomass and species' numbers (or diversity). Therefore, when comparing and drawing conclusions of different bird communities, several community parameters should be estimated. The values of different community parameters are largely affected by different factors, such as the spatial distribution of food resources, general productivity and structure of a habitat, and climatic disturbances.

The density of pairs in a forest bird community declines from the tropical forests to the north boreal areas, to the northern forest limit, from some 1000–2000 pairs per km^2 to about 50–100 pairs per km^2. The information on bird communities in tropical areas is still very poor, but some interesting patterns have emerged. In temperate broad-leaved forests of the northern hemisphere the density of bird pairs in communities is often about 1000 pairs per km^2 or more, so there is no great difference in the total bird density between the tropical and temperate forests. However, the biomass and the number of species in the studied tropical forests are about 5-fold higher than those of the temperate broad-leaved forests. This means that the average local density of a bird species in the tropics is not high. In one study area in Amazonian Peru (11°S) there were about 250 species per km^2, but the most abundant species had a density of only about 10–20 pairs per km^2. In southern Finland, situated in northern Europe in the northern coniferous forest (boreal) zone (60–64°N), the average densities of the most abundant bird species in forests, the willow warbler *Phylloscopus trochilus* and the chaffinch (*Fringilla coelebs*), are on the order of 50 pairs per km^2. There are, of course, species in the tropics which can be locally extremely abundant, but it seems that most of the species are not.

Spatial variation in the bird community composition is much higher in the tropics than in the temperate or boreal areas, whereas temporal varia-

tion is probably larger in the more northern or southern latitudes than in the tropics. For example, forest bird communities in Finland and western Siberia in the boreal coniferous zone consist of largely the same species, although the distance between these areas is ~3000 km. In the tropics there are species with rather limited ranges, and bird communities within the same continent are probably highly different among sites (see Section IV).

IV. TEMPORAL AND SPATIAL SCALES

A. Community Equilibrium

The earlier emphasis in the 1960s and 1970s was that bird communities were at stable equilibrium defined strictly by resources. If variations occurred, they were assumed to closely track the variation of resources. However, during the past 15 years it has been evident that bird communities and populations have large temporal variations both in the short term, from year to year, and in the long term, producing trends. The variation patterns of communities cannot be connected strictly with the variation in resources. A large part of the populations is migratory and thus is affected by several factors other than food resource variation in the breeding grounds. Weather conditions during the migratory period and in the overwintering areas strongly affect the numbers of breeding individuals. A species might not be present in a given year in its breeding grounds although its preferred food is abundant. For example, nomadic owls of boreal areas fluctuate according to cyclic vole populations, but they might be absent in large areas where there are plenty of voles.

Space and time are closely related to each other in the small-scale variation of bird communities: Short-term change in the composition of a bird community at one locality from year to year is similar to the variation in the same year over different localities.

B. Temporal Dynamics

Bird populations and consequently communities vary considerably both in time and space. Temporal variation consists both of short-term year-to-year fluctuation and long-term population trends. The reason behind these temporal variations often remains unclear, although there are several suggestions as to the cause of the variations. Environmental variation, such as differences in weather conditions probably has a profound effect on birds. A large part of bird species is migratory, particularly in temperate and boreal areas, so birds are affected by environmental variation over extremely large areas on earth. Long-distance migrants breeding in the Northern Hemisphere meet environmental variation in three areas: the breeding areas, migratory areas, and wintering areas in the tropics. For example, the numbers of the whitethroat *Sylvia communis* collapsed in England between 1968 and 1969 as a consequence of extreme drought in the Sahel region of Africa, the main wintering area of the species. [*See* SAHEL, WEST AFRICA.]

However, in the northern latitudes tropical migrants do not generally vary in numbers as much as certain nomadic short-distance migrants. Crossbills (*Loxia* spp.), eating seeds of coniferous trees, have huge year-to-year variations in a given area as a consequence of variation in the seed crop of trees. Crossbills can be considered seed specialists, whereas most of the tropical migrants are generalist insectivores.

There is a general tendency for the temporal variation or community instability to increase toward the north. This can be due to more unpredictable weather conditions in the north and huge year-to-year variation in important food resources, such as the seed crop of trees for seed eaters or the number of voles for vole-eating avian predators.

Long-term trends of 86 land birds species are available from Finland in an area of 300,000 km^2 from the mid-1940s to the mid-1970s. During this 30-year period 43% of species had increased in numbers ($\geq$1% per year), 28% had remained stable or fluctuating, and 29% had clearly decreased ($\geq$1% per year). A large part of population changes could be connected with large-scale habitat alteration caused by humans. Also, in North America clear population changes have been observed during the past decades, and tropical migrants have particularly declined in numbers. Climatic change

due to the greenhouse effect will probably alter vegetation zones considerably during the next centuries, which might have a clear effect on the distribution and population trends of bird species.

In general, it has been observed that populations occurring in a given area are highly dynamic and bird communities cannot be considered to be at stable equilibrium. The population dynamic data from tropical areas are very scanty, but the results of John Terborgh and colleagues from Amazonian Peru suggest that at least some birds communities in the tropical forests might be rather stable. However, two factors cause migrations of individuals and consequently variation in communities also in the tropics: (1) There is considerable environmental variation as a result of varying wet and dry seasons, which probably also cause fluctuations in bird populations. (2) A large part of bird communities consists of obligatory frugivores, and these species have been observed to migrate in response to periodic food shortages even in the tropical rainforests. The range of these migrations is, at present, unknown. However, in general, temporal variability of communities in the tropics is probably less than in more northern or southern areas.

C. Spatial Patterns in Communities

Spatial patterns of bird communities refer to two factors: the between-site variation of communities and the spatial scale used in studying communities. The between-site variation of communities is largely affected by variation in landscape structure and dynamics of habitat patches. In general, the more fragmented the landscape, the more variation occurs in communities among sites. The spatial scale used in the study of bird communities is of the utmost importance in drawing conclusions on patterns in communities. If we are to study a community in an area of 1 ha, the results are highly different from those of a study in 1 km^2. There is a stochastic element in the yearly distribution of birds, for which reason a study conducted on a very small scale does not necessarily represent the patterns of a community. On a larger scale the variation of community is clearly smaller and the stochastic patterns are compensated for. The larger variation in a small scale can only be the product of the random yearly redistribution of individuals.

Bird communities in small habitat patches are affected by surrounding landscape. Species of managed landscape intrude into small patches of virgin habitat, and species preferring virgin habitats usually are susceptible to habitat fragmentation: They disappear or occur in low numbers in fragmented virgin habitats. This means that small virgin habitat patches generally reflect largely the avian biota of the managed landscape. Fragmentation implies that species of virgin habitats, such as those occurring in old-growth forests of temperate and boreal areas or in tropical rainforests, decline much more steeply than the area of these habitats: The relationship between the decline of virgin habitats and the decline of species preferring these habitats is not linear. This observation has applications for founding nature reserve networks: The habitat and areal demands of species should be taken into account. [See Virgin Forests and Endangered Species, Northern Owl and Mt. Graham Red Squirrel.]

On the local scale there are areas of population sinks and sources; the variation patterns of populations do differ between different sites. A sink population is highly dependent on the recruited individuals from a source population. On a larger regional scale the variation of populations is lower, as local population variations may be compensated for. On this scale populations form metapopulations (Table I), local populations that interact via individuals moving among populations across habitat types not suitable for the species-specific feeding or breeding activities. "Metapopulation dynamics" refers to the extinction and colonization of local populations in a landscape.

TABLE I
The Hierarchical Organization of a Species (Biological Component) with Corresponding Spatial and Habitat Scales

Biological component	Spatial scale	Habitat scale
Individual	Patch	Within-habitat
Population	Local	Between-habitat
Metapopulation	Regional	Landscape

V. HIERARCHICAL APPROACH

The structure of bird communities should be considered in a hierarchical context in connection with landscape patterns in order to overcome the problems of spatial and temporal variation in bird communities. The patterns observed in a landscape are consequences of disturbances (e.g., fire, storm, and anthropogenic effects), biotic processes (e.g., demography, competition, and predation), and environmental constraints (e.g., weather conditions). These agents should be viewed on different spatial and temporal scales with a hierarchical structure so that events causing a pattern on one scale can be incorporated into a higher level. Generally, low-level events are small and fast, whereas those at a higher level are larger and slower. Patterns of bird communities are sensitive to the spatial scale, and therefore, in studies of bird communities, both higher- and lower-scale phenomena should be taken into account in drawing conclusions on community patterns observed on a particular scale. Different spatial scales can be divided hierarchically (see Table I), for example: (1) the area inhabited by an individual (within-habitat level), (2) local patches occupied by populations of several species (between-habitat level), (3) the region containing metapopulations in a landscape level, and (4) the biogeographic scale covering large areas of several vegetation zones.

At the biogeographic scale events occurring far outside the region studied are essential for bird communities. At this scale processes affecting community patterns have extremely long time scales. The latest glaciation which ended in the Northern Hemisphere about 10,000 years ago has a strong effect on the composition of present bird communities.

Regionally, environmental constraints, such as weather variation, are important for the structure of bird communities. Local effects of habitat structure and alteration should be considered with regional patterns of birds, and processes explaining local between-habitat patterns of communities should be considered at a within-habitat level. By "within-habitat" I mean, for example, the foraging preferences of individuals among tree species. The information for one scale (e.g., local) should be connected to that for another scale (e.g., regional). [See Foraging Strategics.]

Different organisms operate on different scales, for example, a large bird of prey and a small passerine, which poses problems to a community-level approach. Nomadic species may move thousands of kilometers to nest between consecutive years, whereas sedentary passerines may live their entire life in an area of couple of square kilometers or less. Also, an abundant and a rare species in a community may live on highly different scales, and therefore every organism should be considered in connection with species-specific spatial and temporal scales.

One way to overcome the difficulties of the subjectiveness of the study scale is to combine the different spatial scales and to calculate a so-called "fractal dimension," which is based on the observations made on different scales. Also, the landscape patchiness in which birds occur should be analyzed, for example, by taking into account the perimeter–area relationships of patches on different scales of resolution. This is because the larger the patch is, the smaller the perimeter–area relationship will be. Human-caused habitat fragmentation has increased considerably during the past decades, producing smaller habitat patches with higher perimeter–area relationships. The amount of edges has a clear effect on the distribution and abundance of birds. Rates of parasitism and predation are also higher near the edges.

VI. CONSERVATION OF BIRD COMMUNITIES

There are about 9300 bird species on earth. Of these about 1000 (i.e., over 10%) are considered threatened or endangered as a consequence of human activities. In the 1970s only about 250 species were endangered. Most of the threatened species are in the tropics: In both Brazil and Indonesia there are over 100 threatened species. The continent-scale destruction of tropical forests is the main factor affecting the future decrease in avian biodiversity.

In earlier times, during the past 2000 years, extinctions of species (at least some hundreds) occurred mainly on the oceanic islands. For example, half of the original land bird fauna (about 90 species) in New Zealand is now extinct due to human activities. These species included the giant moa *Dinornis maximus* (height, 3–4 m) already hunted to extinction by the polynesians (Maori) before the European settlement. In general, most of the extinct bird species on oceanic islands were flightless: They were adapted to an environment without mammal predators. In addition to direct hunting causing species extinctions on islands, humans have brought with them mammals, such as cats and rats, which have had a devastating effect on the original, often flightless, avifauna of oceanic islands. [See Evolution and Extinction.]

Avifaunas of continents have so far managed better, although several extinctions have been recognized, particularly in the New World. The extinction of the North American passenger pigeon (*Ectopistes migratorius*) is one of the best-known examples, as it was one of the most abundant species on earth. Earlier, it was thought that the massive hunting of this species caused its extinction, but based on the present information, it seems likely that the passenger pigeon became extinct primarily as a result of forest destruction and fragmentation during the nineteenth century.

To preserve avian biodiversity is a difficult task, as birds are mobile. The protection of migratory birds requires both suitable breeding and wintering habitats, often on different continents. Therefore, the protection of bird communities is clearly a global issue: Tropical migrants of northern latitudes do not survive if only their breeding grounds are protected. On the other hand, for the resident birds of the tropics, relatively large areas should be protected, as spatial variation of bird communities in the tropics is high and species' numbers are the highest on earth. The destruction and fragmentation of tropical forests are the major threat to global avian biodiversity. At present we know too little about bird communities in tropical forests.

Whether the study of bird communities in these forests can proceed before these forests are cut remains to be seen. However, time is running out, maybe too quickly.

Glossary

Community Co-occurrence of individuals of several species in time and space.

Equilibrium Communities are at stable equilibrium defined by resources available.

Guild A group of species with similar ecology.

Habitat structure Physiognomy of the habitat.

Hierarchy of communities Effect of different spatial and temporal scales on bird communities, populations, and individuals.

Interspecific competition Negative effect of individuals of one species on the occurrence of individuals of another species.

Intraspecific competition Within-species competition, for example, competition among individuals of different ages or sexes.

Landscape structure Mosaic of habitat patches with different quality.

Spatial scale Scale at which bird communities, populations, and individuals are considered.

Spatial variation Variation of communities between different sites and across different spatial scales.

Temporal dynamics Variation of communities and populations through time.

Bibliography

Cody, M. L., ed. (1985). "Habitat Selection in Birds." Academic Press, Orlando, Florida.

Fretwell, S. D. (1972). "Populations in a Seasonal Environment." Princeton University Press, Princeton.

Keast, A., ed. (1990). "Biogeography and Ecology of Forest Bird Communities." SPB Academic, The Hague, The Netherlands.

Terborgh, J., Robinson, S. K., Parker, T. A., III, Munn, C. A., and Pierpont, N. (1990). Structure and organization of an Amazonian forest bird community. *Ecol. Monogr.* **60,**(2), 213–238.

Virkkala, R. (1991). Spatial and temporal variation in bird communities and populations in north-boreal coniferous forests: A multiscale approach. *Oikos* **62,**(1), 59–66.

Wiens, J. A. (1989). "The Ecology of Bird Communities," Vols. 1 and 2. Cambridge University Press, Cambridge, England.

Bog Ecology

J. B. Yavitt

Cornell University

I. Introduction
II. From Plants to Peat
III. Elemental Dynamics
IV. From Peat to Peatlands
V. Ecology of Bog Plants
VI. Ecological Communities
VII. Carbon Dynamics
VIII. The Historical Record in Peat and Global
Climate Change

Bogs—and closely related fens—occur over roughly 3% of the Earth's land surface. Some of the most extensive ones are in high northern latitudes of central Canada, Great Britain, Fennoscandia (Norway, Denmark, Sweden, and Finland), and Russia (Fig. 1). They are sources of life-sustaining peat (used for fuel) and wildlife for local people and support lucrative agriculture and timber when drained. They also play a role in the world's climate by removing a powerful greenhouse gas, carbon dioxide, from the atmosphere and storing it in the form of peat. The amount of carbon stored in peat is not trivial, and there is considerable concern that global change could induce the release of carbon dioxide (and/or methane) from peat back into the atmosphere, possibly exacerbating climatic warming.

I. INTRODUCTION

Several features distinguish bogs (Gaelic for "soft ground") and fens (Icelandic for "quagmire") from terrestrial and aquatic ecosystems; one of the most conspicuous of these is peat itself. Peat is the residue that remains after incomplete decomposition of shed plant material. It is often acidic, is often waterlogged, and frequently has a low supply of nutrients to support plant growth. Peat formation is a slow process, regenerating less than 1 mm/year. As a result, peat is essentially a nonrenewable resource. Once destroyed, it is virtually lost forever. A second distinctive feature is the nearly constant attendance of water at or above the peat surface—and the demand this places on plants and other organisms for life in a mostly waterlogged environment. [*See* WETLANDS ECOLOGY.]

Bogs occur in—but are not restricted to—maritime and continental climates with long cold winters and with much more rain and snow than water loss though evapotranspiration. Poorly drained land, having less than 1 m of surface relief per kilometer of distance, is most conducive to their development. Not only the quantity of water but also its chemical content is another important consideration. Indeed, all bogs have a low supply of

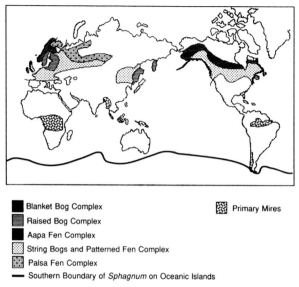

■ Blanket Bog Complex   ▦ Primary Mires
▤ Raised Bog Complex
■ Aapa Fen Complex
▧ String Bogs and Patterned Fen Complex
▨ Palsa Fen Complex
— Southern Boundary of *Sphagnum* on Oceanic Islands

FIGURE I   Worldwide distribution of the main types of mire complexes.

plant nutrients—and many are ombrotrophic (Greek for "rainstorm fed"), meaning that they rely on atmospheric precipitation alone for influx of water and inorganic solutes. Closely aligned to bogs are fens that derive additional inorganic solutes from water having preciously flowed through or over either a mineral soil or a mineral substrate; these are minerotrophic. Bogs and fens frequently intermingle across the landscape, and many bogs actually start out as fens before distinguishing themselves. Europeans use the term "mire" (*myrr* in Old Norse) to embrace all types of bogs and fens and all kinds of vegetation growing in bogs and fens. I also discuss fens in relation to bogs and use "mire" and "peatland" interchangeably throughout this article in reference to peat-forming wetlands.

II. FROM PLANTS TO PEAT

The accumulation of peat on the landscape is ultimately the balance between plant production and incomplete decomposition of shed plant material. Peat is not a single substance, but rather it relects in time and space the collective changes that occur when plant material decomposes by the actions of decomposer organisms (microbial and otherwise).

The plants that are the starting point for peat inevitably must be able to tolerate waterlogged conditions. Mosses are exceptionally adept in this regard, and it is the growth, death, and decomposition of *Sphagnum* mosses (bogmoss) that largely distinguishes peat in bogs. Most bogs also support vascular plants such as sedgelike cotton grass (*Eriophorum vaginatum*) and beak-rush (*Rhynchospora alba*), as well as low-growing shrubs such as bog rosemary (*Andromeda glaucophylla*), leatherleaf (*Chamaedaphne calyculata*), bog laurel (*Kalmia polifolia*), Labrador tea (*Ledum groenlandicum*), and cranberry (*Vaccinium* spp.). Black spruce (*Picea mariana*) occurs in most continental bogs in North America, but it is noticeably absent in maritime regions. In Eurasian bogs the dominant *Sphagnum* species are strikingly similar to those in North America, whereas dwarf shrubs such as heather (*Calluna vulgaris*), heath (*Erica* spp.), crowberry (*Empetrum nigrum*), and cloudberry (*Rubus chamaemorus*) assume more importance. Moreover, Scots pine (*Pinus sylvestris*) replaces the familiar spruce of North American bogs.

One of the most bizarre adaptations to life on a bog is carnivory. This habit of trapping and digesting animals has attracted considerable attention since Charles Darwin's treatise on the subject in 1875. The most common carnivorous plants in bogs are pitcher plants (*Sarracenia* spp. and *Darlingtonia* spp.), with their pitfall traps, along with the sticky leaf surfaces of sundews (*Drosera* spp.).

Fens support much less *Sphagnum* than that found in bogs; rather, it is replaced by fen-indicator mosses (often brown mosses), other sedge and sedgelike plants, grasses, other dwarf shrubs, and other trees. Important indicator species of fens include tussock bullrush (*Scirpus cespitosus*), bog birch (*Betula pumila*), and maybe even northern white cedar (*Thuja occidentalis*), rushes (*Juncus stygius*), reeds (*Phragmites australis*), alders (*Alnus* spp.), willows (*Salix* spp.), and sweet gale (*Myrica gale*).

The decomposition rate of shed plant material in bogs and fens is influenced by the same suite of environmental factors that control decomposition rates in any ecosystem, namely, temperature, available oxygen to support aerobic respiration, nutrients, chemical constituents in the plant material, and nature of the microbial community. There are some situations that are specific to mires, however. For example, water bestows thermal insulation to peat, so waterlogged peat does not undergo such transient fluctuations in temperature as occur in terrestrial ecosystems. Moreover, the slow diffusion of atmospheric oxygen in water limits oxygen availability and favors anaerobic microorganisms such as fermentors, sulfate reducers, and methanogens.

By knowing the starting material for peat and a bit about the decomposition rate of that material, one can quite successfully predict many of its characteristics. Using *Sphagnum* as an example, the peat is probably light to medium brown in color, has a low bulk density of ~0.05 g cm^{-3}, holds several times its dry mass in water, is >85% organic matter by mass, is acidic, and has low concentrations of nutrients that support the growth of plants.

The low nutrient concentration, especially nitrogen, is an inherent feature of *Sphagnum* (1% of dry mass). This low concentration affects growth, by regulating the amount of enzymes available to support photosynthesis, as well as decomposition of dead tissue. One question is, Why does the amount of nitrogen in plant material affect its decomposition rate? Microbial decomposers seem to need more nitrogen to support their metabolic needs than do plants. When the nitrogen content of plant material is too low, then microbial activity is low, which limits that rate of organic matter decomposition, accordingly. There is experimental evidence to support this contention, as *Sphagnum*-dominated sites fertilized with nitrogen have accelerated rates of plant production as well as decomposition rates.

Sphagnum-derived peats also are often acidic. *Sphagnum* acidifies its own environment, affecting plant growth and decomposition of organic matter. Acidification by *Sphagnum* is a natural process, as *Sphagnum*'s composition is largely uronic acid (up to 30% of dry mass). Decomposition of the plant releases acidity (H$^+$) to the environment, while at the same time it ties up basic cations such as calcium and magnesium. The acidic environment, with pH values <4.5 being common, is also toxic to many bacterial decomposers.

The high water-holding capacity of *Sphagnum*-derived peat relates in part to the anatomy of the plant. Unlike vascular plants that have roots to carry out the uptake of water, Sphagnum does not have roots. Instead, plants rely on the transport of water and solutes through elongated conducting cells (hydroids) in their stem, through the free space of the cell walls, by cell-to-cell transfer through intervening walls and cell membranes, or by external capillary spaces. Although some of these mechanisms may be quite effective (e.g., hydroids), freeing some species to rise above the water table, most *Sphagnum* species cannot move water and nutrients long distances throughout the plant. On the other hand, these traits give *Sphagnum* an extraordinary capacity to retain water, being able to hold more than 15 times its own mass in water.

As opposed to *Sphagnum*-derived peat, herbaceous and woody peats dominate in mires with the growth of such plants. Peat derived from sedges and reeds (e.g., sedge peat and reed peat) tends to have a higher bulk density of 0.3 g cm^{-3} and a higher concentration of plant nutrients than *Sphagnum*-derived peat, reflecting the higher nutrient concentrations in fens supporting sedges and reeds. On the other hand, woody peat often has large pieces of wood that can be identified to species even thousands of years after the burial of the wood.

Another facet of peat bears mentioning: compaction of older peat as it becomes buried under the weight of vegetation, newer peat, and possibly even snow in the winter. Experiments by noted ecologist Richard S. Clymo showed the importance of the distinction between new and old peat. Clymo found that organic matter decomposed much more rapidly in surface peat than at depth because oxygen was more readily available to support aerobic microbial activity. When organic matter decomposed to a certain point, it collapsed (compacted), thereby shutting off oxygen transport from the atmosphere to old deep peat. Clymo concluded that decomposition determined the amount of material retained as peat, but compaction determined how the entire peat deposit grew.

The loose configuration of the upper peat allows water to drain through at a relatively rapid rate. If such an arrangement characterized the entire peat deposit, then atmospheric precipitation would eventually drain downward to the water table. However, the rate of water flow (hydraulic conductivity) through smaller spaces within the deeper collapsed peat is several times slower. (In fact, there is a fourth-power relationship between hydraulic conductivity and the size of peat pore spaces.) Thus, the collapsed peat holds water and remains waterlogged, analogous to gelatin "holding" water. Differential rates of hydraulic conductivity between the new and old peat help explain why excess precipitation can run off the surface of a peatland (through the upper peat) without exchanging with the deeper peat. The terms to describe these layers in a peatland are "acrotelm" for the upper layer with high hydraulic conductivity and "catotelm" for the lower, more waterlogged, layer (Fig. 2).

It is important to emphasize that waterlogged peat is not dead or inactive, as was once though; numerous anaerobic bacteria (adapted to life without oxygen) can be quite prevalent in peat. For example, methane is a product of anaerobic microbial activity. Methane is an important greenhouse gas, affecting chemical and radiative properties of the earth's lower atmosphere (the troposphere). Methane released from peatlands alone into the troposphere may account for 20% of the atmospheric methane coming from all sources. Other anaerobic processes produce hydrogen sulfide, which has the characteristic smell of a rotten egg that sometimes permeates peat. Chemical reactions between hydrogen sulfide and various organic compounds produce volatile organic sulfides (carbonyl sulfide and carbon disulfide), which also contribute to the aroma of peat.

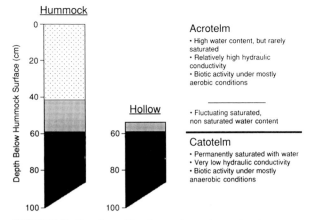

FIGURE 2 Depth profile of acrotelm and catotelm for typical hummock and hollow environments.

III. ELEMENTAL DYNAMICS

The influx and availability of plant nutrients are important in the ecology of bogs and fens, and thus elemental dynamics have received considerable attention. For example, ombrotrophic bogs are influenced by sea spray, soil dust, and air pollution. Sea spray enriches oceanic peatlands in sodium, magnesium, and chloride relative to silica, iron, and aluminium, coming from soil dust. Near agricultural land, soil dust also can contain large amounts of nitrogen and phosphorus used in fertilizer. Air pollution contains much more sulfur, nitrogen, and toxic trace metals such as lead and mercury relative to both sea spray and unpolluted air. The amount of precipitation entering the bog also determines the magnitude of solute deposition. Coastal areas generally receive more atmospheric precipitation than inland areas, so that coastal areas receive the largest amounts of sodium, magnesium, and chloride, whereas inland peatlands receive more silica, iron, and aluminum.

Once deposited, solutes may become concentrated in peat as a result of water evaporation, or they may be taken up by plants, sorbed onto peat surfaces, or leached entirely out of the peat deposit. Evaporation increases solute concentrations, but it does not change the ratio among the dissolved solutes. Uptake removes those solutes with high biotic demand, such as nitrogen and potassium. Once taken up, these nutrients can return to peat in shed plant parts, followed by release (mineralization) as a result of organic matter decomposition. Sorption removes much of the calcium and magnesium from peat pore water, even though plants can draw on adsorbed bases. Leaching is especially important for anions such as nitrate and sulfate that are not readily sorbed by peat.

Inorganic solutes can also be redistributed within the peat by water flow through the acrotelm. Sites losing solutes become more ombrotrophic compared to minerotrophic sites gaining solutes. Thus, within any given peatland, boglike conditions tend to intermingle with fenlike conditions in a predictable manner, entirely as a result of water redistribution (see below). In addition, nutrients can be redistributed in the acrotelm, where organic matter undergoes the most active decay, thereby mineralizing nutrients back into solution. This does not increase the absolute nutrient content of the peat, but it does tend to increase concentrations and the amount available for uptake, sorption, or leaching. Peat transfer from the acrotelm to the catotelm tends to bury nutrients.

The additional solutes entering fens come from either water draining mineral soils located upland of the peat or in water upwelling from mineral soil and sediments beneath the peat. The total amount of inorganic solute that enters depends on the elemental status of the soil or sediment, the residence time of the water in that environment, and the amount of water influx (recharge). Therefore, fens range from ones with slightly higher nutrient concentrations than are found in bogs, supporting so-called poor fen vegetation, to ones with moderate nutrient concentrations, supporting intermediate fen vegetation, and finally ones with relatively high nutrient concentrations and alkaline pH values, supporting rich fen vegetation (Table I).

Finally, disturbances such as fire and erosion can locally alter elemental dynamics. (Incidentally, fire is more common in peatlands than might be expected.) Fire mineralizes nutrients quite rapidly, while at the same time it reduces peat accumulation.

TABLE I

Typical Concentrations of Major Elements in Waters (as Solutes) and in Peat (as Total) of World Mire Types[a]

Mire type	pH	Solute (mg/liter)			
		Ca	Mg	K	Na
Rich fen	7.7	36–80	7–12	0.8–2	5–12
Intermediate fen	4.8	6–24	0.4–5	0.4–1	2–12
Poor fen	4.4	1.2–6	0.2–1.2	0–3	2–7
Bog	3.8	0.5–1	0.5–1.2	0–1	2–4

Mire type	Total (mg/g of peat)			
	Ca	Mg	K	Na
Rich fen	20–200	20–100	1–2	0.1–0.2
Intermediate fen	5–20	2–20	1–4	0.1–0.2
Poor fen	1–10	1–2	1–4	0.1–0.3
Bog	1–10	1–2	<1	0.1–0.3

[a] Data were derived from various studies of German, Swedish, and North American mires.

Plants that survive the fire might benefit from higher nutrient availability immediately following a fire. Erosion seems to occur mostly in the blanket bogs, especially those of Great Britain; the greatest effect is to expose previously buried peat.

IV. FROM PEAT TO PEATLANDS

Whether the growth of peat can be sustained to develop into a mire depends largely on the topography of the environment, with the chemistry and perceived origin of water determining the type of peatland that does develop. Essentially two processes are involved: "terrestrialization" refers to peat that replaces open water or aquatic sediments, usually producing a minerotrophic fen, whereas "paludification" refers to peat that extends laterally across a landscape or above the limit of the local ground water (i.e., resulting in a perched water table fed by atmospheric precipitation falling on it), producing an ombrotrophic bog. Often, the processes are not singular. For example, continual growth of peat on a fen can produce a perched water table, resulting in the progression of a fen to a bog. Indeed, many present-day bogs started out as fens.

Bogs and fens are much more than waterlogged peat deposits with plants growing on them. All mires eventually develop a conspicuous—and relatively stable—undulating surface arrangement of depressions (hollows) and ridges (hummocks). Hummocks are mounds of peat anywhere from 0.25 to several meters across and rising 0.25—1 m above their surrounding surface. Hollows are actual depressions, or they may be a relatively flat matrix (lawn) from which the hummocks rise. How hummocks and hollows form, then persist, has been the subject of much discussion. It now appears that the microtopography of bogs is largely under biotic control. *Sphagnum* species restricted to hummocks are simply more resistant to decomposition than those species restricted to hollow and lawn environments. As an outcome, a hummock rises above the peatland surface because peat preservation is more effective there. Likewise, extensive organic matter decomposition in hollows can cause the peat to collapse, thereby forming open-water "pools."

The arrangement of hummocks, hollows, lawns, and pools on the landscape, among other characteristic features (e.g., floristics and elemental dynamics), is used to describe different mire types. Below I describe some of the more prominent bogs and fens of the world. The lexicon includes English, Finnish, Swedish, and Germanic terms, among others. Figure 3 shows each type. When boglike

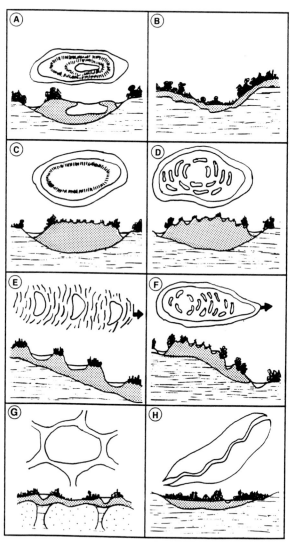

FIGURE 3 Profile and surface features of the main types of mire complexes. For the profiles, stippled areas represent peat growing above mineral substrate. (A) A schwingmoor, (B) a blanket bog, (C) a plateau raised mire, (D) a concentric raised mire, (E) an aapa mire, (F) an eccentric raised mire, (G) a palsa mire (loose stippling represents ice), and (H) a primary mire. [Modified from Crum (1988).]

conditions predominate the mire, it is designated a bog mire complex; examples include blanket bogs and raised bogs. When fenlike situations predominate, it is designated a fen mire complex; examples include aapa fens, patterned fens, schwingmoors, and palsa mires.

Blanket bog complexes occur where the climate is so constantly humid that peat can accumulate over mineral soil on slope up to 20°. Although ombrotrophic situations and *Sphagnum* prevail, sedges typical of fens occur commonly. Despite steep slopes, peat accumulation can be quite impressive, with depths of 1.5–3 m. Under this tremendous burden blanket bogs are susceptible to erosion, slides, and flows, especially with prolonged rains. The best-developed examples of blanket bogs are in upland Scotland, Northern England, Wales, western Ireland, western Norway, and southern Newfoundland.

Raised bog complexes acquire a central mound of peat typically 0.5–10 m above the surrounding mineral soil and 0.5–1.5 km in diameter. From this central dome the sides slope away gradually to a steep edge (rand) that itself contacts a fen water trough (*lagg* in Swedish) surrounding the entire complex. The mire, including the main dome, may be forested. The whole system grows in unison; primary peat laid down in the *lagg* allows the water table to rise and permits the upward growth of peat. Seen from the air, raised mires show spectacular arrangements of open-water pools on the main dome. Plateau mires have a nearly level central dome with irregular pools in a reticulate pattern. Concentric mires have their highest point near the center and pools forming concentric arcs around this point. Eccentric mires have elongated pools (*rimpis* in Finnish) and ridges (*kermis* in Finnish) aligned in arcs, but with the highest point some distance from the center of the dome. Concentric mires almost always develop in flat basins, whereas eccentric mires develop in sloping basins. Sweden, Finland, and continental North America have some of the best examples of raised mires.

Aapa fen complexes are extremely large expanses of peat, developing in wide nearly flat valleys. Besides their large size, ridges and pools accentuate and align themselves 90° from the direction of water flow. The ridges (called strings) are often 0.5–1 m high and 50–400 m long with fenindicator species. These alternate with very wet lawns or open-water pools (called flarks) dominated by sedges and mosses. Aapa fens develop in regions with long cold winters and complete freezing of the surface to reduce evapotranspirational water loss. Northern Finland and north central Quebec have the best examples of aapa fens.

Patterned fen complexes are too large to possess a singular description. On one hand, they are like blanket bogs, covering the landscape. On the other hand, they are like aapa fens with strings and flarks. Fenlike conditions tend to dominate. One striking feature of these mires is the presence of mounds covered by dense stands of black spruce (spruce islands) that are often true bogs. The origin and development of spruce islands are not clear, but underlying peat suggests that they develop from peatland that becomes dry enough in the summer to support shallow-rooted spruce trees. The Hudson Bay lowlands and the Lake Agassiz region of Minnesota have the best examples of patterned fens with raised bog "islands."

Schwingmoor (a German term) describes the filling of small lakes from edges by an encroaching fen carpet of sedges, *Sphagnum,* or reeds. The process starts at the edge, filling the basin inward until finally open water disappears. During development the pioneering mire floats over sometimes substantial water depth. One bounce on this floating peat sends waves traveling away. Sometimes the weight of trees growing on floating peat becomes too great, depressing the surface. This causes waterlogged roots that can kill trees. Standing dead trees along the margins of ponds are testimony to the "success" of those trees. Most people have schwingmoors in mind when they think of bogs. However, they are much less extensive than other types. Their confinement to north temperate and low boreal regions, where people are more common, makes them simply more visible than other types.

One of the best examples is the Everglades of southern Florida. Everglades peat formed largely from incomplete decomposition of plants such as sawgrass (*Cladium jamaicense*) and spikerush (*Eleocharis* spp.). The chemical nature of Everglades peat is quite similar to that of coal in the world's coal deposits, suggesting that coal is the remnant of enormous so-called peat bogs that dominated in the Carboniferous period. Such mires also occur in Sumatra and Sarawak of the Malaysian islands, producing a peat virtually identical to coal.

Palsa mires develop only in high northern latitudes where permafrost is present. They resemble aapa mires, except that they have raised mounds of peat (palsa), 1–5 m in diameter, that rise up to 1 m above the surrounding surface of the fen. The mounds are always associated with thickening ice layers that develop when peat insulates the surface, thereby preventing thaw each summer. The peat can be quite thin on palsas (<40 cm), and it always supports a bog community dominated by *Sphagnum.* The best examples occur in Lapland and in the Canadian Arctic.

V. ECOLOGY OF BOG PLANTS

Plants that prefer to grow in mires face the same essential requirements for life as plants growing anywhere: Water and nutrients (including carbon dioxide) are imperative, while temperatures must be suitable to support metabolism. [*See* PLANT ECO-PHYSIOLOGY.]

As mentioned above, *Sphagnum* retains several times its own mass in water. The high water content can slow the photosynthetic rate, by slowing carbon dioxide diffusion from the atmosphere to the site of assimilation within cells. Thus, the maximum photosynthetic rate for *Sphagnum* often occurs at less than maximum water content. On the other hand, exceptionally dry conditions cause cell death from which recovery is not possible. *Sphagnum* seems to be intolerant of drought through specialized adaptations, but rather it seems to maintain turgor in cells by lifting water through its capillary system above the local water table. This is not true for all mosses. *Polytrichum commune,* which occurs in fens, has a well-developed coating of wax on leaf surfaces to resist water loss. Moreover, *Polytrichum* mosses, dried for several days, can regain more than 50% of their photosynthetic capacity following rewetting, which is not true for most *Sphagnum* species.

In contrast, most shrubs and trees growing in mires have specialized adaptations to prevent water loss; this is ironic, since water is so abundant. However, the evergreen leaves of these plants (i.e., retained for more than a single growing season) show classical adaptations to tolerate drought, being thick and covered by surface waxes or narrow and rolled under. Such leaves are typical of desert plants growing under extreme aridity. In mires, however, such adaptations might be primarily in response to low nutrient conditions. In other words, if nutrients are in short supply, an evergreen leaf is a better investment than the alternative: relying on nutrient uptake, then returning to the peat in leaf litterfall, followed by resorption of nutrients released from the decomposing shed parts. The benefit is that nutrients remain in photosynthetic tissue. The negative side, however, is excessive water loss from sun-exposed leaves during the winter, when uptake of water from cold peat is limited.

While the whole plant deals with limited nutrients, roots must contend with poorly oxygenated water. Oxygen deficiency creates two problems: It decreases respiration (thus, the metabolic rate), and it increases the presence of soluble toxins to plants. To cope with such conditions, vascular plants enhance the transport of photosynthetic and atmospheric oxygen to their roots, thereby supporting root respiration and detoxification. One such way is the evolution of a vast network of open interconnected air spaces throughout plant leaves, stems, and roots. Oxygen transport through these lacunae may be by diffusion along a concentration gradient or by mass flow through pressurized ventilation. In either case oxygen flows from leaves to the roots, with flux of sediment-derived methane and carbon dioxide in the opposite direction. [*See* RHIZOSPHERE ECOPHYSIOLOGY.]

Mire plants also need a broad temperature response to withstand long cold winters and short warm summers. Most plants show positive photosynthetic rates at 0°C, maximum rates at 10–15°C, and a high temperature compensation near 30–40°C. To tolerate low temperature, plants accumulate low- to high-molecular-weight sugars (su-

crose, glucose, and fructose) in cells. Sugars lower the freezing point of liquids, thereby preventing the cells from freezing when air temperature is below 0°C. There is evidence that *Sphagnum* may tolerate slightly higher temperatures at increased concentrations of atmospheric carbon dioxide. The reason for the shift in temperature optimum is not clear, but it has relevance in a world experiencing climatic warming.

VI. ECOLOGICAL COMMUNITIES

Ecological communities have several traits that distinguish themselves from one other, including distinctive species composition, structural appearance, and trophic interactions (who eats whom).

For example, hummocks and dry central areas in raised bogs are frequently dominated by *Sphagnum fuscum,* a major hummock builder, and dwarf shrubs such as Labrador tea and bog laurel grow better there than in wetter places. The height and size of hummocks depend on several factors, such as the particular *Sphagnum* species involved and its relationship to water supply. For example, *S. fuscum* hummocks are broad mounds, whereas *Sphagnum magellanicum* hummocks are much lower. Hummocks rising especially high above the water table—thereby overgrowing the water supply—provide a suitable place for tree growth. In the course of time, trees persist while moss cover may weaken, leading to the development of "treed bogs."

Wet bog communities form in hollows between hummocks and are often dominated by *Sphagnum cuspidatum,* with shrubs being rarer than sedges and sedgelike plants; lawns are often dominated by *Sphagnum papillosum.* Fenlike situations can occur within bogs, indicating local sites of minerotrophy. Any numbers of species fix the limits of fens, and species diversity is usually—but not always—higher than that for bogs. For example, there are bogs in Nova Scotia that harbor as many as 54 species of vascular plants. Pool communities vary in composition as a function of depth to the water and nutrient concentrations within that water. Vascular plants are usually lacking from pools, although aquatic plants such as the water lily (*Nymphaea* spp.) can thrive in deep-water places. In shallower places *Sphagnum fallax,* an emergent form of *Sphagnum,* often dominates.

Many large animals use peatlands, although no species are totally dependent on them for life. These include the woodland caribou (*Rangifer tarandus*), gray wolf (*Canis lupus*), Canada lynx (*Felis lynx*), American mink (*Mustela vison*), and beavers (*Castor canadensis*) in North America. On the other hand, some smaller animals, such as the lemming (*Lemmus sibiricus*) may be found almost exclusively in peatlands. Species lists for birds observed in peatlands are extensive, whereas the diversity of amphibians is limited to very common frogs (*Rana* spp.) and toads (*Bufo* spp.). No doubt the acidic condition of peat is detrimental to amphibians and reptiles.

Invertebrates and microorganisms in mires show increasing specialization for sites, usually as a function of wetness. For example, invertebrate biomass in pools seems to be about 10 times higher than that in the dry bogs. This suggests an aquatic rather than terrestrial affinity by peatland invertebrates. Habitat specificity is also apparent for microbes with aerobic bacteria and fungi in the driest sites and anaerobic bacteria (nitrate- and sulfate-reducing bacteria and methanogenic bacteria) in pools.

One way to study how organisms in ecological communities interact among themselves (i.e., biotic interactions), as well as with their environment (i.e., abiotic interactions), is to construct a food web. Food webs show the flow of energy from primary producers (plants as well as photsynthetic and chemosynthetic microorganisms) through consumers of those plants (herbivores), and finally to secondary consumers (predators). Such studies of mires are surprisingly rare and often incomplete. One detail in this regard is the remarkable observation that almost no animal eats *Sphagnum* directly. The reason for this is not altogether clear, but low nutrient content and/or defensive compounds to prevent herbivory are likely reasons. Hence, the primary consumers in peatlands (e.g., invertebrates) feed on bacterial and fungal decomposers of organic matter, resulting in a detritus based food web.

VII. CARBON DYNAMICS

Bogs and fens are relatively productive ecosystems, with net primary productivity (i.e., the atmospheric carbon taken up through photosynthesis that is incorporated into plant material) ranging from 100 to 350 g of carbon/m^2/year among several sites. In bogs production by *Sphagnum* mosses often predominates, except in treed bogs, where production by trees alone may exceed 300 g of carbon/m^2/year. The contribution by mosses is usually small in fens compared to that produced by vascular plants. Worldwide net primary production of carbon in bogs and fens is ~0.78 Pg/year (1Pg = 10^{15}g), which represents 1.5% of production on the continents. On the other hand, noted ecologist Eville Gorham estimated that the total carbon in peat deposits of bogs and fens worldwide is 455 Pg, representing about 33% of the world pool of soil carbon.

Gorham also estimated that peat accumulates at a rate of ~0.096 Pg/year worldwide. To put this value into some context, ~12% of net primary production becomes incorporated into peat; the remainder is remineralized back to carbon dioxide or methane and released to the atmosphere. Indeed, methane emitted from peatlands to the atmosphere releases ~0.046 Pg/year of carbon—or ~6% of net primary production. For this reason peat accumulation is a slow process.

Conversely, human activity carried out in peatlands accelerates rates of carbon release back to the atmosphere. For example, Gorham estimated that drainage for agriculture induced a net carbon release of 0.0085 Pg/year, whereas a net carbon release of 0.026 Pg/year resulted from combustion of peat for fuel. These releases are certainly much lower than the 5–6 Pg of carbon released to the atmosphere annually from fossil fuel burning, but they are not trivial, on the other hand.

VIII. THE HISTORICAL RECORD IN PEAT AND GLOBAL CLIMATE CHANGE

As fascinating as mires are in their own right, they also chronicle a wealth of information about their own past within the peat deposit. The information occurs in the form of chemicals as well as biological materials preserved in peat. For example, macrofossils are large materials such as wood, seeds, tree needles, animal remains, and even individual *Sphagnum* leaves—or any other material recognized without the aid of a high-powered microscope. Relatively large macrofossils have little chance of being carried any great distance, so they provide information about organisms that grew—or really died—directly on the peatland surface. On the other hand, relatively small microfossils such as pollen can come from quite far off (several kilometers) to complement those from plants growing in the mire, thus providing information about the mire and surrounding areas. The key to understanding mire stratigraphy (i.e, layers or strata of material deposited in peat) (Fig. 4) is recognizing that different species have different fondnesses for different climates and for different levels of minerotrophy. Thus, extracted peat studied *in toto* provides a chronicle—the timing of which can be set by radiocarbon or other dating—of mire development through time, as well as the climate and sup-

ply of chemical bases under which the mire developed.

Interpretation of which factors drive historical changes recorded in peat remains controversial. It seems that the establishment of mires on the landscape occurs in response to climatic changes. For example, the origins of the oldest mires (almost all are fens) in west–central Norway and Scotland date close to the time of deglaciation >8500 years ago, as do the mires in the northernmost region of subarctic Canada. Interestingly, the development of mires in the southern boreal region of Canada started only 5000 years ago, even though the region was deglaciated >10,000 years ago. The most likely explanation for this paradox of longer mire development in more northern regions seems to be that warm dry conditions prevailed in the south during the middle Holocene period that prevented peatland development until the climate changed to colder and more moist about 6000 years ago. In contrast, northern Canada has had a cold moist climate since the early Holocene period.

On the other hand, it appears that developmental changes within most mires are under local biotic or hydrological—or autogenic—control. For example, most present-day bogs started out as rich or poor fens, becoming bogs as peat accumulation moved the peat surface further away from the influence of local groundwater. Furthermore, just as the differential decomposition rate of *Sphagnum* in hummocks than in adjacent hollows allows more peat accumulation in the former site, hollows apparently turn into open-water pools through time as the rising water table swamps plants. For example, the gradual flattening of the mire surface brings the water table to the surface, allowing hollows to expand and coalesce to form broad open-water pools. Therefore, the most well-developed pools occur over the oldest portion of the peatland, while the newest portions show the least microtopographic development.

With the arrival of human-induced environmental change, the "rules" governing mire formation and development may change. The world's mires could experience a temperature increase of 2–4°C. A wetter or drier moisture regimen could accompany this temperature change; the impact on nutri-

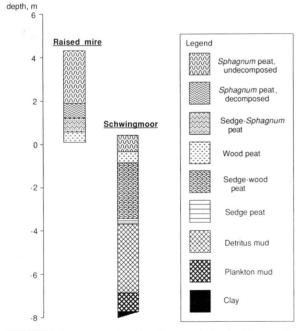

FIGURE 4 Peat stratigraphy for a typical raised bog and a schwingmoor. The former developed by paludification, whereas the latter developed by terrestrialization.

ent and hydrological interactions, as well as fire and erosion, is unclear. Moreover, how these changes will play out for species, populations, and communities is anyone's guess. However, they are likely to affect physical properties such as albedo, surface roughness, and heat balance that have profound feedbacks on both regional and global climates. The anticipated effect on carbon storage is not rosy, because temperature change has a greater effect on the decomposition rate than on the rate of plant production. As a result, the increase in temperature alone might change peatlands from net sinks to net sources for atmospheric carbon dioxide. If wetter conditions prevail, stored carbon may be mobilized as methane, which has a much stronger effect on the radiation budget and chemistry of the atmosphere than carbon dioxide.

However, it is certainly plausible that direct perturbations by humans will swamp any climatic effects. Draining peatlands for agriculture or to extract peat for use as a plant growth medium poses real and immediate threats to peat. Holland has already lost all of its peat bogs, and there is concern that British bogs may be the next to go. Demand for forest products speeds up efforts to drain peatlands and convert them to forest plantations (e.g., Finland). Perhaps there is no need to worry; perhaps the rate of peat accumulation will simply speed up in more northern regions. I doubt, however, that we can rely on such salvation; it seems that human-induced environmental change is far more capable of nibbling away at the majestic mires of the world than it is of ensuring their persistence.

Glossary

Acrotelm The relatively well-oxygenated surface layer of peat in a peatland, characterized by relatively high rates of water flow and organic matter decomposition.

Bog Peat-covered wetlands in which vegetation shows the effects of a high water table, low concentrations of inorganic solutes, and acidic conditions. This definition corresponds to nutrient-poor (ombrotrophic) wetlands.

Catotelm The poorly oxygenated deeper peat in a peatland, with impeded water flow and organic matter decomposition under anaerobic conditions (i.e., in the absence of oxygen).

Fen Peat-covered wetlands with a high water table, but water is enriched by inorganic solutes from upslope soil or sediments. This definition corresponds more to nutrient-rich (minerotrophic) wetlands than to bogs.

Minerotrophic Nourished by mineral water; refers to peatlands that receive inorganic solutes derived from local sediment, soil, or rocks, in addition to those from atmospheric deposition.

Mire A general term embracing all kinds of peatlands and all kinds of vegetation grwoing in peatlands; bogs and fens are examples.

Ombrotrophic Nourished by rain; refers to peatlands that are entirely dependent on inorganic solutes deposited only from atmospheric deposition.

Bibliography

Clymo, R. S. (1987). The ecology of peatlands. *Sci. Prog. (Oxford)* **71,** 593–614.

Crum, H. (1988). "A Focus on Peatlands and Peat Mosses." University of Michigan Press, Ann Arbor.

Damman, A. W. H. (1990). Nutrient status of ombrotrophic peat bogs. *Aquilo Ser. Bot.* **28,** 5–14.

Gore, A. J. P., ed. (1983). "Mires—Swamp, Bog, Fen, and Moor." Elsevier, Amsterdam.

Gorham, E. (1991). Northern peatlands: role in the carbon cycle and probable reponses to climatic warming. *Ecol. Appl.* **1,** 182–195.

Moore, P. D., and Bellamy, D. J. (1974). "Peatlands." Elek (Scientific Books), London.

National Wetlands Working Group (1988). "Wetlands of Canada." Polyscience, Montreal, Quebec, Canada.

Wright, H. E., Jr., Coffin, B. A., and Aaseng, N. E., eds. (1992). "The Patterned Peatlands of Minnesota." University of Minnesota Press, Minneapolis.

Captive Breeding and Management of Endangered Species

Phillip A. Morin

University of California, Davis

Captive management of endangered species for the purpose of conservation is a growing field involving both basic and applied science. The immediate need to conserve far more species than there is space available to house them requires assessment of the minimum population sizes for optimal social, demographic, and genetic factors, as well as development of decision processes for choosing which populations, subspecies, and species are to be captively managed. Finally, management programs differ according to their anticipated duration and goals, whether they be for eventual reintroduction of captive animals to the wild or simply maintaining species in captivity for as long as possible. The problems associated with captive management are primarily associated with the lack of information on natural habitats, behaviors, social structures, and life tables. The need to manage endangered species is, however, immediate and often without room for error, as few or none of the endangered populations remain to start over, and problems arise from founding populations with small numbers of individuals.

I. INTRODUCTION

The human species currently inhabits almost all portions of our planet and numbers more than 5 billion. The World Bank and the United Nations predict that, given our current exponential growth rate, and barring catastrophes such as pandemics and famine, the population will peak and plateau at approximately 11–12 billion about 100 years from now. Presently, over half of the world's forests have been destroyed by human exploitation, and pressures on other biomes such as grasslands and oceans are becoming apparent as grazing lands and fisheries decline. In the next 100–500 years, we will experience a "demographic winter," during which one-quarter to one-half of the earth's estimated 10 million species will become extinct in their natural habitats as a direct result of human population growth and associated increases in exploitation. As an insurance policy against the loss of plants and animals, captive breeding and management of species will be necessary, with the goal that one day the biosphere will recover sufficiently to again support these species in "natural" (though necessarily different from today's) ecosystems.

II. SCOPE OF THE PROBLEM

Species extinction is not a rare event in the earth's history. Paleontologists and biologists estimate that since the Cambrian, species have become extinct at the rate of about 1 per year (though this pattern is far from uniform). In contrast, it is estimated that we will lose approximately 3 to 5 million species in the next 100 years. That translates to about one species extinction every 11–18 minutes. Of course, most of these extinctions will involve species of insects and small plants, the majority of which will never even be described by biologists. The only way that these species will be conserved is by the protection of large land areas that include many habitat types. [See EVOLUTION AND EXTINCTION.]

Of the 10 million extant species, only a small portion (about 2 million) have been formally recognized, and only a small percentage (about 10,000) of those have been ecologically, behaviorally, or genetically (a few thousand) characterized. Many of those that have been studied are not currently "endangered," but it is estimated that the number of known species that will require some sort of captive management by the end of this century will be in the tens of thousands. [See SPECIES DIVERSITY.]

As impending human population pressure will require more and more exploitation of animal and plant habitat, the only practical approach to conserving large species that have historically required and/or spanned wide geographic ranges, and endemic species or varieties, which will almost certainly become extinct in the wild, is maintenance of captive populations. The problem we face is how to carry them through the impending demographic winter. If we are successful, a steady or decreasing human population size and wiser use of renewable resources may allow the reintroduction of species that will have become extinct in their natural habitats.

III. PURPOSE OF CAPTIVE BREEDING

Captive breeding of nondomestic species has become a primary concern for zoos, aquaria, botanical gardens, and research facilities only in the last few decades, as the supply of wild-caught animals dwindles and regulations such as those imposed by the Convention on International Trade of Endangered Species of Wild Fauna and Flora (CITES) prevent trade of endangered species. Replacement of current populations for biomedical use (some endangered primate species, such as chimpanzees, are used in biomedical research because of their similarity to humans; this is a controversial practice, considered by many to be unjustified) and entertainment of the public in display-oriented facilities (i.e., zoos, circuses, and gardens), however, is only one function of captive breeding. There are at least five functions of captive-bred populations of wild animals that facilitate conservation of endangered species. [See CONSERVATION AGREEMENTS, INTERNATIONAL.]

1. Basic Research

Biologists strive to learn about the generalities of nature by observing specific qualities of species and ecosystems. Captive populations can serve as substitutes for wild groups in some studies of genetics, behavior, reproductive biology, sociobiology, and other areas of basic biological research. The use of captive populations reduces the difficulty and expense of studying wild populations and allows for the use of experimental manipulation. Study of different species and strain hybrids of agriculturally important species is important for maintenance of stocks that can resist diseases that plague monotypic crops.

2. Management and Breeding Research

Only about 20% of all mammal species and 10% of bird species have been successfully bred in captivity. Most of the problems surrounding the breeding of wild animals in captivity stem from insufficient knowledge of the natural conditions under which animals breed in the wild. Given the paucity of this kind of information, however, the successes in breeding some species in captivity, and research into the important variables in those successes, can help to develop successful breeding and management plans for other species. For example, successful breeding of several species of hawks and falcons has been useful in the development of a program for the endangered peregrine falcon, and these species have even served as surrogate parents to young peregrine.

3. Demographic and Genetic Reservoirs

Some species are threatened in all or parts of their ranges, or have experienced catastrophes from which they cannot recover without intervention. Captive stocks can serve as reservoirs to infuse natural populations with either numbers or genetic material to help populations recover, or to found new populations where local extinctions have occurred.

4. Species Extinct in the Wild

Captive stocks are, in a growing number of cases, the only immediate opportunity for survival of some species. Captive maintenance of these populations is the only way to preserve options for the future of evolution on our planet. For many, the long-term goal of maintaining species that are extinct in their natural habitats is their eventual reintroduction into the wild.

5. Education

Finally, through the display of endangered species, and wise use of their entertainment value to provide a means of educating the public, captive management of endangered species can help to increase awareness of our ecological predicament. Future socioeconomic goals of species and ecological preservation are the only hope for recovery from the demographic winter. Although large vertebrates make up the majority of species on display in zoos, wild animal parks, and aquaria, they can serve to focus education on the plight of the many more threatened species of plants and animals.

IV. CHOICES OF SPECIES FOR CAPTIVE MANAGEMENT

Conservation biologists have estimated that, among terrestrial vertebrates alone, approximately 2000 species may have to be captively bred to prevent extinction. These include approximately 160 primates, 100 artiodactyls, 100 large carnivores, 800 birds, and several hundred amphibian and rep-

tile species. Estimates for only the United States suggest that approximately 700 species of plants will probably become extinct in the next decade without artificial management, and another 3300 are currently threatened. Also in the United States, at least 24 species of fish have become extinct since the arrival of European settlers, and 63 are now listed as endangered. Considering the difference in relative biodiversity of tropical and temperate regions (such as the United States), and the relative lack of information on particular plant and insect species for which conservation efforts are needed, these world numbers are probably underestimates.

Traditionally, the highest priority for captive management has been given to taxa that are least likely to survive in the wild. One of the primary dilemmas, given limited resources, has been the choice of which taxa to attempt to manage, and which taxonomic unit to preserve. Taxonomy is the science of classifying living things and establishing their relationships in a hierarchical phylogenetic system of nomenclature. Unfortunately, although taxonomic units are discrete and absolute, evolution is a continuous and, in terms of human time scales, gradual process. Various attempts have been made to establish a species definition that can apply to extant and extinct taxa, and to disjunctly distributed populations, but population, semispecies, subspecies, and species have often been defined based on out-of-date taxonomic methods and may not reflect the phylogenetic relationships of populations or the evolutionary fate of those taxa. Thus, after identifying a taxon to be managed, the scope of individual animals to be targeted must be clearly defined. If the primary goal of conservation is to retain as much of the biotic and genetic diversity as possible, then all such groups should be maintained. In reality, each taxon must be evaluated individually to determine the appropriate population or set of populations to manage to give the optimal taxonomic, demographic, and genetic conservation strategy under severe limitations.

A. Limitations

Currently, just over half a million individuals of about 3000 vertebrate species are held in zoos. For long-term propagation, however, it is unrealistic to assume that over half of that number can be maintained at sufficiently large population sizes. As the number of threatened species grows, it is likely that most of the space in zoos will be devoted to captive breeding of endangered species. Still, the number that can in all likelihood be maintained is less than 1000 species (if maintained at a population size of 300; see the following).

Zoos and wild animal parks have limited space, money, and technology. Most zoos are supported by municipal funds and/or public entrance fees. To be accessible to the public, they are typically located in or near cities and must strike a balance between conservation, education, and entertainment. These factors, in the current socioeconomic atmosphere that does not put a high priority on spending money for conservation, place a premium on the space currently available for conservation. In addition, the annual cost (in U.S. dollars) of maintaining a single species in captivity can be nearly as much or more than the cost of maintaining an entire ecosystem (e.g., the projected cost for the California condor recovery plan is more than $1 million annually, more than twice the cost of maintaining the whole Serengeti ecosystem in East Africa). Multiplied by the number of such plans that will potentially be needed, the annual cost could be as high as $10 billion, over twice the present annual budget of the U.S. National Institutes of Health.

B. Priorities

Given the aforementioned limitations on space and money for captive management, how are choices made for captive management of taxa? Historically, animals bred in captivity were chosen on the basis of their usefulness for food or clothing, or because of their aesthetic and entertainment value. "Charismatic megavertebrates" and unusual flowering plants have received the most attention. For conservation, however, a closer look at phylogenetically (evolutionarily) important taxa and those most likely to become extinct in the wild is necessary. Species may be especially vulnerable for many reasons or combinations of reasons; the species charac-

teristics in Table I can sometimes be used to predict which species groups may be threatened, but sometimes the cause is not clear, as in the recent decrease in amphibian species worldwide. Finally, species whose activities are critical to the stability of entire ecosystems should be considered as conservation priorities. Even if whole ecosystems cannot be saved, some species in each system are critical and/or cannot survive without interaction with one or more other species. Emphasis on these species groups may accomplish species preservation in less space and with less money, and provide a greater opportunity for the preservation of biological communities.

C. Diversity

Besides choosing among taxa as outlined here, balancing among taxa to prevent loss of entire gene pools and taxonomic groups must be stressed. Many species are widespread and subdivided into isolated subspecies that may have morphological, behavioral, and genetic differences. Understanding these differences can be very important to the survivability and future evolution of the species, as outbreeding depression or other problems associated with mixing of locally adapted lineages may occur when individuals from different groups are crossed (see Section VI,B). For example, the Tatra

TABLE I
Some Characteristics of Vulnerable Species

Narrow geographic range (endemics)
One or few populations
Small population sizes
Low population density
Large home ranges
Large body size
Low population growth rates
Limited dispersal
Migratory
Low genetic variability
Specialized niche requirements
Characteristically found in stable environments
Form permanent or temporary aggregations
Hunted or harvested by humans

Mountain ibex was successfully reintroduced from Austrian populations when it became extinct in Czechoslovakia. Subsequently, animals from different subspecies in Turkey and Sinai were added to the population. The resulting hybrids rutted in winter instead of early autumn, and their young were born in the coldest part of winter. As a result, the entire reintroduced population in the Tatra Mountains went extinct. Similar outbreeding depression problems have resulted in captivity when cryptic species or subspecies have been bred, either knowingly or unknowingly, by managers. Alternatively, when current genetic and ecobehavioral data contradict formal subspecies designations, such designations should be ignored in the interest of efficient use of available space in maintaining larger captive breeding populations and the corresponding levels of genetic diversity. Again, the evolutionary history and relationships of taxa are of primary importance and must be evaluated individually to make such decisions.

Finally, especially among plants, endemic species and varieties (landraces) are often of scientific and economic value. Prior to the 1950s, when Western agriculture took root in Asia, farmers in Sri Lanka cultivated over 2000 varieties of rice. After high-yield hybrid strains were introduced, that number dropped to 5. Large monotypic crops are especially vulnerable to disease outbreaks, so the availability of resistant strains for interbreeding is important for the future of agriculture, as well as the preservation of species diversity.

V. SPECIFIC MANAGEMENT GOALS

Once decisions have been made to manage a species in captivity, several factors must be considered and appropriate goals set. Haphazard breeding of animals without finite goals and genetic, demographic, and behavioral management will almost without exception result in the extinction of captive populations with limited population sizes. These goals are described next, followed by more detailed descriptions of management plans and the factors that contribute to their design.

A. Time Scale

The demographic winter could last 100 to 500 years. Realistically, though, regaining land for reintroduction of species could be thousands of years away. Most of the land areas that still support wildlife today are those that have been spared development because they are hydrologically and agriculturally marginal, and even they will eventually be exploited. Because these lands will decay after they have been exploited, decades to centuries might pass before they can support wildlife again. So, our function as the guardians of species will continue for millennia. Within the next 200 years, expected advances in reproductive and cryopreservation technologies might provide alternatives to maintenance of large populations in captivity. As a beginning, the Zoological Society of San Diego has begun a "frozen zoo," which maintains cell lines from over 300 species and subspecies of mammals for genetic and cellular biology research. Technology for preserving embryos does not, however, exist for most animal species. Of course, maintenance of some live populations for the preservation of learned behaviors and appropriate demographic profiles will be necessary, but the space constraint will be decreased, and loss of genetic variation greatly slowed or stopped.

For about 85% of plant species, seed banks represent a tremendous potential for long-term storage of viable populations, and botanical gardens, arboretums, and associated nature reserves may be able to cope with at least a portion of the growing population necessary for restocking of seed banks as seeds begin to lose viability, and for those species that do not produce seeds that can be stored.

B. Natural Population Structure and Behavior

The key to successful breeding of many captive species lies in the conditions in which they are kept. Early attempts to breed many species failed because they "failed to adapt to captivity." However, one goal of the conservation of species is to preserve them in their natural state, so adaptation to captivity is not desirable nor necessarily possible. As more is learned about ecology and behavior of species in their natural habitats, it becomes clear that key ecological conditions, such as spatial requirements, diet, environmental needs (light cycles, temperatures, humidity), and housing, as well as appropriate social groupings, are necessary and sufficient to allow adequate captive reproduction. For example, white rhinoceroses rarely reproduced in captivity at the San Diego Wild Animal Park when kept in pairs or small groups, but the population has now grown to the point where it is necessary to use various birth control methods after the animals were maintained in larger groups. Simply providing the appropriate social context was enough to promote natural levels of reproduction.

For long-term conservation purposes, it is necessary to retain appropriate behaviors and family groups as well. Some species require "helpers" in rearing the young, or can only learn appropriate behaviors for rearing their own young if brought up in the appropriate social context. This includes maintaining an optimal age and sex structure by selectively removing animals, and doing so in a way that optimizes demographic and genetic plans and does not disrupt the social dynamics in the group. Although some natural behaviors, such as dispersal and predator escape, cannot be adequately mimicked in captivity, providing an appropriately naturalistic captive environment is often the first step in a successful breeding program, and ensures the best chance of maintaining the evolutionarily significant behaviors of the species.

C. Genetic Variation

The second primary goal of captive propagation, after conservation of biodiversity, is maintenance of genetic diversity. Natural populations of hundreds to millions of individuals, whether genetically linked or isolated, are tremendous storehouses of genetic variation. It is this variation that allows species and populations to adapt when faced with environmental or ecological changes, and to take advantage of new niches that are created by these changes. For long-term preservation of a species, beyond just maintaining the fitness of the captive population, conservation biologists have arbitrarily

set a goal of maintaining at least 90% of the genetic variation for the 200-year time period. This is a number arrived at intuitively as one that will not substantially decrease the ability of a species to evolve, while not requiring huge (>1000) population sizes to be maintained.

Achieving this goal is not easy. First, natural levels of genetic variation are not known for many managed species. Second, what is known usually consists of the variation observed in a small set (usually less than 25) of electrophoretically polymorphic enzyme and blood serum loci (gene products). When one considers that there are approximately 100,000 gene loci in vertebrates, and the corresponding primary gene products may be alternatively spliced in different tissues or different life stages to produce about 1 million secondary gene products, this type of assay can be inadequate because of sampling error (it must be noted, however, that these techniques can often produce substantial and accurate information about the populations and taxa being studied). Further, measures of mean heterozygosity per locus, though indicative of the amount of inbreeding in a population, do not indicate much about the amount of allelic diversity (numbers and frequencies of alleles per gene locus) in a population. Although mean levels of heterozygosity are sometimes positively related to the fitness of populations, allelic diversity is more important for the future ability of a species to evolve; genetic plasticity (alternative combinations of alleles) is the key to a species' ability to adapt to novel and changing environments. New methods in molecular genetics for detecting many alleles are quickly becoming available and will help to determine the extent and rate of loss of allelic diversity in captive populations.

When a captive population is founded by wild-born animals (or plants), provided that the appropriate knowledge is available to breed them in captivity, the number of founding individuals and their relative and absolute reproductive success in the first few generations can drastically affect the distribution and amount of genetic variation. For example, reducing a population to only 10 individuals will result in the loss of only about 5% of the species' natural heterozygosity, but almost 70% of the alleles. Continued reproduction with a small population size will lead to "genetic drift," or random loss of alleles from the population, and increased homozygosity from inbreeding. The methods of genetic management to maximize retention of genetic variability and diversity are discussed next.

D. Minimum Viable Populations

Closely linked with maximizing genetic variation is the maintenance of a minimum viable population (MVP). In the context of long-term conservation, this means the minimum population size that has a 99% chance of surviving for at least 200 years, and includes the maintenance of at least 90% of the genetic variation found in the source population. Decreases in genetic variation can lead to inbreeding depression and genetic drift (called genetic stochasticity) that can cause population fitness to decline, resulting ultimately in extinction; but there are other causes of extinction as well. Small populations are subject to demographic stochasticity, in which chance events in the survival and reproductive success of a finite number of individuals result in random extinctions. One example of such an extinction event is when, by chance, only males remain in the population, as occurred in the now extinct dusky seaside sparrow population. Both captive and wild populations are subject to natural catastrophes, such as floods, fire, and extreme weather, which can abruptly reduce the size of populations (causing a so-called "population bottleneck"), though partitioning of populations into different zoos reduces the chances of an entire species being lost. Finally, although environmental stochasticity (changes in ecological or environmental conditions, or interactions with other species) can be minimized in captivity, it can nevertheless affect populations in the form of parasites and disease outbreaks.

Determining MVPs depends on knowledge of the natural patterns of dispersal and density of each species, as well as sound genetic and demographic management practices designed to minimize the ratio of census to genetically effective population size while optimizing genetic and demographic parameters.

E. Reintroduction

Several captive propagation programs have already realized or partially realized their goal of reintroducing captive-bred populations to the habitat in which they had become extinct. These include the California condor, the last 14 of which were in captivity by 1987. Subsequent breeding to increase the population size has resulted in 75 condors being raised in captivity. This success was due mostly to double-clutching, as California condors raise only one chick every other year if allowed to hatch and raise their own young. By removing the one or more eggs as they were laid, managers induced females to lay more eggs, which were artificially reared (using puppets shaped like condor heads), thereby increasing population growth rate. To optimize the chances of survival after release, researchers used Andean condors, which are not threatened, to determine the methods for reintroduction into the wild that resulted in the greatest survival rates. In 1993, the first 8 captive-bred condors were released into the wild (4 have survived), and the population will continue to be supplied with more birds from the captive population to increase the wild population size and to reduce chances of further inbreeding when the released birds start to reproduce and the population becomes self-sustaining.

One of the first successful reintroductions was that of the Arabian oryx, which has established self-sustaining herds where they were once extinct. Reintroduction of Przewalski's horse, once reduced to 13 captive individuals, into the Gobi Desert is planned. These are examples of species that have been extirpated in the wild by human pressure (hunting and/or loss of grazing lands), but for which suitable habitat still exists. In contrast, most of the species that will need captive propagation to avert extinction are endangered by loss of native habitat, and they will have to be maintained in captivity until suitable habitat is again available. Until then, maintenance of genetic variation, necessary coadapted species (such as specialized pollinators and food sources), and natural behaviors are the primary goals of captive management for eventual reintroduction to the wild.

VI. MANAGEMENT TECHNIQUES

The goals of preserving heterozygosity and allelic diversity can be achieved primarily through two techniques: demographic and genetic management. Demographic management is the use of methods to quickly increase the founding population size to the target size to minimize loss of heterozygosity, combined with equalizing founder contribution and family size to minimize loss of allelic diversity. Genetic management methods are used to minimize inbreeding and outbreeding and to identify genetically important (underrepresented) individuals in the pedigree. Table II summarizes the techniques discussed in the following sections.

A. Demographic Management

Genetic drift, resulting in the loss of allelic diversity and, when accompanied by inbreeding in small

TABLE II
Captive Species Managment Practices[a]

Maximize effective population size (N_e)

Minimize variance in population growth rate

Attain viable population size as soon as possible

Equalize the genetic contributions of founders

Monitor and maintain inherent genetic variability

Reduce inbreeding or purge populations
   of genes responsible for inbreeding depression

Avoid outbreeding depression

Maintain multiple populations (metapopulations)

Avoid selecting for "type" or for domestication

Facilitate natural behaviors, including:
   Dispersal and migration
   Social and breeding structure

Manage interacting species, including:
   Pollinators
   Prey species
   Predators
   Parasites
   Competitors

---

[a] Adapted from Table 8.2 in D. S. Woodruff (1989). *In* "Conservation for the Twenty-first Century" (D. Western and M. Pearl, eds.), pp. 76–88. Oxford University Press, New York.

populations, loss of heterozygosity, is often the greatest threat to captive populations. Most species in captivity have shown some effects of inbreeding depression, usually observed as lower fecundity, higher juvenile mortality, and higher morphological or developmental instability. The most important means of reducing the chances of inbreeding and preserving genetic variation and heterozygosity is rapid growth of a population from founding size to target size. Figure 1 shows the effects of maintaining a population at various target sizes on the loss of heterozygosity. This figure assumes an "ideal" (randomly mating) population and shows the retention of heterozygosity relative to that found in the founding population. In many species, members of one or both sexes disperse among populations, maintaining "gene flow" (mixing of genetic variation among populations). Breeding facilities can simulate dispersal via foster–infant exchange between groups or facilities to mimic more closely a large, "randomly mating" population.

To meet the goal of maintaining at least 90% of the variation found in the wild population, founding population size is also critical. Below a founding size of 6 individuals, there is no possible way to ever reach this goal. As the number of founders increases, the target size of the population necessary to maintain genetic variation decreases. When possible, starting with a founding population of at least 20 unrelated animals will capture enough of the genetic variation to allow the minimum population size target.

The target population size is that which will allow the goal of maintaining 90% of the genetic variation in the wild population, assuming that the captive population is randomly mating and there is no variation in reproductive success (the genetic "effective population size," or N_e). In reality, the conditions of random mating and equal reproductive success are almost never realized, especially when the "population" includes colonies distributed among zoos or other breeding facilities. Thus, the census population size is usually larger than the effective population size. Demographic management aimed at equalizing family size, rapid growth to target population size, and maximizing genera-

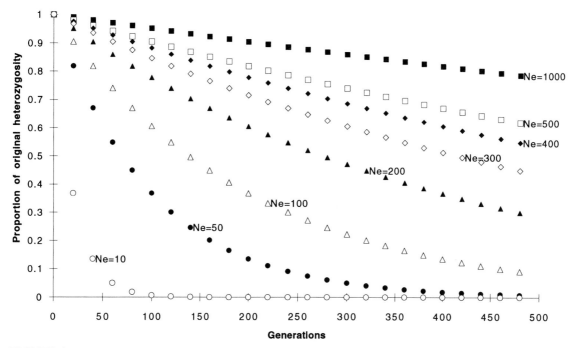

FIGURE I Decline in heterozygosity per generation for different-sized populations. The equation used was $H_t = H_0 e^{-t/2N_e}$, where H_t = the heterozygosity at time t (generations), H_0 = the original heterozygosity (assumed to be 1 for this figure), and N_e = effective population size.

tion time at the target population size will allow for a decrease in the census size to more closely match N_e (Table III).

1. Population Growth

Rapid population growth, especially when a species is first being bred in captivity and the population is small, is often difficult to achieve. Management techniques that facilitate early rapid growth often must include the uses of technologies such as artificial insemination and embryo transfer, where the technology exists (typically only for a few bovid, equid, and primate species), and cross-fostering, hand-rearing, and double-clutching in some species to increase the production of young above that which would be possible in the wild. Birds will often lay second, third, and even more clutches of eggs if their eggs are removed (e.g., the California condor). The removed eggs can then be hatched either artificially or by surrogate parents, including closely related species. Hand-rearing young mammals removed from their mothers will often result in the mothers' quicker return to breeding condition, so that rates of

reproduction can be increased, but may result in incompetent offspring that are unable to reproduce or raise their own young.

2. Generation Time

Generation time is the average time a female takes to produce a reproductive female offspring. Genetic variation and diversity change only when genes are passed from parent to offspring, when inbreeding and genetic drift can cause changes in allele frequency and heterozygosity and the loss of alleles. Once the target population size is reached, increasing generation time can slow the rate of loss of genetic variability. If the rate of loss is decreased, fewer individuals are needed to meet the same goal of maintenance of genetic variability. For example, it is clear from Fig. 1 that a species with a generation time of 20 years (such as the elephant) can be maintained at an effective population size of only 100 individuals, as it will only go through 10 generations in the 200 years, and still lose less than 90% of its genetic variation. Increasing the generation time also can allow better management of family size, so that loss of allelic diversity due to variation

TABLE III
Carrying Capacities Necessary for the Retention of 90% of the Initial Genetic Variance[a]

No. of founders	Length of generation (years)									
	1	2	4	6	8	10	15	20	25	30
6	—	—	—	—	—	—	—	—	—	—
8	—	—	—	—	—	—	—	—	—	—
10	—	—	999	999	999	999	682	392	225	102
12	999	999	999	617	432	321	174	117	80	49
14	999	999	655	399	285	216	122	85	61	40
16	999	999	517	321	229	174	102	71	53	35
18	999	955	450	280	200	154	90	65	48	33
20	999	848	399	253	184	140	84	60	45	32
22	999	783	376	234	170	132	79	57	43	31
24	999	738	355	220	161	124	76	55	42	30
26	999	709	334	212	154	119	73	53	42	30
28	999	682	321	204	151	117	71	52	41	29
30	999	655	315	200	145	115	70	52	40	29

[a] The exponential growth rate per generation is 1. Adapted from Table 6 in M. Soulé, M. Gilpin, W. Conway, and T. Foose, (1986). *Zoo Biol.* **5,** 101–113.

in reproductive success is minimized, and there is less need to cull animals that have achieved their target reproductive output. This, in turn, results in fewer disruptions of social groups and individual relationships in the captive populations.

For plants with seeds that can be stored in seed banks, generations can be greatly extended, and large genetic reserves can be maintained without the need for large growing populations. Stored seeds, however, lose their capacity for germination over time, so seed banks need to be regenerated periodically to ensure a continually viable population. [*See* SEED BANKS.]

3. Demographic Models

Estimates of population growth rates and generation time are based on estimates of age-specific birth and death rates. Demographic models estimate birth and death rates based on the history of individuals. Unfortunately, data are few for small captive populations that have been maintained in captivity for few years relative to their generation time. For species that reproduce rapidly, such as some lemurs, managers must make decisions on when individuals should reproduce and how many offspring they should have long before good es-

timates of birth and death rates for that species are available. New models are being devised by researchers that reduce the need for multiple-generation data sets by using information on other species. Having estimates of the risk of death that a female faces will help managers decide whether they should risk postponing breeding the female so as to increase generation time. Generation time will in turn determine the target size of the captive population. During the growth phase of the population, knowing the probability of survival will optimize decisions on equalizing family size.

4. Limitations to Demographic Management

The biggest limitation to current demographic management is the lack of demographic data on captive species to allow estimates of growth rates and optimum family sizes. These data will continue to accumulate as better records are kept on captive populations, but record keeping of sufficient quality to estimate demographic parameters is a fairly recent addition to most zoo management programs. Mathematical and simulation-based models are being developed that can more accurately predict ranges of growth rates based on limited data, but these are still not widely used by managers. Once a model has been described for a species, implementation is hindered by the need for cooperation among multiple institutions, often scattered worldwide, in implementing the plans; these problems are beginning to be overcome for some species (see Section VII).

B. Genetic Management

Once genetic variation is lost from a captive population, it can only be restored by addition of more founding individuals from the wild, or by genetic mutation. Mutations occur at a rate of about one per genome per generation, or about 10^{-4} per locus per generation; thus, mutations are not adequate to maintain genetic variation in a population of less than about 10,000 individuals. For many endangered species, obtaining more individuals from the wild is neither possible nor desirable (if wild populations are to be maintained as well). Therefore, it is left to efficient management of the current genetic stock to ensure maintenance of genetic variation.

I. Inbreeding and Outbreeding

Inbreeding is defined as the mating between relatives that share greater common ancestry than if they had been drawn at random from a large ($N_e > 1000$) randomly mating population. Outbreeding is defined as random mating in populations large enough to avoid levels of relatedness that would result in what is defined as inbreeding. Inbreeding and outbreeding depression reduce the fitness of breeding groups directly as a result of inbreeding (increased homozygosity) or outbreeding between individuals genetically distinct enough to cause disruption of coadapted gene complexes or changes in traits that affect fitness (as in the example of hybridization of ibex subspecies presented earlier). Because species, and even populations within species, have different population structures and evolutionary histories, their responses to inbreeding and outbreeding will vary substantially. Levels of inbreeding that will produce severe effects in some populations may be normal and have no effects in others.

Inbreeding depression occurs in part because of the presence of recessive deleterious alleles in the population, whose harmful effects are expressed only if the alleles are in homozygous form, so that the increase in homozygosity associated with inbreeding results in expression of more deleterious alleles. In a mammalian species, the number of "lethal equivalents" has been estimated at about 4.6 (range 1.4 to 30.3; median = 3.1). This means that there are an average of 4.6 alleles per genome, each of which, if made homozygous, would result in death of the individual. The average cost of parent–offspring or full sibling mating is 0.33, so mortality would be 33% higher in offspring of these matings than in matings of unrelated parents. Because the range of observed values in mammals is so large and is not clustered around the mean or median, it is unlikely that this information can be used to predict the response to inbreeding of any particular species. Rather, information on natural levels of inbreeding will give the best prediction of whether inbreeding in captivity will result in loss of fitness.

Outbreeding depression can occur when individuals from populations that are genetically distinct are crossed. Although outcrossing among populations can result in a temporary increase in certain characters that can affect fitness, such as size and growth rate (called hybrid vigor, or heterosis), greater genetic distance among populations might have resulted in differences in coadapted gene complexes, local adaptations, and even translocations of genes on chromosomes or changes in the number of chromosomes. Unfortunately, outbreeding depression is often difficult to distinguish from inbreeding depression in captive populations unless cytogenetic analyses or largely disjunct genetic distances in groups of captive individuals can be detected, or if hybrid sterility occurs, indicating some kind of chromosomal or genetic incompatibility between individuals.

2. Pedigree Analysis

Genetic management is most efficiently planned when complete pedigrees are available for the entire population. Retrospective and prospective pedigree analyses allow planning to minimize variation in family size and inbreeding, while maximizing the effective population (relative to census population) size, equalizing founder contribution, and selecting genetically important animals for breeding. Pedigrees can be accurately obtained if animals are housed in pairs or single-male groups, or if genetic data are available for accurate identification of fathers (and mothers, as some species of mammals swap infants, and some birds lay eggs in the nests of other females).

Once a pedigree is produced, it is possible to estimate the current genetic contribution of each founder, the inbreeding coefficient (level of inbreeding of individuals in the population relative to unrelated founders) of all offspring, and the retained genetic variability of the colony as a whole. Inbreeding coefficients indicate how inbred an individual is, and constructing alternative prospective pedigrees will allow evaluation of alternative breeding schemes that will result in lower or higher inbreeding coefficients for each offspring. At the same time, the average relatedness of each individual to all other individuals in the population can be determined from the pedigree and used to balance the reproductive contributions and inbreeding levels in the populations.

3. Founding Population

As mentioned earlier, the size of the founding population has a large and direct impact on the amount of genetic variation and diversity that can be captured and maintained in a captive population. As an example, Table III shows that if a founding population size is increased from 12 to 24 animals, and concomitant increases in the generation time from 6 to 10 years can be made in a vertebrate population, the target population size needed to meet genetic goals is reduced from 617 to 124 animals. This assumes, of course, that all of the founders are able to contribute equally to the target population.

4. Founder Genome Equivalents

Founder contributions in the present population (and under various management schemes) can be determined from pedigrees. Given a pedigree for a population, the number of founder equivalents can be calculated as the number of equally contributing founders that would be expected to produce the same genetic diversity as is found in the population being studied. This means that, although there may have been 20 founders, they may not have all contributed equally, so that there will be fewer founder equivalents than would be possible under optimal management. This analysis can be further refined to determine the number of founder genome equivalents (FGE), which represent the number of equally contributing founders *with no random loss of founder alleles* that would be expected to produce the same genetic diversity as in the population under study. The FGE is directly analogous to the effective number of alleles and is similar to effective population size.

These measures represent the current genetic state of the populations and, in coordination with individual genetic retention analysis, can help to produce the optimal breeding plan to retain the maximum amount of founder genetic variability. Simulation and mathematical models for founder retention are widely used for genetic management, for the purpose of determining how much of each founder's genome is still present in the population and (given that the founder can no longer contribute directly) to set a target for each founder's contribution to the future population. For example, an animal that produced only one offpsring, who then produced many offspring, has contributed at most half of its genome to the population, even though its descendants may represent a larger proportion of the population. Genetic managers would target this founder lineage to represent only half of a founder genome equivalent, in order that the greater genetic diversity contributed by a founder who had many offspring (and therefore higher genetic retention) might be retained in the population.

5. Limitations to Genetic Management

Many captive species are managed based on studbooks, or pedigrees. However, studbooks are often incomplete or incorrect, and only detailed genetic analysis on living animals (and dead animals if, in rare instances, tissue or blood were obtained and preserved) can allow reconstruction of the true pedigree. Even with thorough (and expensive) genetic analysis, pedigrees are often only rough guides for management. For populations founded with individuals of unknown provenance and/or the last of the wild individuals (such as the California condor), levels of relatedness among founders can be a problem, but one that may be avoided to some extent if genetic analysis of relative genetic similarity (e.g., mean numbers of shared alleles) can be determined. These tests are rarely done, as no central genetic facility is available for conservation and management, and genetic analyses are expensive.

Finally, despite the best intentions of genetic managers who have determined which animals should be bred, there are endless logistical problems of transporting gametes or animals between facilities, disruption of social groups, and individual preferences of animals for mates (which can be very strong and unpredictable). Some ways to overcome these difficulties involve artificial insemination, egg transfers, and cross-fostering of young among populations to reduce disruption of social groups while introducing new genetic lines.

VII. WORLD CAPTIVE POPULATIONS

In 1973, the International Species Information System (ISIS) was created to organize and facilitate transfer of basic information on animals in zoos and other facilities worldwide. ISIS now enjoys the participation of 485 institutions in 51 countries and has a total inventory of information on over 720,000 organisms (living and recently deceased). This exchange of information on numbers and relationships of captive animals is the first step in creating worldwide networks for captive breeding programs. ISIS has been instrumental in developing microcomputer-based programs (e.g., VORTEX and SPARKS) for population management and is involved directly with several other organizations that facilitate and/or develop captive breeding programs on a national or international level.

In North America, the most intensive program for management of endangered species is the Species Survival Plan (SSP) program of the American Association of Zoological Parks and Aquariums (AAZPA). This program develops national and international management programs for single species and currently has plans for over 70 species. Worldwide, there are at least 11 such regional programs, and all work in coordination with the Captive Breeding Specialist Group (CBSG), part of the Species Survival Commission (SSC) of the World Conservation Union (IUCN), which facilitates both captive and wild species conservation.

There are currently over 1500 botanical gardens and arboretums worldwide, growing at least 35,000 species of plants (about 15% of the world's flora). Several groups, including the Botanical Gardens Conservation Secretariat (BGCS) of the IUCN, the Consultative Group on International Agricultural Research (CGIAR), and the International Board of Plant Genetic Resources (IBPGR), are involved in coordinating international plant conservation. Groups that specialize in certain agriculturally important species maintain or coordinate maintenance of strains; for example, the International Rice Research Institute (IRRI) has 86,000 rice collections, and the International Maize and Wheat Improvement Center has 12,000 samples of maize and 100,000 samples of wheat. Finally, there are over 50 major seed banks in the world, containing over 2 million collections of seeds. In the United States, the U.S. Food and Drug Administration maintains the National Seed Storage Laboratory.

Although these institutions have made giant leaps forward into a world captive management plan, the logistics and barriers to implementation are still tremendous. Most management plans work on a national level, or at best among several countries that may or may not contain the majority of captive individuals of the species of interest. Currently, there are too few detailed plans to cope with the magnitude of the problems that captive managers face. There is no one method for planning a captive management program, and as a result most plans are put together in an ad hoc fashion, with different organizations taking on different aspects of the problem. Finally, the overemphasis on land mammals and birds still represents an overwhelming bias in the conservation of species; captive management of key ecological species and interacting species must also be emphasized to achieve a balanced approach.

VIII. CONCLUSION

Captive management knowledge, techniques, and programs have made great strides in the last two decades. Theoretical models and empirically derived methods for demographic and genetic management, as well as improvements in reproductive biology and animal care, have resulted in successful captive management programs for many threatened species. Despite this progress, however, we are facing the largest single extinction of species that our planet has ever seen within the next 100 years. With only 3.7% of the land surface in protected areas, the burden of maintaining species, and their ability to evolve, will rest more and more upon captive management. As a last ditch effort, this is the least effective and most expensive conservation method, but a critical one if we are to preserve many of the larger animals and plants that make up the unique biodiversity of this planet.

Acknowledgments

I am very thankful for helpful suggestions on the manuscript by Bela Dornon, Steve Forbes, David G. Smith, Barbara L. Taylor, and David S. Woodruff, and for information supplied by the staff of the Center for the Reproduction of Endangered Species, Zoological Society of San Diego.

Glossary

Critical Official designation of species estimated to have a 50% probability of becoming extinct in 5 years or two generations.
Demography Population structure, growth parameters, and age distributions.

Effective population size (N_e) Number of individuals that would, in an ideal (randomly mating) population, contribute the same amount of genetic variation to the next generation as the population being studied.
Endangered Official designation of species estimated to have a 20% probability of becoming extinct in 20 years or 10 generations.
Founder Wild-caught individual that contributes genetic material to the captive population.
Founder genome equivalent (FGE) Number of equally contributing founders *with no random loss of founder alleles* that would be expected to produce the same genetic diversity as in the population under study.
Heterozygosity Average proportion of observed loci that are heterozygous (two different alleles in an individual) in a population.
Inbreeding depression Decrease in fitness of a population resulting from expression of deleterious recessive alleles as homozygosity increases with inbreeding.
Minimum viable population (MVP) Minimum population size that has a 99% chance of remaining extant for at least 200 years and retains at least 90% of the genetic variation found in the source population.
Outbreeding depression Decrease in fitness of a population resulting from crossing of individuals sufficiently genetically distinct to cause breakdown of coadapted gene complexes or loss of local adaptations.
Phylogeny Evolutionary relationships among taxa.
Vulnerable Official designation of species estimated to have a 10% probability of becoming extinct in 100 years.

Bibliography

Ballou, J., Gilpin, M., and Foose, T., eds. (1995). "Population Management for Survival and Recovery." New York: Columbia University Press (in press).

Primack, R. B. (1993). "Essentials of Conservation Biology." Sunderland, Mass.: Sinauer Associates.

Soulé, M. E., and Wilcox, B. A., eds. (1980). "Conservation Biology: An Evolutionary–Ecological Perspective." Sunderland, Mass.: Sinauer Associates.

Thornhill, N. W., ed. (1993). "The Natural History of Inbreeding and Outbreeding: Theoretical and Empirical Perspectives." Chicago: University of Chicago Press.

Tudge, C. (1991). "Last Animals at the Zoo: How Mass Extinction Can Be Stopped." London: Hutchinson Radius.

Western, D., and Pearl, M., eds. (1989). "Conservation for the Twenty-First Century." New York: Oxford University Press.

Community-Based Management of Common Property Resources

Fikret Berkes
The University of Manitoba

As with many other species, human populations or communities may develop social self-regulatory mechanisms, such as territoriality, in the use of resources on which they depend. Community-based resource management has been important in traditional societies and continues to be significant in the contemporary world. This is because local people who farm, hunt, fish, or simply enjoy an area are more familiar with it than are outsiders; they may have a broader contextual understanding of the environment or a longer time series of observations on it; and local participation in resource management ensures self-interest without which conservation efforts will likely fail. Enabling communities to take care of their environments is one of the nine principles discussed in "Caring for the Earth: A Strategy for Sustainable Living" (see Bibliography).

I. COMMON-PROPERTY RESOURCES

Common-property (or common-pool) resources are nonexclusive by nature and the property of no one. As a class, they share two key characteristics that distinguish them from other resources. First, the exclusion (or control of access) of potential users is problematic. Migratory resources such as fish and wildlife, groundwater, global commons such as the open ocean and the atmosphere pose problems of exclusion. Second, there is a problem of subtractability: each user is capable of subtracting from the welfare of others. As one user pumps water from an aquifer or harvests fish from a lake, there is less resource left for all other users. These two problems often create a divergence between individual and collective economic rationality and lead to the "tragedy of the commons". [*See* TRAGEDY OF THE COMMONS.]

Common-property resources tend to be indivisible, that is, not separable into commodities in time and space, and privatization is often not an option. Rather, these resources require collective decision making, cooperation in resource use, and enforcement of agreed-upon rules among group members. Locally, a community of herders may get together to manage a village grazing commons; regionally, countries of the Baltic Sea basin may manage the Baltic Sea. Individual herders or nations cannot act alone to solve commons problems. This interdependence of users parallels the interdependence of ecosystem components.

Common-property theory combining social science and natural science has emerged only since the mid-1980s. It has been trying to explain why both successes and failures are found with each of private, government, and community-based management systems. Compared to the earlier conventional wisdom that the depletion and degradation of all common-property resources were inevitable, a more comprehensive common-property theory explores possible solutions to the "tragedy of the commons" by focusing on property rights and historical and cultural factors underlying local institutions.

II. TRADITIONAL MANAGEMENT SYSTEMS

The evolution of community-based management systems is possible through biological evolution or cultural evolution. The two mechanisms can act in concert when human populations live in territorial, extended kin-groups. Territoriality may be considered a social self-regulatory mechanism that serves, at the ultimate level of causation, to limit population to available resources. Such self-regulatory mechanisms tend to be more complex in human societies than in nonhuman groups. Many animal populations have territories; human populations may have common-property institutions with rules for access, reciprocity, sharing, social sanctions, and appropriate harvesting behavior and ethics.

Cultural evolution is likely to be more important than biological evolution for community-based resource management. This is because natural selection in human societies has been largely dominated by the human cultures themselves for the last few hundred thousand years. Reciprocity/game theory and coevolution are two approaches used for the study of the evolution of cooperation leading to self-regulatory mechanisms. Numerous examples of traditional management systems, covering a diversity of resources from a wide range of regions and cultures throughout the

world, show that many communities of resource users have been able to exclude other potential users and regulate their own use. This has allowed communities to secure the generation of resources from the ecosystems on which they depend. The societies that were unable to do this disappeared long ago. [*See* ECOSYSTEM INTEGRITY, IMPACT OF TRADITIONAL PEOPLES.]

III. CONTEMPORARY COMMONS

In the contemporary world, nation-states have ultimate jurisdiction over the environment and natural resources. Thus, community-based resource management is never entirely free of its context of government regulation, whether explicit or implicit. Also, there is an important role for government regulation because most natural resources are utilized by several, potentially competing, groups of users. For example, in a mountain watershed ecosystem, the uses of forest, grazing land, water, and agricultural land all affect one another, and the various groups of users are therefore interdependent.

Government resource management systems have become widespread only since the middle of the twentieth century. Scientific resource management is the dominant mode of resource management in Western countries. In many developing countries, however, scientific resource management by government is difficult because of limited scientific knowledge, infrastructure, budget, and enforcement capability. Hence, community-based resource management is often the most realistic option for the sustainable use of resources.

In all parts of the contemporary world, community-based resource management has been documented for a variety of resources. Hierarchically organized, community-based resource management can be effective over large areas and for populations of over 10,000, as in the case of Spanish *huerta* irrigation systems. Some applications of community-based resource management appear to be gaining importance in practice and environmental policy. Examples include conservation projects and local ecosystem rehabilitation by stewardship groups; and bioregionalism, the idea that the scale of human organization should match regions governed by natural, rather than political, boundaries.

IV. CO-MANAGEMENT

Developing a role for user communities in resource management is not easy if there is no cultural background of self-regulation or stewardship ethic. Even with such background, however, community-based management will likely fail if there are unresolved problems of exclusion and subtractability. Government intervention is needed to enforce community-based rights and responsibilities. It is well established in the common-property literature that time-stable systems of community-based management require legal safeguards. All known cases of long-standing and sustainable resource use systems involve some kind of enabling legislation; an example is the "Fisheries Law" of Japan, which delegates management authority for inshore fisheries to village-based fishery cooperative associations.

Common-property theory provides some general guidelines for designing successful community-based resource management: (i) eliminate open-access conditions, (ii) balance resource-use rights of the local population with responsibilities, and (iii) legally protect these rights. Sharing of management responsibility and benefits often requires cooperative management (co-management) arrangements between local groups and governments, especially when more than one group has an interest in a given resource. Such co-management regimes have the potential to combine scientific management with locally evolved management systems, with their stores of traditional ecological knowledge. A key challenge is to learn from time-tested traditional systems to design more effective, joint management institutions that work with the local ecosystem, rather than against it.

Glossary

Coevolution Reciprocal selection; an interrelationship in which two sides change one another continuously by mutual feedback.

Common property resources Class of resources for which exclusion (or control of access) is difficult, and where each user has the potential of subtracting from the welfare of all other users.

Community Social group possessing shared beliefs and values, stable membership, and the expectation of continued interaction.

Cultural evolution Evolution by selection of cultural (rather than biological) traits.

Institutions Codes of conduct that define practices, assign roles, and guide interactions; the set of rules actually used.

Open access Free-for-all; resources owned in common but freely open to any user; Hardin's "unmanaged common."

Property rights Bundle of Rights and responsibilities of individuals or groups to the use of a resource base.

Tragedy of the commons Metaphor formulated by Garrett Hardin to explain the individually rational use of a resource held in common, in a way that eventually brings ruin to the resource and all who depend on it.

Bibliography

Berkes, F., ed. (1989). "Common Property Resources: Ecology and Community-Based Sustainable Development." London: Belhaven.

Bromley, D. W., ed. (1992). "Making the Commons Work. Theory, Practice and Policy." San Francisco: Institute for Contemporary Studies Press.

Feeny, D., Berkes, F., McCay, B. J., and Acheson, J. M. (1990). The tragedy of the commons: Twenty-two years later. *Human Ecol.* **18**, 1–19.

Ghai, D., and Vivian, J. M., eds. (1992). "Grassroots Environmental Action. People's Participation in Sustainable Development." London/New York: Routledge.

Gunderson, L. H., Holling, C. S., and Light, S. S., eds. "Barriers and Bridges to Renewal of Ecosystems and Institutions." New York: Columbia University Press (in press).

IUCN/UNEP/WWF (1991). "Caring for the Earth: A Strategy for Sustainable Living." Gland, Switzerland: IUCN.

Norgaard, R. B. (1994). "Development Betrayed. The End of Progress and a Coevolutionary Revision of the Future." London/New York: Routledge.

Ostrom, E. (1990). "Governing the Commons. The Evolution of Institutions for Collective Action." Cambridge, England: Cambridge University Press.

Conservation Agreements, International

Joel T. Heinen
Florida International University

International conservation agreements are considered to be agreements that focus on the protection of native species or natural areas. The first such agreements were formulated by the United States and Canada to protect migratory birds and parks that crossed the U.S.–Canadian border. Throughout the twentieth century, other bilateral conservation agreements have been made, that is, any agreement signed by two nations. Since the 1970s, a series of very important multilateral agreements have also been formulated. These agreements are open to signature by more than two nations, and those considered here are open to all United Nations members. Much of this article will explore the provisions of four major multilateral conservation agreements.

I. INTRODUCTION

The first international agreement that focused on biotic conservation was the Migratory Bird Treaty Act of 1916, which came into force in 1918. This historic act was a bilateral initiative formulated and signed by the United States and Great Britain acting for Canada. Since this time, numerous countries have formulated bilateral agreements for the protection of migratory species or important natural areas. The first international park was also created by the United States and Canada in the 1930s (Waterton–Glacier International Peace Park), and borderline parks are now quite common in many parts of the world. These bilateral initiatives implicitly or explicitly recognize that many species and ecosystems do not honor political borders; if such biotic resources are to be conserved, international legal and institutional mechanisms must be created. [*See* NATURE PRESERVES.]

This article focuses on more recent multilateral agreements designed to conserve species and natural areas, most of which were formulated under and are maintained by one or another agency of the United Nations. The movement in multilateral environmental policy is much more recent than the bilateral programs just mentioned and is potentially more important for the creation of international awareness and an international legal climate for fostering the protection of important species and natural areas. The most significant works of this type are: The Man and the Biosphere Program, The Convention on Wetlands of International Importance, The Convention on International Trade in Endangered Species of Wild Fauna and Flora, The World Heritage Convention, and the newly formulated United Nations Convention on Biological Diversity.

Each of these will be considered. A great deal has been written and understood about some international conservation agreements (e.g., CITES), whereas much less has been written and understood about others (e.g., MAB). The treatment given here is therefore not equal, but efforts were made in all cases to present the framework and intent for each agreement and to present some important ramifications of each in the general discussion.

II. THE MAN AND THE BIOSPHERE PROGRAM

Unlike the other initiatives described in this article, there is no legally binding convention that covers the creation of biosphere reserves. The Man and the Biosphere Program, commonly known by the acronym MAB, is an international scientific endeavor with the objective of creating a network of biosphere reserves in all ecosystem types throughout the world for the multiple purposes of research, monitoring, training, and conservation. MAB was formulated by the United Nations Scientific, Educational, and Cultural Organization (UNESCO) in 1971. As of 1990, over 100 nations had national MAB programs, and over 70 nations had designated biosphere reserves. [*See* Biosphere Reserves.]

The broad goals of MAB include studying human relationships to and influences on the biosphere, especially for long-term human-induced impacts and conservation for sustainable development, though individual nations have interpreted these general goals in different ways. Biosphere

reserves within individual nations are quite variable in size and management designation. Within the United States, for example, designated biosphere reserves range from about 40,000 km^2 in size (Champlain–Adirondack Biosphere Reserve) to about 6 km^2 in size (Stanislau–Tuolumne Experimental Forest). Units are included that are managed by universities (e.g., The University of Michigan Biological Station), state governments (e.g., New Jersey Pinelands), state/private cooperation (e.g., Champlain–Adirondack), and several different agencies of the federal government (e.g., Everglades National Park and Aleutian Islands National Wildlife Refuge).

In comparison to other initiatives, there are both negative and positive aspects of MAB. Because there is no legal framework in the form of a multilateral convention around which these reserves can be planned and managed, there is a good deal of confusion about the purpose, creation, and maintenance of biosphere reserves. However, this also creates a situation in which individual nations have a great deal of latitude in designating biosphere reserves within their borders, and in designating overall goals with the development of national MAB programs. Individual programs can therefore differ greatly from others, but can address more directly unique national concerns than is the case of other programs that have well-developed and legally binding conventions.

III. THE CONVENTION ON WETLANDS OF INTERNATIONAL IMPORTANCE

The Convention on Wetlands of International Importance, Especially as Waterfowl Habitat is frequently referred to as the Ramsar Convention after the Iranian town in which it was formulated in 1971. This convention entered into force when Greece became the seventh nation to ratify the document (in December 1975); it was the first multinational conservation convention, and it remains the only one that focuses on the protection of one general type of ecosystem. As of 1990, about 60 nations had become party to Ramsar. [*See* WETLANDS ECOLOGY.]

The preamble to the convention instructs parties to recognize the interdependence of humans and the environment, and to consider the ecological functions of wetlands as fundamentally important (e.g., for flood control, nutrient cycling, habitat for migratory waterfowl and commercially important fish). The preamble further espouses the philosophy that wetland loss would be irreparable because such habitats are of great economic, scientific, and recreational value. The parties are also instructed to formulate national policies that decrease the further loss of wetlands and to recognize that migratory waterfowl represent an important international resource because of seasonal movements across national borders. Finally, the preamble instructs parties to agree that national policies and international coordination are both needed to conserve wetlands and their native flora and fauna.

A. The Articles and Amendments of Ramsar

Article 1 defines the types of habitats under consideration. Marshes, fens, peatlands, and marine areas, including those in which low tide does not exceed 6 m, are covered by the convention. Thus wetland ecosystems are considered rather broadly and would include mangrove swamps as well as some types of coral reefs by this definition. Waterfowl are defined as birds that are dependent during at least part of their life cycle on any type of wetland as described here. The parties are instructed in Article 2 to include at least one wetland found in their territory on the List of Wetlands of International Importance, which is kept by The World Conservation Union (IUCN), designated by Article 8 as the bureau responsible for maintaining the convention. The designation of wetlands of international importance is based broadly on ecological, botanical, zoological, limnological, and/or hydrological criteria in Article 2. Article 3 obligates parties to promote the conservation of listed sites through national policies and to inform IUCN of changes in the ecological characteristics of listed sites.

Article 4 promotes the further conservation of wetlands within contracting states by instructing them to establish nature reserves, whether or not the particular sites are listed by the convention, and it requires parties to compensate in area, as much as possible, for listed sites that are removed from the list for "urgent national interests." This article also encourages parties to train personnel for the purposes of conducting research and managing areas to increase waterfowl populations. Article 5 instructs parties to consult with each other about the implementation of Ramsar, an important provision in cases in which listed wetland boundaries cross national borders. Parties are obligated to convene on the conservation of wetlands as the need arises under Article 6, for the purposes of discussing implementation of the convention and any changes to the list.

Articles 7 through 12 are devoted to the bureaucratic and institutional framework of the convention. Representatives of contracting parties must be wetland experts, and each party is given one vote at conferences, as described by Article 7. IUCN is the bureau designated in Article 8 that maintains the list and organizes the conferences described in Article 6. The convention remains open indefinitely for signatures by more parties. As stipulated in Article 9, this can include any member-agency of the United Nations and several other international organizations in addition to individual countries. The convention entered into force four months after the accession of the seventh state member (Greece) as stipulated in Article 10. This article also outlines procedures for amending the convention. Parties have the power to denounce the convention five years after their ratification date, as described in Article 11. The Depositary is required by Article 12 to announce new parties; to deposit all ratification or accession documents, dates of entry, and notifications of denunciation; and to register the convention with the Secretariat of the United Nations.

Since the development of Ramsar, several important amendments were added to address concerns about implementation of certain procedures. For example, an amendment was added to Article 7 during the 1987 Conference of Contracting Parties that stipulated any resolutions and decisions could be adopted by a simple majority of parties. Other resolutions were included in the 1990 conference on financial issues, the framework for implementation, and priorities for attention of the convention.

IV. THE CONVENTION ON INTERNATIONAL TRADE IN ENDANGERED SPECIES

The Convention on International Trade in Endangered Species of Wild Fauna and Flora is commonly known by the acronym CITES. It is perhaps the most important of the older international conservation agreements and the largest in the number of contracting parties, and has probably received more legal and administrative attention than any other such agreement. CITES was originally signed by 85 nations, and 117 had become party as of January 1993. The convention was formed after the United States passed the historic Endangered Species Act of 1973. This national legislation allowed for the listing of American and foreign species and recognized that international trade hindered global species protection. The United States hosted the 1973 Washington Conference that led to CITES.

A preamble and 25 articles are included in CITES, and many additional amendments have been passed since it was formulated. The preamble instructs parties to recognize that wild species are irreplaceable parts of natural systems that warrant protection, that wild species have irreplaceable intrinsic values, that international cooperation is needed to protect species exploited by trade, and that taking appropriate measures is urgent.

A. The Articles of CITES

Article 1 provides a statement of scientific and legal definitions. These are rather broad in the parlance of CITES; a "species," for example, refers to a biological species, subspecies, or separate population, and a "specimen" refers to any animal or plant, whether alive or dead, or recognizable parts or derivatives thereof, that are listed on one of the Appendices of CITES. Definitions are also provided for "trade," "export," "re-export," and the "scientific" and "management" authorities that each party is required to designate. Article 2 lists "The Fundamental Principles of CITES," which define the purposes of the three Appendices. Species listed on Appendix I are those threatened with extinction that are or may become affected by international trade. Those listed on Appendix II are not necessarily threatened with extinction by trade, but could become so if protective measures are not adopted. Species listed on Appendix III are those that are identified by any party as being affected by trade and protected within national jurisdiction. Article 2 concludes that parties are required to agree not to engage in trade in any species listed on any Appendix, except in accordance with the permit system of CITES.

Articles 3, 4, and 5 provide broad legal guidelines for the regulation of trade in parts or specimens of species included on the Appendices, and define the role of the scientific and management authorities of the parties regarding permit requirements for any listed species. Articles 6 and 7 describe in more detail the permits, certificates, and exemptions allowed under Articles 3, 4, and 5; for example, they include exemptions for specimens with traveling zoos and circuses.

Article 8 obligates parties to take certain measures to enforce the convention, including penalties for, and confiscation of, illegally obtained specimens listed on the Appendices. Parties are instructed under Article 9 to designate a Management Authority to grant permits on behalf of that party and a Scientific Authority to provide information on any specimen in question. Parties are further instructed under Article 10 about obligations in cases in which the party engages in trade with nations that are not party, and the CITES Secretariat is obliged under Article 11 to schedule Conferences of Parties at least every two years. The role of the Secretariat is elaborated under Article 12 as arranging conferences, undertaking research, and publishing periodic editions of all Appendices, as well as preparing annual reports and implementation recommendations.

The responsibilities of the Secretariat are further defined under Article 13 to inform parties if they are not in compliance and to instruct parties to respond to such information. Any inquiry is subject to review at the next conference, and any party can make recommendations about noncompliance of another party. Parties may adopt stricter domestic protective measures if it is considered necessary under Article 14, and procedures for amending Appendices I, II, and III are provided in Articles 15 and 16. Procedures for amending the convention itself are provided in Article 17, and procedures for dispute resolution are provided in Article 18.

The remaining seven articles provide information on administrative aspects of CITES. These include signature (Article 19), ratification (Article 20), and accession to the convention (Article 21), which remains open indefinitely. The provisions under which CITES was entered into force are given in Article 22, and procedures for making specific reservations are provided in Article 23. Parties are also permitted to denounce CITES with proper notification under Article 24. Finally, the duties of the Depositary Government (Switzerland) are outlined in Article 25. CITES was witnessed and signed on March 3, 1993, in Washington, D.C., by the original 85 signatories.

V. THE WORLD HERITAGE CONVENTION

The International Convention for the Protection of World Cultural and Natural Heritage, known commonly as the World Heritage Convention, was adopted in Paris in 1972 under the auspices of UNESCO. The convention came into force in 1975 and allows parties to nominate both natural and cultural areas considered to have "outstanding universal value." The World Heritage Convention was formulated with the understanding that many natural and cultural sites possess international importance and some such sites cannot be adequately maintained and financed within some developing nations. The broad purpose of the convention is to designate such sites and to provide financial help where it is needed.

With these purposes in mind, UNESCO maintains a list of cultural and natural sites of international importance (World Heritage Sites) and the World Heritage Trust, which is specifically designated to provide financial resources to developing countries for the management of their World Heritage Sites. Criteria for the listing of both natural and cultural heritage sites are published by UNESCO.

Parts of the preamble and many clauses in the articles of the convention refer specifically to cultural sites. Here we will consider legal aspects of the convention as they relate to the designation of natural sites. Article 2 states that natural heritage sites must include "natural features consisting of physical and biological formations or groups of such formations, which are of outstanding universal value from the aesthetic or scientific point of view." This definition then expanded to include areas that constitute important habitat for endangered species of universal value and outstanding examples of geological formations or other such natural features of outstanding beauty. Within Nepal, for example, Royal Chitwan National Park was nominated under the first criterion, and Sagarmatha (Mt. Everest) National Park was nominated under the second.

Parties to the convention are responsible for nominating sites within their borders as either cultural or natural sites; nominations are then subject to approval by the World Heritage Committee of UNESCO. As of 1990, 113 countries had ratified convention, including most countries in all major regions of the world. Geographic exceptions to ratification include the nations of southern Africa, several in East Africa, a few nations in the Asia/Pacific region, and several nations in the neotropical region. Despite this rather wide geographic coverage, only about one-third of the parties had natural sites listed as of 1990; others had cultural sites only listed (e.g., Egypt and Pakistan), and about 40% of the parties to the convention had no sites listed.

VI. THE UNITED NATIONS CONVENTION ON BIOLOGICAL DIVERSITY

Two major conventions were formulated prior to and during the 1992 United Nations Conference on Environment and Development, which took place in Rio de Janeiro and is commonly referred to as the Earth Summit. The first, The United Nations Framework Convention on Climate Change, relates indirectly to the conservation of biological resources and will not be considered here. The second is the United Nations Convention on Biological Diversity, and it has the potential to become the most important international conservation agreement thus far; it is the most complex and the broadest, and includes a preamble, 42 articles, and two annexes. These are only briefly outlined here; the full text can be found in "Biodiversity Prospecting," published by the World Resources Institute.

The preamble to the convention instructs contracting parties to be conscious of the intrinsic values of biological diversity (hereafter referred to as "biodiversity") and the importance of maintaining biodiversity for evolution and life-sustaining systems. The preamble also maintains that biodiversity conservation is a common human concern, that biodiversity is being reduced significantly, and that nations have both rights over the use of their biodiversity as well as responsibilities to conserve it. The preamble further notes the importance of *in situ* (on site) conservation through the protection of natural ecosystems and the general lack of information about biodiversity in many parts of the world.

A. The Articles of The Convention on Biological Diversity

Article 1 of the convention outlines the objectives, including conservation and sustainable use of biodiversity, as well as sharing the benefits arising from the use of genetic resources. Article 2 defines important terms subsequently used, and Article 3 outlines the overall principle: states have the sovereign right to exploit natural resources in accordance with national laws, and the responsibility to assure that such activities do not cause harm to other states. The jurisdictional scope of the convention is outlined in Article 4, and provisions for cooperation for conservation and sustainable use of biodiversity are described in Article 5. Article 6 instructs parties to develop national strategies for conservation and sustainable use of biodiversity, and to inte-

grate these into relevant cross-sectoral plans. Article 7 instructs parties to identify and monitor components of biodiversity, as well as to identify processes likely to hinder its conservation and sustainable use.

In situ (on site) conservation strategies are outlined in Article 8, which obligates parties to plan, develop, and manage a system of protected areas and to promote environmentally sound development in regions adjacent to these areas. It also includes a clause to preserve the knowledge and innovations of indigenous and local communities. *Ex situ* (off site) conservation strategies are explored in Article 9, which instructs parties to establish and maintain research facilities and to prepare recovery and reintroduction programs in the case of rare species. Article 10 focuses on sustainable use and instructs parties to encourage customary use, adopt measures to avoid adverse impacts, and encourage cooperation between government and the private sector. Article 11 instructs parties to provide appropriate economic and social incentives to promote conservation and sustainable use.

Articles 12 through 14 outline generally the obligations to parties for research and training, public education, and carrying out impact assessments to minimize any adverse impacts. Articles 15 through 18 cover provisions about access to genetic resources, technology transfer, information exchange, and technical and scientific cooperation. The provisions for distribution of the benefits of biotechnology are given in Article 19. Articles 20 and 21 deal specifically with financial matters. The former obligates parties to pay financial support according to their ability, and further obligates developed countries to provide new finances to help developing countries to meet full incremental costs. Article 21 outlines financial mechanisms by which this can be accomplished.

Article 22 stipulates that the convention does not affect rights or obligations of any party that derive from other international agreements, and Article 23 outlines the establishment of a Conference of Parties and the procedures for meetings. The Secretariat is defined in Article 24, and a subsidiary body to provide technical and scientific advice is established by Article 25. This body is to be made up of government representatives competent in relevant fields of expertise. Details about report preparation, dispute resolution, adoption of protocols, and the formulation and adoption of amendments are provided by Articles 26 through 30.

Legal issues regarding voting rights, signature, ratification, accession, entry into force, reservations, withdrawals, and interim financial arrangements are outlined in Articles 31 through 39. Article 40 describes that the interim secretariat is to be provided by the United Nations Environmental Program, and the Secretary-General of the United Nations is to assume the role of Depositary as stipulated in Article 41. Finally, Article 42 stipulates that the original texts published in the six international languages recognized by the United Nations are all equally authentic (i.e., Arabic, Chinese, English, French, Russian, and Spanish).

Two annexes are also included in the document. The first considers identification and monitoring of components of biodiversity, including ecosystems, communities, species, genes, and/or genomes, that are scientifically or economically important. The second outlines procedures for arbitration (17 articles) and conciliation (6 articles).

VII. GENERAL DISCUSSION

International conservation agreements have come very far since the first bilateral initiatives between the United States and Canada. They precipitated in part from the growing awareness of the extent of various environmental problems that were publicized widely in the 1960s and have continued to expand since the early 1970s. As described earlier, there is a great deal more literature about and awareness of the earlier legally binding conventions (Ramsar, The World Heritage Convention, and especially CITES) than about MAB, a scientific program with no legal ratification procedures.

An interesting point of departure here is potential overlap of and conflict between these various agreements. Everglades National Park, for example, is designated as a World Heritage Natural Site under the World Heritage Convention, as a Wetland of International Importance under Ramsar,

and as a Biosphere Reserve under MAB. There is no potential conflict between the two legally binding agreements (Ramsar and World Heritage) as both suggest that the U.S. Park Service maintain as much as possible the ecological integrity of this unique wetland system into perpetuity.

The designation under MAB, however, is perhaps more perplexing. The agreement, though not legally binding, recommends ongoing research in the Everglades about the effects of humans on the system. This is currently a contentious and widely publicized issue in south Florida because of the effects of the sugar industry upstream on water flow, with potential ramifications throughout the Everglades and Florida Bay to the south. It would seem that the legal obligations of the Park Service under Ramsar and the World Heritage Convention would be to ameliorate these effects to the extent possible, but this would necessarily involve private land holdings well outside of the park, over which the Park Service has no jurisdiction under U.S. law. An agreement involving all parties is currently being implemented to address the issue on the regional scale of south Florida. The point here is that the MAB designation almost assumes there are ongoing human effects to study, which is obvious in the case of the Everglades, whereas the other designations almost assume that the system is in some more pristine state and could be managed so indefinitely, which is not the case.

The language of the legally binding conventions in all cases is quite broad (e.g., "maintain to the extent possible"). Virtually all ecosystems worldwide are or have been affected to some degree by humans, and studying this is essential to understanding potential local, regional, and global impacts. Ramsar sites are particularly interesting in this regard. The historical importance of this convention is in its recognition that the utility of many wetlands goes far beyond their national borders, and that waterfowl are international resources owing to their seasonal migrations. The latter had already been addressed prior to Ramsar by several nations in bilateral treaties, but it was the first multilateral convention to address this issue. Ramsar is also significant in that it is the only one to deal with one ecosystem type, and could be used as a model for the formulation of treaties on other ecosystems of international importance, such as tropical rain forests. The framework for this expansion was set in place at the 1992 Earth Summit with the Statement of Principles on Forests, which is not yet legally binding but is likely to become so.

Ramsar may prove especially important for studying potential impacts of global warming. The convention theoretically has the latitude to deal with the multitude of potential global changes through the provisions in Articles 3, 4, 6, and 8, but further amendments may be in order. Additional amendments may be especially important to document climate change at finer scales than is currently possible in the atmospheric sciences, because Ramsar sites are comparatively well described and monitored, and wetlands are expected to be sensitive to minor changes in rainfall patterns and local temperatures and thus may represent important barometers of change.

Another issue to consider with all such agreements is the effect of nonmember states on implementation. CITES has a specific clause outlining the obligations of parties in trading with nonparties, but the fact that there are many nonparties means that there is substantial ongoing trade in endangered species worldwide. This is also a concern with respect to Ramsar; for example, several of the largest countries in South America (e.g., Brazil, Argentina, and Colombia) were not party to the convention as of 1990, nor were about half the nations of Africa. On the Asian continent, coverage is more complete, but several nations with significant freshwater and marine wetlands were also not party as of 1990 (e.g., Bangladesh, Indonesia, Malaysia, and Thailand). These conventions are still relatively new and constantly evolving, and all remain open for signature; thus many of the national gaps may fill with time.

As stated earlier, The United Nations Convention on Biological Diversity is the newest, potentially the most important, and already the largest in terms of the number of signatures. However, the United States, previously the most progressive nation on the issue of international conservation agreements, refused to sign this historic treaty in 1992 under the Bush Administration. The Clinton Administration renegotiated parts of the convention and signed it in 1994.

Within member states, there is also concern about implementation of and compliance with all such agreements. This is not an issue with MAB, because there is no legal convention, but it is recognized as important in other cases. For example, during the adoption of Ramsar in Denmark, concerns were raised regarding issues such as national sovereignty, how Ramsar affected obligations to other treaties, and its effects on local and regional planning issues. Regarding the biological importance and management of Kosi Tappu Wildlife Reserve, Nepal's only Ramsar site, management of the area currently provides very little real protection for migratory waterfowl populations. A great deal of literature has emerged over the past twenty years on the implementation of and compliance with CITES, for example, a survey of the Himalayan fur trade in Nepal, much of which is conducted by Indian merchants with furs obtained in India and sold to Western tourists. Both India and Nepal are parties to CITES. Most of the fur coats encountered were made of species listed on one of the Appendices and protected under national law in one or both countries. The trade continued in part because of a lack of enforcement mechanisms in both Nepal and India.

The World Heritage Convention presents perhaps fewer potential issues with regard to compliance, for it places some responsibility of managing and financing outstanding cultural and natural sites on the international community, yet nations presumably have a strong interest in listing sites and have vested national pride in such places. Many sites are also very popular tourist destinations and therefore important for national and local economies.

Most sites listed thus far are cultural sites, but many World Heritage Natural Sites are important to local residents for other reasons (e.g., extraction of forest products), and protecting global values may at times conflict with local interests. These potential conflicts are currently not addressed by the convention, but may become more important as more sites are listed worldwide. For example, there are various problems with regard to thatch grass extraction and the management of Nepal's Royal Chitwan National Park, a World Heritage Natural Site. The overview and recommendations provided by J. Thorsell in a 1992 IUCN publication outline several general problems with respect to the World Heritage Convention, including scarce funds and the lack of precise criteria to designate natural sites, and the report indicated that many listed sites are threatened in some way.

Because the United Nations Convention on Biological Diversity is new, potential issues of implementation and compliance are only speculative. It is by far the broadest and includes articles and clauses concerning the management of species and natural areas, as well as genetic resources, trade in technologies, monitoring for change, local development issues, and local equity and access concerns. It therefore addresses more issues than all the other previously discussed agreements combined. As such, this convention has the greatest potential for problems in implementation and compliance, but also the greatest potential for achieving major conservation and sustainable development goals around the planet. The prospects for the Biological Diversity Convention are immense and fascinating; time and further study will tell how the provisions of this historic work are implemented within, between, and among nations.

Glossary

Biosphere reserve Any natural reserve recognized under the Man and the Biosphere Program. Such reserves are set aside to study human effects on natural systems.

CITES *See* Convention on International Trade in Endangered Species of Wild Fauna and Flora.

(The) Convention on International Trade in Endangered Species of Wild Fauna and Flora Most commonly known by the acronym CITES, this 1973 multilateral convention was formulated after the United States passed the Endangered Species Act. It is among the largest and most important international agreements, and it is designed to control commercial trade in species in danger of extinction.

(The) Convention on Wetlands of International Importance, Especially as Waterfowl Habitat Commonly known as Ramsar, this was the first major multilateral convention, and it is the only one that focuses on one type of ecosystem (wetlands). It is significant in recogniz-

ing that many biotic resources, because of seasonal movements, cannot be protected within one nation.

Depositary Officially designated government or multigovernmental organization whose responsibility it is to maintain the articles, signatures, and amendments of any convention.

(The) Endangered Species Act First major piece of legislation that afforded special protection to species in danger of becoming extinct in the wild. This was passed in the United States in 1973, and many other nations now have similar legislation.

(The) International Convention for the Protection of World Cultural and Natural Heritage Commonly known as the World Heritage Convention, this agreement is kept by UNESCO for the purposes of recognizing and listing cultural and natural sites of outstanding universal importance. A trust is also maintained to provide developing countries with some finances to maintain their world heritage sites.

International Park Any protected natural area that crosses the border between two nations and is legally protected in both nations, for example, the Waterton–Glacier International Peace Park between the United States and Canada.

IUCN The International Union for the Conservation of Nature and Natural Resources, which has been renamed the World Conservation Union, although it still goes by the abbreviation IUCN. This is a multinational body created in 1948 that compiles lists of rare species and protected natural areas and provides expertise to national governments on the management of biotic resources. IUCN is also the Depositary of the Ramsar Convention.

MAB *See* The Man and the Biosphere Program.

(The) Man and the Biosphere Program International, not legally binding, scientific agreement under UNESCO with the goal of establishing a global network of reserves, called biosphere reserves, for the purposes of studying human effects on natural systems.

Migratory Bird Treaty Act 1918 act signed between the United States and Great Britain acting for Canada that gave legal protection to migratory North American birds. This was the first recognized international conservation agreement of any kind.

Ramsar *See* The Convention on Wetlands of International Importance, Especially as Waterfowl Habitat. Ramsar is the name of the Iranian town in which the convention was formulated in 1971.

UNCED *See* The United Nations Conference on Environment and Development.

UNESCO United Nations Educational, Scientific, and Cultural Organization, headquartered in Paris.

(The) United Nations Conference on Environment and Development More commonly known by the acronym UNCED, and also known as the Earth Summit, this 1992 conference was the largest international conference of any kind and was attended by more Heads of State than any other. The Biological Diversity Convention (*see* The United Nations Convention on Biological Diversity) became open for signature at UNCED, which took place in Rio de Janeiro, Brazil.

(The) United Nations Convention on Biological Diversity 1992 convention that deals comprehensively with the protection of natural areas, species, and genetic materials for the purposes of global sustainable development. As of late 1993, most nations of the world have signed this convention.

World Conservation Union *See* IUCN.

World Heritage Committee Committee appointed by UNESCO whose responsibility it is to determine if nominated sites can qualify as World Heritage Cultural or Natural Sites. If so, the site is placed on a World Heritage List.

World Heritage Convention *See* The International Convention for the Protection of World Cultural and Natural Heritage.

World Heritage Natural Site Any natural site of outstanding global importance that is included on the World Heritage List maintained by UNESCO. Some North American examples are Yellowstone National Park and Everglades National Park.

Bibliography

Anonymous. (1990a). "1990 United Nations List of National Parks and Protected Areas." Gland, Switzerland: IUCN Publications.

Anonymous. (1990b). "Convention on Wetlands of International Importance, Especially as Waterfowl Habitat. Proceedings of the Fourth Meeting of the Conference of Contracting Parties." Gland, Switzerland: Ramsar Convention Bureau.

Favre, D. S. (1989). "International Trade in Endangered Species: A Guide to CITES." Dordrecht, Netherlands: Martinus Nijhoff Publishers.

Fitzgerald. S. (1989). "International Wildlife Trade: Whose Business Is It?" Washington, D.C.: World Wildlife Fund.

Hales, D. F. (1984). The World Heritage Convention: Status and directions. *In* "National Parks, Conservation, and Development: The Role of Protected Areas in Sustaining Society" (J. A. McNeely and K. R. Miller, eds.), pp. 744–750. Washington, D.C.: Smithsonian Institution Press.

Heinen, J. T. (1990). Range and status updates and new seasonal records of birds in Kosi Tappu Wildlife Reserve. *J. Natural History Museum (Nepal)* **11,** 41–49.

Heinen, J. T., and Kattel, B. (1992). Parks, people, and conservation: A review of management issues in Nepal's protected areas. *Population and Environment* **14**(1), 49–84.

Heinen, J. T., and Leisure, B. (1993). A new look at the Himalayan fur trade. *Oryx* **27**(4), 231–238.

Hough, J. (1991). Social impact assessment: Its role in protected area planning and management. *In* "Resident Peoples and National Parks: Social Dilemmas and Strategies in International Conservation" (P. C. West and S. R. Brechin, eds.), pp. 274–283. Tucson: University of Arizona Press.

Kellert, S. R. (1986). Public understanding and appreciation of the biosphere reserve concept. *Environ. Cons.* **13**(2), 101–105.

Koester, V. (1989). "The Ramsar Convention: A Legal Analysis of the Adoption and Implementation of the Convention in Denmark," IUCN Environmental Policy and Law Paper No. 23. Gland, Switzerland: Ramsar Convention Bureau.

Lehmkuhl, J. F., Upreti, R. K., and Sharma, U. R. (1988). National parks and local development: Grass and people in Royal Chitwan National Park, Nepal. *Environ. Cons.* **15**(2), 143–148.

Thorsell, J., ed. (1990). "Parks on the Borderline: Experience in Transfrontier Conservation." Gland, Switzerland: IUCN Publications.

Thorsell, J., ed. (1992). "World Heritage Twenty Years Later." Gland, Switzerland: IUCN Publications, Protected Areas Program.

World Resources Institute (1993). "Biodiversity Prospecting." Washington, D.C.: World Resources Institute.

Conservation Practices—*See* Traditional Conservation Practices

Conservation Programs for Endangered Plant Species

Kent E. Holsinger

University of Connecticut

I. Introduction
II. Identifying Which Plants Are Endangered
III. Identifying the Causes of Endangerment
IV. Conserving Species in Their Native Habitat
V. Conserving Species in Off-Site Collections
VI. An Integrated Approach to Plant Conservation

Conservation efforts for vertebrates have often focused on individual species at risk of extinction. Plant conservation programs must often take a broader approach. The Atlantic coastal forests of Brazil, for example, support one of the most diverse arrays of plants found anywhere in the world. Unfortunately, these once extensive forests have been reduced to less than 5% of their original cover through development for agriculture and housing. Similarly, slash-and-burn agriculture in Madagascar has removed nearly two-thirds of the forest encountered by the first colonists 15 centuries ago, endangering many of the nearly 8000 endemic plant species. And such examples are not limited to the tropics. The Cape floristic province at the southern tip of Africa may have as many as 6500 flowering plant species found nowhere else in the world, but nearly one-third of the native vegetation has already been lost to agriculture and urban development. In southwestern Australia nearly one-quarter of the roughly 3600 species are now classified as rare or endangered. In such places the first priority of plant conservationists is to protect the existing habitat. Merely protecting the habitat, however, is not enough. It is often necessary to focus attention on individual endangered species that would otherwise be lost.

A plant conservation program focused on individual species has three components: (1) identifying the plant species in need of attention, (2) identifying the threats to their long-term persistence, and (3) choosing management tactics that reduce the identified threats. Although rare species are more likely to become extinct than common ones, not all rare plants are endangered. A species should be considered endangered only if, in addition to being rare, it is likely to become extinct in the near future. Declining populations are probably the best indicator of endangerment, but life history characteristics, reproductive biology, and habitat preferences must also be considered. Identifying the factors that threaten a species' persistence also provides clues to the tactics appropriate for conserving it. For many endangered plant species protecting the sites where they occur is not enough to ensure long-term survival. Both habitat manipulations and demographic manipulations may be required. Similarly, protecting and managing existing natural populations is the most important part of any plan to protect an endangered plant, but off-site collections can provide insurance against catastrophes that might wipe out remaining natural populations. Because of the genetic changes likely to accompany long-term cultivation in off-site collections, however, seed banks are preferable to living collections in the off-site component of an integrated conservation plan.

I. INTRODUCTION

Species are now going extinct more rapidly than at any other time in the last 65 million years. Not since the mass extinction that marked the end of the Mesozoic era have so many species become extinct in such a short time. Human activity in tropical rainforests, for example, may have increased the rate of extinction between 1000 and 10,000 times over what it was before. Even more frightening is that the rate of extinction seems to be accelerating. Only 1000 of the 250,000 species of vascular plants known to have been extant in historical times became extinct in the past century, but another 60,000 species may become extinct in the next 50 years. The number of species at risk is so great in tropical latitudes and the threats they face are so severe that the best chance of saving many of them is probably to protect large tracts of natural habitat, hoping that protection of the habitat will ensure the long-term survival of the species found there. There are simply too many species for each one to be the focus of a special conservation effort. In temperate zones, on the other hand, the number of species at risk is lower, and it will often be appropriate to supplement habitat conservation efforts with conservation programs directed to the protection of individual species. [*See* EVOLUTION AND EXTINCTION.]

Although the number of threatened plant species in temperate zones is far smaller than that in the tropics, it is not small. A survey by the Center for Plant Conservation in 1988 identified nearly 700 species of plants in the United States that could become extinct before the year 2000. To put it another way, there are roughly as many plants in the United States in imminent danger of extinction as there are bird species that breed in all of North America north of Mexico. This pattern is repeated even on much smaller geographical scales. The flora of the state of Connecticut, for example, is not particularly rich, having approximately 1600 native species of plants. Nonetheless, nearly one-fifth of these species are recognized as endangered, threatened, or of special concern in state legislation. In fact, almost half of the species in the category of special concern are known only from herbarium records and are presumed to be extinct in the state. Moreover, both the number and the proportion of plants falling in these categories far exceed those for mammals and birds. None of the 60 species of mammals and only 14 of the 114 species of birds that breed in the state (12%) are threatened.

Detailed information on demography, ecology, genetics, and metapopulation dynamics can be enormously useful in choosing management tactics for an endangered species when such information is available. In a few cases, for example, those in which the number of remaining individuals is extremely small, it may even be essential. Given the large numbers of plant species at risk, however, biologists have come to recognize that conservation plans must be made for many plants long before detailed biological studies are complete. Fortunately, much can be done even in the face of our limited knowledge. The plants in the most immediate danger of extinction can be identified from information on the number and size of existing populations combined with observations on habitat specificity. The most important management choices—What can be done to eliminate the immediate threats to persistence? Is a supplementation effort necessary? What habitat management is necessary? Should an off-site seed bank be established?—sometimes require only rudimentary biological information. Intelligent choices can be made for some species based merely on changes in the pattern of distribution and abundance, habitat preference, and life history, information that is often available from specimen labels in a well-curated herbarium. Many other species will require careful investigation, but by focusing efforts on identifying the cause of endangerment, it may still be possible to identify appropriate management tactics quickly. [*See* SEED BANKS.]

II. IDENTIFYING WHICH PLANTS ARE ENDANGERED

The first task facing a conservation program focused on preventing the extinction of plant species is to identify the plants in need of attention. This task can be further divided into two parts: (1) identifying those species that are in the most immediate danger of extinction and (2) determining which of these species deserve the highest conservation priority. Several important principles should be kept in mind while developing the list of species to be included in a conservation program. First, although in some sense all endangered species are rare, not all rare species are endangered. A species should be considered endangered only if, in addition to being rare, it is in immediate danger of extinction or would be in immediate danger of extinction in the absence of protection. Second, although the likelihood of extinction plays an important part in determining conservation priority, many other factors (e.g., the chances of successful restorative action, the cost of restorative action, and taxonomic distinctiveness) will also play a role. The most endangered species are not necessarily the highest priority for conservation, although they do deserve very careful attention. Third, when in doubt, it is always better to regard a species as endangered than to regard it as safe. By the time enough data accumulate for us to be certain of a species' endangerment status, its populations may already have declined so far that only heroic efforts can save it. Fourth, determinations of endangerment status are always provisional. The list of species included in any plant conservation program should be periodically revised as new information comes to light. [*See* PLANT CONSERVATION.]

A. Indicators of Endangerment

Many factors determine how likely it is that a plant species will persist over the long term, but just three variables provide easily accessible indicators (Table I) of how likely extinction is in the near future: (1) geographic distribution, (2) population size, and (3) habitat specificity. Because a single catastrophe (e.g., fire, flood, or hurricane) can destroy an entire population, plant species with only a few existing populations are more likely to become extinct than those with many. The risk associated with a small number of populations is compounded if those populations are also confined to a small geographic area. Similarly, those that occur in large populations are less likely to become extinct than those that occur in small populations. Finally, species confined to one or a few specialized habitats are more likely to go extinct than those that occupy a broader range of habitats, because elimination of just a few sites could eliminate all appropriate habitat for the species. In addition to information on the number and size of existing populations, information on the trends in these variables provides important clues about the degree of endangerment. A species in which both the number of individuals per population and the number of populations is declining is in greater danger than one in which both numbers are stable or increasing.

For the purposes of determining the initial list of species to consider, information on these broad categories of risk is all that is necessary. Species with high ranks on several of the criteria for endangerment are obviously at greater risk of extinction than those with high ranks on few, but the trends in distribution and population size may be more important indicators than the size and extent of existing populations. Species that have always been rare have already demonstrated, by the simple fact of their continuing existence, an ability to cope with the demographic and genetic consequences of rarity. Those that have only recently become rare, however, are likely to lack the ecological adaptations that would allow them to persist. They may suffer significant declines in survivorship and fecundity if populations become very small, and chance failures of reproduction could lead to extinction of the entire population. The rayless layia (*Layia discoidea*), for example, is restricted to two serpentine outcrops in central California, but it has apparently never occurred elsewhere, no popula-

TABLE I

Indicators of Endangerment

Risk of extinction	Number of populations		Geographic distribution	Population size		Habitat requirements
High	Few	Decreasing	Narrow	Small	Decreasing	Specific
↑	↑	↑	↑	↑	↑	↑
Low	Many	Increasing or stable	Wide	Large	Increasing or stable	Broad

tions have been lost, and the existing populations are large and relatively stable. The small whorled pogonia (*Isotria medeoloides*), on the other hand, is found from Maine to Georgia, but all known populations are small and many of them may be declining. Moreover, many populations known to have existed in the past are no longer extant. Thus, *I. medeoloides* appears to be in more imminent danger of extinction than *L. discoidea*, even though *I. medeoloides* has over 10 times as many populations and a much broader geographic range.

B. Indicators of Conservation Priority

Given the large number of plant species at risk and the limited resources that can be used to prevent their extinction, it is not enough to identify which species are in the most danger. Some species will receive attention early in a conservation program while other will wait, and a conservation program must decide which species deserve the most immediate attention. Clearly, the degree of endangerment is an important criterion in making this decision. After all, the objective of a species conservation program is to prevent extinction, and this cannot be done by focusing efforts on species that are not in danger. Despite its obvious importance, however, the degree of endangerment is not the only criterion that should be used in determining conservation priority.

Species that play a unique ecological role or that represent a distinctive evolutionary line contribute more to biological diversity than those that are less distinct. If part of the objective in conserving species is to save as much of the remaining biological diversity as possible, then ecologically or evolutionarily distinctive species deserve a higher prior-

ity than those that are less distinctive. Similarly, protecting plants that are part of a mutualistic association with other endangered species, such as the sundial lupine (*Lupinus perennis*), which is the larval host for the endangered Karner Blue butterfly, does more to conserve biological diversity than protecting plants lacking such associations. Similarly, protecting plants that are good indicators of important and specialized habitats, such as the Tiburon jewel flower (*Streptanthus niger*), a serpentine endemic, may often protect other species restricted to these habitats. The last two considerations can be grouped together in the broad category of ecological importance. In short, there are at least three criteria to be used in determing conservation priority (Table II): (1) degree of endangerment, (2) ecological or evolutionary distinctiveness, (3) ecological importance.

Another factor may also influence the conservation priority given to a species: the chances that a conservation plan will be successful in preventing its extinction. Some have argued that it is better not to spend limited resources on plants that cannot be saved. Because so little is known about the biology of most endangered plants, however, it is almost impossible to say which can be saved and which cannot. Whether or not a conservation plan for a particular plant will succeed cannot be known until it is tried. Thus, no plant should be excluded from consideration simply because the conservation plan may fail. Rather, every plant needing conservation should have a plan developed for it. It is, of course, reasonable to exclude a plant from a particular part of a conservation program if there is good evidence that it will not benefit from that part of the program, but that is no reason to exclude it from other parts of the program from which it

TABLE II
Criteria for Conservation Priority

Priority	Degree of endangerment	Ecological or evolutionary distinctiveness[a]	Ecological importance
High	Likely to become extinct in the immediate future	Very distinct	Important mutualist or good indicator
↑	↑	↑	↑
Low	Not likely to become extinct in the immediate future	Marginally distinct	Not important mutualist or good indicator

[a] Plants scoring high on this scale are often recognized as taxonomically distinct, for examples, monotypic genera, species in monotypic sections. Because of the variation in taxonomic treatments from group to group, however, hierarchical categories should be used as a guide to distinctiveness only for closely related groups.

might benefit. Excluding rare aquatic plants from seed bank collections of endangered plants is reasonable, for example, because there is good evidence that they cannot tolerate long-term storage. But forgoing efforts to protect a critical habitat for them simply because they cannot be included in a seed bank would be a grave error.

III. IDENTIFYING THE CAUSES OF ENDANGERMENT

Once a species has been chosen as the focus of a conservation program, the next step is to identify the most important threats to its continued persistence. In some cases the threat will be obvious, for example, loss of suitable habitat. Even in cases in which the threat is "obvious," however, some time spent analyzing the factors that affect the species' abundance is likely to be informative. It may suggest threats whose effect was obscured by the obvious ones.

The most formal method of identifying a threat is population viability analysis. A complete population viability analysis requires at least two things: (1) a demographic model of the individual populations of the species that describes how population size and age structure changes from one year to the next as a result of reproduction and mortality and (2) estimates of the amount of year-to-year variability of each of the parameters included in the demographic model. In addition, a model of metapopulation dynamics that describes the rates of population extinction, recolonization, and migration among existing populations is required if there is evidence that these rates are substantial. In small populations these demographic models may be combined with genetic models intended to reflect genetic changes that occur, simply by chance, in small populations. With these models in hand, an analysis proceeds by performing a series of computer simulations in which the parameters of the model are used to project changes in population size and age structure, using information about the degree of year-to-year variability to simulate stochastic environmental and demographic effects. The probability of extinction over a specified period is calculated as the fraction of runs in which the modeled population size declined to zero in that time. A viable population (or set of populations, if a metapopulation model was constructed) is often defined as one that has a 95% chance of surviving for 100 years. [See POPULATION VIABILITY ANALYSIS.]

When a complete population viability analysis is possible, great insight into management tactics is possible. By changing the parameters in the model, either individually or in sets, and rerunning the analysis, those components of the life cycle at which management intervention is likely to have a significant impact can be identified. A complete population viability analysis is such a large task; however, it is unreasonable to expect that most endangered plants will ever receive such detailed study. In fact, because of the effort involved, a complete population viability analysis can probably be justified for only a few of the most significant endangered plant species. Although identifying the exact causes of decline will be difficult for many endangered plants, it may still be possible to identify the appropriate stages of the life cycle for management intervention without a complete population viability analysis. Even when a complete population viability analysis is not necessary, however, the focus should be on understanding the factors that regulate the distribution and abundance of individuals.

A. Threats to Persistence in Small Populations

Some populations are at risk of extinction because the average number of individuals that become established each year is smaller than the average number that die. In these populations some deterministic process is responsible for their decline. Even populations in which the mean replacement rate exceeds the mean death rate, however, may have a nonzero probability of extinction if random variation in the demographic parameters can result in a death rate that exceeds the replacement rate in any one year. If several years in which the death rate exceeds replacement succeed one another, the number of individuals left in a population may be so low that it is no longer self-sustaining. A small population is more likely to become extinct than a large one as a result of such stochastic variation, because a shorter period of decline can reduce its number to unsustainable levels. Similarly, populations subject to a large amount of stochastic variation in replacement and death rates are more likely to become extinct than those subject to smaller amounts of variation.

It is useful to distinguish four sources of stochastic variation. *Environmental stochasticity* is a consequence of variation in the physical and biotic environment and is reflected in year-to-year variation in the parameters of a demographic model, for example, rates of reproduction and survival. In some ways *catastrophes* are simply an extreme form of environmental stochasticity, reflecting as

they do occasional extreme changes in the physical or biotic environment. Nonetheless, it is useful to distinguish environmental variation in which a series of bad years is required for extinction (environmental stochasticity). Another reason for distinguishing them is that although some catastrophes cannot be avoided (e.g., floods or earthquakes), others can be (e.g., shopping malls or condominium complexes). To the extent that catastrophic events can be controlled or prevented, extinction probabilities can also be reduced. *Demographic stochasticity* is a manifestation of the chance events that affect which individuals survive and reproduce and which do not. Just as chance events affect how many offspring are produced, they affect the genetic composition of the offspring produced: *genetic drift*. In large populations such chance events have little effect. In small populations, however, these chance events can have a large impact on population size and the genetic composition of the population, even if demographic parameters are unchanged. Moreover, the population may eventually become genetically depauperate, if the rate at which genetic variability is lost through genetic drift exceeds the rate at which it is replaced by mutation.

Few detailed analyses of population viability exist, but those that do agree on one important point. Only in extremely small populations (on the order of 50 or so individuals) do demographic stochasticity and the loss of genetic variability associated with genetic drift represent a significant threat to short-term persistence. In fact, populations must be at least three times as large to buffer the effects of low levels of environmental stochasticity as to buffer the effects of demographic stochasticity or genetic drift. Typically, populations large enough to buffer the effects of environmental stochasticity are more than 20 times as large, that is, more than 1000 individuals. In short, demographic stochasticity and loss of genetic variation pose an immediate threat only in populations that are also at substantial risk from environmental stochasticity.

Loss of genetic variability, unlike demographic stochasticity, may also pose a long-term threat to the viability of a population, because long-term viability depends in part on the amount of genetic variability a population retains. Even here, however, it appears that populations large enough to buffer the effects of environmental stochasticity are little threatened by loss of genetic diversity. There is little, if any, evidence that plant populations with 500–1000 individuals have less adaptively significant genetic variation than larger ones. Furthermore, rare alleles are more likely to be lost than common ones, and adaptation to changed environmental conditions is more likely to result from changes in the frequency of common alleles or from the incorporation of new alleles than from the replacement of common alleles by rare ones. There are two reasons for this. First, most of the genetic variance that can respond to natural selection is found in alleles present in moderate to high frequency that is, common alleles. Second, low-frequency alleles are likely to be lost through genetic drift in just a few generations in small populations, unless they are maintained by selection. In short, loss of genetic variation is unlikely to pose a significant threat to long-term persistence of populations that are large enough to buffer the effects of environmental stochasticity.

B. Inferring the Causes of Endangerment

Demographic stochasticity and the loss of genetic diversity associated with genetic drift are likely to pose a threat only to plant populations that are very small, say, 50–100 individuals. Environmental stochasticity and the possibility of catastrophe serve to compound these threats. Populations large enough to escape the threats of demographic stochasticity and loss of genetic diversity, however, may still be subject to high extinction probabilities associated with environmental stochasticity or catastrophes. In short, all endangered plants are subject to such high extinction probabilities. Moreover, because demographic stochasticity and loss of genetic diversity become important threats only in very small populations, they are unlikely ever to be the causes of endangerment. They are far more likely to be its result. Similarly, chance fluctuations in rates of reproduction and survival (environmental stochasticity) are unlikely to cause a spe-

cies to become endangered unless the number of its populations and the number of individuals within them are both already small. In short, the reasons that a particular species is endangered are more likely to be found in deterministic than in stochastic processes, even though stochastic processes are critically important in determining how large a population must be to remain viable. Thus, in seeking the causes of endangerment, attention should be focused on trying to identify deterministic factors that have changed the balance between reproduction and survival within populations, processes that affected the rates at which new colonies are founded and existing ones become extinct, or catastrophes (including human-caused ones) that eliminated previously existing populations.

I. Patterns of Distribution and Abundance.

The first place to look for clues about the cause of endangerment is to some of the same basic data that are used to identify an endangered species in the first place, namely, its pattern of distribution and abundance. Particularly important is information about *changes* in these patterns. By examining these changes, it may be possible to identify some of the causes of endangerment and suggest additional questions that must be answered to identify others.

a. Are Populations Larger or Smaller Than They Used to Be, or Do They Have about the Same Number of Individuals?

If individual populations are decreasing in size, then demographic processes within populations are at least part of the cause of endangerment. The specific components of the life cycle responsible for causing a decline in population size may be sought by studying patterns of recruitment and survival. If individual populations are stable or their numbers are increasing, however, the cause of endangerment is unlikely to be found in demographic processes operating within populations. It is more likely to be found in processes that reduce the number of existing populations or, in plant species in which extinction and reestablishment of populations play an important role, in changes that diminish the rate at which new populations are founded or that increase the rate at which individual populations become extinct. Analyses of Furbish's lousewort (*Pedicularis furbishiae*), for example, suggest that land use patterns, which affect the frequency and extent of ice scouring and alter the rates of colony extinction and reestablishment, have a greater impact on its long-term viability than within-population dynamics.

b. Are There More Populations or Fewer Than There Used to Be?

When the number of populations of a plant species has declined in historic times, the most common cause is probably habitat modification or conversion associated with human activities, for example, agriculture or construction. Habitat modification is often identified only when it leads directly to the extirpation of previously existing populations. The California fan palm (*Washingtonia filifera*), for example, is threatened by ground water pumping, which can damage or destroy the few remaining desert oases where it occurs. Important as direct habitat modification is, however, long-term changes associated with changing land use patterns may be even more significant in some cases. The running buffalo-clover (*Trifolium stoloniferum*), for example, is apparently threatened because it is adapted to the disturbed habitat characteristic of buffalo wallows, a habitat that has not been available in the eastern United States since European settlement. This example illustrates that changes in the frequency and scale of disturbances may also affect the number of populations. Not only may the number of populations of early-successional plants decline if the frequency of disturbance is reduced, but the number of populations of late-successional species may be reduced if the frequency of disturbance is increased following human settlement. The number of populations may also decline if an important mutualist, for example, a mycorrhizal associate or specialist pollinator, is eliminated from a part of its range.

2. Patterns of Recruitment and Survival.

Once the patterns of distribution and abundance have been examined, further analysis of any popu-

lation whose numbers have been declining is needed. When the size of a population is declining, there are two possible causes: (1) a change in the physcial or biotic environment that results in a decreased recruitment rate, an increased death rate, or both, or (2) a change in the physical or biotic environment that results in a decreased rate of immigration from other populations, an increased rate of emigration from this population, or both.

a. Are New Individuals Being Recruited into the Population?

If few individuals are becoming established in the population, it may be able to sustain itself only through constant immigration. If recruitment is low in all populations of a species, then the species may be doomed to extinction unless the rate at which new populations are formed exceeds the rate at which existing ones become extinct. Thus, low rates of recruitment within a population are always a possible source of endangerment, and some attention should be given to identify the causes. For example, no seedlings have been observed in Chapman's rhododendron (*Rhododendron chapmanii*) or McKittrick's pennyroyal (*Hedeoma apiculatum*), and identifying the conditions necessary for successful recruitment is obviously critical in designing a management plan for these species. In annual plants a low rate of recruitment is evidenced by small (and declining) population sizes. For perennials the presence of a small number of size or age classes (relative to the average life span) is evidence that recruitment may be limiting, although the possibility that recruitment has always been episodic cannot be excluded. If recruitment limitation is suspected as a cause of endangerment, Table III provides a key that can be used to identify those life history stages that are responsible.

b. Are Adults Dying at a High Rate?

Just as low rates of recruitment are always a possible source of endangerment, so also can high rates of adult mortality endanger a population. There are many possible causes of increased adult mortality, for example, successional change, competition with invasive exotics, introduction of new herbivores, and changes in nutrient and water availabil-

TABLE III
Identifying Possible Causes of Low Recruitment Rates

1. Vegetative reproduction possible for example, a stoloniferous grass, a strawberry, or a goldenrod.[a]. 2
   Vegetative reproduction not possible. 3
2. Vegetative offspring being produced . . . . Juvenile mortality
   Vegetative offspring not being produced 3
3. If vegetative reproduction is possible but vegetative offspring are not being produced, there may be many causes, for example, resource limitation, physiological stress, or habitat change leading to inappropriate environmental cues for reproduction.
4. Seedlings observed. . . . . Juvenile mortality
   Seedlings not observed. 5
5. Abundant seed produced. . . . . Seedling mortality
   Seed not abundantly produced. 6
6. Pollinators absent or rare. . . . . Pollinator limitation
   Pollinators abundant. 7
7. If pollinators are abundant but seed set is low, there may be many causes, for example, an insufficient number of male plants in a dioecious species, an insufficient number of incompatibility types in a self-incompatible plant, physiological stress, or resource limitation.

[a] Do not make a choice between these alternatives. Both branches should be followed if they both apply.

ity associated with changes in land use patterns. Which of these causes needs to be considered will depend on the particular species and the habitat in which it is found. Physiological work on the response to environmental factors and transplant experiments may be necessary to identify the precise cause.

3. Metapopulation Dynamics.

Determining the pattern of extinction and recolonization and measuring rates of migration among existing populations are very difficult tasks. In many cases, however it may not be that important. Most endangered plant species will persist only if individual populations are self-sustaining and will perish if they are not. Furbish's lousewort (*P. furbishiae*) is a notable exception. It is restricted to banks of the St. John River in northern Maine. Viable populations are found only in areas where cover is low—early successional habitats that are temporary and disappear over a relatively short period. The creation of new habitat and the destruction of existing populations by ice scouring in spring floods dominates the dynamics of the metapopulation, and changes in the frequency and intensity of ice scouring associated with changes in land use within the watershed will have a significant impact on its long-term viability. Other plants in which colonization depends on environmental disturbance, e.g., those that depend on fire, such as the Florida golden aster (*Chrysopsis floridana*) and the hairy rattleweed (*Baptisia arachnifolia*), are likely to show similar patterns. Thus, if many populations of a species are declining, an examination of metapopulation dynamics is critical, because it may suggest that management on a broader scale than the individual population is required to ensure long-term viability.

4. Life History Characteristics.

Strictly speaking, the information gleaned from patterns of distribution and abundance, patterns of recruitment and survival, and metapopulation dynamics will provide all the information necessary to identify the reasons that any particular plant species is endangered. Nonetheless, different aspects of the life history are important in different species, and these differences may provide a way to identify the causes of endangerment more quickly than would otherwise be possible.

a. Annuals.

The population dynamics of annual plants are dominated by recruitment from seed. The first question to be asked is, Is there a persistent seed bank? If not, the failure of reproduction in any one year is far more serious than it would otherwise be. Members of the genus *Clarkia,* for example, apparently lack a seed bank in natural populations, though seed remains viable for many years in off-site collections. Thus, population sizes in one year are directly related to the number of seed produced in the immediately preceding year. In the genus *Orcuttia,* on the other hand, seedlings commonly emerge after lying dormant in the soil for several years. The presence of this seed bank makes population sizes less immediately dependent on reproduction in any single season. Additional questions to be considered should include, What conditions are necessary for germination and establishment? What proportion of plants set seed every year? If only a few plants are responsible for producing most of the seed, the population is likely to be subject to more severe stochastic fluctuations in abundance than if reproduction is more evenly spread through the population.

b. Monocarpic Perennials, Herbaceous or Woody.

The population dynamics of perennials that flower once and then die, such as the Haleakala silversword (*Argyroxiphium sandwicense* subsp. *macrocephalum*), are also likely to be dominated by recruitment from seed, although high rates of adult mortality may be a further source of endangerment in some species. In addition to the questions posed for annuals, the adult mortality rate should also be measured.

c. Polycarpic Perennial Herbs, Not Reproducing Vegetatively.

Understanding the population dynamics of perennial plants that reproduce many times during their life is much more complicated than understanding the dynamics of annual plants or monocarpic perennials. Without a complete demographic analysis it is impossible to make precise predictions about population dynamics, but it may still be possible to identify some sources of endangerment. Questions to be considered should include, What is the age (size) structure of the population? An even-aged population may reflect either episodic recruitment or a long-term failure of recruitment. Does the rate at which new individuals become established exceed the rate at which adults die?; that is, is the population size increasing, decreasing, or remaining about the same? Although an increase in population size from one year to the next is a poor predictor of long-term population viability, a *trend* of increasing population size or a stable population size is a good indicator of viability. Similarly, a trend of decreasing population size suggests that the population is no longer self-sustaining.

d. Polycarpic Perennial Herbs, Reproducing Vegetatively.

The population dynamics of clonally propagating herbs is similar to that of nonclonal herbs in many respects, except that the importance of reproduc-

tion by seed may be substantially lessened. Clonal propagation allows a population to persist and spread even when sexual reproduction has failed, and it increases the number of ecological individuals. Thus, it may reduce some of the ecological disadvantages of small population size. Of course, it can also reduce the effectiveness of outcrossing in self-incompatible plants if the clones are large enough that most pollen transfer occurs within a clone. For example, an endangered mint from the mountains of southern New Mexico (*Hedeoma todsenii*) occurs in widely separated patches, each apparently of clonal origin, and only a small percentage of flowers set seed. In species in which extinction and reestablishment of populations play an important role, seed production may also be critical to long-term persistence of the population system, even if it plays a small role, if any, in the persistence of individual populations.

e. Woody Shrubs and Trees.

The population dynamics of woody species are poorly understood, partly because of their long life span. What little is known suggests that recruitment is more likely to limit population size than adult mortality. Because of the longevity of their adults, the impact of short-term environmental fluctuations on adult mortality is probably small, and the causes of population decline are more likely to be found in processes that limit recruitment than in those that increase rates of adult mortality. As with polycarpic perennial herbs, a study of age structure can be very informative, although an even-aged population may reflect either episodic recruitment or a long-term failure of reproduction.

IV. CONSERVING SPECIES IN THEIR NATIVE HABITAT

After a species has been chosen for conservation and the threats facing it have been identified, the next task is to develop a management plan to conserve it. The goal of such a plan should be to establish and protect self-sustaining populations in the species' native habitat, using whatever management tactics are appropriate. These tactics may include modification of the disturbance regime, augmentation of existing populations with seed from seed banks or other populations, or the establishment of an off-site seed collection. Whatever the tactics, however, the focus is on establishing and protecting self-sustaining populations, a focus that requires a change in the traditional approach to plant conservation. Animal conservationists have long recognized that population dynamics must be managed to establish and protect self-sustaining populations, but plant conservationists have often been satisfied with protecting sites that support populations of endangered plants without attempting to manipulate the rates of reproduction and survival. Fortunately, plant conservationists are increasingly aware that the decision simply to protect sites is simultaneously a decision not to manage population dynamics, and therefore a decision that must be made consciously. For populations that are stable or increasing in size, deciding only to protect the sites may be appropriate (unless the populations are very small), but for populations in decline, some intervention to halt or reverse the decline is required.

It is useful to distinguish three aspects of population management in native habitats. *Population monitoring* provides information on the status of a population and the effectiveness of management procedures. *Habitat management* aims to affect rates of reproduction and survival indirectly through manipulation of the physical and biotic environment. *Demographic management* aims to affect rates of reproduction and survival by manipulating them directly. The choice of specific management techniques is governed by the causes of endangerment that have been identified, but the overall approach can be easily summarized (Table IV).

- Monitor all populations on a periodic basis to assess their status and the effectiveness of the management program.
- Protect sites where the population occurs.
- Manipulate the habitat to provide a suitable biotic and abiotic environment at sites where populations are in decline.
- Augment existing populations if habitat manipulations alone prove to be unsuccessful,

if there is reason to expect that habitat manipulations alone cannot ensure long-term viability of the population, or if the number of plants is very small.
- Reintroduce populations to sites where they are known to have existed or introduce populations to apparently suitable sites where they are not known to have existed if the number of existing populations is very small or the existing populations are subject to external threats that make their long-term viability questionable.
- Directly manipulate individual reproduction and survival when indirect means of increasing population size have failed or when populations are so small that extinction is probable before more indirect techniques can have an effect.
- Make changes to the management program as information on the effectiveness of each component is accumulated and as new data on the biology of the species being managed becomes available.

A. Population Monitoring

Population monitoring provides data on the status of populations. Comparing the status of a population or set of populations before and after the implementation of a conservation plan provides a way of evaluating the effectiveness of the plan. It may also provide clues about ways in which the plan could be changed to increase its chances of success. The specific data collected during monitoring will depend on the biology of the species and the management techniques being used to conserve it, but at a minimum they should include (1) an estimate of the total population size, (2) an estimate of the number of reproductive individuals, and (3) a brief description of associated soils and vegetation. For species in which habitat protection alone appears sufficient to ensure their long-term viability, no more is needed. For species in which habitat protection is accompanied by active management of habitat characteristics or demographic manipulations, data that allow the success of those techniques to

TABLE IV

Choosing Appropriate Tactics for Management in Native Habitats

Cause of threat	Possible management actions
Loss of early-successional habitat	Manipulate disturbance regimen, for example, controlled burns or moderate grazing Establish new populations in suitable habitat
Loss of late-successional habitat	Exclude incompatible land uses, for example, off-road vehicles Establish new populations in suitable habitat
Excesive adult mortality	Remove invasive exotics Manipulate disturbance regimen Exclude incompatible land uses
Excessive juvenile mortality	Remove invasive exotics Supplement population with propagules drawn from an off-site collection or from other populations
Excessive seedling mortality	Remove invasive exotics Supplement population with propagules drawn from an off-site collection or from other populations Outplant juveniles propagated in an off-site collection
Pollinator limitation	Hand pollinate individuals in population Supplement population with propagules drawn from an off-site collection or from other populations Minimize insecticide use
No seed set or clonal reproduction [a]	Do experiments to determine physiological and ecological requirements for reproduction, for example, reciprocal transplants or manipulation of water, nutrient, and light levels Supplement population with propagules drawn from an off-site collection or from other populations

[a] Where the failure in seed set is not the result of pollinator limitation.

be monitored should also be collected. Similarly, species under active management will need to be monitored more frequently than those in which habitat protection is the primary management tool.

B. Habitat Management

The objective of habitat management for an endangered species is 2-fold: (1) to identify the habitat required to establish a self-sustaining population and (2) to choose the manipulative actions necessary to maintain that habitat. The first step in the process is to protect native populations of endangered species by protecting their habitat. For some species that may be enough. Many serpentine endemics in California, for example, are likely to persist indefinitely, provided that the serpentine outcrops on which they occur are protected. The only management necessary is to provide protection and to monitor the status of the population periodically to ensure against an unexpected decline.

When a population is in decline, however, protecting the land on which it is found is not enough.

Changes in the surrounding physical or biotic environment are often the cause of the decline, and these changes must be reversed if the population is to become self-sustaining. Changes may be the result either of single catastrophic events (e.g., fire, flood, or hurricane) or of long-term changes (e.g., those associated with successional changes in the surrounding vegetation, the introduction of exotic pests, or changes in land use patterns), but whatever their cause, they cannot be reversed without active habitat manipulation. Individuals of the federally endangered Malheur wireweed (*Stephanomeria malheurensis*), for example, are significantly smaller, slower to bolt and flower, and less fecund in plots with cheatgrass (*Bromus tectorum*) than in plots from which this exotic has been removed. In other cases it may be necessary to manipulate the disturbance regimen by altering the frequency of fire or the intensity of grazing or to exclude incompatible land uses, for example, off-road vehicles. Recovery of the large-flowered fiddleneck (*Amsinckia grandiflora*), for example, may depend on the judicious use of selective herbicides to increase the availability of low-competition patches. Monitoring on at least a yearly basis is necessary, not only to assess the status of the population but to measure the effectiveness of the habitit manipulations.

C. Demographic Management

For some populations in decline, even habitat manipulations may be insufficient to reverse the decline. If extinction of these populations is to be prevented, direct intervention to affect the rates of recruitment and survival is necessary. Such intervention may take a variety of forms. Propagules can be introduced into an existing population (augmentation), they may be used to reestablish a previously existing population on a site at which it is now extinct (reintroduction), or they may be used to establish a new population on a site that appears suitable but is not known to have supported a population (introduction).

The appropriate source of propagules for augmentation, reintroduction, or introduction is not always apparent. Reciprocal transplant experiments on both annuals and perennials have consistently shown that individual survivorship and fecundity are greatest when individuals are replanted in the same microhabitat from which they were collected. Theoretical and experimental results also suggest that genetic diffentiation among populations subject to identical selection pressure may often be sufficient to lower average individual viability if individuals are exchanged. Together, these results suggest that propagules for augmentation programs should be drawn from the population being augmented, or from an off-site collection derived from that population. Similarly, propagules for reintroduction should be drawn from an off-site collection derived from the population being reestablished if possible.

When establishing a new population, however, mixing propagules from several populations could provide an advantage. In self-incompatible plants, for example, it is likely to increase the number of incompatibility types present. The three surviving members of the last Illinois population of the lakeside daisy (*Hymenoxys acaulis* var. *glabra*), for example, were all of the same self-incompatibility type. Hybrid seed derived from crosses between Illinois and Ohio plants were used in the reintroduction plan, because this was the only way to ensure representation of the Illinois gene pool in the restored population. Some plant conservationists have gone further and argued that the enhanced genetic diversity expected from a mixed population may allow natural selection to produce locally adapted genotypes, increasing the changes of a successful introduction. Unfortunately, there are few data with which to evaluate this suggestion. We do know that once a hybrid population is established, its hybrid character cannot be eliminated. Given the evidence that transplantation is most successful when individuals are replanted in a microhabitat similar to that from which they were collected, it seems best for a reintroduction or introduction program to use propagules from a single source first. If this fails or if the plant is self-incompatible and no single population can provide enough propagules to ensure a wide array of incompatibility type, then propagules from several populations can be used.

There are many other demographic management techniques that may be required under specific circumstances, for example, outplanting of juveniles when seedling establishment is limiting and hand pollination when the number of pollinators is low or when only a few compatibility types are present. The state of Hawaii, for example, has substantially increased the population size of the Mauna Kea silversword (*Argyroxiphium sandwicense* subsp. *sandwicense*) through an outplanting program. Which of these techniques is appropriate will depend on the species being managed and the threats it faces. Of course, the more deeply involved a manager become in manipulating reproduction and survival of individual plants, the more time and money is required. As a result, direct intervention to ensure reproduction of individual plants can probably be justified for only a few of the most significant endangered plant species.

V. CONSERVING SPECIES IN OFF-SITE COLLECTIONS

The primary focus of an effort to conserve an endangered plant species must be to establish self-sustaining populations in its native habitat. To accomplish this goal, however, it is often necessary to do more than protect the sites where it occurs and manage the habitat on the sites. If it is necessary to augment existing populations, to reestablish those that have been lost, or to establish new ones, a source of propagules other than existing populations may be required. Off-site collection in botanical gardens or seed banks can serve as such a source. They may also serve as insurance against catastrophes that might eliminate existing natural populations. The Malheur wireweed (*S. malheurensis*) become extinct only 20 years after its discovery. Recovery efforts were possible only because a seed stock had been maintained at the University of California at Davis, for research purposes. This off-site collection was the source for all seed that has been used in the reintroduction and recovery effort.

In addition to providing a source of propagules for on-site management, off-site collections can play two additional roles in plant conservation programs. First, the horticultural research necessary for cultivation and display may provide new insight into habitat requirements, insight that can be used to suggest new management practices in natural populations. Second, public display of endangered plants in botanical gardens provides an opportunity to communicate the importance of conservation to the public and to show people the plants that are the focus of our concern.

A. Living Collections versus Seed Banks

The primary purpose of an off-site collection is to provide a source of propagules for augmentation, reintroduction, and introduction efforts. If these collections are maintained as living plants, however, it is virtually impossible to prevent adaptation to the cultivated environment, no matter how carefully a crossing program is designed and no matter how much effort is put into matching garden conditions to those in a plant's native habitat. Even if such adaptation could be avoided, a complex crossing scheme is required to maintain genetic diversity in living collections. Such a crossing program also

requires the assistance of professional geneticists and much time and money.

Fortunately, such complex programs are often unnecessary. Seeds of most temperature zone plants are amenable to long-term storage. In fact, cryogenic storage may permit them to be stored indefinitely. Seed banks reduce the problem of using off-site collections for conservation to (1) obtaining a representative sample of genetic diversity from remaining populations and (2) ensuring continued viability of the seed in the seed bank through periodic germination trials and re-collection as necessary. Living collections are invaluable for display and educational purposes, but they should be used as the source of propagules for augmentation, reintroduction, or introduction only for species with recalcitrant seeds that are not amenable to storage.

B. Collecting Guidelines

There are several problems to be considered in collecting seed for an off-site seed bank: How many populations should be sampled? How many individuals should be sampled per population? How many seeds should be collected from each individual? Under what circumstances is a multiyear collection plan necessary?

The design of an intelligent collecting strategy usually depends on an accurate understanding of the genetic structure of a species, requiring an intense preliminary survey to identify patterns of geographic differentiation. Fortunately, such preliminary surveys are not needed for most endangered species. The largest source of error in the design of a collecting strategy is the failure to collect genotypes that are locally common but rare in the species as a whole. Because most endangered plants occur in a small number of populations, however, the chance of excluding such genotypes from the sample is quite small, provided that collectors use a little common sense about the populations to be included, for example, by collecting from populations that are geographically isolated from one another or that occur in distinctive habitats. For most endangered plants a sample from four or five carefully chosen populations is usually sufficient to provide a representative sample of the adaptively significant genetic diversity in the species.

Similarly, the number of individuals from which seed should be collected is a relatively simple matter. To provide a 95% chance of capturing all alleles with a frequency of 5% or greater within that population, a sample from 50 maternal plants is sufficient in an outcrossing plant, while in a selfer a sample from 100 maternal plants is sufficient. There is a strong law of diminishing returns between the capture of genetic diversity and sample size. The capture of genetic diversity is a logarithmic function of sample size. Thus, we gain as much from our first sample of 10 individuals as we do from the next 90. In short, there is little justification for seeds to be collected from more than 50–100 maternal plants.

Determining the number of seeds to be collected from each plant and when to undertake a multiyear collection effort is also straightforward. The number of seeds collected from each maternal plant should be sufficient to provide a high probability that at least one seed from each maternal plant will be viable. If the total amount of seed required from within a population could endanger that population, the collecting effort should be spread over several years. Similarly, if only a few individuals produce most of the seed in any given year, several small collections from different years are likely to produce a more representative sample of the diversity present than a large sample from a single year.

VI. AN INTEGRATED APPROACH TO PLANT CONVERSION

Most endangered plants are poorly studied. Little is known about their physiology, genetics, or ecology. Fortunately, we need not wait until these studies are done to begin conserving them. Many of the necessary actions are dictated by the simple fact that the plants are endangered. The goal is clear: to establish and maintain self-sustaining populations of endangered plant species in their native habitat. To do that, we first identify those species most at risk of extinction and identify our conservation priorities. We then identify the factors that have caused those species to become endangered. Having identified the causes of endangerment, we guide the populations to recovery through a combination of protecting the sites on which populations occur, managing the habitat to provide a suitable physical and biotic environment for growth and reproduction, and directly manipulating individual growth and reproduction when necessary. As insurance against catastrophic loss of natural populations, we establish an off-site collection that can be used as a source of propagules for augmentation and introduction. The off-site collections also provide the opportunity for public education in the value and importance of plant conservation through the display of living examples.

Each of these approaches is valuable, but each becomes even more valuable when it is used as part of an integrated strategy. Site protection alone may be enough to prevent the extinction of some species. For many, however, persistence can be assured only if their population dynamics are actively managed. Off-site collections provide an additional degree of insurance against unforeseen catastrophes. The chances of preventing extinction are far greater when all appropriate techniques are used than when some of them are not. An integrated approach to plant conservation, in which the protection and management of natural populations are complemented by the creation of off-site collections as a buffer against catastrophe, is likely to prevent the extinction of many endangered plants.

Glossary

Augmentation Introduction of seed or other propagules into an existing population with the intent of increasing the number of individuals in that population.

Demographic model A model that describes the changes in population size and age and size structure from one year to the next in terms of the rates of reproduction (both sexual and asexual) and survivorship. In dioecious plants it may include age and size structure for both males and females.

Demographic stochasticity Random variation in survival and reproduction within a generation that affects population size in the next generation.

Environmental stochasticity Random variation in the physical or biotic environment from one year to the next that is reflected in variation in demographic parameters from one year to the next.

Genetic drift The random change in allelic frequency or allelic composition that occurs as an inevitable result of reproduction in small populations.

Introduction Introduction of seed or other propagules into a site that is not known to have supported a population of the species, with the intent of establishing a new population.

Metapopulation A set of populations that are partially coupled, both demographically and genetically, through extinction/recolonization and migration.

Migration Exchange of individuals between existing populations that results in some degree of demographic and genetic coupling.

Population viability analysis A formal demographic analysis of a specific population or set of populations that provides estimates of the probability that a population will persist for a specified period.

Recruitment rate The rate at which new individuals are incorporated into a population through local reproduction.

Reintroduction Introduction of seed or other propagules into a site that formerly supported a population of the species, with the intent of reestablishing a previously existing population.

Seed bank A collection of seed in long-term storage for conservation purposes.

Bibliography

Bramwell, D., Hamann, O., Heywood, V., and Synge, H., eds. (1987). "Botanic Gardens and the World Conservation Strategy." Academic Press, London.

Falk, D. A. (1992). From conservation biology to conservation practice: Strategies for protecting plant diversity. *In* (P. L. Fiedler and S. Jain, eds.), "Conservation Biology: The Theory and Practice of Nature Conservation, Preservation, and Management" pp. 397–431. Chapman & Hall, New York.

Falk, D. A., and Holsinger, K. E., eds. (1991). "Genetics and Conservation of Rare Plants." Oxford University Press, New York.

Holsinger, K. E. (1992). Setting priorities for regional plant conservation programs. *Rhodora* **94,** 243–257.

Lande, R. (1988). Genetics and demography in biological conservation. *Science* **241,** 1455–1460.

Mace, G. M., and Lande, R. (1990). Assessing extinction threats: Toward a re-evaluation of IUCN threatened species categories. *Conserv. Biol.* **5,** 148–157.

Menges, E. S. (1992). Stochastic modeling of extinction in plant populations. *In* "Conservation Biology: The Theory and Practice of Nature Conservation, Preservation, and Management" (P. L. Fiedler and S. Jain, eds.) pp. 253–275. Chapman & Hall, New York

Rabinowitz, D. (1981). Seven forms of rarity. *In* "The Biological Aspects of Rare Plant Conservation" (H. Synge, ed.), pp. 205–218. Wiley, New York.

Continental Shelf Ecosystems

J. Ishizaka
National Institute for Resources and Environment, Japan

I. Introduction
II. Seasonal Change
III. Physical and Chemical Forcing
IV. Organisms
V. Material/Energy Flow and Human Impacts

The continental shelf area is the marginal sea around the continents, and the average width and depth of the edge are about 75 km and 130 m, respectively. It is one of the most productive ecosystems, and it is suggested that more than 90% of the world fish yield comes from this 7.5% of the ocean surface and 0.15% of the ocean volume. Different kinds of external physical and chemical forcings, from open ocean, land, and atmosphere, give the system its high primary production. The major physical forcing is different for each continental shelf and varies even with area of shelf and with time. The continental shelf ecosystem is important for the global material cycle and energy flow because of its high productivity; however, the variable, open nature of this ecosystem makes it difficult to quantify the material and energy budgets. Human activities directly and indirectly affect the continental shelf and the global feedback effects are still uncertain.

I. INTRODUCTION

The continental shelf area is the marginal sea that is located around the continents, where relatively shallow water extends from the shoreline to the shelf-edge (Fig. 1 and Table I). The landward boundary of the shelf may be an estuary, bay, or lagoon, which are semienclosed areas of brackish water. The oceanic boundary of the shelf can be defined topographically as a shelf-break, where the slope is much steeper than on the shelf area and continuously deepens to the continental slope and continental rise down to the ocean floor. However, the boundary of the ecosystem is difficult to define because the water continues from river to open ocean, and many organisms move back and forth across these areas. The average depth of the shelf-edge is about 130 m, and the average width of the continental shelf is about 75 km. The depth and width vary considerably, with the width ranging from nearly 0 to 1500 km. It is believed that the continental shelf area was formed by continuous sedimentation and erosion processes caused by a sequence of sea-level changes during the last 1 million years and that the present topography formed 8000–14,000 years ago when the sea level was about the depth of the present continental shelf-break. [See LAND–OCEAN INTERACTIONS IN THE COASTAL ZONE.]

The continental shelf area is characterized by a variety of physical and chemical forcings. Physical forcing is represented by tide, major current system, wind mixing, wind-induced upwelling, and

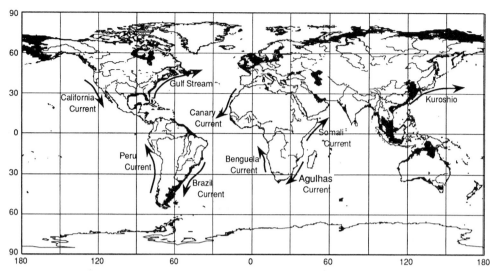

FIGURE 1 Global distribution of the continental shelf (the region less than 150 m deep). Major current systems and rivers that strongly affect the continental shelf ecosystems are also shown.

sea-level change. Chemical forcing includes salinity and other materials that flow in from rivers and that are deposited from the atmosphere by wind and rain. The continental shelf ecosystem is continuously affected by these physical and chemical forcings and is characterized by large spatial and temporal variability.

Productivity on the continental shelf area is much higher than that of the open ocean. The shelf comprises only 7% of the ocean but is responsible for at least 25% of oceanic primary production. This high productivity is caused by the physical and chemical forcing as described in a later section. Because the continental shelf is relatively shallower than the open ocean, interaction between the pelagic (planktonic) community and the benthic community is stronger. However, most of the continental shelf is deep enough so that light cannot reach to the bottom, resulting in less primary production there. Accordingly, the benthic community depends mostly on the production in the upper water column. [*See* OCEAN ECOLOGY.]

II. SEASONAL CHANGE

Primary production of the pelagic community often depends on the light energy and nutrient supply, and these factors change with season in temperate and polar areas (Fig. 2). During winter when

insolation is weak and the water column is strongly mixed, phytoplankton have access to little light energy not only because the original insolation is weak but also because phytoplankton is transported to the deep, dark environment by the water-mixing process. Solar radiation increases in the spring and the water column starts to become stratified, whereupon the phytoplankton biomass increases (often known as spring bloom) using nutrients in the surface water.

Nutrients and biogenic materials in the upper ocean used by phytoplankton eventually drop out of the euphotic zone. Nutrients transported as organic particles recycle in deeper layers or on the sediments. When the water is stratified, nutrient depletion is severe in the upper water column because the nutrients transported do not return. The depletion of nutrients and grazing by increased zooplankton reduce the biomass of the spring phytoplankton bloom. The abundance of phytoplankton remains low at the water surface until autumn, when the stratification weakens and nutrients are supplied from the deeper water. The subsurface maximum of phytoplankton, which is adapted to a low-light environment, often occurs at the base of the nutrient-depleted euphotic zone during the summer.

This general seasonal change is modified by the latitude because of the difference of solar radiation and water temperature. In the polar region, where

TABLE I

Continental Shelves with Major River Discharge, Area, and Annual Primary Production[a]

Latitude	Region	Major rivers	(10^2 m^3 sec^{-1})	Area (10^5 km^2)	Unit production (g C m^{-2} yr^{-1})
		Eastern boundary currents			
0–30	Ecuador–Chile		—	2.7	1000–2000
	Southwest Africa		—	1.7	1000–2000
	Northwest Africa		—	2.6	200–500
	Baja California		—	1.1	600
	Somali Coast	Juba	5.5:5.5	0.6	175
	Arabian Sea	Indus	75.5:86.0	3.8	200
30–60	California-Washington	Columbia	79.6:97.4	1.6	150–200
	Portugal-Morocco	Tagus	3.1:8.9	0.8	60–290
		% Discharge: 3[b]		14.9	Mean: 644.2
		Western boundary currents			
0–30	Brazil	Amazon	1750.0:2005.1	6.0	90
	Gulf of Guinea	Congo	396.4:589.9	3.5	130
	Oman/Persian Gulfs	Tigris	14.5:14.5	4.3	—
	Bay of Bengal	Ganges	116.0:371.8	3.0	110
	Andaman Sea	Irrawaddy	135.6:150.6	4.0	50
	Java/Banda Seas	Brantas	4.0:8.6	6.7	—
	Timor Sea	Fitzroy	1.8:1.8	4.0	100
	Coral Sea	Fly	24.5:62.5	3.2	20–175
	Arafura Sea	Mitchell	3.6:40.3	14.2	150
	Red Sea	Awash	<0.4	1.6	34
	Mozambique Channel	Zambezi	70.8:93.3	2.6	100–150
	South China Sea	Mekong	149.0:363.3	15.7	215–317
	Caribbean Sea	Orinoco	339.3:339.3	3.7	66–139
	Central America	Magdalena	75.0:143.3	4.5	180
	West Florida Shelf	Appalachicola	6.9:6.9	2.1	30
	South Atlantic Bight	Altamaha	3.9:7.2	1.4	130–350
		% Discharge: 62[b]		80.5	Mean: 137.4
		Mesotrophic systems			
30–60	Australian Bight	Murray	7.4:7.7	8.2	50–70
	New Zealand	Waikato	4.1:25.8	2.9	—
	Argentina-Uruguay	Parana	149.0:220.1	10.4	—
	Southern Chile	Valdivia	4.5:44.3	4.0	—
	Southern Mediterranean	Nile	9.5:9.5	2.9	30–45
	Gulf of Alaska	Fraser	35.4:80.8	3.1	50
	Nova Scotia-Maine	St. Lawrence	141.6:178.7	6.7	130
	Labrador Sea	Churchill	15.8:147.5	21.2	24–100
	Okhotsk Sea	Amur	103.0:103.0	7.1	—
	Bering Sea	Kuskokwim	12.8:39.3	11.7	170
		% Discharge: 12[b]		78.2	Mean: 81.9
		Phototrophic systems			
60–90	Beaufort Sea	Mackenzie	97.2:97.2	2.6	10–20
	Chukchi Sea	Yukon	62.0:62.0	6.1	40–80
	East Siberian Sea	Kolyma	22.4:39.8	7.8	—
	Laptev Sea	Lena	163.09:180.2	6.9	—
	Kara Sea	Ob	122.0:300.0	10.1	—
	Barents Sea	Pechora	33.6:78.7	7.3	25–96
	Greenland-Norwegian Seas	Tjorsa	3.6:13.4	2.8	40–80
	Weddell-Ross Seas		—	3.9	12–86
		% Discharge: 11[b]		47.5	Mean: 58.9
		Eutrophic systems			
30–60	Mid-Atlantic Bight	Hudson	3.7:24.6	1.3	300–380

continues

Continued

Latitude	Region		Major rivers (10^2 m^3 sec^{-1})	Area (10^5 km^2)	Unit production (g C m^{-2} yr^{-1})
	Baltic Sea	Vistula	10.1:71.1	3.9	75–150
	East China Sea	Yangtze	220.0:240.3	10.7	—
	Sea of Japan	Ishikari	5.0:22.8	1.7	100–200
	North-Irish Seas	Rhine	25.4:47.8	8.7	100–250
	Northern Mediterranean	Po	14.7:49.0	2.9	68–85
	Caspian Sea	Volga	75.8:75.8	1.5	—
	Black Sea	Danube	65.3:97.9	1.6	50–150
	Bay of Biscay	Loire	8.3:16.7	2.7	—
	Texas/Louisiana	Mississippi	178.0:200.1	2.1	100
			% Discharge: 12[b]	37.1	Mean: 154.2

[a] After J. J. Walsh (1988). "On the Nature of Continental Shelves." San Diego, Calif.: Academic Press.

[b] Total percentage of the freshwater input from the earth's 215 largest rivers (>30 m^3 sec^{-1}) within each of these shelf regions.

light intensity is much weaker and the water column is less stratified than in the temperate region, the maximum biomass may occur during summer. In contrast, in tropical surface waters the water column is typically stratified and nutrients and production are low. These seasonal changes are also modified around the continental shelf area by different physical and chemical forcings.

III. PHYSICAL AND CHEMICAL FORCING

The seasonal changes just described are part of the basic cycle of plankton communities in the ocean. On the continental shelf, additional physical and chemical forcings are important for the enhancement of production and for variability of ecosystem structures. The following sections describe these forcing mechanisms.

A. Estuarine and River Plumes

River and estuarine plumes often extend over the continental shelf or even reach to the open ocean (e.g., the Amazon River and Chesapeake Bay plumes). Because the water of these plumes has lower salinity and density, these plumes overlay the denser water of the continental shelf. The fresh and brackish waters of estuarine and river plumes are important sources of nutrients, organic materials, and other chemical materials for the continental shelf ecosystems. Entrainment of nutrient-rich subsurface water into the surface, less saline waters is another mechanism of enrichment of coastal water

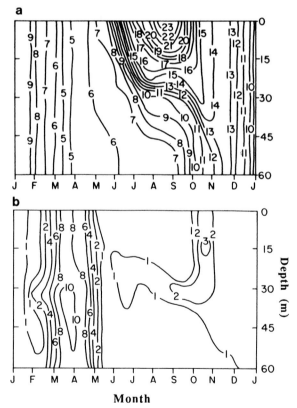

FIGURE 2 Seasonal changes of (a) temperature (°C) and (b) chlorophyll a (μg liter^{-1}). [After J. J. Walsh (1981). *In* "Analysis of Marine Ecosystems" (A. R. Longhurst, ed.), pp. 159–196. Academic Press, San Diego, Calif.]

by these plumes. Discharges of suspended materials from rivers and bays decrease light penetration in the upper water column and sometimes result in lower primary production near the mouth of the rivers and bays. However, phytoplankton, typically euhaline diatoms, increase their biomass by using abundant nutrients in the plume after these sediments settle down and their density peaks at some distance from the river mouth.

B. Tide

The tide is one of the most important hydrodynamic forcings in the continental shelf ecosystem. The relationship between the tidal current and bottom topography induces the mixing of the water column. A combination of strong tidal energy and a shallow bottom induces thorough vertical mixing and sometimes draws nutrients back into the system, but too strong a mixing may decrease the light intensity that is required for phytoplankton production. The transition zone between stratified and mixed conditions corresponds to a high phytoplankton biomass area (Fig. 3), and a distribution of

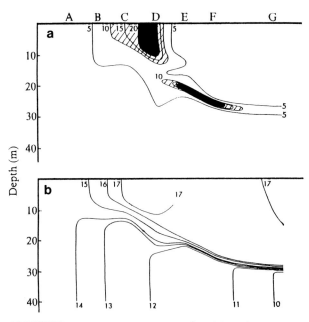

FIGURE 3 (a) Chlorophyll a (μg liter^{-1}) and (b) and temperature (°C) sections through the tidal front. A–G are stations across the front. [After R. D. Pingree *et al.* (1975). *Nature* **258**, 672–677. Copyright 1975 Macmillan Magazines Limited, used with permission.]

tidal energy dissipation, derived from the relation between the mean current speed and water depth, describe well the mixing intensity on continental shelves. This spatial difference is similar to the temporal variability of the temperature spring bloom in which the phytoplankton increase with the increase of stratification and decrease with depletion of nutrients after longtime stratification (cf. Fig. 2). As described in the next section, the distribution of tidal mixing also corresponds well to the distribution of fish larvae and benthic organisms. Another important aspect of tidal motion is the residual current, which is induced by oscillatory motion of tide and which transports materials on the continental shelf region.

C. Internal Wave

When tidal motion or current contacts the bottom topography and the water is stratified, a wave is generated at the boundary between the stratified waters. This wave is called the internal wave because the wave motion is below the surface. Large internal waves can have an amplitude of more than 50 m and a wavelength of 50 km and may propagate with a velocity of 1 m s^{-1}. These large internal waves cause vertical mixing on the continental shelf, supply nutrients to the upper water, and enhance primary productivity. Circulation around the internal waves also creates small-scale divergence and convergence and positive buoyancy materials often converge at the water surface. Floating materials as well as zooplankton that swim upward can be aggregated at the convergent zone. This may be an important mechanism for transport of the zooplankton on the continental shelf.

D. Coastal Upwelling

Wind-induced coastal upwelling is a well-known hydrodynamic factor on the western edges of the continents, such as off Peru, Oregon, California, northwest Africa, and Angola. In these regions, alongshore wind from the equator toward the pole, combined with the effect of earth's rotation (the Coriolis effect), induces an offshore surface current and upward movement of water from the subsurface layer along coast. These regions are highly productive and excellent fishing grounds because of the high nutrient concentrations supplied from the subsurface. These major upwelling zones are fairly large (up to hundreds of kilometers) and the effects of upwelling are seen annually; however, the intensity of upwelling varies spatially because of differences in wind fields and topography and temporally because of the seasonal changes in wind velocity. The variability is important for the biological community, as phytoplankton and fish behavior are often evolutionarily adapted to the variability of these short (days to seasonal) time scale events. Longer-scale wind events, like El Niño, cause far-reaching changes to the ecosystem. Smaller-scale wind-induced upwelling is frequently observed at the various continental shelves and is also important for ecosystem productivity.

E. Western Boundary Current

Western boundary currents, such as the Gulf Stream, Kuroshio, and the Agulhas Current, are major oceanic current systems that flow along the eastern side of continents. Because these strong currents flow along the shelf-break, they have considerable effect on the continental shelf ecosystems. Frontal eddies and other movements of these currents induce sporadic but strong upwelling events and bring nutrients from the subsurface to the outer continental shelf. These upwelling events are one of the major factors that control the primary production on the continental shelves at the eastern side of the continents.

Other than these upwelling events, warm core rings detached from these western boundary currents often approach the coast and cause major exchange between coastal water and open-ocean water. Warm core rings are cyclonic eddies that separate from the poleward extension of the western boundary current, and they contain warm water from the equator side. These rings move westward and frequently contact the poleward part of the eastern continental shelf, where they often introduce plankton communities and fish populations from the open ocean into the coastal area.

IV. ORGANISMS

Organisms living in the continental shelf ecosystem are well adapted to this productive and variable environment. The following sections describe characteristics of different organisms in the continental shelf ecosystem.

A. Phytoplankton

Diatoms are one of the most typical phytoplankton groups on the continental shelf. Diatom blooms are seen in most of the high-productive areas, such as spring bloom, river plume, tidal mixing, and upwelling. These bloom-forming diatoms have high growth rate under high-nutrient and high-light conditions. As described in the previous section, following the well-mixed water conditions of winter, gradual stratification of the water column traps the nutrient-rich water in the euphotic layer. This is a suitable condition for diatom growth, and diatom predominates in the spring bloom until the nutrients are depleted in the upper layer. The transition zone between the tidal mixing area and the stratified area creates a spatially similar condition to that of the spring bloom, and dominance of diatoms is also observed. River plumes and upwelled water also supply nutrients to the euphotic zone, and diatom blooms are typical at a distance from the river mouth or core of the upwelling.

A diatom bloom persists if the water turbulence is high enough to prevent the diatom cells from sinking and to keep high nutrient concentrations in the euphotic zone. The sinking rate of diatom cells increases when nutrients are depleted, and the aggregation of cells observed at the end of a bloom further increases the sinking rate. Some diatom groups also form resting spores that have thicker and heavier silicate shells than the usual vegetative cell, which further increases the sinking rate and protects them from digestion in the gut of zooplankton. The greater sinking rate increases the possibility of cell transport to deeper nutrient-rich areas, and strong turbulence or upwelling may bring these cells up with high-nutrient water to become the seeds of the next bloom. In the deeper coastal upwelling area, it is hypothesized that the faster-sinking diatom cells compensate by having a higher upwelling velocity and so return to the upper water column without settling to the bottom.

Dinoflagellates are another important phytoplankton group that often dominates on the continental shelf area. The conditions necessary for a dinoflagellate bloom are more complicated than for a diatom bloom, and they are not well understood. Because dinoflagellates can swim with their flagella, they form a visible patchiness and the infamous red tide. It is believed that this accumulation results from their motility and hydrographic convergence. Dinoflagellates often become abundant during summertime after the diatom spring bloom, and they increase their biomass in autumn by utilizing the nutrient input that increases with the turbulent condition. It is hypothesized that zooplankton do not graze dinoflagellates much, allowing a gradual accumulation of biomass. The dinoflagellate bloom is also observed in stratified water with a shallow mixed layer. Dinoflagellates can swim between the surface high-light zone and deeper high-nutrient layer and use both light energy and nutrients. Many species of dinoflagellate also form resting cells, which can be found on the bottom of the coastal area, especially in shallow water. The dinoflagellate bloom may be seeded by the resting cells in the bay, after which it moves offshore; however, the bloom has also been observed to start offshore and then move toward shore along the Florida coast.

Other phytoplankton groups may also be important on some occasions. The abundance of small flagellates and coccoid cyanobacteria (*Synechococcus*) is fairly constant across the seasons, but there is a relatively larger abundance during the nutrient-depleted summertime. Large filamentous cyanobacteria (*Tricodesmium*) are seen in the tropical and subtropical ocean and sometimes form a red tide. This phytoplankton can fix gaseous nitrogen, and it may be important for the nitrogen budget of a coastal area. The autotrophic ciliate *Mesodinium rubrum* often forms a red tide in a coastal upwelling area where the upwelled nutrient-rich water is covered with low-nutrient warmer water. The bloom of coccolithophorids, often seen in the North Sea and Gulf of Maine, forms white water that is the result of $CaCO_3$ scales detached from the cells.

B. Zooplankton

Copepods are the most typical zooplankton group, and some species are adapted to the seasonal variation of the continental shelf. Many large copepods in the outer shelf area overwinter in a deeper water layer when the surface layer has low productivity and is highly turbulent, and move back to the upper layer for the warmer, high-productive spring time. Many species on the shallow continental shelf that cannot migrate to deeper layers survive the winter in small numbers or form resting eggs, which sink to the bottom until spring. Many species can also respond to short time scale diatom bloom and increase their population. Zooplankton populations can efficiently graze spring bloom phytoplankton, but more evidence is accumulating that more phytoplankton sink to the bottom and are utilized by benthic organisms without being eaten by zooplankton than was previously believed.

Krill is a well-known grazer and the main prey of whales in the Antarctic. Krill appear to use the hydrographic conditions around the Antarctic continent to keep themselves in the suitable environment for their life cycle stage. It is hypothesized that Krill eggs control their buoyancy so that they are transported to regions where food is abundant when they hatch.

Protozoan zooplankton, such as flagellates and ciliates, are often abundant in water on the continental shelf, but it is still not clear how their biomass is regulated and what qualitative role they play in the food chain. These organisms can grow rapidly, sometimes as fast as the phytoplankton. Bacteria are important food for the protozoan, and the microbial loop, which transfers the dissolved and particulate organic matter to bacteria, microzooplankton, and larger zooplankton, is an important component of marine ecosystems. In the continental shelf ecosystem, the microbial loop is important in the oligotrophic environment during summer and in tropical stratified water, and nutrient recycling occurs in the system and loss of material is minor.

C. Benthic Organisms

Because the continental shelf area is shallower than the open ocean, benthic organisms, which live on the sea bottom, play a significant role for the continental shelf ecosystem. However, the depth of most of the area is usually much deeper than the euphotic zone and so limited primary productivity occurs on the bottom. Thus, the benthic organisms on the continental shelf are mainly heterotrophic and sinking organic materials, from upper-water phytoplankton, coastal seaweed and sea grass, and supplied from land, are the primary source of food. The benthic organisms are important for recycling those materials and for supporting demersal fish population. Because the larvae of many benthic organisms are planktonic, they may also have significant direct effects on the biology and chemistry of the water column.

Bacteria and protozoa (including ciliates) are the most abundant and important benthic organisms, and bacteria sometimes contributes more than 50% of community respiration. The meiofauna are those organisms that pass through 0.5- to 1.0-mm sieves and consist of nematodes, harpacticoids, turbellarians, tardigrades, gastrotrichs, archiannelids, and others. Large macrofauna are retained on 0.5- to 1.0-mm sieves and can be separated into filter, deposit, and browsing feeders by their feeding behavior or into epifauna (living on the sediment) and infauna (living in the sediment) by habitat.

Most of the bottom of the continental shelf is composed of mud and sand and rarely by rock, and the benthic community distributions are closely related to the distribution of sediment particle size. The particle sizes are sorted by current speed above the bottom, and in general coarse sand is found under strong current and fine mud is distributed under weaker current. The particle size is also related to nutritional quality for benthic organisms; fine mud contains more organic matter than coarse sand. Because of the relationships between the physical regime of the water column, particle size, nutritional quality, and other factors, the distribution of the benthic community corresponds to the distribution of physical forcing in the water column. For example, the bottom community of the

tidal mixing area is different from that of a more stagnant area.

The water column hydrography is also important for the planktonic larvae of benthic organisms. For the larvae, it is necessary to settle into good habitat for their less mobile adult stage. It is believed that many organisms are evolutionarily adapted to take advantage of the current system by adjusting the timing for spawning eggs. The larvae of crabs are often concentrated in the internal wave slick, and it is hypothesized that they use the slick to return from offshore to the inshore region.

D. Fish

The continental shelf is known as good fishing ground and more than 90% of the world fish catch is supported by this 7.5% of the world ocean area. This high fish productivity is supported by the high primary production caused by enrichment by rivers, mixing, and upwelling. Many kinds of fish, not only demersal fish but also oceanic fish, spawn their eggs and the fish larvae grow on the continental shelf. The high productivity supplies enough food for both larvae and adult fish, and complex bottom topography and hydrographic structures on the continental shelf create good nursery grounds. Because oceanic fish have superb swimming ability and migrate considerable distances, the continental shelf ecosystem is not a closed ecosystem on the level of fish. [*See* FISH ECOLOGY.]

Although the continental shelf ecosystem is a good nursery ground for fish, diverse fish species have their own diverse behavior, and the behavior is not easily interpreted with relation to the complex environment on the shelf. It is known that herring larvae concentrate on the tidally mixed high-productive and high-turbulent areas on the various continental shelves; however, they are often found in these regions even when the tidal mixing area is not so productive. It is hypothesized that the tidal mixing retains the young fish by hydrodynamics, similar to being retained in an enclosed bay, so that the young fish "remember" the location of the spawning ground.

The coastal upwelling area is known as one of the best fishing grounds in the ocean. The dominant fish are clupeid, including sardines and anchovies, and most of the species appear to feed on both phytoplankton and zooplankton in different growth stages. Adult and juvenile fish can eat zooplankton and diatoms, but first feeding larvae require a high concentration of dinoflagellates and not diatom bloom. Compared to the adult and juvenile, the first feeding larvae cannot swim extensively and tend to avoid strong upwelling regions, where they may be flushed offshore. A rise in temperature and decrease in food during El Niño adversely affect the fish populations in upwelling areas, and subsequently both population size and catch decrease. However, it is still unknown whether large fish die during El Niño or escape by migrating. It seems that these fish are adapted to the large-scale environmental change because the populations survive for years. However, it is also known that the fish species in the upwelling area are radically changed following some of the large El Niño events. The ecosystem off Peru was the world largest single-species fishing ground; however, after 1972, overfishing combined with El Niño effects resulted in a dramatic decline of the anchovy stock (Fig. 4).

V. MATERIAL/ENERGY FLOW AND HUMAN IMPACTS

Because the continental shelf area is the interface between land and ocean and is one of the most productive areas on the earth, it is obvious that it is important for global material/energy flow. However, it is not easy to quantitatively define the material/energy flow in the continental shelf ecosystem, even for one continental shelf area, because of its large spatial and temporal variability. Construction of the carbon flow of a continental shelf ecosystem, as shown in Fig. 4, requires the compiling of a large number of data sets. The material/energy flow of only a few continental shelf areas has been determined, and there is considerable uncertainty even for those. Thus, the quantitative estimate of the material/energy cycle in and through the continental shelf ecosystem is still not clear. A more accurate estimate of the material/

energy flow of this variable environment is probably possible only by the combination of realistic regional numerical models with various data sets. Time series of synoptic spatial observations from satellites, high-resolution time series from mooring buoys, and high-quality observations from ships are now available.

An important, unanswered question is what role does the continental shelf play in anthropogenic carbon flux. It has been hypothesized that the anthropogenic carbon which is fixed as organic materials is exported and buried on the continental slope area as sediment and/or diffused out to the open ocean as dissolved organic matter. It was formerly believed that the high primary production on the continental shelf by diatoms was efficiently used by the pelagic zooplankton community; however, more evidence suggests that this production sinks to the sediment and is used by the benthic community and/or buried in the sediment.

The continental shelf ecosystem is affected by various human activities both directly and indirectly. It is obvious that the area is easily polluted by human waste, because the coastal area is heavily populated. Eutrophication of rivers dumps more nitrogen and phosphorus into the coastal area. The subsequent eutrophication of coastal areas often induced red tides, makes the bottom water and sediment anoxic, and causes fish kills. Nitrogen and phophorus loading shifts the phytoplankton community from the silicate-requiring diatom to non-silicate-requiring species, such as dinoflagellates. Various toxic materials are also discharged into the coastal area with significant direct effects on organisms in the continental shelf ecosystem. Water management on land may also drastically change the river water input and cause a change of density structure and hydrographic and chemical conditions in the coastal area. Unmanaged heavy fishing activity will almost certainly devastate fish populations. [See EUTROPHICATION.]

Other than these direct effects of human activities on organisms and ecosystem on the continental shelf, indirect effects may be also important. Sea-level rise, caused by global warming is one such indirect effect that will flood on large areas of low-land below and widen the continental shelf area. Global

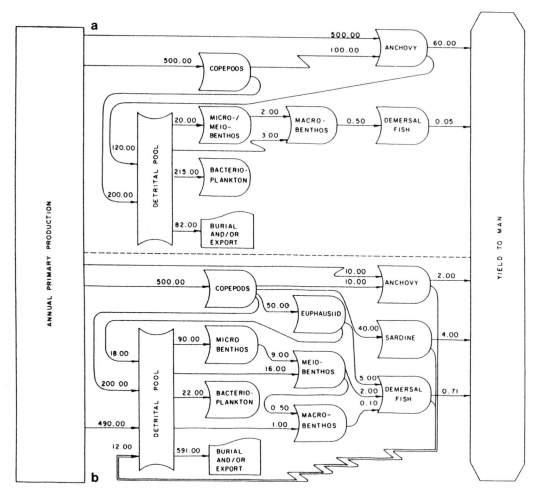

FIGURE 4 Carbon budget of the Peru upwelling ecosystems (a) before and (b) after the anchovy population collapsed in 1972. [After J. J. Walsh (1981). *Nature* **290**, 300–304. Copyright 1981 Macmillan Magazines Limited, used with permission.]

climate change may establish totally different physical and chemical forcings on the continental shelf ecosystems and result in drastic change. Furthermore, those modifications of the continental shelf ecosystems and associated biogeochemical processes may cause changes of global environment as a feedback because the ecosystems play an important role for the global environment; however, presently the feedback effects are completely unknown.

Glossary

El Niño Westerlies induce equatorial upwelling and result in cold surface water at the eastern side of the tropical ocean. Westerlies weaken over a period of several years, whereupon equatorial upwelling weakens. The warmer temperature of surface water in the eastern equatorial ocean affects the earth's climate as well as local fishing activities.

Front Interface between different water masses, where various mechanisms are present. A tidal front is the interface between the tidally mixed water and stratified water. The boundary between a river and estuarine plume and open ocean water is also a front. An upwelling front is the region where the wind-driven surface water moves downward. The front between coastal water and a western boundary current is often visible because of texture differences of the water, and the meander of the current often forms a cyclonic eddy (frontal eddy) that induces upwelling. On a large scale, the western boundary current itself is a large front between low-density water and high-density water.

Red tide Occurrence of a high concentration of phytoplankton (sometimes zooplankton) that makes the color of water visibly different from water without the plankton, where the cell number may reach $10^4–10^6$ cells ml^{-1}. A number of species may form a red tide but often it is dominated by only a few species. Red tides occur in many coastal environments and some of them are natural phenomena; however, eutrophication of coastal water by human activities often causes red tide and fish kills.

Stratification Layering of water by density, temperature, or salinity. The water column is stable when the upper water is less dense than the water below; seawater density is mainly controlled by water temperature and salinity. Water temperature can be changed by a supply of heat by solar energy, loss of heat by longwave radiation, loss or gain of heat through the sea surface by conduction, and loss or gain by evaporation and condensation. Salinity can be changed by a supply of fresh water from rivers and rain and by evaporation. Mixing is caused by wind and by interaction between current and topography. Advection of water with different density also affects localized stratification.

Upwelling An upward current of water; its opposite is downwelling. The velocity of coastal upwelling is on the order of meters per day, much lower than that of the horizontal current, which is on the order of meters per second. Coastal upwelling and equatorial upwelling occur over a large area; however, smaller-scale upwelling is observed in many coastal areas caused by local wind, current, topography, and other physical forcing. Because the temperature of deeper water is lower than that of surface water, and because deeper water contains more nutrients, an upwelling area is characterized by lower temperature and higher nutrients.

Bibliography

Blackburn, T. H., and Sorensen, J., eds. (1988). "Nitrogen Cycling in Coastal Marine Environments." New York: John Wiley & Sons.

Holligan, P. M., and Reiners, W. A. (1991). Predicting the responses of the coastal zone to global change. *Adv. Ecol. Res.* **22,** 211–255.

Holligan, P. M., and de. Boois, H., eds. (1993). "IGBP Global Change Report. 25. Land–Ocean Interactions in the Coastal Zone (LOICZ)." Stockholm: IGBP.

Iles, T. D., and Sinclair, M. (1982). Atlantic herring: Stock discreteness and abundance. *Science* **215,** 627–633.

Mann, K. H., and Lazier, J. R. N. (1991). "Dynamics of Marine Ecosystems: Biological–Physical Interactions in the Oceans." London: Blackwell Scientific.

Mantoura, R. F. C., Martin, J.-M., and Wollast, R., eds. (1991). "Ocean Margin Processes in Global Change." New York: John Wiley & Sons.

Pingree, R. D., Pugh, P. R., Holligan, P. M., and Forster, G. R. (1975). Summer phytoplankton blooms and red tides along tidal fronts in the approaches to the English Channel. *Nature* **258,** 672–677.

Postma, H., and Zijlstra, J. J., eds. (1988). "Ecosystems of the World. Vol. 27. Continental Shelves." New York: Elsevier.

Wollast, R. (1993). Interactions of carbon and nitrogen cycles in the coastal zone. *In* "Interactions of C, N, P and S Biogeochemical Cycles and Global Change" (R. Wollast, M. T. Mackenzie, and L. Chou, eds.), pp. 195–210. New York: Springer-Verlag.

Walsh, J. J. (1981). 6. Shelf–sea ecosystems. *In* "Analysis of Marine Ecosystems" (A. R. Longhurst, ed.), pp. 159–196. San Diego, Calif.: Academic Press.

Walsh, J. J. (1981). A carbon budget for overfishing off Peru. *Nature* **290,** 300–304.

Walsh, J. J. (1988). "On the Nature of Continental Shelves." San Diego, Calif.: Academic Press.

Wroblewski, J. S., and Hofmann, E. E. (1989). U.S. interdisciplinary modeling studies of coastal–offshore exchange process: Past and future. *Prog. Oceanogr.* **23,** 65–99.

Controlled Ecologies

Walter H. Adey

Smithsonian Institution

Controlled ecologies are ecosystems whose boundaries are purposely managed by human actions. Wild ecosystems are not closed; they are open to exchange of energy, materials, and genetic and environmental information. In a controlled ecology, physical boundaries are established either as enclosures of a wild system or as a constructed new system, the living components then being transferred from the wild. The exchange of energy, materials, and information is managed across the boundaries of a controlled ecosystem as a function of its intended use in a wide variety of research, education, and recreation activities. Controlled ecologies have been called experimental ecosystems, microcosms, mesocosms, macrocosms, terraria, and aquaria. Though sometimes called closed ecologies, this term is too restrictive, because very few of the systems included in this category are closed.

I. INTRODUCTION

In the context of this article, ecosystems are environmentally supported and constrained communities of organisms that utilize a primary energy source to self-organize into hierarchically arranged food webs. Such communities progressively degrade the energy that has been stored in organically synthesized chemicals, releasing that energy as low-grade heat. In this process, biogeochemical cycles of elemental building blocks are developed. Multiple species, usually large numbers, are a normal component of wild ecosystems. To varying degrees this is also true of controlled ecologies. [*See* BIOGEOCHEMICAL CYCLES.]

Some researchers have postulated that the coupling of production and consumption is crucial to ecosystem development. Certainly this is true in many small ecosystems, but the concept can only be generalized at larger scale. For example, many ecosystems export a major part of their production to other ecosystems for breakdown. Nutrients from the "consumer ecosystems" are then returned by environmental process to the "producer ecosystems." Thus, although the production-consumption linkage is crucial, it cannot be used as a central element of ecosystem definition.

Wild ecosystems are only partially bounded physically and at all scales are open to some exchange of energy, materials, and information. Though it might be argued that the earth's biosphere is a very large, closed ecosystem, the biosphere receives energy and astronomical information from outside the earth. In addition, the earth's crust and mantle are sources of energy, materials, and information, and also provide short- and long-term materials and energy sinks. Some controlled ecologies may be truly closed for experimental purposes. However, the understanding to be gained from this scenario alone is very limited, as there is no analog in the wild.

The primary energy source for most wild ecosystems is solar energy, although some ecosystems receive chemical energy supplied directly from geological processes (e.g., the communities of midocean rift systems). Many ecosystems are energetically based on photosynthetically produced organic compounds, the production of which is distant in space and time from the "consumer" ecosystem. Wild ecosystems exchange materials, typically water, oxygen, carbon dioxide, nutrient elements, and organic compounds, across their boundaries. They are also affected by external environmental factors such as heat exchange and physical energy introduction (wind, waves, currents). Ecosystems exchange organisms, including reproductive and resting phases, across their boundaries. The genetic information inherent in organism exchange, as well as external environmental and astronomical cycles, provides information that is crucial to the function of ecosystems.

Management of some or all of these factors across an ecosystem boundary for human purposes creates a controlled ecology. To provide controllable ecosystem boundaries, physical walls (glass, plastic, concrete) can be established to create enclosures. More commonly, containers are built specifically for the purpose and organisms and/or communities are transferred to the container. Microcosms are small, typically of laboratory or bench dimensions. Mesocosms are larger, of room or greenhouse size.

Macrocosms are larger than typical human construction and are formed by enclosure. The terms aquarium and terrarium are much more loosely used. The author prefers these terms for systems for which no wild analog exists or whose principal

ecosystem components, for example, environmental limitation, energy supply, hierarchical food webs, species diversity, and biogeochemical cycles, are quite different from those of wild ecosystems. Microcosmology, or alternatively synthetic ecology, is the study of controlled ecosystems, microcosms, mesocosms, macrocosms, aquaria, and terraria.

The limits of what are included in controlled ecologies are poorly demarcated. The spectrum of possibilities ranges from a wild ecosystem from which a single predator species is excluded with a cage to aquaculture systems with a single macrospecies and all energy inputs constrained to operator-added food and mixing. This author prefers a narrower range that includes only those systems in which several kinds of energy, materials, or information transfers are controlled and several hierarchical trophic levels are present. Thus, field exclusion cages for larger grazers and salmon aquaculture in enclosures are not closed ecologies by this definition. Closed Ecological Life-Support Systems (CELSS) for space travel, as currently practiced by national space agencies, are included in this article, but they only marginally meet the definition used here.

A recent extensive bibliography of the field lists 677 references that specifically refer to controlled ecologies. The number published per year is shown in Fig. 1. The earliest references to controlled ecolo-

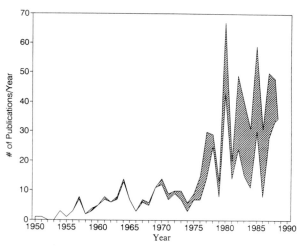

FIGURE I Publications that relate directly to the use of controlled ecologies in basic research (lower line) and applied pollution-related research (upper line).

gies appear in the later 1800s, but until the 1950s work in the field was sporadic. A steady increase in basic research on controlled ecologies occurred through the 1960s and 1970s and reached about 25 papers per year in the 1980s. The use of controlled ecologies to perform pollution research was only a very small percentage of the work in the field prior to the late 1970s. Papers on the effects of polluting compounds and elements using model ecosystems exploded in the 1980s, becoming about equal in number to those on basic research. Standardized microcosms have recently been developed for testing toxic chemicals. Space-related research on CELSS is a small part of the basic research listed.

Controlled ecosystems, as microcosms or "balanced aquaria," have been used in formal biological and ecological education at all levels. In recent years public aquaria, zoological parks, and museums have increasingly attempted to construct and maintain controlled ecosystems for public education in ecological concepts.

Considerable debate over the utility and validity of controlled ecosystems in scientific research has occurred. Where limited hierarchical species diversity was allowed and many boundary exchanges were omitted, analogies were poor and population oscillations and crashes or failures of species populations not characteristic of wild analogs occurred. However, such problems were surprisingly small in number, and success stories are increasing as systems level understanding improves and utilization of controlled ecologies is developed across a broad spectrum of scientific investigation. The application of traditional reductionist scientific methods to wild ecosystems is difficult for several reasons. Most importantly, the major characteristics of the species components (the building blocks) can be defined only in the context of a complex environment of physical and biotic factors. In addition, organic species can adapt and evolve, and symbiotic interrelationships are a basic characteristic of all life from the subcellular to the community level. In short, ecosystems are considerably more than the sum total of their parts and the environmental spectrum that surrounds them. This is also why laboratory research using entire controlled ecosystems as the subject is so important to understanding wild ecosystem and biosphere function as well as human impacts.

Some attempts have been made to establish controlled ecosystems in which all the species building blocks are known, well studied as species, and purposefully injected. These gnotobiotic microcosms, in which all unwanted species are excluded, can be successful to a limited degree. However, their inevitable lack of biotic complexity limits the applicability of most research results to the real world.

In controlled ecosystems, where adequate attention has been paid to boundary exchange parameters and a reasonably large component of species colonization has been allowed, self-organization and a pattern of ecological succession develop. This leads to a climax community and a steady state of materials processing. Some wild ecosystems have predator–prey oscillations and this is also true of some controlled ecosystems. In many wild ecosystems, upper-level predators exert a stabilizing control over lower trophic levels. Because higher predators are usually large and many controlled ecosystems are too small to have the required territory for these animals, human operators must either simulate higher predator effects or tightly manage their influence. Controlled ecosystem research has contributed extensively to a deeper understanding of ecosystem structure and function as well as to our knowledge of the effects of human changes.

II. PHYSICAL PARAMETERS AND ASSOCIATED ENGINEERING

It has become increasingly clear that human engineering (buildings, highways, canals, etc.) can drastically alter ecosystem function at the landscape and biosphere level. Thus, it should come as no surprise that because of the small size of controlled ecologies, engineering and materials design are especially crucial to modeling veracity. This facet of microcosmology has not always been appreciated even in the most sophisticated of designs.

The inner surface of the physical container or enclosure for most laboratory controlled systems has been glass or plastic (vinyl, acrylic, epoxy, and silicone). Metals tend to be strictly avoided to prevent leaching into the ecosystem. In some cases (e.g., rocky shore models), the container surface itself has been either utilized or has had little practical effect. At the other extreme, in small pelagic models that are not gnotobiotic and where the surface-to-volume ratio is high, container surface can seriously impact the veracity of a model. Effectively, an intended pelagic system becomes a permanent tide pool. Likewise, condensation on greenhouse walls can be a serious hazard for flying insects and can also become a condenser for both moisture and dust in terrestrial models. The temperature of controlled ecosystems is often maintained externally. Where internal heating or cooling probes are used, surface area and contamination problems have been an uncontrolled factor.

Solar energy is the basic driving element of most ecosystems. In enclosure controlled ecologies that are open to the atmosphere, solar energy supply may be little altered, though a need to allow for heat flow may be critical. In totally closed systems, typically with glass walls, the solar spectrum will be altered and the resultant energy delivered to the ecosystem will be significantly reduced. Quartz panels and/or supplemental artificial lighting can solve this problem. Many experimental ecosystems are operated with artificial lighting, with metal halide and VHO fluorescent lamps being most commonly employed. Daylight spectra are now available, but sufficient energy to match solar intensity can be achieved only with careful attention to lighting design. Many workers have neglected this problem to the detriment of appropriate primary productivity.

Physical energy inputs, such as wind, waves, tides, and currents, are important structuring and behavioral elements in wild ecosystems. They are also mixing and forcing factors of medium to organism exchange and organism metabolism, including respiration and photosynthesis. Though these inputs are included in the more sophisticated experimental systems, sufficient magnitude is often excluded for practical reasons. In aquatic systems, the required water movement is driven by pumps, many of which are destructive of plankton and swimming larvae, typically providing a major element of selection and environmental limitation to experimental ecosystems. Some pumps, such as Archimedes' screws, paddle wheels (Fig. 2), and to a lesser extent bellows and disc flow types, reduce organism damage. More attention needs to be paid to this aspect of controlled ecosystems, as well as to the materials used in pump construction.

III. CHEMICAL PARAMETERS AND THEIR MANAGEMENT

Model ecosystems established with soils or waters from wild analogs and given appropriate boundary material and energy exchanges, as well as multiple stocking of a broad range of species, will self-organize food webs. The net result will be the self-establishment of biogeochemical cycles as in the wild. However, the disturbance created during construction tends to alter normal population structure and chemical cycling. Biomass or storage organics will be released to bacterial metabolism, and even assuming controlled external inputs, damping of biogeochemical oscillations during start-up is desired. Traditional bacterial filtration, protein skimming, or ozonation, often used in aquarium or aquaculture situations, are self-defeating as stringent limitations to ultimate diversity and function are thereby created. One solution is to operate as an enclosure or with a connection to the wild analog during start-up and early operation, so the larger wild ecosystem becomes a temporary sink and source providing needed start-up stability. In aquatic systems, algal scrubbing (see the next section) can be used to control water chemistry and lock up metabolites in algal biomass for export or temporary storage (Fig. 3).

During experimental manipulations, metered additions of nutrients or other inorganic or organic chemicals can be used to establish desired levels. A variety of chemical or mechanical scrubbing techniques for atmospheres or soils through which gases can be passed are also applicable. In aquatic systems, where plankters are usually a factor, algal turf scrubbing or other aquatic plant technology can be effective in establishing stable levels of oxygen, hydrogen, ion concentration, and nutrients.

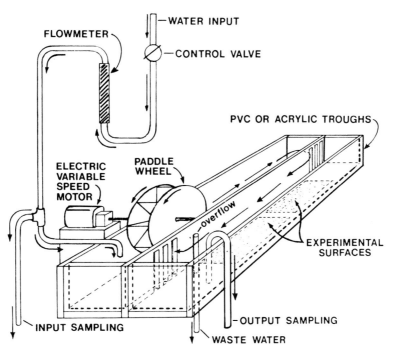

FIGURE 2 Simple freshwater stream microcosms using a paddle wheel to create current without the destructive effects of centrifugal pumps.

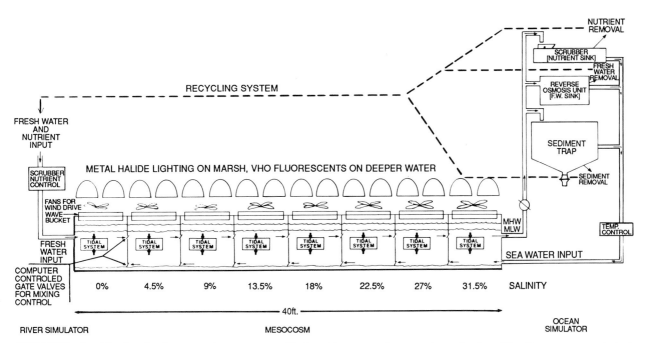

FIGURE 3 Scrubber function in an estuarine mesocosm. Algal scrubbers grow algae at optimum levels and function as ocean-equivalent nutrient sinks. The scrubbers also provide ocean-equivalent high-oxygen, high-pH water.

IV. EXTERNAL EXCHANGE AND MODELING

Manipulation and experimental alteration of internal system elements and function have been a major part of basic research using model ecosystems. However, it is manipulation of external inputs and internal exports that is most fruitful in establishing human relationships to wild ecosystems as well as in achieving steady-state models. Materials or chemical exchange across ecosystem boundaries is a defining element of all wild ecosystems. Developing appropriate mechanisms for achieving these exchanges without disturbing system function is a primary goal of ecological engineering in controlled ecologies. Examples of a variety of energy and materials input and export devices and processes are shown in Figs. 3–5.

Species movement across ecosystem boundaries is also a critical element of both function and control in microcosms and mesocosms. Such species exchange across boundaries can be thought of as materials and energy, and in some cases this exchange is crucial. However, the primary feature of organism exchange is the information carried by the collective genome of the organisms. It is this information, acquired through millions of years of natural selection in ecosystems, that establishes self-organization in food webs and the development of biogeochemical cycling.

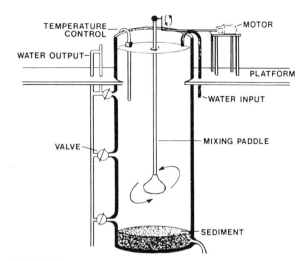

FIGURE 5 One of a series of estuarine, benthic planktonic mesocosms used for many years at the Marine Ecosystems Research Laboratory (MERL) at the University of Rhode Island. This diagram shows standard mixing, heat exchange, and external water exchange devices. To varying degrees these mesocosms use flow-through from the waters of Narragansett Bay.

V. BIOLOGICAL ELEMENTS AND THE LIMITATIONS OF SCALE

In establishing microcosms or mesocosms, species control at the boundaries results in patterns that are similar to those that occur in the wild on newly formed islands or in newly available habitats. The resulting process is a balance between immigration (or addition) and extinction (or loss), and a major element of randomness is present. The basic princi-

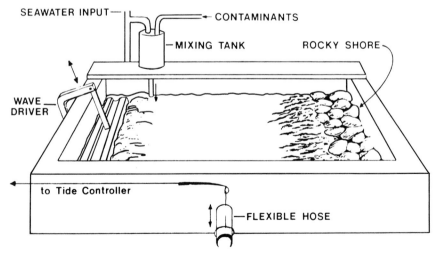

FIGURE 4 Diagram of one of the rocky shore mesocosms at Solbergstrand, Norway, showing tidal and wave energy introduction as well as seawater and polluting input.

ples of island biogeography apply, and the process has been experimentally studied many times. In general, frequent additions from a wild pool over a long period, or broadly open access to controlled ecosystems, will result in a continued increase in species diversity and system complexity to a saturation level. A model can be "supersaturated" but will then lose species when isolated to a level determined by the nature and size of the ecosystem and the veracity of its environmental parameters.

Because of the size and territorial range of some organisms, every ecosystem has its unique scaling problems. A coral reef ecosystem tends to repeat its basic elements at the scale of a few square meters. The largest plants and coral colonies fit within a few square meters, and the largest common predators often remain within a radius of a few hundred meters. A conifer forest, on the other hand, requires at least tens of square meters to a hectare, for its structuring plant elements and some of its higher predators may range over hundreds of kilometers. Fewer compromises and limitations to veracity apply to the modeling of a coral reef as compared to a conifer forest. [*See* CORAL REEF ECOSYSTEMS.]

In general, grazing animals and protists control plants and producer microorganisms, whereas predators open up spaces and increase diversity. Highly successful species are subject to predator explosions and disease epidemics, thereby giving rise to excessive system cycling. The primary scaling limitations to ecosystem modeling tend to lie in higher predators. On the other hand, a voracious protozoan predator will be limited by the size of only the smallest of models, whereas a mature sequoia will be acceptable in only the largest of macrocosms. A major element of stability from oscillation in many ecosystems whether wild or controlled must be supplied by its larger predators or in some cases by larger herbivores. In general, the effects of larger-size organisms, particularly predators, must be offset by the manipulations of the controlled ecology operator.

No generally accepted means of scaling the results of microcosm and mesocosm experiments to wild ecosystems have been devised, although a variety of methodologies have been suggested in the literature, including ratios of volumes, surface area, surface to volume, turnover times (organic and physical), and metabolism. Assuming that physical parameters are accurately provided, sufficient community seeding has been allowed, and boundary conditions are understood and applied, including management of cross-boundary organism exchanges, there is no reason to assume that functional scaling is necessary. Microcosms and mesocosms metabolize at the same levels as their wild counterparts.

VI. SELECTED EXAMPLES OF CONTROLLED ECOSYSTEMS

A 1989 review of the use of aquatic microcosms in ecotoxicological research cited over 90 sets of constructed models and over 30 enclosures since 1962. Some of these systems have been extensively used for a wide variety of basic and applied research [CEPEX (Fig. 6) in over 60 published papers, and

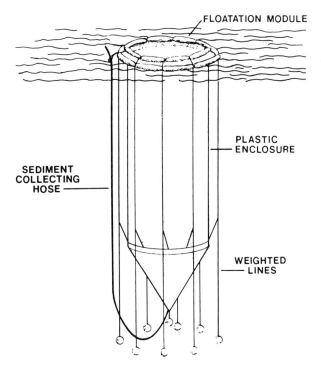

FIGURE 6 CEPEX enclosure used in British Columbia coastal waters to determine plankton responses to pollutants and food web dynamics. [Redrawn from G. Grice and M. Reeve (1982). "Marine Mesocosms," p. 8. Springer-Verlag, New York.]

MERL (Fig. 5) in over 70 published papers]. [*See* ECOTOXICOLOGY.]

A. Aquatic Systems (Freshwater, Marine, and Wetland)

One of the oldest and historically most important methods in biological oceanography, the use of small, light–dark bottles suspended at various depths in the sea and in lakes for determining the primary productivity of the plankton, is in effect a short-term ecosystem enclosure. With all the problems inherent in enclosures, the light–dark bottle technique, utilizing carbon-14, has nevertheless provided our basic understanding of planktonic productivity over the world ocean. Steady-state operation has been demonstrated in similar, but much longer term (to 20 years) materially closed but energetically open flask-size laboratory microcosms. Some of these had very few species (e.g., a lower metazoan, an alga, and microbes). Many microcosms have been developed for examining microscale functions in ecosystems.

At usually a much larger scale, and enclosed by porous or nonporous plastic bags, a number of mesocosms ranging from 1 to 1300 m^3 have been used to investigate food web dynamics, pollution effects, and larval fish recruitment (Fig. 6). Major oceanographic institutions and universities have invested in land-based tanks, up to 2500 m^3 in volume, for similar studies. Enclosures have also been used on mud flats and in terrestrial ecosystems.

A wide variety of raceways have been constructed to simulate freshwater stream ecosystems ranging from microcosm to mesocosm size. Some of these controlled ecologies have been open to natural water flow and others have been closed. In most cases, such stream models have been open to the atmosphere and to natural light (Fig. 2). They have been used extensively for pollution studies, although many of the older units were developed for basic research.

A wide variety of benthic mesocosm models have been constructed, although not in the large numbers characterizing small planktonic models. These have ranged from coral reefs, rocky shores, and mud flats to marsh systems. A 70,000-liter series of gate-connected tanks with tidal control, representing the large Chesapeake Bay estuary from fresh to coastal salinity, is shown in Fig. 3.

B. Terrestrial Systems

In terrestrial ecology, microcosms have not found the widespread utility that they have in the aquatic sciences. The scale of many terrestrial ecosystems is such that realistic models would have to be large and very expensive. Nevertheless, multiple soil-core microcosms on the order of 10 to 20 cm in diameter and 15–30 cm long have been extensively used for grassland and forest floor studies. One study using cores of an old field with soil and plant communities intact (Fig. 7) was able to demonstrate that microcosms (cores) with higher functional complexity, as demonstrated by CO_2 dynamics, were able to retain calcium better than the

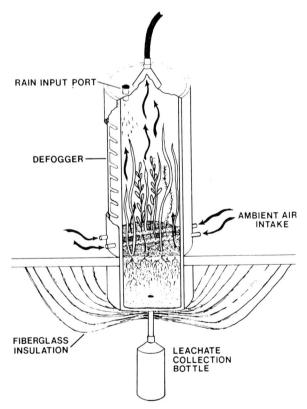

FIGURE 7 Terrestrial microcosm of an old field using soil cores. A series of these were used to demonstrate the relationship between complexity and resistance to heavy metal injection. [Redrawn from P. Van Voris *et al.* (1985). *Ecology* **61**.]

less complex microcosms when treated with the heavy metal cadmium.

C. Human Life-Support Systems (CELSS)

Both NASA and the Russian space program as well as the Biosphere II project in Arizona, U.S.A., have built ecologically oriented systems ultimately intended for human life support in space. The systems built by or for the space agencies have tended to be species-poor with a few higher plant species, a single algal species, and a spectrum of microbes. These systems only very marginally fit the definition used in this article as they have a very minimal trophic structure and species diversity, yet they have partially supported human life for up to at least six months in tests. The Biosphere II project, on the other hand, is a large mesocosm or macrocosm. Although plagued by engineering difficulties, the Biosphere II system supported eight humans in a nearly closed set of ecosystems for two years. The basic ecological difficulties in Biosphere II derive from inadequate solar energy supply due to shading by the enclosing space frame. This problem has been exacerbated by the use of synthetic soils rich in organics (and thus an extraordinary high respiration rate), as well as a failure to include a device, the equivalent of algal turf scrubbers, to manage carbon dioxide fluctuation in the relatively small volume of atmosphere. [See BIOSPHERE RESERVES.]

VII. UTILIZATION OF CONTROLLED ECOSYSTEMS

A. Basic Research

Microcosms and mesocosms have been used to test ecological principles that are difficult to isolate from interconnected factors in wild ecosystems. Basic elements of ecosystem metabolism and patterns of change relative to a wide variety of physical-chemical factors have been repeatedly demonstrated. Also, ecological succession, particularly as it relates to effects of varying patterns of species introduction, nutrient levels, and the relationship between ultimate species diversity and the regularity of environmental factors, as well as the availability of basic metabolites, have all been successfully examined. The role of subsidiary energy sources in mediating the primary energy source (photosynthesis) was repeatedly established in microcosm before its demonstration in the wild. Controlled ecologies have also been used to demonstrate the self-establishment of biogeochemical cycles and the resulting homeostasis during succession. The demonstration of biogeochemical pathways using tracers is considerably easier in microcosms and mesocosms than in the wild because of the lesser dilution and greater environmental control.

B. Pollution Research and the Standardized Microcosm

Single-species tests for the effects of environmental pollutants have become a standard approach, much like the use of laboratory rats and mice. However, correlations of laboratory and wild effects have not been good because individual species react quite differently in the laboratory as compared to the wild. This situation has led to the development of controlled mini-ecosystems for ecotoxicological testing. Extensive tests have been carried out to compare the effects of pollutants in controlled ecosystems and in wild systems, and with few minor exceptions, patterns of effect and fate were identical.

Many of the large planktonic enclosures, although used for a variety of basic and fisheries research, were also used for pollution studies. Some of the benthic mesocosms (e.g., Solbergstrad in Norway) have been used to study the effects of oil pollution. Perhaps the most extensively researched and certainly the longest-lived set of mesocosms is the MERL (Marine Ecosystems Research Laboratory) towers in Narragansett Bay, Rhode Island. Here the number of tests of fates and effects carried out have been truly extraordinary, ranging from heavy metals and fuel oils and their components to a wide variety of toxic organic compounds.

Though the numerous individual publications on this topic should be consulted, a generalization is useful. Most xenobiotic compounds or elements added to an aquatic environment in moderate concentrations cause an initial loss of some species of phytoplankton and zooplankton. The first effect is a drop in primary production and then a strong upswing, sometimes with a phytoplankton bloom due to loss of grazing zooplankters and an increased availability of nutrients. Although there is a temporary drop in diversity, tolerant species can become quite abundant. The pollutant is often taken up by plankton or adsorbed by particulates with a general tendency to sedimentation, and is thus flushed from midwater environments. The pollutant often then affects the benthic community in a similar way, although the results are longer lasting.

Standardized Aquatic Microcosms (SAM) have been developed for predicting the effects of pollutants in wild ecosystems. One such freshwater test with 10 algal species and five animals in a complex food web has a well-defined protocol for colonization, self-organization, and succession for 7 days. After the addition of test chemicals, a monitoring and testing procedure for a total of 63 days is established.

C. Simulation Systems and Computer Modeling

Many researchers using controlled ecologies feel that the strongest understanding of ecosystems can be achieved through the concurrent use of simulation modeling and microcosms or mesocosms. Simulation models are conceptual, using flows of energy or materials (carbon, nutrients) among researcher-defined compartments (grazers, predators, soils, etc.) and from the ecosystem through its boundaries. Following identification of compartments, flows, and controls on flows, mathematical equations are written based on a preliminary understanding of the unit processes involved. These equations are continuously modified as understanding improves and data are collected. Using computers, ecosystem function can be predicted with simulation models and then corroborated by comparison of results with living systems. Because

living models are usually simpler than wild systems, use of the continuum from simulation models to living models to living ecosystems maximizes progress in our understanding of these very complex systems.

Glossary

Analog ecosystem Wild example of a specific controlled ecology; a wild ecosystem against which the success of an ecosystem model, microcosm, mesocosm, or macrocosm is measured.

Biodiversity Term referring to the variability in biological systems. In an ecological context, it is often used to indicate the number of species populations in an ecosystem, although the most acceptable measures include both species number and abundance, that is, species richness.

Biogeochemical cycle Process of serial metabolic use of a nutrient (i.e., any essential chemical element) as it is taken up from the environment in inorganic form and passed through an organism or series of organisms (food web), ultimately being returned to the environment in inorganic form.

Community Collectively, all the species populations of organisms that live in a given locality; the biological units of an ecosystem without consideration of their dynamic (biogeochemical) relationships or their environment. A climax community is the steady state of species composition reached after adjustment to a perturbation.

Food web Feeding or trophic relationships of the species populations in an ecosystem. A highly idealized characterization of a food chain consisting of producers (plants)/herbivores/lower carnivores/upper carnivores/top carnivores is seldom realized. The web concept of a complex of many changing feeding relationships is a more realistic concept both in the wild and in controlled ecologies.

Hierarchical Serial ranking of control that in biological usage has typically referred to trophic structure or the "top to bottom" arrangement of species in a community. More recent usage has tended to view ecosystems as being composed of several different kinds of hierarchies. Attempts have been made to develop a core of ecological theory based on a hierarchy of spatiotemporal scale.

Homeostasis In biology, the ability of an organism to maintain an internal steady state; in ecology, an extension of this concept applies to the ecosystem or landscape level. The Gaia theory proposes that homeostasis operates at the biosphere level. Although it seems clear that homeostasis operates at all levels of ecology, a consensus on its extent has yet to develop among scientists.

Self-organization Individual organisms and species populations have an array of genetic information (the genome) that controls the development and function of the individual. The relationships among different species are also, in large measure, genetically derived. Thus, given the establishment of an appropriate collection of species, self-organization will develop trophic and other relationships that are genetically controlled.

Simulation model Simplified conceptual device, based on quantifiable elements (energy, nutrients), for which mathematical relationships of properties or parameters can be derived. Alterations of elements in a simulation model (e.g., change of solar energy input, change in nutrient concentrations) can be carried out mathematically, often with a computer, to determine the expected resulting variations in the more complex, wild analog or its intermediate controlled ecosystem.

Bibliography

Adey, W., and Loveland, K. (1991). "Dynamic Aquaria: Building Living Ecosystems." San Diego, Calif.: Academic Press.

Beyers, R., and Odum, H. T. (1993). "Ecological Microcosms." New York: John Wiley & Sons.

Folsom, C., and Hanson, J. (1986). The emergence of materially-closed system ecology. *In* "Ecosystem Theory and Application." (N. Polunin, ed.). New York: John Wiley & Sons.

Gearing, J. (1989). The role of aquatic microcosms in ecotoxicologic research as illustrated by large marine systems. *In* "Ecotoxicology: Problems and Approaches" (S. Levin, M. Kelly, and K. Kimball, eds.). New York: Springer-Verlag.

Gillett, J. (1989). The role of terrestrial microcosms and mesocosms in ecotoxicologic research. *In* "Ecotoxicology: Problems and Approaches" (S. Levin, M. Kelly, and K. Kimball, eds.), pp. 367–410. New York: Springer-Verlag.

Gitelson, J. (1992). Biological life-support systems for Mars mission. *Adv. Space Res.* **12,** 167–192.

Lalli, C., ed. (1990). "Enclosed Experimental Marine Ecosystems: A Review and Recommendations." New York: Springer-Verlag.

Santschi, P. (1985). The MERL mesocosm approach for studying sediment–water interactions and ecotoxicology. *Environ. Tech. Lett.* **6,** 335–350.

Schwartzkopf, S. (1992). Design of a controlled ecological life support system. *Bioscience* **42,** 526–535.

Coral Reef Ecosystems

Michel Pichon

Ecole Pratique des Hautes Etudes, Perpignan, France

Coral reefs are highly complex and diverse ecosystems developed in shallow water in the intertropical regions of the oceans. They are the result of the accumulation of limestone secreted by a variety of marine invertebrates, principally scleractinian corals and encrusting coralline algae. The very high species biodiversity found in reefs reflects the large variety of habitats and their biogeographical history. Many reef dwellers are primary producers, either marine plants or invertebrates living in symbiosis with unicellular algae, essentially zooxanthellae. Gross primary production at the scale of the whole reef ecosystem is often very high, but conversely net production is generally close to zero. Coral reefs occur in more than 100 countries and are exploited for subsistence and other local use. They are stressed by natural events such as cyclones, but there is growing concern fueled by evidence of large-scale degradation over the sustained effects of ever-increasing demographic pressure and human-induced impacts.

I. INTRODUCTION

A coral reef is a characteristic, constructional, physiographic feature of tropical seas, mostly established subtidally in the zone of strong illumination. It consists fundamentally of a rigid, calcareous framework made up mainly of the interlocked and encrusted skeletons of reef-building corals, calcareous red algae, and associated organisms.

Reef formations can also be produced by calcareous algae or by invertebrates other than corals, for example, by serpulid polychaetes and vermetid gastropods. In instances where reefs are not built by corals, the resulting build-up may take place in deep water or, when in shallow water, in temperate to cold parts of the world ocean. The present chapter deals exclusively with coral reefs. Coral reefs are the most actively calcifying ecosystems in the world and the resulting accumulation of limestone is enormous. These "bioherms" or "buildups" have the ability to persist throughout geological ages and they represent the most massive bioconstructed feature in the geological record.

Modern-day coral reefs are the most biodiverse of all marine ecosystems because of their topographic complexity at several scales and the variety of environmental conditions encountered. Symbiosis with unicellular algae is widespread among reef invertebrates, and the high diversity of primary producers is a characteristic of the reef ecosystem.

The global surface area covered by coral reefs has been estimated at approximately 617,000 km², a surface area almost entirely lying within the intertropical zone (30°N–30°S latitude and representing 0.17% of the world ocean area). Many developing countries and island nations, most of which are situated in the intertropical zone, consider coral reefs as a resource of considerable importance. Hence the stress placed on the ecosystem must be balanced by appropriate management policies with a conservation perspective for sustainable development strategy.

II. DISTRIBUTION OF CORAL REEFS

The foregoing definition of coral reefs clearly refers to the horizontal and vertical distributional areas: tropical seas and zone of strong illumination. These limits are determined by the ecological requirements of the fundamental reef builders (scleractinian corals) in terms of seawater temperature and solar irradiance.

Although individual species have different tolerance ranges for temperature, most reef-building scleractinian corals live in warm tropical waters, and coral reefs are present only when the surface temperature remains above 18°C, except for short periods of time. With very few exceptions, coral reefs are developed in the parts of the oceans where the surface winter temperature average is not lower than 22°C. Such maritime areas cover the "intertropical" zone and extend north approximately to 35°C latitude and south to 28°C latitude (Fig. 1). The most notable exception is that of the eastern Atlantic, where the above-mentioned temperature regime spans only from 15°N to 10°S, along the

African coast. Coral reefs are best developed when the mean annual temperature is between 25° and 29°C. The importance of their calcareous buildup and their species diversity decreases from the central, approximately equatorial, zone toward their northern and southern latitudinal limits.

Most reef-building scleractinian corals (or "hermatypic" corals) are symbiotic organisms that harbor dinoflagellates in their tissues. As a result, symbiotic (or "zooxanthellate") scleractinian corals have a vertical distribution that is at least in part controlled by the light requirements of the symbionts for a normal photosynthetic activity. Most reef-dwelling species require at least 3% of surface irradiance, but some symbiotic corals still survive at an irradiance of 1% of surface levels. In clear waters, this corresponds to a depth of approximately 100 m, and this value is taken as a global limit for hermatypic corals as a group. Below 40 m, however, there is a rapid decline in species diversity.

III. TYPES AND ORIGIN OF CORAL REEFS

A coral reef structure is essentially a calcium carbonate buildup, made up of the accumulation of coral skeletons and cemented together by other limestone-producing organisms. Live organisms are merely a thin veneer established on the surface of the reef limestone buildup. The upper part of the reef structure is a horizontal surface called the reef flat, the width of which may vary from a few meters up to 2 to 3 km. In areas without tide, the level of the reef flat is just below that of the sea surface (a few centimeters to a few decimeters) and live corals and associated reef organisms are normally submerged. In areas with a significant tide range, the reef flat and the upper slopes are normally submerged, but they may be exposed for up to several hours during low spring tides. Depending on the relationship between the reef buildup and the nearby land, one may distinguish three basic morphological categories of coral reefs: fringing reefs, barrier reefs, and atolls.

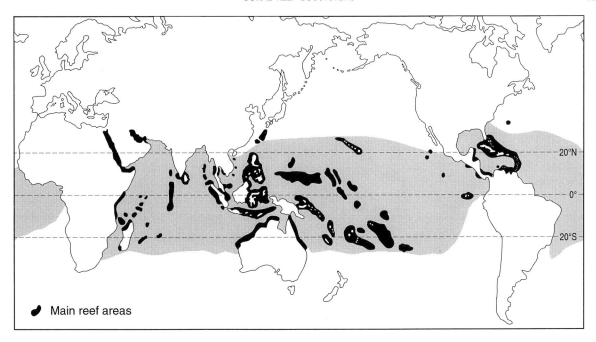

FIGURE I Geographic distribution of coral reefs. The gray-shaded area indicates zones where average surface seawater temperature is 20°C or warmer.

I. Fringing Reefs

A fringing reef is continuous with the shoreline of a landmass (including that of a continental island) (Fig. 2). In a number of instances, a narrow and shallow channel or inshore gutter may be developed between the reef flat and the shoreline.

2. Barrier Reefs

A barrier reef lies some distance offshore from a nonreefal emersed landmass, from which it is sepa-

rated by a lagoon (Fig. 3). The depth of the lagoon is generally 10 m or more and its width may vary from a few hundred meters (oceanic islands) to several tens of kilometers (continental shelves). Fringing reefs abutting on the shoreline may be established in barrier reef lagoons. Barrier reefs are generally close to the shelf margin. A number of "midshelf" reefs may develop in a position intermediate between fringing and barrier reef. The morphology of such reefs may be highly variable and include reefs with sand cays, and low

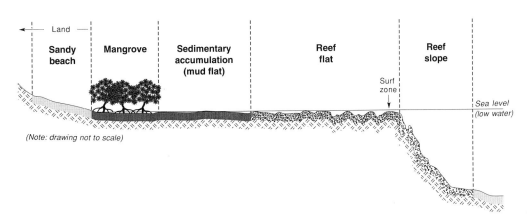

FIGURE 2 Vertical cross section of a fringing reef, with an indication of the major morphological and ecological zones.

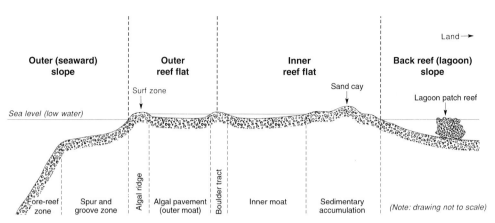

FIGURE 3 Vertical cross section of a barrier reef, with an indication of the major morphological and ecological zones.

wooded islands with development of mangrove vegetation.

3. Atolls

An atoll is an offshore reef formation, broadly annular in shape, surrounding a central lagoon, and devoid of any emersed, nonreefal land. True atolls are typically oceanic formations and do not exist on continental shelves.

Numerous intermediate stages exist between these three basic types, particularly in the central portions of the oceans, for reefs established around volcanic islands are affected by subsidence. This forms the essence of Charles Darwin's theory for the origin of oceanic coral reefs, which is as follows (Fig. 4).

Fringing reefs became established adjacent to the shores of islands produced in the ocean by volcanic activity. When such volcanic islands are affected by subsidence (0.04 mm per year for the Hawaiian Islands), reef growth (averaging up to 3 to 5 mm per year) can keep pace with the sinking of its volcanic basement, while the island decreases in size as it sinks. The fringing reef gradually becomes a barrier reef and, when the island disappears below the surface, an atoll is formed.

It should be noted that atolls display much morphological variability around the basic type: for example, they can be incomplete, tilted, submerged, or almost submerged, filled up, almost filled up, or uplifted.

IV. CORAL REEF MORPHOLOGY

A number of distinct morphological features can be observed in any reef structure. The morphological complexity of a coral reef is highly variable and de-

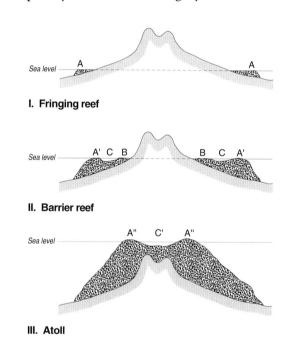

AA	-	Outer edge of the fringing reef
A'A'	-	Outer edge of the barrier reef
A"A"	-	Outer edge of the reef now forming an atoll
BB	-	Shores of the island
CC	-	Channel between the reef and the shores
C'	-	Lagoon

FIGURE 4 The origin of coral reefs and formation of an atoll, according to Charles Darwin's theory.

pends on a number of factors such as environmental parameters, geological and biogeographical history. However, most reefs display some common morphological features, which will be briefly described.

A. Outer Slope

The outer slope is the seaward, submerged part of the reef. The outer slope is often composed of two elements: spur and groove zone and fore reef slope (see Fig. 3). The spur and grove zone is composed of a succession of bioconstructed ridges, roughly perpendicular to the reef edge, and of grooves whose floor is bare rock or partly covered with biodetrital blocks or sediments. The spur and groove zone may extend horizontally seaward for up to 400 m, but its outer (deeper) part may be replaced by a succession of larger buttresses protruding seaward.

The fore reef slope may be more or less steeply sloping, varying from a moderately inclined platform to a dropoff. It is irregularly covered with various types of sessile assemblages and, when it is not too steep, with biodetrital reef sediments.

B. Reef Flat

The reef flat is the whole of the subhorizontal platform constituting the top of a coral reef structure. It may emerge during low-water spring tides. Different types of biodetrital and storm-deposited sediments may accumulate on the reef flat surface, for example, boulder tract (often in the shape of a storm rampart), sand cays, or rubble islands.

When a boulder tract is present, the part of the reef flat situated seaward of the boulder tract is generally called the outer reef flat. In areas with a high level of hydrodynamic energy, the seaward margin of the reef flat is usually represented by a constructional feature, built by encrusting calcareous algae, called the algal ridge. In barrier reefs, the inner (landward) part of the reef may be in part covered by a large accumulation of sand and gravel. When such a barrier reef is close to a large landmass, this sedimentary accumulation of the reef flat may display a dense vegetation of seagrasses. In many fringing reefs, the landward part of the reef flat is followed by a sandflat or mudflat ending, on the shoreline, in a sandy beach that may be preceded by a mangrove curtain (Fig. 2).

C. Lagoon (Inner) Slope

The lagoon slope is the submerged part of the reef sloping more or less steeply from the reef flat to the sedimentary floor of the lagoon. The lagoon slope may be composed of an apron of sediments of various types (from rubble to muddy sand), or may be a coral-built formation. Bioconstructed elements such as pinnacles, knolls, and coral patches may be developed on the lagoon slope or on the lagoon floor.

V. BIOTIC COMPONENTS OF THE CORAL REEF ECOSYSTEM

Coral reefs are the most diverse marine ecosystems. Several hundred species of algae and fishes and thousands of species of invertebrates can live in the coral reef ecosystem, which, by virtue of the high variability of within-reef environmental conditions, offers a large spectrum of different biotopes and niches. The species diversity of most groups (e.g., scleractinian corals, molluscs, crustaceans, and fishes) decreases when one moves away from an area situated in the western Pacific (covering Papua New Guinea, the eastern Indonesian Archipelago, and the southern Philippines), and in particular toward higher latitudes.

A. Algae

A large variety of multicellular algae live in the reef environment. Large fleshy algae, in particular phaeophytes (*Sargassum, Turbinaria, Cystoseira*), are mostly developed on fringing reefs, especially when they are established on the shore of a large landmass. Some of these tall algae may exhibit a clear seasonal development.

Turf algae, a heterogeneous category including juvenile stages of large species and small, mostly filamentous species, are inconspicuous, but present almost everywhere in the reef environment. Turf algae have a high level of weight-specific productivity and, because of their wide distribution, contribute significantly to the overall reef organic production. Also characteristic of reef environments are the endolithic filamentous blue-green and green algae, which lives in the skeleton of scleractinian corals and are able to photosynthesize at very low light intensities.

Calcified algae may be articulated and erect, as is the case for the green algal genus *Halimeda* or the red algal genera *Amphiroa* and *Jania,* or encrusting such as *Porolithon* and *Lithophyllum.* Calcifying algae are not only organic matter producers but also calcium carbonate producers. As such they contribute to the cementation of the reef framework, to the formation of reef rock (encrusting species), and to the production of reef sediments (articulated species).

Blue-green algae (for instance, *Calothrix, Hormothamnion, Microcoleus*) may also be found in the reef environment, where they may exhibit a seasonal development. Blue-green algae are of special significance to the functioning of the reef ecosystem on account of their ability to fix nitrogen (up to 1.8 kg N ha^{-1} day^{-1}).

B. Seagrasses

Seagrasses are only developed when reef sediments contain a high level of organic matter, a situation that is found in reefs established close to large landmasses. Up to 12–15 species can be found in a reef system, where they may form extensive meadows on the inner reef flat sedimentary accumulation. In oceanic reefs, they are very rare or absent. For instance, only one species (*Halophila ovalis*) is found on the lagoon sediment of the Tuamotu atolls. In the Caribbean, *Thalassia testudinum* is common in reef environments. Seagrasses provide a substratum and shelter for a large number of invertebrates and fishes and their net production may be very high (up to 4000 g C m^{-2} yr^{-1}).

C. Sponges

More than 300 species of sponges have been recorded in reef environments. Many of these species, however, are restricted to cavelike habitats, cavities, or crevices. The sponge fauna of reef slope caves includes calcified demosponges (which deposit both silica-as spicules and calcium carbonate) and, in the Indo-Pacific only, is characterized by the presence of pharetronids, a relict group of calcareous sponges. Large-sized non-cave-dwelling species are most abundant on the deep fore reef slope and, when calcified (e.g., *Ceratoporella*) are primary framework builders, especially in the Caribbean. On the reef flat, the sponge fauna is less abundant; however, some cup-shaped species of *Phyllospongia* are conspicuous in areas of rapidly flowing waters. Overall sponges contribute little to reef bioconstruction. Conversely, boring sponges (in particular species of *Cliona*) are active reef biodestructors and sediment producers with rates of bioerosion of up to 3 kg CaCO$_3$ m^{-2} yr^{-1}.

Many sponges contain bacteria and blue-green, or other algal symbionts. Bacteria recycle essential nutrients back to seawater, whereas blue-green symbionts play a role in carbon and nitrogen fixation.

D. Coelenterates

All major groups of benthic coelenterates are well represented in a coral reef ecosystem. These include the hydroids and hydrocorals, the octocorals (the "soft corals" or Alcyonacea, the sea fans or Gorgonacea, as well as the "blue" coral *Heliopora coerulea* and the "organ pipe" coral *Tubipora musica*), the stony (scleractinian) corals, the actinians, zoanthids, and antipatharians.

A large proportion of reef-dwelling coelenterates harbor in their gastrodermis unicellular algae approximately 10 μm in diameter called zooxanthellae. These are a symbiotic stage of the dinoflagellate genus *Symbiodinium.* Reef-building coelenterates have the ability to deposit calcium carbonate (as aragonite), which constitutes the skeleton of the animal or, in most instances, of the colony. This is the case for the hydrocorals Stylasteridae and Milleporidae, as well as for the scleractinians. Because of the massive quantities of calcium carbonate deposited, milleporids and scleractinians, but more particularly the latter, are the primary framework builders of the reef. Although the skeleton of scleractinian corals is composed of up to 99.5% calcium carbonate, it also contains small amounts of magnesium carbonate (up to 1.0%), SiO_2, and traces of $CaSO_4$ and $Ca_3P_2O_8$.

Very few nonzooxanthellate scleractinians are found in the reef environment. These species are generally restricted to the darkest environments (cavities, caves, deep fore reef slope), and with only few exceptions they are solitary or produce only small colonies.

There is a striking difference in the diversity of the scleractinian hermatypic fauna between the Atlantic and the Indo-Pacific provinces. Only 25 reef-building hermatypic genera are recorded in the reefs of the tropical Atlantic versus 86 genera in the Indo-Pacific. Only 7 genera (and probably only one species) of hermatypic scleractinians exist in both provinces.

E. Polychaetes

Polychaetes living in the reef environment are either endobiotic species living in the sedimentary deposits or endocryptobiotic species living in small cavities and crevices of the reef substratum. In addition, a small number of species belonging to the family Serpulidae are sessile on hard substrata. The distribution of species dwelling on the soft bottom is controlled by the sediment particle size distribution and organic matter content. Polychaetes are often the dominant group of the reef invertebrate cryptofauna, with several hundred species represented (belonging to the families Eunicidae, Nereidae, Cirratulidae, Terebellidae, Sabellidae) and a biomass reaching 2 g dm^{-3} (dry weight).

Most polychaetes are active at night, when they swim out of the sediment or hard substrata cavities and crevices and are preyed upon by a large number of reef fishes. They constitute, for instance, up to 82% of the food ingested by holocentrid fishes. Invertebrates may also utilize polychaetes as a primary source of food, for example, the molluscan genus *Conus*.

F. Molluscs

Molluscs, and particularly bivalves and gastropods, are one of the most important groups of reef-dwelling invertebrates because of their high species diversity and biomass. Their importance, however, is often underestimated as a result of the endobiotic life of many species. On the other hand, some species are highly conspicuous by their size (such as the giant clams, *Tridacna* spp.) or, in the case of nudibranchs, by their peculiar shapes and color patterns. Gastropods are the dominant class of molluscs on hard reef substrata (for instance, *Vasum, Turbo,* and *Trochus* on the outer reef flat), whereas bivalves are relatively more important in areas of sediment deposition.

In the atoll reefs of the central Pacific, the average density of molluscs is 4.3 individuals m^{-2}. In atoll lagoons, with little communication with the open sea, the molluscan fauna is very depauperate (only 28 species at Reao, in the Tuamotus), but the number of individuals and biomass may be extremely high.

Molluscs contribute very little to reef framework construction and cementation (with the exception of vermetid gastropods, e.g., *Dendropoma*), but remnants of molluscan shells are a major component of detrital deposits and some species are also active limestone destroyers, further contributing to reef sediment formation.

Molluscs living in the reef environment, especially gastropods, are much sought after by shell collectors. Some species belonging to *Tridacna, Arca,* and *Fasciolaria* are collected and used by humans for food. The black lip pearl shell *Pinctada margaritifera* is used in the cultured pearl industry, whereas the gastropods *Trochus* and *Cassis* were, until recently, intensively collected for mother-of-pearl and cameo handicrafts.

G. Echinoderms

All five classes of echinoderms (echinoids, holothuroids, asteroids, ophiuroids, and crinoids) are found in the reef environment in significant numbers. There is much variation in the echinoderm species composition, however, from one geographic area to the next. Few species have a restricted within-reef distribution and are therefore characteristic of a particular reef biotope. This is the case, however, for the echinoids *Heterocentrotus mamillatus, H. trigonarius,* and *Colobocentrotus* spp. and for the ophiuroids *Ophiarthrum elegans* and *Ophiomastix caryophyllata,* which live on the reef front in a high turbulence environment. The soft substrata of the inner reef flat sedimentary accumulation are often characterized by sea stars belonging to the family Oreasteridae, by echinoid genera such as *Tripneustes* and *Toxopneustes,* and by holothuroids. Crinoids and ophiuroids are more cryptic in their habit than the other classes, and they tend to show a maximum of activity at night.

Although echinoderms contribute to some extent to sediment production (through their test or spicules), they play a more important role in reef biodestruction. In the Indo-Pacific, *Echinometra mathaei* can remove up to 260 g $CaCO_3$ m^{-2} yr^{-1}, and in the Caribbean, *Diadema antillarum* up to 4600 g m^{-2} yr^{-1}. Some echinoid species are among the many important grazers in the reef environment (*Diadema setosa* and *Tripneustes gratilla* in the Indo-Pacific, *Diadema antillarum* and *Lytechinus variegatus* in the Caribbean). They graze mostly on algae, sea grasses, and sponges, but other species (e.g., *Eucidaris thouarsii*) may graze directly upon live corals and could contribute to a limitation of reef growth.

The asteroid *Acanthaster planci* (crown-of-thorns) is a predator on live corals. It may occur at times in plague numbers on reefs and is then responsible for the near-total disappearance of live corals. Some species of echinoderms such as *Tripneustes gratilla* are traditionally and locally used by humans as a source of food. Collecting of holothuroids for the preparation of trepang (beche-de-mer) may reach a commercial scale and may locally contribute to the decrease in densities of *Holothuria atra* in particular.

H. Crustacea

The high zoological diversity of the Crustacea is reflected in the reef environment. However, most groups are largely inconspicuous because of either their small size or cryptic habit. This is less so for a number of stomatopods (though some live in burrows) and decapods, which are generally of a larger size.

Among the decapods, many of the Caridea found in the reef environment (Palaemonidae, Alphaeidae, Hippolytidae) have developed commensal or symbiotic associations with other reef invertebrates (sponges, coelenterates, bivalves, nudibranchs, echinoderms, ascidians) and with fishes. Numerous instances of associations with other invertebrates are found among anomurans (notably within the families Porcellanidae and Galatheidae) and thalassinids (*Upogebia*), although the latter are mostly very effective burrowers, with some species of *Callianassa* able to process up to 3 kg (dry weight) m^{-2} day^{-1} of sediment.

Brachyurans also display a variety of habitats. Many Xanthidae, Hymenosomatidae, and Parthenopidae live in dead coral or in rubble, whereas Calappidae, Leucosidae, and many Portunidae are sand dwellers. Other Portunidae and Hapalocarcinidae are commensals.

Crustaceans play a major role in reef trophodynamics and have representatives at all trophic levels from grazers and browsers (many Xanthidae) to large-sized, very active predators such as the stomatopods (feeding on molluscs and fishes) and many Portunidae.

I. Fishes

Fishes are a particularly conspicuous group of reef dwellers by their body shape, coloration pattern, and behavior. More than 1000 species have been recorded from the reefs of the equatorial western Pacific, and species diversity decreases when moving away from this area of maximum species richness. [*See* FISH ECOLOGY.]

In the Indo-Pacific, a clear within-reef zonation of fish assemblages can be observed. The deep fore reef slope assemblage is very distinct from that of the rest of the reef structure. The latter is relatively homogeneous, although the slope communities (spur and groove zone and inner slopes) have more similarity with each other than with the inner reef flat community (in which Pomacentridae, Apogonidae, and Labridae are the dominant families).

When seagrass beds are developed on the inner reef flat sedimentary accumulation, they harbor a fish assemblage with little affinity to that of the reef structure proper and that of the deep fore reef slope. The surf zone is generally poor in species but is characterized by some Acanthuridae, Pomacentridae, and Blenniidae. Maximum species diversity is reached on the mid outer slope and on the lagoon slope. Overall, the zonation of fish communities is very uniform, in its broad lines, throughout the Indo-Pacific (Table I).

Biomass of reef fishes shows much inter- and intra-reef variability. On the Great Barrier Reef, fish biomass varies from 20 g m^{-2} on the upper slopes to 200 g m^{-2} at the transition between the reef and off reef floor.

Although it has been suggested that grazing by herbivorous fishes is very intense, herbivores do not represent more than 20% of the total number of species and 25% of the number of individuals on the outer reef flat, where their importance is at a maximum. In terms of biomass, herbivores represent less than 20% of the total reef fish biomass.

Herbivores and omnivores feed only during the day, whereas carnivores feed mostly, but not exclusively, at night. On Indian Ocean reefs, 60% of the species and 63% of the individuals feed during the day. Algae (and seagrasses) represent 27% in weight of the food ingested, sessile invertebrates 17%, mobile invertebrates 35%, and fishes 17%. Nocturnal carnivores (representing 40% of the species and 37% of the individuals) feed on fishes representing 21% in weight of the prey ingested and mobile invertebrates amounting to 79% in weight of the prey ingested.

VI. CORAL REEF ZONATION

A. General

Coral reef species are not evenly distributed across the different morphological components of the reef system, but there exist recurring groups of organisms corresponding to different areas. This succession of organismic assemblages across the reef flat and down the slopes results from the combined action of environmental factors, principally hydrodynamic energy, light, sedimentation of fine particles, exposure, and emersion, and of biotic factors, such as interspecific competition and demographic dynamics (reproduction strategies, larval ecology and settlement, growth, fecundity, longevity, mortality) of the many species.

Overall, both in the Indo-Pacific and in the Atlantic, the zonation of coral growth forms appears to be primarily controlled by the three-dimensional gradient of hydrodynamic energy down the slopes and across the reef flat and longshore. From the areas of higher hydrodynamic (swell and waves) energy (the breaker zone) to areas of lower energy, the following succession is conspicuous:

1. algal ridges;
2. *Millepora* and encrusting or short digitate species (*Acropora, Pocillopora*);
3. *Acropora* (branching);
4. massive species (Faviidae, Poritidae).

Several major types of communities are intricately distributed in the reef environment. The spatial complexity of their distribution pattern results

TABLE I
Comparison of Reef Fish Distribution at Tulear (Madagascar) and One Tree Island (Great Barrier Reef)[a]

Location	Ubiquitous species	Species limited to one biotope	Species living on the outer slope only	Species of the deep fore reef slope	Other species
One Tree Island	6.6	47.6	15.4	11.4	19.0
Tulear	8.2	44.0	14.5	6.0	27.3

[a] Values are in percentages of total number of species recorded on each reef.

largely from substratum microtopographic heterogeneity. The main types of communities that can be identified are:

- photophilic sessile communities (dominated by anthozoans and algae);
- skiophilic sessile communities (under the boulders, dead corals, and caves of the outer slope);
- endocryptofaunal communities (including the borers and mobile species living in the small cavities and crevices of the substratum);
- soft substrata communities (with and without macrophytic vegetation); and
- fish communities.

B. Indo-Pacific

In the Indo-Pacific, the broad lines of barrier reef zonation are as follows:

I. Outer Slope

a. The Spur and Groove (or Equivalent) Zone

The spur and groove zone (see Fig. 3) or its equivalent is colonized by species adapted to or tolerant of high hydrodynamic energy conditions: encrusting calcareous algae (mostly *Porolithon*), the hydrozoan *Millepora platyphylla,* and coral species of the genera *Acropora* (with stunted, encrusting, or short digitate growth forms) and *Pocillopora*.

b. The Mid Slope

Diversity is high on the mid slope and coral communities are dominated by branching *Acropora* (e.g., *A. formosa*). Deeper, on the lower part of the mid slopes, these branching species are replaced by massive corals (Faviidae and *Porites* essentially).

c. The Deep Fore Reef Slope

The deep fore reef slope is characterized by reduced light intensity. Some green algae (*Caulerpa, Halimeda*) or red algae (*Galaxaura, Mastophora*) may still be abundant. The scleractinian coral fauna is much reduced; the most conspicuous families are the Agariciidae (*Pachyseris, Leptoseris*) and the Pec-

tiniidae (*Echinophyllia, Oxypora*), with flattened, lamellar colonies followed by a few Mussidae (*Cynarina, Blastomussa, Scolymia*) and some widely distributed species. Other noteworthy invertebrates are the sponges, hydroids, alcyonaceans, gorgonaceans, and antipatharians.

2. Reef Flat

In most reefs established in high hydrodynamic energy environments, a boulder tract is developed and separates the reef flat into an outer reef flat and inner reef flat (see Fig. 3).

a. Outer Reef Flat

An algal ridge, rising up to 1 m above the average level of the reef flat, may be produced on the reef front (outer margin of the reef flat) by the genus *Porolithon*. Even when an algal ridge is not developed, *Porolithon* remains dominant. Other species characteristic of this zone are *Millepora platyphylla,* a few *Acropora* species, and *Pocillopora*. A flat "algal" pavement, the "outer moat" at low tide, is composed of encrusting calcareous algae, often covered by algal turf. This area is swept by strong currents and is mostly barren, with only a few corals (*Hydnophora, Montastrea*), gastropods (*Turbo, Vasum*), sponges (*Phyllospongia*), hydroids (*Aglaophenia*), and clumps of the algae *Caulerpa* and *Chlorodesmis.*

b. Boulder Tract

The boulder tract is an accumulation of blocks of corals that pile up and extend upward towards (and sometimes higher than) the middle of the intertidal zone. Species diversity and substratum cover are reduced, except for the lower, shaded sides of the blocks, which are colonized by sponges, bryozoans, tunicates, and serpulid polychaetes.

c. Inner Reef Flat

Coral assemblages of the inner reef flat show much regional variation but always display a high level of species diversity and substratum cover. When the reef flat is comparatively narrow and gently sloping toward the lagoon, the coral communities, at first with a mixture of species, become more and more dominated by branching to tall branching species of the genus *Acropora*. When a sedimentary accumulation has developed on the inner reef flat, the branching forms (*Acropora*) are gradually replaced by massive corals (Faviidae, *Symphyllia, Porites*), which often develop into a typical "microatoll" shape. Besides the omnipresent turf algae and *Halimeda*, algae are represented by calcareous species, often existing as free "rhodoliths" (*Lithophyllum moluccense*) and, on reefs subjected to terrestrial influence, by the large brown *Turbinaria* and *Sargassum*, often developing seasonally. On the sedimentary accumulation proper, corals are absent and seagrass beds may develop. On fringing reefs, the sedimentary accumulation, more muddy than on barrier reefs, may be partly colonized by mangroves developed in front of the sandy beach of the shoreline.

3. Lagoon Slope

The lagoon slope usually is a sedimentary slope, which may be covered with seagrasses or is dominated by coral assemblages. As hydrodynamic energy levels are lower on the lagoon slope than on the outer slope, encrusting calcareous algae are much less abundant. Scleractinian corals are mostly represented by branching *Acropora*, followed downslope by *Montipora* and massive Faviidae and Poritidae. Overall, scleractinian coral diversity is again very high, often even higher than on the inner reef flat. Sessile invertebrate fauna includes alcyonaceans (*Sarcophyton, Lobophytum*) and, in deeper water, gorgonaceans (*Ctenicella, Junceella*).

Not all the biotic zones referred to here necessarily exist in any one reef. Individual reef zonation mirrors the reef morphological structure, which itself is largely determined by the geological history and the interaction of environmental (and to a certain extent biotic) factors.

C. Atlantic

In the Atlantic, the zonation pattern of the outer slope is in general very similar to that of the Indo-Pacific, if one refers to dominant growth forms (rather than to species, which are different in the two provinces). The general pattern of zonation from the reef front downslope in Atlantic reefs is as follows:

1. An algal ridge (encrusting calcareous algae)/*Millepora alcicornis* zone, which is the equivalent of the algal ridge/*Millepora platyphylla* zone of the Indo-Pacific.
2. An *Acropora palmata*/*Diploria strigosa* zone; although no similar species exist in the Indo-Pacific, this zone corresponds to the encrusting/short digitate zone (Acroporidae/Pocilloporidae) of the Indo-Pacific.
3. An *Acropora cervicornis* zone, which is the exact equivalent of the tall branching *Acropora* assemblages of the Indo-Pacific.
4. A *Montastrea annularis*/*Montastrea cavernosa*/*Porites* zone, which is equivalent to the "massive" species zone of the Indo-Pacific.
5. An *Agaricia* spp. (+ *Helioseris*) zone on the deep fore reef slope, which corresponds to the Agariciidae (*Pachyseris, Leptoseris*) and Pectiniidae (*Echinophyllia, Oxypora*) zone of the Indo-Pacific.

Despite this remarkable similarity in the zonation, it should be noted that gorgonaceans are generally more abundant on the slopes and to a certain extent sponges on the lower slopes in the Atlantic than in the Indo-Pacific. Also reef flats are generally less well developed in Atlantic reefs. The back reef biotopes, which represent the transition between reef "crest" and lagoon floor, include a mixture of sediments (which may bear seagrasses) and coral-dominated communities (but without the components corresponding to the "high hydrodynamic energy" environment).

VII. CORAL REEF PROCESSES

A. Trophodynamics

I. Primary Production

Coral reefs are characterized by a high diversity of primary producers, covering all major taxonomic groups of marine plants (Table II). The importance of symbiosis (between unicellular or filamentous algae and invertebrates, including protozoans, molluscs, and numerous coelenterates) in primary production is also noteworthy and in coral reefs

TABLE II
Primary Production in Coral Reefs

Component	Net primary production (g C m^{-2} day^{-1})	Location
Phytoplankton		
Lagoon	0.12–0.27[a]	Takapoto
Lagoon	0.10–0.42[a]	Tahiti
Benthic diatoms		
Sand (5 m)	0.41[a]	Nosy-Be
Cyanophyceae		
Oscillatoria (lagoon)	0.62	Moorea
Calothrix and *Schizothrix* (reef flat)	1.06–1.39	Enewetak
Endolithic algae		
Ostreobium	0.4–0.6	Curaçao
Corals (with zooxanthellae)	2.63[b]	
Turf algae		
Reef flat	1.2[c]	Curaçao
On coral panels	1.9–2.8	Hawaii
Large fleshy algae		
Mixed community	4.4	Virgin Islands
Mixed community	1.9	Laccadives
Padina tenuis/Hypnea pannosa	0.1–1.5	Central Great Barrier Reef
Encrusting calcareous algae		
Porolithon	0.2–0.5	Hawaii
Chiefly *Porolithon*	0.2	Barbados
Erect calcareous algae		
Halimeda	0.1–2.3	Jamaica
Seagrasses		
Thalassia and *Syringodium*	5.8	Laccadives
Thalassia	0.4–2.7[d]	Barbados

[a] Data obtained by ^{14}C method, providing results intermediate between net and gross production, but assumed to be closer to net production.

[b] Average value expressed per square meter of coral colony surface (and not of reef surface area).

[c] Reported oxygen data converted into carbon units, assuming metabolic quotient equals unity.

[d] Production calculated from leaf parameters; dry weight converted into carbon by assuming 36% carbon content of seagrasses.

reaches an importance unequaled in other marine ecosystems. Overall, coral reef ecosystem gross productivity ranges between 1000 and 5000 g C m^{-2} yr^{-1}. Such values compare favorably with that obtained for other shallow-water tropical marine communities such as seagrass beds.

The sustained existence of flourishing coral reefs such as atolls, with a high gross production in nutrient-poor oceanic waters, has long been considered a paradox. However, corals (and other symbiotic invertebrates) have developed highly efficient internal mechanisms to recycle nutrients between the animal host and the plant symbiont and therefore require little external nutrient supply. Nitrogen fixation (mostly by blue-green algae) is also a significant source of nitrogen in particular in oceanic reefs. Other sources of nutrients include geothermal endo-upwelling and, for reefs affected by the proximity of a landmass, terrestrial runoff and groundwater (and sometimes sewage) discharge.

2. Trophic Webs

Transfer of organic matter from primary producers to higher trophic levels takes place via grazing and detritus formation. Grazing by herbivorous fishes (e.g., Acanthuridae, Scaridae) and invertebrates (echinoids, some molluscs) may account for up to almost 80% of the net primary production. Grazing pressure is therefore higher in reef environments than in nonreef systems, where only about 10% of the net primary production is consumed directly by grazers. Detritus pathways include formation of particulate detrital material, its degradation and alteration by mechanical breakdown, release of dissolved organics, and microbial decomposition. Dissolved organic carbon in reef waters is primarily autochthonous in origin and represents an additional pathway of organic matter transfer to secondary producers, either indirectly by microbial utilization or by direct utilization by a number of invertebrates.

Benthic detritus feeders are particularly abundant in the inner or back reef areas, such as sedimentary deposits or seagrass beds, downstream of the major areas of primary production dominated by algae, as well as in the small cavity network of the inner reef flat, where a large proportion of the cryptofauna is composed of detritus feeders. Suspension feeders are widespread on hard substrata both on the slopes and on the reef flat. They include a large number of coelenterates and sponges and are generally few in numbers in soft-bottom areas, except for a few bivalves (Pinnidae, Lucinidae) in seagrass beds.

Predators are abundant in most reef biotopes and among them the importance of large-sized invertebrates is conspicuous. These invertebrate predators belong mostly to the prosobranch gastropods (Conidae, Terebridae, Naticidae), asteroids, decapods, and stomatopods. Predator fishes include representatives of the families Serranidae, Platycephalidae, Synodontidae, Labridae, Lutjanidae, and Holocentridae. With the exception of the Labridae, most predators are active essentially at night. Overall, trophic webs in a reef ecosystems are extraordinarily complex. A number of organic carbon pathways are still imperfectly identified and the corresponding quantitative fluxes are poorly known. In general terms, it appears that benthic processes are dominant over planktonic processes and that there is a considerable degree of diurnal variation (besides that of photosynthetic activity).

3. Reef Ecosystem Metabolism

As mentioned earlier, gross productivity (P_g) of coral reefs is comparatively high. However, this is balanced against respiratory losses (R) of the whole ecosystem. A positive net production ($P_n = P_g - R$) indicates autotrophy, that is, the reef either increases its biomass without external input of organic matter or exports organic matter to other ecosystems. A negative net production indicates heterotrophy (the reef either decreases its biomass or inputs organic matter). Table III gives productivity values for a number of reef systems or major component communities of reef systems.

It is noted that some reef areas (e.g., algal-dominated communities or outer slopes) might be sources (net producers) of organic matter, whereas others (e.g., rubble or sand deposits) would be sinks. A seasonality in metabolic activity is also observed, the highest P_g/R values being noted in summer. Overall, the whole ecosystem P_g/R is close to unity, a necessity in the case of atolls, if they are to maintain themselves in the absence of significant organic input from the oligotrophic surrounding waters.

This indicates that there is little excess production, in the form of fisheries, that can be extracted by humans from reef systems. Although estimates of maximum sustainable yield for fishes are about 20 tons $km^{-2} yr^{-1}$, reef fish populations are unlikely to withstand commercial fishing pressures of the same order of magnitude as other ecosystems, particularly those in temperate waters.

B. Calcium Carbonate Pathways

I. Reef Framework Construction and Cementation

The calcium carbonate buildup that constitutes the geological mass of the reef is primarily the result of the calcifying activity of scleractinian corals, hydrocorals, and encrusting calcareous algae (primary frame builders). Their constructional activity is completed by that of secondary frame builders and cementers, which bind together the skeletons laid down by the primary frame builders and partly fill the voids. These include encrusting calcareous algae, sessile foraminiferans, vermetid gastropods,

TABLE III

Organic Productivity of Coral Reef Communities and Ecosystems

Location and biotope	Gross production, P_g	Respiration, R	P_g/R
	(g C m^{-2} day^{-1})		
Outer slope			
Puerto Rico	2.0	3.0	0.7
St. Croix	7.1	6.3	1.1
Reef flat			
Enewetak (algal flat)	11.6	6.0	1.9
Enewetak (coral–algal flat)	6.0	6.0	1.0
Rangiroa	2.6	2.5	1.0
Kavaratti	6.2	2.5	2.5
Tulear	19.2	10.9	1.8
Moorea (fringing reef)	7.2	8.4	0.9
Lizard Island	7.5	8.4	0.9
Patch reef			
Lizard Island	9.3	5.5	1.7
Houtman Abrolhos Islands	16.6	17.0	1.0
Lagoon			
Canton	6.0	5.9	1.0
Takapoto	4.0	3.6	1.1
Complete reef system			
One Tree Island	2.3	2.3	1.0
Lizard Island	3.2	3.2	1.0

some sessile bivalves, serpulid polychaetes, and bryozoans. The voids in the primary framework are further filled by carbonate sediment, which is in part produced by bioerosion of the reef rock itself and is in part of biodetrital origin (skeletal fragments).

2. Skeletal Growth

Growth rates of reef-dwelling scleractinian corals are dependent on environmental parameters, in particular light and temperature. Whereas growth is positively correlated with light intensity, the effects of temperature are more complex. On the average, maximum growth is observed at or around 27°C, with lower rates for temperature above or below this value. This imparts a seasonality to coral growth that to date is not fully documented. Average growth rates have been measured for several species and a selection of the results is given in Table IV. Because a variety of methodologies have been employed, the method has been specified for each measurement.

3. Sediment Production

Sediment production in the reef environment may result from a combination of two processes: mechanical breakdown and bioerosion. Mechanical breakdown may affect the bioconstructed primary framework, whether already lithified into reef rock or not, and the skeletons of reef organisms (algae, foraminiferans, corals, bryozoans, molluscs, echinoderms). The quantitative contribution of reef dwellers containing calcium carbonate structures to sedimentation is poorly known. The large foraminiferan *Amphistegina madagascariensis* can contribute up to 500 g m^{-2} yr^{-1} of skeletal debris to reef sediment, and at Glory Be Reef, the green calcareous alga *Halimeda* produces 90 g $CaCO_3$ m^{-2} yr^{-1}. Mechanical breakdown may also induce fragmentation of live corals (particularly branching species) and, in the high hydrodynamic energy areas, removal of large-sized elements of reefs (which are thrown on the reef flat or rolled down the outer slope). This mechanical process leads to the formation of boulder tracts and rubble banks on the reef flat.

TABLE IV
Growth Rates of Reef-Dwelling Scleractinian Corals

Species	Location	Growth rate ($cm\ yr^{-1}$)	Method
Massive species			
Porites lobata	Great Barrier Reef	0.4–0.9	X-radiography
Porites lutea	Enewetak	1.35	X-radiography
Porites lutea	Enewetak	0.3–1.2	X-radiography
Porites lutea	Enewetak	0.9–1.2	X-radiography
Favia speciosa	Enewetak	0.5	X-radiography
Favia speciosa	Moreton Bay	0.6	X-radiography
Favites flexuosa	Bikini	0.8	X-radiography
Montastrea annularis	Barbados	1.9	Real time
Montastrea annularis	Jamaica	0.2–0.7	Real time (alizarin)
Montastrea annularis	St. Croix	0.7–0.9	Real time (alizarin)
Montastrea annularis	Florida	0.9–1.3	Real time (alizarin)
Montastrea annularis	Florida	1.7	X-radiography
Digitate species			
Pocillopora damicornis	Hawaii	0.9–1.5	Real time (alizarin)
Pocillopora damicornis	Panama	3.9	Real time (alizarin)
Pocillopora damicornis	Samoa	2.3	Real time
Pocillopora damicornis	Great Barrier Reef	2.5	Real time
Branching species			
Acropora palmata	Curaçao	8.8	Real time
Acropora palmata	St. Croix	5.9–10.1	Real time (alizarin)
Acropora cervicornis	Barbados	14.4	Real time
Acropora cervicornis	Florida	10.0	Real time
Acropora cervicornis	St. Croix	10.0	Real time (alizarin)
Acropora pulchra	Philippines	18.1	Real time
Acropora pulchra	Carolines	22.6	Real time
Acropora formosa	Samoa	18.5	Real time
Acropora formosa	Great Barrier Reef	8.0–16.6	Real time
Acropora formosa	Phuket	8.5	Real time

Bioeroders destroy carbonate substrata by chemical action, mechanical action, or a combination of both. Although biodestruction affects principally the reef rock (reef framework already cemented and lithified), it contributes as well to the breakdown of encrusting algal deposits, large coral skeletons, and mollusc shells. Bioeroders in the reef environment are found in a large number of groups: bacteria, fungi, algae, sponges, polychaetes, sipunculids, bivalves, echinoderms, and fishes. Their quantitative importance is imperfectly documented. Selected data on bioerosion rates in the reef environment are given in Table V. Sediment produced within the reef system partly accumulates in the voids in the framework, is partly deposited (bottom of the fore reef slope, reef flat, lagoon slope, and lagoon floor), and is partly exported from the system.

4. Carbonate Balance and Reef Accretion

Deposition of carbonates (mostly by calcifying organisms) is compensated by losses, represented by export of particulate (and possibly dissolved) carbonates. The net balance between the two processes represents the reef capacity to maintain itself and for vertical accretion.

Data on net calcification rates in coral reefs are presented in Table VI. This table indicates that there exists a clear zonation of net calcification within a reef system: areas with a very high coral cover may produce more than 10 kg $CaCO_3$ m^{-2} yr^{-1}, whereas sedimentary areas (whether on the reef flats or in lagoons) may produce only 0.5 kg $CaCO_3$ m^{-2} yr^{-1} or less. It is estimated that 1–2% of the reef area calcifies at about 10 kg m^{-2} yr^{-1},

TABLE V
Rates of Bioerosion for Selected Reef Organisms

Organism	Location	Bioerosion ($g\ CaCO_3\ m^{-2}\ yr^{-1}$)
Sponges		
Cliona lampa	Bermuda	250 (up to 3000)
Molluscs		
Tridacna crocea	Great Barrier Reef	140 (max. 3150)
Echinoderms		
Echinometra mathaei	Enewetak	70–260
Echinometra lucunter	Caribbean	3900
Diadema antillarum	Caribbean	4600
Fishes		
Coral and rock browsers	Marianas	425–618
Scaridae	Caribbean	490

TABLE VI
Net Calcification Rates for Various Coral Reef Communities or Mixture of Communities[a]

Location	Biotopes or communities	Rate ($kg\ CaCO_3\ m^{-2}\ yr^{-1}$)
Houtman Abrolhos Island	Coral bank	12
Johnston Atoll	Back reef, heavy coral	9.6
Central Kaneohe Bay	Coral zone	8.8
Rangiroa Atoll	Encrusting coralline pavement	7.5
Johnston Atoll	Lagoon, heavy coral cover	6.4
One Tree Island	Reef flat, coral zone	4.6
Johnston Atoll	Coral/algal pavement	4.4
One Tree Island	Algal pavement	4.0
Enewetak Atoll	Reef flat, coral–algal community	4
Enewetak Atoll	Reef flat, algal turf	4
Lizard Island	Lagoon entrance pinnacle	3.6
Lizard Island	Lagoon reef flat	3.1[b]
Lizard Island	Seaward reef flat	2.7[b]
Kaneohe Bay	Coral algal ocean reef	2.6[b]
Enewetak Atoll	Windward fore reef	1–2
Tulear Barrier Reef	Coral algal reef flat	1.9
Johnston Atoll	Lagoon, reticulated reefs	1.5[b]
One Tree Island	Lagoon, reticulum	1.5[b]
Central Kaneohe Bay	Sand flat	1.2
Fanning Atoll	Lagoon	1
Canton Atoll	Lagoon	0.5
Bahama Banks	Carbonate shoals	0.5
One Tree Island	Open lagoon	0.5
One Tree Island	Reef flat, sand/rubble	0.4
Lizard Island	Sand/algal flats	0.3

[a] All data obtained by the alkalinity depression method.

[b] May represent a mixture of communities with intermediate and low calcification rates.

4–8% at the rate of about 4 kg, and 90–95% at the rate of about 0.8 kg. On the basis of these figures, a coral reef would produce 1.0–1.2 kg $CaCO_3$ m^{-2} yr^{-1}.

Vertical reef growth or vertical net accretion can be inferred from such a figure. With a carbonate (aragonite) density of 2.9 kg dm^{-3} and a porosity of 50%, the average vertical reef growth would be within the range 0.7–0.8 mm yr^{-1}, with values up to 7 mm for the most actively calcifying reef zones. These estimations are not inconsistent with growth rates obtained from ^{14}C data (Table VII), although net vertical accretion rates of more than 10 mm have been reported in the Caribbean.

VIII. NATURAL AND HUMAN IMPACTS

Although coral reefs are sometimes seen as an example of mature, highly diverse, and stable ecosystems, some evidence suggests that the ecosystem equilibrium may be fragile (i.e., could be easily destroyed by disturbances of minor importance).

TABLE VII

Vertical Holocene Reef Growth (Growth Rates Estimated from ^{14}C Dating)

Location	Growth (mm yr^{-1})
Atlantic	
Florida	1.4–4.8
Bermuda	1.2
Jamaica	1.2
Virgin Islands	
Fringing reef	1–12
Bank-barrier reef	1.75–8
Curaçao	1–4
Mexico (Alacran)	1.25–12
Panama (Galeta Pt.)	0.6–3.9
Pacific	
Panama (Gulf of Chiriqui-reef flats)	1.3–4.2
Enewetak	1.2
Great Barrier Reef	
Northern	1.3–14.2
Central	1.9–10.2
Indian Ocean	
Reunion Island	1–6

Throughout their geological history, coral reefs have adapted to and have been shaped by natural events such as rise and fall of sea level and fluctuations of sea surface temperature. At a different spatial and temporal scale, major changes to the structure and functioning of modern coral reefs result from natural events such as cyclones, El Niño (lower sea level and warmer sea surface temperatures in the central Pacific), outbreaks of disease-bearing pathogens causing coral or echinoid mass mortality, bleaching, and coral predator infestations (e.g., *Acanthaster planci*). In the latter case, there is still much uncertainty as to whether outbreaks are a natural phenomenon or are directly or indirectly triggered by human activities.

There is currently widespread concern about the decline in the global conditions or coral reefs, with estimations that 10% of total reef area has already been degraded beyond recovery. It is predicted that another 30% is likely to collapse within the next 10–20 years. To date, the most significant degradations of coral reefs have been through direct human impact, and reefs nearest to large populations (fringing and nearshore reefs) are deteriorating most.

Anthropogenic stress on coral reefs results largely from the fact that socioeconomic forces impart strong pressure for immediate exploitation of all possible reef resources (more so in developing countries), as well as for developmental activities that are likely to introduce disturbances in the reef proper or its environment. Competitive and destructive impacts or activities that can cause stress or damage coral reefs are numerous. They include collecting and fishing (collecting shells and corals by tourists or for commercial purposes, spear fishing, collecting reef fish for aquariums, collecting coral reef resources for local populations, commercial fishing, dynamite used for fishing and for public works, poison used for fishing), pollution (by pesticides and detergents, sediments from land, sewage and eutrophication, storm water runoff disposal, oil pollution, heated effluents from power stations, highly saline effluents from desalination plants, industrial wastes, heavy metals, radioactivity), and disturbances (nuclear effects, dredging and filling activities, land reclamation, constructional activities contiguous to the reef or on the reef/lagoon systems, such as docks, piers, runways, and resorts, recreational impacts such as scuba and snorkel activities, boating and anchor damage, introduction of alien species by accident or for economic purpose).

From both a structural and functional standpoint coral reefs are highly complex and still very imperfectly understood ecosystems. As a result, no reliable means exists to assess how much stress they can tolerate before temporary or permanent degradation starts to occur. Neither is it possible to predict in which direction the ecosystem will evolve when subjected to a given level of a particular stress. These considerations dictate that caution be exercised when dealing with human impacts on the reef ecosystem. Any type of stress and direct or indirect use of the reef resource must be understood as to its possible effects and, if not properly monitored and regulated, may be or become inconsistent with the "conservation for sustainable development" principle.

IX. MANAGEMENT OF CORAL REEFS

Management policies are aimed at a codification of a reasonable level of usage of the reef, while ensuring preservation for the benefit of future generations. As such, they represent a compromise between conservation and exploitation, and between environmental and social (economic and cultural) constraints. The various types of stress or disturbances listed earlier are not uniformly applied to the reefs from either a spatial or temporal standpoint. As a result, certain reefs may be, at certain periods, under conditions of high disturbance pressure, while others will remain totally unaffected for variable periods of time. One of the keys to management for conservation is the maintenance (or the decrease) to an acceptable level of the anthropogenic pressure on the reef ecosystem.

To achieve this goal, management policies include both qualitative and quantitative aspects. The qualitative approach consists of defining the spatial and temporal distribution of the different types of usage (or of activities, disturbances, or stress). This partitioning in time and space of the different types of usage forms the basis of zoning plans whereby specified types of human activities are allowed on certain reefs at certain times. The quantitative approach imposes limitations on the *level* of usage permitted (which is equivalent to the amount of stress allowed) for a given type of activity. Such limitations or constraints apply to:

- yield for living resources subjected to any extractive activity (which in turn may be achieved by limitations placed on boat size and numbers and type of fishing gear, and by the declaration of closure periods);
- levels of all types of pollutants;
- direct impacts resulting from human presence (e.g., limitations of the numbers of waders, snorkelers, divers, boats, and resorts allowed in the area).

The effectiveness of management policies depends on the existence of a legal basis, which in many instances is lagging behind the increasing pressures being applied to reef systems. In many countries, for instance, island nations, the legal basis of management policies has to take into account the special circumstances presented by the traditional rights (in terms of ownership and fishing) of the local populations. Reef management is developed on the basis of predictions or assumptions that, if human users and impacts are maintained or controlled within certain limits, the condition of the managed reef area will be maintained or restored. It is therefore necessary to undertake regular reviews of management policies and practices to assess whether the initial goals and objectives have been achieved. These reviews also provide the opportunity to introduce new policies resulting from changes made in already existing reef usage and to take account of the ever-increasing types of damaging activities resulting from increased development and population growth.

Glossary

Bioherm Fossil reef of biological origin embedded in sedimentary layers of a different lithological nature; the bioherm can be surrounded by an accumulation of biodetrital sediments.

Bleaching Loss of color of corals resulting from the expulsion of the brownish zooxanthellae, leaving the white skeleton of the coral apparent through the translucent tissue. Expulsion of zooxanthellae results from a disruption of the symbiosis, under a variety of stresses such as freshwater runoff, high light intensities including U.V., and higher than normal temperatures. Bleaching events have been recorded mostly since the beginning of the 1980s and may affect large geographic areas or be limited to one reef. If the stress is neither strong nor prolonged, coral may recover over periods of a few weeks to a few months.

Endocryptofauna Small-sized fauna that live in the limestone substrate. It comprises two components: the true borers, also referred to as bioeroders or lithophagic species, and the mobile, opportunistic fauna, including mostly polychaete worms and small crustaceans, which utilize the small cavities, holes, and crevices produced by boring species.

Hermatypic Significantly contributing to the framework of reefs. All hermatypic forms are constructional. The word has often been used to denote the possession of zooxanthellae, however, invertebrates that harbor such symbionts should more properly be called "zooxanthellate."

Skiophilic Shade-loving; includes organisms that are most often sessile and that live in all areas experiencing reduced light intensity, such as dark cavities, cracks, crevices, caves, and the deep fore reef slope.

Zooxanthellate Harboring zooxanthellae. The genus name *Zooxanthella* originally referred to yellow algal cells found in some radiolarians, hydrozoans, and actinians. Subsequently, it was used to designate yellow algal symbionts of diverse taxonomic position. It is now commonly applied to dinoflagellates living in symbiosis with marine invertebrates. The dinoflagellates living in symbiosis with scleractinian corals belong to the genus *Symbiodinium*.

Bibliography

Barnes, D. J., ed. (1983). "Perspectives on Coral Reefs." Canberra, Australia: Brian Clouston.

Dubinsky, Z., ed. (1990). "Coral Reefs," Vol. 25, Ecosystems of the World. Amsterdam: Elsevier.

Fagerstrom, J. A. (1987). "The Evolution of Reef Communities." Chichester, England: John Wiley & Sons.

Guilcher, A. (1988). "Coral Reef Geomorphology." Chichester, England: John Wiley & Sons.

Salvat, B., ed. (1987). "Human Impacts on Coral Reefs. Facts and Recommendations." Papetoai, Moorea, French Polynesia: Antenne Museum E.P.H.E.

Salvat, B. (1992). Coral reefs—A challenging ecosystem for human societies. *Global Environ. Change,* 12–18.

Sheppard, C.R.C. (1983). "A Natural History of Coral Reefs." Poole, Dorset, England: Blandford Press.

Deciduous Forests

James R. Runkle
Wright State Univesity

I. Temperate Zone Deciduous Forests
II. Tropical Deciduous Forests

Deciduous forests are the tree-dominated vegetation type in which most woody taxa shed their leaves in synchrony for part of the year in response to a regularly occurring climatic pattern. Deciduous forests can be divided into two main types: temperate zone deciduous forests, in which leaves are shed during a period of cold which includes some below-freezing temperatures; and tropical deciduous forests, in which leaves are shed during a period of drought.

I. TEMPERATE ZONE DECIDUOUS FORESTS

The deciduous forests of the temperate zone occupy a small section of Chile and large sections of eastern North America, western and central Europe, and eastern Asia. Altogether they occupy approximately 7.0 million km^2, which is equivalent to 1.4% of the total Earth's surface (including water bodies) or 4.7% of the total land area on Earth.

These forests show perhaps the greatest seasonal changes of any vegetation type. During winter, almost all plants are leafless and dormant. During spring, plants come into leaf, with the ground layer leafing out before the canopy. Many perennial herb species which survived the winter as bulbs or rhizomes flower during the month or so before the canopy leaves reduce light levels at the forest floor.

During summer, canopy leaves are dense and little light penetrates to the forest floor (2–5% of full sunlight or even less). The dominant herbs during this time have emerged from buds which survived the winter immediately below the ground surface. Understory species survive the low light levels during the summer by growing in lighter sections of the forest, by reducing respiration and growth to very low levels, by living off energy reserves which had accumulated before the canopy leaves emerged, or by dying back to bulbs or rhizomes. An important source of solar energy for summergreen herbs are sun flecks, where single rays of sunshine have penetrated through the canopy. Some herbs grow slowly due to competition with trees for soil moisture rather than because of competition for light. In autumn, canopy leaves turn color when the dominant summer pigments degrade and then fall. The ground layer may show a second period of increased growth and flowering. [*See* FOREST CANOPIES.]

The relative length of these four seasons (winter, spring, summer, autumn) varies throughout the range of the deciduous forest. For example, in North America the length of the growing season is about 120 days in the north and 250 days in the south.

Relative to other vegetation types, deciduous forests occur in a climate with a warm growing season of 4–7 months containing adequate rainfall and a mild winter lasting 3–4 months. They usually do not occur immediately adjacent to the coast or in the dry interiors of continents. A typical range of precipitation for eastern North America is 125 cm near the Atlantic coast and 85 cm along the grassland boundary of the west, as well as 75 cm in the north, near the Great Lakes, and 150 cm near the Gulf coast. Higher values may be obtained for special topographic features, for example, 200 cm in part of the southern Appalachian mountains. Temperatures vary from −30°C to +40°C.

A deciduous forest is multilayered with the total area of leaves usually 4–8 times the total area of the ground surface. Often the leaves will consist of two or more layers from trees, one from shrubs and one or more from herbs. A mossy ground layer is lacking, though, since it would be covered by the falling leaves.

The normal range of net primary productivity for temperate deciduous forests is 600–2500 g organic matter per square meter of surface area, with an average value of 1200 g/m². This forest type accounts for 4.9% of the world's total annual net primary productivity. The normal range for biomass in deciduous forests is 6–60 kg/m² with an average value of 30 kg/m². This forest type accounts for 11.4% of the world's total biomass.

Although many species have restricted ranges, the forests as a whole have a floristic unity based on several wide-ranging tree genera, common to all three continents of the northern hemisphere. These genera include maple (*Acer*), oak (*Quercus*), beech (*Fagus*), ash (*Fraxinus*), elm (*Ulmus*), and cherry (*Prunus*). Many other tree, shrub, and herb genera also have wide distributions. Some genera, e.g., hickory (*Carya*) and magnolia (*Magnolia*), are found widely distributed in Asia and North America but are absent from Europe. Such distributions are explained in part by the impact of the Pleistocene ice age: because of the orientation of mountain chains, more genera became extinct from Europe than from the other deciduous forest regions.

Compared to the distribution of plant growth forms worldwide, temperate zone deciduous forests have fewer tree, shrub, and epiphyte species and more perennial herbs. These trends occur because of the severity of winter conditions on aboveground plant parts: the species composition of the rest of the world is dominated by the trophics, for which a very high fraction of species are tall plants or epiphytes.

Many animal groups also are common to much of the north temperate zone deciduous forest. Examples include deer, elk, squirrels, mice, lynx, fox, wolves, bears, and many specific groups of birds. The distribution of large animals in particular has been modified by human activity, however.

Trees in the deciduous forest die from several factors. Some die from fungi, insects, or other biological entities while others die of climatic extremes, such as drought, severe cold, ice storms, strong winds, hurricanes, tornadoes, and lightning. Some die of fire, although fire frequency varies greatly from forest to forest. How trees die influences the physiological features of the tree species which dominate an area. If large groups of trees die at the same time, they tend to be replaced by species, termed shade intolerant or understory intolerant, which can grow rapidly in full sunlight. In contrast, if trees die singly then they tend to be replaced by species, termed shade tolerant or understory tolerant, which can persist in more shaded conditions. [*See* Forest Pathology.]

Humans have been modifying the deciduous forest for several thousand years. Although North America's deciduous forest was the one most intact up until a few hundred years ago, native Americans had had a substantial effect on it. Then, with settlement from other parts of the world, the forest was cleared rapidly. For example, Ohio went from almost totally forested in 1800 to 17% forested by 1884. Today, the original forest exists as scattered remnants on private property and parks. Land once cleared for forest has been abandoned from other uses and allowed to return to forest in many places. Humans continue to have a major influence on the forest. Current concerns include air pollution, such as acid precipitation and global warming, and introduced insect pests and diseases. [*See* Forest Stand Regeneration, Natural and Artificial.]

II. TROPICAL DECIDUOUS FORESTS

Tropical deciduous forests are sometimes called tropical seasonal forests and are found in Central America and South America around the edges of the tropical rain forest, in west and central Africa between the tropical rain forest and dry woodlands, in much of India and central southeast Asia, and parts of Australia. Altogether they occupy approximately 7.5 million km^2, which is equivalent to 1.5% of the total Earth's surface (including water bodies) or 5.0% of the total land area.

Their distribution is difficult to define exactly because they intergrade gradually with the tropical rain forest on the wet side and savanna woodlands on the dry side. As rainfall decreases and as seasonal droughts become more pronounced, forests change from evergreen to semievergreen to deciduous. More and more trees become deciduous until the general appearance of the forest changes. The magnitude and duration of the leafless phase varies with precipitation and habitat.

Other aspects of the tropical deciduous forest change during the year. Many tree species flower during the dry season. A dense understory of grass, herbs, and shrubs may develop because the canopy is open at some times of the year. Animal migrations take advantage of changes, moving in and out of the forest according to their needs for flowers, fruit, ground vegetation, or canopy leaves.

The normal range of net primary productivity for tropical seasonal forests is 1000–2500 g organic matter per square meter of surface area, with an average value of 1600 g/m^2. This forest type accounts for 7.1% of the world's total annual net primary productivity. The normal range for biomass in these forests is 6–60 kg/m^2 with an average value of 35 kg/m^2 This forest type accounts for 14.1% of the world's total biomass.

Natural sources of tree mortality may include fire, wind, and landslides. Sometimes trees die singly and sometimes large openings are created when many neighboring trees die at once, e.g., in a hurricane.

Humans have used this forest type extensively because it is productive but accessible for the dry part of the year. Clearance for timber and overgrazing have affected most of these forests, causing them to disappear or become degraded at a rate similar to and sometimes surpassing that of the tropical rainforest.

Glossary

Biomass Weight of living material, including the dead parts of living organisms, such as most of the wood in a living tree.

Epiphyte Plant attached to another plant and supported by it.

Net primary productivity Rate at which energy is accumulated by plants in photosynthesis. The difference between gross primary productivity (total energy fixed by photosynthesis) and respiration (energy used by plants in their metabolism).

Shade tolerance or understory tolerance Relative capacity of a forest plant to survive and thrive in the understory.

Bibliography

Braun, E. L. (1950). "Deciduous Forests of Eastern North America." New York: Hafner.

Ellenberg, H. H. (1987). "Vegetation Ecology of Central Europe." Cambridge: Cambridge Univ. Press. [English translation]

Leigh, E. G., Rand, A. S., and Windsor, D. M. (1982). "The Ecology of a Neotropical Forest: Seasonal Rhythms and Longer-Term Changes." Washington, D.C.: Smithsonian Press.

Miyawaki, A., Iwatsuki, K., and Grandtner, M. M., eds. (1994). "Vegetation in Eastern North America: Vegetation System and Dynamics under Human Activity in the Eastern North American Cultural Region in Comparison with Japan." Tokyo: University of Tokyo Press.

Peterken, G. (1993). "Woodland Conservation and Management." New York: Chapman and Hall.

Rohrig, E., and Ulrich, B. (1991). "Temperate Deciduous Forests," Vol. 7. Amsterdam: Elsevier.

Spurr, S. H., and Barnes, B. V. (1980). "Forest Ecology," 3rd Ed. New York: Ronald Press.

Whitney, G. G. (1993). "From Coastal Wilderness to Fruited Plain." Cambridge, Cambridge Univ. Press.

Deforestation

R. A. Houghton
The Woods Hole Research Center

I. Introduction
II. Rates of Deforestation
III. Local and Regional Effects of Deforestation
IV. Global Effects of Deforestation
V. The Future

I. INTRODUCTION

Deforestation refers to the replacement of forest with nonforest. The process may occur abruptly, when forests are cleared for agricultural land, or more gradually, as a result of unsustainable logging practices. Deforestation began before the development of settled agriculture some 10,000 years ago, but the rate of deforestation today (estimated to be 15 million ha/year) is greater than ever before. Most deforestation is currently occurring in the tropics. The consequences include local and regional effects, such as erosion of soils, siltation of reservoirs, reduced rainfall, elevated temperatures, and increased frequency and severity of floods. The global effects include loss of species and emissions of heat-trapping gases to the atmosphere. Emissions of these gases from deforestation and from the uses of land following deforestation account for 20–25% of the global warming calculated to result from all anthropogenic emissions.

Most deforestation occurs as a by-product of increasing agricultural area: croplands, pastures, or shifting cultivation. Although the fallow period in traditional shifting cultivation allows sufficient time for forests to begin to grow back, these fallow forests are usually cleared again for crop production before the forests have regained their original stature.

Generally, the harvest of wood from forests is not considered deforestation. Under most circumstances forests can be renewably harvested and yet remain as forests. Although a recently logged forest may be temporarily without trees, if the land is managed appropriately, a forest will return. However, there are some forms of logging that amount to deforestation. If harvest is carried out without regard to the sensitivity of the site, or if repeated harvests remove wood more rapidly than growth replenishes it, the net effect of the harvest is equivalent to deforestation. The distinction between logging and deforestation is a fuzzy one, but can be defined such that if a forest has not reestablished after a specified length of time following logging, the land may be called deforested.

Sometimes repeated logging leads to a gradual reduction in the size of the trees, a process defined here as degradation. Although degradation is not strictly deforestation, it may eventually lead to deforestation and has some of the same environmental consequences. Forest degradation is widespread in the tropics as a result of selective logging. Selective logging sounds benign, but the harvest of only a few trees from an area commonly damages and kills about one-third of the remaining trees. Furthermore, although selectively logged forests would recover if allowed to, rates of harvest often exceed rates of regrowth, and the net effect is a reduction in the size of trees. This phenomenon has been observed throughout the tropics, causing the Food and Agriculture Organization (FAO) of the United Nations to list forest degradation as causing a greater reduction in total growing stocks than outright deforestation.

Like deforestation, forest degradation is not new. In parts of the world with long histories of human habitation at relatively high densities, such as in Africa, Asia, and parts of Europe, degradation began long ago. As much as 50% of tropical Asian forests is estimated to have been degraded by 1850, and most of them are degraded at present. That is, the amount of wood per unit of area of forest is less than it would be in the absence of humans.

Degradation of forests may also occur because of air pollution, acid precipitation, or other environmental changes less obvious than logging or clearing. Forest decline in the northeastern United States and Waldsterben in central and eastern Europe are the most well-studied examples of this degradation, although the exact mechanisms of dieback or decline remain to be determined.

In the tropics deforestation and degradation of forests have a number of environmentally deleterious local and regional effects, such as erosion of soils, reduced rainfall, elevated temperatures, reduced capacity of soils to hold water, increased frequency and severity of floods, and siltation of streams, rivers, and dams. In addition, deforestation results in a loss of fuel, shelter, and other resources for local inhabitants. These effects are particularly troublesome for the people of developing nations, a great number of whom depend directly on the land for their survival. The major cause of deforestation, of course, is the need for new agricultural land and greater production of food, both for local consumption and export. One of the tragedies of tropical deforestation is that only about half of the land deforested each year for agricultural purposes remains in agricultural production. The other half is abandoned from production after a few years, having lost its fertility. In many

locations abandoned land may return to forest, but the worldwide increase in degraded lands shows that such reforestation is generally not taking place. Instead, these abandoned agricultural lands often remain degraded, eroded, or waterlogged, with an impoverished low-statured vegetation of woody shrubs. The net effect of deforestation, with subsequent overuse of the land, is a reduction in the capacity of the earth to support human populations.

The effects of deforestation are also global. They include not only the conversion of potentially productive land to land with diminished capacity to support crops, forests, or people, but also the irreplaceable loss of species, and emissions to the atmosphere of heat-trapping trace gases, such as carbon dioxide (CO_2), methane (CH_4), and nitrous oxide (N_2O).

II. RATES OF DEFORESTATION

A. Temperate and Boreal Forests

The highest rates of degradation and deforestation are presently in tropical forests, but in the past major changes occurred in forests of the temperate and boreal zones, and some changes continue. Between 1950 and 1980 there were small increases in the areas of forest in the temperate regions of some countries and small decreases in others, but the total net change throughout these regions was close to zero. Since 1980 rates of clearing through logging in the northwestern United States, western Canada, and the former Soviet Union appear to have increased, while forests in eastern North America and Europe are generally expanding.

In addition to changes in the area of forests, changes are occurring in the stature of forests, or the amount of carbon held in trees and soils. In much of North America and Europe forests are increasing in stature as they recover from earlier logging and from agricultural abandonment more than a century ago. In some areas increased deposition of nitrogen and sulfur from industrial emissions may have enhanced tree growth and stature, at least temporarily. On the other hand, in areas adjacent to heavy industrial activity and in areas affected by air pollution, forest decline has reduced the stature of forests. The frequency of fires in Canada and perhaps elsewhere also seems to have increased during the 1980s, reducing the area and stature of forests there. Whether the change is related to the warming of the 1980s and whether it will continue in the future are, of course, unknown at present. [See FOREST STAND REGENERATION, NATURAL AND ARTIFICIAL.]

B. Tropical Forests

According to the FAO, about 15 million ha of tropical forests were cleared each year during the 1980s, mostly for new agricultural lands. Only 10 years ago the rate of tropical deforestation was estimated to have been about 11 million ha/year (Table I). The rate of deforestation in the tropics has been a topic of considerable uncertainty and controversy. The recent estimate by the FAO is ~50% higher than their estimate for the late 1970s. Part of the increase is real; part of it the FAO attributes to their having underestimated rates in the earlier period. Nevertheless, the FAO acknowledges that the rates have accelerated throughout most of the tropics, although in some Asian countries and Brazil the rates may have declined.

Estimates of deforestation in Brazil have been by far the most variable. Recent estimates range from 8 million ha/year in 1987 to an average of 1.5 million ha/year for the 1980s. The high estimate was based on the number of fires observed by satellite during the dry season of 1987. The spatial resolution of the satellite imagery (Advanced Very High Resolution Radiometer) was coarse, ~1 km, and small fires appeared to occupy an entire 1 km × 1 km square, when in fact they

may have covered only a small fraction of that area. Recent estimates from the Brazilian Space Agency describe a rate that has declined from 1.88 million ha in 1988–1989 to 1.38 million ha in 1989–1990 to 1.11 in 1991. These recent estimates for Brazil are thought to be among the best estimates of deforestation because they are based on high-resolution (80-m) Landsat data. Although the use of satellite data is not without difficulties, the method offers an objective and repeatable approach to the measurement of deforestation. The uncertainty of deforestation rates in other regions of the tropics is surprisingly large given the importance of deforestation to climatic change (see below) and the demonstrated ability of existing satellite data to measure the change in forest area. Systematic measurements of deforestation over major regions of the tropics using satellite data have only recently begun.

III. LOCAL AND REGIONAL EFFECTS OF DEFORESTATION

A. Soils and Food Production

Deforestation in the tropics increases the area of both productive and unproductive land. The statement seems contradicted by the fact that most deforestation is for agricultural land. However, despite this fact, the annual net expansion of agricultural lands is considerably less than the annual net reduction in forest area. For the entire tropics the expansion of croplands between 1980 and 1985 accounted for only 27% of total deforestation. The increase in pasture area accounted for an additional 18% of the area deforested. Fully 55% of the deforestation in this 5-year period was explained by an increase in the FAO category "other land." Although some of this other land is urban land, roads, and other settled lands, these uses are unlikely to have accounted for more than a few percent of the area deforested. Most of the other land seems likely to be abandoned degraded croplands and pastures, lands that no longer support crop or livestock production but that do not revert readily to forest either.

Forests are not converted directly to degraded areas, of course. The transformation of land is from

TABLE I
Rates of Tropical Deforestation (10^6 ha/yr)[a]

Dates	Tropical America	Tropical Africa	Tropical Asia	Total for all tropics
1976–1980	5611	3676	2016	11,303
1981–1990	7400	4100	3900	15,400

[a] Data are from the Food and Agriculture Organization.

forest to agriculture and, subsequently, from agriculture to degraded land. The important point is that only about one-half of the area of tropical forest lost each year actually expands the area in productive agriculture. The other half is only temporarily useful. After a few years it is lost from production, no longer either agriculturally productive or forested.

The fraction of deforestation used to expand the area in agriculture, as opposed to replacing worn-out land, varies among tropical regions. In Africa the expansion of croplands accounted for only ~12% of the net area deforested. Eighty-eight percent of the lost forest area became other land. In tropical Asia only 40% of the net reduction in forests appeared as an expansion of agricultural lands. In Latin America about two-thirds of the reduction in forests could be accounted for by the expansion of croplands and pastures.

Current land use practices are thus eliminating both agricultural land and forests. If agriculture could be made sustainable, rates of deforestation could be halved without reducing the expansion of agricultural lands, and large areas of marginal or degraded lands might be reforested (see Section V).

The loss of fertility with cultivation is not universal. Some tropical soils are rich and continue to support agriculture indefinitely. Many more soils could do so with proper management. Many tropical soils are less fertile than the soils of the glaciated temperate zone because they are older. Furthermore, high rates of precipitation in the humid tropics have weathered the soils and leached out important nutrients. In many tropical forests essential nutrients are in the vegetation rather than in the soil. This is one of the reasons that shifting cultivation works well. The vegetation is burned, leaving nutrients in the ash. These nutrients contribute to the production of food, but only for a few years, after which the land must be left fallow to accumulate nutrients again in vegetation. With a long enough fallow period shifting cultivation is sustainable. Since the 1970s, however, growing numbers of people needing food have increased the pressure on forests and shortened fallow periods. Shifting cultivation is becoming nonsustainable. The greatest increase in degraded lands is in regions where

shifting cultivation is growing most rapidly. [See SHIFTING CULTIVATION.]

B. Climate

Deforestation affects local and regional climates through direct modification of energy and water budgets. The evidence comes from both direct observation and simulations with general circulation models. In general, forests are darker than the grasslands, crops, or bare soils that replace them, and thus forests absorb more of the sun's radiation (or reflect loss). Because the albedo (i.e., reflection) of forests is less, deforestation should lead to a local cooling. The effect is particularly important in boreal forests, where clearing and logging increase the fraction of the land surface covered with snow. The higher albedo of the snow can be expected to cool the region, perhaps preventing the reestablishment of the forest where forests are already near their northern (i.e., temperature) limit.

By the same reasoning, deforestation in the tropics should also decrease surface temperatures, because grasslands are better reflectors of the sun's energy (i.e., they have a higher albedo than forests). Research has demonstrated, however, that the replacement of tropical forests with grassland increases local air and soil temperature, decreases evapotranspiration, and decreases precipitation. The warming from a reduction in evapotranspiration more than compensates for the cooling from the increased albedo. Temperature is increased because evapotranspiration, which is a cooling process, is reduced by deforestation. Evapotranspiration is reduced, in part because less radiant energy is absorbed (more is reflected), in part because atmospheric turbulence (surface roughness) is greater above a forest than above a grassland and hence can evaporate water more rapidly from forests, and in part because the roots of trees generally penetrate to deeper layers of soil than the roots of pastures, and hence have access to more water. If less water is available, the vegetation will respond by closing stomata, thereby increasing the resistance to evapotranspiration and increasing temperatures.

The decrease in evapotranspiration as a result of deforestation has further consequences for the local and regional climate. Evapotranspiration not only cools the surface; it also provides moisture for the atmosphere. If the supply of atmospheric moisture is reduced, precipitation must also be reduced either locally or downwind. In the Amazon basin 50–75% of the precipitation is recycled within the basin. The reduction in precipitation in turn reduces evapotranspiration further and increases temperature because less water is available for the plants.

Increased temperatures result not only from a reduction in evapotranspiration but also from an increase in the hours of sunshine. Less evaporation means fewer clouds and more sunshine. Up to 40% of the increase in annual temperature associated with deforestation may result from an increase in hours of direct sunlight. The other 60% is from reduced evapotranspiration, that is, more energy leaving as sensible heat than as latent heat.

In simulations using mathematical models, when tropical forests of the Amazon were replaced with degraded grasslands (pastures), mean surface temperature increased by ~2.5°C, evapotranspiration was reduced by ~30%, and precipitation was reduced by ~25%. The changes were larger during the dry seasons of the year. In the southern part of the Amazon, the dry season was lengthened. These changes are generally larger that those predicted from an enhanced greenhouse effect (see below). A complete and rapid deforestation of Amazonia might well change the southern part of the region irreversibly. Reductions in rainfall could be large enough that forests, once removed, might not be reestablished. [See MODELING FOREST GROWTH.]

Might such irreversible changes also occur in tropical Africa or Southeast Asia? For Africa, the answer is "yes." There, the elimination of forests at their northern and southern edges, whether induced through human activity or by larger-scale changes in climate, might lead to a further warming and drying in those locations, and to subsequent loss of forests.

In Southeast Asia the feedbacks among forests, land use, and climate are less clear. From one perspective, the monsoon rains are driven by large-scale interactions between the oceans and the atmosphere, and reductions in the area of forests might not be expected to be important. From another perspective, however, different rates of heating between the land surface and the ocean surface are also important in controlling the monsoons, and deforestation might well play a role in changing these rates. Possible links between deforestation and monsoon climates are not well understood, but there are examples of local climatic change attributed to deforestation in India, Malaysia, and the Philippines.

The important point here is that deforestation in tropical regions may be more deleterious to local and regional environments than a global warming brought about, in part, by emissions of greenhouse gases from deforestation. Preservation of forested land helps reduce local and regional environmental variability.

IV. GLOBAL EFFECTS OF DEFORESTATION

The concentration of CO_2 in the atmosphere has increased by more than 25% since the start of the industrial revolution in the mid-nineteenth century. Most of that increase has been a result of the combustion of fossil fuels (coal, oil, and gas), but about one-third of the increase was contributed through deforestation. When forests are cleared and replaced with agricultural lands, the carbon originally held in the trees is released to the atmosphere, either immediately if the trees are burned, or more slowly with the decay of organic matter not burned. Cultivation of forest soil similarly oxidizes the organic carbon held in the soil and releases it to the atmosphere as CO_2. When the area of forest is expanded, or when logged forests regrow, the net flux of CO_2 is reversed. Growing forests withdraw carbon from the atmosphere and accumulate it again in trees and soil.

It is important to recognize, however, that forests are not the "lungs of the world," as is often stated. Young vigorously growing forests withdraw carbon from the atmosphere and store it in wood and soil. At the same time, they release oxygen to the atmosphere. A molecule of oxygen is produced for every molecule of CO_2

consumed. As forests mature, however, the rate of accumulation of carbon (or the production of oxygen) diminishes significantly. Carbon continues to accumulate in soils and trees, but the rate of carbon accumulation in mature forests is low relative to the gross exchanges of carbon and oxygen between forests and the atmosphere ($\sim 100 \times 10^{15}$ g annually). Mature forests are approximately in equilibrium with respect to carbon and oxygen, neither accumulating nor releasing these elements. Mature forests hold a large amount of carbon in their vegetation and soil, and, when deforested, this carbon is released. But left standing, mature forests produce very little oxygen. Whatever oxygen they produce during the day through photosynthesis is largely consumed at night through respiration.

A. Carbon

I. The Global Carbon Cycle

The increasing concentration of CO_2 in the atmosphere and its potential to warm the earth have focused attention on the global carbon cycle. CO_2 has been the most important greenhouse gas emitted from human activities in the past and is expected to be so in the future. Only water vapor is more important, but its concentration is not affected significantly by direct human activity. Indirectly, however, human activity is very important, because the emissions of greenhouse gasses that are under human control will determine the rate of global warming, and the warming itself will increase the concentration of water vapor in the atmosphere (a positive feedback that will accelerate the warming). Among the gases under human control, CO_2 is expected to be most important by a factor of 4. [See GLOBAL CARBON CYCLE.]

The concentration of CO_2 in the atmosphere is controlled by the exchanges of CO_2 with the oceans, terrestrial ecosystems, and reserves of fossil fuel. Components of two of these reservoirs actively exchange carbon with the atmosphere: the earth's vegetation and the surface ocean. Each of these subreservoirs holds about the same amount of carbon as the atmosphere, and each exchanges about 100×10^{15} g of carbon annually with the atmosphere. The fact that these reservoirs exchange carbon rapidly with the atmosphere, and the fact that they each contain about as much carbon as the atmosphere, means that they have the potential to change atmospheric concentration of CO_2 rapidly. For example, the regular seasonal oscillation in CO_2 concentrations is largely the result of terrestrial metabolism (photosynthesis and respiration). The late summer/early fall minimum CO_2 concentration results from the excess of photosynthesis over respiration during the summer growing season. The maximum CO_2 concentration in late spring results from a season in which respiration exceeded photosynthesis.

Each of these subreservoirs that exchanges carbon rapidly with the atmosphere also exchanges carbon with another component that holds even more carbon. The world's soils hold about twice as much carbon as the atmosphere. Soils and vegetation together are the major components of terrestrial ecosystems. Animals, including humans, are also part of terrestrial ecosystems, but animals are negligible in terms of carbon (they hold less than 0.1% of the carbon held in vegetation). The largest reservoir of carbon is the deep ocean, which holds about 50 times more carbon than the atmosphere. Over long periods (thousands of years), the deep ocean can be expected to take up most of the CO_2 emitted to the atmosphere, for in the long term the atmospheric concentration of CO_2 comes into equilibrium with the total amount of carbon in the ocean. Over shorter periods, however, rates of mixing between the surface ocean and the deep are a bottleneck to oceanic uptake. It is this bottleneck that has forced the concentration of CO_2 to rise in response to emissions.

The addition of carbon to the atmosphere each year is, as already mentioned, due to combustion of fossil fuels and to deforestation. The removal from the atmosphere is accomplished by uptake by the oceans and perhaps by uptake by terrestrial ecosystems. For the decade of the 1980s, a simplified equation for the global carbon budget can be written as follows (where the units are in 10^{15} g of carbon/year):

$$\begin{array}{ccccccc}
\text{Atmospheric} & = & \text{Fossil} & + & \text{Deforestation} & - & \text{Oceanic} & - & \text{Unidentified} \\
\text{increase} & & \text{fuels} & & & & \text{uptake} & & \text{sink(s)} \\
3.2 \pm 0.2 & & 5.4 \pm 0.5 & & 1.6 \pm 1.0 & & 2.0 \pm 0.5 & & 1.8 \pm 1.2
\end{array}$$

The unidentified sink(s) indicates that the terms in this global carbon equation are not well enough known at present to balance the equation. The atmospheric increase is known precisely, as are the emissions of carbon from the combustion of fossil fuels. However, neither the oceanic uptake nor the terrestrial release is directly measured. Instead, they are determined with mathematical models which make use of information that can be directly measured. Indirect geophysical analyses based on atmospheric and oceanic data and models implicate terrestrial ecosystems as the unidentified sink, but there are no direct measurements of an accumulation on land. Because the unidentified sink is calculated by the difference from other terms, some of which have considerable uncertainty, its current assignment to terrestrial ecosystems may be revised with further research.

The release of carbon to the atmosphere from deforestation is also calculated rather than measured directly (see below). It is important to note here that this calculated release of carbon from deforestation is actually a net release based on changes in land use; the calculations include both deforestation and reforestation. The basis for this flux is described below.

2. Carbon Stored in Vegetation and Soils of Different Ecosystems

Carbon has been redistributed as a result of human activities. In the eighteenth century $700–800 \times 10^{15}$ g of carbon was held in the earth's vegetation and $\sim600 \times 10^{15}$ g was in the atmosphere. Today, trees, shrubs, and grasses are believed to hold $\sim550 \times 10^{15}$ g (Table II) and the atmosphere contains about 740×10^{15} g. The amount of organic carbon stored in the soils of the earth is $1400–1500 \times 10^{15}$ g of carbon. Most terrestrial carbon is stored in forests. Forests cover $\sim30\%$ of the land surface and hold almost half of the world's terrestrial carbon. If only the carbon in living form is considered (soils are ignored), forests hold $\sim75\%$ of this carbon.

The amount of carbon released to the atmosphere from deforestation depends not only on the rate of deforestation, but also on the amount of carbon held in the vegetation and soil of the ecosystems affected. Forests hold 20–50 times more carbon per unit area in trees than the ecosystems that generally replace them, and this carbon difference is released to the atmosphere as forests are transformed to other uses. Table III compares the relative losses of carbon that result from converting forests to other uses. The losses of biomass range from 100% for permanently cleared land down to 0% for the nondestructive harvesting of fruits, nuts, and latex (extractive reserves). Losses of carbon from the soil may also occur, especially if the soils are cultivated. In the tropics much of the carbon is released immediately through burning. Afterward, decay of soil organic matter, logging debris, and wood products continues to release carbon to the atmosphere, but at lower rates. If croplands are abandoned, regrowth of live vegetation and redevelopment of soil organic matter withdraw carbon from the atmosphere and accumulate it again on land.

To calculate the net flux of carbon from deforestation and reforestation, scientists have documented the changes in carbon associated with different types of land use and different types of ecosystems in different regions of the world. Annual changes in the different reservoirs of carbon (live vegetation, soils, debris, and wood products) determine the annual net flux of carbon between the land and the atmosphere. Because of the variety of ecosystems and land uses, and because the calculations require accounting for cohorts of different ages, bookkeeping models are used for the calculations.

3. Emissions of CO$_2$

The annual amount of carbon released to the atmosphere from deforestation, globally, can be determined by multiplying the annual rate of deforestation by the average carbon content of forests. In fact, the computations are carried out regionally, rather than globally, and they are more complicated. They account for delays in the release of carbon following deforestation. Not all of the carbon is released at once. Some is released slowly

Area, Total Carbon, and Mean Carbon Content of Vegetation and Soils in Major Ecosystems of the Earth in 1980

Ecosystem	Area (10^6 ha)	Carbon (10^{15} g) Vegetation	Soil	Mean carbon content (Mg of carbon/ha) Vegetation	Soil
Tropical evergreen forest	562	107	62	190	110
Tropical seasonal forest	913	116	84	127	92
Temperate evergreen forest	508	81	68	159	134
Temperate deciduous forest	368	48	49	130	133
Boreal forest	1168	105	241	90	206
Tropical fallows (shifting cultivation)	227	8	19	35	84
Tropical woodland	776	28	53	36	68
Tropical grass and pasture	1021	17	49	17	48
Temperate woodland	264	7	18	27	68
Temperate grassland and pasture	1235	9	233	7	189
Tundra and alpine meadow	800	2	163	3	204
Desert scrub	1800	5	104	3	58
Rock, ice, and sand	2400	0.2	4	0.1	2
Temperate cultivated land	751	3	96	4	128
Tropical cultivated land	655	4	35	6	53
Swamp and marsh	200	14	145	70	725
Total	13,648	554	1423		

through decay. Furthermore, the analyses include reforestation as well as deforestation, hence the slow accumulation of carbon in growing forests. The balance of emissions and accumulations determines the net flux to the atmosphere.

For temperate and boreal forests net emissions of carbon from changes in land use (largely logging followed by regrowth in these regions) are thought to be negative. That is, carbon is thought to be accumulating in these regions at a rate that varies, depending on the analysis, between 0 and 1×10^{15} g of carbon. Carbon accumulates in growing forests in these regions and is released from oxidation of products harvested over the last decades.

In the tropics, the net release of carbon from deforestation and regrowth is estimated to have been almost 2×10^{15} g of carbon, mostly as CO_2. Ten countries (Brazil, Indonesia, Myanmar, Mexico, Thailand, Colombia, Nigeria, Zaire, Malaysia, and India), representing all three major tropical regions, contributed ~75% of the total net release. The total global emissions, counting boreal, temperate, and tropical forests, are thus calculated to be between 1 and 2×10^{15} g of carbon/year.

TABLE III

Percentage of Initial Carbon Stocks Lost to the Atmosphere When Forests Are Converted to Different Kinds of Land Use[a]

Land use	Vegetation	Soil
Cultivated land	90–100	25
Pasture	90–100	12
Degraded croplands and pastures	60–90	12–25
Shifting cultivation	60	10
Degraded forests	25–50	<10
Logging	10–50	<10
Plantations	30–50	<10
Extractive reserves	0	0

[a] For soils the stocks are to a depth of 1 m. The loss of carbon may occur within 1 year, with burning, or over 100 years or more, with some wood products.

B. Emissions of CH_4

A small fraction (0.5–1.5%) of the carbon released to the atmosphere during the burning of forests is released as CH_4. The total burning of all forms of biomass, including fires in grasslands, pastures, and forests as well as burning of wood fuels, is estimated to release ~40×10^{12} g of CH_4. Most releases of CH_4 following deforestation, however, result from metabolism rather than from burning. CH_4 is a product of anaerobic respiration. About 80×10^{12} g of CH_4 is released from cattle and other ruminants, and ~60×10^{12} g is released from rice paddy cultivation. The expansion of wetlands through flooding of forests for hydroelectric dams could become a significant new source of CH_4 in the future. Overall, about half of the global emissions of CH_4 result directly and indirectly from deforestation. More than half of this flux, in turn, is a result of tropical deforestation and land use, with a lesser fraction released from outside the tropics.

C. Emissions of N_2O

N_2O is also a biogenic gas emitted to the atmosphere following deforestation. Small amounts of N_2O are released during burning, but most of the release occurs in the months following a fire, especially from new pastures and fertilized croplands. Fire affects the chemical form of nitrogen in soils, and as a result favors a different kind of microbial activity (nitrification). One of the by-products of nitrification is the production of NO and N_2O.

Estimates of the global emissions of N_2O are uncertain. Industrial sources are thought to contribute ~1.3×10^{12} g of N_2O–N/year as a result of fossil fuel combustion and the production of adipic acid (nylon) and nitric acid. Biomass burning is estimated to release $0.2–1.0 \times 10^{12}$ g of N_2O–N/year, and cultivated soils are estimated to release between 0.03 and 3.0×10^{12} g. The role of soils is especially uncertain. They may be both a major sink and a major source of atmospheric N_2O. Fertilized soils may release 10 times more N_2O per unit area than undisturbed soils, and the soils of new pastures may release even higher amounts. Deforestation for tropical pastures may be a major contributor to the global increase in N_2O concentrations.

D. Emissions of Carbon Monoxide

Carbon monoxide (CO) is not a greenhouse gas, but it affects the oxidizing capacity of the atmosphere through interaction with OH, and thus indirectly affects the concentrations of other greenhouse gases, such as CH_4. CO emissions are generally 5–15% of CO_2 emissions from burning, depending on the intensity of the burn. More CO is released during smoldering fires than during rapid burning or flaming. The burning associated with deforestation may thus release $40–170 \times 10^{12}$ g of carbon as CO. In addition, the repeated burning of pastures and savannas in the tropics is estimated to release 200×10^{12} g of carbon as CO. Together, these emissions from the tropics are as large as industrial emissions.

E. Radiative Forcing of Different Greenhouse Gases

The properties of the major greenhouse gases are well enough known that scientists can calculate which ones are likely to be most important in the warming of the earth. The calculation depends on three attributes of each gas: its radiative properties, its atmospheric lifetime, and its emission rate (Table IV). The radiative properties of greenhouse gases depend on their molecular structures. In Table IV the radiative strength, or forcing, of each gas is shown relative to the radiative forcing of 1 kg of CO_2. Thus, 1 kg of CH_4 is about 20 times more effective in heating the earth than 1 kg of CO_2; 1 kg of N_2O is about 290 times more effective than 1 kg of CO_2; and 1 kg of chlorofluorocarbons (CFCs) is thousands of times more effective than 1 kg of CO_2. There are many different CFCs, each with a different radiative forcing. The value of thousands given here is a simplification: 1 kg of the major CFCs is 4000–7000 times more effective than 1 kg of CO_2. [See GREENHOUSE GASES IN THE EARTH'S ATMOSPHERE.]

The second property of greenhouse gases that determines their effectiveness in trapping heat is their atmospheric lifetimes, or the average time a molecule remains in the atmosphere before being either removed or broken down. Except for CH_4, the lifetimes of the major greenhouse gases are long. A molecule of N_2O emitted this year will remain in the atmosphere, heating the earth, for about 130 years on average. Again, the variation

TABLE IV

Characteristics of Greenhouse Gases and Their Relative Contributions to Global Warming over a 100-Year Time Horizon

Gas	1990 emissions $(10^{12}$ g)	Warming effect of an emission of 1 kg relative to that of CO_2	Atmospheric lifetime (years)	Relative contribution over 100 years
CO_2	26,000[a]	1	50–200[b]	61%
CH_4	300	21[c]	10	15%
N_2O	6	290	130	4%
$CFCs$[d]	1	1000s	10s–100s	11%
Others[e]				9%

[a] $26,000 \times 10^{12}$ g of CO_2 = 7×10^{15} g of carbon.

[b] The broad range in lifetime for CO_2 results from uncertainties in the global carbon cycle and from the fact that the removal of CO_2 from the atmosphere depends not only on the amount but on the rate at which CO_2 is emitted to the atmosphere.

[c] Includes the indirect effects to concentrations of other greenhouse gases through chemical interactions in the atmosphere.

[d] The radiative properties and atmospheric lifetimes of specific CFCs are known precisely. Only order of magnitude averages for the more abundant gases are shown.

[e] Principally tropospheric ozone, indirectly generated in the atmosphere as a result of other emissions.

for CFCs reflects the different gases in that category. The large range for CO_2 is a measure of uncertainty in the global carbon cycle. Atmospheric lifetimes are determined from an understanding of the quantities of the gas in the atmosphere and rates of emission or breakdown. Estimates of carbon emissions do not balance with estimates of carbon accumulations, and the imbalance is responsible for much of the uncertainty shown for the atmospheric lifetime of CO_2. The fact that lifetimes differ among gases means that the relative effectiveness of a gas in trapping heat depends in part on the time frame of interest. The time scale shown for Table IV is the next 100 years. Had a shorter time frame been chosen, the relative importance of CH_4 would have been larger. Gases with longer lifetimes are more important in the distant future, because their concentrations do not fall as rapidly as those of short-lived gases.

The third aspect of a gas that determines its effectiveness as a greenhouse gas is the rate at which it is emitted to the atmosphere. The emissions are clearly dominated by CO_2, and indeed, over the next 100 years CO_2 is calculated to account for ~60% of the warming expected under business-as-usual scenarios. Under these scenarios current growth rates in global energy use are assumed to continue. However, most CFCs are phased out of production and become relatively less important despite their long lifetimes.

Table IV gives the three radiative characteristics for the most important greenhouse gases and shows their expected contributions to global warming in the next century if current trends in emissions continue. Over the next 100 years the relative contributions to a global warming are calculated to be 61% for CO_2, 15% for CH_4, 4% for N_2O, and 11% for CFCs. Other gases, principally tropospheric ozone, are expected to contribute ~9%.

The emissions of greenhouse gases from tropical deforestation and from subsequent use of the land are shown in Table V. Summing the emissions and taking into account their relative contributions to radiative forcing in the decade of the 1980s show that deforestation, directly and indirectly, has accounted for 20–25% of the radiatively active emissions globally. Most of these emissions were from the tropics, but significant emissions of CH_4 and N_2O result from land use outside the tropics.

TABLE V
Relative Contribution of Deforestation to the Radiative Effect Calculated for the 1980s

Gas	Annual emissions	Percentage of radiative forcing	Calculated contribution to radiative forcing	
			Total	Deforestation
Carbon dioxide (10^{15} g of carbon)		55		
Industrial	5.6			
Natural	0			
Deforestation	2.0		25%	14%
Total	7.6			
CH$_4$ (10^{12} g of CH$_4$)		15		
Industrial	100			
Natural	150			
Deforestation	250		50%	7.5%
Total	500			
N$_2$O (10^{12} g of N$_2$O)		6		
Industrial	1.3			
Natural	7.2			
Deforestation	2.0		20%	1%
Total	10.5			
CFCs (10^{9} of CFCs)		24		
Industrial	1.0			
Natural	0			
Deforestation	0		0%	0%
Total	1.0			
Totals		100		22.5%

V. THE FUTURE

Knowing the radiative properties, the atmospheric lifetimes, and the emission rates of the various gases, one can calculate not only the relative contributions of different gases to future warming, but also the reductions in emissions required to stabilize concentrations of the gases in the atmosphere. Stabilization of concentrations is important, because warming will continue as long as the concentrations of these gases in the atmosphere increase. The concentrations in turn will continue to increase for as long as the gases are emitted to the atmosphere at current rates, or even at very much reduced rates. Stabilization refers to concentrations, and for stabilization to occur, emissions must be reduced. Stabilization of concentrations at present-day levels would require reductions in emissions, reductions of 60% for the long-lived gases (all but CH$_4$). Reductions of 60% are considerably greater than gov-

ernments are contemplating. However, without large reductions, concentrations will continue to increase for years to come. If emission rates continue to increase, stabilization of concentrations in the future will require even larger reductions. Without strong deliberate action, starting immediately, the world seems destined to a warming and to other changes in climate, more difficult to predict. Even with immediate reductions on the order of 60%, the earth is commited to a warming of another 0.5°C over what has been observed to date.

The steps needed for stabilization are (1) a very large (>60%) reduction in the use of fossil fuels, through increased efficiency of energy use and a much expanded use of renewable energy sources, (2) the elimination of deforestation, and (3) reforestation of large areas of land, either to store carbon or to provide renewable fuels to replace fossil fuels. These steps required to curb global warming are good for a host of reasons independent of climatic

TABLE VI

Potential Annual Fluxes of Carbon (10^{15} g of Carbon/Year) to or from the Atmosphere from Human Activities[a]

Strategy[b]	Fossil fuel	Deforestation	Reforestation[c]	Sustainable harvest of fuel	Total	Potential reduction from present
I	6	2	<0.1	0	8	0%
II	6	0	<0.1	0	6	25%
III	6	2	−1 to −2	0	6–7	10–25%
IV	0	0	0	0	0	100%

[a] Positive values indicate emissions; negative values indicate a removal of carbon from the atmosphere.

[b] (I) This strategy represents 1990 emissions of carbon. (II) No deforestation. (III) Reforestation: 100–200 × 10^{15} g of carbon might be stored in new plantations, in forests protected from further logging and shifting cultivation, and in agroforestry. The accumulations of carbon are assumed arbitrarily to take place over 100 years. (IV) Replacing fossil fuels with wood-based fuels grown sustainably (this assumes that the world energy consumption will not increase substantially above 1990 rates). Combustion of wood fuels emits at least as much carbon to the atmosphere as combustion of fossil fuels. The emissions from wood, however, are balanced by accumulations of carbon in the forests, growing to provide future fuel.

[c] This rate of accumulation of carbon in growing forests only persists while the forests are growing or while new lands are being reforested. Once forests have regrown, in 15–100 years, they continue to hold carbon, but they withdraw it from the atmosphere at greatly reduced rates.

change, and initial steps may even have negative costs. Eventually, the realization of these steps will be difficult and expensive, but the costs of coping with rapid, continuous, irreversible climatic change will also be difficult and expensive.

If current trends in deforestation continue, most tropical forests will be gone in 50–100 years, adding another 120–335 × 10^{15} g of carbon to the atmosphere during their destruction. At the low end of this range, the amount of carbon released will be similar to that released from deforestation over the last 130 years. At the high end of the range, the amount is about equal to the total amount of carbon emitted from both deforestation and combustion of fossil fuels to date. These human activities are currently adding ~8 × 10^{15} g of carbon to the atmosphere each year (strategy I, Table VI).

If, on the other hand, deforestation were stopped, the total emissions of carbon to the atmosphere would be reduced by 2 × 10^{15} g of carbon/year, a 25% reduction (strategy II). Massive reforestation might withdraw from the atmosphere on the order of 1–2 × 10^{15}g of carbon/year (strategy III). If deforestation were stopped and massive reforestation were implemented, the net flux of carbon could be cut almost in half. Reforestation on a scale required to sequester 1–2 × 10^{15} g of carbon/year would require an area of ~1000 million ha, a

little more than the land of the United States. Areas as large as this, which supported forests in the past and are not now in either agriculture or human settlements, may exist. However, while a strategy of reforestation would help stabilize the concentration of CO_2 in the atmosphere, the solution would be temporary. Once regrown, in 40 years or so, forests would continue to withdraw only a small amount of carbon from the atmosphere.

The greatest effect that forest management could have on atmospheric CO_2 would be through the elimination of fossil fuels. If sustainably harvested wood fuels were substituted for fossil fuels (strategy IV, Table VI), net emissions of carbon to the atmosphere could be reduced by 6 × 10^{15} g of carbon/year (75–80%) indefinitely (and by 100% if deforestation were stopped as well). Gross emissions from burning would be approximately balanced by accumulations in forests producing fuel for the future. Management would have to be sustainable. Clearly, this strategy is not going to stabilize CO_2 concentrations by itself. Fossil fuel emissions must be reduced through a number of mechanisms.

If the emissions of greenhouse gases are not reduced substantially, and the earth warms, the warming itself may be lead to deforestation of another kind. One of the effects of a global warming, for example, is likely to be an increased rate of respiration (including decomposition of soil organic matter). Increased emissions of respiratory CO_2 and CH_4, in turn, would increase atmospheric concentrations further, increasing the warming, etc. This may be the significance of the observation that concentrations of CO_2 have lagged changes in global temperature during the last 160,000 years, that is, that temperature itself may force increased emissions of greenhouse gases from terrestrial ecosystems.

There is also the possibility that forests will withdraw more carbon from the atmosphere as CO_2 concentrations increase. CO_2 fertilization is known to occur in short-term laboraory experiments with herbaceous plants and tree seedlings. It has not been demonstrated in forests. If it does apply to natural forested ecosystems, the effect would be a negative feedback, helping to reduce the build-up of CO_2 in the atmosphere. However, even if elevated concentrations of CO_2 could be counted on to stimulate plant productivity and increase the storage of carbon in forest ecosystems, deforestation is shrinking the global area of forests and diminishing the land's capacity to store carbon. The danger of deforestation, locally and globally, is that it reduces the capacity of the earth to support life, and further, that its contribution to a global warming may initiate processes in the earth's climate system that are irreversible for generations.

Glossary

Degradation of forests A reduction in the standing stocks (mass) of woody material per unit area, often accompanied by reductions in productivity and biodiversity as well.

Degradation of land A reduction in the capacity or potential of an ecosystem to produce organic matter. In economic terms degradation can be thought of as a reduction in capital.

Feedback A process that tends to modify the outcome of a cause-and-effect relationship, either enhancing (positive feedback) or diminishing (negative feedback) the effect that would have occurred in the absence of any feedback. In the enhanced greenhouse effect, for example, elevated concentrations of CO_2 in the atmosphere (the cause) warm the surface temperature of the earth (the effect). If the warming, in turn, increases concentrations of CO_2, for example, through increasing rates of respiration, the warming will lead to a further warming (positive feedback). On the other hand, if the warming reduces concentrations of CO_2, for example, through enhanced rates of forest growth (carbon storage), the effect of the warming will tend to reduce the warming (negative feedback).

Radiative forcing The suite of processes that maintains or changes the balance between the energy absorbed by the earth and that emitted by it in the form of long-wave infrared radiation. Enhanced radiative forcing refers to the increase in forcing caused by increased concentrations of radiative gases in the atmosphere, in turn caused by anthropogenic emissions of the greenhouse gases.

Selective logging (also called high-grading) Harvest and removal of only a fraction of the trees, generally the largest or most commercially valuable, from a stand.

Shifting cultivation A rotational form of land use alternating between a short period of cultivation and a longer period of fallow, during which the natural ecosystem tends to reestablish itself.

Bibliography

Houghton, J. T., Jenkins, G. J., and Ephraums, J. J., eds. (1990). "Climatic Changes: The IPCC Scientific Assessment." Cambridge University Press, Cambridge, England.

Myers, N., guest ed. (1991). Deforestation Rates in Tropical Forests and Their Climatic Implications. *Clim. Change* **19.**

National Research Council (1993). "Sustainable Agriculture and the Environment in the Humid Tropics." National Academy Press, Washington, D.C.

Ramakrishna, K., and Woodwell, G. M., eds. (1993). "World Forests for the Future: Their Use and Conservation." Yale University Press, New Haven, Connecticut.

Wilson, E. O. (1992). "Biodiversity." National Academy Press, Washington, D.C.

Desert Biotic History—*See* Packrat Middens, Archives of Desert Biotic History

Desertification, Causes and Processes

David S. G. Thomas
University of Sheffield, United Kingdom

I. Definitional Issues
II. Scope and Scale of Desertification
III. Causes of Desertification
IV. Dryland Dynamics and Desertification
V. Processes of Desertification
VI. Human Processes of Desertification
VII. Desertification: Fact or Fiction?
VIII. Conclusion

Desertification is land degradation in arid, semi-arid, and dry–subhumid areas caused by adverse human impacts. Land degradation is damage to the soil, but also to ground water and to some extent vegetation systems, that reduces sustainability. Damage to soil systems may be through the actual loss of soil by erosion or by internal changes to the physical properties and chemistry of the soil. The environments in which desertification occurs are together called the susceptible drylands, which exclude areas that are naturally extremely desertlike and which therefore have extremely low natural productivity. Desertification is caused by humans, although their damaging activities may themselves be triggered by pressures caused by natural climatic stresses. However, the natural temporal variability of dryland ecosystems, which are a response to the dynamism and variations of dryland climatic regimes, and from which systems would naturally recover, does not constitute desertification.

I. DEFINITIONAL ISSUES

Desertification is an environmental issue with a high profile but which is surrounded by contro-versy and debate. The World Bank has considered desertification to be one of the world's 10 most pressing environmental problems, and since the late 1970s the United Nations Environment Programme (UNEP) has had a branch, based in Nairobi, Kenya, that is devoted to antidesertification activities. Despite this, however, desertification is not always clearly perceived as a scientific or a social issue. Over 100 definitions of desertification have been published over the last two decades. Many differ only in detail, but others are more diverse with regard to where desertification occurs, what causes it, and the very form that it takes. It is not surprising, therefore, that there has been debate about the extent of the desertification problem, its nature, and the actions, whether environmental, social, or political, that might be taken in an attempt to resolve it. [*See* DESERTS.]

The word "desertification" first appeared in 1949, when a French forestry scientist, A. Aubreville, used it to describe the environmental impact of forest clearance in west Africa, ultimately leading, in his opinion, to the creation of an ecological desert. The term was revived and sometimes used interchangeably with desertization in the 1960s and

1970s, when the prolonged major Sahel drought of that time led to a decline in biomass, widespread famine, and major livestock and human deaths in the countries located along the southern fringes of the Sahara Desert. [*See* SAHEL, WEST AFRICA.]

The difficulties being experienced by these developing countries led to the United Nations' holding a major conference on desertification (UNCOD) at Nairobi in 1977. This effectively set an agenda for desertification being seen as an environmental problem with a major social impact. It also effectively placed much of the publicity and antidesertification coordination activities in the hands of various United Nations bodies, such as UNEP, United Nations Development Programme, and the Food and Agriculture Organization. To some extent this led to the politicizing of desertification prior to a full and reasoned scientific understanding of what it entails. This emphasizes just how big a problem desertification has been regarded as, but has contributed to desertification being treated hand in hand with other socially undesirable problems, such as drought and famine. They are not necessarily the same, although certain areas of the world might be particularly liable to their occurrence, perhaps simultaneously. Both drought and desertification may lead to famine, and drought may contribute to people carrying out activities that result in desertification.

Placing desertification high on the agenda of environmental issues has resulted in scientists from a range of disciplines carrying out investigations into the causes, processes, nature, and distribution of the problem. There is as yet no full consensus on the definition, but it can be argued that advances have been made in recent years. These relate in particular to the types of environment where desertification occurs, relationships among desertification and natural environmental and ecosystem dynamics in susceptible regions, the major environmental processes that lead to desertification, and their human triggers.

II. SCOPE AND SCALE OF DESERTIFICATION

It is now widely, but not exclusively, accepted that desertification is a problem that occurs in the world's drylands (Fig. 1). Occasionally, the term is used in a broader spatial context, applied to any environment where land degradation is occurring. Such a lack of exclusivity, however, renders the term of little use when attempting to ascribe causes, establish common processes, or seek solutions. Many scientific definitions, including that accepted by UNEP and utilized at the United Nations Conference on Environment and Development at Rio de Janeiro in 1992, accept this spatial constraint. A further useful refinement is to consider the susceptible drylands—susceptible to experiencing full desert conditions if mismanaged. This therefore excludes hyperarid environments, because it would be difficult to make them more desertlike than they naturally are.

The three environmental categories included in the susceptible drylands differ in their degree of aridity, but all are marked by distinctly seasonal climates, particularly with respect to rainfall occurrence, times of the year (running into several months) when potential evapotranspiration (PET) markedly exceeds precipitation (P), an overall annual excess of PET over P, the likelihood of high year-to-year variability in rainfall totals (interannual variability may be in excess of 50% of mean annual totals), and a liability to drought events. Defined on climatic principles, using an aridity index, the susceptible drylands occupy almost 40% of the earth's land surface (Table I). These areas currently support a human population of about 850 million, both in developing and developed nations and in all of the permanently populated continents (Fig. 1).

Given our earlier definition, desertification is therefore *land degradation in the susceptible drylands.* Land degradation can itself be defined in many ways. What is meant here is essentially a reduction in potential productivity or, as it was expressed in a definition of desertification used at the 1977 UNCOD meeting in Nairobi, a reduction or destruction of the biological potential. This is not the same as a simple reduction in biomass or in species composition, which can be caused by short-term natural or human-induced impacts on dryland ecosystems without any change in the environment's ability to support production. If desertification is defined in this manner, it therefore requires a detrimental change in the biological basis of production

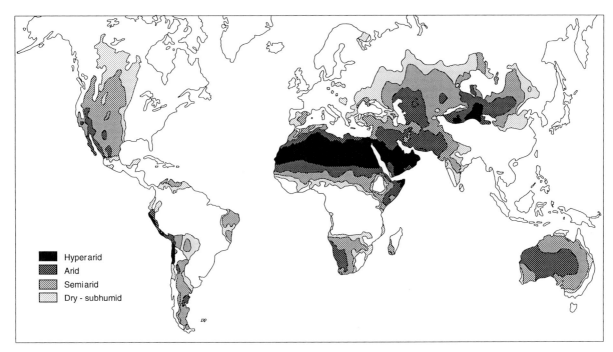

FIGURE I The world's drylands as defined using an aridity index, P/PET (precipitation divided by potential evapotranspiration).

III. CAUSES OF DESERTIFICATION

Understanding why desertification occurs and what causes it is vital if the problem is to be either reduced in its impact or even reversed in places where it has previously occurred. Desertification has been ascribed both to human actions and to climatic events in drylands. While both may play a role, the clear identification of what causes desertification can be severely hampered by the interaction of both anthropogenic and natural events and the confusion of natural short-term cyclic changes in the environment with those that represent long-term degradation. Long-term monitoring of changes in the environment, for example, through the use of satellite imagery, is beginning to shed significant light on the latter point. In this context a major contribution is being made to our understanding of the dynamics of dryland ecosystems, and this in turn is leading increasingly to the point of view that human actions cause desertification but that these actions may, under some circumstances, be triggered by climatic events.

The desertification debate is increasingly being linked to the issue of the sustainability of land use practices in drylands. Therefore, to identify the primary causes of desertification, events and processes must be sought that cause humans to conduct unsustainable activities. Two entwined categories

to take place, which is usually taken to mean a change within the soil or the occurrence of erosion.

TABLE I
Extent of Dryland Zones[a]

	Million ha	% World land area
Dry–subhumid	1294.7	9.94
Semiarid	2305.3	17.72
Arid	1569.1	12.06
Total drylands susceptible to desertification	5169.1	39.72
Hyperarid (very desertlike, therefore not susceptible to desertification)	978.2	7.52

[a] After data of UNEP (1992).

exist: human pressures and environmental pressures.

A. Human Pressures

The most obvious human pressures come from the need to produce more food for growing populations, either by intensifying production from existing agricultural land or by bringing new areas into production. Social anthropologists widely regard traditional dryland societies as being well adapted to their environments, but in effect increasing population numbers and the application of sometimes inappropriate technology and land use practices, imported from more temperate environments, have altered environment-society relationships, especially low-intensity and low-density land use practices. [See GLOBAL ANTHROPOGENIC INFLUENCES.]

Several other human traits are important in increasing vulnerability to desertification. Ignorance of environmental conditions, both directly by farmers, as perhaps was the case in the Dust Bowl years in the American Midwest, and by politicians and agricultural planners in developed and developing world drylands alike, is likely to lead to poor environmental management. Of widespread importance in a developing world context is the incorporation of constituent countries into the world market economy, creating a need to produce commercial crops and products. In Sudan and other Sahel countries the growth of commercial agriculture has displaced subsistence groups from the most fertile and productive areas to desert margin locations. As an example, in Niger millet is now grown 100 km farther north (toward the Sahara) than the official limit of cultivation, increasing the susceptibility to crop failure through drought and the erosion of land where the natural vegetation has been removed for cultivation. In the last two decades civil wars in several eastern Sahel countries have created substantial refugee populations, with people and their formerly productive lands taken out of production. This has then increased the pressure to produce food on areas not directly affected by conflict.

B. Climatic Triggers

Droughts are natural events in drylands. They can trigger desertification either when human actions have lowered natural environmental resistance to drought events, for example, by baring the soil through cultivation practices, or where dought in one region increases the demands for crop production or by livestock in unaffected neighboring areas.

The complexities of drought–desertification links are well illustrated by events during the Dust Bowl years. Drought in the early 1930s in states such as Kansas, Montana, Oklahoma, and Wyoming saw crops fail and wind erosion remove considerable quantities of top soil, lowering potential productivity. However, droughts are now known to be a natural occurrence in this region. What is likely to have caused susceptibility to degradation was the plowing, by settling farmers, of natural grasslands over the previous three decades in an attempt to grow cereal crops. The natural vegetation cover, root mat, and soil structures were destroyed by this, creating a reservoir of potentially deflatable sediment awaiting the meteorological conditions necessary for their mobilization.

Overall, complex social and environmental links lie behind the desertification problem. There is no single cause, but pressures on the environment interact, leading to reduced biological productivity. Once pressures are exerted, the processes of desertification may be self-fueling. Two examples of this are shown in Fig. 2. In the societal feedback cycle land degradation caused by human pressures, perhaps exacerbated by drought, leads to further pressures on the environment which enhance degradation.

In the biophysical feedback cycle, of which two variants were proposed in the 1970s, land degradation triggers a reduction in rainfall that reduces plant cover and increases the potential for wind erosion, causing further and enhanced desertification. In one model, proposed by J. Charney and co-workers in 1975, reduced plant cover and soil degradation increase surface albedo, which in turn reduces the warming of the soil and the lower atmosphere. This causes subsidence of the air, inhibiting

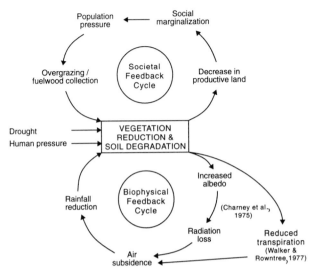

FIGURE 2 Possible feedback cycles of desertification. (Top) Degradation or drought fuels human actions and responses that generate further degradation. (Bottom) Degradation may lead to one of two routes to changes in the atmosphere that lower rainfall. This may reduce plant growth and trigger further drought or degradation by humans.

convective rainfall. The second model, proposed by J. M. Walker and P. R. Rowntree in 1977, suggested that removal of vegetation would lower transpiration to the atmosphere, thereby contributing to atmospheric subsidence and rainfall reduction. While both of these proposed mechanisms have been subject to debate and controversy, they do, together with the suggested societal feedback cycle, illustrate the potential complexities of the mechanisms that interlink and contribute to the occurrence of desertification.

IV. DRYLAND DYNAMICS AND DESERTIFICATION

A realistic understanding of desertification requires the environmental changes that it entails to be correctly placed in the context of the natural characteristics and dynamics of drylands. The seasonality of rainfall regimens, interannual variability, and droughts all have a profound influence on the nature of dryland ecosystems and the manner in which they are adapted to cope with stress and externally driven changes.

Although some plants, such as succulents, display considerable resistance to the disturbance that periodic water shortages impart, other dryland plant communities respond to the stresses caused by events such as drought in a different manner. Grasses (perennial and annual), herbs, and trees are now noted for their resilience, which is the ability of a system to recover from disturbance. In other words, they do not have the ability to resist change but are adapted to recover in the postdisturbance period. Dryland vegetation systems, and probably faunal systems too, are therefore naturally dynamic. From a desertification perspective this means that there are no simple natural equilibrium states against which human-induced changes to vegetation systems can be measured.

When desertification is being considered, it is consequently important to be able to separate natural fluctuations in environmental conditions from those that are more long term and caused by human-induced degradation. Relying on changes in biological systems alone as indicators of the latter can therefore cause difficulties, as was the case with a number of studies reported from Africa in the 1970s. Rather, it is necessary to be able to establish that degradation has occurred to the basis of productivity, which usually means degradation of the soil system.

A. The Problem of the Advancing Sahara

In 1975 a subsequently widely quoted study was produced by H. Lamprey that proposed that the southern margin of the Sahara Desert in Sudan had migrated southward by 100 km in the period 1958—1975, equivalent to an annual advance of 5.5 km. This finding was derived by delimiting the desert boundary in terms of vegetation communities and comparing the findings of field work in 1975 with the analysis of air photographs of the same area dating from 1958. The desert expansion was attributed to the expansion and intensification of human activities and was used as a clear indication of the desertification process (Fig. 3).

Advances in the ecological understanding of desert plant systems and the availability of daily bio-

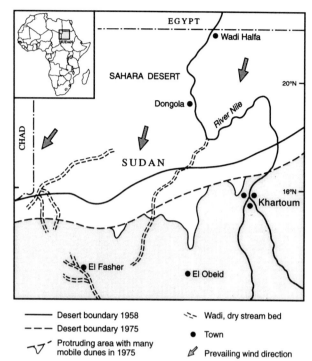

FIGURE 3 Map showing the apparent advance of the Sahara in Sudan from 1958 to 1975. The supposed progressive desert advance due to human-induced desertification was based on changes in vegetation communities. Subsequent studies have shown these to have been due to natural fluctuations in climate, not a progressive southward shift in the desert front.

mass information in the form of a normalized difference vegetation index (NDVI), derived from analysis of National Oceanographic and Atmospheric Administration satellite data, have contributed to a reappraisal of the advancing desert concept. C. J. Tucker and co-workers have used NDVI data to compare the location of the desert boundary in Sudan in the 1984 drought year with the wet year of 1985 and also Sahara-wide vegetation changes from 1980 to 1990. Both of these studies indicate that biomass fluctuated seasonally and annually in response to rainfall. No evidence was found of a simple advancing desert front. On the contrary, the size of the Sahara, as measured by biomass, directly relates to interannual rainfall variability, with both expansions and contractions occurring. This indicates well both the dynamic nature of natural desert conditions and that simple vegetation changes alone cannot be used as an indication of long-term human-induced land degradation.

B. Fluctuating Desert Ecosystems

A number of other studies conducted in the 1980s and 1990s, using detailed field work and high-resolution satellite data, have been produced that show the complexities of unraveling trends in desertification from dryland plant community dynamics. Studies from Sudan have shown no trend at the decade scale for several parameters that might reflect the progressive spread of deserts. Severe impacts of droughts on crop production and biomass were noted, but with subsequent recovery in wet years. In Senegal no evidence of a consistent spread of biomass decline and therefore desert growth has been identified from around boreholes that act as centers for pastoralism. These studies were conducted by Swedish researchers; in the Kalahari region of Botswana, British workers have shown how biomass and species composition changes linked to grazing around boreholes are complex, spatially variable, and probably overridden in their regional impact by the natural changes imparted by rainfall variation and fires.

Overall, evidence indicates that simple analyses of vegetation cover changes may have misinterpreted natural variations in dryland ecosystems with long-term productivity-reducing degradation caused by human activities. To determine the latter, it is necessary to look at processes that primarily lead to changes within the soil system though, which can include certain vegetation and ground water changes that reduce sustainability.

V. PROCESSES OF DESERTIFICATION

Soils in drylands are often thin or skeletal, and formation rates are slow due to limited water availability for weathering and the slow build-up of organic material, except in specific localities such as valley floors. Recovery from the loss of soil by erosion is therefore likely to be slow, and in cases such as salinization through irrigation schemes, salts are likely to remain in the soil for long periods due to high evapotranspiration rates and the absence of a flushing mechanism. In contrast to plant communities, therefore, dryland soils are still believed to have a low resilience to degradation, particularly under the influence of human activities.

A. Biological Processes

There are certain situations in which biological changes can be regarded as reducing long-term productivity in drylands. Long-term sustained grazing pressures may lead to the reduction in seed banks of palatable plant species, such that even if grazing pressure were to be reduced, recovery to pregrazing biomass levels may be inhibited. Encroachment by bush and shrub species onto grazed lands has also been regarded as lowering productivity, but it is now believed that bush encroachment is not permanent, as fires, a natural component of many dryland ecosystems, especially the savannas of Africa, can and do destroy dense shrub vegetation, permitting the reestablishment of grass communities. A lowering of ground water tables, through pumping to supply human and livestock needs, may be a further factor that inhibits biological recovery in human-utilized drylands.

B. Soil System Changes

Soil degradation can be in the form of erosion processes or changes within the soil itself. The former may be relatively easy to recognize, and in susceptible drylands both wind and water erosion can be important. Depending on the local and regional climatic and topographical contexts, these may be favored by vegetation removal, exposing the ground surface to the impact of winds and dryland rainfall. Wind erosion is most likely to occur in areas of low relief and unconsolidated sediments, while areas of steeper topography are more susceptible to water erosion, for example, in the highlands of Ethiopia and Kenya (Fig. 4).

Changes within the soil due to human activities can be more insidious, but a particular threat to productivity. Salinization and alkalinization associated with irrigation schemes (discussed below) are widely cited causes of productivity decline in drylands. Other internal changes relate to waterlogging, also associated with irrigation, and the crusting and compaction of soils, increasingly caused

FIGURE 4 Population growth has led to human habitation and agriculture extending to steep slopes as, for example, shown here in central Kenya. These areas are susceptible to water erosion, especially when cultivation is intensive.

by the application of mechanized cultivation procedures. A more widespread but often less obvious form of internal soil degradation is the loss of nutrients. In some instances this is linked to the actual physical loss of soil by erosion, but in other cases it is due to the clearance of natural vegetation, intensity of cultivation, or lack of application of fertilizers, especially in developing nations.

VI. HUMAN PROCESSES OF DESERTIFICATION

It has been noted above that climatic, environmental, and societal attributes may lead dryland populations to pursue actions, deliberately, accidentally, or through short-term necessity, that lead to stress being exerted on dryland environments and ultimately to the occurrence of desertification. Four main categories of human activities are widely seen as the major causes of a reduction of potential biological productivity in drylands: excessive cultivation, overgrazing, deforestation, and the poor management of irrigation systems.

A. Overcultivation

Overcultivation has sometimes been viewed as the main cause of desertification, especially in areas of the developing world where increasing populations have necessitated attempts to increase yields without the resources available for additional fertilizers. The ensuing nutrient depletion is a serious problem in African drylands, with the World Bank attributing declining yields of staple crops in Sahel nations and parts of South America to this problem. Efforts to increase crop offtake, for example, by reducing the length and frequency of fallow periods, can reduce productivity. Another major aspect of attempts to overcultivate is the importing of farming methods and crops from nondryland environments. Attempts to increase yields through the use of mechanized agriculture can lead to soil compaction, increasing runoff and erosion during intensive dryland storms. Wind erosion can be facilitated by the removal of shelter belts and field boundaries to allow large machinery to be used, for example, in the developing world, in parts of Sudan and Iran, and in the American Midwest. Deep plowing may destroy soil structure and enhance the availability of sediment for erosion. This is a good example of the increase in potential for or susceptibility to desertification.

B. Overgrazing

Pastoralism is an important traditional land use activity in many parts of Africa, Asia, and South America. Overgrazing, both under the influence of these systems and under commercial ranching in the arid parts of Australia, in the western United States, and increasingly in developing countries such as Botswana, has been treated as a major desertification cause. Excessive grazing is seen not only as altering plant community composition but as reducing overall plant cover and therefore increasing the susceptibility to erosion processes. An often cited reason for overgrazing is excessive stocking levels, both under traditional and commercial methods of management. While overgrazing does occur and damage to soil systems can ensue—for example, one Australian report of the late 1970s estimates that irreversible erosion would affect 15% of one area of western Australia if sheep grazing did not stop immediately—there are several reasons that a blanket assumption of pastoralism leading to land degradation in drylands must be questioned. [*See* ECOSYSTEM INTEGRITY, IMPACT OF TRADITIONAL PEOPLES.]

First, changes in plant communities brought on by grazing pressure do not necessarily reduce the long-term productivity of the land, although they can affect the quality of livestock and the potential animal yields in the short term. Second, changes in plant communities in areas where livestock production occurs may be caused not by grazing but, as noted previously, by natural ecosystem variability. Third, where degradation does occur, it may well be very localized, occurring around wells and boreholes. Research using satellite imagery of parts of the Sudan and Senegal shows that these degraded areas remain localized and do not expand into human-made deserts. Fourth, livestock numbers commonly decline naturally during drought periods, creating a postdrought window in which vegetation communities and biomass can recover. This has been observed in Sudan and Botswana.

C. Deforestation

Just as grazing may remove the protection grass communities offer the soil against erosion, so the removal or destruction of wooded areas in drylands can have a similar impact. Woodland clearance can and has occurred on a wholesale basis, with large areas being cleared to facilitate the expansion of agriculture, or in a piecemeal fashion. The latter is increasingly significant in the context of the collection of firewood in African countries. In most sub-Saharan countries fuelwood is the principal source of domestic power, accounting in many cases for over 80% of national fuel use (Table II). [*See* DE-FORESTATION.]

Deforestation does not automatically result in land degradation: If areas cleared for crop production are rapidly cultivated in the postclearance period, erosion may not follow. Over time, however, the problems associated with overcultivation, especially nutrient depletion, may well ensue. Perhaps the greatest threat deforestation offers for desertification is in areas of steep slopes, where the risk of high rates of runoff leading to soil erosion are greatest. Although such areas may be relatively restricted in the context of the world's drylands, the problem can be significant. It can be no coincidence that soil loss is now considerable in the Ethiopian highlands, where there has been a 10-fold reduction in woodland cover in the twentieth century, with less than 5% of the country remaining under tree cover today.

D. Irrigation Systems

Irrigation systems using ground water supplies or diverted river water are an important feature of efforts to cultivate dryland regions. They have been utilized for millennia in some regions, for example, the Mesopotamian area of the Middle East, but in the twentieth century basin-wide drainage and more localized schemes have been an important feature of dryland development. This applies as much to the developed world, including schemes such as the Colorado River Storage Project, which included providing irrigation water as one of its supplementary aims, and the Welton–Mohawk scheme in California, to those in the Nile and Indus basins of Egypt and Pakistan, respectively, which are central to those countries' food production programs.

Land degradation associated with irrigation schemes has been described as one of the few situations in which desertification can clearly be demonstrated, in that human intervention can unequivocally be demonstrated as the root cause of detrimental changes in the relevant soil systems. Irrigation causes desertification through waterlogging, salinization, and alkalinization (i.e., excess sodium accumulation). All of these effects arise through excessive water application to the land; in the case of the last two processes, the high rates of evapotranspiration in drylands lead to precipitation of salts at or near the soil surface. Crop productivity is reduced because most cultivated crops have low salinity tolerances. Problems can be compounded by waterlogging of the soil being enhanced through poorly lined supply canals persistently leaking water, causing ground water tables to be raised excessively. Although these problems are clearly limited to areas where irrigation schemes exist, their impact on national food production can be considerable, especially for dryland countries where cultivable areas are limited.

VII. DESERTIFICATION: FACT OR FICTION?

The disparate nature of the processes that together constitute desertification, the potential confusion with the outcome of natural ecological variability in drylands, and the shear scale and extent of drylands all make gaining reliable data on the scale of the desertification problem a difficult task. However, such data are necessary for planning combative and preventive measures, for land use management, and for estimating food production to meet the needs of dryland populations. As yet there is no simple scientific method for gaining absolute data on the occurrence of desertification. Satellite imagery, even at the highest resolution, is both too

TABLE II

Fuelwood Dependency in Selected African Dryland Countries, 1981[a]

Kenya	74%
Sudan	81%
Cameroon	82%
Nigeria	91%
Ethiopia	93%
Chad	94%
Tanzania	94%
Mali	97%

[a] Values are expressed as a percentage of total energy consumption.

crude to detect many features of soil erosion and inappropriate for identifying many of the processes of soil internal changes. While it is possible to use field studies to determine degradation in individual fields, such data cannot be readily extrapolated to provide regional information.

The United Nations has produced estimates of the global scale of desertification and of its occurrence on a subcontinental scale. Many of the problems mentioned above have arisen, and estimates produced in the 1970s and 1980s have been heavily criticized, not least for their lack of scientific methodology and for confusing natural cyclic changes in vegetation communities with actual desertification. The extent of debate has led, in the late 1980s, to reports and papers suggesting that the problem has at best been overemphasized and at worst used as a scapegoat for famine and social problems caused by droughts and political failings. The term "myth" has been used in conjunction with a number of recent accounts (see the Bibliography).

During the late 1980s UNEP utilized the International Soil Reference Centre in the Netherlands to develop a consistent methodology for the global assessment of soil degradation (GLASOD), including desertification. The approach utilized a geographic information system to handle and analyze data and a consistent methodology permitting interregional comparisons to be made. GLASOD still has many limitations, not the least of which is that much of the initial data input is qualitative and relies on the assessments of individual regional experts. It does, however, provide the opportunity to examine the scale of desertification and its severity in a form that will permit the same approach to be used in future years, allowing assessments of the rate of change to be made. [See SOIL ECOSYSTEMS.]

The GLASOD study estimates (Table III) that about 1030 million ha, or 20% of the susceptible drylands, have experienced degradation processes caused by human activities. This is, however, regarded only to reach strong or extreme levels, meaning that land is unreclaimable without major restorative measures and that the original biotic functions of the soil have been destroyed in less than 4% of the total susceptible dryland area. Water erosion accounts for 48% of the processes of degradation, wind erosion, 39%, and only 10% and 4%, respectively, by *in situ* chemical and physical changes.

VIII. CONCLUSION

There is widespread agreement that the term "desertification" should be restricted to dryland environments. It should also be limited to human-induced degradation imparting changes to the environment that reduce the potential for sustainable production, which primarily comprises changes within the soil system. This use pays regard to developments in understanding the natural dynamics of the biological component of dryland environments. A recent assessment of the scale of deserti-

TABLE III
Estimated Extent of Desertification by Continent[a]

	Susceptible dryland area	Light and moderate desertification	Strong and extreme desertification	Total desertified
Africa	1286.0	245.3	74.0	319.3
Asia	1671.8	326.7	43.7	370.4
Australasia	663.3	86.0	1.6	87.6
Europe	299.7	94.6	4.9	99.5
North America	732.4	72.2	7.1	79.3
South America	516.0	72.8	6.3	79.1
Totals	5169.2	897.6	137.6	1035.2

[a] Values are expressed as millions of hectares. [Data were derived from UNEP (1992).]

fication suggests that it affects ~20% of susceptible drylands, but further monitoring is required both of the processes involved and of the rate of change in its occurrence.

Glossary

Alkalinization Accumulation of sodium salts in the soil.

Aridity Lack of moisture availability. In a climatic context aridity refers to areas experiencing an excess of potential evapotranspiration (PET) over rainfall or precipitation (P) on an annual basis and during many months of the year.

Desertification Land degradation in susceptible drylands caused directly or indirectly by human activities.

Drought Meteorological drought: a period of weeks, months, or years when rainfall is below mean levels; agricultural drought: a period when rainfall is below the growth requirements of specific crops.

Drylands Environments experiencing aridity. Drylands include hyperarid (true desert), arid, semiarid, and dry–subhumid areas (in decreasing order of aridity), defined using climatic indices that consider PET and P.

Land degradation Reduction of the biological potential or productivity of the land, primarily through soil erosion or internal changes in the soil.

Nutrient depletion Loss of soil nutrients, possibly through vegetation clearance but primarily through over-cultivation or soil erosion. Nutrient depletion is an insidious form of soil depletion that can have a marked effect on productivity.

Resilience Ability of an ecosystem to return to a former state after being disturbed.

Resistance Inertia of an ecosystem in the face of disturbance; the ability of an ecosystem to withstand change.

Salinization Accumulation of salts in the soil, often due to the excess application of irrigation water in areas where evapotranspiration rates are high and drainage is poor.

Water erosion Removal and loss of soil through the effects of rainfall and running water. Water erosion may be subtle, through the gradual loss of particles, or more obvious, in the form of rilling and gullying.

Wind erosion Removal of unconsolidated soil and sediment through deflation. Wind erosion can affect particles in the silt and sand size ranges. Silt (dust) particles may be deflated over large distances (from North Africa to the Gulf of Mexico, for example), while sand particles may accumulate around field boundaries or other obstacles close to the source area.

Bibliography

Aubreville, A. (1949). *Climats, forêts, et désertification de l'Afrique tropicale*. Societie d'editions géographique maritimes et colonides, Paris.

Binns, T. (1990). Is desertification a myth? *Geography* **75**, 106—113.

Charney, J. C., Stone, P. H., and Quirk, W. J. (1975). Drought in the Sahara: A biophysical feedback mechanism. *Science* **187**, 434—435.

Grainger, A. (1990). "The Threatening Desert." Earthscan, London.

Hellden, U. (1991). Desertification—Time for a reassessment? *Ambio* **20**, 372—383.

Thomas, D. S. G. (1993). Sandstorm in a teacup? Understanding desertification in the 1990s. *Geogr. J.* **159**, 318–331.

Thomas, D. S. G., and Middleton, N. J. (1994). "Desertification: Exploding the Myth." Wiley, London.

Tucker, C. J., Dregne, H. E., and Newcomb, W. W. (1991). Expansion and contraction of the Saharan Desert from 1980 to 1990. *Science* **253**, 299.

Tucker, C. J., Dregne, H. E., and Newcomb, W. W. (1991). Expansion and contraction of the Sahara Desert, 1980 to 1990. *Science* **253**, 299–301.

UNEP (1992). "World Atlas of Desertification." Arnold, London.

Walker, J. M., and Rowntree, P. R. (1977). The effect of soil moisture on circulation and rainfall in a tropical model. *Quaternary Journal of the Royal Meteorological Society* **103**, 29–46.

Warren, A., and Khogali, M. (1992). "Assessment of Desertification and Drought in the Sudano–Sahelian Region 1985—1991." United Nations Sudano–Sahelian Office, New York.

Deserts

Neil E. West

Utah State University

I. Definitions
II. Distribution of Present Deserts
III. Evolution of Deserts
IV. Climate of Deserts
V. Landforms and Soils
VI. Hydrology of Deserts
VII. Life in Deserts
VIII. Desert Ecosystems
IX. Human Occupation and Use of Deserts

Deserts may be defined by both their apparent sparsity of life and aridity of their climates. Most deserts of the world are located at latitudes 15° to 30° N and S of the equator because of the generally dry air masses there. Deserts have moved as the continents have shifted through geologic time. Plants and animals of deserts have evolved many interesting morphological, physiological, and behavioral adaptations to escape, avoid, or tolerate the heat and drought common to deserts. Desert ecosystems, rather than being durable as commonly thought, are easily disturbed by human activity and are contributing to and being affected by global environmental change.

I. DEFINITIONS

A. Deserts, Defined Biologically

Deserts are commonly thought to be land areas of the world generally lacking in life, human or otherwise. Upon closer and longer inspection, very few places in the world are entirely lifeless. Instead, there are gradients of progressively more life toward the wetter parts of the world's landmasses.

Extreme deserts have no apparent life if seen during a dry period—the usual case. Slightly wetter deserts have perennial vegetation only in ephemeral water courses with augmented soil moisture, a phenomenon known as "contracted vegetation." Semi-deserts have vegetation dispersed across most of the landscape.

Operational definitions of other bordering biomes—grasslands, woodlands, savannas, and tundra—are useful in delineating the desert fraction on the remainder of the world's land mass.

B. Deserts, Defined Climatically

Rather than using biological definitions, an alternate way of defining deserts is by climate. Deserts are areas where aridity dominates; where mean annual precipitation (MAP) is less than about 25 cm, deserts prevail (Fig. 1). However, mean annual temperature (MAT) influences the effectiveness of that precipitation (Fig. 2). As MAT increases, the amount of precipitation needed to maintain similar

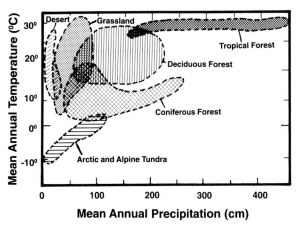

FIGURE 1 Occurrence of biomes in relation to mean annual precipitation and mean annual temperature.

vegetation structure increases, because evaporation and transpiration (water loss from plants) strongly increase with temperature. Holdridge's triangular classification of biomes (Fig. 3) better accommodates the nonlinear interaction between temperature and precipitation, stressing that deserts occur in places where potential evapotranspiration exceeds by 2 to 32 times the actual transpiration plus precipitation.

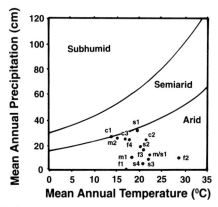

FIGURE 2 Depiction of arid and semiarid moisture province boundaries. Climatic data from some desert localities are superimposed. Mojave desert = m1 (Las Vegas, Nevada), m2 (St. George, Utah), m/s1 (Needles, California, a Mojave–Sonoran transition site); Sonoran desert = s1 (Tucson, Arizona), s2 (Phoenix, Arizona), s3 (Yuma, Arizona), s4 (Brawley, California); Chihuahuan desert = c1 (Socorro, New Mexico), c2 (Ojinaga, Chihuahua), c3 (El Paso, Texas); sites on other continents = f1 (Atacama Desert, Antofagasta, Chile), f2 (southern Sahara Desert, Tessalit, Mali), f3 (northern Sahara Desert, Biskra, Algeria), f4 (Karroo Desert, Beaufort West, South Africa). [From J. MacMahon (1985). *In* "North American Terrestrial Vegetation." Cambridge University Press, Cambridge, England, by permission.]

Soils, fire regime, and land use history also influence whether shrub steppes, grasslands, open woodlands, or savannahs prevail. Excessively rocky, clayish, and/or salty soils select for desert organisms. Excessive fire and livestock grazing favor organisms that are more tolerant of aridity. Polar deserts and tundras can also occur in arid and semiarid regions, respectively, but with much lower MAT (Figs. 1 and 3). [*See* TUNDRA, ARCTIC AND SUB-ARCTIC.]

C. Deserts, Defined Agriculturally

Another way to distinguish deserts is by observing what kinds of agriculture are possible without irrigation. Farmers usually consider any lands that cannot support crop plants as desert. Extreme deserts, or the hyperarid zone, can go for a year or more without rainfall. Except around natural oases, wells, or rivers originating elsewhere, no kind of tillage-based agriculture is possible. Only ephemeral plants occur sporadically.

The arid zone has sparse upland vegetation, usually dominated by thorny shrubs, succulents, and ephemerals. Nomadic livestock use is practicable, but rainfed tillage-based agriculture is impossible. Untilled parts of the semiarid zone are occupied by semidesert shrub steppes and steppes, where a more or less continuous layer of low herbaceous plants with scattered shrubs or trees occurs, which can be used for sedentary livestock production. Rainfed plant crop agriculture is also possible with tillage and seeding, but yields will vary as a function of rainfall.

II. DISTRIBUTION OF PRESENT DESERTS

Because climatic data are usually sparse for arid and semiarid regions, most mapping of deserts and semideserts is based on vegetation and soil boundaries (Fig. 4).

Arid and semiarid lands occur on all continents and collectively comprise about 35% of the earth's surface. An additional 5% of the earth's land area can be categorized as hyperarid; the latter regions

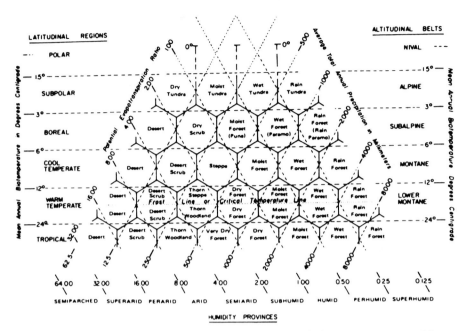

FIGURE 3 Classification of natural life zones after L. R. Holdridge. Note occurrence of deserts to the left side of diagram with axes of mean annual biotemperature (that portion of the year above 0°C), mean total precipitation, and potential evaporation ratios. [From L. R. Holdridge (1967). "Life Zone Ecology." rev. ed. Tropical Science Center, San Jose, Costa Rica.]

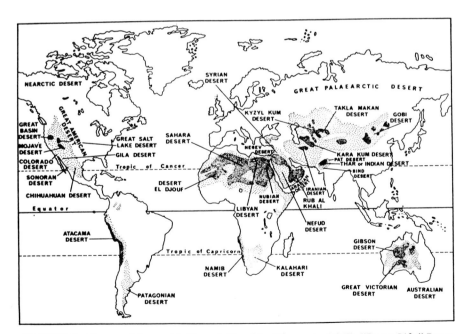

FIGURE 4 Deserts of the world. [From J. Cloudsley Thompson (1965). "Desert Life." Pergamon Press, New York, by permission of author.]

being (in order of decreasing area) the central Sahara and Namib desert areas of Africa; the Hizad, Nafud, and Rub al Khali on the Arabian Peninsula; the Takla Makan and Turfan depressions in Central Asia; the Atacama desert in Peru and Chile; Dasht-eLut of Iran; and small parts of the southwestern United States (Death Valley and Lower Colorado River Valley) and Mexico (Gran Desierto). [*See* SAHARA].

III. EVOLUTION OF DESERTS

Many people intuitively think of deserts as having persisted from ancient times, however, abundant evidence has now accumulated to the contrary. Change has occurred on several time scales. Let us begin within the deeper past. Two great belts of aridity are found between latitudes of about 15° and 30° N and S (Fig. 5A) and are related to general atmospheric circulation resulting from differential

heating of the globe (more radiation impacts near the equator than at the poles) and its rotation (Figs. 5B and 5C). These climatic belts have probably been rather permanent throughout the existence of the planet. The location of the world's landmass, however, has not been constant.

Prior to about 200 million years ago, most of the world's landmass was in one supercontinent called Pangaea. The study of plate tectonics explains how six major (and a number of lesser) plates moved apart, sending the new continents into different climatic zones at different times in geologic history. For instance, Austrialia was once located in a more southern position than now. As it moved northward, a higher fraction of its landmass entered the southern arid belt and the portion that was forested diminished. [*See* DESERTS, AUSTRALIAN.]

Evidence of climatic change since the Cretaceous Age (60 million years before present) can be found in relict floras and faunas (e.g., isolated fishes) that reflect connections to more humid times. Various

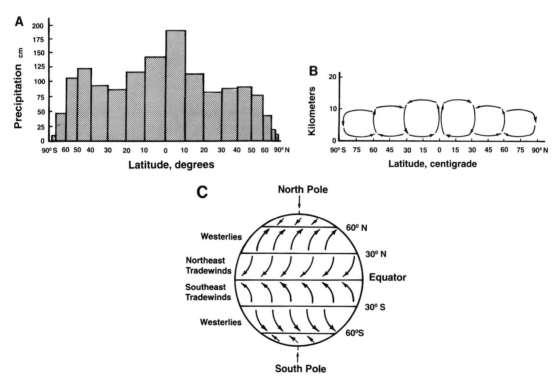

FIGURE 5 (A) Histogram of average annual precipitation versus latitude. (B) Vertical profile of air circulation in relation to latitude. (C) Prevailing wind currents at the earth's surface. [From E. R. Pianka (1974). "Evolutionary Ecology," 2nd ed. Harper Collins, New York, by permission of author.]

physical features also demonstrate that there were once wetter conditions in areas that are now dry. These features include fluvial (river) and lacustrine (lake) deposits such as gravels, clays, and travertine. Reddish color and other soil features are often related to formerly warm, wet conditions. Coverings of drainage systems by sand dunes and other windblown deposits also indicate subsequent aridization, which is a natural process. The Pleistocene (Ice Age), from about 1 million to 12,000 years before present, brought wetter conditions to most desert areas. Pictographs of savanna animals in central Saharan caves confirm that humankind lived through these climatic shifts.

An important question is the degree to which recent human activities may be reducing the capacity of deserts to sustain life and causing semideserts to become more desertlike. This human-caused, permanent reduction in the capacity of the land to produce life has been called "desertification." This issue will be addressed later, after we have discussed some other prerequisite issues.

IV. CLIMATE OF DESERTS

A large number of indices of aridity have been proposed to integrate the singular and interactive effects of precipitation, its amounts, duration, and timing, temperature, wind, relative humidity, and other factors. Such indices typically stress average conditions. Variability in precipitation is as important as the usual amounts. Generally, the lower the mean annual precipitation, the less seasonally dependable the biotic production of an area becomes and the greater the risk of drought (long-term below average amounts of precipitation). In general, deserts occur where air masses usually descend and become warmer, taking up moisture instead of yielding it. This happens in four major circumstances. [*See* Desertification, Causes and Processes.]

The majority of the world's deserts are located in the global "desert belts" in the subtropics (Figs. 4–6). The usually high-pressure atmospheric cells that occur there (Fig. 5B), along with the twist of the Coriolis force, mean that the winds coming

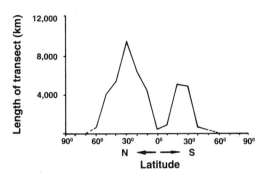

FIGURE 6 Distribution of arid lands in the northern and southern hemispheres, estimated from transects along each 10° longitude; polar arid regions are not included. [From H. E. Dregne (1976). "Soils of Arid Regions." Elsevier, Amsterdam, by permission of author.]

from the northern hemisphere deserts usually flow from the northeast (Fig. 5C). In the southern hemisphere, these winds usually blow from the southeast, creating the "trade winds" that converge near the equator.

The deserts outside the subtropics are due to other causes. Where large mountain ranges intervene in the usual flow of winds off the world's oceans, "rain shadow" deserts are created, because the mountains force the wet air to rise, cooling it, and causing extra precipitation to be dropped on the windward side of the mountains. As the air descends the leeward side of the mountains, it warms up and actually takes on more moisture, making the leeward area drier than it would normally be (Fig. 7). The Great Basin Desert to the lee of the Cascade Mountains and Sierra Nevada in the western United States is an example of such circumstances.

A third cause of deserts is sheer remoteness from the oceans, such that by the time clouds reach there, there is little chance that they will rain. The central Sahara and Gobi are such deserts (Fig. 4). In western Tibet, deserts occur up to about 7000 m in elevation because deep, long-lasting snowpacks cannot exist in such dry air.

A fourth and more minor cause of deserts is dry, descending or stable air masses over the land next to large, cold, upwelling ocean currents on the western edge of continents. These places have the most extreme deserts in the world, particularly the Atacama in Chile and Peru. Fog may penetrate a

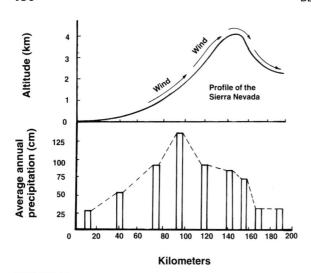

FIGURE 7 Illustration of the rain shadow effect to create the Great Basin Desert to the lee of the Sierra Nevada of eastern California, U.S.A. [From E. R. Pianka (1974). "Evolutionary Ecology," 2nd ed. Harper Collins, New York, by permission of author.]

few miles inland there, but the interiors of such land regions are extremely dry. Some climatic stations in the Atacama have recorded no precipitation for over 150 years of record!

The usual lack of cloud cover and low relative humidity in deserts mean that nights, even at low latitudes, can become very cool. Clear skies allow the escape of reradiated heat, and low humidity minimizes the boundary layer of still air around organisms. One can literally shiver in desert shade because of the latter effect.

V. LANDFORMS AND SOILS

A. Landforms

Most people have the general impression that deserts are typified by vast expanses of moving sand. Actually less than about one-fourth of the world's deserts are covered by sand seas, with mountains occupying the largest fraction of most deserts (Table I). The diversity of desert landforms and soils is very rich and well studied because of the relative ease in examining them, being only sparsely covered by vegetation. Geomorphic processes are the means by which landscapes are sculpted. There are four processes that are important in deserts: weathering, gravitation, fluvial (flowing water), and eolian (wind) movement.

Weathering involves the mechanical and chemical breakdown of rocks into smaller constituent parts. This is an essential first step that allows subsequent wind, water, and gravitational movement to reduce higher, rougher terrain into lower, smoother landforms. Though weathering is relatively slow in desert environments, the abundance of unconsolidated sediments in deserts, especially in the valleys, bears testimony to its ultimate effectiveness.

A peculiar feature of rock weathering in desert environments is called desert varnish, which involves the production of a darkish patina of iron and manganese oxides. Recent investigations have shown that microbes are key players in the production of this varnish; they thrive on infrequent dews. Varnish becomes darker and deeper with age and can thus be used to date the exposure of surfaces.

Another unique feature of soil formation in deserts is subsurface calcium carbonate or silicate deposits. Calcium carbonate layers, usually formed at the maximum wetting front in soils, are called petrocalcic layers, caliche, or kankar. Layers of cemented silicate are called duripans. The time and environmental requirements for formation and breakdown of these deposits are poorly understood. Because they constitute enormous stores of CO_2, the stability of soil carbonates could be very important to greenhouse warming of the earth.

Gravitational forces lead to the downward movement of spalled-off or exfoliated particles. Hydration and solution have been found to be relatively more important than heating and cooling in producing smaller rock particles. Freezing and thawing dislodgement occurs mainly in higher latitudes or altitudes.

Fluvial processes in desert regions are influenced by the sparsity of vegetation, abundance of weathered sediments, and infrequent distribution of precipitation in time and space. Intense storms can easily lead to overland flow, which merges into progressively larger channels and creates ephemeral walls of water carrying huge amounts of sediment. These events resemble mudflows more than the perennial streams of wetter biomes.

TABLE I

Arid Zone Landscapes in Different Regions (Expressed as a Percentage of Area)[a,b]

	Southwest U.S.A.	Sahara	Libya	Arabia	Australia[c]
Mountains	38.1	43	39	47	16
Low-angle bedrock surfaces	0.7	10	6	1	14
Alluvial fans	31.4	1	1	4	
River plains	1.2	1	3	1	13
Dry watercourses	3.6	1	1	1	
Badlands	2.6	2	8	1	—
Playas	1.1	1	1	1	1
Sand seas	0.6	28	22	26	38
Desert flats[d]	20.5	10	18	16	18
Recent volcanic deposits	0.2	3	1	2	—

[a] Percentages given are only approximate, with the degree of accuracy differing between areas.

[b] From R. Cooke, A. Warren, and A. Goudie (1993). "Desert geomorphology." UCL Press, London.

[c] From J. A. Mabbut (1977). "Desert Landforms." Australian National University Press, Canberra. Categories used by Mabbut do not necessarily coincide with those used in other areas; included for comparison only. Remaining data are from Clements, T., R. H. Merriam, R. O. Stone, J. F. Mann, Jr. and J. L. Eyman (1957). "A study of desert surface conditions". Headquarters Quartermaster Research and Development Command Environmental Protection Research Division, Technical Report EP53.

[d] Undifferentiated; includes areas bordering playas.

Another unique feature of desert erosion is creation of underground erosional conduits called "pipes." The ground above these subsurface channels eventually collapses into the pipe, leading to further head-cutting of surface drainage channels, leading to washes, arroyos, or wadis.

Eolian processes are driven by wind. The drier the environment, the more wind dominates in movement of sediments. In addition to the familiar desert dunes, there are other features of desert landscapes that owe their existence to wind. For instance, desert pavement is created from the lag gravel surfaces left after wind and water remove the fines. Blowouts or deflation basins, circular to oval in shape, are another landform caused by wind. Ventifacts (wind-sculpted rocks) leave evidence of the prevailing direction of wind.

The dust from deserts is carried well beyond adjacent landscapes and may even influence the oceans. Some of the best farmlands of the world are composed of deep deposits of silt (called loess). Although the major input of sediment to these regions was from glacial outwash plains, additional depositon comes from windward deserts, for example, the Loessial Plateau of China, downwind of the Takla Makan-Gobi Deserts.

The processes described here lead to distinctive desert landforms (Fig. 8). One landform that is unique to deserts is the pediment (Fig. 9). These shallow-sloped surfaces have a relatively thin and uniform covering of alluvium over bedrock. They are thought to have arisen because rapid head-cutting quickly leads to their isolation from the

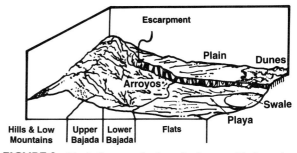

FIGURE 8 Block diagram of a desert landscape with the major geomorphic features that could be found there. [After J. Ludwig (1977). *In* "The Below-ground Ecosystem" (J. Marshall, ed.). Colorado State University, Fort Collins.]

main water flow patterns. Once the pediment surface becomes disconnected from the main drainage, it erodes very little and becomes the most stable part of the landscape where the most mature soils can develop.

Another landform feature unique to deserts is the alluvial fan. These are fan-shaped trains of alluvial materials that are deposited by flash floods emerging from canyons at the base of desert moutains. Over time, they grow convexly in height, broaden, and coalesce to form a continuous apron at the base of mountains or hills that is called a bajada (Fig. 10). Bajadas are the principal water and sediment "transfer" zone in desert landscapes. There is a rapidly diminishing size distribution of the sediment as the debris is carried away on the progressively gentler slope of the fan. The lessening energy of the water drops rocks and gravels near the head of the fan, but carries sands, silts, and clays progressively farther down the slope.

Usually dry lakes, salt pans, or playas (landlocked sediment basins) and plains are major sinks for sediments (Fig. 8). Mountains are the primary source of new sediments and because they are rapidly eroding and poorly protected by vegetation. They are also frequently steep and angular.

Older desert landscapes are more subdued, typically having gently sloping pediments, mesas capped by more resistant rock, and dune fields. Ballenas, which are low, rounded, linear, downsloping hills, form as old alluvial fans are dissected (Fig. 11). Dunes may be either fixed by vegetation or mobile, and come in many sizes and shapes (Fig. 12) caused by differences in wind regime. However, despite much investigation, predictive models are still lacking.

Biologists must understand these landforms because landscape position, surface coverings, and rock fractures to depth greatly influence moisture regimes—the major constraint of life in the desert. Accordingly, habitats can be classified in relation to geomorphic features (Fig. 13). For instance, algae can grow under translucent rocks on hammada; this is technically known as a hypolithic habitat.

B. Soils

A desert biologist cannot fully understand desert life unless he or she understands soils. This is because soil characteristics greatly influence the soil moisture dynamics in deserts, which in turn greatly influence plant growth. Plant growth then influences what animals and microbes can exist there.

Contrary to other biomes, stone-free loamy soils do not support the greatest concentrations of native life in deserts. Moderately rocky and sandy soils in desert regions usually have the greatest diversity and productivity, because water infiltrates faster and deeper on coarser-textured soils and fractured rock than in loams or clays (Fig. 14). The more deeply the water infiltrates, the lower the chance that it will be lost to evaporation, a very powerful force in deserts.

Clay-dominated surfaces can quickly evolve into the complex, rounded topography known as "badlands" because of poor water infiltration and

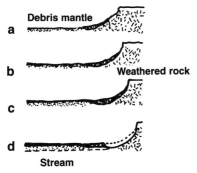

FIGURE 9 Evolution of a pediment: a = youngest to d = oldest landscape. [After C. Twidale (1968). "Geomorphology with Special Reference to Australia." Nelson, Melbourne.]

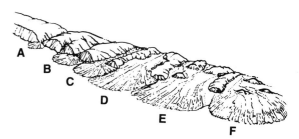

FIGURE 10 Evolution of alluvial fans over time in the Basin and Range Province of North America. A = youngest to F = oldest. [From C. B. Hunt (1973). "Geology of Soils." Freeman, San Francisco, by permission of author.]

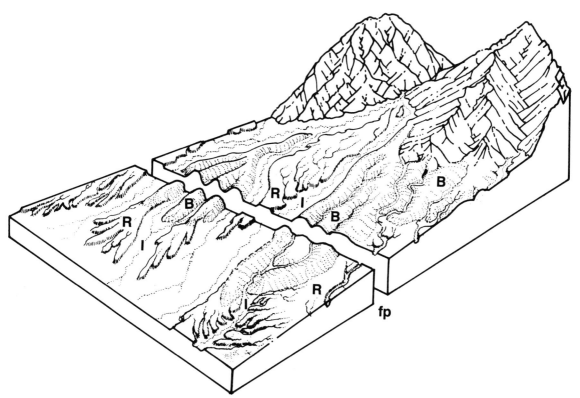

FIGURE 11 Formation of ballenas on old alluvial fans. B = ballenas, R = erosional fan remnants, I = inset fans. [From F. Peterson (1989). "Landforms of the Basin and Range Province." Nevada Agricultural Experiment Station, Reno, by permission.]

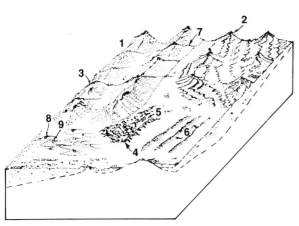

FIGURE 12 Block diagram of aeolian landforms. Dune types: 1 = draa, 2 = oghgroud, 3 = demka, 4 = aklé, 5 = anguie, 6 = seif, 7 = silk (plural, slouk), 8 = barkhan, 9 = linguoid. Large sand sea in background are ergs. [From F. Howard and C. Mitchell (1985). "Phytogeomorphology." John Wiley & Sons, New York, by permission.]

storage and thus little vegetational cover. Silt-dominated surfaces in deserts are blown by wind into isolated hummocks around shrubs or even sinuous stripes.

Limited leaching of salts and low amounts of organic matter accumulation in desert soils result from their general aridity. Soil development is relatively slow because lack of water limits both chemical and biological activity. Without biological activity, there is limited feedback to modify sediments and microclimates. The pioneering conditions found by desert organisms thus remain rather permanent. Only the oldest parts of desert landscapes—the upper pediment surfaces—have strongly developed horizons indicative of "mature" soils. Because most desert landscapes are eroding faster than soils can form, they are mostly poorly differentiated and "young" (Fig. 15).

Desert soils are particularly subject to the formation of crusts and other layers restrictive to passage

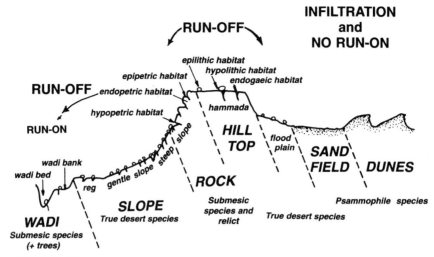

FIGURE 13   Schematic representation of major desert habitats as defined by geomorphology. [From A. Schmida (1981). *In* "Hot Deserts," Vol. 14A, "Ecosystems of the World" (M. Evenari *et al.,* eds.). Elsevier, Amsterdam, by permission of author.]

of water, air, and organisms. These crusts, also known as caps, carapaces, films, veils, or skins, may be largely caused by physical rearrangement of silt and clay particles following wetting. Some crusting is due to slaking caused by sodium or deposits of calcium carbonate or gypsum. Some bacteria, algae, mosses, and lichens may exude mucilaginous materials and thus "glue" soil particles together. These microphytic crusts frequently overlay or intermingle with vesicular (foam soil) horizons. All of these features influence retention and infiltration of liquid water. Whether the influences are positive or negative depends on the combinations of crusting present. Some of the microphytes (e.g., cyanobacteria) may contribute to nitrogen fixation. Nitrogen is usually the limiting element to plant growth once water needs are satisfied.

Because there usually is not enough water to produce continuous plant growth in deserts, scattered shrubs create their own microenvironments (Fig. 16) and cause formation of hummock and mound microtopography around themselves on level areas or striped patterns on slightly sloping topography. In this way "islands of fertility" develop (Fig. 17). Animals and microbes also orient themselves around these spots of more favorable

environment. Their digging and decomposing slowly lead to more soil organic matter over time.

VI. HYDROLOGY OF DESERTS

The major processes in the water budget of deserts are shown in Fig. 18. Most precipitation is lost to evaporation. Because of crusting, a relatively low proportion of rainfall infiltrates to depth in

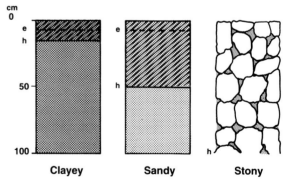

FIGURE 14   Diagrammatic representation of water retention in three desert soils following 50 mm of precipitation. h = lower level of moistening, e = lower level to which soil dries out. Because of evaporation, the clay retains only 50%, the sandy soil 90%, and the stony soil nearly 100% of the water applied to the surface. [From H. Walter (1973). "Vegetation of the Earth," 2nd ed. Springer-Verlag, New York/Berlin, by permission.]

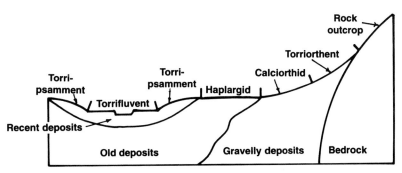

FIGURE 15 Example of soil–landscape relations in an arid region. Steep rocky outcrops have no significant soil cover. Torriorthents are weakly developed. Calciorthids and Haplargids are more developed than the Entisols centered around the youngest deposits in the wadi. [From H. Dregne (1976). "Soils of Arid Regions." Elsevier, Amsterdam, by permission of author.]

the soils. Low soil organic matter limits retention as well as aggregate stability (soil crumb structure) to enhance infiltration and aeration. Stream flow in deserts is very chaotic as it depends on large, rare precipitation events. If most of the water from such flash floods goes into ephemeral lakes, most is lost to evaporation under the generally warm, windy, low relative humidity conditions. Most desert lakes without outlets to the sea become saline sinks and evaporites (salts) form as they dry out.

Because of their poor protection by vegetation yet relatively frequent occurrence of violent convectional storms, the world's semiarid environ-

ments are where the naturally highest rates of water-induced soil erosion occur (Fig. 19). Because much of the water in desert stream channels is hyporheic (flowing below the bottom of the stream channel), the profile of the water table along ephemeral desert stream channels is much different than in wetter regimes (Fig. 20).

Vast amounts of subsurface water are commonly present in desert areas and, if not saline, usually become a dependable source of water for humans. Much of this water is relictual from the wetter times of the Pleistocene. Current use is frequently

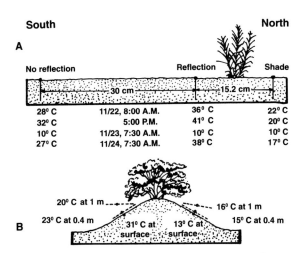

FIGURE 16 Differences in ground temperature in relation to perennial plants in Death Valley, California. [From C. B. Hunt (1973). "Geology of Soils." Freeman, San Francisco, by permission of author.]

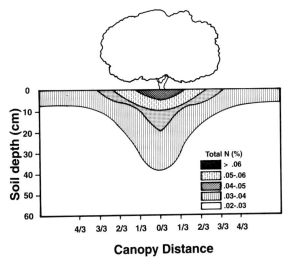

FIGURE 17 Distribution of soil nitrogen around a mesquite tree in the Sonoran Desert. [From N. West and J. Klemmedson (1979). *In* "Nitrogen in Desert Ecosystems" (N. West and J. Skujins, eds.). Stroudsburg, Penn.: Dowden, Hutchinson & Ross, by permission of author.]

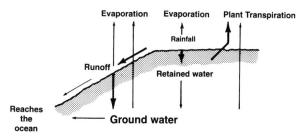

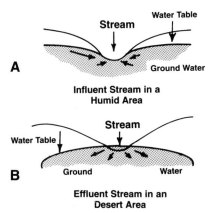

FIGURE 18 Main components of the hydrologic cycle in desert regions. [From H. Walter (1973). "Vegetation of the Earth," 2nd ed. Springer-Verlag, New York/Berlin, by permission.]

much greater than natural replenishment, meaning that some human uses are nonsustainable. Once these fossil waters have been depleted, water will have to be transported from outside the desert region or local saline water will have to be desalinated.

VII. LIFE IN DESERTS

A. In General

The usually dry, hot and windy conditions of deserts, plus their inherent high temporal variability, create a stressful environment for ordinary kinds of organisms. Lack of available water in most places and at most times has led to selection of a wondrous array of plants, animals, and microbes in deserts. Survival centers around a "pulse and reserve strategy," that is, most organisms opportunistically carry on a flurry of activity when conditions are favorable and then go dormant or inactive

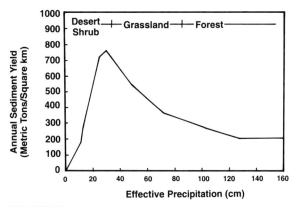

FIGURE 19 Stream sediment yields as related to annual precipitation. [After F. A. Branson *et al.* (1972). "Rangeland Hydrology." Society for Range Management, Denver, Colorado.]

FIGURE 20 Comparison of expected profiles of water tables around stream courses in (A) humid and (B) arid regions.

during the more common times of duress. Few plants can continue photosynthesizing under drought stress. Limited plant growth means that there is limited food and cover for animals and limited amounts of detritus for decomposers.

In addition to the water limitations in all deserts, cold temperatures for part of the year limit biological activity in both high-elevation and high-latitude deserts. Furthermore, the very low winter temperatures there have a more severe impact on the biota than they might in a wetter region because protection of snow cover is less. Cold is no problem in the tropical deserts, but the clear, low-humidity atmosphere there results in high radiation loads during the day, followed by cool nights. The diurnal extremes in temperatures can thus exceed seasonal swings elsewhere.

Other stresses on desert life result from wind, such as the abrasive and suffocating effects of blown sediments. High salinity of soil surfaces and water is another restraint on life.

What is most different about deserts compared to humid ecosystems is the limited proportion of rich habitats in a desert landscape (Fig. 21). The wadi and oasis habitats contribute disproportionately to total biodiversity. Many organisms show marked structural, physiological, and behavioral adaptations to neutralize at least some of the more extreme features of the rest of the landscape.

B. Plants

The most noticeable feature of the perennial plant components of desert vegetation is their usual spar-

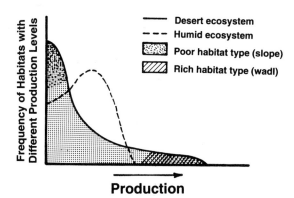

Production

FIGURE 21 Frequency distributions of habitats ordered by relative primary production in desert and humid zones. [From A. Shimida (1986). *In* "Hot Deserts and Arid Shrublands," Vol. 12B, "Ecosystems of the World" (M. Evernari *et al.*, eds.). Elsevier, Amsterdam, by permission of author.]

sity. In hyperarid environments, no perennial plants are found. Only ephemerals can grow when very rare significant rainfalls occur. The remainder of the time they persist as seeds.

As precipitation becomes greater and more dependable in comparatively wetter zones, shrubs and succulents can occupy the whole landscape, but at wide average spacing. However, it should be realized that the large, apparently unoccupied soil patches between the aerial parts of plants may well be occupied by roots. When rain does penetrate the soil, these large, diffuse root systems may exploit the soil so effectively that establishment of other perennials is unlikely.

Desert plants can be classified according to their requirements for water (Fig. 22). Water spenders have no better water use efficiency than plants of humid regions. They escape or avoid drought by

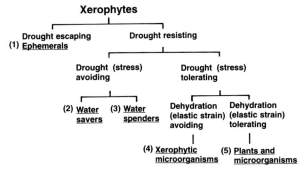

FIGURE 22 System of classification of desert plants based on their needs and utilization of water. [From J. Levitt (1972). "Responses of Plants to Environmental Stress." Academic Press, San Diego, Calif., by permission.]

having bursts of growth during wet periods and have among the highest rates of photosynthesis in the world. Drought-tolerating plants can extract water from much drier soils than plants restricted to mesic (wetter, nutrient-rich) environments. Succulents like cacti take in water rapidly when the soil is moist and store it in their own tissues for later use. These plants usually have crassulacean acid metabolism, which allows them to open their stomates exclusively at night to let in and store CO_2 and thus reduce water losses through transpiraton. Consequently, they can photosynthesize during the day with their stomates closed. Perennial plants frequently grow cork tissue between their vascular structures and so divide themselves into functional segments such that part of the plant can die of drought but allow the remainder to persist.

C. Animals

The diversity of animals in deserts is very high, although population densities for any one species are usually low. Animals find a greater variety of habitats than plants because some habitat is provided by the different plants themselves. Furthermore, because animals can usually move and do not require as much light as plants, they are free to seek spaces and times of activity that are most comfortable and productive.

I. Invertebrates

When desert invertebrates are mentioned, one usually thinks of insects, spiders, and scorpions. Wood lice, millipedes, centipedes, and mites are also common desert dwellers. Desert snails are active when dew is present and humidity is high. Major insect groups in deserts are flies, beetles, butterflies, moths, ants, termites, bees, and wasps, as well as grasshoppers, crickets, and true bugs.

Desert life is easier for most insects and spiders than for vertebrates because, like many desert plants, they have waterproof cuticles that help retain water. Additional invertebrate advantages are short life cycles and resting stages such as diapause and cryptobiosis. Metamorphosis allows the vulnerable stages of invertebrates to be timed for the most favorable seasons. Similar to the seeds of ephemeral plants, many invertebrates persist through unfavorable periods in an inactive stage such as eggs or pupae, and thus avoid many of the rigors of desert life. Like the deciduous woody plants, some invertebrates avoid environmental difficulties during rigorous parts of the year by continuing activity only below ground (e.g., cicadas, termites, and ants).

Insects also adjust their daily activities, mainly based on temperature. For instance, desert nematodes have recently been found to move up and down in desert soil profiles to avoid the heat of day and exploit the more favorable conditions of night.

2. Vertebrates

Reptiles are probably the most characteristic group of desert vertebrates. Lizards and snakes are numerous in the deserts of all continents, and tortoises are found in some deserts. Their scaly skin and plates protect against water loss, but they have to move into shade or below ground to avoid overheating in the summer.

Birds, though diverse in deserts, may be only part-time residents because of food limitations. The general lack of trees and tall shrubs limits bird nesting sites to the ground. Cursorial (running) birds are common in deserts (e.g., ostrich, emu, bustard, roadrunner). Raptors and soaring carrion eaters use the high visibility of desert habitat to their advantage in locating food.

Many desert mammals exhibit extreme modifications of behavior and physiology. For instance, many granivores (seed eaters) and folivores (leaf and stem eaters) have developed renal (kidney) and digestive capabilities such that they can get all of their water needs from their food intake. Many mammals are nocturnal, which decreases water losses. Others spend daytime in soil burrows with higher humidity and lower temperatures. Desert marsupials and ungulates (hoofed animals) may not breed or implant fetuses during droughts. Kangaroos and ungulates also range widely to seek out free water and forage.

Two exceptionally successful desert mammals are the camel and dromedary. Broad feet allow them to move easily over loose sand, and large fat reserves enable them to survive long periods with access to only poor forage. They drink enormous amounts of water when it is available and then go for long periods without drinking. Their fur also helps reduce heat loads.

D. Microbes

Though the diversity of soil microorganisms in deserts is about the same as in other biomes, they are much less abundant and active only following rainfall or dew events. Because their major role is in decomposition, that is, freeing energy in plant and animal biomass and recirculating nutrients, they cannot be overlooked for a full understanding of desert ecosystems.

VIII. DESERT ECOSYSTEMS

Because of the primary limitation of water, deserts have relatively simple plant community structure and low primary productivity (Fig. 23) and standing crops of plants. The intrayear and between-year variabilities are also very high, which lead to relatively limited sources of food for consumers and decomposers, but food webs are usually as complicated as elsewhere. Landscape position and soil texture influence infiltration and use of soil moisture. The complexity of vegetation structure usually declines with lessened water input. Figure

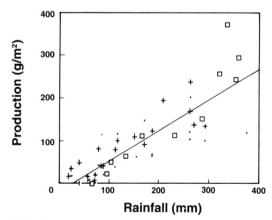

FIGURE 23 Cumulative aboveground net primary production versus cumulative rainfall over 2 years at a desert site in northern Kenya. [From M. B. Coughenour *et al.* (1987). *J. Arid Environ.* **19**, by permission.]

24 shows these interactions for an idealized landscape profile in southern Arizona.

There is comparatively less modification of the macroclimate over the entire landscape by organisms in deserts compared to in most other biomes. However, the islands of fertility centered around large perennial plants create considerable microenvironmental patchiness. Biological feedback to prevent soil erosion and assist soil development is thus much lower than for wetter parts of the world. The chemical and physical features of the environment take precedence over biological modifications of the environment, limiting feedback between trophic (feeding) levels in desert ecosystems (Fig. 25).

Persistent perennial plants are critical to all other life. They maintain their relatively abundant biomass by growing slowly over many years and protecting it against herbivores with spines and chemicals that are poisonous or prevent easy digestion or decomposition.

Granivores are the most abundant kind of consumer in deserts because seeds are the most abundant and dependable source of food. The importance of seeds facilitates interactions among groups of organisms that do not interact elsewhere. For instance, seed-eating birds, mammals, and ants have been found to compete and thus influence plant community structure both collectively and independently. Their digging for seed also influences which plants regenerate.

The dry conditions of deserts favor fungi over bacteria because fungi can continue their activities at soil moisture levels below that required by both bacteria and plants. Termites and mites are very important in breaking down detritus. These, plus insects such as dung beetles, transport detritus below ground, where it can decompose more readily. Direct physical and chemical breakdown of detritus leads to some loss of nutrients from deserts. Levels of organic carbon and nitrogen are usually less than saturated in desert soils.

Because the same species of tolerant or evasive desert organisms return relatively rapidly after disturbance, without intermediate stages characterized by different physiognomy and substantial environmental modification, succession in deserts is called "autosuccession." The concepts of climax and balance are thus not as applicable as for wetter biomes. Wildfires are not part of the true desert environment because of a general lack of continu-

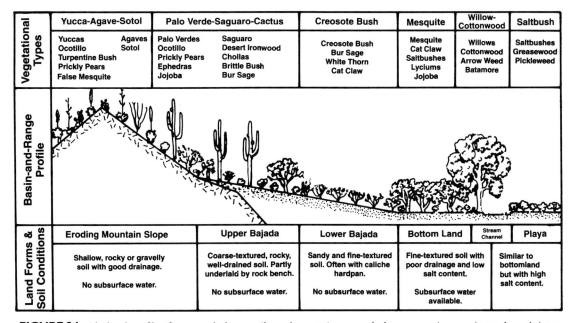

Vegetational Types	Yucca-Agave-Sotol		Palo Verde-Saguaro-Cactus		Creosote Bush	Mesquite	Willow-Cottonwood	Saltbush
	Yuccas Ocotillo Turpentine Bush Prickly Pears False Mesquite	Agaves Sotol	Palo Verdes Ocotillo Prickly Pears Ephedras Jojoba	Saguaro Desert Ironwood Chollas Brittle Bush Bur Sage	Creosote Bush Bur Sage White Thorn Cat Claw	Mesquite Cat Claw Saltbushes Lyciums Jojoba	Willows Cottonwood Arrow Weed Batamore	Saltbushes Greasewood Pickleweed

Land Forms & Soil Conditions	Eroding Mountain Slope	Upper Bajada	Lower Bajada	Bottom Land	Stream Channel	Playa
	Shallow, rocky or gravelly soil with good drainage. No subsurface water.	Coarse-textured, rocky, well-drained soil. Partly underlaid by rock bench. No subsurface water.	Sandy and fine-textured soil. Often with caliche hardpan. No subsurface water.	Fine-textured soil with poor drainage and low salt content. Subsurface water available.		Similar to bottomland but with high salt content.

FIGURE 24 Idealized profile of geomorphology, soils, and vegetation around a low mountain range in southern Arizona. [From L. Benson and R. Darrow (1981). "Trees and Shrubs of the Southwestern Deserts." University of Arizona Press, Tucson, by permission.]

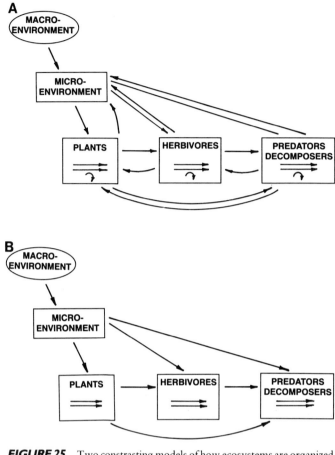

FIGURE 25 Two contrasting models of how ecosystems are organized and function: (A) with much biological feedback in mesic regions or (B) with no feedbacks between trophic levels in deserts. [From I. Noy-Meir (1979/1980). *Israel J. Botany* **28,** pp. 1–19, by permission of author.]

ous fuels. However, semideserts are frequently burned, particularly following wetter than average years.

IX. HUMAN OCCUPATION AND USE OF DESERTS

A. Ancient Use

Because of low productivity and rigorous climates, most humans avoided living and traveling in deserts until the advent of air conditioning. Except for regions where major rivers originated beyond its borders, for example, the Nile, Tigris–Euphrates, and Colorado river valleys, thus allowing irrigated agriculture, human populations were low in desert regions until recently. The ma-

jor aboriginal desert peoples were either hunter-gatherers (e.g., Kung! Bushmen, Paiutes, Australian Aborigines) or pastoral nomads (e.g., Bedouin, Mongols, Uighhurs, Kazaks, Berbers, Navaho). These people moved with spotty rains within the desert proper and retreated into adjacent semiarid lands during prolonged regional droughts. The cultural and behavioral adaptations of the nomads were linked to domestication of ungulates, weaving of textiles for clothing, design of temporary or portable housing, and devlopment of water-preserving structures such as cisterns.

B. Modern Use

The development of nations and modern civilization has created national boundaries, tax collection, education and health services. Most hunter-gatherers and nomads have been forced into sedentary life-styles. Although some of these peoples are now receiving formal education and health services, the environments surrounding their settlements are experiencing unsustainable levels of use, especially from excessive livestock grazing, fuel wood gathering, and water pumping.

Industrialized nations with desert lands have come to regard them as wastelands. Accordingly, "nuisance" activities such as military training and development, as well as nuclear, explosive, and hazardous chemical production, are often centered there. Toxic wastes originating in humid regions are often sent to desert regions for disposal. Only in recent decades have desert nature reserves and parks been established.

People who have not experienced desert living tend to believe that because desert organisms survive rigorous environments, they should be able to withstand human-induced disturbances such as livestock grazing, pollutants, and offroad vehicles. Actually, desert organisms live on a narrow balance point between success and failure. Their reserves of water, energy, and nutrients are often too small to provide for unexpected or multiple disturbances. Biotic communities and soils of deserts are thus unresilient, that is, they do not recover easily or quickly. Slight damage to these ecosystems can easily become permanent. Sustainable management of desert and semidesert lands is thus much more difficult than in wetter parts of the world.

Semiarid regions naturally fluctuate between years of grassland-type climate to years that are more desert like. If populations of livestock built up during wet years remain during droughts, irreversible damage to vegetation and soils will be done. This permanent lowering of the capacity of the land to produce due to human actions is known as desertification. This movement of land capacity through a threshold of permanent change can occur in true deserts, but is more common in adjacent semiarid regions where human expectations are normally higher. Desertification leads to a higher albedo (reflection of radiation) of desert surfaces after its organic components are diminished. This greater albedo may contribute to higher regional aridity and changed spatial and temporal weather patterns. Increased dust storms are also changing the atmosphere and leading to alteration of ecosystems elsewhere, for example, the smothering of coral reefs around the Arabian Peninsula.

If soil erosion, harvest of plants, and nutrient losses are faster than their replenishment, then natural capital becomes squandered and unique organisms and ecosystems are destroyed. The greater the damage, the greater the cost of repair. Such costs will have to be subsidized by wealthier parts of a nation or the world. It behooves us to better understand desert environments and strive to lighten our impact on them so they can exist into perpetuity.

Glossary

Aridization Natural development of deserts through evolution of drier climates.
Bajada Coalesced alluvial fans at the base of desert mountain ranges.
Caliche Petrocalic layers; indurated deposits of calcium carbonate.
Desert pavement Soil surfaces covered by lag gravels and loose stones.
Desert varnish Darkish patina of iron and manganese on the surface of exposed rocks in deserts.
Desertification Human-induced, permanent reduction in the ability of the land to produce life.
Deserts Land areas characterized by dry climates, sparse life, and limited agricultural activity.
Drought A short period of dry weather that can occur anywhere, but is usual in deserts.
Evaporites Salts formed as temporary desert lakes evaporate.
Hammada Stony, upland surfaces.

Bibliography

Cooke, R., Warren, A., and Goudie, A. (1993). "Desert Geomorphology." London: UCL Press.
Dregne, H. E., ed. (1992). "Degradation and Restoration of Arid Lands." Lubbock: International Center for Arid and Semiarid Land Studies, Texas Tech University.
Evenari, M., Noy-Meir, I., and Goodall, D. W., eds. (1985). "Hot Deserts and Arid Shrublands," Vols. 12A and B, "Ecosystems of the World." Amsterdam: Elsevier.
Mabbut, J. A. (1977). "Desert Landforms." Canberra: Australian National University Press.
Schlesinger, W. H., and Reynolds, J. F., Cunningham, G. L., Huenneke, L. F., Jarrell, W. M., Virginia R. A., and Whitford, W. G. (1990). Biological feedbacks in global desertification. *Science* **247**, 1043–1048.

Skujins, J., ed. (1991). "Semiarid Lands and Deserts: Soil Resource and Reclamation." New York: Marcel Dekker.

West, N. E., ed. (1983). "Temperate Deserts and Semideserts," Vol. 5, "Ecosystems of the World." Amsterdam: Elsevier.

West, N. E., (1986). Desertification or xerification? *Nature* **321,** 562–563.

Whitehead, E. E., Hutchinson, C. F., Timmermann, B. N., and Varady, R. G., eds. (1988). "Arid Lands: Today and Tomorrow." Boulder, Colo.: Westview Press.

Whitford, W. G., ed. (1986). "Pattern and Process in Desert Ecosystems." Albuquerque: University of New Mexico Press.

Deserts, Australian

Mark Stafford Smith
CSIRO, Australia

I. Introduction
II. Environment
III. Ecology
IV. Management
V. Future

Nearly three-quarters of the Australian continent is arid or semiarid desert. The flat, well-weathered, and ancient landscape is subject to a climate that is particularly unreliable by global standards. The result is a cascade of ecological implications leading to an unexpectedly high level of vegetation cover, and a greater dominance of social insects and lizards than in most other deserts. Aboriginal people have interacted with this ecology over many millennia, especially through the use of fire. The introduction of domestic and feral animals into the continent 200 years ago precipitated major changes to the ecology and considerable land degradation. A better idea of how to manage this unpredictable environment is now emerging.

I. INTRODUCTION

Australia is the world's largest island nation, and 70% or 5.5 million km^2 of the continent is arid or semiarid; within this area, a core of 4.2 million km^2 (55% of the continent) is arid (Fig. 1). Although Australia as a whole is the second driest continent (after Antarctica), no part of its arid zone receives less than 100 mm annual rainfall on average, and there are no areas that are hyperarid in the sense of the central Sahara. Within the arid core, the Great Sandy, Gibson, Great Victoria, Simpson, and Sturt deserts represent about 1.5 million km^2 and are actually called deserts, in the sense that there is little human use of them. However, this article discusses all the arid lands, which include substantial areas used for grazing. [*See* DESERTS.]

The desert "outback" is a quintessential part of the image of Australia. With Antarctica and the Amazon Basin, it is one of the last great open spaces of the world that is relatively unaltered, yet it is far easier to visit than the others. The isolation of the Australian continent over evolutionary time has allowed the development of a high proportion of endemic species. Furthermore, as this article describes, the continent has an unusual combination of climatic and edaphic factors that affects its ecological functioning in ways that shed light on ecological interactions on other continents. It has a surprising level of perennial plant cover in most areas, partly because of these factors, but also because domestic and feral herds of ungulates only reached inland Australia a century ago. This recent collision with European-based land uses has trig-

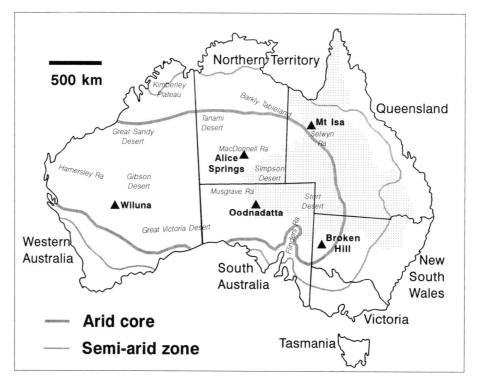

FIGURE I   Map of Australia showing the extent of the arid and semiarid zones, the major "deserts," the main areas of ranges, and other localities mentioned in the text. Stippling shows the approximate extent of the Great Artesian Basin. The boundary of the arid and semiarid zones (thick and thin solid lines) are best defined in terms of rainfall effectiveness; for the arid zone this corresponds to the 250-mm isohyet in the south and the 350-mm isohyet in the north, and for the semiarid zone to about the 375-mm and 750-mm isohyets, respectively. However, the discrimination in common parlance relates more to land use, with grazing being marginal in the arid area and cropping marginal in the semiarid area.

gered radical changes in the deserts, which represent a fascinating management challenge to their occupants.

II. ENVIRONMENT

Inland Australia is vast, flat, ancient, and weathered with well-sorted soils; it is subject to a wide variety of climates, but rainfall is erratic everywhere. Isolated from the other continents since Gondwanaland finally broke up about 50 million years ago, the genetic legacies of the long separation have interacted with the effects of relatively recent aridity. The result is a desert biota with some interesting parallels with other continents, as well as certain unique characteristics.

A. Physical Structure

Inland Australia is generally very flat. The highest point on the continent west of the eastern Great Dividing Range is 150 km west of Alice Springs; here, Mt. Zeil rises to 1500 m from plains that are already at 900 m. Despite their small total area (about 16% of the arid zone), the isolated mountain ranges—such as the MacDonnells, Musgraves, Hamersleys, Selwyns, and Flinders (Fig. 1), as well as smaller outcrops—have important effects on water redistribution.

Apart from the mountain uplands and piedmonts, the main physiographic types are shield deserts (mainly in Western Australia), stony deserts (mainly southern and eastern), riverine and clay plains (scattered throughout, but large areas in western Queensland), and sand deserts (most of the core area). Geologically, the mountains and shield areas are very old, generally being of Precambrian origin; furthermore, some erosional surfaces have been exposed continuously since the Cambrian, some 500 million years ago. Some plateaus represent gently folded Palaeozoic rocks. The Great Artesian Basin—water-holding strata underlying the eastern half of the continent (Fig. 1)—was formed as a result of inundation during the Jurassic and Cretaceous. Over the top of most lower-lying areas is Quaternary alluvium and the wind-blown deposits that form the large sand deserts. Extensive salt deposits have developed during the long history of shallow inundation and exposure; the drying climate of the late Tertiary has left a legacy of extensive salt lakes strung like pearls along ancient, dried-up river courses, or "palaeodrainage lines." [*See* SOIL ECOSYSTEMS.]

Soils include those with uniform, gradational, and texture contrast profiles. Uniform soils include all textures. Most widespread are infertile red siliceous and earthy sands throughout the core of the arid zone. At the opposite extreme are very sticky, fertile cracking clays, often with gilgai micro-relief, on alluvial plains feeding southwest from the eastern half of the continent. Most of the gradational soils are relatively fertile calcareous earths in the southern half of the continent. Texture contrast, duplex soils occur in smaller areas throughout, especially on plains, where they are susceptible to degradation. Inland Australia is dominated by infertile soils, notoriously poor in phosphorus (e.g., total P is 100–300 ppm, available P is 3—22 ppm). The cracking clays, calcareous earths, and alluvial soils may be more fertile. Low phosphorous levels can also limit nitrogen accumulation, and in any case most nutrients are typically held in the top 2–5 cm of soil, where they are highly susceptible to erosion.

Erosion is driven by the two proximate forces of wind and water. Wind erosion has been important in past arid climatic phases during glaciations, giving rise to the large sand deserts dominated by linear dunes, as in the Simpson, Great Sandy, and Great Victoria deserts, with their dune trends reflecting the wind circulation of the time. Wind erosion is less important today, although it can be significant on grazed areas in dry years. Sandy material does not move very far, although fines can be carried many thousands of kilometers.

Water erosion is important only on plains and plateaus surrounding low uplands. However, because these areas are also usually more productive for grazing, erosion patterns have been greatly affected by human use. On these landscapes with low relief, most soil moves only within one "erosion cell" during a single rainfall event: an erosion cell is made up of a source zone, where material is eroding, a transfer zone, through which it passes, perhaps pausing briefly, and a depositional sink zone. Active parts of the landscape can be regarded as a mosaic of interlocking erosion cells at a variety of spatial scales from meters to tens of kilometers, with water movement becoming channeled where gradients increase. Erosion cells can be mapped from satellite data, and it is significant that most landscapes have much larger areas of source zones than of sinks; if a landscape degrades, soil may not move very far, but it may accumulate in relatively small areas, thus burying a significant portion of the productive potential.

B. Climate

Contemporary weather patterns are dominated in northern Australia by tropical influences, and in southern Australia by the passage of frontal systems around the South Pole with a mean periodicity of 6.8 days. This leads to an enormous climatic gradient across the continent (Fig. 2). Rainfall is summer-dominated and cyclonic in nature in the north, which experiences distinct wet and dry seasons without frosts. In the south, rain falls more gently and dominantly in winter, although occasionally cyclonic incursions track across the continent to bring heavy rains in summer. The southern half of the continent experiences significant numbers of frosts. Throughout the continent, mean monthly summer temperatures (January) are around 30°C, with most arid areas experiencing daily maximums of >37.8°C on more than 100 days per year; daily temperature ranges average 10—17°C in both summer and winter. Pan evaporation exceeds rainfall 10-fold in most areas.

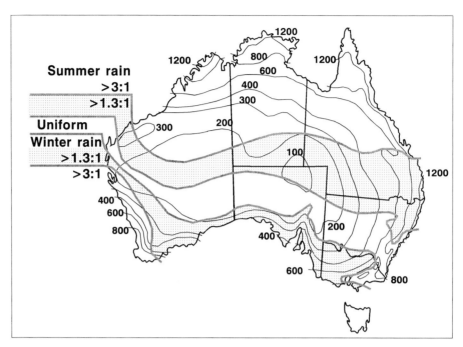

FIGURE 2 Patterns of median annual rainfall and rainfall seasonality in Australia. Note the steep gradients in rainfall around the periphery of the continent, especially in the north. Patterns of rainfall reliability and rainfall effectiveness are generally similar to those of median annual rainfall. (Based on AUSLIG (1992). The Ausmap Atlas of Australia. Cambridge University Press: Canberra, Australia.)

These statistics are relatively consistent from year to year. This is not the case for rainfall; hourly, monthly, and annual totals are severely skewed (e.g., Fig. 3). Rainfall variability in Australia fits with the general global pattern of increasing toward the equator and as annual totals decline. However, it shares an additional influence with India's Thar Desert and southern Africa; the El Niño–Southern Oscillation (ENSO) effect, and other climatic syndromes driven by sea surface temperatures in the Pacific and Indian oceans, causes intermittent periods of severe drought and deluge. For example, the rainfall record at Alice Springs, with an annual mean total of 283 mm (S.D. 144 mm) and median of 263 mm, includes years of 61 mm and 903 mm in its century of weather data; furthermore, the decade of 1958–1967 averaged 178 mm per year, whereas 1968–1977 averaged 435 mm. Perhaps more important is the distribution of periods of drought (Table I); the Alice Springs record includes periods of over a year with no rainfalls exceeding 12.5 mm, and over 5 years with no rainfalls exceeding 50 mm. The resulting variability is remark-

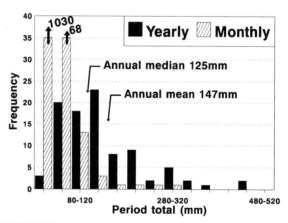

FIGURE 3 Histogram of the sizes of monthly and annual total rainfalls for Oodnadatta, South Australia, for records from 1891 to 1985; each column represents the number of months or years in a 40-mm range (i.e., 0–40 mm, 40–80 mm, etc.). The monthly totals are dominated by small amounts, including zeros of course, but have been as large as 340 mm. The annual totals include whole years with less than 40 mm, but also as high as 460 mm, that is, from less than a quarter to over three times the long-term mean. This greatly skewed distribution is common to many weather stations in Australia.

TABLE I

Length of Periods (in Days) without a Rain of Different Size in the Rainfall Records for Alice Springs, 1872–1985

Event size (mm)	Maximum period	Mean period	SD
12.5	382	62	67
25.0	713	116	120
50.0	1728	288	291

able in global terms and has profound implications for the ecology of arid Australia.

C. Flora

The vegetation communities of the fertile and infertile soils of arid Australia are distinctive (Table II). Of the more fertile soils, the cracking clays are characterized by an open grassland dominated by mitchell grasses (*Astrebla* spp., especially *A. pectinata*); the mobile, self-mulching nature of the soil combines with fire to maintain vast treeless areas. The southern calcareous clays are dominated by low (1–2 m high) chenopod shrubs, most notably saltbush (*Atriplex vesicaria*) and long-lived bluebushes (*Maireana* spp.), sometimes with an open *Acacia* spp. or *Eucalyptus* spp. overstory. Being mainly in the winter-dominant rainfall area, the understory is herbaceous with some short-lived grasses (mainly *Stipa* spp.); there is rarely enough fuel for fire, but most of the shrubs are readily killed if burned. There are also incursions of *Eucalyptus* shrublands and other low woodlands from the semiarid zone (Fig. 4). [*See* GRASSLAND ECOLOGY.]

The infertile bulk of arid Australia is dominated by two vegetation communities, themselves characterized by species of three genera. Australians tend to think of the wattles, *Acacia* spp., as the common plant of the inland; indeed, some 200 species of *Acacia* form the overstory in much of the arid zone. Most dominant is mulga, *Acacia aneura*, which dominates 1.5 million km^2 (20%) of the continent with many variant forms. The acacia shrublands occur mainly on poor red earths and sands, where groved formations around gentle contours sometimes develop ("brousse tigrée"). There is commonly a minor component of many other shrub species, and an understory of perennial grasses in the north or herbaceous ephemerals in the south.

However, "spinifex" is undoubtedly the quintessential plant of the inland, dominant over a third of the continent and a significant component in many other areas, including some with *Acacia* overstory. Spinifex is the common name for the spiky hummock grasses of the genera *Triodia* and *Plectrachne*. Most of the time there is little vegetation in between spinifex hummocks, although a plethora of species appear after rain and fire, partly from seed and to a surprising degree from cryptic,

TABLE II

Main Vegetation Types of the Arid Zone

Formation	Area[a] (million km^2)	Dominant plant genera
Arid		
Shrub steppe	0.43	*Atriplex/Maireana* chenopod shrubs
Acacia shrublands	1.60	*Acacia* shrubs
Hummock grasslands	1.60	*Triodia/Plectrachne* grasses
Arid tussock grasslands	0.50	*Astrebla* grasses
(Salt lakes	0.44	Unvegetated)
Periphery, including semiarid		
Semiarid shrub woodlands		*Eucalyptus/Acacia* trees/shrubs
Mallee woodlands		*Eucalyptus* shrubs
Arid/semiarid low woodlands		Varied, *Eucalyptus*, *Acacia*, *Casuarina*, *Callitris*

[a] The areas are approximate owing to overlaps between zones.

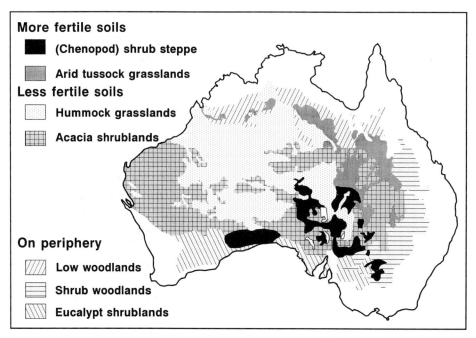

FIGURE 4 Simplified distribution of the major vegetation types of the Australian arid zone and its semiarid periphery. Note the limited contiguous areas of fertile soils. (Based on Williams and Calaby 1985, see bibliography).

dormant rootstock. Spinifex shares many of its soils with the acacia shrublands, and the boundary between the two communities is maintained in many areas by fire. Mulga, like many other longer-lived species of *Acacia,* is intolerant of fire and must recover from seed; the seed is quite short-lived, so that a second fire before establishing plants have reached maturity can locally remove the species. Spinifex recovers from fire at a rate determined by rainfall and temperature; in the south of the continent this takes 20–50 years, in the north only 2–5 years. However, reinvasion of shrubs can occur during a period in which fire is suppressed.

D. Fauna

Without doubt the most prolific animal class is the insects (Table III), with ants and termites preeminent among these. Ants of the genus *Iridomyrmex* dominate to a remarkable degree as elsewhere in Australia, although a range of other genera partition the foraging opportunities to avoid this aggressive group. For example, *Melophorus* spp. emerge as the soil surface temperature rises above about 35°C and other ants hide below ground, and many species, such as *Camponotus* spp., are active at night when the *Iridomyrmex* are relatively subdued. Termites are very widespread, but are particularly dominant in the poorer soil types such as those carrying spinifex grasslands. In these areas, insect predators, such as gryllacridid orthopterans, as well as spiders, are also diverse, taking advantage of the abundant insect prey. Grasshoppers, beetles, and moths are also common. Invertebrates survive either by occupying the more reliable niches in the landscape, like the processionary caterpillars (*Ochrogaster contraria*), or by being mobile, like plague locusts (*Chortoicetes terminifera*), or by having resistant phases, like aquatic crustaceans (e.g., *Triops australiensis*). A significant number inhabit the more reliable environment of perennial roots, such as some 15 species of cicada and the notable "witchetty grubs," which are larvae of several moth species (e.g., *Xyleutes leucomochla*).

Turning to vertebrates, a surprising number of fish species inhabit inland waters, actively swimming upstream during floods to recolonize dried-up waterholes. There are also numerous species of desert-living frogs, some of which actively burrow and cocoon themselves in dry times. Most notable

TABLE III

Approximate Statistics on Some Major Groups of Organisms
in the Arid Zone

Group	No. spp.	No. extinct[a]	No. rare[a]
Insects	Many thousands	?	?
Fish	25	?0	?1
Amphibians	27	0	0
Reptiles	210	0	1
Birds	230	0	31
Mammals	95	11	23
Angiosperm plants	2600	5	34[b]

[a] "Extinct" here means believed completely extinct; "rare," more subjectively, includes species that have shown a major decline, or that have been
lost from the arid areas (e.g., persist only on off shore islands). Numbers
are speculative for insects and approximate for plants. Note the high proportion of mammals lost or at risk.

[b] Approximate numbers of *threatened* species in arid and semiarid areas,
does not include rare species.

among the vertebrates, however, are the reptiles;
there are 40 species of snakes and nearly 200 of
lizards, and spinifex grasslands near Alice Springs
boast the richest recorded lizard fauna in the world.
Although there are large species, such as the sand
goanna, this fauna is dominated by skinks, *Ctenotus*
spp. Farther north in the Tanami Desert, the sand-
swimming legless skinks, *Lerista* spp., are diverse.
Their small body size allows them to show an
astounding degree of niche differentiation, foraging at different times of year or of day, or in different parts of the spinifex landscape, and even in
different parts of individual spinifex clumps.

There are also at least 200 species of birds, including many water birds that find their way to inland
waters after rain. About 40 species are more-or-
less restricted to desert areas: some of these, like
budgerigars and crimson chats, are nomadic,
whereas others, such as the singing honeyeater and
the thornbills, are quite sedentary. The majority
breed seasonally, although their success is strongly
dependent on the quality of season.

Finally, mammals range from the tiny insectivorous planigale (10 g) to 4 species of large herbivorous kangaroo weighing up to 85 kg, and include 18
species of bat. A significant complement of much
larger species is known from Pleistocene records,
when the hippopotamus-sized *Diprotodon* and giant
kangaroos attained 1–2000 kg; many of these seem
to have become extinct in association with a drying
climate, increased fire, and human presence around
30,000 years ago. In the past 200 years, by contrast,
extinctions and range contractions have occurred
primarily among species weighing 50–5000 g;
nearly half the terrestrial species previously present
are extinct or limited to small areas outside the
deserts (Table III). However, with a decrease in
dingo predation and an increase in the provision of
artificial waters, some 7 species of larger kangaroos
(>10 kg) have greatly increased in population in
many regions.

III. ECOLOGY

The key driving forces on the biota are climate,
grazing, and fire. However, these forces are fundamentally modulated by the underlying physical
patterns that vary enormously in space and time.
The result is a cascade of implications for the patterns and process of ecological function in Australian deserts (see List 1).

A. Resource Patterns

Although the landscape is relatively flat and infertile, important spatial patterns have been developed
through sorting over a very long time. The isolated

LIST I
Key Concepts Hypothesized for Arid Australia[a]

1. Rainfall, the principal driving force, is very unpredictable in time and variable in space
2. Unpredictable big rains structure the physical and biotic environments
3. The landscape is ancient, well-sorted, and generally flat and infertile
4. Plant production is highly patterned across the landscape at various scales
5. Soil moisture extremes interact with fertility and fire to constrain feasible plant life-history strategies
6. Low fertility leads to forms of plant production that are indigestible to herbivores
7. Energy is not limiting so carbohydrate is generally plentiful
8. Plentiful, slowly decomposing plant tissue means fire is important
9. Food and nutrients rather than water govern animal life
10. Herbivores in infertile areas must use production that is either poorly digestible or very intermittent
11. Much production in infertile areas goes directly to detritivores, especially termites
12. Localized patches of more reliable production support persistent consumers
13. Social insects are prominent, being able to buffer uncertainty and infertility
14. Reptiles and invertebrates dominate higher-order consumers in poorer landscapes
15. Consumer stability is higher than the unpredictable climate would suggest

[a] After D. M. Stafford Smith and S. R. Morton (1990). *J. Arid Environments* **18**, 255–278.

mountain ranges and even smaller pockets of relief have a profound effect, concentrating runoff at a number of different scales (Fig. 5). Not only is the water supply downstream a little more reliable, but with the water goes soil, seeds, and, most importantly, nutrients. The resulting floodplains, floodouts, and short-lived swamps are critical fertile patches in a sea of poor soils. Even in the more fertile landscapes, where this effect is less pronounced, these run-on areas are still less drought prone.

Equally important are the patterns through time. The occasional years of major rainfall recharge aquifers and sustain water tables under dry riverbeds. They precipitate the growth of plentiful fuel and allow fires to traverse the landscape. In between the good years, the intermittent extreme droughts

kill all but the most tolerant of perennial plants. The ever-changing interplay of unpredictable wet and dry periods means that various life-history strategies are favored at different times and consequently coexist. Different sizes of rainfall events interact with the spatial patterns of runoff and run-on to exaggerate the differential reliability and fertility of the landscape.

B. Biotic Implications

Plants can respond to unreliable resource supplies by seeking to make those resources more reliable in time or in space. To do so in time involves producing a dormant seed stage that can wait for wet times or for bursts of nutrients such as those released by fire; in space, plants must invest in roots to seek out the necessary resources to ensure their survival through harsher times. The Australian arid zone is unexpectedly well-vegetated partly because numerous perennial plants adopt the latter strategy, taking advantage of wet years to establish and grow down to a water table that is relatively reliable; these plants (e.g., *Acacia* spp., *Allocasuarina decaisneana*, *Eucalyptus camaldulensis*, *Grevillea* spp., *Hakea* spp.) may be very long-lived (50–400 years). Other plants adopt shorter-lived strategies: at the extreme is a large ephemeral flora (dominated by Compositae) that appears only after rain or, in some infertile environments like the spinifex grasslands, only after fire and rain. In between, however, are a suite of species of medium longevity (5–20 years), which survive in reasonable years but succumb rapidly to drought (e.g., *Senna* spp., *Eremophila* spp., smaller *Acacia* spp.). As one would expect, these species often have hard-coated seeds that survive well in the soil. On the other hand, one strategy common in other deserts—that of complete succulence as practiced by cacti—is very rare in Australia, probably because of the intermittent occurrence of severe drought.

Although water is the primary limiting factor to growth in deserts, for those perennial plants that do invest in sufficient roots to locate the water table, it is nutrients that become limiting. Light, and hence energy or carbohydrate, is readily available, so many desert plants are surprisingly profli-

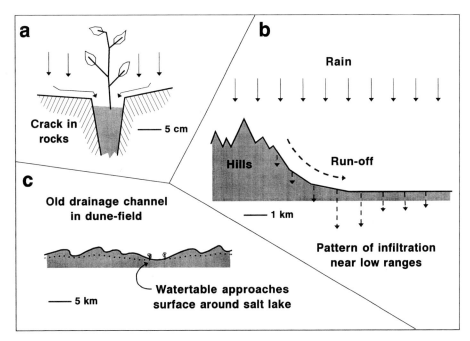

FIGURE 5 The effects of spatial redistribution of water at different scales: (a) local runoff into cracks in rocky areas makes small patches unexpectedly mesic; (b) a similar effect at a landscape scale greatly increases the reliability of run-on areas surrounding ranges of hills, which also receive extra nutrients carried in with the water; (c) water tables recharged by widespread rains may near the surface in old drainage channels even in the most xeric sandy country, thus permitting patches of more reliable production—this is captured by the perennial plants that invest in water-tapping roots, but flows on through all trophic levels.

gate with nectar production (e.g., extra floral nectaries are common in *Acacia* spp.) and in tolerating high loads of sap-sucking insects. Although belowground processes are poorly known, trees probably cope also with heavy loads of belowground herbivory. These food supplies flow through to the fauna, and the populations of perennial plants therefore provide a surprising degree of buffering against the environmental variability, so that consumer populations do not vary as much as might be expected.

Nutrients, however, provide a different story. First, only limited parts of the landscape are fertile. Mammalian herbivores, as well as a particular suite of insects including some grasshoppers, depend on the resulting higher-quality food and focus their activities in these richer patches especially in times of drought. On the broad sweeps of less fertile country, most growth is of such low quality and so well protected by resins that it goes directly into the detritivore pathway, especially to termites. One

result is the diversity of termites; another is the consequent diversity of termite predators, lizards, spiders, and various insect groups.

C. Special Features

Low-fertility soils, unpredictable climates, and redistribution of nutrients occur in many desert landscapes, and co-occur especially in southern Africa. What makes Australia unique in this regard is the geographical extent over which these factors co-occur (about three-quarters of the arid zone) and the fact that they occur in a climate that is only moderately arid. Though this may have led to the *dominance* of termites, ants, and lizards, the *diversity* of this fauna may then be driven by the fact of their small sizes; these allow a far greater niche differentation than is possible for large herbivores like wildebeest and bison, or large predators like lions.

There are other special features of Australia's deserts that are more a result of history, biogeography, or chance rather than of ecological function. These include the presence of marsupial mammals and the dominance, prior to the arrival of domestic stock and rabbits, of many species of plant that were not tolerant of heavy grazing. The lack of intensive grazing and browsing for at least the last 10,000 years may also have resulted in the relatively small numbers of Australian plants that are heavily spiny in comparison to African species.

IV. MANAGEMENT

The underlying ecology and functioning of the Australian desert landscape has been affected by human management for many years, initially at a low density but with profound significance, and recently in a more catastrophic fashion.

A. Aboriginal Land Use

Aboriginal people probably reached the Australian continent about 50,000 years ago. Although not yet proven, it is likely that the hunter-gatherer society combined with climatic changes to increase the frequency of fire, expand grasslands, and reduce the populations of large grazing marsupials. By the time that Europeans arrived in 1788, Aboriginal groups occupied the vast majority of the inland, moving out over the deserts in good times and retreating to reliable water holes in dry times. In some cases the most permanent waters were regarded as sacred and no hunting was normally allowed near them. This may well have been a useful approach to conservation, although the human population itself would have suffered severely in bad droughts, naturally reducing the pressure on the native plants and animals.

As Aboriginal people traveled, they made constant use of fires, for signaling, flushing prey, and clearing areas to walk through; there may also have been deliberate burning to ensure flushes of new growth a year later. The result of this activity was the development of a mosaic of patch burns of different ages; close to their main traveling routes, the patches would be small, farther away they would be larger, being burned less frequently. The net effect was that the whole landscape was subtly

but very effectively managed, and wildfires ignited by lightning would never burn very far. [*See* TRADITIONAL CONSERVATION PRACTICES.]

B. Modern Uses and Abuses

Europeans first settled Australia in 1788; within 50 years sheep and cattle had reached the desert areas, and by 1890 today's extent of desert land use had already been plumbed. With settlers came cats and foxes and, after the 1870s, devastating plagues of rabbits. Other feral animals, including goats, horses, donkeys, camels, and pigs, got out of control in the deserts during this century. At the same time, early grazing practices in almost all regions had two to three times the numbers of stock now considered sustainable on the land; stock numbers built up to unsustainable numbers in sequences of good years, then crashed in extended drought periods, most notably in the 1890s. As the animals died of starvation, land degradation took place. This loss of productivity resulted from ecological changes over several such cycles—palatable perennial vegetation was lost to be replaced by a less reliable annual vegetation, then unpalatable perennial shrubs invaded, further suppressing palatable grasses, and finally, in some areas, significant soil erosion occurred.

Other factors contributed to the changes in productivity. Aboriginal people were settled and ceased their burning practices; grazing reduced fuel levels anyway, and many of the new managers feared fire and tried to suppress it. As a result, the mosaic maintained by Aboriginal practices disappeared. Where the land was too poor for grazing, fuel built up over large areas and there were vast wildfires—up to 30,000 km^2 in one central Australian fire in 1983. Where grazing was occurring, fires were suppressed and a natural control on the balance between grasses and shrubs was removed; shrub seedlings were no longer burned when small, so dense scrubs built up, eventually crowding out the grasses altogether.

A critical factor not adequately appreciated by the new users of the land was that most of the grazing production was coming from very small areas in the landscape—the same fertile patches that were described earlier. These areas, critical for the native mammal fauna and important for Aboriginal people, were also where domestic stock spent most of their time and where feral herbivores such as rabbits, horses, and goats all congregated. It is common for 80% of pastoral production in a paddock to come from less than 40% of the area. The pressure on the more fertile land units was intense, and in many areas they were damaged.

In 1975 it was estimated that two-thirds of the area used for grazing had experienced significant changes in vegetation composition, and vegetation change or erosion was severe enough over a third of these lands to demand active intervention. These figures were very subjective, but, because of the limited fertile areas, they probably underestimate the impact of the changes on productivity and on the overall ecological function. The habitat changes, coupled with the impact of the introduced predators and competition from rabbits, led to the demise of many native mammal species; although no birds are known to have become extinct, many have suffered severe range reductions, and various other animal and plant species are endangered (Table III). [See AGRICULTURE AND GRAZING ON ARID LANDS.]

Today, about 300,000 people occupy the 5.5 million km^2 of the arid and semiarid lands. The main land uses are grazing (ca. 60%), Aboriginal homelands (15%), and conservation reserves (4%); a further 22% is still unoccupied, although ownership is steadily being returned to Aboriginal interests. Much smaller areas are occupied intensively by mining and tourism; however, both mining exploration and recreational four-wheel-driving make very extensive use of the landscape in a less focused way and, with grazing, constitute the major direct human impact on the deserts. There are still large numbers of feral animals (especially rabbits, goats, donkeys, horses, pigs, mice, cats, and foxes), but the populations of the larger species are being reduced; rabbits and foxes are subject to research programs for genetically engineered viral control. A growing issue is that of introduced weeds.

C. Management Principles

Although some degradation is still occurring today, good management principles for coping with the variable environment of Australian deserts are now clearer (see List 2). More than anything this means being prepared to manage the variability through time, and managing the variability in space by protecting the critical parts of the landscape.

The management of the effects of climatic variability on animal production focuses on preparedness for drought and on capitalizing on the opportunities afforded by wet years. Two general approaches to managing stocking rates are possible: to trade in stock to attempt to match grazing mouths with the variable supply of forage; or to operate at low enough stocking rates that they are almost always "safe." The former approach is often more profitable, but harder to do, and far more risky both economically and ecologically. Therefore there is a trend toward lower stocking rates and consequent better reliability and quality of production. Either strategy also requires good control of pest animals and a good understanding of grazing distributions; opportunistic management of exceptional events is essential, such as burning to control the balance between trees and grasses. Additionally, a necessary part of management in such

---

LIST 2

Recommended Suite of Pastoral Management Actions to Cope with Drought: Note That Most Revolve around "Being Prepared"

Before drought
 - plan for flexibility
 - stock lightly
 - maximize per-animal production and quality
 - look after drought reserve country
 - look after areas receiving run-on
 - control pest animals
 - identify important classes of stock to keep
 - set decision dates for staged destocking
 - monitor rainfall prospects, including ENSO forecasts
 - assess economics of destocking strategies
 - plan expenditure so it can be minimized in dry times
 - maintain a cash reserve

During drought
 - destock or agist (move stock to another property) early
 - only graze resilient land types
 - avoid general supplementary feeding
 - maintain control of pest animals
 - protect important conservation habitat

After drought
 - breed back rather than buy in
 - allow pastures to recover before restocking

---

variable environments is adaptive feedback, that is, monitoring the response of the landscape and modifying management accordingly. This applies for both land managers and governments, and systems are being developed that use a mixture of satellite and ground technologies to monitor the vast areas.

Although these approaches may allow the maintenance of enterprise-level sustainability in the marginal grazing areas, pastoralism and other land uses must be integrated to meet the regional goals of ecologically sustainable land management. To maintain the ecological functioning of a region, and to thereby sustain its biodiversity, it is vital to protect the fertile patches. Because these are critical to most productive uses of the land, they are the inevitable focus of land use conflict. Regional planning must therefore not only develop a representative system of reserves that contains important examples of such patches. It must also ensure off-reserve habitat protection in association with surrounding land users, whether pastoralists, tourist operators, or Aboriginal interests. This principle is starting to be applied but still needs considerable development before the ecology of the landscape is fully reflected in our management.

V. FUTURE

The future of inland Australia is important for several reasons. Scientifically, the challenge of understanding and managing the problems posed by the extreme levels of spatial and temporal heterogeneity is fascinating. Such heterogeneity is common to most environments, but normally not so forcefully expressed: consequently science in most parts of the world can often cope by regarding the heterogeneity as noise around a norm. In the desert environment, the noise *is* the norm, and the environmental biology of both ecological function and management derives directly from that variability, as this article has sought to show.

From the management point of view, Australia is lightly populated and fortunate in its technological status. There is therefore the opportunity to develop solutions to fundamental problems that would be hard to tackle in the hurly-burly of land use and population pressures elsewhere; these solutions may then be applied on those other continents. A major example is that of global change, which is likely to push the climatic zones of the Australian deserts (Fig. 2) southward, but also exaggerate the occurrence of extreme wet and dry events still further. The management of these changes is a challenge for all nations with arid areas.

Glossary

Aboriginal people Original hunter-gatherer inhabitants of the Australian continent, who arrived there at least 40,000 years B.P., and who affected the vegetation of the deserts extensively through their use of fire.

Feral animals Various animal species that have been introduced to the isolated continent of Australia, mainly since Europeans arrived 200 years ago, and have since returned to the wild, often with devastating effects on native flora and fauna. The species include horses, donkeys, pigs, goats, cats, foxes, rabbits, rats, and mice, as well as various fish, the cane toad, millipedes, and insects. There are also numerous species of introduced plants that have become weeds.

Land degradation Reductions in the potential productivity of the land caused by human-related activities, in particular grazing of domestic and feral herbivores; the immediate effects vary from modest changes in vegetation composition to severe soil erosion, whereas longer-term effects include the loss of biodiversity and biological function. Logically, land degradation must be defined in terms of a particular land use, whether local or regional. Often used interchangeably with "desertification," although the latter can include climatically induced changes as well as anthropogenic effects.

Life-history strategy Combination of establishment, growth, and reproduction characteristics of plant species (and the equivalents in animals) that identify the ways in which those species interact with their environment; thus in uncertain arid areas, plants tend to either tolerate or avoid extreme conditions through strategies involving major investments in roots (perennial plants) or seeding (ephemerals), but there are many variants on these.

Opportunistic management Management that recognizes that variable climates, such as those of the Australian deserts, mean that most changes are "event-driven," caused by extreme events such as big rains, fires, or droughts, and makes the most of these opportunities for change, rather than attempting to intervene when the ecosystem is not prone to change.

Sustainable Buzzword of the decade with many definitions, "sustainable" has two important uses in relation to land management: *sustainable land use* is the use of a single unit of land in a way that does not cause a loss of future productivity for that land use (i.e., does not cause land degradation); *ecologically sustainable land management* is a regional concept that means maintaining the ecological function of a whole region, including aspects such as water quality and conservation of biodiversity. The latter is not possible on every unit of land (or wheat fields would not be feasible), but because of the large scale of land use, the two concepts are often confused in deserts.

Bibliography

Barker, W. R., and Greenslade, P. J. M., eds. (1982). "Evolution of the Flora and Fauna of Arid Australia." Frewville, South Australia, Australia: Peacock Publications.

CSIRO Division of Soils (1983). "Soils: An Australian Viewpoint." Melbourne, Australia: CSIRO & Academic Press.

Foran, B. D., Friedel, M. H., MacLeod, N. D., Stafford Smith, D. M., and Wilson, A. D. (1990). "The Future of Australia's Rangelands." Melbourne, Australia: CSIRO.

Harrington, G. N., Wilson, A. D., and Young, M. D. (1984). "Management of Australia's Rangelands." Melbourne, Australia: CSIRO.

Heathcote, R. (1983). "The Arid Lands: Their Use and Abuse." London: Longman.

Mabbutt, J. A. (1977). "Desert Landforms." Canberra, Australia: Australian National University Press.

Stafford Smith, D. M., and Morton, S. R. (1990). A framework for the ecology of arid Australia. *J. Arid Environments* **18,** 255—278.

Stafford Smith, D. M., and Pickup, G. (1993). Out of Africa, looking in: Understanding vegetation change and its implications for management in Australian rangelands. *In* "Rethinking Range Ecology: Implications for Rangeland Management in Africa" (R. H. Behnke and I. Scoones, eds.), pp. 196–226. London: Commonwealth Secretariat, Overseas Development Institute and International Institute for Environment and Development.

Van Oosterzee, P. (1991). "The Centre: The Natural History of Australia's Desert Regions." Sydney, Australia: Reed Books.

Williams, O. B., and Calaby, J. H. (1985). The hot deserts of Australia. *In* "Hot Deserts and Arid Shrublands" (M. Evenari, I. Noy-Meir, and D. W. Goodall, eds.), pp. 269–312. Amsterdam: Elsevier.

Diversity Crises in the Geologic Past

Douglas H. Erwin

National Museum of Natural History

I. Introduction
II. Major Diversity Crises
III. Minor Diversity Crises
IV. Impact of Past Diversity Crises
V. Controls on Biotic Diversity

Diversity crises have occurred episodically throughout the history of life, with magnitudes ranging from minor, regional events affecting only a few lineages to extensive, worldwide mass extinctions removing a large part of the earth's biota, events that have often redirected the history of life. Inadequate preservation and sampling of fossils, imprecise taxonomy, and other difficulties have generally forced paleontologists to chronicle changes in the diversity of families and genera instead of species. Such synoptic compilations reveal at least five major mass extinctions over the past 550 million years, some only affecting marine organisms, others both marine and terrestrial ecosystems. Although extinctions reorganize ecosystems and often create significant evolutionary opportunities, postextinction rebounds following major mass extinctions require millions of years, following a lengthy interval of low-diversity, low-complexity ecosystems.

I. INTRODUCTION

Biodiversity crises punctuate the history of life over the past 650 million years (Ma). These sudden changes in the biota were used by geologists in the 1800s to define the periods of the geologic time scale (Fig. 1). Over the past decade the claim that mass extinctions, the end–Cretaceous (K/T) event in particular, were caused by the impact of comets or meteorites has received considerable attention. This hypothesis has, in turn, generated detailed investigations of many global biodiversity crises and evaluation of a variety of other proposed mechanisms, ranging from global cooling and glaciation to oceanic anoxia, massive volcanic eruptions, and marine regressions. In most cases rapid climatic change has been invoked as either a direct agent of extinction or as a correlative process. Stunning advances have been made in understanding these events, as additional well-resolved biostratigraphic data are collected and more rigorous sampling and data analysis procedures are developed. However, inadequate preservation and difficulties distinguishing between correlation and causality continue to plague this area of research. [*See* EVOLUTION AND EXTINCTION.]

Becoming a fossil is a remarkably difficult thing to do. Those species that survive the preservational process and make it into the fossil record are most commonly composed of durable skeletons or shells, rapidly buried, and belong to abundant, broadly distributed species. There are numerous exceptions of course, but such species comprise the bulk of the fossil record. Thus marine species are

TOTAL DIVERSITY

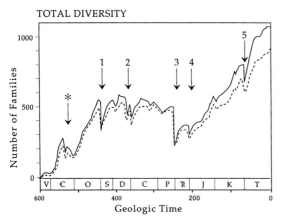

FIGURE I   The number of durably skeletonized marine families through the Phanerozoic. The graph is based on a synoptic compilation of the paleontological literature by J. J. Sepkoski, Jr., of the University of Chicago. The dotted line shows Sepkoski's compilation for 1982 while the heavy line shows the data as of 1993. The five major mass extinctions are shown: (1) the end-Ordovician; (2) late Devonian; (3) end-Permian; (4) end-Triassic; and (5) end-Cretaceous. The asterisk notes the occurrence of several apparent biodiversity crises during the Cambrian. The geologic periods are shown along the bottom of the graph: V, Vendian; C, Cambrian; O, Ordovician; S, Silurian, D, Devonian; C, Carboniferous; P, Permian; Tr, Triassic; J, Jurassic; K, Cretaceous; and T, Tertiary.

more likely to be preserved than species that lived on land (since the land is being eroded and deposited in the oceans) and shallow-water marine species are more likely to be preserved than deeper-water forms. In modern shallow-water communities generally less than half the species have durable skeletons. Not surprisingly, the fossil record of soft-bodied, poorly preserved forms is spotty to nonexistent. However, studies of Pleistocene (1.5 Ma) molluscs in California have demonstrated that the only Recent species missing from the fossil record are small, rare, or have delicate shells, suggesting that most durably skeletonized species may make it into the fossil record. Although detailed studies of vertebrates have not been conducted, the pattern appears to hold true for them as well. The record appears to get progressively worse as one goes back in time, however, as fossils are destroyed by chemical processes and fossiliferous rocks are eroded or heated and metamorphosed. Paleontologists estimate that, overall, less than 10% of the well-skeletonized marine species are preserved in fossil record and that most of these are still awaiting discovery and description. The reliability of the fossil record is further hampered by inadequate sys-

tematic treatment or publication (many groups have not been restudied since early in this century).

These problems, while significant, should not be overestimated. Analysis of the history of biotic diversity is based on synoptic compilations of marine families and genera, vertebrate families and genera, and families of insects. Plants have been summarized at the species level. Paleontologists have sound reasons for believing that such compilations provide a fairly good record of biodiversity over the past 545 million years. The record of well-skeletonized marine genera and families appears to be more than adequate for analyzing changes in biotic diversity, reconstructing community assemblages, identifying changes in biogeographic provinces, and describing large-scale evolutionary trends. The terrestrial record of vertebrate, plant, and insect families, although not as good as the invertebrate marine record, nonetheless appears sufficient to address many of the same questions. Indeed, the fossil record is the only available source on the response of populations, species, and ecosystems to past environmental crises.

The difficulties interpreting the fossil record have an important implication for comparing past diversity crises to current concerns about biodiversity. Many of the species that are currently disappearing have (or, more appropriately, had) small population sizes, limited geographic distribution, and limited preservation potential. Since extinct species with such characteristics have generally vanished without a trace, paleontologists have no direct way of determining how they were affected by past extinction episodes or whether seemingly minor crises actually removed a large fraction of these poorly preserved species. While not belittling the significance of the problems we now face, it is equally important not to overstate the comparisons. At present we appear to be experiencing an interval of intense background extinction instead of mass extinction. [See MASS EXTINCTION, BIOTIC AND ABIOTIC.]

II. MAJOR DIVERSITY CRISES

Five great diversity crises have been identified from the fossil record of marine invertebrates (Fig. 1).

A variety of less significant diversity drops are also evident in Fig. 1. Some of these are geographically or taxonomically restricted events while others are simply less intense global bioevents.

The first known marine biodiversity crisis actually does not appear on Fig. 1 since it occurred over 900 million years ago. The earliest fossils appeared 3.5 billion years ago, but until about 650 million years ago the fossil record of life was limited to the microscopic remains of various single-celled organisms embedded in chert and the sedimentary structures these microbes constructed. Many microbial communities secreted calcium carbonate or trapped passing sedimentary grains. These activities produced finely laminated mats, domes, and columns known as stromatolites. Curiously, many stromatolites are very distinctive and some morphologies are restricted to particular intervals of time. About 900 million years ago the number of different types of stromatolites began a precipitous decline. This diversity crisis is believed to be due to the first appearance of burrowing and grazing animals; today stromatolites are limited to high-salinity bays and other settings.

According to some paleontologists, a series of biodiversity crises occurred during the initial radiation of animals during the late Proterozoic and Cambrian. An assemblage of unusual, soft-bodied animals, most of them probably related to cnidarians, was found in 650–550 million year old marine sandstones in many regions of the world. Since few of the animals in this Ediacara Fauna, as it is known, are found in younger rocks, a diversity crisis may have wiped out most of these animals just before the beginning of the Cambrian explosion of animals.

The Cambrian radiation involved the rapid appearance of essentially all well-skeletonized animals as well as a whole host of poorly skeletonized forms. Many of the common fossils of the Early Cambrian disappeared quickly, suggesting that a mass extinction may have occurred near the end of the Early Cambrian. In any event, there was considerable species turnover near the end of the Early Cambrian. Later, during the Middle and Late Cambrian, trilobites repeatedly flourished in shallow seas, only to be wiped out as the sea level dropped and then expanded again with the next rise in sea level.

The first of the five great Phanerozoic mass extinctions occurred at the end of the Ordovician Period, with the elimination of about 61% of marine genera (Table I). Although only the end-Permian extinction was more extensive, the Ordovician crisis had only a moderate effect on the history of life. Indeed, it probably had less effect on the history of life than any of the other five major crises. This limited impact on marine ecosystems may reflect that although all marine groups were affected, no major groups of animals became entirely extinct, and hence ecological communities were able to reform relatively quickly. Each of the other major extinction events wiped out major components of marine or terrestrial ecosystems and thus were far more disruptive. The end-Ordovician extinction occurred in several steps over 1–2 million years, and apparently began in North America before spreading to Europe. A major glaciation and global cooling event coincides with the extinction interval and, along with resulting changes in sea level and the amount of habitat area, is believed to have triggered the extinction.

A second interval of heightened extinction occurred during the Late Devonian, culminating in

TABLE I

Percentage Extinction of Marine Genera, Estimated Percentage Species Extinction, and Percentage Extinction of Families of Nonmarine Vertebrates during Major Phanerozoic Mass Extinctions

Extinction	Age (Ma)[b]	Generic extinction (%)	Species extinction (%)	Nonmarine families (%)
Early Oligocene	30	—	—	10
Late Eocene	35	15	35 ± 8	—
End-Cretaceous[a]	65	47	76 ± 5	14
Late Cenomanian	90	26	53 ± 7	—
End-Jurassic	146	21	45 ± 8	—
Early Jurassic	187	26	53 ± 7	—
Late Triassic[a]	208	47	76 ± 5	22
End-Permian[a]	250	84	96 ± 2	58
Late Devonian	367	55	82 ± 4	—
Late Ordovician	439	61	85 ± 3	—

Note. Marine generic and species data from D. Jablonski and J. J. Sepkoski, Jr., University of Chicago; nonmarine tetrapod extinction data from M. J. Benton, University of Bristol.

[a] The marine and nonmarine extinction episodes appear to correlate.

[b] Ages in millions of years.

the extinction of some 55% of marine genera; terrestrial ecosystems were still poorly developed and there is no indication of a correlative extinction on land. The available data indicate that this extinction lasted over several million years, but whether there were particular extinction peaks within this interval or a more gradual extinction is not clear. Nonetheless, biodiversity, ecosystem diversity, and biomass all plummeted during this event. Tropical, shallow-water and reef ecosystems were particularly hard hit and coral-dominated reef ecosystems did not really recover until the Triassic (Fig. 2). (Carboniferous and Permian reefs were dominated by an unusual assortment of algae, sponges, and bryozoans; there were few real reef builders during this time.) Thus because of the groups that were eliminated, the Devonian biodiversity crisis had a greater long-term effect than the larger Ordovician event. In Europe the extinction coincides with a series of black limestones and shales known as the Kellewasser horizons. Geochemical analysis of these rocks suggests they were deposited in low-oxygen waters. This extinction may have been caused by an incursion of such dysaerobic waters from the deep ocean onto the shallow marine shelves. Although small glassy spherules have been found near the extinction interval, there is little evidence of an extraterrestrial impact.

The greatest diversity crisis of the Phanerozoic took place 250 million years ago at the close of the Permian. Over a span of 2–3 million years 49% of marine families disappeared and 72% of marine genera. The drop in species diversity is difficult to estimate, but statistical techniques suggest over 90% of marine species became extinct. As pervasive as this event was, the pattern of extinction was far from uniform. Organisms that lived attached to the seafloor (epifaunal sessile forms) and suspension feeders were particularly hard hit. Mobile forms such as gastropods, bivalves, and others that moved or burrowed in the sediment were less heavily affected. Shallow-water forms seem to have been more heavily affected than deeper-water species as well. On land, vertebrates suffered a fairly severe extinction with over 75% of families disappearing, but this was one of several biotic crises during the

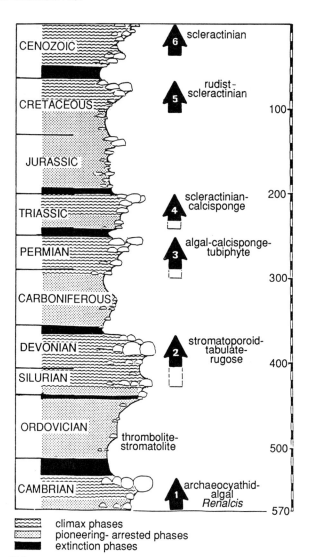

FIGURE 2 The diversity history of reefs through the Phanerozoic. Geologic systems are along the left, showing the development of six different successions of reef development, of which the first five ended with mass extinctions. Each phase of reef development begins with a pioneering phase with the development of occasional patch reefs and leading to a climax phase of well-developed reefs, each characterized by a different set of dominant reef formers. Cambrian reefs are dominated by archaeocyathids (unusual sponges) and the alga *Renalcis,* while stromotoporoids, another unusual group of sponges, and Paleozoic corals predominate in Siluro–Devonian reefs. After a long absence, reefs return during the Permian, largely composed of algae and sponges. Reefs return again during the middle Triassic with scleractinian (modern) corals and calcareous sponges forming the framework. These reefs are eliminated by the end-Triassic mass extinction. A long period of reef development finally climaxes during the mid- and late-Cretaceous with scleractinian corals and the large, odd rudist bivalves. Following the end-Cretaceous crisis, truly modern reefs appear. Note that during many periods of geologic time that environments conducive to reef development (for example, warm, shallow tropical platforms) were present but reefs did not develop because good reef-forming taxa were not present. [From Copper (1988).]

Permian and Triassic. Other than an interesting change in pollen and a brief increase in the abundance of fungal spores, plants show little evidence of mass extinction across the Permo–Triassic boundary. The quality of the plant record is not high, however, and further study may reveal more extinctions. More interestingly, however, insects experienced the only major extinction event in their history, triggering a major transition from several archaic forms to the groups that dominate today. Taken together, the terrestrial groups indicate that the cause of the extinction affected both land and the oceans.

The Late Permian and Early Triassic was an interval of pronounced physical change. Virtually all of the continents, other than those which presently comprise east Asia, were united into a single land mass called Pangea. A gradual withdrawal of the seas led to the exposure of much of Pangea during the latest Permian, with resulting climatic instability. This was followed in the very latest Permian and Early Triassic by a very rapid rise in the sea level by perhaps 180 m. In addition, the largest known flood basalt province erupted in Siberia at the Permo–Triassic boundary and may have caused severe climatic disruptions, including both rapid cooling and warming. Finally, studies of shifts in carbon, oxygen, sulfur, and strontium isotopes have revealed sharp changes in ocean chemistry across the Permo–Triassic boundary, although the interpretation of these shifts is not unambiguous.

Understanding the cause of this extinction is difficult because none of the proposed causes seem capable of causing such a massive extinction. There is no evidence of extraterrestrial impact, which restricts our attention to those mechanisms that affect both oceanic and terrestrial ecosystems. It thus appears likely that a multitude of causes were involved in the extinction, including the loss in habitat area and climatic effects of sea level drop, oceanic anoxia, and the impact of the Siberian volcanism.

In many ways the evolutionary consequences of this diversity crisis are more interesting than the extinction itself. The end-Permian mass extinction eliminated the major marine community types that

had dominated since the Ordovician. In the aftermath of the extinction a variety of new forms appeared, leading to the establishment of new community types in which active burrowers and predators were far more common. In fact, modern marine ecosystems really owe their origin to the end-Permian mass extinction.

The end-Permian crisis was so pervasive that it was followed by a 5 million year period of low diversity before marine communities recovered, but by the middle of the Triassic, marine ecosystems seemed fully rejuvenated. Reefs flourished again in tropical environments, and marine diversity had rebounded. But just when marine and terrestrial ecosystems seemed to be well on their way to recovery, the end-Triassic mass extinction intervened, eliminating about 23% of both marine and vertebrate families. Cephalopods were particularly hard hit and many of the marine groups that disappeared were Paleozoic holdovers that had just survived the end-Permian mass extinction. Several groups of marine vertebrates disappeared. Some investigators feel that there were several phases to this extinction, but most evidence points to a single episode, although not necessarily a catastrophic event. The cause of this extinction is uncertain. Some evidence of an extraterrestrial impact has been found near the extinction horizon, but the relationship to the extinction remains unclear. As with the earlier extinctions, the cause appears to lie in climatic change and a drop in sea level. In the aftermath of this extinction the earliest dinosaurs radiated, coming to dominate terrestrial ecosystems during the Jurassic and Cretaceous Periods.

The extinction of the dinosaurs has long captured the interest of the public and paleontologists alike. Discovering that the culprit was an extraterrestrial impact has made the end-Cretaceous mass extinction the best-known of all the mass extinctions and the most intensively studied. The discovery in 1980 that rocks spanning the Cretaceous/Tertiary boundary in Gubbio, Italy, had anomalously high levels of the element iridium sparked a revolution in the earth sciences. Iridium is rare on the earth but is far more common in comets and meteorites, which suggests that the iridium may have been deposited after the impact of a large extraterrestrial object. Although some geologists have argued that high iridium concentrations may occur through volcanic activity or chemical processes, there is additional evidence supporting an impact at the K/T boundary. Geologists have now found evidence of an extraterrestrial impact at more than 100 boundary sections around the world; the apparent impact site is located in Mexico. The Chixulub crater is 300 km in diameter and lies buried in the northern Yucatan peninsula.

Prior to 1980 most geologists had rejected catastrophist explanations, particularly those involving extraterrestrial events, since they appeared difficult to test and hence unscientific. The recognition that high iridium concentrations was an indicator of at least some types of impact provided a means to test such ideas and gave new credence to catastrophist explanations.

Granted that an impact occurred at the K/T boundary, but does the evidence suggest that this was the primary cause of the extinction? It is entirely possible that the impact simply coincided with an ongoing extinction and was at best a contributory extinction mechanism. If this were the case then the fossil record should indicate that extinctions began well before the impact horizon. The fossil record is sufficiently spotty, however, that for many species even a very catastrophic extinction appears gradual. Hence paleontologists must examine the record of species with exceptional preservation. The constant rain of calcareous and siliceous organisms on the ocean floor provides such a record and provides persuasive evidence for a very rapid extinction. Other evidence does suggest that some species may have begun disappearing before the impact, but the major cause of the extinction appears to be the impact. But how did the impact cause the extinction? Geologists continue to debate this point, but the impact would have produced a dust cloud that would have circled the globe after several months, blocking out sunlight and cooling the globe. Additionally, acid rain, unchecked wildfires, and the direct effects of the blast (tidal waves, earthquake, etc.) probably exacerbated the extinction.

About 47% of marine genera disappeared during this extinction, but only 14% of vertebrate families.

Most marine groups experienced some degree of extinction. Many skeletal microfossils, like foraminifera, were very hard hit by the extinction, suggesting that oceanic ecosystems were particularly disrupted. Of the marine invertebrates, the large rudistid and inoceramid bivalves disappeared well before the latest Cretaceous. The ammonite cephalopods, another important Cretaceous group, seem to have been decimated by the impact, although exceedingly careful collecting was required to establish this pattern.

On land, large animals suffered the most, although freshwater species (turtles, crocodiles, etc.) and scavengers and detritivores were among the least affected. Mammals did not experience much extinction, yet the dinosaurs were completely wiped out (except for the birds of course). Unlike the end-Permian extinction, when plants were seemingly not affected by the extinction, there was a dramatic change in plants at the K/T boundary. Fern spores became far more common for a brief period, as they do today after a severe forest fire, and early Cenozoic floras were quite different from those of the late Cretaceous.

The creative effects of mass extinctions were again on display in the aftermath of the K/T extinction. Both flowering plants and mammals rapidly diversified and established themselves as the predominant groups on land. Marine ecosystems had been less disrupted so no major changes were evident. Fish expanded, replacing the ammonite cephalopods, and scleractinian corals took over the reef-building role that had been temporarily occupied by rudistid bivalves.

As revolutionary as the impact scenario was, another offshoot of the hypothesis was even more radical: that mass extinctions had occurred about every 26 million years over the past 250 million years (the precision of the data is too poor to reveal if the cycle continues into the Paleozoic). The periodicity hypothesis was based on statistical analysis of the data shown in Fig. 1. Although the pattern is not evident in this figure, the analysis suggests fairly regular spacing of extinction peaks from the end-Permian into the Miocene, and includes both the end-Triassic and K/T episodes as well as smaller events shown in Table I. Curiously, this pattern roughly coincides with a number of known impact craters (although most are too small to have caused significant extinction) and the eruption of extremely large volcanic eruptions known as flood basalts. Since the cycle includes the K/T event and geologists have been unable to identify an earth-bound process that could produce such a cycle, attention has focused on periodic extraterrestrial bombardment as a cause. There are many difficulties with this hypothesis as well, and some paleontologists continue to question the reliability of the extinction data.

This discussion of diversity crises concludes with two smaller Cenozoic events that are also of some interest: the Eocene/Oligocene boundary and the disappearance of many large vertebrates during the Pleistocene.

The Eocene/Oligocene fits into the periodic extinction cycle and there is some evidence for an extraterrestrial impact. However, current evidence suggests that the extinctions covered a period of perhaps 10 million years, and thus involves more than an impact. Calcareous and siliceous microplankton underwent a series of extinctions, as did some shallow-water marine invertebrates, particularly molluscs. On land, plants provide evidence for at least two periods of global warming followed by global cooling, with the final cooling event being particularly severe. Vertebrates show a marked drop in diversity from the mid- to late Eocene, largely among crocodiles, turtles, and several mammal groups. Many of the mammals that disappeared in North America were considered "archaic" forms and were replaced by species better adapted to the less forested terrain that appeared at this time. Thus there is considerable evidence that the extinctions and faunal replacements were intimately linked to climatic changes. Of particular interest are suggestions that the climate changes were linked to the initial development of ice in Antarctica and the changing patterns of oceanic circulation caused by continental drift. Finally, it is worth remembering that much of our ability to reconstruct these complex patterns of change, particularly on land, reflects the relatively short interval since the Eocene. Had this occurred during the Paleozoic, paleontologists would have far greater difficulty discriminating among the several extinction pulses.

The earth experienced numerous episodes of extensive continental glaciation, global cooling, and drops in sea level during the Pleistocene. Although cooling and glaciation have been invoked as causing earlier extinction episodes, the extinctions associated with glaciation were relatively minor. The major late Pleistocene extinction focused on large terrestrial mammals. In North America, 37 genera of large mammals disappeared, South America lost 46 genera, Europe lost 13 genera, and an uncertain number were lost in Africa and Asia. The extinctions occurred at different times on different continents, at times at least roughly coincident with the first appearance of modern humans in each area. The association between the appearance of humans and extinction led to one of the two most common explanations for these extinctions: the overkill hypothesis. According to this hypothesis the spread of humans into new regions led to overhunting and to the eventual elimination of the major prey species. In regions where humans had long been established, the extinction may reflect the development of new hunting techniques. In support of this hypothesis, the disappearance of many birds on Pacific islands also tracks the migration of groups across the Pacific. The alternative mechanism for the extinctions is climatic and environmental change associated with the end of the last glaciation. The difficulty, however, is that continental glaciation has waxed and waned for several million years with no apparent effect on the biota. It is not clear why such environmental changes would only cause extinction at the close of the last glacial episode. Some regional marine extinctions did occur during the Pleistocene, but these were relatively minor. For example, many molluscs (and presumably other marine invertebrates) along the Atlantic seaboard became extinct during cooler episodes because they could not migrate far enough south (Florida was as far south as they could move). In contrast, no such extinctions occurred along the Pacific coast because species could migrate north and south with little difficulty.

III. MINOR DIVERSITY CRISES

As dramatic as these major mass extinctions appear, they account for less than 10% of species extinctions during the Phanerozoic. Most species disappeared during the intervals between major events, as so-called background extinctions. Some became extinct in limited diversity crises that affected only a few groups or a limited region, others simply disappeared in ones and twos. Does this suggest some dichotomy between mass and background extinctions? Does the selectivity of mass extinctions somehow differ qualitatively as well as quantitatively from background extinctions? Some research suggests that during mass extinctions some groups may have disappeared not because they were poorly adapted during background times, but simply because the regions or environments in which they lived suffered preferential extinction. Normally, species extinction rates are inversely correlated with geographic range: the greater the species range the lower the extinction rate. Similarly, generic extinction rates are generally correlated with the number of species per genus (species richness), the safety in numbers principle. During a number of mass extinctions, particularly the K/T episode, these principles may fail and since species richness does not buffer against extinction, geographic range of the genus is advantageous. Since the latter is not a species-level property it is not an adaptation and cannot be selected for. In other words, some data suggest that mass extinctions are qualitatively as well as quantitatively different than background extinctions in that the characteristics that buffer against extinction during normal times fail during at least some mass extinction events. If this is so, it raises the question of the long-term efficacy of "normal" evolutionary processes, currently a contentious topic among evolutionary biologists.

David Raup of the University of Chicago has even proposed that most extinction horizons, mass extinctions as well as most smaller events, are due to extraterrestrial impacts. In his view smaller extinctions reflect smaller objects; larger extinctions reflect the impact of larger objects. If this hypothesis is true, it suggests that the history of life may be more punctuated than previously believed. As yet there is no evidence from the rock record to support this idea. Nonetheless, it does illustrate how thoroughly the impact hypothesis has revolu-

tionized our thinking about catastrophic extinctions.

IV. IMPACT OF PAST DIVERSITY CRISES

The history of past diversity crises teaches us several important lessons about the history of life. First and foremost is the recognition that both background and mass extinctions are simply a fact of life. Some level of species loss is inevitable and is actually beneficial because it helps keep the evolutionary process going. Second, mass extinctions are most commonly due to a set of environmental stresses that push species and ecosystems beyond their ability to adapt. Climatic change often seems to be implicated in these episodes, which suggests that as humans continue to modify the climate, increased rates of extinction can be expected. At the same time, the response of ecosystems (or, in many cases, the lack of response) to Pleistocene climatic fluctuations is a reminder of the resilience of ecosystems as well. Decreased environmental heterogeneity also appears linked to reduced biotic diversity.

The biotic recoveries following mass extinctions often have an important impact on macroevolution. As noted earlier, the end-Ordovician extinction did not eliminate any major groups, and thus had little impact on the composition of Silurian marine communities. In contrast, the end-Permian mass extinction was so extensive that it broke the hold of many groups and liberated a variety of taxa that had been of minor significance during the Paleozoic. This triggered the formation of an entirely new set of communities during the Mesozoic. In many ways the composition of a coastal tide pool today reflects which groups lived and died during this extinction. On land, the post-Triassic recovery allowed the dinosaurs, previously a minor group, to diversify rapidly. Similarly, after the K/T extinction, mammals radiated even more extensively. New evidence suggests that the success of flowering plants, which first appeared during the Cretaceous, was also a produce of the extinction.

V. CONTROLS ON BIOTIC DIVERSITY

Global diversity is frequently divided into three components: within-community species richness (α diversity), differentiation between communities within a region (β diversity), and differentiation between regions or biotic provinces (γ diversity). Together these three produce the global diversity patterns discussed earlier, and thus changes in any of the three could be responsible for variations in global diversity. During mass extinctions all three elements are generally affected, but there is far more to global diversity than occasional mass extinctions (Fig. 1). Particularly noteworthy is the extensive Cambrian–Ordovician radiation, the Silurian–Permian diversity plateau, and the consistent increase in diversity from the Triassic to the Recent. Although Fig. 1 is based on marine families, vertebrates and plants exhibit similar trends. What is the relative contribution of α, β, and γ diversity to these global trends?

Within-habitat diversity more than doubled during the Paleozoic, with much of the increase coming from a greater number of types of animals filling a particular ecologic role. During the Cambrian, trilobites and a few worms were the only active shallow burrowers that removed food from the sediment (deposit feeding). Today, trilobites are long extinct, but species representing a host of other groups occupy the same ecological role, including bivalves, other worms, echinoids, holothuroidian echinoderms, and others. Some of the increase in diversity may have been generated by dividing the available resources more finely, but testing this in the fossil record is generally difficult. When all these factors are taken into consideration, however, increases in α diversity only seem to account for some 50% of the increase in global diversity between the Cambrian and the present.

γ diversity appears to be closely related to the number of biotic provinces through time. As the continents split and coalesce the number of provinces changes as well. During the lower Paleozoic the number of provinces was quite low (less than one-fourth of today's number). The number rose toward the end of the Paleozoic, but then plunged as a result of the end-Permian extinction and the collision of virtually all the continents during the Permian to form the supercontinent of Pangea. Since the Late Triassic, Pangea has split apart, creating the Atlantic Ocean and dramatically increasing the number of marine biotic provinces to 30–35. Because the oceans are north–south, instead of girdling the equator, continental positions increase the temperature gradient between the equator and the poles, raising biotic diversity. Taking all of γ diversity into account still does not account for the other 50% of the Phanerozoic increase in global diversity.

The development of new community types is more poorly studied than α or γ diversity, but appears to account for the remaining increase in diversity, including the development of a variety of reefs and other organic buildups and the construction of communities in new environments. A number of important questions remain about the controls on biotic diversity and the impact mass extinction has on this diversity.

Glossary

Background extinction Relatively constant disappearance of low numbers of species.

Diversity Number of taxa present during an interval of time.

Mass extinction Major losses of biodiversity that are global in extent, involve many clades, and occur in a geologically brief interval of time; may involve the marine or terrestrial realms, or both.

Phanerozoic Geologic interval of the last 545 million years, with visible animal fossils and divided by major extinction episodes into the Paleozoic era (545–250 Ma), Mesozoic era (250–65 Ma), and Cenozoic era (65 Ma–today).

Bibliography

Briggs, D. E. G., and Crowther, P. W. (1990). "Paleobiology: A Synthesis." London: Blackwell Scientific Publications.

Copper, P. (1988). "Ecological succession in Phanerozoic reef ecosystems: Is it real?" *Palaios* **3,** 136–152.

Donovan, S. K., ed. (1989) "Mass Extinctions." New York: Columbia Univ. Press.

Erwin, D. H. (1993). "The Great Paleozoic Crisis: Life and Death in the Permian." New York: Columbia Univ. Press.

Jablonski, D. (1995). Extinctions in the fossil record. *In* "Estimating Extinction Rates" (J. H. Lawton and R. M. May, eds.). Oxford: Oxford Univ. Press.

Raup, D. M. (1991). "Extinction: Bad Genes or Bad Luck?" New York: Norton.

Signor, P. W. (1990). The geologic history of diversity. *Ann. Rev. Ecol. Syst.* **21,** 509–539.

Walliser, O. H., ed. (1994) "Global Bio-Events and Event Stratigraphy." Berlin: Springer-Verlag.

Duties to Endangered Species

Holmes Rolston III
Colorado State University

Whether humans have duties *to* endangered species is a significant theoretical and urgent practical question. Few persons doubt that we have some obligations *concerning* endangered species, because persons are helped or hurt by the condition of their environment, which includes a wealth of wild species, currently under alarming threat of extinction. The U.S. Congress, deploring the lack of "adequate concern [for] and conservation [of]" species, has sought to protect species through the Endangered Species Act. Congress has also entered into a Convention on International Trade in Endangered Species. The United Nations has negotiated a Biodiversity Convention, signed by over 100 nations. Taking or jeopardizing endangered species (at least the listed ones) is illegal and, many think, immoral.

But these might be all obligations *to* persons who are benefited or harmed by species as resources. Is there a human duty directly *to* species, in addition to obligations that humans have to other humans, fellow members of their own species? This would be part of an interspecific environmental ethics, and involves a challenging mix of science and conscience. An answer is vital to the more comprehensive question of the conservation of biodiversity, how humans can achieve a sustainable relationship to the natural world.

I. ETHICAL DUTIES AND BIOLOGICAL SPECIES

A rationale for saving species that centers on their worth to humans is anthropocentric, in which species have instrumental values; a rationale that includes their intrinsic and ecosystemic values, those values they may have in themselves or in their functions in ecosystems, in addition to or independently of persons, is naturalistic. Some say there are no duties *to* endangered species, only duties *to* persons. The preservation of species, by the usual utilitarian account, is commended only insofar as human beings have or might have interests at stake. This includes duties to future human beings, duties derived from our stewardship role as keepers of the planet for later people. Any duties concerning species will then be a matter of finding out whatever human values are at stake with the loss of species and of applying classical duties to persons to protect these values. [*See* MASS EXTINCTION, BIOTIC AND ABIOTIC.]

Persons have a strong duty not to harm others (called a duty of nonmaleficence) and a weaker, though important, duty to help others (called a duty of beneficence). Many endangered species—which ones we may not now know—are expected to have agricultural, industrial, and medical benefits. Relatively few plants have been tested for their usefulness. Loss of the wild stocks of cultivars leaves humans genetically vulnerable, so it is prudent to save native materials. According to this reasoning, the protection of nature is ultimately for the purpose of the enlightened exploitation of nature. Norman Myers urges "conserving our global stock."

Where they are not directly useful, wild species may be indirectly important for the roles they play in ecosystems—as part of the human life support system. they are "rivets" in the airplane, the Earthship in which we humans are flying, and one ought not to pop rivets in people's planes. The loss of a few species may have no evident results now, but the loss of many species imperils the resilience and stability of the ecosystems on which humans depend. The danger increases exponentially with subtractions from the ecosystem, a slippery slope into serious troubles. Even species that have no obvious or current direct value to humans are part of the biodiversity that keeps ecosystems healthy. [*See* ECOLOGICAL ENERGETICS OF ECOSYSTEMS.]

On the benefit side, there are less tangible benefits. Species that are too rare to play roles in ecosystems can have recreational and aesthetic value—even, for many persons, religious value. Species can be curiosities. The rare species fascinate enthusiastic naturalists and are often key scientific study species. They may serve as indicators of ecosystem health. They provide entertainment and new knowledge, regardless of their stabilizing or economic benefits. They can be clues to understanding natural history. Destroying species is like tearing pages out of an unread book, written in a language humans hardly know how to read, about the place where we live. This is called the Rosetta stone argument (named after the famous obelisk found at the town of Rosetta in Egypt in 1799, which enabled the deciphering of forgotten languages of the ancient past). Humans need insight into the full text of natural history. They need to understand the evolving world in which they are placed. It is safe to say that, in the decades ahead, the quality of life will decline in proportion to the loss of biotic diversity, although it is sometimes thought that one must sacrifice biotic diversity to improve human life.

Following this logic, humans do not have duties to the book, the stone, or the species, but to ourselves—duties both of prudence and education. Such anthropogenic reasons are pragmatic and impressive. They are also moral, since persons are benefited or harmed. But are there also naturalistic reasons? Can all duties concerning species be analyzed as duties to persons? Many endangered species have no resource value, nor are they particularly important for the other reasons given above. Beggar's-ticks (*Bidens* spp.), with their stick-tight seeds, are a common nuisance weed through much of the United States. However, one species, the tidal shore beggar's-tick (*Bidens bidentoides*), which differs little from the others in appearance, is increasingly endangered. It seems unlikely that it is either a rivet or a potential resource to humans. So far as humans are concerned, its extinction might be good riddance.

Are there completely worthless species—not good for anything at all? If so, is there any reason or duty to save them? Are the humanistic reasons exhaustive? A primary environmental ethics answers that species are good in their own right, whether or not they are any good for humans. The duties-to-persons-only line of argument leaves deeper reasons untouched. The deeper problem with the anthropocentric rationale is that its justifications are submoral and fundamentally exploitive and self-serving, even if subtly so. This is not true intraspecifically among humans, when out of a sense of duty an individual defers to the values of fellow humans. But it is true interspecifically, since, under this rationale, *Homo sapiens* treats all other species as "rivets," resources, study materials, or entertainment.

Ethics has always been about partners with entwined destinies. But it has never been very convincing when pleaded as enlightened self-interest (that one should always do what is in one's intelligent self-interest), including class self-interest, even though in practice altruistic ethics often need to be reinforced by self-interest. To value all other species only for human interests is rather like a nation's arguing all its foreign policy in terms of national self-interest. Neither seems fully moral.

Nevertheless, those who try to articulate a deeper environmental ethic often get lost in unfamiliar territory. Natural kinds, if that is what species are, are obscure objects of concern. Species, as such, cannot be directly helped or hurt, although individual tokens of the species type can be. Species, as such, don't care, although individual animals can care. Species require habitats, embedded in ecosystems that evolve and change. Ninety-eight percent of the species that have inhabited Earth are extinct, replaced by other species. Nature doesn't care, so why should we? All of the familiar moral landmarks are gone. We have moved beyond caring about humans, or culture, or moral agents, or individual animals that are close kin, or can suffer, or can experience anything, or are sentient. Species are not valuers with preferences that can be satisfied or frustrated. It seems odd to say that species have rights, or moral standing, or need our sympathy, or that we should consider their point of view. None of these elements has figured within the coordinates of prevailing ethical systems.

In fact, ethics and biology have had uncertain relationships. An often-heard argument forbids moving from what *is* the case (a description of scientific facts) to what *ought to be* (a prescription of moral duty); any who do so commit, it is alleged, the naturalistic fallacy. On the other hand, if species are of objective value, and if humans encounter and jeopardize such value, it would seem that humans ought not destroy values in nature, at least not without overriding justification producing greater value. We might make a humanistic mistake if we arrogantly take value to lie exclusively in the satisfaction of our human preferences. What is at jeopardy and what are our duties?

II. THE THREAT OF EXTINCTION

Although projections vary, reliable estimates are that ~20% of Earth's species may be lost within a few decades, if present trends go unreversed. These losses will be about evenly distributed through major groups of plants and animals in both developed and developing nations, although the most intense concerns are in tropical forests. At least 500 species, subspecies, and varieties of fauna have been lost in the United States since 1600. The natural rate would have been about 10. In Hawaii, of 68 species of birds unique to the islands, 41 are extinct or virtually so. Half of the 2200 native plants are endangered or threatened. Covering all states, a candidate list of plants contains over 2000 taxa considered to be endangered, threatened, or of concern, although relatively few of these have been formally listed. A candidate list of animals contains about 1800 entries. Humans approach, and in places have even exceeded, the catastrophic rates of natural extinction spasms of the geological past.

Throughout the Endangered Species Act, from the title onward, the mood is one of danger. The Act laments the irretrievable extinction of any species, climaxing in a "no-jeopardy" clause. That clause has proved the toughest part of the Act, where nearly all of the litigation has arisen. This instructs all federal agencies to take whatever action is necessary "to insure that actions authorized, funded, or carried out by them do not jeopardize the continued existence of such endangered species or threatened species." Existence is at stake, both of species and of habitats that are critical for them.

It may be though that, although the terms are maleficent, humans are not in jeopardy; only the plants and animals are. Humans have important, but not life-jeopardizing, benefits to be gained from saving species. Congress did want to protect values at stake to the nation and its people. Yet, in the snail darter case (Tennessee Valley Authority versus Hill), the U.S. Supreme Court found in the Act "repeated expressions of congressional concern over what it saw as the enormous danger presented by the eradication of *any* endangered species." Although Congress has not said that humans have duties to species (such an ethical judgment might not be the prerogative of Congress), the Court insisted "that Congress intended endangered species to be afforded the highest of priorities." All of this suggests considerable peril, and responsibility proportionate to the peril.

Nor is this simply an Act for the utilitarian conservation of important economic resources. Congress declared that species have "esthetic, ecological, educational, historical, recreational and scientific value" but refused to put on the list that value that has since become the one most often given: economic value. Rather, revealing what Congress thought was at stake, economic value is sharply set opposite to these others. Congress laments "economic growth and development untempered by adequate concern and conservation," and it has consistently refused to allow the economic benefits or costs of the preservation of a species to be one of the criteria that determine whether it is listed. Since economic concerns must sometimes be considered, Congress in the 1978 amendments authorized a high-level interagency committee to evaluate difficult cases, and, should this committee deem fit, to permit economic development at the cost of extinction of species that impede such development. But it clearly places a high burden of proof on those who wish to put species at peril for development reasons.

III. QUESTIONS OF FACT: WHAT ARE SPECIES?

There are problems at two levels: one is about facts (a scientific issue—about species), one is about values (an ethical issue—involving duties). It is difficult enough to argue from an *is* (that a species exists) to an *ought* (that a species ought to exist). Matters grow worse if the concept of species is troublesome to begin with, and there are several differing concepts of species within biology. Perhaps any concept is arbitrary, conventional—a mapping device that is only theoretical. Darwin wrote, "I look at the term species, as one arbitrarily given for the sake of convenience to a set of individuals closely resembling each other." Is there enough factual reality in species to base duty there? [*See* SPECIATION.]

No one doubts that individual organisms exist, but are species discovered? Or made up? Indeed, do species exist at all? Systematists regularly revise species designations and routinely put after a species the name of the "author" who, they say, "erected" the taxon. If a species is only a category

or class, boundary lines may be arbitrarily drawn, and the species is nothing more than a convenient grouping of its members, an artifact of the classifier's thoughts and aims. Some natural properties are used--reproductive structures, bones, teeth, or perhaps ancestry, genes, ecological roles. But which properties are selected and where the lines are drawn are decisions that vary with systematists.

Botanists are divided whether *Iliamna remota,* the Kankakee mallow in Illinois, and *Iliamna corei* in Virginia, which are both rare, are distinct species. Perhaps all that exists objectively in the world are the individual mallow plants; whether there are two species or one is a fuss about which label to use. A species is some kind of fiction, like a center of gravity or a statistical average. Almost no one proposes duties to genera, families, orders, and phyla; biologists concede that these do not exist in nature, even though we may think that two species in different orders represent more biodiversity than two in the same genus. If this approach is pressed, species can become something like the lines of longitude and latitude or like map contour lines, or time of day, or dates on a calendar. Sometimes endangered species designations have altered when systematists decided to lump or split previous groupings. To whatever degree species are artifacts of those doing the taxonomy, duties to save them seem unconvincing.

There are four main concepts of species: (1) morphological, asking whether organisms have the same anatomy and functions; (2) biological (so-called), asking whether organisms can interbreed; (3) evolutionary, asking whether organisms have the same lineage historically; and (4) genetic, asking whether they have a common genome. But these concepts are not mutually exclusive; organisms that have enough common ancestry will have a similar morphology and function; they will be able to interbreed, and they can do so because they have similar genomes.

All of these concepts combine for a more realist account than the artifact-of-taxonomy subjectivist account. A species is not just a class that taxonomists decide on; it is a living historical form (Latin: *species*), propagated in individual organisms, that flows dynamically over generations. Species are dynamic natural kinds, historically particular lineages. A species is a coherent ongoing natural kind expressed in organisms that interbreed because that kind is encoded in gene flow, the genes determining the organism's morphology and functions, the kind shaped by its environment. In this sense species are objectively there as living processes in the evolutionary ecosystem—found, not made by taxonomists. Species are real historical entites, interbreeding populations. By contrast, families, orders, and genera are not levels at which biological reproduction takes place. So far from being arbitrary, species are the real evolutionary units. This claim—that there are specific forms of life historically maintained in their environments over time—is not fictional, but, rather, seems as certain as anything else we believe about the empirical world, even though at times scientists revise the theories and taxa with which they map these forms.

Species are more like mountains and rivers, phenomena that are objectively there to be mapped. The edges of such natural kinds will sometimes be fuzzy, to some extent discretionary. We can expect that one species will modify into another over evolutionary time, often gradually, sometimes more quickly. But it does not follow from the fact that speciation is sometimes in progress that species are merely made up, instead of found as evolutionary lines articulated into diverse forms, each with its more or less distinct integrity, breeding population, gene pool, and role in its ecosystem. It is quite objective to claim that evolutionary lines are articulated into diverse kinds of life. What taxonomists do, or should do, is, as Plato said, to "carve nature at the joints."

G. G. Simpson concluded, "An evolutionary species is a lineage (an ancestral-descendant sequence of populations) evolving separately from others and with its own unitary evolutionary role and tendencies." Niles Eldredge and Joel Cracraft insist, with emphasis, that species are *"discrete entities in time as well as space."* As convincing an account as any finds that species, though not individual organisms, are another natural kind of historical *individual,* each a unique event in natural history, and that species names are proper names. The various criteria for defining species (recent descent,

reproductive isolation, morphology, and distinct gene pool) come together at least in providing evidence that species are really there. What survives for a few months, years, or decades is the individual animal or plant; what survives for millennia is the kind as a lineage. Life is something passing through the individual as much as something it possesses on its own. Even a species defends itself; that is one way to interpret reproduction. The individual organism resists death; the species resists extinction through reproduction with variation. At both levels, biological identity is conserved over time.

IV. QUESTIONS OF DUTY: OUGHT SPECIES BE SAVED?

Why ought species be protected? Beyond a humanistic set of answers, when we confront the objective history of speciation and evolution of species, is there some nonhumanistic reason to save endangered species? One reply here is that nature is a kind of wonderland. As curiosities and relics of the past, even species that are presently not good for anything in particular can be given an umbrella protection by saying that humans ought to preserve an environment adequate to match their capacity to wonder.

But nature as a wonderland introduces the question of whether preserving resources for wonder is not better seen as preserving a remarkable natural history that has objective worth—an evolutionary process that has spontaneously assembled as its products millions of species. Valuing speciation directly, however, seems to attach value to the evolutionary process (the wonderland), not merely to subjective experiences that arise when humans reflect over it (the wonder). It will be better, beyond our pragmatic self-enlightened strategies for conservation, beyond our obligations to other humans, beyond even our wonder, to know the full truth of the human obligation, to have the best reasons for saving species, as well as the good ones.

We might say that humans of decent character will refrain from needless destruction of all kinds, including destruction of species. Vandals destroying art objects do not so much hurt statues as do they cheapen their own character. By this account the duty to save endangered species is really a matter of cultivating human excellences. It is philistine to destroy species carelessly; persons of character will not do it. It is uncalled for. But such a prohibition seems to depend on some value in the species as such, for there need be no prohibition against destroying a valueless thing. Is there not here some insensitivity to a form of life that (unlike a statue) has an intrinsic value that places some claim on humans?

Why are such insensitive actions "uncalled for" unless there is something in the species itself that "calls for" a more appropriate attitude. If the excellence of character really comes from appreciating something wonderful, then why not attach value to this other, so full of wonder? It seems unexcellent—cheap and philistine—to say that excellence of human character is what we are after when we preserve these endangered species. We want virtue in the human beholder that recognizes value in the endangered species. Excellence of human character does indeed result, but let the human virtue come tributary to value found in nature. An enriched humanity results, with values in the species and values in persons compounded—but only if the loci of value are not confounded.

A naturalistic account values species and speciation intrinsically, not as resources or as a means to human excellence. Humans ought to respect these dynamic life forms preserved in historical lines, vital informational processes that persist genetically over millions of years, overleaping short-lived individuals. It is not *form* (species) as mere morphology, but the *formative* (speciating) process that humans ought to preserve, although the process cannot be preserved without some of its products, and the products (species) are valuable as results of the creative process. An ethic about species sees that the species is a bigger event than the individual organism, although species are always exemplified in individual organisms. Biological conservation goes on at this level too, and in a sense this level is more appropriate for moral concern, since the species is a comprehensive evolutionary unit but the single organism is not. [*See* SPECIES DIVERSITY.]

A consideration of species is both revealing and challenging, because it offers a biologically based counterexample to the focus on individuals—typically sentient animals and usually individual persons—that has been so characteristic in Western ethics. As evolution takes place in ecosystems, it is not mere individuality that counts. The individual represents (re-presents) a species in each new generation. It is a token of a type, and the type is more important than the token. Though species are not moral agents, a biological identity—a kind of value—is here defended. The dignity resides in the dynamic form; the individual inherits this, exemplifies it, and passes it on. The evolutionary history that the particular individual has is something passing through it during its life, passed to it and passed on during reproduction, as much as something it intrinsically possesses. Having a biological identity reasserted genetically over time is as true of the species as of the individual. Respecting that identity generates duties to species.

When a rhododendron plant dies, another one replaces it. But when *Rhododendron chapmanii*—an endangered species in the U.S. Southeast—goes extinct, the species terminates forever. Death of a token is radically different from death of a type; death of an individual, different from death of an entire lineage. The deaths of individual rhododendrons in perennial turnover are even necessary if the species is to persist. Seeds are dispersed and replacement rhododendrons grow elsewhere in the pinewood forest, as landscapes change or succession shifts. Later-coming replacements, mutants as well as replacements, are selected for or against in a stable or changing environment. Individuals improve in fitness and the species adapts to an altering climate or competitive pressures. Tracking its environment over time, the species is conserved, modified, and continues.

With extinction, this stops. Extinction shuts down the generative processes, a kind of superkilling. This kills forms (*species*)—not just individuals. This kills "essences" beyond "existences," the "soul" as well as the "body." This kills collectively, not just distributively. To kill a particular plant is to stop a life of a few years, while other lives of such kind continue unabated, and the possibilities for the future are unaffected; to superkill a particular species is to shut down a story of many millennia and leave no future possibilities.

A species lacks moral agency, reflective self-awareness, sentience, or organic individuality. Some are tempted to say that specific-level processes cannot count morally. But each ongoing species defends a form of life, and these forms are, on the whole, good kinds. Such speciation has achieved all the planetary richness of life. All ethicists say that in *Homo sapiens* one species has appeared that not only exists but ought to exist. A naturalistic ethic refuses to say this exclusively of a late-coming highly developed form and extends this duty more broadly to the other species—although not with equal intensity over them all, in view of varied levels of development.

The wrong that humans are doing, or are allowing to happen through carelessness, is stopping historical gene flow in which the vitality of life is laid, and which, viewed at another level, is the same as the flow of natural kinds. A shutdown of the life stream is the most destructive event possible. Although all specific stories must eventually end, we seldom want unnatural ends. The difference between natural extinction and human-caused extinction is something like that between death by natural causes and murder. Humans ought not to play the role of murderers. The duty to species can be overridden, for example, with pests or disease organisms. But a prima facia duty stands nevertheless. [*See* EVOLUTION AND EXTINCTION.]

What is wrong with human-caused extinction is not just the loss of human resources, but the loss of biological sources. The question is not, What is this rare plant or animal good for? But, What good is here? Not, Is this species good for my kind, *Homo sapiens*? But, Is *Rhododendron chapmanii* a good of its kind, a good kind? True, we are censuring insensitivity in persons, but we are appreciating an objective vitality in the world, one that precedes and overleaps our personal or cultural presence. To care directly about a plant or animal species is to be quite nonanthropocentric and objective about botanical and zoological processes that take place independently of human preferences.

Never before has this level of question been faced. Previously, humans did not have much power to cause extinctions, or knowledge about what they were inadvertently doing. But today humans have more understanding than ever of the natural world they inhabit, of the speciating processes, more predictive power to foresee the intended and unintended results of their actions, and more power to reverse the undesirable consequences. Increasingly, we know the natural histories of flora and fauna; we find that, willy-nilly, we have a vital role in whether these stories continue. The duties that such power and vision generate no longer attach simply to individuals or persons but are emerging duties to specific forms of life.

A consideration of species strains any ethic fixed on individual organisms, much less on sentience or persons. But the result can be biologically sounder, although it revises what was formerly thought logically permissible or ethically binding. When ethics are informed by this kind of biology, it is appropriate to attach duty dynamically to the specific form of life. The species line is the more fundamental living system, the whole, of which individual organisms are the essential parts. The species too has its integrity, its individuality; and it is more important to protect this than to protect individual integrity. The appropriate survival unit is the appropriate level of moral concern.

V. SPECIES IN ECOSYSTEMS

A species is what it is inseparably from the environmental niche into which it fits. A species is what it is where it is. Particular species may not be essential in the sense that the ecosystem can survive the loss of individual species without adverse effect. But habitats are essential to species, and an endangered species often means an endangered habitat. Species play lesser or greater roles in their habitats. Integrity in the species fits into integrity in the ecosystem. The species and the community are complementary goods in synthesis, parallel to, but a level above, the way the species and individual organisms have distinguishable but entwined goods. It is not preservation of *species* that we wish, but the preservation of *species in the system*. It is not merely *what* they are, but *where* they are that we must value correctly.

This limits the otherwise important role that zoos and botanical gardens can play in the conservation of species. They can provide research, a refuge for species, breeding programs, aid on public education, and so forth, but they cannot simulate the ongoing dynamism of gene flow over time under the selection pressures in a wild biome. They only lock up a collection of individuals; they amputate the species from its habitat. The species *can* only be preserved *in situ;* the species *ought* to be preserved *in situ*. That does move from scientific facts to ethical duties, but what ought to be has to be based on what can be.

Neither individual nor species stands alone; both are embedded in an ecosystem. Plants, which are autotrophs, have a certain independence that animals and other heterotrophs do not have. Plants need only water, sunshine, soil, nutrients, and local conditions of growth; animals, often mobile and higher up the trophic pyramid, may range more widely, but in this alternate form of independence depend on the primary production of plants. Every natural form of life came to be what it is where it is, shaped as an adaptive fit, even when species acquire a fitness that enables them to track into differing environments. (A problem with exotic species, introduced by humans, is often that they are not good fits in their alien ecosystems.) The product, a species, is the outcome of entwined genetic and ecological processes; the generative impulse springs from the gene pool, defended by information coded there. But the whole population or species survives when selected by natural forces in the environment for a niche it can occupy.

In an ethic of endangered species, we want to admire the evolutionary or creative process as much as the product. This involves regular species turnover when a species becomes unfit in its habitat, goes extinct, or tracks a changing environment until transformed into something else. On evolutionary time scales species too are ephemeral. But the speciating process is not. Persisting through vicissitudes for 2.5 billion years, speciation is about as long-continuing as anything on earth can be.

VI. NATURAL AND HUMAN-CAUSED EXTINCTIONS

It might seem that for humans to terminate species now and again is quite natural. Species become extinct all the time in natural history. But although extinction is a quite natural event, there are important theoretical and practical differences between natural and anthropogenic (human-caused) extinctions. Artificial extinction, caused by human disturbance or encroachments, is radically different from natural extinction. In natural extinction a species dies out when it has become unfit in habitat, and other existing or future species appear in its place. There are replacements. Such extinction is normal turnover in ongoing speciation. Although harmful to a species, extinction in nature is seldom an evil in the system. It is rather the key to tomorrow. The species is employed in, but abandoned to, the larger historical evolution of life.

By contrast, artificial extinction typically shuts down future evolution because it shuts down speciating processes dependent on those species. One opens doors, the other closes them. Humans generate and regenerate nothing; they only dead-end these lines. Relevant differences make the two as morally distinct as death by natural causes is from murder. Anthropogenic extinction differs from evolutionary extinction in that hundreds of thousands of species will perish because of culturally altered environments that are radically different from the spontaneous environments in which such species evolved and in which they sometimes go extinct. In natural extinction nature takes away life, when it has become unfit in habitat, or when the habitat alters, and typically supplies other life in its place. Natural extinction occurs with transformation, either of the extinct line or related or competing lines. Artificial extinction is without issue.

From this perspective, humans have no duty to preserve rare species from natural extinctions, although they might have a duty to other humans to save such species as resources or museum pieces. No species has a "right to life" apart from the continued existence of the ecosystem with which it cofits. But humans do have a duty to avoid artificial extinction.

Over evolutionary time nature, though extinguishing species, has provided new species at a higher rate than the extinction rate, hence the accumulated global diversity. There have been infrequent catastrophic extinction events, anomalies in the record, each succeeded by a recovery of previous diversity. Although natural events, these extinctions so deviate from the normal trends that many paleontologists look for causes external to the evolutionary ecosystem—supernovae or collisions with asteroids. Typically, however, the biological processes that characterize Earth are both prolific and with considerable powers of recovery after catastrophe. Uninterrupted by accident, or even interrupted so, they steadily increase the numbers of species.

An ethicist must be circumspect. An argument might commit what logicians call the genetic fallacy to suppose that present value depended on origins. Species judged today to have intrinsic value might have arisen anciently and anomalously from a valueless context, akin to the way in which life arose mysteriously from nonliving materials. But in an ecosystem, what a thing is differentiates poorly from the generating and sustaining matrix. The individual and the species have what value they have to some extent inevitably in the context of the forces that beget them. There is something awesome about an Earth that begins with zero and runs up toward 5–10 million species in several billion years, setbacks notwithstanding. Were the sole moral species, *Homo sapiens,* to conserve all Earth's species merely as resources for human preference satisfaction, we would not yet know the truth about what has been, is, or ought to be going on in biological conservation.

VII. RESPECT FOR RARE LIFE

Duties to endangered species will be especially concerned with a respect for a rare life. Such respect must ask about the role of rarity in generating respect, if this differs from a more general respect for common life. Rarity is not, as such, an intrinsically valuable property in fauna and flora, or in human experiences (even though people take an interest in things just because they are rare). Certain diseases are rare, and we are glad of it. Monsters and other sports of nature, such as albinos, are rare, and of no particular intrinsic value for their rarity (curiosities though they sometimes become). Indeed, if a species is naturally rare, that initally suggests its insignificance in an ecosystem. Rarity is no automatic cause for respect. Nevertheless, something about the rarity of endangered species heightens the element of respect, and accompanying duty.

Naturally rare species, as much as common or frequent species, signify exuberance in nature; each species presents an actual unique expression of the prolific potential driving the evolutionary epic. A rare species may be barely hanging on, surviving by mere luck, and we have already noticed that there is no duty to save species going extinct naturally. But a rare species may be quite competent in its niche, not at all nearing extinction if left on its own; it is only facing extinction when made artificially more rare by human disruptions. The rare flower is a botanical achievement, a bit of brilliance, an ecological problem resolved, an evolutionary threshold crossed. The endemic species, perhaps one specialized for an unusual habitat, represents a rare discovery in nature, before it provides a rare human adventure in finding it.

Rhododendron chapmanii is a particular evolutionary achievement. Though rare, it is a satisfactory fit, well placed in its niche in the transition zone between the dry longleaf pine forests and the moist *Cyrilla* thickets. Millions of years of struggle lie behind it; the results of that history are now genetically coded within it. Rare species—if one insists on a restricted evolutionary theory—are random accidents (as are the naturally common ones), resulting from a cumulation of mutations. But this mutational fertility generates creativity, and, equally by the theory, surviving species must be satisfactory fits in their environments. Sometimes they live on the cutting edge of exploratory probing; sometimes they are relics of the past. Either way they offer promise and memory of an inventive natural history. Life is a many-splendored thing; extinction of the rare dims this luster. From this arises the respect that generates a duty to save rare lives.

This respect for life is sometimes expressed in terms of rights. Aldo Leopold, advocating respect for the fauna and flora on the landscape, says "A land ethic of course cannot prevent the alteration, management, and use of these 'resources,' but it does affirm their right to continued existence, and, at least in spots, their continued existence in a natural state." They "should continue as a matter of biotic right." Charles S. Elton, an ecologist, reports a belief that he himself shares: "There are millions of people in the world who think that animals have a right to exist and be left alone." This appeal to a biotic right must be taken as evidence of the strength of conviction that there are duties to species. Nevertheless, many philosophers have concluded that the vocabulary of rights, though useful rhetorically, is not the most appropriate category of analysis for values at the species level. "Rights" is best developed as a category for protecting personal values; rights are not objectively present in the natural world. But endangered species are objectively valuable kinds, good in themselves; they do have their own welfare. Respect for life ought to be directly based on this value.

The seriousness of respect for rare life is further illustrated when the idea approaches a "reverence" for life. As noticed earlier, when the U.S. Congress declared that species have multiple values, it left economic value off the list. Another notable omission is religious value. Congress would have overstepped its authority to declare that species carry religious value. Nevertheless, for many, Americans and others around the globe, this is the most important value at stake. Species are the creation itself, the "swarms of living creatures" (biodiversity) that "the earth brought forth" at the divine imperative; "God saw that it was good" and "blessed them." Noah's ark was the aboriginal endangered species project; God commanded, "Keep them alive with you."

God's name does not appear directly in the Endangered Species Act but nevertheless occurs in connection with the Act. The high-level interagency committee may permit human development at the cost of extinction of species. In the legislation this committee is given the rather nondescript name "The Endangered Species Committee," but almost at once it was nicknamed "the God Committee." The name mixes jest with theological insight and reveals that religious value is implicitly lurking in the Act. Humans are trustees of creation and ought to "play God" with extreme care. Any

who decide to destroy species take, fearfully, the prerogative of God. When one is conserving life, ultimacy is always nearby. Extinction is forever; and, when danger is ultimate, absolutes become relevant. The motivation to save endangered species can and ought to be pragmatic, economic, political, and scientific; deeper down it is moral, philosophical, and religious. Species embody a fertility on Earth that is sacred.

This genesis is, in biological perspective, spontaneous and autonomous; and biologists find nature to be prolific, whether or not the God question is raised. Whether the conviction rises to a reverence for life or not, the respect for life in jeopardy becomes intense. Life is the peculiar value on our planet, among the rare phenomena in the universe, indeed, not yet elsewhere known. Natural history is a vast scene of birth and death, sprouting, budding, flowering, fruiting, passing away, passing life on. Biologists know, better than others, that Earth has brought forth the natural kinds exuberantly over the millennia. Ultimately, there is a kind of creativity in nature demanding at least that one spell nature with a capital N, if one does not pass beyond nature to detect some deeper sacred presence.

Biologists today are not inclined, nor should they be as biologists, to look for explanations in supernature, but biologists meanwhile find a nature that is superb! This commands a deep respect. Science, many think, eliminates from nature any suggestions of teleology, but it is not so easy for science to dismiss genesis. What has managed to happen on Earth is startling by any criteria. Ernst Mayr concludes, "Virtually all biologists are religious, in the deeper sense of the word, even though it may be a religion without revelation. . . . The unknown and maybe unknowable instills in us a sense of humility and awe." "And if one is a truly thinking biologist, one has a feeling of responsibility for nature, as reflected by much of the conservation movement." If anything at all on Earth is sacred, it must be this enthralling creativity that characterizes our home planet.

Species are a characteristic expression of the creative process. The swarms of species are both presence and symbol of forces in natural systems that transcend human powers and utility. Generated from earth, air, fire, and water, these fauna and flora are an archetype of the foundations of the world. Earth is a fertile planet, and in that sense the genesis on Earth is the deepest valuational category of all, one classically reached by the concept of creation. Many find it impossible to be a conservation biologist without a respect for life. Whatever biologists may make of the mystery of life's origins, they almost unanimously conclude that the catastrophic loss of species that is at hand and by our hand is tragic, irreversible, and unforgivable. That generates duties to endangered species.

On the scale of evolutionary time, humans appear late and suddenly, a few hundred thousand years on a scale of billions of years, analogous to a few seconds in a 24-hour day. Even more lately and suddenly, they increase the extinction rate dramatically, in this one century in several thousand years of recorded history. What is offensive in such conduct is not merely senseless destabilizing, not merely the loss of resources, but the maelstrom of killing and insensitivity to forms of life. What is required is not prudence, but principled responsibility to the biospheric Earth. Only the human species contains moral agents, but conscience ought not be used to exempt every other form of life from consideration, with the resulting paradox that the sole moral species acts only in its collective self-interest toward all the rest.

Several billion years' worth of creative toil, several million species of teeming life, have been handed over to the care of the late-coming species in which mind has flowered and morals have emerged. On the humanistic account, such species ought to be saved for their benefits to humans. On the naturalistic account, the sole moral species has a duty to do something less self-interested than count all the products of an evolutionary ecosystem as human resources; rather, the host of species has a claim to care in its own right. There is something Newtonian, not yet Einsteinian, besides something morally naive, about living in a reference frame in which one species takes itself as absolute and values everything else relative to its utility.

Glossary

Anthropogenic extinction Extinction caused by human-introduced causes, as distinguished from natural extinction.

Catastrophic extinction Extinction at extremely high rates, differing from normal rates, at unusual periods in natural history.

Instrumental value Value as a means to an end. Species have instrumental value for humans if they have medical, industrial, agricultural, recreational, or other uses.

Intrinsic value Value that is inherent in something, without necessary reference to its instrumental value. Intrinsic value in species claims that natural kinds are good in themselves, whether or not they are useful to humans.

Natural extinction Extinction that takes place due to natural causes, without human causes, as has occurred throughout evolutionary history.

Naturalistic fallacy An alleged fallacy when one argues from statements of fact in premises to statements of duty in conclusions, from descriptions to prescriptions.

Bibliography

Council on Environmental Quality and Department of State (1980). "The Global 2000 Report to the President." U.S. Government Printing Office, Washington D.C. [Includes projections of impending losses of species.]

Darwin, C. (1872/1968). "The Origin of Species." Penguin, Baltimore, Maryland.

Ehrlich, P., Ehrlich, A. (1981). "Extinction." Random House, New York. [A comprehensive study of extinctions, with the rationale for saving species.]

Eldredge, N., Cracraft, J (1980). "Phylogenetic Patterns and the Evolutionary Process." Columbia University Press, New York.

Elton, C. S. (1958). "The Ecology of Invasions by Animals and Plants." Wiley, New York.

Ghiselin, M. (1987). Species concepts, individuality, and objectivity. *Biol. Philos.* **2,** 127–144.

Hull, D. L. (1978). A matter of individuality. *Philos. Sci.* **45,** 335–360.

Leopold, A. (1968). "A Sand County Almanac." Oxford University Press, New York.

Mayr, E. (1982). "The Growth of Biological Thought." Harvard University Press, Cambridge, Massachusetts.

Mayr E. (1985). How biology differs from the physical sciences. *In* "Evolution at a Crossroads" (D. J. Depew and B. H. Weber, eds.). MIT Press, Cambridge, Massachusetts.

Myers, N. (1979a). "The Sinking Ark." Pergamon, Oxford, England. [A comprehensive study of species losses worldwide.]

Myers, N. (1979b). Conserving our global stock. *Environment* **21**(9), 25–33.

Norton, B. G., ed. (1986). "The Preservation of Species." Princeton University Press, Princeton, New Jersey. [Oriented to policy and philosophical issues.]

Opler, P. A. (1977). The parade of passing species: A survey of extinctions in the U.S. *Sci. Teach.* **44,** 30–34.

Rolston, H., III (1985). Duties to endangered speceis. *Bioscience* **35,** 718–726.

Rolston, H., III (1988). Life in jeopardy: Duties to endangered species. *In* "Environmental Ethics," Temple University Press, Philadelphia.

Simpson, G. G. (1961). "Principles of Animal Taxonomy." Columbia University Press, New York.

U.S. Congress (1973). Endangered Species Act of 1973, 87 Stat. 884, Public Law 93-205.

U.S. Fish and Wildlife Service (1990). Review of plant taxa for listing as endangered or threatened species. *Fed. Regist.* **55,**(35), 6184–6229.

U.S. Fish and Wildlife Service (1991). Animal candidate review for listing as endangered or threatened species. *Fed. Regist.* **56,**(225), 58804–58836

Ecological Energetics of Ecosystems

S. K. Nisanka and M. K. Misra
Berhampur University, India

I. Introduction
II. Source and Subsidy of Energy in Ecosystems
III. Utilization of Energy in Ecosystems
IV. Energy Flow in Ecosystems
V. Energy Losses
VI. Energy Budget of Ecosystems
VII. Energy Conservation

The functional interrelations between living organisms and their habitat constitute an ecosystem. Energy is the driving force for an ecosystem as each and every activity of an organism requires energy, and the amount of output of such activities depends on the energy available to and mobilized by the organism. The ultimate source of energy for ecosystems or for the biosphere as a whole is the sun. The flow of energy through ecosystems, along with the pattern and efficiency of energy utilization is an important ecological process and comprises the field of ecological energetics. Green or photosynthetic plants are first to absorb solar energy in an ecosystem and transfer it to animals in the form of food. Hence there is wide variation in energy-flow patterns within different ecosystems depending on the availability of energy, the primary production (photosynthesis) pattern, the energy-capturing efficiency of plants, energy losses at different steps of energy transfer, ecosystem structure, and various interrelations between and within biotic and abiotic components.

I. INTRODUCTION

Any system can be defined as a set of interdependent units having its own boundary. The units may be designated as subsystems. A system operates by extracting necessary inputs from the surroundings. Figures 1a and 1b are simple models showing a system and interdependent subsystems. Any signal external to the system to which the system reacts is an input. The output may be any attribute that is transmitted to the environment as a response to the input.

An ecosystem is basically an energy-processing system consisting of living organisms (biotic component) and the nonliving environment (abiotic component). Living organisms are responsible for the functioning of the system, whereas the abiotic inputs are energy, oxygen, carbon dioxide, nutrients from soil, and rain. The nonliving environment consists of soil, air, water, and so on. In an ecosystem, both the biotic and abiotic components are influenced by one another, thus both are in continuous interaction with each other and are necessary for the survival (existence) and maintenance of the ecosystem. The biotic components of an ecosystem are producers, consumers, and decomposers. [*See* EQUILIBRIUM AND NONEQUILIBRIUM CONCEPTS IN ECOLOGICAL MODELS.]

Producers refer to all green plants, ranging from phytoplankton to giant-sized trees, that are capable of preparing their own food (organic) by the pro-a

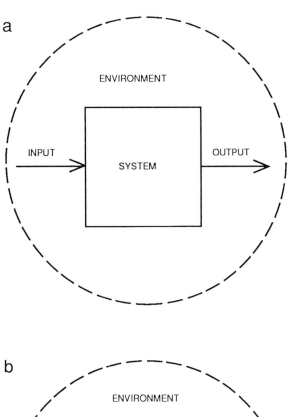

b

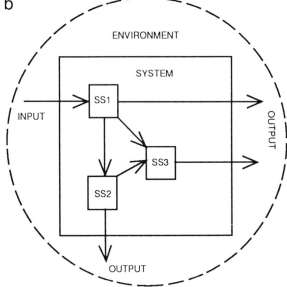

FIGURE I Diagrammatic representation of an open system with inputs from and outputs to the environment (a) and subsystems within a system (b). SS1, SS2, and SS3 are three subsystems interacting with each other.

cess of photosynthesis. Producers are also called autotrophs as they are self-dependent. Consumers are chiefly animals, or those who feed on green plants or producers directly or indirectly. They include both herbivores (directly feeding on plants)

and carnivores (feeding on herbivores). Decomposers are microorganisms (bacteria and fungi) and macroorganisms (insects) that are responsible for releasing the nutrients of dead organisms back to the system. Both consumers and decomposers constitute the heterotrophic component, as they depend on the food stored by autotrophs.

Ecosystems may be as large as an ocean or uninterrupted tracts of forests or as small as a pond or a ditch. Other examples of ecosystems are grasslands, deserts, tundras, swamps, rivers, lakes, estuaries, oceans, and mountains, or human-made ecosystems such as croplands, orchards, plantations, aquaria, a village, and a town. In a grassland ecosystem the biotic components are the grasses and herbs (producers), grasshoppers, rats, hares, and other grazing animals (herbivores or first-order consumers), lizards, snakes, and frogs (carnivores or second-order consumers), and other decomposing organisms present in the soil, water, and air. In a pond ecosystem the producers are phytoplankton and other aquatic plants; the first-order consumers are young sunfish, tadpoles, and aquatic insects; and second-order consumers are frogs, adult sunfish, basses, and herons. Irrespective of the types of ecosystems, the role of the components is uniform in all ecosystems, that is, producers utilize sunlight and manufacture carbohydrates out of CO_2 and water; primary consumers feed on these stored food in plants; and secondary consumers feed on the herbivores. When they die, these organisms reach the substratum, where decomposers start their work to release the nutrients locked up in the dead organisms in the form of various organic compounds back to the substratum or atmosphere.

Energy is defined as the ability to do work. An organism requires energy as an input for all its activities. The output (work) depends on the availability and mobilization of energy. The sun is the ultimate source of energy, which is transmitted and transformed by the biotic components. The different processes of energy exchange in an ecosystem are radiation, absorption, conduction, convection, evaporation, and reradiation (Fig. 2). Such transmission and transformation of energy can be explained by the first law of thermodynamics:

$$Q = E + w \qquad (1)$$

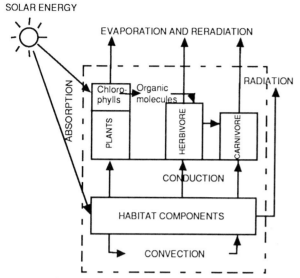

FIGURE 2 Physical and chemical processes involved in exchange of energy in an ecosystem.

where Q = energy input into the ecosystem, E = change in the energy content of the ecosystem, and w = work done (output) by the ecosystem. The existence of life or ecosystem in the dynamic form requires energy exchange or a transfer process, which is provided by the flow of energy through ecosystems.

II. SOURCE AND SUBSIDY OF ENERGY IN ECOSYSTEMS

Solar energy is the driving force of the ecosystem and is also responsible for circulation of other inputs within the system. The solar radiation reaching the earth's surface provides both heat and photochemical energy. Heat energy is responsible for atmospheric warming, production of air and water currents, and the water cycle. The photochemical energy is trapped by green plants and is utilized to produce biochemical compounds such as carbohydrates, proteins, and fats. Energy from the sun comes to earth in the form of electromagnetic waves as a result of continuous processes of nuclear fission and fusion taking place inside it. This reaction, being responsible for the conversion of hydrogen atoms into helium atoms, produces a broad spectrum of radiant energy. The energy that is re-

leased from the sun into space each second is equivalent to about one million times that of the earth's past deposits of coal, natural gas, and petroleum before they began to be consumed by humans.

The total energy released from the sun does not reach earth's surface, only 56% of it. Of the rest, 34% is reflected back into the atmosphere itself and 10% is captured by the ozone layer, water vapor, and outer atmospheric gases. Again, of the total energy coming to earth's surface, only a small fraction (1 to 5%) is utilized by green plants for photosynthesis, while the rest (95 to 99%) is reflected by shining/bright plant surfaces and absorbed as heat by ground vegetation or water.

Solar energy reaches earth's surface in the form of electromagnetic radiation, which may be characterized by wavelengths. These wavelengths range from 100 to 10,000 nm and consist of various rays of the electromagnetic spectrum (Fig. 3). Out of these, about 4% is ultraviolet (UV) rays (100 to 380 nm), 44% is visible light (380 to 760 nm), and the remaining 52% is infrared or longwave

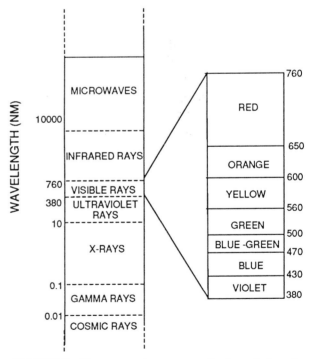

FIGURE 3 Electromagnetic spectrum of solar radiation; the visible rays are magnified to show the colors at different wavelengths. All wavelengths are in nanometers. (The intervals between wavelengths are not to scale.)

radiation (>760 nm). The visible radiation consists of a spectrum of six colors within the range 380 to 760 nm (Table I). But the photophysiological processes of plants utilize radiation of 280 to 800 nm wavelength, which includes ultraviolet and near-infrared rays. Plants efficiently absorb blue (430–500 nm) and red (650–760 nm) light as they contain energy sufficient to initiate electronic transitions called excitation and thereby carry on photochemical reactions. Electromagnetic radiation contains energy in the form of photons or quanta, which can be considered as small units or packets of energy. The energy of a photon (E) is inversely proportional to the wavelength (λ):

$$E = \frac{h}{\lambda} \quad (h = \text{Planck's constant}) \quad (2)$$

The energy of a photon or quantum can be measured in calories (cal) or joules (J). From this relationship of energy and wavelength it is clear that the energy of a photon at shorter wavelengths is higher than the energy of a photon of longer wavelength. Thus blue light (radiaton) contains more energy than red light.

It has been estimated that around $64 \times 108 \, \text{J m}^{-2}$ of solar energy enters the earth's atmosphere annually. But this amount varies by geographical location or latitude, as represented in Table II. However, the daily input of solar energy to the upper layer of an ecosystem containing green plants varies within 418 to $1674 \, \text{J cm}^{-2} \, \text{min}^{-1}$, with a mean range of ap-

TABLE I

Distribution of Wavelengths and Photon Energy in the Visible Spectrum

Color	Wavelength (nm)	Photon energy (kJ mole quanta^{-1})
Ultraviolet (UV)	100–380	—
Violet	380–430	299.33
Blue	430–470	274.23
Blue-green	470–500	261.67
Green	500–560	249.12
Yellow	560–600	224.02
Orange	600–650	198.91
Red	650–760	178.66
Infrared (IR)	760+	—

TABLE II

Latitudinal Variation of Solar Insolation in the Northern Hemisphere

Latitude (°N)	Solar energy (kJ m^{-2} year^{-1})
0–20	723.85×10^4
20–40	682.01×10^4
40–60	476.99×10^4
60–80	305.44×10^4

proximately 1255 to $1674 \, \text{J cm}^{-2} \, \text{min}^{-1}$ for temperate areas. Furthermore, solar radiation also fluctuates within a wide range regarding strata and season. In terrestrial ecosystems, it is reduced from upper stratum to lower due to changes in plant canopy.

A. Energy Subsidy

Besides solar energy, an ecosystem may also be supported by external sources of energy that lessen its internal self-maintenance. This is called energy subsidy or auxiliary energy and is responsible for higher outputs (productivity) of an ecosystem. Examples of such sources of energy subsidy are faster wind, warmer temperature, greater rainfall, running water with extra nutrients for agricultural crops, fossil fuels used in equipment for irrigation and other agricultural operations, tides in an estuary, and so on. When estimated in terms of energy inputs, these sources can be compared with solar energy.

III. UTILIZATION OF ENERGY IN ECOSYSTEMS

A. Energy Utilization by Green Plants

Photons of solar radiation are the chief and only significant source of ecosystem energy. This energy is fixed by green plants, via photosynthesis, which is defined as a process of conversion of solar energy into chemical energy. This process begins with absorption of radiant energy by green parts of plants containing several pigments, such as chlorophyll a, chlorophyll b, and carotenoids. Chlorophylls a and b are responsible for absorption of red light (650–760 nm) and blue light (400–500 nm); carotenoids also absorb in the blue region of the spectrum. When a photon with appropriate energy hits the pigment, an electron from an inner orbit gets excited and jumps to the nearest outer orbit. If the electron then drops back to the inner orbit by losing its excited state, then energy is liberated. Under excited conditions the pigments may react in many ways depending on their electronic configuration as well as on the energy status of the radiation. The energy of the excited electron may be emitted as heat and radiant energy or it may be channeled into photochemical or photophysiological reaction. These chemical reactions include many energy exchange processes involving ATP (adenosine triphosphate) and other energy-rich compounds, and finally lead to formation of carbohydrates using carbon dioxide and water. Thus a translation of energy from the electrons into carbohydrate bond energy occurs, which is diagrammed in Fig. 4.

During the process of photosynthesis, oxygen gas is liberated to the atmosphere and the simple carbohydrates produced undergo further metabolic processes to produce other essential organic compounds like proteins, lipids, and nucleic acids. These organic compounds are further elaborated into leaves, stems, roots, tubers, fruits, seeds, and other tissues in plants. Thus, energy stored in the various organic compounds is either utilized by the plants themselves or remain as such. The latter form is the major food source of all other organisms of the earth.

B. Primary Production

Storage or accumulation of energy by green plants is called primary production, as this is the first and

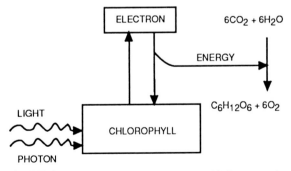

FIGURE 4 Electron-mediated conversion of light energy into chemical ($C_6H_{12}O_6$) bond energy in green plants.

the primary step of energy accumulation in the ecosystem. The rate of this production is expressed as primary productivity. By definition, primary productivity of an ecosystem is the rate at which radiant energy is stored by photosynthetic activity by producer organisms (mainly green plants) in a kind of organic matter that can be used as food.

Practically, all the sun's energy absorbed and assimilated by plants is not retained as organic matter. Some of it is lost through the physiological processes of respiration, excretion, and reproduction in the plant body. This loss also takes place in animal bodies, meaning that all the energy that a herbivore takes in from plant(s) is not accumulated in the body. This loss may be through carbon dioxide, water, nitrogenous compounds, and so on. Therefore, in the case of plants, primary productivity should be divided into two categories: *Gross Primary Productivity* (GPP) and *Net Primary Productivity* (NPP). Gross primary productivity is the rate of total assimilation or total photosynthesis, including the organic matter (energy) used up or lost in respiration. Net primary productivity is the rate of accumulation of organic matter or energy in plant tissues, excluding the respiratory utilization by plants. In other words, this is called "apparent photosynthesis" or "net assimilation." In ecosystems this net production is usually consumed by grazing animals or other heterotrophic consumers. Therefore, the rate of accumulation of organic matter after consumption may be called *Net Community Productivity* (NCP).

From an ecological energetics point of view the first law of thermodynamics may be modified as

$$GPP = NPP + R \qquad (3)$$

or

$$NPP = GPP - R \qquad (4)$$

where GPP = gross primary productivity, NPP = net primary productivity, and R = respiratory loss. In this equation, GPP is the energy input, NPP is the energy stored, and R is expenditure of energy to do work. This equation can be applied to individual organisms as well as to the ecosystem as a whole. From this equation it is clear that net primary productivity is nothing but the rate of increase in stored energy with time. Sometimes this net productivity is mistaken for biomass, which in fact refers to the total stored energy content at any one time or at the time of measurement. Hence net productivity can be defined as rate of biomass accumulation with time. This equation can also predict the status of an ecosystem when the three parameters (GPP, NPP, and R) are measured.

Let us assume three probable conditions as follows:

1. GPP = R,
2. GPP < R, and
3. GPP > R.

In the first case, NPP = 0 [according to Equation (3)], which indicates that in the ecosystem there is no accumulation of energy and the ecosystem just maintains itself. In the second case, respiratory losses exceed gross production, resulting in a negative NPP, which indicates a state of deterioration of the ecosystem. In the third case, however, the value of NPP is positive, indicating accumulation of biomass (or energy) in the ecosystem.

C. Measurement of Primary Productivity

Primary productivity is the change in energy or biomass per unit area per unit time. The common units for expression of primary productivity are grams per square meter per day ($g \, m^{-2} \, day^{-1}$) or kilograms per hectare per year ($kg \, ha^{-1} \, year^{-1}$). Primary productivity of any ecosystem is of great value for both human beings and the rest of the animal kingdom. Hence, its estimation/measurement is also of great value in assessing the state of an ecosystem. For example, a farmer should keep an account of crop production, a forester has to estimate the wood production, and a fishery scientist should take into account the algae production in a culture pond. Primary productivity of ecosystems is measured by various methods, such as (1) harvest method, (2) measurement of oxygen evolution, (3) measurement of Co_2 absorption, (4) estimation of pigments, (5) use of radioactive carbon

elements, (6) measurement of litter or disappearing raw materials, and (7) measurement of caloric value or energy content.

The suitability of these methods varies from ecosystem to ecosystem. In the study of primary productivity, however, determination of caloric value or energy contents of organic matter accumulated in ecosystems is most extensively used. Caloric value of any material is the amount of energy (calories or joules) liberated by burning one unit of the material. Energy content of any material is usually measured by an instrument called a calorimeter or commonly known as an oxygen-bomb calorimeter. The basic principle of this method is to measure the heat generated by combustion of one gram of the compressed powdered material in the presence of oxygen gas. The instrument includes a bomb having facilities for inletting and outletting oxygen and sample materials, a water bath surrounding the bomb, electrical connections from outside to inside the bomb for ignition, and a very sensitive Beckmann's thermometer (sensitivity 0.01°C). The calculation is based on the formula

$$V = \frac{W(\Delta t - \Sigma c)}{G}$$

where V = caloric value ($cal \, g^{-1}$),
W = water equivalent (constant) of the instrument,
t = increase in temperature after burning (°C),
c = correction values for the acid formed and combustion or ignition wire and thread, and
G = dry weight of the sample material (g).

The total amount of stored energy in an ecosystem can be calculated by multiplying the energy contents of all sample materials with their respective dry weights. This will give an estimation of total net production in terms of energy. This method is most important in ecological energetics as it has its applicability in calculating steps in energy flow and values of ecological efficiency in ecosystems.

D. Range of Primary Productivity

The efficiency of plants or plant communities to fix energy in an ecosystem depends on the ratio of net primary production to gross primary production. This ratio generally ranges from 0.4 to 0.8. Plants like large algae, grasses, and phytoplankton, which do not maintain a high supporting biomass, are found to be most efficient. The productivity of any ecosystem is regulated by environmental inputs—carbon dioxide, water (or rainfall), temperature, oxygen, and nutrients from soil or any habitat. This is also influenced by chlorophyll content, leaf surface area, and intensity of solar radiation. Therefore, productivity of ecosystems varies widely due to variations in these factors. The highest productivity (1000 to $3500 \, \mathrm{g \, m^{-2} \, year^{-1}}$) among terrestrial ecosystems is found in tropical forests with high rainfall and warm temperature.

Sometimes biological productivity is confused with yield or industrial production. In the latter case the reaction ends with the production of a certain amount of material, whereas in the former there is continuity in the process in time and therefore it needs a time unit for complete expression. In any case, productivity generally refers to the richness of an ecosystem.

E. Secondary Production

Secondary production deals with the process of energy accumulation at the consumer (herbivore) level and the rate of energy accumulation at this level is called secondary productivity. Unlike primary productivity, secondary productivity cannot be divided into "gross" and "net" amounts as the role of consumers is only to utilize food energy already produced with respiratory losses and to convert it into different forms of tissues. However, meat or flesh of animals, milk, dung, and any other animal product or residue are a few examples of secondary production.

F. Ecological Efficiency

Ecological efficiency or energy-capturing efficiency may be defined as the ratio of energy output to the energy input at any level when expressed as a percentage. But the performance of an individual plant or especially the plant community (primary producers) of an ecosystem is best evaluated in terms of Energy-Capturing Efficiency (ECE) or Efficiency of Energy Conversion (EEC). This is also known as photosynthetic efficiency and is actually the efficiency of plant(s) in converting solar energy into chemical energy, which fluctuates with variations in many internal and external factors that influence the process of photosynthesis. Energy-capturing efficiency is the ratio of energy stored in the plant to the total solar radiation available, expressed as a percentage:

Energy-capturing efficiency
$$= \frac{\text{Energy captured/time/area}}{\text{Available solar radiation/time/area}} \times 100 \quad (5)$$

Energy stored per unit area per unit time can be determined by determining the total energy content or caloric value. For practical purposes, available solar energy is considered to be half of the total solar radiation. Therefore, Equation (5) may be written as

Energy-capturing efficiency (%)
$$= \frac{\text{Energy captured/time/area}}{\frac{1}{2} \text{ solar radiation/time/area}} \times 100 \quad (6)$$

In agricultural ecosystems of different regions of the world, energy-capturing efficiency ranges between 1% and 5%. Under water scarcity, extreme fluctuation in temperature, nutrient deficiency, and careless practices this value may be as low as 0.1%. Similarly, with proper irrigation, fertilization, and other better management practices these values may be increased up to 10–12% in crop field ecosystems for a short period. However, theoretically it is estimated that the primary production potential of vegetation is limited within 20–30%. However, under natural conditions this varies within 1 to 5%. The other terms related to ecological efficiency between and within trophic levels are discussed in Section IV.

IV. ENERGY FLOW IN ECOSYSTEMS

A. Food Chain

Energy (solar energy) fixed by green plants flows in the ecosystem in the form of food through various consumer organisms along a series of steps of eating and being eaten—this is called a food chain. Food chains link consumer organisms starting from herbivores to decomposers. There are basically two types of food chains:

1. Grazing food chains linking green plants (producers), herbivores (consumer I), and carnivores (consumer II, III . . .); and
2. Detritus food chains linking dead organic matter, predators, detritus-feeding organisms, and microorganisms (decomposers).

In the body of the consumers, the consumed energy is partly utilized for maintenance, growth, and reproduction and partly discharged from the body as feces and urine. In the latter case, energy contents are transferred to the detritus food chain in which soil acts as an important medium. Figures 5a and 5b represent two simple food chains. Energy transfer along a food chain is not 100% because at each step a large amount (around 90%) of potential energy is wasted as heat. Therefore, in nature the maximum number of links in a food chain is limited to four or five, and the fewer the links the greater is the availability of energy.

B. Food Web

The food of an organism (whether herbivore or carnivore) is subject to change according to its availability. Some kind of food is shared by varieties of consumers especially at the beginning of the food chain (consumer I level). For example, in a grassland ecosystem, grasses and herbs are eaten

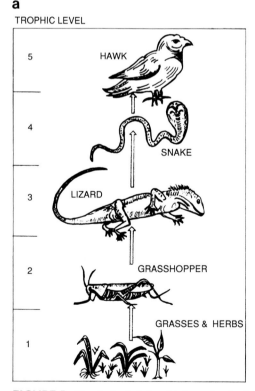

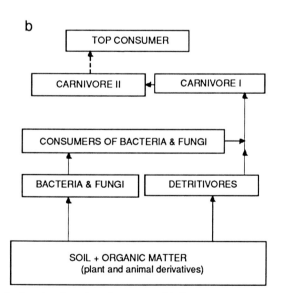

FIGURE 5 Simple food chains. (a) A grazing food chain showing the role and position of different organisms in a grassland ecosystem. Trophic level 1 = producer; 2 to 5 = consumers; 2 = herbivore or primary consumer; 3 to 5 = carnivores and secondary, tertiary, and quarternary or top consumers, respectively. (b) A general model of a detritus food chain linked to a grazing food chain.

by hares, grasshoppers, and field mice; grasshoppers are eaten by garden lizards; hares, field mice, and lizards can be eaten by snakes; hares, lizards, mice, and snakes can be eaten by hawks. Similarly, in a marshland ecosystem, marsh plants are consumed by fishes, birds, and other grazing mammals; these consumers are then consumed by many predators. This implies that food chains do not exist in isolated sequences, rather they are interlinked with one another. Such food chains collectively form a food web (Fig. 6).

C. Trophic Levels

In natural ecosystems, steps in a food chain are occupied by specific organisms and are called trophic levels (Fig. 5a). Thus, all organisms obtaining food from the same source in the food chain belong to one trophic level. In any ecosystem, green plants (producers) occupy the first trophic level; herbivores (primary consumers or consumer I or first-order consumers) belong to the second trophic

level; carnivores (secondary consumers or consumer II or second-order consumers) belong to third, and so on.

D. Role of Different Components of a Food Chain

Producers, herbivores, carnivores, omnivores, decomposers, parasites, scavengers, and saprophytes are different components of a food chain and play their specific roles in energy flow within the ecosystem. Herbivores start the process by converting the stored energy of plant tissues into their own body tissue (animal tissue), the food or source of energy for the next higher trophic level. Herbivores such as deer, goat, sheep, rabbit, hare, and cow feed on plant materials and possess adaptive features of their alimentary canal, such as different kinds of teeth (cutting, tearing, and grinding), complicated stomach, long intestine, well-developed cecum, and association of symbiotic microorganisms in the digestive tract. First-order consumers are flesh

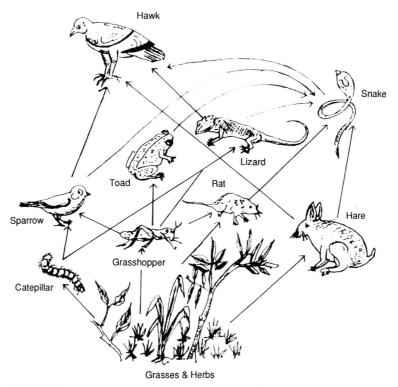

FIGURE 6 Food web of a grassland ecosystem involving many interwoven food chains.

eaters, which are habituated to catching living prey, tearing their flesh with the help of claws, beaks, and canine teeth, and digesting them. In a food chain, some consumers, such as a hawk in the grassland ecosystem and a tiger in the forest ecosystem, are not eaten further—they are designated as top-level consumers. Humans, along with other animals (red fox, crow, white-footed mouse, dog, cat), feed on both plants and animals. They are called omnivores and their position in the food chain is not constant.

The role of decomposers is opposite to that of producers, as the former help to release the nutrients of dead organisms into soil and atmosphere and "run" the nutrient cycle. Decomposers include both macroorganisms (mites, earthworms, crabs, and other insects) and microorganisms (fungi and bacteria). Plants use the soil nutrients released by decomposers. Parasites depend on living plants and animals for food and energy, scavengers are animals that feed on dead organisms, and saprophytes are plants that feed on organic matter.

Energy-flow patterns vary greatly in grazing and detrital food chains. The detrital food chain is common to all ecosystems and is the major path of energy flow in terrestrial ecosystems, as consumption of plant biomass by grazing animals is comparatively low. In aquatic ecosystems, on the contrary, the role of grazing animals (the meaning of grazing refers to direct consumption of plants or plant materials) is dominant in the energy flow. However, during transformation of energy through the grazing food chain, availability energy is reduced to 10% as 90% is lost in the system. For example, if 10,000 kJ of plant energy is consumed by herbivores, about 1000 kJ goes into herbivore tissue, 100 kJ to first-order carnivore production, 10 kJ to second-order carnivores, and only 1% will reach third-order consumer (Fig. 7). Hence, higher trophic level organisms get a lower amount of energy.

E. Analyses of Energy Flow

For analyses of energy flow, aquatic ecosystems are preferable to terrestrial ecosystems because of their smaller range of temperature fluctuations, distinct and clear-cut boundary, and slower dissipa-

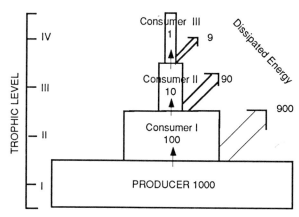

FIGURE 7 Dissipation of energy at different trophic levels during transfer of energy (according to the 10% law). Figures indicate only fractions of the total (i.e., 1000).

tion of evolved gases. It is a very difficult task to estimate energy flow by summing up population energy at different trophic levels, because certain organisms have dual positions. However, this is the only practical approach and it may be attempted in three ways: the ecosystem approach, the food chain approach or individual approach, and the population approach.

1. Ecosystem Approach

In an ecosystem, gross and net energy production along with energy loss by respiration are estimated for all trophic levels. It is observed that stored energy (NPP) decreases at successive trophic levels owing to higher respiratory losses and exports. In every ecosystem, the least energy reaches the top consumers not only because of these losses but also because of the larger size and smaller number (population) of these organisms. For example, in a pond ecosystem the number of big fishes and in a grassland the number of hawks will be least among the organisms. Flow of the least energy to this trophic level indicates that there is no trophic level beyond them in the ecosystem.

2. Food Chain Approach

The food chain approach to energy-flow analyses is easier than the ecosystem approach and better explains the demarcation of energy resources used by individual species. In this case, energy moves into only one species at each trophic level and flow of energy along other directions is not considered. Gross and net production along with respiratory and other losses are estimated for a single species of a particular trophic level. Additional energy losses due to interconnected or overlapped food chains can also be easily estimated. Findings through this approach reveal that at the producer level, respiratory losses from gross production are of lower percentage than at the consumer level. For example, in a Michigan old field, respiratory losses are 15% of gross productivity at the producer level and around 97% at the consumer level. As a result, there are drastic reductions in available energy at each level.

3. Population Approach

Population-level analyses of energy involve estimation of energy losses by taking many environmental factors and adaptive features of organisms into consideration. The external factors include rate at which energy is transferred to predator organisms, food resources, and ambient temperature, and the internal factors are body temperature and growth rate. In this approach, estimations of gross, net, and respiratory energy are calculated for generally one year. The reason for considering environmental and body temperature is the greater influence of these two factors in energy flow. For example, warm-blooded animals (homeotherms) lose more assimilated energy by respiration than do cold-blooded animals (poikilotherms), as the former utilize much energy to maintain their body temperature. Respiratory loss also varies with age of organisms, as in young animals the growth rate is higher due to higher rate of biomass accumulation and lower rate of respiratory loss and in old animals the reverse holds.

F. Classification of Ecosystems Based on Energy Flow

Energy flow determines the structural and functional status of an ecosystem. Based on the pattern of energy flow, ecosystems can be divided into four types:

1. unsubsidized ecosystems,
2. nature-subsidized ecosystems,
3. human-subsidized ecosystems, and
4. fossil-fuel-powered ecosystems.

I. Unsubsidized Ecosystems

Unsubsidized ecosystems are natural ecosystems such as ponds, lakes, grasslands, and forests that are operated by solar energy. They produce the essential resources of life (food, fodder, fuel, fiber, medicine). In such ecosystems the annual energy flow ranges from 4.17 to 41.7 MJ m^{-2}.

2. Nature-Subsidized Ecosystems

Nature-subsidized ecosystems are also natural ecosystems but are supported by energy subsidies such as wind, rain, evaporation, and tidal water in addition to solar energy. These ecosystems possess higher productivity as discussed earlier. A few examples are tidal estuaries, tropical rain forests, and coral reef ecosystems. The annual energy flow in these ecosystems is also very high (41.7 to 166.8 MJ m^{-2}).

3. Human-Subsidized Ecosystems

Human-made ecosystems like plantations, crop fields, agriculture-based villages, and aquacultures are powered by both the sun (a natural source) and subsidiary energy sources (auxiliary energy) such as fertilizers, pesticides, human and draught animal energy, and fossil fuels for agricultural machinery. Human efforts to increase the productivity of such ecosystems through these energy subsidies are remarkable. In these ecosystems the annual energy flow is as high as in nature-subsidized ecosystems (41.7 to 166.8 MJ m^{-2}).

4. Fossil-Fuel Powered Ecosystems

Urban ecosystems (towns, cities, industries) are human-made ecosystems that are fully dependent on fossil fuel (gasoline, diesel, coal, natural gas) and other nonrenewable energy resources. These ecosystems are in fact responsible for the depletion of natural resources and the creation of environmental pollution. The annual energy flow in these ecosystems is very high, ranging between 417 and 12540 MJ m^{-2}.

G. Energy-Flow Models

A model is an assumptive structure consisting of sequentially related hypotheses. It is used as a testing device to explain some process or phenomenon. The energy-flow pattern in ecosystems is also better explained with the help of models. There has been an evolution of energy-flow models since they were first proposed and developed by Raymond Lindeman in 1942, based on the principles of thermodynamics and the "trophic dynamic concept." This is a very simple mathematical model in the form of the equation

$$\frac{\Delta\Lambda n}{\Delta t} = \lambda_n - \lambda_{n'} \tag{7}$$

where $\Delta\Lambda n$ = rate of change of energy content of any trophic level, Δt = change in time, λ_n = energy content of a trophic level, and $\lambda_{n'}$ = energy content of the next higher trophic level.

This model [Equation (7)] states that the rate of change in energy content at any trophic level is equal to the rate at which the energy is received by the trophic level without taking the energy loss into account. The dynamic nature of a trophic level along with continuous receipt and loss of energy is indicated in this model. But here, the total amount of energy transferred from one trophic level to the next, is considered as true productivity assuming that this production is equivalent to the gross primary production at the producer level and assimilation at the consumer levels.

For a clear-cut explanation of energy flow many ecologists have proposed structural models such as the box and pipe model, the simple two-channel model, the Y-shaped double-channel model, and so on.

Box and pipe models can explain energy-flow patterns through individuals (Fig. 8a), populations (Fig. 8b), and the ecosystem as a whole (Fig. 8c). In the individual model it has been shown that out of the total food energy consumed (C), a part is assimilated (A) and a part is lost in the form of feces and urine (FU). The assimilated energy is utilized for respiration (R), for storage in the form of new tissues (B), and for growth and reproduction of the organism. In the population model (Fig. 8b), besides the aspects exhibited in the individual model, the additional features are flow of energy from biomass to predators and parasites and inflow (gain) and outflow (loss) of energy. Ecosystem models are designed by fitting the population boxes into many trophic levels (Fig. 8c).

Box and pipe modeling of energy flow has become widespread since the 1960s and is proposed by eminent ecologists like H. T. Odum and E. P. Odum. These models follow the thermodynamic laws. The boxes and pipes indicate the trophic levels and energy flow, respectively. The ecosystem boundary shows inflows and outflows, which are said to balance each other, and during its flow energy at each trophic level is partly dispersed as heat, and thereby the two laws of thermodynamics are satisfied. For constructing the ecosystem boundary in an energy-flow model, natural features such as the shore of a lake, the bank of a river or pond, the edge of a forest, or arbitrary limits like the pathway around a field and partitions of crop fields are chosen. Furthermore, in natural ecosystems, grazing and detritus food chains need to be indicated separately, which are exhibited in Figs. 9 and 10.

The grazing model, known as the Y-shaped double-channel model, has a practical applicability as it separates grazing circuits from organic detritus circuits, herbivorous consumption of food from saprophytic utilization of dead organic matter, and the role of phagotrophic macroconsumers from that of decomposers or microconsumers (Fig. 9). The special features of the detritus model (Fig. 10) are the transfer of energy beyond producers along two ways—the distinction of organisms feeding on living materials (biophages) and those feeding on nonliving or dead living materials (saprophages)—fractionalization of decomposers, and accommodation of import and export of energy.

In a practical sense, modeling of energy flow in any ecosystem or unit of area is possible as long as the import (or inflow) and export (or outflow) of energy are taken into account. This makes it easier to model energy flows in political units such

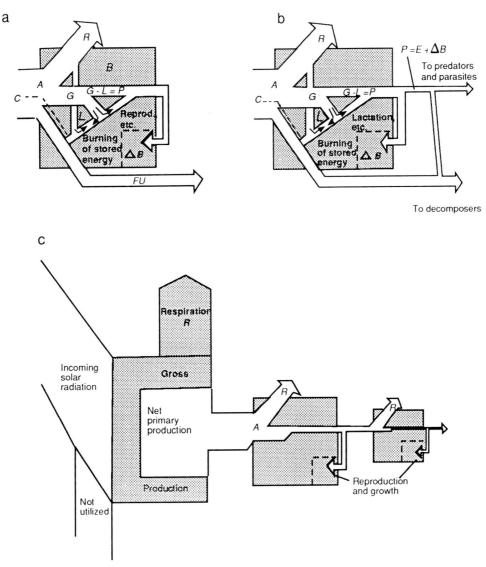

FIGURE 8 Box and pipe models of energy flow through individual organism (a), population (b), and ecosystem (c). C = energy ingested; A = energy assimilated; FU = energy lost through feces, urine, gas, and other rejecta; P = secondary production; R = respiratory loss; B = biomass; ΔB = change in biomass; G = growth expenditure; and L = loss of energy. [After R. L. Smith (1980). "Ecology and Field Biology," 3rd ed. Harper & Row, New York.]

as a village, a town, or a city, which can be studied as ecosystems from an energy standpoint. The limits of the ecosystem boundary can be represented by following political or revenue maps. For example, in an agriculture-based village or a forest-based tribal village, the limitations of revenue boundary will be an erroneous approach because intervillage dependence on energy resources is possible. In this case the boundary can be the area of dependence

of inhabitants (human and domestic animals) for the extraction of resources. Two such models are depicted in Figs. 11 and 12. The salient features of these models are: an agriculture-based and a forest-based village are considered as ecosystems by taking the boundaries of the revenue area and the area of natural resources (cropland, plantations, and natural vegetations) in the former and the area of dependence within the reach of village inhabitants in

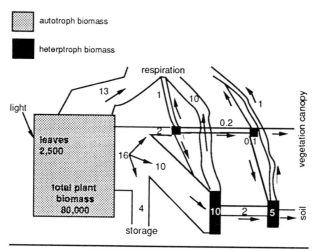

standing crop biomass in kilocalories / m^2
energy flow in kilocalories / m^2 / day

FIGURE 9 A Y-shaped or double-channel energy-flow model showing separation of grazing food chain (vegetation canopy) from detritus food chain (soil) in a forest ecosystem. [After E. P. Odum (1971). "Fundamentals of Ecology," 3rd ed. Saunders, London.]

the latter; (2) energy flow through all the subsystems (animal husbandry, agriculture, human population, and other economic activities) in kind of food, fodder, fuel, and other biomass resources is utilized for other purposes; and (3) imports and exports are also emphasized to show the balance of energy.

V. ENERGY LOSSES

Trophic-level studies of energy flow accounts for much of the flow within organisms. So far it has been discussed that respiration is the major avenue of energy loss from organisms (plants and animals) and the rest of the energy is utilized within the

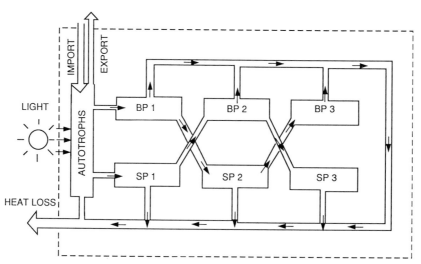

FIGURE 10 Two-channel energy-flow model showing import (inflow) and export (outflow) of energy. BP = biophages or organisms feeding on living organisms and SP = saprophages or organisms feeding on dead organic matter.

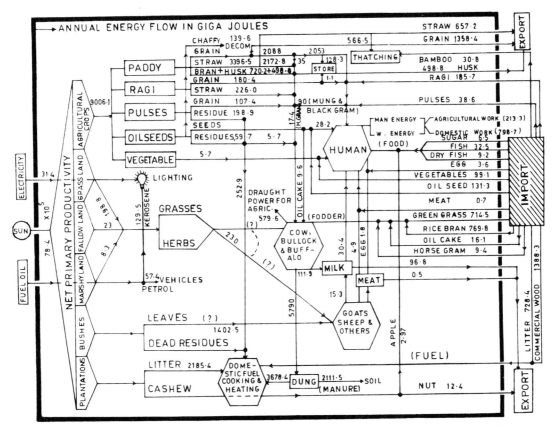

FIGURE 11 Energy-flow model within an Indian agriculture-based village ecosystem. Note: Paddy = rice grain; ragi = finger millet; pulses = legumes; chaffy = unutilized crop biomass; agriculture residue for fuel (252.9 GJ) = legume and sesamum residues; groundnut residue is used as fodder; paddy straw (2739.3 GJ) is used for fodder (2172.8 GJ) and thatching (566.5 GJ); rice grain (2088 GJ) is consumed by humans (2053 GJ) and cattle (35 GJ); legumes include mung (*Phaseolus aureus*) and blackgram (*P. mung*) consumption by humans (90 GJ) and horsegram (*Dolichos biflorus*) by cattle (17.4 GJ); human energy expenditure for agricultural (213.3 GJ) and domestic (798.7 GJ) activities is shared by men and women (W. energy). [Reprinted from *Biomass* **23**, S. K. Nisanka and Misra, 165–178, Copyright (1990) with kind permission from Elsevier Science Ltd., Kidlington, U.K.]

organisms. But there are a series of steps of energy loss starting from energy consumption until it is utilized or stored in the body. As a result, only a part of consumed energy is assimilated and utilized by the organism, and the rest is liberated or returned to the environment by one way or another. A few steps involved in these losses or unutilized energy are (1) unavailable energy from food, (2) unused surplus energy, (3) leftover energy, (4) defecation, and (5) respiration.

Within an ecosystem, herbivores or the primary consumers depend on the producers for their food. This does not mean that all the stored energy of these plants is utilized by the herbivores. Some plant parts may be rejected because they are toxic, inpalatable, or inaccessible or because of the natural removal of

net energy production from one ecosystem to another. An example of the former case may be non-consumption of roots by maximum herbivores, and removal of green plants (producers) from salt marshes by tides exemplifies the latter. Forest litter lasts longer when there is a lack of proper consumers. This is an example of a faster rate of production than that of consumption, which results in surplus energy in an ecosystems. Unused surplus energy can also be generated when a prey–predator relationship in a forest is affected by fluctuations in prey populations. The percentage of predation decreases when the density of prey population is high as well as low. At lower prey density, predation pressure is lower due to greater expenditure of the predator's energy

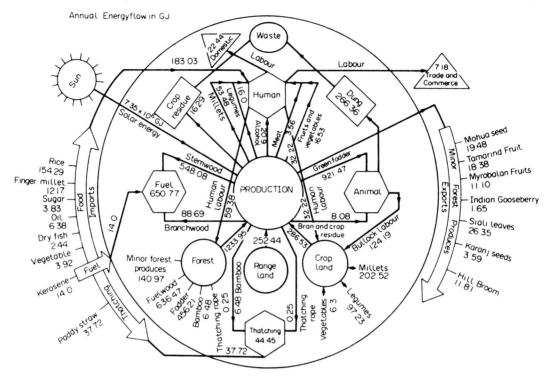

FIGURE 12　Energy-flow pattern within a forest-based village ecosystem. [Reprinted from *Biomass and Bioenergy* **4**(1), S. P. Nayak, S. K. Nisanka, and M. K. Misra 23–34, Copyright (1993), with kind permission from Elsevier Science Ltd., Kidlington, U.K.]

to catch the prey. On the other hand, at higher prey density, the lower percentage of predation is due to the smaller population of predators. However, this results in an increase in prey population at the concerned trophic level, the former being responsible for unused surplus energy.

In natural ecosystems, sometimes consumer organisms harvest more food than what they can physically consume. This results in a form of energy loss called "left over" energy. For example, underground storage of grains by rodents goes unconsumed even though the storage is made to be consumed at a time of future food scarcity. Similarly, many food items contain nonedible (discardable) and indigestible fractions, which are a considerable percentage of available food energy and known as nonassimilated energy. Seeds of many fruits, fibers and cellulose of plant materials, skin, fur, bones, and exoskeletons of many prey animals are a few examples of such energy losses. So, energy that is assimilated can be used to build up net production or partly be lost through respiration and other associated ac-

tivities, such as maintenance of body temperature, replacement of worn and torn tissues, interaction with other animals and plants, courtship behavior and reproduction, and other homeostatic mechanisms. Energy losses within an ecosystem with reference to each trophic level can be estimated by assessing all of these forms.

VI. ENERGY BUDGET OF ECOSYSTEMS

The energy budget of an ecosystem is the overall estimation of energy available, energy flow to different components, and energy loss. To prepare the energy budget of any ecosystem one should determine total available solar energy (or solar radiation), total assimilation by green plants (or total energy content of plant materials), and flow to different trophic levels. Practically, in energy budgets the net energy production is determined and the values are used to manipulate gross energy production based on the assumption that 30% of net production is lost during respiration, which is added to net production to obtain the gross energy production [Equations (3) and (8)]:

$$R = NPP \times 0.3 \qquad (8)$$

However, respiratory losses can also be determined by taking sample organisms in ecosystems. Energy budgets for various ecosystems are depicted in Tables III, IV, and V.

In aquatic ecosystems, energy-capturing efficiency is said to be less than that of terrestrial ecosystems. In the Silver Springs ecosystems in Florida (Table III), out of total solar energy only 24.1% is absorbed and only 1.22% is stabilized in the form of gross production. In the tropical grass-land ecosystem dominated by *Aristida,* gross production of energy at the producer level comes to 0.52% of total solar energy and net energy production is 0.40% (Table IV). This energy budget is restricted to the producer level only and therefore may be designated as an energy budget of a community that explains the flow of net energy into different component parts such as roots, standing dead parts, and litter along with losses due to litter and root decomposition. The losses are 65.69% of net production and 0.26% of total solar energy. The energy budget of a village ecosystem can be prepared by measuring the production, consumption, export, import, and storage of energy (Table V). The balance of energy in such ecosystems is represented by

TABLE IV
Energy Budget of *Aristida* Community in a Grassland Ecosystem[a]

Parameter	Total energy received or transferred (MJ m^{-2} year^{-1})	Percentage of total solar energy
Sunlight	60.16×10^2	100
Available energy	30.08×10^2	50
Gross energy production	31.502	0.52
Respiration	7.270	0.12
Net energy production	24.232	0.40
Net production to standing dead matter	13.260	0.22
Net production to roots	8.223	0.14
Net production to litter	2.749	0.04
Standing dead matter to litter	10.878	
Total loss from litter and roots	15.918 (65.69% of net production)	

[a] From M. K. Misra and B. N. Misra (1989). *Folia. Geobot. Phytotax.* **24,** 25–35.

$$P + I = C + S + E \qquad (9)$$

where P = production, I = import, C = consumption, S = storage, and E = export.

TABLE V
Energy Budget of an Agricultural Village Ecosystem (Village Area = 125 ha)[a]

Parameter	Energy (GJ year^{-1})
Solar insolation	78.4×10^5
Production (net yield in all kinds of food, fodder, fuel, and other utilizable biomass)	18,763
Consumption (food consumption by human beings, fodder consumption by domestic animals, fuel consumption and utilization in other forms)	20,451
Storage (seeds for sowing next year's crop)	129
Exports (selling products outside the village)	2,642
Imports (purchase of commercial items and collection of noncommercial items from outside ecosystem)	4,441

[a] Reprinted from *Biomass* **23,** S. K. Nisanka and Misra, 165–178, Copyright (1990), with kind permission from Elsevier Science Ltd.. Kidlington, U.K.

TABLE III
Energy Budget of a Silver Springs, Florida, Ecosystem[a]

Parameter	Percentage of total available energy
Solar input	100
Not absorbed	75.9
Absorbed	24.1
Not stabilized	22.88
Gross energy production	1.22
Respiratory loss	0.70
Net energy production	0.52

[a] From H. T. Odum (1957). *Ecol. Monogr.* **27,** 27–55.

In such budgets, because of the overlapping of many subsystems, only the dominant components such as agricultural production, grassland production, and human and domestic animal consumption are taken into account.

VII. ENERGY CONSERVATION

Biological conservation entails the responsible management of all life-forms to ensure the healthy functioning of diverse ecosystems. One benefit to humans may be sustainable resources for the present generation while maintaining the potential to meet the needs of future generations. From a socioeconomic but also ecological point of view, sustainable development requires the integrated conservation and management of resources. Hence, energy conservation can be considered as one of the major aspects of ecosystem management, especially in rural and urban ecosystems where human beings are the chief consumers.

A. Energy Resources

Energy conservation means the conservation of energy resources. Therefore, considering the processes and methods of energy conservation, let us first discuss the various energy resources. The two broad categories of energy resources are renewable and nonrenewable.

I. Renewable Resources

Renewable resources are reproducible, inexhaustible, and mostly biological in origin. Biomass energy, solar energy, wind energy, water energy (or tidal energy), geothermal energy, and hydroelectricity belong to this category.

Biomass, a conventional renewable source of energy, refers to all materials obtained directly or indirectly from plants. This includes wood and other plant parts, dung, biogas, and agricultural residues. Because these resources originated from plants, they can be renewed through plantations or afforestation (plantation of forest trees). Biogas is produced using animal dung and other organic wastes and it is very economic to produce. Biomass

meets 85% of the rual energy demand and 51% of the urban poor energy needs in the world.

Solar energy, wind energy, tidal energy and geothermal energy are some of the forms of renewable energy that can augment power in specific areas in a decentralized manner. Solar energy has been discussed in preceding sections. It is harnessed through photovoltaic devices to produce heat and electricity for cooking, lighting, domestic irrigation, and other uses. Solar cookers are now in wider use in developing countries for domestic cooking and are responsible for shifting the demand away from other diminishing energy resources.

Wind energy is highly useful in remote areas, saves fossil fuels, and produces energy that is locally available. Wind energy is converted to mechanical energy and electric energy by windmills. Hence, this energy causes no pollution or environmental degradation. In coastal areas, wind flows at high velocity and therefore in such areas wind energy can meet the local energy demands. Windmills are widely used in countries like India and Holland.

Tidal energy is generated by the rise and fall of sea level due to tides and is sufficient to rotate a turbine that can produce electricity. This is possible through small tidal power plants, which have been installed in China, Russia, France, and other countries. Geothermal energy is derived from the high temperature of underground waters and is converted into electricity. This can also be used for creating refrigeration.

2. Nonrenewable Resources

Nonrenewable resources are physical resources like fossil fuels (oil, coal, natural gas), hydroelectricity, and nuclear power. Fossil fuels actually contain solar energy as they are the extracts of forest trees that were buried millions of years ago and were converted into coal, oil, and gas under pressure and temperature. These fossil fuels account for 90% (gasoline 45%, natural gas 20%, coal 25%) of the global commerical energy demand. Atomic energy from atoms of radioactive elements like uranium (^{238}U) can be produced in atomic reactors. The generation of energy from 1 kg of natural uranium is estimated to be equivalent to that of 35,000 kg of coal. Nuclear energy is widely used in developed countries like

Germany, France, the United States, Russia, and England and to a lesser extent in developing countries. Modern civilization presently relies on these fuels, which are going to be exhausted one day. Hence they are called nonrenewable resources and are also responsible for causing more environmental pollution than the renewable forms of energy.

B. Energy Demand

The demand for energy today is growing rapidly and almost doubled in the last 14 years. The developed countries consume much more energy than the developing countries. For example, the United States consumes 33% of the world's energy production and has 6.25% of the world's population, whereas India with 15% of the world's population consumes only 1.5% of the world's energy production. However, in developing countries the energy sources are mainly biomass or nonconventional renewable types, whereas in developed countries it is mostly of nonrenewable type. In developing countries, per capita energy consumption was 36 GJ, and in developed countries this rate was 218 GJ during 1987. The rate of energy production has not kept up with the rapid growth of population, industrialization, and urbanization in developing countires.

C. Energy Waste and Conservation

The major necessary steps toward conservation of energy are the consumption of nonrenewable resource at a minimum rate and the continuous reproduction of renewable resources to balance the growing energy demand. The former step involves reducing energy use and waste and modification of devices for the consumption of renewable sources. This is possible by avoiding the unnecessary use of vehicles and machinery on one hand and manufacturing fuel-efficient devices on the other. Whether the consuming unit is a machine or a motor vehicle or a domestic stove, it should be fuel-efficient in the sense that energy consumption should be at a minimum to yield maximum output. In rural sectors of developing countries, defective stoves are used owing to lack of availability, awareness, and skill, and thus energy waste is very high.

Conservation of biomass resources is possible through the conservation of forests of firewood trees and replanting of fast-growing and high-energy species. Tree plantations are often located on wastelands or nonagricultural lands using fast-growing and suitable species like *Acacia, Casuarina,* and *Prosopis.* These trees have high caloric value and high energy-capturing efficiency are resistant to various environmental stresses, and require little management cost. Biomass resources like firewood can also be modified into more efficient forms through modern technologies like briquetting, gasification, and liquification.

The burning of cattle dung in rural areas is said to be an ecologically damaging practice because it is more valuable as a source of organic biofertilizer. However, a wiser use would be for the production of biogas, which can yield fuel for cooking and lighting and fertilizer in the form of slurry. Biogas can also be produced from human excreta and sewage sludge.

Glossary

Ecosystem System composed of living organisms and the nonliving environment with which they interact continuously, for example, a pond, an aquarium, a forest.

Energy budget Estimation of total inputs, outputs, and demand of energy in an ecosystem.

Energy capturing efficiency (ecological efficiency) Ratio of total energy accumulated by green plants per unit time and unit area, expressed as percentage of total available solar radiation; percentage of energy in biomass (organic matter) produced into biomass by the next higher trophic level.

Energy conservation Management of energy resources and utilization for a sustainable benefit and maintenance of their potential for future needs.

Energy flow One-way transfer of energy through an ecosystem, including the way in which energy is converted and used at each trophic level within an ecosystem.

Energy subsidy External supply (from outside ecosystem) of energy that decreases the energy need for internal self-maintenance and increases the output of an ecosystem.

Food chain and food web Food chain is a sequence of transfers of food energy among organisms belonging to different trophic levels. The complex and interlocking series of food chains form a food web.

Primary productivity Rates of assimilation (gross) or accumulation (net) of energy and nutrients by autotrophs (green plants).

Surplus energy Extra or unused energy generated in an ecosystem due to faster rate of energy production and slower rate of consumption.

Trophic level (food level) Level at which food energy is transferred from one organism to another; the position of an organism in the food chain.

Bibliography

Dash, M. C. (1993). "Fundamentals of Ecology." New Delhi: Tata/McGraw–Hill.

Misra, M. K., and Misra, B. N. (1989). Energy structure and dynamics in an Indian grassland. *Folia. Geobot. Phytotax.* **24,** 25–35.

Nayak, S. P., Nisanka, S. K., and Misra, M. K. (1993). Biomass and energy dynamics in a tribal village ecosystem of Orissa, India. *Biomass and Bioenergy* **4**(1), 23–34.

Odum, E. P. (1971). "Fundamental of Ecology," 3rd ed. London: Saunders.

Odum, H. T. (1957). Trophic structure and productivity of Silver Spring, Florida. *Ecol. Monogr.* **27,** 27–55.

Smith, R. L. (1980). "Ecology and Field Biology," 3rd ed. New York: Harper & Row.

Ricklefs, R. E. (1990). "Ecology," 3rd ed. New York: W. H. Freeman.

Ecological Energetics of Terrestrial Vertebrates

Gwendolyn Bachman
University of California at Los Angeles

Sandra Vehrencamp
University of California at San Diego

I. INTRODUCTION

Ecological energetics is the study of energy flow through ecological systems. In living organisms, energy flow occurs whenever one organism eats another or when one form of energy is transformed into another form. In terrestrial ecosystems, energy flow begins with the capture of light energy by green plants, and continues through the bodies of herbivores that eat the plants and the carnivores that eat the herbivores. Ecological energetics can be examined at various levels of ecological organization: the individual, the population, the community, and the ecosystem. The primary data on energy flow are usually gathered at the individual level by quantifying energy intake and the efficiency with which it is used for maintenance, activity, and the production of new biomass. This information is combined into an energy budget, which changes during the annual cycle as energy demands and availability fluctuate. Individual effects can then be summed across populations and species to estimate energy flow at the community and ecosystem levels. This article deals primarily with energetics at the level of the individual organism but identifies important aspects of this approach that

are used for model building and analysis of energy flow at higher ecological levels. Furthermore, discussion will focus on the ecological energetics of terrestrial vertebrates such as birds, mammals, and, to a lesser extent, reptiles. The general principles, however, are applicable to all organisms. General information is summarized and a few specific examples are presented with the intention of comparing the energetics of different kinds of vertebrate animals and indicating the relative impact of different variables on energy flow through an individual.

A. What Is Energy?

Energy is the capacity to do work. There are various forms of energy, some of which are critical to or characteristic of living organisms. Kinetic energy is a form of energy that is possessed by moving objects. This includes animal activity as well as heat or thermal energy, and light. The energy in a photon of light is the basis of all energy flow in most terrestrial ecosystems by virtue of the ability of green plants to capture and use light energy. Potential energy may be stored in an object as a result of its location or physical state. An important form of potential energy is chemical energy stored in molecules as a result of the bonding arrangement of atoms.

Energy can be transformed from one form to another. For example, the potential energy in the chemical bonds of a molecule of glucose (sugar) can be converted to heat energy or movement. Light energy can ultimately be converted to chemical energy when it is absorbed by specific molecules in plants (e.g., chlorophyll) and in the photoreceptors of animals (e.g., rhodopsin). Energy transformations are a key feature of energetics and they are governed by the laws of thermodynamics. Of particular relevance is the observation that different forms of energy can be arranged along a quantitative scale of disorder, called entropy, from highly organized energy to random energy, and that every process or transformation involving energy leads to an increase in entropy. Light and chemical energy are highly ordered forms of energy that can be used by organisms. In living systems, most energy ultimately becomes transformed to heat, which is a random form of energy that cannot be further transformed by organisms.

The most commonly used measurement unit of energy is the joule. One joule is the energy required to raise one kilogram of mass to a height of 10 cm in earth's gravity. One joule is equal to 0.239 calories. The calorie is a measure of energy that only applies to heat.

Animals obtain chemical energy by eating other animals or plants. Much of the energy contained in consumed food is passed to the chemical bonds of a molecule called ATP (adenosine triphosphate) by way of an oxidation process called respiration. Respiration is a tightly regulated sequence of biochemical reactions in which oxygen is used to release the chemical potential energy of organic compounds. For example, the production of ATP from a carbohydrate such as glucose sugar ($C_6H_{12}O_6$) proceeds according to the reaction

$$C_6H_{12}O_6 + 6O_2 + ADP + P_i$$
$$\rightarrow 6CO_2 + 6H_2O + ATP + heat$$

The products of respiration are carbon dioxide, water, ATP (from ADP, adenosine diphosphate), and heat released during the reactions. The reactions associated with respiration occur in mitochondria, organelles found in most cells of the body. The potential energy stored in ATP can later be made available to the cells to perform work.

B. Energy Budgets

Studies of ecological energetics at the individual level are based on the concept of an energy budget (Fig. 1). An animal's energy budget is an accounting of all energy expenditures with reference to the energy gained in food. In the analysis presented here, the energy budget will focus primarily on chemical potential energy, however, the energy exchanges associated with heat gain and loss can also be formulated into an energy budget. The energy in food is used for a variety of functions, including digestion and assimilation of food, general maintenance, growth, reproduction, sustaining activity required for foraging, predator de-

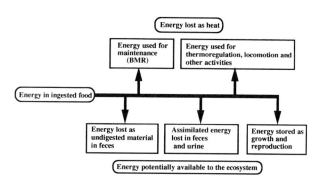

FIGURE I A simple energy budget illustrating energy flow through an individual. The allocation of energy to different components of the budget is shown in the approximate order of use from left to right. Only the energy remaining after essential functions such as digestion, maintenance, and activity have been fulfilled can be used for growth and reproduction. Components of the budget that represent energy lost as heat are shown above the horizontal line, and components that are potentially available to the ecosystem are shown below the horizontal line. Most of the energy in feces is in undigested food, though secretions and cells are also a component of feces and contain potential energy. Not all the energy that is assimilated, or that passes into the body across the walls of the digestive tract, is completely metabolized. Urinary excretions include assimilated but unmetabolized molecules as well as "nitrogenous waste" from the metabolism of proteins and other nitrogen-containing compounds. Nitrogenous waste is in the form of molecules of urea (mammals) or uric acid (birds and reptiles). Much of the energy lost as feces and urine is available to detritivores, including microorganisms; energy stored in body tissues or offspring is available to predators in the next trophic level.

fense, and behavioral components of reproduction. Because the amount of energy available for use is limited by the amount of energy taken in and the extent to which an animal is willing to use its energy reserves, the distribution of energy to each function must be allocated in a manner that will maximize its survival and reproductive success. Increasing the allocation of energy to one component may reduce the energy available to one of the other components, so flexibility and trade-offs in the allocation of energy are common. If an animal's mass and body composition are not changing, the energy budget is balanced: the energy gained in food is sufficient to cover all expenditures. At different times of year or different life stages, an animal's energy budget may be balanced, positive, or negative. Weight gain occurs when the energy budget is positive. Weight loss occurs when the energy budget is negative and the energy requirements are met in part by using energy stored as body tissues.

The relative amount of energy that an animal allocates to each component of its energy budget varies according to its role in the community and factors such as diet, body size, thermoregulatory strategy, sex, age, and physical properties of the environment. Only the energy captured in tissues by the growth and reproduction of the individual, and the energy in excreted material, are potentially available as food to other organisms in the community or ecosystem. Energy allocated by animals to digestion, maintenance, and activity is dissipated as heat, which is unusable by most organisms in the community. The energy budget of an individual thus determines the energy flow to higher levels of ecological organization. The following discussion will present details on each component of the energy budget.

II. ENERGY INTAKE

All animals obtain energy by the consumption or ingestion of food. Energy intake is a measure of the potential chemical energy contained in the ingested food. Different types of food, such as carbohydrate, fat, and protein, contain different amounts of potential energy, so an accurate estimate of en-

ergy in food requires measuring the proportion of these components in the diet. The energy content for 1 g of each food type is as follows: carbohydrate, 17.6 kilojoules (kJ); fat, 39.4 kJ; protein, 17.9 kJ.

A. Digestible Energy Intake

Some of the food that is eaten cannot be digested and therefore passes through the animal to be eliminated as feces. The energy that remains is called digestible energy intake. The digestibility of a diet is a function of the food types in the diet, and the morphological and physiological adaptations of the animal. One way to compare differences in digestible energy intake is to measure digestive efficiency:

$$\text{Digestive efficiency} = (\text{energy consumed} - \text{energy in feces})/\text{energy consumed}$$

Table I summarizes the relative digestibilities of different food types. The digestive efficiency of individual herbivores varies substantially according to the amount of plant fiber consumed. Animals that are specialized for a particular diet will often have adaptations that can increase their ability to extract the energy from the food they eat.

TABLE I
Digestible Energy Intake Measured as Digestive Efficiency[a]

Diet	Digestible energy intake (%)
Woody plants	≤30
Young leaves, flowers	30–60
Seeds, fruits	80–90
Meat	85–95
Insects	70–80

[a] Digestive efficiency is the percentage of total energy intake that is digested and absorbed across the membranes of the digestive tract. Energy that is not retained is excreted as feces. Many plant parts contain material that is not easily digested by vertebrates. Seeds with relatively strong or fibrous hulls may have digestive efficiencies as low as 60–70%. Insects have chitinous exoskeletons that are difficult to digest. If the diet contains primarily mature insects instead of larvae, the digestive efficiency may be as low as 50%. The values presented here apply in general to all vertebrates.

The supportive tissues of a plant, such as stems, trunks, and even most roots, are made of relatively indigestible forms of carbohydrate such as cellulose. Vertebrates do not produce cellulases, which are enzymes that can break down these structural carbohydrates. Consequently, vertebrates consuming woody material or coarse grasses may have a low digestive efficiency. To enhance the digestion and absorption of energy from such a diet, some vertebrate species have developed symbiotic relationships with microorganisms that do have cellulases. Ruminants such as cows and antelope are a well-known group of animals that exhibit this symbiosis, though herbivorous birds and reptiles also have symbiotic microorganisms. Ruminants have a specialized stomach morphology to accommodate their symbiotic microorganisms, which enable the ruminant to digest 50–75% or more of the plant energy they ingest. Some herbivores such as rabbits and horses have cellulose-digesting microorganisms lower in their digestive tract, where the ability to absorb nutrients and thus the benefits of microbial action are limited. Rabbits overcome this limitation by reingesting feces, thereby taking advantage of the predigestion provided by the microorganisms to achieve a digestive efficiency of approximately 69%.

Selectively browsing animals eat young leaves or flowers that contain more digestible carbohydrates. However, these plant parts may also contain compounds that reduce their digestibility or otherwise deter predation. For example, tannins in leaves reduce the ability of the herbivore to access the protein in the diet, and plant toxins such as cyanide will deter predation by making animals ill. Herbivores may select a variety of plants and plant parts to avoid ingesting too much of any one toxin. Fruits and seeds are the easiest plant products to digest. They are rich in energy in the form of very accessible sugars, lipids, and protein and digestive efficiencies are high for both. These differences in digestibility can also be detected as the diet of a single animal changes. For example, a small rodent such as a vole can digest 65 to 80% of the energy in green vegetation. On a higher-quality diet of grain, the vole can digest 84 to 94% of the energy it consumes.

The digestive efficiency for a carnivore eating pure meat can reach 99%, although a lower efficiency may be more typical as indigestible bone, hair, or feathers are often consumed as well. The chitinous exoskeleton of insects is also resistant to digestion. As mature insects are likely to have more chitin than larval forms, the efficiency of a diet of larval insects may be higher than that of a diet of mature insects.

The digestive efficiencies presented in Table I apply to reptiles as well as birds and mammals. However, the rate of digestion by reptiles is affected by their body temperature. After ingesting a meal, a reptile may select a warm area of its environment that will enable it to elevate its body temperature. The selected or preferred temperature, approximately 36°C or more, can promote the reptile's energy intake rate by reducing the time it takes for food to pass through the digestive tract.

B. Metabolizable Energy Intake

Once food has been digested or broken down, the smaller particles cross the membranes of the stomach and intestine and enter the body. Not all of this assimilated energy is available to the animal (Table II). Proteins are composed of amino acids, which contain amino groups that must be excreted as urea (mammals) or uric acid (birds, reptiles).

TABLE II

Metabolizable Energy Intake as a Percentage of Total Energy Intake for a Horse (nonruminant), Ox, and Sheep (ruminants) with Different Diets[a]

Species	Diet	Metabolizable energy intake (%)	Metabolizable energy intake (less methane gas) (%)
Horse	Hay and grain	60.5	59
	Straw	35	32
Ox	Hay and grain	71.3	65.1
Sheep	Grain	80	69.4
	Straw	37.2	30.5

[a] The metabolizable energy intake in the left column is total energy less the energy lost in feces and urine. The values in the column on the right are also reduced by the energy lost in methane gas production.

The energy contained in these compounds, though digested and absorbed, is not available to the consumer. Metabolizable energy intake is the energy that is actually retained and available for use by an animal.

Metabolizable energy intake
 = digestible energy intake
 − (energy in urea or urate
 + energy in other unoxidized molecules)

Other compounds are also excreted in urine. For example, oils in pines and eucalyptus are easily digested, but are not oxidized and thus are excreted.

Microorganisms in the digestive tract of herbivores produce various amounts of methane gas. In nonruminant animals, the energy lost as methane can represent less than 1% of the digestible energy of their diet. Because of the activity of their symbiont microorganisms, ruminants may lose nearly 10% of their digestible energy intake as methane gas. The energy in gases produced during digestion is released into the atmosphere.

C. Essential Nutrients

Animals need to acquire nutrients as well as energy from their diet. If the concentration of an essential nutrient is low in a particular food, then the animal may not be able to process enough food over a period of time to meet its nutrient requirements. A food item may then be included in a diet primarily for its nutrient content, which may conflict with diet choices that provide the maximum energy intake. Carnivores, because they eat other animals, generally obtain an adequate supply of both energy and nutrients. Herbivores, however, may face trade-offs between nutrients and energy content. The moose is a well-studied example of an herbivore that adjusts its diet to accommodate both energy requirements and sodium requirements. For the moose, aquatic plants are a concentrated source of sodium, but are a dilute source of energy. Other plants are higher in energy, but are too low in sodium to satisfy sodium requirements. Another example of nutrient requirements affecting diet

choice is provided by Costa's hummingbird. These nectar feeders need a daily supply of both energy and protein. Nectar is a concentrated energy source but provides insufficient protein. Therefore, 2–12% of a hummingbird's foraging time must be spent catching insects.

III. ENERGY USE

As discussed in the foregoing, of the total amount of energy that is consumed by an animal only the metabolized energy, or metabolizable energy intake, can be used by the animal. The energy budget equation can now be rewritten to simplify the following discussion of energy use:

Energy metabolized
 = Energy used for maintenance (basal metabolism)
  + Energy used for thermoregulation
   + Energy used for activity
    + Energy used for production

A. Basal Metabolism

Basal metabolism is the minimal energy expenditure required to sustain essential body functions such as the maintenance of organ and cell operations, blood flow, muscle tension, and body temperature. For birds and mammals, the basal metabolism is operationally defined as the minimum energy expenditure of a nonstressed, postabsorptive (nondigesting) animal, resting in a thermally neutral ("comfortable") ambient temperature. The basal metabolic rate (BMR) is measured in a laboratory setting so that other types of expenditures can be eliminated, that is, the animal must be resting to remove effects of activity and the animal is fasted to exclude expenditures associated with digestion. BMR is typically estimated by the measurement of oxygen consumption over a period of time. As illustrated previously, oxygen is used during respiration to convert the chemical potential energy in food to usable energy in the form of ATP. The amount of oxygen required to produce a given amount of energy depends on whether fat, protein, or carbohydrate is oxidized. Therefore, an estimate of the metabolic rate also requires an estimate of the substrate being oxidized: often, a mix of substrates is assumed. For birds and mammals, measures of BMR are taken under conditions where no additional energy is used to maintain body temperature. However, the metabolic rate of reptiles and other ectotherms changes with temperature (see the following section on thermoregulation). Therefore, "basal" metabolic measurements in animals such as reptiles are often taken at temperatures characteristic of the animal's habitat or determined by the question being asked by the investigator. The term standard metabolic rate (SMR) refers to the minimum maintenance expenditures of these animals at the designated ambient temperature, and in that respect it is equivalent to BMR.

There is a positive relationship between the body size of an animal and its basal metabolic rate. A larger mass requires more energy to maintain (Fig. 2). In proportion to body mass, however, the BMR of larger animals is lower than that of smaller animals. The typical relationship between BMR and body mass is a function of an animal's weight in kilograms raised to the 0.75 power. Another way to visualize this relationship is to note that 1 g of a larger animal requires less energy to maintain than 1 g of a smaller animal. A generally accepted explanation for this relationship has yet to be found. [*See* ECOLOGY OF SIZE, SHAPE, AND AGE STRUCTURE IN ANIMALS.]

As depicted in Fig. 2, the basal metabolic rates of animals can be affected by taxon independently of body size. For example, marsupial mammals have lower BMRs than placental mammals. Marsupial mammals are those in which the young develop largely within an external pouch, whereas for placental mammals the gestation of young occurs via a well-developed placenta. Passerines (small perching birds) have BMRs that are twice as high as those of equivalently sized mammals. The BMR of passerines is at least 10 times higher than that of an equivalently sized reptile.

Because BMR is a measure of minimum energy expenditure, it provides a baseline against which

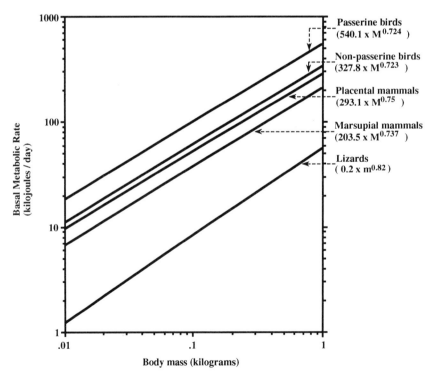

FIGURE 2 The relationship between basal metabolic rate and body mass in kilograms (M) or grams (m) for different vertebrate groups. The SMR data for lizards assume a body temperature of 37°C. The equations on the right side of the figure give the regression equation for each line in units of kilojoules per day.

other types of energy expenditures can be compared. All other expenditures require energy in addition to BMR, thus increasing total metabolism above BMR. In the following discussion of energy use, expenditures will be expressed as multiples of BMR (or SMR for reptiles) or as a percentage increase over BMR, and the units of BMR will be kilojoules per day. This convention provides a standard index of the relative cost of each type of expenditure and the impact of different variables on the energy budget, and permits comparisons among different sizes and categories of animals.

Animals rarely function at their basal metabolic rate in nature. The total daily energy expenditure of most vertebrates under free-living natural conditions is approximately 3 times BMR. Field metabolic rate (FMR), expressed as kilojoules per day, is often measured using the doubly labeled water technique. This technique uses the differential loss of water "labeled" with oxygen and hydrogen isotopes to estimate energy use. The isotopes (deuterium or tritium, and oxygen-18) are injected into the animal and equilibrate with other, nonlabeled atoms within hours, after which a sample of body fluid is taken and the animal is released into its habitat. The animal is recaptured a day or more later, and a second sample is taken. Both fluid samples are analyzed for their isotope concentrations. As illustrated earlier, during respiration, oxygen is released as carbon dioxide and water, whereas hydrogen is released only in water. The difference in the rate of loss of hydrogen and oxygen isotopes is a function of the loss of carbon dioxide. As with the measurement of metabolic rate by oxygen consumption, the relationship between the amount of carbon dioxide released and the energy used depends on the energy substrate (carbohydrate, protein, or fat). The doubly labeled water method can provide an estimate of the average daily energy expenditure. Along with behavioral observations, it can also be used to estimate the energy requirements of specific behaviors and activities. Most mammals exhibit daily energy expenditures ranging from 2.4 to 3.5 times BMR. Daily expenditures of birds mostly fall within a similar range, although they tend to be no greater than 3 times BMR. Changes in activity or physiological state can alter

these approximations in birds and mammals. For example, hibernating animals have extremely low metabolic rates, whereas in some species reproduction can increase metabolism up to 6 times BMR or more. In summer, when reptiles are active, they too have daily energy expenditures ranging from 2 to 3 times SMR. However, in winter, when temperatures are cooler, reptiles have lower daily energy expenditures of 2 times SMR or less.

B. Thermoregulation

Thermoregulation is the regulation or control of body temperature. Vertebrates exhibit two strategies for regulating body temperature that differ in the source of heat used. Ectotherms, such as reptiles, amphibians, and fish, are unable to generate and retain sufficient metabolic heat to maintain a high and constant body temperature. Their thermoregulatory strategy primarily consists of absorbing and losing heat energy from and to their environment. (Fig. 3). Their body temperatures, therefore, vary in direct relationship to the ambient temperature (Fig. 4A). Endotherms, such as birds and mammals, are able to generate and retain substantial amounts of heat from their own metabolic processes in addition to heat exchange with their environment. Endotherms maintain a high and relatively constant body temperature (Fig. 4A), regardless of the ambient temperature, by varying their metabolic rate (Fig. 4B). Endotherms have a thermoneutral zone (TNZ), a range of ambient temperatures over which they can maintain a constant body temperature without the use of energy above BMR. The TNZ is bounded by an upper critical temperature and a lower critical environmental temperature. Within the TNZ, body temperature is regulated with virtually cost-free mechanisms such as sleeking or fluffing the fur or feather insulation layer and dilating or constricting blood vessels in the skin to increase or decrease heat transfer from the body core to the skin surface. Above and below the TNZ, more energy must be expended to maintain body temperature. In low environmental temperatures (below the lower critical temperature), an endotherm's metabolic heat production will increase. Only at very low tempera-

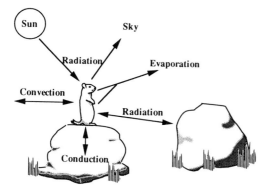

FIGURE 3 Four modes of heat exchange between an animal and its environment. Conduction is the exchange of the kinetic energy of heat through direct contact between an animal and another object (such as the substrate). The extent and direction of conductive heat exchange depends on the temperature gradient and the amount of surface area in contact. Convection is similar to conduction but involves the movement of liquid or gas (air), which enhances heat exchange. Convective heat loss varies in proportion to the wind speed, temperature difference, and the exposed surface area of the animal. Wind speeds are relatively low at the ground surface, so small animals living there may be minimally affected by convection. Radiation is the absorption or emission of electromagnetic radiation. Radiant heat is absorbed from the sun and other objects and is lost to the sky and objects. Solar radiation can be a significant source of energy for ectotherms. Evaporative water loss is the loss of energy from the animal as it evaporates water from its surface. At 30°C, the evaporation of 1 g of water removes 2.43 kJ of energy. In warm environments, evaporative water loss can be an effective mechanism for keeping cool.

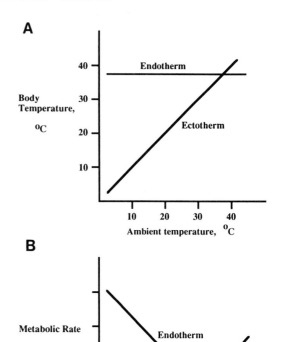

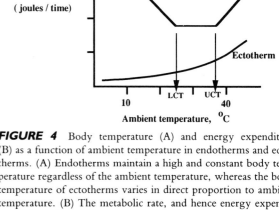

FIGURE 4 Body temperature (A) and energy expenditure (B) as a function of ambient temperature in endotherms and ectotherms. (A) Endotherms maintain a high and constant body temperature regardless of the ambient temperature, whereas the body temperature of ectotherms varies in direct proportion to ambient temperature. (B) The metabolic rate, and hence energy expenditure, of ectotherms increases exponentially when their body temperature is driven up by ambient temperature. Endotherms, on the other hand, regulate their body temperature by adjusting their metabolic rate. Within the thermoneutral zone endotherms can maintain their body temperature without the use of energy above BMR. The thermoneutral zone is bounded by a lower critical temperature (LCT) below which metabolic heat production must be used to keep warm. The upper critical temperature (UCT) defines the upper end of the endotherm thermoneutral zone. Because heat exchange depends on temperature gradients, the upper critical temperature is often near an endotherm's body temperature.

tures will shivering be evident. At moderately cold temperatures, extra heat production is provided by cellular mechanisms collectively referred to as nonshivering thermogenesis. Above the upper critical temperature metabolic options for unloading excess heat are limited. Evaporative water loss is the primary physiological mechanism for heat dissipation. Adaptations for heat loss in both endotherms and ectotherms include adjustments in behavior and use of cooler portions of the habitat.

The higher basal metabolic rate of birds and mammals compared to reptiles is associated with higher body temperatures. Normal body temperatures for various passerine bird species are around 40–41°C. Placental mammals have body temperatures in the range of 36–38°C, whereas those of marsupial mammals are lower, near 35 or 36°C. The additional energy expenditures incurred when temperatures are below or above the thermoneutral zone comprise thermoregulatory costs. In their natural environment, a substantial portion of an endo-

therm's daily energy expenditure may be devoted to thermoregulation. For example, a small bird with a 41°C body temperature may use energy at a rate of approximately 0.5 times BMR for thermoregulatory costs (above BMR), which is approximately 16% of its average daily energy expenditure.

Ectotherms such as reptiles are unable to control their body temperature by generating extra metabolic heat. However, selective use of sun, shade, and warmer and cooler areas of their habitat allows ectotherms to regulate their body temperature within a few degrees around a preferred body temperature. The preferred body temperatures of most reptiles fall within the range of 34–40°C. The preferred body temperatures of reptiles seem to provide them with increased metabolic efficiency. For example, at these temperatures, reptiles attain their maximum endurance and running speed.

Although ectotherms do not actively increase their metabolism for the purpose of increasing body temperature, their energy expenditures are affected by their thermoregulatory behavior. The energy expenditure of a reptile increases and decreases with its body temperature (Fig. 4B). The rate of chemical reactions, including the reactions of metabolism, are positively related to temperature. The "Q_{10}" for a reaction describes how the reaction rate changes with a 10°C change in temperature. For most biological reactions, the rate will approximately double or triple with a 10°C increase in temperature (Q_{10} = 2 or 3). Behavioral thermoregulation in reptiles involves some energy expenditure in the form of movement or locomotion to get to an area with an appropriate temperature, as well as the cost associated with maintaining a basking posture. The extent of these costs can be demonstrated in the marine iguana. At night, when the iguana's body temperature is near 22°C, it has a metabolic rate of 0.75 kJ per hour. When it is resting during the day, with a body temperature of 35°C, it has a metabolic rate of 2.8 kJ per hour. After foraging in the water, the iguana must warm itself from 28°C to approximately 36°C. While the iguana is basking it is using energy at a rate equal to 1.9 times the expected SMR at 35°C. Half of this energy goes to the metabolic rate of resting at 35°C, and half goes to the cost of modest movements and postures necessary to bask effectively as well as modest increases in various physiological processes associated with changes in blood distribution relative to the skin surface. Overall, the energy associated with basking is 37% of the iguana's daily energy expenditure.

The thermoregulatory cost for an endothermic animal is affected by body size, shape, thickness of fur or feather insulation, and additional determinants of the upper and lower critical temperatures of its thermoneutral zone. The lower critical temperature is generally higher in smaller endotherms than in larger ones. Larger organisms have relatively less surface area for their volume and can carry thicker insulation than smaller organisms. This effectively reduces the lower critical temperature for larger endotherms and lowers the rate of increase in metabolic rate with decreasing temperature. The Arctic fox provides an extreme example with a lower critical temperature of −40°C due in large part to its thick insulating layer of fur. Small endotherms such as mice have a limited ability to insulate themselves, as well as a relatively large surface area, so heat loss in cold climates can be rapid. The gradient between the mouse's body temperature and a cooler environmental temperature does not have to be very great before the mouse must seek shelter or increase its metabolism to keep warm.

The cost of thermoregulation can be greatly modified by environmental variables that affect heat exchange between the animal and its habitat (Fig. 3). Heat exchange follows a gradient from warmer temperature to a cooler temperature, therefore conduction and convection can affect heat flow to or from an animal. For example, if an animal's body temperature is above ambient temperature, a wind will increase heat lost by convection. Similarly, if body temperature is below ambient temperature, convection can increase heat gain. Radiation also affects heat exchange, for example, a desert lizard can use radiative heat gain from the environment to keep its body temperature relatively stable at 40°C, even when air temperatures are near 30°C. All of these positive and negative effects on the temperature perceived by the animal can be combined with air temperature into a measure of operative temperature. The operative temperature is typically measured with a hollow body form, usually made of copper, covered with an animal skin. This taxidermic mount, with a temperature-measuring device inside, integrates all the passive thermal effects discussed previously to provide a single temperature that the animal would attain at equilibrium if it lacked metabolic heat production and metabolic heat loss mechanisms (i.e., evaporative water loss). The operative temperature provides a closer predictive relationship between temperature and metabolic rate under field conditions than would air temperature alone.

The advantages of endothermy include the ability to remain active, to be alert for predators, and to forage and reproduce over a wide range of ambient (or operative) temperatures. In this respect, ectotherms are at a disadvantage as their activity level is dictated by their body temperature. However, ectotherms are able to save energy by lowering their body temperature on a regular basis at night and during periods of inactivity. The economy of the ectotherm life-style allows them to survive in habitats where relatively little energy is available, and by having reduced thermoregulatory costs they are able to divert more energy to reproduction. Only a few endotherms are able to achieve energy savings by reducing thermoregulatory costs when they use torpor, either on a daily basis or seasonally (hibernation).

Daily torpor and hibernation are both states of reduced energy expenditure associated with reductions in body temperature, activity, metabolism, and food consumption. Torpor appears to be an adaptation for escaping periods of environmental stress usually associated with limited food or water availability typically associated with low temperatures. Some desert vertebrates become torpid during the hotter summer months, which results in a reduction in water lost through respiration.

Daily torpor is characterized by bouts of torpor lasting only a few hours in any 24-hr period and a shallow reduction in metabolism, during which body temperatures are reduced by only a few degrees. It is frequently observed in small endotherms such as mice and hummingbirds when exposed to a sudden and unexpected drop in temperature, conditions in which short-term energy demands are greater than the animal's energy stores. Entry into torpor may be influenced by an animal's physiological state. Animals in reproductive condition enter torpor less readily.

Long periods of torpor during winter are collectively called hibernation. During hibernation, bouts of torpor may last as long as 2 weeks, with only a few hours of normal body temperature and metabolism in between. Hibernation is an annually "programmed" event that occurs at approximately the same time each year, typically lasting from late fall to spring. The cues that trigger hibernation are not clear, but photoperiod as well as a reduction in ambient temperature seem to play an important role. Hibernation is not common in species that weigh much more than 5 kg as these larger animals are usually able to tolerate longer periods of starvation or are able to migrate to more hospitable wintering areas. When hibernating, the body temperature of a small endotherm is regulated to be only a few degrees above ambient temperatures, resulting in typical body temperatures ranging from 0 to 15°C. If the body temperature drops below a minimum temperature "set-point," metabolic heat production will increase. The thermoregulatory cost savings during hibernation is substantial. Hibernating bats and ground squirrels have energy expenditures of at least 1/30th to 1/40th of what they would have if they maintained normal body temperatures.

As an alternative to becoming torpid, some animals huddle together when ambient temperatures are low. Huddling reduces the surface area that each individual presents to a cold environment. The savings increases with group size. For example, a group of three voles will each see a 30% savings on the energy associated with thermoregulation at a given environmental temperature, whereas individuals in a group of five can each save 45%.

Hibernation is observed in reptiles (ectotherms) that live at higher altitudes in temperate regions. As with endotherms, reptile hibernation is associated with cold temperatures and is a mechanism for saving energy when food and growth opportunities are limited. The metabolic rate of a hibernating reptile is lower than what the body temperature alone would dictate. The reduced metabolism is not simply due to the Q_{10} effects associated with a drop in body temperature. However, unlike birds and mammals, the hibernating reptile does not regulate body temperature around the lowered point because mechanisms to rewarm without external heat are limited.

C. Activity

The energy dissipated by activity involving skeletal muscles is lost as heat and work and on average may account for 20–40% of the normal daily energy expenditure, or a rate of energy use approximately equal to 0.8 times BMR. Even postural differences can affect energy expenditures. In mammals, standing may increase metabolism by 10–20% over lying down. Large mammals reduce the postural cost of standing with skeletal specializations that permit them to lock the joints in their legs, thereby minimizing the use of muscle tension.

1. Locomotion

The cost of locomotion is related to an animal's body size as well as its mode of locomotion. Calculating the "cost of transport," the most efficient energy expenditure per kilogram per kilometer, is a convenient way to compare different modes of locomotion and different animals. This cost includes both the energy required for maintenance (BMR) as well as the additional energy required to move. The cost of transport decreases as body size increases for running and flying (Fig. 5) and swimming. In part, this is due to the lower per gram energy requirements for larger animals. However, the cost of locomotion is reduced even more for larger animals owing to the increased efficiency of longer limbs and muscles. For example, in running mammals of different sizes, all muscle tissue operates at a similar efficiency, which can be thought of as work produced for energy used. The longer limbs of larger animals give them a longer stride length for a given amount of effort relative to a smaller animal, which has to move its legs faster to travel at the same rate. Therefore, at a given velocity, the energy expenditure of a smaller animal is greater per gram than that of a larger one. Given that the BMR per unit mass is greater in smaller animals, the cost of locomotion adds to the disproportionately larger relative (per unit mass) energy requirements of a small animal. For any running animal, the energy expenditure increases linearly with velocity.

The cost of locomotion is also affected by the terrain that is encountered. Soft substrates like sand, mud, or snow can increase energy expenditures substantially. For example, a deer walking through deep, soft snow can experience a doubling in the energy required to move a given distance.

Moving uphill also increases energy expenditure. Although the cost of transport on level ground is relatively high for small animals, the energy required to run uphill is the same for animals of all sizes, and is about 27 J per kilogram moved 1 m vertically. Therefore, a small mammal running uphill or over rough terrain experiences a minimal impact on its already high cost of transport. For example, moving vertically at 2 km per hour requires a 23% increase in the metabolic rate of a running mouse, relative to a 630% increase for a horse.

Flapping flight in both birds and bats can use energy at a rate of 10 to 14 times BMR, compared to a 2 to 4 times BMR requirement for running in mammals. Birds much larger than 10 kg do not depend on flapping flight to a large extent. Instead, larger birds glide and soar, activities with rates of energy expenditure that are proportionally similar to that of running. Though flight is metabolically expensive, it is an economical way to cover ground. For a given distance, a flying bird can use as little as 1% of the energy used by a similar-sized running mammal. Unlike running, the relationship between the velocity of flapping flight and energy expenditure is parabolic. Slow flight requires deeper and faster wing beats to provide sufficient lift and thus is more expensive than flight at an intermediate velocity. When flying very rapidly, the cost of flight increases exponentially with velocity owing to the increase in air resistance.

2. Foraging

Foraging can account for a substantial proportion of the energy used in activity and of course results in the acquisition of food.[ ] The energetic cost of foraging depends on the style of feeding, that is, the kind and amount of locomotor activity. For example, birds that catch insects in flight may spend up to 38% more energy in obtaining food than birds with more sedate feeding styles that require less flight. Similarly, most lizards fall into one of two foraging strategy categories: those that actively search for prey and those that sit and wait to ambush prey. The daily energy expenditure of a lizard with an active foraging mode is 3.1 times SMR, but while active the rate of energy use can be 12

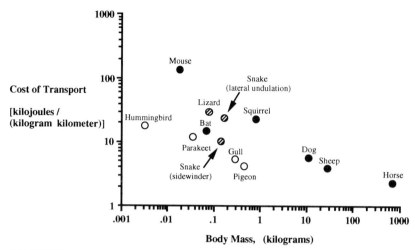

FIGURE 5 The energetic cost of transport as a function of body size for different animals. The solid circles are running mammals and a flying bat. The open circles are birds in flight. The striped circles are for reptiles: a running lizard and two modes of snake locomotion, sidewinding and lateral undulation. Within a given mode of locomotion, the cost of moving one kilogram of mass over one kilometer is greater for smaller animals. The cost of transport includes maintenance costs as well as net transport costs. For a given body mass, it is more costly to run than to fly despite the fact that per unit time, flight uses more energy. This is partly because a runner has to spend more time to cover a given distance and thus spend more energy on maintenance during that time.

times SMR. A lizard with an ambush style of foraging has an overall daily energy expenditure of 2.2 times SMR, with a nearly equivalent rate of energy use while foraging of 2.8 times SMR. Much of the additional energy expenditure associated with foraging in reptiles is a consequence of the increased body temperature that they must attain to increase neural and muscular activity. The increase in body temperature alone will increase a reptile's SMR, but the differences illustrated here show that activity also affects their energy expenditures. [*See* FORAGING STRATEGIES.]

3. Digestion

In the context of energetics, "activity" in the broadest sense refers to energy-consuming events ranging from the cellular level (e.g., biosynthesis, secretion, and neural processing) to the more overt activity associated with skeletal muscle contraction (movement, locomotion). Some activity at the cellular level accounts for part of the basal, or minimal, requirements for energy discussed earlier. However, activity at this level can increase measurably above basal in some instances, for example, the increase in metabolic rate associated with digestion.

There is an increase in energy expenditure associated with digestion equivalent (on average) to approximately 10% or more of the daily energy expenditure, or roughly 0.3 times BMR. Digestion can account for as much as 30% of the daily energy expenditure in ruminants. This energy requirement is not reflected in the previously discussed measures of digestibility. The energetic cost of digestion is due to an increase in cellular activity as enzymes are secreted and nutrients are transported across intestinal membranes into the body, as well as the muscular contractions of the stomach and intestine, which increase when food is present. Typically, the estimated cost of digestion also includes the energy required to synthesize and finally store the food energy in body tissues. The extent of the energy used during these digestive processes depends on the composition of the food and comes from the metabolizable energy intake component of the energy budget. However, because the amount of energy expended during digestion depends on the amount and kind of food being processed, the cost is usually expressed as a percentage of the energy in the food. For example, approximately 6% of the potential energy in a relatively digestible carbohydrate (such as a starch) is required to fully digest and store it. This cost is in addition to the energy required to access such molecules when they are locked up behind sturdy plant cell walls (see earlier section on digestible energy intake). Approximately 13% of the energy in fat is required to digest it. Nearly 30% of the energy in protein is required to fully process ingested protein. This high cost of digestion is due to the production of enzymes, urea, or other waste products, and storage molecules. The variability of the energy used during digestion makes it necessary that animals be in a fasted state when BMR is being measured.

D. Production

Production is the increase in biomass resulting from both the increase in an individual's mass as well as the addition of new individuals to the population by reproduction. Generally, the energy requirements for maintenance metabolism, thermoregulation, and activity take precedence over those for production. The total amount of energy retained in tissues, as well as the efficiency with which they are produced, is a critical feature of energy flow in ecosystems.

1. Tissue Growth

An increase in body mass is characteristic of young animals as well as fully grown adult individuals. Young, growing animals are depositing mostly muscle, organ, and bone mass, with relatively small amounts of fat. However, increases in the mass of adult animals are characteristically due to increases in body fat. The energetic cost of growth depends mostly on the amount of fat and protein being deposited. One gram of protein has an energy value of 17.9 kJ. The biosynthetic activity required to deposit a gram of protein requires an additional 28 kJ. Because muscle tissue is approximately 21% protein (the remainder being mostly water), a total of at least 10 kJ is required to increase muscle mass by 1 gram. In contrast, 1 g of fat has an energy value of 39.7 kJ. Approximately 14.8 kJ is required to process the fat and deposit it into adipose tissue, which is approximately 90% fat. Therefore the to-

tal cost of increasing adipose by 1 g is minimally 49 kJ.

Adipose tissue, or fat, is the primary energy storage tissue in vertebrates because of its overall economy. Though more energy is required to deposit fat, the efficiency with which protein is deposited (as a percentage of the energy provided by protein) is approximately 40%, whereas the efficiency of fat deposition is nearly 73%. One gram of adipose tissue provides approximately 10 times more energy than 1 gram of muscle tissue, making adipose a more economical energy storage tissue for mobile vertebrates. Energy storage increases in anticipation of a period of food shortage or a period of intense energy demands. Small birds and mammals, when experiencing cool night temperatures, will deposit a few grams of fat during the day for use overnight. Female vertebrates will increase fat stores for use at some point during reproduction, when energetic demands may exceed the limits of energy intake, or conflicting demands limit foraging time. The increases in fat stores for reproduction are under endocrine (hormonal) control, and in some vertebrates are influenced by environmental cues such as photoperiod, rainfall, or the availability of green plants. Some of the most dramatic examples of mass changes are seen in animals that hibernate for long periods and in long-distance migrants. A doubling in mass by fat deposition prior to hibernation or migration is not uncommon in these animals.

Ultimately, the energy for growth comes from the metabolizable energy intake of the animal. Growth efficiency is typically expressed as the percentage of metabolizable energy intake converted to new tissue. In mature animals, growth efficiency is close to 10% for nonfat tissues, and approaches 50% for fat deposition. In young, rapidly growing endotherms, growth efficiency may be close to 50% even though nonfat tissues are being deposited. During rapid growth, more of the ingested energy is retained as tissues rather than lost for maintenance, especially thermoregulation.

2. Reproduction

Energy used for reproduction includes the cost of activity associated with mate and territory acquisition and the feeding of dependent young, as well as the energy required for gestation or egg or sperm production. Reproduction is a seasonal event, during which the metabolic expenditures of mature males and females may increase an additional 1 or 2 times BMR, or more.

The energy expenditure of males during reproduction is generally limited to activity costs because the energy for sperm production is trivial. For example, in birds, the additional cost per day of growing testes is 0.005 times BMR, and 0.008 times BMR for the growth and maintenance of sperm. In contrast, male house martins, which must fly to catch insects, experience an increase in energy expenditure from 2.8 times BMR during incubation up to 3.1 times BMR when feeding nestlings. Similarly, swallows may have a daily energy expenditure of nearly 5 times BMR when feeding nestlings as a consequence of increased flying required to catch sufficient insect prey. In some species, male display behavior can have a measurable impact on the energy expenditure during the breeding season. For example, the strut display of the male sage grouse may be as energetically intense as flight (10 times BMR). A very active male that displays for 2–3 hours each morning exhibits a total daily expenditure that is 4 times BMR compared to 2 times BMR for a nondisplaying male. For male mammals, reproduction is not necessarily a time of maximal energy expenditure. In some ground squirrel species the energy expenditure of the male during mating early in summer is only 2.5 times BMR. This is not substantially different from the energy expenditure during the rest of the summer when they are fattening prior to hibernation. Often, the real cost of reproduction in males is not energetic but rather the increased injury and predation risks to which their activity may expose them.

Female birds and mammals devote more energy to the products of reproduction (eggs and fetus) than do males (sperm). The cost of growing and maintaining the mammalian oviduct is nearly insignificant. The reproductive tract of female birds, however, is more complex and requires more energy to grow and maintain than that of mammals. In birds, the growth of ovary and oviduct requires an increment of approximately 0.4 times BMR. Depending on the number and size of the eggs, the daily cost of

egg laying can be an additional 0.3 to 2.0 times BMR. During egg laying, 70% or more of a female's metabolizable energy intake above maintenance requirements is used to produce egg mass.

Incubation requires the transfer of heat energy from the parent to the eggs, yet the overall impact of incubation on energy expenditure is estimated to be slight. Maximally, an increase in metabolism equal to 0.3 times BMR can be associated with the cost of warming the eggs, but this is offset by the reduction in activity and the increase in thermal insulation provided by the nest. One estimate suggests that the energy expenditure of 1 hour of incubation is equivalent to 2 minutes of flight. Energy expenditures overall may be reduced for the incubating parent as activity levels are reduced. However, foraging time is also reduced, which may place demands on energy reserves.

Gestation and lactation have a large impact on the energy expenditure of the female mammal. The rate of energy expenditure during gestation ranges from 2 to 4 times BMR. Lactation is one of the more energetically stressful periods in the life of a female mammal. The metabolic rate during lactation peaks at nearly 7 times BMR just prior to weaning in a variety of small mammals, and lactation is approximately 65% of the total cost of reproduction for a female. During gestation and lactation, food consumption and digestive assimilation abilities increase in some mammals. In small rodents, for example, food consumption may increase as much as 50% during gestation, and up to 150% during lactation. During gestation, most of this energy is required to maintain maternal tissues, including those supporting the fetus, and contributes to maternal fat deposition. Approximately 13% of the metabolizable energy intake is converted to fetal tissue. The efficiency of converting metabolizable energy intake to milk is approximately 67%. However, it can be as efficient as 85% if the energy source is maternal tissues, such as the fat deposited during and prior to gestation. This efficiency is important, because the energy cost of milk production can be 3 times BMR in mammals with large litters.

Ectotherms, lizards in particular, have reproductive expenses that are similar to those of birds and mammals when viewed as a proportion of their annual average for SMR. The overall expenditures for males may be an additional 0.3 times SMR and 0.6 times SMR for females. However, in some species, the male and female expenditures over the reproductive season are much more similar. The more impressive aspect of ectotherm reproduction is the efficiency with which they can convert energy intake into biomass. This production efficiency is near 1–4% for endotherms, but estimates of 20% or more have been reported for vertebrate ectotherms. This is because much of the energy intake in endotherms is used for maintenance of body

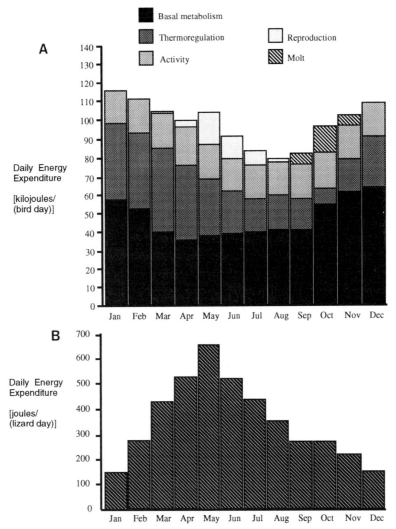

FIGURE 6 Monthly changes in daily energy expenditures of a house sparrow (A) and lizard (B). For the bird, the thermoregulatory and BMR costs consistently account for the greater part of the total energy expenditure and are greatest in winter and spring. The energy devoted to activity (outside of reproduction) is farily constant throughout the year. Molt in birds requires energy. Food intake, especially the intake of protein, increases during this time. The cost of molt includes the energy contained in the feather, as well as the cost of creating the proteinaceous feather structure. Total energy expenditure is highest in the winter months and the months that span the period of reproduction. The energy expenditures of the 2- to 3-g lizard shown in (B) illustrate that the cool ambient temperatures of the winter months reduce the ectotherm's expenditures relative to those of the nonhibernating endotherm. The costs of reproduction for the lizard are included in the energy expenditures for the months of March through May, and account for an average of 33% of the energy expenditure for these months.

temperature. Ectotherms do not incur such an extensive thermoregulatory cost because they use energy from their environment for thermoregulation. Therefore, reptiles are able to divert a greater portion of their energy intake into growth and reproduction and, from this perspective, are more efficient than endothermic animals.

IV. ANNUAL ENERGY BUDGET

The energy requirements for most terrestrial vertebrates vary over the course of a year primarily due to fluctuations in temperature and reproductive state. These fluctuations may be especially apparent in the temperate zone. As shown in Fig. 4, ambient temperature is important for endotherms as well as ectotherms.

The annual energy budget of the adult house sparrow (Fig. 6A) illustrates a number of features that apply in general to all vertebrates. The BMR of the house sparrow is higher in the winter months. This is not necessarily characteristic of all endotherms, but serves to illustrate the fact that BMR is not always static. In house sparrows as with some other vertebrates, BMR varies with annual or even geographical climate—the BMRs of more northern populations are higher than those in the south. A higher BMR may be a consequence of a greater body mass, as well as overall greater enzyme and hormonal activity. The growth and/or replacement of scales, fur, or feathers is characteristic of all vertebrates. Energy is sequestered in these structures, and energy is required to rearrange ingested proteins to form them. Energy allocated to activity and reproduction varies over the course of a year. The amount allocated to activity will vary with a species' particular habits. For example, migration will increase the energy expenditure for activity, whereas hibernation will virtually eliminate activity costs. Additionally, some vertebrates do not reproduce every year. In endotherms, thermoregulatory costs may be a substantial component of daily energy expenditure, though this too will vary with habitat and species.

For ectothermic animals, body temperature and energy expenditures tend to vary directly with temperature. The average daily energy expenditure for a small lizard living in the temperate zone is diagrammed in Fig. 6B. Whereas a bird's daily energy expenditure, and especially BMR, is higher in the cooler months of the year (Fig. 6A), the lizard's metabolism is reduced, reflecting their different thermoregulatory strategies.

Glossary

Digestible energy intake Digestion is a mechanical and chemical process in which large particles of food are broken into smaller molecules that can be transported across the membranes of the digestive tract and into the body. The digestible energy intake is the energy value of the food that is retained within the body, that is, the energy that is absorbed. It is calculated to be the energy content of food minus the energy content of undigested material in the feces. Also called assimilated energy.

Ectotherm Animals that use an external source of heat for body temperature regulation, in contrast to endotherms. These animals include reptiles, amphibians, fish, and invertebrates. Ectothermic vertebrates are able to maintain a high and constant body temperature during daylight hours only through the selective use of their habitat, which determines heat gain or loss. Their options for regulation are limited at night, when their body temperatures are typically lower.

Endotherm Animals (birds and mammals) that use heat produced internally to control body temperature, in contrast to ectotherms. Endothermic vertebrates typically maintain a high and constant body temperature despite changes in the environmental or ambient temperature.

Energetics Study of energy and its transformations into various forms. The amount of energy flow, energy retention, and the energy transformations in a system.

Energy Capacity to perform work. In biological systems, this work may range from creating the temporary molecular bonds that ultimately result in animal locomotion to forming molecules that themselves store energy.

Energy budget Description of the energy obtained by food and the uses to which this energy is allocated.

Metabolic rate Total energy expenditure of an organism over a given period of time. Often, the metabolic rate is expressed as the daily energy expenditure, or the energy used over the course of a single day. The basal metabolic rate (BMR) is the minimum rate of energy use in a normal, endothermic vertebrate animal at rest and not expending additional energy on thermoregulation or digestion. The standard metabolic rate (SMR) is the rate of energy use for an ectothermic vertebrate resting at its preferred temperature.

Metabolism Collective biochemical reactions that take place in a living organism. Metabolism consists of energy-consuming or energy-absorbing reactions, as well as energy-releasing reactions.

Metabolizable energy intake Fraction of the energy in the food that is eaten that is used or stored in an animal. The metabolizable (or metabolized) energy intake is the energy in food minus the energy of all excreted material. Unused energy is excreted from the body as feces, urea or uric acid, molecules that have been digested but remain unusable and are therefore excreted by the kidneys, and methane gas, especially in ruminant vertebrates.

Respiration Cellular respiration is the name given to the oxidation (oxygen-consuming) reactions that capture the energy in food by transferring it to the molecule adenosine triphosphate (ATP).

Thermoneutral zone Range of ambient or environmental temperatures over which an endotherm can maintain body temperature without additional energy expenditure. Within the thermoneutral zone, the ambient temperature may rise or fall, but there will be no perceptible change in the metabolic rate or body temperature of the animal. Heat gain or loss is effected by adjusting fur or feather fluffing, blood flow patterns to warmer or cooler regions of the body, and modest changes in posture. Also known as the thermal neutral zone.

Thermoregulation Control of body temperature. Both endotherms and ectotherms thermoregulate. However, the endotherm uses a great deal of energy for thermoregulation, whereas the ectotherm uses heat passively gained from or lost to the environment to control body temperature.

Bibliography

Blaxter, K. (1989). "Energy Metabolism in Animals and Man." Cambridge, England: Cambridge University Press. (Individual level, quite factual and covering most variables in energy budget.)

Carey, C., Florant, G. L., Wunder, B. A., and Horwitz, B., eds. (1993). "Life in the Cold: Ecological, Physiological, and Molecular Mechanisms." Boulder, CO: Westview Press. (Based on a recent symposium, focus is primarily on issues relating to endotherm hibernation.)

Grodzinski, W. (1985). Ecological energetics of bank voles and wood mice. *Sympos. Zool. Soc. London* **55,** 169–192. (This study presents the energetic impact of these species on their environment. Examines seasonal body condition, digestive abilities, metabolism, thermoregulation, and energy budgets.)

Nagy, K. A. (1983). Ecological energetics. *In* "Lizard Ecology: Studies of a Model Organism" (R. B. Huey, E. A. Pianka, and T. W. Schoener, eds.), pp. 24–54. Cambridge, Mass.: Harvard University Press. (Complete energetic study on a lizard population.)

Pimm, S. L. (1982). "Food Webs." London: Chapman & Hall. (General overview, ecosystem level.)

Prosser, C. L. (1991). "Environmental and Metabolic Animal Physiology." New York: Wiley–Liss. (Part 2 of a two-volume set in comparative physiology.)

Ricklefs, R. E. (1989). "Ecology," 3rd ed. New York: W. H. Freeman. (Ecology with an energetic viewpoint.)

Robbins, C. T. (1993). "Wildlife Feeding and Nutrition," 2nd ed. San Diego: Academic Press. (Individual level, digestion, maintenance, production.)

Schmidt-Nielsen, K. (1984). "Scaling: Why Is Animal Size So Important?" Cambridge, England: Cambridge University Press. (General metabolism, locomotion, thermoregulation.)

Tomasi, T. E., and Horton, T. H., eds. (1992). "Mammalian Energetics: Interdisciplinary View of Metabolism and Reproduction." Ithaca, N.Y.: Comstock Publ. Associates. (Overview, many references.)

Townsend, C. R., and Calow, P., eds. (1981). "Physiological Ecology: An Evolutionary Approach to Resource Use." Oxford, England: Blackwell Scientific. (Includes some topics not covered, easy to read.)

Ecological Restoration

Sheila M. Ross
University of Bristol

I. Introduction
II. Purposes of Ecological Restoration
III. System Establishment
IV. Postrestoration System Management
V. Philosophy for the Success of Future Restorations

Ecological restoration is the practice of artificially recreating natural or seminatural habitats. Worldwide we see a range of degrees of habitat disturbance and damage, and few ecosystems could be considered as anthropogenically unmodified. Thus, "natural" habitats are taken here to mean ecosystems whose disturbance and damage are minimal. "Natural" habitats may be taken to mean ecosystems that are as undisturbed and undamaged as possible. Several different terms are used for practices of renewing damaged land. *Ecological restoration* may only be used where the aim is to return land to semi-natural or natural vegetation. Damaged land does not necessarily have to be restored to the same ecosystem that was present on the site before destruction. Thus, woodland may have been cleared for mining operations, but could be subsequently restored to a lake habitat. Other terms used for repairing land damage include *reclamation,* where land may be returned to a new use, *rehabilitation* may be used if the improvements are more aesthetic than ecological, whereas *revegetation* is used for situations where the original vegetation has been destroyed and new vegetation growth is being encouraged. While restoration is an umbrella term used for all attempts to upgrade land, ecological restoration is reserved for those activities seeking to recreate a sustainable ecosystem that mimics its undisturbed analogue. Restoration covers a wide range of practices and a wide range of timescales, from highly technical, often long-term, attempts at faithful reproduction of complex ecosystems such as woodlands and grasslands, to short-term landscaping projects that produce habitats with less scientific credibility, but which are more popularly appealing. For land to require ecological restoration, the original ecosystem must already have been destroyed or degraded. Again, several interchangeable terms are used to describe degraded land, including *derelict land,* a term used specifically for land abandoned from its previous use. Terms such as *damaged, disturbed, degraded, spoiled,* or *destroyed* can be used to refer to the degree of habitat perturbation or pollution. Ecological restoration is not restricted to terrestrial environments but is equally applicable to freshwater, estuarine, and marine habitats.

The three characteristics of ecological restoration projects vary greatly:

i. Reason for restoration. Reasons range from the restoration or repair of a degraded or polluted site, to the creation of a completely new habitat, perhaps to replace a site lost by development, or the creation, piece by piece, of a new ecosystem for education or research.

ii. Degree of habitat restoration. This can vary from attempted wholesale ecosystem replacement from scratch, to the simple importation of key individual species, designed to increase diversity of an existing, degraded ecosystem.

iii. Techniques of restoration. Two groups of techniques are involved in ecological restoration: techniques designed to alter environmental conditions, such as soil or hydrology, and techniques designed to produce the correct biological diversity and ecological integration.

I. INTRODUCTION

Earliest attempts to restore degraded natural habitats or to recreate natural habitats such as woodlands and meadows lay in landscaping, frequently in the urban context and often with a horticultural instead of an ecological orientation. The scope for both restoring and diversifying natural habitats and for building new habitats from scratch has widened considerably over the last 10 to 20 years. This has been due to more numerous and more diverse environmental problems to be solved, more field experience of both successful and unsuccessful habitat restorations, improved ecological knowledge of the integrated soil, environment, and biological processes, and new climates of planning philosophy.

To successfully restore damaged land to its former habitat or to create a new ecosystem if the former habitat is unknown, a good understanding of the growth requirements of plants and of soil–plant interactions is necessary. The basic requirements for plant growth are adequate supplies of light, carbon dioxide, water, and nutrients. Since light and carbon dioxide are usually present in excess in restoration sites, restorers must direct their attention to the soil moisture and chemical conditions. In modern ecological restoration projects, the need to marry the understanding of ecological principles and processes with practical horticultural and silvicultural techniques is also clearly appreci-

ated. As understanding of all three disciplines has improved over time, so too has success in habitat reconstruction and restoration. It is important to understand how the land was damaged, to identify critical conditions that are likely to be detrimental for plant growth, and to devise management techniques for mitigating these conditions. Many disciplines and professions may be involved in ecological restoration, starting with site engineers to modify physical conditions, followed by landscape architects, agronomists, horticulturalists, and ecologists to modify soil and plant conditions. In ecological restorations of freshwater and marine habitats, similar integrations of knowledge are necessary, this time tapping the disciplines of hydrology, limnology, and ecology. [See ECOLOGICAL ENERGETICS OF ECOSYSTEMS.]

The ecological restoration of a site may have a variety of goals (political, educational, scientific, replacement of lost heritage), with a range of different associated approaches, methods, and management. There are many differing objectives of habitat restoration, ranging from those that attempt to recreate convincing replicas of ancient and complex habitats, to those that aim to create simpler habitats with less scientific credibility, but more immediate popular appeal. These two different objectives are, sadly, often confused and this adds doubt to assessing the success or failure of a restoration project. [See RESTORATION ECOLOGY.]

For each restoration project, a priori criteria should be defined so as to determine how successful the restoration had been. The temporal basis for postrestoration assessment criteria depends to some extent on the type of habitat being restored and should examine changes in both environmental conditions and in species diversity over time. The success of temperate woodland restoration must be assessed over a much longer time period than for grassland restoration. Natural ecosystems are characterised by *stability* and *resilience,* which describe, respectively, their abilities to resist perturbations and to recover successfully from them. These should also be the long-term characteristics of anthropogenically restored habitats. [See FOREST STAND REGENERATION, NATURAL AND ARTIFICIAL.]

II. PURPOSES OF ECOLOGICAL RESTORATION

It is possible to identify at least four different types of reasons for ecological restoration. First, there are increasing demands worldwide for more "natural" and low maintenance landscaping of urban and civil engineering projects such as construction projects and roadside verges. A second impact of development and construction is the need for habitat replication or duplication in the near vicinity of an ecosystem(s) consumed or destroyed by the development. Third, there is the increasingly important requirement for reclamation of natural habitats on sites that have been degraded or destroyed by pollution, mining, industrial dereliction, agricultural abandonment, or natural disaster. The fourth reason for ecological restoration is to build functional ecosystems designed for education or to provide systems for process research, in which interacting processes and ecological theories can be tested. Although the bulk of attention in the past has focused on the reclamation of mining and industrial derelict land, a much greater interest in urban natural landscaping, in habitat removal and relocation, and in the reconstruction of ecosystems for research has greatly broadened the scope of ecological restoration activities.

A. Natural Landscaping

Modern solutions to landscaping of housing estates, road construction, and city developments are tending to reject more formal horticultural designs in favor of simplified, seminatural vegetation. These approaches may adopt minimum restoration practices, with "assisted" natural colonization, often using pioneer and ground cover species, while attempting to create diverse tall herb communities with colorful flowers that are aesthetically pleasing. This type of landscaping seeks to establish seminatural vegetation communities which in some way resemble the seminatural original, although not necessarily recreating their full diversity. Some authors have suggested that simpler restorations, or habitat creations, could be thought of as political habitats, aiming to educate or spread propaganda

for wildlife conservation, particularly in cities and on the verges of urban motorways. Simple reconstructions will normally diversify through time as invasion and colonization processes take place.

B. Habitat Duplication

Where valued ecosystems are destined for destruction during roadbuilding or construction projects, ecologists now plan "rescue" operations in advance. These programs either (i) recreate the ecosystem elsewhere locally or (ii) move and relocate the entire ecosystem. In such cases, the entire habitat is moved, usually in intact pieces, from the *donor site* to a nearby *receptor site* whose soil and environmental conditions are suitable. Such procedures are possible using modern earth-moving equipment. While this solution to habitat destruction has been attempted for grassland and heathland ecosystems, where large turfs can be moved with their original soil intact, transplantation of woodlands has generally been restricted to moving individual mature trees instead of relocating whole woodland ecosystems.

C. Restoration of Degraded and Derelict Land

The destruction of land by mineral exploitation, coal extraction, and other industrial uses has been the target for ambitious ecological restoration work for several decades. Mining activities completely alter the physical and chemical characteristics of slopes and soils. Frequently these sites suffer from high concentrations of potentially toxic metals and often serious chemical deficiencies. Under these conditions, plant establishment, survival, and growth can be seriously hindered. Soil management to remove or at least minimize the plant growth limitations is the first step in habitat restoration. To ecologically restore sites that have been severely altered both physically and chemically is often too great a challenge in the short term so other land uses, such as amenity open spaces, recreation playing fields, or picnic sites, are used until further site amelioration can occur. Further site amelioration often depends on time instead of a carefully prescribed management. *Progressive restoration* may be used to describe the stage by stage site operations and revegetation procedures designed to reclaim and restore quarry and mine sites where dumping may still be taking place or where vegetation colonization is sequenced over time.

The physical and chemical problems associated with different types of degraded and derelict land are shown in Tables Ia and Ib, respectively. These summaries allow soil scientists and agronomists associated with ecological restorations to plan their ameliorative management programs. In many cases, the modification of adverse site conditions, e.g., reduction of soil erosion or prevention of acid runoff, has a higher initial priority than the replacement of original species, which can only take place after an acceptable degree of environmental amelioration has been achieved.

D. Habitat Reconstruction for Education and Research

Constructing a replica ecosystem from scratch can provide additional information about gaps in

knowledge, effects of time and space, synchronization of processes, and the integration and linkages between component parts of the system. The recreated ecosystem becomes a massive hardware model, or outdoor experiment, for testing theories about how the ecosystem works. Famous longer-term experiments of this type include the temperate forest at the University of Michigan Biological Station and the prairie grassland at the University of Wisconsin arboretum. Exponents of these experiments term their work *restoration ecology* or *synthetic ecology* instead of ecological restoration. [*See* GRASSLAND ECOLOGY.]

III. SYSTEM ESTABLISHMENT

Successful ecological restorations depend very heavily on four main criteria: (i) careful appraisal of initial site and soil conditions and development of prescribed amelioration programs; (ii) clear understanding of site and soil requirements of the desired ecosystem to be restored, based on knowl-

TABLE Ia
Physical Problems Associated with the Ecological Restoration of Differently Disturbed Sites[a]

Site and substrate	Physical problems[b]			
	Texture and structure	Stability	Water supply	Surface temperature
Coliery spoil	---	---/o	-/o	o/ + + +
Strip mining	---/o	---/o	--/o	o/ + + +
Fly ash	--/o	o	o	o
Oil shale	--	---/o	--	o/ + +
Iron ore mining	---/o	--/o	-/o	o
Bauxite mining	o	o	o	o
Heavy metal waste	---	---/o	--/o	o
China clay waste	---	--	--	o
Acid rocks	---	o	--	o
Calcareous rocks	---	o	--	o
Sand and gravel	-/o	o	o	o
Coastal sands	--/o	---/o	-/o	o
Land from sea	--	o	o	o
Urban wastes	---/o	o	o	o
Roadsides	---/o	---	--/o	o

[a] Modified from Bradshaw and Chadwick (1980).
[b] Deficiency, ---, severe; --, moderate; -, slight; adequate, o; excess, +, slight; + +, moderate; + + +, severe.

TABLE Ib

Chemical Problems Associated with the Ecological Restoration of Differently Disturbed Sites[a]

| Site and substrate | Nutrients | | | | | | pH | Ion exchange capacity | Heavy metals | Other toxins |
	N	P	K	Ca	Mg	Na				
Coliery spoil	---	---	o	---/o	-/+ +	+	---/o	--	o	o
Strip mining	--	---	-	---/o	-/+ +	+	---/o	-	o	o
Fly ash	---	-	o	o	o	o/+	+/+ + +	-	o/+ +	boron
Oil shale	---	-	-	--	+ +	o/+ +	--/o	o	o	o
Iron ore mining	---	---	-	-/+	-/+	o	o	o	o	o
Bauxite mining	--	--	-	o	o	o	o	o	o	o
Heavy metal waste	---	---	-	---/+ +	-/+ +	o	---/+	--	+/+ + +	o
China clay waste	---	---	o	-	-	o	-	---	o	o
Acid rocks	---	---	o	--	-	o	-	--	o	o
Calcareous rocks	---	---	-	+ +	-	o	+	o	o	o
Sand and gravel	--/o	-/o	-/o	-/o	-/o	-/o	-	o	o	o
Coastal sands	--	-	o	o	o	o	o	-	o	o
Land from sea	-	-	-	-/+	o	+ + +	-/+	o	o	o
Urban wastes	-	-	o	o	o	o	o	-	o/+	various
Roadsides	--	--	-	-/+	o	o/+ +	-/o	-	o/+	o

[a] Modified from Bradshaw and Chadwick (1980)

[b] Deficiency, ---, severe; --, moderate; -, slight; adequate, o; excess, +, slight; + +, moderate; + + +, severe.

edge of undisturbed systems; (iii) postrestoration monitoring programs that identify shortfalls in amelioration, vegetation establishment, and ecosystem management programs; and (iv) postmonitoring remedial management to rectify shortfalls. The approach taken for ecological restoration depends on whether the goal is to create an ecosystem from scratch on a more or less virgin site or whether the restoration strategy is to improve existing, but degraded, habitats. Both approaches are briefly discussed here.

A. Ecological Restoration on Virgin Sites

The restoration of virgin sites is usually associated with the reclamation of construction sites, derelict land, or mining wastes. Once site environmental conditions have been assessed, four major steps are required for the development of an ecological restoration program. In chronological order, step 1 involves site preparation, which may include topographic regrading or improving site hydrology; step 2 involves soil amelioration, in which soil physical and chemical conditions are modified; step 3 is the development of a suitable seed mixture and its establishment; while step 4 involves the importation of transplants, which may be trees, shrubs, or locally rare species.

Some commonly encountered site and soil conditions that require amelioration prior to successful habitat restoration are illustrated in Fig. 1. Only once site and soil conditions have been suitably restored to specifications typical of undisturbed ecosystems should vegetation restoration begin. After vegetation establishment, its development should be regularly monitored. Depending on the outcome of monitoring, further site/soil modification and/or planting may be deemed necessary, together with a routine system of management or maintenance practices. This scheme for ecological restoration is illustrated in Fig. 1. The soil and vegetation establishment stages in ecological restoration are addressed next. [See SOIL ECOSYSTEMS.]

I. Soil Requirements

For most ecological restoration projects, soil physical conditions are modified first. Achieving the cor-

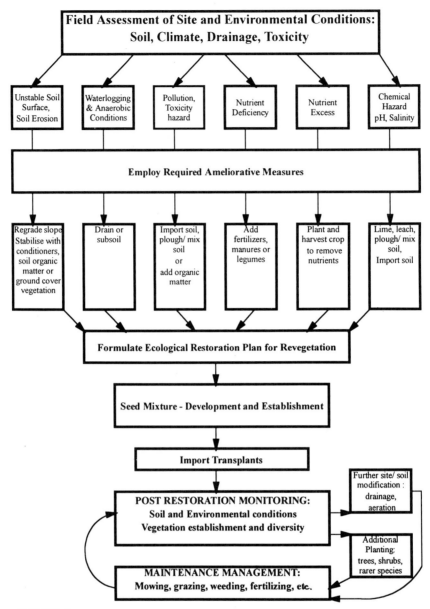

FIGURE I Stages in the ecological restoration of damaged and derelict land, illustrating possible management practices for site amelioration.

rect soil drainage state with suitable soil hydrological characteristics is often the most serious problem in transported soils, imported soils, mixed soils, or in soils that are replaced on site after stockpiling. The alteration of soil chemical properties can take place after site and soil physical conditions are acceptable. Rectifying nutrient deficiencies is a simpler task than ameliorating nutrient excesses or toxicities, but in all cases, identifying the degree of the deficiency or toxicity problem requires much

laboratory analyses. In addition, it is imperative that remedial management does not create or exacerbate nutrient imbalances. Techniques for dealing with some of these problems are discussed next.

a. Soil Hydrology, Drainage, and Soil Physical Conditions

Restoration of site and soil hydrological conditions is often problematic, particularly after mining or construction work, where site topographic alteration, soil stockpiling, or soil compaction has occurred. Compaction is usually caused by the heavy earth-moving equipment used in topographic regrading and landscaping. A further difficulty is due to the disruption of soil structure. When soil aggregates break down, their component soil particles wash through the soil, blocking up the soil pores. The relative proportions of macropores (large pores >0.2 mm) and micropores (small pores < 0.2 mm) in the soil are important since macropores provide routing for fast water transmission while micropores are responsible for soil water retention and good aeration in the soil matrix. Artificially recreating the original distribution of soil macropores and micropores is virtually impossible. The best that can be achieved is to promote the recreation of good soil structure by adding soil organic matter and synthetic soil conditioners and stabilizers or to wait for colonizing soil organisms and the roots of vegetation to help develop good soil aggregation over time.

Soil compaction can be rectified by *subsoiling,* where tines or heavy metals blades are dragged through the soil to the required depth. In addition, *rotavation* is used to mix and loosen surface soil layers, whereas *harrowing* is used to break up clods of earth and dense soil structures. At this stage in ecological restoration, it may be important to protect the soil surface against soil erosion, using organic matter, mulches made of leaf litter, straw, or branches, or encouraging ground cover plants. Mulches and ground cover vegetation provide three important forms of protection at the soil surface where fragile new seedlings must root: (i) thermal insulation, reducing the range of temperatures in surface soils; (ii) moisture retention, reducing the amount of evaporation from the bare soil surface; and (iii) a physical barrier to hold the surface soil in place and to intercept the energy of raindrop impact and splash that would otherwise disrupt surface soil particles and move them downslope.

Alleviating waterlogging in compacted soils by employing artificial drainage systems or by breaking up subsurface compaction to encourage vertical percolation of water is only one aspect of adverse soil hydrological conditions. Drought can also be a problem for vegetation establishment in coarse-textured substrates such as sand and gravel deposits or china clay quartz waste which have poor soil moisture retention capacities. In such circumstances, the addition of soil organic matter can help to improve soil structure and to improve water-holding properties.

b. Soil Chemistry, Fertility, and pH

Chemical toxicities and deficiencies can cause problems in soils of restoration projects. Three main forms of toxicity affect restoration work: acidity, salinity, and heavy metal concentrations. Naturally generated acidity, such as the production of dilute sulfuric acid during the oxidation of iron sulfide, pyrite, in colliery spoil is extremely difficult to ameliorate. Even when extremely high levels of ground limestone are applied to such soils, the effects are transient because of the large reserves of pyrite available for further oxidation. A similar acidity problem is typical in sites reclaimed from the sea. These sites also suffer from high salinity and sodium hazard which can be reduced over time through natural leaching. Heavy metals are associated with derelict sites around former mines and mine spoils. When concentrations of heavy metals in soil are higher than trace quantities, nearly all are toxic to plants, particularly lead, cadmium, zinc, and copper. The first symptom of metal toxicity in plants is the inhibition of root growth, which can seriously affect water uptake. Plants suffering metal toxicity are thus very susceptible to drought. Methods of mitigating the toxic effects of metals range from importing uncontaminated topsoil to entirely cover the affected site, to selecting and establishing plants that have some degree of metal tolerance.

Table Ib shows that most types of derelict land suffer from nutrient deficiencies, especially nitrogen. Nitrogen deficiency is the most likely cause of plant establishment and growth problems on derelict and degraded sites. The ecological restoration of such sites generally requires nutrient inputs, either in the form of inorganic fertilizers or as organic manures. Natural nitrogen fixation processes, involving free-living bacteria in soil, help to build up levels of soil nitrogen with time after restoration. Legumes may also be included in revegetation programs to increase the soil nitrogen economy of the system through symbiotic nitrogen fixation. The ecological restoration of agricultural land provides a quite different challenge since previously fertilized fields usually show soil nutrient concentrations in excess of levels required by the ecosystems to be restored.

In ecological restoration projects, ecologists must use their knowledge of the nutrient requirements of different plants and plant associations to decide whether macronutrients (N, P, K, S, Ca, Mg, and Na) and/or micronutrients (Fe, Mo, Mn, Zn, Cu, Bo, Cl, and Co) must be added to the soil. Many of the habitats we wish to conserve, such as heathlands and species-rich grasslands, are typically associated with infertile soils. For restoring these ecosystems, care must be taken to ensure that initial and subsequent soil nutrient levels are not too high, otherwise ubiquitous, fast-growing and nutrient-demanding weed species can successfully compete to the detriment of the less-demanding, perhaps slower-growing, native species we wish to encourage. Reduced species diversity is often seen in habitats with excess nutrients. In addition, experiments in which fertilizers have been applied to natural ecosystems cause reduced species diversity. Difficulties exist in attempting to recreate high species diversity grassland on sites that have previously been heavily fertilized for agriculture. Repeated annual applications of agricultural fertilizers cause soil phosphorus levels to build up over time as quite large proportions of the added phosphates form insoluble calcium, iron, and aluminium compounds. These phosphorus compounds may become slowly available to plants over the long term and hence influence the species composition of the restored system over the longer term. Residual soil phosphorus can become a more serious and long-term problem than high concentrations of soil nitrogen, which can be used up by the plants in the relatively short term of around one to two growing seasons.

Two main options are available to artificially impoverish soil nutrient funds, by managing the system so that (a) nutrient exports are greater than nutrient imports and/or (b) nutrients are sequestered in unavailable pools, where circulation is blocked or reduced to a low level. These options can be carried out using three management strategies: (i) managing the site to promote natural processes, such as natural succession, that will impoverish nutrient supply; (b) indirect nutrient removal using some form of continuous cropping with regular harvesting or grazing; or (c) direct soil nutrient pool removal, such as topsoil stripping.

In addition to inherited soil nutrients, aerial pollution can progressively elevate soil fertility levels, as well as altering soil pH and acidity. Anthropogenic emissions to the atmosphere, particularly of the oxides of nitrogen and sulfur during the combustion of fossil fuels, has resulted in increased inputs of these elements to the soil via rainfall. Nitrate–nitrogen concentrations in rainfall have increased substantially and, at the same time, the increased acidity of rainfall has resulted in elevated levels of hydrogen ions added to the soil, causing soils to become increasingly acidic, particularly those derived from noncalcareous substrates and parent materials.

2. Vegetation Establishment

At the revegetation stage of a restoration project a good understanding of ecological theory is vital since a potentially immense number of species and/or management options are available, particularly for species-rich habitats. Choices of species for revegetation rely on our understanding of their ecological strategies. Grime classified plants into three strategic types: (i) stress-tolerants, those that are unresponsive to environmental changes but are extremely sensitive to damage; (ii) ruderals, those that colonize frequently disturbed fertile habitats, and (iii) competitors, those species that strongly compete and eventually monopolize fertile sites where there is low disturbance. An equally important ecological principle is that there are no species that can exploit sites which combine low productivity and high levels of disturbance. Thus, no flowering plants can survive in the heavily trampled pathways of recreational grasslands and uplands.

The vegetation establishment plan for most ecological restorations involves several stages and techniques. Most start with herbage seeding, using a mixture of grasses and herbs. A second stage may employ turf transplants and/or shrub and tree transplants. At this stage, rarer species may be imported, either seeded or as transplants. In the woodland habitat, establishment by seed is not a practical on-site option since the species that persist under the canopy tend to produce few, heavy seeds which give rise to seedlings whose survival in the gloom of the forest floor is good but which do not flourish in bright light and fluctuating temperatures typical of restoration sites. Seeds of temperate woodland species may also possess a chilling requirement before germination can occur. In general, woodland reconstruction is best achieved using transplants, with seeding of early succession understory species, and late succession species added as transplants at a later date.

a. Choice of Seeding Composition

Soil seedbanks, or seed stores, may contain long-persistent seeds of unwanted weeds that are strong competitors, with abilities for fast invasion and dominance. If the ecosystem to be rehabilitated is different from the previous vegetation, care must be taken at the vegetation establishment stage to ensure that weed regrowth from the seedbank is suppressed and/or that a seeding mixture is chosen that ensures a balance of species that are fast germinating and persistent. In the early stages of revegetating many sites, legumes are included in the sward to enhance nitrogen fixation. Legumes are only valuable if they possess the correct strain of symbiotic *Rhizobium* bacterium. In the restoration of derelict and mine spoil sites, and in soils where the legume has not previously grown, legumes must be inoculated with the appropriate *Rhizobium*. Nurse crops can also be sown in the initial establishment phase to aid and protect seedling growth. Nurse crops are usually grasses, sown at low density to allow good light levels at the ground surface. [*See* SEED BANKS.]

Before choosing a seeding mixture, it is important to know the species characteristics important for successful establishment. Agricultural grasses with competitive ruderal or competitive strategies establish more successfully than stress tolerators and species typical of calcareous pasture, arable land, mire, and woodland. Species requiring seed chilling, those that are slow to establish a population of fast-growing individuals, and late-flowering species all appear to be disadvantaged in restored swards. Given that commonest species initially out compete other less common but habitat-typical species, most successful attempts to create the original species diversity involve at least two different phases of vegetation establishment. In the first place, ecologically appropriate common species should be sown, ideally in the autumn. In the second phase, at least 1 year later, additional species, which may be more difficult to establish, should be sown in specially created vegetation gaps or as established plants of root cuttings.

A range of commercially available seeding mixtures are widely available, including "species-rich meadow," "wildflower," "woodland," and "shade" seed mixtures. These may satisfy basic establishment criteria, with later management used to modify the species mix. Wild, native grasses and herbs can be seeded using mowings laid directly on the soil surface. A similar technique, using branches or chipped branch mulch, can be adopted for seeding shrubs and trees. Wild seed either can be collected at the right time of year or can be sown in carefully collected and transported topsoil. Seeds may be present only in the immediate surface soil layer, as is the case with *Calluna* moorland and some woodland types. In such cases, only the top few centimeters of soil should be scraped off and replaced in the later stages of soil restoration. Soil transfer is much more valuable than just a seeding source. Transported native soil also contains other components of the soil ecosystem, including mycorrhizae, other fungal spores, soil invertebrates, and bacteria. All of these improve soil conditions for ecosystem development. Transported soil also contains vegetative parts of plants, such as bulbs (e.g., *Allium ursinum* and *Endymion non-scriptus*) or rhizomes (e.g., *Arum* and *Circaea*) which contribute significantly to vegetation regrowth.

Hydraulic seeding techniques, called hydroseeding, in which the seed mixture is suspended in an aqueous solution and sprayed onto the soil surface, are advantageous in sites where soil instability and soil erosion are a problem. This is the case with many derelict sites and mine spoils, particularly where the spoil has been graded into steep dumps. Some of the advantages of hydroseeding techniques are due to the possibility of producing liquid cocktails of seeds, nutrients, lime, soil stabilizers, dilute organic slurries, and mulching materials which, when sprayed onto the site, not only sows the seed, but glues it to the soil surface, provides nutrients, conditions the pH, helps in preventing soil erosion, helps in ameliorating soil temperatures, and retains soil moisture. In addition, the spraying machine acts from up to 60 m away and does not compact the soil to be revegetated.

b. Species Transplants

Once a site has successfully established a sward, turf transplanting from nearby, or importing nursery-grown shrub and tree seedlings, can provide reliable and fast methods of speeding up the acquisition of perennials. Transplanting nursery material is also the most successful method of establishing metal-tolerant species to metal-contaminated restoration sites. While a very large range of tree and shrub species could provide planting stock for reclamation sites, ecological restorations should choose native species, ideally using material germinated from seed collected locally. Four types and maturities of transplants may be used: bare-rooted seedlings, usually 2–3 years old, and around 15–30 cm tall; tube seedlings, usually 1 year old and around 10 cm tall; container seedlings, usually 2–3 years old and 15–30 cm tall; or standards, which are young trees, around 1–2 m tall. Tube and container seedlings benefit from having their roots protected by soil, whereas bare-rooted seedlings require more careful planting and cannot tolerate drought. Taller standards are a more expensive option, but can provide an immediate visual effect around buildings.

B. Ecological Restoration of Degraded Sites

During natural ecosystem regeneration after perturbation, a restored site may experience gradual changes in species composition as a result of invasion and colonization. Although actively debated for many years, the classical ecological theory of succession provides a basis for ecological restorations since most restoration plans adopt phased planting schemes and rely on invasion to boost species diversity. An example *recovery succession* for an abandoned old field might follow several stages: the early abandonment phase of "weed" growth is replaced by grassland, then shrubland, and finally forest. The whole process may take tens of years. Theories of primary successions, vegetation sequences developed from scratch on bare soil surfaces, suggest the development over time of a series of different vegetation associations, or communities. At each stage, the site becomes less suitable for existing plants and more suitable for invasion by the next group of species. The theory of secondary successions is used for development on abandoned sites that already contain seeds or vegetative material representing the full range of successional stages. Both theories are relevant for ecologically restoring perturbed habitats. In such cases, restoration may have two main goals: to increase species diversity and to increase the speed of successionary changes. These goals may be achieved by importing transplants, particularly shrub and tree species, and by using artificial seeding techniques to diversify the herbage; a kind of assisted succession. Slot seeding techniques, using machinery that cuts a slot in the surface soil or turf, then inserts seed into the slot, can be used to implant swards after initial establishment or to upgrade and diversify degraded habitats. Such techniques are used to insert N-fixing legumes, such as clover, into agricultural pastures.

C. Ecological Restoration of Aquatic Habitats

Two main types of damage to aquatic systems require restoration: (i) the eutrophication, or nutrient enrichment, of rivers and lakes, often caused by the overuse of agricultural fertilizers, and (ii) water pollution caused by emissions of waste materials or agrochemicals such as herbicides. In both cases the first step in restoration is to remove the source or cause of pollution, followed by management designed to clean up the watercourse. Only when water quality has been restored can plants and animals be successfully reintroduced. One of the principle methods of reducing high levels of phosphorus and other nutrients in water is to grow and harvest algae or macrophytes in the water body. Ecologists are now importing techniques from wastewater treatment to improve water quality in lake and wetland environments, including the use of reedbeds, composed of the common reed *Phragmites communis*. [*See* EUTROPHICATION.]

IV. POSTRESTORATION SYSTEM MANAGEMENT

A. Time and Soil/Vegetation Succession

Invasion of species, through natural dispersal mechanisms, goes hand in hand with planned vegetation establishment to increase species diversity in restoration sites. This method of increasing species diversity depends on nearby available seed sources, suitable dispersion processes, and time. Following the scheme outlined in Fig. 1, the success of restoration and species accumulation should be monitored periodically after the establishment phase to assess how successfully introduced species are being retained and how quickly new species are being gained.

Monitoring should aim to assess soil conditions which also change with time after restoration. In particular, nitrogen-fixating species play an important role in gradually improving site fertility for the colonization of the very poorest soils, such as china clay spoils, sand dunes, glacial moraines, and iron ore and coal shale spoil. Developing vegetation adds to topsoil organic matter and fertility by bringing nutrients such as P and K up from the subsoil and depositing them at the soil surface in litterfall. Nutrients deposited from the atmosphere are also accumulated by plants at the soil surface. Improved soil organic matter input in litterfall and the activities of both plant roots and burrowing animals, including soil organisms, help stabilize soil structure. Progressive improvements in soil fertility thus accompany primary succession and community development after restoration. Com-

munity complexity and biomass are likely to increase concurrently.

Improved fertility and invasion of new species over time may result in differential competition, perhaps even to the extent of excluding some of the species restoration sought to restore. Species that grow faster, taller, and more aggressively compete successfully. Good examples of this include fast-growing birch (*Betula pendula*) or sycamore (*Acer pseudoplantanus*) in *Calluna* heathland, the rampant colonization of bracken (*Pteridium aquilinum*) in heather moorland, or the growth of rank grasses such as cocksfoot (*Dactylis glomerata*) in short grass turf. In particularly nutrient-poor communities, such as heather moorland, it may be important to prevent the invasion of nitrogen-fixing plants such as gorse (*Ulex europaeus*) or broom (*Cytisus scoparius*). On the other hand, nitrogen accumulators should be encouraged in the early establishment phases of nutrient-demanding communities, such as woodlands.

It is essential to ensure that aggressive species are not included in the establishment seed mixture. In seeding grassland, this means omitting vigorous species such as ryegrass (*Lolium perenne*) or red clover (*Trifolium pratense*). Excluding aggressors at the establishment phase may be impossible if seedbank or topsoil materials are being used. Where species invasion and excessive competitive growth are a threat to community diversity, ecosystem management practices must be introduced to retain the balance of environmental conditions and desired species composition. Apart from manual weeding, which might be used for removing birch and sycamore from heathland and the selective use of herbicides, three main management techniques that can be adopted are mowing, grazing, and burning.

B. Mowing and Grazing

The main difference between mowing and grazing as management options is that mowing is nonselective. There is also the potential for more controllable nutrient removal in mowed clippings. Mowing schemes should be tailored to allow flowering and seeding/fruiting of the species that are to be encouraged and to prevent seeding in species that are to be discouraged. The choice of grazing regime, namely the type of animal, the stocking rate, and the period of grazing, might depend on stock availability and their husbandry as much as the selectivity of their grazing. Cattle are less selective grazers than sheep, choosing most tall, rank herbage. During the summer months, deer, which browse a wide range of grasses and other plants, might provide a management option in the uplands.

C. Burning

Fire is a less expensive, but potentially more severe, option to the prevention of succession than grazing. Burning is a particularly appropriate form of management for heathlands and some forestry systems where regeneration is stimulated by fire. Burning leads to losses of nitrogen and sulfur in smoke, but these nutrients can be subsequently accumulated with community development. [*See* FIRE ECOLOGY.]

V. PHILOSOPHY FOR THE SUCCESS OF FUTURE RESTORATIONS

A. Role of Monitoring Programs

Short- and long-term objectives of ecological restorations may differ. In the short term, particularly for complex ecosystems such as woodlands, full species diversity and sustainability should not be expected. These are long-term goals. Realistically, the success of establishment and sustainability in the first 2 or 3 years of habitat restoration should be assessed against a priori criteria. Example criteria for grass meadow restoration could include (i) successful short-term establishment of the more common species (e.g., the top 75% of the sown species over the initial 2 to 5 years), (ii) longer term persistence of environmental characteristics required for ecosystem maintenance (e.g., soil physical, hydrological, and chemical characteristics over an initial 5- to 10-year period, and possibly over the longer term), or (iii) successful longer term reintroduction of less numerous species (e.g., the lower 25% of introduced species over 5 to 15 years). It is crucial

to allocate adequate financial support for postrestoration monitoring since remedial system management may depend on identifying and characterizing system change.

B. Roles of Research and Restoration Ecology

The basic understanding of plant growth requirements and soil–plant interactions is an essential precursor to successful ecological restorations. However, further knowledge of plant interactions, particularly in complex field situations, not in simplified, laboratory, or glasshouse experiments, is necessary for ecosystem maintenance and management in the longer term. Existing experience is dominated by ecological restorations of grasslands and prairies. Even in these systems, the complex interactions between plants and animals and between these organisms and their environment, which make up a stable, resilient, and self-sustaining system, are not well understood. In particular, the environmental controls on interspecific competition in different ecosystems require attention. This is especially important in circumstances of changing environmental conditions: increased soil nutrient levels due to air pollution, increased ozone levels and ultraviolet radiation receipt, or increased temperature and altered precipitation patterns predicted by global climatic change. In such changing conditions, the ecosystem composition also changes. Other particularly applicable research for ecological restorations includes studies into regenerative traits of species, including seed weight, number, and longevity, together with controls on germination and establishment. The biological strategies of species and the role of biological associations, symbiotic and otherwise, in species persistence require much further study.

The field experimental scale of restoration ecology is perhaps one of the only ways that we can glean answers to the complex questions outlined earlier. Bradshaw's observation that "what ultimately survives in nature is what is best suited to its environment" points the way to ecological restorations that mimic natural habitats. These restorations are more likely to produce resilient and self-sustaining solutions to landscape rehabilitation. John Ewel summarized these views by suggesting that "restoration is the ultimate test of ecological theory." Thus, the success or otherwise of restoration projects serves to illustrate our proficiency in ecology.

Glossary

Ecology Derived from *oikos* (house) and *logos* (study) to mean the study of living organisms in relation to their environments.

Ecosystem Integrated association of plants, animals, and environmental conditions that are self-sustaining.

Hydroseeding Method of spraying plant seeds onto restored land, using aqueous solutions of seed mixtures, often containing fertilizers, lime, organic matter, and soil stabilizers or conditioners.

Mulch Layer of organic or artificial materials laid on the soil surface to control soil temperatures, to protect the soil from moisture loss, or to protect the soil surface from soil erosion.

Nutrient cycling Stores and fluxes of plant nutrients, primarily the macronutrients (N, P, S, K, Ca, Mg, and Na), in an ecosystem. The main stores are in vegetation biomass and in the soil, whereas the main fluxes are mineralization in the soil, uptake by plants and animals, translocation within the biomass, litterfall and excretion, and organic matter decomposition in the soil.

Primary succession Sequential development over time of vegetation and organisms colonizing a virgin or bare site.

Secondary succession Sequential development over time of vegetation and organisms colonizing an abandoned site that already contains seeds and vegetative material typical of the full range of successional stages.

Soil amelioration Soil management designed to minimize any physical or chemical limitations to plant growth.

Soil fertility Fund and balance of plant-available nutrients present in the soil, both macronutrients (N, P, S, K, Ca, Mg, and Na) and micronutrients (Fe, Mo, Mn, Zn, Cu, Bo, Cl, and Co).

Bibliography

Bakker, J. P., ed. (1989). *Nature Management by Grazing and Cutting. On the ecological significance of grazing and cutting regimes applied to restore former species-rich grassland communities in the Netherlands.* Geobotany 14, 400pp. Kluwer Academic Publishers, Dordrecht, The Netherlands.

Bradshaw, A. D. (1983). The reconstruction of ecosystems. *J. Appl. Ecol.* **20,** 1–17.

Bradshaw, A. D. (1984). Land restoration: Now and in the future. *Proc. Roy. Soc. London B* **223,** 1–23.

Bradshaw, A. D., and Chadwick, M. J. (1980). "The Restoration of Land." Oxford: Blackwell Scientific Publications.

Bradshaw, A. D., Goode, D. A., and Thorp, E., eds. (1986). "Ecology and Design in Landscape." Oxford: Blackwell Scientific Publications.

Buckley, G. P., ed. (1989). "Biological Habitat Reconstruction." London: Belhaven Press.

Cairns, J., ed. (1988). "Rehabilitating Damaged Ecosystems." Boca Raton, FL: CRC Press.

Grime, J. P. (1979). "Plant Strategies and Vegetation Processes." Chichester, UK: John Wiley & Sons.

Jordan, W. R., Gilpin, M. E., and Aber, J. D. (1987). "Restoration Ecology: A Synthetic Approach to Ecological Research." Cambridge: Cambridge Univ. Press.

"Restoration Ecology," (Vol. 1, 1993). Cambridge, MA: Blackwell Scientific Publications.

Ecology, Aggregate Variables

Gordon H. Orians
University of Washington

An *aggregate variable* is a unit created by merging into it a number of smaller units that share certain features in common. Aggregate variables are erected because the human mind can think clearly about complex systems only if the number of variables is greatly reduced. We reduce the number of variables to a manageable number in three ways. *Stratification by velocity* eliminates variables by treating as constants those processes whose rates of change are very slow relative to others. Which processes are fast or slow depends on the problem of interest. The chemical constants of marine ecologists are the fast processes of marine geochemists. *Aggregation* groups together a number of different entities into a larger, composite unit. The larger the unit, the fewer the features shared by its members. Third, entities may be merged into some *macrodescriptor,* such as diversity, biomass, or productivity. The products of both the latter two methods are treated here as aggregate variables. How and why aggregate variables are formed and used in ecology is the subject of this article.

I. HOW ARE AGGREGATE VARIABLES SELECTED?

Aggregate variables, if they are to be useful, must be defined in ways that derive from and are consistent with the properties of their constituents. This requires that they have sufficient internal cohesion and shared traits that the aggregate groups make sense as ways of describing a complex reality. Without these conditions, any aggregate variable is unlikely to be useful in deducing the properties of interactions within and among different groups. All aggregate variables sacrifice information by emphasizing or ignoring particular details of relationships. What should be sacrificed or emphasized depends on the purposes to which aggregate variables are to be put. Therefore, considerable attention needs to be given to what is gained and lost by their use.

Because we use aggregate variables for many purposes in ecology, we construct them in diverse ways. Ecologists need aggregate variables that categorize the physical environment, which is minimally affected by the organisms themselves but which profoundly affects organisms. Because organisms have been molded over long time periods by evolutionary processes, phylogenetic and adaptive aggregates of several types are useful in many circumstances. Present-day interactions result in products that are also usefully studied with the help of aggregate variables. [See ECOLOGICAL ENERGETICS OF ECOSYSTEMS.]

II. AGGREGATE VARIABLES THAT DESCRIBE THE PHYSICAL ENVIRONMENT

A. Climate

The most pervasive physical variable—climate—has long been used as a basis for constructing aggregate variables. Indeed, because of the great importance of climate for the lives of organisms, biologists were at the forefront in devising climatic classification systems. The most general aggregate terms used to describe climates are based on temperature (tropical, temperate, arctic), wetness (arid, mesic, humid), and degree of seasonality (continental, maritime). Such terms are used frequently in the ecological literature as very broad, imprecise descriptors of climate. Biologists developed more precise classification systems designed to categorize climates in ways judged to be meaningful to the distribution of organisms and ecological communities. For example, C. W. Thornthwaite classified climates according to the seasonal availability of water in the soil. L. R. Holdridge developed a classification of climate using mean annual biotemperature (calculated from adjusted monthly mean temperatures), potential evapotranspiration, and total annual precipitation. H. F. Baily et al. described climates according to the difference in monthly mean temperature between the warmest and coldest month, the percentage of precipitation that falls in winter, and the mean annual temperature. Some of these climate classifications have been used as the basis for classifications of vegetation. Biologists have even gone so far as to use vegetation to infer and, hence, categorize, climates.

B. Soils

Soil formation is influenced by a variety of physical, chemical, and biological processes. The characteristics of a soil are determined by climate, the rock from which it was formed, the vegetation growing on it, local topography, and the length of time the soil has been developing. Soils are classified in several ways on the basis of their physical and chemical characteristics. As a soil forms, the sizes of its particles are generally reduced and chemical changes results in the formation of distinct layers, called *horizons*. A widely used, simple system classifies horizons into four major categories, one of which has two prominent subdivisions. In descending order from the soil surface, they are:

O: a layer primarily composed of dead organic matter (litter);

A_1: a layer of partly decomposed organic material (humus) mixed with mineral soil;

A_2: a region of extensive leaching of minerals from the soil;

B: a region with little organic material that may contain minerals and oxides of iron and aluminum leached out of the A_2 horizon;

C: weakly weathered material similar to the parent rock; in dry regions, calcium and magnesium carbonates accumulate here and may form a hard, impenetrable layer.

Soils are also classified on the basis of the particle sizes of which they are composed. These particles, which range over more than a billionfold in diameters from molecules to stones, are divided into eight categories (Table I). Any particle less than 2 μm

TABLE I
Classification of Soil Particle Sizes

Particle diameter (μm)	Designation
2000	Very coarse sand
200	Fine sand
20	Silt
2	Coarse clay
0.2	Fine or colloidal clay
0.02	Fine colloidal clay
0.002	Ultrafine clay

in effective diameter but larger than a small molecule is called a clay, regardless of its chemical or mineralogical composition. An important classification of soils is based on the percentage of particles belonging to the three main size categories—clay, silt, and sand. A *loam* is a mixture of sand, silt, and clay in which all three categories are well represented. [*See* SOIL ECOSYSTEMS.]

The soil-forming processes are continuous but the resulting soils are classified into broad categories known as great soil orders (Table II). These orders are further subdivided into subgroups, families, and series. More than 10,000 series have been identified, described, and classified. These fine divisions are important for local and regional land classifications and for determining the best uses to which the soils can be put.

C. Fresh Water

Ecologists find it useful to divide fresh waters into two large categories—flowing (*lentic*) and stationary (*lotic*). Lakes are classified by their mode of origin (glacial, river oxbow, volcanic, impoundments) and by the state of oxygenation and thermal stratification of their waters. Lakes that are permanently stratified are called *meromictic*. The

TABLE II
United States Department of Agriculture Classification of Soils

Soil order	Dominant characteristics
Entisol	Very young soils with simple profiles
Inceptisol	Fine-textured; only moderate horizon development; brown forest soils
Aridisol	Desert soils low in humus; often have clay horizons
Mollisol	Black prairie soils rich in humus
Spodosol	Light gray A_2 horizon rests on a black and reddish B horizon high in extractable iron and aluminum; often called Podzols
Alfisol	Shallow penetration of humus; high base content; well-developed horizons
Ultisol	Strong clay translocation, intensely leached, low base content; usually in warm climates
Oxisol	Highly weathered tropical and subtropical soils; often called Latisols and laterites
Histisol	Bog and peat soils high in organic matter

oxygenated part of a lake in which the rate of photosynthesis exceeds the rate of use of O_2 is called the *trophogenic* (euphotic) zone. That part in which rate of use of O_2 exceeds its rate of generation by photosynthesis is called the *tropholytic* zone. [*See* LIMNOLOGY, INLAND AQUATIC ECOSYSTEMS.]

In regions with marked seasonal temperature changes, the peculiar thermal properties of water generate patterns of vertical layering in lakes. Water is most dense at 4°C and expands at temperatures both above and below that level. During winter, very cold water floats on top of water at 4°C. As a lake warms in spring, surface waters become less dense than the colder water below and sink until the entire lake reaches a temperature of 4°C. At that point, the lake is very unstable and winds readily cause a complete turnover of the water column, bringing nutrients to the surface and oxygen to deeper waters. As surface waters continue to warm, they float on top of the colder water, which remains at 4°C. The transition from warmer to cold water gradually lowers during summer, but may never reach the bottom of large, deep lakes at high latitudes. This thermal stratification is recognized in the aggregate terms *epilimnion* (warm surface water) and *hypolimnion* (deep cold water). The zone between these regions where temperatures change rapidly over short vertical distances is called the *metalimnion*. The *thermocline* is a plane through the metalimnion at the depth of the inflection point on the temperature–depth curve.

Zones in ponds and lakes are also recognized by features other than temperature. The shallow water around the shore, usually occupied by rooted vegetation, is called the *littoral* zone. Open water constitutes the *limnetic* zone. The bottom below the littoral zone is the *profundal* zone.

D. Marine Waters

The oceans are divided into regions based on the depth of water and distance from the surface or substrate. The zone between high and low tide is the *littoral zone*. The continental shelf below the littoral zone is the *sublittoral zone*. Still deeper are the *bathyal* and *abyssal zones* of the deep ocean and the *hadal zone* of the deepest trenches. The open water above the sublittoral zone is the *neritic zone*. Classification of marine environments into these zones highlights the importance of tidal fluctuations to organisms at ocean margins, the great differences between open water and substrate as an environment in which to live, and the differences between communities living in the zones where photosynthesis is possible and those dependent on the settling to the bottom of materials produced in the photic zone. Another useful division is between the open ocean and coastal estuaries that are strongly influenced by the inflow of fresh water and, hence, may change dramatically in salinity seasonally or during tidal cycles. [*See* OCEAN ECOLOGY.]

III. EVOLUTIONARY AGGREGATES

Evolutionary processes are reflected in biologists' choices of aggregate variables in classification systems of organisms and their adaptive characteristics.

A. Taxonomic Aggregates

The universally used Linnaean system of classifying organisms is a hierarchical one in which organisms are aggregated into increasingly larger taxonomic units, the principal ones being species, genus, family, order, class, phylum, and kingdom. This system serves two major functions. One is to reflect phylogenetic history by grouping organisms according to their presumed common ancestries. The other is to serve as a stable naming and information retrieval system. The systems commonly used today are compromises between these two goals, partly for historical reasons, but also because both functions cannot be accommodated without compromise within a single hierarchical system. The value of taxonomic aggregates for organizing and communicating information about biological diversity is amply demonstrated by the fact that the system persisted through a transition from a special creationist to an evolutionary view of the world. [*See* SYSTEMATICS.]

B. Adaptive Aggregates

Every one of the many millions of species is biologically unique and has unique relationships with other organisms and the physical environment. Description of these myriad traits and developing hypotheses about them would be impossible without the use of aggregate variables. Aggregates are used to describe form, functioning, and ecological relationships.

A widely used system for classifying plant life-forms was proposed by the Danish plant ecologist C. Raunkiaer on the basis of the position of their buds. He distinguished five major life-forms according to how well the buds are protected during unfavorable periods (Table III), and he showed that the frequencies of those forms in local floras were strongly correlated with climate. The unfavorable season that concerned Raunkiaer was winter, but his system works well in regions when the stressful season is dry rather than cold. More complicated systems for classifying plant life-forms have been developed but the simplicity of Raunkiaer's system has resulted in its continuing popularity.

Other aggregate units group organisms according to their physical tolerances. Plants that are tolerant of drought are called *xerophytes;* those tolerant of high salt levels are called *halophytes*. The range of salinity that an organism can tolerate is described by terms such as *euryhaline* (having a broad tolerance range). Comparable terms aggregate organisms according to their temperature tolerances. These terms are useful for comparing the

TABLE III
Raunkiaer's Classification of Plant Life-forms

Life-form	Characteristics
Phanerophytes	Trees and large shrubs; buds exposed to the physical environment on tips of branches
Chamaephytes	Small shrubs and herbs; buds close to the ground
Hemicryptophytes	Buds at ground surface, protected by soil and decomposing leaves
Cryptophytes	Buds buried in soil
Therophytes	Lack persistent buds; plants survive unfavorable season as seeds

responses of species to environmental conditions and for identifying adaptive properties that are common to many species.

IV. FUNCTIONAL AGGREGATES

Some ecological classifications are similar to taxonomic classifications in that they aggregate organisms by the number of interacting units. At lower levels these units correspond to those in taxonomic systems—individual, deme, population, and species. At more inclusive levels, the classifications diverge. The standard ecological units are community, ecosystem, biome, and biosphere. These aggregate units are used to define and delimit the spatial scales of systems being investigated. By themselves these terms do not specifically determine boundaries but, in combination with clearly stated objectives, they provide criteria for selecting useful boundaries. A *community* is the totality of species of organisms in the focal region or the members of some particular subgroup, such as the lizard community. An *ecosystem* includes both the organisms and the physical environment of the focal area. A *biome* is a biotic community that is characterized by a distinctive physical structure formed by its dominant plants. Generally a biotic community is not recognized as a biome unless it covers a broad geographical area. The definition of a biome is vague and somewhat arbitrary and, consequently, ecologists do not agree on the number of biomes that should be recognized. The most general of the commonly recognized types are tundra, boreal forest (taiga), temperate deciduous forest, tropical wet forest, tropical dry forest, tropical savannah, temperate grassland, hot desert, cold desert, and chaparral (broad-leaved sclerophyll). The *biosphere,* the sum total of all ecosystems on earth, is often referred to as the global ecosystem.

Another criterion for establishing ecological aggregates is the energy source of organisms. This basis for classification is very useful because of the central role that energy plays in ecological systems. First introduced by C. S. Elton in 1927 and brought into clear focus by R. I. Lindeman in 1942, *trophic levels* are a widely used method of grouping organisms by their general source of energy. The commonly used trophic levels are photosynthesizers (producers), herbivores (eaters of plant tissues), primary carnivores (eaters of herbivores), secondary carnivores (eaters of primary carnivores), omnivores (animals that feed at more than one trophic level), and detritivores (decomposers) that feed on the dead remains of other organisms. A finer subdivision of a trophic level is a *guild,* a group of organisms exploiting the same general energy source in a similar manner, such as the guild of foliage-gleaning insectivorous birds. Every trophic level consists of a variety of organisms differing in size, population dynamics, specific sources of energy, habitat, and motility. These organisms may interact little with other members of the aggregate and very differently with organisms in other trophic levels. Elephants and grasshoppers may both be herbivores in the same African savannah but they share little else in common. In addition, many animals do not feed entirely within a single trophic level, making the concept a difficult one to define operationally. The concept remains useful despite these problems as a basis for comparing ecological communities with respect to their general trophic structure and energy-processing properties.

Energy-processing traits of communities are also described using macrodescriptors that highlight the flow patterns of energy among species or aggregates of species. The most important of these macrodescriptors are *food chains* (representations of passages of energy through populations in a community) and *food webs* (representations of the various paths of energy flow through populations in a community). The concept of a food web has generated many studies designed to detect patterns in the structures of such webs, and to identify the causes of any similarities that are uncovered. Without the macrodescriptor, such studies would not have been conceived, much less undertaken.

Another group of macrodescriptors focuses on the number of species in ecological communities. A simple list of the species present in a community defines its *species richness.* For a variety of reasons, it is often useful to weight species according to their abundance, size, energy consumption, or some other ecologically relevant attribute. Such weighted measures of species richness are called *species diversity*. Both of these aggregate concepts have spawned a rich literature focused on measuring species richness and diversity and seeking insights that can be derived from the comparative study of those measures.

V. PROCESS AGGREGATES

Static descriptors of energy utilization in ecosystems are supplemented by process-oriented descriptors that focus on quantities of energy flowing through the system. The most general of these aggregates is *productivity*, the rate of capture of energy by the system. The result of productivity is production. *Gross production* refers to the total energy intake, whereas *net production* refers to intake minus maintenance costs. The latter is the energy available to the next trophic level. Production is also divided by trophic level into *primary* (by photosynthetic and chemosynthetic organisms) and *secondary* (by all other trophic levels) *production*. Ratios of production between successive trophic levels can be compared and expressed as production efficiency.

The outcomes of these processes are also described with still another important macrodescriptor—*biomass,* the weight of living material in a population, community, or ecosystem. Biomass is usually expressed on a per unit area basis in terrestrial environments and on a per unit volume basis in aquatic environments. Biomass is the basic unit for comparing the quantities of materials found in different communities and ecosystems.

In addition to processing energy, organisms use mineral nutrients that, unlike energy that has a unidirectional flow, can be and are recycled within individual organisms and ecosystems. The study of nutrient cycling occupies an important place in ecology because the amounts of elements and their movement between compartments provide a convenient and relatively easy to measure index to the flow of energy, and because levels of nutrients regulate primary production in many ecosystems. Important macrodescriptors related to nutrient use are the *recycling index,* the percentage of a nutrient

that is recycled within an ecosystem, and *nutrient use efficiency,* the production per unit of nutrient cycled. The former measure has no counterpart among energy aggregates, but the latter is the mineral equivalent of trophic efficiencies.

In addition to measuring the quantities of energy flowing through ecosystems and the cycling of nutrients, ecologists are also interested in how those patterns change over time as communities change and how the patterns respond to perturbations of the system. An important macrodescriptor in ecology is *succession,* the replacement of populations in a habitat through a regular progression to a relatively stable state. Successions can be classified according to whether they start with an organism-free state (*primary succession*) or from a disturbed community with many species in it (*secondary succession*). The presumed relatively steady end point of succession is called the *climax.* Much controversy has surrounded the utility of climax as an aggregate variable because no truly stable end point is ever reached, and how stable is stable enough to make the concept useful depends on many different perspectives and goals.

Ecologists have long perceived a need for macrodescriptors that categorize types of stability despite the difficulties attending any attempts to measure and define stability. Many different systems are currently in use but the categories shown in Table IV encompass the major types that ecologists have found useful. The classification illustrates an important point, namely, that concepts of stability are useful even if no constant state exists. Indeed, the ways that systems respond to perturbations to return (or not return) to approximately their former states are very useful characteristics to define and measure.

VI. HOW WELL ARE AGGREGATES WORKING IN ECOLOGY?

It is impossible to conduct science in any field without the use of aggregate variables, but aggregate variables need to be erected carefully because they inevitably and powerfully channel thinking. Scientific questions are framed using aggregate terms

TABLE IV
Categories of Ecological Stability

Category	Characteristics
Constancy	The lack of change in some parameter of a system; carries no causal connotations
Persistence	The survival time of a system or some component of it
Inertia	The ability of a system to resist external perturbations
Elasticity	The speed with which a system returns to its former state
Amplitude	The region over which a system returns to its former state
Cyclic stability	The property of a system to cycle or oscillate around some central point or zone
Trajectory stability	The property of a system to move toward some final end point or zone

and the nature of the questions asked by scientists are molded by the aggregate variables with which they have been taught to think. The very act of erecting an aggregate variable implies a judgment about the importance of interactions among entities being aggregated. For example, guilds are identified and studied because members of a guild are thought to influence one another more strongly than they influence members of other guilds. This supposition can, of course, be tested observationally and experimentally, but aggregate variables tend to acquire a life of their own whether or not their utility has been subjected to testing.

Aggregate variables are key components of the conceptual bridge between microecology—the study of individuals and small groups of individuals and their interrelationships—and macroecology—the study of larger ecological systems. Traffic across these bridges is still rather intermittent. Concepts of community energetics and competition theory have, until recently, been little influenced by the underlying microtheory of foraging behavior. Notions of trophic efficiency and complexity of energy pathways are based on still different assumptions. As yet we do not know how our views of energy flow processes at community levels will be influenced by developments in microecology. New aggregate variables are certain to be developed as relationships become clearer, and some

currently popular variables are likely to be abandoned. However, the history of science suggests that our abilities to predict these changes are likely to remain poor. The best we can do is to be alert for the need for changes and to constantly question the utility of the aggregates we are using. If we do not challenge them, they will control us without our being aware of it.

Glossary

Biome Major division of the ecological communities of earth; characterized by distinctive vegetation.

Community Ecologically integrated group of species inhabiting a given area.

Ecosystem Organisms of a particular habitat, such as a pond or forest, together with the physical environment in which they live.

Food web Diagram of the set of food links between species in a community that indicates which ones are the eaters and which are eaten.

Guild Group of organisms that share a common food source and may have strong interactions with one another.

Horizon Vertical layer of a soil distinguished by its physical and chemical properties.

Macrodescriptor Index or term describing some aggregate feature of a group of organisms.

Succession Gradual, sequential series of changes in the species composition of a community following a disturbance; with time it may reach a relatively stable state (climax) in which species composition changes slowly or not at all.

Trophic level Group of organisms united by obtaining their energy from the same part of the food web of a biological community.

Bibliography

Bailey, H. P., Simpson, B. B., and Vervoorst, F. (1977). The physical environment: The independent variable. *In* "Convergent Evolution in Warm Deserts" (G. H. Orians and O. T. Solbrig, eds.), pp. 13–49. Stroudsburg, Penn.: Dowden, Hutchinson & Ross.

de Candolle, A. P. A. (1874). Constitution dans le règne végétal de groupes physiologiques applicables à la géographie ancienne et moderne. *Arch. Sci. Phys. Natur.* (Geneva). **50,** 5–42.

Holdridge, L. R. (1967). "Life Zone Ecology." San José, Costa Rica: Tropical Science Center.

Köppen, W. (1884). Die Wärmezonen der Erde, nach Dauer der Heissen, Gemässigten und Kalten Zeit, und nach der

Wirkung der Wärme auf die Organische Welt betrachter. *Meterol. Z.* **1,** 215–226.

Levins, R. (1977). The search for the macroscopic in ecosystems. *In* "New Dimensions in the Analysis of Ecosystems" (G. S. Innis, ed.). pp. 213–222. La Jolla, CA: Society for Computer Simulation.

Lindeman, R. I. (1942). The trophic-dynamic aspect of ecology. *Ecology* **23,** 399–418.

MacMahon, J. A., Phillips, D. L., Robinson, J. V., and Schimpf, D. J. (1978). Levels of biological organization: An organism-centered approach. *BioScience* **28,** 700–704.

Orians, G. H. (1975). Diversity, stability and maturity in natural ecosystems. *In* "Unifying Concepts in Ecology" (W. H. van Dobben and R. H. Lowe-McConnell, eds.), pp. 139–150. The Hague: W. Junk Publishers.

Raunkiaer, C. (1934). "The Life Forms of Plants and Statistical Plant Geography." Oxford, England: Clarendon Press.

Rigler, F. H. (1975). The concept of energy flow and nutrient flow between trophic levels. *In* "Unifying Concepts in Ecology." (W. H. van Dobben and R. H. Lowe-McConnell, eds.), pp. 15–26. The Hague: W. Junk Publishers.

Thornthwaite, C. W. (1948). An approach to a rational description of climate. *Geogr. Rev.* **38,** 55–94.

Ecology and Genetics of Virulence and Resistance

P. Schmid-Hempel and J. C. Koella
ETH Zürich, Switzerland

I. Introduction
II. Evolution of Virulence and Resistance
III. Variation and Genetics
IV. Coevolution
V. Ecology and Epidemiology

A parasite's success depends on its growth within the host and its transmission to and establishment on other hosts. The resulting damage to the host is usually summarized as the parasite's virulence. To reduce this damage, hosts mount defenses against parasitic infection; they develop resistance against the parasite. Virulence and resistance result from a number of quite different processes in the complex interplay of host and parasite. The details of this interplay largely determine the ecology and evolution of the host–parasite interaction. It has become the accepted view that, contrary to the conventional wisdom, natural selection will lead to avirulent parasites only under a limited set of conditions. The degrees of virulence and resistance are important for the ecological dynamics of the interaction and have ramifications for entire ecosystems. Consideration of the ecology of virulence and resistance is therefore important for applied questions such as predicting the success of an immunization or biological control program. Underlying these evolutionary and ecological questions are the variability of virulence and resistance. Virulence and resistance vary at different scales; genetic variation in these traits is ubiquitous and is based on various genetic systems, from gene-for-gene interactions to polygenic inheritance.

I. INTRODUCTION

A parasite's survival depends on its successful transmission from one host to another. The wide range of parasite life cycles necessarily leads to different ways of achieving success, but one feature seems to be held in common: the production of a large number of transmissible forms of the parasite. To achieve this, microparasites (i.e., parasites such as viruses, bacteria, and protozoa) rely on extensive replication within their host to increase their density and thus enhance their probability of being transmitted, whereas macroparasites (i.e., parasites such as flukes and tapeworms) rely on individual growth and survival within their host and on the production of a large number of offspring that leave the host to infect others. In enhancing their own success, parasites reduce their host's chances of survival and reproduction. Thus, host–parasite relationships are characterized by conflicting interests of the two parties, each increasing its own fitness at the cost of the other's. This conflict of interest forms the basis for many evolutionary and ecological aspects of the host–parasite interactions, including virulence and resistance. [See PARASITISM, ECOLOGY.]

Resistance usually refers to the result of physiological processes in the host that reduce the number, fertility or survival of parasites. In a more ecological view, resistance refers to the result of any process that reduces the negative effects of the parasite on its host. This view measures resistance not in terms of parasite density but in terms of host survival and reproduction, that is, as components of host fitness. Processes leading to resistance can include reproduction at an earlier age, so that the probability of leaving some offspring is increased, or reallocation of energy and nutrients from growth to defense, so that the probability of host survival is increased. However, the concept of resistance normally excludes aspects of behavior and ecology that reduce the chances of contact or establishment in the first place. Such processes include avoiding infected conspecifics, selecting sleeping sites where animals have a low probability of becoming infected, or choosing food items that are unlikely to be intermediate hosts of potential parasites.

Most of our knowledge of the underlying mechanisms of host resistance is restricted to plants and, among animals, vertebrates, in which the immune system mechanisms for recognizing non-self and eliminating pathogens have been well studied. Vertebrate immune systems have the particular ability to remember past infections and to quickly mount a massive defense against a new infection by the same or a sufficiently similar parasite. Most knowledge for arthropods comes from research on the biological control of pest species by means of parasitic organisms. Typically, parasites inside the body cavity of insect hosts become encapsulated by blood cells and are thereby eliminated. Some research has concentrated on snails that are important vectors for human diseases such as bilharzia. But overall, our present-day knowledge of resistance mechanisms is depressingly meager.

Virulence refers to the result of processes induced by the infection that lead to a reduction in some component of the host's well-being. Sometimes this is measured as the rate of multiplication or the density of the pathogen within the host—quantities that are assumed to relate to the amount of damage done—but more often, in particular by ecologists, virulence is measured as the reduction in host survival and reproduction. Thus, virulence is measured in terms of the fitness of its host, although it is considered to be a property of the parasite. Virulence can be determined by a number of different processes at various stages of the infection. For example, harm could be associated with penetration of the host skin, depletion of host resources, damage or destruction of host tissues, production of toxins by the parasite, or the suppression of the host's immune response rendering the host more susceptible to other pathogens. More specifically, most of the symptoms of bilharzia are caused by the reaction of the human host to the eggs of the parasite, *Schistosoma,* trapped in host tissue. *Plasmodium falciparum,* which causes a severe form of malaria, induces clumping of erythrocytes in the bloodstream and can thus lead to dangerous thromboses. Ecologists have extended this view of virulence to allow for parasite-induced behavioral changes that increase the probability of successful transmission for the parasite but put the host at risk. Acanthocephalan parasites, for example, *Plagiorhynchus cylindraceus,* change the behavior of their intermediate host, a wood louse, in a way that makes it more likely to be eaten by their final host, a songbird. Malaria parasites induce their mosquito vectors to change their feeding ecology, so that the mosquito must probe more often to obtain a blood meal, and thus increase the rate of transmission of the parasite from mosquito to humans but put the host at a higher risk of being killed during feeding.

To summarize, infection by parasites is associated with more or less characteristic symptoms that result from the host–parasite interaction. There is an increasing awareness that symptoms cannot be analyzed apart from considerations of virulence and resistance. For example, coughing in patients infected with influenza viruses is beneficial for the parasite, because droplets loaded with virus particles are spread and can infect new hosts. On the other hand, fever aids in suppressing multiplication of many parasites, because they cannot survive at high temperatures. Fever may thus be considered as an aspect of resistance of the host as well as an aspect of the parasite's virulence. These examples show that, to understand the evolution and adaptive value of virulence, one must consider not only the negative consequences of the parasitic infection on the host, but also its consequences for parasite biology.

Virulence and resistance result from the interaction of host and parasite in a given environment. Host, parasites, and their life-styles are extremely diverse. Our present knowledge is far from complete and mostly restricted to cases of medical or economical importance. In the following, we will review several fundamental questions about the evolution, ecology, and genetics of host–parasite interactions, referring mostly to microparasites and macroparasites, and give an overview of the meaning of virulence and resistance, with special reference to animal hosts.

II. EVOLUTION OF VIRULENCE AND RESISTANCE

The conventional wisdom of host–parasite coevolution maintains that host–parasite systems will evolve toward commensalism. This view assumes that virulent parasites will destroy their host population, and thus themselves as well, and will therefore be replaced in time by less virulent strains. Virulence is thus considered a primitive, not yet adapted form of parasitism. In support of this view are several cases of parasites being more virulent when they infect a new, exotic host than when they infect their normal host. For example, in regions of Africa where trypanosomiasis is common, wild ruminants acquire only mild infections, whereas imported cattle experience infections that usually end fatally if untreated. However, as in many similar cases, the observations do not allow one to distinguish whether the development of mild infections over evolutionary time is due to a decrease of the virulence of the parasites or an increase of the

resistance of the hosts. Furthermore, some parasites seem to have achieved a stable level of virulence. The smallpox virus, for example, seemed to have reached an intermediate level of virulence that has remained fairly constant over the past thousand years before it was eradicated.

Apart from such conflicting interpretation of the available data, the conceptual advance in evolutionary biology over the past two decades has raised severe criticism about the conventional wisdom. In particular, selection of reduced virulence by its consequences on parasite persistence would require that parasites evolve for the benefit of the group or the species. However, as modern concepts suggest, such group selection is unlikely to be a potent force for most of the traits of organisms and is usually overridden by individual selection. This is because the genetic material of the most successful individual parasites will be represented in future generations, regardless of the long-term fate of the host–parasite relationship. Ecologists and evolutionary biologists have therefore analyzed the evolution of virulence from the perspective of individual selection by asking: What benefit will individual parasites gain from being virulent? These analyses have demonstrated that, though individual selection may lead to avirulent parasites, it is likely that an intermediate level of virulence or ever-increasing virulence will arise. An illustrative way of showing this is to consider the equation describing the basic reproductive rate, R_0, of a directly transmitted pathogen:

$$R_0 = \frac{\beta N}{\alpha + \mu + v},$$

where β denotes the parasite's transmission rate, N the density of susceptible hosts, μ the background mortality rate, α the parasite-induced mortality rate (virulence), and v the recovery rate of infected hosts. The basic reproductive rate describes the number of secondary infections caused by a single infected host, that is, the rate of spread of a parasite through a population of susceptible hosts, and can thus be thought of as a measure of the parasite's fitness. The parasite can maintain itself only when $R_0 \geq 1$, and the larger R_0 the higher its fitness.

Therefore, natural selection will favor the level of virulence that maximizes R_0. If β and v are not related to virulence (α), then parasite fitness is indeed maximized by minimum virulence ($\alpha \to 0$) and the interaction will tend toward avirulence. However, if β or v is related to virulence, other outcomes are possible, so that it is possible that an intermediate value of virulence maximizes R_0 (see Fig. 2). Which outcome arises depends critically on the exact details of the host–parasite interactions.

There is only one well-studied case in natural populations where these relationships have been analyzed in any detail: the interaction between the European rabbit introduced into Australia and the myxoma virus introduced to control it. Shortly after the introduction of the virus in 1950, the rabbit showed a 99.8% case mortality and nearly went extinct. But evolution in the virus toward reduced virulence was rapid, so that case mortality decreased to an apparently stable equilibrium of about 70% (Fig. 1). The reason for this evolution of intermediate virulence seems to be that virulence is related to transmission in such a way that the basic reproductive rate is maximized at intermediate virulence: too high a virulence kills a rabbit rapidly and too low a virulence allows the host to recover quickly, both resulting in low rates of transmission (Fig. 2).

Though other host–parasite systems are less well studied with respect to the evolution of virulence, many observations suggest relationships between virulence and rate of transmission (Table I). Sleeping sickness in humans may illustrate the point: the density of parasites circulating in the bloodstream is assumed to be associated with the probability that a biting tsetse fly will pick up the pathogen, and hence with the probability of transmission to a new host. The density of parasites is also associated with the clinical symptoms of the disease, and hence with virulence. There is also some evidence for an association of virulence with traits that are important at other stages of the parasitic infection. In trematode parasites of beetles, for example, lines with high virulence are less effective in establishing the infection than lines with low virulence. In the myxoma virus, high degrees of virulence are associated with low rates of host recovery (Fig. 3).

The idea that virulence evolves to the rate that maximizes basic reproductive rate leads to several predictions, only some of which have been tested to date. First, when a host individual is infected by more than one parasite genotype, a clone that rapidly reaches a high density will have a high

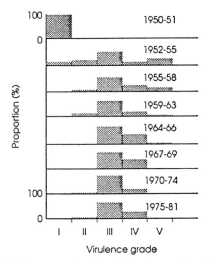

FIGURE 1 The proportions in which various grades of myxoma virus have been found in wild populations of rabbits in Australia over 30 years. The grades refer to case mortality rates. [Figure 16, R. M. Anderson and R. M. May (1983). *Proc. Roy. Soc. London Ser. B* **219**, 281; reproduced with permission.]

FIGURE 2 The relationship between basic reproductive rate, R_0 (a measure of parasite fitness), and virulence, α, as estimated from data for various grades of myxoma virus in Australian rabbits. The figure shows that maximum fitness of the parasite is achieved with intermediate virulence. [Figure 2, R. M. Anderson and R. M. May (1982). *Parasitology* **85**, 411; reproduced with permission.]

TABLE I

Examples of the Relationship between Host Pathology (a Consequence of Virulence) and Transmission for Some Human Pathogens

Disease	Effects
Diarrheal diseases: Viruses, bacteria, protozoa	Transmission facilitated through contact with fecal material
Respiratory diseases: Viruses, bacteria, mycobacteria	Transmission spreads via respiratory secretions, coughing
Diseases of the central nervous system: Rabies, trypanosomiasis	Transmission facilitated by behavioral changes (biting), host death, impaired grooming, sluggishness against biting vectors
Skin infections: ringworm, leprosy, onchocerchiasis	Symptoms related to accumulation of transmission stages in skin for direct contact with new host
Diseases with animal, soil, or plant reservoirs: Tapeworms, anthrax, Lyme disease, tetanus, botulism, mycoses	Host pathology has little effect on pathogen transmission; pathogens are often extremely virulent
Diseases with a carrier state: Hepatitis B, typhoid	Long-lived, minimally symptomatic hosts increase transmission probability
Diseases with long prepatent infectious period: AIDS	Overt host pathology impedes transmission, e.g., with decreased social contacts; not the case during the time when no pathological effects are visible

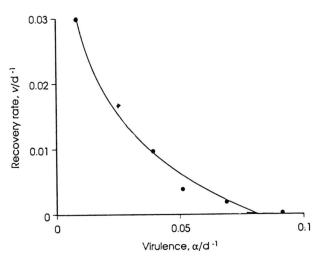

FIGURE 3   The relationship between virulence of myxoma virus and recovery rate of the rabbit host. [Figure 1, R. M. Anderson and R. M. May (1982). *Parasitology* **85**, 411; reproduced with permission.]

probability of transmission but will kill the host before a slowly replicating genotype has grown to a density ensuring its transmission. Therefore, rapidly growing, virulent parasites will outcompete slowly growing ones, so that multiple infections should often select for increased levels of virulence. In this case, although the parasite with the highest basic reproductive rate will be favored, the actual values of R_0 will generally be lower than if the parasites infected their hosts singly. Empirical demonstrations for these predictions are hard to come by.

Second, vertical transmission, that is, transmission of the parasite from an infected host to its offspring, should select for lower virulence than horizontal transmission (i.e., to other hosts in the same generation), because a vertically transmitted parasite must ensure that its host survives long enough to produce offspring. This prediction has been tested for two systems: for parasitic nematodes that infect fig wasps and for filamentous bacteriophage that infect the bacterium *Escherichia coli*. Fig wasps lay all of their eggs into a single flower of a fig tree, where the eggs hatch and the larvae develop into adults. The nematodes' life cycle is closely associated with that of their host. Immature nematodes infect newly emerged wasps and are carried to the next fig. After the wasp has laid her eggs, adult nematodes emerge from the wasp and lay their eggs into the same fig. This life cycle of wasps and nematodes leads to variation in the potential of horizontal transmission of parasites. In some species, only one wasp lays her eggs into a flower. In these species, the parasites must necessarily be transmitted vertically. As expected, this condition is associated with relatively low virulence. In species where several wasps lay their eggs together into a single flower, parasites have ample opportunity for horizontal transmission to offspring of another host. In these species, virulence of the nematodes is high. In the experiment involving filamentous phage and bacteria, two forms of the parasite were used: an avirulent form that permitted a high growth rate of the bacterium but was not infectious, and a deleterious form that was capable of infecting other bacteria. When bacteria were cultured so that no infectious spread was allowed and all parasite transmission was therefore vertical, the avirulent phage prevailed, whereas when conditions allowed for horizontal spread, the selective advantage of the avirulent form was lost.

Third, one effect of parasite virulence may be to immobilize hosts. Thus, if virulence is high, rate of contact among hosts can be reduced, reducing the rate of transmission of directly transmitted pathogens from one host to another. Parasites transmitted by a vector do not suffer from such a cost. Therefore vector-transmitted parasites might be expected to be more virulent than directly transmitted ones. Some data on human diseases support this idea. It is also suggested by the observation that virulence can be artificially increased when a parasite is experimentally passed to the next host, thus bypassing the association between virulence and transmission rate.

These examples lend support to the currently prevailing view that virulence is an evolved character by which host damage is selected because of its consequences for transmission of the parasite. This contrasts three alternative hypotheses: (1) the conventional wisdom that parasites evolve to low virulence so that they persist; (2) that host damage is an unselected, accidental consequence of parasite biology, for example, an inevitable by-product of how parasites must enter a host, develop, or reproduce; and (3) that virulence can be favored by selection within a host, independently of its effect on transmission, so that mutants are selected to grow very rapidly or to invade different organs and tissues. Examples for this may be meningitis and cerebral malaria, where the most virulent strains of the parasites invade the central nervous system from where they cannot be transmitted. In the latter two hypotheses, virulence would persist despite its potential negative effects on host *and* parasite.

In contrast to the evolution of virulence, the evolution of resistance poses less controversy: it is clear that hosts should evolve some degree of resistance against virulent parasites to counteract their damage. The main question is: What level of resistance should be adopted? A key element in these discussions is the cost of resistance, that is, the negative association between resistance and other traits affecting the host's fitness. If this cost is too high, it is likely that a host population will not evolve complete resistance. Costs of resistance are ubiquitous. For example, they have been observed in bacteria where, in the absence of their parasitic phage, resistant genotypes replicate slower than sensitive ones. In snails that act as intermediate hosts for the trematode *Schistosoma,* the snails that are resistant against infection by trematodes have, in absence of the parasite, lower fecundity than the susceptible ones. In humans, resistance against malaria is partly due to a genetically determined defect of the red blood cells, the sickle cell trait. Heterozygous individuals have the advantage of some resistance with no known negative effect, but homozygous individuals, though resistant against malaria infection, die at an early age. Costs of resistance are also suggested by the existence of autoimmune diseases such as multiple sclerosis, where the immune system turns against the host itself. Generally, with limited resources organisms have to "decide" whether to allocate energy and nutrients to further growth, survival, and reproduction or to defense against parasites. For example, malnutrition often has negative effects on the immunocompetence of mammals.

III. VARIATION AND GENETICS

Variation in virulence and resistance is ubiquitous, but its causes are rarely understood. In statistical terms, the total phenotypic variation in resistance and virulence can be partitioned into genotypic variation (i.e., differences due to different genotypes), environmental variation (i.e., differences due to differences in the environments), and variation due to genotype–environment interactions (e.g., when certain genotypes are particularly successful in certain environments). Although all of these components affect the ecological dynamics of a host–parasite relationship, the evolutionary process can lead to changes in virulence and resistance only if there is sufficient genotypic variability of the traits for selection to work on. Such genotypic variation has been found in animals and plants wherever it has been looked for, including such diverse cases as geographical variation in resistance of snails against trematodes, variability in resistance of humans against malaria or of mosquitoes to viruses, variation in the virulence of trypanosomes infecting bumblebees, and variation in aphids being affected by parasites (Table II). The details of this variation can be considered along two dimensions: first, according to the number of genes involved in determining resistance and virulence, and second, according to whether resistance against a parasite species is general, directed against all parasite genotypes, or is genotype specific, so that a host individual is resistant against only a limited number of parasite genotypes.

Where the genetic analysis has been done, resistance has often been found to be due to a single gene. Thus, in humans a particular gene locus, the locus coding for the sickle cell trait, has a strong effect on resistance against the most lethal species of the malaria parasites, and a different gene locus, that coding for glucose-6-phosphate-dehydrogenase, has a strong effect on resistance against other malaria species. Resistance of the mosquito vector against malaria is coded for by a single, dominant gene. Other examples include the resistance of mosquitoes against some nematode species and the resistance of the laboratory mouse against the measles virus. On the other hand, resistance of the snail *Biomphalaria glabrata* against its trematode parasites and resistance of wheat against the attack by the Hessian fly are under the control of several genes. Such multigenic resistance has repeatedly been found in other plant defenses against herbivores.

In some cases, resistance appears to be general, so that resistant individuals are resistant against all genotypes of the parasite species, for example, the resistance of humans and of mosquitoes against malaria parasites. The more common case seems to be genotype-specific resistance: a given individual of the snail *Biomphalaria glabrata* is resistant to only some genotypes of trematodes, whereas other snails may be susceptible to these but resistant to other genotypes.

A particular case of such genotype-specific resistance, the gene-for-gene hypothesis was first proposed by H. H. Flor in 1956. The standard gene-for-gene hypothesis proposes first that resistance against a given parasite and virulence toward a given host are each under the control of a single gene; second, that for each gene controlling resistance in the host there is a specific gene controlling virulence in the parasite; third, that resistance in the host is dominant to susceptibility; and fourth, that virulence in the parasite is recessive to aviru-

TABLE II
Examples of Genotypic Variation in Virulence and Resistance Characteristics in Animals

Host	Parasite	Characteristic
Snail (*Biomphalaria glabrata*)	Trematode (*Schistosoma mansoni*)	Susceptibility of host under genetic control, with a few genes controlling resistance
Bumblebee (*Bombus terrestris*)	Trypanosome (*Crithidia bombi*)	Bee colonies within a population vary in susceptibility
Pea aphid	Fungus (*Pandora neoaphidus*) Parasitic wasps (*Aphidius evi*)	Variation in susceptibility among aphid clones; negative correlation in resistance against fungus vs. resistance against wasps
Sheep	Nematodes	ADA (adenosine-deaminase)- associated immunocompetence
Mosquitoes (*Aedes, Anopheles*)	Nematodes, Malaria agent (*Plasmodium*)	Host resistance coded by a dominant gene
Laboratory mouse	Measles virus	Host resistance coded by a dominant gene
Laboratory mouse	Leishmaniosis (*Leishmania donovani*)	Acquired resistance under recessive genetic control
Humans, domestic animals	Toxoplasmosis (*Toxoplasma gondii*)	Small number of virulent strains within parasite population
Humans	Malaria	A single gene in sickle cell anemia

lence (Table III). This mode of inheritance of resistance and virulence is common in agricultural systems, for example, for rusts on flax, smuts on wheat, nematodes on potato, insects on wheat, and bacteria on legumes. However, its applicability to natural systems of hosts and parasites has to date been tested only rarely. A practical implication of this mode of inheritance is that plant breeders have been able to counter a new pathogen type by selecting single or a few genes in host plants. Unfortunately, relying on single genes for resistance can lead to rapid counteradaptation by the parasites, so that resistance breaks down. Plant breeders have begun to move away from this approach by trying to select for mechanisms of resistance determined by a combination of several genes (polygenic resistance).

Whereas these cases are within the realm of classical genetics, it was discovered in the late 1950s that plasmids play a special role in the genetics of antibiotic resistance in bacteria. Plasmids are autonomously replicating, circular DNA molecules residing inside bacteria. They are capable of carrying resistance genes in addition to genes that enable them to be transferred to other bacteria through cytoplasmic bridges. Hence, plasmids act as transfer factors that can quickly spread resistance in unorthodox ways through the host population. Since then, it has been found that transfer factors can also account for instances of host resistance against viral bacteriophages. The picture is complicated by the fact that resistance genes can transpose within the genome, that is, shift to a different place, and thus become accessible to other transfer factors that may spread resistance to a different set of hosts.

The foregoing examples consider the genotypic variability of a single component of resistance or virulence, for example, the potential of the parasite to become established on its host, the growth rate of parasites within their host, or the lethality of the parasites to their host. What is only rarely studied, however, is how these individual components are related to one another. In the few examples where their relationship has been studied, genotypic correlations among the components, have been found. Lines of trematodes in the beetle *Tribolium,* for example, can be selected for higher lethality, with the correlated response that infectivity to hosts is reduced. A similar phenomenon is that resistance to different parasite species may be genetically associated with one another. Resistance of the pea aphid, for example, against the fungus *Pandora neoaphidus* and the parasitic wasp *Aphidius evi* are negatively correlated, so that as resistance against the fungus increases, so does susceptibility against the wasp. New techniques of molecular typing, for example, the polymerase chain reaction (PCR), will allow the analysis of genetic variation in unprecedented detail, and will certainly lead to dramatic change in the study of the ecology and genetics of virulence and resistance in the future.

IV. COEVOLUTION

That resistance and virulence have a genotypic basis sets the stage for various aspects of the evolution of hosts and parasites. Moreover, parasites depend entirely on their host, whereas the host may still survive and reproduce even when parasitized. Host–parasite interactions are therefore characterized by asymmetric selection imposed on the two parties as well as by conflicts of interest. This makes it likely that continuous and rapid evolution in virulence and resistance occurs, and that a common history of change leading to mutual adaptations (coevolution) can be found.

A common pattern of coevolutionary dynamics, namely, frequency-dependent selection of hosts

TABLE III

The Pattern of Compatibility in a Single-Locus Gene-for-Gene Interaction[a]

Pathogen genotype	Host genotype		
	AA	Aa	aa
BB	Incompatible	Incompatible	Compatible
Bb	Incompatible	Incompatible	Compatible
bb	Compatible	Compatible	Compatible

[a] Compatibility refers to establishment and growth of the pathogen on a host. A = a dominant host gene conferring resistance to the pathogen; a = a recessive gene conferring susceptibility; B = a dominant gene conferring avirulence; and b = recessive gene for virulence.

and parasites, was described by the British biologist J. B. S. Haldane around 1950. Haldane, stimulated by observations on the variability of resistance in wheat cultivars against fungal infections, reasoned that the fitness of a genotype depends on the relative frequency of antagonistic genotypes in the population. Because pathogens are typically more numerous and have shorter generation times than their hosts, they can adapt rapidly to common host genotypes, so that the fitness of a host genotype decreases as it becomes more frequent in a population. Thus, initially a resistant mutant of the host will be successful, but as it becomes more frequent, many parasites will be able to overcome its resistance and render it susceptible. In this type of situation, the gene frequencies in host and parasite constantly change as parasites select against common host genotypes and favor rare ones. Because rare host genes have an advantage, pathogen-mediated selection can prevent their disappearance from the population and therefore genetic variability in a host population can be maintained, as Haldane envisioned for the biochemical diversity of the processes that convey host resistance against pathogens. He was the first to connect asymmetric selection, genetic polymorphism, and rapid evolution in a modern way. This view may help to explain many instances of polymorphism, for example, why the part of the mammal genome called the "major histocompatibility complex," which codes for proteins involved in immune reactions, is typically characterized by high degrees of polymorphism. Similarly, many different strains of a pathogen can be maintained in the same host population, as the example of viruses in humans demonstrates.

Explicit modeling of host–parasite coevolution, mostly based on gene-for-gene interactions, started in the 1950s and 1960s. Because such models involve nonlinear relationships, earlier attempts were limited by the necessity to perform a large number of numerical calculations. With the rapid development of computers and their increasing speed, modeling efforts resumed in the late 1970s and 1980s. Such models show that various outcomes of the coevolutionary process are possible, depending on the exact details of the genetic system, on the costs of resistance and virulence, or on the

time delays in frequency-dependent selection. In the simplest cases, gene frequencies fluctuate in regular oscillations, but in more realistic models they fluctuate irregularly over time, today recognized as instances of deterministic chaos (Fig. 4). As a spatial consequence, it is expected that local adaptations should occur (see Table II).

Because of this "cyclical" selection, pathogens create a continuously deteriorating environment for the host, putting the host under constant pressure to "keep running" and "escape" its enemies. This process has been named the "Red Queen" metaphor by L. Van Valen in 1973, after the charac-

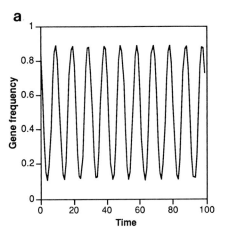

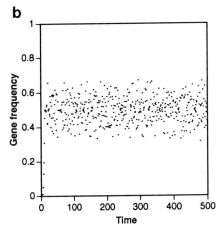

FIGURE 4 The trajectory of gene frequencies in the host (resistance) in model analyses of host–parasite coevolution. Frequencies are between 0 and 1 (100%). In the simplest cases, the dynamics settles to a stable cycle (a). More realistic models produce chaotic fluctuations in the frequencies (b) that can maintain genetic diversity for long periods of time. Genes at low frequencies are "protected" from extinction because they have a fitness advantage under negative frequency-dependent selection.

ter in Lewis Carroll's novel "Through the Looking-Glass" who must keep on running to stay in the same place. A very effective way for the host to escape a continuously deteriorating environment is with sexual reproduction. Therefore, as repeatedly emphasized by W. D. Hamilton, parasites can provide the selection pressure required for the maintenance of sexual reproduction. The problem with sexual reproduction is that, because of the necessity to invest energy in males or in male function, a sexual female can propagate her genes at only half the rate of an asexual female. Thus, a strong selection pressure is required so that sexual reproduction is maintained. Following Hamilton's and other people's ideas, many scientists today believe that this selection pressure is maintained by the coevolutionary dynamics of a host with its parasites. Several questions, however, remain open about the role of parasites in the maintenance of sexual reproduction in the host. First, there is little evidence for the required conditions of frequency dependence in natural populations, though it has been demonstrated in the short-lived perennial grass *Anthoxanthum odoratum*. Second, few studies have yet been able to demonstrate cyclical selection by pathogens in natural populations of hosts. But pathogens could also help to maintain sexual reproduction in another way: populations are often genetically structured because of their social organization, leading to more intense interactions with sibs than with unrelated individuals. Among the clearest examples are colonies of social insects like bees and ants, groups of social carnivores like lions, groups of primates, and human families. Because of the close interaction within groups, pathogens are often more likely to be transmitted within than among groups. This can be seen in the trypanosome *Crithidia bombi* infecting colonies of bumblebees and the common cold and smallpox in human populations. The more closely related the individuals within groups, the more intense transmission within groups is expected to be. Therefore, a mother can profit by reproducing sexually, thus diversifying and reducing transmission among her offspring. Some indirect evidence for this advantage of sexual reproduction is found in plants, where recombination is highest in those species whose seeds are aggregated within fruits and therefore disperse as groups of sibs.

V. ECOLOGY AND EPIDEMIOLOGY

The study of the ecology of host–parasite interactions has been extensive in the fields of biological control and epidemiology of human diseases. These analyses have normally assumed virulence and resistance to be fixed within the ecological time scale and thus not to evolve as a result of the ecological interaction. There is an increasing awareness, however, that the details of the interaction and their microevolutionary changes matter. For example, the dynamics of susceptible populations of the plant *Silene alba* is demonstrably different from that of resistant populations, when an anther smut disease is present. Furthermore, ecologists have also become aware that the genetical structure of host and parasite populations is important for predicting the course of the interaction. Quite generally, a major effect of variation is that the interaction is stabilized, provided the variation is not too large (too few hosts killed for regulation) or too small (too few hosts escape and can reproduce). To understand the population dynamics of hosts and parasites it is therefore often necessary to consider the distribution of parasites within and among hosts. Genetic markers are an invaluable tool for this work. For example, isozyme analyses have demonstrated that several fungal strains may reside in a single lesion of the host plant, or that single parasite genotypes may be spreading over distances of tens of meters and infecting neighboring host trees.

The ecology if the interaction between microparasites and their hosts has been analyzed in mathematical terms ever since D. Bernoulli in 1760 devised the first model of the epidemiology of smallpox. Epidemiological analysis is based on the "mass action principle" introduced by W. H. Hamer and R. Ross early in this century, that is, the assumption that the net rate of spread of the infection is proportional to the product of the density of susceptible hosts times the density of infected hosts. This was extended by W. O. Kermack and A. G. McKendrick, who also postulated the existence of an epidemic threshold, that is, the minimum necessary host population size that will allow a disease to spread and be maintained. Such models suggest that, depending on virulence and resistance properties, the ecological dynamics of the host–pathogen interaction can assume any of several outcomes: stable equilibrium points, stable cycles, or chaotic population fluctuations. Again, the mathematical analysis requires sufficient computing power and is therefore expected to expand considerably in the near future.

In parallel to this theoretical work, experiments since the early work of T. Park on microsporidia infecting beetles in the 1940s have repeatedly demonstrated that parasites and pathogens are indeed able to regulate host populations in laboratory settings. Empirical evidence suggests that in some cases parasites may be able to inflict substantial additive mortality rates on natural host populations and affect their dynamics, as in seals, ungulates in the Serengeti, red grouse, or water mites. Parasites can also affect the outcome of competition between species and affect their geographical distribution, as suggested for malaria in monkeys, nematodes infecting deer and moose, brucellosis in elks and bison in North America, or rinderpest in African bovines. But it is not yet clear to what extent parasitism in the wild is an important ecological process rather than being an occasional, compensatory source of mortality. Hence, virulence and resistance may very well be under strong selection and evolve rapidly, but the ecological consequences in the wild might be limited.

Nevertheless, virulence and resistance are central for applied ecology and should be considered when designing, for example, biological control or immunization programs. The theoretical framework for both of these questions can be mainly attributed to R. M. Anderson and R. M. May. The underlying concept is based on the threshold density needed to maintain the pathogen. For a directly transmitted disease, this threshold density, N_T, can be calculated from the condition that the basic reproductive rate must be greater than one (see earlier):

$$N_T = \frac{\alpha + \mu + v}{\beta}.$$

Thus, virulent parasites are associated with large threshold densities of their hosts: virulent pathogens kill their host so fast that a very large host population is needed to perpetuate the infection. Estimates for this threshold are available for several human diseases. For measles this threshold is on the order of 500,000 people.

The broad objective of biological control is to reduce the abundance of a pest with its pathogens. Mathematical models show that this is possible only if the pathogen is sufficiently virulent. In particular, for directly transmitted pathogens of insects, which have no acquired immunity, the pathogen must kill its host at a rate that is larger than the recovery rate of the host, that is, $\alpha > v$. If this condition is not satisfied, the host population continues to grow, albeit at a diminished rate, until other regulatory processes limit its growth. But even when the condition is fulfilled, a pathogen cannot in general eradicate its host; once the host is driven to sufficiently low densities, it will be below the threshold for maintenance of the pathogen. This leads to the question: What is the degree of virulence that leads to the largest depression of the host's abundance? One can show that the lowest density of a host can be achieved with intermediate virulence. Too low a virulence will have no effect on the host, too high a virulence leads to a high threshold, so that regulated populations still have a high density. This result contrasts the implicit assumption of many control programs that the most virulent pathogens are the best. Although this may be true for programs in which pathogens are repeatedly released, it is not true for programs that aim at a self-perpetuating system.

The final goal of an immunization program is to eradicate the pathogen. Models show that, for directly transmitted pathogens, this is possible if a proportion, p, of the host population is immunized, provided

$$p > 1 - \frac{N_T}{N}.$$

This follows from the observation that, in such a population, the number of susceptibles is reduced from N to $N(1-p)$. The condition that the density of susceptibles must be smaller than the threshold density gives the equation. This requirement for eradication of a pathogen is rather stringent. Even pathogens with low intensities of transmission, for example, scarlet fever or mumps in various localities in the United States, require that about 80% of the population become vaccinated, whereas pathogens with more intense transmission, for example, measles or whooping cough, can require that more than 90% of a population be vaccinated. Clearly, such proportions are not realistic in many areas of the world. Even more disturbing is that, because p depends on population size (N), the increasing level of urbanization will require that an ever-increasing proportion of the human population be vaccinated so that today's level of protection is maintained.

Such studies on the ecology of host–parasite systems do not consider the potential of the parasites to evolve other degrees of virulence or of the hosts to change their resistance. From work on other aspects of the interactions between hosts and parasites, however, it is clear that we should expect such evolution (see earlier). For example, it can be expected that as urban populations grow, pathogens will exploit this increase in number of hosts by increasing their virulence. To resolve the question of how such evolution affects our predictions for the outcome of control programs will be a major challenge for ecologists, population geneticists, and evolutionary biologists alike.

Glossary

Epidemiology Study of host–parasite dynamics. Epidemiology seeks to understand the spread of diseases and to devise rational measures for control. It is based on processes traditionally investigated in ecology and population biology.

Frequency-dependent selection Selection regime in which the strength of selection depends on the relative frequency of the genotype in the population. Parasites are assumed to exert negative frequency-dependent selection, because they can quickly adapt to their hosts and thus select against common genotypes.

Gene-for-gene interaction In this interaction, the host is assumed to possess specific genes for resistance that

match specific genes for the virulence of the parasite in a one-to-one fashion. This contrasts with polygenic and quantitiative inheritance, where virulence and resistance are determined by many genes, each with a small contribution.

Resistance Result of processes in the host that mitigate the effects of parasitic infection, usually measured in terms of host survival or fecundity. Resistance is a consequence of host–parasite interaction but characterizes the ability of the host to withstand parasitism.

Transmission Transmission occurs when a parasite propagule passes from one host to the next. Transmission can be direct, for example, through contact among hosts, or indirect, for example, by means of a vector such as biting fly. It is a crucial element in the life cycle of the parasite.

Variation, genotypic, phenotypic Variation is a statistical measure for the dispersion of values around a mean (e.g., standard deviation, standard error). Phenotypic variation is observed when traits, such as body size, vary among individuals of a population. Genotypic variation accounts for the heritable plus in some cases non-heritable (e.g., dominance and epistasis) component of phenotypic variation, that is, it relates to variation in the genetic information and its interactions present in individuals of a population. Genotypic variation can be made visible by techniques such as enzyme electrophoresis and DNA sequencing, or determined through statsistical methods as used in quantitative genetics.

Virulence Result of processes associated with parasitic infection, establishment, growth, development, and transmission, which damage the host. Virulence is usually measured in terms of host survival and fecundity, but occasionally also as rate of replication of the pathogen. It is associated with the host–parasite interaction but characterizes the negative effect of the parasite on the host.

Bibliography

Anderson, R. M., and May, R. M. (1991). "Infectious Diseases of Humans." Oxford: Oxford University Press.
Ewald, P. W. (1993). The evolution of virulence. *Sci. Amer.* **268** (4), 56–62.
Rollinson, D., and Anderson, R. M., eds. (1985). "Ecology and Genetics of Host–Parasite Interaction." London: Academic Press.
Thompson, J. N., and Burdon, J. J. (1992). Gene-for-gene coevolution between plants and parasites. *Nature (London)* **360,** 121–125.
Wakelin, D., and Blackwell, J., eds. (1988). "Genetics of Resistance to Bacterial and Parasitic Infection." London: Taylor & Francis.
Williams, G. C., and Nesse, R. M. (1991). The dawn of Darwinian medicine. *Q. Rev. Biol.* **66,** (1), 1–22.

Ecology of Mutualism

John F. Addicott
University of Alberta

I. INTRODUCTION

Mutualism occurs when the expected lifetime reproductive success of individuals in each of two or more species increases as a result of their interaction. It is a reciprocally beneficial interaction. This contrasts with exploitation (predation, parasitism, parasitoids, diseases, etc.) in which individuals of one species are harmed by their interaction while individuals of the other species benefit, competition in which both species are harmed, and facilitation (= commensalism) in which only one species benefits but the other is not affected.

The dispersal of seeds by ants is an example of mutualism. Many species of violets (*Viola* spp.) produce seeds to which is attached an elaiosome, a structure rich in lipids. Ants collect seeds at the parent plant, return the seeds to their nest, remove and consume elaiosomes, and discard the seeds at the nest site. This interaction is usually mutually beneficial. Ants obtain an important source of nutrition and violets may experience lower rates of predation by rodents, higher germination or growth success on the nutrient-rich ant nests, or

decreased competition from other plants. There is a cost, however, to the production of elaiosomes, and in the absence of ants, plants producing elaiosomes would be at a disadvantage relative to plants that did not.

Seed dispersal by ants is just one of many kinds of mutualism. Other examples include: pollination of plants by animals; nitrogen fixation by bacteria associated with leguminous plants; ant–plant interactions (e.g., acacias and acacia ants); flagellate–termite associations; lichens; cleaning symbioses (e.g., tick birds and large mammals); endozoic algae and invertebrates (e.g., "green" *Hydra*); invertebrate "commensals" (e.g., hermit crabs and sea anemones); damsel fish and anemones; endophytic fungi and grasses; mycorrhizal fungi and flowering plants; gut bacteria of ruminants; luminescent bacteria of fish; associations of ants and homoptera; phoretic mites and burying beetles; ambrosia beetles and fungi; ant (or termite) fungus gardens; intracellular bacteria in the guts of homoptera; mutual metabolic dependencies of microbes; and intracellular bacteria of tube worms at hydrothermal vents. There is frequently an asymmetry of size and mobility between mutualistic species, and it is therefore often convenient to refer to mutualists as hosts and visitors.

Identifying interactions as actually or potentially mutualistic is an important process, particularly because many interactions once thought to be strictly parasitic have been shown subsequently to be mutually beneficial. But simple classification tends to impose artificial boundaries on the set of continua of directions and intensities of interactions among species, thereby obscuring some very interesting ecological and evolutionary processes, such as (1) interactions between individuals of two species can shift along these continua depending on the biotic and abiotic environment at a particular place and time (see Section III,A); (2) not all individuals of a mutualistic species will necessarily interact with their mutualistic partners in the same way (see Section VII); (3) species may interact mutualistically with respect to one set of traits but not another, or at one life-history state but not another; and (4) interactions can shift along these continua in the course of evolution (see Section I,A). [See PARASITISM, ECOLOGY.]

Interactions vary along many other continua besides the degree of benefit, including the degree of association (symbiotic/casual), permanence of association (permanent/temporary), specificity of association (specialized/generalized) (see Section I,A), necessity of the interaction (obligate/facultative), and mechanism of the interaction (see Section II).

A. Mutualism versus Symbiosis

The concepts of symbiosis and mutualism are distinct. Symbiosis is the relatively long-term association of individuals in which one individual lives on or in another. Symbiotic interactions span the full range from parasitic, exploitative interactions to reciprocally beneficial interactions, whereas mutualism spans the full range from permanent, obligate symbioses to those in which there is never any physical contact between individuals of different species. An example of a symbiotic mutualism is the association between the dinoflagellate *Symbiodinium microadriaticum* and reef-building corals. The

dinoflagellates live within the cells of corals. The nitrogenous wastes of the corals benefit the dinoflagellates and the photosynthates of the dinoflagellates benefit the corals. Most pollination systems are only weakly symbiotic mutualisms, because pollinators, such as hummingbirds, visit plants for only a matter of seconds. There are, however, some pollination systems, such as those between figs and fig wasps or between yuccas and yucca moths, in which all or most of the insect's larval/pupal life is spent symbiotically within the host plant's tissues. Some mutualisms involve no physical contact between individuals of the two species, such as in multispecies foraging associations of birds in which individuals benefit through increased foraging efficiency or increased predator detection rates. [See PLANT–ANIMAL INTERACTIONS.]

B. Mutualism versus Cooperation

Mutualism and cooperation are terms that have been used interchangeably, but this can lead toward thinking of mutually beneficial interactions as being altruistic, with one organism "helping" another. However, for many kinds of mutualism it is much more appropriate to think of mutualism as mutually beneficial exploitation, in which individuals of each species act in their own self-interest. Although there is selection for enhancement of traits or behaviors that facilitate an interaction between mutualists, there is also ample evidence that individuals attempt to increase their own fitness, even though this may cause a decrease in the fitness of their potential mutualistic partner. For example, many animal-pollinated plants produce floral nectar, which attracts birds, bats, bees, moths, or other animals to visit the flowers. The structure of the corolla usually protects the nectar against exploitation by casual visitors, and forces appropriate visitors to contact the anthers and stamens as they attempt to collect nectar. However, some visitors, such as carpenter bees, routinely bypass the normal entry to the flower, bite a hole at the base of the flower, and obtain nectar without contacting the anthers or stamens (see Section VII). Once this occurs, appropriate visitors, such as hummingbirds or bumblebees, may also collect nectar at the base of the flower, thereby failing to collect or transfer pollen. This shift in behavior is more consistent with visitors acting as optimal foragers than with visitors acting to help or cooperate with their mutualistic partners.

C. Measuring Benefits

In theory, measuring mutualistic benefits is easy. One must simply determine whether individuals that are associated with potential mutualists produce on average more offspring than do unassociated individuals. Alternatively, for nonsymbiotic interactions, if the host species were removed or reduced in density, would the visitor density decrease, would its per capita rate of population growth decline, or would the range of habitats occupied shrink? In practice, measuring mutualistic effects is difficult, and there is no single protocol that works for all systems. In very few cases has there been experimental verification that both species in a putative mutualism actually benefit. In most cases it is simply assumed that mutual benefit occurs, based on the nature of the interaction. For example, the benefits to ants from the various interactions with homopterous insects (e.g., aphids and membracids) or plants have almost never been measured.

Expected lifetime reproductive success may be difficult or impossible to measure, particularly in symbiotic mutualisms, and as a consequence, correlates of fitness are measured, such as rate of carbon fixation, nutrient transfer rates, or growth rates. This accounts for the physiological approach to mutualism adopted by most microbial ecologists, in contrast to the population biology approach usually adopted by plant and animal ecologists.

Obligate symbiotic systems are those in which each partner requires the other to survive or reproduce. These would appear to be mutualistic by definition. For example, leaf-cutter ants cultivate fungi, which in the absence of leaf-cutter ants quickly succumb to predators or competitors, and which are rarely if ever found outside of their association with leaf-cutter ants. But, does dependence equate to benefit? There is no satisfactory answer to this question.

II. DIVERSITY OF MUTUALISTIC SYSTEMS

Mutualism, like exploitation and competition, arises from many qualitatively different kinds of interactions among organisms. When considering other ecological processes, such as population dynamics, evolution, or nutrient cycling, it is therefore dangerous to group all mutually beneficial interactions under a single, unified concept of mutualism. This is analogous to exploitative interactions, in which there are very different population dynamics for host/parasite, predator/prey, host/parasitoid, and plant/grazer systems. The diversity of mutualistic systems is illustrated by the qualitatively different ways in which individuals of one species can benefit individuals of another (Table I), whether the benefit is reciprocal (mutualism) or unidirectional (facilitation). The number of pos-

TABLE I

Types of Benefits among Individuals of Different Species in Mutualistic and Facilitative Interactions

Direct benefits
  Nutrient transfer
  Energy transfer
  Provide habitat
  Dispersal of gametes
  Dispersal of individuals to abiotically favorable environments
Indirect benefits
  Modified predator–prey interactions
    Protection from predation
      Feed on predator
      Deter feeding by predator
      Common defense
    Increase prey availability
  Modified competitive interactions
    Decrease competition on mutualist
      Niche differentiation
      Deter competition
    Increase effects on competitor
    Feed on competitor
  Modified mutualistic interactions
    Attract mutualists
      Simultaneously
      Sequentially

sible pairs of types of benefits is large, although not all possible combinations of benefits occur.

There are two classes of benefit, direct and indirect. Direct benefits are those that involve just the species in the mutualism and their abiotic environment. These usually involve at least some degree of symbiosis. By far the most common form of benefit is the direct provision of energy or nutrients by one organism to another, such as plants producing nectar upon which insects, birds, and mammals feed, and mycorrhizal fungi mobilizing inorganic nutrients, particularly phosphorus, for their associated flowering plants. The dispersal of gametes as a mutualistic benefit occurs almost exclusively in pollination systems. One of the benefits for plants of their interaction with fruit/seed dispersers is that seeds are moved to locations that are intrinsically favorable, such as nutrient hot spots. Finally, one organism may simply provide habitat for another, such as scallop shells for sponges or hollow acacia thorns for acacia ants.

Indirect benefits are those that involve at least one other species besides the mutualists in order for the benefit to arise. In the absence of that third species, no benefit is possible. The most common form of indirect benefit occurs through protection from predators, which arises by deterring or feeding on predators. Many of the interactions between ants and plants involve these kinds of benefits, such as acacia ants deterring herbivores from feeding on acacias. Another benefit, usually found only in facilitation (= commensalism), occurs when one species makes prey more readily available, such as predaceous fish driving schools of prey fish to the surface, where marine birds can feed on them. Benefits can also arise through the amelioration of competitive effects, or by increasing the competitive ability of one's partner. Some ant/seed dispersal systems are effective because ants remove seeds to ephemeral habitats, such as fallen logs, where competition intensity is low. In the ant/acacia system, acacia ants effectively increase the competitive ability of their host acacia, because the ants chew on the new growth of plants in the vicinity of acacias.

The benefits that one organism provides another are usually qualitatively different for the host and visitor. For example, plants provide nectar and other energetic rewards to insects and other animals in return for transfer of gametes, and ants provide protection to plants from herbivores in return for energetic rewards. There are, of course, exceptions to this, as occurs in the mutual metabolic dependencies of bacteria, or multispecies foraging groups of birds or mammals. The basic reason for the asymmetry is that an organism with a completely different genome, particularly one in a different phylum or kingdom, is capable of solving "ecological" problems that another organism is not capable of solving on its own. Similar organisms are less likely to be able to provide qualitatively different solutions. In the interactions between ants and plants, the behavioral flexibility of ants can provide behaviorally responsive solutions to many different herbivores and plant competitors, whereas plants can provide only structural or chemical defenses.

A. Intraspecific Beneficial Interactions

Mutualism and beneficial intraspecific interactions are separate but related concepts. At low population densities, the most common beneficial intraspecific interaction involves grouping (flocks, schools, etc.), which results in more effective predator defense, predator detection, or simple dilution of the impact of predation (the selfish herd principle). Alternatively, groupings may increase the success of foraging, as occurs in many fish and mammals, or increase the probability of finding mates. These beneficial effects of increased density at low density are referred to as Allee effects. Many mutualisms are relatively simple multispecies extensions of Allee effects. For example, multispecies schools of tropical fishes often occur as a means of overcoming defense of resources by territorial fishes. Such schools are likely to be effective whether they are composed of single or many species. Other multispecies associations are based on more effective predator detection, for example, where one species has a particularly effective sense of sight and the other an effective sense of smell.

B. Specificity and Multiple Interactions

Direct interactions among two individuals of different species (e.g., pollination systems, or mycorrhizal fungi and flowering plants) can be sufficient for mutualism to arise, and the interaction can be highly specific. This is the case in the pollination interaction between figs and fig wasps in which each species of fig is pollinated by a single species of fig wasp, and each species of fig wasp pollinates only a single species of fig. But simple one-on-one mutualisms are relatively rare, even when benefits are direct. Most pollination systems are less specific than the fig/fig wasp interaction. For example, a given species of hummingbird will visit many species of plants.

Mutualisms based on indirect benefits are usually more complex, because a third species must be present for the interaction to occur. In the absence of herbivores, for example, there may be no benefit to plants with extrafloral nectaries (EFN) from being tended by ants. Complexity may result from a hierarchy of interactions. For example, termites have symbiotic gut flagellates, which in turn have symbiotic bacteria that allow the termites to survive on a diet of wood.

Other mutualistic systems become complex because they involve two or more mutualistic interactions. For example, the ant–acacia interaction in Central America includes ants, acacias, insect and mammalian herbivores, plants competing with acacias, fire, birds, and capuchin monkeys. Acacia ants protect acacias from a variety of herbivores. They also form an area free of other vegetation around their acacia tree, which decreases competition with other plants for light, water, and soil nutrients, as well as minimizing the impact of fire on acacias (and their ants!). Thus, the interaction involves a minimum of four biotic components and one abiotic component. The system is even more complex, however, because acacia trees with their potent ants facilitate the successful nesting of birds in the face of nest predation by capuchin monkeys.

When there are multiple interactions in one mutualistic system, there is the potential for one mutualistic interaction to interfere with another. For example, plants that have EFNs scattered throughout the foliage may attract ants that protect the plant from herbivores. However, the same plant may also harbor aphids or other homoptera that could be tended by ants. If the aphids are restricted to a few large colonies, ant foraging could shift to the aphids from the EFNs and thereby decrease the effectiveness of defense against herbivores.

C. Other Forms of Mutualism

I. Advantages of Being Consumed

An intriguing, but controversial, interaction is one in which there is an apparent advantage to plants from being consumed by herbivores. Plants in the genus *Ipomopsis,* when grazed by mammalian herbivores early during the flowering season, have greater lifetime reproductive success than ungrazed individuals. This counterintuitive result has generated considerable debate. Other apparent mutualisms closely resemble human agriculture, in that the mutualistic partner is cultivated and consumed, with other consumers or competitors of the mutualistic partner being excluded. Examples of this include the fungus gardens of leaf-cutter ants and the algal "meadows" of territorial limpets.

2. Food Chain "Mutualisms"

Net positive interactions among species in a community can arise indirectly from strong direct negative interactions (exploitation, competition). These indirect positive effects arise in a variety of ways, each of which depends on particular kinds of strengths and symmetries of interactions among species. An example of a system displaying net positive indirect interactions involves seed-eating rodents and seed-eating ants in desert environments. Although ants and rodents may be competitors for seeds over short time periods, rodents may actually be beneficial for ants over longer time periods. This depends on rodents preferentially foraging on seeds of large-seeded plants, which decreases the competitive effects on the small-seeded plants upon which ants preferentially feed.

III. REGULATION OF MUTUALISTIC SYSTEMS

The benefits of an interaction for an organism often depend on an organism incurring some costs, whether these be metabolic, developmental, or behavioral. Floral nectar can form a significant part of a plant's energy budget, but it is the floral nectar that provides the primary benefit to pollinators for visiting the flowers and leads to the benefits of pollen removal and pollen deposition. Similarly, to achieve dispersal of their seeds into favorable sites for germination, pine trees lose a high proportion of their seeds to the birds that cache their seeds. The costs and benefits of an interaction are conditional on the biotic and abiotic environment within which the interaction occurs, and the costs and benefits may also be regulated by the organisms themselves.

A. Conditionality of Interactions

Interactions take place in an ecological context of the biotic and abiotic environments, and when these environments change in space or time, interactions may shift. If the benefits that individuals receive as a result of an interaction decrease but the costs remain unchanged, then the costs may outweigh the rewards and the interaction may become at least temporarily detrimental. The interactions between mycorrhizal fungi and plants illustrate this phenomenon. Mycorrhizal fungi make soil nutrients, particularly phosphorus, more readily available to the plants, at the cost to the plants of photosynthates transferred to the fungi. Under conditions of low soil nutrients, the interaction is mutually beneficial. However, under high soil nutrients, plants can obtain adequate soil nutrients directly from the soil, and the performance of individuals with mycorrhizae is worse than that of individuals without mycorrhizae. Effectively the interaction shifts from mutualism to parasitism.

Another form of conditionality is density dependence of costs and benefits. In some cases there is positive density dependence, with larger groups of mutualists attracting proportionately more partners. This occurs in some ant–homopteran interactions and is an example of an Allee effect. In other ant–homopteran interactions, there is classical negative density dependence, with larger aphid populations receiving less benefit from the interaction with ants.

B. Regulation of Costs and Benefits

What control do individuals have over the costs and benefits of their interaction with individuals of another species? In some cases there is a linkage between costs and benefits. For example, the costs to ants from interactions with plants with extra floral nectaries or homoptera are basically energetic costs of searching and guarding. There is a minimum level of activity needed to find new homopteran populations or EFN plants, but once they are found, activity costs increase in proportion to benefits.

In other cases the costs to individuals are fixed, presumably in response to average levels of benefits. Most flowers produce nectar regardless of whether pollinators are available, and the same is true of the food bodies used to reward ants, such as the Beltian bodies of *Acacia* or the Müllerian bodies of *Cecropia*. An unusual case involves plants in the genus *Piper*, which produce food bodies only in the presence of tending ants.

In mutualisms that have arisen from parasitic interactions (see Section V), what prevents the visitor from overexploiting the host? Is it restraint on the part of the visitor or does the host regulate the system? In the pollination/seed predation mutualism between yuccas and yucca moths, yucca moths lay their eggs in the yucca pistil. The primary cost to yuccas from the interaction is that yucca moth larvae feed on the seeds that were fertilized by the pollen transferred by adult yucca moths. Because most yuccas are not pollen limited (see Section VI,A), yuccas selectively abscise those flowers with the greatest numbers of yucca moth larvae. This effectively regulates the costs to yuccas.

IV. POPULATION CONSEQUENCES OF MUTUALISM

Theoretical study of the population dynamics of mutualism began in the 1920s and 1930s along with theoretical studies of exploitation and competition. However, models of mutualism did not receive much additional attention until the 1970s, when

Robert May included mutualism in some of his theoretical work. Since then there have been numerous studies of the population dynamics of mutualism. The overriding conclusion of these models is that because mutualism involves the potential for positive feedback, some intrinsic or extrinsic mechanism must exist that either limits the density of one of the mutualists or limits or decreases the benefits that species experience as a result of their interaction.

Limitation of mutualistic benefits arises commonly when mutualism affects just one part of the life history of an organism. For example, successful pollen transfer is but one factor that has the potential of affecting plant population dynamics. Resource limitation of seed production places an upper limit on seed production in many plants, and in many cases successful seedling establishment is only loosely related to seed production. Therefore, plant population dynamics may be quite independent of the efficacy or density of the plant's pollinators. In some cases the rate of population growth might be affected by mutualism, but the final density of the plant population may be set by other factors. Pinyon pine trees dispersed by birds, such as Clark's nutcrackers and pinyon jays, illustrate this phenomenon. Following the last Pleistocene glaciation, pinyon pines revegetated the isolated mountain ranges of the Great Basin in western North America far faster than would have been possible without the birds. It is questionable, however, whether the bird–pine mutualism determines the density of pinyon pine trees in a mature stand.

V. MUTUALISM AND EVOLUTION

A. The Origin of Mutualistic Systems

How does mutualism arise? Given the number of different mutualistic benefits and the vast array of mutualistic systems, we must expect that mutualism has arisen in many different ways. There are, however, two broad classes for the origin of mutualism. In one set, mutualism arises from antagonistic interactions (exploitation, competition) or in a broader context from environments in which antagonistic interactions are prevalent. In the other set, mutualism arises from interactions in environments that are environmentally stressful. Other mutualisms are probably not evolved at all, at least with their present partners. As humans have moved species among continents, we find novel pairs of species interacting mutualistically. Introduced ants may protect plants with EFNs. This emphasizes the fact that mutualisms need not be coevolved systems.

1. Mutualism from Antagonistic Interactions

One of the processes favoring the development of mutualism from exploitation is the evolution of avirulence, in which the severity of the effect of an exploiter on its victim is diminished through evolution. The conditions for this are restrictive, but far from rare. When there is vertical transmission of the exploiter from parent to offspring, or when the interactions are inevitable in the lifetime of individuals, then exploiters that have a decreased effect on their hosts will be more successful than those that are more virulent. This can drive host–parasite (host–disease) systems toward neutral interactions. Whether these systems will cross the boundary from parasitic to mutualistic then depends on chance events: Can the exploiter provide a resource that is in short supply (e.g., a metabolic by-product), increase the efficiency of a service (e.g., pollen transport), or deter predation (e.g., endophytic fungi and grasses) for its host?

Evolution of the victims in exploiter–victim systems may also lead to mutualism. In seed predation systems, seeds that survive breakage and digestion during gut passage are likely to germinate in high-nutrient microsites. Selection for increased seed hardness accompanied by food rewards for the exploiter, such as the pulp around many bird-dispersed fruits, could quickly lead to mutualism. Alternatively, the antipredator defense strategy of mast fruiting can lead eventually to mutualism. If seeds are plentiful in some years but absent in others, and if the seed predators are relatively dependent on these seeds, then the seed predators will be likely to exhibit seed-caching behavior. This occurs in many birds, rodents, and ants. If the seeds are stored in places where they can germinate, as occurs with various pines and Corvidae (e.g., nut-crackers and jays), mutualism can result.

2. Mutualism in Marginal Environments

The other major pattern for the evolution of mutu-alism is for it to arise in the context of marginal or stressful environments. In many environments, major nutrients may be limiting, such as nitrogen or phosphorus, or sources of energy may be lim-iting. Under these circumstances, associations be-tween individuals of different species may arise as forms of mutual metabolic dependency. One of the harshest terrestrial environments are rock faces, which are deficient in water and available nutrients. These surfaces have been effectively colonized by lichens, which are an algal/fungus mutualism.

B. Mutualism and Localization of Benefits

All mechanisms for the evolution of mutualism depend at least in part on the localization of benefits among partners. In the models for cooperation de-veloped by R. Axelrod and W. D. Hamilton and elaborated by others, the key features are partner fidelity or partner choice: individuals must have a high probability of continuing to interact in order for the localization of benefits to exist. In the con-text of the evolution of avirulence in host–parasite systems, vertical transmission of the parasite from parent to offspring effectively achieves partner fi-delity.

On a broader scale, D. S. Wilson has argued that intrademic group selection can lead to increases in net positive interactions among species. The key feature of such interactions is again that the benefits of an interaction are shared not by an entire popula-tion but by a localized subset of the population, known as a trait group. Even though a trait group does not persist from generation to generation, there is still the potential for the evolution of mutu-alistic interactions.

C. Mutualism and Specialization

Mutualistic systems vary from highly specific one-on-one interactions to highly generalized interac-tions in which hosts interact with many visitors and visitors interact with many hosts (see Section II,B). The degree of specificity appears to be largely a function of the origin of the mutualistic systems. In pollination systems, those that evolved from host–parasite interactions, such as the figs and fig wasps or the yuccas and yucca moths, are relatively specific. Those that evolved from more "graz-erlike" exploiters, such the hummingbirds or bees, tend to be much more generalized.

Another factor affecting the specificity of interac-tions is the community context in which an interac-tion occurs. If there is a rich community with many species potentially interacting with the mutualists, it may be difficult or impossible for specificity to arise, simply because both host and visitor are con-tinually interacting with other species, there is low partner fidelity, and the potential for coevolution between species leading to specificity is thereby limited.

Another aspect of specificity is a function of the kind of mutualistic system. Consider the compari-son between pollination systems and seed dispersal systems. Pollination systems should be relatively specific, because it is advantageous for plants to have very directed dispersal of their pollen. Pollen that is transferred to other species of plants is lost and pollen that arrives from other species of plants may be detrimental (see Section VI,A). Alterna-tively, in seed dispersal systems, diversity of seed dispersal mechanisms may be critical. There is un-likely to be a best location for dispersal. Instead, what is important is the broadest possible seed shadow that samples the full range of locally avail-able environmental variation and leads to few areas of high densities of seeds that could be found and exploited by seed predators.

Finally, in the comparison between mutualistic and antagonistic systems, we expect greater speci-ficity among antagonists. The reasons for this are based on assumptions about the success of species as a function of their genetic structure. In antagonistic systems, victims are more likely to be successful if their offspring are different from themselves, making it more difficult for the exploiters to track them. This should lead to relative specificity of host and parasite. Alternatively, mutualists are more likely to be successful if they produce phenotypes similar to themselves; if the interaction worked well in one generation, then perpetuate it in the next.

VI. MUTUALISM IN THE CONTEXT OF COMMUNITIES

A. Competition for Mutualists

The services or resources provided by a host or visitor may be the basis for competition, both intra- and interspecific. Intraspecific competition for mutualists is one of the essential features leading to population stability of mutualistic systems, at least in the short term. Consider, for example, homoptera being tended by ants. As more homoptera colonize host plants surrounding an ant nest, the ability of ants to tend these homoptera may become limiting and ants tend those homoptera closest to their nest. When ants remove other herbivores from plants, "green islands" can occur, in which plants close to ant nests are more effectively protected from herbivores than plants far from ant nests.

Interspecific competition for mutualists may also be important and can lead to nonrandom assemblages of species. In plant/pollinator systems, coexisting plant species are frequently nonrandom with respect to flower size, flowering time, or flowering syndrome.

B. Mutualism for Mutualists

Not all interspecifc interactions among suites of mutualists need be competitive, however. If one host species does not support a population of visitors on its own, then the presence of two or more host species might be able to. This can take the form of simultaneous or sequential beneficial interactions. A population of visitors may be able to stay in a given location, if two species can sequentially provide the necessary resources. For example, in pollination systems, one plant species may provide nectar/pollen resources early in the season and another in the middle of the season. The early species

benefits the later species by attracting and maintaining pollinators in the vicinity. Ironically, those individuals that overlap in flowering may experience competition, either through exploitative or interference mechanisms, but as a whole there is mutualism at the population level. Alternatively, a population of visitors may stay in a given location if two or more host species simultaneously provide sufficient resources even though a single host species could not. This pattern is implicated for populations of plants in western North America that have tubular red flowers and are pollinated by migratory hummingbirds.

C. Keystone Mutualists

Species that have an overriding effect on the structure of a community or the functioning of an ecosystem are called keystone species. Most examples of keystone species are predators that prevent dominant competitors from monopolizing resources. However, there are also keystone mutualists. The strongest case for keystone mutualists comes from tropical forests, in which it is argued that trees producing nectar during periods of nectar scarcity maintain populations of pollinators and seed dispersers that are important for the pollination and dispersal of other species of plants during the remainder of the year.

VII. CHEATING AND TAKING ADVANTAGE

Mutualistic resources (e.g., nectar) or services (e.g., protection) are also subject to exploitation by individuals that do not provide reciprocal benefits. This is known as cheating, and it can arise in two ways. Species that are not involved in the mutualism may consistently fail to provide reciprocal benefits. Some plants consistently fail to provide nectar, but are nonetheless pollinated, because they mimic the floral colors, odors, or shapes of flowers that do provide rewards. Such floral mimicry, as in mimicry in general, depends on mimetic flowers being relatively rare. Other examples occur among both the fig wasps and yucca moths, where there are groups of species closely related to the pollinators that consistently fail to provide pollination services, yet they exploit the fig and yucca systems.

The other kind of cheating involves individuals of species that normally interact mutualistically. For example, some plants may fail to produce nectar or some pollinators may fail to transfer pollen. Again, such behavior cannot be too common without the risk of the system breaking down. In the yucca system, individual yucca moths can fail to transfer pollen even when they are carrying pollen.

VIII. SIGNIFICANCE OF MUTUALISM

The significance of mutualism is still a matter of some speculation and a function of the level at which the question is posed. Mutualism is given very little consideration in comparison to competition, predation, or environmental heterogeneity as an organizing process for communities or in population dynamics. As in the debate about the importance of competition in natural communities, however, the critical question is not whether, but where, when, and why mutualistic interactions play critical roles in the distribution and abundance of organisms.

Mutualism does have a critical role with respect to many ecosystem functions, the occupation of marginal environments, the exploitation of marginal resources, and the diversity of life. For example, nitrogen-fixing symbiotic mutualisms affect the community/ecosystem-level processes of succession. Particularly in primary successional sequences, such as those following glaciation, early-successional plants are often associated with nitrogen-fixing bacteria, which eventually increases soil nitrogen, thereby allowing later-successional species to enter the community.

One of the most important features of mutualism is that it can allow organisms to occupy habitats, or utilize resources, that are marginal with respect to the availability of resources. The algal–coral symbiosis allows the high production and subsequent high diversity that occur in tropical reefs, which exist in the midst of relatively low productivity environments. Many resources are difficult to utilize, either because they are hard to break down (e.g., cellulose) or because they are low in nutrients (e.g., plant sap). Mutualisms are involved in enabling insects or vertebrates to utilize these resources.

With respect to the diversity of life, mutualisms have played pivotal roles in the evolution of the kingdoms. The serial endosymbiotic theory, championed by Lynn Margulis, is essentially a hypothesis of mutualism becoming obligate, leading to diversification of major groups. At another level, the diversity of land plants has been ascribed to the role of insect pollinators as agents of diversification.

Thus, although mutualism has played a minor role in the development of population and community ecology, mutualistic systems exert major influences on ecosystem process and the diversity of life. The challenge for ecologists is to move beyond the description of mutualistic systems and to develop a more comprehensive understanding of the origin and consequences of mutually beneficial interactions among species.

Glossary

Allee effect Positive relationship at low population densities between expected lifetime reproductive success of individuals of one species and its own population density.

Cheating Utilization of mutualistic resources or services by individuals of a normally mutualistic species or by individuals of a nonmutualistic species without providing mutualistic services or resources in return.

Conditional mutualism Mutualistic interaction that varies as a function of the environment, density of the species, age of the individuals, and so on.

Direct benefits Mutualistic benefits between individuals of two species that do not depend on the presence of any other species (e.g., individuals of one species provide food, such as nectar, to individuals of another species).

Facilitation (= commensalism) Interaction among species that is beneficial for the individuals of one species, while individuals of the other species are neither harmed nor benefited.

Food chain mutualism Indirect positive association among species that arises from strong exploitative or competitive interactions with other species in a community

Indirect benefits Mutualistic benefits between individuals of two species that arise only in the presence of a third species (e.g., individuals of one species deter predators from feeding on individuals of another species).

Keystone mutualists Mutualistic species whose removal from a community has an extraordinarily large effect on community structure and function.

Mutualism Interaction among species that is beneficial for the individuals of each species, where benefit is measured by the expected lifetime reproductive success of individuals with and without the other species.

Symbiosis Individuals of different species living together in close association.

Bibliography

Addicott, J. F. (1984). Mutualistic interactions in population and community processes. *In* "A New Ecology: Novel Approaches to Interactive Systems" (P. W. Price, C. N. Slobodchikoff, and B. S. Gaud, eds.), pp. 437–455. New York: John Wiley & Sons.

Addicott, J. F. (1985). On the population consequences of mutualism. *In* "Community Ecology" (T. J. Case and J. Diamond, eds.), pp. 425–436. New York: Harper & Row.

Boucher, D. H., ed. (1985). "The Biology of Mutualism: Ecology and Evolution." London: Croom Helm.

Boucher, D. H., James, S., and Keeler, K. H. (1982). The ecology of mutualism. *Annu. Rev. Ecol. Systemat.* **13,** 315–347.

Bronstein, J. L. (1994). Our current understanding of mutualism. *Quart. Rev. Biol.* **69**(1), 31–51.

Cushman, J. H., and Beattie, A. J. (1991). Mutualisms: Assessing the benefits to hosts and visitors. *Trends in Ecol. Evol.* **6**(6), 193–195.

Douglas, A. E., and Smith, D. C. (1989). Are endosymbioses mutualistic? *Trends in Ecol. Evol.* **4**(11), 350–352.

Margulis, L. (1992). "Symbiosis in Cell Evolution," 2nd ed. San Francisco: W. H. Freeman.

Margulis, L., and Fester, R., eds. (1991). "Symbiosis as a Source of Evolutionary Innovation." Cambridge, Mass.: The MIT Press.

Smith, D. C., and Douglas, A. E. (1987). "The Biology of Symbiosis." London: Edward Arnold.

Thompson, J. N. (1982). "Interaction and Coevolution." New York: John Wiley & Sons.

Ecology of Size, Shape, and Age Structure in Animals

W. A. Calder

University of Arizona

I. Introduction
II. Scaling Fundamentals
III. Animal Requirements
IV. Toxicity, Drug Dosages, and Turnover
V. Population Biology
VI. Community Structure
VII. Implications for Monitoring
VIII. Shape
IX. Age Structure
X. Conclusion

Many of the million known animal species are facing habitat loss through degradation and fragmentation, overexploitation, and/or pollution, with some showing serious population declines. For most, we lack the natural history data necessary for prudent conservation, and urgently need ways to approximate, even if only crudely, their requirements and tolerances. Within a class, mammals, birds, reptiles, etc., the quantitative details are set more by size than by any other characteristic. Hence size can be a powerful starting point for analyzing the lives of animals.

I. INTRODUCTION

As environmental problems ranging from pollution to depletion of biological diversity and other resources become increasingly apparent, the biological knowledge of animal species becomes crucial, both for preservation of each species per se

(e.g., snow leopard) and for use as indicator species representing the health of an ecosystem (e.g., northern spotted owl.). With basic life history information far from complete for most species, we must utilize those characteristics of species that are available and capable of providing a preliminary picture of how animals live and what their needs may be. Of all characteristics of a species, body size is one of the most influential in quantitatively affecting physiological and behavioral responses to the physical and biological environment, needs for food and space, reproduction, life span, and population–age structure. For example, when compared with a 200-kg "elk" (wapiti, red deer *Cervus elaphas*), a 20-g deer mouse (*Peromyscus maniculatus*) has heart and breathing rates about 10 times as fast, eats 10% of its body mass per day (vs 4% for elk), can produce 12 offspring annually (vs 1–2), and lives, on average, one-sixth as long as the elk, occupying an area 1/430 or less of that claimed by an elk.

II. SCALING FUNDAMENTALS

Quantitative expressions of such size dependencies commonly appear in the form of allometric power functions of body mass M:

$$Y = aM^b,$$

where Y is the physiological, morphometric, or ecological variable, e.g., metabolic or exogenous chemical turnover rate, annual fecundity, life expectancy, and susceptibility to anthropogenic perturbations or detection in censusing. The scaling exponent b shows the proportional effect of size, represented visually by the slope of a regression correlating Y and M values on logarithmic scales (Fig. 1). The Y intercept of such a graph is the coefficient a, which is characteristic of the class (or other taxon) being analyzed, e.g., higher for metabolism of mammals and birds than for reptiles and amphibians. b is independent of the units of M used (g, kg), while the numerical value of a depends on units of M.

Usually derived via least-squares regressions of log-transformed data on standard PC spreadsheets with math functions and general statistics (e.g., Lotus 123, Quatro-pro, etc.), these correlations can serve several purposes:

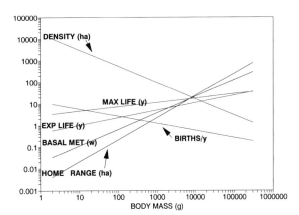

FIGURE I Generalized relationships between body size and six basic ecological or ecophysiological characteristics of eutherian mammals. Maximum density and home range regressions are based on herbivorous (primary consumer) animals only, because carnivores are rarer and require much larger home ranges. Food requirements, drug dosages, and toxic limits are scaled in parallel with basal metabolic rates.

A. Heuristic

Knowing the general trend helps us to identify patterns useful in anticipating requirements and problems. For example, larger animals tend to be fewer and to reproduce more slowly, so they are more prone to extinction than smaller species in the same habitats. Since carnivores are fewer in numbers than their prey species, large carnivores (e.g., snow leopards) are especially vulnerable compared to large herbivores. Allometric relationships represent only empirical descriptions, not cause-and-effect relationships. Nevertheless, persistent scaling patterns are proving increasingly fruitful for explorations of possible theoretical implications and insights into the evolution of life histories. [*See* EVOLUTION AND EXTINCTION.]

B. Baseline

Variability can be quantified as a "departure" from the average size-adjusted value which the regression predicts. When a species exhibits a particularly elevated or depressed value relative to the baseline, this may represent a special adaptation or a phylogenetic legacy which must be taken into consideration in species recovery plans. These "departures" usually occur in suites, so that, for example, low metabolism may be associated with slow reproduction and longer life span.

C. Preliminary Predictions

Log–log plots minimize the visual impact of variability, so apparently neat regressions may give crude predictions at best. However, if the data base includes several orders of magnitude in body size, Y variability can be swamped, yielding strong correlations useful to anticipate ecological relationships until specific data can be obtained. Generalized life history "invariants" or "design constants" can be derived from two independent regressions, e.g., for life span or metabolic rates and annual birth rates to obtain relationships between the two. Since each regression incorporates errors, caution is urged for the uncritical use of ratios so obtained.

III. ANIMAL REQUIREMENTS

Basal metabolism and food requirements (kg/day), energy (kJ/day), and oxygen (liters/day) have been known to scale as the 3/4 power of body mass since the work of Max Kleiber and Samuel Brody in the early 1930s. This means that an increase in body size by 10,000 (10^4; e.g., mouse to elk) would require a 1000-fold increase (10^3) in energy turnover per day. In contrast to metabolism, the increase in spatial requirements (i.e., home range) would be directly proportional to this body size increase (10,000-fold increase).

IV. TOXICITY, DRUG DOSAGES, AND TURNOVER

Toxic or therapeutic dosages, e.g., metabolic poisons, act in proportion to metabolic rates, so should be administered per $kg^{3/4}$, but these have been commonly and erroneously scaled up from experimental animals to humans (or other target species) in units per kilogram of body mass. Suppose that 10 mg of substance "X" produces a certain effect in 20-g mice (0.5 mg/g), and we want to know how much would be needed to produce the same effect on a 70-kg human (3500 times mouse mass). Had the substance been administered at the 0.5-mg/g dosage for the experimental mice, the human subject would have received 3500 times 10 mg = 35 g, over seven times the proper metabolic equivalent ($kg^{3/4}$) dose of 4.55 g [(70,000 g/ 20 g) 0.75 × 10 mg]! The turnover time of an entire body reserve of a substance such as fat or water, or of an administered drug, would scale in proportion to body mass/rate or $M^{1.0}/M^{0.75} = M^{0.25}$.

V. POPULATION BIOLOGY

Smaller mammals (and birds) generally have larger litters (or clutches) and shorter gestation (or incubation) periods, resulting in higher annual fecundity. Mortality rates are also higher and life spans are shorter in small animals. Hence they tend to be quicker to colonize or rebuild populations, and to develop pesticide tolerance. Furthermore, their populations tend to fluctuate more than for larger species.

VI. COMMUNITY STRUCTURE

There are more species of small animals than of large animals, and within each species, the small animals tend to be more abundant, generally occurring at higher population densities than large animals. Larger populations, coupled with more rapid reproduction, make smaller species less prone to extinction. However, the generalizations about more of both species and individuals of smaller sizes seem not to hold at the small extreme for a particular taxonomic class of animals.

VII. IMPLICATIONS FOR MONITORING

Concern for the preservation of biological diversity and ecological function has led to considerable effort in censusing and monitoring populations of sensitive, rare, and indicator species. The reliability of acquired data depends on an understanding of animal characteristics that may be revealed from scaling relationships. For example, techniques for bird censuses usually include detection by songs and calls. The sound power of birds correlates strongly ($r^2 = 0.861$) with body size (M in grams):

$$mW \text{ sound output} = 0.042 \, M^{1.14}$$

This parallels the area requirements of birds, as suggested by the similar scaling ($r^2 = 0.629$):

$$\text{hectares within home range or territory} = 0.33 \, M^{1.17}$$

This means that in a census relying significantly on bird song, larger birds with louder voices claiming larger territories are likely to be detected at greater distances than smaller birds. Counts from linear transects or point counts represent different areas, limiting the validity of comparisons to year-to-year changes on a species-by-species basis. [*See* BIRD COMMUNITIES.]

Diversity is expressed as species richness (total number of species present) or as one of several diversity indices weighted for proportionate contributions. According to the Shannon–Wiener index, for example, diversity is greatest if there are the same number of individuals from each species. However, population density is size dependent, being higher in smaller species (except the very smallest). Hence an even distribution is not to be expected in nature, so measurement of diversity may need methodological rethinking. [*See* SPECIES DIVERSITY.]

VIII. SHAPE

Animal shape appears to reflect adaptations for food habits, type of locomotion, and habitat. A close correlation between size and metabolism of mammals leaves little variation to be accounted for by shape. To a first approximation, the surface areas of mammals (or birds) are simple multiples ($12.3 : 6 = 2.05$ and $12.3 : 4.8 = 2.56$, respectively) of areas of cubes or spheres of the same body volume (Fig. 2). Bodies of mammals, and most other multicellular animals, are cylindrical,

so even before considering appendages, they would be expected to have greater surface areas than of cubes or spheres. The surface areas of long-tailed weasels (*Mustela frenata*), with extended cylindrical bodies, are 15% greater than pack rats (*Neotoma* spp.) of equal body mass. J. H. Brown and R. C. Lasiewski (1972; *Ecology* **53**, 939–943) associated this with standard metabolic rates 21 to 58% higher than the pack rats.

IX. AGE STRUCTURE

For birds and mammals, respectively, average life expectancy increases more with size ($M^{0.46 \text{ to } 0.35}$) than does maximum longevity ($M^{0.14 \text{ to } 0.20}$), so larger species have a better chance of attaining senior citizenship and potential longevity. This provided opportunity to learn from elders, favoring social behavior and structure (herding) and its survival benefits. Another implication of the effects of size on natality and life expectancy is that a larger proportion of the population of small mammals is young, giving a population–age distribution similar to that of a developing country (Fig. 3). Hence the sample of mice or other small animals, drawn from a lab colony for drug or toxicity studies, is more likely to be young, perhaps proving inappro-

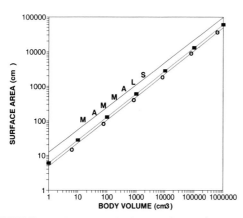

FIGURE 2   Surface areas of spheres, cubes, and mammals increase with body size, in a roughly parallel fashion (which means that size has similar effects, but from progressively greater starting proportions). A cube has 25% more surface than a sphere of the same body volume, whereas a mammal of the same volume has over twice the surface area of the sphere.

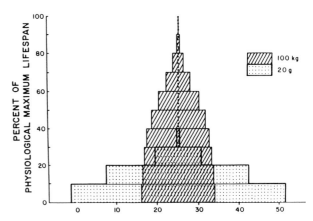

FIGURE 3   Population–age distributions predicted from scalings of maximum longevities and mean life expectancies of a small and a large mammal species show similarity in appearance to those for a developing and a developed country, respectively. This is because of a common underlying basis in natality and mortality rates. [From Calder (1984). "Size, Function, and Life History," with permission.]

priate if the drug is being developed for problems of the elderly.

X. CONCLUSION

The study of biological scaling began as a technique for empirical interspecific comparisons, but is now finding both theoretical and practical applications. Whether an environmental study is related to toxicity, monitoring of population size, or planning for recovery of endangered species or ecosystems, the consequences of body size need to be considered.

Glossary

Allometric Of different measures or proportions, as contrasted with isometric (of same measures or proportions). Pictures of a mouse and an elephant drawn to the same absolute size differ, with the elephant having proportionately thicker legs and body relative to length (L). If thickness of limbs had the same proportions relative to length (isometric; $L_1 = k \, L_2$), an enlarged picture of a mouse would appear similar to that of an elephant. Because volume (V) and mass (M) are products of the cubes of linear dimensions, isometry would also be expressed by a scaling of $M^{0.33}$ for body dimensions or $M^{1.00}$ for volume or mass functions. Metabolic rates, which scale as $M^{0.75}$, are allometrically scaled.

Correlation A statistical expression of the association between two variables, e.g., between body mass ("weight") and territory size. This does not prove that either one depends on the other, as in causation, only indicates that if one is greater, so is the other.

Fecundity Number of offspring born, usually expressed per annum.

Home range Average area of habitat used for a season of residence by an animal; does not include space through or over which the animal migrates.

Least-squares regressions An iterative fit that minimizes the sum of squared deviations of individual data points from the line of best fit.

Life expectancy Mean or average life spans in an animal population, which usually range widely between extremes of juvenile mortality due to predation and disease and to attainment of maximum longevity. Life expectancy often differs between sexes, so is expressed separately.

Linear transects Counts of individuals of one or more species along a standardized route; related to, but does not directly express, population densities.

Maximum longevity Oldest recorded age in a species or population, considered to represent the physiological maximum if based on an adequate data base.

Log-transformation of data Using logarithms of original measurements from a wide body size range makes it possible to see all data points on a reasonably scaled range and tends to convert curved plots to more interpretable straight lines.

Phylogenetic Pertaining to evolutionary relationships and origins.

Point counts A sampling or monitoring technique employing repeated counts from standard points.

Population–age distribution Proportions of a population by age groups (population structure). A growing population is characterized by having a high proportion of juvenile or young individuals, whereas a stable population has a higher proportion of mature and elderly individuals.

Scaling Proportionate or power function increases or decreases accompanying body size differences.

Shannon–Wiener diversity index One of several adjustments of biological species counts for differences in relative abundances, with the intent that a species represented by only one or two individuals with little ecological role is not counted equally with more abundant, and therefore influential, species.

Sound power Rate of sound energy flow at a point 1 m from the emitter.

Territory Any area defended by an animal to the exclusion of others who might compete; usually conspecifics of the same sex.

Turnover time Time in which an amount equal to total stock or reserves is used and replaced.

Bibliography

Calder, W. A. (1984). "Size, Function, and Life History." Harvard University Press.

Charnov, E. L. (1993). "Life History Invariants: Some Explorations of Symmetry in Evolutionary Ecology." Oxford University Press.

Dedrick, R. L. (1986). Interspecies scaling of regional drug delivery. *J. Pharm. Sci.* **75,** 1047–1052.

Demment, M. W., and Van Soest, P. J. (1985). A nutritional explanation for body-size patterns of ruminant and nonruminant herbivores. *Am. Nat.* **125,** 641–672.

Holling, C. S. (1992). Cross-scale morphology, geometry, and dynamics of ecosystems. *Ecol. Monogr.* **62,** 447–502.

Lindstedt, S. L., and Jones, J. H. (1987). Symmorphosis: The concept of optimal design. *In* "New Directions in Ecological Physiology" (M. E. Feder, A. F. Bennett, W. W. Burggren, and R. B. Huey, eds.), pp. 289–304. Cambridge: Cambridge Univ. Press. England.

Maurer, B. A., and Brown, J. H. (1988). Distribution of energy use and biomass among species of North American terrestrial birds. *Ecology* **69,** 1923–1932.

McMahon, T. A., and Bonner, J. T. (1983). "On Size and Life." New York: Scientific American Books.

Peters, R. H. (1983). "The Ecological Implications of Body Size." Cambridge: Cambridge Univ. Press.

Schmidt-Nielsen, K. (1984). "Scaling: Why Is Animal Size so Important?" Cambridge: Cambridge Univ. Press.

Ecophysiology

P. C. Withers
The University of Western Australia

Ecophysiology melds the study of how organisms function at an organ and tissue level (physiology) with the relationships between organisms and their physical and biological environments (ecology). Such study of how both internal and external constraints are balanced to enable organisms to survive and reproduce in their particular environment allows the distribution and abundance of organisms to be correlated with their physiological capabilities and causative inferences to be made, particularly in extreme environments. Temperature, energy, water, and solute balance are often major physiological constraints to the distribution and abundance of organisms.

I. INTRODUCTION

Ecology is largely concerned with how animals and plants interact with their external environment, and physiology largely deals with how animals and plants sustain the differences between the external, extracellular, and intracellular environments. Ecophysiology involves the interactions between the external and extracellular environments. [See PLANT–ANIMAL INTERACTIONS.]

One of the central questions that ecologists seek to answer is how and why organisms are found in their particular habitats. The interests of early ecologists often included the physical factors, such as temperature, rainfall, moisture, light, salinity, pH, nutrients, and oxygen, that might limit the distribution of particular organisms, and this established a close relationship between ecology and physiology. Although modern ecologists tend to focus more on the interactions between individuals (population ecology) or between species (community ecology), there is a current resurgence in the combination of physiology and ecology to explain the distribution, abundance, and habits of organisms.

An important approach for ecophysiological studies in attempting to explain the distribution, abundance, and habits of organisms through their physiology has been a comparative method, that is, to compare individuals of a species in different habitats, or compare related species in different habitats, or compare unrelated species in similar habitats. Any correlated differences in the physiology and ecology of the species are presumed to be adaptive and to provide a mechanistic and causative explanation for both physiological and ecological traits. This fruitful comparative approach to ecophysiology has recently been modified by the incorporation of phylogenetic history and use of general linear regression models. Species generally have not originated explosively from a single common ancestral form, but have arisen along a typically dichotomously branching phylogeny. It is important and necessary to interpret the comparative physiology and ecology of species in the context of their phylogenetic history. Linear regression analysis is an important analytical tool for comparative analyses, particularly allometry (scaling). The normally used (model I) regression analysis is only one subset of a more general structural relations regression model; other subsets are major axis regression and reduced major axis regression. Use of the most appropriate regression model is necessary for the correct biological interpretation of the comparative approach.

II. EXCHANGE PROCESSES AND BUDGETS

Maintenance of the internal environments of organisms (homeostasis) involves exchange, for example, of water and solutes between the environment and the organism (Fig. 1). Exchange occurs across the body surface, either the outer integument or the internal gut tube or respiratory structures. Under steady-state conditions, inward movement is equal and opposite to outward movement, but nonsteady-state conditions of storage or depletion often occur in organisms, at least for short periods of time. The primary exchanges considered for plants are of energy (including light for photosynthesis), water, solutes, and gases (O_2 and CO_2), and those for animals are of energy, heat (including light), water, solutes, and gases (O_2 consumption and CO_2 release). The avenues of gain and loss can be itemized and combined to form a budget, which accounts for all avenues of gain, storage, and loss (Table I). There must be conservation of material

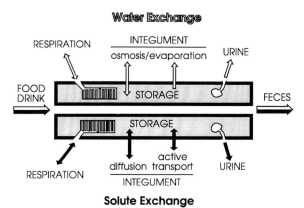

FIGURE I Schematic representation of the routes for exchange between an animal and its environment, using water and solutes as an example. Avenues of intake include food and water, and exchange across the integument and respiratory surface; avenues of loss include urine and feces, and exchange across the integument and respiratory surface. There can be addition to or depletion of body stores.

and energy in biological systems, and so budgets must balance.

III. ENVIRONMENTAL EXTREMES

Numerous physical and biological variables constitute the external environment in which organisms must not only survive, but also reproduce. The ecophysiology of all organisms is affected by the exchange of heat, water, solutes, and energy with

TABLE I

Generalized Budget for Exchange between an Organism and Its Environment, Showing Avenues for Intake, Storage, and Loss

In
Food
Drink
Integument surface
Respiratory surface
Storage
Body fluids/urine
Body depots (e.g., fat pads)
Out
Urine
Feces
Integument surface
Respiratory surface

their environment, but exchange is often a spectacular constraint for organisms in severe environments. These situations illustrate, *in extremis,* mechanisms fundamental to all organisms but not so highly developed in organisms from less severe environments. Sometimes, extreme environments require unique adaptations. Consequently, ecophysiologists have intensively studied organisms from hot and desiccating deserts, cold polar and alpine environments, hypoxic environments, hyperosmotic water bodies, and high-pressure submarine depths.

Desert environments are characterized by extremely high temperatures and low water availability. Temperatures, especially surface soil temperatures, can greatly exceed the lethal thermal limits of most animals and plants, and so special adaptations are required to either avoid or tolerate these extremes. The scarcity of water requires special adaptations to acquire it, to store it, to limit its loss, and to tolerate temporary water imbalances. [*See* DESERTS.]

Polar and alpine environments have extremely low (subfreezing) temperatures, at least seasonally, which require special adaptations of ectothermic animals to avoid freezing or tolerate freezing of some body water, and of endothermic animals to either minimize loss of body heat or hibernate. Water is generally present at least as snow or ice, but liquid water may not be readily available. [*See* NORTHERN POLAR ECOSYSTEMS.]

Hypoxic aquatic environments often occur in the tropics, and at depth in the oceans, because of O_2 depletion by decay in stagnant water. Hypoxic terrestrial environments occur at high altitudes, because of the decrease in atmospheric pressure. Intertidal animals may periodically experience hypoxia because of emersion (if they are water breathers) or immersion (if they are air breathers).

Hyperbaric (high-pressure) environments occur at great depths under water; the pressure increases by 1 atmosphere for every 10 m of water depth. Moderate pressure is a severe physiological problem for animals with an air space, such as a swimbladder, and the very high pressures of abyssal environments (e.g., 1000 atm) exert physiological effects in solutions.

Hyperosmotic bodies of water exceed the osmotic concentration of seawater (about 1100 mOsm liter^{-1}) and can even approach or exceed salt saturation. This is exceedingly hostile to aquatic life because of the double danger of osmotic loss of body water and influx of salts. Drinking is the only significant source of water for these animals, but it incurs an exceptionally high salt load.

IV. WATER AND SOLUTE BALANCE

Water balance is essential to all organisms, as water is the biological solvent for inorganic and organic solutes of their body fluids and is the supporting fluid of the intracellular and extracellular spaces. The water content of organisms is typically 60 to 80% of body mass. It is also important for organisms to maintain solute balance; major solutes include inorganic ions (Na^+, Cl^-, K^+) and organic solutes (amino acids, proteins, and sometimes urea). It is essential for all organisms that their body water content and solute composition be maintained within limits. Only the inactive life stages of a few animals (e.g., cysts of the brine shrimp *Artemia*) and plants (dormant seeds) are able to survive the loss of most body water.

The exchange of water between an aquatic organism and its environment is passive and governed by osmosis. The main sources of water are food and drink, absorption across the body surfaces (skin, gut, respiratory organs), and to a minor extent metabolic water (Fig. 1). Avenues of water loss are the skin, respiratory organs, via feces, and through urine. Solutes can be obtained in the food or by drinking the medium, and absorbed across the gut, or eliminated by the gut in the feces or by specific excretory organs as urine. Solute exchange occurs across the body surfaces and specialized excretory organs. Unlike water, many solutes can be directly transported across epithelia by active transport; solutes can be absorbed from low external concentrations (e.g., 1 μM or less) or excreted into high external concentrations (e.g., 1 M or more). Organisms must remain in steady-state water exchange over extended periods, but may be transiently in positive or negative balance by using body fluids or specific stores as a temporary buffer to water imbalance.

For terrestrial organisms, liquid water is not necessarily present in the environment, and so water gain is often predominated by preformed water (in the food) and metabolic water. A few arthropods are able to absorb water vapor from subsaturated air. Avenues of water loss are predominated by evaporation from the integument or respiratory surfaces, water lost in the feces, and water excreted in urine. The primary avenue for solute gain is the food, and sometimes drink. The body fluids can be a short-term buffer of solute imbalances, and some organs can temporarily sequester or release solutes (e.g., bone, insect fat body).

Drinking is an unnecessary avenue of water intake for freshwater animals and is avoided; passive water gain by osmosis greatly exceeds that required to replenish water lost by elimination and excretion. Drinking is an important source of water for marine animals to replace that lost by elimination and as urine. This has only a minor perturbing effect on solute balance for animals that have similar osmotic and ionic concentrations as seawater, but it is a major perturbation to solute balance for teleost fish and animals that live in hypersaline environments. Terrestrial animals typically drink when free water is available, although amphibians rely on cutaneous water uptake and do not drink.

The body surface is often extremely impermeable to water and solutes and limits passive water and solute exchange. A major role of the integument is active uptake (freshwater animals) or excretion (hypersaline animals) of solutes, particularly NaCl. Specialized integumentary regions with dense aggregations of salt pumps occur in fish and amphibians (gills, skin) and crustaceans (neck gland, phyllopod appendages, gills) to absorb or excrete NaCl.

Most animals have specialized, tubular excretory organs that eliminate solutes in urine. The urine of some animals (e.g., marine crustaceans) has the same osmotic concentration as the body fluids although the specific solute concentrations may differ dramatically, but in many animals it can be considerably more dilute and of very different solute concentrations than the body fluids. The urine formed by the kidney of fishes, amphibians, and reptiles is always either more dilute or the same osmotic concentration as blood; the ratio of osmotic concentration of urine to blood (U/P ratio) is ≤ 1.

The excretory organs of only a few animals can produce a urine that is more osmotically concentrated than their body fluids. Most notable is the hindgut complex of some insects and the kidney of some birds and mammals. The urine produced by the Malpighian tubules of insects generally has the same osmotic concentration as the hemolymph, although the urine that is ultimately voided may be considerably modified by the hindgut to a lower or higher osmotic concentration. In birds and mammals, the urine initially formed by the kidney is also the same concentration as blood, but the urine (U) can be subsequently modified to be either more dilute or more concentrated than plasma (P), when voided. In birds, the maximal U/P ratio varies from about 1.4 in the emu to over 5 in the Savannah sparrow. There is a general tendency for arid-adapted or salt marsh birds to have a high U/P, but there is not such a good correlation between habitat aridity and U/P_{max} for birds as there is for mammals, because the osmotic-concentrating role of the kidney in birds is compromised by the subsequent modification of urine in the rectum and hindgut. Urates are precipitated in the hindgut, and the voided excreta has a very low water content. However, the fluid reabsorbed from the hindgut is always of a higher osmotic concentration than the lumen contents, and so the conservation of water from excreta is accompanied by retention of salts. Many marine and terrestrial reptiles and birds have salt glands to eliminate excess solutes.

In mammals, the urine may be considerably more concentrated (e.g., 4000–9000 mOsm liter^{-1}) than plasma (about 350 mOsm liter^{-1}); the maximal U/P ratio may be as high as 25 in some desert-adapted rodents. This urine-concentrating capacity is due to the structural organization of the kidney; major blood vessels and urine-forming structures (glomeruli) are in the outer cortex, and hairpin loops of Henle and collecting ducts are in a central medullary cone (or multiple medullary cones in some species) that establishes an increasing interstitial osmotic gradient along the medullary cone toward the tip. There is a strong correlation between maximal urine concentration, relative kidney structure, and habitat aridity for mammals. The maximal urine concentration (mOsm liter^{-1}) is related to the relative medullary thickness (RMT; the ratio of the thickness of the medulla to the overall thickness of the kidney): $U_{max} = 680\,RMT - 696$. Mammals from desert environments typically have kidneys with a higher relative medullary thickness than kidneys of mammals from mesic environments. For example, the arid-zone (northwestern Australia) dasyurid marsupial *Pseudantechinus macdonnellensis* has a kidney with a more predominant medulla and a much higher RMT than does the mesic dasyurid *Antechinus swainsoni* (eastern Australia; Fig. 2).

A few terrestrial arthropods can absorb water vapor from subsaturated air, at a relative humidity greater than a critical value (RH$_{crit}$) at which water gained by active uptake is in balance with water loss by evaporation. This water vapor uptake occurs despite the biological difficulty of coupling energy expenditure with the movement of water vapor against an extreme water potential gradient. For example, absorbing water vapor into solution from a relative humidity of 90% is equivalent to transporting water against a 14 Osm liter^{-1} concentration gradient! Nevertheless, some isopods are able to absorb water vapor from a high relative humidity (RH$_{crit}$ = 87–93%), and some insects and ticks can do so even at considerably lower relative humidities (RH$_{crit}$ = 45–75%).

V. ENERGY BALANCE

Energy exchange is critical to the survival of organisms. For plants, energy is absorbed as solar radia-

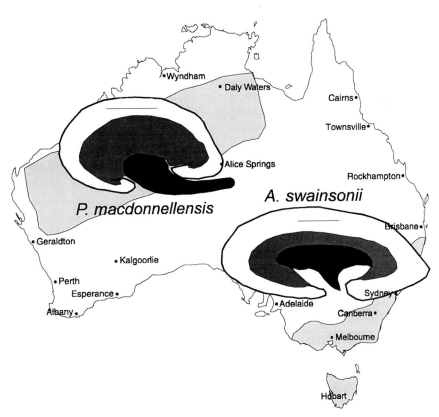

FIGURE 2 Schematic representation of kidney structure and geographic range for dasyurid marsupials from the arid zone (*Pseudantechinus macdonnellensis*) and mesic zone (*Antechinus swainsonii*) of Australia. Shaded regions of kidney are the inner and outer medulla; scale bar is 2 mm. [Data from B. Brooker and P. Withers. (1994) Kidney structure and renal indices of dasyurid marsupials. *Aust. J. Zool.* **42**, 163–176.]

tion and converted by photosynthetic pathways into organic chemicals. Chemical energy is a primary energy source for the chemisynthetic microorganisms of deep-sea vent ecosystems. For animals, the predominant source of energy is ingested food. Energy balance, like the balance of all materials between an animal and its environment, is constrained by the principle of mass balance—all of the energy (or matter) entering an organism must be ultimately lost to the environment, although there may be temporary imbalances between gain and loss due to storage or depletion of body reserves.

Metabolic rate is the amount of energy utilized by an organism per unit time. Metabolism is usually sustained through the oxidation of ingested carbohydrate, lipid, or protein with oxygen. The aerobic oxidation by cellular metabolism of glucose illustrates well the concept of metabolism at the cellular level:

glucose oxidation: $C_6H_{12}O_6 + 6O_2 \rightarrow 6CO_2$
$$+ 6H_2O + 2854 \text{ kJ mole}^{-1}$$

The oxidation of 1 mole of glucose by 6 moles of oxygen forms 6 moles of carbon dioxide and 6 moles of water, and releases 2854 kJ of energy as heat. Metabolic rate is usually expressed as the rate of heat release (Q_h; kJ hr^{-1}), the rate of oxygen consumption (VO_2; ml O$_2$ hr^{-1}), or the rate of carbon dioxide production (VCO_2; ml CO$_2$ hr^{-1}). Similarly, the aerobic oxidation of palmitic acid and alanine illustrate cellular metabolism for lipids and protein:

palmitic acid oxidation: $[CH_2]_{15}COOH + 23O_2$
$$\rightarrow 16CO_2 + 16H_2O + 9791 \text{ kJ mole}^{-1}$$

alanine oxidation: $NH_2CHCH_3COOH + 3O_2$
$$\rightarrow 2\tfrac{1}{2}CO_2 + 2\tfrac{1}{2}H_2O + \tfrac{1}{2}CO[NH_2]_2 + 580 \text{ kJ}$$
$$\text{mole}^{-1}$$

The aerobic oxidation of glucose and palmitic acid forms only CO_2 and H_2O as waste products, but the oxidation of amino acids also forms nitrogenous waste products (e.g., urea as mentioned earlier, but also ammonia, uric acid, and other nitrogenous molecules). For organisms, a complex mixture of substrates is oxidized and the overall stoichiometry of metabolism is more complex.

Oxygen is essential for aerobic metabolism, but it is sometimes not freely available, either because it is scarce or absent in the environment (e.g., hypoxic or anoxic water or mud, gut fluids), or because respiratory gas exchange is temporarily reduced (e.g., emersion of aquatic animals or immersion of terrestrial animals), or because it cannot be transported sufficiently rapidly into the cells (e.g., during strenuous activity). There are a variety of pathways for anaerobic metabolism during O_2 unavailability, but these have a low energy yield of 5–35% of aerobic metabolism and often involve the formation of an end product that accumulates and ultimately limits the anaerobic metabolic pathway. Examples of anaerobic metabolism are the formation of propionate, methylbutyrate, and methylvalerate by microorganisms, acetate, acetaldehyde, and ethanol by some fish, lactate by most vertebrates, and alanopine, strombine, octopine, and lysopine by various molluscs.

The major determinants of the metabolic rate of an organism are cellular maintenance requirements, activity, body temperature, body mass, and phylogeny. The metabolic energy expenditure of animals is apportioned among a variety of functions. Generally, there is a minimal energy expenditure, the standard metabolic rate (SMR; for ectotherms), or the basal metabolic rate (BMR; for endotherms) required for basic cellular requirements. Additional energy is expended on various activities, such as feeding, predator avoidance, locomotion, digestion, growth, and reproduction; endothermic animals also expend considerable energy on heat production or heat loss in order to maintain a constant body temperature.

A variety of cellular processes presumably contribute to the minimum (standard or basal) energy requirements of organisms, for example, maintenance of membrane integrity, sustaining ion gradients and transporting various molecules across membranes, protein synthesis, and heat production. Despite the intuitive concept of minimal metabolic expenditure of organisms for cellular maintenance, it is apparent that this minimal maintenance cost can be depressed by some organisms during inactivity. For example, the metabolic rate of some arid-zone frogs can be reduced to only 20% of their standard metabolic rate during estivation (summer dormancy); a metabolic depression of similar magnitude has also been reported for estivating salamanders and lungfish, snails and crustaceans, and for some O_2-deprived animals. The metabolic rate of some endothermic mammals and birds is dramatically reduced during hibernation, or torpor, but generally the marked reduction in metabolic rate can be ascribed to the relaxation of body temperature regulation and the effect of reduced body temperature on metabolic rate. However, an intrinsic tissue metabolic depression by unknown cellular processes may contribute a further reduction in metabolic rate, particularly in small hibernating mammals and birds.

Temperature has a profound acute effect on the metabolic rate of organisms. A 10°C increase in body temperature generally increases the metabolic rate of ectothermic organisms by about 2.5 times; that is, the Q_{10} is about 2.5. Consequently, there is an exponential relationship between ambient temperature (= body temperature) and metabolic rate for ectotherms. There is a similar relationship between body temperature and metabolic rate for endotherms, with a Q_{10} of about 2.5. However, endotherms generally precisely regulate body temperature (at 35° to 40°C) regardless of the ambient temperature. There is a complex relationship between ambient temperature and metabolic rate for endotherms (see the following discussion). Especially ectotherms, but also endotherms, show complex long-term responses to chronic temperature change, both in the laboratory ("acclimation") and in nature ("acclimatization"). In general, the acclimatory response tends to counteract the acute Q_{10} effect and restore metabolic rate toward the initial rate prior to temperature change (partial acclimation), return the metabolic rate to the preacclimatory value (perfect acclimation), or even overcompensate (overacclimation). Alternatively, there may be no acclimatory response, or the acclimatory response may exacerbate rather than negate the Q_{10} effect (inverse acclimation).

The body mass of organisms has a profound influence on their metabolic rate. The metabolic rate of the smallest organisms, weighing only a few picograms (10^{-9} g), is immensely lower than that of the largest organisms, weighing many gigagrams (10^6 g). However, there is not a simple proportionality between metabolic rate and body mass; in general, metabolic rate increases in larger organisms to a lesser extent than would be expected from the increase in mass. This scaling, or allometry, of metabolic rate (MR) as a function of body mass (M) is most readily analyzed as a power function of the form

$$\text{MR} = \mathbf{a}\, M^{\mathbf{b}} \quad \text{or} \quad \log_{10} \text{MR} = \mathbf{a}' + \mathbf{b} \log_{10} M,$$

where $\mathbf{a}$ is a constant of proportionality and $\mathbf{b}$ is the mass exponent; a double-logarithmic relationship where $\mathbf{a}'$ is the intercept (and equal to $\log_{10} \mathbf{a}$) and the mass exponent $\mathbf{b}$ is the slope of the linear relationship provides a mathematically more convenient manner of expressing this relationship. If there were a direct proportionality between metabolic rate and body mass, then $\mathbf{b}$ would be 1, but in general the value of $\mathbf{b}$ is closer to 0.75 or 0.67, reflecting the smaller increment in metabolic rate than mass. A value of 0.67 is expected if there is a geometric dependence of metabolic rate on some aspect of surface area (e.g., the surface area of the respiratory gas exchange surface), because surface area $\propto M^{0.67}$. The intraspecific mass exponent, that is, $\mathbf{b}$ determined for a single species of organism, is commonly about 0.67, perhaps reflecting geometric similarity or developmental changes in metabolic rate with increasing body mass. There is no commonly accepted explanation for why the interspecific value of $\mathbf{b}$ is often observed to be 0.75 or thereabouts for interspecific comparisons, that is, comparing the metabolic rates of different species. Nevertheless, the metabolic rate of organisms is often expressed per mass$^{0.75}$ to eliminate the confounding effect of body mass variation, and many other physiological parameters are often expressed per mass$^{0.75}$ to remove effects of "metabolic intensity."

The metabolic rate of organisms is also related to their taxonomy, or phylogeny. There appear to be three levels, or "grades," of metabolism among organisms; unicellular organisms have a substantially lower standard metabolic rate than multicellular organisms of similar mass, and endothermic animals have a higher standard metabolic rate than ectothermic animals of similar mass. The different metabolic "grade" of unicellular and multicellular organisms has been related to the smaller size and consequently higher surface area for metabolic exchange of cells from multicellular organisms. The different metabolic "grade" between ectotherms and endotherms has been related to the greater need for cells of the latter to produce heat for thermoregulation, and has been correlated with a higher cellular mitochondrial content. Phylogenetic correlates of metabolic rate are also apparent for narrower taxa, such as within mammals (three "grades": monotremes, marsupials, placentals) and birds (three "grades": ratites, nonpasserines, passerines), and even among these subdivisions of taxa (e.g., a number of "grades" are recognized among various placental mammals, and among various nonpasserine birds).

Field metabolic rate (FMR) has been directly measured for a wide variety of vertebrates and some invertebrates. For ectothermic lizards, there is a clear allometric relationship between FMR and body mass, with a mass exponent of about 0.80. The FMR is about 6.2 times the standard metabolic rate at 20°C, and 2.5 times that at 30°C. For mammals and birds, the mass exponent is about 0.60 to 0.80 and the FMR is about 2 to 3 times the basal metabolic rate.

VI. HEAT BALANCE

The body temperature of an organism is determined by the balance between the various avenues of heat gain and loss. Body temperature is an extremely important variable because physiological and ecological processes are invariably influenced by temperature, and consequently many animals and also some plants expend considerable time or energy to precisely regulate their body temperature. Avenues for heat exchange include conduction, convection, radiation, a change in state of water, and metabolic heat production. For an organism with a constant body temperature, the heat gain is exactly equal to the heat loss; if heat gain exceeds heat loss, there is storage of thermal energy and the body temperature increases; if heat gain is less than heat loss, there is a depletion of thermal energy and the body temperature declines. Most conductive, convective, and radiative heat exchange occurs across the external body surface, although some heat exchange can occur across the gut and respiratory surfaces, and via loss of urine or feces.

Conduction is heat exchange through contact with a solid substratum, for example, via the feet in contact with the ground, the body adpressed against the soil or a rock, or roots in soil. Conductive heat exchange depends on the temperature difference $(T_b - T_a)$ between the organism (T_b) and solid materials (T_a) that it contacts and the thermal conductivity (k) of the materials, and is inversely dependent on the distance for heat transfer, x: $Q_{cond} = k \, (T_b - T_a)/x$.

Convection is heat exchange through the flow of a fluid (either air or water). Convective heat exchange depends on the temperature difference between the organism and surrounding fluid and the thermal conductivity of the fluid (k_f), and is inversely dependent on the thickness of the surrounding fluid boundary layer (δ): $Q_{conv} = k_f \, (T_b - T_a)/\delta$. Convection is "free" if the fluid movement is passively induced by the temperature gradient between the organism and the fluid and is "forced" if the fluid flow is externally generated (e.g., by wind).

All objects radiate heat energy; the energy of emission is $\sigma \, \varepsilon \, A \, T^4$ J sec^{-1}, where A is surface area in m^2, T is surface temperature in degrees Kelvin, σ is the Stefan–Boltzmann constant $(5.67 \times 10^{-8}$ J sec^{-1} K$^{-4})$, and ε is the surface emissivity (typically 0.9–0.95). Net radiative heat exchange (Q_{rad}) is the relative gain of radiative heat from the surroundings $(\sigma \, \varepsilon_{sur} \, A \, T_{sur}^4)$ compared to loss from the organism $(\sigma \, \varepsilon_b \, A \, T_b^4)$.

Evaporation of water from an organism decreases its temperature whereas condensation of environmental water on an organism warms it. Evaporation (E) of water absorbs about 2400 J g^{-1}; the melting of ice releases about 334 J g^{-1}. Rarely in biological systems, heat is released by condensation or freezing. Heat is continuously produced by metabolism (MHP); this is inconsequential for ectotherms but plays a major role in thermal balance for endotherms. Heat can also be stored in the body (S).

The overall equation for heat balance relating metabolic heat production to the various avenues of heat loss or gain is

$$MHP \pm k\,(T_b - T_a)/x \pm k_f\,(T_b - T_a)/\delta \\ + \sigma\,\varepsilon_b\,\varepsilon_{sur}\,A\,(T_{sur}{}^4 - T_b{}^4) \pm E \pm S = 0$$

The thermal balance of an organism is thus exceedingly complex and correspondingly difficult to determine by measuring all parts of the heat balance equation. However, it can be easily measured empirically using an appropriate model of an organism; the equilibrium temperature of such a model is the operative environmental temperature (T_e) and is equal to body temperature of the organism in that environment (with no evaporative heat loss). The use of operative temperature models can readily indicate the theoretical range of body temperatures available to organisms.

The heat balance of ectothermic organisms is dominated by the influence of their thermal environment, especially fluid thermal convection for aquatic ectotherms and radiation for terrestrial ectotherms. The T_b of aquatic ectotherms is almost invariably the same as T_{water}, but it is possible for many aquatic ectotherms to behaviorally thermoregulate by selecting water at their preferred temperature. Terrestrial ectotherms are removed from the constraints of an environment with a high thermal conductivity, and consequently many precisely thermoregulate by exploiting solar radiative heat to elevate T_b significantly above T_a.

The heat balance of endothermic mammals and birds is dominated by metabolic heat production, and the heat balance equation can be simplified to $MHP = k\,(T_b - T_a)/x = C\,(T_b - T_a)$, where C is the thermal conductance, a physical/physiological parameter reflecting insulation. Other metabolically active, well-insulated animals, such as moths, are also endothermic. For mammals and birds, there is a linear relationship between metabolic rate and air temperature, at least at low T_a; the slope of this relationship is $-C$. At intermediate T_a, MHP plateaus at the basal metabolic rate. At high T_a, metabolic rate rises as a consequence of a slight

elevation in T_b (and a Q_{10} effect) and due to the metabolic cost of heat-dissipative processes such as sweating or panting. This complex relationship between metabolic rate and T_a for endotherms is unlike the simple exponential relationship observed for SMR of ectotherms.

Some endotherms occasionally abandon the strict metabolic constraints of endothermy when cold-stressed and food-deprived (hibernation, torpor) or heat-stressed and water-deprived (estivation). Body temperature declines, often to the ambient temperature, and this provides a considerable energy savings. However, thermoregulation is not necessarily abandoned; most hibernating endotherms will maintain T_b above a minimal level, which may reflect a minimal T_b below which arousal from torpor is not possible (e.g., about 15°C for some dasyurid marsupials) or a minimal temperature to avoid freezing of body tissues (e.g., 0°C).

For exceedingly large organisms, the concepts of ectothermy and endothermy become obfuscated by their high thermal inertia and low thermal conductance, and these organisms exhibit gigantothermy. For example, large turtles can significantly elevate their T_b above T_{water}, despite the high thermal conductivity of water, by their moderate metabolic heat production while swimming. Physical and physiological modeling of thermal balance for large dinosaurs (e.g., *Edmontosaurus* at 3.6 metric tons) suggests that T_b might have been as little as 2°C above T_a if activity were minimal and heat dissipation maximal, or more than 45°C above T_a if metabolic rate was maximal and heat dissipation minimal. Even smaller dinosaurs such as *Deinonychus* (75 kg) could have had a T_b as much as 10°C above T_a, but much smaller dinosaurs (e.g., *Compsognathus* at 2 kg) would always have had T_b close to T_a.

Body temperature is one of the most important proximate factors determining the metabolic capacity of organisms; there is a physiological performance curve determined by T_b. For animals, one important consequence of metabolic capacity is the speed and endurance of locomotion (Fig. 3). Many studies of the effects of body temperature on physiological capacity for sprint speed and endurance

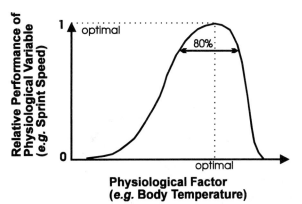

FIGURE 3 Hypothetical example of a physiological performance curve, relating the change in relative value for a physiological parameter (such as locomotor velocity or endurance) with an environmental variable (such as temperature or humidity). Note the optimal relative value (1, or 100%) at an optimal value for the environmental variable. Ranges of the environmental variable can be defined to encompass given relative values of the physiological parameter (e.g., 80% of maximal relative value).

have indicated an optimal temperature, often similar to the preferred body temperature, below which performance (e.g., sprint speed) declines to suboptimal levels and above which performance rapidly declines. Similar effects of temperature on performance have been reported for digestion rate and efficiency, muscle contraction velocity, hearing, and maximal metabolic rate. There is a range of temperatures for which physiological performance is quite high (e.g., 80% of maximal, T_{80}).

VII. PHYSIOLOGICAL LIMITS TO ORGANISMAL DISTRIBUTION

The distribution of organisms is undoubtedly affected by many physical and biological factors, and population and community ecology can only be fully understood by studying the interactions of these factors. For any specific factor, there is an optimal value and a narrow reproductive range over which the organism is able to survive and sustain a continuous population, a broader survival range over which an individual organism is able to live indefinitely, and a still wider geographic range over which an individual organism can survive though not indefinitely; outside this geographic

range, the organism cannot survive at all (Fig. 4). Liebig's "Law of the Minimum" states that the distribution of an organism is determined by that environmental factor for which it has the lowest tolerance. However, the survival and reproduction of a species may be influenced by a number of different factors. For example, the distribution of many organisms might be determined primarily by temperature and food availability, but other factors such as light and humidity might have lesser effects. Extension of this argument to multiple ("n") dimensions yields multidimensional reproductive, survival, and geographic regions. If restricted to physiological parameters, then the n-dimensional hyperspace defines the "physiological niche."

Organismal distribution may be determined by nonphysiological limitations, for example, by evolutionary and zoogeographic history and limits to dispersal, by habitat selection, or by interrelations with other organisms (e.g., competition, predation, allelopathy). Nevertheless, distributions of some organisms are clearly influenced by physical factors, such as temperature, light, moisture, energy, or specific nutrients. One role of ecophysiology is to define these potential physical and physiological limitations to organismal distribution. Temperature and water are probably the two most limiting physical factors that affect the distribution of organisms.

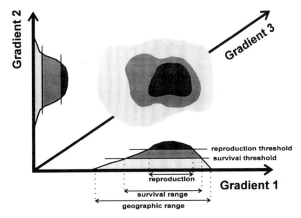

FIGURE 4 Schematic representation illustrating in three dimensions the concept of an n-dimensional hyperspace for survival and reproduction, survival, and geographic range of organisms, defined by axes representing physiological responses to environmental variables. The one-dimensional response curves are shown for two of the axes.

Temperature affects virtually all physiological and ecological processes, and it has been used to explain the distribution and abundance of numerous aquatic and terrestrial organisms. Whether organisms are susceptible to freezing, for example, can determine their latitudinal and altitudinal distributional limits. The northern distribution of loblolly pines is correlated with the winter temperature (and rainfall); low winter temperatures decrease root water uptake. The distribution of many aquatic insects has been linked to water temperature; low temperatures limit both metabolism and growth, whereas high temperatures increase metabolism but reduce growth. Differences in the thermal physiology of tree frogs are associated with distributional differences and variation in breeding pattern; locomotor performance is better at low temperatures for more northern species, which are earlier breeders.

Water availability and solute concentration may limit the distributions of both aquatic and terrestrial organisms. Salinity limits the distribution of many aquatic animals. For example, low salinity tolerance is greater for freshwater killifish than for estuarine killifish. For terrestrial organisms, rainfall, ambient humidity, and soil water content can influence their distribution. The distribution of woody plants is often associated with soil moisture content, although temperature and wind also have confounding effects. Probably not many terrestrial animal distributions are directly determined by water, but may be indirectly determined by plant distributions. For example, the distribution of red kangaroos in Australia is limited to the <400 mm rainfall contour, which determines the distribution of arid-zone grasses.

Energy can be an integrative variable for studying organismal distributions. For example, energy balance is influenced by the combined effects of many physical factors (temperature, salinity) and biological factors (predation, competition). Energy balance contributes directly to the survival and reproduction of organisms; it provides a simple index of fitness; and it can indicate reproductive potential, because organisms generally only reproduce using the energy surplus after other energy-requiring needs are met. For example, the northern limit to the distribution of many American passerine birds is correlated with the average minimum January air temperature isotherm, at which the calculated metabolic rate is 2.5 times the basal metabolic rate ($\approx$FMR).

VIII. PHYSIOLOGICAL INFLUENCES ON ORGANISMAL ABUNDANCE

The abundance of a species is generally related to its distribution, being high in the middle of its range and progressively declining toward the periphery. The analysis of factors affecting abundance is similar to that for distribution. The most favorable part of the n-dimensional hyperspace of physical and biological factors often also defines the region of highest abundance; that part where individuals survive but fail to reproduce defines a surrounding region of lower abundance, relying on immigration to sustain the population; the limits to geographic range and lowest abundance are defined by the survival limits to the n-dimensional hyperspace.

It is more difficult to explain the absolute abundance of a species, that is, how many organisms there actually are in a particular space. One approach is the intuitive relationship between abundance and body size; small animals clearly can have a higher abundance than large animals. A more analytical approach is to assume that energy balance is a limiting factor and relate this to the relationship between body mass and metabolic rate. The maximum amount of energy available to a species (E; kJ m^{-2} day^{-1}) limits the maximum abundance of organisms (N; organisms per square meter) according to the body mass (M; g), the allometry of resting metabolic rate (RMR; $\mathbf{a}\,M^{0.75}$), and the field metabolic rate increment (I_{field}; $\times$RMR): $N = E/[I_{\text{field}}\,\mathbf{a}\,M^{0.75}]$. This approach sets the upper bound to the abundance of organisms, but the realized abundance depends on many other complex biological factors, such as the number of species competing for the energy at this particular trophic level and their relative effectiveness at harvesting the available energy.

In practice, very abundant organisms may be energy limited in this manner, but uncommon and rare organisms are not.

The abundance of all species in a community can be similarly analyzed as a body mass effect. In general, studies of the body mass and abundance for animals from various assemblages suggest that abundance scales with body mass to about the -0.75 power (standard linear regression model) to -0.95 power (reduced major axis model), at least for the more abundant species at any particular mass. Because metabolism scales with body mass to the $+0.75$ power, then the total metabolism of a species in an assemblage is predicted to scale with body mass to the 0.00 power if abundance $\propto$ mass$^{-0.75}$, that is, total metabolic expenditure should be independent of mass. However, the efficiency of biomass conversion is size dependent, $\propto$mass$^{0.20}$, so the species energy use $\propto$ mass$^{0.95}$; if abundance $\propto$ mass$^{-0.95}$, then again total metabolic expenditure is predicted to be independent of mass. Both of these approaches suggest that there may be an "energetic equivalence" such that all abundant species in an assemblage use an equal amount of energy. However, mass is not a good predictor of abundance for all species; the slope depends on which regression model is used; and there are major energy losses between trophic levels. Whether there is any general energetic importance to abundance and body size in assemblages, communities, and ecosystems is as yet unresolved.

Glossary

Acclimation Gradual readjustment of a physiological variable after an acute change, for example, in response to a temperature change.

Allometry (scaling) How a variable alters with respect to body mass; often expressed by a power curve of the form aMass^b, where **a** is the allometric constant and **b** is the allometric exponent.

Basal metabolic rate (BMR) Lowest metabolic rate measured for an endotherm, within the zone of thermoneutrality.

Comparative method Rigorous conceptual and methodological procedure for examining the way in which closely related organisms have radiatively adapted to different environments, and the way in which distantly related organisms have adapted to similar environments.

Ectotherm Organism whose body temperature is determined primarily by thermal equilibrium with its environment.

Endotherm Organism whose body temperature is determined primarily by endogenously produced heat from metabolism.

Homeostasis Occurrence of constant conditions, either as a simple consequence of constancy of environmental conditions or through physiological regulation.

Metabolic depression Reduction in metabolic rate below the normal minimal level (BMR or SMR) in response to cold and food deprivation (hibernation), heat and water deprivation (estivation), or hypoxia.

Metabolic rate Sequence of biochemical reactions used to degrade (catabolize) complex organic molecules; for animals, the metabolic rate is generally measured as the rate of heat production, oxygen utilization, or carbon dioxide production.

Physiological performance Relationship between any physiological variable, such as locomotor velocity or endurance, and a physiological factor, such as temperature or salinity; there is generally an optimal value for the physiological factor at which the variable is maximized.

Standard metabolic rate (SMR) Minimal metabolic rate measured for an ectotherm, at a specified temperature.

Bibliography

Bakken, G. S. (1992). Measurement and application of operative and standard operative temperatures in biology. *Amer. Zool.* **32,** 194–216.

Beuchet, C. A. (1990). Body size, medullary thickness, and urine concentrating ability in mammals. *Amer. J. Physiol.* **258,** R298–R308.

Blackburn, T. M., Brown, V. K., Doube, B. M., Greenwood, J. D., Lawton, J. H., and Stork, N. E. (1993). The relationship between abundance and body size in natural animal assemblages. *J. Anim. Ecol.* **62,** 519–528.

Dunson, W. A., and Travis, J. (1991). The role of abiotic factors in community organization. *Amer. Nat.* **138,** 1067–1091.

Hall, C. A. S., Stanford, J. A., and Hauer, F. R. (1992). The distribution and abundance of organisms as a consequence of energy balances along multiple environmental gradients. *Oikos* **65,** 377–390.

Harvey, P. H., and Pagel, M. D. (1991). "The Comparative Method in Evolutionary Biology." Oxford, England: Oxford University Press.

Huey, R. B. (1991). Physiological consequences of habitat selection. *Amer. Nat.* **137,** S91–S115.

Nagy, K. A. (1987). Field metabolic rate and food requirement scaling in mammals and birds. *Ecol. Monog.* **57,** 111–128.

Root, T. (1988). Energy constraints on avian distributions and abundances. *Ecology* **69,** 330–339.

Spotila, J. R., O'Connor, M. P., Dodson, P., and Paladino, F. V. (1991). Hot and cold running dinosaurs: Body size, metabolism and migration. *Modern Geol.* **16,** 203–227.

Withers, P. C. (1993). "Comparative Animal Physiology." Phildelphia: Saunders College Publishing.

Ecosystem Integrity, Impact of Traditional Peoples

David W. Steadman

New York State Museum

I. Introduction
II. Common Themes across Changing Cultures
III. Continental Ecosystems
IV. Island Ecosystems
V. Relating the Past to the Present

Humans, wherever we live and whatever our life-style, are important participants in ecosystems. We have an unparalleled ability to affect plant and animal communities directly through gathering, chopping, digging, burning, hunting, fishing, and trapping. We also can alter the very air, water, soil, sediment, and bedrock upon which plants and animals depend. This article reviews some of the impacts that traditional peoples have had on ecosystems across the world. As we will see, traditional peoples know a great deal about the habitats that they occupy and modify. Now that modern technology has exerted its influence virtually everywhere, it is appropriate to pause and study the times and places where tools were made of stone, wood, shell, and bone.

I. INTRODUCTION

Many of today's environmental problems seem overwhelming. How can 5.6 billion people ever deal with the situations we have created? If we are to face a problem squarely and attempt to solve it, there is nothing so important as a realistic under-standing of the problem, no matter how sobering that might be. Some of the figures associated with the biodiversity crisis, for example, can elicit a deep sense of loss (Edward Wilson estimates that roughly 27,000 species become extinct each year) or an inescapable feeling of ignorance (the 1.4 to 1.8 million species that have been described may represent only about 10% of the actual number that exists). Because human impact now reaches all corners of the globe, it probably is impossible to preserve examples of ecosystems that are completely unaffected by our activities. Nevertheless, such preservation is a worthy long-term goal, and pursuing it depends on knowing what is natural and what is not. [*See* BIODIVERSITY.]

The purpose of this article is to provide a historical and multicultural perspective on the impact of human activities on natural ecosystems. We have been conditioned to think of environmental impact as a product of the industrial age. Though it is true that virtually every environmental problem has been worsened by the past 200 years of industrial and technological development, the earth was far from pristine in preindustrial times. We are learning more and more that anthropogenic environmental changes have played important roles in both the rise and fall of traditional cultures. [*See* GLOBAL ANTHROPOGENIC INFLUENCES.]

By studying the environmental impact of traditional peoples, we can begin to estimate the composition and function of ecosystems at two critical junctures: before any human contact, and at the time of Western contact (when all previous human impact had been due to traditional peoples). Such knowledge is important for long-term projections of ecosystem composition, stability, and integrity, as well as to help us understand to what extent various ecosystems might recover from human disturbances.

II. COMMON THEMES ACROSS CHANGING CULTURES

From a global perspective, exact definitions are difficult for "traditional peoples" or "native peoples." For example, pre-Columbian American Indians were traditional because their culture was unaffected by Europeans. They were native as well in a broad sense, although it should be remembered that large movements of prehistoric peoples (such as the arrival of Navaho and Apache in the Southwest) displaced earlier native peoples with mutually unintelligible languages. Plains Indians of the 1800s who hunted and butchered bison with horses, guns, and steel knifes were native but certainly not fully traditional.

Precise definitions of native (= indigenous) versus traditional peoples are most difficult in much of Africa and Eurasia. Human occupation there is measured in hundreds of thousands of years (Fig. 1) and thus includes evolutionary intermediates between early forms of *Homo* (ca. 1–1.5 million years ago) "anatomically modern humans," which had evolved sometime between 100,000 and 40,000 years ago. In the Americas, New Guinea, Australia, Oceania, Madagascar, and other islands, the first human occupation was by anatomically modern humans. Lowered sea levels caused by continental ice sheets aided the human colonizations of the Americas, New Guinea, and Australia.

Archaeological evidence (especially from bone assemblages) suggests that people in Africa and Eurasia have hunted large mammals for at least 100,000 years. This early human predation focused primarily on small or weak individuals rather than the largest and most dangerous animals. Then, about 40,000 years ago, there was a major advance in hunting proficiency, perhaps coinciding with the evolutionary emergence of modern people. For the first time, systematic and organized predation of large, powerful, and dangerous animals became an important part of human subsistence. The first appreciable human impact on vertebrate faunas may have occurred at this time.

The development of agriculture during the past 10,000 years has had an enormous impact on the natural ecosystems of both temperate and tropical areas. Though the task was not pursued initially with a farming future in mind, the big-game hunting of preagricultural peoples helped to pave the way for a more sedentary life-style that involved growing much of your own food. Getting rid of most large wild mammals makes it easier and safer to grow and protect crops, and the attraction to crops of certain remaining wild animals renders them easier to hunt. Domestication of a few species of large mammals reduces human dependency on wild mammals for fat, protein, bones, sinew, and hide. Clearing the land, often through burning, may temporarily remove high-quality habitat for some large mammals, further keeping their populations in check. (Fire, a most powerful tool employed by all traditional peoples, had been used to concentrate animals for hunting long before it was used to clear land for agriculture.) If keystone species were eliminated (see the following example), hunting and habitat modification could affect ecosystems far beyond just the direct removal of a few species.

The impact of traditional agriculture on natural ecosystems is not limited to the reduction or loss of some species and the propagation of others. Additional impact, known as "landscape changes," include alterations of local topography, pedology, and hydrology. The upslope erosion and downslope deposition of soils and other sediments, often associated with deforestation or overgrazing, are

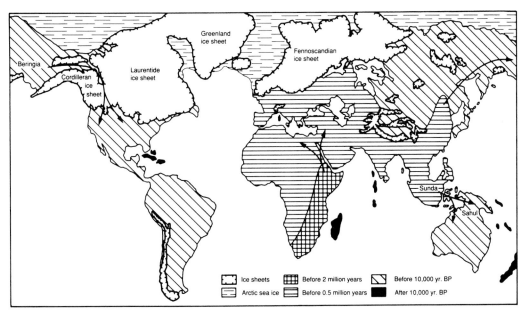

FIGURE 1 Chronological summary of human colonization of the earth. [From N. Roberts (1989). "The Holocene: An Environmental History," Fig. 3.7. Basil Blackwell, Oxford, England.]

important aspects of landscape change, as are alterations of water supply (channeling, flooding, draining, etc.). Regardless of climate, much early agriculture developed in fertile alluvial soils along major rivers. As a result, the natural biotas of riparian areas have been disturbed disproportionately for thousands of years.

Of course not all traditional peoples, even in modern times, used agriculture. Indigenous hunter-fisher-gatherer (h-f-g) peoples existed in a great variety of habitats on all continents (except uninhabited Antarctica) at the time of European contact. Ethnographic studies of h-f-g peoples show that gathering of wild plants often contributes more to subsistence than hunting or fishing. In many cases this may be because the most easily obtained animals already have been overexploited. The movements of h-f-g peoples may be due to necessity as much as any conscious effort to conserve resources. Even though tropical forests, for example, might appear to be lush and teeming with life, the relatively few species of plants and animals regularly used by h-f-g peoples often have scattered distributions, forcing people to travel widely to find enough to eat.

Cross-cultural contacts have occurred for millennia among various peoples from Europe, western Asia, northern Africa, and occasionally beyond. To distinguish traditional from nontraditional peoples in these areas would require arbitrary definitions related to the development and diffusion of technology. European culture and technology did not disperse worldwide, however, until the seafaring explorations of the Portuguese, Spanish, Italians, French, Dutch, and British in the fifteenth through eighteenth centuries. The ensuing geographical explosion of rapidly advancing technologies signals this period as the end of traditional life-styles across most of the world. For the first time, many hundreds of distinct cultures in sub-Saharan Africa, Central and Southeast Asia, New Guinea, Australia, Oceania, and the Americas were exposed to "Western" objects, ideas, and people. These complex events, the very last of which are occurring today, were difficult for the newly contacted peoples. Opportunities to improve wealth, power, and technology often were offset by loss of cultural identity, pressure to change religion, disease, warfare, enslavement, and even genocide.

The last areas to support diverse cultures with little if any Western influence were the most remote parts of Amazonia, sub-Saharan Africa, and Melanesia, where many peoples were unaffected until the late nineteenth or well into the twentieth century. As we approach the twenty-first century, even these regions are inhabited by few if any peoples whose way of life is not influenced in some way by Western technology or thought.

Traditional peoples know a great deal about the ecosystems they inhabit. A detailed knowledge of nature is related to survival and to the ability to extract diverse natural resources. That is why traditional peoples are sought out today by pharmaceutical companies, organismal biologists, ecotourists, and government officials who recognize the economic, social, and scientific importance of their knowledge. Very often, traditional peoples not only can identify and name virtually every local species of bird, mammal, and macroscopic plant, they also can tell you about the behavior and ecology of these species. This is just the sort of information that leads to successful h-f-g activities, which include the avoidance of dangerous species.

Traditional peoples do not live "harmoniously" with the environment. The skillful use of tools and fire has set humans apart from other animals for many tens of millennia. The relatively small environmental impact often attributed to traditional peoples seems to be related mainly to technology. It is more difficult to clear forests with stone axes than with bulldozers and chainsaws. It is more difficult to dig a ditch with a deer scapula hafted to a stick than with a backhoe. It is more difficult to kill animals with spears than with guns. And, it is more difficult to overfish the oceans with dugout canoes and handlines than with driftnet trawlers. In either case, however, the person involved ends up cutting down a tree, digging a ditch, killing an animal, or catching a fish.

This is not intended in any way to belittle or criticize traditional peoples, who engage in these activities, like so many Westernized people, as a means of subsistence. Every person requires and deserves food, clothing, and shelter, all of which is derived from the environment. Humans, just like all organisms, live within, not outside of, their environment. The reason why traditional peoples might seem to have relatively little influence within ecosystems is because their per capita consumption of natural resources is low by Western standards. When low per capita consumption is combined with a low population density, the impact of traditional peoples, especially those with limited or no agriculture, may indeed be relatively small. Small does not equal insignificant, however, especially when it leads to overexploitation of critical human resources (firewood, soil, water, etc.) in marginal environments, or to the loss of keystone species.

The elimination of keystone species can have dramatic consequences even in habitats that would seem, on first inspection, to be intact. For example, Barro Colorado Island in Panamá became an island when the surrounding forest was flooded about 100 years ago to form the Panama Canal. In spite of protection from logging and hunting, the waterbound isolation of forested Barro Colorado Island has led to the absence there of the two largest local carnivores (jaguar and puma). Cocha Cashu in Amazonian Perú is another protected rain forest area with little direct human impact today. Unlike Barro Colorado, Cocha Cashu is inhabited by jaguars and pumas, which are reported by John Terborgh and Louise Emmons to consume annually about 8% of the standing crop of terrestrial mammals weighing 1 kg or more. The absence of these large carnivores on Barro Colorado is believed to be why its population densities of medium-sized mammals (opossum, armadillo, rabbit, agouti, paca, and coati) are from 2 to more than 10 times greater than those at Cocha Cashu. Four of these medium-sized mammals are seed predators; their high numbers may lead to changes in the density and diversity of certain trees in the forest. The large populations of opossum, coati, and perhaps other medium-sized carnivores result in high levels of predation on birds that nest in the understory. Because some of these birds are important seed dispersal agents, their decline could also have long-term impacts on species composition of forest trees. These situations are complicated by the different size classes of seeds and mammalian and avian granivores, frugivores, and carnivores. Assuming that Barro Colorado Island and Cocha Cashu continue to be protected, it will be fascinating to compare the plant and animal communities of these two tropical forests a century from now to learn the overall impact of top-level predators. [See KEYSTONE SPECIES.]

The principle of keystone species relates directly to our central topic—the impact of traditional peoples on ecosystem integrity. The integrity of the Barro Colorado ecosystem has been weakened by the loss of large mammals. Much the same process occurs throughout the Neotropics in areas where the habitat is still largely intact but, during the past human generation or two, the local people have shifted from traditional subsistence to a more Western way of life that includes guns and chainsaws. In such newly developing regions of the Neotropics, I have always noticed that large species (crocodilians, eagles, currasows and other gamebirds, jaguar, puma, deer, peccary, tapir) are among the first to be eliminated or become scarce and wary. To study the Ocellated Turkey in 1977, James Stull, Stephen Eaton, and I went to Tikal National Park in northern Guatemala because the turkeys there were remarkably common and tame after years of relative insulation from human hunting and habitat modification. Through most of its range, this tasty bird is hunted, resulting in scattered flocks of very shy birds.

Tree kangaroos provide another instance where seemingly low levels of human impact have dramatic results. In New Guinea's remote Bewani Mountains, Jared Diamond has noted that tree kangaroos are nocturnal, shy, and arboreal in any area that is even very sparsely settled by people. What would seem to be almost insignificant hunting pressure (one visit per hunting party per forested valley per several years) has nearly eliminated the tree kangaroo, New Guinea's largest surviving terrestrial mammal.

III. CONTINENTAL ECOSYSTEMS

A. Human Predation

To build upon the themes introduced so far, let us compare the late Quaternary losses of large mammals in Africa, Europe, the Americas, and Australia. (The Pleistocene fossil record of Asia is too poorly known, especially in chronology, to be analyzed with much accuracy.) Unlike in earlier geological times, the extinction of mammals in the late Pleistocene is very biased toward large species (≥ 44 kg). Most species (or species lineages) of small mammals have survived the past 100,000 years.

Humans evolved in Africa, alongside other species of large mammals. Thus African large mammals experienced predation from humans as we were evolving anatomically and behaviorally. As a result, African large mammals have, to varying extents, adapted to and survived human predation. In Europe, humans and other large mammals also share a long history, but one that is more sporadic because of Pleistocene glaciation. In the Neolithic of Europe, there are hundreds of sites where human activity is associated directly with the remains of extinct large mammals. Human dwellings made of mammoth bones are among the more astonishing of these sites. Cave paintings may be the most beautiful.

Only seven genera of large mammals (an elephant, horse, camel, deer, and three bovids) were lost in Africa during the late Pleistocene (Table I). Although more were lost in Europe than in Africa, the European losses (a dhole, two hyenas, sabertooth, mammoth, elephant, horse, two rhinoceroses, hippopotamus, deer, and four bovids) were much less than those that occurred so rapidly at or near the end of the Pleistocene in the Americas and Australia. These differences (Fig. 2) are related to the evolution and dispersal of preagricultural humans (see Fig. 1).

The entire world experienced major changes in climate and vegetation during the Pleistocene, as glacial intervals ("ice ages") were succeeded by warmer interglacials, then cooled again to glacial intervals. Ten glacial/interglacial cycles of varying intensity are recorded for the past million years. During the last glacial to interglacial transition, the late Pleistocene to Holocene transition from about 14,000 to 8000 years ago, an overall warming trend resulted in major changes in the latitudinal, elevational, or edaphic ranges for most species of plants. Because individual species of plants seem to have responded independently rather than as entire communities, the plant communities that developed during the Holocene were not exactly like those that existed farther north or south or at lower elevations during the late Pleistocene. The distributions

TABLE I

Extinct versus Living Genera of Late Pleistocene Large (>44 kg) Mammals[a]

Continent	Extinct	Living	Total	% extinct	Major period of extinction
Australia	17	3	20	85	30–15 ka
South America	46	12	58	80	12–10 ka
North America	31	12	43	72	12–10 ka
Europe[b]	15	14	29	52	18–9 ka
Africa	7	42	49	14	18–9 ka

[a] Modified from P. S. Martin in P. S. Martin and R. G. Klein, eds. (1984). "Quaternary Extinctions: A prehistoric Revolution." University of Arizona Press, Tucson; and N. Roberts (1989). "The Holocene: An Environmental History." Basil Blackwell, Oxford, England. ka = 1000 years B.P.

[b] Excluding Mediterranean islands.

of species of animals also changed at this time, again somewhat independently rather than as intact communities. The great majority of species of plants and animals, however, had withstood similar dramatic changes in climate during the previous glacial–interglacial cycles. The last transition was different because of humans.

About 40,000 years ago, during times of glacially lowered sea levels, people walked from New Guinea into previously uninhabited Australia. Between 30,000 and 15,000 years B.P., all species of marsupials larger than the living grey kangaroo became extinct in Australia. The 13 genera that became extinct (Table I) represent a marsupial "lion," three wombats, a marsupial "tapir," four diprotodons, and eight kangaroos. The late Pleistocene loss of large mammals in Australia precedes the time of major changes in climate and vegetation. The arrival of h-f-g humans is the only likely explanation of the extinctions.

At the end of the Pleistocene in the Americas, nearly all species of amphibians, reptiles, and small mammals survived, whereas many large mammals became extinct. In particular, a fantastic variety of herbivores was lost in both North America and South America (Table I). In South America the extinct genera of mammals include nine glyptodonts, ten ground sloths, two litopterns, two toxodonts, capybara, sabertooth, bear, four gomphotheres, three horses, two peccaries, five camels, and four deer. The extinct genera from North America feature two glyptodonts, four ground sloths, two bears, two sabertooths, cheetah, beaver, two capybaras, two mastodons, mammoth, horse, tapir, two peccaries, three camels, two deer, pronghorn, and four bovids.

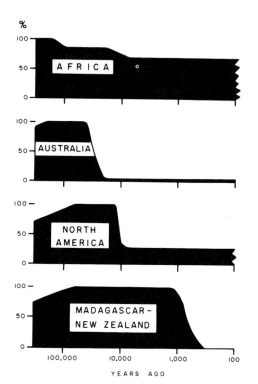

FIGURE 2 Percentage survival of large (≥44 kg) mammal genera in the late Quaternary. [From P. S. Martin and R. G. Klein, eds. (1984). "Quaternary Extinctions: A Prehistoric Revolution," Fig. 17.8. University of Arizona Press, Tucson.]

The collapse of the large herbivorous mammal communities in the Americas led to the demise of certain dependent species such as large carnivorous mammals, carrion-feeding birds, and dung beetles. Eleven thousand years ago, when ground sloths, mammoths, mastodons, horses, tapir, camels, and others still existed across North America, the currently endangered "California" Condor lived as far away from California as Florida and New York. Many other large scavenging birds thrived because of the abundant variety of carrion provided by herds of large animals, much as in modern African game parks.

Why were virtually all plants and nearly all small vertebrates in the Americas able to withstand the end of the Pleistocene, while large mammals and their associates were not? About 11,200 years ago, right in the heart of the Pleistocene to Holocene transition, North America's first people rapidly spread from Beringia and Alaska across the continent. Within only hundreds of years, descendants of these big-game hunters dispersed through Central and South America as well. Their spears were tipped with stylistically distinctive projectile points, the most characteristic of which (Clovis points; Fig. 3) have been found with the bones of extinct mammoths and mastodons at sites that have been radiocarbon dated to within a century or two of 11,000 years B.P.

The "Pleistocene overkill theory" contends that the early big-game hunters were the primary (if not sole) cause of the collapse of American large mammal communities. The overkill concept has been championed by Paul Martin, Jared Diamond, and other prominent ecologists who appreciate the skill and tenacity of traditional hunters. Scientists opposed to the overkill theory (who tend to favor climate change as the sole or main cause of late Pleistocene extinctions) have yet to explain why American large mammal communities that survived so many glacial–interglacial cycles would suddenly collapse 11,000 years ago, exactly when people first arrived in America.

Regardless of the cause, this collapse must have had a serious impact on the ecosystems that once included the extinguished grazers and browsers, and their predators and commensals. Considering the Barro Colorado example of what happens with the loss of only two keystone species, the demise of so many large mammals 11,000 years ago must be regarded as the single most important direct impact that prehistoric humans ever had on New World ecosystems.

With most large mammals extinct, some peoples shifted to a more generalized diet, whereas others specialized in hunting bison, which survived into the Holocene. Study of bones from prehistoric sites younger than 11,000 years B.P. corroborates this dietary diversity (except for the well-organized bison kills on the Great Plains). Bones from North American archaeological sites several thousand to several hundred years old often represent 20 to 50 species of vertebrates that range in size from frogs and songbirds to bear, deer, elk, and moose. These sites often record various birds and mammals from localities that are well outside of their known post-Columbian distribution. Intertribal trading of animals may account for some of these range extensions, although many if not most reflect formerly indigenous populations. Such species include Trumpeter Swan, Swallow-tailed Kite, Whooping Crane, Sandhill Crane, Long-billed Curlew, Carolina Parakeet, Ivory-billed Woodpecker, Common Raven, Fish Crow, rice rat, Allegheny wood rat, and puma.

B. Agriculture

The initial development of agriculture occurred during the latest Pleistocene to early Holocene transition, about 12,000 to 8000 years B.P. An especially important region was the Near East, where there is early evidence of domesticated wheat, barley, rye, peas, lentils, flax, cattle, sheep, goats, pigs, and dogs. In East Asia, newly domesticated species in the early and middle Holocene included rice, bottle gourd, foxtail millet, common millet, water chestnut, mulberry, chicken, pig, water buffalo, and zebu. The first use of finger millet, oil palm, sorghum, and perhaps yams occurred in Africa.

Rice, which feeds much of the world today, can be grown on dry fields or in irrigated plots. The cultivation of rice has had a profound environmental impact in Asia for many millennia, directly

FIGURE 3 Fluted projectile points (Clovis variety) from various North American sites, original specimens and accurate casts. (A) Wapanucket site, eastern Massachusetts, (B) Vail kill site #1, northwestern Maine, (C) Lamb site, western New York, (D) Naco site, Arizona, (E) Vail site, northwestern Maine, (F) Anzick site, Montana, (G) Mueller site, central Illinois, (H) Lamb site, western New York. (Photograph by R. M. Gramly)

through deforestation and landscape changes and indirectly by stimulating human population growth. In the well-watered lowland plains of greater Southeast Asia, the native forests have long been felled and the resulting barren land flooded by water whose levels are controlled by dams, levees, and channels. Most upland forests also have been lost as swidden (= shifting, or slash and burn) cultivation of various crops has intensified to feed growing human populations. Other upland areas have been terraced for centuries to support irrigated rice.

The prehistoric dispersal of rice has been documented, especially by Peter Bellwood and colleagues, by radiocarbon dating the rice grains or husks embedded in early pottery, and by the occurrence of stalks, grains, husks, or siliceous phytoliths of rice in dated cultural sediments. The earliest known cultivation of rice is in the middle and lower Yangzi Valley of central China at about 8000 years

B.P., whence it spread to southern China, northern Thailand, and Taiwan by 6000–5000 years B.P. Cultivated rice was in northern India, central Thailand, Borneo, and the Philippines by 4500–4000 years B.P., after which it continued to spread through the equatorial islands of Southeast Asia (Indonesia). Except for the Mariana Islands of Micronesia, rice was not cultivated prehistorically in Oceania, New Guinea, or Australia.

The earliest records of bananas, sugarcane, taro, and perhaps yams are from New Guinea. Excavations in highland New Guinea have revealed human disturbance of swamps at about 9000 years B.P. (a ditch 2 m wide, 1 m deep, and 450 m long) and complex water management systems for swamp cultivation (presumably taro) at about 6000 years B.P. Pollen records show significant forest clearance by about 5000 years B.P.

In the Americas, domesticated bottle gourd, chili pepper, avocado, beans, squash, maize, dog, and llama/alpaca all existed at or before 6500 years B.P. Maize, probably the most important traditional New World food crop, has been traced to wild or early domesticated forms with much smaller cobs from southern México about 7000 years B.P. The more sedentary existence that came with agriculture was accompanied by a reduced dependence on wild plants and animals. This transition probably was gradual, especially with shifting cultivation rather than long-term use of the same plots. In fact, hunting, fishing, and gathering remained a significant part of subsistence in most of the New World long after the domestication of plants and animals. Even at the time of European contact, North American Indian farmers regularly hunted deer, elk, moose, bison, black bear, turkey, and many smaller species.

Chaco Canyon, in northeastern New Mexico, provides a clear example of prehistoric overexploitation of critical natural resources. This story has been told by Julio Betancourt, Thomas Van Devender, and colleagues, who studied radiocarbon dated plant macrofossils in ancient packrat middens. Spectacular buildings were constructed at Chaco Canyon from about A.D. 900 to 1150 by American Indians known as the Anasazi. This construction involved the use of 200,000 beams from large coniferous trees (ponderosa pine, subalpine fir, and blue/Englemann spruce) brought from as far away as 75 km, Chaco Canyon being at slightly too low an elevation to support these alpine conifers. When the Anasazi arrived, the area around Chaco Canyon was a pinyon–juniper woodland. These two small conifers provided firewood for the Anasazi, who farmed the narrow band of floodplain soil near their buildings.

As time progressed, the Anasazi had to go 15 km or more from Chaco Canyon just to gather firewood, the area nearby having been transformed from a pinyon–juniper woodland to a nearly woodless desert scrub. The removal of pinyon, juniper, and riparian trees produced soil erosion, increased runoff, and probably a lowered water table in Chaco Canyon. This made it difficult to continue irrigated farming of the valley bottom. Stressed agriculture and lack of firewood led to the Anasazi's abandonment of Chaco Canyon more than 500

years ago. Scattered junipers grow at Chaco Canyon today, but no pinyon. The incised streambed would be virtually impossible to modify for agriculture. Because of human activities that began more than 1000 years ago, Chaco Canyon remains a poor place to live.

IV. ISLAND ECOSYSTEMS

The relatively small land areas of most islands result in small populations of organisms that therefore are more vulnerable to habitat modification, direct human predation, and the pathogens, predators, or competitors introduced by people. It is well known that a disproportionate number of species have become extinct on islands during the past several centuries. In the past decade or so, we have learned that an even larger number of human-caused extinctions occurred on islands in prehistoric times. On several Caribbean islands, prehistoric cultural features and artifacts are clearly associated with the bones of extinct populations and species of lizards, snakes, birds, bats, and rodents. On Madagascar, a huge island renowned for its spectacular endemic biota, archaeological sites have yielded the butchered bones of many large extinct species that did not survive into recent centuries, such as tortoises, elephantbirds, many kinds and sizes of lemurs, an aardvark, and a small hippopotamus.

One way to evaluate the environmental impacts of traditional peoples on oceanic islands is to study the biotic history of islands that never were inhabited in pre-Columbian times. The human history of the Galápagos Islands, for example, begins with brief explorations by the Spanish and British in the sixteenth and seventeenth centuries, before which the Galápagos biota truly was pristine. Knowing that extinction of indigenous populations and species seems to be a regular consequence of human occupation of islands, we can look at the pre-Columbian biotic history of the Galápagos to determine the level of natural or "background" extinction (i.e., nonanthropogenic extinction). Biogeographic theory predicts that extinction (often offset by colonization to yield an "equilibrium" number of species) is a part of an island's biotic history. Knowing this rate of natural extinction would allow us to comprehend the severity of the human-caused extinction that we witness today.

In the Galápagos, 34 populations or species of tortoises, lizards, snakes, bats, rodents, hawks, owls, mockingbirds, and finches are known to have become extinct during the past 8000 years. The Holocene vertebrate fossil record from the Galápagos consists of 15 sites from five different islands, with a total of nearly 500,000 identified bones. This record reveals only three extinct populations or species that are not known to have survived to the period of human contact. At least 31 of the 34 taxa that became extinct died out during the past 200 years, after the arrival of people. Thus the rate of natural extinction for reptiles, birds, and mammals in the Galápagos is at least two orders of magnitude less than the posthuman rate. [*See* GALAPAGOS IS-LANDS.]

Birds and mammals of the Galápagos Islands are an excellent example of the extreme tameness of animals that evolved in the absence of humans and other mammalian predators. Not surprisingly, the tameness of insular species leads to great vulnerability with the arrival of people and other nonnative predators (rats, cats, dogs, mongoose, stoats, pigs, etc.). The people who first arrived in the Galápagos were westernized rather than traditional. How important is this distinction? From the standpoint of damaging the integrity of indigenous ecosystems, the arrival of traditional peoples tends to be a more "gentle" event than the arrival of more technologically advanced peoples. Nevertheless, as is apparent from the Caribbean and Madagascar, and as we now will see in Polynesia, the arrival of any group of humans on a previously uninhabited island is likely to have major environmental implications.

Nine to ten thousand species of birds are believed to exist today. The prehistoric human colonization of Oceania (Melanesia, Micronesia, Polynesia) resulted in the loss of as many as two or three thousand species of birds. In other words, the world's avifauna would be 20 to 25% richer today had islands of the Pacific remained unoccupied by humans. Because of such pervasive extinction, biogeographic analyses of Pacific island avifaunas are of limited value unless they consider extinct species along with the relatively few that survive.

The arrival of Polynesians (the Maori) in temperate New Zealand about 1000 years ago initiated the extinction of at least 34 species of birds (moas, pelican, waterfowl, eagle, harrier, rails, aptornithids, owlet-nightjar, xenicids, and raven). In the Hawaiian Islands, the arrival of Polynesians nearly 2000 years ago led to the loss of at least 62 species (petrel, ibises, waterfowl, eagle, harrier, rails, owls, crows, honey eaters, thrush, and finches). These numbers do not include the extirpation or severe reduction in size of individual island populations of extant species.

Most of the prehistoric, human-caused extinction of Polynesian birds did not involve the isolated, highly endemic avifaunas of New Zealand and Hawaii. Three examples will be discussed, the first being Huahine (77 km^2, 16°45'S, 151°W), an island near Tahiti in French Polynesia. Recently Dominique Pahlavan and I studied bird bones from the 1000-year-old Fa`ahia archaeological site on Huahine. The bones, which represent birds that were killed for their flesh and feathers, increase the seabird and landbird fauna of Huahine from the historically known 4 to 15 species and from 7 to 18 species, respectively (Table II). Extinct species of landbirds include a rail, two doves, two parrots, and a starling, whereas a heron, rail, dove, two pigeons, and warbler are extirpated. The occurrence of so many extinct or extirpated species of birds at Fa`ahia suggests that this site represents a very early period of human occupation on Huahine, probably no more than 500 years after people settled on this previously undisturbed island.

The second example is the small island of Mangaia (52 km^2, 21°55'S, 157°55'W) in the Cook Islands. Unlike on Huahine, the data from Mangaia include more than 100 archaeological and paleontological sites that document a variety of prehistoric environmental changes. In 1989 and 1991, Patrick Kirch and I excavated archaeological deposits on Mangaia that produced rich sequences of cultural material associated with bones of 5 species of seabirds (shearwater, petrel, storm petrel, booby, and tern) and 13 species of landbirds (four rails, sandpiper, six pigeons, and two parrots) that no longer occur on the island. Even though this site represents nearly 1000 years of human occupation, the bones of most extinct or extirpated birds were confined

TABLE II
Loss of Birds During the Past 1000 Years on Huahine, French Polynesia[a]

	Archaeological record	Modern record
SEABIRDS		
Shearwaters, petrels		
Puffinus pacificus	X	—
Puffinus nativitatis	X	—
Puffinus lherminieri	X	—
Pterodroma rostrata	X	—
Pterodroma alba	X	—
Pterodroma arminjoniana	X	—
Tropicbirds		
Phaethon lepturus	X	X
Boobies		
Sula leucogaster	X	—
Sula sula	X	—
Frigatebirds		
Fregata minor	X	—
Fregata ariel	X	—
Gulls, terns		
★*Larus* new sp.	X	—
Anous stolidus	X	X
Anous minutus	—	X
Gygis alba	X	X
LANDBIRDS		
Herons		
Egretta sacra	X	X
Ardeola striata	X	—
Ducks		
Anas superciliosa	—	X
Rails		
★*Gallirallus* new sp.	X	—
Porzana tabuensis	X	—
Pigeons, doves		
Gallicolumba erythroptera	X	—
★*Gallicolumba nui*	X	—
Ptilinopus purpuratus	X	X
★*Macropygia arevarevauupa*	X	—
Ducula galeata	X	—
Ducula aurorae	X	—
Parrots		
Vini peruviana	—	e
★*Vini vidivici*	X	—
★*Vini sinotoi*	X	—
Swifts		
Collocalia leucophaea	—	e
Kingfishers		
Halcyon cf. *tuta*	X	X
Warblers		
Acrocephalus caffer	X	e
Starlings		
★*Aplonis diluvialis*	X	—
Total species	29	11
Combined total species	33	
Total seabirds	14	4
Combined total seabirds	15	
Total landbirds	15	7
Combined total landbirds	18	
Total modern landbirds (without e)	—	4

[a] Combined total = modern plus prehistoric; asterisk = extinct species; e = historic record, now extirpated. Updated from D. W. Steadman (1989). *J. Archaeol. Sci.* **16**, 177–205.

to strata more than 700 years old, suggesting that these species were depleted if not exterminated early in the human history of Mangaia.

The loss of so many birds on Huahine, Mangaia, and hundreds of other oceanic islands was not due solely to human predation. Other contributing factors, still operative today, were predation from prehistorically introduced mammals (rats, dogs, pigs) habitat change wrought by deforestation and agricultural development, and perhaps introduced avian diseases. On Mangaia, studies by John Flenley, Stuart Dawson, Frances Lamont, and Joanna Ellison show that the forested interior of Mangaia began to be burned and logged shortly after people arrived about 1600 years B.P. (Fig. 4). The radiocarbon-dated lake sediments reveal decreases in the pollen of trees (forest indicators) and increases in the spores of ferns (indicators of disturbance, especially by fire). These changes were accompanied by increased rates of sediment influx because upland soils and subsoils eroded following deforestation. Increased concentrations of iron, silicon, and aluminum coincided with this major erosional event, which deposited inorganic clays in the lake basin.

The same events that were destructive to indigenous Mangaian forests had direct benefits for the early Mangaian people. The infilling of lake basins allowed the efficient cultivation of taro, an aroid whose starchy tuber still is the most important food crop in much of Polynesia. Not unlike rice cultivation in Southeast Asia, long-term plots for taro were developed on Mangaia through building and maintaining networks of dikes, levees, and channels that controlled the flow of water and influx of sediment in Mangaia's basins (Fig. 5). This rich, relatively reliable source of food fueled increases in the human population, which further stressed the remaining forest biota and which led to violent rivalries over access to prime taro fields.

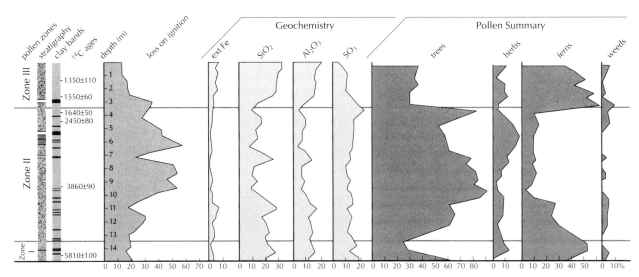

FIGURE 4   Chronostratigraphic summary of geochemistry and pollen from lake sediments on Mangaia, Cook Islands. [From P. V. Kirch, J. R. Flenley, D. W. Steadman, F. Lamont, and S. Dawson (1992). *Nat. Geogr. Res. Explor.* **8,** 166–179.]

The third Polynesian example is Easter Island (Rapanui), a rather large (171 km^2) southeast Pacific island (27°9'S, 109°26'W,) renowned for its great isolation and massive stone sculptures. In 1991, Claudio Cristino F., Patricia Vargas C., and I excavated a small trench at the coastal site of Ahu Naunau. We recovered more than 1400 artifacts of obsidian and basalt from throughout the deposit, which consisted of a basal clay overlain by calcareous sand. Radiocarbon dates on wood charcoal ranged from 900 ± 60 years B.P. at the sand–clay contact to 660 ± 80 years B.P. near the top of the deposit.

The vertebrate assemblage of 7310 identifiable bones was dominated by porpoise (40% of all bones), Pacific rat (32%), and fishes (23%). We recovered no bones of pigs, dogs, or lizards, and only a single human bone. The nonnative chicken,

FIGURE 5   Modern swamp taro cultivation on Mangaia, Cook Islands. Note the treeless hills in the background.

so common in late prehistoric Rapanui middens, represented only 0.4% of the bone assemblage. The abundance of porpoise bones suggests that the prehistoric Rapanui had seaworthy dugout canoes for harpooning porpoises offshore until about 600 years B.P. The absence or scarcity of porpoise bones in Easter Island's late prehistoric sites (<500 years B.P.) probably reflects the lack of such canoes during the late prehistoric period, as mentioned in the earliest European ethnographic accounts of the eighteenth century. Prehistoric deforestation, documented by John Flenley and colleagues in paleobotanical studies similar to those just described from Mangaia, had eliminated the raw material (large trees) needed to make dugout canoes.

Bones of indigenous birds (5%) were found throughout the deposit. They included 14 species of seabirds, 9 of which no longer live on Easter Island or its offshore islets. Combined with previously unstudied bones excavated in the 1970s and 1980s, the prehistoric seabird fauna of Easter Island stands at 22 species (Table III), 14 of which are extirpated (albatross, fulmar, prion, five petrels, two shearwaters, storm petrel, tropicbird, and two terns; six of these nest today only south of 30°S latitude). The 1991 excavations also produced seven fragmentary bones of landbirds that seem to represent 6 extinct species (heron, two rails, two parrots, barn owl). The loss of at least 20 species of birds is only one aspect of the environmental degradation of Easter Island, which also includes extinctions of a palm tree and land snails. Such data help to interpret changes in prehistoric subsistence and ethnobiology of the Rapanui people, as well as to elucidate the biogeographic affinities of the Easter Island biota.

Even though most species of birds were lost on all three islands, why were a few more able to survive on Mangaia and Huahine than on the larger Easter Island? The answer seems to lie in the "user-friendly" topography of Easter Island, which consists of three coalesced shield volcanoes with gentle slopes. Mangaia and Huahine, on the other hand, have large areas that are difficult to traverse and virtually impossible to cultivate. The outer perimeter of Mangaia consists of creviced, pinnacled limestone with little soil, whereas much of the interior

TABLE III

Loss of Seabirds During the Past 1000 Years on Easter Island[a]

	Prehistoric record	Modern Status	Modern breeding range (°S)
Albatrosses			
Diomedea sp.	X	e	01–51
Petrels, shearwaters			
Fulmarus glacialoides	X	e	53–68
Pachyptila vittata	X	e	36–65
Pterodroma macroptera/ lessoni	X	e	34–50
Pterodroma ultima	X	e	15–27
Pterodroma externa	X	e	24–33
Pterodroma heraldica	—	b	08–27
Pterodroma neglecta	—	b	20–33
Procellaria sp.	X	e	36–50
Puffinus carneipes	X	e	31–42
Puffinus nativitatis	X	b	00–27
Puffinus griseus	X	e	32–50
Procellariidae new sp.	X	e	?
Storm petrels			
Nesofregetta fuliginosa	X	e	00–27
Tropicbirds			
Phaethon rubricauda	X	B	00–27
Phaethon lepturus	X	e	00–24
Frigatebirds			
Fregata minor	X	b?	00–27
Boobies			
Sula dactylatra	X	b	00–27
Terns			
Sterna fuscata	X	b	00–27
Sterna paradisaea	X	e?	—
Sterna lunata	—	b?	00–27
Procelsterna cerulea	X	b	00–33
Anous stolidus	X	b	00–33
Gygis candida	X	b	00–27
Gygis microrhyncha	X	e	08–10
Totals	22	B = 1, b = 8–10 e = 13–14, combined = 25	

[a] B = breeds today on Easter Island itself; b = breeds today on islet(s) offshore from Easter Island; e = extirpated on Easter Island and all offshore islets. From author's unpublished data.

of Huahine consists of very steep ("knife-edge") hills of eroded volcanic rocks. As a result, prehistoric deforestation was more complete on Easter Island.

The prehistoric and ongoing extinction of so many populations and species of birds on South Pacific islands has consequences far beyond losing the birds themselves. For example, the loss of so many populations and a few entire species of seabirds undoubtedly has impacts on marine food webs, in which seabirds are top consumers. For landbirds, most of the extinct species were omnivores, frugivores, granivores, or nectarivores. Indigenous plants, particularly trees and shrubs, must have depended on these birds for pollination or seed dispersal. Today, many South Pacific trees and shrubs seem to have no natural means of intra- or interisland dispersal. The exact implications of these situations are rich topics for future research.

V. RELATING THE PAST TO THE PRESENT

No ecosystems on earth function today with complete independence from human influence. Our activities, especially those of the past 200 years, have altered some physical, chemical, or biotic aspects of even the ecosystems that are most isolated from direct human habitation, such as those of Antarctica, the deep oceans, or the highest mountains.

Many terrestrial ecosystems of the Americas, sub-Saharan Africa, Southeast Asia, New Guinea, Australia, and Oceania seemed initially to Westerners to be nearly free of significant human impact. We now know that this was not the case. Unable to return to a time when at least some ecosystems functioned without human influence, we should use historical and ethnographic environmental data to help evaluate current situations, with a goal of preserving ecosystems that at least approach a natural state. The long-term data provided by studies of prehistoric human impact allow us to study plant and animal communities in time intervals of potential evolutionary significance (millennia) rather than the seasons, years, or at best decades that characterize most ecological studies.

As one example, let us consider the birds that existed in North America at European contact. (The North American avifauna already had lost at least 20 to 40 species at the end of the Pleistocene, mainly because of the extinction of so many large mammals on which they depended.) In the past 200 years, at least 5 species have been lost (Great Auk, Labrador Duck, Passenger Pigeon, Carolina Parakeet, Ivory-billed Woodpecker). Two others, the Eskimo Curlew and Bachman's Warbler, are either extinct or very nearly so. At least 6 other species persist today only in dangerously small, localized populations (California Condor, Greater and Lesser Prairie Chickens, Whooping Crane, Red-cockaded Woodpecker, and Kirtland's Warbler). Even if every New World country maintains a reasonable level of political, economic, and environmental stability, some of these 6 species are likely to die out in the next 200 years, as are others known today to be in decline (such as Piping Plover, Spotted Owl, Common Nighthawk, Red-headed Woodpecker, Sedge Wren, Loggerhead Shrike, Cerulean Warbler, Golden-cheeked Warbler, and Henslow's Sparrow).

We should shudder at the prospect that temperate North America may continue to lose species of birds at rates of 20 to 30 per millennium. For if we extrapolate very far into the future, there may be hundreds fewer species of North American birds to face the changing climates and habitats that will come with the next ice age. On a smaller time scale, many species of North American trees today face the projected northward movement of their ranges because of global warming. Unlike in the past, these species must accomplish their range shifts without the Passenger Pigeon, until a century ago the most abundant consumer and disperser of their seeds.

The biotic world as we know it is much impoverished by the activities of ourselves and our ancestors. Reversing this trend is the premier challenge of the twenty-first century.

Glossary

Community The organisms that inhabit an ecosystem. May be divided into various taxonomic or ecological categories, such as the amphibian community or the frugivore community.

Ecosystem The organisms that live in a particular habitat (forest, salt marsh, coral reef, etc.), and the physical aspects of the environment (bedrock, soil, water, air, climate, etc.) upon which the organisms depend.

Ecosystem collapse Alteration of an ecosystem to the point where it ceases to function as the natural entity it once was. Examples of human-caused ecosystem collapses would be clear-cutting a tropical rain forest for conversion to pasture; the introduction, establishment, and dominance of alien species of plants and animals on an island, leading to the loss of native species; draining a wetland for conversion to agriculture; or toxic pollution and overfishing of a lake accompanied by loss of native species and establishment of alien species. Naturally caused ecosystem collapses, which today are less frequent than anthropogenic ones, would include catastrophic geological or climatic events, such as an island biota wiped out by volcanic eruption or a lake drained by seismic activity.

Ecosystem integrity Natural quality or wholeness of an ecosystem. An ecosystem functions with full integrity when all of its biotic and physical components exist without human influence, and therefore variations in any aspect of these components are due to natural processes.

Extinction Worldwide loss of all individuals of any taxonomic category (population/subspecies to phylum); most often used for species. The Passenger Pigeon (*Ectopistes migratorius*) and Steller's sea cow (*Hydrodamalis gigas*) are examples.

Extirpation Local (as opposed to global) loss of any taxonomic category, but most often used for species and subspecies; usually implies human involvement in the loss. Examples include the Wapiti or American elk (*Cervus canadensis*) in eastern North America and the African lion (*Panthera leo*) in the Mediterranean region.

Holocene Geochronological epoch that represents the last 10,000 years.

Keystone species Species whose loss or addition perturbs an ecosystem in trophic levels other than its own, resulting in changes in community structure and perhaps even the physical nature of the ecosystem.

Paleoecology Study of past ecosystems, using a broad approach that combines geology, archaeology, paleontology, and climatology with modern principles and methods of ecology.

Pleistocene Geochronological epoch preceding the Holocene and following the Pliocene; begins about 2 million years ago and ends 10 thousand years ago. The earth's most recent series of glacial intervals ("ice ages") occurred during the Pleistocene.

Quaternary Geochronological period that represents the past 2 million years (= Pleistocene + Holocene). Quaternary is used frequently in paleoecological studies because the time interval sampled often crosses the Pleistocene–Holocene boundary.

Radiocarbon dating Method to date organic materials based on the decay of radioactive carbon (^{14}C) that was incorporated into the organism when it was alive; effective only for materials more than 200 years old and less than 50,000 to 35,000 years old. A radiocarbon date usually is expressed in "years before present" (years B.P.), with A.D. 1950 designated as "present."

Bibliography

Betancourt, J. L., Van Devender, T. R., and Martin, P. S., eds. (1990). "Packrat Middens: The Last 40,000 Years of Biotic Change." Tucson: University of Arizona Press.

Denslow, J. S., and Padoch, C., eds. (1988). "People of the Tropical Rain Forest." Berkeley: University of California Press.

Diamond, J. (1992). "The Third Chimpanzee." New York: Harper Collins.

Flood, J. (1989). "Archaeology of the Dreamtime: The Story of Prehistoric Australia and Its People." New Haven, Conn.: Yale University Press.

Kirch, P. V., Flenley, J. R., Steadman, D. W., Lamont, F., and Dawson, S. (1992). Prehistoric human impacts on an island ecosystem: Mangaia, Central Polynesia. *Nat. Geogr. Res. Explor.* **8,** 166–179.

Martin, P. S., and Klein, R. G., eds. (1984). "Quaternary Extinctions: A Prehistoric Revolution." Tucson: University of Arizona Press.

Nitecki, M. H., and Nitecki, D. V., eds. (1986). "The Evolution of Human Hunting." New York: Plenum Press.

Pielou, E. C. (1991). "After the Ice Age." Chicago: University of Chicago Press.

Roberts, N. (1989). "The Holocene: An Environmental History." Oxford, England: Basil Blackwell.

Steadman, D. W. (1989). Extinction of birds in eastern Polynesia: A review of the record, and comparisons with other Pacific island groups. *J. Archaeol. Sci.* **16,** 177–205.

Steadman, D. W. (1991). Extinction of species: Past, present, and future. *In* "Global Climate Change and Life on Earth" (R. L. Wyman, ed.), pp. 156–169. New York: Routledge, Chapman, & Hall.

Steadman, D. W., Stafford, T. W., Jr., Donahue, D. J., and Jull, A. J. T. (1991). Chronology of Holocene vertebrate extinction in the Galápagos Islands. *Quaternary Res.* **36,** 126–133.

Wilson, E. O. (1992). "The Diversity of Life." Cambridge, Mass.: Belknap Press.

Ecotoxicology

A. W. Hawkes, T. R. Rainwater, and R. J. Kendall
Clemson University

Numerous definitions for ecotoxicology have been developed over the years. The broadest interpretations define ecotoxicology as the study of environmental contaminants and their effects on ecosystems. Ecotoxicology incorporates the scientific methodology utilized in the fields of toxicology and ecology to test hypotheses investigating ecosystem response to environmental contaminants.

ronment from anthropogenic (human-made) and naturally occurring contaminants. The ultimate goal of ecotoxicology is to understand the relationship between contaminants and the ecosystem such that the ecological risk can be identified. In turn, reasonable policies concerning management of environmental contaminants may be developed. [*See* GLOBAL ANTHROPOGENIC INFLUENCES.]

I. INTRODUCTION

Ecotoxicology is a specialized area of study within the broader discipline of environmental toxicology. Environmental toxicology is an extremely interdisciplinary field, drawing from the subdisciplines of ecotoxicology, biochemical toxicology, behavioral toxicology, analytical toxicology, and ecological modeling. Ecotoxicology can be further subdivided into aquatic and terrestrial ecotoxicology. Determining the extent of chemical-induced stress on an ecosystem involves identifying compounds of concern, defining the compound's chemical fate (transport, transformation, and degradation), and measuring ecological effects on terrestrial and aquatic systems at risk. Interest in ecotoxicology is derived from the desire to develop accurate assessments of potential risk to the envi-

II. COMPOUNDS OF CONCERN

Compounds of ecotoxicological concern include any human-made or natural compound that may affect ecosystem integrity. The most common compounds of concern include pesticides (insecticides, herbicides, fungicides, etc.), heavy metals (mercury, lead, cadmium, etc.), oil/petroleum products, and domestic and industrial waste. These contaminants may be released into the environment from "point" (e.g., pesticide applications, industrial discharge) or "nonpoint" (e.g., agricultural or urban runoff, atmospheric deposition) sources. The fate of a contaminant is determined by its physical and chemical properties, and identifying the fate of the compound(s) of concern provides information for determining the distribution in various environmental compartments (air, water, soil, sediment, or biota).

III. MEASURING ECOTOXICOLOGICAL EFFECTS

Measuring ecotoxicological effects involves determining pollutant impact on the structure and function of an ecosystem. Although ecotoxicologists are primarily interested in the effects of environmental contaminants on populations, communities, and ecosystems, they must first understand the impacts of these substances at the organismal level. Laboratory studies are conducted to determine dose–response relationships and to pinpoint the biochemical and physiological mechanisms by which a given compound exerts its effect. This information is then used to design and conduct extensive field investigations. Pen studies, mesocosm (experimental ecosystem) studies, and full-scale field studies are all employed to help determine the overall ecosystem impact of the contaminant(s) in question. Intensive sampling and analysis of both biotic (plant and animal) and abiotic (soil, water, sediment, and air) matrices are essential for determining the direct (e.g., mortality) and indirect (e.g., response to habitat alteration) effects of a contaminant. Typical ecotoxicological end points evaluated are designed to characterize contaminant-induced population level effects and impacts on ecosystem diversity, productivity, resistance, and resilience.

IV. TERRESTRIAL ECOTOXICOLOGY

Terrestrial ecotoxicology focuses on the effects of environmental contaminants on terrestrial ecosystems. Early terrestrial ecotoxicology dealt primarily with exposure and impacts of toxic compounds to birds and mammals. Mammals have often been the focus of toxicity testing because they are considered sentinels to human health hazards. Much attention has also been directed toward the effects of toxic substances on birds. For example, during the 1940s and 1950s, ingestion of lead shot by water-

fowl and exposure of some species to the pesticide DDT were associated with declines of certain avian populations. However, concern has shifted to a broader spectrum of biota. Birds, mammals, reptiles, amphibians, insects and other invertebrates, microorganisms, plants, and soils are all studied to determine how chemical insult to a certain trophic level may influence the productivity and health of the whole ecosystem.

V. AQUATIC ECOTOXICOLOGY

Aquatic ecotoxicology focuses on the effects of environmental contaminants on aquatic ecosystems, ranging from the smallest stream to oceans. This field is extremely complex because of the behavior and fate of different compounds in water and sediments. As in terrestrial ecotoxicology, the physical and chemical properties of a particular environment will dictate contaminant bioavailability to various ecosystem components. Although historically aquatic ecotoxicology studies focused primarily on the impacts of contaminants to fish, a wider variety of biota are studied today. For example, in addition to fish, test species include benthic (sediment-dwelling) and other aquatic invertebrates, reptiles, amphibians, aquatic plants, and algae. [See POLLUTION IMPACTS ON MARINE BIOLOGY.]

It is important to note that aquatic and terrestrial systems are in constant interaction and together make up a larger ecosystem. Food chain transfer of DDT and associated metabolites (breakdown products) is an example of aquatic/terrestrial interactions. DDT was used extensively after its introduction in the mid-1940s, and although it was primarily applied to terrestrial systems, its persistence and widespread use resulted in the transport of toxic metabolites to aquatic systems. As a result, populations of several species of fish-eating birds (bald eagle, osprey, etc.) declined due to eggshell thinning associated with DDT exposure.

Glossary

Bioaccumulation Process by which chemicals are taken up by organisms from water, food, or other sources containing the chemical. If the chemical is persistent and is selectively retained by the organism, the organism may accumulate high concentrations of the chemical in the body.

Biomagnification Process by which tissue concentrations of bioaccumulated chemicals increase as the chemical passes up through two or more food chain (trophic) levels.

Diversity In a community, the combination of richness (number of species) and abundance (numbers of individuals) of various species.

Ecology Study of the structure and function of ecosystems; the totality or pattern of relations between organisms and their environment.

Ecosystem Interacting system consisting of the community and its associated nonliving environment.

Productivity Rate at which radiant energy is converted by photosynthetic and chemosynthetic activity of producer organisms (chiefly green plants) to organic substances.

Resilience Ability of an ecosystem to recover from disturbance.

Resistance Ability of an ecosystem to maintain its structure and function in the face of disturbance.

Risk assessment Process of hazard identification, dose-response evaluation, exposure assessment, and risk characterization designed to estimate the probability of increases of disease or death based on exposure to a substance.

Toxicology Scientific study of the detection, occurrence, properties, effects, treatment, mechanisms, and regulation of toxic substances.

Bibliography

Calow, P., ed. (1992). "Handbook of Ecotoxicology." Oxford, England: Blackwell Scientific.

Forbes, V. E., and Forbes, T. L. (1994). "Ecotoxicology in Theory and Practice" (M. H. Depledge and B. Sanders, eds.), Chapman & Hall Ecotoxicology Series 2. London: Chapman & Hall.

Kendall, R. J., and Akerman, J. (1992). Terrestrial wildlife exposed to agrochemicals: An ecological risk assessment perspective. *Environ. Toxicol. Chem.* **11,** 1727–1749.

Kendall, R. J., and Lacher, T. E., Jr. (1994). "Wildlife Toxicology and Population Modeling: Integrated Studies of Agroecosystems." Boca Raton, Fla.: Lewis Publishers.

Peterle, T. J. (1991). "Wildlife Toxicology." New York: Van Nostrand Reinhold.

Endangered Species—*See* Duties to Endangered Species
Environmental Chemistry—*See* Aspects of the Environmental Chemistry of Elements

Environmental Radioactivity and Radiation

Thomas F. Gesell

Idaho State University

I. Introduction
II. Natural Radioactivity and Radiation
III. Human-Made Environmental Radioactivity
IV. Effects of Radiation on Humans
V. Effects of Environmental Radiation on Plants and Animals
VI. Conclusions

The study of environmental radioactivity broadly includes the sources, transport, and distribution of radioactivity and radiation in the environment. Environmental in this case refers to the natural environment, communities, and homes, but not workplaces or medical establishments. Concern about adverse health effects in humans often stimulates environmental radioactivity studies, but there is also considerable interest from disciplines such as geophysics, oceanography, atmospheric science, archeology, and biology. The two traditional, major categories of environmental radioactivity are natural and human-made, but alteration of the natural radiation environment by humans is receiving increased attention. The latter category is called technologically enhanced natural radiation.

I. INTRODUCTION

Experience with natural environmental radioactivity and its risks preceded the discovery of radioactivity. Mines in central Europe had been exploited for their heavy metals since medieval times. The atmospheres of these mines were unknowingly so radioactive that the miners developed a fatal lung disease that was later diagnosed as lung cancer. It was not until nearly 400 years later that the air of the mines was found to contain high concentrations of the radioactive gas radon.

Some radioactive substances were used even before it was known that they were radioactive. The gas mantle, which was developed in 1885, used the incandescent properties of thorium–cerium oxide to greatly increase the luminosity of gaslight. Uranium oxide has long been used to provide a vivid orange color in ceramic glazes. Other oxides of uranium and thorium have also been used as glazes and for tinting glass. These uses brought natural radioactivity into proximity with humans.

Shortly after the discovery of radioactivity in 1896, natural radioactivity began to be exploited for its supposed benefit to health. The popularity of radioactive mineral waters in spas continues to this day at places such as Saratoga Springs in New York State and the spas of Europe, Japan, and South America. Radioactive Brazilian beaches and the high radon concentrations of old mines in Austria and the United States have long attracted tourists. Many of these visitors believe that exposure to natural radioactivity can cure arthritis, general debility, and a variety of other diseases. The potential beneficial effect of low levels of radiation is called radiation hormesis, a subject that remains controversial among most scientists.

In the early 1920s and continuing up to about 1940, radioactive substances, particularly radium and radon, found a place in the medical faddism of the period. Radium was injected intravenously for a variety of ills but, far from being cured, many patients later developed bone cancer. Devices that were sold for home use made it possible to add radon to drinking water, and radioactive poultices were prescribed for arthritic joints.

Another early experience with natural radioactivity that had unfortunate consequences was the use of radium in luminous paints. A slight amount of radium added to a suspension of zinc sulfide caused the material to glow. Thousands of timepieces, compasses, and instruments were painted with the mixture during and immediately after World War I, with no precautions to protect the employees. Many cases of aplastic anemia and bone cancer developed among the factory workers who ingested radium while engaged in applying these luminous paints prior to about 1940. By then hygienic and safety practices prevented further injuries from occurring.

Some information about environmental radioactivity was available before World War II, but little was known outside the few, highly specialized laboratories that could make measurements of radioactivity. The world inventory of radioactive materials was confined to those found in nature, and less than 1 MBq (mega-becquerel) of artificial radioactivity produced in particle accelerators.

The 1939 discovery that energy contained within the atomic nucleus can be released by fission (splitting) brought a new dimension to environmental radioactivity. Knowledge of nuclear fission coincided with the outbreak of World War II, and the first application of nuclear energy was for military purposes. During World War II, large plutonium-producing reactors and processing facilities near Hanford in Washington State were responsible for the first major radioactive releases to the environment. Contamination was released to a lesser extent at other major U.S. nuclear research and production centers. The early Hanford studies on the behavior of various radionuclides in the environment are classics in the field and demonstrate the caution one must adopt in discharging radioactive substances to the environment.

By far the largest share of radioactivity released to the environment by human activities has been through the detonation of nuclear weapons. Radionuclides produced in the nuclear explosions in various parts of the world soon permeated the atmosphere, the soils, and the food chains. Widespread apprehension began to develop, first in certain scientific circles and later among the public. Protracted studies of the survivors of the bombings in Japan, radium ingestion, medical uses of radiation, and radon in uranium mines have provided data on the long-term effects of radiation. [See GLOBAL ANTHROPOGENIC INFLUENCES.]

The most obvious long-range benefit from fission was the promise of cheap power. A half century since the nuclear reactor was first demonstrated in 1942, reactors are producing a large fraction of the electricity used in many countries of the world, including France (73%), Belgium (59%), Sweden (52%), Hungary (48%), South Korea (48%), Switzerland (40%), Taiwan (38%), Spain (36%), Bulgaria (34%), Finland (33%), Czechoslovakia (29%), Germany (28%), Japan (24%), the United States (22%), and the United Kingdom (21%). Energy demands are expected to increase and the reserves of fossil fuel will become smaller. Unless dramatic breakthroughs occur in renewable energy sources, nuclear energy is likely to play an increasing role in civilian economies. Routine releases from nuclear power stations, as well as from the mining, milling, enrichment, and processing of uranium, make small contributions to environmental radioactivity. However, several accidents, most recently the accident at Chernobyl in the Ukraine, have resulted in large environmental releases.

The widespread interest in the subject of environmental radioactivity resulted in acceleration of research in trace substance behavior in the environment. The concern that began to pervade the scientific community about contamination by toxic chemicals in the mid-1960s was to some extent the result of ecological knowledge that was obtained from studies of fallout from nuclear weapons tests. Those studies raised many difficult questions that at first seemed unique to the subject of environmental radioactivity. What are the ecological pathways by which these substances reach humans? Do they accumulate so that they can result in unforeseen ecological injury? Are there synergistic effects with other environmental pollutants? By the late 1960s, the same questions could be asked about insecticides, food additives, fossil fuel combustion products, trace metals, and other nonradioactive pollutants of the environment. In many respects, the pioneering studies of environmental radioactivity provided the tools by which more general problems of environmental pollution could be understood.

A. Quantities and Units

There are thousands of different radioactive nuclides and several different kinds of radiation. Fortunately, potential harm to biota (including humans) can be specified by using one or more special radiation quantities. These quantities are (1) energy deposited per unit mass (absorbed dose), (2) absorbed dose modified by a factor that accounts for the variable effects of different types of radiations (equivalent dose), and (3) equivalent dose modified by factors accounting for the sensitivities of different tissues and organs (effective dose). Effective dose is the dose to the whole body that is equivalent, in terms of risk, to larger doses to smaller portions of the body. The modern unit of absorbed dose is the gray (Gy, one joule per kilogram) and the modern unit of equivalent dose and effective dose is the Sievert (Sv). The amount of radioactivity is specified in terms of the number of transformations occurring per unit time. More detail on these terms and the older dose terms is given in the Glossary.

II. NATURAL RADIOACTIVITY AND RADIATION

Natural radioactivity originates from extraterrestrial sources and from radioactive elements in the Earth's crust. About 70 radioactive nuclides have been found. The radioactivity of the Earth includes the primordial radionuclides, whose half-lives are sufficiently long that they have survived since their creation, the secondary radionuclides that are derived by the radioactive decay of the primordials, and radionuclides that are continuously produced by bombardment of stable nuclides by cosmic rays, primarily in the atmosphere.

A much larger number of radioactive isotopes than now exist were produced when the matter of which the universe is now formed first came into being several billion years ago, but most of them have decayed out of existence. Radionuclides with half-lives of less than about 100 million years have become undetectable in the 40 or so half-lives since their creation, whereas radionuclides with half-lives greater than a billion years are readily identifiable today. Over much of the terrestrial surface of Earth, the natural radioactivity varies only within fairly narrow limits. In some localities, however, there are wide deviations from normal levels because of abnormally high soil concentrations of radioactive minerals.

The naturally occurring radionuclides can be divided into those that occur singly (Tables I and II) and those that are components of three chains of radioactive elements. The uranium series (Table III) originates with ^{238}U, the thorium series (Table IV) originates with ^{232}Th, and the actinium series (Table V) originates with ^{235}U. Each of Tables I through V provides the isotope, half-life, and principal radiations. Tables I and II also give typical concentrations. The three chains of radioactive elements and ^{40}K account for much of the background radiation dose to which humans are exposed. Of the 22 identified cosmogenic nuclides (Table I), only four, ^{14}C, ^{3}H, ^{22}Na, and ^{7}Be, are of any consequence from the perspective of dose to humans. Only two of the 17 nonseries primordial nuclides (Table II), ^{40}K and ^{87}Rb, are of consequence.

A. Uranium

The uranium normally found in nature consists of three isotopes having mass numbers 234, 235, and 238. Uranium-238 comprises 99.28% and is usually in equilibrium with ^{234}U, which comprises

TABLE I

Radionuclides Induced in the Earth's Atmosphere by Cosmic Rays

Radionuclide	Half-life	Major radiations	Target nuclides	Typical concentrations (Bq kg^{-1})		
				Air (troposphere)	Rainwater	Ocean water
^{10}Be	2,500,000 y	β	N, O			2×10^{-8}
^{26}Al	720,000 y	β^+	Ar			2×10^{-10}
^{36}CL	300,000 y	β	Ar			1×10^{-5}
^{80}Kr	213,000 y	K X-ray	Kr			
^{14}C	5730 y	β	N, O			5×10^{-3}
^{32}Si	~650 y	β	Ar			4×10^{-7}
^{39}Ar	269 y	β	Ar			6×10^{-8}
^{3}H	12.33 y	β	N, O	1.2×10^{-3}		7×10^{-4}
^{22}Na	2.60 y	β^+	Ar	1×10^{-6}	2.8×10^{-4}	
^{35}S	87.4 d	β	Ar	1.3×10^{-4}	$7.7-107 \times 10^{-3}$	
^{7}Be	53.3 d	γ	N, O	0.01	0.66	
^{37}Ar	35.0 d	K X-ray	Ar	3.5×10^{-5}		
^{33}P	25.3 d	β	Ar	1.3×10^{-3}		
^{32}P	14.28 d	β	Ar	2.3×10^{-4}		
^{38}Mg	21.0 h	β	Ar			
^{24}Na	15.02 h	β	Ar		$3.0-5.9 \times 10^{-3}$	
^{38}S	2.83 h	β	Ar		$6.6-21.8 \times 10^{-2}$	
^{31}Si	2.62 h	β	Ar			
^{18}F	109.8 m	β^+	Ar			
^{39}Cl	56.2 m	β	Ar		$1.7-8.3 \times 10^{-1}$	
^{38}Cl	37.29 m	β	Ar		$1.5-25 \times 10^{-1}$	
^{34m}Cl	31.99 m	β^+	Ar			

0.0058%. Uranium-235, the parent isotope of the actinium series, comprises 0.71%.

Uranium is found in all rocks and soils. The typical range of concentrations in normal rocks and soils is listed in Table VI but many exceptions exist. Igneous rocks contain concentrations of the order of 3 ppm, about 100 times greater than the ultrabasic igneous rocks but considerably less than the phosphate rocks of Florida and southeastern Idaho and neighboring areas. Phosphate deposits contain as much as 120 ppm of uranium and have been considered as a commercial source of uranium. The high uranium content of phosphate rocks is reflected in correspondingly high uranium concentrations in commercial phosphate fertilizers and in slag from elemental phosphorous production plants. Uranium series radionuclides are present in coal at concentrations similar to those in soil and are emitted to the environment with the fly ash when coal

is burned. The quantities released are detectable but small in comparison to natural levels.

Because uranium occurs in soils and fertilizers, the element is present in food and human tissues. On average, the annual intake of uranium from all dietary sources is about 5 Bq. The intake of uranium from tap water is negligible by comparison. The skeleton is estimated to contain about 25 μg of uranium, equivalent to about 0.3 Bq. The dose to the skeleton, which receives a higher dose from uranium than any other organ, is about 3 μSv y^{-1}.

B. Radium-226

Radium-226, a member of the uranium series, and its decay products are responsible for a major fraction of the dose received by humans from the naturally occurring internal emitters. Table III indicates that ^{226}Ra is an alpha emitter that decays, with a

TABLE II

Nonseries Primordial Radionuclides

Radionuclide	Half-life (y)	Major radiations	Typical crustal concentration (Bq kg^{-1})
^{40}K	1.26×10^9	β, γ	630
^{50}V	6×10^{15}	γ	2×10^{-5}
^{87}Rb	4.8×10^{10}	β	70
^{113}Cd	$>1.3 \times 10^{15}$	Not reported	$<2 \times 10^{-6}$
^{115}In	6×10^{14}	β	2×10^{-5}
^{123}Te	1.2×10^{13}	X-rays	2×10^{-7}
^{138}La	1.12×10^{11}	β, γ	2×10^{-2}
^{142}Ce	$>5 \times 10^{16}$	Not reported	$<1 \times 10^{-5}$
^{144}Nd	2.4×10^{15}	α	3×10^{-4}
^{147}Sm	1.05×10^{11}	α	0.7
^{152}Gd	1.1×10^{14}	α	7×10^{-6}
^{174}Hf	2.0×10^{15}	α	2×10^{-7}
^{176}Lu	2.2×10^{10}	e^-, γ	0.04
^{187}Re	4.3×10^{10}	β	1×10^{-3}
^{190}Pt	6.9×10^{11}	α	7×10^{-8}
^{192}Pt	1×10^{15}	α	3×10^{-6}
^{209}Bi	$> 2 \times 10^{18}$	α	$<4 \times 10^{-9}$

half-life of 1622 years, to ^{222}Rn, with a half-life of 3.8 days. The decay of radon is followed by the successive transformations of several short-lived decay products that emit alpha, beta, and gamma rays. After six decay steps, ^{210}Pb is produced, which has a half-life of 22 years. Finally, stable ^{206}Pb is produced. Radium does not directly add significantly to the gamma activity of the environment, but does so indirectly through its gamma emitting decay products.

Radium-226, like its parent ^{238}U, is present in all rocks and soils in variable amounts. Igneous rocks tend to contain higher concentrations than sandstones and limestones. Reported mean values are, for example, 16 Bq kg^{-1} in limestone and 48 Bq kg^{-1} in igneous rock. Elevated concentrations present in phosphate industry by-products used in construction result in increased gamma radiation exposure rates in structures.

The radium content of surface waters is low (4–19 Bq m^{-3}) compared to most groundwaters. Dissolved radium sorbs quickly to solids and does not migrate far from its place of release to groundwater.

Radium is chemically similar to calcium and is absorbed from the soil by plants and passed up the food chain to humans. Variability in radium concentrations in soil and in the way it is transferred from soil to plants and animals results in considerable variability in the radium content of foods. The average can be very roughly estimated as 37 mBq kg^{-1}. Daily average intakes of radium by humans have been estimated to range from 30 to 60 mBq d^{-1}.

In an early attempt to estimate the radium content of food, Brazil nuts were found to be much more radioactive than other foods, ranging between 10 and 260 Bq kg^{-1}. This is apparently due to the tendency of the Brazil nut tree (*Bertholletia excelsa*) to concentrate barium, which has chemical properties similar to radium. The average annual effective dose equivalent to human bone from deposited radium is about 70 μSv.

C. Thorium-232 and Radium-228

Thorium-232 is present in soil in amounts similar to uranium (see Table VI) and makes a similar contribution to the external gamma-ray dose. However, because of its relative insolubility and low specific activity, ^{232}Th is present in biological materials only in insignificant amounts. In humans, thorium is found in the highest concentrations in pulmonary lymph nodes and lungs, suggesting that the principal source of human exposure is inhalation of suspended soil particles. Although the thorium descendant, ^{228}Ra, frequently occurs in soil and water in approximately a 1 : 1 ratio to ^{226}Ra, there is little information about its occurrence in foods or in human tissues. The data that do exist suggest that under normal circumstances the ^{228}Ra content of food, water, and human tissues is from one-half to one-fourth that of the ^{226}Ra content.

D. Radon-222 and Radon-220 (Thoron)

When ^{226}Ra decays by alpha emission, it is transmuted to its decay product ^{222}Rn, an inert gas with a half-life of 3.8 days. As an inert gas, ^{222}Rn is much more mobile than the parent radium and readily enters the gas in soil voids and subsequently

TABLE III
Uranium Series

	Nuclide		Historical name	Half-life	Major radiations
	^{238}U		Uranium I	4.47×10^9 y	α, <1% γ
	^{234}Th		Uranium X$_1$	24.1 d	β, γ
	^{234m}Pa		Uranium X$_2$	1.17 m	β, <1% γ
	^{234}Pa		Uranium Z	21.8 y	β, γ
	^{234}U		Uranium II	244,500 y	α, <1% γ
	^{230}Th		Ionium	7.7×10^4 y	α, <1% γ
	^{226}Ra		Radium	1600 y	α, γ
	^{222}Rn		Emanation radon	3.8 d	α, <1% γ
	^{218}Po		Radium A	3.05 m	α, <1% γ
	↙↘				
^{214}Pb (99.98%)			Radium B	26.8 m	β, γ
		^{218}At(0.02%)	Astatine	2 s	α, γ
	^{214}Bi		Radium C	19.9 m	β, γ
	↙↘				
^{214}Po (99.98%)			Radium C'	164 μs	α, <1% γ
		^{210}Tl (0.02%)	Radium C"	1.3m	β, γ
	^{210}Pb		Radium D	22.3 y	β, γ
	^{210}Bi		Radium E	5.01 d	β
	↙↘				
^{210}Po (~100%)			Radium F	138.4 d	α, <1% γ
		^{206}Tl (0.00013%)	Radium E"	4.20 m	β, <1% γ
	^{206}Pb		Radium G	Stable	None

TABLE IV
Thorium Series

	Nuclide		Historical name	Half-life	Major radiations
	^{232}Th		Thorium	1.4×10^{10} y	α, <1% γ
	^{228}Ra		Mesothorium I	5.75 y	β, <1% γ
	^{228}Ac		Mesothorium II	6.13 h	β, γ
	^{228}Th		Radiothorium	1.91 h	α, γ
	^{224}Ra		Thorium X	3.66 d	α, γ
	^{220}Rn		Emanation thoron	55.6 s	α, <1% γ
	^{216}Po		Thorium A	0.15 s	α, <1% γ
	^{212}Pb		Thorium B	10.64 h	β, γ
	^{212}Bi		Thorium C	60.55 m	α, γ
	↙↘				
^{212}Po (64%)			Thorium C'	0.305 μs	α
		^{208}Tl (36%)	Thorium C"	3.07 m	β, γ
	^{208}Pb		Thorium D	Stable	None

TABLE V

Actinium Series

Nuclide		Historical name	Half-life	Major radiations
^{235}U		Actinouranium	7.038×10^8 y	α, γ
^{231}Th		Uranium Y	25.5 h	β, γ
^{231}Pa		Protoactinium	2.276×10^4 y	α, γ
^{227}Ac		Actinium	21.77 y	$\beta, <1\% \gamma$
⤢⤡				
^{227}Th (98.62%)		Radioactinium	18.72 y	α, γ
	^{223}Fr (1.38%)	Actinium K	21.8 m	β, γ
^{223}Ra		Actinium X	11.43 d	α, γ
^{219}Rn		Emanation actinium	3.96 s	α, γ
^{215}Po		Actinium A	1.78 ms	$\alpha, <1\% \gamma$
⤢⤡				
^{211}Pb (~100%)		Actinium B	36.1 m	β, γ
	^{215}At (0.000235)	Astatine	~0.1 ms	$\alpha, <1\% \gamma$
^{211}Bi		Actinium C	2.14 m	α, γ
⤢⤡				
^{211}Po (0.273%)		Actinium C′	0.516 s	α, γ
	^{207}Tl (99.73%)	Actinium C″	4.77 m	$\beta, <1\% \gamma$
^{207}Pb		Actinium D	Stable	None

escapes to the atmosphere. This mobility and the fact that ^{222}Rn decays to charged, metallic descendants mean that ^{222}Rn makes the largest average contribution to human radiation dose. Similarly, ^{224}Ra, which is a descendant of ^{232}Th, decays by alpha emission to ^{220}Rn, known as thoron in the older literature. Radon-220 has a half-life of only 54 seconds, which reduces its ability to escape to the atmosphere and so lessens its contribution to dose.

Reports from several countries suggest that the average concentrations of ^{222}Rn in outdoor air may be taken normally to be in the range of 4 to 19 Bq m^{-3}, but locations far from the soil sources, such as Antarctica, show markedly lower concentrations. Maximum concentrations are observed in the early hours and the lowest values are found in the late afternoon, when the concentrations are about one-third those of the morning maxima. The variations at any given locality are dependent on meterological factors that influence both the rate of emanation of the gases from the earth and the rate of dilution in the atmosphere. Changes in barometric pressure can "pump" radon from the soil.

Because the decay products of ^{222}Rn and ^{220}Rn are electrically charged when formed, they tend to

TABLE VI

Ranges and Averages of the Concentrations of ^{40}K, ^{232}Th, and ^{238}U in Typical Rocks and Soils

Material	Potassium-40		Thorium-232		Uranium-238	
	% K (total)	Bq kg^{-1}	ppm	Bq kg^{-1}	ppm	Bq kg^{-1}
Rock (range)^a	0.3–4.5	70–1500	1.6–20	7–80	0.5–4.7	7–60
Continental crust (average)	2.8	850	10.7	44	2.8	36
Soil (average)	1.5	400	9	37	1.8	22

^{a} Examples of materials exceeding the ranges can be found, but quantities are relatively small.

attach themselves to inert dusts that are normally present in the atmosphere, thus endowing the ordinary dusts of the atmosphere with radioactivity. The dose from the radon series is delivered primarily by the decay products attached to dust that deposits in the lung, although the small unattached fraction of decay products also contributes. When air that contains radon or thoron in partial or total equilibrium with their decay products is inhaled, the inert gases will largely be exhaled immediately. However, a fraction of the dust particles will be deposited in the lung. With each breath, additional dust will be deposited. When radon is in equilibrium with its decay products, the total energy dissipation in the lungs from the decay products is about 500 times greater than that derived from radon itself. Based on models, the dose to the basal cells of the bronchial epithelium could be as high as 20 mSv yr^{-1} from the radon decay products normally present in the outdoor atmosphere.

In confined spaces, especially those bounded by radon-emitting materials, radon concentrations can be orders of magnitude higher than outdoors. Examples include underground mines, especially uranium mines, caves, and human-made structures.

A surprising development has been that the concentration of ^{222}Rn (and the radon descendants) is so high in many homes that the potential risks are far greater than that from many other pollution hazards. The problem exists mainly in residential dwellings because the radon originates primarily from the soil, which has its greatest effects on one- or two-story buildings. The building materials themselves are a minor source of radon compared to soil except where the materials contain relatively high concentrations of radium. Elevated concentrations of radon are often found within structures located on land with high radium concentrations and soil permeability. The elevated radium can be natural or from various mineral wastes used as fill or in construction.

Radon can enter the indoor atmosphere in several ways, including diffusion from construction materials or diffusion from soil through breaches in the foundation. However, there is evidence that diffusion of radon from soil is a minor source compared to the movement of soil gases directly through the foundation because of slight pressure differentials that can result from barometric changes, temperature differentials, or wind velocity. Water supplies usually make a small contribution to the indoor radon concentration, but can be the predominant source in areas where the radon content of groundwater is unusually high. Other minor sources are unvented burning of natural gas or propane, which also contain radon.

Many studies of indoor radon involving small numbers of dwellings have been conducted and larger, more systematic studies are under way. The data suggest that indoor radon levels in the United States are log-normally distributed with a geometric mean of about 37 Bq m^{-3} and a geometric standard deviation of about 3. Using these values, approximately 10% of houses could be expected to be over the U.S. Environmental Protection Agency guideline of 150 Bq m^{-3} (4 pCi liter^{-1}).

E. Polonium-210

Polonium-210 is another important member of the ^{238}U decay chain (Table III). It occurs in the atmosphere from radon decay and also deposits on vegetation. Two population groups have elevated internal deposits of ^{210}Po, smokers and extreme northern ethnic populations that consume a large proportion of reindeer meat. The increased levels in smokers, about a factor of two greater than those in nonsmokers, are believed to result from deposition of ^{210}Po on growing tobacco leaves. Dose estimates to various pulmonary tissues from ^{210}Po from cigarette smoke range from 20 to 850 μSv. Reindeer and caribou feed on lichens that absorb trace elements, including ^{210}Po, from the atmosphere. Lapps living in northern Finland have ^{210}Po levels about 12 times higher than those of residents of southern Finland, where more typical European dietary regimes prevail.

Production of elemental phosphorous from the ore results in the emission of naturally occurring ^{210}Po to the atmosphere, an example of technologically enhanced natural radiation. This emission has been regulated by the U.S. Environmental Protection Agency for some years. Polonium-210 does not occur to any significant degree in domestic water supplies.

F. Potassium-40

Of the three naturally occurring potassium isotopes, only ^{40}K is unstable, having a half-life of 1.3×10^9 years and emitting beta and gamma radiation. Potassium-40 occurs to an extent of 0.012% in natural potassium, imparting a specific activity of approximately 30 kBq kg^{-1} potassium. The typical range of concentrations of ^{40}K in normal rocks and soils is listed in Table VI but exceptions exist. Total potassium concentrations in the Earth's crust are usually lowest for limestones and highest for granites. The potassium content of soils of arable lands is modified by using fertilizers. It is estimated that about 110 TBq of ^{40}K is added annually to the soils of the United States in fertilizer (1 TBq = 10^{12} Bq). Potassium-40 is an important contributor, with the uranium and thorium series, to the external gamma-ray field at the Earth's surface.

Because of its relative abundance and its energetic beta emission (1.3 MeV), the radioactivity of ^{40}K is the predominant radioactive component in normal foods and human tissues. It is important to recognize that the potassium content of the body is under strict homeostatic control and is not influenced by variations in environmental levels, either natural or human-made. Consequently, the concentration of ^{40}K within the adult body is constant, approximately 37 kBq. This isotope delivers a dose of about 200 μSv y^{-1} to muscle and other soft tissues and about 150 μSv y^{-1} to bone.

G. Radionuclides Induced by Cosmic Radiation

Several radionuclides that exist on the surface of the earth and in the atmosphere are continuously produced by the interaction of cosmic rays with atmospheric nuclei. The most important of these in terms of radiation dose to humans and other organisms are tritium (^3H) and ^{14}C. The properties of the cosmogenic isotopes and the extent to which they have been reported in various media are listed in Table I.

Carbon-14 is formed at a rate of 1400 TBq y^{-1} when ^{14}N atoms capture a cosmic-ray neutron and emit a proton. The natural rate of production is believed to have been unchanged for at least 15,000 years before initiation of thermonuclear weapons testing in 1954. Carbon-14 of natural origin is present in the carbon of all biota at the historically constant amount of 0.22 kBq kg^{-1} of carbon. After death, the ^{14}C equilibrium is no longer maintained, and the ratio of ^{14}C to ^{12}C diminishes at a rate of 50% every 5600 years, which makes it possible to use the ratio of these two radioisotopes to estimate the age of organic materials.

Human industrial activities have been altering the ^{14}C concentration in the biosphere. Nuclear weapons testing and to a lesser extent other nuclear activities have increased ^{14}C concentrations. On the other hand, extensive burning of fossil fuels such as coal, oil, and natural gas have released massive amounts of carbon into the atmosphere in which essentially all of the ^{14}C has decayed. This dilution with stable carbon reduces ^{14}C concentrations, a phenomenon known as the Seuss effect.

The ^{14}C content of the body is approximately 3.7 kBq, but the dose is small because the ^{14}C beta particles are of low energy. It is estimated that the dose from ^{14}C is 24 μSv y^{-1} to the skeletal tissues of the body and 5 μSv y^{-1} to the gonads.

Tritium, a radioactive isotope of hydrogen, is formed from several interactions of cosmic rays with gases of the upper atmosphere. Existing in the atmosphere principally as water vapor, tritium precipitates in rain and snow. Like ^{14}C it is produced in the nuclear industry and nuclear weapons testing, and this has increased the atmospheric content of tritium. The natural production rate of tritium over the surface of the earth is estimated to be about 0.19 atom cm^{-2} s^{-1}, corresponding to a steady-state global inventory of about 1 EBq (10^{18} Bq).

The natural concentration of tritium in lakes, rivers, and potable waters was reported to have been 0.2 to 0.9 kBq m^{-3} before nuclear weapons testing. The annual absorbed dose from tritium of natural origin is estimated to be about 0.01 μSv. The concentration of tritium in lakes, rivers, and potable waters reached about 150 kBq m^{-3} in

1963, as a result of atmospheric nuclear weapons testing, and has declined since to about 2 kBq m^{-3}.

The other nuclides formed from cosmic-ray interactions with the atmosphere may be potentially useful as tracers for studying atmospheric and water transport mechanisms, but they make no significant contribution to radiation dose.

H. Natural Radiation from Terrestrial Sources

The terrestrial sources of gamma radiation are ^{40}K and nuclides of the ^{238}U and ^{232}Th series. If the concentrations of these three nuclides in soil are known, the dose rates can be readily estimated using mathematical models. Dose rates can also be measured directly with ion chambers or indirectly with gamma-ray spectroscopy, if the contribution from cosmic radiation is accounted for properly. Gamma-ray measurements can be made from low-flying aircraft as well as from the ground.

The external gamma radiation from radionuclides in the Earth's crust is influenced by the kind of rock over which the measurements are made because of the variation of radioactivity with rock type. The actual doses to people cover a narrower range because most people live on soil rather than rock. Soils tend to be less variable in their radioactive content because the igneous rocks that are high in radioactive content weather more slowly and, therefore, contribute less radionuclides to soils than the softer sedimentary rocks. Soil moisture and snow cover affect the dose rate by shielding the gamma rays, introducing temporal variations amounting to as much as 30%. The current estimate of average dose equivalent rate from terrestrial sources in North America is 280 μSv y^{-1}, based primarily on analysis of extensive airborne gamma-ray spectroscopy measurements. Depending on location, this quantity can vary by a factor of three in either direction. The higher values are found in the Rocky Mountain region; the lower values are found on the coastal plain.

I. Cosmic Radiation

Primary radiations that originate in outer space and impinge on the top of the earth's atmosphere con-

sist of 87% protons, 11% alpha particles, 1% heavier nuclei, and about 1% electrons of very high energy. An important characteristic of the cosmic radiations is that they are highly penetrating, with a mean energy of about 10^4 MeV and maximum energies of as much as 10^{13} MeV. The primary radiations predominate in the stratosphere above an altitude of about 25 km. These radiations originate outside the solar system and only a small fraction is normally of solar origin. However, the solar component becomes very significant outside the atmosphere following flares associated with sunspot activity.

The interactions of the primary particles with atmospheric nuclei produce electrons, gamma rays, neutrons, and mesons. At sea level the mesons account for about 80% of the cosmic radiation, electrons about 20%, and neutrons less than 1%. Only 0.05% of the primary protons penetrate to sea level. As altitude increases, neutrons and protons contribute a larger share of the dose.

The dose from cosmic radiation is markedly affected by altitude. Outdoors the annual cosmic ray dose is about 290 μSv at sea level. For the first few kilometers above the earth's surface, the cosmic ray dose rate doubles for each 2000-m increase in the altitude. However, during the first 1000 m, the total dose rate actually decreases with altitude above the surface, because attenuation of the gamma rays from terrestrial sources diminishes more rapidly than the increase in cosmic radiation. Residents of Denver (altitude 1600 m) receive nearly twice the dose at sea level. In Leadville, Colorado (altitude 3200 m), the residents receive about 1250 μSv y^{-1} from cosmic rays, more than four times the annual dose at sea level.

The average external annual effective dose equivalent from cosmic radiation at ground level in North America is estimated to be 270 μSv y^{-1}. This estimate includes an adjustment for the shielding effect and occupancy of structures.

Doses are much higher at aircraft altitudes, exceeding 15 μSv h^{-1} at the upper limit of high-performance aircraft (25,000 m). Once or twice during the 11-year solar cycle, a giant solar event may deliver dose equivalents in the range 10–100 mSv h^{-1}, with a peak of as high as 50 mSv during the first hour. During a flare in February 1956, dose rates exceeding 1 Sv h^{-1} existed briefly at altitudes as low as 10 km.

Because of the effect of altitude, the passengers and crew of high-flying aircraft are subject to additional dose from cosmic rays, much of it from neutrons. A commercial transcontinental flight will result in a dose of about 50 μSv per round trip and aircraft crew are estimated to receive incremental doses of several mSv y^{-1} above that received at sea level.

When astronauts travel into outer space they are exposed to the intense radiation of the two belts of trapped electrons, the primary cosmic radiation particles and the radiation from solar flares. The Apollo missions lasted from about 6 to 12 days and resulted in average crew doses of 1.6–11.4 mGy. The Skylab missions lasted from 28 to 90 days, and resulted in doses of 16–77 mGy. During the first 25 Shuttle flights, the individual crew doses ranged from about 0.06 to 6 mGy. The doses would be much higher in the event of a solar flare. It has been estimated that the dose from a flare that occurred in July, 1959, could have been between 400 and 3600 mGy. Space radiation is undoubtedly an important constraint on long-term space travel.

J. Technologically Enhanced Natural Radiation

Aerospace activities are only one way in which the dose from nature is increased by technological developments. The dose can also be increased by use of building materials that have high levels of natural radioactivity and, as discussed earlier, by living in houses in which radon and its decay products accumulate. It has also been mentioned that the dose recieved from the radon decay products, especially ^{210}Po, can be increased by the practice of smoking cigarettes. Other ways in which the dose from natural sources can be increased include burning natural gas (which may contain radon), by exploiting phosphate minerals (which are often associated with uranium), and through the mining and milling of uranium, including the injudicious disposition of uranium mill tailings. These alterations of the natural radiation environment have been called technologically enhanced natural radiation. Exposures resulting directly from the mining, milling, and other processing of uranium for nuclear fuel are treated later as part of the nuclear fuel cycle.

K. Areas Having Unusually High Natural Radioactivity

There are places in the world where the levels of natural radiation exposure are abnormally high. In many places the radioactivity of springs and sands has been exploited for their alleged curative powers. Elevated concentrations of radioactive minerals in soil have been reported in Brazil, India, and China.

In Brazil, the radioactive deposits are of two distinct types: The monazite sand deposits along certain beaches in the states of Espírito Santo and Rio de Janeiro and the regions of alkaline intrusives in the state of Minas Gerais.

Monazite is a highly insoluble rare earth mineral that occurs in beach sand with the mineral ilmenite, which gives the sands a characteristic black color. The external radiation exposure levels on these black sands range up to 50 μSv h^{-1}, and people come from long distances to relax on the sands. The most active of the Brazilian vacation towns is Guarapari, where some major streets have radiation levels as high as 1.3 μSv h^{-1}, more than 10 times the normal background. Similar radiation levels are found inside some buildings. Most of the approximately 60,000 inhabitants of these regions are exposed to abnormally high radiation levels, but fewer (about 6600) are exposed to more than 5 mSv y^{-1}.

In the state of Kerala, on the southwest coast of India, the monazite deposits are more extensive than those in Brazil, and about 100,000 persons inhabit the area. The dose from external radiation is, on average, similar to the exposures reported in Brazil (5–6 mSv y^{-1}), but individual exposures exceeding 30 mSv y^{-1} have been reported.

A distinctly different source of exposure to natural radioactivity exists near the city of Araxa in the Brazilian state of Minas Gerais, where the more productive soil contains both uranium and thorium. Food grown in this area contains relatively high amounts of ^{228}Ra and ^{226}Ra but the number of exposed persons is few. Of the 2000 or so persons who live in this area, about 10% ingest radium in amounts that are 10 to 100 times greater than normal.

A unique anomaly located near Pocos de Caldas, also in the state of Minas Gerais, is the Morro do Ferro, a hill that rises about 250 m above the surrounding plateau. Near the summit is a near-surface ore body that contains about 3×10^7 kg of thorium and an estimated 10^8 kg of rare earth elements. The ambient gamma radiation levels near the summit of the hill range from 10 to 20 μSv h^{-1} over an area of about 30,000 m^2. The flora from this hill have absorbed so much ^{228}Ra that they can readily be autoradiographed.

Studies have been undertaken of the exposures of rats living underground on the Morro do Ferro. Of particular interest is the dose to these rodents due to inhalation of ^{220}Rn, which was found in the rat burrows in concentrations up to 3.7 MBq m^{-3}. The dose equivalent to basal cells of the rat bronchial epithelium was estimated to be in the range of 30 to 300 Sv y^{-1}. The external radiation dose to the rats was estimated to be between 13 and 67 mSv y^{-1}. Of 14 rats trapped and sacrificed for pathological study, none was observed to show any radiation effects. This is of little significance, as the Morro do Ferro is a relatively small area and if rats were affected by this exposure, they could be replenished rapidly from the surrounding normal areas.

About 73,000 persons live in an area of monazitic soils in Guangdong Province, China, where the external radiation exposure is about 3.3 mSv y^{-1}. The radiation levels are between three and four times normal. The ^{226}Ra body burdens in the high background area were reported to be 10 Bq, about three times higher than in the control area.

L. The Natural Reactor at Oklo

Much to the surprise of the scientific world, it was discovered in 1972 that the site of the open-pit uranium mine called Oklo in the West African republic of Gabon was the fossil remains of a 2-billion-year-old natural reactor. Although con-temporary uranium contains only 0.7% of the fissionable isotope ^{235}U, this nuclide has a half-life of only 700 million years compared to the 4.5-billion-year half-life of ^{238}U. The percentage of ^{235}U present in the ore was thus very much greater 2 billion years ago, and the conditions for initiating fission reactions apparently existed. Criticality was sustained for more than 1 billion years, during which time an estimated 15,000 megawatt-years of energy was released by the consumption of 6000 kg of ^{235}U. Studies of the migration of the fission products and transuranic elements produced during this period may have implications for radioactive waste management.

M. Summary of Human Exposures to Natural Radiation

The annual effective dose equivalent received by persons living in North America where natural radioactivity is within normal limits is given in Table VII. Radon-222 and its short-lived decay products contribute more than half the total effective dose equivalent. Although the dose from radon can be extremely variable, individual doses to most people from the other components of natural radiation are within a factor of three of the average values in Table VII.

III. HUMAN-MADE ENVIRONMENTAL RADIOACTIVITY

Although small quantities of radioactivity were produced in particle accelerators before the discovery and exploitation of nuclear fission, essentially all the human-made radioactivity in the environment today results from the fission of uranium and plutonium in nuclear reactors or nuclear explosive devices. When the heavy nucleus of uranium or plutonium is split in two, the resultant fission products are almost always radioactive, with half-lives ranging from fractions of a second to hundreds of years. Besides fission products, neutrons are absorbed by stable elements, such as in reactor components or coolants, creating radioactivity through the process of activation.

TABLE VII
Annual Estimated Average Effective Dose Equivalent (EDE) Received by an Individual in the United States from Natural Radiation

Source	Average individual EDE (μSv)	Average individual EDE (mrem)
Inhaled (radon and decay products)	2000	200
Other internally deposited radionuclides (^{40}K, ^{210}Po)	390	39
Terrestrial radiation	280	28
Cosmic radiation	270	27
Cosmogenic radioactivity (^{14}C)	10	1
Total (rounded)	3000	300

A. The Nuclear Fuel Cycle

The production of nuclear energy is based mainly on the fission of ^{235}U. Uranium-238, which is the most abundant isotope of uranium, is not readily fissionable. Following an intensive exploratory program conducted by the U.S. government, the uranium industry grew rapidly during the post-World War II period, serving both civilian and defense needs.

Uranium ore is mined in both open pits and underground workings and is then shipped to mills that produce uranium concentrate. The output of these mills is transported to uranium refineries, where the concentrates are converted into uranium compounds having a high degree of chemical purity. From there, some uranium goes to isotopic enrichment plants, where the ^{235}U content is increased; the remainder is used for natural uranium reactor fuel.

I. Uranium Mining

Uranium mining takes place mainly in underground workings and open pits. Uranium mining by *in situ* underground leaching (solution mining) has also been demonstrated and may become more important in the future.

Radon and its decay products are discharged from the mines to the general environment. The required ventilation rates for the mines vary from 0.5 to over 100 m^3 s^{-1}, and the discharged air contains radon in concentrations that range from 0.7 to 26 kBq m^{-3} of air. For comparison, the natural emission rate of 1 km^2 of the earth's surface is about 22 kBq s^{-1} of ^{222}Rn. The estimated maximum individual doses to the general public from the radioactive emissions from uranium mines are estimated to be the largest of any component of the nuclear fuel cycle.

2. The Uranium Mills

Uranium ore is transported from the mines to mills, where the uranium concentrate (yellowcake) is produced. Although more than 95% of the uranium is removed from the ore by milling, most of the radioactive decay products in the uranium series remain with the tailings contained in slurries that are discharged into holding areas. Because the U.S. mills are located in the arid regions of the Southwest, the impounded slurries dry rapidly and become growing mounds of the residue. Approximately 2 × 10^{11} kg of uranium mill tailings have been accumulated at 52 active and inactive sites, all but one in the western states. The piles of mill tailings are potential sources of environmental problems because of (1) emanation of ^{222}Rn, (2) dispersion by wind and water, and (3) the historical use of the tailings in building construction.

The increased radon flux from the tailings piles is due to the higher ^{226}Ra content of the tailings, which may contain 10,000 Bq kg^{-1} compared to 40–80 Bq kg^{-1} for normal soils. Because the area covered by the tailings piles, usually on the order of 10 to 100 acres, is small compared to the areas not covered by tailings piles, the piles themselves do not make a significant contribution to the concentration of Rn in the general environment. Their influence is localized and can be detected only within about 1 mile. The EPA has established a ^{222}Rn emission limit of 0.75 Bq m^{-2} s^{-1} (20 pCi m^{-2} s^{-1}) and requires that the disposal method be designed to provide "reasonable assurance" that radon emissions will not exceed that limit, averaged over the disposal area for 1000 years.

Materials from the tailings piles can be dispersed into the environment either by impoundment failure or erosion by flowing water or wind action. Impoundment failure can be minimized in the foreseeable future by application of available engineering methods, and if cover is provided for the piles, erosion by wind or water can be avoided in the semiarid regions in which the tailings piles are located. However, there can be no assurance that these protective features will be effective for the hundreds of thousands of years during which the radioactivity remains elevated.

In past years, liquid tailings were allowed to seep into nearby streams. The ^{226}Ra concentration in the Colorado River below Grand Junction was 1100 Bq m^{-3} compared to 11 Bq m^{-3} upstream. The San Miguel River below Uravan, Colorado, where a mill was located, contained 3200 Bq m^{-3} compared to 180 Bq m^{-3} in water immediately upstream of the mill. The Animas River in southwestern Colorado serves as a public water supply for the cities of Aztec and Farmington, and the water is also used for irrigation. In 1955, the radium concentration below Durango, where a mill was located, was found to be 120 Bq m^{-3} compared to 7.5 Bq m^{-3} upstream. That considerable concentration was taking place in the stream biota was shown by the fact that plants below Durango contained 24,000 Bq kg^{-1} compared to 220 Bq kg^{-1} above Durango. Stream fauna below Durango contained 13,000 Bq kg^{-1} compared to 220 Bq kg^{-1} above the mill.

It was found that using the then-existing International Commission on Radiological Protection recommendations as a guide, consumers of untreated river water received about three times the maximum permissible daily intake of radium. This included food grown on land irrigated with river water and the water consumed. About 61% was due to the radium content of food resulting from the contaminated irrigation water. Steps were taken by the mill operators to correct this problem, and by 1963 the radium content of the Animas River sediments had been reduced to three times the background level as compared to several hundred times background several years earlier. The Animas River experience illustrates the danger, if precautions are not taken, of contamination of potable water by radium and other radionuclides and toxic chemicals present in tailings piles.

Airborne dust measurements downwind of three unstabilized tailings piles have shown clear relationships between the distance from the tailings piles and the concentrations of uranium and ^{210}Pb. Air concentrations were well below permissible levels in two cases, but approached recommended upper limits at a distance of about 300 m in the third case.

3. Use of Mill Tailings for Construction

Uranium mill tailings have been used in construction in several areas of the United States, most notably Grand Junction, Colorado, where 3300 structures were affected. This practice, discontinued in 1966, resulted in increased exposure to both gamma radiation and radon. Congress provided financial assistance to the state of Colorado to limit the radiation exposures that existed because of the use of tailings for construction purposes. The cleanup in Grand Junction was performed in accordance with guidelines issued by the Office of the Surgeon General.

4. Refining

The mill concentrates in the United States are sent to any of several locations in which the uranium is converted to either the metal or some intermediate uranium compounds. Refining operations involve the mechanical processing of dry powders of uranium compounds, which can result in the discharge of uranium dust to the environment. Present-day plants are equipped with filtration equipment that effectively removes the uranium dust. The high monetary value of uranium precludes the possibility of its being discharged to the atmosphere in large quantities for sustained periods of time. However, the hastily constructed plants during World

War II had insufficient control over dusts contained in exhaust air, and relatively large amounts of uranium were discharged to the outwide atmosphere. Nevertheless, naturally occurring uranium is so abundant in the environment that the element was undetectable above the natural background at moderate distances from the plants.

Thirty-one plants, laboratories, and storage sites were involved with production of uranium from ore during the World War II era. Operation of those plants was ended by the mid-1950s, and they were cleaned up according to the then-existing understanding of what constituted adequate decontamination. As time went on, more conservative criteria were developed and many sites were found to be above the limits suitable for unrestricted access. In 1974, the Atomic Energy Commission, the predecessor of the Department of Energy, initiated a program of decontamination that is currently under way.

Much uranium goes through a process of isotopic enrichment to increase the proportion of ^{235}U. Environmental regulation, strict accountability procedures, and the economic value of the enriched uranium preclude the likelihood of widespread environmental contamination from enrichment plants. Fuel-element manufacture is carried on at many government and private facilities. Again, the relatively high cost of the uranium and the requirements for strict accountability make it unlikely that significant environmental contamination will occur from these plants, but the possibility of accidents cannot be discounted. Uranium chips are pyrophoric, and if fires or explosions occur, more than normal amounts of activity may be released.

5. Nuclear Reactors

The first nuclear reactor was operated briefly by Enrico Fermi and his associates in Chicago on December 2, 1942, less than 4 years after the discovery of nuclear fission. Now, about 1000 land-based reactors have been built and operated in various parts of the world and about 200 naval vessels are powered by nuclear reactors. After a 15-year period of growth for the nuclear industry until the mid-1970s, the ordering of additional nuclear capacity in the United States came to a sudden stop owing to several interacting economic and political factors. Despite setbacks, nuclear power production is still a major industry.

Contemporary reactors, with only a few exceptions, use either natural uranium or uranium in which the amount of isotope 235 has been enriched. The ^{235}U enrichment of the fuel used in most U.S. civilian power reactors is about 3%. Because of several difficulties with using uranium metal, the fuel material is uranium dioxide in most present-day reactors. The fuel may be fabricated as rods, pins, plates, or tubes, protected by a cladding whose function is to prevent the escape of fission products and protect the fuel from the eroding effect of the coolant.

Fission is caused by the capture of a neutron by the nucleus of a fissionable atomic species. Because more than one neutron is released in the process, a multiplication of neutrons may be achieved. This process allows additional atoms of uranium to be split, which in turn will yield additional neutrons to continue the multiplicative fission process. Enormous amounts of heat and radiation are released by fission in large reactors.

The radionuclides that accumulate in reactors are primarily those produced by fission within the reactor core and, secondarily, are the activation products formed when traces of corrosion products and other impurities contained in the coolant-moderator undergo neutron bombardment in passing through the core. In this way, radionuclides of elements such as chromium, cobalt, manganese, and iron are produced. The fission product inventory is very much larger than the inventory of corrosion products, but the nature of reactor operation is such that the corrosion products may be present in greater quantities in aqueous wastes.

Light-water reactors during normal operation produce both gaseous and liquid wastes. Although sophisticated control systems are used, some radioactivity is discharged directly to the environment. These wastes originate from both the radionuclides produced by activation of elements in the primary coolant and fission products. The fission products may originate from traces of uranium present on the surfaces of fuel elements or other reactor components, but the major source is leakage or diffusion through the fuel cladding. The radionuclides in the coolant system exist in gaseous form, dissolved solids, and suspended solids. The exact composition of the liquid and airborne radioactive emissions from light-water reactors will vary from reactor to reactor depending on construction materials and fuel condition. The noble gas radionuclides, tritium, and volatile nuclides such as iodine are more likely to be released than the others.

Emissions from operating power reactors in the United States are routinely reported by the U.S. Nuclear Regulatory Commission. There has been a gradual reduction in both liquid and gaseous emissions from the power reactors due to consistent improvement in fuel quality and effluent control. Doses to populations have been reviewed by the National Council on Radiation Protection and Measurements. The emissions have generally resulted in insignificant doses to the general population.

Besides releases of radioactivity from routine operations, reactor accidents can contribute to environmental radioactivity. The severity of a reactor accident in which core damage occurs is dependent on the extent to which radioactivity, mainly as ^{131}I and ^{137}Cs, is released to the environment. Fourteen reactor accidents involving core damage have occurred in the 50-year history of the nuclear energy program, but only two, the 1957 accident at Windscale in the United Kingdom and the 1986 accident at Chernobyl, resulted in the release of important quantities of radioactivity. Both of these reactors used graphite to moderate or slow down neutrons rather than the water used in U.S. commercial power reactors. Graphite exhibits some undesirable physical properties when exposed to large quantities of neutrons and is combustible. The Windscale reactor released 740 TBq of ^{131}I during the accident. The Chernobyl reactor had the disadvantage of having been built without some of the protective features that are routinely adopted in Western countries. The Chernobyl accident released 270,000 TBq of ^{131}I and important quantities of many other radionuclides. The much-publicized accident at Three Mile Island in the United States released only 0.6 TBq of ^{131}I.

6. Reprocessing Spent Reactor Fuel

When reactor fuel elements have reached the end of their useful lives, only a small percentage of the ^{235}U will have been consumed in fission. It is possible to reprocess the fuel to recover the usable uranium and stabilize the waste fission products in preparation for waste management. The large inventories of radioactivity, together with the nature of the fuel-reprocessing methods, present opportunities for major environmental contamination unless strict procedures are followed to avoid release of radioactive substances to the vicinity of the plant. Chemical explosions are conceivable, and because fissionable materials are processed, it is possible for critical masses to be assembled accidentally. However, as with reactors, modern plants have been designed to minimize the probability of a serious accident and to mitigate the consequences should an accident occur.

From World War II until 1966, all fuel reprocessing in the United States was performed at government-owned centers of atomic energy development and production at Hanford, Washington; Oak Ridge, Tennessee; Savannah River, South Carolina; and the National Reactor Testing Station near Idaho Falls, Idaho. These plants were constructed to meet military needs and were not intended for the processing of civilian fuel.

Through the Hanford Dose Reconstruction Project, it has become apparent that environmental emissions from the Hanford reprocessing plants were quite large during the early years of operation. Iodine-131 emissions apparently exceeded 30,000 TBq. Studies are currently under way to better define the doses received by the surrounding populations and to determine if adverse health effects can be observed.

Although there was an apparent need for commercial fuel reprocessing, the initial attempts by private industry to meet the needs of the projected nuclear power development were failures for a combination of technical, economic, and political reasons. The first and only privately owned nuclear fuel-reprocessing plant in the United States, Nuclear Fuel Services Inc., began operation in 1966 in

West Valley, New York, with a daily processing capacity of 1 metric ton of low-enriched uranium oxide fuel. Other privately owned plants were constructed by General Electric Company in Morris County, Illinois, and the Allied Chemical Corporation at Barnwell, South Carolina, but these plants never operated.

The plant in West Valley encountered licensing and economic problems that resulted in the 1972 shutdown of the plant and the subsequent abandonment of the venture by its owners. The West Valley plant is now being cleaned up by DOE and the state of New York under the Congressionally mandated West Valley Demonstration Project.

Any incentive to restart civilian fuel reprocessing was eliminated in 1977 when U.S. policy prohibited civilian fuel reprocessing to prevent the diversion of plutonium and the possible uncontrolled proliferation of nuclear weapons. The question of whether this was a justifiable policy became moot because lessened requirements for nuclear power reduced the economic incentive to recycle fuel. Nevertheless, there is some reprocessing of power reactor fuel in Europe.

Other aspects of the nuclear fuel cycle, including spent fuel storage, low-level waste management, high-level waste management, and transportation, potentially contribute to radiation dose but are not discussed in detail here. The estimated doses from these activities, where available, are given in Table VIII.

7. Summary of Doses due to the Nuclear Fuel Cycle

Table VIII is a summary of the estimated doses from the various components of the nuclear fuel cycle. The average effective dose equivalent from the entire nuclear fuel cycle to a U.S. resident, 0.55 μSv, is seen to be trivial in comparison to an average natural background level of 3000 μSv. However, the potential maximum individual dose of 2600 μSv, associated with uranium milling, is comparable with background. Most of the dose from the nuclear fuel cycle is due to to mining and milling of uranium and transportation. It is worth noting that most of the transportation dose is associated with the transportation of uranium ore instead of nuclear waste.

B. Nuclear Weapons

The first test of a nuclear weapon occurred in Alamagordo, New Mexico, in 1945 and about 450 devices were tested in the atmosphere subsequently. The United States, the former U.S.S.R., and the United Kingdom tested in the atmosphere until 1962. India, France, and the Peoples Republic of China tested through 1980. This testing injected fission and activation products directly into both the lower and upper atmospheres, resulting in both local and global fallout of radioactivity. Global fallout has resulted in small radiation doses to large

TABLE VIII

Annual Estimated Individual Maximum and Average Effective Dose Equivalent (EDE) Received by an Individual in the United States from the Nuclear Fuel Cycle

Fuel cycle component	Maximum likely individual EDE (μSv)	Average individual EDE (μSv)	Average individual EDE (mrem)
Mining	610	0.38	0.038
Milling	2600	0.1	0.01
Conversion	32	0.00012	0.000012
Enrichment	4	0.004	0.0004
Fabrication	7	0.0016	0.00016
Nuclear power plants	6	0.0192	0.00192
Low-level waste	10		
Transportation	200	0.05	0.005
Total (rounded)		0.55	0.055

numbers of persons and effects, if any, are masked by the normal incidence and variation of radiogenic diseases such as cancer.

Although testing was always done in remote areas, some populations were exposed to significant local or regional fallout. In 1954, a group of 67 natives of the Marshall Islands on the atoll of Rongelap were inadvertently exposed to large fallout doses from a thermonuclear explosion at Bikini following an unexpected wind shift. Excess cases of thyroid abnormalities, including hypothyroidism, nodules, and seven malignant tumors, had occurred in this population at the end of about 27 years. Studies of populations exposed to fallout from the Nevada Test Site indicated a small increase in thyroid disease in persons exposed as children. The relatively high doses responsible for these effects were delivered primarily by short-lived isotopes of iodine.

The average total dose accumulated by the year 2000 by persons who were alive through the entire atmospheric testing period has been estimated to be about 2000 μSv. Dose rates from fallout today are due mostly to ^{137}Cs and ^{90}Sr and are about 10 μSv per year, quite small in comparison to the 3000 μSv (average) from natural background. Unless atmospheric testing is resumed, the dose rate will continue to decline as the radioactivity decays and becomes less available to the biosphere.

Table IX provides a summary of average doses to humans from the various environmental sources. Clearly natural sources, especially radon, are by far the largest contributors to environmental radiation dose. Other sources, such as medical X-rays and nuclear medicine procedures, also contribute dose (500–600 μSv y^{-1}) but these are not considered environmental sources.

IV. EFFECTS OF RADIATION ON HUMANS

More is known about the effects of ionizing radiation exposure than about the effects of any other of the many noxious agents that have been introduced artificially into the environment. Reports of radiation injury began to appear in the literature within a year of the discoveries of X-rays and radioactivity in 1896. Groups that were significantly exposed in the early years included the early X-ray and radium workers, radium luminous dial painters, radiology patients, Japanese atomic bomb victims, fallout recipients, uranium miners, and workers involved in radiation accidents. Animal studies were initiated and genetic mutations due to X-rays were reported in 1927.

The early exposures provided knowledge of the acute effects of high levels of exposure. Relatively minor effects occur at doses less than 1 Sv. Over 1 Sv, effects include sterility, damage to blood-forming tissues, and, in the case of very high levels of exposure, a complex of symptoms that came to be known as the acute radiation syndrome. About 50% fatalities would be expected to occur in the

TABLE IX

Annual Estimated Average Effective Dose Equivalent (EDE) Received by an Individual in the United States from Environmental Radiation Sources

Source	Average individual EDE (μSv)	Average individual EDE (mrem)
Inhaled radon and decay products	2000	200
Other natural sources (see Table VII)	950	95
Fallout from nuclear weapons testing	10	1
Miscellaneous	0.6	0.06
Nuclear fuel cycle (mining, milling, enrichment, and fabrication)	0.53	0.053
Nuclear fuel cycle (power plants)	0.02	0.002
Total (rounded)	3000	300

range of 4 to 5 Sv. As the whole-body dose approaches 10 Sv, the fatalities would reach 100%. Except for local fallout from nuclear weapons and events such as the nuclear reactor accident at Chernobyl, environmental radiation does not approach levels required to induce acute effects.

The delayed effects of radiation may not appear for several decades after exposure and can result either from massive doses that have caused nonfatal prompt effects or from relatively small exposures repeated over an extended time with no visible effect. The delayed effects that develop in the exposed individual, most often cancers including leukemia, are called somatic effects to differentiate them from genetic effects that occur in the progeny of the exposed person. Radiation injury can also occur in the developing fetus, and these are considered teratogenic rather than genetic effects.

The relationship between dose and the incidence of cancer has been the subject of much scientific research and debate. At doses above about 0.5 Sv there are clear relationships between dose and cancer incidence that are approximately linear or quadratic over most of the sublethal range. Below 0.5 Sv the human data are nonexistent, or equivocal, so models, supported by animal data, are used to extrapolate to the possible effects at environmental levels. Relying primarily on data from the Japanese atomic bomb survivors, the U.S. National Academy of Science has estimated that continuous exposure of populations to the whole-body portion of natural background radiation (1000 μSv y^{-1}) may cause as much as 3% of cancer incidence in the United States. This value is almost certainly an upper bound because no adjustment was made for dose rate. The Japanese survivors received dose at a very high dose rate whereas environmental levels produce very low dose rates.

An apparently linear relationship between radon decay product exposure and incidence of lung cancer has been observed among uranium miners. Although exposure conditions are different to some extent, it is believed by many scientists that the risk estimates based on experience with miners are applicable to the general population exposed to indoor radon. When the risk coefficients were applied to the currently estimated exposure distributions, it was estimated that there may be annually 5,000 to 10,000 cases of lung cancer due to ^{222}Rn exposure. In 1986 The U.S. Environmental Protection Agency estimated that the number could range from 5000 to 20,000 but in 1992 placed the estimate at 7000 to 30,000. Because there are about 130,000 cases of lung cancer per year in the United States, exposure to environmental ^{222}Rn and its decay products may be responsible for 4–23% of the total number of reported cases. These estimates should be regarded as speculative because of the uncertainties involved and the confounding effects of smoking.

Epidemiological studies of human populations exposed to elevated natural radiation have been performed at several locations in the world. The U.S. National Academy of Sciences has summarized these studies and concluded that no increase in the frequency of cancer has been documented in populations residing in areas of high natural background radiation.

Risk rates are very uncertain at low doses and the Academy states that below an exposure of 100 mSv, the excess risk may be zero. Nevertheless, it is common for a linear, nonthreshold model to be used to estimate risk at very low doses.

The assumption that the dose–response relationship is linear and without threshold has important implications for risk assessment and formulation of public policy. The absence of a threshold implies that there is no such thing as an absolutely safe level of exposure. Every increment of dose above zero, however small, results in an increment of risk as well.

A dilemma that also arises from the assumption of linearity and the absence of a threshold is that the risk to individuals can be very small, but a finite number of cancers can be calculated if a sufficiently large population is exposed. A lifetime risk of one in a million is negligible to an individual, but if the world's population of 4.5 × 10^9 persons is exposed to a one in a million risk, 4500 radiation-induced cancers can be calculated. The question of whether acceptable dose should be established on the basis of individual or collective risk cannot be answered by science alone and most involve moral, ethical, and political values.

V. EFFECTS OF ENVIRONMENTAL RADIATION ON PLANTS AND ANIMALS

Species on earth have evolved within natural environmental radiation levels. Although individual organisms may have some low probability of being affected by radiation, the success of populations and communities is presumably not seriously affected by natural radiation. One can even speculate that environmental radiation provides a slight increase in genetic variability, perhaps making it easier for species to adapt to changing environmental conditions.

Environmental radiation doses to plants and animals can be increased by human-made radioactivity because of normal operation of nuclear facilities or, infrequently, because of catastrophic accidents such as Chernobyl. It has been generally assumed that if humans protect their own environment (air, water, land, food) sufficiently to protect themselves then natural systems would automatically be protected.

This assumption was recently examined in detail by the International Atomic Energy Agency. The known effects of radiation on aquatic and terrestrial species, populations, and ecosystems were reviewed. It was found that although some individual organisms might be affected, measurable effects on populations were unlikely at levels of 1 mGy per day for the most sensitive terrestrial species and 10 mGy per day for aquatic species. Most national and international authorities recommend that human-caused environmental radiation doses to humans in the general population be limited to a maximum of 1 mSv per year. Using this value as a starting point, models were used to examine likely doses to plants and animals in natural systems. It was found that doses to natural organisms were unlikely to exceed 1 mGy per day under the prevailing radiation protection standards. The model calculations were conservative so actual doses to natural species are likely to be a factor of 10 or 100 smaller. The current level of protection afforded to humans from routine radioactivity releases appears adequate to protect natural biota as well.

The nuclear reactor accident at Chernobyl illustrates what can happen to natural species and systems following massive releases of radioactivity to the environment. Populations of sensitive species such as rodents were reduced within several kilometers of the plant and evidence of radiation effects was found in surviving members of rodent populations. Coniferous forests were visibly affected, showing loss of needles, suppression of new growth, and abnormal growth. Deciduous trees were less affected and returned to normal after two years. Mollusk communities in the Chernobyl plant cooling pond declined. There were no observed effects on the composition of plant communities but genetic effects in plants were observed. Insect population structure and dynamics were not affected except for aphids, which declined in size and in number of species.

The major effect on populations was indirect. Because of evacuation of humans from the 30-km-radius zone around the plant, there was less disturbance of wildlife and unharvested crops provided additional food and cover. Populations of birds and mammals, including game birds, wild boar, fox, and wolf, actually increased. No suppressing effects of radiation on populations of large animals were observed.

VI. CONCLUSIONS

The living environment continually receives small doses of radiation from both natural and human-made sources. The radiation is received from external sources and from radioactivity incorporated into tissues. The levels associated with natural background in normal regions and emissions from nuclear facilities complying with recommended international limits amount to no more than a few milliSieverts per year. There are no known, deleterious effects on populations or ecosystems at these levels. There may be deleterious effects on individual members of populations, including humans, at these levels, but there is no direct evidence. Risks at these levels are projected from observations at much higher levels, generally 0.5 Sv and above. There is no epidemiological evidence for increased human cancer frequency in areas of high natural background radiation.

Individuals of any species can certainly be affected deleteriously by sufficiently large amounts of radiation, such as that from the testing or use of nuclear weapons or nuclear accidents. Local populations and ecosystems can also be affected, at least for some period of time.

Glossary

Absorbed dose Physical quantity equal to the energy deposited by ionizing radiation to unit mass of a substance, often tissue. The traditional unit of dose is the rad, equal to 100 ergs g^{-1}. The modern SI unit is the gray (Gy), equal to 1 joule kg^{-1}. One Gy equals 100 rads. Standard metric multiples and submultiples are often used.

Dose equivalent Dose adjusted by a quality factor (or radiation weighting factor) for the relative biological effectiveness of the type of radiation. In the newest recommendations of the International Commission on Radiological Protection, the term equivalent dose is used instead of dose equivalent. The traditional unit of dose equivalent is the rem, equal to dose in rads times the unitless quality factor. The modern SI unit of equivalent dose is the Sievert (Sv), equal to the dose in Gy times the radiation weighting factor. One Sv equals 100 rems. Standard metric multiples and submultiples are often used.

Effective dose equivalent Dose equivalent adjusted by tissue weighting factors to account for the individual radiosensitivites of the various organs. Effective dose equivalent, recently shortened by the International Commission on Radiological Protection to "effective dose," is intended to be a single numerical index of detriment to a human that takes into account the delivered energy, the efficacy of different kinds of radiation, and the radiosensitivity of various organs. Effective dose is expressed in rem or Sv because the weighting factors are unitless.

Standard metric multiples and submultiples are often used.

Radiation In this context radiation is restricted to ionizing radiation and includes electromagnetic radiation and directly and indirectly ionizing particles. Ionizing radiation has sufficient energy to remove electrons from atoms or molecules and create an ion pair.

Radioactivity Term used to describe unstable atoms that spontaneously transform into more stable states, resulting in the formation of new nuclei and the emission of one or more of several kinds of ionizing radiation. The traditional unit of radioactivity is the curie (Ci), equal to 3.7×10^{10} transformations per second (approximately equal to the activity of 1g of ^{226}Ra). The modern SI unit is the becquerel (Bq), equal to one transformation per second. Standard metric multiples and submultiples, that is, milli (m), micro (μ), mega (M), and so on, are often used.

Bibliography

Eisenbud, M. (1987). "Environmental Radioactivity." San Diego, Calif.: Academic Press.

International Atomic Energy Agency (1992). "Effects of Ionizing Radiation on Plants and Animals at Levels Implied by Current Radiation Protection Standards," Technical Report Series No. 332. Vienna: International Atomic Energy Agency.

National Academy of Science–National Research Council (1990). "The Health Effects of Exposure to Low Levels of Ionizing Radiations." Washington, D.C.: National Academy Press.

National Council on Radiation Protection and Measurements (1987). "Ionizing Radiation Exposure of the Population of the United States." Bethesda, Md.: National Council on Radiation Protection and Measurements.

United Nations Scientific Committee on the Effects of Atomic Radiation (1988). "Sources, Effects, and Risks of Ionizing Radiation." New York: United Nations.

Environmental Stress and Evolution

A. A. Hoffmann
La Trobe University,
Australia

P. A. Parsons
University of Adelaide,
Australia

Stresses are defined as environmental factors that cause changes in organisms that are potentially injurious and that may cause a drastic reduction in reproductive output and ultimately death. Increasing evidence indicates that stressful conditions have had a profound influence on the evolution of organisms in the fossil and living biota. The effects of stresses on evolutionary change can be seen in examples of natural selection that involve observations on organisms under adverse conditions arising from climate changes or human influences. Stresses can also have indirect evolutionary effects by reducing the stability and size of populations and by affecting the expression of genetic variation. An understanding of adaptation to stress is important because this process determines the potential for organisms to evolve in response to future global environmental changes. This emphasis leads to a reassessment of conservation strategies.

I. INTRODUCTION

Environmental stresses include climatic factors such as extremes of temperature and humidity, as well as food shortages and environmental toxins such as pollutants and pesticides. Stresses may have direct effects on organisms by causing death or severely reducing reproductive output. In addition, stresses may have indirect effects and exert their influence via biotic interactions. For instance, stresses may weaken organisms and thereby influence their susceptibility to diseases or predation.

Numerous studies have documented drastic effects of environmental stresses on natural populations. Examples include the marked changes in marine populations following the 1982–1983 El Niño event that involved an increase in sea temperature and decrease in nutrient availability. This event led to drastic reductions in the populations of corals and algae, as well as a decline in numbers of seabirds, seals, and sea lions because of food shortages. Drastic effects of drought, hot and cold weather, and toxins have been recorded for many populations of animals and plants. [*See* ECOTOXICOLOGY.]

Organisms have several ways of minimizing the effects of environmental stresses. Animals may *evade* a stress by moving away from it. In addition, both plants and animals may evade a stress by becoming dormant or quiescent. If evasion does not occur, a number of physiological and biochemical changes can take place to allow organisms to *resist* the effects of a stress. These include the degradation of proteins, as well as the synthesis of specialized proteins that help to protect cells against damage. In this way, organisms may acquire stress resistance and become *acclimated*. Acclimation may be triggered by environmental changes other than stressful conditions, allowing organisms to increase their resistance before the onset of a stress.

Much evolutionary change occurs as a consequence of natural selection that may be intense under stress. A number of genetic changes are possible as organisms adjust to stressful conditions. Individuals that are more successful at evading these conditions may be favored, perhaps because they enter dormant stages before the onset of a stress or because they are able to move away from it. In addition, individuals may be favored because they have a higher level of stress resistance, perhaps as a consequence of an increased ability to become acclimated. Populations can therefore undergo a range of evolutionary changes as they become adapted to stressful conditions. [*See* EVOLUTION AND EXTINCTION.]

As well as having direct effects on evolutionary processes, stresses may also have indirect effects. Because stresses often cause mortality or reproductive failure in populations, drastic reductions in population size can occur. A consequence of decreasing population size is an increase in the effects of *genetic drift*, which is a random process that can lead to the loss of rare alleles from a population. As a result, populations exposed to repeated stresses may become genetically less variable. This has important implications if the lost genes are important for survival.

Evolutionary studies of stress have focused on three levels: (1) the effects of stress on the genetic composition of extant populations in order to understand evolutionary processes operating currently in populations; (2) comparisons of populations or related species occupying different environments to determine how stress responses have evolved; and (3) the fossil record in order to relate morphological changes in fossils to known and inferred climatic stresses.

II. NATURAL SELECTION AND ENVIRONMENTAL STRESS

Table I shows six methods of demonstrating natural selection in populations involving the impact of environmental stresses. The first method is the most commonly used and involves correlating variation in traits with environmental changes. There are several ways to examine such correlations. One approach is to associate variation in a trait with variation in an environmental stress over a large geographical area. Any association may reflect the selective effects of the stress on the trait. For instance, temperature stress has often been associated with geographical variation in the body

TABLE I

Methods Used for Detecting Natural Selection in Traits Associated with Responses to Environmental Stresses

Method	Examples
Correlating traits with environmental factors	
1. Correlate variation in a trait over large areas	Geographic changes in body color of mammals and invertebrates
2. Correlate trait variation over short distances	Heavey metal resistance in plants and heat resistance in invertebrates
3. Follow changes in traits during stressful periods	Changes in the body size of birds under drought or in severe winters
Other methods	
4. Extrapolations from laboratory fitness tests	Relate fitness estimates for enzyme polymorphisms such as lactate dehydrogenase in killifish to gene frequencies in nature
5. Predicting changes in traits from a prior knowledge of energetics, physiology, or behavior	Associate wing polymorphisms in insects with the permanency of their habitats
6. Predicting mean trait values from optimality models	Predict the clutch size of birds exposed to fluctuating environmental conditions

size and body color of animals. A limitation of this approach is that associations may not reflect causal relationships between a trait and an environmental factor. For instance, variation in a trait may be associated with humidity, but selection may be exerted by factors correlated with humidity, such as temperature.

A second method is to examine environmental variation over short distances, which makes it easier than the first method to identify casual factors. For example, considering heavy metal resistance in plants, those growing on mine tailings differ genetically from conspecifics growing on adjacent soil not contaminated by heavy metals. Resistant plants are often only a few meters from susceptible plants, suggesting that strong selection associated with heavy metals underlies and maintains differences between adjacent areas.

A third method is to look at changes in an environmental factor over time, for traits tracking the environmental change are likely to be under selection. This approach is particularly powerful when populations are monitored at times of environmental stress. For instance, in Darwin's finches in the Galapagos Islands, body size and other morphological traits have been monitored at the time of a severe drought that resulted in 85% mortality in the finch population. Large birds with large beaks were favored during the drought because they could crack the large, hard seeds that formed the main food supply in this period. [See BIRD COMMUNITIES.]

There are also several methods that do not depend on directly associating variation in a trait with environmental factors. The fourth method in Table I involves measuring the fitness of individuals with different genotypes under stressful laboratory conditions. Fitness differences are related to the distribution of genotypes in nature. This approach has often been used to demonstrate selection on specific genetic polymorphisms. For instance, killifish have different morphs of the enzyme lactate dehydrogenase, and these differ in metabolic characteristics. As a consequence, one morph performs better in laboratory fitness tests than others at low water temperatures, whereas another enzyme morph performs better at high temperatures. These fitness differences can, in turn, be related to the frequency of the lactate dehydrogenase morphs in natural populations exposed to different temperatures.

A fifth method involves making predictions about how a trait changes during stress exposure, on the basis of prior knowledge about energetics, physiology, or behavior. For instance, if a phenotype decreases energy demand, it is likely to be favored when there is a food shortage. Wing polymorphisms in insects can be interpreted from this perspective. Many insect species include individuals with and without functional wings. Winged individuals evade food shortages in a habitat by flying to another habitat. However, the production of functional wings is energetically expensive, and insect flight muscles take up a lot of body space that could be used for egg production. These energetic considerations suggest that winged individuals will only be favored in temporary habitats. This prediction has been verified in studies on water striders.

Finally, in a sixth method, natural selection may be demonstrated indirectly by assuming that a trait has been under selection for a long time, so that the common phenotype in a population should represent the "optimal" one for a particular set of ecological conditions. Models can be set up to predict optimal phenotypes, and predictions can be compared to phenotypes in natural populations. For instance, if birds are periodically exposed to stressful seasons that reduce their ability to successfully rear their young, optimality models predict that the clutch size of the birds should be smaller than the number of progeny they can rear in a favorable season. This prediction has been verified in some bird populations living in fluctuating environments.

Conclusive demonstrations of natural selection will often depend on using a combination of these methods. If evolution is to proceed via natural selection, variation in a trait needs to have a heritable component. Phenotypic changes resulting from selection would otherwise not be passed on to the next generation.

III. GENETIC VARIATION UNDER ENVIRONMENTAL STRESS

Natural selection acts on heritable variation between organisms, and the expression of this variation depends on the environment to which organisms are exposed. Environmental stresses can have a number of direct effects on this variation as outlined in Table II. These have been detected by looking at the effects of stresses on mutation rates, recombination, developmental stability and the *heritability* (degree of genetic determination) of a trait.

One way the environment can influence genetic variation is by influencing the rate of recombination. Many experiments in eukaryotic organisms, including *Drosophila,* nematodes, tomatoes, and several fungi, indicate that recombination increases at temperatures above and below normal culture temperatures as well as at extremes of other stresses. Recombination results in the production of new combinations of alleles in the progeny generation. Some combinations may be better adapted to stressful conditions than allelic combinations of the parents. Organisms may therefore be at an advantage if they have an increased recombination rate under stressful conditions, thus increasing their chance of producing progeny that are better adapted than themselves.

Environmental stresses can also increase mutation rates. Well-known stresses that increase rates are ionizing radiation and toxic chemicals. In addition, mutation rates can be increased by other stresses, including extreme temperatures, intensive insolation, and extreme humidities. In general, mutation rates tend to increase with temperature, as demonstrated in *Drosophila,* microorganisms, fungi, and plants. However, cold shock has been shown to increase mutation in *Drosophila melanogaster,* suggesting that temperature extremes are important rather than temperature per se. Some of this increase in mutation may be associated with mobile genetic elements known as *transposons* that insert into DNA. It is possible that mechanisms increasing mutation rates under stress are adaptive, because they allow organisms to generate genetically variable offspring in response to environmental changes. There is evidence that mutation rates are under genetic control and can be altered by natural selection.

The increase in mutation rate under stress is thought to be random, so that mutations are as likely to occur at loci under selection as those not under selection. Recently, however, the random nature of this process has been challenged, with the proposal that stress increases mutation rates specifically at loci under selection. Most of the evidence for nonrandom mutation has come from bacteria and is controversial. Cells are exposed to a starvation stress because they carry a mutation in a biosynthetic pathway. New mutations are detected by renewed growth following reversion of the original mutation. The high frequency at which these mutations occur suggests nonrandom mutation. However, it has been argued that these experiments lack adequate controls to exclude the possibility that random mutation rates are being increased by stress. Moreover, there is still no plausible mechanism for mutations being directed by the environment.

TABLE II
Direct Effects of Environmental Stresses on Heritable Variation

Effect	Evolutionary consequence
Increase in recombination rate	Results in more genetically variable progeny after sexual reproduction
Increase in mutation rate	
Random	Leads to production of mutants at all loci, some of which may be adapted to stressful conditions
Directed	Increases production of mutants only at loci under selection by stress
Production of abnormal phenotypes	Creates potential for selection on unusual phenotypes
Increase in fluctuating asymmetry	Leads to selection for individuals with low fluctuating asymmetry
Increase or decrease in heritability of traits	Influences speed at which a trait can respond to natural selection

As well as increasing mutation and recombination rates, adverse conditions can also increase the incidence of abnormal phenotypes. Examples include the development of vertebrae in snakes and the arrangement of scales on the heads of lizards. Several studies have shown that the phenotypic variability of *Drosophila* is increased under adverse conditions. Unusual phenotypes involving changes in wing venation patterns and the morphology of the thorax region can be generated by exposing immature stages to high temperature or chemical stresses. The incidence of these abnormalities can be increased by selection, so that the abnormal phenotype is eventually expressed in the absence of the environmental stress. Some researchers have suggested that a similar process could result in major phenotypic changes in evolutionary lineages, such as the appearance of limbs or changes in segmentation patterns.

Another phenotype that changes under stress is the *fluctuating asymmetry* (FA) of bilateral characters that are measured on both sides of a body. FA is the absolute difference between the right and left sides of the body. FA has been studied for many characters in a variety of animals, including the number of scales of lizards and number of bristles of flies, the length of veins on the wings of insects, and the length of mammalian bones. It has often been suggested that FA provides a measure of the stability of an organism's development. As the level of stress during development increases, FA increases, so that it is higher in populations exposed to adverse conditions than in those under benign conditions. When averaged over several bilateral characters, low levels of FA have been equated with high fitness, because organisms with high fitness are expected to be better buffered against stress during development.

Finally, environmental stress may influence the heritability of traits in populations. Heritability often increases with the level of stress, although decreases in heritability have also been reported, particularly for yield in crop plants. Because heritability determines the rate at which a population can respond to selection, the speed of adaptation to an environmental change may depend on the stress level experienced by a population. This makes it difficult to predict responses to selection in one environment from heritability estimates in another environment.

IV. GENES UNDERLYING EVOLUTIONARY CHANGES UNDER STRESS

There are two views on the nature of genetic variation underlying adaptation to environmental stresses. The first of these proposes that adaptation normally occurs via many genes, each with a small effect on a trait under selection. These genes are known as *polygenes* or *minor* genes. The alternative view is that adaptation proceeds by the spread of a few genes with a large effect on a trait. These are known as *major* genes.

If selection favors a new optimum value as a consequence of an environmental change, the evolutionary response that will occur is expected to take place via minor genes, as long as the new optimum is not far away from the initial optimum (A in Fig. 1). The new optimum can be readily attained by genetic changes involving minor genes, because evolution involves a shift in phenotype that does not extend outside the normal range of the ancestral population. This means that there are many genotypes consisting of minor genes that can attain the new optimal value.

Once selection becomes intense, major genes are thought to be more likely to contribute to adapta-

resistance to environmental stress

FIGURE I Distribution of stress resistance in a population and two selection intensities for increased resistance. A indicates selection for resistance within the normal phenotypic range likely to favor minor genes and B indicates selection outside the normal range likely to favor rare major genes.

tion that will require a phenotypic change that is outside the normal range of a population (B in Fig. 1). Major genes can more readily produce phenotypes outside this range than minor genes. In addition, major genes are less likely to be diluted out of a population when susceptible individuals migrate into the population and mate with resistant individuals. These considerations explain why insecticide resistance selected in laboratory experiments at a low intensity of selection tends to involve minor genes, whereas major genes are involved in the field, where selection is very intense and resistant individuals must have phenotypes well outside a population's normal range of resistance.

Such arguments may also account for the simple genetic basis underlying other cases of adaptation to stresses imposed by human activities. In plants, resistance to heavy metals associated with mine tailings seems to have a fairly simple genetic basis, perhaps because adaptation has involved intense selection of plants from ancestral populations that contain only a few individuals capable of surviving on contaminated soils. Herbicide resistance in weeds has also been associated with major genes. The genetic basis of responses to other environmental stresses, including those arising from climatic changes, is not known, although major genes may determine much of the variation in resistance to some climatic stresses, including high temperatures in *Drosophila* and bacteria.

V. LIMITS TO EVOLUTIONARY CHANGE

There are limits on the extent to which populations can adapt to changing environmental conditions. This is apparent from the extinction of populations and species that often follow environmental changes. Limits to adaptation are also evident at the margins of species distributions, which often arise because species are unable to adapt to conditions beyond their margin.

One reason why adaptation may be limited is because of trade-offs. Genotypes with a high fitness in one environment may have a low fitness in a different environment, resulting in a trade-off be-

tween environments. In addition, trade-offs may arise between traits rather than environments, in which case genotypes with a high level of performance for one trait have a low performance for another trait. Trade-offs between environments are expected because the same genotype is not likely to have a high fitness in all environments. For instance, increased resistance to stressful environments where resources are limiting may be attained by a low rate of metabolism, but this can decrease fitness in favorable environments when resources are plentiful because high rates of metabolism increase mating success and reproductive output. In addition, trade-offs between traits related to fitness are expected because organisms have only a finite amount of resources. If resources are devoted to a task such as continued survival, less is available for reproduction and mating.

Five methods for detecting trade-offs between environments are listed in Table III. The first method is to artificially select for increased fitness

TABLE III
Methods for Detecting Genetic Trade-offs between Environments

Method	Examples
1. Artifically select for increased performance in one environment, score correlated response in another environment	*Drosophila* selected for increased resistance to desiccation or starvation have decreased fecundity under favorable conditions
2. Allow natural selection in the laboratory or field in one environment, score performance in another environment	Plants that have evolved resistance to heavy metals perform more poorly in uncontaminated soils
3. Compare the performance of some members of a strain or family in one environment, and others of the same strain/family in a different environment	Cultivars of barley that perform well under moist conditions perform poorly under drought conditions
4. Measure the fitness of genotypes at a locus in different environments	Different types of hemoglobin in deer mice are favored at high and low altitudes
5. Correlate variation in a trait with a known genetic basis to environmental variation	In Darwin's finches, large birds are selected at times of drought, whereas small birds are selected during wet periods

under one set of conditions and examine fitness under different conditions after several generations of selection. This has been used in *Drosophila* studies on stress resistance. Selection for increased resistance to dry and starving conditions has been associated with decreased early fecundity under optimal conditions, indicating a trade-off between survival under stress and reproductive output in a favorable environment.

In the second method, trade-offs are detected by natural selection in the laboratory or field. Organisms are exposed to one set of conditions for many generations before examining changes in performance under different conditions. This approach has been used to test for trade-offs in bacteria and plants. Plants selected under stressful environmental conditions often perform more poorly under favorable conditions. One example is heavy metal resistance: populations adapted to grow in soils with high concentrations of heavy metals often have relatively lower fitness when grown in uncontaminated soils.

A third method for detecting trade-offs between environments is to test individuals from the same strains or families in different environments. If trade-offs exist, strains or families performing well in one environment will perform relatively poorly in another environment. Cultivars of agricultural plants and breeds of animals are often used in such comparisons.

The final two methods are to look at the effects of stress on specific loci or specific traits. When one allele at a locus is favored in a stressful environment, and another allele is favored in an optimal environment, variation at this locus may contribute to a trade-off between environments. In the same way, one extreme of a trait may be favored under one set of conditions and the opposite extreme under different conditions. For instance, in Darwin's finches in the Galapagos Islands, large body size is favored during periods of drought as mentioned earlier, but small birds are favored in wet years because they can handle small soft seeds more efficiently. Body size therefore underlies a trade-off between environments.

Although trade-offs may constrain evolutionary change over a short time scale, their long-term evolutionary consequences can only be examined by comparisons of related species. If trade-offs act as constraints, species with relatively high fitness in one environment should have relatively low fitness in another environment. For instance, environmental temperature influences the body temperature of lizards, and this influences their ability to sprint. Sprint speed is related to fitness in lizards because it affects their ability to escape predators and to forage. Lizard species that attain maximum sprint speed at high temperatures can tolerate higher temperatures than those species attaining maximum speed at lower temperatures. This suggests that high temperature resistance is constrained by optimal temperatures for sprinting. In contrast, there is no association between the highest and lowest temperatures at which lizards can sprint, suggesting that resistance to these extremes evolves independently and is not constrained.

Evolutionary constraints may also be caused by factors other than trade-offs, and this is particularly apparent at the margins of species ranges. There are a number of reasons why evolutionary change may be constrained at margins. First, there may be a constraint at the physiological or biochemical level that prevents genes arising that would permit adaptation to conditions beyond a margin. Second, overall levels of genetic variation may be reduced at species margins because of the small size of marginal populations. This can lead to inbreeding or to the loss of favorable genes by genetic drift. A third possibility is that genes enabling range expansion do arise, but these are diluted in populations at margins because of a continuous influx of genes from other populations. Finally, an expansion of species ranges may be constrained because adaptation requires simultaneous genetic changes in a number of traits, so that individuals with all the requisite genes only rarely occur.

Indirect evidence suggests that the margins of some animals and plants are limited by physiological constraints because genes allowing further adaptation do not arise. For instance, some transplant experiments with plants indicate that individuals cannot survive conditions beyond margins. In addition, trade-offs may be important at borders. Genotypes with a high fitness under stressful conditions often have a low fitness under favourable conditions. These genotypes may not increase in frequency if stressful conditions in marginal populations occur only sporadically. However the relative importance of trade-offs, constraints, and other factors will not become clear until detailed genetic studies on traits that are ecologically important are carried out in marginal populations.

VI. ENVIRONMENTAL STRESS, EXTINCTION, AND DIVERSIFICATION

Extinctions occur when species are subjected to biological and physical stresses not present in their prior evolution, and when there is insufficient time for adaptation. Extinctions are an important component of the evolutionary record because the living species on earth represent a very small proportion of the species that have ever existed. A few brief periods of extinction have been so severe that exceedingly large numbers of species were eliminated. Five of these extreme periods are referred to as mass extinction events. In the most extreme of these events at the end of the Permian Period, 250 million years ago, around 96% of all living species were eliminated. For the most recent and most studied mass extinction event, 73 million years ago on the boundary of the Cretaceous Period and the Tertiary (the K-T boundary), around 65–70% of species were eliminated.

Though there is much debate concerning events behind mass extinctions, major changes in global climate are likely to be involved. For instance, temperature changes and sea level changes are both associated with marine invertebrate extinctions, but temperature changes (particularly a temperature decrease) are a more proximal force. Other stresses that may lead to extinctions include changes in ocean salinity and anoxic conditions, especially in shallow marine waters.

Groups of organisms are expected to differ in their susceptibility to extinction. In particular, organisms that can resist stresses are likely to be less susceptible. For instance, major extinctions at the K-T boundary did not occur in insects, which have a variety of mechanisms for overcoming stresses, including quiescence, diapause, and the evasion of stressful habitats by migration. During this extinction event, diatoms survived relatively unscathed compared to other marine plankton groups, apparently because they can evade stress by forming a resting spore during periods of nutritional depletion. Adaptation to conditions of reduced primary production may be particularly important in surviving mass extinction events. Resting cells therefore provide a survival mechanism when the conditions of the planktonic environment become lethal for more active cells. For bivalves at the K-T boundary, extinctions were concentrated among animals with starvation-susceptible feeding modes, active life cycle stages, and high energy budgets. The likelihood of extinction may therefore be reduced if organisms possess inherent resistance to environmental stresses.

Environmental changes have also been associated with periods of evolutionary diversification. At the start of the Cambrian period, 600 million years ago, the remains of complex and diverse organisms suddenly appeared in the fossil record. A great burst of morphological innovation occurred that produced many phyla, classes, and orders. At this time, there were changes in ocean chemistry, including the release of substantial quantities of phosphorus into shallow waters. The influx of phosphorus is important because it normally acts as a limiting factor in marine productivity. There may have been evolutionary opportunities once this resource was no longer limiting. This reduction of environmental stress by the appearance of increased resources could therefore have provided a trigger for a period of evolutionary divergence.

Environmental stress influences the rates of evolution detected in fossil lineages. At one extreme are the living fossils in which observable evolutionary change is arrested over vast time periods that often include several mass extinction events. In many cases, living fossils persist in harsh environments. For example, stromatolites occur in supersaline regions of searing heat and low precipitation, and horseshoe crabs can function in water varying widely in salinity and can endure great swings in temperature and oxygen levels. Stress levels are so high in such environments that predators and competitors cannot survive, allowing for the persistence of these organisms.

On the other hand, there are many cases where evolutionary changes have been associated with climatic changes, suggesting that environmental stresses may directly or indirectly underlie evolution. Environmental changes have been linked to periods of evolutionary change in invertebrates such as bivalves and ammonites, as well as in small and large vertebrates.

Severe and persistent environmental stresses may therefore restrict major evolutionary changes as in the case of living fossils. Evolution may be more likely under moderate stress levels where resources are not limiting, and there is supporting evidence for this notion. Along marine gradients, disturbance levels tend to be high in onshore habitats and lower in offshore habitats toward the deep ocean. Furthermore, food tends to be more abundant in onshore habitats, with organic levels decreasing toward the ocean floor, and oxygen levels fall with depth from the surface. The expectation therefore is for evolutionary change in the onshore habitats that are disturbed and where resources are sufficient to provide energy above that required for maintenance and survival. In support of this, higher taxa of benthic marine invertebrates tend to originate in such habitats. They then expand offshore, where communities tend to be more archaic.

In plants, disturbed environments with moderate stress levels may also promote evolutionary change. Archaic plant taxa occur throughout the Phanerozoic eon, extending from less than 590 million years ago, in wet, acidic, oxygen-poor, and nutrient-poor swamps and other wetlands, whereas evolutionary innovations are characteristic of uplands and floodplains where environmental conditions are more variable. Vascular plants, which started as weedy, fugitive species, also arose in variable, disturbed environments.

When individual fossil lineages are considered, those exposed to less extreme physical perturbations appear to show the greatest degree of evolutionary change. For example, in trilobites from the Ordovician Period (around 470 million years ago), relatively continuous morphological evolution,

known as *phyletic gradualism,* occurs in lineages from environments that showed some fluctuations. However, under more extreme fluctuations, there was little or no morphological change most of the time. Large changes occurred only occasionally in these environments, so the pattern of evolutionary change consisted of periods of little change interspersed with major morphological shifts, in contrast to more gradual changes under less extreme fluctuations.

Evolutionary patterns in the fossil record may therefore be associated with the magnitude of environmental fluctuations and the degree to which organisms experience stress. Diversification may be favored when organisms experience a sudden release from stresses in a resource-rich environment. Phyletic gradualism may be characteristic of moderately stressed environments, whereas morphological stasis may be characteristic of highly stressed environments.

VII. PREDICTING FUTURE EVOLUTIONARY CHANGE

Although there is uncertainty about future climatic trends, substantial and extremely rapid increases in world temperature of 2–4°C have been postulated in the next century. Because a 1°C rise in temperature is known to have a substantial effect on ecological systems, the predicted temperature change is expected to impose severe stress on many organisms. In addition to being associated with a global temperature increase, human activities are having other effects on natural habitats, particularly via the destruction of rain forests. Rapid deforestation of the Amazonian region has led to a significant increase in surface temperature, a decrease in evapotranspiration, and an extended dry season. These changes impose direct stresses on organisms as habitats become increasingly marginal and many populations face extinction. [See ATMOSPHERE–TERRESTRIAL ECOSYSTEM MODELING.]

In line with evidence from fossil studies, extinctions are unlikely to occur randomly. Considering rain forests, species diversity is likely to be reduced and converge toward that of depauperate faunas in heat- and desiccation-stressed habitats as found in some outlying pockets and at rain forest margins. These habitats are characterized by generalist and widespread species that are resistant to climatic extremes. Examples of such changes are already evident from unusual climatic events. For instance, following a severe El Niño drought in a neotropical forest in Panama, there were catastrophic declines in moisture-loving and rare plant species. A fall in biodiversity is therefore likely following stressful conditions, concomitant with a spread of some common species. [See BIODIVERSITY, PROCESSES OF LOSS.]

The extent to which extinction can be avoided will depend on the ability of populations to adapt to changing conditions. Theoretical models have been constructed in which the continued survival of a population depends on evolutionary responses in a single trait. Not surprisingly, these models show that the extent to which a population can track an environmental change depends on the genetic variation in the trait and the strength of selection. The maximum rate of environmental change that a population can tolerate without going extinct decreases with decreasing population size, decreasing genetic variance, and increasing variability in the environment.

These models are probably still too simple to predict the likelihood that natural populations will avoid extinction, particularly as adaptation to future environmental changes is likely to require changes in more than one trait. Organisms will have to adapt to the combined effects of different stresses because chemical pollution, fluxes of ultraviolet (UV-B) irradiation, and other stresses are likely to occur together with temperature increases. Synergistic effects occur between temperature stress and pollutants as well as between these factors and biotic stresses such as those imposed by pathogens, suggesting the need for complex adaptive responses. [See ATMOSPHERIC OZONE AND THE BIOLOGICAL IMPACT OF SOLAR ULTRAVIOLET RADIATION.]

The potential for many populations to evolve in response to rapidly changing conditions appears limited, as indicated by the common tendency for widespread species to track climates rather than undergo range extension independent of climate. Nevertheless, there are several cases in historic times of species extending their range beyond that predicted from climatic factors. In addition, adaptive changes have been documented in populations that have colonized areas with different climates, such as in rabbits introduced into Australia. Because some populations can adapt to changing climatic conditions, it is important for populations to maximize levels of genetic variation that are useful for adapting to future environmental changes.

Populations from ecological margins are likely to have been selected for environmental stress resistance, and so are expected to contain genotypes that are useful in countering stressful periods. In contrast, the energetic costs commonly associated with increased stress resistance mean that genes that increase resistance will be at lower frequencies in populations from favorable environments. Conserving populations from marginal habitats may therefore be one useful strategy for maintaining resistance genes to enhance future survival.

Taking into account these factors, widespread species appear to have the greatest chance of surviving and perhaps adapting to climatic change, and therefore have the least risk of extinction compared with species having a more restricted distribution. The rate of postulated climatic change suggests the initial extinction of stress-sensitive species, followed by climate tracking and some adaptation in more resistant widespread species.

Glossary

Acclimation Increase in resistance to a stress in response to prior nonstressful conditions.
Adaptation Process of genetic change in a population, owing to natural selection, whereby the average of a character with respect to specific function becomes improved.
Genetic drift Random process of change in gene frequencies in a population because of sampling from generation to generation.
Locus Position of a gene on a chromosome.
Natural selection Differential survival and/or reproduction of classes of entities that differ in one or more hereditary characteristics.
Phenotype Character arising from interaction between genotype and the environment.

Polymorphism Presence of different forms of a trait or gene in a population.

Recombination Process in which combinations of alleles arise from crossing-over between chromosomes during meiosis.

Stress Any environmental factor that causes a potentially injurious restriction in growth, reproduction, or survival of organisms.

Bibliography

Grant, B. R., and Grant, P. R. (1989). Natural selection in a population of Darwin's finches. *Amer. Natur.* **133,** 377–393.

Hoffmann, A. A., and Parsons, P. A. (1991). "Evolutionary Genetics and Environmental Stress." Oxford, England: Oxford University Press.

Holt, R. D. (1990). The microevolutionary consequences of climatic change. *Trends in Ecol. Evol.* **5,** 311–315.

Huey, R. B., and Kingsolver, J. G. (1993). Evolution of resistance to high temperature in ectotherms. *Amer. Natur.* **142,** S21–S46.

Karieva, P. M., Kingsolver, J. G., and Huey, R. B., eds. (1992). "Biotic Interactions and Global Change." Sunderland, Mass.: Sinauer.

Parsons, P. A. (1993). Stress, extinctions and evolutionary change: From living organisms to fossils. *Biol. Rev.* **68,** 313–333.

Equilibrium and Nonequilibrium Concepts in Ecological Models

Donald L. DeAngelis

The University of Tennessee

The question of whether ecosystems are more appropriately described and modeled as equilibrium systems or as nonequilibrium systems is a central one in ecology. Ecological theory was long dominated by the idea of a "balance of nature"; that is, that the interactions of species populations create a stable community. Correspondingly, theoretical models of ecological communities have commonly been based on sets of differential equations for species populations that are analyzed at equilibrium. However, the relevance of much theory based on the study of mathematical equilibrium has been challenged by both theoretical and empirical results. Natural ecological systems show persistence, but often not constancy of the sort implied by stable mathematical equilibria. Recent theoretical and empirical research suggests that the apparent persistence of many ecological communities may derive from causes that include spatial extent and heterogeneity, as well as disturbance regimes that prevent competitive exclusion. Results based on equilibrium theory, though still useful in many cases, must be used with an understanding of their limitations.

I. INTRODUCTION

As long as humans have philosophized, they have argued whether permanence or change is more fundamental. During the last two centuries science has shown that over very long time scales the earth has been characterized by change. The climate and physiography of the earth have changed and bio-

This work was sponsored by the National Science Foundation's Ecosystems Studies Program under Interagency Agreement 40-689-78 with the U.S. Department of Energy under Contract De-AC05-84OR21400 with Martin Marietta Energy Systems, Inc.

logical systems have evolved. As a consequence, ecological communities have undergone continuous change on geological time scales.

Ecologists, however, are usually concerned with shorter time scales of years and decades. On such time scales we are more likely to see nature as unvarying. From one year to the next we generally observe the same species of plants, birds, and insects in about the same numbers in our backyards. Still, systematic observations show that this constancy is only approximate at best. For example, major outbreaks of some insect species and diebacks of some plant species occur at frequencies high enough to be noticed. Are these fluctuations merely stochastic deviations from an overall "balance of nature" or equilibrium, or are they symptoms of an underlying nonequilibrium nature of ecological systems? This is a basic question that involves many aspects of ecological theory. [*See* INSECT DEMOGRAPHY.]

II. THE CONCEPT OF STABLE EQUILIBRIUM

Until recently, ecology was dominated by the central concept of a stable equilibrium among species of a community, at least as a state toward which an ecosystem tends. Of course, dynamic aspects were recognized as well. Ecological succession is a dynamic process, and it refers to the observation that following a disturbance, an ecological system tends to go through a number of transient or nonequilibrium stages, dominated by different flora and fauna. But in the classic description of this process, a final state is approached in which a community of climax species exists in stable equilibrium. [*See* POPULATION REGULATION.]

The equilibrium point of view of ecological communities was formalized in mathematical models. Such models typically start with a set of differential equations (or difference equations) for an ecological community, one equation for the rate of change of number density of each species population, N_i, where i identifies a particular species, in terms of the number densities of other species populations in the community:

$$dN_i(t)/dt = F_i(N_1, N_2 \ldots, N_m) \quad (i = 1,2, \ldots, m). \tag{1}$$

These equations are analyzed close to equilibrium. This can be done by linearizing the equations around the one or more equilibrium points of the system, where an equilibrium point $(N_1^\star, N_2^\star, \ldots, N_m^\star)$ is the point at which

$$dN_i/dt = 0 \ (i = 1,2, \ldots, m). \tag{2}$$

The linearized equations are represented as a matrix equation,

$$d\mathbf{n}(t)/dt = A\mathbf{n}(t), \tag{3}$$

where A is an $m \times m$ matrix the terms of which have the form $a_{ij} = \partial F_i/\partial N_j^\star$, and $\mathbf{n}$ is the vector of perturbations of population densities about the equilibrium. The stability properties of the system close to the equilbrium point can be found by analysis of the matrix A.

A stable ecological community would correspond to a stable A-matrix having eigenvalues whose real parts are all negative. Mathematically, this is called "neighborhood" or "Lyapunov" stability, or stability for very small perturbations. For a single population, such stability means that a perturbation in population number density above the equilibrium value, $N_1^\star$, would lead to a lower per capita population reproductive rate and/or a higher per capita population mortality rate. The opposite would occur for a perturbation to number densities below the equilibrium value. These are called "density-dependent" population regulation effects, which tend to return the population to its equilibrium value. In the idealized stable ecological community, any perturbation in the population level of one species above its equilibrium value would engender increased resource limitation, increased predation, increased competition, or some combination of these effects that would tend to return the population to its equilibrium value.

Mathematical ecologists such as Robert MacArthur (1930–1972) believed that the study of such equilibrium systems would be the best approach for generating new hypotheses. Configurations of species and parameter values of equations of type (1) that produced stable model systems were expected to give insights into the nature of ecological communities. Questions of various types were asked of these models. For example, how similar can competing species be and still coexist, and what sort of structure will food webs tend to have? This research program, which can be called the "equilibrium theory of ecological communities," is the basis for much of the model analysis in theoretical ecology. It is exemplified in many of the research papers and reviews in theoretical ecology during the 1960s and 1970s. However, theorists realized the limitations of equilibrium theory and pressed theoretical development in the direction of accounting for such things as the stochastic environmental disturbances that were prevalent in natural systems. [*See* SPECIES DIVERSITY.]

III. CHALLENGES TO THE EQUILIBRIUM CONCEPT

Ironically, the study of systems of equations such as (1), based on the idea of equilibrium ecological communities, gave results that led to a radical questioning of the usefulness of equilibrium theory and models. These results came from the study of abstract community models with an arbitrary number of species, which gave some surprising answers. On the basis of the mathematical properties of matrices, it was shown that the equilibrium points of large, highly connected food web models tend to be unstable; that is, their **A**-matrices are likely to have at least one positive eigenvalue, a likelihood that increases with size and the relative number of connections between species in the web. This seemed to fly in the face of the conventional wisdom of ecology that the complexity of ecological systems gives them stability.

Surprising results were also found for simple ecological systems. In simple models of one predator and one prey, realistic feeding functions often led to predictions that the system should be unstable. This raised the possibility that many systems, such as phytoplankton–zooplankton interactions, are inherently unstable. Analysis of models, especially those with three or more species, showed that mathematical "chaos" commonly occurred even though the equations were deterministic. Chaos in dynamical systems is characterized by model population levels varying in a way that is indistinguishable from random fluctuations and by an arbitrarily small difference in an initial population level leading to a large difference in the model population at a later time.

Another mathematical result of the study of equations of type (1) implies that some number m of species could not coexist in competition for a number of distinct resources, k, where $k < m$. This type of result was also shown experimentally under controlled conditions for closely related organisms, such as different species of flour beetles competing for a single resource. However, in natural systems, the observations are different. Many species of phytoplankton coexist in the pelagic areas of oceans and lakes (the "paradox of plankton" discussed by the ecologist and limnologist G. Evelyn Hutchinson). The fact that all the species appear to compete for light and a small number of limiting nutrients would, on the basis of theoretical predictions from equilibrium theory, implies that only a small number of these species could coexist. Thus, the results of the models themselves indicated that, rather than regulate the community of species populations at stable equilibrium, the assumed biotic interactions would more likely lead to system instabilities and competitive displacement of all but a few species.

Many empirical ecologists criticized the equilibrium theory from a different point of view. They challenged the idea, embodied in theoretical equilibrium models, that density-dependent biotic interactions are consistently strong enough to regulate population levels and determine the structure of ecological communities. The absence of strong regulating forces, such as competition and resource limitation, would mean that well-defined community equilibria would not occur and that species populations in the community would exhibit a great deal of random fluctuation in response to environmenal disturbances and demographic stochasticity. The main arguments of these critics were that studies of many species reveal few unambiguous examples of the types of mechanisms that were assumed to regulate populations and communities, such as resource limitation on species populations or of competition between species under natural conditions. In addition, in many species, at the local level in space at least, populations undergo frequent and drastic fluctuations. This critique is based on the studies of a number of particular taxa. For example, specialists on insects have noted that many populations of insect taxa show no apparent sign of being near equilibrium. Instead, populations are constantly going to extinction on patches of habitat, which are then repopulated by new immigrants of that species from other patches. The same has been asserted of grassland avifauna, coral reef fishes, stream fish communities, and various other communities.

These empirical challenges have not gone unanswered. Other ecologists, by analyzing other data or the same data but with different criteria (such as giving higher weight to population studies conducted over relatively long time periods), have responded with evidence that density-dependent regulation is quite common in populations and that competition and other biotic interactions stabilize, or at least influence, the structure of many communities. The safest conclusion from all of these studies at present is that the role of biotic interactions in structuring and regulating communities may occur in some situations, but are not universally strong and may, in fact, be difficult to demonstrate convincingly in many systems.

In summary: first, the analysis of mathematical models based on equilibrium theory typically reveals instabilities that undermine the idea that ecological systems exist in a stable balance. Second, empirical data do not universally indicate the presence of strong density-dependent biotic mechanisms that are postulated in equilibrium models of populations and communities and suggest that population fluctuations may be more frequently the result of demographic and environmental stochasticity.

IV. PERSISTENCE: AN ALTERNATIVE TO STABLE EQUILIBRIUM POINTS

A variety of both theoretical and empirical results, therefore, indicate difficulties in the use of equilib-

rium theory for describing ecological systems. Ecologists recognized the need for a broader perspective and focused on another aspect of stability, namely, persistence, as a more relevant concept for ecological systems than neighborhood stability. Persistence is a weak form of stability, meaning merely that the species components of an ecosystem will persist over long periods of time, regardless of how strongly population numbers may fluctuate. This is similar to the mathematical concept of global stability, which states merely that a variable is bounded, not that it necessarily approaches a constant value.

In view of the fact that communities show a considerable degree of persistence, but often do not show a constancy that would imply stable equilibrium in the neighborhood sense, a number of alternatives to the simple equilibrium model of ecological communities were formulated during the 1970s and 1980s. The basic ideas of three of these can be briefly summarized as follows. (1) Regulation occurs, but is only strong at very high population levels and is weak at typical ranges of population density. (2) Ecological communities are truly nonequilibrium systems at the small spatial scale (e.g., a habitat patch), but fluctuations in the local subpopulations roughly balance out over the larger spatial scale. (3) Ecological communities are maintained in nonequilibrium states by disturbances and a cessation of disturbances would result in a loss of species.

V. DENSITY-VAGUE REGULATION

The first of these alternatives to equilibrium theory has been called "density-vague regulation." In this view there is no evidence that population changes are tightly governed by population density, at least at intermediate densities; however, density-dependent mortality or reproduction will keep populations from growing too large or will tend to prevent population extinction at very low densities.

Models have been made of the dynamics of density-vague populations subjected to perturbations. The mathematics used in these models might be, for example, that of stochastic Markov chains rather than deterministic differential equations. Instead of having a stable equilibrium, a population can, at best, exhibit θ-persistence. The population is said to be θ-persistent if its size, $N(t)$, obeys the inequality

$$| N_0 - N(t) | < \theta \text{ for all } t > 0, \qquad (4)$$

where N_0 and θ are constants. In words, exogenous disturbances can buffet a population over the range from $N_0 - \theta$ to $N_0 + \theta$, but as those bounds are approached, density-dependent mechanisms come into play and resist further changes in the population size.

According to the density-vague concept, species populations are generally not at levels at which they compete strongly with each other. Competitive exclusion would not be common in such situations and populations would not generally be limited by resources. Only when populations grow to large sizes such that they exceed their carrying capacities would they experience density-dependent regulation, usually resulting in a population crash.

VI. SPATIAL EXTENT AND HETEROGENEITY CAN STABILIZE SYSTEMS OVER LARGE SPATIAL SCALES

A second possible approach to explaining community persistence without the equilibrium model is the invocation of spatial extent and heterogeneity. This view accepts the idea that instabilities due to biotic interactions and chance fluctuations and extinctions due to variability in the environment are likely to affect populations in local areas. When populations are spread out over an extensive area, interactions may be too weak for strong instabilities to occur everywhere. Only small parts of a population may be exposed at a given time to unfavorable events in the abiotic environment, a situation often referred to as "spreading of risks."

The importance of scale in the interpretation of equilibrium and nonequilibrium systems was given a strong impetus by the classic experiments on scale mites by C. B. Huffaker. He studied the dynamics of the herbivorous six-spotted mite, *Eotetranychus sexmasculatus,* which infests oranges, and its preda-

tor *Typhlodromus occidentalis.* Huffaker found that the predator–prey interactions on a single orange quickly led to exinction of first the prey and then the predator. Clearly, the two species on the small spatial patch of the orange constitute a nonequilibrium system. When a large number of oranges were used and separated by distances that retarded but did not prevent movement between them, the extinctions continued at the local level, but recolonizations followed. As a result, the system as a whole persisted much longer.

Mathematical models were developed to explicitly simulate systems with spatially distributed habitat patches. In particular, Huffaker's system of herbivorous and predatory mites on a spatially heterogeneous array of oranges was simulated by R. Hilborn by means of a computer model with 50 habitat patches or cells. Both predators and prey in the model were able to disperse from cell to cell, and the predators were assumed to eliminate any prey on a cell that they reached. It was found that the relative effectiveness of dispersal of the two species determined whether the two species would persist or not. Increasing the number of cells helped to prolong persistence.

A computer simulation model was used by H. Caswell to study the coexistence of two competitors and a mutual predator in a multicell system. The predator slowed the rate of elimination of the weaker competitor. Although no assumptions concerning equilibrium were made in this model, systems with a large number of cells persisted indefinitely and increased cell numbers resulted in reduced demographic stochasticity.

Models have been developed specifically to study the long-term persistence of endangered species when no mechanisms of population regulation were present. For example, S. Hubbell modeled rare tree species by means of a stochastic birth-and-death model in which several species competed for a limited number of spatial sites. No assumptions on regulation were made; the population sizes simply drifted. The model showed that for population sizes starting with thousands of individuals, all species populations could coexist for hundreds of generations. Thus no assumptions of regulation around an equilibrium point need to be made to guarantee long-term persistence if there is sufficient spatial extent.

Other theoretical models have been developed to simulate the internal dynamics of each cell, as well as the migration between cells. The local dynamics were governed by deterministic equations that were nonequilibrium, so that the populations in the cells were transient. When migration between cells is episodic rather than continuous, the overall region may be stable, although the isolated cells are unstable.

The foregoing models are all computer models, but simpler mathematical models have also been developed to predict the effects of spatial extent and heterogeneity. These models, called patch occupancy models, were developed primarily during the 1970s. In patch occupancy models there were two or more species and a landscape that was assumed to consist of a large number of habitat patches that could be occupied by the species. Colonization of an unoccupied patch occurred randomly as a function of the percentage of occupied patches. The internal dynamics of the patches were not explicitly described, just the probabilities of successful occupancy or displacement of another population if colonists of one type landed in a patch. The mathematical explanation for a simple example is as follows. Suppose that x is the fraction of patches that have only prey on them and that y is the fraction of patches where both the predator and prey are present. The rate of colonization of empty patches is assumed to be proportional to the number of patches occupied by prey times the fraction of empty patches $(1-x-y)$, that is, $ax(1-x-y)$. The rate of transition of prey-only patches is assumed to be proportional to the product of the prey-only patches and the predator–prey patches, bxy, while the extinction of prey and predators is assumed to proceed at a rate ky, where a, b, and k are constants. Thus, the equations for the system are

$$dx/dt = ax(1 - x - y) - bxy \qquad (5)$$

$$dy/dt = bxy - ky. \qquad (6)$$

This model has the equilibrium solution $x^\star = k/b$, $y^\star = a(1 - x^\star)/(a + b)$.

In these models, it is assumed that on the large spatial scale a deterministic equilibrium exists [i.e.,

Eqs. (5) and (6) have an equilibrium solution], whereas nonequilibrium conditions exist in the local cells. Another concept, however, is that increased spatial scale does not produce a true equilibrium, but merely prolongs the persistence of a particular system. A class of models designed for computer simulation of competitive or predator–prey systems over multicelled regions chooses migration transfers by Monte Carlo random number generation. The possibility of stable equilibrium over large spatial domains is not obvious, and these studies focus on persistence at the large scale.

The nonequilibrium concept is implicit in some of the current theories of landscape or regional ecology. For example, in the study of forest communities, the idea of succession toward a stable steady state has given way to the "mosaic concept" of the landscape. When a gap is created in a forest, a process of succession takes place in which species replace other species through time. Finally, only one or a few large canopy trees are left. These die and the process begins again. Looked at from the point of view of the specific spatial location, this is a nonequilibrial process. However, over a large section of forest, the relative abundances of plant species and age-classes hardly show any change. Another view of the landscape, with special pertinence to animal populations, is that it is a mosaic of source and sink areas, that is, subregions (sources) in which there is a net surplus of production of organisms of a particular species and other subregions (sinks) in which there is a net loss of organisms of that species. Source–sink conceptions of the landscape incorporate nonequilibrium ideas and stabilization by means of spatial heterogeneity. One subregion may produce a surplus in population every year, so that surplus has to emigrate to surrounding areas, where they may survive but do not reproduce well enough to create a net surplus.

VII. DISTURBANCE REGIMES CAN STABILIZE SYSTEMS

A third general proposal for the long-term persistence of communities in the absence of equilibrium involves the action of disturbances. In the pure theory of equilibrium systems, it is characteristic to think of disturbances as exogenous "spoilers" that prevent the system from achieving perfect equilibrium. However, the study of mathematical models showed that strong biotic interactions had a strong tendency toward elimination of all but a few species.

The idea that disturbances may be vital in promoting coexistence of competing species goes back to Hutchinson, who was puzzled by the high diversity of phytoplankton competing for the same basic resources of nutrients and light. Hutchinson surmised that disturbances would interrupt the process of competitive displacement. An example where the occurrence of disturbance was shown to be critical for coexistence of populations was discussed for competing plant species. If disturbances went away, one plant would take over. A general explanation of the role of different levels of disturbances is as follows. Consider a community of competing species with a variety of life cycle strategies ranging from early-successional types (good colonizers and fast growers) to late-successional types (slow colonizers and growers but better competitors). When subjected to very high disturbance (mortality) rate, the ecosystem is dominated by the early-successional types. When subjected to low disturbance rates, the good competitors eliminate most other species. At some intermediate disturbance rate there is a maximum in species diversity. This is sometimes referred to as the intermediate disturbance hypothesis.

As early as the 1940s ecologists had already suggested that terrestrial plant community dynamics is a nonequilibrium process, in which patches of land change through succession and are occasionally reset to early stages by disturbances. Similar ideas were proposed for phytoplankton communities, intertidal algal species, tropical rain forests, tropical reefs, stream fish communities, host–parasitoid systems, and carrion-breeding communities.

A number of interesting models have been developed from these ideas. Three different types can be distinguished, where (1) disturbances affect population numbers directly, (2) disturbances affect the resources that populations use, and (3) there are

endogenously created fluctuations in consumers and resources.

A model that incorporates disturbances of the first type was developed for competition for space among bivalves, tube-building polychaetes, and small crustaceans. Simultaneous persistence of these species occurred only when recruitment rates fluctuated relative to each other. A similar model is the lottery competition model used to study competing fish populations on coral reefs. Models based on the intermediate disturbance hypothesis were applied to a variety of systems: phytoplankton–herbivore systems, herbivorous diptera and algae on rocky shores, filter-feeding insects in small streams, stream benthic insects, herbaceous communities in old fields, and tropical forests among them.

The second type of disturbance affects the resources of competing species. Mathematical models and computer models of various numbers of competitors examined the stabilizing influence of perturbations. Both deterministic environmental fluctuations, such as applied periodic variations in resources in certain frequency ranges, and random fluctuations were shown to be capable of maintaining coexistence of competing species that would otherwise go to extinction. The third type of disturbance, endogenous variations in population sizes due to population instabilities, was also shown through models to maintain the persistence of a system in which competitive displacement would take place otherwise.

VIII. IS THE EQUILIBRIUM OR NONEQUILIBRIUM DESCRIPTION MORE FUNDAMENTAL?

The dichotomy between equilibrium and nonequilibrium may in part be merely an issue of spatial and temporal scales. If one conducts measurements of ecological communities on sufficiently large spatial scales that aggregate over many habitat patches, the populations should tend to have greater constancy than would be found from measurements on a single patch. Similarly, the longer a record of measurements one has of a population, the more likely its numbers will appear to fluctuate about a well-defined equilibrium. One can use models based on equilibrium theory [of equation type (1)] as good phenomenological descriptions of the measurements of these systems.

However, the fact that the nonequilibrium description converges to an equilibrium description on certain spatial and temporal scales is not a completely satisfactory solution to the question of which description is more valid or basic. The ways in which some observed population and community phenomena are interpreted depend critically on which point of view one is operating with. For example, populations in a community often show large fluctuations. This could be interpreted from the nonequilibrium or density-vague point of view as indicating a lack of population regulation, meaning that equations of the form (1) with density-dependent regulation about an equilibrium are not appropriate. On the other hand, as was discussed earlier, equations (1), if unstable because density-dependent biotic interactions are destabilizing, can produce mathematical "chaos" in population levels that is indistinguishable from stochasticity. Thus, ironically, from the observation of apparently random fluctuations in populations it could be argued that the population obeys the equilibrium description of equations (1), but that the density-dependent biotic interactions are so strong that chaotic instability results.

It follows that if one accepts the broad definition of "the equilibrium theory of ecological systems" as encompassing all systems for which the set of equations (1) are appropriate, regardless of whether these equations are stable at their equilibria or not, then systems with large fluctuations could still fit under this theory.

The debate between equilibrium and nonequilibrium descriptions may not be easy to settle merely from time-series measurements of population numbers but may require more careful studies of the life cycles of organisms and their detailed interactions. It is obvious that the species that we observe are ones that have persisted for long periods of time, so mechanisms that promote persistence (i.e., survival of at least some offspring in every generation to reproductive age and successful re-production) have been selected. It may often be the case that these mechanisms also favor some degree of constancy of population numbers in some species. Other species, however, have opportunistic life cycles adapted to rapid growth when or where environmental conditions are favorable and just as rapid decline when conditions change. Constancy of numbers in either time or space is therefore sufficient but not necessary for a species to persist over long time periods. Hence, natural selection may sometimes select for stability, but not necessarily.

Over sufficiently long time periods, even those ecological systems that appear most stable must change, through either climatic change or the appearance of new invading species. Thus, in a fundamental sense, "equilibrium" is an ideal that never exists. It does not exist even in the physical systems of the earth except under artificial conditions. It certainly does not exist for ecological systems.

One can formulate equlibrium equations and they may be good approximations in some circumstances. They seem to hold better at large spatial scales and over certain time windows. Used appropriately, models based on the equilibrium assumption may be quite helpful in understanding ecological systems. However, on small spatial and temporal scales, equilibrium models may be misleading and may produce incorrect conclusions. Many nonequilibrium models of dynamics at the spatial patch level and for short-time transient dynamics are now being developed and these may hold the key to future progress in understanding ecological systems.

Glossary

Competition Use of resource by one individual that depletes the amount of that resource available to other individuals, either of the same (intraspecific) or of other species (interspecific).

Demographic stochasticity Fluctuations in populations due to the fact that births and deaths occur with some degree of randomness.

Density dependence Influence on the rate of change of a population or of any process (e.g., reproductive rate, mortality rate, growth rate, age at maturation) within the population that depends on the density of the population.

Density-vague regulation Concept that at typical levels of a population, density-dependent effects are negligible, though they may be important at high densities.

Disturbance Event that causes a sudden change in the level of a population or of a process within the population.

Ecological community Group of species that coexist in the same place and that interact with each other in some defined way.

Environmental stochasticity Fluctuations in environmental conditions that lead to fluctuations in populations.

Food web Set of species that interact with each other through feeding (consumer–resource) relationships and, in so doing, pass energy through the ecosystem.

Intermediate disturbance hypothesis Hypothesis that an ecological community has higher diversity of species when subjected to an intermediate level of disturbance than to a higher or lower level.

Negative feedback Tendency of a system to counteract externally imposed changes.

Population regulation Assumed occurrence of feedback mechanisms (e.g., increased mortality rate and/or decreased reproductive rate when the population increases) that keep a population level within certain bounds.

Stability Tendency for a system to return to an equilibrium following a small disturbance (neighborhood stability), to return following a large disturbance (global stability), or for the system to remain within certain bounds though not necessarily converging to an equilibrium point (persistence).

Bibliography

DeAngelis, D. L., and Waterhouse, J. C. (1987). Equilibrium and nonequilibrium concepts in ecological models. *Ecol. Monographs* **57,** 1–21.

Diamond, J., and Case, T. J., eds. (1986). "Community Ecology." New York: Harper & Row.

Hassell, M. P., Latto, J., and May, R. M. (1989). Seeing the wood for the trees: Detecting density dependence from existing life-table studies. *J. Animal Ecol.* **58,** 883–892.

Hastings, A. (1988). Food web theory and stability. *Ecology* **69,** 1665–1668.

Pimm, S. L. (1991). "The Balance of Nature?" Chicago: University of Chicago Press.

Rahel, F. J. (1990). The hierarchical nature of community persistence: A problem of scale. *Amer. Natur.* **136,** 328–344.

Sinclair, A. E. G. (1989). The regulation of animal populations. *In* "Ecological Concepts" (M. Cherrett, ed.), British Ecological Society Symposium. Oxford, England: Blackwell Scientific.

Stiling, P. D. (1987). The frequency of density dependence in insect host–parasitoid systems. *Ecology* **68,** 844–856.

Eutrophication

W. T. Edmondson
University of Washington

Lakes with relatively large supplies of nutrients are called eutrophic (well nourished) and those with poor supplies are oligotrophic (poorly nourished). The categories are not sharply separated and are connected by intermediates, called mesotrophic. Eutrophication is a process by which bodies of water are made more eutrophic by an increase in the nutrient supply. The term is used mostly in connection with lakes, but can be applied to rivers and to estuaries, bays, sounds, or other partly enclosed marine waters.

The term is based on the observation that, for natural reasons, lakes are supplied with nutrients at a given rate from their surroundings, mostly dissolved in the water of inlets. Eutrophic lakes, with a rich supply, tend to develop and maintain higher concentrations of nutrients than others and are more productive biologically, generating large populations of algae and other organisms.

I. INTRODUCTION

A widespread, serious, but usually not dangerous environmental problem is produced by an increase in the delivery of nutrients to lakes or other bodies of water that were previously in acceptable condition. The additional nutrition increases the rate of production of photosynthetic organisms, most notably phytoplankton, and the maximum amount that can be present at any time. It encourages the growth of particular types that can cause unpleasant conditions, mostly cyanobacteria (blue-green "algae"). Expensive treatment of water supplies may be required because of taste, odor, and possible toxicity. Under some conditions these organisms produce toxic materials that can be fatal to warm-blooded vertebrates that drink the water. Because of the increase in abundance of phytoplankton, a lake can become less transparent, having a cloudy appearance, and submerged rooted vegetation may be eliminated from deep water by shading. [*See* LIMNOLOGY, INLAND AQUATIC ECOSYSTEMS.]

II. RELATION BETWEEN PRODUCTION AND NUTRIENT SUPPLY

The rate at which algae can be produced is based on the availability of nutrients to them, and that depends on the concentration in the water. At low concentrations the rate of absorption of nutrients is limited by the rate of encounters of ions with cell surfaces. If the concentration is increased, the rate of encounters increases and therefore the rate of absorption and rates of synthesis of cell material and of cell division increase. At a high concentration the rate of absorption is limited not by encounters, but by the rate of transport of ions through the cell surface, and further increase of nutrient does not increase production. At some times of year, absorption by a growing phytoplankton population may reduce the concentration of one or more nutrients below the minimum needed by that population for further growth.

The concept of nutrient supply (also called nutrient loading or input or income) to a lake is fundamental to understanding biological production. The income of a given nutrient during a period of time is calculated by multiplying the volume of water entering the lake during that time by the concentration of the nutrient. A given quantity can enter in high concentration in a small volume of water or in low concentration in a larger volume. The consequences for the lake are quite different. The entering water generally displaces an equal volume of lake water and mixes with the rest. If the entering water has the same concentration as that of the lake there is no change, and concentration-dependent processes proceed at the same rate. If the entering water has a higher concentration, the concentration in the lake will increase and so can the rate of production and the quantity of organisms produced. However, doubling the water input and halving its concentration will result in delivery of the same amount of nutrient, but the concentration produced in the lake will be less, and the resulting production will be less.

III. CONCEPTS OF EUTROPHICATION

In practice, the term eutrophication has come to be used in at least four different ways.

Meaning 1

The word eutrophication refers to an increase in the rate of supply of nutrients to a lake over the preexisting rate. This is the meaning most closely associated with the etymology, history, and application of the word, and it is widely accepted. The best-known examples of eutrophication have been caused by pollution with sewage effluent, raw or treated, which has very high concentrations of nutrients. Some changes have been caused by drainage from agricultural land or suburban land development. An essential feature of eutrophication is that the content of nutrients in the lake increases.

Sometimes a distinction is made between natural and artificial eutrophication. Truly natural eutrophication appears to be rare. It would be an increase in nutrient loading by a natural process such as the exposure of nutrient-rich deposits, deposition of volcanic ash, a rearrangement of drainage systems by earthquakes, or by increased release from soils resulting from plant succession after a natural disturbance.

Meaning 2

Sometimes one or more consequences of eutrophication by Meaning 1 are taken to be eutrophication itself. The immediate effect of eutrophication by Meaning 1 is an increase in the average concentration of nutrients in a lake. Because lakes with high concentrations of nutrients tend to be biologically more productive than those with low concentrations, sometimes the term eutrophication has been applied to an increase in productivity or in the abundance of organisms. This preempts a meaningful usage of the word for something quite different. The abundance of organisms can increase for reasons having nothing to do with nutrition, as when a lake is relieved of an input of toxic substances. In the long term, undisturbed lakes can show a steady increase in production and population density resulting from morphological changes as they fill in and from succession in the aquatic community, without a continuing increase in external nutrient supply. "Aging" is not eutrophication.

Meaning 3

Eutrophication in this meaning is defined by the ambiguous phrase "nutrient enrichment." This can be interpreted as identical to Meaning 1, but a widespread usage is quite different: the fact that a lake is continuously receiving nutrients through its inlets or other sources is described as "nutrient enrichment." In Meaning 1 this is described as "nutrient income" or "nutrient loading." In Meaning 3 it is called "eutrophication." An increase in input, called "accelerated eutrophication" in Meaning 3, is equivalent to the eutrophication of Meaning 1. However, delivery of a large amount of nutrients over time by a continuous large input of water with a low concentration of nutrients does not create a eutrophic condition, nor make a lake highly productive.

Meaning 4

Some of the material that enters a lake leaves through the outlet. The rest accumulates in the sediment on the bottom, so that the total nutrient content of the original lake basin including the mud steadily increases over time. This increase is called by some "eutrophication." The difficulty here is that only a thin layer of sediment is in active exchange with the water. Most of the sediment is not a functional part of the ecosystem, and therefore has nothing to do with the state of nutrition or productivity of the lake. It forms the paleolimnological record of past conditions.

The first two meanings described here concern different but closely related aspects of the control of productivity of aquatic ecosystems. Meaning 1 is based on the rate of delivery of nutrients. To use this meaning as the basis for a useful formal definition, it is necessary to specify how the rate of supply is defined. It is usually given as mass of nutrient delivered to the lake per unit time, often the year. To permit comparison of lakes of different sizes it can be given per unit area of lake. The volume of the lake must also be taken into account; see Section VI. Meaning 1 implies a net increase in nutrient content of the lake. Meaning 2 is based on the *effects* of a change in the nutrient content. This meaning is frequently presented in textbooks, sometimes combined with Meaning 1.

IV. CONSEQUENCES OF EUTROPHICATION

The most obvious direct effect on a lake of a net increase in nutrition is an increase in the abundance of phytoplankton. The presence of dense populations of phytoplankton causes a conspicuous reduction in transparency, widely regarded as environ-

mental deterioration. Some of the organisms have other properties that cause unpleasant or economically damaging conditions. Most objectionable are cyanobacteria, which can form dense floating masses, produce odors in the water, and may cause problems with the filtration systems in drinking water treatment plants. Many cyanobacteria produce materials that are toxic to warm-blooded vertebrates and have been responsible for widespread deaths of cattle, dogs, and other animals that have drunk water with substantial quantities of the organisms. This has been a considerable economic problem in semiarid parts of western North America. It is becoming clear that toxic conditions are much more frequent than is indicated by the observed deaths of animals. No human deaths have been reported, but cases of temporary illness are known.

The types of algae that grow attached to solid surfaces likewise increase and can form thick coatings on stones, plants, docks, and boats. This has caused difficulty in Lake Tahoe, on the California–Nevada border, still one of the clearest lakes in the world.

The large rooted higher plants (macrophytes) are affected in different ways by eutrophication. The submerged species with weak root systems that absorb nutrients from the water increase with the enhanced nutrient supply. Large quantities can become annoying by entangling boat propellers and swimmers. Eurasian milfoil (*Myriophyllum spicatum*), which has invaded North America, is a noteworthy example. Plants with well-developed root systems get much of their nutrition from the mud, and so recycle nutrients that otherwise would be permanently buried. A dense stand of perennial species of such plants does not necessarily signal enhanced productivity, as it may consist of growth accumulated over some years. In eutrophied lakes that produce large amounts of phytoplankton, the penetration of light is reduced so much that the depth inhabited by macrophytes is reduced, even though their density in shallow water may be increased, and they may carry a large load of attached algae.

An increase in photosynthetic organisms provides food for and an increase in the abundance of certain invertebrates and fish. The amplified biological production leads to increased deposition of dead organic matter onto the bottom, which supports a large population of scavenging animals and heterotrophs, such as bacteria, which use dissolved oxygen. The benthic populations increase, and the species composition changes in a characteristic way. Typically, oxygen concentrations in the deep water of eutrophic lakes are low during summer, and some kinds of fish cannot survive there. Eutrophied lakes may produce more coarse shallow-water fish than formerly, but the species that require cold, oxygenated water may disappear.

Because the abundance of phytoplankton is very responsive to eutrophication, and transparency is strongly affected by it, much use has been made of transparency as an index of the trophic state of lakes. However, conditions other than production can affect phytoplankton. This is well illustrated by Lake Washington, in the state of Washington, which had two major increases in transparency associated with changes in phytoplankton (Fig. 1). The first followed the diversion of sewage effluent, resulting in a marked decrease in phytoplankton. The second followed an increase in the abundance of the small crustacean *Daphnia,* an effective consumer of algae; there was no change in phosphorus input at that time.

Estuaries and other semienclosed marine waters can show effects of eutrophication comparable to those discussed here. As with lakes, some areas are eutrophic for natural reasons, and some may be enriched by human activity. The differences between marine and freshwater systems come about partly because of differences in the character of the biota, especially the life history of the primary producers. Massive seaweeds are classified as algae, but have effects more like those of freshwater macrophytes. Cyanobacteria are not dominant in seawater, but dinoflagellate and diatom populations can become very dense. Diatoms have given trouble in nutrient-rich areas because they clump together on the gills of fish, preventing effective respiration. Some kinds of algae eaten by shellfish produce toxins which are harmless to the shellfish, but cause fatal paralytic poisoning of the people who eat them.

V. NUTRIENTS EFFECTIVE IN EUTROPHICATION

The term eutrophication is best used for macronutrients that form a substantial fraction of the structure of cells and that can limit abundance when

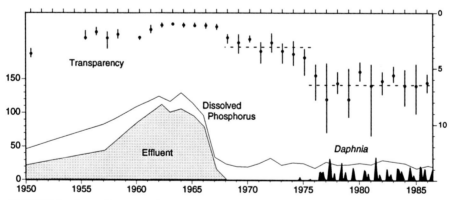

FIGURE I   Eutrophication and recovery of Lake Washington from 1950 through 1986. Top: transparency as measured by the Secchi disc, right scale, depth in meters from the surface. The bars show the range during summer (July–August) and the dots show the mean. In Lake Washington, changes in transparency during summer are controlled mostly by the abundance of phytoplankton. Below, left: annual input of dissolved phosphorus (metric tons per year). Use numbers on left scale. The shaded area shows the amount contained in secondary sewage effluent. Note how closely transparency follows the phosphorus supply until 1976, when suddenly the transparency increased. Below, right: abundance of *Daphnia* in top 10-m layer, monthly means of individuals per liter. Again, use numbers on left scale.

the concentration is low. Depending on geological conditions, the most frequent limitation in temperate regions in the Northern Hemisphere is by phosphorus, with nitrogen a close second. Marine systems are often limited by nitrogen rather than by phosphorus. Phosphorus is an essential part of the structure of chromosomes and cell walls, and nitrogen is part of all amino acids, the basic unit of proteins. Sulfur is essential but has not been demonstrated to limit production in lakes even when scarce. Carbon, a component of all organic compounds, rarely limits phytoplankton growth on a long-term basis because of the availability of carbon dioxide from the air. Several trace metals have been cited as causing eutrophication (Meaning 1), but generation of algal nuisances by addition of such metals alone has never been demonstrated. Enrichment with iron binds phosphate, thus reducing production from an otherwise rich medium. Copper has been used to kill algae in lakes. Excess zinc reduces algal productivity.

A nutrient is said to be limiting when its concentration is so low that the growth of an organism that requires it is less than it would be with a higher concentration. Limiting nutrients for phytoplankton can be identified by bioassay experiments in which samples of lake water are enriched with nutrients singly and in combination and incubated in illuminated chambers at controlled temperatures. The samples may contain the natural population existing at the time of sampling or may be filtered and inoculated with a pure culture of a test species. In either case, the effect is judged by whether the abundance or photosynthesis increases, and how much. The rate of photosynthesis may increase temporarily after the addition of any one of a number of trace metals, but a significant increase of population size will not necessarily follow. That depends on an adequate supply of structural elements. For sustained growth, often only one of the major nutrients is limiting, meaning that when it is exhausted to the point of limitation, others are still present in adequate concentration to form major structures of the cells. The content of phosphorus in phytoplankton varies with the external concentration and supply, and the amount stored may be enough to support several cell divisions in the absence of phosphorus in the water. At times, the addition of either phosphate or nitrate may increase production.

VI. PREDICTION OF EFFECTS OF CHANGES IN NUTRIENT LOADING

Considerable effort has gone into developing ways to predict the results of attempts to prevent or ameliorate the effects of eutrophication, especially in lakes. An effective way to start is to calculate the effect of changes in the input of a limiting nutrient on the potential concentration of that nutrient in the lake. The problem is complicated by the fact that the way in which nutrient supply is converted to phytoplankton is affected by several chemical and physical conditions. For example, differences in the thickness of the epilimnion will affect the density of the phytoplankton population and its exposure to light. Not all organisms produced stay in the productive zone. Some are eaten, but others sink to the bottom, removing the nutrients they contain. Also, if oxygen is exhausted from the hypolimnion, chemical changes occur at the sediment surface of the mud that release phosphate that otherwise would remain tightly bound. This increases the recycling of phosphorus with consequent increased productivity. The phosphorus thus released from sediment is sometimes called internal loading and is added to the external loading in calculations of total loading. However, it should be realized that the released phosphate was measured earlier in the external load, and without the change in oxygen, it would remain in the sediment. In other words, the effectiveness of a given input of phosphorus is increased by anaerobic conditions in the hypolimnion.

For practical purposes it is more useful to calculate consequences of changed loading that are more obvious than phosphorus concentration, such as phytoplankton abundance or a subjective evaluation of condition. Systems have been developed for identifying the effect of changes in loading by combining empirical data on lake condition with theoretical models of photosynthetic production and nutrient dynamics. They provide a prediction of the amount of nutrient loading required to change the condition of a lake of given dimensions and water loading from "not acceptable" to "acceptable."

Such predictions refer to potential changes, for the actual abundance and kind of phytoplankton are affected by conditions in addition to ambient nutrient concentration, such as toxic effluents or grazing by zooplankton.

VII. RESTORATION AND PROTECTION AGAINST EUTROPHICATION

Various techniques have been developed to improve polluted lakes or protect them against deterioration. The most direct way is to prevent the polluting material from reaching the body of water in the first place. This can be done by removing the material from effluent before it enters a sewer or by abandoning the use of the polluting substance. To counter eutrophication in waters where phosphorus is a limiting element, emphasis has been on limiting the use of household phosphate-based detergents.

When such measures are not possible, other methods must be used. Some well-known successful cases of control of eutrophication by domestic wastes have followed diversion of the wastes. Lake Washington responded promptly to diversion of the secondary sewage effluent that had been entering at nearly 76,000 m^3 per day. This procedure is acceptable only when the new recipient is not damaged or can be sacrificed. For chemical and physical reasons, the transfer of Lake Washington's sewage effluent did not cause deterioration of Puget Sound. However, although Lake Monona in Wisconsin was improved when sewage was diverted to the next lake downstream, Waubesa, the latter then began to experience nuisance blue-green algae blooms. In regions where phosphorus is the main limiting element, removal of phosphorus from wastes during treatment can be beneficial. Lake Zürich in Switzerland benefited by precipitation of phosphate from sewage effluent with ferric chloride.

Usually little can be done to control the nutrient supply of a naturally eutrophic lake in a nutrient-rich watershed, but passing the water of phosphate-rich inlets through a phosphorus removal process was beneficial in at least one case. The practicality of such a method depends on the local circumstances and the quantity of water to be treated.

Other protective procedures involve treatment of the lake itself rather than the input. A more practicable method of reducing nutrient concentration than its removal is by the addition of water when a large supply of nutrient-poor water is available. Dilution had favorable effects in Green Lake (Seattle) and Moses Lake (central Washington). It is important to realize that this method does not reduce the quantity of nutrients entering the lake.

Another approach is to limit recycling of phosphorus by adding alum, which forms a layer on the bottom and binds phosphate, removing it from the water and reducing its release from the mud.

A different method takes advantage of the fact that during periods of thermal stratification in eutrophic lakes, high concentrations of nutrients are released by decomposition of organic material deposited from above and accumulate in the hypolimnion. During mixing periods these materials are redistributed throughout the whole lake and used in subsequent production. Before the mixing period, the nutrients can be intercepted by a large siphon pipe laid from the deep water up through the outlet of the lake. Thus the outlet, instead of draining epilimnetic water only, carries some hypolimnetic water, and removes nutrients from the lake that otherwise would have recycled. When the input of phosphorus cannot be limited sufficiently to benefit the lake, or when phosphorus is not the dominant limiting element, some of the symptoms produced by eutrophication can be partly alleviated by a variety of techniques. These may involve oxygenating the hypolimnion by pumping air or oxygen through pipes arranged to distribute the oxygenated water into the hypolimnion without breaking down stratification. In other cases, lakes have been deliberately destratified by bubbling air freely into the bottom water, which is easier to do than aerating the hypolimnion. Whether the effect is beneficial depends on several factors, including the chemical conditions in the hypolimnion at the time of mixing. It may even have undesirable effects by eliminating a cold-water refuge for fish.

The understanding and control of practical problems of lake management require knowledge of the concepts of aquatic ecology. There is reciprocal benefit when the application of those results to a practical problem is accompanied by study of the consequences. Not only does that provide a check on the effectiveness of the procedure, but it can improve the scientific understanding of the lake system.

Glossary

Algae Collective name for a group of diverse groups of mostly microscopic organisms. Current usage tends to limit the term to eukaryotic organisms.

Cyanobacteria (blue-green bacteria) Group of prokaryotic microorganisms capable of photosynthesis in the presence of oxygen; also called blue-green algae.

Hypolimnion See thermal stratification.

Macrophytes Rooted higher plants.

Phytoplankton Component of plankton composed of photosynthetic organisms. It is composed of species of algae and cyanobacteria.

Plankton Community of microorganisms and small invertebrates in the open water of lakes or marine waters that is relatively independent of the bottom for completing life cycles.

Production Organic material synthesized from components. Two major levels can be distinguished: primary production is based on photosynthesis using the carbon dioxide system as a source of carbon, and secondary production is the synthesis of material from organic compounds that are digested and assimilated by animals (consumers). Other types can be distinguished, for example, microbial production by heterotrophic bacteria and chemosynthetic processes in special habitats. The word productivity is often used to indicate rate of production.

Thermal stratification Typically in lakes in temperate regions during summer a layer of warm water (epilimnion) floats on top of the deeper dense cold water. The deepest part (hypolimnion) is separated from the eplimnion by an intermediate layer (metalimnion).

Bibliography

Cooke, G. D., Welch, E. B., Peterson, S. A., and Newroth, P. R. (1993). "Restoration and Management of Lakes and Reservoirs." Boca Raton, Fla.: Lewis Publishers.

Edmondson, W. T. (1991). "The Uses of Ecology: Lake Washington and Beyond." Seattle: University of Washington Press.

Elser, J. J., Marzolf, E. R., and Goldman, C. R. (1990). Phosphorus and nitrogen limitation of phytoplankton growth in the freshwaters of North America: A review and critique of experimental enrichments. *Can. J. Fish. Aquatic Sci.* **47**, 1468–1477.

Harper, D. (1992). "Eutrophication of Freshwaters. Principles, Problems, Restoration." New York: Chapman & Hall.

Likens, G. E., ed. (1972). "Nutrients and Eutrophication, the Limiting-Nutrient Controversy," Special Symposia, Vol. I, *Amer. Soc. Limnol. Oceanogr.* Lawrence, Kan.: Allen Press.

Sutcliffe, D. W., and Jones, J. G., eds. (1992). "Eutrophication: Research and Application to Water Supply." Ambleside, England: Freshwater Biological Association.

Vollenweider, R. A. (1976). Advances in defining critical loading levels for phosphorus in lake eutrophication. *Mem. Ist. Ital. di Idrobiol.* **33**, 53–83.

Everglades, Human Transformations of a Dynamic Ecosystem

Lance H. Gunderson

University of Florida

I. Introduction
II. The Natural System
III. Human System
IV. History of Water Management
V. A Sustainable Future

The Everglades is a large wetland ecosystem situated in the subtropics of Florida. The landscape of open marshes, dark green tree islands, alligators, and wading birds is unique to north America. Efforts to preserve the natural resources of the Everglades during this century have been at the leading edge of conservation movements. During the same period, the human population in and around the Everglades ecosystem has increased dramatically, resulting in a variety of demands on and uses of a unique ecosystem. It is the struggle to sustain the multiple, sometimes competing, uses that define the Everglades at the end of the 20th century. Both the natural and human systems are dynamic, changing in both space and time. The characteristics of these changes contribute to the singularity of the ecosystem and provide the focus for this article.

I. INTRODUCTION

The physiographic region of the Everglades can be defined in explicit dimensions in space and time.

The Everglades system originated about 5000 years ago, when a dramatic rise in sea level created the wetland conditions that persist to date. Prior to intensive human development a little more than a century ago, the freshwater wetlands of the Everglades covered approximately 10,500 km². The system extended about 210 km along the north–south axis, bordered by Lake Okeechobee to the north and Florida Bay to the south (Fig. 1). The widest east–west dimension was about 77 km, from the higher Atlantic coastal ridge on the east to the big cypress swamp on the west. Although these measures define the geographical dimensions of the system, processes that influence the system vary across a wide range of spatial and temporal scales.

This article describes three facets of this dynamic system; the natural components, the human components, and the recent history of attempting to manage the water resources for a variety of human needs. The first section describes the spatial and temporal domains of the biotic and abiotic components of the natural (nonhuman) system. The sec-

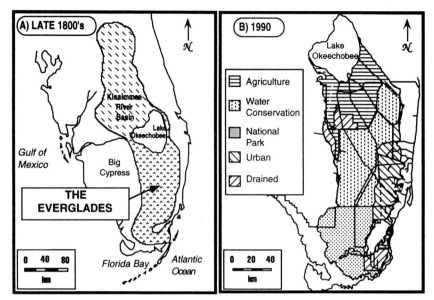

FIGURE I   (A) The historic drainage basin of the Everglades during the late 1800s. The Everglades was the southern third of a hydrologic system that included the Kissimmee River Basin and Lake Okeechobee. (B) Land uses of the historic Everglades ecosystem in 1990, including agriculture in the north, water conservation areas in the center, urban development along the east, and the national park and preserve in the south.

ond section describes the human-dominated system, especially how the wetland complex has been transformed for various land uses. The third section describes one view of the history of water management, as it has progressed during this century, dealing with the crises and surprises of this dynamic system.

II. THE NATURAL SYSTEM

The primary abiotic components of the Everglades include the geologic features, topography, soils, hydrology, fires, freezes, and severe storms. All of these operate at various scales in space and time. The geologic, topographic, and edaphic features are slow variables; changing on the order of centuries and at broad spatial scales. The dominant processes of hydrology, fires, and storms all operate at middle or mesoscale ranges; fluctuating on time scales of decades and covering areas of tens to hundreds of kilometers. The interaction of these variables across space and time scales provides the theater within which the biotic components of vegetation and fauna operate. [*See* WETLANDS ECOLOGY.]

A. Abiotic Components

The key abiotic components that configure the Everglades ecosystem include (1) the edaphic features of geology, soils, and topography; (2) the subtropical climate of hot summers, mild winters, and distinctive wet and dry seasons; and (3) the role of disturbance events of severe storms and fires. Each will be discussed in turn.

The geologic features under the Everglades are limestones formed by shallow marine accumulations during the Pleistocene era. Three surficial formations are recognized. The Fort Thompson formation underlies the northern Everglades and is composed of marine and freshwater marls beds interleaved with limestone and sandstone. The Anastasia formation is found in the northeastern Everglades and is characterized by sandy limestone and calcareous sandstone. The surficial feature of the southern Everglades is Miami limestone, composed of oolitic and bryozoan facies.

The soils of the Everglades are biogenic Holocene sediments, categorized as peats, mucks, and marls. The oldest soils in the Everglades are approximately 5500 years old, dating back to the most recent transgression of sea level. Peats and mucks are histosols, named by the dominant recognizable plant remains from which the soils are derived, and accumulate under extended periods of inundation. The marls are a calcitic mud, produced by the re-precipitation of calcium carbonate from saturated water during photosynthesis by blue-green algae. Land use conversions and water management practices have altered the hydrology and accelerated oxidation of the peat soils. In portions of the agricultural areas in the northern Everglades, as much as 10 to 12 feet of soil has been lost.

The topography of both the bedrock and the soil surface is flat, characterized by almost no relief with extremely low gradients. The maximum elevations in the northern Everglades are approximately 5.3 m above the national geodetic vertical datum (NGVD). The elevational gradient is mostly north to south, with an average slope of 2.8 cm/km. The surficial elevation is associated with the underlying bedrock structure and accumulations of organic sediments. The topography varies at two spatial scales. The macrotopography (measured in linear distances of kilometers) is associated with bedrock features and defines the broad swales and differences in inundation and soil types. Microtopographic variation is due to the combined effects of smaller scale processes such as peat accretion, soil oxidation, chemical solution of bedrock, and soil consumption by fire.

The climate of southern Florida is subtropical, characterized by high temperatures and rainfall during the summer months and mild dry winters. The area has also been classified as a subtropical moist forest type because of the high annual rainfall and moderate annual biotemperatures. Mean monthly temperatures average between 18° (January) and 24°C (August). Frosts are rare (average less than 3 freeze days per year) yet influence the distribution of tropical vegetation and agricultural activities. Severe storms (hurricanes and tropical storms) can occur annually during the late summer months, but those that dramatically affect the biota tend to recur at intervals greater than a decade.

Fires are a critical mesoscale process in shaping the landscape of the Everglades. Fires not only consume the standing vegetation, but during severe droughts can burn the organic soils, thereby lowering the surface elevation and changing the type of vegetation. Fire sizes vary with return intervals; the largest fires (up to 75 km^2) occur about once a decade, while smaller fires occur annually. The patterns of fires and their impacts are related inversely to the hydrologic regime; the wetter the conditions, fewer fires tend to burn whereas the drier hydrologic conditions result in the largest, most severe fires. The factors that influence the hydrologic regime are the subject of the next section. [See FIRE ECOLOGY.]

The hydrology of the wetland complex is also a result of many processes operating at various scales. Sea level, bedrock topography, and soils are the variables that operate over broad ranges and change on time scales of centuries. These slower, broader variables provide the context within which the faster variables operate. The dominant fluctuation of the ground and surface waters is an annual cycle, but cycles of 10 to 12 years are noticed in surface water stages and flows. Rainfall is the primary input to the Everglades system, with minor amounts added from overland flows from the Kissimmee River basin and Lake Okeechobee to the north (Fig. 1A).

Rainfall over the Everglades exhibits both spatial and temporal variability. Rainfall varies seasonally; 85% generally falls during the wet season between May and October. Annual total rainfall over the system averages 130 cm, with measured extremes ranging from 95 cm in 1961 to 270 cm in 1947. The coastal region receives on the average 30 to 35 cm more than the interior marshes.

The rainfall patterns can be related to different processes which influence the timing and amount of precipitation; as shown by the atmospheric hierarchy in Fig. 2. The summer rainy season is attributed to convective thunderstorms, which are linked to land–sea breeze patterns. During the summer, insolation results in differential heating of the air over the land mass compared to air over the water. The heated air over the land rises, creating low pressure and establishing a pressure gradient along which maritime air flows toward the center of the Florida peninsula. The moisture-laden air rises, cools adiabatically, condenses, and forms convective thundershowers. This process has been described as the "rain machine." Some authors suspect that rainfall totals have decreased because drainage and development have altered the net radi-

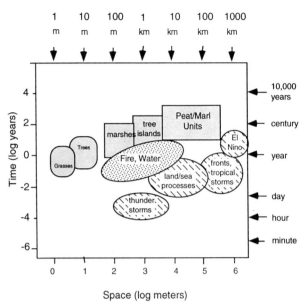

FIGURE 2 Spatial and temporal domains of critical ecosystem components in the Everglades. The upper shaded squares represent the domains within a vegetation hierarchy of individual plants, associations, and landscape units. The crosshatched circles represent the atmospheric hierarchy, the primary sources of water input. The mesoscale processes of fire and surface water conditions are in intermediate positions and hence integrate the vegetation and atmospheric hierarchies.

ation budget (increased reflectance due to a higher albedo of developed areas) which in turn decrease the rate of convection, but this has yet to be substantiated. Statistically higher rain amounts measured during September have been explained by the greater incidence of tropical cyclones during this month. Rain during the winter dry season is associated with the passage of cold fronts that pass on the average every 7 days. The interannual variation in rain totals are thought to be linked to global processes such as tropical cyclones, winter frontal systems, and El Nino southern oscillation (Fig. 2).

The abiotic factors of geologic features, soil types, sea level, hydrology, and periodic disturbances of fires, freezes, and severe storms form the stage within which the biota operate. The unique features of vegetation and fauna that define the Everglades are described in the next sections.

B. Vegetation

The Everglades region is located in the transition zone between temperate and tropical areas. The flora of the Everglades has many representatives from these two areas, but also has elements that are from regions further away (aliens or exotics) as well as taxa that are endemic to the region. The native tropical species are primarily from the Antillean–Caribbean region of the neotropics. Propagules of these species are thought to have crossed the saltwater barrier to Florida carried either by hurricanes, migrating birds, or for those able to tolerate a period of exposure to salt water, by ocean currents. The temperate species range northward into the coastal plain of the southeastern United States. The endemic taxa consist of species and subspecies that have evolved in unique habitats of southern Florida and are found nowhere else in the world. The exotic flora was, for the most part, introduced into southern Florida for ornamental and horticultural purposes. A few of these introduced species (especially trees of the genera *Melaleuca, Casuarina,* and *Schinus*) were so well-adapted to conditions that they have become aggressive invaders.

The vegetation of the Everglades can be described as a hierarchy composed of individual plants, plant associations, and landscape units (Fig. 2). The flora consists of about 850 species; dominated by grasses and sedges. These herbaceous and woody species aggregate into less than 20 plant associations; given names such as tree islands, marshes, and forests. The plant associations are found in one of three landscape groupings: the upland complex of pine and hardwood forests, the peat wetland complex, and marl wetland associations. The landscape groupings represent combined edaphic and hydrologic features.

The upland complex includes pine forests, tropical hardwood hammocks, and areas of newly created or disturbed land. The uplands are found along the eastern edge of the Everglades on the Atlantic coastal ridge and extend into the Everglades National Park as Long Pine Key. Tropical hardwood hammocks occur within the Everglades, but cover relatively little area. Distributions of these two forest types are due not only to effects of fire on the species composition and structure, but also to substrate differences. Pine forests once dominated the Atlantic coastal ridge. Communities of sand pine, *Pinus clausa,* were found over the deep, white sands in the northern areas of the ridge (in and around West Palm Beach). From northern Dade County and extending south into the area now in Everglades Park, outcrops of oolitic limestone are the dominant surficial feature and support a rockland vegetation composed of a matrix of pine forests enclosing hardwood hammocks. Approximately 10% of the original rockland forests is conserved as park area, most of which is in Everglades National park. Small outcrops of bedrock support areas of hardwood hammocks and are scattered throughout the wetland areas of the Everglades proper.

The native freshwater wetlands of the Everglades are composed of both forested and nonforested communities. All communities are described by the dominant species or groups. The forested wetlands include bayheads, willow heads, and cypress forests. Ponds, open water sloughs, sawgrass marshes, and wet prairies comprise the nonforested or graminoid wetlands.

The broad-leafed hardwood associations of the Everglades are also referred to as tree islands. These clumps of hardwood trees are taller than the surrounding marsh and appear as dark-green "islands" amid a sea of grass. Tree islands are mostly swamp forests, although some small areas of higher elevation support tropical hardwood vegetation. The larger tree islands within the Everglades are shaped like an elongated tear drop, generally oriented with the main axis parallel to the main axis of flow. The tear-drop shape may be a result of a bedrock bump near the upstream end of the island, which interacts with the flow pattern to form this shape. Smaller tree islands in the northern Everglades have a circular shape that appears to be related to the mode of formation whereby chunks of peat break loose and form floating islands or "batteries" and are colonized by hardwoods.

Bay trees dominate the tree islands, hence the name bayhead is often applied. Canopy species include red bay, *Persea borbonia,* and swamp bay, *Magnolia virginiana,* as well as dahoon holly, *Ilex cassine;* pond apple, *Annona glabra;* and wax myrtle, *Myrica cerifera.* Other less common speices include willow, *Salix caroliniana;* and strangler fig, *Ficus*

aurea. A dense shrub layer is generally found beneath the canopy, composed primarily of cocoplum, *Chrysobalanus icaco,* but can also include smaller individuals of the just-mentioned overstory species. Other plants in the shrub stratum include buttonbush, *Cephalanthus occidentalis,* and the large leather fern *Acrostichum danaeifolium.* Some areas of the understory are devoid of ground cover because of the dense shade of the overstory.

Willow heads composed of the southeastern coastal plain willow, *S. caroliniana,* are found on sites with a history of severe soil disturbance, such as following a peat fire, lumbering, farming, or alligator excavation. These stands are generally monotypic, with willow as the abundant dominant woody plant. Other less common associates include phragmites, *Phragmites australis,* sawgrass, and flag, *Thalia geniculata.* Herbaceous vines such as *Sarcostemma clausum, Mikania scandens,* and *Ipomoea sagittata* are commonly found. Willow may be more widespread because of changes in hydrology and the associated impacts of dry-season fires.

Cypress forests are relatively minor features of the Everglades, occurring in the southern Everglades and along the western border with the Big Cypress area. Two types of forests occur, domes and cypress prairie, and both are dominated by pond cypress, *Taxodium ascendens.* Cypress prairies are also called dwarf or hat rack cypress. The cypress trees have a stunted growth form and are widely spaced, with a grassy or prairie-like understory composed of sawgrass, muhly grass, *Muhlenbergia filipes,* and other herbs and grasses.

Prior to conversion to agriculture, a large forest dominated by pond apple, *Annona glabra,* with some willow and elderberry, *Sambucus simpsonii,* was located on the southern rim of Lake Okeechobee. Although this forest is gone, small stands of pond apple still occur throughout the Everglades.

Sawgrass is the common, characteristic plant species of the freshwater Everglades system. Sawgrass, not a grass as the common name implies, is a rhizomatous, perennial sedge. The plant is well adapted to the conditions of flooding and burning that occur in the Everglades. Although capable of surviving variable water depths from dry soil to flooding of the lower portions of the plant, sawgrass is killed if high water levels are prolonged. Sawgrass also has low nutrient requirements, a characteristic which promotes its dominance in the oligotrophic waters of the Everglades.

Wet prairies are generally described by the dominant plant found in the associations such as *Eleocharis, Rhynchospora,* or maidencane flats. Common emergent aquatic plants that are found in these wet prairies include spike rush, *Eleocharis cellulosa,* beak rush, *Rhynchospora tracyi;* maidencane, *Panicum hemitomon;* arrowhead, *Sagittaria lancifolia;* and pickerel weed, *Pontederia lanceolata.* At least 25 taxa occur in these associations, but generally spike rush, beak rush, and maindencane dominate. Submerged aquatics include ludwigia, *Ludwigia* spp., and bladderworts, Utricularia spp. Periphyton is a conspicuous and important feature of these associations. As with sawgrass marshes, wet prairies have been invaded by *Melaleuca* within the last 50 years.

Slough communities encompass all associations of floating aquatic plants and are found on the lowest, wettest sites in the Everglades. Dominant macrophytes include white water lily, *Nymphaea odorata;* floating hearts, *Nymphoides aquatica;* and spatterdock, *Nuphar advena.* The remainder of the flora of these associations is composed of submerged aquatics and periphyton. The common submerged aquatics are the bladderworts *Utriculara foliosa* and *U. biflora.* The submerged aquatics provide structure for periphyton and are a key component in what is described as the periphyton mat complex.

Wet prairies over marl substrates occur in the southern Everglades on the east and west margins of Shark and Taylor Sloughs where bedrock elevations are slightly higher and hydroperiods are shorter. Most of the marl prairies in Everglades Park are dominated by two species: muhly grass and sawgrass. Other species that can be locally dominant include *Schoenus nigricans, Aristida purpurascens, Schizachyrium rhizomatum,* and *Eragrostis elliottii.* Beak rush, *R. tracyi,* is common in the lower, wetter areas of the marl prairies. The association is diverse; over 100 species are found. The majority of these are herbaceous plants, yet comprise less than 1% of the ground cover.

C. Animals

The majority of Everglades fauna is comprised of taxa derived from the southeastern coastal plain of the United States. All of the fish species and most of the reptiles, amphibians, and mammals have temperate affinities. The terrestrial invertebrates and lepidopterans are from the neotropics. Most of the wading birds are widespread throughout the tropics as well. Although temperature affiliates dominate the fauna, a few West Indian taxa such as flamingos, American crocodile, mangrove cuckoos, and white crown pigeons contribute to a unique blend of animals. Endemism is low and, if present, is at the subspecific level.

The vertebrate fauna of the Everglades is dominated by birds, fishes, and reptiles. Of the 161 species of native Everglades vertebrates, 70 are breeding birds. Herons and hawks are the most speciose groups. The fish and reptile groups each have 30 species with sunfish, killfish, snakes, and turtles as the most abundant. Carnivores are the dominant group of the 17 mammal species, whereas frogs comprise 10 of the 14 amphibians. The fauna of the Everglades is considered depauperate because of the age of the wetland system, limited habitats for temperate taxa, and the Everglades being at the far end of a peninsula.

Many changes have been noted in the fauna during this century. Most notable are the increase in nonnative species, the endangerment and extirpation of other species, and the dramatic changes in nesting populations of wading birds. Estimates of several hundred species of animals have naturalized along the southeastern coast of Florida; of those a few dozen (primarily fish taxa) exploit altered niches with the Everglades system. At least 17 taxa have been recognized as federally endangered or threatened; most notable are the Florida panther, Cape Sable seaside sparrow, and wood stork. The most dramatic changes in the fauna have been observed with the nesting population of wading birds. During the period from the 1920s to the 1960s, the Everglades provided the primary nesting area for populations in the southeastern United States. Since the 1960s, although the numbers vary annually, only 5 to 10% of the populations that once nested continue to use the area. A number of hypotheses are proposed to explain the loss, including a spatial decrease of early wet season habitat due to development, less water flow through the system, alteration of hydrologic regimes necessary for successful feeding and breeding, and that other sites in the southeast may be better. Despite the loss of nesting, the Everglades still continues to provide an important feeding area for wintertime and transient populations. The history of human use that has both negatively impacted the fauna and is currently attempting to restore the system is described in the next section.

D. Ecosystem Dynamics

The Everglades ecosystem operates on time scales of 1 year and about a decade. The annual cycle is tuned to the summer rainy season followed by a dry winter and spring. The pulse of summer rains create vast flooded areas and trigger aquatic productivity. In this mode, the system acts as a vast solar collector, fixing solar energy. After the rains cease, the system slowly dries out and the wetted area shrinks thereby concentrating local abundance of aquatic organisms. These concentrations in the lower wetter sloughs and alligator holes (small, open water ponds maintained by alligator activity) provide the feeding opportunities for the large wading bird populations. The annual cycle is overlain by longer cycles, characterized by years of higher (flood) and lower (drought) rainfall. The longer term droughts liberate nutrients, allow peat consuming fires, change the nature of the plant and animal communities. The floods drive terrestrial mammals to uplands, increase numbers of aquatic organisms and also change plant and animal communities.

Other broader and longer term changes such as sea level state, and the impact of tropical storms influence the plant and animal communities. Generally, these processes have more of an effect on types close to the coast. For example, Hurricane Andrew in 1992 devastated the mangrove forests south of the Everglades, but had an almost negligible effect on the freshwater marshes.

III. HUMAN SYSTEM

A. History of Use

Evidence of human habitation in southern Florida dates back well over 10,000 years, prior to the existence of the vast wetland ecosystem. The first written accounts, from the 16th century, describe the fierce coastal native Indian tribes and the peaceful Mayami tribes who lived around Lake Okeechobee. These early Americans probably burned the Everglades and used the area for hunting and fishing purposes.

The name Everglades first appeared on British maps in the early 1800s, probably a contraction of 'Never a glade,' descriptive of the large treeless expanses. With the expansion of European-derived settlers throughout the southeastern coastal plain, native Americans translocated from the Carolinas to southern Florida. The term "Seminole," which is the name for the major tribes of current native populations that persist today, means runaway. The Seminoles describe the Everglades in the term "Pahayokee," which loosely translates into "grassy lake," again descriptive of a nonforested wetland system. These native Americans used the elevated tree islands for home sites and for the cultivation of crops, as well as hunted and fished throughout the system. The United States fought a series of wars with the Seminoles during the mid 1800s, restricting their territory to a few reservations through south Florida. One remains within the Everglades proper, where the Miccosukee Indians still retain land use rights.

The latter part of the 19th century marked the first influx of white settlement and attempts at "reclamation" of the wasteland known as the Everglades. Soon after Florida became a state in 1845, early settlers and their governments embarked on programs to drain the Everglades for habitation and agriculture. Buckingham Smith was commissioned by the U.S. Senate to reconnoiter the Everglades for development potential. In 1850, under the Swamp and Overflowed Lands Acts, the federal government deeded over 19,000 km² (7500 mi²) to the state, including the Everglades. The Florida legislature established the Internal Improvement Fund, whose board was to sell and improve these swamp lands through drainage. Attempts at manipulation of the water were ineffective in the 1800s, as the magnitudes of the variations in hydrology were far greater than the minor control structures could handle.

By 1900, initial colonization of the coastal regions east of the Everglades was underway. The population of the Palm Beach, Broward, and Dade counties in southern Florida in 1900 was 28,000. By 1920, the major land uses now found in southern Florida had started. Urban development occurred along the railroad line down the east coast. Agriculture was developing in the peat lands south of Lake Okeechobee. Conservation of the natural resources had begun with the formation of Royal Palm State Park in 1917 in the southern Everglades.

During the period from 1920 through 1990, the spatial extent of these land uses grew, in large around these three general loci. In the 1940s, 283,000 had of the northern Everglades was designated as the Everglades agricultural area. By the mid 1940s water conservation areas were designated in the central regions of the glades to manage water resources for multiple purposes. Conservationists work started during the 1920s came to fruition in 1935 with the establishment of the Everglades National Park in the southern Everglades, although the park was not formally dedicated until 1947. The park area was increased in 1989 to 1.4 million acres by the addition of the Northeast Shark Slough. Urban development along the east coast has followed the exponential increase in population and resulted in the drainage and colonization of former wetland areas. As of 1990, about 5 million people inhabited the area. The current configuration of the Everglades ecosystem depicting agricultural areas in the north, water conservation areas in the central areas, and the Everglades National Park in the south is shown in Fig. 1B.

Through the past century, the spatial extent of the historic Everglades ecosystem has been slowly whittled away, to the degree that perhaps one-half of the original system has been irrevocably converted to specific land uses. As of 1990, the historic Everglades ecosystem was partitioned into at least five major use types: 32% of the historic Everglades is in areas designated for water management, 27% in agriculture, 20% for preservation of natural resources, 12% had been developed for urban purposes, and 9% remains as drained, undeveloped lands. It is estimated that only half of the original land area of the Everglades is still in native vegetation types and that certain landscape types, including a large pond apple forest in the north as well as marshes and cypress forests in the east, are gone. The remaining natural areas have probably been hydrologically altered and their future viability is largely dependent on water management actions. Even though half of the Everglades wetland has been converted to other land uses and corresponding impacts have led to the endangerment of many animals, none have been extirpated. Unlike the adjacent uplands, where almost all of the habitat is gone and at least 10 species have been exterminated.

B. Current Management Institutions

The federal operations or participation in water management in the Everglades has been historically entrusted to the U.S. Army Corps of Engineers. The Corps has jurisdiction (House Document 643) for the design and construction of the water management infrastructure. Everglades National Park (Park) and Big Cypress National Preserve are managed by the U.S. National Park Service. The Arthur Marshall National Wildlife Refuge is managed by the U.S. Fish and Wildlife Service, whereas other branches of this agency are responsible for endangered species enforcement. The U.S. Environmental Protection Agency has maintained a relatively low profile in Everglades matters, not withstanding its review of COE permitting in wetland development.

The state of Florida has a number of agencies that have management responsibilities of various portions of the system. The South Florida Water Management District (SFWMD) operates and maintains the water management system to meet goals of flood protection, water supply, and environmental mandates. The Department of Environmental Protection retains supervisory authority over the SFWMD and has exercised that oversight most frequently on water quality matters. The Florida Game and Freshwater Fish Commission manages for wildlife habitat as well as the harvest of game and fish.

A number of nongovernmental organizations are involved in the politics and management of the Everglades. Agrobusiness concerns are represented by the Florida Sugarcane League and the Florida Fruit and Vegetable Association. The Everglades Coalition, Friends of Everglades, and The Florida Defenders of the Environment are all structured around conservation interests.

IV. HISTORY OF WATER MANAGEMENT

In the past century, the Everglades ecosystem has been transformed from a vast subtropical wetland into a highly managed, multiple-use system as a result of one of the largest public works projects in the world. This transformation was not a linear process, but was characterized by turbulence and punctuated change. For the most part, the interplay between humans and their increasing control over the system was driven by a series of events that were perceived as crises that threatened exploitation of the resources. Each crisis precipitated actions (or a series of actions) that resulted in a reconfiguration and the emergence of a new system. Crises appear to have followed two pathways: those created by external environmental events and those created by human activities.

One type of crises arises from larger scale processes and are perceived as local surprises. One example is the occurrence of tropical cyclones (storms and hurricanes). These storms occur at irregularly spaced, rather long intervals, as perceived by a human observer at a point in south Florida. In the Atlantic each year, tens of hurricanes occur. For the most part, these events are described statistically; the probability that a hurricane will strike along a 80-km segment of south Florida is about 15%. The environmental crises in the Everglades take the form of either too much (flood events) or too little rainfall (droughts) over the system.

The other type of crises occur over a longer period of time. These events tend to be endogenous to the system and reflect a chronic problem defined by a slower variable. One example is the slow water quality degradation that suddenly leads to dramatic shifts in dominant taxa. This class of crises appears to originate as a result of human involvement and development of natural resources. Another way of casting the problem is that humans have accelerated the rate of natural processes that were much slower and may have occurred over a much longer time span. In the Everglades these types of crises have been associated with agricultural activities, primarily in the form of soil loss and water pollution. Other slow variables that have created crises as they change include human myths, institutions and technologies that fail, and conceived operating criteria that have altered hydrologic cycles in Everglades National Park.

The crises perceived in the Everglades, both environmental and human induced, have resulted in at least four major eras of water management. Flooding impacts (natural and man induced) have been a recurrent and dominant predicament that shaped most of the water management infrastructure in the Everglades during the first part of the century. The first two eras were a result of flooding from high rainfall events. The third era was related to drought events and the fourth era resulted from attempts to rectify latent or previously unattended problems.

The earliest settlers were intent on reclaiming land "lost" to natural flooding in order to farm the rich muck soils. Early attempts at drainage were able to control water levels during average water conditions. Periods of dry years allowed for expansion, until the next wet year. The approach was to dig canals and drain the land as fast as possible; a strategy labeled "Cut 'N Try" was coined by Governor Napolean Bonaparte Broward. However, these attempts at drainage were unable to cope with the full variation in climatic regimes that would inevitably occur. Flood crises occurred in

1903 as a result of high rains and again in 1926 and 1928 as severe hurricanes.

The earliest era persisted until 1947, when over twice the normal amount of rainfall fell on South Florida, severely impacting the coastal communities. The acute flooding resulted in the implementation of a widespread plan to avoid this type of flooding in the future. This massive control plan, developed by the U.S. Army Corps of Engineers, called for the creation of specific land-use areas (agriculture, water conservation, and national park) spanning over 46,800 km^2 (18,000 mi^2) and the water management infrastructure (2240 km. of canals and levees, pumping stations with 3.8 billion liters/day capacity, and requisite water regulation schedules) needed to regulate flood waters. Proposed structures on the planning maps were colored green and once construction was completed, the lines were colored red. This era of water management, dubbed "Turning Green Lines to Red," lasted from 1947 through the early 1970s.

The crisis that prompted the next era of water management was a drought in 1971, the worst in 40 years. By the early 1970s, the population of southern Florida topped the 2 million mark and sugarcane production had tripled following the communist takeover in Cuba. The low rainfall coupled with increased urban and agricultural water demands prompted concern for an adequate supply of water to meet these needs. A number of serious problems arose, in correcting observed deficiencies in a water management system that had been operational for about a decade. Another problem arose in trying to retool a system that was designed for flood control to meet water supply concerns. The situation was exacerbated by the difficulty inherent in decisions involving trade-offs among water-use categories. This era of water management is referred to as "No easy answers," a paraphrase from Jack Maloy, then the executive director of SFWMD. Maloy advised his board that all the easy answers had been used up and that only difficult choices (many requiring changes in social behavior) were ahead in trying to solve competing objectives of water management.

The current era of water management (repairing the Everglades) beginning in 1983 is attempting to restore and revitalize the natural values system. A record drought in 1981 was followed by 2 wet years in 1982 and 1983. These events fostered a number of activities seeking to reverse the downward trend in the natural system, mitigate interactions among land uses, and attain goals of sustainability for the entire system, although much of the inherent pathologies still exist. Adaptive management efforts to restore the hydrologic and natural values of the system are underway. Efforts to learn from management and in developing composite policies indicate that key features of the resilient landscape can be restored, even though half of the system has been converted to other uses.

V. A SUSTAINABLE FUTURE

Marjory Stoneman Douglas taught the world that "there are no other Everglades." This statement provided the foundation for the conservation of the unique biologic values of this ecosystem that have been recognized both nationally and internationally. During the past century, half of the historic Everglades has been converted to other human uses and 90% of the nesting wading birds have been lost. In contrast to these bleak facets of human development, however, positive signs are emerging; restoration efforts are underway to recover these biologic features and to preserve pristine features for future generations. The importance of the natural values of the Everglades coupled with creatively adaptive management can define a sustainable future even with the dramatic transformations that have occurred in this dynamic ecosystem.

Glossary

Ecosystem restoration Reestablishment or recreation of lost attributes by active manipulation of critical processes.
Hydrology Study of the occurrence and distribution of water.

Water Management Control of water in systems through the use of physical structures (canals, levees, pumps, gates, and weirs) and operational critera.

Wetland Ecological systems where the soils are saturated for some time during the year.

Bibliography

Craighead, F. C., Sr. (1971). "The Trees of South Florida," Vol. I. Coral Gables, FL: University of Miami Press.

Davis, J. H., Jr. (1943). "The Natural Features of Southern Florida, Especially the Vegetation, and the Everglades." Tallahassee, FL: Florida Geological Survey.

Davis, S. M., and Ogden, J. C., eds. (1994). "Everglades; The Ecosystem and Its Restoration." Delray Beach, FL: St. Lucie Press.

Douglas, M. S. (1978). "The Everglades; River of Grass." Miami, FL: Banyan Books.

Egler, F. E. (1952). Southeast saline Everglades vegetation, Florida, and its management. *Vegetatio* **3**(4–5), 213–265.

Gleason, P. J., ed. (1984). "Environments of South Florida: Present and Past II," 2nd Ed. Coral Gables, FL: Miami Geological Society.

Gunderson, L. H., and Loftus, W. F. (1993). The Everglades. *In* "Biodiversity of the Southeastern United States" (W. H. Martin, S. C. Boyce, and A. C. Echternacht, eds.). New York: John Wiley & Sons.

Harshberger, J. W. (1914). The vegetation of south Florida, south of 27 degrees 30' north, exclusive of the Florida Keys. *Trans. Wagner Free Inst. Sci.* **7**, 49–189.

Hendrix, G., and Morehead, J. (1983). Everglades National Park: An imperiled wetland. *Ambio* **12**(3–4), 153–157.

Leach, S. D., Klein, H., and Hampton, E. R. (1971). "Hydrologic Effects of Water Control and Management of Southeastern Florida." Coral Gables, FL: United States Geological Survey.

Light, S. S., Wodraska, J. R., and Joe, S. (1989). The southern Everglades: evolution of water management. *Natl. Forum* **69**, 11–14.

Loftus, W. F., and Kushlan, J. A. (1987). Freshwater fishes of southern Florida. *Bull. Florida St. Mus. Biol. Sci.* **31**(4), 147–344.

Long, R. W., and Lakela, O. (1971). "Flora of Tropical Florida. Coral Gables, FL: University of Miami Press.

Loveless, C. M. (1959). A study of the vegetation of the Florida Everglades. *Ecology* **40**(1), 1–9.

Myers, R. L. (1983). Site susceptibility to invasion by the exotic tree *Melaleuca quinquenervia* in south Florida. *J. Appl. Ecol.* **20**, 645–658.

Myers, R. L., and Ewel, J. J. (1987). "Ecosystems of Florida." Orlando, FL: University Presses of Florida.

Ogden, J. C. (1978). Freshwater marshlands and wading birds in south Florida. *In* "Rare and Endangered Biota of Florida," (H. Kale, ed.). Gainesville, FL: University Presses of Florida.

Parker, G. G., Ferguson, G. E., and Love, S. K. (1955). "Water Resources of Southeastern Florida with Special Reference to Geology and Groundwater of the Miami Area." Washington, D.C.: U.S. Geological Survey.

Robertson, W. B., Jr. (1959). "Everglades: The Park Story." Coral Gables, FL: University of Miami Press.

Robertson, W. B., Jr., and Kushlan, J. A. (1984). The southern Florida avifauna. *In* "Environments of South Florida: Present And Past: Memoir 2" (P. J. Gleason, ed.), 2nd Ed., pp. 219–257. Coral Gables, FL: Miami Geological Society.

Walters, C. J., Gunderson, L. H., and Holling, C. S. (1992). Experimental policies for water management in the Everglades. *Ecol. Appl.* **2**(2), 189–202.

Evolution and Extinction

Eviatar Nevo
University of Haifa

I. THE BIG QUESTIONS AND OUR PRESENT WORLD PICTURE

The most fundamental question ever asked by humans relates to our origin and place in the universe. This question has been answered in a multidisciplinary way by all sciences in the all-embracing theory of Cosmic Evolution, comprising physical, chemical, biological, and human interlinked stages of evolution. The major feature of the universe and life is that they are critically explorable by humans: mind reflecting on its own evolution from lifeless to living matter, from featureless simplicity to increasing complexity. The universe is not static, small, and young, as conceived in the prescientific world belief. By contrast, science indicates that it is a large, old, evolving, ever-dynamic universe. The major discovery of humans is that the universe and its components, that is, time, space, matter, galaxies, stars, planets, life, and humans themselves, are interrelated historical entities that are constantly evolving. Many gaps and puzzles exist in our knowledge and major mysteries still remain, such as the origin of galaxies, life, and consciousness. Thus, a healthy degree of skepticism is in order, for we are using scanty evidence to draw a grand world picture.

The basic features of nature comprise change and evolution from simple physical and biological entities to complex derivatives, climaxing, on our planet, into astounding biodiversity and the extraordinary complexity of the human brain. Remarkably, this cosmic evolution, whose physical and chemical foundations are the cradle of all life, is explicable by materialistic, explorable, lawful patterns and processes. The evolutionary theory of the universe relates time, space, energy, matter, and life into a consistent, comprehensive, theoretical, and lawful framework beginning with the astronomical Big Bang explosion 15 ± 4 billion years ago (Ba) and culminating, from our local planetary perspective, in life and humans. A grand holistic continuum of cosmic evolution emerges, leading from lifeless matter to life and mind.

I will first briefly set the cosmic stage and then give an overview of the evolutionary patterns and processes of life on our planet, including biological diversification and extinction.

II. ORIGIN, STRUCTURE, AND EVOLUTION OF THE UNIVERSE

According to the general theory of relativity, there must have been a universal state of infinite density about 15 ± 4 Ba, at the Big Bang, which would have been an effective beginning of time of our present universe. By combining quantum mechanics with general relativity, space and time together might form a finite four-dimensional space without singularities and boundaries. This idea could explain the large-scale uniformity of the universe and the smaller-scale departure from it resulting in galaxies, stars, planets, life, and humans. The standard Big Bang model is the best current cosmology explaining the origin and evolution of the universe. It assumes that our universe is expanding according to Hubble's law and the dynamics of the expansion is described by Einstein's general relativity theory. Theories of the expanding universe are rapidly converging with unified theories of force and matter at the level of subatomic particles.

In a tiny fraction of a second after the Big Bang, the universe expanded dramatically by a factor of 10^{30}, or more, and is still inflating and forming cosmic structure involving clusters and superclusters of hundreds of billions of galaxies in the visible universe. Our own Milky Way galaxy alone contains more than 100 billion suns, many of which apparently have planetary systems. However, there are many billions of similar galaxies throughout the observable universe and billions of galaxies lie beyond our cosmic horizon.

The early universe was very hot, dense, and perhaps also irregular. The irregularity gradually decayed. Within minutes after the Big Bang, some nuclear reactions occurred. All the helium in the universe was then synthesized. The cosmic background radiation discovered in 1964 is appar-

ently a residual vestige of the Primeval Fireball of this early era, filling all space. During the expansion of the universe it cooled, eventually condensing into galaxies and quasars and showing cosmic evolution. The galaxies clustered and fragmented into stars. As the first generation of stars were born and died, heavy elements, such as carbon, oxygen, silicon, and iron, were gradually evolving through nucleosynthesis. As stars evolved into red giants they ejected matter that condensed into dust grains. New stars formed from clouds of gas and dust. In at least one such nebula, the cold dust collapsed into a thin disc surrounding the star. Dust grains accumulated into larger bodies that grew in size by their gravitational attraction, forming the diverse array of bodies, from asteroids to giant planets, that constitute our solar system. On the size-fitting planet Earth, both appropriate hydrosphere and atmosphere set the stage for the evolution of life and humans. We are children of the stars.

The Big Bang model for the expanding universe yielded a set of interpretations and successful predictions with no well-established empirical contradictions. It is reasonable to conclude that this standard cosmology has developed into a mature and believable model providing the physical and chemical theater for the evolution of life on Earth.

Evolutionary biology is the science studying the history of life on our planet. Two major ideas unify life and humans with nature: the ideas of biological and molecular evolution. [*See* MOLECULAR EVOLUTION.]

III. BIOLOGICAL EVOLUTION

The idea of biological evolution was substantiated by Charles Darwin (1859) and his followers. Life and humans were not created. They originated ultimately, given the proper environment, from inanimate matter, and evolved gradually from a primitive common ancestor throughout 4 billion years of chemical and biological evolution, generating all biodiversity and, of course, humans (Fig. 1). The evolutionary mechanism advanced by Darwin was natural selection, operating on heritable variability among individuals, thus forming biological adaptations in changing environments. This mechanism of Darwinian evolution, through genetic individuality, replication, mutation (i.e., variation), and natural selection resulting in environmental adaptation, might have been operating since the origin of life. At the primitive level of molecular evolution it was shown experimentally in artificial RNA by Manfred Eigen. Later, complex adaptive systems presumably adapt to and on the edge of chaos, according to the emerging science of complexity. [See BIODIVERSITY.]

The origin of life, from the standpoint of genetics, was the appearance of the first replicating molecule (which could metabolize and transmit information) on earth presumably about 4 Ba. Future progressive evolution involved improving homeostasis and complexity by optimizing structural and functional diversity. Likewise, biodiversity generally increased slowly and gradually, set back occasionally by mass extinctions. Admittedly, scientists are having a hard time agreeing on the greatest mystery ever, the origin of life: on when, where, and how life first emerged on Earth from lifeless matter. Organic chemicals could have been delivered by impacts or synthesized in the atmosphere, tidal pools, or deep-sea hydrothermal vents. These earliest organic chemicals combined stepwise to form more complex organic compounds, including nucleic acids, such as the dual ribozyme and the replicating and catalytic RNA, which generated proteins.

If a self-replicating "naked gene" originated from simpler organic molecules and multiplied, it, or its progenitors, might have started molecular evolution that led to cellular life. Impacts and greenhouse effects, caused by CO_2 derived from volcanoes, prevented life from evolving until 4 Ba. Around then, precellular forms that consumed molecules from natural organic "soup" may have evolved either to early cellular animallike forms or to precellular chemosynthetic forms that did not leave fossils, giving rise to chemosynthetic cellular prokaryotes. Later, photosynthetic bacteria resembling blue-green algae (cyanobacteria) emerged, forming dense stromatolite mounds on shallow sea margins. These primitive organisms, based on carbon bio-

chemistry and generating oxygen through photosynthesis, gave rise to all descendant complex life-forms, including humans (Fig. 1).

IV. MOLECULAR EVOLUTION

The second major idea unifying human and life with nature is the finding that the instructions (i.e., blueprint) for forming an organism are encoded chemically (Fig. 2). A flourishing current theory states that life is derived via the RNA (ribonucleic acid) world. RNA molecules could have evolved from ribose and other organic compounds by "learning" to copy themselves. They may have started both as information and catalytic molecules, but then began to synthesize proteins that served as both catalysts and the building material of organisms. The proteins helped the RNA to replicate and synthesize proteins more effectively. They also helped the RNA make double-stranded versions of itself that evolved into DNA (deoxyribonucleic acids), possibly the most important substance in nature, substantiating life and transmitting its genetic plan over generations. DNA took over in evolution, using RNA to make proteins, which in turn helped DNA make copies of itself and transfer its genetic information to RNA and then to proteins and on to the building up of organisms and their progeny (Fig. 2).

The chemical language of life, conveying genetic information for prescribing organisms, consists then of the two closely related families of giant molecules, the nucleic acids RNA and DNA. The latter is a long, ladderlike, double-stranded, twisted molecule like a spiral staircase consisting of a sugar–phosphate polymer backbone, with four nucleotide side molecules involving only four bases, or "letters": adenine, thymine, guanine, and cytosine in DNA, and the same nucleotides, except with uracyl replacing thymine, in RNA. DNA is the repository of the genetic code and RNA is a single-stranded molecule now primarily executing DNA's plans, producing proteins on cytoplasmic ribosomes by its stored information, analogous to computer-stored data (Fig. 2). DNA's two-stranded double helix can separate and replicate by

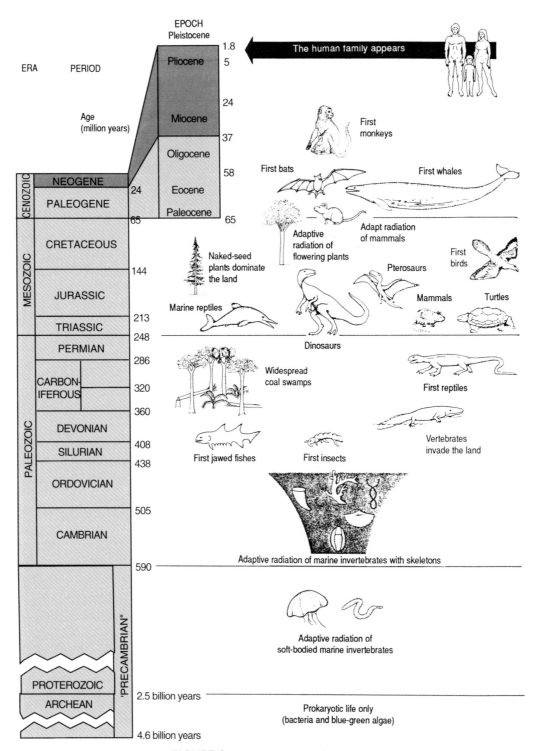

FIGURE I The evolution of life on Earth.

each strand complementing itself. Thus, DNA can both metabolize (produce proteins) and replicate, combining growth and reproduction.

Cloning and sequencing techniques now allow us to characterize genes directly instead of having to deduce their properties from their effect. This is a new genetics climax in the plan to obtain the complete DNA sequence of the human genome, though this goal will be achieved in the first decades of the next century. Small "model" genomes, such as the polio virus, are about 5000 letters, or nucleotides (0.005 megabases, Mb), long; the bacterium *Escherichia coli* (4.7 Mb), yeast (14 Mb), and the fruitfly *Drosophila* (165 Mb). These "model" organisms are better scaled to existing technology. The yeast genome contains genes with functions common to all eukaryotic cells and those of simple multicellular organisms. However, vertebrates, including humans, differ in their morphology and development. Therefore, the ideal small genome of minimum size and complexity with maximum homology to the human genome may be that of the pufferfish *Fugu rubripes* (400 Mb). This genome is 7.5 times smaller than the 3000 Mb of the human genome, but having a similar gene repertoire makes it the best model genome for the discovery of human genes (for update information see the "Genome" issue, *Science*, 30 September, 1994).

The 7 billion nucleotides are packed in 46 chromosomes in each of the nuclei centered in each of the trillions of cells comprising our bodies. This genetic information generates (through messenger RNA on cytoplasmic ribosomes) a second universal complementary chemical language of proteins. It consists of 20 amino acid building blocks, where 3 nucleotides in DNA code for a single amino acid molecule (Fig. 2). Proteins build the living body and provide its biochemical machinery. A length of DNA nucleotides, coding for hundreds of amino acids that comprise a single protein, is called a gene. Our 46 chromosomes in each nucleus consist of some 100,000 genes that prescribe us.

Remarkably, both the genetic (RNA and DNA) and metabolic (protein) languages, as well as the energetic currency, ATP (adenosine triphosphate), are universal to all life-forms: to microorganisms, fungi, plants, animals, and humans. Can there be a better demonstration of human unity with all life? This outstanding molecular uniformity transforms the Darwinian biological revolution into the material chemical language of life, that is, into the domain of molecular evolution. Furthermore, RNA comprises both genetic and metabolic (catalytic) ability, laying to rest the long debate over which of life's characteristics originated first: metabolism or the genetic code.

There are no substances in nature as important as the nucleic acids, RNA and DNA. These nucleic acids provided the genetic and metabolic basis of the evolutionary process generating the billions of different life-forms that have occupied the Earth since life emerged some 4 Ba. The evolutionary uniqueness of DNA and RNA resides in their dual nature, which embodies the capacity to transmit across generations both *constancy* and *diversity*. Constancy is retained through faithful DNA replication, whereas diversity is generated by the capacity to mutate, recombine, and form an infinite number of interchangeable chemical forms. Thus, a single molecule explains in one grand synthesis the constancy of inheritance and its diversity. The latter provides the material basis of evolution by natural selection.

V. UNITY AND DIVERSITY IN ORGANIC NATURE

The fascination of life resides in its unity and diversity. Unity is expressed by the chemical nature of life involving information transfer by the genetic code (Fig. 2), energy transfer by polyphosphates, structure and function through proteins, and cellular and tissue structures. Beyond these universals, nature consists of astounding biodiversity, involving millions of extant biological species and billions of extinct ones, most of which evolved and became extinct in the last 600 million years (My), that is, in Phanerozoic time. Clearly, extinction of species has been almost as common as origination. Only 1.5 to 1.8 million species have been documented to date, including unicellular organisms such as viruses, bacteria, algae, fungi, and protozoans, and multicellular organisms, comprising all living or-

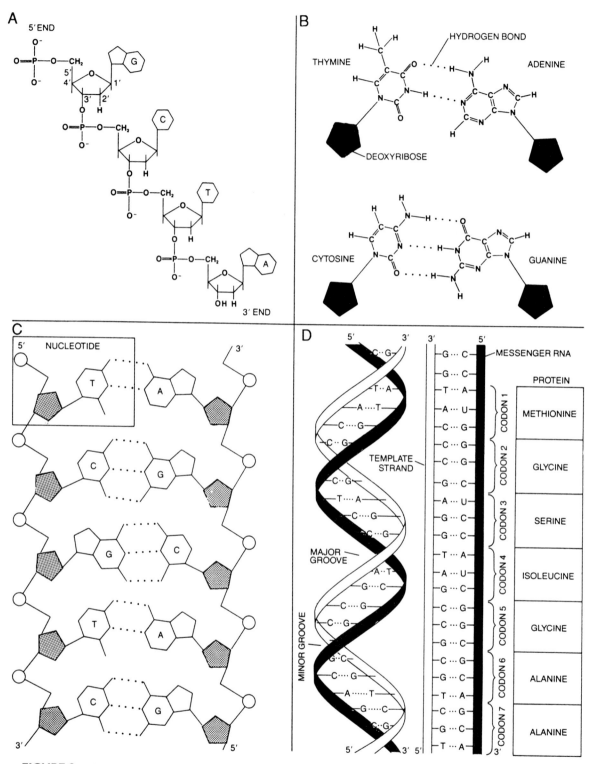

FIGURE 2 The structures of DNA, RNA, and proteins. (A) DNA backbone made of sugar and phosphate groups with attached bases; (B) nucleotide structure; (C) two-dimensional DNA structure; and (D) three-dimensional DNA structure, its transcription into a strand of messenger RNA, and the translation of mRNA into proteins. [From G. Felsenfeld (1985). *Scientific American,* October, p. 46].

ganisms. However, most extant species, involving several to dozens of millions of species (the figure is enigmatic and may differ by an order of magnitude, that is from 10 to 100 millions of species), are still unknown to science. The resolution of the enigma of how many species existed in the past and currently inhabit the Earth could highlight the evolutionary process itself and provide a sound basis for biological conservation and environmental management.

The astounding biodiversity in nature at the organismal level is based on the astronomic capacity for variation at the molecular level of RNA, DNA and proteins. Organic nature's response to environmental heterogeneity, and to the ever-changing physical and biotic stresses, was always the evolution of diversity. This is abundantly clear by observing the progressively increasing biodiversity through time from Precambrian times about 3.8 Ba, as well as in space, across the drifting continents, despite global and regional mass extinctions through the history of life on our planet (Figs. 1 and 3A–3C). The following discussion highlights the patterns and processes in the history of life, involving both diversification and extinctions.

VI. LIFE'S HISTORY: PRECAMBRIAN ORIGINS

The earth originated about 4.6 Ba, with other planets of the solar system, from a whirling cloud of dust. Radioactive dating unravels the earliest history of the Earth and the evolutionary dynamics of the emerging continents. During the first billion years of the Archean Eon of Earth history (see Fig. 1 for Earth geological eras, periods, and epochs), continents were small, but already drifting, volcanism was widespread, and radioactive elements were abundant in the lithosphere, releasing heat at a high rate. The oldest known intact rocks on Earth, the Acasta gneiss in northern Canada, are from 4 Ba, and the earliest zircon crystals, from Mount Narryer, Australia, date to 4.1 to 4.2 Ba. By the beginning of the Proterozoic Eon, 2.5 Ba, the Earth had cooled, larger continents were forming, and plate tectonic processes that operated al-

ready in Archean times continued. Banded iron formations, which formed in the presence of oxygen, accumulated in the sea about 2.5 and 1.8 Ba. Life originated apparently around 4 Ba and consisted first of prokaryotes (cells without nuclei) comprising bacteria and blue-green algae (Fig. 1).

VII. THE FOSSIL RECORD: OVERVIEW OF LIFE'S EVOLUTION

A. Prokaryotes

The fossil record represents a unique repository of information about the constant and dramatic change of organic evolution and increasing biodiversity over time, due to changes in physical (geologic, climatic) and biotic (competition, parasites, predation) diversity and stress. It reveals evolutionary diversification and extinction patterns and rates, whose dates are determined radioactively. Unequivocal evidence indicates that the first cells were prokaryotes, cells without a central nucleus containing DNA, which existed at least 3.5 Ba, in the Archean Eon, and prevailing for 1–1.5 My till the early Proterozoic Eon. This is demonstrated by stromatolite, threadlike bacterial mounds, and filamentous microfossils resembling modern cyanobacteria, or blue-green algae, from Greenland, Australia, and South Africa. The earliest prokaryotes survived presumably by anaerobic metabolism, obtaining energy by chemosynthesis and fermentation. The blue-green algae and other photosynthetic bacteria, following the anaerobes, enriched the atmosphere with oxygen as a by-product by harnessing sunlight energy and storing it chemically through photosynthesis (Fig. 1).

Photosynthetic organisms apparently existed even earlier, 3.8 Ba, as demonstrated by the Isua banded-iron sedimentary rocks from Greenland. These are the oldest terrestrial rocks indicating the existence of liquid water, a prerequisite of life. Even the carbon content of the rocks may indicate that they involved photosynthetic organisms that utilized a certain ratio of carbon isotopes (^{13}C to ^{12}C) characterizing life. If life indeed existed 3.8 Ba as cellular forms, then its origin may date back to around 4 Ba, though fossilized cells disappeared because of strong heating during rock metamorphism. Thus, remarkably, life appeared on Earth several hundreds of millions of years after its birth as soon as the environment became suitable. However, the buildup of atmospheric oxygen may date to 2.3 to 2.2 Ba, as shown by the increasing abundance of stromatolites. This oxygen presumably set the stage for the evolution of more complex organisms, that is, the eukaryotes (Fig. 1).

B. Eukaryotes

Eukaryotes (advanced organisms, both unicellular and multicellular, with larger cells involving nuclei, chromosomes, and organelles) appeared in the early Proterozoic Eon, more than 1.4 Ba. Single-cell eukaryotes arose presumably through symbiosis by the union of two cells. The eukaryote organelles, that is, the genetic nucleus, energy-producing mitochondria, and energy-harnessing chloroplasts responsible for photosynthesis, may have all started by symbiosis as prokaryotes engulfed by other prokaryotes. The symbiosis of prokaryotes led to the evolution of novel and high-energy sources. Prokaryotes reproduce asexually by simple fissioning. The additional evolution of nuclei and chromosomes provided the structures needed for sexual reproduction and meiosis. Thus, eukaryotes "evolved" mitochondrial respiration, sex, and recombination, major emergent evolutionary innovations that increased genetic diversity and speeded up evolution, leading to multicellularity. The biochemistry invented by early life persists today not only in bacteria but also in the most complex lifeforms composed of quadrillions of cells. Life's genetic library exploded in size but retains its original chemical linguistics (Figs. 1 and 2).

C. Multicellular Organisms

Eukaryotes paved the way to multicellularity after 3 billion years of unicellular "gestationary" evolution that transformed the planet atmosphere. Only about 1 Ba, multicellular complex organisms evolved, as suggested by fossil and molecular bio-

logical evidence. These gave rise to five kingdoms of living organisms: Monera, (all prokaryotes, that is bacteria and relatives), Protista (single-celled organisms with nuclei, the consumers among them are protozoans), Fungi (mushrooms, lichens, and their relatives), Plantae (Metaphyta, multicellular algae and higher plants), and Animalia (Metazoa, multicellular animals). During the first 3 billion years of life's history microorganisms prevailed in the oceans, with multicellular organisms appearing only in the last billion years (Fig. 1).

D. The Late Precambrian Radiation and Extinction

Trace fossils of tubes of wormlike, multicellular, burrowing animals in late Precambrian times occur only in rocks less than about 1 Ba. A global (Europe, North America, Australia, Africa) adaptive radiation of soft-bodied marine invertebrates, increasing in both complexity and variety (medusae, sea pens, annelid worms, and the first arthropods), occurred in the late Precambrian, originating about 800 million years ago (Ma). The best known is the Ediacara fauna of southern Australia (670 Ma), discovered in 1974 (Fig. 1). However, they seem to be a dead end and largely unrelated directly to extant living organisms. Their widespread presence is as mysterious as their disappearance (see later discussion). Other Precambrian lineages led to the great diversity of plants and animals in water, on land, in air, and just now in space.

E. The Cambrian Explosive Radiation and Paleozoic Life

The dramatic and seemingly explosive evolutionary radiation of marine invertebrates with skeletons (trilobite crustaceans, brachiopods, graptolites, bryozoans, etc.) occurred in the early Cambrian, almost 600 Ma, when most of the 26 major extant phyla appeared and others had been exterminated. These include unique, ancient, durable body plans (Bauplanes) surviving until today, such as radial symmetry (e.g., anemones, consisting of two layers of tissue), bilateral symmetry (e.g., flatworms with three primary tissue layers), and the coelomates with three body layers and a cavity in the middle layer. The latter involve segmented worms (annelids), pentamerally symmetric creatures (echinoderms, sea stars, sea cucumbers, and starfish), and bilaterally symmetric coelomates, including arthropods (insects, spiders, and crustaceans), the mollusks, the vertebrates, and others. Thus, with the initial decline of stromatolites about 1 Ba, the major eukaryote radiation and probably the first metazoans appeared between 1000 and 900 Ma. The Ediacaran faunas representing the diploblastic grades evolved extensively during the period 580–550 Ma, followed by the Cambrain faunas representing the triploblastic grades (Fig. 1).

Phyla are distinguished by characters reflecting the oldest and deepest levels of evolutionary association and are hierarchically divided into classes, orders, families, genera, and species. The times of origin of the phyla and their relations remain obscure, but must have taken place in late Precambrian times more than 900 Ma, despite their spectacular explosion due to the possession of hard skeletons in the early Cambrian 600 Ma. The most spectacular assemblage of Cambrian fossils comes from the Burges shales in British Columbia. This assemblage contains many unique body plans that flourished early in the Cambrian, becoming extinct later. During Cambrian and Ordovician times, hundreds of class-level taxa originated.

Early Paleozoic biota were in the sea. The first vertebrates appeared about 500 Ma in the sea, among a rich invertebrate fauna. Colonization of land occurred about 400 Ma, involving plants, fungi, invertebrates, and vertebrates. The amphibians were the first land vertebrates that became associated with water bodies for reproduction. About 300 Ma the breeding pond was "enclosed" inside the amniotic egg, which is bounded by a hard envelope. This innovation liberated the reptiles, the descendants of amphibians, from their water dependence, thus completing the trek toward land colonization. Dinosaur reptiles prevailed in water, land, and air during the entire Mesozoic era, together with naked-seed plants (gymnosperms and cycads) (Fig. 1).

F. Life's Evolution in the Mesozoic and Cenozoic

The mammals, which originated in Triassic times, more than 200 Ma in the early Mesozoic era, radiated dramatically in Cretaceous time, becoming predominant and colonizing most habitats in the Cenozoic (Fig. 1). Placental mammals, including primates, originated 65 Ma, the monkeys 50 Ma, and the apes 30 Ma. The human family appeared 6–7 Ma, and the genus *Homo* a mere 1.6 Ma. Archaic *Homo sapiens* appeared 200,000–300,000 years ago (Ya), and modern *Homo sapiens* showed up 100,000 Ya. The agricultural revolution occurred 10,000 Ya, the Industrial Revolution about 200 Ya, and the communication revolution in the last several decades, and is still on the rise.

In 1953 humans deciphered the genetic code that prescribes all organisms. This chemical language encoded in DNA and RNA is universal to all organisms and registers genetically the history of life on Earth in the genomes of viruses, bacteria, multicellular organisms, and humans. This is a uniquely dramatic testimony to the unified evolutionary origin of all organisms from a common ancestor in early Precambrian, Archaic times. This history is condensed in humans in 46 chromosomes contained in each of the trillions of cells comprising our bodies. In recent years, we have learned to engineer the genetic code. At present, the human genetic code is being actively unraveled in the Human Genome Project, possibly the climax of all current biological research programs (see "Genome" issue in *Science*, 30 September 1994). An understanding of the chemical language of life allows us to genetically manipulate organisms, an immense, sobering power that has begun to dramatically change medicine, agriculture, and industry through genetic engineering of plants and animals by transgenic operations.

If we condense the history of life into one year to illustrate the time line of major evolutionary events, then life originated on January 1; vertebrates appeared on November 16, with land colonization on November 25; the human family appeared on December 31 at 6 a.m.; australopithecids

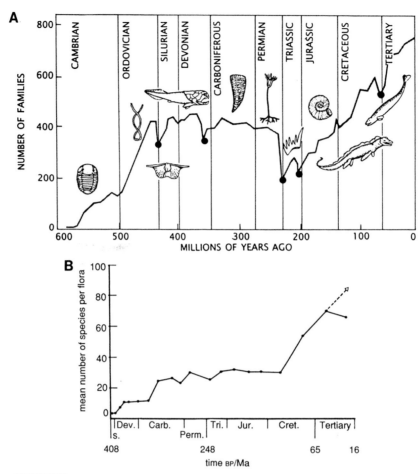

FIGURE 3 (A) Biodiversity trajectory of families of marine organisms in the Phanerozoic. (B) Biodiversity trajectory of plant species in the Phanerozoic. (C) Biodiversity trajectory of tetrapod families in the Phanerozoic. 1–6 indicate beginnings of extinctions. I, II, and III indicate 3 family assemblages that succeeded each other through geological time. [(A) from E. O. Wilson (1989). *Scientific American* **261**(3), 60–66; (B) from Maynard Smith in Chaloner and Hallam, 1989, after Knoll, 1986; (C) from M. J. Benton (1989). *Philos. Trans. Roy. Soc. London Ser. B* **325**, 369–386.]

evolved on December 31 at 1:00 p.m. and *Homo sapiens,* our own species, appeared on the same day at 11:55 p.m.; the agricultural revolution arrived at 11:58:41; and human cultural history occurred in but fragments of the last few minutes.

VIII. BIODIVERSITY: DIVERSIFICATION AND EXTINCTION PATTERNS

Large-scale evolutionary change over time clearly has various patterns, rates, tempos, and modes.

The previous section dealt with qualitative evolutionary change in life's history, deduced from the fossil record, from the simplest Precambrian organisms to the extant complex organisms. Before overviewing the causes leading to variation in the trajectory of life's large-scale evolutionary change, I will first describe briefly the general quantitative evolutionary patterns of diversification and extinction.

A. Diversification: Origination and Speciation

The history of global biodiversity over evolutionary time is one of a generally increasing diversity,

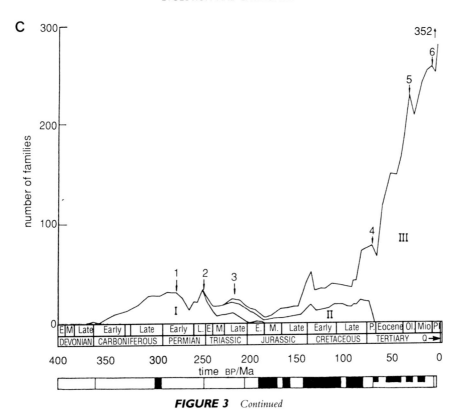

FIGURE 3 *Continued*

through speciation by either cladogenesis (i.e., fragmention of populations into reproductively isolated and ecologically compatible species in space) or anagenesis (i.e., phyletic evolution over time), in both marine (Fig. 3A) and terrestrial (Figs. 3B, 3C) organisms. The biodiversity trajectory of marine organisms indicates an initial experimental period followed by a swift rise in family number in early Paleozoic times, about 600 Ma. It remained roughly constant for the remaining 200 million years of the Paleozoic era and finally displays a slow but steady climb through the Mesozoic and Cenozoic eras to diversity's present all-time high (Fig. 3A). The rise in biodiversity was set back by five massive extinction episodes during the end-Ordovician, end-Devonian, end-Permian, end-Triassic, and end-Cretaceous periods (Figs. 3A and 4). The last of these is by far the most famous, because it ended the age of dinosaurs and opened ecological opportunities to the extensive mammalian adaptive radiation in the Cenozoic, leading, among other lineages, to the evolution of humans. A series of smaller extinctions brings the total to at least 12 extinction

events, 9 of which stand out above the regular background extinction levels (Fig. 4).

The diversity of terrestrial plants (Fig. 3B) and tetrapods (Fig. 3C) increased similarly from the Devonian to the Permian. Plant diversity remained roughly constant during the Mesozoic, then began to increase in the late Cretaceous and continued to do so during the Tertiary (Fig. 3B). The rapid radiation of "modern" tetrapod groups—frogs, salamanders, lizards, snakes, turtles, crocodilians, birds, and mammals—was hardly affected by the famous end-Cretaceous extinction event. Rodent peri-Mediterranean diversification patterns in the Neogene (24–3 Ma) show three peaks of high origination rates, two in the Miocene (17.5 and 11.5– 11 Ma) and one in the early Pliocene (4.2–3.8 Ma).

B. Insect Diversity in the Fossil Record

Insects possess a surprisingly extensive fossil record. Their diversity exceeds that of preserved vertebrate tetrapods through 91% of their evolutionary history [see Figs. 1–5 in Labandeira and Sepkoski (1993);

see Bibliography.] The great diversity of insects was achieved not by high origination rates but rather by low extinction rates comparable to the low rates of slowly evolving marine invertebrate groups. The great radiation of modern insects began 245 Ma and apparently was not accelerated, at least at the familial level, by the expansion of angiosperms during the Cretaceous period. The basic trophic machinery of insects was in place nearly 100 My before angiosperms appeared in the fossil record, and may relate to the evolution of seed plants in general rather than to angiosperm evolution.

C. Extinction

"Mass extinction" is the term used in paleobiology to describe the relatively short interval of geological time when large and diverse segments of the world's biota underwent extinction, involving global reduction in diversity and biomass, as revealed in the fossil and geochemical record. (Figs. 3A, 3C, and 4; Table I). Both perspectives provide significant results but they must be interpreted carefully to avoid potential pitfalls. Historically, five mass extinctions have been identified in the fossil record as having been the greatest crises in the history of life during the Phanerozoic: they occurred in the end-Ordovician (~440 Ma), end-Devonian (~360 Ma), end-Permian (~250 Ma),

end-Triassic (~215 Ma), and end-Cretaceous (65 Ma). In addition, four additional global events are interpreted as mass extinction phenomena, especially, those in the late Precambrian (650 Ma), end-Cambrian (500 Ma), Eocene–Oligocene (32 Ma), and end-Pleistocene (11,000 Ya) (Fig. 4 and Table I). [See DIVERSITY CRISES IN THE GEOLOGIC PAST.]

Background extinction describes a spectrum of smaller events, resulting from abiotic and biotic stresses leading to the regular disappearance of taxa. In general, Phanerozoic patterns of phytoplankton radiation and extinction parallel those for skeletonized marine invertebrates. In plants, extinctions tend to follow innovation. Major mass extinctions among tetrapods took place in the early Permian, late Permian, early Triassic, late Triassic, late Cretaceous, early Oligocene, late Miocene, late Pliocene, and Holocene (Fig. 3C). Many of these events appear to coincide with the major mass extinctions among marine invertebrates.

D. Patterns of Extinctions Across Life's History

Extinction may be episodic at all scales, with relatively long periods of stability alternating with short-lived extinction events (Figs. 3A–3C, 4, and 5). Most extinction episodes are biologically selective, that is, the victims and survivors are not ran-

TABLE I
Principal Extinctions and Probable Cause[a]

Extinction event	Probable cause
Late Pleistocene	Postglacial warming plus predation by humans.
Eocene to Oligocene	Stepwise extinction associated with severe cooling, glaciation, and changes of oceanographic circulation, driven by the development of the circum-Antarctic current.
End-Cretaceous	Bolide impact producing catastrophic environmental disturbance.
Late Triassic	Possibly related to increased rainfall with implied regression.
End-Permian	Gradual reduction in diversity produced by sustained period of refrigeration, associated with widespread regression and reduction in area of warm, shallow seas.
End-Frasnian (Devonian)	Global cooling associated with (causing?) widespread anoxia of epeiric seas.
Late Ordovician	Controlled by the growth and decay of the Gondwanan ice sheet following a sustained period of environmental stability associated with high sea level.
Late Cambrian	Habitat reduction, probably in response to a rise in sea level, producing a reduction in number of component communities.
Late Precambrian	Complex, including widespread regression, physical stress (restricted circulation and oxygen deficiency), and biological stress (increased predation, scavenging, and bioturbation).

[a] Based on S. K. Donovan (1989). "Mars Extinctions: Processes and Evidence," Chaps, 4–12. Belhaven Press, London.

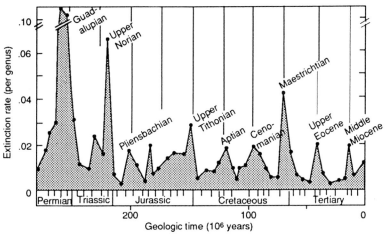

FIGURE 4 Record of percent extinction (per million years) computed from the records of 9773 genera of marine fossil animals (with the addition of the 26-My periodicity in best-fitting position). Marks along the abscissa indicate standard stages and are not sampling intervals; dots are placed at the centers of the 51 sampling intervals. [From Raup D. M., and Sepkoski, J. J. (1988). *Science* **241**, 94–96.]

dom samples of the preextinction biota. Analyzing survivorship of lineages could highlight the proximal ecological and physiological causes of extinction (Table I). The data base for extinction analysis is the distribution in space and time of about 250,000 known marine fossil species, an extremely small sample of past life. This is true because of the negligible probability of preservation and discovery of all species, as well as the problematics of the morphospecies as compared with the biological species. Thus, many sibling, morphologically indistinguishable good biological species occur in nature on the one hand, and much geographic variation within species could be wrongly interpreted as representing different species. Clearly, a better resolution derives at higher taxonomic levels (genera, families, and orders) (Figs. 3A, 3C, and 4).

Several factors may contribute to extinction resistance: broad ecogeographic and genetic spectrum, large population size, dispersal ability, and habitat selection. However, these may be more effective in background than in mass extinctions. Tropical biotas and specialist species are the first victims of extinction.

I. Precambrian Extinctions

The end-Precambrian (~650 Ma) witnessed apparently a widespread and early global extinction of the Ediacara fauna, due to combined abiotic and biotic stresses. The latter derived from sea re-

gression and emergence of continental shelves associated with plate tectonics and climatic changes involving glaciation, anoxia (i.e., oxygen deficiency), hypersalinity, scavenging, predation, and bioturbation, all leading to the extinction of many lineages of medusoids, algae, annelid worms, and "protoarthropods."

2. Paleozoic Extinctions

a. End-Cambrian (~500 Ma)

Global mass extinctions are registered in the fossil record (Fig. 3A). Thus, at the end of the Cambrian nearly two-thirds of the 60 existing families of crustacean trilobites became extinct (Fig. 5). Major biogeographic and ecological reorganization, due to a sea level rise, led to progressive reduction in component communities or biofacies, causing a nonrandom extinction and selective survival of trilobite families. Those families, represented in upper-slope refuges, had a higher survival probability, regardless of their species richness. This increases clade survival during background extinction, supporting D. Jablonski's idea of differential causation in background as compared with mass extinctions.

b. End-Ordovician (~440 Ma)

Marine organisms, including trilobite and graptolite families, disappeared in the late Ordovician, amounting to 20% loss of diversity (Figs. 3A and

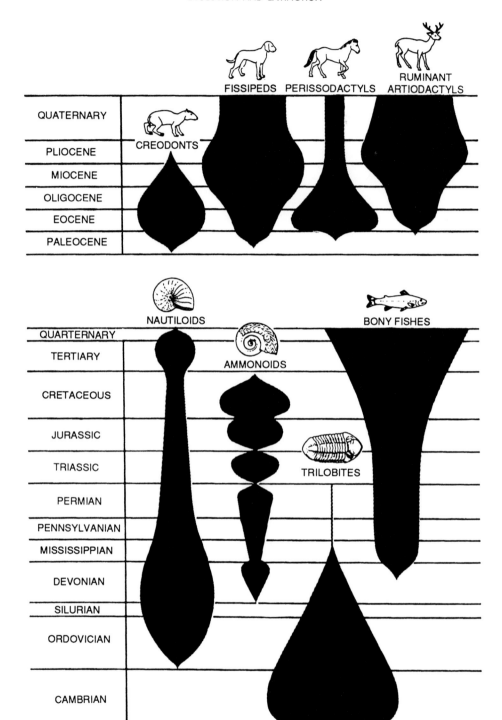

FIGURE 5 Ecological replacement of some taxa in the Phanerozoic. [From N. D. Newell (1963). *Scientific American*, February, p. 187.]

5). The end–Ordovician extinction involved two peaks of extinction separated by hundreds of thousands of years, characterized by the growth and decay of the Gondwanan ice sheet. This promoted widespread environmental perturbations, particularly temperature and sea level changes, which caused niche destruction and extinction. The major factors involved sea level and climatic changes were associated with chemical sea changes and anoxia. These caused the extinction of tropical, climatically stabile, stenotopic species and their replacement by cosmopolitan, generalist eurytopic species (see replacement of trilobites by nautiloids in Fig. 5). Extinction was selective among biotas (tropics versus temperate), communities (trilobites, corals, and brachiopods), niches, and species. The end–Ordovician extinction provides one of the best examples of ecological collapse. It was followed by a reestablishment of the ecological structure, setting the stage for the early Silurian radiation and the diversification of new Silurian communities.

c. End–Devonian (~360 Ma)

In the late Devonian as many as 21% of all families, 50% of all genera, and 70% of all species disappeared, involving reefal, peripheral, and shallow-water organisms, especially corals, stromatoporids, tentaculids, and brachiopods (Figs. 3A and 5). The ultimate cause of the end–Devonian extinction remains enigmatic. Tectono-glacial, paleogeographic, and bolide impact hypotheses have been proposed, each predicting a resultant temperature decline. The terrestrial ultimate causes seem more substantiated than the bolide hypothesis. The extinction lasted 3 My and was associated with lethal temperature decline as a proximate cause, creating widespread anoxia in the surface waters across the planet. This caused the global decimation of low-latitude tropical reef ecosystems and of warm-water shallow marine faunas. It was combined with higher survival of high-latitude faunas, deep-water faunas, and terrestrial faunas and floras. At the local and regional level, additional complications were added.

d. Early Carboniferous (~340 Ma)

The middle Mississippian blastoid (phylum Echinodermata) extinction event was a rapid, habitat-specific extinction. Blastoids presumably became rare worldwide or absent in shallow-water environments after the extinction. Onshore–offshore habitat shifts have been recognized as an important historial trend among marine benthos. The blastoids diversified immediately after this extinction and repopulated shallow-water habitats after a period of diminished diversity and abundance.

e. End–Permian (~250 Ma)

The end–Permian was the greatest global extinction in Earth's biota terminating the Paleozoic great experiment in marine life during an interval of intense climatic, tectonic, and geochemical change (Figs. 3A and 4). Decimation of 77–96% of marine and terrestrial species occurred in the late Permian. Marine life was devastated, experiencing a reduction of 57% of families and 96% of species, with crinozoans (98%), anthozoans (96%), brachiopods (80%), and bryozoans (79%) suffering the greatest extinction. Other severely affected groups were the cephalopods, corals, ostracods, and foraminiferans, all predominantly tropical groups or members of the reef-building community. Likewise, 75% of amphibian families and more than 80% of reptile families also disappeared. Such an enormous reduction in the biota may suggest a catastrophic event, but geochemical and faunal evidence favors a gradual process of reduction in species diversity. The most plausible causal mechanism seems to be climatic change, with a sustained period of global refrigeration, accompanied by widespread regressions and reduction of warm, shallow seas on the western and eastern margins of Pangaea. Notably, the main suborders of the decimated taxa survived the Permian to carry over into the Triassic.

3. Mesozoic Extinctions

a. End–Triassic (~215 Ma)

Multiphase late Triassic extinctions occurred in Europe involving scallops, crinoids, ammonoids, bryozoans, conodonts, and reef-building organisms (Figs. 3A and 4). Most ammonite families became extinct, except one that gave rise to the scores of families of Jurassic and Cretaceous times (Fig. 5). Likewise, primitive reptiles and amphibians were replaced by the expanding dinosaurs (Fig.

3C). End-Triassic extinctions were caused primarily by terrestrial causation facies changes, and there is no need for invoking extraterrestrial impacts. Ultimate tectonic, proximate sea level regression causing anoxia, and the loss of reefal facies, leading to reduced habitable area, best explain marine invertebrate extinction at this time. Two Jurassic mass extinctions occurred on a regional, not a global, scale, as shown by A. Hallam. They can be related to severe reductions in habitat area caused by regression of the epicontinental sea or by widespread anoxia.

b. End-Cretaceous (~65 Ma)

In the late Cretaceous (K-T), mass marine and terrestrial extinctions, extending over several million years, eliminated the ammonites, dinosaurs, and marine and flying reptiles, together with numerous families of corals, rudistid bivalves, echinoids, planktonic foraminifera, and belemnites, and also brought declines in marsupial mammals (Figs. 3A, 3C, 4, and 5). Plants, primarily tropical and subtropical lineages, also became selectively extinct (20–70% of the species, depending on latitude). Plant megafossils at the K-T boundary, most strongly in the Northern Hemisphere and less so in the Southern Hemisphere, express ecological catastrophe, some climatic selective extinction of broad-leaved evergreen species as compared with deciduous foliage, and long-term vegetational restructuring from numerous refugia (Fig. 3B). Although 50% of marine genera died out completely, and probably 60–75% of the species, many families of cephalopods and nautiloids survived with only minor evolutionary modifications (Fig. 5). This is also true for most bony fish and tetrapods. Dinosaurs, the symbols of the K-T extinction, died out at different times in different places rather than suddenly and simultaneously worldwide, sometimes surviving the iridium anomaly and sometimes coexisting with Tertiary-type mammals. Nevertheless, the demise of the dinosaurs opened the door to the explosive radiation of placental mammals, leading ultimately, among other lineages, to human evolution.

4. Cenozoic Extinctions

Global gradual extinctions, due to climatic change embracing both marine and terrestrial faunas and floras, occurred in the mid-Eocene–Oligocene transition in five steps over 10 My (40, 38, 36, 33.7, and 30.5 Ma), when the world changed from a "hothouse" to a "coldhouse" (Figs. 4 and 5). The major extinction event was caused by proximal cooling, glaciation, and changes in oceanographic circulation at the middle to late Eocene transition (about 40–41 Ma). Warm-humid tropical plants and animals were severely reduced. Lesser extinctions took place at the end of the Eocene (34 Ma), although extinction among taxa was selective. In the early Oligocene (about 33.5 Ma), there was significant cooling and increase of ice volume, which resulted in major changes in land floras and possibly the "Grand Coupure" immigration event in European land faunas. Major Antarctic glaciations marked the middle Oligocene (about 30 Ma), though most aboveground organisms were cold-adapted survivors, and there were relatively few extinctions.

These Eocene–Oligocene extinctions were clearly related to climatic change rather than to extraterrestrial impacts, triggered by the thermal isolation of Antarctica as Australia drifted northward and allowed the development of deep, cold, bottom water. In addition, the world Cenozoic orogenies (Andean, Laramide, and Alpine–Himalayan), which started differentially in the Paleogene across the planet, but were reinforced in the Neogene, complemented the global trend of increasing cooling, drought, and seasonality. The Cenozoic ecological theater of open country biota, starting in the Eocene–Oligocene transition, resulted then from global climatic change, extensive sea regressions, and mountain formation. The subterranean ecological zone was opened for the global convergent evolutionary experiment of adaptive radiation of small mammals underground on all continents, thus avoiding the harsh aboveground Oligocene climate. Likewise, big running mammals radiated in the extensive open biota.

5. Miocene and Pliocene Extinctions

Two important rodent extinctions, among other mammals, followed immediately dynamic originations (11.5–11 Ma and 4.2–3.8 Ma). The most important rodent extinction occurred in the middle to late Miocene boundary (11.5–11 Ma) (Fig. 4).

At the Miocene–Pliocene boundary and during the early Pliocene, the faunal turnover seems to increase, thus decreasing mean species duration.

6. Late Pleistocene, Holocene, and Present Extinctions

Dramatic worldwide extinctions occurred during the late Pleistocene and Holocene periods since 50,000 Ya, affecting primarily large mammals (Fig. 6). Although a global phenomenon, late Pleistocene extinctions in northern Eurasia and North America were highly variable in their severity in different regions. In North America, Asia, and Australia, but not in Africa, many of the large herbivores and carnivores became extinct rapidly, between 12,000 and 6000 Ya, with a maximum rate around 8000 Ya, when the climate had become milder and the glaciers were shrinking. South America lost approximately 46 out of 58 mammalian genera (i.e., 80%); Australia lost 15 out of 16 genera (94%); North America lost 33 out of 45 genera (73%) and at least 19 genera of birds vanished; Africa lost only 7 out of 49 genera (14%); Africa south of the Sahara lost only perhaps 2 out of 44 genera (5%); and Europe lost 7 out of 24 genera (29%). Asia also endured only a few extinctions. In general, the larger the animal, the more it was at risk of extinction. Changes in geographic ranges, reassembling into new communities affected, beside extinctions, also the relative composition of world's terrestrial biota. Marine extinctions, however, were insignificant.

IX. THE CAUSES OF BIOLOGICAL DIVERSIFICATION

Biodiversification in the Phanerozoic, in both marine and terrestrial plants and animals, displays generally increasing diversity, not only for species but also for genera and families (Figs. 1, 3A–3C and 7). The general pattern exhibits the rise and fall of families over geological time. J. J. Sepkoski described three basic patterns of marine fauna represented by Newell in spindle diagrams (Fig. 5): *Cambrian Paleozoic fauna:* highest biodiversity of trilobites and inarticulate brachiopods; *post-Cambrian Paleozoic fauna:* highest biodiversity in the Paleozoic (articulate brachiopods, rugose corals, cephalopods, crinoids, and many others); and *Mesozoic and Cenozoic fauna:* dramatic increase in biodiversity after the Paleozoic (bivalves, gastropods, echinoids, teleost fish, and modern fauna) (Figs. 1, 3A–3C, 5, and 7).

Three models of biotic diversification attempted to explain these patterns globally: (1) the *logistical model,* following the ecological theory of island biogeography, where the probabilistic rates of family origination and extinction are diversity-dependent (formulated for islands by R. H. MacArthur and E. O. Wilson, for continents by M. Rosenzweig, and for the planet by J. J. Jr Sepkoski, and (2) the *multiphase logistical model,* which focuses the Competitive interaction between successive evolutionary faunas (Fig. 5). Calculations by L. Van Valen and A. Hoffman based on Sepkoski's compendium of marine fauna in the Phanerozoic show that diversification may be diversity-dependent during low (as in the early Cambrian and Triassic periods) and high (as in most of the Paleozoic beginning in the late Ordovician) biodiversity levels.

Global rates of speciation seem to be independent of diversity. Van Valen, Maiorana, and Hoffman suggested that while the probabilistic rates of family origination and extinction in the Phanerozoic decrease, the family diversity increases with geological time (Figs. 3A–3C). There is no significant intercorrelation between either family extinction or origination rate and diversity. The rates seem to be diversity-independent. Finally, (3) the *"lithospheric complexity" model* is another nonequilibrium model for global biodiversity developed by Joel Cracraft, suggesting that speciation and species extinction are determined by environment (climate gradients and barriers, and topographic barriers to migration, etc.). The biosphere in this model can never reach equilibrium and is in a constant flux due to the incessant geological–climatic change, despite mass extinctions that temporarily reverse the trend. Clearly, this model is diversity-independent in contrast to other models, including the famous "Red Queen" hypothesis of L. Van Valen. The Red Queen hypothesis visualizes a never-ending evolution of taxa due to biotic stresses and interactions only, even in the absence of any change in the physical environment. Clearly,

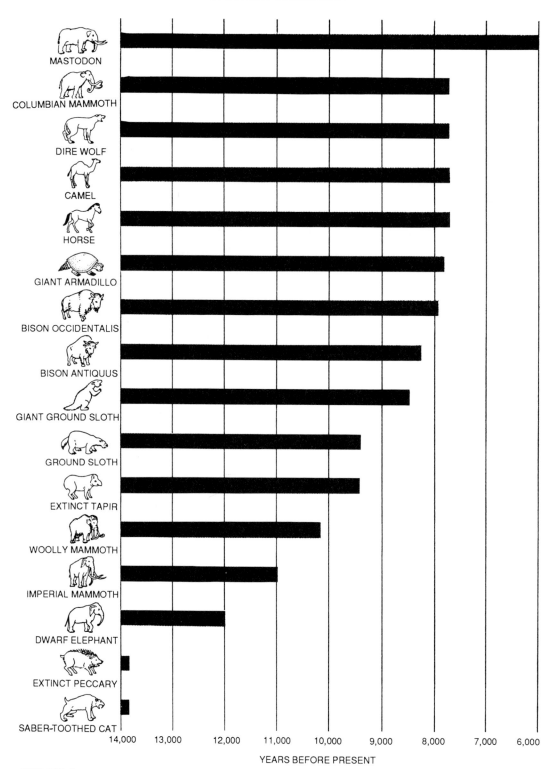

FIGURE 6 Extinction of large mammals in the Upper Pleistocene and Holocene. [From N. D. Newell (1963). *Scientific American,* February, p. 190.]

the dichotomic either/or approach was always wrong in evolutionary studies. To argue that either abiotic or biotic change only predominates is unrealistic. As usual in biology, interactions of complex factors drive evolution. Our present major ignorance is in quantifying the relative importance of the various evolutionary interacting abiotic and biotic forces in the evolutionary process, which involves both diversification and extinction.

X. THE CAUSES OF EXTINCTIONS

A. Overview: Patterns and Theories

Speculation on the causes of extinctions in the geological past dates back to the early days of geology in the early nineteenth century when Georges Cuvier's catastrophism model (1822) was replaced by Charles Lyell's gradualism and W. Whewell's uniformitarianism models. However, little thought has been given to the possibility that mass extinctions might show some regular cyclic pattern. Furthermore, suggestions of driving mechanisms for major extinction events varied from terrestrial, such as gradual or abrupt changes in climate and/or sea level, to sudden extraterrestrial large-body impacts.

The study of mass extinctions only recently became one of the major areas of evolutionary biology. The main factors for this are the spectacular and provocative hypothesis that (1) the latest Cretaceous mass extinction (K-T) was caused by an impact of a huge bolide; (2) mass extinctions are periodic, with a periodicity of approximately 26 million years since the end of the Permian, and are caused by comet showers triggered by an unseen solar companion; and (3) the biotic effects of mass extinctions are qualitatively different from all other phenomena in the history of life on Earth.

Extinction is an evolutionary and ecological problem. All plant and animal species that have been driven to extinction were primarily affected by environmental changes leading to background extinction. Species become extinct because they are unable to cope with rapidly changing environments, being constrained by their own slow evolu-

tion, or because their ecological niches disappear and their adaptive repertoire is no longer relevant to the new environments. Or, because of biotic factors (parasites, diseases, and competitors). Both causes, either singly or in combination, can lead to species extinction individually and/or massively. Extinctions are biologically not random, either taxonomically or ecologically. For this reason, extinction must play an important role in the evolution of life in a Darwinian selective sense.

Prediction of the extinction of populations and species requires ecological, demographic, genetic, and evolutionary information. Loss of genetic diversity due to genetic drift or other factors in small populations can diminish future adaptability to a changing environment. Resistance to background ("normal") extinction may be ineffectual during mass extinction. In general, but with many exceptions, groups surviving great extinctions were generalists and conservatives, supporting the idea of "survival of the unspecialized," which was already recognized by Darwin. According to Jablonski, mass extinctions tend to remove not only more clades, but different clades from those lost during times of background extinction. Traits conferring resistance during mass extinctions (e.g., broad ecogeographical range at the clade level) are, according to Jablonski, poorly correlated with traits that enhance survival and diversification during background times. Removal or reduction of dominant groups during mass extinctions provides opportunities for diversification of taxa that had been minor constituents of the preextinction biota. Thus, evolution can proceed in directions not predictable from the background extinction preceding the mass extinction.

Extinction is most pronounced in animals. Each of the three successive land floras, the lower vascular plants (mosses and ferns), gymnosperms, and angiosperms, evolved rapidly followed by a long period of stability (Fig. 7). Extinction may be caused by either or both abiotic (physical) and biotic causes, and/or their interaction, which may predominate. Extinction can operate on individuals, species, higher taxa, and even entire ecosystems.

Background extinction, that is, the ongoing process of the normal rate of replacement of one species

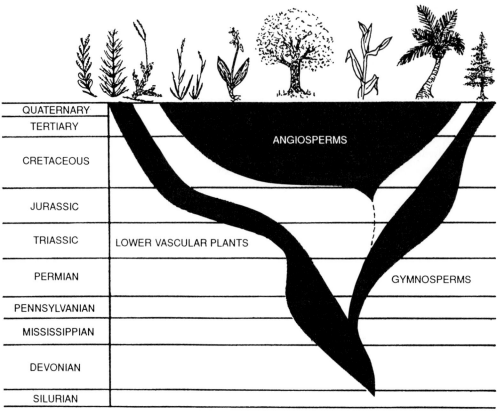

FIGURE 7 Biodiversity trajectory of land plants in the Phanerozoic. [From N. D. Newell (1963). *Scientific American,* February, p. 183.]

by another, known as Van Valen's law, may be caused by physical and biotic factors or their combination. However, their relative importance may vary in space and time and is largely unknown. Recent introduction of predators, competitors, pathogens, and parasites into new areas, primarily islands, either small (e.g., in the Pacific Ocean or West Indies) or large (e.g., Australia), emphasizes the relative importance of biotic factors under relatively slowly evolving environments. However, abiotic factors, such as climatic changes, fluctuations of sea level, and transgression–regression phenomena, occurred throughout evolutionary history and more recently due to the dramatic changes of the global environment resulting from anthropogenic influences.

Transgression-dependent anoxic events have been associated with several extinctions in end-Ordovician, end-Devonian, end-Triassic, end-Jurassic, Cenomanian (mid-Cretaceous), and end-

Cretaceous times. The idea of an association of extinction with anoxia–eustatic cycles is generally supported, as claimed by Hallam. Although we have clues as to the combinatorial factors causing specific extinctions (Table I), such as in the late Permian, the relative importance and interaction of the cause(s) in specific extinctions are largely unknown, and no simple current theory of community ecology provides a satisfactory explanatory model. Such a theory is expected to emerge from the integration of current hypotheses and new empirical insights.

Mass extinctions are regarded by some as caused by one single factor, usually the same factor for all extinctions. This could be an extraterrestrial impact, an extraordinary volcanic eruption, or a major paleogeographical, oceanographic, or climatic change, but it is always just one single driving force that is conceived as being the ultimate cause of all extinctions during the time interval interpreted as a mass extinction. D. M. Raup and J. J. Sepkoski suggested in 1984, primarily on the stratigraphic ranges of marine families, that the fossil record of the past 250 My is regularly spaced in geological time and shows a 26-My periodicity caused by climatic changes, meteoritic impact, and other signals in environmental history that may corroborate periodicity. The case for or against extraterrestrial causes of extinctions depend on further clarification of the timing of extinction in the stratigraphic record. However, recent data appear to be more against than in support of "the cause" as a single factor, and also question the periodicity idea.

B. Evaluation of Current Extinction Theories

In-depth statistical analysis challenges the 26-My periodicity. J. A. Kitchell and D. Penna showed in 1984 that the best-fitting time-series model is a stochastic autoregressive model that displays a pseudoperiodic behavior with a cycle length of 31 My for the past 250 My. Periodicity in a time series is insufficient evidence that a periodic external force is causally responsible. Cyclical-like genetic cycles and supercycles caused by cyclical selection (V. Kirzhner, A. Korol, Y. Ronin, and E. Nevo) demonstrated that causality should be evaluated with independent evidence. The periodicity lengthens and weakens when the analysis is extended to the entire Phanerozoic. The history of the probability of extinction for the entire Phanerozoic, based on time-series analysis, has decreased uniformly over Phanerozoic time, whereas the inertia or stability of the biotic system after the late Permian has increased. Moreover, in 1987 S. M. Stigler and M. J. Wagner challenged the 26-My periodicity conclusions on statistical grounds. Important questions concerning mass extinctions include, among others, proximate and ultimate timing (terrestrial, solar, cometary) causes, periodicity, and organismal response. Whether a mass extinction is periodic, pseudoperiodic, or random, and whether the signal is simple or complex, earthbound or extraterrestrial, are now under intensive critical and

stringent testing in several laboratories, including broad sampling of evidence for the cause(s).

The supposed 26-My periodicity of extinction since the mid-Permian has also been challenged by the work of C. Patterson and A. B. Smith with fish and echinoderms, which together make up about 20% of the available marine Phanerozoic data analyzed by Sepkoski. Only 25% of these fish and echinoderm extinctions are real, reflecting disappearance of a monophyletic group. The remaining 75% is noise, chiefly involving "extinctions" of nonmonophyletic groups, mistaken dating, and families containing one species only. They concluded that periodicity in their sample is a feature of the noise component, not of the signal. Similarly, family appearances and extinctions in the ammonoids, summarized by M. R. House in 1989, are significantly correlated with temperature and sea level fluctuations, associated with anoxic or low-O_2 events, supporting the species–area theory. The tetrapod record is largely equivocal regarding the theory of periodicity of mass extinctions. Overall, the tetrapod data are suggestive, but by no means conclusive, evidence against periodicity.

Ammonoides lived for about 320 My, from the early Devonian to late Cretaceous. They experienced major extinctions at the end of the Devonian, Permian, Triassic, and Cretaceous, and smaller extinctions primarily in the Paleozoic. Extinctions were preceded by decline in diversity and followed by low diversity and individual abundance. Character novelties were gradually elaborated and diversified as in other taxa. Innovation appears steadier than the rather more spasmodic and irregular extinctions. Earthbound, nonperiodic environmental and paleogeographical changes best explain the evolution and extinction of ammonoids.

Hallam suggested in 1987 that the famous end-Cretaceous (K-T) mass extinctions were not a geologically instantaneous event and were selective in character, with a high proportion of both terrestrial and marine groups surviving with little or no change. Likewise, he has forcefully argued in 1989 the case for sea level change as a dominant causal factor in mass extinction and radiation of Phanerozoic marine invertebrates, on both small and large scales. In many cases, the major extinction factor was apparently not regression, but the spreading of anoxic bottom water associated with the subsequent transgression. The sea level–extinction relation was also formulated ecologically.

An ecological mass extinction model involving mantle–core interactions, at the end of the Paleozoic and Mesozoic, appears plausible. This may involve magnetic field reversal patterns, together with changes in sea level, volcanicity, and climate, causing mass extinctions. Our minds should always be open to alternative scenarios, but we must also entertain a healthy degree of skepticism and beware of simplistic, single-factor models. To what extent the hypothesized large bolide impact was responsible for the K–T extinction, or was merely a contributing or coincidental factor to an already deteriorating environment, is currently being tested and debated by paleontologists. Clearly, the linking of biostratigraphy and chemostratigraphy has great potential for unraveling the evolutionary history of life on Earth.

These features are incompatible with the original L. W. Alvarez hypothesis in 1980 of the K–T extinction being caused by a single asteroid impact that produced a world-embracing dust cloud with devastating environmental consequences. By analysis of physical and chemical evidence from the stratigraphic record, Hallam and colleagues have shown that a modified extraterrestrial model in which stepwise extinctions resulted from encounter with a comet shower is less plausible than one intrinsic to the Earth. The latter involved significant disturbance in the mantle, causing plate tectonics and increased volcanism. Sea level fall and rise, and concomitant increased volcanicity, would also have caused seasonal extremes of temperature on the continents to increase, thus augmenting environmental stress on the dinosaurs. Increased volcanism may cause devastating atmospheric consequences, including acid rain, reduction in the alkalinity and pH of the surface ocean, global atmospheric cooling, and ozone layer depletion, all leading to selective ecological disaster among terrestrial plants and animals.

C. Climatic Causes of Extinction: Patterns and Theory

Climatic diversity is a major determinant of biotic diversity. Climate change might have contributed to both background and mass extinctions. A direct influence for climate on Phanerozoic extinctions has been argued by S. M. Stanley. Slowly changing boundary conditions can cause stepwise or geologically abrupt responses in climate models and result in slow to rapid transition between climate and states. These could cause extinction and/or origination events, that is, negative or positive climatic selection, locally, regionally, or globally. Theoretical and empirical results support the concept of geologically abrupt climate change, that is, those lasting thousands to hundreds of thousands or millions of years. Climate–life transitions often coincide in Earth history. If biotic turnover is indeed significantly affected by climate change, ecosystems may be more sensitive to forcing during early stages of evolution from an ice-free to a glaciated state, as recurred several times in evolutionary history (e.g., Precambrian, Ordovician, Devonian, Eocene–Oligocene, and Pliocene–Pleistocene). Thus, climate instability represents an important mechanism for causing stepwise to rapid environmental change and biotic turnover, in addition to other factors, terrestrial or extraterrestrial.

Long-term climatic change occurred from a "hothouse" to a "coldhouse," or from ice-free Earth in the mid-Cretaceous (100 Ma) to a bipolar glacial state (antarctic and arctic), with periodic glacial expansion into northern midlatitudes in the Cenozoic. These lead to global climatic heterogeneity and seasonality, documented by the oxygen isotope record. Likewise, there has also been significant increase in cooling and aridity during the last 30–40 My, climaxing in the Pleistocene glaciations. This started with the Eocene–Oligocene extinction, the most significant biotic turnover in Cenozoic biota, and continues in our current biota and ecosystems and their catastrophic change and destruction by human activities. Three of the Cenozoic biotic extinctions coincide with climatic change of glaciation and cooling: 31–40 Ma

(Eocene–Oligocene), 10–14 Ma (middle Miocene), and 2.4–3.0 Ma (Pliocene–Pleistocene). Smaller extinctions are also correlated with the O_2 record. Other extinctions are associated with ocean anoxic conditions, and changes in organic carbon, which affected atmospheric CO_2 levels.

Climate model experiments suggest that the long-term trend over the last 100 My may be largely determined by climate changes induced by plate tectonics and mountain formation (orogenies, i.e., earthbound). Similarly, plate tectonics-induced changes in the seasonal cycle may have triggered the Ordovician and Carboniferous glaciations. The climate change in the late Precambrian coincided with the breakup of a super-continent, analogous to the breakup of Pangaea in the Mesozoic. Over the past half-billion years, the face of the Earth has changed markedly. The pieces of Gondwana assembled during the period 540–300 Ma at the South Pole and united to become the supercontinent Pangaea. The latter moved northward across the equator and eventually broke up to form the familiar continents we find today. These developments may have been linked with an increase in atmospheric CO_2, due to enhanced seafloor spreading rate. This may demonstrate a close global link between geological and climatic changes affecting biotic evolution.

Remarkably, two additional climate–extinction events are well correlated in the Phanerozoic: the late Ordovician (440 Ma) and late Devonian (360 Ma), coinciding with glaciations following long intervals of ice-free conditions. In general, increasing glaciations are associated with sea level fall. The biggest of all extinctions in the late Permian (250 Ma) has often been related to changes in sea level, but glaciation and salinity changes associated with extensive evaporite formation may have been substantial, implicating thermohaline instabilities as a contributing mechanism. The late Triassic extinction occurred during a peak in Phanerozoic evaporite formation. Finally, a significant evolutionary event, the expansion of soft-bodied metazoans in the late Precambrian (after 670 Ma), followed the most widespread phase of late Precambrian glaciation.

Even the celebrated K-T extinction may have involved terrestrial causes such as a general fall in sea level, which might have affected seasonality, and the decrease in ocean productivity, independent from a hypothesized impact. Preliminary evidence indicates that ecosystems may be more sensitive to the climatic stress of cooling (e.g., mid-Paleozoic and mid-Cenozoic) during the initial stages of environmental change. Later, more extensive glaciations had a lesser effect on biota, possibly because of resistance and increasing adaptation of organisms to stress through natural selection. Though multiple causes may contribute to extinction events, terrestrial, geological-climatic, and consequently biotic stresses may override extraterrestrial factors in relative importance.

Two complementary causes seem to be responsible for the Upper Pleistocene–Holocene extinction: climate and humans ("prehistoric overkill"). Climate change, correlating with late Pleistocene and Holocene extinctions, occurred on global, regional, and local scales, as is evident by palynological data. Vegetational shifts occurred across the globe from tundra and boreal conifer forests, from about 18,000 years before present (B.P.) to present closed-canopy deciduous forest starting about 10,000 years B.P., following postglacial warming. During this time the fauna changed drastically, primarily by large terrestrial mammal extinctions. Although a global phenomenon, late Pleistocene extinctions were most severe in North America, South America, and Australia, and moderate in northern Eurasia. In Africa, where nearly all of the late pleistocene "megafauna" survived to the present century losses were slight, but become alarming at present, when apes, elephants, and big game are under very severe wild hunting by humans ("human overkill") (Fig. 6). Climate change today in the composition and chemistry of the atmosphere involves devastating changes of greenhouse gases (carbon dioxide, methane, nitrous oxide, and chlorofluorocarbon-11[D], decreased stratospheric ozone concentrations, and increased ultraviolet input. The consequent increases in ultraviolet radiation, alarming global warming, and pollution are likely to affect drastically biodiversity, ecosystems, agriculture, and human health. Climate model simulations confirm a climatic causation (involving increased seasonality) for the biotic turnover of flora and vertebrate fauna, as described in the foregoing.

Climatic extinctions in the Pleistocene–Holocene times involved mammoths, mastodonts, horses, camels, sloths, and peccaries, among others (Fig. 6). The disappearance of large herbivores led to the extinction of large carnivores and scavengers dependent on them. This model of climatic extinction was designated "coevolutionary disequilibrium" by R. W. Graham and E. L. Jr. Lundelius (1984). The change from Pleistocene to Holocene environments, with major reorganization of vegetational communities, was certainly very unfavorable for large mammals. Each species is considered to have responded independently to the climatic and vegetational changes at the end of the Pleistocene, thus destroying that delicate balance and resulting in extinctions, climaxing at our times.

Across the planet the addition of a new predator, *Homo sapiens,* complemented and aggravated climatic change in causing the terminal Pleistocene extinction. Large herbivores (e.g., ground sloths) became extinct, even though their supposed habitat remained. Artefacts were found associated with 5 of 37 genera of large vertebrates that disappeared (mammoths, mastodonts, horses, camels, and giant tortoises; (Fig. 6), and worldwide extinction of megafauna is roughly correlated with the first appearance of humans in North America. On islands, almost all recent extinctions of species derive from human activity. Victims include New Zealand's moas, Madagascar's giant lemur, and many bird species on Hawaii and other tropical Pacific, Atlantic (West Indies), and Indian oceanic islands. Rarity of species is the precursor of extinction. Small environmental effects can sometimes cause large ecological changes that lead to extinctions.

The modern crisis of extinction is approaching the catastrophic proportions of late Cretaceous times. Current estimates of extinction rates underestimate the actual values by a large factor, particularly in the tropics. The human demand for space, hunting, fishing, lumbering, grazing, farming, road building, deforestation, industrialization, use of insecticides, pesticides, and herbicides, effect of introduced species (biological invasions) and habitat destruction all lead to accelerating destruction of ecosystems, biodiversity, endemic species, and genetic diversity. Pollution, fires, insect outbreaks, disease vectors, alien grasses, and species migration increase the causes of mass destruction, potentially climaxing in the elimination of the delicately balanced and fragile tropical rain forest. [*See* GLOBAL ANTHROPOGENIC INFLUENCES.]

Humans are primarily responsible for the extinction of birds and mammals during the last 2000 years, most significantly on oceanic islands, but also in continental North and South America, Africa, and Australia. The African savannah's big game and tropical forest fauna and flora are under increasing pressure. Both biodiversity and genetic diversity are currently being reduced at an alarming rate, causing the extinction of populations, species, ecosystems, and biotas. At present, the rate of extinction is accelerating and it will likely continue to increase because of the human population explosion. This growth directly led to global changes caused by excessive predation, introductions of competitors and pathogens, habitat destruction (primarily the deforestation of the tropical rain forest), and extensive land conversion for agriculture and industrialization. The distinctly human driving forces of change, or anthropogenic stresses, primarily the catastrophic population explosion from the present 5 billion people to possibly 10 billion in the next century, severely threaten the future of both the biosphere and humans. A contrasting cornucopian economic view assumes that greater numbers of people will help us to solve environmental problems by modern techology and economy. Ecological predictions, however, clearly indicate that future human survival depends on stopping our population explosion and conserving the biosphere.

D. Challenges to Mass Extinction: Simplistic Theories

A. Hoffman and others have recently challenged the periodicity and single-cause models, suggesting that many phenomena designated traditionally as mass extinctions are in fact clusters of extinction episodes roughly associated in geological time, within periods of a few million years (see Table I). He and others suggested that different causes led to the latest Ordovician, late Devonian, mid-Cretaceous, latest Cretaceous, and late Eocene–Oligocene extinctions. Thus, different environmental causes, and coincidental combinations thereof, might lead to extinctions and so they can hardly be considered as individual events. Even the largest of all global extinctions, the Permo-Triassic, attributed by Hoffman to a single factor and characterized as a noncatastrophic extinction (i.e., due to atmospheric and ocean depletion and enrichment of O_2, respectively, causing marine nutrition deficiency), may not be an exception; it involved climatic, tectonic and geochemical change. As rightly emphasized by D. H. Erwin, few complex events stem from a single cause. More common is a complex web of· causality, a web that can be difficult to untangle, and the end-Permian extinction is no exception. The most plausible explanation according to Erwin would appear to be a three-phase model combining elements of several mechanisms. The extinction began with the loss of habitat area as the regression dried out many marine basins, converting the two-dimensional coastlines of the mid-Permian to more linear coasts. The increased exposure of Pangaea as the regression progressed exacerbated climatic instability. This instability, coupled with the effects of continuing volcanic eruptions and an increase in atmospheric carbon dioxide (with some global warming), led to increasing environmental degradation and ecological collapse. Furthermore, Hoffman believes that much of the evidence for the 26-My period mass extinction derives from arbitrary, ambiguous, and imprecise decisions concerning geochronometry, that is, he questions the absolute dating of stratigraphical boundaries, the culling of the data base, and the definition of mass extinction as opposed to background extinction. In his view, rapid but apparently staggered appearance of major new taxa, such as at the Cambrian–Ordovician boundary, elevated taxonomic pseudoextinctions. Thus, the apparent periodicity of mass extinctions may result from stochastic processes, and periodicity theory may overdramatize the patterns that are truly obtainable from the fossil record.

Carbon isotope ratios in marine carbonate rocks have been shown to shift at some of the time boundaries associated with extinction events, for example, the K-T and Ordovician–Silurian. The Permian–Triassic boundary, the greatest extinction event of the Phanerozoic (Fig. 3A), is also marked by a large carbon ratio depletion. New carbon isotope results from sections in the southern Alps show that this depletion did not actually represent a single event, but was a complex change that spanned perhaps a million years during the late Permian and early Triassic. These results suggest that the Permian–Triassic extinction may have been in part gradual and in part "stepwise," but was not in any case a single catastrophic event. Both the carbon isotope shifts and the chemical events (including an iridium anomaly) may have causes related to a major regression of the sea.

According to Jablonski, a substantial and rapid fall of sea level seems to provide the best correlation with marine invertebrate mass extinction episodes throughout the Phanerozoic, presumably due to reduction in the neritic habitat area. Climatic deterioration has been claimed as the proximal cause in a number of cases, for example, in the global Eocene–Oligocene and regional Pliocene–Pleistocene extinctions. Clearly, associations between climatic deterioration and extinctions do not always reveal cause–effect relationships. However, if many such associations occur across phylogeny, the causality may become more robust and plausible. One of the major unresolved problems and future challenges in extinction theory is the biological selectivity of victims and survivors.

XI. THE EVOLUTIONARY PROCESS

The evolutionary process is based on several major driving forces and their interactions, including mutation (in the broad sense), recombination, migration, natural selection, and stochastic processes. Unfortunately, we know very little even today about their relative importance in the evolution of natural populations and the role of evolutionary constraints. Clearly, however, genetic diversity in nature is the basis of the evolutionary process through its twin processes of speciation and adaptation (i.e., in successful originations), but it also may substantially affect the probability of extinction. The evidence derived from extant natural populations of plants and animals, at the local, regional, and global scales, is best explained by ecological heterogeneity and change in space and time. These involve physical (climatic, geological, hydrological) and biotic (pathogenic, predatory, and competitive factors) changes and stresses. It is plausible to assume that the interaction of abiotic and biotic ecological heterogeneity and stress are also the major determinants of background and mass extinctions, according to the extent and severity of the stress.

Natural selection, in its various forms and in combination with stochastic factors, appears to be a major differentiating and orienting force of evolutionary change at the molecular (genotypic) and organismal (phenotypic) levels, locally, regionally, and globally. This holds primarily at the individual level, but also at higher, group selection levels (species and higher taxa) in the taxonomic hierarchy all the way up to whole biota. Likewise, it presumably assumes a major role in originations and extinctions across individuals, species, ecosystems, and biota. Critical testing and analyses, in nature and the laboratory, are still greatly needed to substantiate this conclusion. Theory should attempt to cope with the accumulating evidence at the single and multilocus genetic structure of protein and DNA levels and their interface with developmental biology and organismal evolution, as well as in origination and extinction. The explosive accumulation of genetic maps (locus maps of complex genomes) will substantially highlight evolutionary patterns and processes. Clearly, evolutionary success, in overcoming environmental stresses and even global changes, depends at least partly on the genotypic and phenotypic (morphological, physiological, and behavioral) resources available during ecological crises at all scales, local, regional, and global.

Microevolution appears to extend smoothly into macroevolution. Alternative views have been voiced by leading paleobiologists, including Gould, Eldredge, Stanley, and Vrba, who argue for a hierarchical model in which macroevolution is decoupled from microevolution. The aforementioned basic evolutionary forces appear to operate at all levels. However, selection at higher taxonomic and ecosystematic levels may be important in evolution. The instability of arctic and subarctic species has been advocated by C. S. Elton. In contrast to intuition, as emphasized by R. May, stability becomes less likely in complex ecosystems like the tropical rain forest. Complex ecosystems are composed of numerous food chains involving primary producers (diversifying according to light, temperature, water, and nutrients) and secondary consumers (herbivores, predators, and parasites). Hence, these complex ecosystems are vulnerable to slight, domino effect changes, because they may be at the transition phase between order and chaos. This was emphasized by S. A. Kauffman in his 1993 discussion on the origins of order and the roles of self organization and selection in evolution.

XII. CONCLUSIONS

The universe unfolds a large-scale evolution of order emerging from chaos and primordial uniformity. The expanding and inflating universe, following the Big Bang 15 Ba, evolved in a hierarchical cosmic order of galaxies, quasars, stars, planets, and life. The emergence of life and later consciousness from inanimate matter are wonders equal in significance to the cosmic birth. Organisms evolved from star stuff and the atoms comprising their bodies were born in primordial nuclear fires at the birth of time or were derived from dying suns. The solar system was born about 4.6 Ba. Life's origin is still a scientific mystery, although its evolution from inanimate matter is highly plausible and decipherable. Overwhelming geochemical, biochemical, biological, and fossil evidence reveals that life originated about 3.8 or possibly 4 Ba. Based on replication, metabolism, and transmission of information over generations, life evolved relentlessly over time at varying tempos.

Overall, the evolution of life displays an increase in biodiversity and complexity of organisms, ecosystems, and biota. Although originations proceeded to increase biodiversity, background regional and global mass extinctions recurred through the Earth's history, always followed by innovation and diversification. Originations and extinctions, that is, the stuff of evolution, are associated primarily with terrestrial abiotic (geological, climatic, environmental) and biotic (parasites, diseases, predators, competitors) factors. Extraterrestrial impacts (asteroids, comets, bolides) might have affected the evolution of life, but their relative importance may be secondary, and at any rate are currently hotly debated. The disentangling and assessing of the relative importance of terrestrial and extraterrestrial factors may prove hard to resolve, but the paramount importance of climate–life interactions is indisputable. Likewise, though several massive extinctions clearly recurred in the Phanerozoic, the predominant background and mass extinction of species, ecosystems, and biotas, analogous to individual mortality, has operated through evolution, opening ever-new ecological niches for life's evolutionary experimentation, innovation, and emergence.

The dogged evolution of life from simple lifeless objects to complex living organisms is based on a combination of stochastic and deterministic factors. However, from molecular to organismal evolution, and from the original few life-forms that evolved from inanimate matter to our current all-time high biodiversity and complexity, evolution has been primarily dominated by natural selection despite its strong interaction with stochastic processes.

Cosmic evolution, on physical, biological, and cultural (human) levels, appears to display a trajectory from simplicity to complexity. This trajectory toward increasing complexity depends on self-organizing, materialistic, physical, chemical, and biological laws. Cosmic evolution seems to have been continuous ever since the Big Bang, about 15 Ba to the present. The universe may either proceed to expand relentlessly or crush and recycle, depending on cosmic conditions. As long as the sun provides planet Earth with free energy, and unless extraterrestrial, astronomical, terrestrial, or human forces destroy the biosphere, life's evolution will keep ticking without any obvious target except that of better survivorship against all odds and expansion by adaptive radiations into all open ecological niches.

Evolution is based on a unified chemical language and genetic code, generating an infinite biodiversity of individuals, populations, and species. Thus, unity and diversity are the cornerstones of ever-increasing biological evolution. Only humans can instill meaning in this ongoing game of survival. Only humans can understand the past and present and attempt to responsibly plan and control the foreseeable future.

The origin of the mind from living matter, together with the origin of life from lifeless matter and matter's origin from radiation after the Big Bang explosion, are the three highlights of universal history. The mind reveals evolutionary insights, wonders, and mysteries and manifests the evolution of biodiversity within a unified cosmic evolution, proceeding from chaos to increasing order and diversity. Mind's cosmic awareness closes the circle of thinking matter consciously and scientifically examining and reflecting on its own origins and past, present, and future evolution. Though cosmic evolution may seem comprehensible but meaningless, the relentless search for origins and evolutionary patterns and processes, as well as for beauty and harmony, is the mind's meaningful creation. A human, as suggested by David Bohm, is "a microcosm of the universe." Understanding and seeking truth, wondering and enjoying the beauty and mysteries of the macrocosmos and microcosmos, may be the only meaningful creations generated by our mind.

As so aptly said by Albert Einstein, "The most beautiful experience we can have is the mysterious. It is the fundamental emotion which stands at the cradle of true art and true science." Moreover, as cogently written by physicist Steven Weinberg, "The effort to understand the universe is one of the very few things that lifts human life a little above the level of farce, and gives it some of the grace of tragedy." Unquestionably, a unique and deep grace, wonder, and mystery inhere to a fragment of the universe that aspires to comprehend the whole. In a sense, the universe came to know itself. Cosmic evolution indeed displays the most dramatic holistic saga ever: the lawful, stepwise, and gradual evolution of energy and simple matter into the immense complexity of life, and the slowly yet ever-increasing level of biodiversity since 4 Ba, with new forms replacing old ones (background extinction), occasionally set back by mass extinctions. The climax of that Primeval Fireball is the evolution of thinking-matter, brain-mind, capable of reflecting on the origin and evolution of life and our place in the universe.

Glossary

Hubble's law The reddening of light from a star that is moving away from us due to the Doppler effect, which states that velocity is proportional to distance, i.e., the change in the observed frequency of a wave due to relative motion of source and observer. This leads to Hubble's law, which states that velocity is proportional to distance.

Infinite density The density of the universe at the Big Bang when density and the curvature of space-time would have been infinite. The Big Bang is the singularity at the beginning of the universe, that is the point in space-time at which the space time curvature becomes infinite.

Irregularity The early state of the universe, which shows no definite order or shape.

Nucleosynthesis, or nucleogenesis The theoretical process(es) by which atomic nuclei could be created from possible fundamental dense plasma.

Symbiosis An association of dissimilar organisms to their mutual advantage.

Bibliography

Chaloner, W. G., and Hallam, A. (eds.). (1989). Evolution and extinction. *Philos. Trans. Roy. Soc. London Ser. B* **325**, 241–488.

Darwin, C. (1859, 1872). "On the Origin of Species by Means of Natural Selection," 1st and 6th eds. London: Murray.

Davies, P. (1988). "The Cosmic Blueprint." New York: Simon & Schuster.

Donovan, S. K. (1989). "Mass Extinctions: Processes and Evidence." London: Belhaven Press.

Fautin, D. G., Futuyma, D. J., and James, C. (1992). Special section on global environmental change. "Annual Review of Ecology and Systematics." Palo Alto, Calif.: Annual Reviews, Inc. pp. 1–235.

Fedoroff, N., and Botstein, D. (1992). "The Dynamic Genome." New York: Cold Spring Harbor Laboratory Press.

Forey, P. L. (1981). "The Evolving Biosphere." Cambridge: Cambridge University Press.

Gesteland, R. F., and Atkins, J. F. (eds.) (1993). "The RNA World." New York: Cold Spring Harbor Laboratory Press.

Gillespie, J. H. (1991). "The Causes of Molecular Evolution." Oxford: Oxford University Press.

Grant, V. (1991). "The Evolutionary Process," 2nd ed. New York: Columbia University Press.

Hawking, S. H. (1988). "A Brief History of Time. From the Big Bang to Black Holes." New York: Bantam Books.

Hoffman, A. (1989). "Arguments on Evolution." New York: Oxford University Press.

Hoffman, A. A., and Parsons, P. A. (1991). "Evolutionary Genetics and Environmental Stress." Oxford: Oxford University Press.

Kauffman, S. A. (1993). "The origins of order: Self organization and selection in evolution." Oxford: Oxford University Press.

Kirzhner, V. M., Korol, A. B., Ronin, I., and Nevo, E. (1994). Cyclical behavior of genotype frequencies in a two-locus population under fluctuating haploid selection. *Proc. Natl. Acad. Sci. U.S.A.* **91**, 11432–11436.

Labandeira, C. C., and Sepkoski, J. J., Jr. (1993). Insect diversity in the fossil record. *Science* **261**, 310–315.

Levinton, J. (1988). "Genetics, Paleontology, and Macroevolution." New York: Cambridge University Press.

Mallove, E. F. (1987). "The Quickening Universe." New York: St. Martin's Press.

Margulis, L. (1982). "Early Life." Boston: Science Books International.

Margulis, L. (1993). "Symbiosis in Cell Evolution." New York: W. H. Freeman & Company.

Mayr, E. (1976). "Evolution and the Diversity of Life." Cambridge: Harvard University Press.

Mayr, E. (1988). "Toward a New Philosophy of Biology: Observations of an Evolutionist." Cambridge, Mass.: Harvard University Press.

Nevo, E. (1988). Genetic diversity in nature: Patterns and theory. *Evol. Biol.* **23**, 217–246.

Nevo, E. (1991). Evolutionary theory and processes of active speciation and adaptive radiation in subterranean mole rats, *Spalax ehrenbergi* superspecies, in Israel. *Evol. Biol.* **25**, 1–125.

Nevo, E. (1993). Adaptive speciation at the molecular and organismal levels and its bearing on Amazonian biodiversity. *Evol. Biol.* **7**, 207–249.

Nitecki, M. H. (ed.) (1984). "Extinctions." Chicago: The University of Chicago Press.

Nitecki, M. (ed.) (1990). "Evolutionary innovations." Chicago: The University of Chicago Press.

Prothero, D. R., and Berggren, W. A. (1992). "Eocene-Oligocene climatic and biotic evolution." New Jersey: Princeton University Press.

Raup, D. M., and Jablonski, D., eds. (1986). "Patterns and Processes in the History of Life." Berlin: Springer-Verlag.

Simpson, G. G. (1944). "Tempo and Mode in Evolution." New York: Columbia University Press.

Simpson, G. G. (1953). "The Major Features of Evolution." New York: Columbia University Press.

Stanley, S. M. (1986). "Earth and Life through Time." New York: W. H. Freeman & Company.

Weinberg, S. (1977). "The First Three Minutes: A Modern View of the Origin of the Universe." New York: Basic Books.

Wilson, E. O., and Peter, F. M., eds. (1988). "Biodiversity." Washington, D.C.: National Academy Press.

Evolutionary History of Biodiversity

Philip W. Signor
University of California

The variety of organisms extant on earth has fluctuated dramatically throughout geological history. These changes in diversity reflect the interactions of physical and biological processes over the vast expanse of geological time. The biosphere was dominated by bacteria and, later, eukaryotic protists for the first 3 billion years of life's history. The earth's biota was devoid of Metazoa or Metaphyta until approximately 600 million years ago, when the first animals appeared. The numbers of marine animal families, genera, and species increased rapidly with the first appearance of animals, and then fluctuated for the next 200 million years around levels of 20 to 50% of modern diversity. Following a mass extinction at the end of the Paleozoic era, marine animal diversity increased rapidly to the peak levels observed today. Terrestrial animals and vascular plants appeared much later than marine animals, and their subsequent diversification roughly parallels the diversity of marine Metazoa. Superimposed upon this gross pattern of diversification are occasional mass extinctions, marked by the sharp reduction of marine and terrestrial diversity.

I. INTRODUCTION

Life on earth is monophyletic; the known forms of life share a common ancestor. At some point in the earth's ancient history there existed a single species from which the multitudes of today's species are descended. The trajectory of that diversity, from a single species in the early history of the earth to the profusion of species present in modern habitats, has been the subject of considerable analysis. This article summarizes the history of organic diversity through geological time as it is currently understood. [See BIODIVERSITY.]

Paleontologists have long wrestled with the history of biodiversity. In the last century, progressionists clashed with uniformitarians over the possibility of trends in biodiversity. The diversity controversy continued episodically into this century and reemerged as a vigorous debate that raged through the 1970s over the history of marine faunal diversity in geological time. One side to the debate argued that strong increases in diversity were evident in the fossil record despite the well-known biases of the fossil record. The opposite view held that diversity has been at an equilibrium through much of the past 550 million years (the Phanerozoic, or age of visible life); the apparent fluctuations of diversity reflected only sampling effects. Despite the acknowledged presence of powerful biases in the fossil record, a majority of paleontologists now agree on the general outlines of diversity, as determined from the fossil record. Supporters of the faunal equilibrium have abandoned their former position and, for the most part, accept that the fossil record of marine organisms reflects, more or less accurately, the history of diversity within the marine realm.

The only source of information on the history of diversity in evolutionary time (beyond the 10,000 years of human history) is the fossil record. Unfortunately, the quality of that record varies from habitat to habitat and among taxonomic groups. The resolution of the record also varies according to the taxonomic level of the analysis. The suite of problems associated with inferring biological pattern from the fossil record are typically lumped together under the rubric "sampling problems." These sampling problems have injected considerable uncertainty into attempts to reconstruct the history of diversity. The tendency of sampling effects to mask biological patterns has led to intensive analysis of bias in the fossil record.

This article employs the geological time scale for recording the history of diversity. The time of visible life, the Phanerozoic, is divided into three eras: the Paleozoic (550–245 Ma), Mesozoic (245–65 Ma), and Cenozoic (65 Ma to today). (Ma is a standard abbreviation indicating age in millions of years before the present.) Each era is composed of periods. The Paleozoic includes 6 periods: Cambrian, Ordovician, Silurian, Devonian, Carboniferous, and Permian. The Paleozoic is preceded by the Ediacaran (or Vendian) period, a short interval characterized by unusual, soft-bodied faunas. The Mesozoic is composed of the Triassic, Jurassic, and Cretaceous. Finally, the Cenozoic is composed of the Tertiary and the Quaternary.

One final introductory note: What is this thing called the fossil record? There is no tangible "fossil record," although most paleontologists write and speak as though it not only exists but they can lay

their hands on it at a moment's notice. The fossil record is actually used by paleontologists in two very different senses. One common usage is the sum total of current human knowledge about fossils. The second usage is broader and encompasses not only current knowledge but the additional information to be gained from fossils, yet undiscovered, preserved in existing sedimentary rock. This latter definition is adopted in this article.

II. SAMPLING THE FOSSIL RECORD

The vast majority of extinct species were never preserved as fossils or, if they were preserved, were subsequently lost through erosion or metamorphosis of sedimentary rock. It is difficult to estimate what fraction of all species is forever lost to human knowledge. Estimates of the proportion of ancient marine species preserved and subsequently described from the fossil record vary between 0.1 to 0.001 of the total number of extinct marine skeletogenous (skeleton-forming) species. The proportion of terrestrial species or species lacking mineralized tissues that have been recovered as fossils must be much lower.

Marine organisms have a greater likelihood of being preserved as fossils because they live in an environment where the net accumulation of sediment is more frequent, hence the burial of animals or animal remains is more likely. Terrestrial environments are more likely to be dominated by erosion or nondeposition. Freshly dead organisms that remain exposed to the elements are subject to physical and biological degradation. Rapid burial can increase the odds of preservation by halting or ameliorating the damage.

However, the ocean floors are not a benign environment for fossils. The sediments of the deep sea are constantly being destroyed by subduction of oceanic crust. For this reason, the useful marine fossil record accumulates only on the continents during times of relatively high sea level. When sea level is low, the only continental accumulation of marine sediments is on the extreme margins and slopes of the continents, where the sediments cannot be studied unless they are transported or uplifted. Thus, periods of low sea level will usually generate gaps in the marine fossil record.

The survivorship of sedimentary rock and the fossils preserved in them is a function of geological time. Much more sedimentary rock, together with the fossils entombed therein, survives from younger times than from older times (Fig. 1). In general, longer periods are represented by larger volumes of sedimentary rock, and younger rocks tend to overlay older rocks. The exposed area and volume of sedimentary rock of the eleven periods of the Phanerozoic are strongly correlated with the

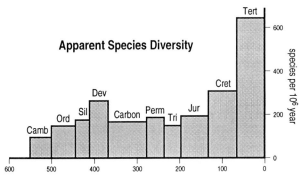

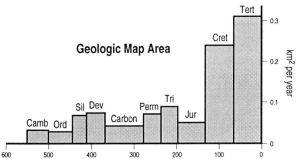

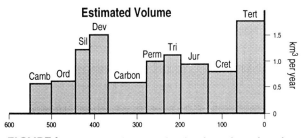

FIGURE I Species, rock area, and rock volume through geological time. The numbers of described species as of 1970 (apparent species diversity) are strongly correlated with rock area and rock volume. The geological map area of sedimentary rock differs from the estimated volume because younger rocks usually overlie the older sedimentary rock. All three histograms are normalized to the length of the respective geological periods.

numbers of described species from each period, reflecting at least in part the control of sedimentary rock on diversity. Interpretations of the fossil record must allow for these biases.

Analysis of young (Pleistocene) fossil deposits indicates that the majority of the skeletogenous marine species are incorporated into the fossil record. The very low frequencies of ultimate recovery of species from the fossil record therefore probably reflect the operation of metamorphism, diagenesis, and erosion that destroys sedimentary rock or the fossils entombed within the rock. Alternatively, an optimist might hope that the fossils are there, awaiting discovery. But this seems unlikely after two centuries of paleontological endeavor.

Possession of a skeleton, teeth, or other mineralized tissues greatly enhances the chances of preservation. Soft tissues are vulnerable to destruction by predators, scavengers, bacteria, and other biological agents. Consequently, the likelihood of soft tissues being preserved is very low. Even mineralized tissues are most often destroyed before they can be preserved. The biases created by the varying potential for preservation of different types of tissue are well illustrated by the fossil record of terrestrial vertebrates. Most fossil vertebrates are known only from teeth or from teeth and small portions of the skull and jaw. Not surprisingly, teeth are the hardest, most resistant parts of the vertebrate body. Elements of the postcranial skeleton are uncommon and complete skeletons are quite rare.

Nevertheless, there are a number of spectacular fossil deposits that preserve soft-bodied fossils in great abundance, such as the Cambrian Burgess Shale, the Jurassic Solnhofen Limestone, and the Eocene Green River Shale. These unusual deposits, or *Lagerstätten,* provide insights into the nature and diversity of ancient communities but are too infrequent to allow a comprehensive reconstruction of the history of soft-bodied biotas. Therefore, the most detailed efforts to reconstruct diversity in geological time have focused on marine skeletogenous organisms.

Far more species are known from the modern world than from the fossil record. Nevertheless, the large numbers of described fossil species make direct counts of species difficult and unwieldy. Furthermore, tabulations of species are especially susceptible to the sampling biases outlined earlier. To circumvent these difficulties, paleontologists have usually employed higher taxa (genera, families, or orders) to infer diversity at the species level, for there are fewer higher taxa, making them easier to count. More importantly, higher taxa are easier to "discover" because one need only recover a single constituent species from the fossil record to establish the presence of the genus, family, or order. This technique has provoked debates over the utility of higher taxa as surrogates for species diversity. Clearly, higher taxa are not created on the basis of diversity; they reflect evolutionary relationship. But the presence of 100 genera does suggest fewer species than, for example, 500 genera. The fidelity of genera, families, and orders as surrogates for species diversity remains to be demonstrated.

III. TAXONOMIC DIVERSITY

A. Marine Realm

In a series of studies published over two decades, J. J. Sepkoski, Jr., has explored the history of taxonomic diversity in geological time. In one of his earliest papers, he demonstrated that the number of metazoan orders increased rapidly following the appearance of Metazoa in the late Proterozoic. The number of orders reached a plateau around 130 in the Ordovician, and subsequently fluctuated near that value for the next 450 million years (Fig. 2).

In subsequent papers, Sepkoski reviewed the history of metazoan diversity at the familial level, where a very different pattern emerged. Family diversity increased very quickly during the Cambrian and Ordovician, where upon it reached a plateau. This plateau was maintained throughout the Paleozoic. At the end of the Paleozoic, the number of families was severely reduced but eventually rebounded. The rediversification of families continued, with minor setbacks, until the Recent, when the number of families stood at approximately twice the number extant in the Paleozoic (Fig. 2).

Sepkoski documented a similar pattern in the numbers of marine genera over geological time.

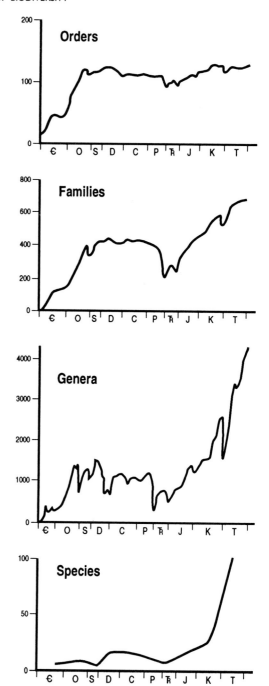

FIGURE 2 Taxonomic diversity of marine animals through geological time. The numbers of orders, families, and genera represent actual tabulations of taxa (based on work by J. J. Sepkoski, Jr.). The curve for the numbers of species is an estimate.

This Herculean task involved the compilation of the geological ranges of more than 35,000 genera. The data suggest a rapid initial diversification following the appearance of Metazoa, reaching a Paleozoic diversity plateau in the Ordovician. The number of genera declined sharply following the Paleozoic, but rediversified nearly continuously for the remainder of the Phanerozoic, reaching a modern level between three and four times higher than the Paleozoic plateau (Fig. 2).

Direct counts of species diversity (Fig. 1) suggest a pattern that is similar to familial and generic diversity through time. But it is difficult to interpret direct counts because of the pervasive influence of sampling effects. Efforts to estimate the numbers of species by modeling the influence of sampling effects have produced patterns similar to Sepkoski's tabulation of genera (Fig. 2). [*See* SPECIES DIVERSITY.]

The variations in pattern between the different levels of the taxonomic hierarchy have prompted considerable discussion about which level best represents genuine biological phenomena. If the number of extant species is the desired variable, it could be argued that the number of genera is most likely to reflect changes in underlying species richness. Alternatively, one might argue that species and genera are relatively susceptible to sampling effects, and the numbers of families provide a proper balance between biological diversity and sampling biases. Interestingly, the vascular plants also show a divergence between different levels of the taxonomic hierarchy, but a divergence that is dissimilar to that observed among marine animals. In the former case, because the number of species were actually tabulated, it can be determined that the numbers of species are more closely represented by genera than by families.

In any event, the tabulations of marine animal taxa are dominated by skeletogenous species, and a further assumption is needed to extrapolate these results to global marine biodiversity. It must be accepted that the proportion of skeletogenous species in marine communities has been relatively constant through geological time. This assumption is certainly false because the earliest animal faunas were almost exclusively soft-bodied (as with the Vendian, or Ediacaran, faunas). In modern marine communities, skeletogenous species constitute on average one-third to one-half of the total number of animal species. Evidence from the Cambrian

Burgess Shale suggests that early faunas included a smaller proportion 'of skeletogenous species. Whether this reflects unusual conditions in Burgess communities or temporal trends in the proportion of skeletogenous species is not known.

B. Terrestrial Animals

The history of terrestrial vertebrate diversity is reminiscent of marine diversity through geological time. Terrestrial vertebrates first appeared much later than marine animals, in the Late Silurian epoch, and diversified rapidly, reaching a plateau of around 25 orders. Following the extinction of dinosaurs at the end of the Cretaceous, the number of orders increased more than threefold to the levels observed today. This tremendous increase is entirely a function of the simultaneous and prodigious radiations of birds and mammals on separate continents; reptile diversity actually declined during this period and amphibian diversity increased only slightly (Fig. 3).

Recent research has clarified the diversification of insects, now the most diverse class of animals on earth. The fossil record of insects is surprisingly good and reflects a sustained diversification extending over much of insect history (Fig. 3). The number of insect families increased substantially through the Mesozoic and Cenozoic in a pattern

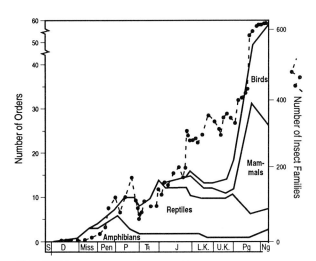

FIGURE 3 Taxonomic diversity of terrestrial animals through geological time. The curves for vertebrates represent the numbers of orders. The curve for insects represents families.

reminiscent of families of marine animals. Curiously, this pattern shows no obvious linkage between the diversification of insects and angiosperm plants, their frequent hosts. Previous workers had suspected a strong coevolutionary relationship between the two groups and had suggested that the diversification of angiosperms had triggered or amplified the radiation of insects. Instead, the two radiations were largely independent.

C. Vascular Plants

Tabulations of the global taxonomic diversity of vascular plants are not available, but a number of analyses of floras from the Northern Hemisphere have been completed. These analyses indicate that the history of vascular plant evolution reflects two phases of diversification. The first phase followed immediately after the appearance of vascular plants in the fossil record. In the second diversification, the numbers of vascular plant taxa increased substantially in the latest Mesozoic and Cenozoic (Fig. 4). There was a twofold increase in the number of families and a sixfold increase in the numbers of species. The increase in overall vascular plant diversity was entirely a reflection of the diversification of angiosperms.

To summarize, it is apparent that the diversity of animal and vascular plant species has increased substantially through geological time, although the precise magnitude of that increase is uncertain. Marine metazoans, terrestrial vertebrates and insects, and vascular plants all show large increases in taxonomic diversity in the late Mesozoic and Cenozoic. These increases are not synchronous, however, and could result from different processes.

IV. ECOLOGICAL COMPLEXITY

The changing numbers of species in global tabulations undoubtedly reflect underlying trends in the numbers of species within ancient communities. But changing diversity within individual communities or habitats might be swamped by variations induced by plate tectonics and other factors. Direct analysis of changes in alpha (within-habitat) and beta (between-habitat) diversity is necessary to dis-

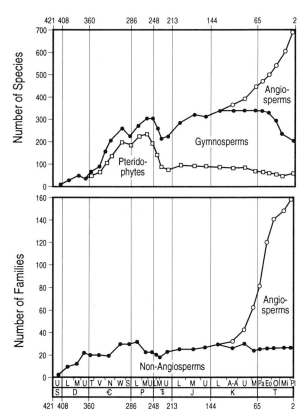

FIGURE 4 Taxonomic diversity of vascular plants. The curves indicate the tabulated diversity of vascular plants in the Northern Hemisphere.

cern trends in ecological complexity over geological time.

A. Within-Habitat Diversity

The number of species present within marine benthic communities has apparently increased, in a stepwise fashion, through time. Nearshore, physically stressed communities show little change in alpha diversity over time (Fig. 5), but nearshore variable and offshore communities apparently doubled their numbers of constituent species in the Cenozoic. This increase in the accommodation of species was not accomplished by finer partitioning of resources, but by adding new guilds, or trophic roles, to existing communities.

Even more than the tabulation of global species richness, which did include the occasional soft-bodied fossil, this analysis is based exclusively on marine skeletogenous benthic metazoans. Consequently, this analysis is predicated on the assump-

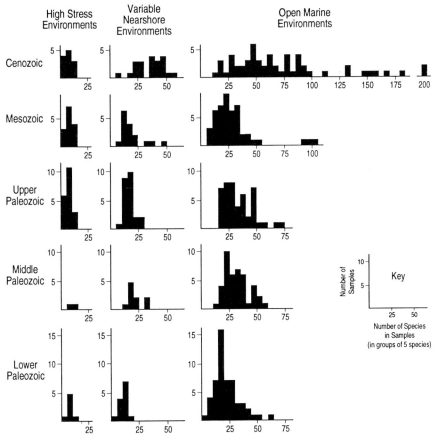

FIGURE 5 Within-habitat species richness of marine invertebrates in geological time. Variable nearshore environments and open marine environments show substantial increases in the Cenozoic (based on work by R. Bambach).

tion that skeletogenous animals have remained a constant proportion of the species in marine benthic communities through geological time. The large proportion of molluscs, especially gastropods and bivalves, in Cenozoic marine communities undoubtedly reflects the extraordinary radiations of those two invertebrate classes. It is not known if these organisms were replacing soft-bodied species or were creating entirely new ecological roles within communities. Nevertheless, the composition of faunas preserved in the Cambrian Burgess Shale, which includes abundant fossils of soft-bodied organisms, suggests that there has been an increase over geological time in the proportion of skeletogenous organisms in benthic communities. That increase, however, was small in comparison to the large increase in within-habitat species diversity documented in Fig. 5.

A second and larger potential source of bias might be found in the way fossil communities are sampled. Older communities are invariably found in lithified sediment. Fossils occurring within rock can be sampled only by tedious methods of breaking, dissolving, sectioning, or otherwise penetrating the rock. Younger faunas are found with some frequency in unconsolidated sediment, which can be easily disaggregated (sometimes even sieved) and the fossils removed. It might be coincidental, but the increase in within-habitat diversity occurs at about the time when faunas preserved in unconsolidated sediment become common.

B. Between-Habitat Diversity

More recent work indicates that communities in adjacent marine habitats have become more dissimilar through geological time. Early in the Paleozoic, species had relatively broad ecological ranges. With the passing of time, the ecological range of species was increasingly limited and the composition of adjacent communities became more dissimilar. This pattern suggests increasing ecological specialization through time. Such specialization increases the numbers of species present within biogeographic provinces.

V. CONTROLS ON DIVERSIFICATION

The available evidence indicates that the global diversity of marine animal species was maintained at a level that was less than half of modern diversity until the end of the Paleozoic, followed by a steady increase in diversity to modern levels. The remaining question is the process (or processes) responsible for controlling global diversity of marine animals. Certainly, within-habitat (alpha) and between-habitat (beta) diversity contribute to global diversity. But most authors attribute the primary role in controlling global diversity to plate tectonics. At times in the geological past when the cratons were gathered together into a single landmass, there were relatively few biogeographic provinces. In contrast, when the cratons were widely separated, there were a large number of provinces, each mostly composed of species unique to that province. Marine species diversity, therefore, tracks the numbers of provinces, which in turn are controlled by plate tectonics.

The diversity of terrestrial vertebrates is less closely linked to plate tectonics. Certainly, the presence of unique faunas, such as the Cenozoic marsupial faunas of Australia and South America, directly reflects geographic isolation resulting from plate tectonics. Yet the total numbers of taxa are not closely tied to the development of separate biogeographic provinces. There is no simple relationship between terrestrial vertebrate taxonomic richness and plate tectonics.

As noted previously, the diversification of insects was long thought to parallel and reflect the relatively recent radiation of angiosperms. However, the most recent data indicate that the radiation of

insects preceded the diversification of angiosperms. Also, it appears that the various trophic roles of the insects became established before the evolution of angiosperms. This trophic diversity probably fueled the radiation of the insects.

The processes controlling the diversification of vascular plants are, not surprisingly, a contentious subject. Biogeography apparently plays a much smaller role in plant diversification than for marine animals. The breakup of Pangea early in the Mesozoic led to no noteworthy increase in diversity. Instead, the size of landmasses and habitat diversity, especially the area of the highlands, appear to be the major controls on plant diversity.

VI. MASS EXTINCTIONS

Mass extinctions are unique phenomena of the fossil record: the geologically sudden elimination of large proportions of the earth's biota. Compared to the great extinctions of the past, extinctions caused by human activity over the past 100,000 years seem puny by comparison. [*See* MASS EXTINCTION, BIOTIC AND ABIOTIC.]

A. Diversity Crises of the Past

There are five major mass extinctions in the history of life: the Ashgillian (end of the Ordovician), the Frasnian-Famennian (Late Devonian), the terminal Permian (Permo-Traissic), the Norian, and the Maastrichtian mass extinctions (Fig. 6). Other,

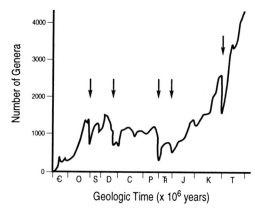

FIGURE 6 Five major mass extinctions substantially reduced species diversity on earth. Recovery from mass extinction can be prolonged, even in terms of geological time.

lesser extinctions occur with greater frequency. To date, no obvious general mechanism for mass extinction has been identified. Considerable evidence has accumulated to suggest that an asteroid impact was responsible for the Cretaceous extinction event, but evidence for a similar causation of earlier extinctions is less convincing. There is limited evidence for an impact in conjunction with the Frasnian-Famennian extinction, but little or no evidence associated with the other extinctions.

B. Lessons from Mass Extinctions

After each mass extinction, diversity has rebounded and the earth's biota has been replenished. However, the recovery of diversity is a process that takes millions of years. For example, it required nearly 90 million years for the Mesozoic fauna to rebound to the levels of Paleozoic diversity prior to the great Permian mass extinction. Of course, that rebound was hampered by the intervening Triassic (Norian) extinction, nevertheless, the rebound clearly required an extended period of time. The obvious conclusion from this observation is that any losses in diversity cannot be recovered within time intervals subject to human attention. Furthermore, any later gains in diversity will necessarily be new life-forms; extinct forms are lost to the world forever. For these reasons, decisions to kill off species or even to risk extinction must be made with the greatest caution.

VII. SUMMARY

Taxonomic diversity has been a dynamic variable over geological time, fluctuating in response to plate tectonics, mass extinction, and evolution within the biosphere. Other physical and biological processes undoubtedly contributed, but it is difficult to demonstrate cause and effect in ancient and complex biological systems. Marine animals, terrestrial vertebrates, insects, and vascular plants all show substantial increases in diversity in the late Mesozoic and Cenozoic, although the increases were not completely synchronous and might have resulted from different processes. Modern faunas and floras encompass the largest numbers of taxa in the history of the biosphere.

Glossary

Apparent diversity Numbers of species, genera, and families that are actually collected and described from the fossil record. Apparent diversity is usually far lower than actual or true diversity.

Geological time scale Temporal framework of stages, epochs, periods, eras, and eons employed by geologists to represent the vast expanse of geological time.

Lagerstätten Noun borrowed from German that translates roughly as a mother lode. The term is applied to unusual fossil occurrences, especially those characterized by exceptional preservation of mineralized tissues or soft tissues.

Mass extinction Substantial reduction in global taxonomic diversity occurring in a geologically brief interval.

Phanerozoic Time of visible life, conventionally employed to include the Paleozoic, Mesozoic, and Cenozoic eras. There has been some debate over inclusion of the Ediacaran or Vendian period in the Phanerozoic, yet most geologists prefer to mark the beginning of the Phanerozoic at the beginning of the Paleozoic.

Sampling effects Casually employed in paleontological study to include the many ways that the known fossil record differs from the actual communities of the past. In this sense, sampling effects include the differential preservation of organisms, systematic biases in the formation and preservation of sedimentary rock, and the way that the fossil record is collected and studied.

Bibliography

Bambach, R. K. (1983). Ecospace utilization and guilds in marine communities through the Phanerozoic. *In* "Biotic Interactions in Recent and Fossil Benthic Communities" (M. J. S. Tevesz and P. L. McCall, eds.), pp. 719–746. New York: Plenum.

Benton, M. J. (1990). The causes of the diversification of life. *In* "Major Evolutionary Radiations" (P. D. Taylor and G. P. Larwood, eds.), pp. 409–430. Oxford, England: Clarendon.

Erwin, D. H. (1993). "The Great Paleozoic Crisis: Life and Death in the Permian." New York: Columbia University Press.

Labandeira, C. C., and Sepkoski, J. J., Jr. (1993). Insect diversity in the fossil record. *Science* **261**, 310–314.

Sepkoski, J. J., Jr. (1991). Diversity in the Phanerozoic oceans: A partisan review. *In* "The Unity of Evolutionary Biology" (E. C. Dudley, ed.), Vol. 1, pp. 210–236. Eugene: Discorides Press.

Signor, P. W. (1990). The geologic history of diversity. *Annu. Rev. Ecol. Systematics* **21,** 509–539.

Tiffney, B. H., and Niklas, K. J. (1990). Continental area, dispersion, latitudinal distribution and topographic variety: A test of correlation with terrestrial plant diversity. *In* "Biotic and Abiotic Factors in Evolution" (W. Allmon and R. D. Norris, eds.). Chicago: University of Chicago Press.

Valentine, J. W. (1990). How good was the fossil record? Clues from the California Pleistocene. *Paleobiology* **15,** 83–94.

Evolutionary Taxonomy versus Cladism

Patricia A. Williams
Virginia State University

Biological taxonomy is the theory and practice of classifying organisms or taxa. Taxa (singular, taxon) are groups of organisms sharing certain characteristics, groups that taxonomists name and put into categories. For its nomenclature and method of categorization, modern taxonomy looks to the classification system developed by Carolus Linnaeus (1707–1778), a system that employs Latin binomial nomenclature and arranges categories in an inclusive hierarchy of nested sets, a system designed to reflect the creative plan of a rational God. In taxonomy, contention arises over taxonomic methods, groups, names, and categories. Some contentions develop into rival schools. In contemporary taxonomy, evolutionary taxonomy and cladism constitute two such schools. Cladists are evolutionists who claim that they employ logical rigor and objective rules in their taxonomic system. Evolutionary taxonomists are evolutionists who say that they reflect as much about nature as possible in their taxonomic system, but that nature is not always rational. Both schools have much in common. Both employ the Linnaean system of classification, and both use similar analytical methods to ascertain the relationships of organisms and taxa in order to construct evolutionary histories (phylogenies). In large part, their differences arise from their disparate ways of imposing the highly rational Linnaean system on the products of a nonrational evolution. To delve into their differences, this article first examines the Linnaean system of classification. Next it discusses the problems raised for the Linnaean system by the fact that organisms and taxa are the products of evolution rather than the creations of a rational deity. Finally, it presents cladism and evolutionary taxonomy as two modern solutions to the difficulties inherent in the attempt to classify the products of evolution rationally.

I. FOUNDATIONS OF MODERN BIOLOGICAL TAXONOMY

During most of his life the founder of modern biological taxonomy, Carolus Linnaeus, believed that God created each species separately according to a rational plan and that species have not changed essentially since their creation. In these beliefs, he was a person of his age. Most people who thought about the subject, and therefore most taxonomists, held similar beliefs. They believed that a rational God created an ordered world according to a design, much as a human being might design and then construct a house. Once designed, species did not change essentially, although they produced variations on the plan owing to environmental influences and the effects of habit. The design for the organic world, and therefore that world itself, was primarily atemporal and static.

In Linnaeus's time, to engage in biological taxonomy was to discover the natural system, the system of nature that God had designed. The nested categories that Linnaeus found were thought to reflect the deity's plan. Categories above the species level, such as genus, family, order, class, phylum, and kingdom, constituted the categories of God's thought, reflecting the order of His mind and His wisdom. This does not imply that all biological classification systems were of this sort or that all taxonomists sought to know the mind of God. Many classifications were simple identification keys designed for specialized purposes such as the recognition of medically efficacious herbs. Such specialized systems exist today as well. But Linnaeus's system, like that of many of his scientific peers, proposed to discover the natural order and, in doing so, to reflect the mind of God and the plan of His creation. [See SYSTEMATICS.]

Also like his educated contemporaries, Linnaeus had studied the logic of Aristotle (384–322 BCE) in which classification is by dichotomous division. For example, a personal library might be divided first into books about science and books on the humanities. Books on science might be further divided into those on physical science and those on biological science. Such division is often described as an exclusive or divisional hierarchy, for each category excludes objects in the other, as science excludes the humanities.

Linnaeus claimed that he was using the method of dichotomous division in his classification. However, Linnaean scholars believe that he used it only partially and classified more artfully, basing his classifications on his wide knowledge of organisms. In using methods other than dichotomous division, Linnaeus moved toward the position that Aristotle presents in his biological works. In these,

Aristotle ridiculed dichotomous division as a method for classifying organisms. He thought that organisms should be classified in reference to their essences, and he thought that essences could not be characterized by single attributes, a simplicity encouraged by dichotomous division.

Aristotle knew, and later taxonomists rediscovered, that classification by dichotomous division is likely to produce unnatural taxa, especially if all organic attributes are given equal weight. The sex of sexually reproducing organisms has to be taken into account as to developmental stages. Separating organisms into those that are larvae and those that are adults clearly would not cut nature at the joints. Equally unnatural divisions would be created if attributes like legged and not-legged were used early on in the classification, and then legged were divided into two- versus four- versus six-legged creatures, a division that would place people with birds and separate them from other mammals. Aware of these difficulties, many taxonomists in the eighteenth and early nineteenth centuries sought "affinities" or God's plan, not dichotomies, in their classifications.

Despite its internal tensions, the Linnaean system had great success, a success not enjoyed by other proposed systems of biological classification. It is the system used today. It is characterized by its Latin binomial nomenclature—people are *Homo sapiens*—and by its inclusive hierarchy of nested sets.

The Linnaean hierarchy has been successful because it serves many of the needs of a system of general classification. Its nested sets make it particularly useful as a memory and information-retrieval system. To know that something is an animal is already to know much about it; to know that it is a mammal is to know much more. The same characteristics aid in prediction and generalization. The system also works well because it can absorb new information without great disruption. When a new species is added, genera are not generally disrupted; certainly kingdoms rarely change.

The Linnaean system also served a heuristic function: by organizing taxa into a hierarchy of nested sets, it helped Charles Darwin (1809–1882) to recognize that contemporary organisms are the products of common descent and thus to develop his theory of evolution by natural selection. Ironically, Darwin's work undermined the theoretical basis for the Linnaean hierarchy as the plan of a rational deity.

II. FOUNDATIONS AND THE PROBLEMS OF EVOLUTION

Darwin was not the first person to write about evolution, for his grandfather and others had published books on the subject prior to his work. However, Darwin was the first person to explain convincingly how evolution might occur and, therefore, he was the first person to persuade a large number of educated people that evolution is a fact. The mechanism that Darwin posited is natural selection. Biologists today agree that evolution is a fact. They disagree about the role played by natural selection, about whether evolution occurs mostly by natural selection or whether other mechanisms might be equally important or more important. Currently, this is a lively area of discussion. Thus, a classification based on evolutionary information employs important facts, not debatable and changing theories, as some taxonomists think. [See EVOLUTION AND EXTINCTION.]

In the early and middle nineteenth century, evolution was a revolutionary idea. The idea that species change fundamentally, that they evolve from one another, that ancestral taxa differ markedly from contemporary taxa, and that contemporary taxa are related through descent, is a radically different idea from the concept that species and higher taxa come to be because of the rational plan of a creator and remain essentially unchanged. In taxonomy, the evolutionary origin of taxa requires that distant ancestors find their unique places in classifications, places not needed when they were considered to be essentially the same as living taxa. And it means that the organic world is not likely to be rationally organized. Indeed, one thing that evolution demonstrates is that there is no rational plan underlying organic nature. This lack of rational plan poses tremendous problems for the development of a classification system that is itself rational.

Three areas raise special difficulties: organisms appearing to be similar are not necessarily descended from a close common ancestor; speciation occurs in a variety of ways; and whether a population is a species cannot always be clearly determined, in principle. These problems require elaboration. [See SPECIATION.]

First, whatever theory of classification or organic origins taxonomists hold, as a practical matter they must classify organisms according to similar and distinguishing attributes, whether these be attributes of physiology, anatomy, chemistry, behavior, or habitat. But as taxonomists long before Darwin knew, simply comparing similar attributes is insufficient to give a good classification, so taxonomists search for meaningful similarities, for homologous attributes, attributes that evolutionists define as derived from the nearest common ancestor. Similarity of attributes that mimic homology but that are independently acquired by two or more taxa is called "homoplasy."

Homoplasy has three sources. First, it may develop by convergence due to similarity of function. For example, the similar shape of whales and fish is a homoplasy of convergence because the function of rapid movement through fluid media is enhanced by having a torpedolike shape. Second, it may develop due to genetic similarity because of parallel evolution in taxa that are distantly related. The presence of stalked eyes in some flies is an example. Third, attribute loss (character reversal) may occur, making existing species resemble distant ancestors in striking ways. The taxonomist must determine whether similarities are due to homology or to homoplasy, not an easy task in practice.

Another reason why similarity does not indicate genealogy is mosaic evolution, that is, different rates of evolution in the same taxa for different structures, organs, or behaviors. Mosaic evolution occurs frequently between sexes of the same species, sometimes producing differences so great that different sexes of the same species have been mistakenly relegated to different genera. Mosaic evolution is particularly prominent in species that metamorphose having egg, larval, and adult stages. Metamorphosis is common in insects, amphibians, and parasites. The taxonomist who emphasizes only one stage can produce radically different classifications from the taxonomist who focuses on a different stage.

Second, speciation, the evolution of a new species, occurs in a variety of ways, commonly when a small population (a "founder") becomes isolated from its large, parent population and undergoes rapid change due to genetic and environmental pressures. Perhaps the most studied example is the evolution of the fruit fly in the Hawaiian island chain. During and after speciation by the founder, the parent population remains virtually unchanged. Over the course of geologic time, a parent population may bud off many small populations that later form separate species. This method of speciation is frequently referred to as the "founder principle." An illustration of the founder principle is found in Fig. 4 near the end of this article.

Populations may also speciate by hybridization, a method especially common in plants. In this process, two different species, or parts of them, hybridize to form a new species. In hybridization, the parent species may continue to exist, or one or both of them may be absorbed in the hybridization.

Speciation may also occur by splitting when a species becomes divided by geographic or other means, and the two halves become separate species over time. In splitting, the parent species ceases to exist.

Third, it is not always clear, in principle, whether a population is a species, a problem compounded by the fact that biologists seem unable to agree on what a species is. Evolutionary taxonomists tend to adhere to the biological species concept, which defines a species as a population that is reproductively isolated from other populations in nature, a definition that tends to be ahistorical. On the other hand, cladists prefer the evolutionary species concept, which emphasizes the evolutionary history of lineages and considers a species to be a single lineage of ancestor–descendant populations that maintains its identity over time and has its own evolutionary tendencies and historical fate.

Reproductive isolation is marked by the failure of populations to interbreed satisfactorily for successive generations when they are in geographical proximity. Successful laboratory experiments in cross-fertilization do not affect what it means to be a species. Two populations that cross-fertilize in the laboratory may be good species in nature, for their habits may be so different as to reproductively isolate them from one another, as with two species of flowering plant, one of which produces pollen in the spring and the other in the autumn. Reproductive isolation is central to the biological species concept and is a major corollary of the evolutionary species concept, for to maintain their separate identities, tendencies, and fates, lineages must exhibit considerable reproductive isolation in nature.

Under these definitions of species, ambiguous speciation occurs in four distinct cases. First, populations may form a partial ring (ring species) such that each of the adjoining populations successfully interbreeds with its neighbors, but the end populations cannot interbreed with one another. Across this gap, the end populations may be two good species, sufficiently reproductively isolated to maintain their own evolutionary tendencies; consider the adjoining populations, however, the entire ring may constitute one species, evolving its own history.

Allopatric species constitute a second case. These are populations that are geographically isolated from one another yet still very similar. If they were not geographically isolated, they might successfully interbreed and so be considered one species—or they might not, in which case they would be two good species with separate identities.

A contrasting case is found in sister species. These are species in geographical proximity that are so similar as to be virtually indistinguishable by taxonomists who examine specimens, yet they do not interbreed in nature. Because for practical reasons most taxonomy is done in museums and laboratories rather than in the field, two sister species can easily be mistaken for a single species. The problem is compounded by trying to take their evolutionary histories into account, histories written in fossils, for it is more difficult to recognize separate species among fossils than it is among living organisms. When taxonomists examine fossils, they can neither retrieve intact specimens nor observe breeding habits, even in principle.

Fourth, many species are extinct, known only by their fossils. Being extinct, their members do not currently interbreed. Being fossils, much material is missing from the specimens, especially soft parts, which decay rapidly or served as a predator's meal. Paleontologists decide whether populations are good species by examining morphological similarities and differences among fossilized remains and by noting geographical location and geographical age. Information is usually scarce. As techniques for extracting DNA from fossils improve, important information will become available about genealogical relations among fossils, but DNA is not a panacea for indicating speciation, as the close chemical resemblance between human beings and chimpanzees makes clear.

In sum, evolution is complex and messy. There is no rational plan. Accidents occur. Some are vitally important in evolutionary history, particularly those that seem to have caused the great extinctions, but smaller accidents play a role as well. Developing a rational classification despite these facts is not an easy matter.

One response of some taxonomists has been to eschew this entire body of knowledge and to try to classify by overall similarity alone. This endeavor is known variously as "phenetics," "numerical taxonomy," "patterned cladism," and/or "reformed cladism." If taxonomists want to reflect the diversity and relations among the taxa around them, that is, if they want a natural system, classification by overall similarity is inadequate. The existence of sexual dimorphism among almost all sexually reproducing species is itself an insurmountable barrier, and more barriers are erected by the other difficulties posed by the fact of the evolutionary origin of taxa.

In contrast to this position, evolutionary taxonomy and cladism deliberately employ evolutionary information. Both agree that classifications should reflect phylogeny, the tracing out of lines of descent, among groups of organisms. Evolutionary taxonomists and cladists both wrestle with the problems that nature poses, some of which have been adumbrated here. However, they go about it differently. The remainder of this article will discuss accomplishments and challenges in each of

these schools of biological taxonomy. Because cladism claims to be the most logical of the two, it will be treated first.

III. ONE SOLUTION: CLADISM

Of the claims made by cladism, three are basic. First, cladism claims that its method of attribute analysis is the best means for reconstructing phylogenies. Because many evolutionary taxonomists have adopted cladistic methods of analysis, this claim is not in serious dispute except in the use of fossil evidence, evidence that some cladists repudiate.

Second, cladism claims to be the most logical method of phylogenetic analysis and of taxonomic classification. Both of these claims are seriously disputed. Cladistic logic is based on dichotomous division, a method that Aristotle thought was unsuited to the classification of organic nature and that Linnaeus found problematic in practice. However, cladistic methods are both somewhat different and also more sophisticated than those available in previous centuries, so this claim deserves careful examination. In the area of taxonomic classification, critics of cladism protest that cladistic methods do not give classification; they merely give phylogeny.

Third, cladism claims that its method of dichotomous division will reproduce the Linnaean hierarchy because the two hierarchies—divisional and Linnaean—are logically equivalent. This claim is that of Willi Hennig (1913–1976), the founder of cladism. It has some considerable confusion. Thus, there are disputes regarding each of the three fundamental claims made by cladism.

Least in dispute is cladism's method of attribute analysis and, through that analysis, its ability to reconstruct genealogical relationships among closely related taxa. Cladists use two methods that proved reliable long before the advent of cladism. First, they use as many attributes as possible, including chemical analysis (DNA comparison) where available. A sign of taxonomy's increasing employment of chemical analysis has been the establishment of high-technology laboratories at institutions that long considered classical analysis to be best: the Smithsonian's Museum of Natural History, the American Museum of Natural History in New York, the Field Museum in Chicago, and natural history museums at major centers in the European Union. Second, cladists use weighted attributes, a technique advocated by Aristotle, Linnaeus, Darwin, Hennig, Ernst Mayr (1904–), and many others.

Cladists make taxonomy more sophisticated by emphasizing two other methods. First, analysis is based on synapomorphic attributes, not on pleisomorphic ones. Synapomorphic attributes are those that are recently derived genealogically, shared by two or more taxa, and assumed to have been inherited from the nearest common ancestor, not from earlier ancestors. For taxonomic purposes, synapomorphies serve to unite a lineage. Pleisomorphic attributes are those derived from ancient ancestors. The former help taxonomists locate speciation events; the latter do not. Second, cladism uses out-group comparison to discover which attributes are synapomorphic and which are pleisomorphic. In out-group comparison, three related groups are compared. If two have an attribute that the other lacks, the attribute is a synapomorphy for the sister groups; if all three have it, it is a pleisomorphy for the sister groups. Clearly, the terms are relative: an attribute that is a pleisomorphy in one group may be a synapomorphy in another.

There are other techniques for locating nearest common ancestor and sister groups as well. However, no matter how elegant, none can eliminate the difficulties caused by the fact of evolution, difficulties like convergence, parallel characters, reversed characters, and mosaic evolution. No matter how rigorous their logic, taxonomists need to have wide experience with a group of organisms to make sound taxonomic judgments. Nonetheless, cladistic analysis has done much to make taxonomy more objective than it was before Hennig published.

More in dispute is cladism's method of dichotomous division. Contention arises over whether the method of dichotomous division is too rigid to be the basis for the natural system that Darwin, Hennig, and Mayr all seek. Dichotomous division does not accurately reflect natural speciation events. In nature, the most common method of speciation

is by the founder principle. Speciation by hybridization also occurs and is particularly common among plants.

Cladism's method of dichotomous division cannot reflect these natural events, for cladism insists that the original population becomes extinct at each branch point in a cladogram, a line drawing representing a speciation event. It maintains that splitting alone occurs, and that all divisions should be resolved into dichotomous ones. Cladism's method captures only a type of speciation thought to be relatively rare in nature. Thus, a case can be made that the logical methods of cladism significantly distort natural events and therefore cannot be the basis for a natural system of classification, a system designed to reflect nature as accurately as possible.

One of the most troubling aspects of cladism's insistence on dichotomous division is that ancestor–dependent relationships are not represented in cladograms and, because they are not, fossils tend to be discarded from cladistic classifications. In a system whose foundation is phylogeny, this is strange; in one that embraces the evolutionary species concept, a concept originally designed by George Gaylord Simpson (1902–1984) to capture the temporal dimension of evolution, it is bizarre. Even when the definition of "phylogeny" is limited to branching relationships, the narrow definition of "phylogeny" that cladism prefers, it is odd, for sister-groups branch from their nearest common ancestor, and it is the unrepresented ancestor that gives the synapomorphy.

This peculiarity in cladism has two sources. The first source is the fact that the Linnaean hierarchy was not designed to represent ancestors, for the continuity of God's plan did not require that they be represented. Because the hierarchy has no obvious place for ancestors, the tendency is to leave them out of Linnaean classifications. The second is Hennig's conflation of the divisional hierarchy with the Linnaean hierarchy. Hennig thought that the two hierarchies are logically equivalent; however, they are not.

Dichotomous division produces a divisional hierarchy. Divisional hierarchies are exclusive hierarchies, as cladism's insistence on finding synapomorphies and rejecting pleisomorphies makes clear. Cladistic sister-groups exclude other groups. Tellingly, other groups are called "out-groups."

In contrast, the Linnaean hierarchy is an inclusive hierarchy, a hierarchy of nested sets. A species is included in its genera, a genera in its order, an order in its class, and so on. The divisional hierarchy of cladistic methodology and the Linnaean hierarchy of cladistic classification are logically different.

Their difference can be illustrated graphically. Figure 1 represents three different versions of the Linnaean hierarchy. A set of nested boxes (a) best illustrates the inclusive relationships of this hierarchy. Part (b) shows the same relationships represented in a line drawing, with I being the most inclusive category and a, b, c, d, e, and f being encompassed by all the higher categories. An inclusive hierarchy may also be represented in the form of an indented outline (c). Figure 2 represents a divisional hierarchy by line drawing, practically the

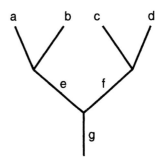

FIGURE 2 An exclusive, divisional hierarchy formed by dichotomous division. In this tree diagram, all the categories (a–g) are equal; none includes any other. a and b formed when e split; c and d when f split. e and f formed from the division of g. If this drawing represented biological relationships, a–g would all be species; no higher categories are present.

only way to illustrate it. Here a–g are equal; none is included in the other. The groups are formed by dichotomous splitting, g giving rise to e and f, e to a and b, and f to c and d. In biological taxonomy, a–f in Fig. 1 would represent species, 1–3 genera, and I, perhaps, a kingdom. In Fig. 2, a–g represent species; no higher categories are present. The line drawings of Figs. 1 and 2 are not logically equivalent. Figure 2 is a phylogenetic tree; Fig. 1 is a Linnaean classification.

The lack of logical equivalence between the hierarchies poses two problems for cladism. The first is how to turn cladograms, representing sister-group relationships, into phylogenetic trees, representing ancestor–descendent relationships. One answer to this problem is to note that the construction of hypothetical ancestor–descendent relationships is necessary in order to find sister-groups, and that, although this information is deleted from cladograms, it is available if a cladist wants to construct a tree. As this answer suggests, cladograms cannot be turned directly into trees, but cladists can construct trees if they employ additional information to do so.

The second problem is how to turn trees into classifications. As noted by Darwin, Mayr, and others, genealogy alone does not give classification. A genealogy is a divisional hierarchy, not a Linnaean one. Somehow, the taxonomist must cut up the trees and collapse its branches into nested sets. Early in his work, Hennig uses the axe of time to cut the tree, chopping it at significant geological

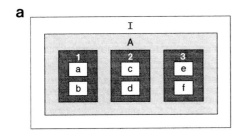

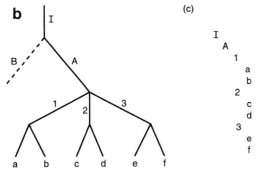

FIGURE I An inclusive hierarchy such as the Linnaean hierarchy can be represented in a number of equivalent ways. In the figure, (a), (b), and (c) all represent the same inclusive relationships. Part (a) is a set of nested boxes, (b) a type of tree, and (c) an indented outline. In all three, I is the most inclusive category, encompassing all the others. A includes 1–3 and a–f. 1 includes a and b, 2 includes c and d, and 3 includes e and f. If these illustrations represented biological relationships, only a–f would be species. The other alpha numbers would symbolize higher categories: 1–3 might be genera, A perhaps a class, and I possibly a kingdom.

ages, then designating extant taxa as species and ancestral species as genera, families, orders, and so on, as one plunges through geologic time. In practice, this approach turns out to be unworkable, and because it conflates ancestral species with higher categories, it is logically untenable as well.

One solution is to revert to pure Linnaeanism. Linnaeus did not recognize fossils, and because he thought that ancestors were essentially like living forms, he classified only living forms. Later in his work, Hennig suggests that fossils be shunned and ancestors not be represented or, if represented, that they be classified as addenda to the Linnaean hierarchy. Many cladists follow these practices.

Such an answer rejects ancestor–descendent relationships and fossils and also fails to rank taxa into Linnaean categories. Anticipating such problems, Darwin suggests that geneaology will give nested sets by the collapse of the branches of trees, but that rank must be given by the amount of difference in various taxa. Thus he refers to his famous genealogical tree diagram in the fourth chapter of *On the Origin of Species* (1859) and suggests cutting it up according to the degree of modification distinguishing taxa, taxa equally related by genealogy. To Darwin, degree of modification gives rank. Evolutionary taxonomy embraces Darwin's solution.

IV. ANOTHER SOLUTION: EVOLUTIONARY TAXONOMY

Evolutionary taxonomy is similar to cladism in significant ways. It emphasizes phylogeny and has adopted cladistic analysis, agreeing that such analysis best delineates relationships among existing species. In emphasizing phylogeny, evolutionary taxonomy acquires all the difficulties that stem from the fact of evolution, including the importation of evolutionary information into the Linnaean hierarchy.

However, in substantial respects, evolutionary taxonomy also differs from cladism. The two most important differences are, first, the evolutionary taxonomy is willing to forego logical rigor in order to reflect the nonlogical facts of nature and, second,

that it defines "monophyly" differently from cladism, a distinction of importance because both rest their classifications largely on their respective views of monophyly.

Unlike cladism, which often tries to ignore the difficulties for classification imposed by evolution, evolutionary taxonomy seems to embrace the complications. It recognizes all the problems. Particularly, it recognizes that the founder principle of speciation does not produce splitting, and certainly not dichotomous splitting in which the parent population goes extinct. It uses ancestor–descendent relationships and recognizes fossils, although Mayr seems ambiguous about them. And it recognizes that phylogenetic trees do not give classification.

Partly as a result, evolutionary taxonomy is logically less rigorous than cladism claims to be, for it employs three criteria for classifying that are not logically related. First, it uses phylogeny: members of the same taxa must have a common ancestor. Second, it employs genetic similarity, which, before the chemical revolution, had to be inferred from morphology and ecology: if two taxa have a nearest common ancestor, but one taxon has diverged significantly in morphology or ecology from its sister taxon, the divergent taxon will be given a different rank. Third, evolutionary taxonomy counts numbers of species: if a taxon is highly speciose, it will be given a higher categorical rank than a taxon that is less numerous.

Evolutionary taxonomy accepts the first two criteria because of the way it defines "monophyly." It defines a taxon as monophyletic if all its members are derived from the nearest common ancestor, the traditional definition initiated by Ernst Haeckel (1834–1919). Haeckel used the concept of monophyly to separate evolutionary relationships from the nonevolutionary ones proposed by his opponents who thought that organisms had several points of independent origin. Cladism, in contrast, defines a group as monophyletic only if it includes all the descendents of the nearest common ancestor.

Because monophyly has acquired these two different meanings, several new terms have been introduced into taxonomy to try to clarify the situation, although the new terms are not always used consistently or by all taxonomists. Cladism's "monophyletic" has become "holophyletic," whereas Haeckel's "monophyletic" has become "paraphyletic." Groups without a nearest common ancestor are "polyphyletic." These relationships are illustrated in Fig. 3. Both evolutionary taxonomy and cladism agree that polyphyletic groups are unacceptable in a classification. Evolutionary taxonomy allows paraphyletic groups; cladism allows only holophyletic ones.

Evolutionary taxonomy claims that allowing paraphyletic groups lets it illuminate several aspects of nature that holophyletic classifications obscure. First, evolutionary taxonomy can display morphological gaps that occur in nature, for example, emphasizing the difference between human beings and other primates.

Against this view, it has been argued that, although nature now has such gaps, the gaps are artificial because they are the consequence of extinctions and that, if taxonomists possessed complete phylogenetic information, there would be no gaps. Being artificial, morphological gaps should not be represented in a natural classification. However, not only is this a somewhat curious use of "artificial"—extinctions are natural events—but this argument overlooks the effects of mosaic evolution. At least some gaps occur because one phylogenetic line evolves and speciates rapidly into a new ecological zone while its sister line remains static. If evolutionary trees are drawn with nature's pen, crossing geologic time, then mosaic evolution creates gaps even without extinctions (see Fig. 4).

Some of the morphological and ecological gaps that have appeared in the classifications of evolu-

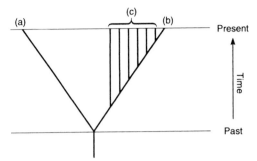

FIGURE 4 This figure illustrates the founder principle and it shows how gaps can appear without extinctions. (a) and (b) are two lineages that speciated by dichotomous splitting sometime in the past and diverged from one another in about equal measure. (a) has not speciated further; (b) has speciated frequently. (c) are all founder species whose parent is (b). (b), the ancestor of (c), continues to exist. (b) and (c) are not sister-groups, but an ancestor and its descendents. Common attributes cluster together in taxa (c) and (b). There are no extinctions; nonetheless, between this cluster and (a), a large attribute gap exists.

tionary taxonomists reflect neither phylogenetic relationships nor genetic ones. Because one rationale behind the criterion of divergence has been that it captures underlying genetic dissimilarities, and genetic relationships can now be chemically tested to some extent, chemical comparisons can be used to test evolutionary classifications. Where these methods have been used, some evolutionary classifications have been shown to be misguided, if the genetic rationale is employed. For example, evolutionary taxonomy has recognized reptiles and birds as two distinct groups, placing crocodiles in their morphologically obvious place with the reptiles. However, chemical comparisons appear to show that crocodiles are more closely related to birds than they are to other reptiles. If on more complete analysis this turns out to be the case, then evolutionary taxonomy's methods have developed a mis-

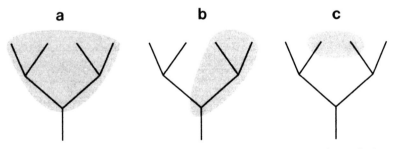

FIGURE 3 The shaded taxa are (a) holophyletic, (b) paraphyletic, and (c) polyphyletic.

leading classification if the criterion of genetic divergence supplies the underlying rationale for classification by morphological and ecological divergence.

Dividing groups because they are rich in species and giving higher rank to some taxa than would be given if the taxon were less speciose reflect neither phylogeny nor genetic similarity. It seems to have become a practice in evolutionary taxonomy because it is a practical aid for the taxonomist. Employing a practicality criterion would seem to be contrary to evolutionary taxonomy's search for a natural system.

However, even as a natural system, evolutionary taxonomy may be inconsistent if it maintains the rationale of genetic foundations. Evolutionary taxonomy recognizes the founder principle, and part of that principle is that founder populations undergo a genetic revolution in the process of speciation. Because founders frequently differ in appearance and ecology from the parent population, this sounds like a reasonable assumption. However, for the most part it is untested speculation, and the bird–crocodile case suggests that it may be false. Genetic similarity may closely track phylogenetic relationships. If this turns out to be the case, paraphyletic classifications will not give information about genetic divergence.

Nevertheless, they would give information about morphological and ecological divergence, and this may be information that is desirable in a general classification. For example, chimpanzees and human beings are very closely related genetically but are divergent in their ecology, morphology, and other characteristics. A classification that did not indicate this would obscure important information about the relationship between the two species. This point is a general one: a classification that ignores ecological relationships will be less informative than one that contains such information. It will also be less useful to the environmental biologist.

In summary, systems such as cladism—systems rooted in phylogeny yet rejecting morphological and ecological divergence as criteria for ranking—lack a logical method of ranking, for phylogenetic trees are not logically equivalent to the Linnaean hierarchy. It appears that phylogenetic classification systems either must do without a logical method of ranking or rank by criteria that are not strictly holophyletic. Such is the quandary in which taxonomic classification finds itself today. Perhaps, in the end, taxonomists will find that they want a method of classification that includes evolutionary divergence, a method that captures the fact that this article will never be read by a chimpanzee.

Glossary

Founder principle Principle that the founding population of a new species will undergo a genetic revolution due to the fact that it contains only a fraction of the genetic variation of the parental population.

Holophyletic group Group that contains all the descendants of its most recent common ancestor.

Homology Attribute of two taxa derived from the nearest common ancestor.

Mosaic evolution Different rates of evolution in the same group of organisms for different attributes.

Paraphyletic group Group derived from the nearest common ancestor, but not containing all the descendants of the common ancestor.

Phylogeny Inferred line of descent of a group of organisms.

Pleisomorphy Ancestral state of organic attributes.

Polyphyletic group Group derived from two or more ancestral sources.

Synapomorphy Organic attribute found in two or more taxa, present in the nearest common ancestor but not in earlier ancestors, and, for taxonomic purposes, serving to unite a lineage.

Bibliography

Brooks, D. R., and Funk, V. A. (1990). Phylogenetic Systematics as the Basis of Comparative Biology." Washington, D.C.: Smithsonian Institution Press.

Brooks, D. R., and McLennan, D. (1991). "Phylogeny, Ecology, and Behavior." Chicago: University of Chicago Press.

Gibbons, A. (1991). Systematics goes molecular. *Science* **251**, 872–874.

Mayr, E. (1994). Systems of ordering data. *Biology and Philosophy* (in press).

Mayr, E., and Ashlock, P. D. (1991). "Principles of Systematic Zoology," 2nd ed. New York: McGraw–Hill.

McDade, L. (1990). Hybrids and phylogenetic systematics: Patterns of character expression in hybrids and their implications for cladistic analysis. *Evolution* **44**(6), 1685–1699.

Scott-Ram, N. R. (1990). "Transformed Cladistics, Taxonomy and Evolution." Cambridge, Mass.: Cambridge University Press.

Sober, E. (1993). Experimental tests of phylogenetic inference methods. *Systematic Biol.* **42,** 85–89.

Soltis, P., Soltis, D., and Doyle, J., eds. (1991). "Molecular Systematics of Plants." New York: Chapman & Hall.

Williams, P. A. (1992). Confusion in cladism. *Synthese* **91,** 135–152.

ISBN 0-12-226731-1

90018